T5-CWR-618

CORROSION HANDBOOK

THE
CORROSION HANDBOOK

edited by

HERBERT H. UHLIG, PH.D.

ASSOCIATE PROFESSOR OF METALLURGY IN CHARGE OF THE CORROSION LABORATORY, MASSACHUSETTS INSTITUTE OF TECHNOLOGY, CAMBRIDGE, MASSACHUSETTS

and sponsored by

THE ELECTROCHEMICAL SOCIETY, INC.

NEW YORK, N. Y.

NEW YORK · JOHN WILEY & SONS, INC.
LONDON · CHAPMAN & HALL, LIMITED
1948

Copyright, 1948
BY
JOHN WILEY & SONS, Inc.

All Rights Reserved

This book or any part thereof must not be reproduced in any form without the written permission of the publisher.

Third Printing, June, 1951

PRINTED IN THE UNITED STATES OF AMERICA

PREFACE

The tax imposed by corrosion of metals on industry, the community, and the nation is an appreciable proportion of our national income. The costs have been not only in replacement of metals, the waste of which we can afford less and less as our ore reserves are depleted, but also in damage to products by contamination, shutdown of production, loss of efficiency, and in psychological factors attending imminent failure or explosion of corroding equipment. Unlike some forms of taxation, however, the tax leveled by corrosion can be reduced effectively by application of science and technology. Efforts in this direction have only barely begun. Vast opportunities are available to improve performance of metals in corrosive environments, and to reduce to a less staggering figure the annual toll exacted by metal deterioration. It is on such a premise that the subject of corrosion justifies its bid for attention and support in the fields of science and engineering.

This handbook, as its name implies, is a condensed summary of corrosion information, including within its scope a cross section of scientific data and industrial experience. It is logical that discussion should be concerned primarily with corrosion protection, and the behavior of metals and alloys in environments at ordinary and elevated temperatures. Some chapters are devoted briefly to current theories of corrosion and to corrosion testing, but obviously in any single book covering so broad a field, the latter subjects are treated less prominently than the former.

Emphasis is on quantitative information. Qualitative data leaving the reader uncertain as to proper interpretation have been minimized. By listing actual corrosion rates under precisely defined exposure conditions, the reader will have, so it is hoped, more adequate basis for sound judgment in using the information. Effective in this respect are some unpublished data supplied by several authors. Such data, until the *Corrosion Handbook* appeared, found no ready medium for publication; yet the information has practical importance.

During organization of the book many gaps of information were uncovered which will probably be made more prominent by the present correlation and classification of data. It is hoped that, in addition to serving as an authoritative reference, this book will stimulate appropriate experimental work directed toward supplying what is now missing.

As far as practicable, patented processes, formulas, and coatings are so indicated in the references, but the absence of patent reference does not necessarily imply that a means or method is in the public domain. Neither should it be assumed that the mention of any patented process or proprietary formula is in any sense a recommendation for its application or use.

The attempt was made in expressing corrosion rates to transpose most data to inches penetration per year (ipy) or milligrams per square decimeter per day (mdd). It is expected that these particular units will find favor in the corrosion literature, both because they have already had widespread use, and also because some standardization is of great advantage to a ready familiarity with orders of magnitude.

To those making their first acquaintance with the subject, it should be pointed out that corrosion rates reported in these units do not necessarily imply a total time of exposure confined to days in the case of mdd, or to years in the case of ipy. The rates are used in the same sense as miles per hour. For example, a plane is said to travel an average of 100 miles per hour, and yet may have been in the air either a matter of minutes or a matter of hours. In flight, its velocity may have been greater or less than 100 miles per hour at any single moment. Corrosion rates likewise represent an average over the time of exposure and may or may not include a rapid initial rate and a less rapid final rate. In general, therefore, extended extrapolation of corrosion rates is not justified unless the characteristics of the corrosion reaction with respect to time are well known. It is for this reason that whenever possible the total time of exposure accompanies the corrosion rates reported in this book.

Units converted from English to metric system, or the reverse, take into account approximate reproducibility of the measurements. For example, the boiling point of water at 100° C can be precisely interpreted as 212° F, but a furnace temperature of 1000° C is not always regulated to a precision justifying a reported temperature of 1832° F. It is probably better expressed as 1830° F, in which case the temperature is listed as 1000° C (1830° F).

The sign of potentials conforms with the convention adopted by The Electrochemical Society and National Bureau of Standards and is also that used in other countries. Accordingly, a positive potential refers to a noble or cathodic potential, whereas a negative potential applies to an active or anodic potential. Ambiguity is avoided by using the words *anodic* or *cathodic* in place of or parallel with *positive* or *negative*.

Plans for this book grew out of informal conversations at The International Nickel Company's annual corrosion inspection at Kure Beach, North Carolina, and the early Gibson Island corrosion conferences sponsored by the American Association for the Advancement of Science. These plans in part led to organization of the Corrosion Division within The Electrochemical Society in 1942. One of the first objectives of the Corrosion Division was the publication of a convenient reference volume covering the entire field of corrosion, to bring together, in effect, much of the information scattered broadly throughout the scientific and engineering literature. This book is the first response to that objective. Royalties from its sale will be used by The Electrochemical Society to support similar activities in corrosion.

The times during which the project was organized and prosecuted were not the most favorable for an undertaking of this kind, but postponement seemed less justified than an attempt to do our best under the circumstances. Perhaps the demand for a practical summary of corrosion information in a postwar period acutely aware of conservation, as well as required economies in industry, justified our decision to go ahead.

Much encouragement and help during organization and planning of this book came from R. M. Burns who, as president of The Electrochemical Society, gave his full and active support to the project. He and F. L. LaQue generously contributed time and advice during the metamorphosis of an idea into a 1200-page book.

Grateful acknowledgment is due many reviewers of chapters who, besides reading the manuscripts and checking information, often assisted still further by contributing additional data. Appreciation is expressed to The General Electric Company for providing the necessary facilities in the early organization of the book. I am pleased to acknowledge the assistance of W. D. Robertson in some of the proofreading, and, finally, the help of my wife in many routine tasks associated with editing a book of this kind.

CAMBRIDGE, MASSACHUSETTS
July 21, 1947

HERBERT H. UHLIG
EDITOR

MEMBERS OF EDITORIAL ADVISORY BOARD

R. M. BURNS
Bell Telephone Laboratories,
Murray Hill, N. J.

U. R. EVANS
Cambridge University,
Cambridge, England

W. H. FINKELDEY
Singmaster and Breyer,
New York, N. Y.

F. L. LAQUE
The International Nickel Company,
New York, N. Y.

R. J. MCKAY
The International Nickel Company,
New York, N. Y.

R. B. MEARS
Carnegie Illinois Steel Corporation,
Pittsburgh, Pa.

F. N. SPELLER
Consultant,
Pittsburgh, Pa.

L. G. VANDE BOGART
Crane Company,
Chicago, Ill.

G. H. YOUNG
Mellon Institute,
Pittsburgh, Pa.

LIST OF CONTRIBUTORS

	PAGE
ALMEN, J. O.	590

Research Laboratories Division, General Motors Corp., Detroit, Mich.

ANDERSON, E. A. 331, 722, 862, 1023
  Research Division, The New Jersey Zinc Co., Palmerton, Pa.
ANDREWS, A. I. 872
  Department of Ceramic Engineering, University of Illinois, Urbana, Ill.
AUSTIN, J. B. 630, 1122
  Research Laboratory, U. S. Steel Corp., Kearny, N. J.
BALKE, CLARENCE W. 60, 320, 620, 720
  Fansteel Metallurgical Corp., North Chicago, Ill.
BARTON, G. B. 119
  Research Department, American Smelting and Refining Co., Barber, N. J.
BASH, F. E. 997
  Driver-Harris Co., Harrison, N. J.
BINDER, W. O. 182, 640
  Union Carbide and Carbon Research Laboratories, Inc., Niagara Falls, N. Y.
BLUM, WILLIAM 803, 970, 1030
  National Bureau of Standards, Washington, D. C.
BORGMANN, CARL W. 953
  University of Colorado, Boulder, Col.
BRENNER, ABNER 803
  National Bureau of Standards, Washington, D. C.
BROWN, M. H. 747
  Experimental Station, E. I. DuPont de Nemours and Co., Wilmington, Del.
BROWN, R. H. 481
  Aluminum Research Laboratories, Aluminum Company of America, New Kensington, Pa.
BRYAN, W. T. 201
  The Duriron Co., Inc., Dayton, Ohio
BULOW, C. L. 69, 85, 96, 551, 747, 1015
  Bridgeport Brass Co., Bridgeport, Conn.
BURGHOFF, H. L. 965
  Chase Brass and Copper Co., Waterbury, Conn.
BURNS, R. M. 609
  Bell Telephone Laboratories, Murray Hill, N. J.
CAMPBELL, W. E. 916
  Bell Telephone Laboratories, Murray Hill, N. J.
CLAPP, WILLIAM F. 433
  William F. Clapp Laboratories, Duxbury, Mass.
CLARK, C. L. 664
  Timken Roller Bearing Co., Canton, Ohio
COLLINS, LEO F. 538
  Chemical Engineer, Detroit, Mich.

LIST OF CONTRIBUTORS

	PAGE
COMPTON, K. G.	1006, 1023

Bell Telephone Laboratories, New York, N. Y.

COPSON, H. R. ... 569, 1009
Research Laboratory, The International Nickel Co., Inc., Bayonne, N. J.

DARSEY, V. M. ... 867, 871
Parker Rust Proof Co., Detroit, Mich.

DENISON, I. A. ... 1038, 1048
National Bureau of Standards, Washington, D. C.

ECKART, RICHARD J. ... 884
The Sapolin Co., Inc., New York, New York

EDWARDS, JUNIUS D. ... 857
Aluminum Research Laboratories, Aluminum Company of America, New Kensington, Pa.

ELDREDGE, G. G. ... 905, 1083
Shell Development Co., Emeryville, Calif.

EVANS, U. R. ... 3
Cambridge University, Cambridge, England

EWING, SCOTT P. ... 601
Research Laboratory, The Carter Oil Co., Tulsa, Okla.

FAIR, W. F., JR. ... 898
Mellon Institute, Pittsburgh, Pa.

FIELD, B. E. ... 294
Union Carbide and Carbon Research Laboratories, Inc., Niagara Falls, N. Y.

FINK, COLIN G. ... 103, 606
Division of Electrochemistry, Columbia University, New York, N. Y.

FOOTE MINERAL CO. ... 347
Philadelphia, Pa.

FRANKS, RUSSELL ... 150, 188
Union Carbide and Carbon Research Laboratories, Inc., Niagara Falls, N. Y.

FRASER, O. B. J. ... 1052
Research and Development Division, The International Nickel Co., Inc., New York, N. Y.

FRIEND, W. Z. ... 253, 266, 281
Research and Development Division, The International Nickel Co., Inc., New York, N. Y.

FUGASSI, J. P. ... 1126
Department of Chemistry, Carnegie Institute of Technology, Pittsburgh, Pa.

GEIGER, G. F. ... 729, 1000
Research and Development Division, The International Nickel Co., Inc., New York, N. Y.

GLEASON, C. A. ... 545
Scovill Manufacturing Co., Waterbury, Conn.

GONSER, BRUCE W. ... 323, 829
Battelle Memorial Institute, Columbus, Ohio

GURRY, R. W. ... 630
Research Laboratory, U. S. Steel Corp., Kearny, N. J.

GWATHMEY, ALLEN T. ... 33
School of Chemistry, University of Virginia, Charlottesville, Va.

LIST OF CONTRIBUTORS

	PAGE
HADLEY, RAYMOND F.	466
Susquehanna Pipe Line Co., Philadelphia, Pa.	
HARING, H. E.	978
Bell Telephone Laboratories, Murray Hill, N. J.	
HAWKINS, G. A.	511
Department of Mechanical Engineering, Purdue University, Lafayette, Ind.	
HECHT, MAX	520
Consultant, Pittsburgh, Pa.	
HIERS, G. O.	207, 747
National Lead Co., Brooklyn, N. Y.	
HUNTER, M. A.	683
Department of Metallurgical Engineering, Rensselaer Polytechnic Institute, Troy, N. Y.	
HUTCHINS, L. H.	433
Woods Hole Oceanographic Institution, Woods Hole, Mass.	
JENKS, R. H.	61
Revere Copper and Brass, Inc., Rome, N. Y.	
KENDALL, V. V.	1058
National Tube Co., Pittsburgh, Pa.	
KERR, S. LOGAN	597, 993
Consulting Engineer, Philadelphia, Pa.	
KNAPP, B. B.	1077
Research Laboratory, The International Nickel Co., Inc., Bayonne, N. J.	
KROLL, W. J.	56, 252, 314, 329, 330, 331, 378, 380
Consulting Metallurgist, Albany, Ore.	
LAQUE, F. L.	194, 383, 747, 1060
Research and Development Division, The International Nickel Co., Inc., New York, N. Y.	
LARRABEE, C. P.	120, 1043
Research Laboratory, Carnegie-Illinois Steel Corp., Vandergrift, Pa.	
LEACH, ROBERT H.	314, 718
Handy and Harman Company, Bridgeport, Conn.	
LEE, JAMES A.	353, 354, 359, 365, 372, 723
Chemical and Metallurgical Engineering, McGraw-Hill Publishing Co., New York, N. Y.	
LOANE, C. M.	559, 1034
Research Department, Standard Oil Co., Whiting, Ind.	
LOGAN, K. H.	446
National Bureau of Standards, Washington, D. C.	
LOOSE, W. S.	218, 670, 864
Dow Chemical Co., Midland, Mich.	
MAXWELL, H. L.	143, 747
Engineering Department, E. I. DuPont de Nemours and Co., Inc., Wilmington, Del.	
MCKINNEY, D. S.	1126
Department of Chemistry, Carnegie Institute of Technology, Pittsburgh, Pa.	
MEARS, R. B.	20, 39, 493, 617, 747
Research Laboratory, Carnegie-Illinois Steel Corp., Pittsburgh, Pa.	
MILEY, H. A.	11
Radiation Laboratory, Massachusetts Institute of Technology, Cambridge, Mass.	

LIST OF CONTRIBUTORS

	PAGE
MUDGE, W. A.	675

    Research and Development Division, The International Nickel Co., Inc., New York, N. Y.

NELSON, HARLEY A. 1071
    Technical Department, The New Jersey Zinc Company, Palmerton, Pa.

PARTRIDGE, E. P. 520
    Hall Laboratories, Inc., Pittsburgh, Pa.

PEARSON, JOHN M. 923
    Physical Research and Development Laboratory, Sun Oil Co., Chester, Pa.

PHAIR, R. J. 1023
    Bell Telephone Laboratories, New York, N. Y.

POPE, ROBERT 935
    Bell Telephone Laboratories, New York, N. Y.

POWERS, P. O. 890
    Battelle Memorial Institute, Columbus, Ohio

PRATT, W. E. 747
    Worthington Pump and Machinery Corp., Harrison, N. J.

PRAY, H. A. 1016
    Battelle Memorial Institute, Columbus, Ohio

REDFIELD, ALFRED C. 1111
    Woods Hole Oceanographic Institution, Woods Hole, Mass.

REINHARD, C. E. 331
    Research Division, The New Jersey Zinc Co., Palmerton, Pa.

RHINES, FREDERICK N. 621
    Metals Research Laboratory, Carnegie Institute of Technology, Pittsburgh, Pa.

RUDER, W. E. 638, 667
    Research Laboratory, General Electric Co., Schenectady, N. Y.

SAMPLE, C. H. 1002, 1034
    The International Nickel Company, Inc., New York, N. Y.

SANDS, G. A. 747
    Electrometallurgical Co., New York, N. Y.

SCHEIL, M. A. 174
    A. O. Smith Co., Milwaukee, Wis.

SCHNEIDER, W. K. 902
    Mellon Institute, Pittsburgh, Pa.

SCHROEDER, W. C. 520
    U. S. Bureau of Mines, Washington, D. C.

SEAGREN, G. W. 430
    Mellon Institute, Pittsburgh, Pa.

SODERBERG, GUSTAF 837, 845
    Graham, Crowley and Associates, Inc., Jenkintown, Pa.

SOLBERG, H. L. 511
    Department of Mechanical Engineering, Purdue University, Lafayette, Ind.

SPELLER, F. N. 496, 506
    Consultant, Pittsburgh, Pa.

STEARN, R. F. 1103
    Research Laboratory, The International Nickel Co., Inc., Bayonne, N. J.

STRADER, JAMES E. 323, 829
    Battelle Memorial Institute, Columbus, Ohio

LIST OF CONTRIBUTORS

	PAGE
TRACY, A. W.	105, 110

The American Brass Company, Waterbury, Conn.

UHLIG, H. H. 20, 125, 165
  Corrosion Laboratory, Department of Metallurgy, Massachusetts Institute of Technology, Cambridge, Mass.

VANDE BOGART, L. G. 747
  Research and Development Laboratories, Crane Co., Chicago, Ill.

VOSBURGH, F. J. 348, 723
  National Carbon Co., Inc., New York, N. Y.

WALDRON, LEO J. 970
  Federal Housing Authority, Washington, D. C.

WARNER, J. C. 905, 1126
  Department of Chemistry, Carnegie Institute of Technology, Pittsburgh, Pa.

WESCOTT, BLAINE B. 578, 987
  Gulf Research and Development Co., Pittsburgh, Pa.

WESLEY, W. A. 481, 817, 959
  Research Laboratory, The International Nickel Co., Inc., Bayonne, N. J.

WHIRL, S. F. 520
  Power Stations Department, Duquesne Light Co., Pittsburgh, Pa.

WILKINS, R. A. 61
  Revere Copper and Brass, Inc., Rome, N. Y.

WISE, E. M. 112, 299, 699
  The International Nickel Co., Inc., New York, N. Y.

WISSLER, W. A. 57, 618
  Union Carbide and Carbon Research Laboratories, Inc., Niagara Falls, N. Y.

YOUNG, G. H. 441, 878
  Mellon Institute, Pittsburgh, Pa.

CONTENTS

Glossary of Terms Used in Corrosion. xxv
Abbreviations xxxiii

Section I. Corrosion Theory

An Outline of Corrosion Mechanisms, Including the Electrochemical
Theory 3
    by U. R. Evans
Fundamentals of Oxidation and Tarnish 11
    by H. A. Miley
Passivity 20
    by H. H. Uhlig and R. B. Mears
Effect of Crystal Orientation on Corrosion 33
    by Allan T. Gwathmey

Section II. Corrosion in Liquid Media, the Atmosphere, and Gases

Aluminum and Aluminum Alloys 39
    by R. B. Mears
Beryllium 56
    by W. J. Kroll
Cadmium, cross reference to 57
Chromium, cross reference to 57
Cobalt Alloys 57
    by W. A. Wissler
Columbium 60
    by Clarence W. Balke
Copper 61
    by R. A. Wilkins and R. H. Jenks
Copper Alloys
  Copper-Zinc Alloys 69
    by C. L. Bulow
  Copper-Nickel Alloys 85
    by C. L. Bulow
  Copper-Tin Alloys 96
    by C. L. Bulow
  Copper-Silicon Alloys 105
    by A. W. Tracy
  Copper-Beryllium Alloys — Beryllium Copper 110
    by A. W. Tracy
Gold and Gold Alloys 112
    by E. M. Wise

INDIUM AND INDIUM ALLOYS 119
    by G. B. Barton
ATMOSPHERIC CORROSION OF IRON 120
    by C. P. Larrabee
IRON AND STEEL 125
    by H. H. Uhlig
IRON ALLOYS
    CHROMIUM-IRON ALLOYS 143
        by H. L. Maxwell
    CHROMIUM-NICKEL AUSTENITIC STAINLESS STEELS 150
        by Russell Franks
    PITTING IN STAINLESS STEELS AND OTHER PASSIVE METALS 165
        by H. H. Uhlig
    STRESS CORROSION CRACKING IN STAINLESS ALLOYS 174
        by M. A. Scheil
    16% CHROMIUM–2% NICKEL STAINLESS STEEL 182
        by W. O. Binder
    CHROMIUM-MANGANESE AND CHROMIUM-MANGANESE-NICKEL STAINLESS
    STEELS 188
        by Russell Franks
    NICKEL-IRON ALLOYS 194
        by F. L. LaQue
    SILICON-IRON ALLOYS 201
        by W. T. Bryan
LEAD AND LEAD ALLOYS 207
    by G. O. Hiers
MAGNESIUM AND MAGNESIUM ALLOYS 218
    by W. S. Loose
MOLYBDENUM 252
    by W. J. Kroll
NICKEL 253
    by W. Z. Friend
NICKEL ALLOYS
    NICKEL-COPPER ALLOYS 266
        by W. Z. Friend
    NICKEL-CHROMIUM ALLOYS 281
        by W. Z. Friend
    NICKEL-MOLYBDENUM AND NICKEL-MOLYBDENUM-IRON-(CHROMIUM)
    ALLOYS 294
        by Burnham E. Field
PLATINUM GROUP METALS AND ALLOYS 299
    by E. M. Wise
RHENIUM 314
    by W. J. Kroll
SILVER AND SILVER ALLOYS 314
    by Robert H. Leach
TANTALUM 320
    by Clarence W. Balke
TIN 323
    by Bruce W. Gonser and James E. Strader

CONTENTS

TITANIUM	329
by W. J. Kroll	
TUNGSTEN	330
by W. J. Kroll	
URANIUM	331
by W. J. Kroll	
ZINC	331
by E. A. Anderson and C. E. Reinhard	
ZIRCONIUM	347
by Foote Mineral Company	
NON-METALLIC MATERIALS AND SEMI-METALS	
INTRODUCTION	348
CARBON AND GRAPHITE	348
by F. J. Vosburgh	
CHEMICAL STONEWARE	353
by James A. Lee	
GLASS AND VITREOUS SILICA	354
by James A. Lee	
PLASTICS	359
by James A. Lee	
PORCELAIN	365
by James A. Lee	
NATURAL AND SYNTHETIC RUBBER	365
by James A. Lee	
WOOD	372
by James A. Lee	
BORON	378
by W. J. Kroll	
SILICON	380
by W. J. Kroll	

Section III. Special Topics in Corrosion

CORROSION BY SEA WATER	
BEHAVIOR OF METALS AND ALLOYS IN SEA WATER	383
by F. L. LaQue	
PAINTS AND ORGANIC COATINGS FOR SEA-WATER EXPOSURE	430
by G. W. Seagren	
MACRO-ORGANISMS IN SEA WATER AND THEIR EFFECT ON CORROSION	433
by William F. Clapp	
ANTI-FOULING MEASURES	441
by G. H. Young	
CORROSION BY SOILS	446
by K. H. Logan	
CORROSION BY MICRO-ORGANISMS IN AQUEOUS AND SOIL ENVIRONMENTS	466
by Raymond F. Hadley	
FUNDAMENTAL BEHAVIOR OF GALVANIC COUPLES	481
by W. A. Wesley and R. H. Brown	
HOT AND COLD WATER SYSTEMS	496
by F. N. Speller	

DEACTIVATION AND DEAERATION OF WATER	506
by F. N. Speller	
CORROSION BY HIGH-TEMPERATURE STEAM	511
by G. A. Hawkins and H. L. Solberg	
BOILER CORROSION	520
by Max Hecht, E. P. Partridge, W. C. Schroeder, and S. F. Whirl	
CORROSION OF STEAM CONDENSATE LINES	538
by Leo F. Collins	
CONDENSER CORROSION	545
by C. A. Gleason	
CORROSION BY LUBRICANTS	559
by C. M. Loane	
EFFECT OF MECHANICAL FACTORS ON CORROSION	
STRESS CORROSION	569
by H. R. Copson	
CORROSION FATIGUE	578
by Blaine B. Wescott	
FRETTING CORROSION	590
by J. O. Almen	
CAVITATION EROSION	597
by S. Logan Kerr	
CORROSION BY STRAY CURRENTS	601
by Scott P. Ewing	
CORROSION-RESISTANT ANODES	606
by Colin G. Fink	
CORROSION OF LEAD AND LEAD-ALLOY CABLE SHEATHING	609
by R. M. Burns	

Section IV. High-Temperature Corrosion

ALUMINUM AND ALUMINUM ALLOYS	617
by R. B. Mears	
CHROMIUM, cross reference to	618
COBALT ALLOYS	618
by W. A. Wissler	
COLUMBIUM	620
by Clarence W. Balke	
COPPER AND COPPER ALLOYS	621
by Frederick N. Rhines	
IRON AND STEEL	630
by J. B. Austin and R. W. Gurry	
IRON ALLOYS	
ALUMINUM-IRON ALLOYS	638
by W. E. Ruder	
CHROMIUM-IRON, AUSTENITIC CHROMIUM-NICKEL-IRON, AND RELATED HEAT-RESISTANT ALLOYS	640
by W. O. Binder	
CHROMIUM-SILICON-IRON ALLOYS	664
by C. L. Clark	

CONTENTS xxi

CHROMIUM-ALUMINUM-IRON ALLOYS 667
    by W. E. Ruder
MAGNESIUM AND MAGNESIUM ALLOYS 670
    by W. S. Loose
NICKLE AND NICKEL-COOPER, NICKEL-MANGANESE, AND RELATED HIGH-NICKEL
    ALLOYS 675
    by W. A. Mudge
HIGH-NICKEL CHROMIUM-(IRON) ALLOYS 683
    by M. A. Hunter
NOBLE METALS AND THEIR ALLOYS 699
    by E. M. Wise . .
SILVER 718
    by Robert H. Leach
TANTALUM. 720
    by Clarence W. Balke
ZINC 722
    by E. A. Anderson
NON-METALLIC MATERIALS
    INTRODUCTION 722
    CARBON AND GRAPHITE 723
        by F. J. Vosburgh
    REFRACTORIES 723
        by James A. Lee
OTHER METALS, cross references to 724

Section V. High-Temperature Resistant Materials. . 729
    by G. F. Geiger

Section VI. Chemical Resistant Materials . . . 747
by L. G. Vande Bogart, Chairman, and C. L. Bulow, M. H. Brown, G. O. Hiers,
F. L. LaQue, H. L. Maxwell, R. B. Mears, W. E. Pratt, G. A. Sands

Section VII. Corrosion Protection

METALLIC COATINGS
    ZINC COATINGS ON STEEL 803
        by William Blum and Abner Brenner
    NICKEL AND CHROMIUM COATINGS 817
        by W. A. Wesley
    TIN COATINGS 829
        by Bruce W. Gonser and James E. Strader
    CADMIUM COATINGS. 837
        by Gustaf Soderberg
    LEAD COATINGS 845
        by Gustaf Soderberg
INORGANIC COATINGS
    CHEMICAL CONVERSION COATINGS
        ANODIC TREATMENT OF METALS 857
            by Junius D. Edwards

CHROMATE COATINGS ON ZINC	862
by E. A. Anderson	
CHROMATE COATINGS ON MAGNESIUM ALLOYS	864
by W. S. Loose	
PHOSPHATE COATINGS	867
by V. M. Darsey	
CHEMICAL OXIDE COATINGS	871
by V. M. Darsey	
VITREOUS ENAMELS	872
by A. I. Andrews	
ORGANIC COATINGS	
INTRODUCTION	878
by G. H. Young and Collaborators	
PAINTS AND VARNISHES	884
by Richard J. Eckart	
LACQUERS AND BAKING ENAMELS	890
by P. O. Powers	
BITUMINOUS COATINGS	898
by W. F. Fair, Jr.	
RUBBER AND RUBBER-LIKE COATINGS	902
by W. K. Schneider	
INHIBITORS AND PASSIVATORS	905
by G. G. Eldredge and J. C. Warner	
TEMPORARY CORROSION-PREVENTIVE COATINGS—SLUSHING COMPOUNDS	916
by W. E. Campbell	
FUNDAMENTALS OF CATHODIC PROTECTION	923
by John M. Pearson	
APPLICATION OF CATHODIC PROTECTION	935
by Robert Pope	

Section VIII. Corrosion Testing

LABORATORY TESTS	
GENERAL DISCUSSION OF LABORATORY CORROSION TESTING	953
by Carl W. Borgmann	
TOTAL IMMERSION TESTS	959
By W. A. Wesley	
ALTERNATE IMMERSION TESTS	965
by H. L. Burghoff	
SALT-SPRAY TEST	970
by William Blum and Leo J. Waldron	
POTENTIAL MEASUREMENTS AND ELECTROCHEMICAL TESTS	978
by H. E. Haring	
TESTS FOR CORROSION FATIGUE	987
by Blaine B. Wescott	
TESTS FOR DAMAGE BY CAVITATION	993
by S. Logan Kerr	
HIGH-TEMPERATURE TESTS	997
by F. E. Bash	
GALVANIC COUPLE TESTS	1002
by C. H. Sample	

CONTENTS

HIGH-HUMIDITY AND CONDENSATION TESTS 1006
    by K. G. Compton
TESTS FOR STRESS CORROSION CRACKING 1009
    by H. R. Copson
TESTS FOR BRASS 1015
    by C. L. Bulow
TESTS FOR STAINLESS STEELS 1016
    by H. A. Pray
STEAM TEST FOR ZINC 1023
    by E. A. Anderson
TESTING OF ORGANIC COATINGS 1023
    by K. G. Compton and R. J. Phair
TESTS FOR METAL COATINGS 1030
    by William Blum
TESTS FOR CORROSION BY LUBRICANTS 1034
    by C. M. Loane
SOIL CORROSION TESTS 1038
    by I. A. Denison
FIELD AND SERVICE TESTS
  ATMOSPHERIC EXPOSURE TESTS 1043
    by C. P. Larrabee
  SOIL EXPOSURE TESTS 1048
    by I. A. Denison
  CHEMICAL PLANT EQUIPMENT TESTS 1052
    by O. B. J. Fraser
  WATER PIPE SERVICE TESTS 1058
    by V. V. Kendall
  SEA-WATER CORROSION TESTS 1060
    by F. L. LaQue
  FIELD TESTING OF PAINTS 1071
    by Harley A. Nelson
PREPARATION AND CLEANING OF SPECIMENS 1077
    by B. B. Knapp
STATISTICAL METHODS 1083
    by G. G. Eldredge

Section IX. Miscellaneous Information

ILLUSTRATIONS OF TYPICAL FORMS OF CORROSION 1103
    by R. F. Stearn
CHARACTERISTICS OF SEA WATER 1111
    by Alfred C. Redfield
HIGH-TEMPERATURE EQUILIBRIA FOR OXIDATION AND CARBURIZATION OF IRON . 1122
    by J. B. Austin
GENERAL TABLES 1126
INDEX 1163

GLOSSARY OF TERMS USED IN CORROSION

Prepared jointly by the Editorial Advisory Board for the *Corrosion Handbook* and Subcommittee V of the American Coordinating Committee on Corrosion. The members of Subcommittee V of A.C.C.C. are:

W. E. Campbell, Chairman	A. W. Tracy
Marc Darrin	H. H. Uhlig
Robert Pope	C. J. Walton
R. F. Stearn	J. C. Warner

GLOSSARY

Aeration Cell (Oxygen Cell). An electrolytic cell, the emf of which is due to a difference in air (oxygen) concentration at one electrode as compared with that at another electrode of the same material.

Aggressive Carbon Dioxide. Free carbon dioxide in excess of the amount necessary to prevent precipitation of calcium as calcium carbonate.

Anaerobic. Free of air or uncombined oxygen.

Anion. A negatively charged ion of an *electrolyte*,* which migrates toward the *anode* under the influence of a potential gradient.

Anode. The electrode of an electrolytic cell at which *oxidation* occurs. In corrosion processes, usually the electrode that has the greater tendency to go into solution. Typical anodic processes are *anions* giving up electrons; metal atoms becoming *ions* in solution or forming an insoluble compound of the metal; and the oxidation of an element or group of elements from a lower to a higher valence state.

Anode Corrosion Efficiency. The ratio of the actual corrosion of an *anode* to the theoretical corrosion calculated from the quantity of electricity which has passed.

Anodic Polarization. That portion of the *polarization* of a cell which occurs at the *anode*.

Anolyte. The electrolyte of an electrolytic cell adjacent to the *anode*.

Cathode. The electrode of an electrolytic cell at which *reduction* occurs. In corrosion processes, usually the area that is not attacked. Typical cathodic processes are *cations* taking up electrons and being discharged, oxygen being reduced, and the reduction of an element or group of elements from a higher to a lower valence state.

Cathodic Corrosion. *Corrosion* resulting from a cathodic condition of a structure, usually caused by the reaction of alkaline products of electrolysis with an amphoteric metal.

Cathodic Polarization. That portion of the *polarization* of a cell which occurs at the *cathode*.

Cathodic Protection. Reduction or prevention of *corrosion* of a metal surface by making it cathodic, for example, by the use of sacrificial anodes or impressed currents.

Catholyte. The *electrolyte* of an electrolytic cell adjacent to the *cathode*.

Cation. A positively charged *ion* of an *electrolyte*, which migrates toward the *cathode* under the influence of a potential gradient.

Caustic Embrittlement. *Embrittlement* of a metal resulting from contact with an alkaline solution.

Cavitation Erosion. Damage of a material associated with the formation and collapse of cavities in the liquid at a solid-liquid interface.

Chalking. The development of loose removable powder at or just beneath a coating surface.

Checking. The development of slight breaks in a coating which do not penetrate to the underlying surface. Checking may be described as visible (as seen by the naked eye) or as miscroscopic (as seen under magnification of 10 diameters).

Chemical Conversion Coating. A protective or decorative coating produced in situ by chemical reaction of a metal with a chosen environment.

---

* Italicized words are defined elsewhere in the Glossary.

GLOSSARY

Coefficient of Corrosion. The reciprocal of *anode corrosion efficiency*. A term used in applied *cathodic protection*.

Concentration Cell. An electrolytic cell, the emf of which is due to a difference in concentration of the electrolyte or active metal at the *anode* and the *cathode*.

Concentration Polarization. That portion of the *polarization* of a cell produced by concentration changes resulting from passage of current through the *electrolyte*.

Contact Corrosion (Crevice Corrosion). *Corrosion* of a metal at an area where contact is made with a material usually non-metallic.

Corrosion. Destruction of a metal by chemical or electrochemical reaction with its environment.

Corrosion Fatigue. Reduction of fatigue durability by a corrosive environment.

Corrosion Fatigue Limit. The maximum repeated stress endured by a metal without failure in a stated number of stress applications under defined conditions of *corrosion* and stressing.

Couple. A pair of dissimilar conductors in electrical contact.

Couple Action. See *Galvanic corrosion*.

Cracking (of Coating). Breaks in a coating which extend through to the underlying surface. Observation under a magnification of 10 diameters is recommended where there is difficulty in distinguishing between cracking and *checking*.

Crazing. A network of checks or cracks appearing on a surface.

Critical Humidity. The relative humidity above which the atmospheric corrosion rate of a given metal increases sharply.

Deactivation. The process of prior removal of the active corrosive constituents, usually oxygen, from a corrosive liquid by controlled corrosion of expendable metal or by other chemical means.

Decomposition Potential (or Voltage). The practical minimum potential difference necessary to decompose the *electrolyte* of a cell at a continuous rate.

Depolarization. The reduction of counter-emf by removing or diminishing the causes of *polarization*.

Deposit Attack. Corrosion occurring under or around a discontinuous deposit on a metallic surface.

Dezincification. Corrosion of a zinc alloy, usually brass, involving loss of zinc, and a residue or deposit in situ of one or more less active constituents, usually copper.

Differential Aeration Cell. See *Aeration cell*.

Drainage. Conduction of current (positive electricity) from an underground metallic structure by means of a metallic conductor.

1. *Forced Drainage.* Drainage applied to underground metallic structures by means of an applied emf or sacrificial anode.
2. *Natural Drainage.* Drainage from an underground structure to a more negative (more anodic) structure, such as the negative bus of a trolley substation.

Electrochemical Equivalent. The weight of an element or group of elements oxidized or reduced at one electrode of an electrolytic cell by the passage of a unit quantity of electricity. It is generally expressed in grams per coulomb.

Electrolysis. The production of chemical change in an *electrolyte* resulting from the passage of electricity.

Electrolyte. A chemical substance or mixture, usually liquid, containing *ions* which migrate in an electric field.

Electrolytic Cleaning. The process of degreasing or descaling a metal by making it an electrode in a suitable bath.

Electromotive Force Series (Emf Series). A list of elements arranged according to their *standard electrode potentials,* the sign being positive for elements whose potentials are cathodic to hydrogen and negative for those anodic to hydrogen. (This convention of sign, historically and currently used in European literature, has been adopted by the Electrochemical Society and by the National Bureau of Standards, and is employed in this book. The opposite convention of G. N. Lewis has been adopted by the American Chemical Society.)

Electronegative Potential. A potential corresponding in sign to those of the active or anodic members of the *Emf Series*. Because of existing confusion of sign in the literature, it is suggested that "anodic potential" be used whenever "electronegative potential" is implied. (See *Electromotive Force Series [Emf Series]*.)

Electropositive Potential. A potential corresponding in sign to potentials of the noble or cathodic members of the *Emf Series*. It is suggested that "cathodic potential" be used whenever "electropositive potential" is implied. (See *Emf Series.*)

Embrittlement. Severe loss of ductility of a metal or alloy.

Erosion. Destruction of metal or other material by the abrasive action of liquid or gas. Usually accelerated by the presence of solid particles of matter in suspension and sometimes by *corrosion.*

Exfoliation. Scaling off of a surface in flakes or layers.

Film. A thin, not necessarily visible layer of material.

Fogged Metal. Metal, the luster of which has been sharply reduced by a *film* of corrosion products.

Fretting Corrosion. *Corrosion* at the interface between two contacting surfaces accelerated by relative vibration between them of amplitude high enough to produce slip.

Galvanic Cell. A cell made up of two dissimilar conductors in contact with an *electrolyte* or two similar conductors in contact with dissimilar electrolytes. More generally, a galvanic cell converts energy liberated by a spontaneous chemical reaction directly into electrical energy.

Galvanic Corrosion. *Corrosion* associated with the current of a *galvanic cell* made up of dissimilar electrodes. Also known as couple action.

Galvanic Series. A list of metals and alloys arranged according to their relative potentials in a given environment.

Graphitization (Graphitic Corrosion). *Corrosion* of gray cast iron in which the metallic iron constituent is converted into corrosion products, leaving the graphite intact.

Hydrogen Embrittlement. *Embrittlement* caused by entrance of hydrogen into the metal, as for example through pickling or cathodic polarization.

Hydrogen Overvoltage. *Overvoltage* associated with the liberation of hydrogen gas.

Impingement Attack. *Corrosion* associated with turbulent flow of a liquid. For some metals the action is considerably accelerated by entrained bubbles in the liquid.

Inhibitor (Applied to Corrosion). A chemical substance or mixture which, when added to an environment usually in small concentration, effectively decreases *corrosion.*

GLOSSARY

Intercrystalline Corrosion. See *Intergranular corrosion*.
Intergranular Corrosion. Preferential *corrosion* at grain boundaries of a metal or alloy. Also called intercrystalline corrosion.
Internal Oxidation. The precipitation of one or more oxides of alloying elements beneath the external surface of an alloy as a result of oxygen diffusing into the alloy from an external source. Also known as subscale formation.
Ion. An electrically charged atom or group of atoms.

Liquation. The process of separating a fusible substance from one less fusible by heat.
Local Action. Corrosion caused by *local cells* on a metal surface.
Local Cell. A cell, the emf of which is due to differences of potential between areas on a metallic surface in an electrolyte.
Long-Line Current. Current (positive electricity) flowing through the earth from an anodic to a cathodic area which returns along an underground metallic structure. Usually used only where the areas are separated by considerable distance and where the current results from *concentration cell* action.

Matte Surface. A surface with low specular reflectivity.
Metallizing. The process of spraying a surface with a metal.
Metal Replacement. The deposition of a metal from a solution of its ions on a more anodic metal accompanied by solution of the latter metal. Also called "Immersion Plating."
Mill Scale. The heavy oxide layer formed during hot fabrication or heat treatment of metals. Especially applied to iron and steel.

Noble Metal. A metal which in nature occurs commonly in the free state. Also a metal or alloy whose corrosion products are formed with a low negative or a positive free energy change.
Noble Potential. A potential substantially cathodic to the standard hydrogen potential.

Open Circuit Potential. The measured potential of a cell from which no significant current flows in the external circuit.
Overvoltage. Difference between the potential of an electrode at which a reaction is actively taking place and another electrode at equilibrium for the same reaction.
Oxidation. Loss of electrons by a constituent of a chemical reaction.

Parting. The selective corrosion of one or more components of a solid solution alloy.
Parting Limit. The maximum concentration of a more noble component in an alloy, above which *parting* does not occur within a specific environment.
Passivator. An *inhibitor* which appreciably changes the potential of a metal to a more cathodic value.
Passive-Active Cell. A cell, the emf of which is due to the potential difference between a metal in an active state and the same metal in a passive state.
Passivity.
Definition 1. A metal active in the *Emf Series* or an alloy composed of such metals is considered passive when its electrochemical behavior becomes that of an appreciably less active or *noble metal*.
Definition 2. A metal or alloy is passive if it substantially resists corrosion in an environment where thermodynamically there is a large free energy decrease

associated with its passage from the metallic state to appropriate corrosion products.

Patina. A green coating consisting principally of basic sulfate and occasionally containing small amounts of carbonate or chloride, which forms on the surface of copper or copper alloys exposed to the atmosphere a long time.

pH. A measure of hydrogen ion activity defined by

$$pH = \log_{10} \frac{1}{a_{H^+}},$$

where a_{H^+} = hydrogen ion activity = the molal concentration of hydrogen ions multiplied by the mean ion activity coefficient.

Pickle. A solution or process used to loosen or remove corrosion products such as scale and tarnish from a metal.

Pitting Erosion. See *Cavitation erosion*.

Pitting Factor. The depth of the deepest pit resulting from corrosion divided by the average penetration as calculated from weight loss.

Polarization. The production of counter-emf by products formed or by concentration changes resulting from passage of current through an electrolytic cell.

Prime Coat. A first coat of paint originally applied to improve adherence of the succeeding coat but now frequently containing a corrosion *inhibitor*.

Reaction Limit. The minimum concentration of an alloy component below which appreciable attack of the alloy takes place in a given environment, but above which the alloy is corrosion resistant.

Reduction. Gain of electrons by a constituent of a chemical reaction.

Relative Humidity. The ratio, expressed as a percentage, of the amount of water present in a given volume of air at a given temperature to the amount required to saturate the air at that temperature.

Rusting. *Corrosion* of iron resulting in the formation of products on the surface consisting largely of hydrous ferric oxide.

Scaling. The formation at high temperatures of partially adherent layers of corrosion products on a metal surface.

Season Cracking. Cracking resulting from combined *corrosion* and internal stress. A term usually applied to *stress corrosion cracking* of brass.

Self-corrosion. See *Local action*.

Slushing Compound. A non-drying oil, grease, or similar organic compound which, when coated over a metal, affords at least temporary protection against *corrosion*.

Spalling. The chipping or fragmenting of a surface or surface coating caused, for example, by differential thermal expansion or contraction.

Standard Potential (Standard Electrode Potential). The reversible potential for an electrode process when all products and reactants are at unit activity on a scale in which the standard potential for hydrogen is zero.

Stray Current Corrosion. *Corrosion* caused by current through paths other than the intended circuit or by any extraneous current in the earth.

Stress Corrosion. *Corrosion* of a metal accelerated by stress.

Stress Corrosion Cracking. Cracking resulting from the combined effect of *corrosion* and stress.

Subscale Formation. See *Internal oxidation*.

Tarnish. Discoloration of a metal surface due to formation of an adherent continuous *film* of corrosion products.

Tuberculation. The formation of localized corrosion products scattered over the surface in the form of knoblike mounds.

Underfilm Corrosion. Corrosion that occurs under lacquers and similar organic films in the form of randomly distributed hairlines (most common) or spots.

Weld Decay. Corrosion notably of austenitic chromium steels at specific zones away from a weld.

ABBREVIATIONS

absolute	abs.	milligram	mg
alternating current	a-c	milligrams per square deci-	
ampere	amp	meters per day	mdd
atmosphere	atm.	milliliter	ml
		minute	min
boiling point	b. p.		
British thermal unit	Btu	natural convection	nat. convect.
		normal	N
centigrade	C		
centimeter	cm	ounce	oz
cubic	cu	ounces per square foot	oz/sq ft
cubic centimeter	cc		
		page	p.
decimeter	dm	parts per million	ppm
direct current	d-c	pound	lb
		pounds per square inch	psi
electromotive force	emf		
		relative humidity	R. H.
Fahrenheit	F	revolutions per minute	rpm
foot	ft	room temperature	R. T.
hour	hr	second	sec
		specific gravity	sp. gr.
inch	in.	square	sq
inches penetration per year	ipy	specimen	spec.
kilogram	kg	temperature	temp.
		temperature of environment	temp. envir.
logarithm (base 10)	log		
logarithm (natural)	ln	weight	wt.
melting point	m. p.	year	yr
milliampere	milliamp		

CORROSION THEORY

AN OUTLINE OF CORROSION MECHANISMS, INCLUDING THE ELECTROCHEMICAL THEORY

U. R. Evans*

DEFINITIONS

The word *corrosion* denotes destruction of metal by chemical or electrochemical action; a familiar example is the rusting of iron. Comminution by mechanical agencies — for instance, the grinding of iron to dust — may conveniently be termed *erosion*. Some of the most dangerous types of metallic wastage, however, result from processes that are partly chemical and partly mechanical. Purely chemical corrosion, although starting rapidly, often becomes slow as soon as an obstructive layer of corrosion product has been formed upon the metallic surface. If, however, this corrosion product is continually being cracked by bending, or removed by scraping or other mechanical operation, corrosion will continue unchecked at its original rapid rate. Familiar examples of this *conjoint action* are provided by corrosion fatigue breakage, and the impingement attack by bubbles passing into condenser tubes or by vacuum cavities generated from ships' propellers.

This chapter provides a short account of the mechanism of corrosion of metals immersed in liquid. No attempt is made to discuss oxidation at high temperatures or the corrosion of metals exposed to the atmosphere, which will be dealt with by other writers in the chapters which follow.

MECHANISMS OF REACTION

Direct Attack

Since the reaction of most metals with oxygen and/or water would result in a diminution of free energy, it may well be asked why metals do not quickly disappear in service. The commonest reason has already been suggested, namely, that a film of solid corrosion products tends to isolate the metal from the corrosive agency, sometimes before the film has attained visible thickness. Rapid destruction may be expected where the immediate corrosion product is soluble in any liquid present. Parsons[1] studied the behavior of several metals towards iodine dissolved in various organic solvents, and found that whenever the iodide of the metal under test was soluble in the organic liquid employed, corrosion continued apace; where the iodide was neither soluble nor peptizable, corrosion was slowed down by a film of solid iodide. Quantitative examination of the attack on silver by a chloroform solution of iodine, carried out by Evans and Bannister,[2] indicated that the rate of attack fell off with time as a film of silver iodide (insoluble in chloroform) developed, being at any moment inversely proportional to the film thickness.

Two-Stage Attack

Since most metallic oxides (except those of the alkali metals, alkaline earth metals, and thallium) are sparingly soluble in water, it might be expected that metals (other

---

* Cambridge University, England.

[1] L. B. Parsons, *J. Am. Chem. Soc.*, **47**, 1830 (1925).
[2] U. R. Evans and L. C. Bannister, *Proc. Roy. Soc.* (A), **125**, 370 (1929).

than the exceptions mentioned) would remain almost unattacked by distilled water containing oxygen. In general, attack by high-purity water and oxygen is indeed slow; Bengough, Stuart, and Lee[3] have shown how attack on zinc by water and oxygen falls off with time. But in some cases a metal may enter the liquid as one oxide or hydroxide, being precipitated as a less soluble oxide or hydroxide at a distance from the metal, so that a protective layer of the first (more soluble) oxide is not maintained. For instance, iron placed in distilled water under conditions of limited oxygen supply may enter the liquid as ferrous hydroxide, and be precipitated as the less soluble hydrated ferric oxide (yellow rust) at a slight distance from the surface. Such rust being formed out of physical contact with the metal will be non-protective (Fig. 1). If the supply of oxygen is sufficient to cause the hydrated ferric oxide to be formed in physical contact with the metal, attack will be "stifled." Mears and Evans,[4] in a statistical study of the behavior of drops of distilled water on iron surrounded by different mixtures of oxygen and nitrogen, found that the proportion of drops causing attack fell off steadily as the oxygen content was increased, although those drops which did produce attack acted more rapidly at high oxygen concentrations than at low ones.

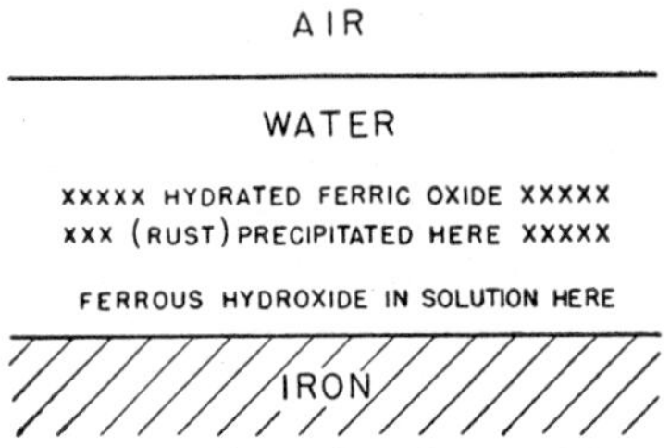

FIG. 1. Rusting of Iron in Pure Water. (The diagrams for this chapter were prepared by R. S. Thornhill.)

Somewhat analogous results had been found by Forrest, Roetheli, Brown, and Cox,[5] in their work on iron specimens rotating in water containing oxygen. They found that at low oxygen concentrations, where the corrosion product was loose, granular black magnetite, corrosion was roughly proportional to the oxygen concentration. At high concentrations, where the corrosion product was a gelatinous hydrated ferric oxide precipitated in close contact with the metal, so as to provide a clinging protective film, the corrosion rate was lower than this rule would predict.

The corrosion products formed in the first and second stages need not always differ in state of oxidation. Our knowledge of the behavior of lead in water containing oxygen[6] rather suggests that lead dissolves in such water as lead oxide or hydroxide, and is precipitated as basic carbonate by the small amount of carbon dioxide invariably present. If this second product is formed in physical contact with the lead, corrosion is slow; if precipitated at a distance, it is more rapid.

ELECTROCHEMICAL ATTACK

A common case in which a corrosion product is precipitated at a distance from the metal, thus failing to stifle further corrosion, occurs when the mechanism of attack is electrochemical. In one sense almost any chemical action can be regarded as electrochemical, being accompanied by the transfer or displacement of electrons. In this chapter, however, the term *electrochemical attack* will be applied only to cases

[3] G. D. Bengough, J. M. Stuart, and A. R. Lee, *Proc. Roy. Soc.* (A), **116**, 449 (1927).
[4] R. B. Mears and U. R. Evans, *Trans. Faraday Soc.*, **31**, 527 (1935).
[5] H. O. Forrest, B. E. Roetheli, R. H. Brown, and G. L. Cox, *Ind. Eng. Chem.*, **22**, 1197 (1930); **23**, 350, 650, 1010, 1012 (1931).
[6] J. F. Liverseege and A. W. Knapp, *J. Soc. Chem. Ind.*, **39**, 30T (1920). O. Bauer and G. Schikorr, *Mitt. deut. Materialprüfungsanstalte.*, **28**, 67 (1936). O. Heckler and H. Hanemann, *Z. Metallkunde*, **30**, 410 (1938). G. Schikorr, *Korrosion u. Metallschutz*, **16**, 181 (1940). E. A. G. Liddiard and P. E. Bankes, *J. Soc. Chem. Ind.*, **63**, 39–48 (1944).

connected with spatially separated anodic and cathodic areas, so that corrosion is accompanied by electric currents flowing for perceptible distances through the metal.

Consider, for instance, an article (Fig. 2) consisting partly of iron and partly of copper immersed in sodium chloride solution below air. A current will pass between the iron as anode and the copper as cathode, its strength being determined by the rate of arrival of oxygen, the cathodic stimulator, at the copper surface. The two reaction products, ferrous chloride at the anode and sodium hydroxide at the cathode, will meet in the middle of the liquid, yielding ferrous hydroxide, which will take up further oxygen to produce hydrated ferric oxide (yellow rust). The final result is, therefore, a combination of iron, oxygen, and water to give hydrated iron oxide. It should be noted, however, that the iron passes into the liquid at one point, the oxygen is taken up at a second, and the rust appears at a third, so that the precipitation cannot stifle further attack. It is the fact that solid corrosion product often appears at a distance from the seat of corrosion which renders electrochemical action so dangerous. In those cases where electrochemical action would yield a sparingly soluble body as the direct anodic product (e.g. lead in a sulfate solution), the attack will stifle itself at the outset

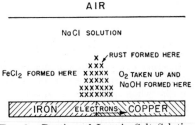

Fig. 2. Rusting of Iron in Salt Solution at Junction with Copper.

The presence of a second metal is not needed for electrochemical action. Iron covered with mill scale suffers severe attack at small interruptions in the scale when placed, say, in salt water (Fig. 3). Here the mill scale acts as cathode and the iron exposed at the discontinuity as anode, and the attack is made more intense by the fact that the anode is small compared to the cathode. The large amount of oxygen reaching the large cathodic area will usually permit a fairly high current to flow, and since its effect is concentrated on the small anodes, the attack will produce serious pitting at breaks in the scale.

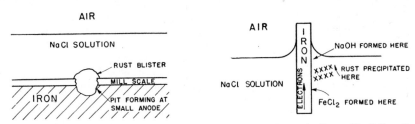

Fig. 3. Pitting at Break in Mill Scale.

Fig. 4. Rusting of Iron Partially Immersed in Salt Solution.

Even the presence of mill scale is unnecessary for electrochemical action. A piece of bright iron partially immersed in sodium chloride solution (Fig. 4) will at the outset develop numerous anodic and cathodic areas interspersed all over the surface, owing to local chemical or physical differences in or on the metal. However, since the oxygen needed for the main cathodic reaction will readily be replenished only near the water line, the cathodes on the lower part of the specimen will cease to function, and these parts will become wholly anodic. Since the total cathodic current must be equal to the total anodic current, the part near the water line will become predominantly cathodic, and rust will be precipitated at the level where the cathodic and anodic zones meet.

Currents set up through differences of oxygen supply are generally called *differential aeration currents*. There are, of course, many other factors that can set up potential differences between different parts of a metal, and these have been investigated by Mears and Brown.[7] Prolonged discussion regarding the electrochemical mechanism of corrosion has taken place between the Laboratories at Teddington and Cambridge (England), leading in 1938 to an agreed statement[8] by four authors, two from each laboratory.

PROOF OF THE ELECTROCHEMICAL CHARACTER OF CORROSION

In order to ascertain whether electric currents are really flowing over the surface of corroding metals in quantity sufficient to account, in the sense of Faraday's Law, for the whole attack, several series of experiments have been carried out in the author's laboratory. Hoar[9] studied the action of potassium chloride solution on partially immersed plates of a steel that showed reproducible boundaries of the corroding area. In some experiments, he cut the specimens along the boundaries, protected the edges with a wax mixture, and passed known currents between two segments, noting the relation between current and the potential of the cathodic segment. This relation enabled him, by measuring the potential of uncut specimens immersed in the same liquid, to know what current was passing, and he found that this current was almost exactly equivalent to the corrosion that the loss of weight showed to be taking place.

Later Agar[10] developed a different method for measuring electric currents flowing on zinc partly immersed in sodium chloride solution. He traced out equipotential lines as contours on a chart, and calculated the current from the distances between these lines, which will be crowded together when the current is high and far apart when it is low. The current measured was sufficient to account for the whole of the loose corrosion product formed. A small amount of adherent matter formed above the water line seems to have a slightly different mode of origin.

Previously Thornhill had carried out several researches with *dielectrodes*. In his work on partially immersed zinc,[11] the dielectrode consisted of two silver-silver chloride electrodes, each provided with a tubulus. The open end of one tubulus was pressed against any point on a corroding zinc specimen which it was desired to examine; the end of the other tubulus remained a short distance from the zinc surface. The two electrodes were joined to a sensitive galvanometer. Clearly, in such a case, if the point were anodic, the galvanometer would register a current in one direction, whereas, if it were cathodic, the deflection would be in the opposite sense. Thornhill found that corrosion was observed only at those points that the electrical instrument showed to be anodic; at cathodic or neutral points, no attack took place. Although purely qualitative, this investigation pointed very definitely to the fact that the corrosion of zinc partially immersed in sodium chloride was solely of an electrochemical character.

A dielectrode can, however, furnish quantitative results. Thornhill[12] measured the

---

[7] R. B. Mears and R. H. Brown, *Ind. Eng. Chem.*, **33**, 1001 (1941).
[8] G. D. Bengough, U. R. Evans, T. P. Hoar, and F. Wormwell, *Chem. Ind*, p. 1043 (1938).
[9] T. P. Hoar and U. R. Evans, *Proc. Roy. Soc.* (A), **137**, 343 (1932).
[10] J. N. Agar, unpublished work, summarized by U. R. Evans, *J. Iron Steel Inst.*, **141**, 221P (1940).
[11] R. S. Thornhill and U. R. Evans, *J. Chem. Soc*, p. 2109 (1938).
[12] R. S. Thornhill and U. R. Evans, *J. Chem. Soc.*, p. 614 (1938).

currents flowing around a scratch line made on iron wetted with sodium bicarbonate solution of a concentration chosen to cause rusting along the scratch but not elsewhere. He found that currents were actually flowing between the scratch line as anode and the uncorroded region around as cathode, and somewhat exceeded the strength needed to account for the measured corrosion, assuming that the iron was entering the solution solely as ferrous salts. It was concluded that the electrochemical oxidation of ferrous to ferric ions contributed appreciably to the current flowing.

A very interesting study of corrosion around scratch lines was carried out on aluminum.[13] As is well known, the corrosion of aluminum usually starts at small scratches or blemishes representing weak places in the invisible oxide film. By coating all the weak places on one piece of aluminum with a wax mixture, and then covering up the whole surface except the weak places on a second piece connected through an ammeter to the first, Brown and Mears measured the current passing between the weak places and the film-covered surface, and found it to be the electrical equivalent of the corrosion actually proceeding. They also obtained correlation of corrosion rate with the current flowing between naturally occurring pits and unattacked areas on aluminum surfaces.

MECHANISM OF CORROSION OF IRON FULLY IMMERSED FOR LONG PERIODS IN WATER OR SALT SOLUTIONS

In the cases studied above, it is fairly certain that corrosion is connected with electric currents passing through relatively large and stable anodic and cathodic areas. In the case of a horizontal iron surface kept fully immersed for long periods, the situation is less certain. Bengough and his colleagues[14] have measured the corrosion of horizontal iron or steel disks over periods of several years. In the case of steel immersed in N potassium chloride solution, they found that during the first 500 to 600 days the metal is covered with a thin layer of black magnetite, overlaid by a loose mass consisting mainly of hydrated ferric oxide, that can easily be shaken off without affecting the rate of corrosion. Later a gradual change occurs to a more coherent form that cannot be shaken off, and this transformation is accompanied by a steady fall of corrosion rate. Since many of Bengough's specimens ultimately became corroded over the whole surface, it is clear that corrosion is not connected with large stable anodic and cathodic areas. One possible view is that large anodic and cathodic areas do exist, but that their polarity changes, so that every point on the metal becomes anodic in turn, thus explaining why the whole surface ultimately suffers attack. Another possible explanation is that the mechanism is not electrochemical at all, but consists of a two-stage oxidation of the form suggested above. A third explanation has been put forward by Herzog,[15] who found that iron placed in soft or distilled water also became covered with a film of magnetite that was fairly adherent, but not sufficiently continuous to prevent attack on the iron below, and was in turn covered with a layer of hydrated ferric oxide (yellow rust). Herzog believes that the magnetite forms the cathode of the corrosion couple, the iron below

---

[13] R. B. Mears and R. H. Brown, *Trans. Electrochem. Soc.*, **74**, 495 (1938); **81**, 455 (1942). See also similar work on steel, magnesium, and stainless steels by L. J. Benson, R. H. Brown, and R. B. Mears, *Trans. Electrochem. Soc.*, **76**, 259 (1939); R. H. Brown and R. B. Mears, *Trans. Faraday Soc.*, **35**, 467 (1939).

[14] G. D. Bengough, J. M. Stuart, A R. Lee, and F. Wormwell, *Proc. Roy. Soc.* (A), **116**, 425 (1927); **121**, 88 (1928); **127**, 42 (1930); **131**, 494 (1931); **134**, 308 (1931); **140**, 399 (1933). G. D. Bengough and F. Wormwell, *Rept. Corrosion Committee Iron Steel Inst.*, **3**, 123 (1935); **4**, 213 (1936); *J. Iron Steel Inst.*, **131**, 285 (1935).

[15] E. Herzog, *Bull. soc. chim.*, **3**, 1530 (1936); **5**, 187 (1938).

it being the anode and suffering attack. The loose yellow rust acts as depolarizer (cathodic stimulator), being presumably reduced to hydrated magnetite, which may either be reoxidized by the oxygen dissolved in the solution, or may lose water, reinforcing the film of anhydrous magnetite. Herzog supports his views by special experiments in which it was found that contact with magnetite does greatly accelerate the rate of corrosion of iron by water; contact with rust produces a much smaller effect.

MECHANISM OF ATTACK BY ACIDS

The attack by non-oxidizing acids will proceed rapidly even in the absence of oxygen. The cathodic reaction here consists in the elimination of hydrogen according to the equation

$$2H^+ + 2e^- = H_2$$

instead of the absorption of oxygen

$$\tfrac{1}{2}O_2 + H_2O + 2e^- = 2(OH^-)$$

which is the principal cathodic change in nearly neutral solutions. Both cathodic changes lead to a rise in pH value; but the first is accompanied by liberation of hydrogen, and the second by absorption of oxygen. The anodic reaction may in either case be written (assuming the metal to be divalent, as is usually the case)

$$M = M^{++} + 2e^-$$

In the hydrogen-evolution type of attack, the cathodes, instead of being large areas, will usually be small points of low hydrogen overpotential. The distinction between the cathode operating in two types of attack has been emphasized by Brennert.[16] Palmaer[17] considers that when acid attacks cast iron or a malleable casting the grains of graphite act as cathodes, although Thiel and Eckell[18] prefer the view that the furrows along the edges of the flakes, rather than the flakes themselves, constitute points favorable to hydrogen evolution. Hoar and his colleagues[19] have obtained evidence that, in the action of organic acids on steel, cementite acts as cathode, since steel containing massive cementite is more susceptible to such acids, *ceteris paribus*, than steel free from this constituent. However, the sulfur content of the iron has a greater influence on attack by acids, since hydrogen sulfide stimulates the anodic reaction. Hoar and his colleagues showed that if sufficient copper is present to lock up the sulfur as relatively stable copper sulfide, susceptibility to acid is greatly diminished.

In the attack by oxidizing acids, such as nitric acid, the cathodic reaction consists in the reduction of the oxidizing agent — a change which is sometimes autocatalytic.[20] In such cases there is no reason to believe that there are distinct anodic and cathodic areas. Probably the anodic and cathodic portions of the change proceed alternately at the same point or concurrently at points exceedingly close together.

Where corrosion by an acid would lead to a sparingly soluble salt, the attack will generally tend to stifle itself. The resistance of lead to sulfuric acid, of silver to hydrochloric acid, and of magnesium to hydrofluoric acid is connected with the low solubilities of lead sulfate, silver chloride, and magnesium fluoride. But sometimes

---

[16] S. Brennert, *Trans. Electrochem. Soc.*, **76**, 231 (1939).

[17] W. Palmaer, *The Corrosion of Metals* (Svenska Bokhandelscentralen AB), **1**, 176 (1929); **2**, 14 (1931).

[18] A. Thiel and J. Eckell, *Z. Electrochem.*, **33**, 385 (1927).

[19] T. P. Hoar, D. Havenhand, T. N. Morris, and W. B. Adam, *J. Iron Steel Inst.*, **133**, 239P (1936); **140**, 55P (1939); **144**, 133P (1941).

[20] H. J. T. Ellingham, *J. Chem. Soc.*, p. 1565 (1932); U. R. Evans, *Trans. Faraday Soc.*, **40**, 120 (1944).

the velocity of dissolution of a compound may be more important than its equilibrium solubility. The surprising resistance of certain trivalent metals to nitric acid is connected with the *slow* rate of dissolution of sesquioxides. Films of ferric oxide, if once isolated from the metallic basis, will remain undissolved even in dilute sulfuric acid for hours,[21] although the same films, if present on iron dipped into the same acid, are dissolved within a minute. In the latter case the cell Fe : Acid : Fe_2O_3 is set up at every discontinuity in the film, and causes cathodic reduction of Fe_2O_3 to the ferrous state with immediate dissolution. The destruction of (visible) films of ferric oxide on iron can be avoided by any precaution that will prevent reduction to the ferrous state, such as anodic treatment, or the presence of an oxidizing agent (e.g., chromic acid) in the sulfuric acid. Now nitric acid is both oxidizing agent and acid. At high concentrations the oxidizing nature prevails, so that iron becomes passive, through the formation of an oxide film; at lower concentrations, the acidic nature prevails, and iron is violently attacked by dilute acid, although the same acid produces only a slow attack on aluminum — a metal without divalent compounds.

SOLUBLE INHIBITORS

It has been stated that electrochemical attack will stifle itself, just like any other type, when the solution is capable of yielding a sparingly soluble substance as the direct *anodic* product. The deposition of a sparingly soluble *cathodic* product will also slow down corrosion either by shutting out oxygen needed for the commonest cathodic reaction, or by "poisoning" spots favorable for other cathodic reactions, such as the production of hydrogen. Thus there are two separate classes of chemicals that can be introduced into waters to diminish corrosion. They are called *anodic* and *cathodic inhibitors,* according as they stifle the anodic or cathodic reaction.

Anodic inhibitors are efficient, and if added in sufficient quantity will practically prevent the corrosion of iron altogether. If added in insufficient quantity, they will often — for electrochemical reasons set forth elsewhere[22] — diminish the area affected more rapidly than they diminish the total corrosion, and thus actually increase the intensity of attack (the corrosion per unit area of the part affected). Thus anodic inhibitors may be dangerous, particularly in places containing corners or crevices where replacement is difficult, or points where sludge or debris is likely to collect. Typical anodic inhibitors include sodium carbonate, phosphate, silicate, or chromate. The first three lead to sparingly soluble iron compounds as direct anodic products. A chromate, however, will interact with any soluble ferrous salt, throwing down a mixture of ferric and chromic (hydrated) oxides. An iron surface that has been exposed to air for some time will already carry a film of ferric oxide, which, however, will be too discontinuous to confer protection against any ordinary natural water. If introduced into a water containing a chromate, the discontinuities will be "repaired" with the ferric-chromic oxide mixture, thrown down in close proximity to the metal, so that no loose rust will be formed.* Thus the film stripped by the iodine method from iron exposed to air for a long period before being introduced into potassium chromate solution will contain only a little chromium, whereas that formed on iron freshly abraded just before its introduction will contain a larger quantity, as shown by the analytical studies of Hoar.[23]

---

* For a different view of the action of chromates, see *Passivity*, p. 31.

[21] U. R. Evans, *J. Chem. Soc.*, p. 481 (1930); *Nature*, **126**, 130 (1930).
[22] U. R. Evans, *Trans. Electrochem. Soc.*, **69**, 213 (1936); E. Chyzewski and U. R. Evans, *Trans. Electrochem. Soc.*, **76**, 215 (1939).
[23] T. P. Hoar and U. R. Evans, *J. Chem. Soc.*, p. 2476 (1932).

Cathodic inhibitors never prevent corrosion so completely as anodic inhibitors, but are safer, being less likely to intensify attack if added in insufficient amount. The cathodic inhibitor that makes possible the use of steel pipes for conveying hard, natural waters is calcium bicarbonate. Such waters coat the pipe with calcium carbonate, owing to the rise in pH produced by the cathodic reaction. This chalky deposit is often subsequently converted to a clinging, protective type of rust by interaction, *in situ*, with anodically produced iron salts. But the stifling of attack will not occur if carbonic acid is present in excess of the amount needed for stabilizing the calcium bicarbonate,[24] since such an excess will prevent any immediate deposition of calcium carbonate. Since the quantity of carbonic acid needed for stabilization varies with the *cube* of the $(HCO_3)^-$ concentration,[25] it follows that a natural hard water that contains no "excess" or carbonic acid as it is pumped (for instance) from a chalk well, may sometimes come to contain such excess after softening by the base-exchange process.[26] It may thus lose its power of stifling corrosion. From the point of view of corrosion, base-exchange softening is, in general, less desirable than lime-soda softening.

The intense corrosion often set up by water containing inhibitors at crevices where metal surfaces approach one another is generally attributed to the fact that inhibitors (which are slowly used up in stifling corrosion) will not be readily replenished at remote corners.[27] Carius[28] considers that crevices, as such, displace the potential in the anodic direction, but his views have been criticized by Müller,[29] and later by Werner,[30] who has repeated and extended Carius' experiments, obtaining results that suggest that crevice corrosion is generally connected with differential aeration and other concentration cells.

INHIBITIVE PIGMENTS

Obviously the introduction of a freely soluble pigment, like sodium chromate, into a protective paint would be futile, since it would rapidly be washed away by the rain. On the other hand, a completely insoluble chromate (if such a substance existed) would probably fail to inhibit. Theory suggests the choice of a pigment with intermediate properties. Thus Sutton and Le Brocq[31] consider that, for the protection of magnesium alloys, strontium chromate strikes "a happy balance between having a reserve of chromate undissolved and a useful amount in solution." It is often considered that the vehicle chosen to carry an inhibitive pigment should not be so completely waterproof as to prevent the inhibitor passing into solution. Mühlberg,[32] for instance, states that inhibitive pigments, like red lead and zinc chromate, give better results in linseed oil than in media based on dehydrated castor and tung oils, which have lower water-absorption values.

The specific inhibitive nature of certain lead pigments used in paints — quite apart from water exclusion — has been demonstrated in laboratory experiments. Lewis

[24] J. R. Baylis, *J. Am. Water Works Assoc.*, **27**, 220 (1935). W. F. Langelier, *J. Am. Water Works Assoc.*, **28**, 1500 (1936). F. E. de Martini, *J. Am. Water Works Assoc.*, **30**, 85 (1938).
[25] G. Bodländer, *Z. Phys. Chem.*, **35**, 23 (1900).
[26] U. R. Evans, *Journées de la lutte contre la corrosion* (Special number of *Chimie et industrie*, Paris), p. 492C (1938). *Chem. and Ind.* (London), **19**, 867 (1941).
[27] R. B. Mears and U. R. Evans, *Trans. Faraday Soc.*, **30**, 417 (1934).
[28] C. Carius, *Ber. über die Korrosionstagung*, **5**, 61 (1935).
[29] W. J. Müller, *Ber. über die Korrosionstagung*, **5**, 71 (1935).
[30] M. Werner, *Chem. Fabrik*, **13**, 120 (1940).
[31] H. Sutton and L. F. Le Brocq, *J. Inst. Metals*, **57**, 223 (1935).
[32] G. Mühlberg, *Kolloid Beihefte*, **52**, 277 (1941).

and Evans[33] found that iron specimens shaken with air and water containing these pigments (no oil being present) suffered less corrosion than similar specimens shaken with air and water containing inert powders. The passivating power of these compounds has been clearly demonstrated by the potential measurements of Burns and his colleagues.[34]

FUNDAMENTALS OF OXIDATION AND TARNISH
H. A. Miley*

INTRODUCTION

Clean surfaces of many metals exposed to air oxidize quickly. The initial rapid reaction rate decreases as protective films are formed. Oxide films formed in air at ordinary temperatures are generally invisible, but when formed at higher temperatures they may pass through several orders of well-known interference colors. Prolonged heating at high temperatures produces thick surface layers of oxide or scale.

The usual tarnish films developed in air at room temperature are composed of mixtures of oxides and sulfur compounds, as illustrated by cuprous oxide and cuprous sulfide on copper, and by silver sulfide, silver sulfate, cuprous oxide, and cuprous sulfide on coin or sterling silver. However, when a pure metal is exposed to only one reactant, a tarnish film consisting of a single compound may be formed.

GROWTH OF OXIDE AND TARNISH FILMS

Permeability of the Surface Products

The rate of reaction of metals with formation of corrosion product coatings depends in most instances upon the relative permeability of the coating to the reactants. A porous corrosion product film is less protective than one non-porous. Whether or not corrosion product films are porous apparently depends on the relative volume of the corrosion product compared to the volume of the metal consumed in forming it. Pilling and Bedworth[1] showed that for oxidation if the ratio Md/mD (where M is the molecular weight of the oxide and D its density, m is the atomic weight of the metal multiplied by the number of metal atoms in the oxide formula, and d is the metal density) is *greater* than unity, the oxide coating is *protective;* when *less* than unity, it is *non-protective*. Metals accordingly can be divided into two groups.

1. One group is composed of the lighter metals with porous oxides smaller in volume than the equivalent metal consumed in producing the oxides. These metals at constant temperature oxidize at a rate nearly constant with time. Examples are calcium (CaO) and magnesium (MgO), which have ratios of Md/mD equal to 0.64 and 0.79, respectively. Data for magnesium showing a linear rate of oxidation with time are given in Fig. 1, p. 670.

2. Aluminum and the heavier metals with non-porous oxides of greater volume than the equivalent metal consumed make up the second group. Many metals of this group oxidize at a rate inversely proportional to the square root of the time or, as

* Radiation Laboratory, Massachusetts Institute of Technology, Cambridge, Mass.

[33] K. G. Lewis and U. R. Evans, *J. Soc. Chem. Ind.*, **53**, 25T (1934). Cf. J. E. O. Mayne, *J. Soc. Chem. Ind.*, 1946 (in press).

[34] R. M. Burns, H. E. Haring, and R. B. Gibney, *Trans. Electrochem. Soc.*, **69**, 169 (1936); **76**, 287 (1939).

[1] N. B. Pilling and R. E. Bedworth, *J. Inst. Metals*, **29**, 534 (1923).

commonly expressed, according to the parabolic equation. (Under some conditions they oxidize proportionally to the logarithm of the time.) Examples are nickel (NiO),

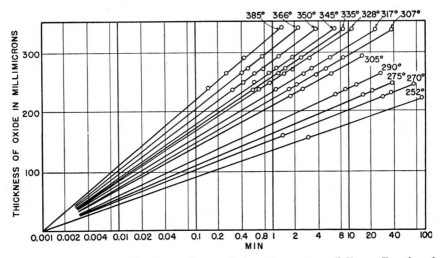

Fig. 1. Thickness of Oxide Films on Iron at Various Temperatures (° C) as a Function of Time. (G. Tammann and W. Köster, Z. anorg. allgem. Chem., **123**, 196 [1922].)

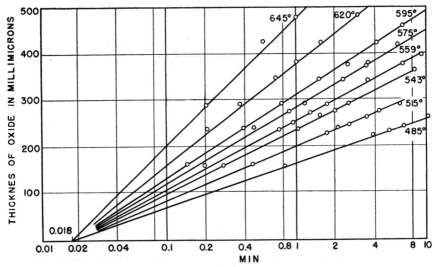

Fig. 2. Thickness of Oxide Films on Nickel at Various Temperatures (° C) as a Function of Time. (G. Tammann and W. Köster, Z. anorg. allgem. Chem., **123**, 196 [1922].)

copper (Cu_2O) and chromium (Cr_2O_3), which have ratios of Md/mD equal to 1.60, 1.71, and 2.03, respectively.[2] The parabolic relation with time for the oxidation rates

[2] A table of these ratios is presented by U. R. Evans in *Metallic Corrosion Passivity and Protection*, p. 122, Edward Arnold and Co., London, 1937.

FUNDAMENTALS OF OXIDATION AND TARNISH

of copper is shown in Fig. 2, p. 623, and the logarithmic relation for oxidation rates of iron and nickel in Figs. 1 and 2 of this chapter.

CRACKING AND SPALLING

Although a coating may be protective when first formed, a limited thickness is often reached at which the coating cracks. In some instances further heating serves to fill in the cracks or, after the film has become sufficiently thick, appreciable protection may exist despite the cracks. In other instances, spalling or fragmenting of the coating occurs because the locked-up stresses in the oxide eventually exceed the binding forces of the coating. On cooling or heating the oxidized metal, differential contraction or expansion of oxide and metal also may cause spalling, thereby exposing the underlying metal to fresh attack.* Spalling occurs each time the metal is cooled and may continue for some hours after room temperature is reached.[3] For this reason intermittent heating and cooling of a metal usually results in rates of reaction higher than those for continuous heating.

EFFECT OF METAL PROPERTIES†

The orientation of the base metal has an appreciable effect on the rate of oxidation of metals although there is as yet no agreement on measured rates for the various crystal faces. Tammann[4] suggested that different oxidation rates of crystal faces is explained by a fixed orientation relation between the lattice of the metal and that of the oxide film. Lustman and Mehl,[5] however, found that whereas a relation of metal to oxide orientation exists, the relation of orientation to rate of oxidation is complex and is temperature dependent. (See also *Effect of Crystal Orientation on Corrosion*, p. 33.)

The lattice structure of a metal apparently can also influence the rate of oxidation. It was shown that oxidation rates of face-centered cubic (γ) iron differ inherently from rates for body-centered iron (α) when extrapolated to the same temperatures.[6, 7] Data for the oxidation of iron by carbon dioxide showing a break at the approximate temperature of the phase transformation are presented in Fig. 3. If it is assumed that the nature and thickness of an oxide or tarnish film control the rate of reaction, it is logical that orientation and lattice structure of a metal in some way effect the composition, structure, electrical conductivity, or other property of the protective coating. There is some evidence, however, that the reaction rate may be controlled by transfer of electrons from metal to film, which, in turn, is affected by the electron configuration of the metal.[8] For example, discontinuities in the slopes of curves representing oxide film thickness formed at given temperatures on Fe-Ni alloys plotted with per cent nickel coincide with the Curie temperatures of the alloys. The subject requires further study.

---

* Tendency to spall is in some degree dependent on the shape of the metal surface. Spalling of iron oxide on iron exposed to high-temperature steam is less pronounced on a concave surface than on a surface either convex or flat. See p. 518. EDITOR.

† This section is by the Editor.

[3] W. Hessenbruch, *Zunderfeste Legierungen*, p. 60, Berlin, 1940.

[4] G. Tammann, *Stahl u. Eisen*, **42**, 615 (1922).

[5] B. Lustman and R. Mehl, *Trans. Am. Inst. Mining Met. Engrs.*, **143**, 246–267 (1941).

[6] A. Portevin, E. Pretet, and H. Jolivet, *Rev. Met.*, **31**, 221 (1934).

[7] K. Fischbeck and F. Salzer, *Metallwirtschaft*, **14**, 753 (1935).

[8] N. Mott, *Trans. Faraday Soc.*, **35**, 1175 (1939); **36**, 472 (1940). B. Lustman, *Trans. Electrochem. Soc.*, **81**, 359 (1942).

Composition[*]

The composition of a metal is a sensitive factor in determining the oxidation or tarnish rate. The alloy components that preferentially oxidize such as aluminum in copper or brass, or silicon and chromium in iron, even though in small concentration, can effectively reduce the rate.[†] It is thought in these instances that the oxide first formed is enriched in these particular components and is therefore more effective than the atom fraction of components in the metal would lead one to suspect. The oxide layer high in alloy elements appears to form an important barrier to the diffusion of reactants. In the process of reaction the ions of the base metal usually diffuse outward through the inner protective layer to react with the non-metallic components

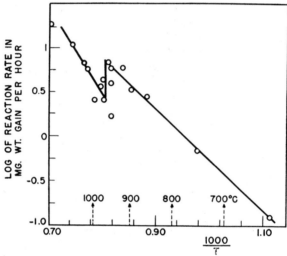

FIG. 3. Effect of $\alpha \rightarrow \gamma$ Transition on the Reaction Rate of Iron with Carbon Dioxide Saturated with Water. (K. Fischbeck and F. Salzer.)

of the coating, but at a rate much less than occurs in the absence of this layer. The presence of an inner layer enriched in the alloying elements that produce oxidation resistance has been demonstrated by chemical analyses of the oxide coatings formed on iron alloys both in air[9] and in steam.[10]

Surface Conditions

The thickness of the air-formed oxide films referred to the apparent area[‡] varies with the surface conditions. The surface conditions, in turn, vary from the relatively

[*] This section is by the Editor.

[†] The minor constituents should possess: (1) a value of $Md/mD > 1$; (2) an oxide of high electrical resistivity; (3) small-sized ions compared with those of the base metal.

It should be noted that small amounts of alloying constituents have little or no effect on the tarnish rate of silver (L. E. Price and G. J. Thomas, *J. Inst. Metals*, **63**, 29 [1938]) or copper (W. H. Vernon, *Trans. Faraday Soc.*, **23**, 113 [1927]) at ordinary temperatures probably because restricted diffusion in the metal lattice prevents concentration of the alloy constituents in a surface-product film. Private communication from W. E. Campbell, Bell Telephone Laboratory, Murray Hill, N. J.

[‡] The apparent area as measured by a ruler placed on the surface is independent of the roughness of the surface. The local or contour area takes into account all irregularities such as the corrugations of the abraded surface.

[9] L. B. Pfeil, *J. Iron Steel Inst.*, **119**, 530 (1929). A. White, C. Clark, and C. McCollam, *Trans. Am Soc. Metals*, **27**, 125 (1939).

[10] G. Hawkins, J. Agnew, and H. Solberg, *Trans. Am. Soc. Mech. Engrs.*, **66**, 291 (1944).

stress-free surfaces of evaporated metal to the corrugated "shattered" metal surfaces on specimens vigorously abraded with a coarse abrasive.

The initial film formed in pure air at ordinary temperatures on an annealed film-free copper surface may be 10° A (10^{-7} cm) thick,[11, 12] whereas it may be as much as eight times this amount for a copper specimen machined or abraded. The latter film is probably formed at the high local surface temperatures produced by abrasion, which may reach several hundred degrees centigrade.[13] The initial oxidation rate at room temperature of such an abraded surface is of the same order as for evaporated metal. The thickness of the oxide film on an abraded metal surface may therefore be influenced by (1) stresses induced by abrasion, (2) increased area (caused by corrugations, shattered metal, and cracks), and (3) superficial oxidation by reason of high surface temperatures produced by abrasion.

When some metals are alternately oxidized and reduced several times, the surface is found to be much more active in further oxidation tests.[14] This activation of metal surfaces is attributed to the dispersal of the reduced metal in a larger volume (the volume that was occupied by the oxide), providing a "spongy" layer with a greater surface area exposed to the attacking oxygen. The reduction of the initial oxide on any metal, therefore, as part of a cleaning process, may leave the surface more active with respect to subsequent oxidation or tarnish.

In general, smooth surfaces such as a polished surface, oxidize more slowly than such rough surfaces as are obtained by machining, abrading, or etching. This was indicated from oxidation experiments on copper[15] and on iron.[16]

Effect of Environment

The presence of sulfur-containing gases and moisture appreciably increases the oxidation rates of copper,[17] iron (see *Iron and Steel*, p. 635), and many other metals and alloys in air. Moisture in hydrogen sulfide is necessary for tarnish of silver.* The mechanism of behavior of these constituents is not well known.

THE EQUATIONS OF OXIDATION AND TARNISH

There are three equations by which common metals are known to oxidize and tarnish under ordinary conditions: (1) the linear; (2) the parabolic; and (3) the logarithmic equations.

The *linear equation* is generally expressed by

$$y = K_1 t + A_1 \tag{1}$$

the *parabolic equation* by

$$y^2 = K_2 t + A_2 \tag{2}$$

and the *logarithmic equation* by

$$y = K_3 \log (A_3 t + c) \tag{3}$$

where y = film thickness, t = time, and K, A. and c are constants. The history of the

---

* Apparently oxygen as well as moisture is necessary. (S. Lilienfeld and C. E. White, *J. Am. Chem. Soc.* **52**, 885 [1930].)

[11] A. H. White and L. H. Germer, *Trans. Electrochem. Soc.*, **81**, 305 (1942).
[12] W. E. Campbell and U. B. Thomas, *Trans. Electrochem. Soc.*, **76**, 303 (1939).
[13] F. P. Bowden and K. Ridler, *Proc. Royal Soc.* (A), **154**, 640 (1936).
[14] C. N. Hinshelwood, *Proc. Royal Soc.* (A), **102**, 318 (1922).
[15] B. Lustman and R. Mehl, *Trans. Am. Inst. Mining Met. Engrs.*, **143**, 246-267 (1941).
[16] M. Day and G. Smith, *Ind. Eng. Chem.*, **35**, 1098 (1943).
[17] W. H. Vernon, *Trans. Faraday Soc.*, **23**, 113 (1927).

discovery of these equations and a discussion of the constants in their equations have been given elsewhere.[18]

The linear and parabolic equations were proposed by Pilling and Bedworth[19] corresponding to porous and non-porous film growth. Tammann and Köster[20] expressed the logarithmic equation in the exponential form $t = Ae^{by} - A$, where A and b are constants. Vernon, Akeroyd, and Stroud[21] were the first to express it in the form of Eq. 3 above. The two expressions become equivalent when c of Eq. 3 is made equal to 1, so that $y = 0$ when $t = 0$. For thick films, Bangham and Stafford[22] obtained an approximation to this equation.

Effect of Temperature*[23]

The Arrhenius Equation† has been found in many cases to describe the change of rate with temperature. Accordingly if the logarithm of the velocity constant K is plotted with $1/T$, where T is the absolute temperature, a straight line is obtained. (See Fig. 3. The lattice transformation presumably accounts for the discontinuity in slope.) There are some recorded instances, however, where this linearity does not apply.[24, 25]

The Electrolytic Theory of Oxidation and Tarnish

The electrolytic theory, based largely on the work of Wagner,[26] proposes that oxidation and tarnish are electrochemical phenomena. It was once thought that oxidation and tarnish represented direct chemical reaction whose rate was controlled by simple diffusion of reactant atoms (as by Fick's Law) through a non-porous corrosion-product film. Experiment has shown, however, that reaction may occur by *metal ions* diffusing outward through the film as well as by the less frequent diffusion of *reactant ions* inward. Wagner showed, for example, that in the formation of silver sulfide on silver, the movement of silver through the tarnish layer is more pronounced than movement of sulfur. Usually the diffusion of the larger of two ions involved in the reaction is negligible.

A growing film (Fig. 4) can be compared to a current-producing cell with the metal-film interface (as anode) supplying cations and electrons for outward diffusion and the attacking substance-film interface (as cathode) supplying the anions for inward diffusion. The film acts as both the internal and external circuit of a closed cell. Then,

* This section is by the Editor.
† The Arrhenius Equation is the following:

$$\ln K = - \frac{\text{(Heat of Activation)}}{RT} + \text{Constant}$$

where
K = specific reaction rate,
R = gas constant,
T = absolute temperature.

[18] H. A. Miley, *Trans. Electrochem. Soc.*, **81**, 391 (1942). U. R. Evans, *Corrosion, Passivity and Protection*, pp. 121–128, Edward Arnold and Co., London, 1937.
[19] N. B. Pilling and R. E. Bedworth, *J. Inst. Metals*, **29**, 534 (1923).
[20] G. Tammann and W. Köster, *Z. anorg. allgem. Chem.*, **123**, 216 (1922).
[21] W. H. J. Vernon, E. I. Akeroyd, and E. G. Stroud, *J. Inst. Metals*, **65**, 301 (1939).
[22] D. H. Bangham and J. Stafford, *Nature*, **115**, 83 (1925).
[23] See also discussion by J. S. Dunn, *Proc. Roy. Soc.* (A), **111**, 203 (1926), and by E. A. Gulbransen, *Trans. Electrochem. Soc.*, **83**, 301 (1943).
[24] C. A. Siebert and C. Upthegrove, *Trans. Am. Soc. Metals*, **23**, 187 (1935).
[25] B. Lustman and R. Mehl, *Trans. Am. Inst. Mining Met. Engrs.*, **143**, 246–267 (1941).
[26] C. Wagner, *Z. physik. Chem.*, **21B**, 25 (1933); **32B**, 447 (1936); **40B**, 455 (1938); (with K. Grünewald), *Trans. Faraday Soc.*, **34**, 851 (1938).
See summary by L. E. Price, *Soc. Chem. Ind.*, **56**, 769 (1937).

using Ohm's Law:

$$I = \frac{E_o}{R + r} \quad (4)$$

where R = the electronic resistance of the film, r = the electrolytic resistance of the film, and E_o = the emf of the cell in volts as provided by the free-energy decrease of the film-forming reaction $M^+ + X^- = MX$, where M is the metal and X is the

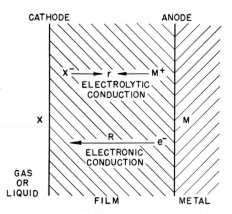

FIG. 4. A Cell Representing an Oxide or Tarnish Film on a Metal (M) Reacting with a Non-metal (X) as Follows: $M + X \to MX$.

oxidizing or tarnishing agent. The general equation expressing rate of film growth in terms of current I is

$$\frac{dy}{dt} = BI \quad (5)$$

where y = film thickness, t = time, and B is a constant.

Derivation of Parabolic Rate Constant.* The major triumph of Wagner's theory was the successful calculation of oxidation and tarnish rates, using independent physical-chemical data. The following derivation of the rate constant is for the case where resistivity of the film is independent of the reactant pressure, and polarization effects can be neglected.

Let κ be the measured specific conductivity of the film substance and n_1, n_2, and n_3 the mean cation, anion, and electron transference numbers, respectively. Let σ be the specific resistivity of the film substance. Then $\sigma = 1/\kappa$.

The Wagner cell comprises essentially an electronic resistance and an electrolytic resistance in series (Fig. 4). If σ_e represents the electronic resistivity and σ_i the electrolytic resistivity, the following relation holds for the oxidation tarnish cell:

$$\sigma = \sigma_e + \sigma_i \quad (6)$$

In measuring the resistivity of the film substance, however, these two resistances are in parallel. Remembering that the transference number of an ion is the fraction it carries of total current I passing through the cell, and letting E be the voltage drop

---

* This section is by the Editor. The derivation, with slight modifications, is based on that given by T. P. Hoar and L. E. Price in *Trans. Faraday Soc.*, **34**, 867 (1938).

across a centimeter cube of film substance produced by flow of such current,

$$\sigma_e = \frac{E}{n_3 I} = \frac{\sigma}{n_3} = \frac{1}{n_3 \kappa} \tag{7}$$

and

$$\sigma_i = \frac{E}{(n_1 + n_2)I} = \frac{1}{(n_1 + n_2)\kappa} \tag{8}$$

Therefore, from Eq. 6,

$$\sigma = \frac{1}{n_3 \kappa} + \frac{1}{(n_1 + n_2)\kappa} \tag{9}$$

and, since $n_1 + n_2 + n_3 = 1$,

$$\sigma = \frac{1}{n_3(n_1 + n_2)\kappa} \tag{10}$$

The total resistance R of a film having area A and thickness y is given by

$$R = \sigma \frac{y}{A} \tag{11}$$

or

$$R = \frac{1}{n_3(n_1 + n_2)\kappa} \frac{y}{A} \tag{12}$$

The rate of film growth must be equivalent to the current I flowing through the cell. Then, if w equivalents of film substance form in t seconds,

$$\frac{I\,dt}{F} = dw \tag{13}$$

where F is the Faraday, or

$$\frac{dw}{dt} = \frac{I}{F} = \frac{E_o}{FR} \tag{14}$$

$$= \frac{E_o n_3 (n_1 + n_2)\kappa A}{Fy} \tag{15}$$

where E_o is the emf of the cell, the value of which is derived either from potential measurements of appropriate cells or from values of free energy for the film reaction.

Equation 15 is equivalent to the equation:

$$\frac{dw}{dt} = \frac{kA}{y}$$

where k is equal to $\dfrac{E_o n_3 (n_1 + n_2)\kappa}{F}$.

A value of k for copper oxidizing at 1000° C and 100-mm oxygen pressure was calculated, making use of the following experimental data: $(n_1 + n_2) = 4 \times 10^{-4}$, $\kappa = 4.8$ ohm-cm, and E_o for the cell $Ag(O_2)/borax/Cu_2O$, Cu is 0.367 volt at 1000° C and 1 atm., and calculated by Nernst's equation to be 0.311 volt at 100-mm pressure. The resultant calculated value of k is 6×10^{-9} equiv. cm^{-1} sec^{-1}, which is in excellent agreement with the value 7×10^{-9}, as observed by Feitknecht.[27] A similar calculation applied to the reaction of silver with sulfur at 220° C gave a value for the reaction rate

[27] W. Feitknecht, Z. Electrochem., **35**, 142 (1929); **36**, 16 (1930).

constant of 2 to 4 × 10^{-6} equiv. cm^{-1} sec^{-1}, as compared with the observed value of 1.6 × 10^{-6} equiv. cm^{-1} sec^{-1}.

Equation 15 above can be transformed into the familiar parabolic equation, letting M represent the gram-equivalent weight of film substance and d its density. Then

$$w = \frac{yA\,d}{M} \qquad (16)$$

and

$$\frac{dw}{dt} = \frac{Ad}{M}\frac{dy}{dt} \qquad (17)$$

Substituting dy/dt for dw/dt in Eq. 15 and integrating, one obtains:

$$y^2 = \frac{2E_o n_3 (n_1 + n_2)\kappa M}{Fd} t + \text{Const} \qquad (18)$$

The Logarithmic Equation

Many metals oxidize according to the logarithmic rather than the parabolic equation, especially in the lower temperature range. Satisfactory reasons are not known. Some attempts have been made to derive the logarithmic equation by introducing polarization effects into Wagner's model,[28] by assuming obstruction of diffusion paths in the film,[29] and by introducing quantum mechanical concepts of electron and ion emission from metal surfaces combined with the "tunneling effect" across energy barriers.[30]

Equations Obeyed by Turns

In some cases film growth may obey each equation in turn. For example, the growth of a non-porous film may follow OM (Fig. 5) while the linear law is obeyed, MF while the parabolic law is in control, and GQ after the conditions become favorable to the logarithmic law. It will obey neither law in the interval of transition (F to G).

In the tarnishing of silver[31] by iodine dissolved in organic liquids, a straight-line relationship was approximated for thin films, and for films of greater thickness, the parabolic equation was obeyed in most cases. Still thicker films might possibly provide conditions favorable to the logarithmic equation.

It has been shown[32] that copper at 180° to 300° C (356° to 572° F) can oxidize parabolically and change to a logarithmic rate of growth.* It is clear that such a parabola could not start at the origin since this would require an infinite rate of growth at the beginning.†

Equations Applied to Porous and Non-porous Films. The preparabolic period, Ot_1, may be greatly extended for porous films, and greatly suppressed for the more protective films. The critical thickness, represented by N, decreases rapidly with

---

* In such a case, the logarithmic equation is $y = k \log [a(t + t') + 1]$, where $t' = OL$ (Fig. 5), c having been set equal to 1.

† The preparabolic curve is not always a straight line, but it always has a common tangent with the parabola at the point of transition M of Fig. 3.

[28] H. A. Miley, *Trans. Electrochem. Soc.*, **81**, 391 (1942). O. H. Hamilton and H. A. Miley, *Trans. Electrochem. Soc.*, **81**, 413 (1942).

[29] U. R. Evans, *Trans. Electrochem. Soc.*, **83**, 335 (1943).

[30] N. Mott, *Trans. Faraday Soc.*, **35**, 1175 (1939); **36**, 472 (1940). B. Lustman, *Trans. Electrochem. Soc.*, **81**, 359 (1942).

[31] U. R. Evans and L. C. Bannister, *Proc. Roy. Soc.* (A), **125**, 370 (1929).

[32] A. L. Dighton and H. A. Miley, *Trans. Electrochem. Soc.*, **81**, 321 (1942).

temperature, and varies with all the factors that affect the perviousness of the films. At 80° C (176° F), oxide films grew logarithmically[33] on monocrystalline copper beyond about 40 Å (40 × 10⁻⁸ cm), and at room temperature[34] films of almost atomic dimensions grew logarithmically on evaporated copper specimens.

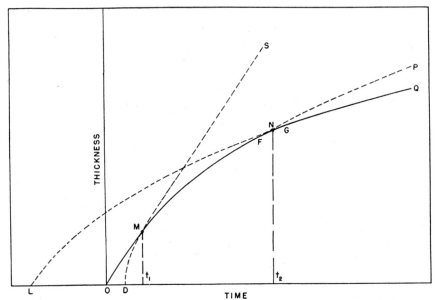

Fig. 5. A Hypothetical Diagram Showing How a Metal May Oxidize or Tarnish so as to Obey the Three Equations by Turns.

PROTECTION AGAINST TARNISH

In light of the electrolytic theory, electronic conduction is essential to film growth. Therefore films of very high electric resistance should provide good protection. Accordingly, films of alumina and beryllia developed on silver or copper[35] have yielded a high measure of protection against tarnish. The rapid oxidation of aluminum to a limiting thickness, after which the film grows very little, is possibly due to the very high specific resistance of the film.[36]

PASSIVITY

H. H. Uhlig* and R. B. Mears†

DEFINITION

Passivity is a property of a metal commonly defined in two ways. One of these is based on a change in the electrochemical behavior of the metal, and the other on its corrosion behavior. These definitions may be stated as follows:

* Corrosion Laboratory, Department of Metallurgy, Massachusetts Institute of Technology, Cambridge, Mass.
† Development Division, Aluminum Company of America, New Kensington, Pa. Present address: Research Laboratory, Carnegie-Illinois Steel Corp., Pittsburgh, Pa.

[33] B. Lustman and R. F. Mehl, *Trans. Am. Inst. Mining Met. Engrs.*, **143**, 246–267 (1941).
[34] A. H. White and L. H. Germer, *Trans. Electrochem. Soc.*, **81**, 305 (1942).
[35] L. E. Price and G. J. Thomas, *J. Inst. Metals*, **63**, 21–28, 29–57 (1938).
[36] N. F. Mott, *Trans. Faraday Soc.*, **35**, 1175 (1939).

Definition 1. A metal active in the EMF Series, or an alloy composed of such metals, is considered passive when its electrochemical behavior becomes that of an appreciably less active or noble metal.

Definition 2. A metal or alloy is passive if it substantially resists corrosion in an environment where thermodynamically there is a large free energy decrease associated with its passage from the metallic state to appropriate corrosion products.

By either definition, to passivate a metal is to make it passive by either physical or chemical treatment. This may be accomplished, for example, by exposure to oxygen, to an oxidizing solution, or by anodic polarization.

Definition 1 is more specific than definition 2, the latter having the disadvantage of defining as passive many common metals like zinc and iron exposed to air or immersed in deaerated water. However, definition 1 does not cover all cases commonly considered examples of passivity in which active metals prove highly resistant to corrosion. It approximates the classical meaning of passivity as first discussed by Schönbein and Faraday. The latter[1] cautioned against confounding the particular state (passivity) of iron when it is inactive in HNO_3 with the inactive state of amalgamated or pure zinc in dilute H_2SO_4. "The distinction is easily made," Faraday stated, "by the contact of platinum with either in the respective acids," the iron cell producing little or no current, whereas the zinc cell develops current normally associated with zinc.

Despite this, the term broadened with use to include many cases of corrosion rate decrease without regard to galvanic potentials or anodic behavior. In many circles today, definition 2 is the one commonly implied.

Cases under definition 1 include iron in contact with passivators, for example, solutions of HNO_3 (sp gr > 1.4), $NaNO_2$, $K_2Cr_2O_7$, Na_2FeO_4.[2] Also included are practically all the transition metals of the Periodic Table and passive alloys, for example, the stainless steels.

Cases under definition 2 include those of definition 1 plus a multitude involving chemical conversion coatings, such as lead in H_2SO_4, zinc or aluminum in Na_2SiO_3, iron in conc. H_2SO_4, magnesium in H_2O or HF ($> 2\%$), silver in HCl, etc. Also included are metals immersed in inhibited pickle baths and tarnished copper. It is generally accepted that examples of this category represent passivity in which relatively thick films afford physical protection in the manner of a paint film. Metals of this classification, in general, when passivated undergo only slight change of potential. A good example is the slight change in potential of iron whether immersed in an acid electrolyte or the same electrolyte containing a pickling inhibitor.[3,4] In some cases for which data exist, e.g., silver covered with a AgCl film in dilute chloride solutions,[5] lead covered with $PbSO_4$ in dilute sulfate solution,[6,7] or iron covered with $Fe(OH)_2$ in 3% NaCl free of dissolved oxygen,[8] the potential change is accounted for by Nernst's equation for reversible potentials: $E = E_o - \frac{RT}{nF} \ln$ (ion activities). Potentials of metals under definition 1, on the other hand, undergo an appreciable change on becoming passive, and are not accountable by the same equation. For example, iron in

---

[1] M. Faraday, *Experimental Researches in Electricity*, Vol. II, p. 243, University of London, 1844.

[2] L. Gmelin, *Handbuch der anorgischen Chemie*, 8th Ed., No. 59, Pt. A, pp. 279–298 (1929); 327–329 (1930), Verlag Chemie.

[3] R. M. Burns, *Bell System Tech. J.*, **15**, 20 (1936).

[4] H. H. Uhlig, *Ind. Eng. Chem.*, **32**, 1493 (1940).

[5] Lewis and Randall, *Thermodynamics*, p. 407, McGraw-Hill Book Co., New York, 1923.

[6] *Ibid.*, p. 422.

[7] R. M. Burns, *Bell System Tech. J.*, **15**, 36 (1936).

[8] H. Uhlig, N. Carr, and P. Schneider, *Trans. Electrochem. Soc.*, **79**, 111 (1941).

0.01 N $K_2Cr_2O_7$ solution becomes over 0.5 volt more noble compared with its potential in tap water. (See Fig. 2, p. 981.)

HISTORY

The first recorded observation that iron is unattacked by conc. HNO_3 is frequently ascribed to J. Keir[9] in 1790. However, parallel observations were recorded by C. F. Wenzel[10] in 1782 and by several others.[11]

The word passivity was first used by Schönbein[12] in 1836 to describe the behavior of iron immersed in conc. HNO_3 or when anodically polarized. This peculiar inertness of iron in concentrated nitric acid has been the subject of much study and discussion up to the present time. Hisinger and Berzelius[13] were the first to point out that anodic polarization aided in the occurrence of passivity, while Wetzlar,[14] at an early date, emphasized the electrochemical nature of the process and the fact that there are several degrees of passivity. Schönbein called this phenomenon to Faraday's attention[15] and subsequently Faraday[16] made extended investigations along these lines.

Iron passivated by chemical media or by anodic polarization is only temporarily passive. The fact that passivity of iron is stabilized by alloying with 12% or more of chromium was accidentally discovered by H. Brearley in 1912 and independently by B. Strauss and E. Maurer.[17] It has since been found that other alloying constituents of iron such as molybdenum or nickel can act similarly and that passivity in copper-, nickel-, or cobalt-base alloys can also be either stabilized or increased through alloying.

GENERAL PROPERTIES OF PASSIVE METALS AND ALLOYS

Passivity is not an absolute property like melting point or heat of fusion. A metal may possess variable degrees of passivity. The degree of passivity is measured by (1) galvanic potential* or (2) reaction or corrosion rate.

In general, oxidizing conditions favor passivity while reducing conditions destroy it or cause activity. For example, anodic polarization passivates iron whereas cathodic polarization activates, or destroys passivity in iron. Reducing solutions in general act in the same manner as cathodic polarization. Iron in contact with a more noble metal (corresponding to anodic polarization) is more easily passivated, but, in contact with a less noble metal, cathodic polarization occurs and passivity is more difficult or impossible to attain.

Halogen ions, particularly chloride ions, are destructive to passivity and usually

---

* Only for metals and alloys passive by definition 1. A more noble potential of a metal can result from either passivation, or from anodic polarization through local action currents of a heterogeneous surface containing more-noble phases. Adequate interpretation of potential measurements as evidence of passivity must take into account the latter possibility.

[9] J. Keir, *Phil. Trans.*, **80**, 359 (1790).
[10] C. F. Wenzel, *Lehre von der Verwandschaft der Körper*, p. 108, Dresden, 1782.
[11] See L. Gmelin, *Handbuch der anorgischen Chemie*, 8th Ed., No. 59, Part A, p. 314, Verlag Chemie, 1930.
[12] C. Schönbein, *Pogg. Ann.*, **37**, 590 (1836).
[13] W. Hisinger and J. Berzelius, *Gilb. Ann.*, **27**, 275 (1807).
[14] G. Wetzlar, *J. Schw.*, **49**, 470 (1827); **50**, 129 (1827).
[15] See letter from Schönbein to Faraday, published in *Phil. Mag.*, **9**, 53 (1836).
[16] M. Faraday, *Experimental Researches in Electricity*, Vol. II, p. 243, University of London, 1844.
[17] E. Thum, *The Book of Stainless Steels*, 2nd Ed., pp. 1–5, American Society for Metals, 1935.

should be avoided whenever it is essential that passivity be retained. In metals or alloys stable with respect to passivity in air (e.g., chromium, aluminum, or stainless steels) passivity is broken down by chloride ions at points or local areas. The local areas of active metal become anodes in cells, the cathodes of which are large areas of passive metal. The considerable emf of such cells produces active corrosion at the anodes causing pitting. (See *Pitting in Stainless Steels and Other Passive Metals*, p. 165.) With metals that are amphoteric, like aluminum or zinc, alkaline solutions can also destroy passivity.

Elevated temperatures are destructive to passivity (below 100° C for iron in nitric acid). Magnetic fields are reported to destroy passivity, but the experimental evidence is inconclusive since the magnetic fields often induce secondary effects that are damaging, such as increased temperature (for varying fields), mechanical shock, or physical effects due to magnetostriction.

THEORIES OF PASSIVITY

GENERALIZED FILM THEORY

Many theories of passivity have been proposed. The earliest seems to be Faraday's,[18] who apparently considered the passivity of iron in nitric acid to be attributable to the formation of an oxide film. For instance, he states: "Why the superficial film of oxide which I suppose to be formed when the iron is brought into the peculiar state by voltaic association, or occasionally by immersion alone into nitric acid, is not dissolved by the acid, is, I presume, dependent upon the peculiarities of this oxide and of nitric acid of the strength required for these experiments; . . ."

This oxide film theory of Faraday's has been generalized and, as the so-called film theory, is the one most widely held at the present time. Evans[19] has stated it as follows: "Most cases of passivity . . . appear to be attributable, directly or indirectly to a protective film, although not always an oxide film." Similar views have been expressed by Hedges,[20] by Glasstone,[21] and by Mears.[22] Hedges states: "During the last few years, the production of protective oxide or similar films has come to be regarded almost universally as the true cause of all types of passivity. . . ." Glasstone writes as follows: ". . . if iron is placed in nitric acid, there will be rapid attack at first . . . with the result that a thin adherent film of ferric oxide, or a basic salt, is formed which protects the metal from further attack. The passivity produced in certain metals, particularly the chromium-iron alloys, e.g., stainless steel, on exposure to air, is probably also due to similar oxide films."

This generalized film theory embraces the special cases where a film of oxygen rather than oxide causes passivity. Langmuir[23] pointed out that a film of adsorbed oxygen on tungsten does not react with hydrogen at as high a temperature as 1200° C, although the compound WO_3 is reduced by H_2 at 500° C. If the adsorbed film was weakened by evaporation or through reduction of oxygen pressure in the space surrounding the tungsten, gaps occurred in the adsorbed layer, and hydrogen suddenly and rapidly reacted with the oxygen. The mechanism for the inertness of adsorbed oxygen was ascribed to satisfaction of valence forces by the surface atoms of tungsten. The mechanism of passivity in metals like chromium was thought to be similar.

---

[18] M. Faraday, *Experimental Researches in Electricity*, Vol. II, p. 243, University of London, 1844.
[19] U. R. Evans, *Metallic Corrosion, Passivity and Protection*, p. 4, Edward Arnold and Co., London, 1937
[20] E. S. Hedges, *Protective Films on Metals*, p. 112, D. Van Nostrand Co., New York, 1937.
[21] S. Glasstone, *Text Book of Physical Chemistry*, p. 1011, D. Van Nostrand Co., New York, 1940.
[22] R. B. Mears, *Trans. Electrochem. Soc.*, **77**, 288 (1940).
[23] I. Langmuir, *J. Am. Chem. Soc.*, **38**, 2273 (1916); *Trans. Electrochem. Soc.*, **29**, 260 (1916).

Tammann[24] later suggested that a film of adsorbed oxygen, rather than the oxide, is responsible for the passivity of iron, nickel, cobalt, and chromium.

Several other theories have been proposed,[25] most of them in opposition to the oxide film theory. A recent theory that attempts to correlate and combine many of these with some aspects of the generalized film theory is discussed next.

The Electron Configuration Theory

The first qualitative suggestion of a relation between electron configuration of atoms and passivity was made by Russell[26] (1925), also by Swinne[27] (1925) and Sborgi[28] (1925). A theory accounting quantitatively for critical passivity concentrations in alloys was outlined by Uhlig and Wulff[29] in 1939 and extended by Uhlig.[30]

The latter theory, called the Electron Configuration Theory,[31] centers about the fact that metals passive according to definition 1 are largely those in the transition groups of the Periodic Table. The atoms of these elements according to the electron theory of metals (e.g., chromium, nickel, cobalt, iron, molybdenum, and tungsten) are characterized by incomplete inner shell (d electron) energy levels and by unfilled d energy bands in the metallic state. The Electron Configuration Theory assumes that these unfilled energy states tend to fill with electrons in the same sense that closed atomic shells are the tendency in chemical compounds. The state of passivity is ascribed to the condition of unfilled d bands in the metal or alloy, and the active state to the situation that, in effect, fills the d band with electrons. Thus adsorbed oxygen or adsorbed oxidizing substances accompany maximum passivity because they are electron absorbers with no tendency to supply electrons to atoms of the metal surface. On the other hand, interstitial hydrogen (by pickling or cathodic polarization) or certain classes of alloying elements in adequate concentration supply electrons and thereby favor the active over the passive state.

In Cr-Fe alloys (stainless steels), for example, the constituent chromium is naturally passive and is considered to impart this property to iron by electron sharing resulting from the strong tendency of chromium to absorb electrons. Iron can apparently be transformed to the passive state by sharing of at least one extra electron per iron atom. This assumption is supported by the fact that active iron as anode in an electrolyte, e.g., sodium sulfate, corrodes as Fe^{++}, but from stainless steel in which iron is passive, it corrodes with one electron less, namely as Fe^{+++}. Chromium with 5 vacancies in the $3d$ shell of the atom has a corresponding number of vacancies in the $3d$ band of the metal. With a stronger tendency than iron to absorb electrons, it can share at least 5 electrons or can passivate 5 iron atoms. This proportion corre-

---

[24] G. Tammann, *Z. anorg. Chem.*, **107**, 104, 236 (1919).

[25] "The Reaction Velocity Theory": Le Blanc *Lehre der Elektrochemie*, 6th Ed., p. 299, Leipzig, 1914. O. Sackur, *Z. Elektrochem.*, **10**, 841 (1904); **12**, 637 (1906).

"The Hydrogen Solution Theory": E. Grave, *Z. phys. Chem.*, **77**, 513 (1911). G. Schmidt and W. Rathert, *Trans. Faraday Soc.*, **9**, 257 (1914). E. Müller, *Z. phys. Chem.* (A), **181**, No. 2, 89 (1937). E. Müller and V. Cupr, *Z. Elektrochem.*, **43**, 42 (1937).

"The Allotropic Change Theory": F. Schönbein, *Pogg. Ann.*, **44**, 70 (1838). F. Finklestein, *Z. phys. Chem.*, **39**, 91 (1902). A. Smits and A. Aten, *Z. phys. Chem.*, **90**, 46 (1915). A. Smits, *Theory of Allotropy*, Longmans, Green and Co., New York, 1922. A. Smits, *Trans. Faraday Soc.*, **19**, 772 (1924).

[26] A. Russell, *Nature*, **115**, 455 (1925); **117**, 47 (1926). A. Russell, D. Evans, and S. Rowell, *J. Chem. Soc.*, **1926**, 1877.

[27] R. Swinne, *Z. Elektrochem.*, **31**, 422 (1925).

[28] U. Sborgi, *Atti Lincei* (6), **1**, 315, 388 (1925); *Gazz.*, **56**, 532 (1926).

[29] H. H. Uhlig and J. Wulff, *Trans. Am. Inst. Mining Met. Engrs.*, **135**, 494–521 (1939).

[30] H. H. Uhlig, *Trans. Electrochem. Soc.*, **85**, 307–318 (1944).

[31] For criticism of the theory, see discussion of papers: footnotes 29 and 30. Also R. Landau, *Trans. Electrochem. Soc.*, **81**, 559 (1942) and U. R. Evans, *Trans. Faraday Soc.*, **40**, 129 (1944).

sponds to 16.7 atomic per cent or 15.7 weight per cent chromium which is in satisfactory agreement with the observed critical minimum amount of chromium (about 12%) necessary to produce so-called stainless Cr-Fe alloys. If iron is alloyed with chromium in amount greater than about 84 to 88 atomic per cent, the available electrons of iron more than fill the d band of the alloy, with the result that the alloy is no longer passive.

The mechanism for the ennobled potential and reduced reactivity of passive metals and alloys is considered to reside largely in the surface-metal atoms, although in alloys electron interaction (sharing) between components enters as a major factor in determining passivity, and this occurs *throughout* the metal. It is the electronic configuration of the surface metal atoms, however, that in final analysis determines whether an approaching substance capable of reaction *will react or not*. If it reacts, a protective layer of reaction product may form on the surface, accounting for passivity by definition 2, but not for the passivity with which the Electron Configuration Theory is concerned (definition 1). If the substance does not react, it will *adsorb* on the metal surface because of existing affinities. In so doing it satisfies secondary valence forces of the metal surface and displays in itself considerably less chemical affinity for substances impinging on the adsorbed layer. The metal surface is therefore less reactive and by this mechanism is considered passive. This explanation is similar to that proposed by Langmuir for the chemically inert behavior of oxygen adsorbed on tungsten.

Many substances are capable of activated adsorption on metals and hence each can produce some degree of passivity. Oxygen is one of the most common, but, with some metals, H_2O very likely can also function. Carbon monoxide adsorbed on 18-8 stainless steels is effective in this respect even in hydrochloric acid where adsorbed oxygen fails.[32]

The condition that determines whether a substance capable of reaction will react or adsorb appears to be related to the work function (w.f.) of the metal (a measure of the heat of evaporation of electrons from a metal) and the heat of sublimation (ΔH_s) of the metal.[33] When oxygen, for example, approaches a metal surface it will either extract an electron from the metal surface and become adsorbed, or a metal atom will be dislodged from its position in the lattice to form an oxide. The preferred process is that requiring approximately the lesser amount of energy, or more accurately the greater decrease of free energy. Hence, since the state of adsorption corresponds to passivity, the condition for passivity is approximately given by:

$$\frac{\text{w.f.}}{\Delta H_s} < 1 \text{ and for reaction by } \frac{\text{w.f.}}{\Delta H_s} > 1.$$

The reasonableness of this relation is shown in Table 1. Of the elements for which data exist and that are recognized as being passive by definition 1, the calculated ratios of w.f./ΔH_s are largely equal to or less than unity, whereas other elements, not passive by the same criteria, have ratios greater than one.

Alloying, which affects both the work function and the sublimation energy, can induce passivity or activity, therefore, depending on the approximate ratios of these quantities at any alloy composition. It is by this general outline of corresponding properties that the electron configuration of the alloying elements is thought to determine the nature of metal surface behavior.

According to the Electron Configuration Theory, therefore, metals passive by

[32] H. Uhlig, *Ind. Eng. Chem.*, **32**, 1490 (1940).
[33] H. H. Uhlig, unpublished.

TABLE 1

	Heat of Sublimation, ΔH_s, 25° C, Electron-Volts	Work Function, Volts*	Work Function/ΔH_s
K	0.93	2.24	2.4
Na	1.12	2.37	2.1
Ca	1.85	3.09	1.7
Mg	1.56	3.60	2.3
Zn	1.35	3.7–4.2	2.7–3.1
Hg	0.63	4.50	7.1
Pb	2.01	3.8	1.9
Al	2.92	3.6–4.4	1.2–1.5
Ag	3.0	3.9–4.7	1.3–1.6
Cu	3.54	4.80	1.36
Mn	3.02	3.76	1.25
Pt	5.40	6.3	1.17
Ni	4.26	4.1–5.0	0.96–1.2
Fe	4.20	4.2–4.7	1.0–1.1
Cr	3.87	4.35	1.1
W	8.8	4.52	0.51
Mo	6.75	4.12	0.61
C	8.6	4.82	0.56

* These are either the thermionic work function or the photoelectric threshold potential, considered identical for a given metal. The determinations of work function are not always precise and hence a range of experimental values is given for some metals.

definition 1 can be considered covered with a film, but this film is largely monatomic or monomolecular* and does not include as part of the essential mechanism of passivity any thicker films of oxides or other compounds. Iron in nitric acid is passive presumably because of an unstable higher oxide of iron produced by momentary reaction. The oxide is strongly electron absorbing, and *adsorbed* on the surface, in agreement with the early proposal of Bennett and Burnham.[34] On this basis, a film of Fe_2O_3 isolated from passive iron can be considered as evidence only that the higher oxide decomposed to this end product.

PASSIVITY IN ALLOYS

The major components of passive alloys within definition 1 are the transition metals. It is found that a metal normally passive in air can often passivate another metal if the two form a solid solution alloy over an adequate range of composition. Multi-phase alloy systems can be passive, in general, only if the galvanic potentials of the electrically conducting phases are of the same order of magnitude (e.g., the passive two-phase Fe-Mo system of greater than 55% molybdenum); otherwise galvanic action destroys passivity.

Passivity usually initiates at a minimum alloy composition that is critically and uniquely defined for a large number of different environments. The critical composition may shift, however, or entirely disappear in some types of active chemical media. For example, a minimum of 12% chromium is the critical concentration necessary to

* The monomolecular film is distinct from the measurably thick oxide surface layers that are protective to metals at elevated temperatures. In the latter instance, the oxide layer is a barrier to diffusion of reactants and impedes the reaction by this process rather than by reduction of surface chemical affinities. The mechanism of high temperature corrosion is discussed in *Fundamentals of Oxidation and Tarnish*, p. 11.

[34] C. Bennett and W. Burnham, *Trans. Electrochem. Soc.*, **29**, 217 (1916). W. Bancroft and J. Porter, *J. Phys. Chem.*, **40**, 37 (1936).

stabilize passivity of iron in the atmosphere and in water, but in non-oxidizing acid salt solutions, such as $FeSO_4$, the critical concentration may shift to 15 or 20% chromium. Furthermore, the alloy is not passive in hydrochloric acid, no matter how high the chromium concentration.

Tammann[35] differentiated between parting limits (see p. 118) characteristic, for example, of Au-Ag alloys, and passivity limits or critical alloy concentrations for passivity. He considered the phenomena not the same. The correspondence pointed out by others between critical alloy composition for passivity and the $n/8$ rule previously suggested by Tammann[36] for parting or reaction limits was attributed to coincidence.

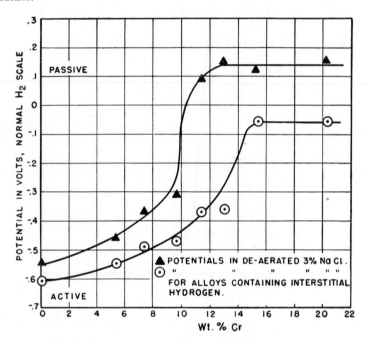

Fig. 1. Potentials of Cr-Fe Alloys. (According to H. Uhlig, N. Carr, and P. Schneider *Trans. Electrochem. Soc.*, **79**, 111 [1941].)

A similar shift of potential occurs at 12 to 16% Cr in 1 N $FeSO_4$ (B. Strauss, *Stahl u. Eisen*, **45**, 1201 [1925]) and above 15% Cr in 0.1 N H_2SO_4 (G. Tammann and E. Sotter, *Z. anorg. Chem.*, **127**, 257 [1923]).

Five electron vacancies in the $3d$ shell of chromium permit 1 chromium atom to share a maximum of 1 electron each from 5 iron atoms. Predicted minimum chromium content for passivity is at 16⅔ atomic per cent or 15.7 weight per cent Cr.

THEORY OF PASSIVITY IN ALLOYS

Since the advent of the Cr-Fe stainless steels, the mechanism of passivity in the many passive alloys has commonly been attributed to a protective oxide film. The effect of alloying chromium with iron is considered by some to impart to the alloy an impermeable, tenacious film of oxide similar to that hypothesized for pure

[35] G. Tammann, *Z. anorg. Chem.*, **169**, 151 (1928).
[36] G. Tammann, *Lehrbuch der Metallkunde*, 4th Ed., p. 440, L. Voss, Leipzig, 1932.

chromium. Tammann[37] and others ascribed passivity of stainless steels to a layer of oxygen rather than oxide.

The viewpoint of the Electron Configuration Theory with respect to passivity in Cr-Fe alloys has already been discussed. The potential behavior of these alloys and the effect of interstitial hydrogen, as shown in Fig. 1, are accounted for by the theory. Figure 2 presents corrosion data, characteristic of the Cr-Fe alloys, illustrating that the critical minimum chromium concentration for passivity as measured by corrosion

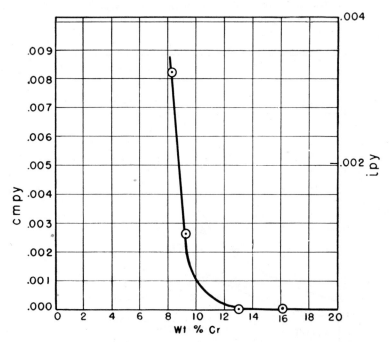

FIG. 2. Corrosion of Cr-Fe Alloys in Intermittent Water Spray, Room Temperature. (W. G. Whitman and E. Chappell, *Ind. Eng. Chem.*, **18**, 533, 1926.)

A similar corrosion minimum occurs for 30% nitric acid between 8% and 18% Cr, depending on the temperature (J. Monypenny, *Stainless Iron and Steel*, p. 348, Chapman and Hall, London, 1931) but no minimum occurs for HCl or H_2SO_4.

Predicted minimum chromium content for passivity is at 15.7% Cr.

rate is also at about 12% chromium.

The theory has been successful in accounting for passivity in several other alloy systems, starting with the atomic configurations of the constituent atoms and assuming a mechanism of electron sharing as in stainless steels. The potential behavior of Mo-Fe alloys (Fig. 3) has been considered in this light, as have corrosion data of the three-component Mo-Ni-Fe system (Fig. 4). The critical nickel content of Ni-Fe alloys is predicted to occur at 34% nickel (33⅓ atomic per cent) whereas data in dilute sulfuric acid show a passive limit at about 40% nickel. (See Fig. 5, also Fig. 1, p. 195.) This agreement is considered satisfactory. The maximum iron content permitting retention of corrosion resistance in a Mo-Ni-Fe alloy containing

---

[37] G. Tammann, *Stahl u. Eisen*, **42**, 577 (1922).

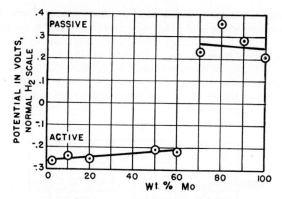

FIG. 3. Potentials of Mo-Fe Alloys in 0.1 N H_2SO_4. (According to G. Tammann and E. Sotter, *Z. anorg. Chem.*, **127**, 257 [1923].)

Five vacancies in the $4d$ shell of molybdenum permit 1 molybdenum atom to share a maximum of 1 electron each from 5 iron atoms. According to the phase diagram, a solid solution phase exists up to 9 weight per cent molybdenum (5.5 atomic per cent). It contains less than the critical amount of molybdenum; hence it is not passive and, in turn, imparts activity by galvanic action to the two phase-region below 55% molybdenum. Above 55%, a compound containing more than the critical molybdenum content in contact with a solid solution phase, essentially all molybdenum, accounts for passivity in this region.

Predicted minimum molybdenum content for passivity is at 55 weight per cent.

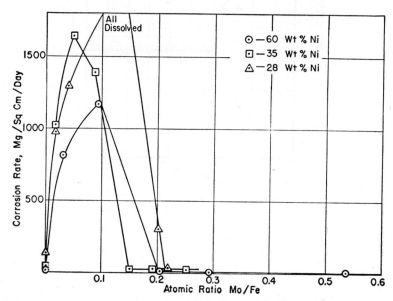

FIG. 4. The Corrosion Rates of Mo-Ni-Fe Alloys in 29.8% H_2SO_4 at the Boiling Point as Affected by the Ratio of Mo to Fe. (According to M. Schmidt and L. Wetternick, *Korrosion u. Metallschutz*, **13**, 184 [1937].)

Five vacancies in the $4d$ shell of molybdenum indicate that one molybdenum atom can impart passivity to five iron atoms in solid solution alloys.

20% molybdenum is predicted to be 35%. Confirmation is obtained by data of Fig. 3, p. 296 in dilute sulfuric acid which show that the maximum occurs at 30% iron.

In the Cu-Ni system, magnetic and specific heat data lead to the conclusion that the d band of electronic energy levels for the alloy is filled at 60 atomic per cent copper and above. Passivity, therefore, according to the theory, should be a property of alloys containing 40% and more of nickel. Pitting, a result of passive-active cells, and biofouling data in sea water (Fig. 6) (restricted release of toxic copper ions by passive alloys encourages fouling) as evidence of the passive state confirm this predicted critical composition.

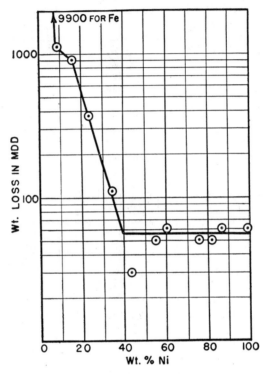

FIG. 5. Corrosion of Cast Ni-Fe Alloys in 10% Sulfuric Acid, Room Temperature. (According to Y. Utida and M. Saito, *Science Reports*, Tohoku Imperial University, **14**, 295 [1925].)

Two vacancies in the 3d shell of nickel permit 1 nickel atom to share a maximum of 1 electron each from 2 iron atoms.

Predicted minimum nickel content for passivity is at 33⅓ atomic per cent or 34 weight per cent.

THEORY OF PASSIVATORS

Passivators are a special class of inhibitors of corrosion (see *Inhibitors and Passivators*, p. 905). In low concentrations they appreciably change the galvanic potential of a metal in a more noble direction. Potassium chromate added to cooling waters is an example.

Pigments used in priming-paint coats usually contain passivators. Zinc chromate is one such pigment, largely insoluble, but soluble enough in water diffusing through the paint film to bring an appreciable concentration of chromate ions to the metal surface. This small concentration passivates the base metal. An appreciably less soluble chromate, for example lead chromate, does not have this property.

The action of chromate ions, from one viewpoint, is to form a thin protective film on the surface composed of an insoluble iron compound such as iron chromate or a ferric-chromic oxide mixture (see p. 9). According to the Electron Configuration Theory, chromate ions form an adsorbed layer on iron. The chromate film shares

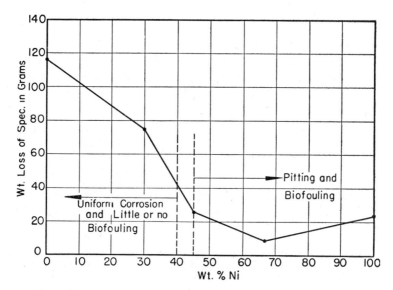

Fig. 6. The Behavior of Cu-Ni Alloys in Sea Water. (According to F. L. LaQue, *J. Am. Soc. Naval Engrs.*, **53**, 29 [1941] [and with W. Clapp] *Trans. Electrochem. Soc.*, **87**, 103 [1945].)

The overall region where maximum pitting and biofouling occur is evidence of the passive state. The d band of the alloy, as indicated by physical data, is filled at 60 atomic per cent copper or unfilled at 40 atomic per cent nickel and above. Predicted minimum nickel content for passivity is at 40 atomic per cent or 38 weight per cent.

electrons from surface iron atoms and satisfies secondary valence forces, but does not disrupt the metal lattice; wherefore the metal surface is less reactive and is more noble in the galvanic series.

Red lead in priming coats is also a passivator.[38] Chemically speaking, it is plumbous orthoplumbate so that the passivating ion analogous to CrO_4^{--} is probably PbO_4^{--}. The mechanism whereby it passivates iron may be considered to be the same as that of chromates.

Sodium nitrite used as a passivator in oil pipe lines (see p. 914) renders iron several tenths volt more noble than iron in distilled water or in a similar concentration of sodium nitrate.[39] The nitrite ion is oxidizing in nature and hence acts like chromates or other oxidizing passivators to reduce corrosion.

[38] R. Burns and A. Shuh, *Protective Coatings for Metals*, p. 306, Reinhold Publishing Corp., New York, 1939.
[39] H. H. Uhlig, unpublished observations.

Passivators for stainless steels, among which are Fe^{+++}, Cu^{++}, NO_3^-, appreciably reduce corrosion of these alloys in sulfuric or nitric acids.[40] By the chemical film theory, they are considered to form or alter the protective oxide or other-type compound surface layer. According to the Electron Configuration Theory, these ions absorb or share electrons from the surface metal atoms whether or not an oxide layer is present, thereby reducing the tendency for hydrogen to enter the lattice to destroy passivity. At the same time they are probably adsorbed on the surface, and in this state reduce reactivity of the surface metal atoms in the same manner as does adsorbed oxygen when stainless steels are exposed to air.

GENERAL DISCUSSION OF PRESENT THEORIES OF PASSIVITY

The Generalized Film Theory and the Electron Configuration Theory of passivity have much in common, and it is more than probable that they are different expressions of the same basic idea. Adherents of the film theory generally admit that the electron configuration of the metal or alloy must affect its corrosion behavior, just as it is considered to affect the other chemical properties. Cathodic hydrogen is considered to be an activator by either theory. However, according to the film theory its action is by direct reduction of the protective film, whereas, according to the Electron Configuration Theory, it first adsorbs on or dissolves in the metal and changes the electron configuration of the metal atoms, which, in turn, alters the nature of adsorbed surface films.

Chloride ions destroy passivity according to the first theory by penetrating pores of the film or colloidally dispersing it, whereas, according to the second theory, these ions preferentially adsorb at favored areas and alter the electron configuration of metal atoms on the surface so that reaction can occur.

The Electron Configuration Theory in its present form applies only to instances of passivity listed under definition 1, which with few exceptions involve the transition metals and their alloys. Aluminum is the outstanding case of a non-transition metal that fits into this definition of passivity, but as yet it does not find an adequate place in the theory. This is not surprising in view of the fact that aluminum, as Hume-Rothery expresses it,[41] "possesses many abnormal properties," and little is known about its electronic structure in the metal lattice. U. R. Evans pointed out[42] a similarity of aluminum to the transition metals in that both resist anodic corrosion. He suggested that the partial ionization of the aluminum atom in the metal lattice, as for the transition elements, may contribute to this behavior. In other words, the electronic structure peculiar to aluminum as well as the nature of the protective oxide film appear to play a part in its corrosion behavior.

All cases of passivity which fall only under definition 2 are adequately explained, for the most part, by the physical protection of metal oxides or other type chemical compounds.* Any change in surface chemical forces or secondary valence forces of the metal by these films is considered to be relatively unimportant, whereas for the metals under definition 1, whether or not such chemical compounds cover the surface,

---

* An exception is the relative inertness of pure or amalgamated zinc in dilute sulfuric acid, passive by definition 2. The film which stifles reaction in this case is supposedly an adsorbed hydrogen layer.

[40] J. H. G. Monypenny, *Stainless Iron and Steel*, pp. 343–373, Chapman and Hall, London, 1931.

[41] W. Hume-Rothery, *The Structure of Metals and Alloys*, p. 26, Institute of Metals Monograph and Report Series No. 1, 1936.

[42] U. R. Evans, *Metallic Corrosion, Passivity and Protection*, p. 357, Edward Arnold and Co., London, 1937.

the alteration of valence forces by adsorbed films, sometimes aided by alloying elements, is of first importance according to the Electron Configuration Theory.

The generalized film theory, therefore, is a satisfactory working theory for all cases of passivity. The Electron Configuration Theory supplements this by dealing more specifically with the details of mechanism in one group of passive metals and alloys (definition 1), about which most of the lengthy discussion of passivity in the past years has centered.

EFFECT OF CRYSTAL ORIENTATION ON CORROSION
Allan T. Gwathmey*

INTRODUCTION

A metal surface is a complex of crystal faces, edges, corners, boundaries, and disturbed layers. The properties of a surface are a composite of the properties of these many types of structure. In order to understand the behavior of a metal, therefore, the properties of these different surface structures must be understood.

Although it is impossible at the present time to predict in advance the crystal face which under a given set of conditions will be the most resistant to corrosion, a knowledge of the principles and experimental facts bearing on the effect of crystal face is helpful in obtaining a better understanding of the mechanism of corrosion.

Principles Involved

The driving force of a reaction is dependent on the change in free energy in going from the initial to the final state. Since the arrangement of atoms in a surface varies with the crystal face, the free energy of a solid surface varies with face. The free energy of a surface also depends on the medium in which it exists. Calculations made of the free surface energy of several ionic crystals[1] and of tungsten[2] indicate that the ratio of free energies between two faces may in some cases be as high as a factor of 5.

The many varied crystal forms in which solids exist as a result of growth and of etching suggest that the free energy varies with face. Contact potential,[3] thermionic[4] and photoelectric[5] emissions, and rupture strength[6] have been found to vary with face.

Free energy changes influence the forms which crystals will attain only when equilibrium is reached. In most cases the time required to establish this equilibrium is so long that the form having the lowest energy is not always reached in practice. The speed with which a reaction takes place is dependent on the energy of activation, which is the energy barrier through which the reactants must pass. There are few theoretical calculations of the energy of activation; however, in the adsorption of hydrogen on nickel the energy of activation has been shown mathematically to be a minimum for the (110) plane,[7] and therefore the rate of adsorption should be greatest

---

* School of Chemistry, University of Virginia, Charlottesville, Va.

[1] J. E. Lennard-Jones and P. A. Taylor, *Proc. Roy. Soc.* (A), **109**, 476–508 (1925); G. E. Boyd, "Surface Chemistry," *Publication* 21 of the American Association for the Advancement of Science, 128–140 (1943).
[2] R. Smoluchowski, *Phys. Rev.*, **60**, 661–674 (1941).
[3] H. E. Farnsworth and B. A. Rose, *Proc. Nat. Acad. Sci.*, **19**, 777–780 (1933).
[4] R. Smoluchowski, *Phys. Rev.*, **60**, 661–674 (1941).
[5] H. E. Farnsworth and R. P. Winch, *Phys. Rev.*, **58**, 812–819 (1940).
[6] F. Seitz, *The Physics of Metals*, pp. 165–167, McGraw-Hill Book Co., New York, 1943.
[7] G. Okamoto, T. Horiuti, and K. Hirota, *Sci. Pap. Inst. Phys. Chem. Res.*, Tokyo, **29**, 223 (1936); A. Sherman and H. Eyring, *J. Am. Chem. Soc.*, **54**, 2661–2675 (1932).

on this plane. In agreement it has been found experimentally that the rate of catalytic reaction between hydrogen and ethylene on nickel films is greatest when the (110) plane of nickel is parallel to the surface.[8] Thus for adsorption it is expected that the free energy change, which determines the equilibrium, and the energy of activation, which determines the rate, will vary with crystal face.

EXPERIMENTAL EVIDENCE FOR THE VARIATION OF CORROSION WITH CRYSTAL FACE

The most striking evidence for the variation of corrosion rate with crystal face is obtained in experiments with large single crystals on which the different faces are easily identified. A sphere is the most convenient form for the crystal since all possible faces appear parallel to the surface at some point on the sphere. Using interference colors[9, 10] and ellipticity of polarized light[11] as a measure of the thickness of oxide films, it has been found that the rate of oxidation of copper crystals in air over a temperature range of 80°–550° C(175°–1020° F) varies greatly with face, the ratio of the rates for two faces in some cases being as high as 3.5. The face of greatest rate varies among investigators, possibly owing to different methods of preparing the surfaces. The rates of oxidation of iron, nickel, lead, and zinc crystals in air at elevated temperatures also vary with face.[12, 13]

The acceleration of corrosion in aqueous media due to the presence of oxygen has been shown to vary with face. For example, when a single-crystal sphere of copper is immersed in water at room temperature through which carbon dioxide and hydrogen are bubbled and air is excluded, no etching takes place.[14] When carbon dioxide and air are passed through the water, definite crystal facets parallel to the (111) planes are developed in certain regions. Copper is also preferentially etched at room temperature by acetic acid containing dissolved air, while no appreciable etching takes place when air is excluded and an atmosphere of hydrogen maintained. Oxidizing acids, such as dilute nitric acid, etch copper preferentially with face. The relative rates of etching on various faces of copper crystals by acids containing hydrogen peroxide are given in the table below, the rate of attack on the (111) face being taken as unity.[15]

RELATIVE RATES OF CORROSION OF DIFFERENT FACES OF COPPER CRYSTALS

	Crystal Face		
	(111)	(100)	(110)
0.3 N HCl — 0.1 N H$_2$O$_2$	1	0.90	1.0
0.3 N Acetic acid — 0.1 N H$_2$O$_2$	1	0.90	0.55
0.3 N Propionic acid — 0.1 N H$_2$O$_2$	1	1.28	1.33

For iron crystals a slight preferential etching takes place with acetic acid in an atmosphere of hydrogen, but in an atmosphere of air the rate is greatly increased. Iron is etched preferentially by nitric and sulfuric acids. Both alkalies and salts have

---

[8] O. Beeck, A. E. Smith, and A. Wheeler, *Proc. Roy. Soc.* (A), **177**, 62–90 (1940).
[9] G. Tammann, *J. Inst. Metals*, **44**, 29–73 (1930).
[10] A. T. Gwathmey and A. F. Benton, *J. Phys. Chem.*, **46**, 969–980 (1942).
[11] B. Lustman and R. F. Mehl, *Trans. Am. Inst. Mining Met. Engrs.* **143**, 246–267 (1941).
[12] R. F. Mehl and E. L. McCandless, *Trans. Am. Inst. Mining Met. Engrs.*, **125**, 531 (1937).
[13] A. T. Gwathmey and H. Leidheiser, unpublished results, 1942–1944.
[14] A. T. Gwathmey and A. F. Benton, *Trans. Electrochem. Soc.*, **77**, 211–222 (1940).
[15] R. Glauner and R. Glocker, *Z. Krist.*, **80**, 377–390 (1931).

been found to etch metals preferentially. For example, a 10% solution of sodium hydroxide produces hexagonal pits on the basal plane of a zinc single crystal. A strong solution of ammonium persulfate in 15 hr converts a cylindrical crystal of copper into a twelve-sided prism with sides parallel to the (211) planes.[16] Hot lubricating oils in an atmosphere of air etch many of the common metals preferentially with crystal face, and hot gases reacting on metal surfaces often cause rearrangements with the development of special facets.[17] The electrodeposition of copper on a single crystal of copper and the electroetching of a single crystal of copper vary greatly with face at low current densities.[18]

It appears probable that all surface reactions between crystalline solids and liquids vary preferentially with face if the reactions are made to proceed slowly enough. Where the rate of etching is great, the tendency of the various faces to react preferentially may be hidden; for rapid diffusion at the sharp projections produced by preferential etching will increase the attack at these points, and slow diffusion in the valleys will decrease the attack there, the overall result being to produce a smoother surface. Thus copper may be electrolytically polished at high current densities, and a fairly smooth surface may be obtained by treating copper with concentrated acids. Fitting the reacting molecule structurally into the lattice of the solid surface seems to play an important part in surface reactions, because a certain critical angle between the facet developed by etching and the geometric surface of the solid is often required.

Since the shearing of metals varies greatly with crystal plane, the opening up of large cracks along slip planes, into which reacting gases and liquids may diffuse, is a factor which varies with crystal face and may affect *corrosion fatigue*. Most of the evidence, such as the above, for the variation in reaction with face has been obtained with large single crystals, but similar results are obtained in the case of polycrystalline materials when care is taken in the preparation of the surface. The differential etching of ordinary metallographic specimens of pure metals for examination under the microscope is an example of the effect of grain orientation on chemical attack.

There are several methods whereby the orientation of surface crystals in a polycrystalline material might be controlled in order to obtain the most resistant surface to corrosion. Mechanical working and the addition of special agents in electrodeposition[19] produce an orientation of surface crystals. For example, in the electrodeposition of copper, the addition of thiourea to the bath produces an orientation of the surface grains with the (100) faces parallel to the surface.[20] Sufficient studies have not yet been made on the control of orientation and its effect on corrosion in order to make practical use of these possibilities. A few studies have been made.[21]

[16] C. H. Desch, *The Chemistry of Solids*, p. 74, Cornell University Press, Ithaca, N. Y., 1934.
[17] A. T. Gwathmey and H. Leidheiser, unpublished results, 1942–1944.
[18] A. T. Gwathmey and A. F. Benton, *Trans. Electrochem. Soc.*, **77**, 211–222 (1940).
[19] C. S. Barrett, *The Structure of Metals*, pp. 381–442, McGraw-Hill Book Co., New York, 1943.
[20] F. L. Clifton and W. H. Phillips, *Metal Finishing*, **40**, 457 (1942).
[21] V. Caglioti, *Atti IV congr. naz. chim. pura applicata*, 442–446 (1933); R. Glauner and R. Glocker, *Z. Metallkunde*, **20**, 244–247 (1928).

CORROSION IN LIQUID MEDIA, THE ATMOSPHERE, AND GASES

ALUMINUM AND ALUMINUM ALLOYS[1]

R. B. MEARS*

Alloy 2S, sometimes known as commercially pure aluminum, contains about 99.0–99.3% aluminum. The rest of the alloy is made up mainly of iron and silicon with minor amounts of copper. Purer aluminum, containing up to about 99.95% aluminum, is also available in some quantity, and a small amount of very pure metal, over 99.99% aluminum, has been produced by electrolytic refining.

TABLE 1. THE COMPOSITIONS AND MECHANICAL PROPERTIES OF SOME WROUGHT ALUMINUM ALLOYS

Alloy and Temper	Composition	Typical Tensile Strength,* psi	Typical Yield Strength,* psi	Typical Elongation,* % in 2 in.
2S–O	"Commercially pure" aluminum	13,000	5,000	35
2S–H	" " "	24,000	21,000	5
3S–O	Al + 1.2% Mn	16,000	6,000	30
3S–H	Al + 1.2% Mn	29,000	25,000	4
17S–T	Al + 4.0% Cu + 0.5% Mn + 0.5% Mg	62,000	40,000	20
24S–T	Al + 4.5% Cu + 0.6% Mn + 1.5% Mg	68,000	46,000	19
52S–O	Al + 2.5% Mg + 0.25% Cr	29,000	14,000	25
52S–H	Al + 2.5% Mg + 0.25% Cr	41,000	36,000	7
53S–W	Al + 0.7% Si + 1.3% Mg + 0.25% Cr	33,000	20,000	22
53S–T	Al + 0.7% Si + 1.3% Mg + 0.25% Cr	39,000	33,000	14
61S–W	Al + 0.25% Cu + 0.6% Si + 1.0% Mg + 0.25% Cr	35,000	21,000	22
61S–T	Al + 0.25% Cu + 0.6% Si + 1.0% Mg + 0.25% Cr	45,000	39,000	12
56S†	Al + 5.25% Mg + 0.1% Mn + 0.1% Cr	....	....	....
Alclad‡ 3S–O	A duplex product made of a 3S core with a coating of 72S (Al + 1% Zn) on one or both sides.	16,000	6,000	30
Alclad‡ 3S–H	A duplex product made of a 3S core with a coating of 72S (Al + 1% Zn) on one or both sides.	29,000	25,000	4
Alclad§ 17S–T	Duplex products with cores of the alloys indicated (17S–T or 24S–T) and coatings on one or both sides of aluminum.	55,000	32,000	18
Alclad§ 24S–T	Duplex products with cores of the alloys indicated (17S–T or 24S–T) and coatings on one or both sides of aluminum.	64,000	43,000	18

* Based on sheet specimens 1/16 in. thick.
† Not available in sheet form.
‡ A similar alloy is called Clad.
§ A similar alloy is called Pureclad.

* Development Division, Aluminum Company of America, New Kensington, Pa. Present address: Research Laboratory, Carnegie-Illinois Steel Corp., Pittsburgh, Pa.

[1] See also R. J. McKay and Robert Worthington, *Corrosion Resistance of Metals and Alloys*, Chapter VI, Reinhold Publishing Corp., New York, 1936. E. H. Dix and R. B. Mears, *Mech. Eng.*, **58**, 786 (1936). *Aluminum in the Chemical Industry*, booklet published by Aluminum Company of America, 1944.

Pure aluminum has a relatively low strength, and its use is, therefore, rather limited. However, the resistance of pure aluminum to attack by most acids and many neutral solutions is higher than that of aluminum of lower purity or of that of most of the aluminum-base alloys.

The commercial alloys of aluminum generally contain one or more of the following elements: copper, silicon, manganese, magnesium, or zinc. Such additions materially increase the strength of the material or have other desirable effects.

TABLE 2. THE COMPOSITIONS AND MECHANICAL PROPERTIES OF SOME CAST ALUMINUM ALLOYS

Alloy and Temper	Composition	Typical Tensile Strength, psi	Typical Yield Strength, psi	Typical Elongation,* % in 2 in.
		Sand Castings		
43	Al + 5% Si	19,000	9,000	3.0
195–T4	Al + 4.5% Cu	31,000	16,000	8.5
214	Al + 3.8% Mg	25,000	12,000	9.0
356–T4	Al + 7.0% Si + 0.3% Mg	28,000	16,000	6.0
B214	Al + 1.8% Si + 3.8% Mg	20,000	13,000	2.0
220–T4	Al + 10% Mg	46,000	25,000	14.0
		Permanent Mold Castings		
43	Al + 5% Si	24,000	9,000	6.0
A108	Al + 4.5% Cu + 5.5% Si	28,000	16,000	2.0
356–T4	Al + 7.0% Si + 0.3% Mg	32,000	18,000	9.0
		Die Castings		
13	Al + 12% Si	33,000	18,000	1.8
43	Al + 5% Si	29,000	13,000	3.5
81	Al + 7% Cu + 3% Si	32,000	24,000	1.3
218	Al + 8% Mg	38,000	23,000	5.0

* Based on round specimens ½ in. diameter for sand castings and permanent mold castings, and on round specimens ¼ in. diameter for die castings.

The compositions and mechanical properties of some of the common aluminum-base alloys[2] are given in Tables 1 and 2. The wrought alloys and some of the cast alloys are available in different tempers identified by characteristic symbols. Thus, for wrought products, the soft or annealed temper is identified by an O, the fully work-hardened temper by an H, and various amounts of work hardening by the symbols ¼H, ½H, or ¾H. The strengths of certain alloys can be raised by appropriate thermal treatments. The symbol W for wrought alloys or T4 for castings indicates that the alloy has been given a solution heat treatment followed by quenching. After a solution heat treatment of this type, the properties of some of the alloys (such as 17S or 24S) alter rapidly as the alloy stands or "ages" at room temperature. This change in properties is termed age hardening or aging. The age hardening temper of these alloys is indicated by a T. Some of the other alloys must be aged at a temperature somewhat above room temperature. This is termed artificial aging or precipitation heat treatment. The artificially aged tempers of these alloys are indicated by a T for wrought products or by a T6 for castings.

[2] See also *Alcoa Aluminum and Its Alloys*, booklet published by Aluminum Company of America, 1944

Alclad alloys are duplex wrought products, supplied in the form of sheet, tubing, and wire, which have a core of one aluminum alloy and a coating, on one or both sides, of aluminum or another aluminum alloy. Generally, the core composes 90% of the total thickness with a coating comprising about 5% of the thickness on each side. The coating is metallurgically bonded to the core over the entire area of contact. In the most widely used Alclad materials, the coating alloys are selected so that they will be anodic to the core alloys in most natural environments. Thus the coating will electrolytically protect the core where it is exposed at cut edges, rivet holes, or at scratches. Such Alclad alloys are usually more resistant to penetration by neutral solutions than are any of the other aluminum-base alloys.

AQUEOUS MEDIA

Effect of O_2, Temperature, and pH

Oxygen. Aluminum-base alloys are relatively insensitive to the concentrations of oxygen present in most aqueous solutions. In general, high concentrations of dissolved oxygen tend to stimulate attack somewhat, especially in acid solutions, although this effect is less pronounced than for most of the other common metals. (See Table 3.) Carbon dioxide or hydrogen sulfide, even in high concentrations, appears to have a slight inhibiting action on the effect of aqueous solutions on aluminum alloys. Water solutions of sulfur dioxide have some etching action on aluminum, but even so aluminum resists the action of such solutions better than copper or steel. (See Table 4.) Water solutions of hydrogen chloride are strongly corrosive to aluminum. Hydrogen and nitrogen are without effect, except as they influence the oxygen content.

Table 3. EFFECT OF GAS ATMOSPHERE ON THE CORROSION RATE OF ALUMINUM (2S-½H)

Tests run on three specimens. Those below are averages.
Immersed area of specimens — 37.8 sq cm (5.85 sq in.).
Volume of solution — 150 ml.
Temperature — room.
Rate of aeration — 15 ml per min.

Solution	Gas	Duration of Test	Corrosion Rate	
			ipy	mdd
1.0% sodium carbonate	Oxygen bubbled through solution	18 hr	0.17	320
" " "	Air " " "	" "	0.17	320
" " "	Nitrogen " " "	" "	0.18	338
0.1% sodium hydroxide	Oxygen bubbled through solution	6 hr	0.65	1220
" " "	Air " " "	" "	0.64	1200
" " "	Nitrogen " " "	" "	0.63	1190
0.1% hydrochloric acid	Oxygen bubbled through solution	48 hr	0.066	124
" " "	Air " " "	" "	0.036	68
" " "	Nitrogen " " "	" "	0.006	11
" " "	Air over quiescent solution	" "	0.007	13

Temperature. At low temperatures (4° C [40° F] or below), the action of most aqueous solutions is much slower than at room temperature. However, in many solutions, increase in temperature above about 80° C (180° F) results in a decrease in rate of attack. Thus a temperature of 70°–80° C (160°–180° F) is likely to result in more severe corrosion than temperatures of 20° C (70° F) or 100° C (212° F).

pH. There is no general relationship between pH and rate of attack. The specific ions present largely influence the behavior. Thus most aluminum alloys are inert to strong nitric or acetic acid solutions but are readily attacked in dilute nitric, sulfuric, or hydrochloric acid solutions. Similarly, solutions with a pH as high as 11.7 may not attack aluminum alloys, provided silicates are present; but, in the absence of

TABLE 4. RESISTANCE OF ALUMINUM TO WATER SOLUTIONS OF SEVERAL GASES

Metal	Carbon Dioxide* and Water		Sulfur Dioxide,† Air, and Water		Hydrogen Sulfide‡ and Water	
	Av. Wt. Loss, grams	Av. ipy §	Av. Wt. Loss, grams	Av. ipy §	Av. Wt. Loss, grams	Av. ipy §
Aluminum(2S)	0.0003	0.00004	0.150	0.0498	0.002	0.00028
Copper	...	...	0.681	0.0701	0.237	0.01030
Steel	0.2153	0.00977	8.583‖	1.02‖	1.366	0.06800

* Metal specimens 1 × 4 × 1/16 in. (2.5 × 10.2 × 0.16 cm) were partially immersed (to a depth of 2 in.) (5.1 cm) in distilled water through which carbon dioxide and air were bubbled. The total period of exposure was 342 hours at room temperature.

† Metal specimens 1 × 4 × 1/16 in. thick were partially immersed (to a depth of 2 in.) in distilled water through which air and sulfur dioxide were bubbled. The total period of exposure was 135 hours at room temperature.

‡ Metal specimens 1 × 4 × 1/16 in. were partially immersed (to a depth of 2 in.) in distilled water through which hydrogen sulfide was bubbled. The total period of exposure was 320 hours at room temperature.

§ This calculation was based on the assumption that all corrosion was confined to the immersed areas of the specimens.

‖ Steel specimen corroded completely through at the water line.

silicates, attack may be appreciable at a pH as low as 9.0. In chloride-containing solutions, generally less action occurs in the near-neutral pH range, say 5.5 to 8.5, than in either distinctly acid or distinctly alkaline solutions. However, the results obtained will vary somewhat, depending on the specific aluminum alloy under consideration.

FRESH WATERS

Aluminum-base alloys are not appreciably affected by distilled water[3] even at elevated temperatures (up to 180° C [350° F] at least). Furthermore, distilled water is not contaminated by contact with most aluminum-base alloys. For this reason, there is a fairly extensive and satisfactory use of aluminum alloy storage tanks, piping valves, and fittings for handling distilled water.[4] Aluminum alloys such as 2S, 3S, 52S, 53S, and 61S are those most widely used for storage tanks and piping, while alloys such as 43 and B214 are used for cast valves and fittings.

Because natural fresh waters differ so widely in their composition and behavior, it is extremely difficult to make generalizations regarding the resistance of aluminum-base alloys to their action. Most commercial aluminum-base alloys show little or no general attack when exposed to most natural waters at temperatures up to 180° C (350° F) at least. However, certain waters may cause a severe localized attack or pitting. Pitting is of most importance where the metal section thickness is small,

[3] H. V. Churchill, *Ind. Eng. Chem.* (Analytical Ed.), **5**, 264 (1933).
[4] H. V. Churchill, *Chem. Met. Eng.*, **46**, 226 (1939).

ALUMINUM AND ALUMINUM ALLOYS

since the rate of attack at the pits generally falls off with increasing time of exposure. In general, the time necessary to perforate aluminum alloy sheet 0.040 in. (0.10 cm) thick or greater is prolonged, as attested to by the wide and successful use of aluminum tea kettles.

The Alclad products are much more resistant to perforation by pitting than are the other aluminum alloys. Therefore, wherever the characteristics of a specific water are not known in advance, it is safer to employ aluminum alloys such as Alclad 3S.

SEA WATER

Of the aluminum alloys in common use, those which do not contain copper as a major alloying constituent are resistant to unpolluted sea water. As in other natural waters, any attack which does develop in sea water is likely to be extremely localized (pitting). Therefore, rates of attack calculated from weight change data have little value. Measurement of change in tensile strength is the most widely used criterion. Typical results[5] are given in Table 14, p. 410.

STEAM CONDENSATE

Condensate from steam boilers, if free from carry-over of water from the boiler, is similarly inert to aluminum-base alloys. Thus, either wrought or cast aluminum alloys are used successfully for steam radiators or unit heaters. Where aluminum alloys are used, it is desirable to install suitable traps in the steam lines, since entrapped boiler water, especially if alkaline water-treating compounds are employed, may be corrosive.

ACIDS

Acid mine waters are corrosive to the aluminum-base alloys. The extent of attack depends upon the specific composition of the water. Some use of aluminum pipe has been made in soft coal mines for handling acid mine waters. It has been found that pipe of aluminum alloy 3S greatly outlasts bare or galvanized steel pipe in this application.

Many aluminum-base alloys are highly resistant to *nitric acid* in concentrations of about 80 to 99%. Alloys such as 2S, 3S, and 61S have received the widest use for handling nitric acid of these concentrations. Nitric acid of lower concentrations is more active. Figure 1 illustrates the relationship between rate of attack and acid concentration for commercially pure aluminum (2S).

Dilute *sulfuric acid* solutions, up to about 10% in concentration, cause some attack on aluminum-base alloys, but the action is not sufficiently rapid at room temperature to prevent their use in special applications. In the concentration range of about 40 to 95%, rather rapid attack occurs. In extremely concentrated or fuming acid, the rate of attack drops again to a very low value (Fig. 2).

The action on aluminum (2S) of solutions containing sulfuric acid, nitric acid, and water is illustrated in Fig. 3. It will be noted that aluminum is most resistant to solutions dilute in both acids, or high in nitric acid concentration, or in 100% sulfuric acid.

Hydrofluoric, hydrochloric, and *hydrobromic acid* solutions, except at concentrations below about 0.1%, are definitely corrosive to aluminum alloys. The rate of attack is greatly influenced by temperature (Fig. 4).

Both *perchloric acid* and *phosphoric acid* solutions in intermediate concentrations

---

[5] See also R. B Mears and R. H. Brown, *Trans. Soc. Naval Architects and Marine Engrs.*, **52**, 91-113 (1944).

definitely attack aluminum. Dilute (below 1%) phosphoric acid solutions have a relatively mild, uniform etching action that makes them useful for cleaning aluminum surfaces.

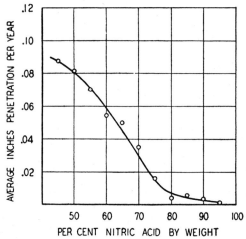

FIG. 1. Action of Nitric Acid on 2S-H Aluminum (90-day tests, room temperature).

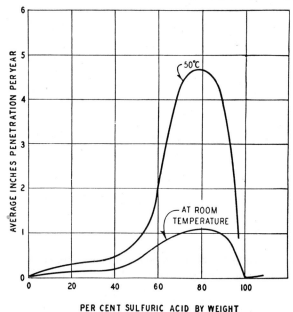

FIG. 2. Action of Sulfuric Acid on 2S Aluminum (24-hour tests).

Boric acid solutions in all concentrations up to saturation have negligible action on aluminum alloys.

Chromic acid solutions in concentrations up to 10% have a mild, uniform etching action. Mixtures of chromic acid and phosphoric acid have practically no action on a wide variety of aluminum alloys, even at elevated temperatures. Such mixtures are used for quantitatively removing corrosion products or oxide coatings from aluminum alloys.

Most organic acids are well resisted by aluminum alloys at room temperature. (See Fig. 5.) In general, rates of attack are highest for solutions containing about 1 or 2%

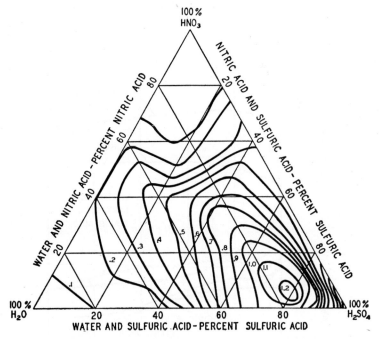

FIG. 3. Action of Mixtures of Nitric and Sulfuric Acids on 2S Aluminum (24-hour tests, room temperature; contours labeled in ipy).

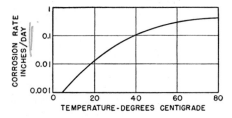

FIG. 4. Effect of Temperature on Corrosion Rate of 53 S-T Aluminum in 10% HCl.

of the acid. *Formic acid, oxalic acid,* and some organic acids containing chlorine (such as *trichloroacetic acid*) are exceptions and are definitely corrosive. Equipment made of aluminum alloys, such as 2S or 3S, is widely and successfully used for handling acetic, butyric, citric, gluconic, malic, propionic, and tartaric acid solutions.

Aluminum alloys also have a high resistance to the action of uncontaminated natural fruit acids. Contamination of these substances by heavy metal compounds may cause them to become corrosive. In contrast to this, the addition of sugar to fruit acids causes them to become even less corrosive.

Alkalies

Solutions of *sodium hydroxide* or *potassium hydroxide* in all but the lowest concentrations (less than 0.01%) rapidly attack aluminum and its alloys. Attack by the very dilute caustic solutions can be inhibited by suitable corrosion inhibitors (such as silicates or chromates; see pp. 914–915), but in more concentrated solutions none of the usual inhibitors is very effective. The alloys of aluminum containing more than about 4% magnesium are somewhat more resistant to attack by alkalies[6] than are the other aluminum-base alloys.

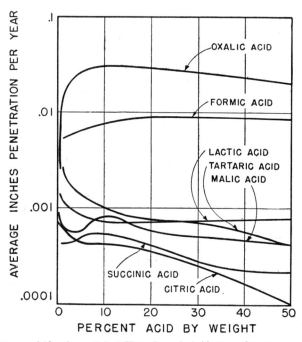

Fig. 5. Resistance of Aluminum (3 S-½H) to Organic Acids (1-week test, room temperature).

Lime or *calcium hydroxide* solutions are also corrosive, but the maximum rate of attack is limited by the low solubility of these materials.

The aluminum-base alloys are highly resistant to *ammonia*[7] and *ammonium hydroxide*. The alloys that contain appreciable magnesium tend to be even less affected by ammonium hydroxide solutions than the other aluminum alloys.

The *amines* generally have little or no action on aluminum alloys. However, a few of the most alkaline do cause definite attack.

Salt Solutions

Neutral or nearly neutral (pH from about 5 to 8.5) solutions of most inorganic salts have negligible or mild action on aluminum-base alloys at room temperature. This is true for both oxidizing and non-oxidizing solutions. Any attack that does

[6] L. J. Benson and R. B. Mears, *Chem. Met. Eng.*, **49**, 88 (1942).
[7] J. R. Willard and R. B. Mears, *Refrig. Eng.*, p. 2 (December, 1940).

occur in such solutions is likely to be highly localized (pitting) with little or no general corrosion. Solutions containing chlorides are likely to be more active than other solutions. The simultaneous presence of salts of the heavy metals (especially copper) and chlorides may be very detrimental.* Table 5 lists rates of attack of several types of salt solutions on commercially pure aluminum.

Distinctly acid or distinctly alkaline salt solutions are generally somewhat corrosive. The rate of attack depends on the specific ions present. In acid solutions, chlorides generally stimulate attack greatly. In alkaline solutions, silicates or chromates greatly retard attack.

TABLE 5. EFFECT OF SALT SOLUTIONS ON ALUMINUM (2S)

Size of specimen — partially immersed area = 6 sq in. (15.2 sq cm).
Volume of solution per specimen — 150 ml.
Velocity — quiescent.
Aeration — nat. convect.
Temperature — room.
Duration of test — 6 weeks.

Salt	Concentration, %	Average Depth of Attack, ipy
Ammonium sulfate	1	No appreciable attack
	5	" " "
Magnesium sulfate	0.0001	0.0000
	0.001	0.0000
	0.01	0.0004
	0.1	0.0003
	1.0	0.0002
	10.0	0.0003
Sodium nitrate	0.0001	0.0009
	0.001	0.0010
	0.01	0.0014
	0.1	0.0012
	1.0	0.0000
	10.0	0.0001
Sodium sulfate	0.0001	0.0006
	0.001	0.0003
	0.01	0.0001
	0.1	0.0000
	1.0	0.0000
	10.0	0.0000
Sodium thiosulfate	1	No appreciable attack
	5	" " "

EFFECT OF VELOCITY

In the case of neutral solutions, the velocity of the solution, up to about 6 meters per sec (20 ft per sec), seems to have little effect on the rate of attack. In some cases increased movement of the liquid may actually reduce attack by assuring greater uniformity of environment. However, increases in velocity apparently decrease the variation in pH that can be tolerated without special erosive attack occurring.

* Mercury salt solutions are likely to be extremely corrosive. Mercury is deposited which then acts as described on p. 618.

Galvanic Effects

Aluminum is anodic to many of the other common metals in many types of aqueous solutions. This means that galvanic attack is likely to occur on aluminum articles exposed in aqueous solutions in contact with parts made of dissimilar metals.[8] Contact with copper or copper-base alloys often results in very severe galvanic action on the aluminum parts of the couple. Contact with steel often causes galvanic action on the aluminum, but in certain solutions and in some natural waters this action may be reversed so that attack of the steel is accelerated and the aluminum is protected. Contact between stainless steel and aluminum in sea water or other saline solutions usually results in less galvanic action on the aluminum than does contact of aluminum with steel.[9] However, no cases of reversal of this action, such as occurs with ordinary steel, are known.

Cadmium and aluminum are similar in potential in many aqueous solutions. Consequently, usually little galvanic action results when aluminum is coupled to cadmium. Zinc is anodic to aluminum in most neutral or acid solutions; hence in such solutions contact with zinc results in protection of the aluminum articles.[10] In alkaline solutions, the potentials reverse so that in these media contact with zinc can cause accelerated attack of aluminum.

Mechanical Factors

Constant stress below the yield strength does not affect the rate of attack of most commercial aluminum alloys under normal conditions of use.[11] Alloys rendered susceptible to intergranular attack by undesirable thermal treatments may show accelerated attack when stressed. Cyclically applied stresses in combination with exposures to corrosive environments (corrosion fatigue) may result in increased attack. The frequency and magnitude of the cyclic stresses and the particular alloy and environment to which it is exposed will all influence the results obtained.

Vibration[12] or agitation of a liquid in contact with aluminum alloys will generally increase attack of the metal over that in quiescent exposure.

In many types of exposure, *cold work* does not appreciably affect the resistance to corrosion of a wide variety of aluminum-base alloys. In solutions of the non-oxidizing acids, however, cold work seems to stimulate attack to some extent and also indirectly stimulates attack of the aluminum alloys containing over about 5% magnesium. With the latter alloys, severe cold work increases the tendency for a magnesium-aluminum constituent to precipitate selectively from solid solution. On exposure to certain media, selective attack of this constituent then occurs.

Metallurgical Factors

Since corrosion is an electrochemical phenomenon, it might be expected that alloys composed of one homogeneous phase or of two or more phases, all of which had very similar solution (galvanic) potentials, would be more resistant to attack than alloys composed of two or more phases with widely different solution potentials. This expectation is generally fulfilled.[13] Thus pure aluminum or single-phase alloys

---

[8] R. H. Brown, *Am. Soc. Testing Materials Bull.* 126 (January, 1944).
[9] R. B. Mears and R. H. Brown, *Ind. Eng. Chem.*, **33**, 1001 (1941).
[10] R. B. Mears and H. J. Fahrney, *Trans. Am. Inst. Chem. Eng.*, **37**, 911 (1941).
[11] R. B. Mears, R. H. Brown, and E. H. Dix, Jr., *Symposium on Stress-Corrosion Cracking of Metals (1944)*, p 323, American Society for Testing Materials and American Institute of Mining and Metallurgical Engineers, Philadelphia and New York, 1945.
[12] R. B. Mears and L. J. Benson, *Ind. Eng. Chem.*, **32**, 1347 (1940).
[13] R. B. Mears and R. H. Brown, *Ind. Eng. Chem.*, **33**, 1001 (1941).

of aluminum and magnesium or aluminum and silicon are all relatively resistant to attack. Al-Cu alloys heat-treated and quenched to retain the copper in solid solution are much more resistant to attack than are similar alloys treated so that the copper precipitates out of solution as a constituent,[14] $CuAl_2$, which differs in solution potential[15] from the matrix solid solution and may cause intergranular corrosion.* Al-Mn alloys (such as 3S or 406) are highly resistant to corrosion, although the manganese constituent is present as a separate phase. The reason for this is that the manganese constituent has a solution potential very similar to that of the matrix.

ORGANIC COMPOUNDS

Aqueous solutions of the organic acids and amines have already been discussed. Aqueous solutions of organic chemicals having a substantially neutral reaction are generally without appreciable action on aluminum-base alloys, unless these solutions are contaminated with other substances, particularly chlorides and heavy metal salts.

At room temperature or slightly above, most organic compounds, in the absence of water, are completely inert to aluminum-base alloys. This is true for organic-sulfur compounds as well as for other organic compounds. At elevated temperatures some organic compounds, such as *methyl alcohol* and *phenol*, become definitely corrosive, especially when they are completely anhydrous.

GASES

Most gases, in the absence of water and at or near room temperature, have little or no action on aluminum-base alloys. In the presence of water, the acid gases, such as HCl and HF are corrosive, and wet SO_2 also has some action (see p. 42). *Hydrogen sulfide* or *ammonia*, either in the presence or absence of water and at room temperature or slightly above, has negligible action on aluminum-base alloys.

Halogenated hydrocarbons, such as *dichlorodifluoromethane, dichlorotetrafluoromethane,* and *monochlorodifluoromethane,* are almost completely inert to aluminum.[16] However, *methyl chloride*[16] and *methyl bromide* are corrosive and should not be used in contact with aluminum-base alloys.

PROTECTIVE MEASURES IN AQUEOUS MEDIA

The most appropriate protective measures vary with the conditions of exposure and the service requirements. Attack by nearly neutral waters or dilute neutral salt solutions can usually best be avoided by the use of:

1. Alclad alloys.[17]
2. Protective surface-conversion or anodic coatings.[18]
3. Cathodic protection by zinc attachments.[19]
4. Appropriate chemical inhibitors.[20]
5. Organic coatings.[21]

* For further discussion of heat treatment and intergranular corrosion, see p. 54.

[14] F. Keller and R. H. Brown, *Aluminum Res. Lab. Tech. Paper* 9 (1943).
[15] E. H. Dix, Jr., *Trans. Am. Inst. Mining Met. Engrs.* (Inst. Metals Div.), **137**, 19 (1940).
[16] J. R. Willard and R. B. Mears, *Refrig. Eng.*, **40**, 381 (1940).
[17] E. H. Dix, Jr., *Mining and Met.*, p. 397 (1927).
[18] J. D. Edwards, *Trans. Electrochem. Soc.*, **81**, 341 (1942). (See *Anodic Treatment of Metals*, p. 857).
[19] R. B. Mears and H. J. Fahrney, *Trans. Am. Inst. Chem. Engrs.*, **37**, 911 (1941). R. H. Brown and R. B. Mears, *Trans. Electrochem. Soc.*, **81**, 455 (1942).
[20] R. B. Mears and G. G. Eldredge, *Trans. Electrochem. Soc.*, **83**, 403 (1943).
[21] J. D. Edwards and R. I. Wray, *Ind. Eng. Chem.*, **27**, 1145 (1935). J. D. Edwards and R. I. Wray, *Ind. Eng. Chem.* (Analytical Ed.), **7**, 5 (1935). H. J. Fahrney and R. B. Mears, *Chem. and Met. Eng.*, **49**, 86–89 (1942).

In definitely acid or definitely alkaline solutions, methods 4 and 5 are generally the most effective methods of protection. The avoidance of harmful effects may also be classed as a protective measure. Among the most common harmful effects are galvanic action, resulting from direct contact between aluminum and a dissimilar metal, such as copper, and indirect galvanic effects resulting from contact between aluminum and solutions containing reducible compounds of heavy metals. In some cases a suitable design or construction[22] will prevent serious corrosion even though no other factors are altered. Similarly, since the various alloys of aluminum differ widely in behavior, the selection of the most suitable alloy is important.

In a limited number of cases, corrosion can be prevented by removing some minor constituent from the contacting liquid or gas. For instance, copper compounds, which may make a water corrosive to aluminum, can be removed by passing the water through a tower packed with aluminum chips. Finally, the use of suitable periodic cleaning procedures[23] may be highly beneficial in specific cases.

SOIL CORROSION

The extent of attack that occurs on aluminum alloys buried underground varies greatly, depending on the soil composition and climatic conditions. In dry, sandy soils corrosion is negligible. In wet, acid or alkaline soils, attack may be severe.

TABLE 6. SOIL BURIAL TESTS OF FIVE YEARS DURATION* WITH ALUMINUM ALLOY SPECIMENS

Alloy	Well-Drained Soil			Marshy Soil		
	Max. Depth of Attack,† Inches	Per Cent Change in Tensile Strength‡	Remarks	Max. Depth of Attack, Inches	Per Cent Change in Tensile Strength	Remarks
2S-½H	0.0017	−1	Mild general etching	0.0280	−7	Pitted
52S-½H	0.0007	+1	" " "	0.0140	0	"
53S-T	0.0007	0	" " "	0.0150	0	"
53S-T, Alrok #13 coated.	0.0006	0	" " "	0.0006	0	Mild general etching
53S-T, Alumilite #204 coated.	0.0003	0	" " "	0.0002	+2	" " "
17S-T	0.0380	−20	Severe pitting	0.0310	−41	Severely pitted
Alclad 24S-T	0.0013	0	Mild general etching	0.0028	−1	Generally etched
Steel	0.0640	−27	Completely perforated at 3 spots	0.0190	−17	Pitted

* Specimens in the form of panels 3 × 9 × 0.034 in. thick were buried to a depth of 2 ft in soil at the property of the Aluminum Research Laboratories in New Kensington, Pa.
† Depth of attack determined by microscopic examination of cross sections.
‡ Change in tensile strength determined by machining tensile specimens from the panels after exspoure and comparing their strength with that of unexposed tensile specimens of the same materials.

Results[24] of soil corrosion tests in two locations are summarized in Table 6. In both these locations, panels of the various alloys were buried in clayey soil of the Aluminum Research Laboratories' property in New Kensington, Pa. One location

[22] R. B. Mears and J. R. Akers, *Proc. Am. Soc. Brewing Chemists*, Fifth Annual Meeting (1942).
[23] J. R. Akers and R. B. Mears, *Soap and Sanitary Chemicals*, **17**, 25 (1941).
[24] From unpublished work by G. G. Eldredge and R. B. Mears.

was in relatively well-drained soil and the other was in a marshy area less than 100 ft away. In the well-drained soil, attack on all the aluminum-base alloys, except 17S–T, was mild after five years. The 17S–T was severely attacked, although to not so great an extent as was the steel.

In the marshy soil, maximum depths of attack on all the uncoated aluminum-base alloys, except Alclad 24S–T, were appreciable and of the same order as on ordinary steel, although the relative loss in tensile strength was definitely less for most of the aluminum-base alloys than for steel. In the case of the Alclad 24S–T, the attack that occurred was all confined to the coating, as would be expected. Chemical dip and sulfuric acid anodic coatings were definitely protective to 53S–T (and presumably to the other aluminum alloys). Steel was corroded less severely than 17S–T in the marshy exposure. Apparently the reason for this is that in the marshy exposure less oxygen is present in the soil, which is beneficial to the steel but not to 17S–T. This same reversal in behavior of 17S–T and steel has been noted in sea water tests, where steel corrodes faster when exposed at tide level than when continuously immersed; the reverse is true for 17S–T.

ATMOSPHERIC CORROSION

OUTDOOR EXPOSURES

The aluminum-base alloys as a class are highly resistant to normal outdoor exposure conditions. The alloys containing copper as a major alloying constituent (over about 1%) are somewhat less resistant than the other aluminum-base alloys, whereas the Alclad alloys are generally the most resistant. Results[25] of typical outdoor exposure tests are given in Table 7. The data given in the table are all based on exposure of machined tensile specimens, 0.064 in. (0.16 cm) thick. Changes in strength would be smaller if panels had been exposed and machined into specimens after exposure. Also, if the specimens had been thinner, obviously the losses would be greater; if they had been thicker, the losses would have been smaller. This effect of thickness is especially pronounced in the case of aluminum-base alloys, since the rate of attack greatly decreases with increasing time of exposure. This "self-stopping" action has been mentioned before[26] and can be readily seen from the data in Table 7. Note, for example, that the loss in tensile strength for 17S–T exposed at Point Judith is nearly as great (9%) after one year as after ten years (12%).

All the specimens referred to in Table 7 were freely exposed to the outdoor locations. If they had been partially sheltered, the rate of attack would have been somewhat greater; if they had been largely sheltered, very little attack would have occurred. Apparently, in the case of aluminum-base alloys, periodic exposure to rain is beneficial probably because the rain washes off corrosive products that settle out of the air. This view is supported by the behavior of the aluminum alloy specimens at Georgetown, British Guiana. The exposure site of Georgetown was immediately adjacent to the Demerara River about 1½ miles from its mouth. The climate is warm, and there are two wet seasons and two dry seasons each year. The total rainfall each year is about 100 in. This exposure proved to be one of the mildest of the group. Evidently, free exposure to rain is not harmful but, on the contrary, appears to be actually beneficial.

The gases ordinarily found in industrial atmospheres have little effect in accelerating the corrosion of aluminum-base alloys. Settled carbon particles may accelerate

---

[25] See also Reports of Committee B–3, Sub. VI, *Proc. Am. Soc. Testing Materials.*
[26] E. H. Dix, Jr., *Proc. Am. Soc. Testing Materials*, **33**, Pt. II, 405 (1933).

CORROSION IN LIQUID MEDIA, ATMOSPHERE, GASES

TABLE 7. RESULTS OF OUTDOOR EXPOSURE TESTS ON ALUMINUM–BASE ALLOYS

All alloys were exposed as machined tensile specimens, 0.064 in. thick.
The data are all averages of at least three specimens, generally from several lots of material.

% Change in Tensile Strength

Alloy and Temper	Years at Point Judith, R. I.				Years at New Kensington, Pa.				Years at Georgetown, British Guiana				Years at Edgewater, N. J.			
	1	2	4	10	1	2	5	10	1	2	5	10	1	2	5	10
2S–O	−1	−2	−2	−7	−1	−1	−2	−4	0	0	−2	−1	−7	−2	−1	−4
2S–H	−1	−3	0	....	+1	−1	..	−4	0	0	0	−1	0	−1	0	−4
3S–O	−2	−1	−1	−4	−1	−2	..	−5	+1	−1	−2	0	−6	−2	−2	−4
3S–H	−1	−6	0	....	+2	−2	−1	−4	−2	0	−1	0	0	−1	−4	−2
17S–T	−9	−11	−13	−12	−4	−4	−5	−8	−2	−3	−2	−1	−5	−4	−7	−6
24S–T	−7	−10	−12	....	−5	−4	−6	..	..	..	..	..	..	..	..	..
52S–½H	−1	−1	−1	....	−1	−3	..	..	..	..	..	..	..	..	..	..
53S–T	−2	−6	−5	....	−1	−2	..	..	..	..	..	..	..	..	..	..
Alclad 17S–T	...	−3	−1*	−3	..	..	..	−1	+3	+2	0	+1	+2	+2	+1	+1
Alclad 24S–T	0	−1	+1	....	0	−1	0	..	..	..	..	..	..	..	..	..

Alloy and Temper	Years at Miami, Fla.			Years at St. Louis, Mo.			Years at Golden Gate, Calif.		Years at Wilmington, N. C.			Years at Pitcairn Island, British Oceania	
	1	2	4	1	2	4	1	2	1	2	4	1	2
2S–O	−1	+2	...	−2	−1	−1	0	−4	−5	−4	−9	0†	−2†
2S–H	...	...	...	...	...	...	...	...	...	...	...	...	...
3S–O	0†	−2†	...	−1†	−3†	−2†	...	...	−2†	−4†	−5†	...	...
3S–H	...	...	...	...	...	...	...	...	...	...	...	...	...
17S–T	−1	−2	...	−2	−3	−5	−5	...	−13	−12	−21	...	...
24S–T	−2	−2	−3	−3	−5	−6	−6	−6	−12	−14	−20	...	...
52S–½H	...	...	...	+2	0	−1	0	0	−2	−1	−3	−1	+2
53S–T	+1	+1	...	−2	−2	−5	+2	...	−3	−4	−9	+1	+1
Alclad 17S–T	...	...	...	...	...	...	...	...	...	...	...	...	...
Alclad 24S–T	0	0	+2	+2	...	0	...	...	...	...	...	+2‡	0‡

* Exposed for 5 yr.   † ½H temper.   ‡ RT temper.

corrosion by galvanic action. Under outdoor atmospheric exposure conditions, this factor is of secondary importance even in intensely industrial regions, such as Pittsburgh and Altoona. Sulfur compounds, such as H_2S, have no specific effect in accelerating the tarnishing or corrosion of aluminum alloys. In many cases, settled pools of water containing hydrogen sulfide or carbon dioxide will cause less action than similar pools from which these gases are absent. However, the highly acidic nature of water containing dissolved SO_2 or SO_3 causes it to become somewhat corrosive.

INDOOR EXPOSURES

The effects of indoor exposures differ greatly, depending on the exposure conditions. Exposure indoors in homes or offices ordinarily causes, at most, only a mild surface dulling of aluminum-base alloys even after prolonged periods of exposure. In damp locations, especially where there is contact with moist insulating materials, such as wood, cloth, and paper insulation, attack may be more appreciable. In factories or chemical plants, fumes or vapors incident to the operations being conducted may cause a definite surface attack. However, in most indoor atmospheres where pools of contaminated water do not remain in prolonged contact with aluminum alloys, or where extended contact with moist, porous materials is avoided, no appreciable loss of mechanical properties through corrosion will occur. In particular, aluminum alloys are highly resistant to warm, humid conditions where there is appreciable moisture condensation so long as contact with porous materials is avoided. Bare aluminum alloy panels have been used in constructing humidity cabinets that operate just above the dew point at 50° C (125° F). After five years' use, there is no corrosion other than a mild surface staining.

Galvanic Action

Aluminum-base alloys are anodic to many of the other common structurally used metals and alloys. Thus, if aluminum-base articles are exposed outdoors or in moist locations in contact with parts made of dissimilar metals, galvanic attack of the aluminum surfaces adjacent to the dissimilar metal is likely to occur. Galvanic action is much more pronounced in marine or seacoast atmospheres than in rural or industrial locations.

Contact with copper or copper-base alloys causes more pronounced galvanic attack than contact with most other metals. In rural or industrial locations, contact with ordinary steel does not generally cause a very pronounced acceleration in rate of attack of aluminum-base alloys (especially the Al-Cu alloys such as 17S–T and 24S–T). In seacoast locations, attack may be appreciably accelerated. Contact with stainless steel is usually even less harmful than contact with ordinary steel. Cadmium has about the same solution (galvanic) potential as aluminum. Therefore, contact with cadmium usually results in negligible galvanic action. Zinc is definitely anodic to most aluminum-base alloys under most conditions of exposure and tends to protect adjacent aluminum areas cathodically.

Magnesium and its alloys are definitely anodic to the aluminum alloys. Thus contact with aluminum increases the corrosion of magnesium. However, such contact is also likely to be harmful to aluminum, since magnesium may send sufficient current to the aluminum to cause *cathodic corrosion*. Cathodic corrosion, as mentioned above, is most likely to be encountered in seacoast locations. Certain aluminum-base alloys (such as 56S or 220) are less affected by contact with magnesium than are the other aluminum alloys. For this reason 56S rivets have been extensively employed in assembling magnesium alloy structures.

In designing outdoor structures, it is often necessary to combine dissimilar metals in the structure. Suitable protective methods are available* that, if adopted, will greatly reduce the risk of galvanic corrosion.

Use of Clad Alloys

Because of mechanical considerations, it is not always possible to use the most corrosion-resistant aluminum alloy for a specific application. The availability of Alclad alloy products has gone a long way toward solving this problem. The Alclad alloys have strengths nearly as high as those of the core alloy and yet

* To be published.

have excellent resistance to corrosion. Alclad alloys are now supplied as sheet, plate, wire, rod, and tubing.

In assemblies, the regions most susceptible to corrosion are the joints. Where Alclad sheet (or plate) is riveted or spot-welded to shapes of a non-clad alloy, the coating on the Alclad sheet (or plate) will usually tend electrolytically to protect adjacent areas of the shapes. In aircraft, where strong lightweight construction receives great emphasis, the skin sheet is normally made of unpainted Alclad 24S-T sheet and the stiffening members are formed from Alclad 24S-T sheet or are made from 14S-T or 24S-T extrusions. The extrusions are commonly anodically coated and painted before assembly, but no paint covering is considered necessary on the Alclad sheet even for aircraft that fly over the ocean.

In some cases other mechanical factors, such as formability or hardness, may be of great importance in the selecting of an appropriate alloy for a specific application. Aluminum alloys such as 2S, 3S, or 52S, in the softer tempers, are readily formable and are also highly resistant to corrosion. If greater strength is required, alloy 61S should be considered. This alloy combines good formability (in the W temper) with relatively high strength and good resistance to corrosion.

Metallurgical Factors

Metallurgical factors are often of importance in influencing rates of corrosion. They are probably best known in the case of the Al-Cu alloys of the duralumin type. Such alloys contain about 4% copper (alloys 17S-T and 24S-T). This amount of copper is soluble in solid aluminum at elevated temperatures (above 480° C [900° F]) but is not entirely soluble at room temperature. After fabrication, such alloys are commonly heat-treated at about 490° C (920° F) in order to effect solid solution of the copper in the aluminum. They are then immediately quenched in cold water, which retains the copper in solution. Upon standing at room temperature, after quenching, the hardness and strength of the alloys increase, approaching maximum values after aging about 4 days. It is generally assumed[27] that this age hardening is caused by the precipitation of some $CuAl_2$ constituent from the Al-Cu solid solution. However, the precipitate particles, if present, are in a very finely divided state and, when in this condition, the alloys, from a corrosion standpoint, behave as if they were substantially single-phase alloys; that is, they are relatively resistant to corrosion. It is in this quenched and room-temperature-aged condition that they are generally used and, as such, are susceptible only to pitting corrosion with no selective attack at grain boundaries. However, if the alloys are quenched more slowly from the heat-treating temperature, that is, quenched in boiling water instead of cold water, they become susceptible to selective grain boundary attack (intergranular corrosion). This type of attack is attributed to the selective precipitation of relatively large particles of the $CuAl_2$ constituent at the grain boundaries.[28,29] The Al-Cu solid solution adjacent to the grain boundaries becomes depleted in copper, since the copper is precipitated out of solution. This depleted zone is definitely anodic to the Al-Cu solid solution of the main body of the grain and also to the precipitated particles of $CuAl_2$. Consequently the depleted zone corrodes out, giving an intergranular form of attack.

Somewhat similar results occur if the rapidly quenched Al-Cu alloy is heated

---

[27] W. L. Fink, *J. Applied Phys.*, **13** (No. 2), 75–83 (1942).
[28] E. H. Dix, Jr., *Trans. Am. Inst. Mining Met. Engrs.* (Inst. Metals Div.), **137**, 19 (1940).
[29] R. B. Mears and R. H. Brown, *Ind. Eng. Chem.*, **33**, 1003 (1941).

(artificially aged) to a somewhat elevated temperature (above 120° C [250° F]) for a critical period of time.[30] This heating also causes the alloy to become susceptible to intergranular attack. However, if the heating is carried out for a sufficiently extended period of time, the susceptibility to intergranular attack again disappears, probably because substantially all the copper has precipitated out of solid solution and, therefore, the zones adjacent to the grain boundaries are no more depleted in copper than are the other areas in the grain boundaries.

For many of the other aluminum-base alloys, metallurgical factors have relatively little effect on resistance to corrosion. Alloys such as 2S, 3S, 52S, 53S, and 61S are relatively insensitive in this respect.

Protective Measures

The methods used for protecting aluminum-base alloys from atmospheric effects are:

1. Alloy selection.
2. Joint-sealing compounds.
3. Anodic and surface conversion coatings.
4. Organic coatings.
5. Metal spray coatings.
6. Electrodeposited coatings.

The first method, alloy selection, is generally the most important. As indicated above, aluminum-base alloys, such as 2S, 3S, 52S, 53S, Alclad 3S, Alclad 17S-T, and Alclad 24S-T, are highly resistant when freely exposed to most natural environments. They will all discolor or darken appreciably in most outdoor exposures but will suffer no structurally appreciable changes in properties unless exposed in relatively thin sections (below about 0.030 in. [0.076 cm] thick).

Joints, depressions, or other areas that pocket moisture and dirt are more susceptible to corrosion than regions freely exposed to the atmosphere. Most plastic or semi-solid joint-sealing compounds that conform and are firmly adherent to adjacent metal surfaces are highly effective in preventing special attack in these regions. Some of these joint-sealing compounds that contain soluble inhibitors (generally chromates) are particularly suitable.

Anodic coatings (see p. 857), particularly those applied in a sulfuric acid electrolyte and suitably sealed, are highly effective in preventing discoloration or surface staining of the aluminum-base alloys mentioned above. In addition, aluminum members that are used architecturally are more readily cleaned of atmospheric dirt if they have been anodically coated.

Organic coatings are widely used on aluminum-base alloys. As with other metals, the method of surface preparation employed before painting is of great importance. The surface should be grease-free and should generally be treated with a phosphoric acid type cleaning compound (or a chromic acid solution) prior to painting. As a priming coat, zinc chromate pigmented paints are probably the most satisfactory. The primer should be followed by one or two coats of a good finish paint. Where protection from corrosion is the main result desired, aluminum pigmented finish coats are generally the most satisfactory. The anodic or surface conversion coatings make excellent paint bases. Therefore, for finishes of the highest protective value, such coatings are often applied prior to priming and painting.

---

[30] R. B. Mears, R. H. Brown, and E. H. Dix, Jr., *Symposium on Stress-Corrosion Cracking of Metals* (*1944*), p. 323, American Society for Testing Materials and American Institute of Mining and Metallurgical Engineers, Philadelphia and New York, 1945.

Metal-spray coatings have received limited use in protecting aluminum-base alloys. If coatings of metals cathodic to aluminum are used, severe galvanic attack at pores in the coatings is likely to develop. Thus spray coatings of copper, brass, nickel, iron, tin, or chromium are not generally used on aluminum-base alloys. Spray coatings of cadmium or zinc do not have this disadvantage, since they either have the same solution potential as the aluminum alloy base material or else are anodic to it. To some extent, coatings of these latter metals compete with Alclad aluminum alloys. Wherever a suitable Alclad alloy product can be obtained, it is generally better to use it rather than to apply a zinc or cadmium spray coating. The Alclad product will probably be cheaper than the metal-sprayed product and will also generally be at least as resistant to attack.

Electrodeposited coatings are generally applied to aluminum-base alloys in order to achieve certain decorative effects or to obtain some special property, such as ease of soldering or abrasion resistance. Such coatings are not used in general to give protection from corrosion. As is the case with spray coatings, electrodeposited coatings of the metals cathodic to aluminum will result in severe undermining of the coating and attack of the aluminum alloy base at any discontinuity in the coating when the article is exposed in a corrosive environment. In mild exposures such coatings may have the advantage of retaining a bright surface appearance when aluminum itself would discolor.

BERYLLIUM

W. J. Kroll*

This element has properties in common with magnesium, aluminum, and silicon. It is easily attacked by mineral acids, including hydrofluoric and nitric acids, and slowly forms soluble berylliates in strong alkalies. Despite the fact that its oxide has a high melting point, it does not improve the heat-resisting properties of alloys and steels to any degree of commercial importance.† The metal has been rolled, but it is brittle in the cold state.

The corrosion data given below refer to crystalline powder, the properties of which vary greatly with the physical state and fineness of the material. The rolled metal is probably more resistant to attack.

Beryllium is slightly tarnished in air. Oxidation is slight up to 400°–500° C (750°–930° F). At 800° C (1470° F) the oxide forms more readily and burning starts at 1200° C (2200° F). The behavior is analogous in oxygen, but temperatures are lower.

The nitride forms in the presence of nitrogen at 500° C (930° F). HCN and CO also react, the latter forming a carbide. Hydrogen does not react with beryllium.

Cold and hot water do not attack the metal, but steam reacts at elevated temperatures. Molten alkalies react explosively.

Fluorine reacts at room temperature. Chlorine, bromine, and iodine react at elevated temperatures, as do NO_2, H_2S, and the elements sulfur, selenium, tellurium, boron, arsenic, and phosphorous.

Concentrated nitric acid, when cold, does not attack the metal, but reacts violently when hot. Hydrochloric acid of any concentration dissolves beryllium rapidly, as

---

* Consulting Metallurgist, Albany, Oregon.

† It is stated that as little as 0.001% beryllium added to A. S T. M. No. 4 magnesium alloy (6% Al, 0.2% Mn, 3% Zn) raises the approximate ignition temperature from 580° C (1080° F) to over 800° C (1470° F). (L. Carapella and W. Shaw, *Metals and Alloys*, **22**, 415 [1945].) Editor.

does hydrofluoric acid. Concentrated sulfuric acid reacts slowly, but the dilute acid more rapidly. Acetic, citric, tartaric acids react, but the action stops after a time. Most of the heavy metal chlorides in aqueous solution are reduced. Caustic solutions such as 10% KOH dissolve the metal slowly.

GENERAL REFERENCE

GMELIN, L., *Handbuch der anorganischen Chemie* (26), 8th Ed., pp. 68-71, Verlag Chemie, 1930.

CADMIUM

See *Cadmium Coatings*, p. 837.

CHROMIUM

See *Nickel and Chromium Coatings*, p. 824.

COBALT ALLOYS

W. A. WISSLER*

Cobalt is so similar in chemical properties to nickel, and at the same time so much more costly, that it is seldom, if ever, used as a pure metal. It cannot be classed as highly corrosion resistant.

A series of cobalt-base alloys containing chromium and either tungsten or molybdenum are produced commercially.† In general, the chromium content ranges from 25 to 35%, the tungsten from 0 to 20%, and the molybdenum, if used, up to 6%. Carbon is present in amounts of 0.20 to 2.50%. Iron, manganese, silicon, and nickel are always present in small amounts as impurities, but may be added to confer special properties.

The particular advantage gained by the use of cobalt as the base instead of iron is that the resulting alloy possesses a high degree of red hardness, or, in other words, remains much harder at elevated temperatures than alloys of similar composition

TABLE 1. COMPOSITION OF COBALT-BASE CORROSION-RESISTANT ALLOYS*

	No. 8	No. 6	No. 12	No. 4	No. 1	Star J
Chromium	30	30	30	30	31	32
Tungsten	2	4.50	8	14	13	17
Carbon	0.35	1.10	1.30	0.60	2.50	2.40
Cobalt	Bal.	Bal.	Bal.	Bal.	Bal.	Bal.
Hardness, Rockwell C	35	40	48	50	54	61

* Haynes Stellites.

having an iron base. This property makes it highly suitable for resistance to wear, which is usually accompanied by high local surface temperatures, and also for use in valves and similar parts operating at elevated temperatures. However, this same property makes forging and rolling operations very difficult, and only the softer grades can be worked. In Table 1, a number of standard commercial alloy grades are arranged in the order of increasing hardness. Of these, No. 8 and No. 6 can, with some difficulty, be forged, rolled, and machined.

* Union Carbide and Carbon Research Laboratory, Niagara Falls, N. Y.
† Haynes Stellite, registered trade mark, Haynes Stellite Co., Kokomo, Ind.

Although these alloys have often been called stainless metals because they retain their luster in the atmosphere, they are not commonly used solely for resistance to corrosion. Their high cost, together with the difficulty of working, make them uneconomical for such uses, unless abrasive conditions are also present. Since the alloys are hard and wear resistant, there are a number of applications in which they are used to withstand the combination of wear and corrosion. A typical example is the No. 4 grade, which is being used as castings for dies and other tools used in the forming and handling of filling mixtures of the ordinary electric dry cell. It is estimated that the life is from four to eight times that of the steels formerly used.

TABLE 2. WEIGHT LOSSES OF CAST Co–Cr–W ALLOYS (STELLITES) IN ACIDS, SALTS, AND NaOH

Weight Loss in mdd*

Size of specimen — 2.5 × 5.0 × 0.6 cm (1 × 2 × ¼ in.).
Surface preparation — unknown.
Velocity and aeration — nat. convection.
Duration of test — 15 hr.

Corrosive Medium	Wt %	Temp.	Alloy Grade, No. 6	Alloy Grade, No. 1
Sulfuric acid	10	Room Temperature	2.4	x†
		Boiling	38,870	1,330
Nitric acid	10	Boiling	x	21.0
	Conc.‡	"	12,040	1,540
Hydrochloric acid	10	Room Temperature	1,260	140
	10	Boiling	12,326	20,090
	Conc.	Boiling	1,680,000	2,520
Chlorine in water	Saturated	Room Temperature	1,820	x
Phosphoric acid	38	Boiling	9.8	119
Acetic acid	10	Boiling	16.8	x
	Conc.	"	21.0	x
Ferric chloride	10	Room Temperature	665	x
		Boiling	12,040	501
	30	Room Temperature	3,570	x
		Boiling	37,250	7,000
Cupric chloride	10	Room Temperature	567	x
		Boiling	6,230	17,010
Ferric sulfate	10	Room Temperature	7.0	x
		Boiling	11.9	x
Sodium hydroxide	10	Boiling	249	70
Sodium bichromate	20	Boiling	7	x

* These corrosion rates do not include pitting, concerning which information is not available. Of the media listed above, pitting might be expected in ferric and cupric chlorides and in Cl_2 sat'd water. Refer to *Pitting in Stainless Steels and Other Passive Metals*, p. 165. EDITOR.

† x — no measurable loss.

‡ Conc. — commercial concentrated.

Dilute aqueous solutions containing oxygen, including potable waters, do not in general affect the alloys, but water containing chlorine and hypochlorites may slowly produce corrosion accompanied by pitting. The alloys are attacked by solutions of strong acids, alkalies, and some heavy metal salts, typical examples of which are given in Table 2.[1]

Erosion tests have been made to establish the comparative value of various alloys

[1] *Properties of Haynes Stellite Alloys*, Haynes Stellite Company, Kokomo, Ind.

for use as steam turbine blades. Samples of commonly used metals were fastened to a rotating fixture in such a way that at a peripheral speed of 1200 ft per sec (366 meters per sec) they met and passed through a ⅛-in. jet of water. The erosion of samples sheathed with No. 6 alloy, as measured by loss in weight, was 0.03% per min, and less than that of 12% Cr (0.2%), tantalum (0.35%), 5% Ni steel (0.4%), and Nitralloy (0.1%).[2]

Very little attack occurs on exposure to mine waters, sea water,* and boiler waters at ordinary temperatures. Conditions in boilers and especially superheaters that allow sodium salts to accumulate and fuse will cause corrosion of these alloys, used, for example, as valve facings.

ATMOSPHERIC CORROSION

The reflectivity of polished surfaces of these alloys is about 65% that of silver.[3] This degree of reflectivity is unaffected by exposure to both inland and marine atmospheres. In severe industrial atmospheres, darkening by a dirt film would probably take place as with other materials. It is expected that luster in this instance could be restored simply by cleaning off the dirt.[4]

As the grade usually employed, No. 8, is reasonably resistant to surface abrasion, it has been widely used for reflectors of arc lights and similar applications. In one test, it was found that two reflectors, after having been exposed 18 months to the marine atmosphere of Kure Beach, N. C., showed an average reflectivity of 62.3% before being cleaned and 63.5% after being cleaned.[4]

ALLOY OF 65% COBALT, 30% CHROMIUM, 5% MOLYBDENUM

The 65% Co–30% Cr–5% Mo alloy† is one of the softer alloys used for dentures and inserts employed in repairing or replacing bones of the human body. Surgical parts inserted in the body are allowed to remain in place permanently. Dentures and surgical appliances have shown no pitting or loss of luster after years of service.

Corrosion tests simulating body and mouth conditions have been made at about body temperatures (37.5° C, 99° F) for 30 days in aqueous solutions containing 10% sugar, 10% lactic acid, vinegar, and ammonium sulfide. No weighable loss or pitting occurred. In dilute hydrochloric acid (5.8 ml of concentrated acid to 500 ml of water or about 0.014 N) the loss was 0.59 mdd over a 30-day period.[5] "Passivating" with nitric acid avoids an initial corrosion loss which may occur in some media.

When galvanically coupled with aluminum, copper, nickel, tin, dental amalgam, and silver in 1% NaCl, the corrosion rate of the Co-Cr-Mo alloy is unaffected, at least in short-time tests. No weighable loss occurs for either electrode of couples with gold or 18-8 stainless steel immersed in 1% NaCl for 4 days.[5]

* In sea water, Stellites are susceptible to contact corrosion and pitting. See pp. 417–418.
† Vitallium.

[2] C. Richard Soderberg, "Turbine Blade Erosion," *Elec. J.*, **32**, 533 (December, 1935).
[3] W. Coblentz and R. Stair, *J. Research Natl. Bur. Standards*, **2**, 343 (1929).
[4] Private Communication, F. L. LaQue, International Nickel Co., New York, N. Y.
[5] A. W. Merrick, *Cobalt-Chrome Alloys in Prosthetic Dentistry*, published by Austenal Laboratories, New York, N. Y.

COLUMBIUM

Clarence W. Balke*

Columbium is highly resistant to corrosion only at ordinary temperatures. An increase of temperature to 100° C (210° F) results in a greatly increased corrosion rate. It is much more active than tantalum and much more susceptible to hydrogen embrittlement.

The metal used in the tests reported below was in the form of thin annealed sheet. It contained a small amount of tantalum, and traces of titanium, iron, and carbon.

CORROSION IN AQUEOUS MEDIA

The results of laboratory tests with the common acids and alkalies are presented in Table 1. Hydrofluoric acid, or a mixture of HF and HNO_3, dissolves the metal

Table 1. CORROSION OF COLUMBIUM IN VARIOUS MEDIA

Size of specimen — sheet 0.2 mm (0.008 in.) thick
Area of surface — 26 sq cm (4 sq in.).
Immersion — 75% of surface in liquid; 25% in air.

Solution	Temperature		Duration of Test, days	Corrosion Rate		Condition of Specimen at End of Test
	°C	°F		mdd	ipy	
HCl, 20%	21	70	82	0.025	0.0000	No change
HCl, conc.	21	70	82	0.6	0.0001	Slight etch, not embrittled
HCl, conc.	100	212	67	23.4	0.004	Brittle
HNO_3, conc.	100	212	67	0.0	0.0000	No change
Aqua regia	22	72	6	0.0	0.0000	No change
H_2SO_4, 20% by vol.	21	70	3650	0.02	0.0000	No change, total loss = 0.7%
H_2SO_4, 25% by vol.	21	70	3650	0.03	0.0000	No change, total loss = 1.0%
H_2SO_4, conc., 98%	21	70	3650	0.56	0.0001	Partial embrittlement; loss = 18.3%
H_2SO_4, conc.	50	122	67	4.8	0.0008	Brittle
H_2SO_4, conc.	100	212	32	113.1	0.019	Brittle
H_2SO_4, conc.	150	302	2	1247.0	0.213	Brittle
H_2SO_4, conc.	175	347	1	8320+	1.42+	Completely dissolved
H_2SO_4, conc. $+CrO_3$	100	212	42	46.4	0.008	Pitted and brittle
H_3PO_4, 85%	21	70	82	0.07	0.0000	No change
H_3PO_4, 85%	100	212	31	19.3	0.0033	Brittle
Tartaric acid, 20%	22	72	82	0.0	0.0000	No change
Oxalic acid, 10%	21	70	82	3.3	0.0006	Brittle
NH_4OH	21	70	82	0.0	0.0000	No change
Na_2CO_3, 20%	100	212	50	7.4	0.0013	Brittle
NaOH, 5%	21	70	31	6.6	0.0011	Action at surface of liquid
NaOH, 5%	100	212	5	108.6	0.019	Brittle
KOH, 5%	21	70	31	44.2	0.008	Action at surface of liquid
KOH, 5%	100	212	5	274.4	0.047	Brittle
H_2O_2, 30%	21	70	61	1.1	0.0002	Oxide film, not brittle

* Fansteel Metallurgical Corporation, North Chicago, Ill.

completely. In many cases, even where the weight loss is small, the metal is rendered brittle. In addition to the substances listed in the table, numerous sodium salts, lactic, acetic, and perchloric acids, 5% phenol solution, and various acid inks have been tested at 22° C (72° F) for a period of 82 days. In these cases there was no change in weight or deterioration of the specimens.

ATMOSPHERIC CORROSION

In 1931, plates of twelve metals were fastened to a marble slab with Bakelite pins, and exposed on the outside of a factory building at North Chicago, Illinois. A near-by chemical plant frequently contaminates the air with chlorine, HCl, and HF. Examined thirteen years later, the tantalum showed no tarnish; the columbium was still bright but showed a slight tarnish. Tungsten and then molybdenum showed more color and tarnish in that order. The remainder of the metals were black and appreciably corroded.

COPPER

R. A. WILKINS* AND R. H. JENKS*

Electrolytic tough pitch copper is the type of copper most widely used in the United States. However, there are a variety of commercial coppers and certain modified forms of electrolytic copper that are also of industrial significance.

There are several types of fire-refined copper characterized by the presence of moderate quantities of minor impurities, notably arsenic and silver. There are synthetic Lake coppers usually comprising electrolytic copper to which arsenic or silver has been added in small amounts. In addition, there are the oxygen-free coppers where the refining and casting procedure has been altered to provide a material substantially free of oxygen and possessing high electrical conductivity as distinguished from the deoxidized coppers in which oxygen has been substantially removed by the use of a deoxidizing element, such as phosphorus.

Table 1 lists typical compositions of some of the commonly encountered commercial coppers and Table 2 some physical and mechanical properties.

All these coppers have varying characteristics, particularly with respect to certain physical properties. However, this variation does not extend in any broadly significant way to the matter of corrosion resistance; hence this discussion is generally applicable to all the types of copper enumerated above.[1]

AQUEOUS MEDIA

The electrode potential of copper in both aerated and air-free aqueous solutions is cathodic to the reversible hydrogen electrode, except in very strong acid solutions (above 5 N). Ordinarily the reduction of H^+ ions is not possible as a cathodic reaction, and in the absence of oxygen no local action process is available to provide metal solution. Therefore, generally speaking, copper is not appreciably corroded in the absence of oxygen. In aerated natural waters low in anions and dissolved carbon dioxide, a protective film of cuprous oxide and cupric hydroxide is formed on the copper which appreciably retards its rate of corrosion.

* Research and Development, Revere Copper and Brass, Inc., Rome, N. Y.
[1] J. L. Gregg, *Arsenical and Argentiferous Copper*, pp. 27–40, Chemical Catalog Co., New York, 1934.

TABLE 1. TYPICAL COMPOSITIONS OF COPPER

	Oxygen-Free High Conductivity	Electrolytic (Remelted Cathodes)	Prime Lake	Natural Arsenical Lake	Casting
Copper	99.98	99.953	99.92	99.88	99.44
Silver	0.002	0.0018	0.03	0.03	0.01
Lead	0.0004	0.001	0.0000	0.0000	0.05
Bismuth	Trace	0.0000	0.0000	0.0000	0.01
Arsenic	0.0008	0.0000	0.0020	0.0400	0.02
Antimony	0.0028	0.0009	0.0000	0.0000	0.05
Selenium and tellurium	0.0031	0.0026	0.0000	0.0000	.....
Iron	0.0015	0.0038	0.0020	0.0020	0.38
Nickel	0.0016	0.0028	0.0015	0.0015	0.15
Sulfur	0.0025	0.0026	0.0015	0.0015	0.002
Oxygen	0.0000	0.0315	0.0430	0.0450	Trace
Silicon	0.0015	.....	.....	.....	.....

TABLE 2. TYPICAL PHYSICAL AND MECHANICAL PROPERTIES OF THE COPPERS

Property	Material			
	Electrolytic Copper	Phosphorized Copper	Arsenical Copper	Argentiferous Copper (4–12 oz Ag per ton)
Melting point — °C	1,083	1,083	1,080	1,083
°F	1,981	1,981	1,976	1,981
Density — lb/cu in.	0.322	0.323	0.323	0.322
grams/cc	8.91	8.94	8.94	8.91
Coefficient of thermal expansion — 20° to 300°C	1.77×10^{-5}	1.77×10^{-5}	1.77×10^{-5}	1.77×10^{-5}
Electrical conductivity — % IACS	101	85	50	101
Thermal conductivity — gram cal/(sec)(sq cm)(°C/cm) at 20°C	0.934	0.81	0.40	0.934
Modulus of elasticity — psi	1.7×10^7	1.7×10^7	1.7×10^7	1.7×10^7
Tensile strength* — psi { Hard	50,000	50,000	52,000	50,000
Soft	32,000	32,000	33,000	32,000
Yield strength* — psi { Hard	45,000	45,000	46,000	45,000
Soft	10,000	10,000	10,000	10,000
Elongation in 2 in.* { Hard	6	6	6	6
Soft	45	45	45	45
Rockwell Hardness* { Hard	50 B	50 B	54 B	50 B
Soft	40 F	40 F	42 F	40 F

* For strip 0.040 in. thick.

NATURAL WATERS

The rate of corrosion of copper in water largely depends on the rate of diffusion of oxygen to the metal-liquid interface. Any factor which changes the solubility of oxygen in a solution will of necessity change the corrosion rate. It would be erroneous to assume, however, that the corrosion rate is directly proportional to the dissolved oxygen since the proportionality probably exists only if the chemical and physical characteristics of the solution either prevent the formation of films or permit the deposition of exclusively non-protective types. In solutions wherein protective metallic compounds are capable of existing, oxygen performs a dual role — promoting corrosion by its reaction at cathodic areas and retarding it by formation and maintenance of protective oxygen-bearing primary films and secondary corrosion layers. Thus a limited amount of dissolved oxygen at the copper surface may constitute a more corrosive condition than a more ample supply.[2]

Natural fresh waters generally promote the formation of protective coatings on copper. Actual rates of corrosion are usually very low; consequently copper is widely used in water lines, water tanks, and heat exchangers.

With very soft waters (rain water and some mineral waters) characterized by low

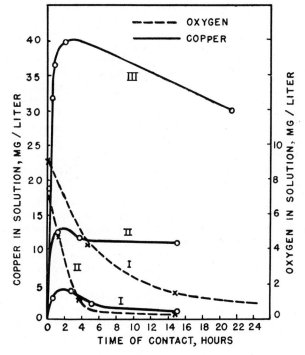

FIG. 1. Copper and Oxygen Contents of Tap Water Containing Varying Amounts of Added Carbon Dioxide, After Standing for Different Periods of Time in Copper Pipes at 20° C.

I ... 45 mg CO_2/liter
II ... 180 mg CO_2/liter
III ... 210 mg CO_2/liter

permanent hardness in combination with high carbon dioxide and oxygen content, the rate of corrosion may become excessive[3] (Fig. 1). The carbonic acid formed, even though a weak acid, prevents the formation of the protective films ordinarily developed on copper. The activity of carbon dioxide in such waters can of course be reduced or prevented by appropriate treatment of the water. (See *Hot and Cold Water Systems*, p. 496.)

[2] R. A. Wilkins, E. S. Bunn, and W. Lynes, *Proc. Am. Petroleum Inst.*, Tenth Midyear Meeting, Sec. III, **21**, 46–57 (1940).
[3] L. Tronstad and R. Veimo, *J. Inst. Metals*, **66**, 17–32 (1940).

Other constituents, such as acids, metallic chlorides, ammonia, and hydrogen sulfide, occasionally present in natural waters may also accelerate corrosion.

SEA WATER*

Copper withstands sea water corrosion as satisfactorily as any commercially available metal, provided that relative velocity with respect to the water is low. It is widely used for marine applications. Generally copper is corroded at a low, uniform rate, amounting to approximately 0.002 ipy. Analysis of the corrosion products on copper exposed to sea water shows them to be usually a mixture of cupric oxychloride, cupric hydroxide, basic cupric carbonate, and calcium sulfate. Cuprous oxide also occurs, typically, as a distinct layer adjoining the metal. The protective value of such a scale formed on copper is generally much greater in stagnant or slowly moving water than in rapidly moving waters.

An interesting and traditional use of copper in contact with sea water is exemplified by the sheathing of wooden vessels for the dual purpose of protecting the wooden construction and eliminating the fouling of the bottom by marine growths. Copper in this application possesses adequate resistance to the corrosive attack of sea water, but at the same time there is slow solution of copper sufficient to prevent the growth of barnacles and similar marine organisms. The function of copper in this respect is recognized by the inclusion of copper compounds in marine anti-fouling paints. (See *Anti-fouling Measures,* p. 441.)

STEAM CONDENSATES

As with natural waters, carbon dioxide in combination with oxygen are the constituents that render steam condensates corrosive to copper. With boiler water that is properly treated chemically or is naturally non-corrosive, no significant corrosion of copper tubes occurs on the steam side. The primary consideration in the selection of a heat-exchanger tube is the corrosiveness of the cooling water under the conditions of velocity and temperature anticipated, rather than the action of condensate.

Where condensate is fed back to the boilers, conditions must be maintained that will prevent appreciable solution of copper in the condenser, regardless of whether such solution is of significance with respect to the life of condenser tubes. Copper so dissolved can redeposit on boiler tubes, resulting in galvanic corrosion of the tubes with serious consequence.†

It should be recognized, however, that the use of copper in the form of condenser tubes is limited, and that where the usual alloys of copper are utilized, this particular type of corrosion difficulty is minimized.

ACIDS, SALTS, AND BASES

Copper is substantially resistant to the chemical attack of a large number of acids, salts, and bases. Attack by mineral and organic acids is dependent largely upon the presence of an oxidizing agent in solution.[4,5] Solutions of non-oxidizing acids containing little or no oxygen have little corrosive effect regardless of concentration.

---

* Refer also to discussion beginning p. 393 and Table 3, p. 394.

† This has also occurred with galvanized hot water tanks coupled to copper pipes .(L. Kenworthy, *J. Inst. Metals,* **69,** 67 [1943].)

[4] R. P. Russell and A. White, *J. Ind. Eng. Chem.,* **19,** 116–118 (1927).

[5] G. H. Damon and R. C. Cross, *J. Ind. Eng. Chem.,* **28,** 231 (1936).

Acids of inherent oxidizing capacity, such as nitric, sulfurous, concentrated sulfuric, and acids carrying metallic salts readily susceptible of chemical reduction, for example Fe^{+++}, rapidly dissolve copper. Figure 2 shows the effect of dissolved oxygen on the corrosion of copper by dilute acids at room temperature.[4] *

Copper is resistant to alkaline solutions† but not to those containing ammonium hydroxide or substituted ammonium compounds or cyanides.

Copper has excellent resistance to most non-oxidizing salt solutions but low resistance to oxidizing salt solutions. Low concentrations of ferric, stannic, mercuric, cupric, and other ions that are readily susceptible of reduction, as well as complex or substituted ammonia ions, are a source of danger when present in otherwise non-oxidizing solutions.[6] Acid chromate solutions corrode copper rapidly.

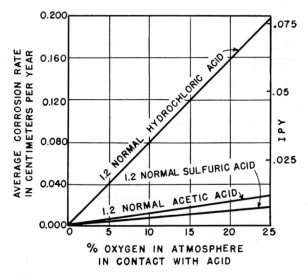

FIG. 2. Effect of Dissolved Oxygen on the Corrosion of Copper by Dilute Acids. (Temp. ... room; duration of test ... 24 hours. [In O_2-free runs, duration was 2 days to 2 weeks.] O_2-N_2 mixtures bubbled through acids 10 bubbles per minute.)

Potential Behavior. Copper, when immersed in aerated aqueous solutions, tends to establish a potential which shows little variation with time or pH if a stable protective film is formed. When protective films, such as cuprous oxide, are not formed, the potential is generally displaced toward more active values.[7] Figure 3 shows the effect of time on the electrode potential of copper in various solutions of HCl at 25° C (77° F).[8]

---

* Data for copper immersed in acetic acid are given in Table 6, p. 108.

† Reported rates in hot concentrated caustic soda up to 400° C (750° F) range from 1 to 120 mdd. Refer also to Table 7, p. 108. EDITOR.

[6] R. J. McKay and R. Worthington, *Corrosion Resistance of Metals and Alloys*, pp. 406–430, Reinhold Publishing Corp., New York, 1936.

[7] N. D. Tomashov, *Compt. rend. (Dokl.) acad. sci. U. R. S. S.* (N. S.), **23**, (7) 649–652 (1939). (In English.)

[8] O. Gatty and E. C. R. Spooner, *The Electrode Potential Behaviour of Corroding Metals in Aqueous Solutions*, pp. 182–252, Clarendon Press, London, 1938.

ORGANIC COMPOUNDS AND GASES

The resistance of copper to attack by organic compounds is varied by the role which complex anions and cations play. Many of the chlorinated and halogenated hydrocarbons react with copper. In many cases the action is accelerated by moisture, free acids, free chlorine, or by decomposition products formed by exposure to sunlight and heat.

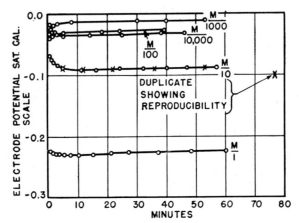

FIG. 3. Potentials of Copper in HCl Solutions of Various Concentrations.

Copper is widely used in systems employing various organic refrigerants containing halogen atoms. This use is entirely successful and devoid of difficulties arising as a consequence of corrosion. The complete absence of moisture is essential to successful operation in such applications.[9]

At room temperature, dry fluorine, chlorine, bromine, and iodine do not corrode copper appreciably, but are corrosive when wet. Ammonia and substituted ammonias are corrosive.

Sulfur and some sulfur compounds have a strong tendency to corrode copper.

IMPINGEMENT ATTACK

Impingement attack (erosion-corrosion) occurring at high water velocities will frequently cause failures where copper otherwise is not susceptible to failure. When impingement attack is anticipated, means of reinforcement or protection should be employed, or preferably an alloy of copper should be used (see pp. 76, 93, 100, 393) which possesses adequate resistance to this type of corrosion.[10]

GALVANIC COUPLING

When copper and a metal more active in electrochemical potential are coupled in an electrolyte, or when solutions of copper salts come in contact with such a metal, corrosion of the metal may be stimulated. In exposure to the milder types of aqueous solutions, such as rain water and other fresh waters, dilute salt solutions, and moist

---

[9] J. W. Mellor, *Comprehensive Treatise on Inorganic and Theoretical Chemistry*, Vol. 3, pp. 69–104, Longmans, Green and Co., New York, 1941.
[10] G. D. Bengough and R. May, *J. Inst. Metals*, **32**, 81–269 (1924).

atmospheres, the galvanic attack that results when copper is coupled with copper alloys, or with nickel, lead, and tin is usually negligible. Copper when coupled with iron and zinc has a tendency to accelerate attack of these metals.

Contact of copper with aluminum, magnesium, or zinc under corrosive conditions should be avoided. In industrial applications, galvanized steel or iron has been subject to failure as a consequence of couples involving copper.[11]

MECHANICAL AND METALLURGICAL FACTORS

Corrosion Fatigue

Corrosive types of fresh water may reduce the fatigue durability of copper perceptibly. As might be expected, the corrosion fatigue of copper is usually more marked in salt water or other strongly corrosive solutions than in fresh water. Copper is peculiarly sensitive to the so-called notch effect when in the hard condition. This sensitivity is drastically reduced when in the soft or annealed condition.[12]

Effect of Arsenic

There is some evidence that arsenic as it occurs in Lake copper, or in copper to which it is intentionally added, exerts a moderately beneficial effect in improving the resistance to certain types of corrosion, notably that of sea water. The presence of arsenic also reduces the tendency of copper to scale at high temperatures.

Effect of Cold Work

The varying commercial tempers in which copper is supplied are the consequence of varying degrees of cold work. With respect to corrosion, the temper so induced in copper is of no significance, except as tempered copper might be coupled with annealed copper. In this event, slight potential differences exist which can, under proper conditions, lead to preferential corrosion of the hard metal.

SOILS

Copper is resistant to the corrosive action of most soils. However, soils with a high content of organic matter or alkaline soils, in which the ratio of chlorides and carbonates to sulfate is high, may be corrosive. Copper should not be embedded directly in cinders or in tidal marshes where it may be subjected to attack by sulfur compounds.

ATMOSPHERIC CORROSION

Copper in exposure to the atmosphere slowly develops a thin, protective coating or patina, but otherwise is resistant to atmospheric attack. The green patina developed on copper consists essentially of basic copper sulfate except at the seaboard, when it is accompanied by more or less copper oxychloride. The stable basic sulfate is chemically the same as the mineral brochantite, which is extremely resistant to atmospheric action. Sulfur compounds derived from products of combustion

---

[11] U. R. Evans, *Metallic Corrosion Passivity and Protection*, p. 529, Edward Arnold and Co., London, 1937
[12] D. J. McAdam, Jr., and R. W. Clyne, *J. Research Natl. Bur. Standards*, **13**, 527 (1934).

(especially sulfur dioxide) and disseminated by wind are necessary agents in the development of the green patina. At relatively long distances from urban areas or the sea the formation of the green patina ceases. After a long period of exposure, some 70 or 80 years, the composition of the patina becomes stable or mineralized, after which no further change takes place.[13]

With copper, as with other materials, the lack of proper provision for expansion and contraction of the metal with temperature changes in conjunction with mild atmospheric corrosion can lead to corrosion fatigue failure. Adequate provisions for such expansion and contraction and the use of sufficiently heavy copper should always be made, particularly in the design of structures utilizing extended areas of metal.

While normal atmospheres, or even industrial atmospheres, do not present conditions which of themselves will cause corrosion of copper to the point of failure, there may be complicating conditions arising from the nature of the structure that lead to active corrosive attack. For example, a "line" corrosion can occur where water is retained by a heavy solder seam or held by capillarity along a shingle line, with consequent establishment of an oxygen concentration cell and corrosion of the copper. An overhanging structure leading to dripping on a copper surface may result in a combination of erosion and corrosion, particularly in industrial areas where the dripping water is likely to be contaminated with combustion products.

TABLE 3. TYPICAL USES OF COPPER IN CHEMICAL EQUIPMENT

Material	Equipment
Phenolic resins	Stills, fractionating columns, condensers
Soda pulp	Caustic evaporators
Copper nitrate	Pumps
Copper sulfate	Piping, storage
Potassium hydroxide	Evaporator tubes
Sodium hydroxide	Evaporator tubes
Acids	
Acetic	Stills
Citric	Filters, piping
Fatty	Piping, stills
Gallic	Piping
Lactic	Evaporators
Oxalic	Evaporators
Phthalic	Stills
Salicylic	Evaporators, piping
Tannic	Autoclaves
Alcohols	
Amyl	
Benzol	
Ethyl	
Ethylene glycol	
Glycerol	
Isoamyl	Coils, kettles, pumps, stills, storage
Isobutyl	
Isopropyl	
Methyl	
Propyl	
Formaldehyde	Stills, reactors, storage
Glycerol	Condenser tubes

[13] W. H. J Vernon and L.Whitby, *J. Inst. Metals*, **42**, 181–197 (1929).

COPPER-ZINC ALLOYS*

C. L. Bulow†

Compositions

Brass is an alloy consisting of a solid solution of zinc in copper. The brasses range from about 45% copper up to almost pure copper, with accompanying noticeable variations in the physical properties.

There are hundreds of modifications of the brasses,[1] most of them consisting of the addition of 0.1% to approximately 5.0% of fifteen different elements, added singly or in combination. The elements and the more commonly used concentrations are tabulated below. Not all these elements are commonly added to commercial Cu-Zn alloys, however.

	Percentage	
Lead	0.1 to 12	
Aluminum	0.1 to 3.0	
Tin	0.5 to 6.0	
Nickel	0.5 to 10	(sometimes up to 30)
Iron	0.1 to 2.0	
Silicon	0.1 to 2.0	
Manganese	0.05 to 5	(sometimes up to 25)
Phosphorus	0.01 to 0.10	
Arsenic	0.01 to 1.0	
Antimony	0.01 to 0.1	
Gold	0.5 to 1.0	
Bismuth	0.1 to 3.0	
Vanadium	0.1 to 0.5	
Tungsten	0.1 to 2	
Chromium	0.05 to 0.5	

Many of these added elements render the alloys unsuitable for cold working, but may increase the tensile strength of castings. Some castings, for example, have tensile strength values as high as 100,000 psi. The physical properties of several of these alloys are described in the A.S.T.M. Standards on Copper and Copper Alloys. Properties of the common brasses and condenser tube alloys are listed in Table 2, p. 556.

DEZINCIFICATION

Many Cu-Zn alloys corrode in major part by a process called dezincification. (See Glossary for definition.) This occurs in local areas (plug type, Fig. 11B, p. 1108) or in broad areas (uniform layer type, Fig. 11A, p. 1108). As the name implies, zinc is lost from the alloy, leaving as a residue, or by a process of redeposition, a porous mass of copper having little mechanical strength. Soft waters especially may lead to corrosion failures from localized dezincification of the brasses containing much zinc, such as Muntz metal (60% Cu, 40% Zn), non-inhibited aluminum brass (76% Cu, 22% Zn, 2% Al), and yellow brass (67% Cu, 33% Zn) containing no dezincification inhibitor. Red brass (85% Cu, 15% Zn), Admiralty metal (70% Cu, 29% Zn, 1% Sn, 0.05% As or Sb) and arsenical aluminum brass (76% Cu, 22% Zn, 2% Al, 0.05% As), generally resist dezincification in these corrosive waters, which explains their widespread use as piping materials. Where conditions are quite severe (i.e., where the carbon dioxide

---

* The reader should also consult *Condenser Corrosion*, p. 545.
† Bridgeport Brass Co., Bridgeport, Conn.
[1] William Campbell, "A List of Alloys," *Proc. Am. Soc. Testing Materials*, **30**, Pt. 1, 336–397 (1930).

content of the water is high) the coating of these alloys with tin gives added protection.

In the process of localized dezincification, porous plugs of copper may form *in situ* up to approximately 0.2 in. in depth within one or more years. Usually when plug-type dezincification occurs in pipe lines the copper plugs or localized zinc-depleted areas develop on the bottom side of horizontal runs of piping and in a random manner on vertical sections of piping and where the pipe screws into a fitting. This condition indicates that debris of one sort or another settling on brass surfaces or the presence of crevices (such as threaded joints) aids to initiate this localized corrosion. Frequently, plug-type dezincification in heavy-walled yellow-brass pipes, may perforate the pipe wall, whereupon the water oozes through the porous plug of copper to the exterior. The water may evaporate, leaving on the surface of the pipe a deposit of mineral matter in the form of a small nodule. Sometimes the deposit will form a stalactite as long as an inch. In most instances, if these nodules or stalactites on the pipe line are not disturbed, the leaking may stop and the pipe continue to give service for many years after perforation of the pipe wall. Dezincification, therefore, is not always a serious type of corrosion; the seriousness depends upon the rate of dezincification and the function of the brass part. This type of corrosion is most troublesome when it occurs in threaded sections because the strength of the thread is lost or the fitting "freezes."

If the rate of uniform-layer-type dezincification proceeds at a rate of 0.001–0.003 ipy, a Muntz metal condenser tube with an initial wall thickness of 0.065 in. (0.165 cm) may have a service life of 20 to 40 years. This length of life of Muntz metal tubes is commonly obtained in the Great Lakes region and in the midwestern states.

Reduction of Dezincification through Alloying

A reduction in the zinc content of the alloy decreases its susceptibility to dezincification. Brasses, for example, containing more than 85% copper are practically immune. The addition of tin or arsenic (also antimony* and phosphorous) to the brasses containing more than 15% zinc usually is quite effective in slowing up or inhibiting the dezincification reaction in fresh waters and in sea water. A few examples are Admiralty metal (1% tin), Naval brass (¾% tin), arsenical aluminum brass (0.04% arsenic), and arsenical Muntz metal (¼% arsenic). These are appreciably more resistant than the parent Cu-Zn alloys free of the inhibiting alloy additions.

AQUEOUS MEDIA

Dissolved Oxygen and Other Gases

The corrosion of copper-base alloys in aqueous solutions is influenced considerably by the concentration of dissolved oxygen, carbon dioxide, hydrogen sulfide, sulfur dioxide, and other gases. Numerous investigations have revealed that concentration of dissolved oxygen in aqueous solutions controls the rate of corrosion. Damon and Cross[2] clearly demonstrated that there is a close relation between the sulfuric acid concentration, oxygen concentration, and the corrosion rate of copper alloys. Tests conducted on copper in sulfuric acid solutions (at room temperature) freed of

---

* It is stated that bismuth (although related chemically to antimony) accelerates dezincification of Muntz metal. (W. B. Price and R. W. Bailey, *Trans. Am. Inst. Min. Met. Engrs.*, **147**, 136 [1942].)

[2] G. H. Damon and R. C. Cross, *Ind. Eng. Chem.*, **28**, 231–233 (1936).

air by boiling showed practically no corrosion at the end of 6 months.[3] (See also Fig. 2, p. 65.)

Sulfur dioxide dissolved in water is more active than dissolved oxygen since it acts as a mild oxidizing acid in a manner similar to nitric acid.

The presence of carbon dioxide in aqueous solutions appreciably increases the corrosion rate. The corrosion products whch usually form on copper-base alloys are soluble in water containing carbon dioxide and may not have an opportunity, therefore, to stifle the corrosive action.

The presence of hydrogen sulfide in aqueous solutions (fresh and salt waters) accelerates the corrosion of some copper-base alloys by the formation of a voluminous non-protective (though only very slightly soluble) corrosion product. High-zinc brasses are more resistant to hydrogen sulfide than copper or the red brasses.

Natural Fresh Waters

The corrosiveness of natural fresh waters depends, in a complicated matter, upon the concentration of mineral matter, gases, organic matter, and debris. The sum of all the cations and anions gives the total mineral content of the water, which will vary from approximately 0.003% in fresh waters to 3.5% in sea water.

The Cu-Zn alloys generally show very good resistance to most types of unpolluted fresh water. The corrosion rates for these materials, in the absence of dezincification, average 0.0001 to 0.001 ipy. Non-scale-forming fresh waters containing aggressive carbon dioxide produce higher corrosion rates of high-zinc brasses often accompanied by dezincification.

Mineral Scales. Dissolved salts in aqueous solutions vary considerably in their influence on the corrosion rate of copper-base alloys. For example, certain waters will deposit thin or thick mineral scales of variable composition on the metal surfaces with which these waters come into contact. (See discussion of saturation index, p. 502.) The most common scale found in water-heating coils, condenser tubes, and heat-exchanger tubes, consists essentially of calcium carbonate ($CaCO_3$) formed by the decomposition of calcium bicarbonate, $Ca(HCO_3)_2$, at elevated temperatures. These scales frequently account for considerable stifling of corrosion.

Unfortunately, the build-up of a thick mineral scale on the inside of a heater tube or heat exchanger tube seriously interferes with heat transfer. This loss in heat transfer may result in considerably higher temperatures on the metal surfaces facing the fire or hot vapors, which in turn lead to more rapid oxidation of the metal. Such failures are described as resulting from "overheating" or actual "burning."

Sea Water

The corrosion rates of Cu-Zn alloys in sea water, on the average, range from as low as 0.0003 up to 0.004 ipy. (See *Behavior of Metals and Alloys in Sea Water*, p. 393, particularly Table 5.) The outstanding exception to this is the relatively high corrosion rate (as much as 0.0085 ipy or higher) of Muntz metal, which suffers from rapid dezincification. This high rate of attack occurs with little or no change in the dimension of the Muntz metal part. The corrosion can be diminished considerably by the addition of either tin or arsenic to produce Naval brass or arsenical Muntz metal, respectively. The addition of these elements lowers the corrosion rate to 0.003 ipy or less.

Where dezincification is not a factor the corrosion rates for the alloys containing more than 65% copper are quite low. Over many years the 65-35 brass, 70-30 brass,

---

[3] Unpublished data of Bridgeport Brass Company.

and Admiralty metal (70% Cu, 29% Zn, 1% Sn) have been widely used in the manufacture of condenser tubing. Under some conditions of service, corrosion proceeds at rates averaging 0.001 to 0.005 ipy. The 0.065-in. gage condenser tubes last approximately 10 to 20 years. Occasionally considerably higher corrosion rates are encountered under debris which lodges in the tubes. Under such conditions the 65 to 75% copper brasses may fail from plug-type dezincification at rates as high as 0.050 to 0.100 ipy. Admiralty metal subjected to the same conditions suffers similarly.

Effect of Temperature. Figure 1 shows the effect of temperature on the rate of corrosion of Admiralty metal in a 3% sodium chloride solution using an alternate immersion test.[4] In this instance, the depth of corrosion was measured in the pitted

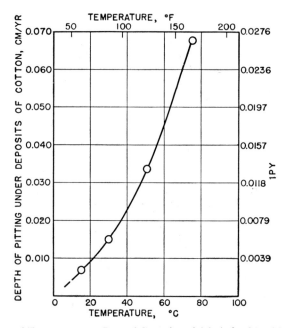

FIG. 1. Effect of Temperature on Rate of Corrosion of Admiralty Metal in 3% Sodium Chloride Solution.

areas which developed under a deposit of cotton (to produce an oxygen concentration cell). It is apparent that a temperature rise of about 20° C (35° F) will double the rate of corrosion. Because of this influence of temperature, it is widely recognized that heat-exchanger equipment operating at too high a temperature may fail prematurely.

The higher temperature of cooling waters (salt water or sea water) during the summer time contributes towards more rapid corrosion. Instances have been reported where new condenser tubes installed during the winter season have given longer life than tubes installed during the summer. This difference in performance occurs because of difference in the type of protective film formed. The corrosion product film which forms on metal in cold water is generally thinner and more continuous than one which forms in warm water.

[4] Unpublished data of Bridgeport Brass Company

In warm and hot water, the susceptibility towards pitting generally increases substantially. As a result, the useful life of a metal part or heat exchanger tube may be considerably shorter than that indicated by the overall corrosion rate as evaluated by loss in weight or loss in tensile strength. Figure 2 shows the effect of seasonal variations in temperature and water composition upon the rate of corrosion.[5] The increased corrosion coincided with the period when the water gave off an unpleasant odor (hydrogen sulfide and other gases) and when the normal green corrosion products

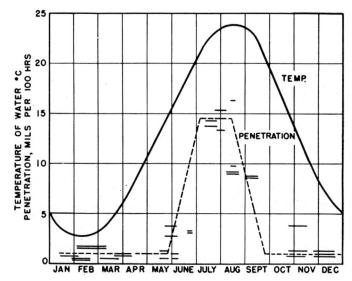

FIG. 2. Results of Jet Tests Showing Effect of Seasonal Variations on Corrosion Rate of Admiralty Metal, Muntz Metal, and 66-33 Brass in Sea Water. (Morris.) (Averaged over three-year period.)

which formed during the winter season changed into black copper sulfide. In some instances, 1 to 15 ppm of hydrogen sulfide have been found where rapid corrosion has occurred in copper-zinc condenser tubes and cast bronze impellers during the summer season.

A similar effect of temperature is encountered in non-scale-forming fresh waters.

BOILER WATER AND CONDENSATES

The comments which have been made in connection with the corrosion resistance of Cu-Zn alloys towards fresh waters and the discussion pertaining to the effect of temperature are of particular importance in considering the corrosiveness of boiler waters. Generally the copper-base alloys are not used in construction of boilers except as they are used for hot water heaters where relatively little corrosion occurs with copper or red brass. When excessively high temperatures (over 300° C [550° F]) or corrosive soft waters are encountered (high in dissolved CO_2), corrosion rates as high as 0.025 ipy have been observed.

Steam condensates of adequately treated boiler water lead to very low corrosion rates of the Cu-Zn alloys, the corrosion rates generally being under 0.0005 ipy. With

[5] A. Morris, *Trans. Am. Inst. Mining Met. Engrs.*, **99**, 274 (1932).

untreated water the rates are increased appreciably by condensates containing oxygen, carbon dioxide, and sometimes ammonia. Instances have been observed where 30 ppm of ammonia in a steam condensate caused deep grooving of unstressed brass and stress corrosion cracking of stressed brass.

Mine Waters

The composition and pH of mine waters vary considerably, depending upon the nature of the coal, ore, rock, and soil through which the water is seeping. Such waters may be very corrosive, even when the dissolved oxygen content is low, owing to the presence of ferric ion (Fe^{+++}) which acts as an oxidizing agent. The corrosion rates of the Cu-Zn alloys and the casting bronzes under these conditions are considerably higher than the pH value alone would indicate. (See "Oxidizing Salts" below.)

Dissolved Salts, Acids, and Bases

Non-Oxidizing Salts.

Acid salts. The acid non-oxidizing salts (the alums and certain metal chlorides such as magnesium chloride) which hydrolyze to yield dilute acid solutions behave essentially the same as dilute solutions of the corresponding acids. The corrosion rates of the Cu-Zn alloys range from 0.002 to 0.060 ipy.

Halide salts. In neutral salt solutions such as sodium chloride the corrosion rates of the brasses range from 0.001 to 0.005 ipy at room temperature, and at rates of 0.005 to 0.075 ipy at 75° C (170° F). (Refer also to Table 5, p. 396.) In the temperature range of 20° to 75° C (70° to 170° F) the order of increasing corrosiveness of sodium halide solutions towards Cu-Zn alloys is as follows: sodium fluoride (very low), sodium chloride, sodium bromide, and sodium iodide (most corrosive). Ammonium chloride attacks Cu-Zn alloys at rates of approximately 0.050 ipy under stagnant conditions and at considerably higher rates when aerated or agitated.

Alkaline salts. The alkaline salts, such as silicates, phosphates, and carbonates, at room temperature attack the Cu-Zn alloys at rates of 0.002 to 0.005 ipy. The addition of carbon dioxide to sodium carbonate solutions may increase the corrosiveness to 0.010 to 0.020 ipy. The alkali cyanides ($2 N$) attack the Cu-Zn alloys at rates ranging from 0.025 to 0.076 ipy at room temperature.[6]

Oxidizing Salts.
The oxidizing salts, such as sodium dichromate and the parent acid, chromic acid, corrode Cu-Zn alloys at high rates. The salts of polyvalent metals, such as ferric, cupric, mercuric, and stannic salts, corrode the Cu-Zn alloys at rates as high as 0.025 in. per *day*. The salts of metals more noble than copper, such as mercury and silver, corrode copper and Cu-Zn alloys at approximately the same rate.

The presence of a film of liquid mercury (produced by reaction with mercury salts) on brasses which are stressed internally or by applied loading may suffer from intercrystalline cracking. The liquid mercury penetrates the metal preferentially along the grain boundaries. The reaction with mercurous nitrate is used as a test for internal stresses in brass. (See *Tests for Stress Corrosion Cracking*, p. 1013.)

The alkaline hypochlorites attack Cu-Zn alloys at rates ranging from 0.003 to 0.030 ipy.

Acids.

Nitric acid. The Cu-Zn alloys show very high corrosion rates in $8 N$ nitric acid (about ¼ in. per *day*). $2 N$ nitric acid at room temperature corrodes the brasses at approximately 0.5 ipy (⅜ in. round spec., 32-day test period).

---

[6] Unpublished data of Bridgeport Brass Company.

HYDROCHLORIC ACID. Hydrochloric acid is one of the most corrosive of the nonoxidizing acids when in contact with Cu-Zn alloys. The corrosion rates in dilute stagnant hydrochloric acid solutions may range from 0.0015 to 0.025 in. per *day*. Water lines tests (show effect of oxygen) conducted at room temperature in $2N$ hydrochloric acid solution showed corrosion rates ranging from 0.03 to 0.06 in. per day. At elevated temperatures and high acid concentrations the corrosion rate is still higher and may proceed with the liberation of hydrogen.

SULFURIC ACID. Sulfuric acid attacks all the Cu-Zn alloys less rapidly than hydrochloric acid. The average rate of corrosion in stagnant solutions averages about 0.003 ipy in a 10% sulfuric acid solution at room temperature. At the water line in such a solution the corrosion rate may be as high as 0.50 ipy. The presence of an oxidizing agent such as ferric sulfate or sodium dichromate accelerates the overall corrosion rate approximately two to three hundredfold. Where the material is only intermittently exposed to the acid, red brass, copper, leaded bronzes, and the aluminum bronzes are frequently used for tie rods on pickling tanks containing 2 to 15% sulfuric acid.

In the absence of oxygen the corrosion rate of the Cu-Zn alloys in dilute sulfuric acid solutions is practically nil. Hot concentrated sulfuric acid solutions rapidly attack the Cu-Zn alloys.

SULFUROUS ACID. Sulfurous acid solutions are actively corrosive towards Cu-Zn alloys. The corrosion rates may range from 0.01 ipy at room temperature to 1.0 ipy at higher temperatures; and, in addition, pitting may occur. The corrosiveness of sulfurous acid is due to the fact that this acid acts as an oxidizing agent.

PHOSPHORIC ACID. The corrosion rates at room temperature in relatively pure phosphoric acid solutions range from 0.001 to 0.20 ipy, depending upon the degree of aeration. Raising the temperature will increase the corrosion rate as much as tenfold to a hundredfold. The presence of oxidizing salts also increases the corrosion rate appreciably.

ORGANIC ACIDS. Acetic acid and similar organic acids attack the Cu-Zn alloys at rates ranging from 0.001 to 0.030 ipy in quiet solutions at room temperature with limited aeration. Increased aeration and elevated temperatures may increase the rate of corrosion a hundredfold. Velocity of acetic acid relative to a brass surface increases the corrosion rate.

Citric acid as found in orange juice attacks the Cu-Zn alloys at a rate of approximately 0.002 ipy. While this corrosion rate is low it is sometimes considered too high for foodstuffs because of the accumulation of metal ions.

FATTY ACIDS. Fatty acids rapidly attack the Cu-Zn alloys. For example, in heat exchangers the corrosion rate ranges for 0.010 to 0.050 ipy. In some instances this rate of corrosion may not be objectionable, but where it is desirable to obtain colorless fatty acids the brasses are not used. Pure oleic acid in the absence of oxygen does not attack the Cu-Zn alloys.

ALKALIES. Alkalies attack the Cu-Zn alloys at rates of approximately 0.002 to 0.020 ipy at room temperature under stagnant conditions. Increased aeration and elevated temperatures increase the corrosion rates from 0.020 to 0.070 ipy. Tests conducted in hot caustic liquors containing 10 to 50% sodium hydroxide and 5 to 16% sodium chloride showed that the corrosion rates for copper and Cu-Zn alloys ranged from 0.017 to 0.024 ipy.[7] Further work conducted in sodium hydroxide solutions containing wood extracts (black liquor) from a soda pulp mill indicated that the presence of

---

[7] *Proc. Am. Soc. Testing Materials*, **35**, Pt. 1, 161, 164, 165 (1935).

organic material in the sodium hydroxide solution increased the corrosion rate appreciably.

AMMONIUM HYDROXIDE. Ammonium hydroxide corrodes Cu-Zn alloys very much more rapidly than the alkali hydroxides. A $2N$ solution of ammonium hydroxide corrodes brasses at rates ranging from 0.072 to 0.240 ipy at room temperature.[8] The presence of applied or residual stresses in the Cu-Zn alloys renders them highly susceptible to stress corrosion cracking in this medium. (This is discussed further under "Stress," opposite page.)

ORGANIC COMPOUNDS AND SOLUTIONS

Formaldehyde dissolved in water (37% by weight) with 10 to 15% methanol corrodes Cu-Zn alloys at rates ranging from 0.002 to 0.001 ipy.[8] The aldehydes such as benzaldehyde and butyraldehyde attack the Cu-Zn alloys fairly rapidly in the presence of air at room temperature and with greater rapidity on warming. This reaction will proceed in the presence or absence of toluene or ethyl acetate. In the absence of air the corrosive reaction does not take place.[9]

Anti-freeze solutions consisting of methanol, ethanol, or ethylene glycol with and without corrosion inhibitors corrode the brasses at rates ranging from 0.00002 to 0.0025 ipy.[10]

Dry carbon tetrachloride is not corrosive towards the Cu-Zn alloys. The presence of moisture results in a water layer which is acid and corrosive. The brasses containing more than 20% zinc in contact with moist carbon tetrachloride dezincify at rates ranging from 0.001 to 0.050 ipy. The organic chlorides such as ethyl chloride and methyl chloride attack the Cu-Zn alloys at very low rates (less than 0.001 ipy) in the absence of moisture. Therefore the brasses are widely used in the handling of chlorinated compounds.[11]

The organic bromides behave in a manner similar to that of the organic chlorides.

The fluorinated organic compounds are practically without effect. The copper alloys are widely used in the construction of refrigeration equipment using the fluorinated hydrocarbons (e.g., CF_2Cl_2 ["Freon 12"]) as refrigerants.

VELOCITY

In many instances, only a slight movement of the liquid is required to start the rapid localized corrosion of Cu-Zn alloys at such areas as tube and pipe bends, screens, and where the cross-sectional area of the piping system is suddenly reduced (water box to tubes in heat exchangers). In heat exchanger and condenser tubes, the movement or turbulence of water at the inlet ends of tubes frequently leads to localized corrosion, commonly called inlet end corrosion. When sand, debris, and gas bubbles are present the effect of velocity becomes even more serious. For example, when much entrained air is present in the fluid stream, pits that are undercut at the forward end may form at those points where the gas bubbles impinge on the metal surface. This corrosion is usually called *impingement corrosion* or *impingement attack*. Arsenical aluminum brass, 70-30 cupro nickel containing iron, certain aluminum bronzes, and the high-tin bronzes show good resistance to this type of attack,

---

[8] Unpublished data of Bridgeport Brass Company.
[9] Tenney L. Davis and Walter P. Green, Jr., *J. Am. Chem. Soc.*, **62**, 3014–3015 (1940).
[10] D. H. Green, H. Lamprey, and E. E. Sommer, *J. Chem. Educ.*, **18**, 488–492 (1941).
[11] Karl S. Willson, Walter O. Walker, William R. Rinelli, and C. V. Mars, *Chem. Eng. News*, **21**, 1254–1261 (1943).

especially in sea water and other saline solutions. The unmodified binary Cu-Zn alloys do not stand up well in sea water and certain fresh waters under such conditions. (See "Copper and Copper Alloys," p. 393, and *Condenser Corrosion*, p. 545.)

The effect of velocity on corrosion by wet steam is discussed under the heading "Steam," p. 83.

GALVANIC COUPLING

Generally, the coupling of copper-base alloys to one another does not lead to a significant increase in corrosion rate. A problem may arise with the galvanic coupling of a copper alloy to a less noble metal, such as aluminum, steel, and zinc, attended by increased corrosion of the latter metals. The electrical resistivity of fresh water is quite high, so that if galvanic corrosion occurs, the attack is localized near to or at the junction of the two metals, which at ordinary temperatures is not usually severe. In solutions of salts, acids, and bases with higher electrical conductivity than fresh waters, deep pitting may occur at the junction as well as some distance away.

In practice, the coupling of copper alloys with iron is quite common in equipment handling fresh water and slightly corrosive liquids. For example, copper and brass tubes are used in conjunction with thick steel tube sheets in heat exchangers where river or lake water is used for cooling purposes. Steel tubes joined by bronze brazing are used in fresh waters. Steel and copper are coupled together in locomotive boilers.*

Copper alloy rivets in aluminum in contact with salt water or marine atmosphere may cause considerable trouble. Cadimum plating of copper alloy aircraft parts is commonly used to prevent this. Steel piping in contact with brass or bronze valves may fail at the points of contact, especially in the threaded area.

MECHANICAL FACTORS

STRESS

Effect of Environment. The effect of applied or internal stresses upon copper-base alloys varies considerably, depending upon the composition of the alloy and the corrosive environment. Season cracks (the result of internal stress) may follow a zigzag pattern in a piece of cold-worked metal, such as a drawn cup or tube, eventually forming a network of cracks. These cracks penetrate the metal and may finally allow pieces of metal to drop out. Sometimes hard-drawn brass tubing, or rod, will split open longitudinally. Laboratory tests reveal that stress corrosion cracks develop at right angles to the direction of applied stress.

Experience over the years indicates that the presence of ammonia and nitrogen-bearing materials (amines, ammonium sulfate, etc.) plays an important part in producing many stress corrosion failures encountered in service. The contamination of a condensate on cold-stressed brass condenser tubes or basement water pipes by ammonia or ammonium compounds from impure steam or from the products of combustion (gas or coal-fired furnaces and water heaters) may lead to stress corrosion cracking. Stress corrosion rarely occurs in warm or hot tubes, plates, etc., because they are generally free from condensate. Severely stressed brass does not crack if exposed to air from which the common contaminating substances have been removed.[12] Considerable protection against cracking is apparently obtained if the stressed brass is covered with a continuous, non-porous, electrolytic deposit of nickel.

---

* In some waters, a copper or brass, although not corroding at a rate in itself serious, may cause copper to deposit out at remote areas of iron or galvanized iron exposed to the same water. This may result in severe attack of the iron or zinc through localized galvanic contact with the deposited copper. (See *Hot and Cold Water Systems*, p. 504, and *Copper*, p. 64.) EDITOR.

[12] H. Moore, S. Beckinsale, and C. E. Mallinson, *J. Inst. Metals*, **25**, 35–152 (1921).

Applied Stresses. Where applied stresses are involved there is relatively little difference between alloys containing 60, 70, 80, and 90% copper, according to the stress corrosion data of Fig. 3. These data indicate that alloys in the critical range of 64 to 65% copper should be avoided where applied stresses and a corrosive atmosphere are encountered in service.

Experience has shown that the 90-10 and 85-15 brasses are more resistant than high-zinc brasses to stress corrosion cracking when the stress is applied.

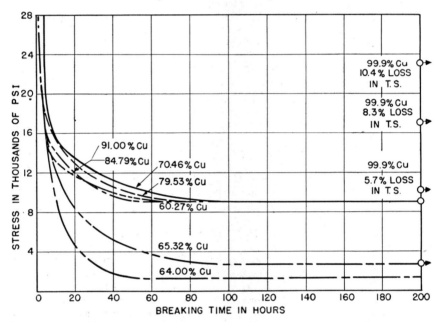

Fig. 3. Relation of Stress to Breaking Time of Copper-Zinc Alloys in Moist Air Containing Ammonia. (Time of test, 200 hours; T.S. = Tensile Strength.)

Residual Stresses. In a general way, the resistance of Cu-Zn alloys to season cracking or to stress corrosion cracking from residual stresses increases progressively from 65 to 100% copper. Most service failures of Cu-Zn parts are directly related to the shape of the article and the amount of strain or deformation resulting from internal or locked-up stresses. Consequently, if stress corrosion cracking occurs in brass articles made by plastic deformation, such as cold heading, drawing, and spinning, an improvement in resistance can be expected by increasing the copper content (Fig. 4).

Effect of Stress Relief Anneal. The effect of various stress relief annealing temperatures on hard-drawn brass tubing (68% Cu, 31.5% Zn, 0.5% Pb) were tested by alternate immersion (22.5 sec immersed. 22.5 sec in air) in a 15 N ammonium hydroxide solution at room temperature.[13] The percentage loss of tensile strength and the

---

[13] C. L. Bulow, *Symposium on Stress-Corrosion Cracking of Metals (1944)*, American Society for Testing Materials and American Institute of Mining and Metallurgical Engineers, Philadelphia and New York, 1945

percentage loss in weight were plotted against time as shown in Figs. 5 and 6. The number of stress corrosion cracks which appeared in the specimens decreased progressively as the annealing temperature rose. No noticeable difference could be detected between the hard-drawn specimens and those stress-relief-annealed at 100° C (212° F). Microscopic examination showed that the cracks were mainly transcrystalline.

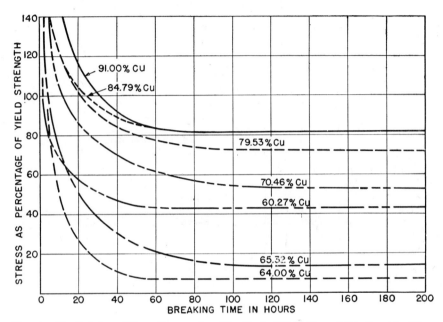

FIG. 4. The Relation of Stress Expressed as "Per Cent of Yield Strength" to Breaking Time of Copper-Zinc Alloys in Moist Air Containing Ammonia.

Figure 6 shows that no significant relation was found between the corrosion weight losses and temperature of heat treatment. It should be noted (1) that the 300° C (570° F) stress relief anneal resulted in an improvement as indicated by the change in the slope of the curve for per cent loss in tensile strength, and (2) that all the physical properties were essentially the same as those of the original hard-drawn tubes except per cent elongation. At 350° C (660° F) and 550° C (1020° F) partial and complete recrystallization resulted, as shown by microscopic examination and change in physical properties. No stress corrosion could be detected, as indicated by the merging of tensile loss and weight loss curves and absence of cracking. These results indicate that the most effective stress relief anneal for this class of alloys is one in which the temperature approaches the sharp change in slope of the physical property curves. Replotting the data of Fig. 5 for the 16- and 24-hour exposures shows that the relief of stresses for 1-hour annealed occurs over a narrow temperature range and apparently is not complete until recrystallization begins (Fig. 7).

Various modifications of Muntz metal (Naval brass, brazing rod, free-turning brass, forging brass, etc.) and *some* of the nickel silvers, such as A. S. T. M. Grade A and B, show about the same degree of susceptibility to stress corrosion as that

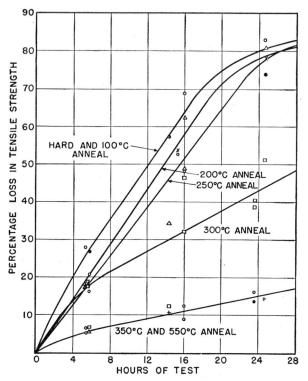

Fig. 5. Effect of Alternate Immersion in Ammonium Hydroxide on Tensile Strength of Annealed Low Lead Content Brass Tubing. (Time of anneal, one hour.)

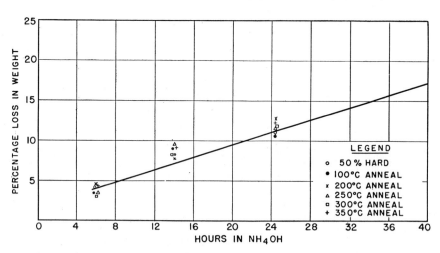

Fig. 6. Weight Loss of Hard and Annealed Low Leaded Brass Tubing Alternately Immersed in 15 N Ammonium Hydroxide.

of the copper alloys discussed. In general, however, the addition of nickel to a brass to produce nickel silver slightly increases resistance to stress corrosion.

METALLURGICAL FACTORS

Metallurgical factors governing stress corrosion and dezincification have already been discussed in this chapter.

In general there is no direct relation between service life and microstructure such as fine grain size versus coarse grain size, or annealed structures versus cold-worked

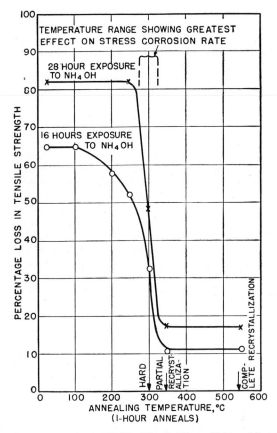

FIG. 7. Loss in Tensile Strength of 50% Hard Leaded Brass Tubing After Alternate Immersion in 15 N NH$_4$OH as Affected by Temperature of Prior Anneal and Time of Exposure.

structures. This statement refers particularly to single-phase copper-base alloys in aqueous solutions under conditions where stress corrosion is not a factor.

Muntz metal, containing both alpha and beta phases, in some environments may suffer from preferential corrosion of the beta phase. This corrosion may occur in the non-scale-forming fresh waters and almost always in sea water. The rate is quite markedly diminished by the presence of a dezincification inhibitor such as arsenic.

In the process of dezincification of Muntz metal the front line of attack consists of corrosion of the beta phase, with the alpha phase later undergoing dezincification where it is in contact with the porous copper left by the dezincified beta phase.

PROTECTIVE MEASURES

In addition to the protective measures already suggested, both chromium and nickel plating of brass plumbing goods serve the double purpose of being protective and decorative.

Crevice corrosion occurring between threaded fittings and pipes is offset by the use of a compound in the joint or by the use of soldered fittings or by brazing, all of which prevent liquids from getting into the crevice.

In some localities a trace of copper picked up by the water as it passes through red-brass pipe or copper tubing may lead to the "green staining" of porcelain wash basins and tubs. The use of tin-lined tubing or pipe has been successfully employed to combat green staining. A coating of tin only 0.1 mil thick is frequently very effective. Staining can also be reduced by water treatment. (See *Hot and Cold Water Systems*, p. 496.)

SOILS*

Light, sandy, clay-bearing soils usually are only slightly corrosive to copper and the Cu-Zn alloys. Sodium chloride, calcium chloride (used to melt ice around railroad switches), human or animal refuse (source of ammonium compounds), and gypsum greatly increase the corrosiveness of all soils. The presence of cinders in a soil leads to dezincification in those alloys containing more than approximately 20% zinc. If there is a question concerning the corrosive nature of a soil, copper or brass tubing should be covered with a water-repellant coating.

Tidal-marsh soils, which are characterized by a deficiency of oxygen, extreme acidity, sulfides, and the presence of sea water, are very corrosive to the high-copper brasses. Muntz metal and high-zinc brass withstand the action of the sulfide-bearing soils very well. Tin-coated copper has more corrosion resistance than uncoated copper.

GASES

Halogens. When the halogen gases fluorine, chlorine, bromine, as well as hydrogen chloride gas, are perfectly dry and held at or below room temperatures they are practically non-corrosive towards Cu-Zn alloys. When moisture is present, if in only small quantities, chlorine, bromine, and their compounds may be quite corrosive.

At elevated temperatures chlorine, bromine, and iodine react directly with considerable rapidity.

Hydrogen Fluoride. Anhydrous hydrogen fluoride is resisted quite well by the Cu-Zn alloys, but the presence of water or oxygen leads to rapid attack. Copper is extensively used in the manufacture and use of hydrogen fluoride.

Sulfur Dioxide. Sulfur dioxide in air begins to show its corrosive effect on the Cu-Zn alloys when a definite concentration is exceeded (approximately 0.9%) and when the humidity rises above a critical value of approximately 70%; when this occurs free sulfuric acid appears on the corroding metal surface. In distilleries the most rapid corrosion of copper alloys usually occurs on the top plates of the rectifiers

* Refer also to *Corrosion by Soils*, p. 452.

where sulfur dioxide gas collects (the corrosion product being essentially basic copper sulfate). This type of corrosion has been diminished by aeration to lower the sulfur dioxide concentration.[14]

Hydrogen Sulfide. Hydrogen sulfide reacts with the Cu-Zn alloys even when the humidity is low. The rate of the corrosive reaction is increased by a rise in temperature and decreased by thin films of moisture. A Cu-Zn alloy containing more than 30% zinc resists the action of hydrogen sulfide considerably better than the alloys containing lower concentrations of zinc. Wet vapors of hydrogen sulfide at 100° C (210° F) corrode Muntz metal, Naval brass, or Admiralty metal at a rate of 0.002 to 0.003 ipy and copper and red brass at a rate of 0.048 to 0.062 ipy.[15]

Carbon Dioxide. Carbon dioxide at ordinary temperatures is without effect on the Cu-Zn alloys in the absence of moisture. In the presence of moisture and oxygen, relatively mild corrosion occurs with the formation of bright green copper carbonate.

At red heat carbon dioxide produces a thin zinc oxide film on the surface of Cu-Zn alloys.

Carbon Monoxide. Carbon monoxide is without effect on Cu-Zn alloys at room temperature or elevated temperatures. Since alloy steels are attacked by carbon monoxide, high-pressure plant equipment used for handling carbon monoxide is frequently lined with copper or copper alloys.[16]

Nitrogen. Nitrogen at room temperature and at elevated temperatures is without effect on copper and Cu-Zn alloys. Brass in commercial nitrogen may tarnish at elevated temperatures, but this is caused by the presence of a trace of oxygen.

Ammonia. Copper has been reported as reacting with liquid ammonia in the presence of oxygen.[17] Dry ammonia at room temperature does not react with copper, but in the presence of traces of water vapor and oxygen rapid corrosion results.

Steam. The action of wet steam traveling at high velocities (3000 to 4000 ft per sec or 900 to 1200 meters per sec) is considerable. The Cu-Zn alloys corrode at rates of 0.001 to 0.050 ipy or more, depending upon the number and size of the water droplets, the steam velocity, and the composition of the steam. A number of investigations indicate that the size of droplets of water in the steam is of great importance. The wet steam impingement attack of metals is found in all parts of equipment handling steam, such as steam turbines, blades, condenser tubes, valve seats and disks, and piston rings in steam engines.

Figure 8 is a photomicrograph of the steam impingement corrosion produced on a hard-drawn copper tube (brass behaves similarly) located near the throat of a steam condenser. The steam entered the condenser from the turbine, which was located immediately above, struck a deflector plate, and then glanced off the side of the copper tube. The roughened surface consists of a large number of copper cones tilted in one direction. Sometimes the tips of the cones may not develop and the tube surface is then covered with a large number of small cones with rounded tops (hemispheroids).

Numerous attempts have been made to eliminate this type of corrosion of condenser tubes and other non-ferrous parts. One procedure consists of removing as much moisture as possible from the steam as it leaves the last wheel of the turbine, so that a fairly dry steam reaches the condenser tube or other alloy parts.

---

[14] Y. K. Raghunatha, *Int. Sugar J.*, **41**, 438–439 (1939).
[15] Unpublished data of Bridgeport Brass Company.
[16] Dudley M. Newitt, *The Design of High Pressure Plant and the Properties of Fluids at High Pressures*, Clarendon Press, Oxford, England, 1940.
[17] E. C. Franklin, *The Nitrogen System of Compounds*, A. C. S. Monograph Series, 1935.

84 CORROSION IN LIQUID MEDIA, ATMOSPHERE, GASES

Since this cannot always be realized the alternative consists of installing materials more resistant to impingement attack in the area where steam impingement occurs.

In steam valves, steam impingement corrosion may be very severe and is commonly referred to as *wire drawing*. Corrosion of brass valve seats or disks may take the form of a single deep clean groove or a cavity or a number of grooves or cavities. Under these conditions, the attack can be prevented to a considerable extent by streamlining the interior passages of the valve and providing traps for the removal of moisture before the steam enters.

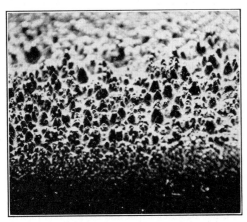

FIG. 8. Photomicrograph of Copper Tube Showing Tilted Cones Formed by Steam Impingement Corrosion. (Tops of cones represent original surface of the tube. Magnification × 10.)

In dry steam or wet steam (of good quality) at low velocities copper and Cu-Zn alloys generally give very satisfactory service. The average corrosion rates are less than 0.0001 ipy.

The copper and Cu-Zn alloys are seldom used for steam service above 260° to 290° C (500° to 560° F). This is due to the fact that their mechanical properties diminish quite markedly above these temperatures.

ATMOSPHERIC CORROSION

The rate of corrosion indoors and outdoors in a given locality depends very largely upon the local atmospheric pollution. Since the number of possible contaminants of the atmosphere is large, it is frequently very difficult to predict the precise performance of a given material.

It has been observed that many copper or Cu-Zn structures, such as church steeples, window roofs, and trim, have the green patina developed more perfectly on the windward side, especially if this side is directed toward a distant town or towards the ocean. Analyses of typical inland patinas show that they consist essentially of basic copper sulfate, which may contain as much as 2.75% basic copper carbonate, with more or less wind-borne material. The green basic sulfate is practically insoluble in water and offers much protection to the metal before the final protective compound of maximum stability is obtained.[18,19]

[18] W. H. J. Vernon and L. Whitby, *J. Inst. Metals*, **42**, 181 (1929); **44**, 389 (1930).
[19] W. H. J. Vernon, *J. Inst. Metals*, **52**, 93 (1933).

Because of the relatively high corrosion resistance of Cu-Zn alloys, they normally are not protected by lacquer, paint, rubber, or other coatings except where it is desirable to preserve their color or resist some specific type of corrosion. Rubber coatings are used to overcome corrosion and erosion by the highly polluted atmosphere and the dust and cinders coming from stacks of locomotives at high velocities in railroad tunnels.

The results of tests run in various types of atmospheres over a period of 10 years by the A.S.T.M.[20] reveal that, while the corrosion rates of Cu-Zn alloys are very low, the rates vary considerably from place to place throughout the United States. The Cu-Zn alloys showed corrosion rates (based on loss in weight) ranging from 0.00005 to 0.00015 ipy in industrial and semi-industrial atmospheres, from 0.000015 to 0.00005 ipy on the sea coast, and from 0.000004 to 0.000030 ipy in rural atmospheres.

COPPER-NICKEL ALLOYS

C. L. Bulow*

COMPOSITIONS

The Cu-Ni alloys consist of solid solutions of nickel in copper, where copper is the predominating alloy constituent.† The commercial alloys range from 54% copper to almost pure copper. Various modifications are possible by the addition of small quantities of iron, manganese, tin, zinc, and aluminum.

The most widely used alloy is the 70% Cu–30% Ni alloy employed, for example, in the construction of Navy surface condensers where high velocity sea water is used for cooling purposes. In the form of condenser tubing the 80-20 Cu-Ni alloy is also used to some extent. Copper-nickel clad steel plate and tubes are used for parts of small high-duty heat exchangers to resist certain corrosive conditions. The 60-40 Cu-Ni and the 54-45-1 Cu-Ni-Mn alloys are widely used as resistance wires (Constantan).

The physical properties of the Cu-Ni alloys are quite similar to those of the corresponding brasses. (See p. 556.) Mechanical properties are maintained to a large degree with elevation in temperature. (See latest A.S.M.E. Boiler Code.)

AQUEOUS MEDIA

GENERAL

Corrosion of the Cu-Ni alloys in aqueous solution is influenced by the concentration of dissolved oxygen and other gases in a manner similar to that discussed under *Copper-Zinc Alloys*, p. 76. The presence of hydrogen sulfide appears to accelerate corrosion of the 70-30 Cu-Ni alloy more than it does the 70-30 Cu-Zn alloy.

FRESH WATERS

Since the corrosiveness of natural fresh waters depends, in a complicated manner, upon the concentration of mineral matter, gases, organic matter, and debris, it is difficult to predict how the Cu-Ni alloys will behave in every detail. In the main, the alloys show very good resistance to both polluted or unpolluted fresh water. The 70-30 and 80-20 Cu-Ni alloys (which sometimes contain as much as 5% zinc, 1% tin, or 1% iron) show corrosion rates of less than 0.001 ipy.

* Bridgeport Brass Co., Bridgeport, Conn.
† Alloys in which nickel is the predominating component are discussed under *Nickel-Copper Alloys*, p. 266
[20] *Proc. Am. Soc. Testing Materials*, **44**, 224–239 (1944)

At elevated temperature, the rate of corrosion of the 70-30 Cu-Ni alloy in aqueous solution increases in approximately the same manner as the Admiralty metal described in Fig. 1 of *Copper-Zinc Alloys*, p. 72. Sometimes there appears to be a marked tendency for the pits to be wide and shallow rather than deep and narrow, but there are exceptions. Preferential loss of nickel, similar to dezincification sometimes encountered in brasses containing more than 15% zinc, is rare in 70-30 Cu-Ni alloy.

Boiler Water and Condensates

70-30 Cu-Ni alloy has been used in boiler water preheaters handling steam condensate and boiler feed waters with good results. The corrosion rates usually average less than 0.003 ipy.

Short time tests (22 hours) conducted[1] in steam condensate at 70° C (160° F) containing carbon dioxide and air at 35 pounds gage pressure gave rates shown in Table 1.

Table 1. CORROSION OF Cu-Ni ALLOYS IN STEAM CONDENSATE

Alloy		Concentration of CO_2 in Entering Gas, %	Corrosion Rate, ipy
Cu	Ni		
55	45	30	0.0172
70	30	30	0.0117
100	..	30	0.0235
55	45	70	0.0140
70	30	70	0.0104
100	..	70	0.0071

Steam condensate containing a trace of acetic acid (pH 5.9) corroded 70-30 Cu-Ni at a rate of 0.0078 ipy during an exposure of 11 days.[1]

Comparative data for 80-20 Cu-Ni in steam condensate containing 9.8 ml CO_2 plus 3.2 ml O_2 per liter are given in Fig. 2, p. 542.

Sea Water

70-30 Cu-Ni alloy is widely used in Naval surface condensers. 80-20 Cu-Ni has been used to a smaller extent. See "Copper and Copper Alloys" in *Behavior of Metals and Alloys in Sea Water*, p. 393, for information concerning the performance of the Cu-Ni alloys under varying conditions.

Effect of Temperature

In aqueous solutions, the effect of temperature on the rate of corrosion of 70-30 Cu-Ni is similar to that discussed in detail in connection with Admiralty metal in *Copper-Zinc Alloys*, p. 72. Under adverse conditions, a temperature rise of the corroding solution of about 20° C (35° F) will approximately double the rate of corrosion in 3% sodium chloride solution. This relation appears to hold true in the temperature range of 0° to 75° C (32° to 170° F). Some waters, when heated, will deposit a thick mineral scale of variable composition on the metal surface. The most common scale found in water-heating coils and heat exchanger tubes consists essentially of calcium carbonate. Frequently, when such a scale forms on a metal surface, considerable stifling of the corrosive action occurs.

The build-up of a thick mineral scale on the inside of the heat exchanger tube may seriously interfere with heat transfer. This loss in heat transfer may result in con-

[1] Data from International Nickel Co.

siderably higher metal temperatures on alloy surfaces facing the hot vapors, which in turn, may lead to rapid oxidation of the metal.

INORGANIC ACIDS

Hydrochloric Acid. Hydrochloric acid is one of the most corrosive of the non-oxidizing acids when in contact with Cu-Ni alloys. For example, the corrosion rate of 70-30 Cu-Ni in 2 N HCl at 25° C (75° F) may range from 0.09 to 0.3 ipy, depending upon the degree of aeration. An increase in the temperature, acid concentration, or velocity of the acid increases the corrosion rate still more. Hydrobromic acid behaves similarly.

Specimens of 80-20 Cu-Ni submerged in stagnant solutions of HCl solutions at room temperature[2] corroded at the following rates (500-hour exposure):

Concentration of HCl, %	Corrosion Rate, ipy
1	0.012
10	0.031

Sulfuric Acid. The corrosion rates of 70-30 and 60-40 Cu-Ni alloys average about 0.003 ipy in a 10% sulfuric acid solution at room temperature; at the water line the corrosion rate may be as high as 0.3 ipy (probably the maximum rate to be expected in aerated solution). The presence of an oxidizing agent, such as ferric sulfate or sodium dichromate may accelerate the corrosion rate as much as three hundredfold. In the absence of oxygen and oxidizing agents the corrosion rate of 70-30 Cu-Ni in dilute sulfuric acid solution is practically nil.

Tests conducted[2] in 6% sulfuric acid solutions at 85° C (185° F) in the presence of other substances gave the following rates of corrosion for 55-45 Cu-Ni:

Solution	Corrosion Rate, ipy
4% $FeSO_4$ + 6% H_2SO_4	0.263
4% $FeSO_4$ + 1% $CuSO_4$ + 6% H_2SO_4	0.520
Aerated 6% H_2SO_4	0.128
H_2 saturated 6% H_2SO_4	0.011

Hydrofluoric Acid. Laboratory tests conducted on specimens of 70-30 Cu-Ni alloy two-thirds immersed in solution contained in a closed bomb[2] gave the following rates of corrosion:

Concentration of HF, %	Temperature	Length of Exposure, hr	Corrosion Rate, ipy
38	110° C (230° F)	48	0.036
98	38° C (100° F)	86	0.002

Laboratory corrosion tests carried out in conjunction with the hydrofluoric acid alkylation process[3] in essentially anhydrous acid and aqueous solutions led to the conclusion that the Cu-Ni alloys are satisfactory for use with hydrofluoric acid. The results of tests are recorded in Table 2. Tests conducted in duplicate in aqueous hydrofluoric acid solutions containing 40 to 50% acid at 85° C (185° F) for 18 days showed that 80-20 Cu-Ni, 70-30 Cu-Ni, 70-25-5 Cu-Ni-Zn, and other copper-base alloys corroded at rates ranging from 0.0006 to 0.0040 ipy.

Chromic Acid. Chromic acid plating solution (400 grams chromic acid and 4 grams sulfuric acid per liter) at room temperature (stagnant solution) corroded

[2] Data from International Nickel Co.
[3] M. E. Holmberg and F. A. Prange, *Ind. Eng. Chem.*, **37**, 1030–1033 (1945).

TABLE 2. CORROSION OF Cu-Ni ALLOYS IN ESSENTIALLY ANHYDROUS HYDROFLUORIC ACID

Alloy Cu	Ni	Temperature	Length of Exposure	Corrosion Rate, ipy
80	20	26°–38° C (80°–100° F)	33 days (2 tests)	High 0.0052 Low 0.0051 Average 0.0052
70	30	26°–38° C (80°–100° F)	8–33 days (7 tests)	High 0.0070 Low 0.0007 Average 0.0020
70	30	54° C (130° F)	22 days (1 test)	0.0003
70	30	87° C (190° F)	8 days (1 test)	0.01

70-30 Cu-Ni at a rate of 0.0891 ipy.[4] The boiling solution rapidly attacks this alloy.[5]

Nitric Acid. The Cu-Ni alloys show very high corrosion rates in concentrated nitric acid solutions. For example, 70-30 Cu-Ni corrodes at a rate as high as 0.25 in. or more per *day* in 50% nitric acid solutions at room temperature.

Sulfurous Acid. Aqueous solutions saturated with sulfur dioxide (sulfurous acid) may corrode the Cu-Ni alloys at rates of approximately 0.10 ipy or more. Frequently localized corrosion or pitting occurs under the black corrosion product coating which forms.

Phosphoric Acid. Phosphoric acid does not corrode the Cu-Ni alloys as rapidly as sulfuric acid. The corrosion rate at room temperature ranges from 0.001 to 0.20 ipy, depending upon the degree of aeration. The presence of oxidizing salts (such as ferric salts often found in crude phosphoric acid) increases the corrosion rate.

Aerated 8.4% phosphoric acid solutions at room temperature[6] gave the following rates of corrosion (8-hour exposure):

Alloy Cu	Ni	Corrosion Rate, ipy
55	45	0.0199
67	33	0.0199
75	25	0.0234
80	20	0.0229
90	10	0.0209
100	..	0.0310

Specimens of 70-30 Cu-Ni alloy installed for 30 days[7] in the outlet of a tank handling 75% phosphoric acid at 70° C (160° F) showed a uniform corrosion rate of 0.028 ipy.

ORGANIC ACIDS

Acetic Acid. Acetic acid and similar organic acids attack the Cu-Ni alloys at rates ranging from 0.001 to 0.025 ipy at room temperature, depending upon the degree of aeration. An increase in the concentration of acetic acid apparently does not have so marked an effect on increasing the corrosion rate as is encountered with other types of acids. An increase in temperature may or may not increase the corrosion rate, depending upon its influence on the solubility of oxygen in the solution.

[4] Data from International Nickel Co.
[5] Unpublished data, Bridgeport Brass Co.
[6] Data from International Nickel Co.
[7] Unpublished data, Bridgeport Brass Co.

COPPER–NICKEL ALLOYS

Tests conducted[8] in quiet unaerated solutions of a number of organic acids at room temperature for 500 hours gave rates of corrosion listed in Table 3.

TABLE 3. CORROSION OF Cu-Ni ALLOYS IN ORGANIC ACIDS

Alloy		Acid	Corrosion Rate, ipy
Cu	Ni		
75	25	10% Acetic	0.0010
50	50	10% Acetic	0.0021
80	20	10% Acetic	0.0011
80	20	1% Acetic	0.0008
80	20	5% Citric	0.00078
80	20	2% Citric	0.0025
75	25	7% Citric in blackberry syrup	0.000065
80	20	5% Lactic	0.00092
80	20	5% Malic	0.00065
80	20	2% Tannic	0.00021
75	25	2% Tartaric	0.0016
80	20	5% Tartaric	0.00075

Tests conducted in an evaporator containing 10 to 22% lactic acid at 55° C (130° F) showed that 70-30 Cu-Ni corroded at a rate of 0.057 ipy (360-hour exposure).

Fatty Acids. The fatty acids under service conditions attack the Cu-Ni alloys at somewhat higher rates than the organic acids (such as acetic acid and citric acid) because the temperature and the degree of aeration generally are higher.

Tests conducted[8] in a copper-lined (wood) splitting tank containing a mixture of approximately 60% fatty acid, 39% water, and 1.17% sulfuric acid which was violently mixed with steam at 100° C (212° F) gave the following rates of corrosion:

Alloy		Corrosion Rate, ipy	Location of Test Specimens
Cu	Ni		
80	20	0.0026	Submerged near top of tank
70	30	0.0024	" " " " "
80	20	0.0070	Bottom of tank
70	30	0.0073	" " "

70-30 Cu-Ni specimens installed at the liquid interface in a mixing tank containing a mixture of oleic acid with 1 to 5% methyl alcohol and water at 100° C (210° F) for 38 days corroded at a rate of 0.0035 ipy with no pitting.

ALKALIES

Ammonium Hydroxide. Ammonium hydroxide corrodes the Cu-Ni alloys much more rapidly than do the metal hydroxides. Solutions of $2N$ ammonium hydroxide corrode the alloys at rates ranging from 0.010 to 0.2 ipy at room temperature.[9] An increase in nickel content increases their corrosion resistance. The 70-30 Cu-Ni alloy is very resistant to stress corrosion cracking in this medium.

Results of short-time tests[10] at 30° C (85° F) in 3.5% ($1N$) ammonium hydroxide solution under varying conditions of aeration and motion are given in Table 4.

[8] Data from International Nickel Co.
[9] Unpublished data, Bridgeport Brass Co
[10] Data from International Nickel Co.

TABLE 4. CORROSION OF Cu-Ni ALLOYS IN 3.5% (1 N) AMMONIUM HYDROXIDE *

Alloy		Corrosion Rate, ipy			
Cu	Ni	Submerged Unaerated, 96 hours	Submerged Aerated, 96 hours	Alternate Immersion, 24 hours	Spray, 30 days
55	45	0.00010	0.000034	0.00022	0.0000081
70	30	0.00021	0.000027	0.00039	0.0000086
100	..	0.0320	0.093	0.009	0.0068

* H. S. Rawdon and E. C. Groesbeck, Bureau of Standards, *Tech. Paper* 367, 409–446 (1928).

Sodium Hydroxide. The 70-30 and 55-45 Cu-Ni alloys corrode at rates of less than 0.0002 ipy in 1 to 2 N NaOH. The degree of aeration in these dilute solutions usually is without appreciable effect. An increase in temperature may increase the corrosion rate twofold or threefold.

Tests conducted[11, 12] in hot caustic liquors containing 10 to 50% NaOH and 5 to 16% NaCl showed that 70-30 Cu-Ni corroded at a rate of approximately 0.0002 ipy; 83-17 Cu-Ni corroded at a rate of 0.005 ipy.

Other tests conducted in NaOH solutions containing wood extracts (black liquor) from a soda pulp mill indicated that the presence of organic material increased the corrosion rate appreciably. Under these conditions 70-30 Cu-Ni corroded at a rate of 0.0011 ipy and 83-17 Cu-Ni corroded at a rate of 0.011 ipy.

Potassium Hydroxide. Tests conducted for 6 hours[13] in 95% KOH at 140° to 160° C (285° to 320° F) gave the following corrosion rates:

Alloy		Corrosion Rate, ipy
Cu	Ni	
78	22	0.00007
70	30	0.00002

70-30 Cu-Ni alloy continuously immersed for 3360 hours in an aqueous solution containing 25.2% KOH, 37.8% potassium isobutyrate, 5.5% K_2S, 1.9% potassium mercaptides, and 2.1% K_2CO_3 at approx. 140° C (285° F) with no aeration corroded at a rate of 0.015 ipy. The higher than usual corrosion rate in this instance is due to the presence of sulfides and mercaptides in the alkaline solution.

SALT SOLUTIONS

Non-Oxidizing Salts. The *acid non-oxidizing salts* (alums and certain metal chlorides, such as magnesium chloride and calcium chloride) which hydrolyze in water behave essentially the same as dilute solutions of the corresponding acids. The corrosion rates of the Cu-Ni alloys range from 0.001 to 0.060 ipy at room temperatures, depending upon the extent of aeration and the acidity.

In *neutral salt* solutions, such as sodium chloride, the corrosion rates of the Cu-Ni alloys containing 30% or more of nickel range from less than 0.0002 to 0.002 ipy in 6% solutions at room temperature, depending upon the degree of aeration. At 75° C (170° F) the corrosion rate may be as high as 0.012 ipy.[14]

Tests conducted[15] in several salt solutions showed the rates of corrosion listed in Table 5.

[11] *Proc. Am. Soc. Testing Materials*, **34**, 225–227 (1934).
[12] *Proc. Am. Soc. Testing Materials*, **35**, 161–165 (1935).
[13] Data from International Nickel Co.
[14] Unpublished data, Bridgeport Brass Co.
[15] Data from International Nickel Co.

TABLE 5. CORROSION OF Cu-Ni ALLOYS IN SALT SOLUTIONS

Salt Solution	Alloy		Corrosion Rate, ipy	Test Conditions
	Cu	Ni		
10% Sodium sulfate	80	20	0.0033	Quiet unaerated immersion for 500 hours at room temperature
10% Sodium nitrate	80	20	0.000013	Quiet unaerated immersion for 500 hours at room temperature
10% Potassium nitrate	80	20	0.000011	"
20% Magnesium fluosilicate	70	30	0.060	Unaerated immersion for 24 hours at 130° F at 23 ft per min
5.8% Sodium choride	55	45	0.00024	Unaerated immersion for 96 hours at room temperature
" " "	70	30	0.00034	"
" " "	55	45	0.00082	Immersed in aerated solution for 96 hours at room temperature
" " "	70	30	0.00062	"
" " "	55	45	0.00037	Alternate immersion for 24 hours at room temperature
" " "	70	30	0.00007	"
" " "	55	45	0.000052	Spray for 30 days at room temperature
" " "	70	30	0.000052	"

The *alkaline salts,* such as sodium silicate, phosphate and carbonate, attack the Cu-Ni alloys at rates ranging from 0.0001 to 0.0015 ipy. An increase in temperature does not increase the corrosion rate appreciably; frequently at the higher temperatures lower corrosion rates may be obtained.

Alkaline cyanide solutions attack the Cu-Ni alloys at appreciably higher rates; for example, 2 N sodium cyanide[16] will corrode these alloys at rates ranging from 0.02 to 0.10 ipy at room temperature.

The alkaline sulfides also attack the Cu-Ni alloys at considerably higher rates. For example, boiling solutions of varying concentrations in 5-hour tests corroded 70-30 Cu-Ni as follows:

Alkaline Solution	Corrosion Rate, ipy
5% Na_2CO_3, 0.5% Na_2S, 5% dye solution	0.0186
5% Na_2CO_3, 5.0% Na_2S, 5% dye solution	0.225

Oxidizing Salts. Oxidizing salts, such as sodium dichromate, corrode the Cu-Ni alloys at rates ranging from 0.0001 to 0.01 ipy. The addition of an acid, such as sulfuric, to the dichromate solution may increase the corrosion rate several hundredfold, depending upon the acid concentration and the temperature.

Short-time tests conducted at 30° C (85° F) for 96 hours in 4.9% potassium dichromate solutions and spray under varying conditions of aeration and motion showed the results given in Table 6.

Copper sulfate solution (5%) containing ¼% H_2SO_4 was found[17] to corrode 80-20 Cu-Ni at a rate of 0.0011 ipy at room temperature (500-hour exposure).

Solutions of the polyvalent metals, such as ferric, mercuric, and stannic salts, corrode the Cu-Ni alloys at rates as high as 0.003 in. per *day*. The salts of metals more noble than copper, such as the nitrates of mercury and silver, corrode the alloys at approximately the same rate with the displacement of mercury or silver on the metal surface. The silver layer which builds up on the metal surface sometimes

[16] Unpublished data, Bridgeport Brass Co.
[17] Data from International Nickel Co.

TABLE 6. CORROSION OF Cu-Ni ALLOYS IN 4.9% $K_2Cr_2O_7$*

Alloy Cu	Ni	Corrosion Rate, ipy	Test Procedure
55	45	0.000086	Immersed, unaerated
70	30	0.000086	" "
55	45	0.000086	Immersed aerated
70	30	0.000034	" "
55	45	0.000070	Alternately immersed
70	30	0.000018	"
55	45	0.0000081	Sprayed
70	30	0.0000081	"

* H. S. Rawdon and E. C. Groesbeck, Bureau of Standards, *Tech. Paper* 367, 409–446 (1928).

peels off and the irregular film then favors localized corrosion and pitting. The film of liquid mercury produced by reaction with mercury salts may penetrate the alloy along the grain boundaries if the metal is internally or externally stressed.

Alkali hypochlorites attack the Cu-Ni alloys at rates ranging from 0.003 to 0.050 ipy, depending upon the concentration of chlorine.

AQUEOUS ORGANIC COMPOUNDS

Formaldehyde dissolved in water (37% by weight) with 10 to 15% methanol corrodes 70-30 Cu-Ni at approximately 0.0006 ipy.[18]

The organic chlorides, such as ethyl chloride, methyl chloride, and carbon tetrachloride, attack the Cu-Ni alloys at low rates (less than 0.001 ipy) in the absence of moisture. The presence of moisture and higher temperatures may lead to corrosion rates as high as 0.020 ipy.

The results of tests[19] conducted on 55-45 Cu-Ni in a number of chlorinated organic compounds are tabulated in Table 7.

TABLE 7. CORROSION RATES OF Cu-Ni ALLOYS IN CHLORINATED SOLVENTS

Corrosive Medium	Corrosion Rate, ipy
Anhydrous carbon tetrachloride at 25°–30° C	0.000008
Partially immersed with water layer at 25°–30° C	0.000057
Anhydrous carbon tetrachloride at 77° C (boiling)	0.00047
Carbon tetrachloride with water layer at 77° C (boiling)	0.0020
Anhydrous ethylene dichloride at 25°–30° C	0.000016
Partially immersed with water layer at 25°–30° C	0.000028
Anhydrous ethylene dichloride at 84° C (boiling)	0.000075
Ethylene dichloride with water layer at 84° C (boiling)	0.0041
Anhydrous chloroform at 25°–30° C	0.000011
Partially immersed with water layer at 25°–30° C	0.000028
Anhydrous chloroform at 61° C (boiling)	0.00037
Chloroform with water layer at 61° C (boiling)	0.00230

Specimens of 70-30 Cu-Ni under test for 99 days in a condenser handling chlorinated solvents and steam at 60° to 90° C (140° to 195° F) corroded at a rate of 0.0314 ipy.

[18] Unpublished data, Bridgeport Brass Co.
[19] Data from International Nickel Co.

70-30 Cu-Ni[20] in an ethyl acetate still handling a mixture of ethyl acetate, ethyl alcohol, and traces of sulfuric and hydrochloric acids at 80° to 100° C (175° to 210° F) corroded at a rate of 0.033 ipy (32-day test).

A long-time test (one year) in an esterification still operating at 70° to 170° C (160° to 340° F) and handling a mixture of acetic, propionic, and butyric acids (49%), ethyl, butyl, and amyl alcohols (50%) and 1% sulfuric acid showed that 70-30 Cu-Ni corroded at a rate of 0.00086 ipy.

Monoethanolamine[20] in a carbon dioxide stripping tower (CO_2 stripped from H_2) at 82° to 104° C (180° to 220° F) corroded 70-30 Cu-Ni at a rate of 0.00185 ipy (no pitting developed during 100-day exposure).

Velocity

Sometimes localized corrosion of the Cu-Ni alloys occurs at those points where the corrosive liquid is in motion, such as in tube and pipe bends, screens, and in areas where the cross-sectional area of the piping system is suddenly reduced. When another phase is present in the fluid stream, such as sand, various debris, and gas bubbles, the effect of velocity becomes more serious.

Under service conditions appreciable differences in the performance of the Cu-Ni alloys have been encountered. This, in some instances, may be due to differences in alloy composition since laboratory and field tests show that the addition of iron to the Cu-Ni alloys considerably improves their impingement and general corrosion resistance to sea water. See "Effect of Alloy Additions," p. 94.

Galvanic Coupling

Generally the coupling of Cu-Ni alloys to other copper-base alloys does not lead to a significant change in the corrosion rate. The seriousness of coupling to a less noble metal depends not only upon the potential difference in a given corrosive environment, but also upon the resistivity of the corrosive liquid, and the relative areas of the metals involved. (See *Fundamental Behavior of Galvanic Couples*, p. 481.) The resistivity of fresh water is high; therefore, if galvanic corrosion occurs, the attack is usually localized quite near to or at the junction of the two dissimilar metals. In some waters stifling of corrosion may occur owing to the formation of a relatively high-resistance mineral scale on the cathodic surface.

The coupling of Cu-Ni alloys to other copper-base alloys or iron is fairly common in equipment handling fresh water and slightly corrosive liquids. For example, Cu-Ni heat exchanger tubes are used in conjunction with steel or brass tube sheets where river or lake water is used for cooling purposes.

Galvanic Couples to Be Avoided. The coupling of Cu-Ni alloys to aluminum in contact with sea water or marine atmospheres should be avoided. Steel piping in contact with Cu-Ni alloys may fail at the point of contact (especially in threaded areas) in sea water, saline or acid solutions, and occasionally in some fresh waters. The coupling of Cu-Ni alloys with zinc, lead, or tin in sea water and other solutions of low resistivity may be expected to lead to accelerated corrosion of the less noble metals. Where it is necessary to join Cu-Ni alloys for use in sea water it is advisable to use silver solder rather than soft (Pb-Sn) solder.

Mechanical Factors

Stress. Externally applied or internal stresses usually have no detrimental effect upon the Cu-Ni alloys when exposed to environments which may lead to stress

[20] Data from International Nickel Co.

corrosion cracking in some susceptible Cu-Zn alloys. The Cu-Ni alloys (especially 70-30 Cu-Ni) have replaced the stress corrosion susceptible alloys, such as Admiralty metal and aluminum brass, when unfavorable operating conditions proved troublesome. (This usually consists of an environment containing a small amount of ammonia or amines.)

Vibration. The Cu-Ni alloys show good resistance to corrosion fatigue in fresh and saline waters. These alloys are generally considered among the best of the copper-base alloys for application where vibration is present.

Metallurgical Factors

The Cu-Ni alloys are single-phase solid solution alloys. Sometimes they show a cored structure, but there is no evidence to show that such a structure has any practical detrimental effect on corrosion resistance.

Corrosion tests conducted in sea water[21] clearly showed that the size of the grain within commercial limits did not affect the corrosion resistance of the alloy.

Other tests also conducted in sea water[22] showed that hardness (produced by cold rolling) has no influence on the corrosion rate of 70-30 Cu-Ni in sea water.

Effect of Alloy Additions. Some recent tests[23] in sea salt solutions, indicated quite conclusively that the addition of iron to the Cu-Ni alloys containing from 5% to 30% nickel improved impingement corrosion resistance. The tests indicate that as the nickel content of the alloys decreases from 30% to 5%, more iron is required for optimim corrosion resistance; for example, 0.5% Fe for the 30% Cu-Ni; 0.60% for the 20% Cu-Ni; 0.75% for the 10% Cu-Ni; and 1% for the 5% Cu-Ni. Other tests, based on loss in weight,[24] conducted in sea water showed that the addition of 0.4% iron to 70-30 Cu-Ni lowered the corrosion rate from 0.00216 to 0.00033 ipy.

GASES

Halogen Gases

The halogen gases, fluorine, chlorine, bromine, and their hydrogen compounds, when perfectly dry at room temperature or lower, are practically non-corrosive toward the Cu-Ni alloys. Traces of moisture will considerably increase attack. At elevated temperatures, chlorine and bromine rapidly react with the Cu-Ni alloys. Iodine vapors are also reactive.

Sulfur Dioxide

Sulfur dioxide when perfectly dry is only slightly corrosive toward the Cu-Ni alloys. When the humidity rises above the critical value of approximately 60 to 70% a considerable increase in corrosion occurs.[25]

Hydrogen Sulfide

Hydrogen sulfide reacts with the Cu-Ni alloys at rates ranging from 0.001 to 0.050 ipy. The range in corrosion rates appears to depend primarily upon the concentration of water vapor and the temperature.

[21] Maurice Cook and I. Boodson, *Trans. Faraday Soc.*, **38**, 391–394 (1942).
[22] C. L. Bulow, *Trans. Electrochem. Soc.*, **87**, 319–352 (1945).
[23] A. W. Tracy and R. L. Hungerford, *Proc. Am. Soc. Testing Materials*, **45**, 591–612 (1945).
[24] C. L. Bulow, *loc. cit.*
[25] W. H. J. Vernon, *J. Inst. Metals*, **48**, 121 (1932).

CARBON DIOXIDE

Carbon dioxide gas at ordinary temperatures is without effect on the Cu-Ni alloys in the absence of moisture. In the presence of moisture and oxygen relatively mild corrosion will occur.

STEAM

In dry steam or wet steam (of good quality) at low velocity the Cu-Ni alloys usually corrode at rates less than 0.0001 ipy. Under severe conditions (corrosive steam or high velocity), especially where the direction of the steam flow changes, the corrosion rate may be increased considerably. Rates as high as 0.050 ipy have been encountered.

Results[26] obtained from a 180° bend test showed that 70-30 Cu-Ni exposed to superheated steam (150° C [300° F] and 6 kg per sq cm [85 psi]) did not suffer from corrosion fissuring as did the other alloys tested. Higher temperatures, however, produced corrosion fissuring.

MOLTEN METALS

Molten metals, such as solder, tin, aluminum, lead, and zinc, rapidly attack the Cu-Ni alloys. Molten mercury attacks the 70-30 Cu-Ni alloy at a very much lower rate. If the metal, however, is stressed in tension, there is great danger that the mercury will penetrate the metal along the grain boundaries. 70-30 Cu-Ni alloy stressed in tension in contact with molten solder has been reported as being only slightly susceptible to intercrystalline penetration and weakening.[27]

ATMOSPHERIC CORROSION

The rate of atmospheric corrosion indoors and outdoors in a given locality depends very much upon local atmospheric pollution. Since the number of possible contaminants of the atmosphere is large, it is frequently very difficult to predict the performance of a given material. Tests run in various types of atmospheres over a period of 10 years by A.S.T.M.[28] reveal that while the corrosion rates of the Cu-Ni alloys are very low, the rates vary considerably from place to place throughout the United States. The results of these tests are shown in Table 8.

TABLE 8. ATMOSPHERIC CORROSION OF Cu-Ni ALLOYS

Test Site	Rate of Corrosion Based on Loss in Wt., ipy	
	75 Cu, 20 Ni, 5 Zn	70 Cu, 29 Ni, 1 Sn
New York, N. Y. (Industrial or semi-industrial atmosphere)	0.0000876	0.0000796
Altoona, Pa. (Industrial or semi-industrial atmosphere)	0.0000855	0.0000796
Sandy Hook, N. J. (Marine atmosphere)	0.0000410	0.0000435
Key West, Fla. (Marine atmosphere)	0.0000117	0.0000139
La Jolla, Calif. (Marine atmosphere)	0.0000121	0.0000106
State College, Pa. (Rural atmosphere)	0.0000138	0.0000137
Phoenix, Ariz. (Rural atmosphere)	0.0000035	0.0000030

In unusual atmospheres such as those encountered in railway tunnels, the Cu-Ni alloys stand up well.[29]

[26] J. F. Saffy, *Compt. rend.*, **186**, 1116–1118 (1928).
[27] G. Wesley Austin, *J. Inst. Metals*, **58**, 173–192 (1936).
[28] *Proc. Am. Soc. Testing Materials*, **44**, 224–239 (1944).
[29] Sidney C. Britton, *J. Inst. Civil Engrs.* (London), **5**, 65–72 (1941).

At Passamaquoddy, Maine (marine atmosphere), the 70-30 and 55-45 Cu-Ni alloys corroded at rates of 0.000031 and 0.000024 ipy respectively (9-month exposure).[30]

COPPER-TIN ALLOYS

C. L. BULOW*

COMPOSITIONS

The bronzes are essentially solid solutions of tin in copper. Phosphorus is commonly used as a deoxidizer in the casting of these alloys. Because of the residual phosphorus (0.01 to 0.40%), these Cu-Sn alloys are known as the Phosphor Bronzes.

The most common wrought bronzes contain 1 to 10% tin and seldom more than 10% tin. Most of the wrought Cu-Sn alloys, when in the completely homogenized condition, are single-phase alloys having a microstructure similar to the alpha brasses. Above 5% tin, it becomes increasingly difficult to homogenize them because of the inverse segregation of a tin-rich constituent called *delta*. The excellent cold-working properties, tensile properties,[1] resilience, and good resistance to fatigue and corrosion of the Cu-Sn alloys account for their wide use in the form of marine hardware, diaphragms, springs, snap switches, fuse clips, socket and plug electrical contact points, flexible tubing, stranded cables, pole line hardware, bolts, nuts, screws, wire screens, and welding rods.

The Cu-Sn alloys containing more than 8 to 10% tin usually are used in the form of castings. These cast bronzes are employed as bearings, bells, bushings, gears, instruments, rings, pump parts, gun metal, piston rings, valve bodies, whistles, statuary, elbows and fittings, slides, and hardware.

The homogenized wrought Cu-Sn alloys, being single-phase alloys, corrode in a uniform manner, and the corroded surfaces are very smooth. The wrought Cu-Sn alloys may show the presence of *delta* phase, and the cast Cu-Sn alloys show a very marked dendritic structure. Under some conditions, especially in acid solutions, the corrosion proceeds in an irregular manner, with the result that the grains stand out in relief and the metal is quite rough.

The hardness of the wrought Cu-Sn alloys appears to have no appreciable influence on their corrosion rate in sea water, fresh water, and in the atmosphere.

There are numerous modifications of the Cu-Sn alloys,[2] most of which deal with variations in the concentration of tin, zinc, lead, and phosphorus. About twelve other elements have been added singly or in combination, depending upon the application of the alloy, but it is seldom that more than four other elements are added. A summary of compositions is tabulated in Table 1.

AQUEOUS MEDIA

FRESH AND SALT WATERS

The corrosiveness of natural fresh waters depends in a complicated manner upon the concentration of mineral matter, gases, organic matter, and debris. (See *Copper-Zinc Alloys*, p. 69.) For this reason it is difficult to predict how the Cu-Sn alloys will behave in every detail. In the main, both the cast and wrought Cu-Sn alloys show

---

* Bridgeport Brass Co., Bridgeport, Conn.
[30] Data from International Nickel Co.
[1] R. A. Wilkins and E. S. Bunn, *Copper and Copper Base Alloys*, pp. 266–290, McGraw-Hill Book Co., New York, 1943.
[2] William Campbell, "A List of Alloys," *Proc. Am. Soc. Testing Materials*, **30**, Pt. 1, 336–397 (1930).

good corrosion resistance to fresh waters and sea water. Some of the commonly used alloys are: 88% Cu, 8% Sn, 4% Zn; 86.5% Cu, 6% Sn, 1.5% Pb, 4% Zn; 92% Cu, 8% Sn; 95% Cu, 5% Sn; 88% Cu, 10% Sn, 2% Zn; 85% Cu, 5% Sn, 5% Zn, 5% Pb; and 88% Cu, 4% Sn, 4% Zn, 4% Pb.

TABLE 1. RANGES OF VARIOUS COMPOSITIONS OF Cu-Sn ALLOYS

Percentage

Copper	60.0 –99.5	
Tin	0.5 –35.0	
Zinc	0.5 –15.0	(sometimes up to 30)
Lead	0.5 –15.0	
Phosphorus	0.01– 3.0	
Cadmium	0.5 – 1.0	
Nickel	0.10–15.0	
Iron	0.05– 4.0	
Silicon	0.05– 2.0	
Aluminum	0.5 – 2.5	
Arsenic	0.5 – 2.0	
Antimony	0.1 – 8.5	
Cobalt	5.0 ⎫	
Platinum	10.0 ⎪	
Tungsten	10.0 ⎬	not common
Manganese	3.0 ⎪	
Bismuth	0.5 ⎭	

The Cu-Sn alloys containing more than 5% tin and a small amount of other elements, such as lead and zinc, are widely used in the construction of marine hardware and other fittings. Where the water velocity is high, it is desirable to have a tin content in excess of 5%. The Cu-Sn alloys are considered the best of the copper-base alloys for marine use. These alloys in sea water usually corrode at rates less than 0.002 ipy.[3]

An increase in the tin content of the Cu-Sn alloys up to 10% produces a progressive increase in the corrosion resistance to sea water.[4] The presence of large quantities of lead in the Cu-Sn alloys in some instances has led to a noticeable increase in the corrosion rate. For example, alloys of 92% Cu and 8% Sn and of 75% Cu, 10% Sn, and 15% Pb corroded at rates of 6.5 and 22.3 mdd (0.001 and 0.003 ipy), respectively, in salt harbor water.[5] Additional data on the performance of the Cu-Sn alloys in sea water are presented in Tables 6, p. 398, and 9, p. 400.

ACIDS

Nitric Acid. The Cu-Sn alloys show very high corrosion rates (0.025 in. or more per *day*) in concentrated nitric acid solutions. At first the rate is high, but with time it is lowered appreciably owing to the formation of white metastannic acid on the surface. The amount of protection given by the coating of metastannic acid varies with the concentration of the acid, extent of agitation, abrasion, and temperature.

Hydrochloric Acid. In contact with the Cu-Sn alloys, hydrochloric acid is one of the most corrosive of the non-oxidizing acids. The presence of tin in the alloy appears to increase the corrosion rate at room temperature. For example, copper corrodes at a rate of approximately 5500 mdd (0.9 ipy) in aerated $2N$ hydrochloric

---

[3] C. L. Bulow, *Trans. Electrochem. Soc.*, **87**, 127–160 (1945).
[4] Unpublished data, Bridgeport Brass Co., Bridgeport, Conn.
[5] *Proc. Am. Soc. Testing Materials*, **35**, 160–166 (1935).

acid solution at room temperature, but 95% Cu–5% Sn alloy corrodes at a rate of 11,800 mdd (1.9 ipy).[6] *

Sulfuric Acid. An alloy of 95% Cu and 5% Sn in stagnant 10% sulfuric acid solution at room temperature corrodes at rates ranging from 0.001 to 0.003 ipy. At the water line the corrosion rate may rise to a value as high as 0.015 ipy. The casting bronzes (89% Cu, 7% Sn, 4% Zn; and 87% Cu, 10% Sn, 3% Sb) in strongly agitated 5% sulfuric acid solutions corrode at rates as high as 0.060 ipy.

Water line tests[6] conducted in 10% sulfuric acid solution at room temperature revealed the beneficial effects of tin added to copper. These tests showed that copper corroded at a rate of 372 mdd (0.06 ipy), and 5% Sn-Cu alloy corroded at a rate of 60 mdd (0.01 ipy); the respective tensile losses were 37 and 6.7%. The more uniform corrosion which occurred with the 5% Sn-Cu alloy is shown in Fig. 1.

The presence of an oxidizing agent such as dissolved oxygen, ferric sulfate, or sodium dichromate will considerably accelerate the corrosion of all bronzes.

The presence of a corrosion inhibitor, such as benzyl thiocyanate, has a pronounced retarding effect on the rate of corrosion as shown in Table 2.

At temperatures above 80° C (175° F), solutions containing more than 60% sulfuric acid rapidly corrode Cu-Sn alloys.

Sulfurous Acid. Sulfurous acid solutions such as aqueous solutions saturated with sulfur dioxide corrode the Cu-Sn alloys at generally lower rates than sulfuric acid. Sometimes pitting may develop at a rate as high as 0.1 ipy. The hot sulfite solutions used in the sulfite process for making paper corrode the casting bronzes at approximately 0.060 ipy, and at room temperature the corrosion rates range from 0.015 to 0.030 ipy. The casting bronzes (84% Cu, 9% Sn, and 7% Pb, etc.) are commonly used in the construction of sulfite cellulose digesters.

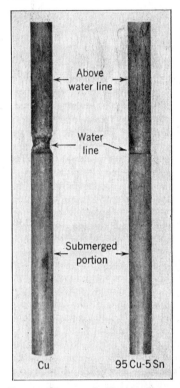

FIG. 1. Photographs Showing Beneficial Effect of Tin in 95% Cu–5% Sn Alloy Compared with Copper in Water-Line Corrosion. (Test in 10% sulfuric acid.)

Phosphoric Acid. Phosphoric acid solutions do not corrode the Cu-Sn alloys as rapidly as sulfuric acid. The corrosion rates at room temperature range from 0.001 to 0.020 ipy, depending upon the degree of aeration. An increase in temperature may increase the corrosion rate to as high as 0.060 ipy. The presence of oxidizing salts, such as ferric salts, often found in crude phosphoric acid, will increase the corrosion

---

* These rates appear to be lower under different test conditions. In 3 N hydrochloric acid (approx. 10%) using alternate immersion tests (1½ min immersed, 1½ min out) the corrosion rate for copper was found to be 0.10 ipy; for 5% Sn-Cu alloy it was 0.13 ipy; and for 10% Sn-Cu alloy 0.15 ipy. (Private communication from A. W. Tracy, American Brass Co., Waterbury, Conn.)

[6] Unpublished data, Bridgeport Brass Co., Bridgeport, Conn.

rate still further. The Cu-Sn alloys are considered among the best of the copper-base alloys for use in contact with phosphoric acid solutions.

Acetic Acid. Acetic acid (0 to 40% and 80 to 99%) and similar organic acids attack the Cu-Sn alloys at rates ranging from 0.001 to 0.025 ipy at room temperature, depending upon the degree of aeration. The effect of temperature on the corrosion rates depends upon its influence on the solubility of oxygen in solution.

TABLE 2. INHIBITING EFFECT OF BENZYL THIOCYANATE IN 10% SULFURIC ACID SOLUTION

Quiescent Solutions, 25° C (77° F)

Metal	10% Sulfuric Acid	10% Sulfuric Acid plus 0.05% Benzyl Thiocyanate	Ratio of Corrosion Rate in Non-Inhibited Acid/ Inhibited Acid
99.95% Copper	102 mdd (0.0164 ipy)	2.35 mdd (0.00038 ipy)	43.4
95% Cu, 5% Sn	34 mdd (0.0055 ipy)	2.46 mdd (0.00040 ipy)	13.8

Fatty Acids. Fatty acids at temperatures usually encountered in practice attack the Cu-Sn alloys at somewhat higher rates than the organic acids, such as acetic and citric acids at normal temperatures. The corrosion rate depends very markedly upon the temperature and the degree of aeration.

ALKALIES

The Cu-Sn alloys usually corrode at rates less than 0.010 ipy at room temperature in 1 to 2 N sodium hydroxide solutions. The degree of aeration in these dilute solutions usually is without appreciable effect. For example, 95% Cu–5% Sn alloy corrodes at a rate of 0.0016 ipy in a stagnant solution and at a rate of 0.0020 ipy when aerated.

Tests conducted[7,8] in hot caustic liquors containing 10 to 50% sodium hydroxide and 16 to 50% sodium chloride showed that the 92% Cu–8% Sn alloy corroded at a rate of approximately 0.020 ipy. Other tests conducted in sodium hydroxide solution containing wood extracts (black liquor) from a soda pulp mill indicated that the presence of organic material in the sodium hydroxide solution increased the corrosion rate appreciably.

Ammonium Hydroxide. Ammonium hydroxide corrodes the Cu-Sn alloys more rapidly than the alkali hydroxides. Ammonium hydroxide of 2 N concentration corrode the Cu-Sn alloys at rates ranging from 0.050 to 0.10 ipy at room temperature.[9]

SALT SOLUTIONS

Non-Oxidizing Salt Solutions. The non-oxidizing salts, such as the tartrates, sulfates, and nitrates of sodium, potassium, magnesium, and calcium, are only slightly corrosive towards the Cu-Sn alloys. The average corrosion rate in stagnant solutions at room temperature is less than 0.0002 ipy. Occasionally, corrosion rates up to 0.010 ipy may be encountered, depending upon the extent of aeration, acidity, and the temperature.

HALOGEN SALTS. In neutral salt solutions, such as sodium chloride, the corrosion rates of the Cu-Sn alloys containing more than 5% tin range from less than 0.001

---

[7] *Proc. Am. Soc. Testing Materials*, **34**, Pt. 1, 225–227 (1934).
[8] *Proc. Am. Soc. Testing Materials*, **35**, Pt. 1, 161–164 (1935).
[9] Unpublished data, Bridgeport Brass Co., Bridgeport, Conn.

to 0.0035 ipy in 2 N solutions at room temperature, depending upon the alloy and the degree of aeration. A 95% Cu–5% Sn alloy in 2 N sodium chloride solution corroded at a rate of 0.0026 ipy at 25° C (77° F), and at a rate of 0.0041 ipy at 75° C (170° F).

Magnesium and calcium chlorides which hydrolyze to acid solutions behave essentially the same as dilute solutions of the corresponding acid. If the acid produced by hydrolysis is not neutralized, corrosion rates up to 0.020 ipy may be anticipated.

Ammonium chloride is more corrosive than the alkali chlorides. Corrosion rates as high as 0.16 ipy may be encountered in 2 N solutions at room temperature.

ALKALINE SALTS. Alkaline salts, such as silicates, phosphates, and carbonates of sodium and potassium, at room temperature attack the Cu-Sn alloys at rates less than 0.002 ipy. The addition of carbon dioxide to sodium carbonate solutions may increase the corrosion rate to 0.005 ipy.

Alkali cyanide solutions attack the Cu-Sn alloys at appreciably higher rates than the above-mentioned alkaline salts. For example, 2 N sodium cyanide solutions will corrode these alloys at a rate of approximately 0.035 ipy at room temperature.[10]

Oxidizing Salts. Sodium dichromate solutions corrode the Cu-Sn alloys at the very low rate of approximately 0.0002 ipy. The addition of acid may increase the corrosion rate as much as several hundredfold, depending upon the acid concentration and the temperature.

Solutions of the polyvalent metals, such as ferric, mercuric, and stannic salts, corrode the Cu-Sn alloys at rates as high as 0.025 in. per *day*. The salts of metals more noble than copper, such as the nitrates of mercury and silver, corrode the alloys at approximately the same rate, with the displacement of mercury or silver on the metal surface. The silver layer which builds up on the metal surface sometimes peels off, and the irregular film then favors localized corrosion and pitting. If the alloy is tensile-stressed, the film of liquid mercury produced by reaction with mercury salt may penetrate the metal along the grain boundaries, causing failure.

Alkali hypochlorites attack the Cu-Sn alloys at rates ranging from 0.005 to 0.095 ipy, depending upon the concentration of chlorine.

VELOCITY

Sometimes movement of the corrosive liquid may start rapid localized corrosion of the Cu-Sn alloys containing less than 5% tin. Sea water moving at high velocity is best resisted by the Cu-Sn alloys containing more than 5% tin, or preferably more than 8% tin, such as the high-tin alloys used in bronze pump impellers. When another phase is present in the fluid stream, such as sand, debris, and gas bubbles, the effect of velocity may become serious.

GALVANIC COUPLING

Generally, the coupling of Cu-Sn alloys to other copper-base alloys does not lead to a noticeable change in the corrosion rate. In sea water and 3% sodium chloride solutions, the potential difference between the Cu-Sn alloys and other copper-base alloys generally is less than 0.050 volt with the Cu-Sn alloys slightly cathodic to the other alloys.

The coupling of Cu-Sn alloys with other copper-base alloys and iron is fairly common in equipment handling fresh water and slightly corrosive liquids. The coupling of Cu-Sn alloys to other copper-base alloys is common in equipment in contact with sea water and saline solutions.

[10] Unpublished data, Bridgeport Brass Co., Bridgeport, Conn.

Galvanic Couples to Be Avoided. Soldering of the Cu-Sn bronze alloys to be exposed to sea water or saline solutions should be avoided. Under these conditions, the tin is anodic and is susceptible to highly localized corrosion, especially along the contact boundary of the two metals. For the same reason, the coupling of Cu-Sn alloys with zinc, lead, or aluminum in sea water and other solutions of low resistivity may be expected to cause accelerated corrosion of the less noble metals. Where it is necessary to join Cu-Sn alloys for use under such conditions, it is advisable to use silver solder rather than soft (Pb-Sn) solder.

When parts are joined by welding, a bronze welding rod alloy should be used. Steel piping in contact with Cu-Sn alloys may fail at the point of contact in sea water and other solutions of low resistivity. Sometimes, much difficulty can be avoided by inserting a sacrificial nipple of steel over the joint. The use of brazed connections may prove to be less troublesome than threaded connections.

EFFECT OF MECHANICAL FACTORS

STRESS

The Cu-Sn alloys are considerably more resistant to stress corrosion cracking than the Cu-Zn alloys. Hard-drawn (cold-worked) alloys — such as 98% Cu, 2% Sn; 99% Cu, 1% Sn; 99.5% Cu, 0.5% Sn; 97.5% Cu, 1.5% Sn, 1.0% Si; and 98.5% Cu, 0.6% Sn, 0.9% Cd — have been used extensively during the past 30 years for (1) trolley and power transmission wires, (2) cable and wires used in catenary construction holding power transmission wires, and (3) associated pole line hardware, nuts, and bolts. All these services subject the metal to conditions which may produce stress corrosion cracking in susceptible alloys of other compositions.

VIBRATION

The Cu-Sn alloys show good resistance to corrosion fatigue in air, fresh water, and sea water. The wrought Cu-Sn alloys are generally considered among the best of the copper-base alloys for applications subject to vibration. These alloys compare favorably with the best stainless steels.[11] For this reason, the Cu-Sn alloys are widely used in the construction of springs, spring-type electrical contacts, and diaphragms.

ORGANIC COMPOUNDS

Ethyl alcohol dissolved in water is only slightly corrosive to the Cu-Sn alloys. The corrosion rate is approximately 0.0001 ipy or less. Methyl alcohol solutions may be slightly more corrosive because of the usual presence of other compounds.

The organic chlorides, such as ethyl chloride and carbon tetrachloride, in the absence of moisture corrode the Cu-Sn alloys at rates less than 0.0001 ipy. The presence of moisture and higher temperatures may lead to corrosion rates as high as 0.050 ipy.

GASES

HALOGEN GASES

Fluorine, chlorine, bromine, and their hydrogen compounds when perfectly dry at room temperature (or lower) are practically non-corrosive toward the Cu-Sn alloys. Traces of moisture will increase the corrosiveness of these gases and their

---

[11] H. J. Gough and D. G. Sopwith, *J. Inst. Metals*, **60**, 143 (1937).

compounds considerably. At elevated temperatures, chlorine, bromine, and iodine react with the Cu-Sn alloys with considerable rapidity, forming volatile tin halides.

Oxygen

Oxygen at room temperature does not visibly corrode the Cu-Sn alloys. Cu-Sn alloy castings and wrought parts are widely used in the manufacture of fittings for use in oxygen gas lines and compressed air lines where it is desirable to keep corrosion at a minimum.

Sulfur Dioxide

Sulfur dioxide when perfectly dry is only slightly corrosive toward the Cu-Sn alloys. In the presence of moisture a considerable increase in corrosion occurs. (See "Sulfurous Acid," p. 98.)

Hydrogen Sulfide

Hydrogen sulfide reacts with the Cu-Sn alloys even when the humidity is low. Wet vapors at 100° C (210° F) corrode the alloys at a rate of approximately 0.050 ipy.

Carbon Dioxide

Carbon dioxide gas at ordinary temperatures is without effect on the Cu-Sn alloys in the absence of moisture. In the presence of moisture and oxygen, relatively mild corrosion occurs with the formation of bright green corrosion products.

Steam

In dry steam or wet steam (of good quality) at low velocity, the Cu-Sn alloys usually corrode at rates less than 0.0001 ipy. Under severe conditions (corrosive steam or high velocity) the corrosion rate may be considerably higher. In some instances, rates as high as 0.035 ipy have been encountered. The Cu-Sn alloys are seldom used for steam service above 300 psi at 260° to 290° C (500° to 560° F) since the mechanical properties decrease quite markedly above these temperatures.

ATMOSPHERIC CORROSION

The rate of atmospheric corrosion indoors and outdoors in a given locality depends very much upon local atmospheric pollution. Since the number of possible contaminates of the atmosphere is large, it is frequently very difficult to predict the performance of a given material. Ten-year tests by A.S.T.M.[12] in various types of atmospheres reveal that while the corrosion rates of the Cu-Sn alloy tested (92% Cu, 8% Sn) are very low, they varied considerably from place to place throughout the United States. The results of these tests are shown in Table 3.

The color of the tarnish or corrosion product developing on the Cu-Sn alloy depends upon the atmosphere. In sulfur-bearing industrial atmospheres greenish-black and black corrosion films form. In coastal regions bright green corrosion films form. In dry rural regions, such as in Arizona, reddish-brown films predominate.

In unusual atmospheres, such as those encountered in railway tunnels, the Cu-Sn

[12] *Proc. Am. Soc. Testing Materials*, **44**, 224–239 (1944).

alloys stand up well. Railway signal wires made of 98% Cu–2% Sn alloy exposed simultaneously with galvanized steel wires in a railway tunnel were still serviceable after 40 months of exposure, whereas the galvanized wires had been replaced six times during that period.[13]

Hard-drawn bronze wire (99.05% Cu, 0.91% Sn) 0.066 in. (0.17 cm) in diameter exposed for 2 years on a roof of a building in South Kensington, London, was corroded at a rate of 0.000245 ipy.[14]

Periodic Formation of Corrosion Products*

Studies of the corrosion products of tin bronzes have shown the products to have decided periodicity. A cross section of the corrosion layer on a 90% Cu–10% Sn alloy readily reveals distinct alternate layers of red cuprite and green malachite.[15] On old bronzes, as many as seventy-five such layers have been observed.

The basis for this phenomenon was at first attributed to the fluctuating seasons,

Table 3. ATMOSPHERIC CORROSION OF 92% Cu–8% Sn ALLOY

Test Site	Type of Atmosphere	Rate of Corrosion Based on Loss in Weight, ipy
Altoona, Pa.	Industrial or semi-industrial	0.000058
New York, N. Y.	Industrial or semi-industrial	0.000072
Sandy Hook, N. J.	Sea coast	0.000050
Key West, Fla.	Sea coast	0.000031
La Jolla, Calif.	Sea coast	0.000081
State College, Pa.	Rural	0.000031
Phoenix, Ariz.	Rural	0.000006

in particular when the metal article lay buried in the soil, where the corrosion velocity would respond to fluctuations in the temperature, moisture content, salt content, etc., of the soil. However, there were, among the bronze articles examined, some that had lain buried for hundreds of years in locations where fluctuations in temperature and moisture were practically zero. Accordingly, a more generally applicable interpretation of the observations was needed.

Several laboratory specimens were corroded under constant temperature and constant humidity conditions, the corrosion products of which showed a decided laminated structure, that is, alternate thin layers of red oxide of copper and malachite. The thickness of these layers was approximately 0.03 to 0.08 mm (7 to 20 mils). Since new layers of cuprite (primary product) are formed next to the metal surface, the products of corrosion, cuprite and malachite, must move or migrate away from the surface of the metal.

On the basis of direct tests, it was established that cuprite is the primary product in the corrosion of copper and that malachite is formed from the cuprite. This latter step requires the presence of CO_2 and, of course, moisture. It was concluded that the periodicity observed in the corrosion products of copper and bronze is due to the fluctuating supply of CO_2 in the surrounding moist air or moist soil.

* This topic is discussed by C. G. Fink, Columbia University, New York, N. Y.
[13] S. C. Britton, *Engineering*, **152**, 78–80 (1941).
[14] J. C. Hudson, *J. Inst. Metals*, **44**, 409–422 (1930).
[15] See, for example, Fig. 25 in "Microscopic Study of Ancient Bronze and Copper," by Colin G. Fink and E. P. Polushkin, *Trans. Am. Inst. Mining Met. Engrs.*, Inst. Metals Div., **122**, 90–120 (1936).

TABLE 1. PHYSICAL CONSTANTS OF SOME TYPICAL Cu-Si ALLOYS*

	Everdur 1010	Everdur 1015	Everdur 1000	Olympic Metal A	Olympic Metal B	Duronze I	Duronze II	Cusiloy	Hereuloy 418	Hereuloy 419	Hereuloy 420	Hereuloy 421	PMG 3	PMG 10
	%	%	%	%	%	%	%	%	%	%	%	%	%	%
Copper	95.80	98.25	94.90	96.0	97.5	97.6	97.0	95.0	96.25	97.75	96.0	98.25	98.2	95.6
Silicon	3.10	1.5	4.00	3.0	1.5	1.0	3.0	3.0	3.25	2.00	3.0	1.5	1.2	3.2
Manganese	1.10	0.25	1.10									.25		
Zinc				1.0	1.0						1.0			
Tin						1.4		1.0	0.5	0.25				
Iron								1.0					0.6	1.2
Melting point, °C	1019	1055	1000	1027	1060	1040	1025		993	1021	1019	1055	1060	1020
Density	8.539	8.74	8.15	8.58	8.75	8.795	8.55		8.54	8.72	8.54	8.74	8.78	8.55
Coefficient of linear expansion $\times 10^5$, °C	1.73	1.67	1.73	1.73	1.67	1.67	1.67		1.73		1.73			
Electrical conductivity, I. A. C. S.,† %	6.0	8.5	5.6	7.5	12.0	13	6.5		8.1	10.9	6.7	12.0	15	6.5
Thermal conductivity, % that of copper	8.4	14.0	7.2	8.6	14.0	20	9		9.3	10.6	8.4		21	
A. S. T. M. type	A, C	B		A, C	B	B	A, C	A, C	A, C	B	A, C	B	B	A, C

* *A. S. M. Handbook*, p. 1421, 1939.
† International Annealed Copper Standard.

COPPER-SILICON ALLOYS

A. W. TRACY*

There are three general types of silicon bronzes containing about 1½, 3, or 4% silicon and small additions of manganese, zinc, iron, or tin. The silicon bronzes were developed for structural and engineering uses requiring metals with high strength and fabrication possibilities combined with corrosion resistance equal to that of copper. Typical analyses and physical constants of several silicon bronzes are given in Table 1.

The approximate tensile strengths of several types of Cu-Si alloys at different tempers are given in Table 2.

TABLE 2. TENSILE STRENGTHS OF Cu-Si ALLOYS — SHEET AND STRIP*

Temper	Tensile Strength, psi	
	Types A and C[†]	Type B[†]
0.070-mm anneal	52,000– 58,000	38,000–45,000
0.040-mm anneal	55,000– 64,000	40,000–50,000
Quarter-hard	62,000– 72,000	42,000–52,000
Half-hard	71,000– 81,000	47,000–57,000
Hard	87,000– 97,000	60,000–70,000
Extra-hard	99,000–108,000	67,000–76,000
Spring	105,000–113,000	71,000–79,000

* A. S. T. M. Designation B 97-44.
† See Table 1.

Many of the data given in this chapter were obtained with the 3.10%[†] and 1.5% Si[‡] alloys, but such data will be applicable to other silicon bronzes of similar silicon content.

AQUEOUS MEDIA

NATURAL WATERS

Silicon bronzes offer excellent resistance to corrosion by natural waters. The 3% silicon bronzes are used extensively for hot water tanks for both industrial and domestic service. The amount of oxygen in ordinary potable or industrial water is not a serious factor, as the metal will build up a protective film. However, the simultaneous presence of oxygen and free carbon dioxide will definitely increase the corrosion rate, particularly at temperatures above 60° C (140° F). The corrosion rate in such waters, however, is usually not so important a consideration as the discoloration of the water by the pick-up of traces of copper. For this reason, hot tin-coated silicon bronzes are sometimes used for fabrication of water tanks.

Sea Water.[§] Silicon bronzes offer excellent resistance to corrosion by sea water, providing the velocity of flow is less than 5 ft (1.5 meters) per sec. Both the 1.5% silicon and the 3% silicon bronzes have been used extensively for fastenings on boats where they are protected from contact with high-velocity sea water, but are in contact with salt-water-soaked wood. The silicon bronzes may be considered resistant

* Technical Department, The American Brass Co., Waterbury, Conn.
† Everdur 1010.
‡ Everdur 1015.
§ Refer also to *Behavior of Metals and Alloys in Sea Water*, p. 393; in particular Table 4, p. 395.

to fouling by marine organisms under most conditions. The 3% silicon bronze exposed at half tide to clean sea water in Long Island Sound for 2 years corroded at a rate of 2.4 mdd (0.0004 ipy) and showed no measurable pitting. The results of a laboratory test on silicon bronzes in sea salt solution are given in Table 3.

TABLE 3. SILICON BRONZES TOTALLY IMMERSED IN 2.5% TURKS ISLAND SEA SALT SOLUTION

Size of specimen — 20.3 × 1.6 × 0.13 cm (8 × 0.625 × 0.050 in.).
Surface preparation — pickled.
Velocity — turbulently stirred.
Aeration — probably saturated with air.
Temperature — 50° C (120° F).
Duration of test — 8 months.

Material	Composition			Corrosion Rate	
	Copper, %	Silicon, %	Manganese, %	mdd	ipy
Copper	99.98			6.2	0.001
3% Silicon bronze	95.88	2.94	0.90	11.8	0.002
4% Silicon bronze	94.26	4.59	1.12	11.4	0.002

MINE WATER

Mine water is an actively corrosive solution because of its acidity and ferric sulfate content, the latter acting as a cathodic depolarizer. Laboratory tests on a 3% silicon bronze in mine water have given corrosion losses from 10 to 1000 mdd (0.002 to 0.17 ipy) depending on concentration, temperature, and velocity of the mine water.

TABLE 4. EFFECT OF INCREASING SILICON CONTENT ON CORROSION RESISTANCE OF SILICON BRONZES

Type of test — alternate immersion, 1½ min in air; 1½ min in solution.
Size of specimen wires — 0.47 cm (0.187 in.) diam × 20.3 cm (8 in.) long.
Surface preparation — pickled.
Aeration — probably saturated with air.
Temperature — 60° C (140° F) — sulfuric acid tests; 21° C (70° F) — hydrochloric acid tests.
Duration of test — 216 hr.

Silicon, %	Manganese, %	10% Hydrochloric Acid				10% Sulfuric Acid			
		Wires 6 B & S Nos. Hard		Wires Annealed, 750° C, ½ hr		Wires 6 B & S Nos. Hard		Wires Annealed, 750° C, ½ hr	
		mdd	ipy	mdd	ipy	mdd	ipy	mdd	ipy
0.57	0.22	1705	0.277	1535	0.249	3030	0.492	2760	0.448
0.78	0.27	1450	0.235	1290	0.209	3015	0.489	2895	0.470
1.14	0.40	1310	0.216	1085	0.179	3065	0.505	2500	0.412
1.61	0.51	1205	0.198	845	0.139	2720	0.448	1980	0.326
2.12	0.68	1255	0.207	725	0.119	2395	0.394	1345	0.221
2.56	0.87	1155	0.194	655	0.110	2720	0.458	1465	0.246
3.04	1.07	1240	0.209	705	0.119	2360	0.397	1100	0.185
3.17	0.89	1190	0.200	725	0.122	2360	0.397	1240	0.209
4.01	1.21	1190	0.208	775	0.126	2240	0.391	895	0.156

Acids

The corrosion of silicon bronzes by acids depends on many factors, such as concentration, degree of ionization, temperature, velocity, and degree of saturation with oxygen. Tests reported in Table 4[1] show the effect of increasing silicon content on the corrosion resistance of silicon bronzes in 10% sulfuric acid and in 10% hydrochloric acid. The ratio of silicon to manganse in these alloys was maintained at approximately 3 : 1. The tests were made by alternate immersion consisting of a 3-min cycle in which the test specimens were immersed 1½ min and in the air 1½ min.

Laboratory tests on three types of silicon bronzes in sulfuric and hydrochloric acids are given in Table 5. The specimens were annealed at 750° C (1380° F) for ½ hour.

TABLE 5. SILICON BRONZES TOTALLY IMMERSED IN SULFURIC AND HYDROCHLORIC ACIDS

Size of specimens — 5.1 × 2.5 × 0.13 cm (2 × 1 × 0.050 in.).
Surface preparation — pickled.
Velocity of solutions — nat. convect.
Aeration — none.
Duration of test — 48 hr.

	Copper....95.70% Silicon.....2.96% Manganese..1.07%		Copper.95.82% Silicon . 3.07% Zinc ... 1.05%		Copper.95.50% Silicon . 3.07% Zinc ... 0.96% Tin 0.38%	
	mdd	ipy	mdd	ipy	mdd	ipy
Sulfuric acid, 25° C (77° F)						
3% by weight	16	0.0027	15	0.0025	16	0.0027
10% "	14	0.0023	14	0.0023	14	0.0023
25% "	8	0.0014	10	0.0016	10	0.0016
70% "	4	0.0007	4	0.0006	5	0.0008
Sulfuric acid, 70° C (158° F)						
3% by weight	42	0.0070	39	0.0066	44	0.0073
10% "	15	0.0026	35	0.0058	33	0.0055
25% "	22	0.0037	20	0.0034	23	0.0038
70% "	5	0.0008	8	0.0013	3	0.0005
Hydrochloric acid, 25° C (77° F)						
3% by weight	23	0.0039	21	0.0036	23	0.0039
10% "	23	0.0039	21	0.0036	21	0.0035
20% "	18	0.0031	18	0.0031	20	0.0034
35% "	123	0.0207	113	0.0190	111	0.0186
Hydrochloric acid, 70° C (158° F)						
3% by weight	183	0.0307	195	0.0327	175	0.0294
10% "	137	0.0230	140	0.0235	105	0.0177
20% "	238	0.0401	242	0.0406	246	0.0413
35% "	1608	0.2701	1280	0.2148	1670	0.2803

The corrosion rates of 3% silicon bronzes and copper in several concentrations of acetic acid are given in Table 6. In dilute acetic acid, the 3.5% Si-Cu alloy is more resistant than copper.

The silicon bronzes are not sufficiently resistant to nitric acid except where the acid concentration is very low, that is, of the order of 1%.

[1] H. A. Bedworth, *Trans. Am. Inst. Mining Met. Engrs.*, **83**, 160–174 (1929).

TABLE 6. EFFECT OF ACETIC ACID ON 3% SILICON BRONZE

Size of specimen — 5.1 × 2.5 × 0.13 cm (2 × 1 × 0.050 in.).
Surface preparation — pickled.
Velocity of solutions — nat. convect.
Aeration — containers open to air.
Duration of test — 7 days.
Temperature — 21° to 24° C (70° to 75° F).

Acetic Acid	Copper ....94.93% Silicon..... 3.51% Manganese. 0.96%		Copper	
	mdd	ipy	mdd	ipy
10% by weight	1.2	0.0002	4.2	0.0007
25% "	9.5	0.0016	8.3	0.0014
50% "	11.9	0.0020	10.7	0.0018
75% "	23.8	0.0040	16.7	0.0028
99.5% "	76.1	0.0128	80.9	0.0136

ALKALIES

Silicon bronzes are satisfactorily resistant to alkalies, except at high concentration and high temperature. Like copper, they are not suitable for use with ammonium hydroxide solutions. The results of some tests in sodium hydroxide solution are given in Table 7.

TABLE 7. EFFECT OF SODIUM HYDROXIDE ON SILICON BRONZE

Size of specimen — 5.1 × 2.5 × 0.13 cm (2 × 1 × 0.050 in.).
Surface preparation — bright-rolled and degreased.
Velocity of solution — nat. convect.
Aeration — containers open to air.
Temperature — 60° C (140° F).
Duration of test — 4 weeks.

Metal				30% Sodium Hydroxide	
				mdd	ipy
Copper				17	0.0028
	Copper, %	Silicon, %	Manganese, %		
Silicon bronze	98.32	1.21	0.22	15	0.0025
Silicon bronze	95.50	3.00	1.34	11	0.0019

SALT SOLUTIONS

The chlorides are more actively corrosive than sulfates, but the silicon bronzes are quite suitable for both types of salts. Ammonium chloride is definitely corrosive because of the activity of the chloride ion and because of the formation of soluble complex copper corrosion compounds. The 3% silicon bronze has been used in ammonia-recovery equipment to handle ammonium sulfate solutions acidified with sulfuric acid. The effect of sea water on silicon bronzes has already been mentioned (p. 105). The effect of brines encountered in the purification of salt from salt wells is given in Table 8.

Table 8. EFFECT OF SALT BRINES ON 3% SILICON BRONZE AND COPPER

Size of specimen — 5.1 × 2.5 × 0.013 cm (2 × 1 × 0.050 in.).
Surface preparation — pickled.
Velocity of solutions — nat. convect.
Aeration — containers open to air.
Duration of test — 4 weeks.

Composition of Brines	Copper .... 95.81% Silicon .... 2.99% Manganese . 0.93%		Copper	
	mdd	ipy	mdd	ipy
Brine from wells, 25.7% salt in solution				
Room temperature	2.1	0.00035	1.9	0.00031
Boiling*	6.0	0.00100	6.0	0.00096
Salt slush, 26.5% salt in solution				
Room temperature	1.3	0.00022	1.4	0.00022
Boiling*	2.8	0.00047	2.7	0.00044
Brine from Grainers, saturated				
Room temperature	1.8	0.00030	2.8	0.00045
Boiling*	4.3	0.00087	3.8	0.00079

* Solutions allowed to cool each night.

Silicon bronzes, similar to copper, are corroded rapidly by oxidizing salts, such as ferric chloride, ferric sulfate, and acid chromates.

NON-AQUEOUS CORROSION

Water-free organic compounds are generally non-corrosive to silicon bronzes at ordinary temperatures. The organic chlorides may be found corrosive to these alloys at boiling temperatures if moisture is present. Under such conditions carbon tetrachloride is the most severely corrosive of the common organic chlorides.

GASES

Silicon bronzes are resistant to corrosion by dry gases such as chlorine, bromine, fluorine, hydrogen fluoride, hydrogen sulfide, hydrogen chloride, sulfur dioxide, and ammonia. All these gases are corrosive in the presence of moisture.

STEAM

Silicon bronze containing 3% silicon, tested in steam at 125 to 150 psi for 78 weeks corroded at the rate of 3.5 mdd (0.0006 ipy) (based on weight loss). There has been some indication that the 3% silicon bronze is corroded intergranularly by high-temperature steam.

ATMOSPHERIC CORROSION

The silicon bronzes have the same order of resistance to corrosion by normal atmospheres as pure copper. The corrosion resistance of sheet metal specimens exposed vertically for 10 years in several atmospheres is shown in Table 9.[2]

[2] W. H. Finkeldey, *Proc. Am. Soc. Testing Materials*, **43**, 137–154 (1943).

110 CORROSION IN LIQUID MEDIA, ATMOSPHERE, GASES

TABLE 9. EFFECT OF ATMOSPHERES ON SILICON BRONZE

Copper, 95.95%; silicon, 2.88%; manganese, 1.13%

Size of specimen — 22.9 × 30.5 × 0.09 cm (9 × 12 × 0.035 in.).
Surface preparation — pickled.
Duration of test — 10 yr.
Note: Corrosion products removed by 10% sulfuric acid dip at end of test.

Test Location	mdd	ipy
Altoona, Pa., industrial	0.32	0.000053
New York, N. Y., industrial-marine	0.36	0.000061
Sandy Hook, N. J., marine	0.27	0.000045
Key West, Fla., marine	0.23	0.000039
La Jolla, Calif., marine	0.42	0.000071
State College, Pa., rural	0.11	0.000019
Pheonix, Ariz., dry, rural	0.04	0.000007

Silicon bronzes are definitely more resistant to *stress corrosion cracking* (season cracking) than brass on exposure to polluted atmospheres.

COPPER-BERYLLIUM ALLOYS

Beryllium Copper

A. W. TRACY*

Beryllium copper was developed as a high-strength, heat-treatable alloy having corrosion-resistant properties equivalent to those of copper. Beryllium copper usually contains 1.70 to 2.25% beryllium, and often about 0.35% nickel and the remainder copper. The following discussion is based largely on experience with the alloy containing approximately 97.50% copper, 2.15% beryllium, and 0.35% nickel. The nickel is added principally to prevent grain growth on heat treating and to increase ductility.[1] It probably has little if any effect on corrosion resistance.

Physical constants and tensile properties of the alloy are given in Tables 1 and 2.

The Be-Cu alloys find widest use for springs, diaphragms, bearing parts, valves, etc., where high strength and resistance to shock and fatigue are required in addition to good corrosion resistance. The resistance of beryllium copper to various atmospheric conditions is thus more important than its resistance to actively corrosive solutions. As a result, the corrosion data for this alloy are quite meager. In general, however, it may be said that the corrosion resistance of beryllium copper is equal to or slightly better than the corrosion resistance of pure copper in most environments.

Like copper, beryllium copper is not readily corroded intergranularly, but highly stressed beryllium copper diaphragms have been cracked when exposed to moist ammonia and air.

In general, the alloy may be considered to be resistant to selective corrosion. However, halogen gases at slightly elevated temperatures are known to have attacked the beryllium content selectively.

* Technical Department, The American Brass Co., Waterbury, Conn.
[1] H. F. Silliman, *Ind. Eng. Chem.*, **28**, 1424–1428 (1936).

TABLE 1. PHYSICAL CONSTANTS OF BERYLLIUM COPPER

Melting point, °C	955
Density, grams/cc	8.22
Coefficient of linear expansion	$1.77 \times 10^{-5}/°C$ ($0.98 \times 10^{-5}/°F$)
Electrical conductivity, I. A. C. S.,* %	18–25
Thermal conductivity (% that of copper)	21.5–26

* International Annealed Copper Standard.

TABLE 2. TENSILE PROPERTIES OF BERYLLIUM COPPER*

Material Not Precipitation-Hardened

	Tensile Strength	Yield Strength, min, psi	Elongation in 2 in., min, %
Annealed	80,000	...	35.0
¼ Hard	73,000	...	10.0
½ Hard	80,000	...	5.0
Hard	95,000	...	2.0

Material Precipitation-Hardened

AT	150,000	90,000	5.0
¼ HT	160,000	92,000	3.5
½ HT	170,000	93,000	2.0
HT	180,000	95,000	1.0

* A. S. T. M. Designation B 120 — 41T.

Sea Water

Beryllium copper offers better resistance to impingement corrosion in rapidly flowing sea water than copper, but the impingement resistance is not comparable to that of the alloys regularly used for condenser tubes such as Admiralty, aluminum brass, or cupro-nickel. (See *Condenser Corrosion*, p. 545.) Corrosion tests made in clean sea water in Long Island Sound indicate that the alloy is more resistant than phosphorus deoxidized copper (Table 3).

TABLE 3. BERYLLIUM COPPER IN SEA WATER IN LONG ISLAND SOUND

Size of specimens — $20.3 \times 1.6 \times 0.13$ cm ($8 \times 0.625 \times 0.050$ in.).
Surface preparation — pickled.
Velocity of sea water — tidal flow, specimens at half tide.
Aeration — probably saturated with air.
Temperature — 0° C (32° F) to 21° C (70° F) (water temperatures).
Duration of test — 10 months, June–April.

Metal	Composition, %				Corrosion Rate	
	Copper	Phosphorus	Beryllium	Nickel	mdd	ipy
Phosphorus deoxidized copper	99.94	0.017	....	....	5.6	0.0009
Beryllium copper	97.56	....	2.00	0.35		
800–4–275*					2.3	0.0004
800–300†					2.3	0.0004

* Quenched from 800° C, rolled 4 B & S Nos. hard, precipitation heat-treated at 275° C for 2 hr.
† Quenched from 800° C, precipitation heat-treated at 300° C for 3 hr.

In a potential test in sea water, beryllium copper was anodic to phosphorus deoxidized copper by 50 millivolts when both metals were immersed, but within 24 hours the potential was reversed and beryllium copper was cathodic by 50 millivolts.

ACIDS

An accelerated test indicates that beryllium copper is less resistant than phosphorus deoxidized copper to corrosion in 10% sulfuric acid and more resistant in 10% hydrochloric acid. The results of the test are given in Table 4.

TABLE 4. BERYLLIUM COPPER — ALTERNATE IMMERSION IN ACID SOLUTIONS

Size of specimens — 20.3 × 1.6 × 0.13 cm (8 × 0.625 × 0.050 in.).
Surface preparation — pickled.
Alternate immersion cycle — 1½ min in solution, 1½ min in air.
Aeration — probably saturated with air.
Temperature — 10% sulfuric acid — 60° C (140° F).
10% hydrochloric acid — 21°–24° C (70°–75° F).

	Copper, %	Phosphorus, %	Beryllium, %	Nickel, %	10% Sulfuric Acid, 60° C (140° F)		10% Hydrochloric Acid, 21°–24° C (70°–75° F)	
					mdd	ipy	mdd	ipy
Phosphorus deoxidized copper	99.94	0.017	....	....	3780	0.600	453	0.073
Beryllium copper 800-4-275*	97.56	.....	2.00	0.35	4880	0.852	332	0.058
800-300†					5115	0.892	332	0.058

* Quenched from 800° C, rolled 4 B & S Nos. hard, precipitation heat-treated at 275° C for 2 hr.
† Quenched from 800° C, precipitation heat-treated at 300° C for 3 hr.

GOLD AND GOLD ALLOYS

E. M. WISE*

Pure gold has a low chemical affinity for most elements, and its corrosion resistance arises from this fact, rather than from the formation of passive films. On electrolysis, however, a gold anode may become passive. Thermochemical data useful in estimating the domain within which corrosion can be anticipated are presented by Bichowsky and Rossini.[1]

Gold is resistant to alkalies, salts, and many acids, including sulfuric, hydrochloric, and nitric acids, the last-named acid having a specific gravity not over 1.46 and free from traces of halogens. Gold, however, is rapidly attacked by chlorine and bromine, and generally by iodine.

USES OF GOLD AND GOLD ALLOYS

PURE GOLD

The rather moderate melting point of gold, 1063° C (1945° F), its softness, as well as its vulnerability to halogens, restrict its use in the pure form. Pure or nearly pure

---

* The International Nickel Co., Inc., New York, N. Y.

[1] F. R. Bichowsky and F. D. Rossini, *The Thermochemistry of the Chemical Substances*, pp. 296–298, Reinhold Publishing Corp., New York, 1936.

gold is occasionally used in chemical equipment as a lining, and at one time found considerable use in pans for concentrating sulfuric acid. It is also used as a solder for platinum, and sometimes as a decorative and partially protective electroplate in laboratory equipment. Thin electroplates of gold like those of other metals tend to be porous. In addition to its use for jewelry and other appearance items, it was used until recently for coinage in alloys containing 90% or more of gold. Pure and slightly alloyed gold in the form of leaf about 6 millionths of an inch thick has numerous uses for signs and decoration in both indoor and outdoor exposure.

Platinum and Palladium Alloys

In its most important chemical application, namely for rayon spinnerets, a gold alloy containing 30% of platinum is one of the alloys employed. The 20% Pd-Au alloy had limited use for spinnerets, although it is believed that a ternary alloy containing gold, platinum, and palladium is much more suitable than the binary alloy. In recent years, considerable replacement of gold-base spinnerets by high-platinum alloys has occurred. The 90% Au–10% Pt alloy,[2] as well as the alloys containing 20 to 30% palladium, remainder gold, have been used for laboratory chemical ware as substitutes for platinum, and possess higher melting points and greater hardness than pure gold.

Silver Alloys

The simple Au-Ag alloys are solid solutions at all temperatures, are all very soft, and find little direct application, aside from the restricted decorative use of the 75% Au–25% Ag alloy constituting "green gold." Some small use has been made of the 70% Au–30% Ag alloy and the 10 to 20% Au, remainder Ag, alloys for electrical contacts, as well as the harder 70% Au–6% Pt–24% Ag[3] alloy, now replaced almost wholly by palladium. Alloys containing 20 to 30% gold, remainder silver, are very important in assaying because the silver can be quantitatively extracted, as will be discussed later.

Copper Alloys — Ordering

Copper is an extremely effective hardener for gold. It is employed for hardening gold coins and is a component of many more complex alloys. Occasionally, the 75% Au–25% Cu alloy (18 carat*) of reddish color is employed for decoration on jewelry.

The 25% Cu alloy, which closely approximates 50 atomic per cent Cu-Au, or the compound AuCu, is a solid solution at high temperatures. However, below about 430° C (805° F) a rearrangement occurs, in which the random arrangement of atoms in the simple face-centered cubic lattice is replaced by a slightly tetragonal ordered structure in which the gold and copper atoms have a regular arrangement. This change causes an alteration in the mechanical and electrical properties, but its effect

---

* The *carat* is the proportion of gold in an alloy in twenty-fourths. It is calculated on the following basis:
$$\frac{\text{Wt. \% Gold} \times 24}{100}$$
The *fineness* of gold is the gold content in parts per thousand.

Gold-clad or laminated stock, frequently employed for jewelry, is designated by the ratio of weight of the gold alloy to the total weight of the material. For example, 1/10 14-carat stock would contain 1/10 × 14/24 or 5.8% gold by weight.

It should also be noted that Troy weight is standard in the precious metal industry. The Troy ounce is equal to 1.097 avoirdupois ounces and is equivalent to 31.1 grams.

[2] K. W. Frohlich, *Chem.-Ztg.*, **66**, 161–163 (1942).
[3] E. B. Craft and J. W. Harris, U. S. Patent 937,284 (1909).

upon corrosion resistance is not decisive. Harder[4] presents a detailed review of the views concerning this transformation and the ordering at a composition corresponding to $AuCu_3$. The AuCu composition is very troublesome to work, and for this and other reasons finds little practical use, although it is of great scientific interest.

COPPER AND SILVER ALLOYS

Alloys in which part of the copper is replaced by silver are much more generally useful, and are the basis of the yellow and greenish-yellow golds employed in the jewelry industry.[5,6] Most of these alloys respond to age hardening. In the high-carat alloys, this is due largely to the Au-Cu ordering reaction. In the low-carat golds, age hardening may be due to a phase based on $AuCu_3$, or to vestiges of the Ag-Cu eutectic which extends out into the ternary field.[7] Some of the low-carat alloys also contain zinc, which produces age hardening phases.[8]

As much as 5% nickel and some zinc are added to produce the 10- and 12- carat pink or tan alloys used for eyeglass frames. Alloys containing Au, Ag, and Cu, and generally palladium and platinum, are standard for dental restorations because of their good corrosion resistance, high strength, and age hardenability. Their properties in both wrought and cast form have been studied in detail.[9,10,11,12]

Pen nibs of Au-Ag-Cu alloys are subjected to several varieties of corrosive agents in the ink and also to tarnishing; hence a 14-carat alloy is required for this application. Even at this level, stress cracking or stress corrosion cracking has, at times, been a problem.

NICKEL ALLOYS

The Au-Ni alloys are solid solutions at high temperatures, but separate into nearly pure gold and nearly pure nickel on slow cooling. In that state they are easily attacked by nitric acid. The solid-solution alloys are hard, and the 75% Au–25% Ni alloy is white but has little use. Related alloys containing zinc, and generally copper, form the basis of the 18-carat, 14-carat, and lower-carat white golds which are extensively used for jewelry. Many of these alloys undergo complex and incompletely understood transformations on heating at low temperatures.[8]

CORROSION CHARACTERISTICS OF PURE GOLD

Some indication of the behavior of gold toward a number of corrosive media is summarized in Tables 1 and 2, and additional data for temperatures above 100° C (210° F) are given in the section on high-temperature behavior (p. 699).

Pure gold is not attacked by *oxygen* or by *sulfur, sulfur dioxide,* or *selenium;* but at high temperatures *tellurium* reacts with it.[13] There is, however, some evidence of

[4] O. Harder, *A. S. M. Metals Handbook*, pp. 1496–1502, 1939.
[5] F. E. Carter, *Trans. Am. Inst. Mining Met. Engrs.*, **78**, 786–802 (1928).
[6] L. Sterner-Rainer, *Z. Metallkunde*, **18**, 143 (1926).
[7] R. F. Vines and E. M. Wise, "Age Hardening Precious Metal Alloys," *Am. Soc. Metals Symposium on Age Hardening*, 190–226 (1939).
[8] E. M. Wise, *Trans. Am. Inst. Mining Met. Engrs.*, **83**, 384–403 (1929).
[9] W. Souder and G. Paffenbarger, *Natl. Bur. Standards (U. S.)* Circ. C-443 (1942).
[10] R. L. Coleman, *J. Am. Dental Assoc.*, **12**, 520 (1925); *Dental Cosmos*, **69**, 1007 (1922).
[11] E. M. Wise, W. C. Crowell, and J. T. Eash, *Trans. Am. Inst. Mining Met. Engrs.*, **99**, 383–407 (1932).
[12] E. M. Wise and J. T. Eash, *Trans. Am. Inst. Mining Met. Engrs.*, **104**, 277–303 (1933).
[13] L. Nowack and J. Spanner, *Die Korrosion Met. Werkstoffe*, by O. Bauer, O. Kröhnke, and G. Masing, Vol. II, p. 823, S. Hirzel, Leipzig, 1938.

the existence of a minute oxide film, and that it produces mild passivity.* A film sufficient to prevent amalgamation is said to develop on exposure to H_2S,[14] and it is known that alkaline sulfides attack gold, presumably in the presence of oxygen. Gold is extremely resistant to *sulfuric acid,* and the slight corrosion that does occur

TABLE 1. BEHAVIOR OF GOLD IN VARIOUS MEDIA

Medium	Temp.	Rating	Medium	Temp.	Rating
Sulfuric acid	R. T.	A	Fluorine	R. T.	A
	100° C	A		100° C	A
Sulfuric (fuming)	R. T.	A	Chlorine—dry	R. T.	C
Persulfuric acid	R. T.	A	" —moist	R. T.	D
			Chlorine water	R. T.	D
Selenic acid	R. T.	A	Bromine — dry — liquid	R. T.	D
	100° C	A	Bromine water	R. T.	D
			Iodine	R. T.	B
Nitric acid, 70%	R. T.	A	Iodine in KI	R. T.	D
	100° C	A	Iodine in alcohol	R. T.	C
Nitric acid (fuming)	R. T.	B			
			Ferric chloride solution +HCl	R. T.	B
Aqua regia	R. T.	D			
			Sulfur	100° C	A
Hydrofluoric acid, 40%	R. T.	A	Selenium	100° C	A
			Hydrogen sulfide — moist	R. T.	A
Hydrochloric acid, 36%	R. T.	A			
	100° C	A	Sodium sulfide + oxygen	R. T.	C
Hydrobromic acid, sp. gr. 1.7	R. T.	A	Potassium cyanide + oxygen	R. T	D
Hydriodic acid, sp. gr. 1.75	R. T.	A			
			As anode where acid halogens or CN are involved	R. T	C–D
Perchloric acid	R. T.	A			
	100° C	A	Mercury and other molten metals		D
Acetic acid		A			
Citric acid		A			
Tartaric acid		A			
Hydrogen cyanide solution + oxygen		C			
Phosphoric acid	100° C	A			

R. T. — room temperature.
A — little or no attack.
B — slight and generally acceptable.
C — considerable corrosion. (Metal is not useful except for small laboratory equipment)
D — rapid corrosion.

above 250° C (480° F) appears to be dependent upon the availability of oxygen. A hot *mixture of sulfuric and nitric acid* rapidly attacks gold,[15] as do hot mixtures of sulfuric acid and oxides of the heavy metals.[16]

The reactivity of the halogens toward gold is a minimum when they are extremely dry, but attack occurs at moderately elevated temperatures, generally

* Since the free energy of formation for Au_2O_3 is positive (18,710 gram-calories per mole) it appears more likely that any such film would consist of adsorbed oxygen. EDITOR.
[14] W. Skey, *Chem. News,* **22,** 282 (1870); **23,** 277 (1871).
[15] V. Lenher, *J. Am. Chem. Soc.,* **26,** 552 (1904).
[16] J. W. Mellor, *A Comprehensive Treatise on Inorganic and Theoretical Chemistry,* Vol. 3, p. 529, Longmans, Green and Co., London, 1923.

forming rather volatile compounds. *Iodine* becomes active at about 50° C (120° F),[17] *chlorine* is active at 80° C (175° F),[18] and *fluorine* becomes active at 300° C (570° F). Liquid *chlorine* is very active at 100° C (212° F).[19] * *Bromine* is perhaps the most reactive halogen with respect to gold.

TABLE 2. THE CORROSION RATES OF GOLD IN VARIOUS MEDIA*

Volume solution — 25 ml.
Area of specimen — 12.9 sq cm.
Aeration — nat. convect.
Velocity — quiescent.

Corrosive Medium	Temp.	Duration of Test	Wt. Loss, mdd
Hydrofluoric acid	R. T.†	....	0.0
5% Hydrochloric acid, intermittent exposure	Boiling	....	2.2
10% Phosphoric acid	100° C	5 hr	2.4
Acetic acid (glacial)	100° C	4 hr	1.4
10% CuSO$_4$	100° C	5 hr	0.0
Chlorine, dry‡	R. T.	23 hr	7.6
Chlorine, moist‡	R. T.	....	865
Saturated chlorine in water	R. T.	18 hr	1510
Bromine, dry‡ (liquid)	R. T	21 hr	1770
Bromine, moist‡	R. T.	21 hr	672
Saturated bromine in water	....	23 hr	1750
Iodine, dry	R. T.	....	Trace
Iodine, moist	R. T.	....	0.0
5% Iodine in alcohol	R. T.	24 hr	12.0
Aqua regia	22° C	....	Rapid attack
Aqua regia	100° C	....	Rapid attack
H$_2$S, moist	R. T.	45 days	0.0

* R. H. Atkinson, A. R. Raper, and A. Middleton, *Platinum Metals Laboratory Report*, Mond Nickel Co., London, 1937.
† R. T. — room temperature.
‡ Reaction products removed prior to weighing.

The presence of moisture increases the activity of the halogens with the exception of fluorine (the latter will of course react with water) so that gold is unserviceable in their presence. Moist iodine has been reported to be active at room temperature, but possibly this depends upon the presence of light, as according to the experiments reported in Table 2[20] no reaction was observed for this reagent. Iodine in KI solution is extremely corrosive, with rates as high as 25,000 mdd[21] being quoted. Bromine behaves similarly to iodine in this regard. *Ferric chloride* plus HCl and air is corrosive, particularly hot, and the unstable oxyhalogen acids are also corrosive, as might

* The approximate corrosion rates of gold in dry Cl$_2$ determined for short-time tests are: 0.03 ipy at 120° C (250° F), 0.06 ipy at 150° C (300° F), 0.12 ipy at 175° C (350° F) and 1.2 ipy at 205° C (400° F). In dry HCl at 980° C (1800° F) gold corrodes at a rate of 0.05 ipy in short-time tests. The suggested upper limit for continuous service in dry HCl is 875° C (1600° F). Private Communication, M. H. Brown, E. I. Du Pont de Nemours Co.

[17] L. Nowack and J. Spanner, *Die Korrosion Met.Werkstoffe*, by O. BAUER, O. KRÖHNKE, and G. MASING, Vol. II, p. 816, S. Hirzel, Leipzig, 1938.

[18] Julius Meyer and W. Aulich, *Z. angew. Chem.*, **44**, 21 (1931).

[19] J. W. Mellor, *A Comprehensive Treatise on Inorganic and Theoretical Chemistry*, Vol. 3, 527, Longmans, Green and Co. (London), 1923.

[20] R. H. Atkinson, A. R. Raper, and A. Middleton, *Platinum Metals Laboratory Report*, Mond Nickel Co., London, 1937.

[21] British Patent, 358, 383.

be expected. Mixtures of hydrochloric, hydrobromic, or hydriodic acids with nitric acid are extremely active, the first combination (aqua regia) being widely used to effect solution of gold. *Hydrofluoric acid plus nitric acid,* however, is not corrosive.

In the absence of air or other oxidizing components solutions of pure HCl and its homologues are not corrosive to gold, but on heating at high temperatures, probably under air pressure, HCl becomes quite reactive, according to Ogryzlo.[22]

Gold is attacked by a solution of HCN containing oxygen, and is more rapidly attacked by solutions of NaCN or KCN containing air or other suitable oxidizing agents. Use is made of this fact in extracting gold from certain of its ores.[23,24] A mixture of 10% KCN and 10% $(NH_4)S_2O_8$[25] is widely used for etching gold alloys in metallography.

Anodic Corrosion

A gold anode in an alkaline cyanide solution is attacked, and in suitable solutions a very smooth deposit of gold occurs at the cathode. For these reasons, and also because of the low noble-metal ion concentration in such solutions, the cyanide bath is normally used for gold plating.[26]

As an anode, gold dissolves quite readily in acid-chloride solutions, but at sufficiently high current densities it may become passive. When considerable silver is present as an alloy, solution may become slow, even at lower current densities. The superposition of alternating current, as in the Wohlwill[27] Process for gold refining,[28] insures rapid solution, despite the presence of silver.

CORROSION CHARACTERISTICS OF GOLD ALLOYS

Additions of silver, and particularly copper, lower the nobility of gold, as might be expected; but the effect on the acid resistance is very slight at room temperature down to about 50 atomic per cent gold (which is about 65 weight per cent gold in the case of the silver alloys and about 75 weight per cent for the copper alloys). Below this the activity toward strong corrosive agents increases rapidly, as studied in detail by Tammann[29,30,31] and his coworkers. For weaker corrosive media a generally less sharply defined limit slightly above 25 atomic per cent can be observed, as indicated in Table 3. While these reaction limits offer some guidance in specifications of alloys, complete reliance cannot be placed upon them in practice.

The effect of variations in composition on the corrosion behavior of 10-carat jewelry alloys has been presented by Jarrett.[32]

---

[22] S. P. Ogryzlo, *Econ. Geol.*, **30**, 400 (1935).
[23] Robert Peele, *Mining Engineers' Handbook*, 3rd Ed., Vol. II, 33–07, 33–10, John Wiley and Sons, New York, 1941.
[24] Ernst Beyers, *J. Chem. Met. Mining Soc.*, South Africa, **37**, 37–89; 148–152 (1936).
[25] M. C. Jewett, *Special Report*, Research Laboratory of the Wadsworth Watch Case Co., Dayton, Ky., 1923.
[26] W. Blum and G. B. Hogaboom, *Principles of Electroplating and Electroforming*, 2nd Ed., pp. 362–372, McGraw-Hill Book Co., New York, 1930.
[27] German Patent 207, 555 (1908).
[28] H. J. Creighton and W. A. Koehler, *Principles and Applications of Electrochemistry*, 2nd Ed., Vol. II, *Applications* (by Koehler), pp. 179–181, John Wiley and Sons, New York, 1944.
[29] G. Tammann and E. Brauns, *Z. anorg. Chem.*, **200**, 209 (1931).
[30] G. Tammann, *Die Korrosion Met. Werkstoffe*, by O. Bauer, O. Kröhnke and G. Masing, Vol. II, pp 788–793, S. Hirzel, Leipzig, 1938.
[31] G. Masing, *Trans. Am. Inst. Mining Met. Engrs.*, **104**, 16 (1933).
[32] Tracy Jarrett, *Trans. Am. Inst. Mining Met. Engrs.*, **137**, 447–455 (1940).

Table 3. REACTION LIMITS*

Corrosive Medium	Cu-Au Solid Solutions		Cu-Ag-Au Solid Solutions (Atomic Ratio, Cu to Ag = 1)	Ag-Au Solid Solutions	
	At. % Au	Wt. % Au	At. % Au	At. % Au	Wt. % Au
PdCl$_2$	24.5–25.5	50.2–51.5	26–30	.....	.....
Pd(NO$_3$)$_2$	.....	.....	20–24	24.5–25.5	37.2–38.5
PtCl$_2$	24.5–25.5	50.2–51.5	.....	24.5–25.5	37.2–38.5
(NH$_4$)$_2$S$_2$	24.5–25.5	50.2–51.5	20–24	32	46.5
Na$_2$S$_2$	.....	.....	.....	27–32	40.3–46.2
Na$_2$S	22	46.7	.....	27	40.3
HNO$_3$ (sp. gr. 1.3)	.....	.....	35–40	48.0–49.0	62.8–63.7
H$_2$SO$_4$	49–50	74.5–75.5	.....	50	64.7
HgCl	No limit	No limit	.....	.....	.....

* Below these concentrations at room temperature, appreciable attack occurs. "Reaction limit" is here used instead of "parting limit" because complete parting may not occur in all cases.

Parting

The resistance of gold to chloride-free nitric acid of not too high concentration[33] is the basis for the "parting" operation in assaying for gold.[34] In this step an essentially Au-Ag alloy, usually containing 20 to 30% gold, rolled to a thin strip, is treated with nitric acid (1 part nitric to 3 parts water). After action ceases the sample is treated with 1-1 nitric acid to remove any residual silver. The residue is nearly pure gold. If the alloy is too low in gold, or is treated initially with too strong acid, the gold does not remain in a coherent strip, and so mechanical losses may result. Treatment with nitric acid above a density of 1.46 will produce appreciable loss of gold by solution.

Tarnishing

Tarnishing due to H$_2$S, elemental sulfur, or other non-oxidized sulfur compounds is of considerable importance from an aesthetic standpoint and in use in electrical contacts. In the normal dental and jewelry alloys, 14-carat and above, the corrosion resistance is generally sufficient for normal applications, and such alloys usually withstand a spot test with nitric acid.* The tarnishing problem generally becomes more acute in the low-carat jewelry alloys, and the corrosion resistance of the latter is not always adequate. These lower-carat alloys are generally vulnerable to the nitric acid spot test.

The effect of various additions on the tarnish resistance of the multi-component alloys is not simple. For instance, white-gold jewelry alloys containing rather large amounts of nickel have relatively good resistance to indoor atmospheric tarnishing, and the replacement of a portion of the gold in the dental-type alloys by palladium rather sharply improves the tarnishing behavior. For indoor atmospheres the rate of tarnishing of the Au-Ag alloys drops rather steadily from a high value for silver

---

* If a drop of concentrated nitric acid placed on the alloy does not discolor and leaves no spot, the alloy is said to withstand the nitric acid spot test.

[33] G. Tammann and E. Brauns, *Z. anorg. Chem.*, **200**, 209 (1931).

[34] O. C. Shepard and W. F. Dietrich, *Fire Assaying*, 1st Ed., 75–84, McGraw-Hill Book Co., New York, 1940.

to a negligible value for 70% gold. The effect of composition on the tarnish rate under outdoor exposures is uncertain, as there has been little occasion to study gold alloys in such exposures. In the case of gold leaf, not less than 95% gold is considered necessary to avoid tarnishing. In tropical climates 18-carat alloys may be required in jewelry to withstand service conditions.

INDIUM AND INDIUM ALLOYS

G. B. Barton*

Indium has been available in commercial quantities for such a comparatively short time that its corrosion characteristics have been studied for only a few applications. Its physical characteristics of low melting point (155° C, 311° F), low strength, and extreme softness (1 Brinell scale) make it useful for special applications. To date, it has found most widespread use for plating engine bearings for protection against corrosion. Patents have been granted for plating on silver or alloying with silver to improve resistance to tarnish, particularly by sulfur and hydrogen sulfide.

Of the commonly used metals, tin most nearly resembles indium in behavior in aqueous acids, neither being very resistant, especially in the presence of dissolved oxygen. However, indium, not being amphoteric, shows excellent resistance to the action of alkaline solutions. Dissolved oxygen accelerates the rate of attack, particularly in acids. Since indium is generally placed between cobalt and cadmium in the electromotive series, it can be surmised that its galvanic behavior would resemble that of these metals.

Some tests were carried out to determine the relative rates of attack by various acids and alkalies.[1] The results are tabulated in Table 1.

Since lubricating oils at times contain organic acids which are formed by breakdown of the oil, experiments were conducted to obtain data on the accelerated rate of attack of pure indium. A solution of 5% by volume of oleic acid in a Penn base S. A. E. 10 motor oil was used. It is to be emphasized that this concentration of acid is much higher than would ordinarily be found in lubricating oils. Two specimens of indium similar to those used in the aqueous tests were suspended on glass hooks in 1.1 liter of the acid-oil solution at room temperature. Air was bubbled through the solution continuously. At the end of 12 days there was no change in weight. The specimens were then placed in a loosely stoppered flask containing 500 ml of the same acid-oil mixture and placed in an oven at 136° ± 3° C (277° ± 5° F). At the end of 2 weeks the specimens showed a loss of 7.65 mdd and were covered with a brown tarnish film.

Experiments with indium-plated bearings have been reported.[2,3] Since the indium plating rapidly diffuses into the surface layer of the underlying metal at operating temperatures, these results are actually those of an alloy. Others[4] have reported on the action of used lubricating oil on Pb-In diffusion alloys. In all cases they reported that indium greatly improved the resistance of the bearings or bearing alloys to corrosion by lubricating oils.

Indium shows good resistance to tarnish in indoor atmospheric exposure. It resists

---

* Central Research Laboratory, American Smelting and Refining Co., Barber, N. J.

[1] Unpublished data, American Smelting and Refining Co.
[2] C. F. Smart, *Trans. Am. Inst. Mining Met. Engrs.*, Inst. Metals Div., **128**, 295 (1938).
[3] L. Raymond, *J. Soc. Automotive Engrs.*, **50**, 533 (1942).
[4] J. M. Freund, H. B. Linford, and P. W. Schutz, *Trans. Electrochem. Soc.*, **84**, 65–69 (1943).

TABLE 1. CORROSION RATES OF 99.9% INDIUM IN ACIDS AND ALKALIES

Size of specimen — 26 × 76 × 2.35 mm (1 × 3 × 0.084 in.).
Surface — 0.44 sq dm.
Surface preparation — etch in 1 : ⅛ : 4 HCl, HNO₃, H₂O for 30 sec or until grain pattern is visible
Temp. — room (approx 30° ± 5° C).
Aeration and velocity — air-lift pump kept solution in slow circulation.
Duration — 5 days in 1% solutions and 7 days in 5% solutions.
Vol. of solution — 1.5 liters.
Total immersion.
The 1% solutions changed every 2 days; 5% solutions every 4 days.

Solution	Composition	pH	Wt. Loss, mdd	Appearance
HCl	1%	....	123	Bright
HNO₃	1%	....	2920	Bright with grain boundaries etched
H₂SO₄	1%	....	176	Bright
Acetic	1%	2.1	130.3	Gray tarnish
"	5%	....	172.9	" "
Citric	1%	2.15	21.8	Bright
"	5%	....	122.2	"
Oxalic	1%	1.45	117.6	Very light tarnish
"'	5%	....	98	Bright
Lactic	5%	....	151.5	Very light tarnish
Succinic	5%	....	122.8	" " "
Na₂CO₃	5%	....	.7 gain	" " "
NaOH	5%	....	2.5 "	Light gray film

oxidation at temperatures up to a little above the melting point, but at higher temperatures is oxidized to form In_2O_3. Sulfur and H_2S in the atmosphere are virtually without effect.

Two specimens similar to those used in the immersion tests, 26 by 76 by 2.35 mm (1 by 3 by 0.084 in.), were placed in a salt-spray chamber for 6 weeks. The chamber was operated at 35° ± 2° C (95° ± 4° F), using 3% NaCl. Upon removal, the specimens were found to be partially covered with a hard yellow film, the color of In_2O_3, only slightly soluble in acetic acid. A gain in weight of 35.5 mg (1.93 mdd) was noted.

Indium alloys have been little investigated. It has been reported[5,6] that the addition of indium to tin reduced the tendency of tin to corrode in aqueous media.

ATMOSPHERIC CORROSION OF IRON

C. P. LARRABEE*

The rate of atmospheric corrosion is dependent upon: A. The length of time moisture is in contact with the surface. B. The extent of pollution of the atmosphere. C. The chemical composition of the iron or steel.

That factors A and B are both of great importance is seen in the results of tests of the Iron and Steel Institute (British).[1] Samples of the same iron exposed in a

---

* Research Laboratory, Carnegie-Illinois Steel Corp., Vandergrift, Pa.

[5] C. G. Fink, E. R. Jette, S. Katz, and F. J. Schnettler, *Trans. Electrochem. Soc.*, **75**, 463–469 (1939).
[6] G. Derge, *Trans. Electrochem. Soc.*, **75**, 469–470 (1939).
[1] J. C. Hudson, Official Investigator to the Corrosion Committee. *Paper* 10, p. 49. Iron and Steel Institute, 4 Grosvenor Garden, London, S.W. 1, 1943.

similar manner lost one hundred times as much weight in the humid industrial atmosphere of Frodingham, England, as in the dry semi-tropical atmosphere of Khartoum, Egypt. Corrosion of unprotected steel is also very severe in marine atmospheres, especially where the metal is subjected to periodic wetting by salt water or spray. The relative corrosion rate is small in unpolluted inland atmospheres in the arctic, temperate, or tropic zones.[1]

Variations in the compositions of irons or steels (factor C) may play as important a part as atmospheric environment. This is illustrated by the fact that in the industrial atmosphere of Kearny, N. J., a structural steel with 0.04% residual copper lost 13.8 grams per sq dm after 8 years of exposure, whereas a similar specimen of a steel containing 11.4% chromium lost only 0.16 gram per sq dm. A very thin and adherent red rust formed on the latter specimen.

There are two common methods of decreasing the atmospheric attack on ferrous metals when the alloy content is below that of the stainless grades. The first is to cover the metal with a protective coating, either metallic, such as zinc, tin, lead, or nickel and chromium, or non-metallic, such as paint. The second is to add alloying elements to iron which by promoting the formation of a more dense, adherent, and hence protective rust, cause a much slower attack. Only the second method of decreasing atmospheric attack is considered in this chapter. (For discussion of first method, see *Corrosion Protection,* beginning p. 801.)

Since the corrosion rates of steels which remain continuously moist vary greatly, depending upon the particular conditions of exposure, the following discussion is intended to apply only where the rust can dry periodically.

COPPER STEELS AND IRONS

The most extensive attempt to determine the relative atmospheric corrosion resistance of commercial sheet products was started in 1916 by the American Society for Testing Materials, and has continued to the present time.[2] It is noted in these reports that the rust on ferrous materials containing 0.1 to 0.3% copper is more dense and adherent than that on similar products of lower copper content.

In Fig. 1 are shown graphically the results of two of these tests on corrugated sheets of 22 gage (0.79 mm, 0.031 in.) exposed 30° to the horizontal. Examination of the chart shows a large difference in the corrosiveness of the atmospheres at Pittsburgh, Pa., and Annapolis, Md. However, at both locations, all materials containing copper have a longer life (years to visible perforation) than similar low-copper products. At Pittsburgh, the open hearth copper steel remained without visible perforation nearly three times as long as the average low-copper open hearth sheet. Owing to the combined effect of phosphorus and copper, this ratio is increased in Bessemer steels (averaging 0.10% phosphorus), with the result that the copper-Bessemer steel sheets were the most resistant in the test. Attention is called to the fact that the addition of 0.3% copper does not improve the resistance of the wrought iron as much as it does the open hearth or Bessemer steels. The same is true, but to a lesser extent, of the materials designated by the A.S.T.M. as open hearth irons. The Pittsburgh test was discontinued after about 6 years (1923), but that at Annapolis was still in progress in 1946.

Phosphorus (0.01 to 0.3%) apparently has a very marked beneficial effect upon the

---

[2] *Proc. Am. Soc. Testing Materials,* 1916–

atmospheric corrosion resistance of copper steel.[3,4,5] It has a more noticeable effect when present in amounts greater than 0.05%, but it is seldom present in large amounts in open hearth copper steel on account of the adverse effect upon cold-forming qualities.

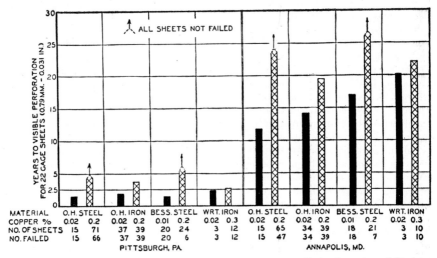

FIG. 1. Data on Effects of High and Low Copper on Atmospheric Corrosion of Various Steels and Irons. (From tests of Subcommittee III, Committee A–5, A. S. T. M.)

LOW–ALLOY STEELS
(Less than 10% total alloy content.)

Economic reasons limit the practical additions of suitable alloying elements. Nevertheless, loss-of-weight determinations show that relatively small amounts of many elements, singly or combined, confer appreciable atmospheric corrosion resistance on steels. Space does not permit a detailed exposition of the effect of each individual alloying element, but French and LaQue[6] cover the subject quite thoroughly. Experience has shown that it is necessary to expose each steel with its particular combination of alloying elements in the type of atmosphere in which it will be used in order to determine with exactness its relative corrosion rate. Data from tests in industrial atmospheres cannot be applied to marine atmospheres, or vice versa. This is evident from an examination of Fig. 2. It is seen that the loss in weight of the copper steel during the first 0.5 and 1.5 years is considerably less in marine than in industrial atmosphere, but from the shape of the curves it is probable that the corrosion losses in the marine atmosphere will eventually be higher than in the industrial atmosphere. It is also apparent that, although the 3.5% nickel steel loses

[3] K. Daeves, "Die Weiterentwicklung witterungsbeständiger Stähle," *Arch. Eisenhüttenw.*, **9**, No. 1, pp. 37–40 (1935).

[4] V. V. Kendall and E. S. Taylerson, "A Critical Study of the A. S. T. M. Corrosion Data on Uncoated Commercial Iron and Steel Sheets," *Proc. Am. Soc. Testing Materials*, **29**, Part 2, 204–219 (1929).

[5] C. H. Lorig and D. E. Krause, "Phosphorus as an Alloying Element in Low Carbon, Low Alloy Steels, III," *Metals and Alloys*, **7**, 69 (1936).

[6] H. J. French and F. L. LaQue, *Alloy Constructional Steels*, Part VIII, pp. 140–206, American Society for Metals, Cleveland, Ohio, 1942.

considerably less weight in marine than in industrial atmosphere, the opposite is true of the more complex Ni-Cu-Si-P steel.

The effects of different alloying elements in promoting corrosion resistance are not always cumulative, although such additive properties are often observed. Phosphorus increases the atmospheric corrosion resistance of many low-alloy steels, especially in an industrial atmosphere. As with copper steels,[7] phosphorus in amount

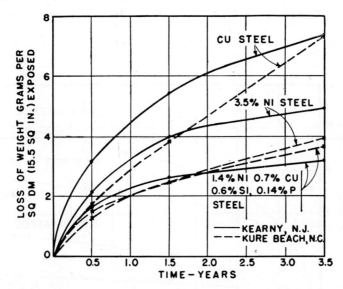

FIG. 2. Time-Corrosion Curves of Three Steels in Marine and Industrial Atmospheres. (Specimens 10.2 by 15.2 cm [4 by 6 in.].)

greater than 0.05% is usually necessary to cause a worth-while improvement.

Visual inspection of rusted specimens of alloy steels shows that the rust is more adherent and of finer texture than even that on copper steel, and thus presumably less permeable. However, as stated above, such resistant films are formed only when the exposure conditions permit the rust to become dry part of the time. Such conditions exist in the A.S.T.M. and similar tests. Under continuously moist conditions the corrosion rates of plain and low-alloy steels are more nearly alike.

Under conditions of test as outlined in *Atmospheric Exposure Tests*, p. 1043, many combinations of alloying elements in steels produce rust films that improve the corrosion resistance appreciably over that of copper steels. In Fig. 3 are shown typical time-corrosion curves for open hearth copper steel, Bessemer copper steel, 3.2% nickel structural steel, and a "band" in which lie the losses in weight of three typical low-alloy, high-strength, corrosion-resistant steels which were among the first of this type introduced in 1934–1935. Analyses of these steels are given in Table 1.

In order to obtain losses in weight below the "band" in Fig. 3 (higher corrosion resistance), it is necessary to add alloying elements in amounts disproportionate to their value. This is due to the fact that, when steels are exposed as referred to above, a loss in weight of about 3 grams per sq dm of most alloy steels is used up in supply-

[7] C. H. Lorig and D. E. Krause, "Phosphorus as an Alloying Element in Low Carbon, Low Alloy Steels, III," *Metals and Alloys*, **7**, 69 (1936).

TABLE 1. ANALYSES OF STEELS IN FIG. 3

	C	Mn	P	S	Si	Cu	Ni	Cr	Mo
O. H. Cu Steel	0.02	0.39	0.006	0.018	0.005	0.20	0.00	0.07	....
Bessemer Cu Steel	0.10	0.40	0.112	0.059	0.018	0.21	0.003	0.03	....
3.2% Ni Struct. Steel	0.19	0.53	0.016	0.022	0.009	0.07	3.23	0.10	....
Low-Alloy Steels									
Ni-Cu Steel	0.05	0.36	0.054	0.016	0.008	1.14	1.99	0.01	....
Cu-Ni-Mn Steel	0.09	0.86	0.008	0.025	0.019	1.41	0.95	0.03	0.09
Cr-Si-Cu-P Steel	0.09	0.24	0.154	0.024	0.80	0.43	0.05	1.07	....

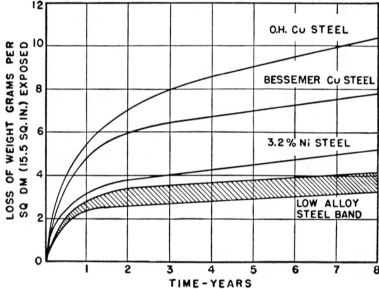

FIG. 3. Time-Corrosion Curves for Steels in Industrial Atmosphere. (Specimens 10.2 by 15.2 cm [4 by 6 in.].)

ing metal for a protective oxide film. Therefore, the corrosion rates shown by the lower limit of the band in Fig. 3 appear to be about the optimum that can be obtained economically in this type of steel. It has been shown[8] that under these conditions of exposure approximately 40% of the loss occurs on the skyward side of the specimen.

CAST IRON

As cast iron is usually employed in very thick sections, little work has been done on its atmospheric corrosion resistance. However, Friend[9] gives data which show

---

[8] C. P. Larrabee, "The Effect of Specimen Position on Atmospheric Corrosion Testing of Steel," *Trans. Electrochem. Soc.*, **85**, 297 (1944).

[9] J. N Friend, "Final Report on the Relative Corrodibilities of Various Commercial Forms of Iron and Steel," Iron and Steel Institute, *Carnegie Scholarship Memoirs*, **18**, 61–71 (1929).

that, in a 6-year exposure test in an industrial atmosphere, the corrosion rate of cast iron containing 1% phosphorus was about the same as that of copper steel.

IRON AND STEEL

H. H. UHLIG*

INTRODUCTION

The electrochemical theory of corrosion was proposed as a result of observations on the behavior of iron in aqueous media.[1] This theory, now with overwhelming evidence in its support, proposes that corrosion of metals is largely accomplished by the action of a network of short-circuited electrolytic cells on the metal surface. Metal ions go into solution at the anodes of these cells in amount chemically equivalent to the reaction at the cathodes.

At anode areas of an iron surface, the following reaction takes place:

$$Fe \longrightarrow Fe^{++} + 2e^- \qquad (1)$$

The rate of this reaction is found to be dependent upon the rate of the cathode reaction; hence the corrosion rate is "cathodically controlled."

Either of two reactions is typical of the cathode areas:

$$H^+ \longrightarrow \tfrac{1}{2}H_2 - e^- \qquad (2)$$

$$2H^+ + \tfrac{1}{2}O_2 \longrightarrow H_2O - 2e^- \qquad (3)$$

Reaction 2 is fairly rapid in acids, but is very slow in alkaline or neutral media. It can be speeded by dissolved oxygen as in reaction 3, a process called *depolarization*. The cathodic reaction rate, and hence the overall corrosion rate for case 3, is proportional to the rate of diffusion of oxygen to the metal surface. The diffusion rate, in turn, is proportional to the concentration of dissolved oxygen in the aqueous environment.

AQUEOUS MEDIA

EFFECT OF DISSOLVED OXYGEN

At ordinary temperatures, oxygen and moisture are the basic factors necessary for the corrosion of iron in neutral or near-neutral media. Both must be present simultaneously, because oxygen alone or water free of dissolved oxygen does not corrode iron to any practical extent.

Iron corrodes in natural waters according to Eqs. 1 and 3 at a rate usually proportional to the concentration of dissolved oxygen.† Water in contact with iron continues to cause corrosion only until the dissolved oxygen is consumed. This fact is made use of in deactivation equipment employing expendable iron sheet to react with dissolved oxygen before water enters the iron piping of a building. (See *Deactivation and Deaeration*, p. 506.)

Calculations show that if all the dissolved oxygen in neutral water saturated with air is used up to corrode iron, with Fe_2O_3 as end product, metal is penetrated to the

---

* Corrosion Laboratory, Department of Metallurgy, Massachusetts Institute of Technology, Cambridge, Mass.

† The rate is not proportional if the oxygen concentration is high. See "Effect of High Oxygen Concentration," p. 127.

[1] W. R. Whitney, *J. Am. Chem. Soc.*, **25**, 394 (1903).

extent of 0.0025 cm for every sq cm of surface per liter of water (0.0006 in. per sq in. iron surface per gal water).

Although Eq. 3 is adequate to explain most of the corrosion that occurs in neutral and alkaline media, and Eq. 2 that which takes place in acids, the two processes in practice are seldom clearly one or the other. Dissolved oxygen accelerates corrosion of iron by non-oxidizing acids, and it is known that some hydrogen is liberated when iron corrodes in aerated neutral media. It is stated that 3% of the total corrosion of iron in neutral tap water at 82° C (180° F) can be ascribed to hydrogen evolution (Eq. 2); the remainder is due to oxygen consumption (Eq. 3).[2]

In the absence of dissolved oxygen, the corrosion rate, already minimized by the slow rate of reaction 2, is still further retarded by the formation of an alkaline corrosion product film on the surface. This film, probably hydrous ferrous oxide, forms a saturated solution in water, the pH of which is reported to be 8.2 to 9.6 with some evidence that the lower value is more probable.[3,4] In the presence of dissolved oxygen, this or a similar film slows up diffusion of oxygen to the metal surface, as discussed below under "Effect of pH," p. 129.

Corrosion in Absence of Dissolved Oxygen. Effect of Bacteria. Iron can corrode rapidly in the absence of oxygen if sulfate-reducing bacteria are present. These bacteria are commonly found in deep wells, in soils, and in sea water. The bacteria, by their metabolic processes, reduce dissolved sulfates to sulfides in the course of which they are able to depolarize cathodic areas of iron. Corrosion, thereupon, proceeds as rapidly as bacterial action permits. Galvanized pipe carrying cold water has failed from this source within two years time.

When sulfate-reducing bacteria are responsible for observed corrosion, sulfide can be identified in the iron corrosion products, and often H_2S accumulates in the aqueous environment as corrosion progresses. The bacteria do not thrive in aerated or chlorinated waters. This subject is discussed in further detail under *Corrosion by Micro-organisms in Aqueous and Soil Environments*, p. 466.

Corrosion Caused by Differential Aeration. Oxygen as described previously is a chief factor in the general corrosion of iron immersed in water, but in addition it can cause localization of attack through so-called differential aeration cells. These cells are formed whenever iron is in contact with aerated solution at one place and with oxygen-deficient solution elsewhere. The oxygen-deficient areas become anodes and therefore suffer proportionately more corrosion. Differential aeration cells commonly account for localized corrosion or pitting under partially protective rust layers on iron, in open crevices, or whenever aerated solution is shielded from portions of an iron surface.

Differential aeration also accounts for corrosion at the *water line*.[5] In the presence of chlorides, localized attack of this kind is stimulated by insufficient addition of inhibitors such as chromates, sodium phosphate, or sodium silicate. The remedy is to add more inhibitor the higher the chloride concentration. (See *Inhibitors and Passivators*, p. 905.)

Water line attack may occur in hard waters. The meniscus area of partially immersed iron is freely exposed to oxygen of the air; hence iron in this region tends to become cathodic to regions elsewhere. Alkaline cathodic reaction products form

---

[2] F. N. Speller, *Corrosion, Causes and Prevention*, pp. 33–34, McGraw-Hill Book Co., New York, 1935.

[3] R. Corey and T. Finnegan, *Proc. Am. Soc. Testing Materials*, **39**, 1242 (1939). (Contains a critical summary of the literature.)

[4] W. Whitman, R. Russell, and G. Davis, *J. Am. Chem. Soc.*, **47**, 70 (1925).

[5] U. R. Evans, *Metallic Corrosion, Protection and Passivity*, p. 178, p. 296, Edward Arnold and Co., London, 1937.

which cause precipitation of insoluble magnesium and calcium compounds. The insoluble compounds shield a portion of the iron from aerated solution so as to induce localized attack at the water line.

Effect of High Oxygen Concentration. Although increasing oxygen concentration at first accentuates corrosion of iron in water, as one expects from Eq. 3, it is found that beyond a critical concentration the corrosion rate may drop again to a low value. In moving distilled water the maximum corrosion rate of mild steel (0.06 ipy) appeared at 16 ml of oxygen per liter and fell to a value much lower (0.005 ipy) at 24 ml of oxygen per liter.[6] (Velocity 15 ft/min; 48-hour test, 30° C.)

This is clearly an example of iron passivated by oxygen, as shown by a potential of 0.1 to 0.4 volt (hydrogen scale) for iron in water containing the higher oxygen concentration, as compared with -0.4 to -0.5 volt in water normally saturated with air[7] (6 ml of oxygen per liter at 25° C). Further evidence in this direction is shown by the fact that presence of chloride ions (3.5% NaCl) in solution effectively retards passivation, and iron then corrodes at increasing rates for oxygen concentrations above those for which the rate decreases in distilled water.[8]

EFFECT OF DISSOLVED CARBON DIOXIDE

Carbon dioxide is not necessary, as once thought, to initiate the corrosion reaction in natural waters or in the atmosphere. The small amount of carbon dioxide normally present in hard waters affects the rate of corrosion largely by determining whether or not a protective layer of insoluble compound (e.g., calcium carbonate) will precipitate on an iron surface. Waters containing carbon dioxide in excess of that required to keep calcium carbonate in solution (aggressive carbon dioxide) are more corrosive than those containing just enough or less carbon dioxide. (See *Hot and Cold Water Systems*, p. 496.)

Carbon dioxide in appreciable concentration definitely accelerates the corrosion of iron. Dissolved carbon dioxide, or carbonic acid, is similar to any acid that reacts with iron evolving hydrogen. This fact is of practical importance in the corrosion of steam return lines, where carbon dioxide expelled from boiler waters dissolves in the condensate to form carbonic acid at above-room temperatures. (See *Corrosion of Steam Condensate Lines*, p. 538.)

The corrosion rates of mild steel (0.15% carbon) in various concentrations of dissolved carbon dioxide or carbon dioxide plus oxygen at 60° C (140° F) and 90° C (195° F) are given in Fig. 1.[9] Carbon dioxide is not as corrosive as oxygen in equal concentrations. At 60° C (140° F) for example, 4 ml of oxygen per liter produces seven times more corrosion in a given time than a solution containing 4 ml of carbon dioxide per liter. At gas concentrations of 20 ml/liter at the same temperature oxygen is ten times more corrosive than carbon dioxide.

The corrosion rate in distilled water with only oxygen present increases about twofold between 60° and 90° C and about 2.6 fold with only carbon dioxide present. For dissolved mixtures of the two gases, the temperature dependence of rate is intermediate between these two values. It should be recognized that when pitting accompanies corrosion, the temperature dependence of rate of metal penetration may be appreciably higher than these values.

[6] E. Groesbeck and L. Waldron, *Proc. Am. Soc. Testing Materials*, **31**, Part II, 279–291 (1931).

[7] O. Bauer, O. Kröhnke, and G. Masing, *Die Korrosion metallischer Werkstoffe*, Vol. 1, p. 209, S. Hirzel, Leipzig, 1936.

[8] F. G. Frese, *Ind. Eng. Chem.*, **30**, 83–85 (1938).

[9] G. Skaperdas and H. Uhlig, *Ind. Eng. Chem.*, **34**, 748 (1942).

Effect of Temperature

In a closed system, when corrosion is controlled by diffusion of oxygen, the rate measured by weight loss increases approximately double its value for every 30° C (60° F) rise in temperature.[9] In an open vessel, the corrosion rate in water increases

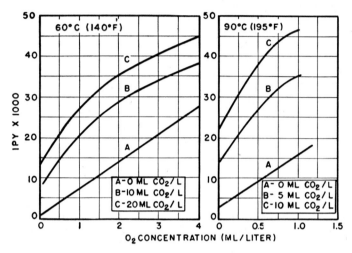

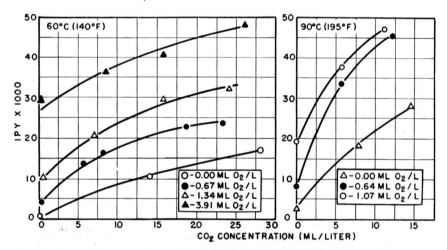

Fig. 1. Corrosion of Mild Steel (0.15% Carbon) as a Function of Dissolved CO_2 and O_2 Concentration. (5-hr test; velocity, 45 cm [0.025 ft] per min; spec. size, 6.3 × 2.5 × 0.318 cm [2.5 × 1 × ⅛ in.].)

with temperature up to about 80° C (175° F), then falls to a very low value at the boiling point.[10] The falling off of corrosion above 80° C is related to decreased solubility of oxygen in water at the higher temperatures.

[10] F. N. Speller, *Corrosion, Causes and Prevention*, p. 153, McGraw-Hill Book Co., New York, 1935.

When corrosion is by hydrogen evolution the rate increase is more than double for a 30° C temperature rise.

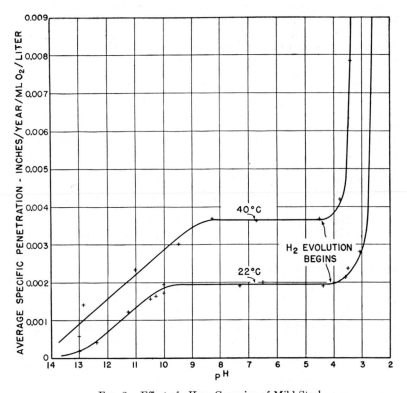

FIG. 2. Effect of pH on Corrosion of Mild Steel.

EFFECT OF pH

The effect of hydrogen ion activity (pH) of water on corrosion of iron at 22° and 40° C (72° and 104° F) is shown in Fig. 2.[11] Specimens of mild steel were exposed to water containing an average of 5 ml of oxygen per liter. NaOH and HCl were added to produce the alkaline and acid ranges of pH, respectively.

At values of pH greater than needed for hydrogen evolution (pH = 4) the results of Fig. 2 can be explained by a protective layer of hydrous ferrous oxide on the iron surface formed by the initial corrosion reaction. Regardless of the actual pH of water between pH 4 and 9.5, the surface of iron is always in contact with an alkaline saturated solution of hydrous ferrous oxide (pH = approx. 9.5*). Corrosion continues as rapidly as oxygen can diffuse through the protective layer, the layer being continually renewed by the corrosion process. Since the corrosion product film next to the iron is essentially unchanged by external conditions within the above range of pH, the corrosion rate is not altered except by change in dissolved oxygen.

* Lower values have been reported. See R. Corey and T. Finnegan, *Proc. Am. Soc. Testing Materials*, **39**, 1242 (1939).

[11] W. Whitman, R. Russell, and V. Altieri, *Ind. Eng. Chem.*, **16**, 665 (1924).

At pH 9.5 (22° C), however, further increase in alkalinity extends its effect to the iron surface and either decreases the surface reaction rate (the iron becomes passive) or decreases permeability of the corrosion product layer to oxygen (by decreasing the solubility of hydrous FeO*). In either instance the corrosion rate is expected to decrease, as is observed.

Within the acid region (pH < 4), on the other hand, the alkaline corrosion product layer is dissolved and the acid reacts directly with iron. For acids that are not totally dissociated into their component ions (weak acids), the pH at which hydrogen is evolved will shift to a higher (less acid) value. For example, carbonic acid at room temperature causes hydrogen evolution at a pH of 6 and phenol solutions at a pH of 7, whereas with sulfuric or hydrochloric acid the value of pH must reach 4. It appears that total acidity rather than pH of an acid is the controlling factor.

At a concentration of sodium hydroxide equal to about 1 gram-equivalent per liter (4%) the corrosion rate of iron is nearly zero and the potential of iron on the hydrogen scale is about 0.1 volt. As the concentration of sodium hydroxide is raised to 16 gram-equivalents per liter (43%) the potential drops to the very active value of -0.86 volt. In this concentration region iron corrodes with formation of soluble sodium ferrite ($NaFeO_2$).† In an oxidizing environment sodium ferrate (Na_2FeO_4) may form. The corrosion rate corresponding to the more active potential in concentrated sodium hydroxide is measurably higher, but is still within the range of 0.0001 to 0.004 ipy.[12]

Corrosion in Natural Waters

Soft waters are more corrosive than hard waters. Hard waters tend to precipitate an insoluble layer, e.g., $CaCO_3$, particularly on the cathodic surfaces of corroding iron, which impedes diffusion of oxygen to the metal. Information as to whether a given water will precipitate a protective surface layer can be estimated from analyses of the water followed by calculation of the so-called saturation index. (See *Hot and Cold Water Systems*, p. 502.)

The corrosion rate of mild steel immersed in quiet soft water saturated with air ranges from 0.002 to 0.006 ipy. The occurrence of pitting may produce maximum rates of penetration as high as ten times these rates. The extent of pitting with time, and factors influencing pitting, are approximately the same in fresh waters as in sea water. These are discussed on p. 388.

The overall corrosion rate of mild steel in natural *sea water* is approximately 0.005 ipy. (See *Behavior of Metals and Alloys in Sea Water*, p. 383.)

Effect of Dissolved Salts

Non-Oxidizing Salts. Salts of the *alkali metals* added to aerated water at first increase the corrosion rate. The increase is caused by two factors: (1) The electrolytic

---

* The formation of a more protective oxide layer is also properly called passivation (definition 2, p. 21). However, a supplementary mechanism of passivity accounting for a decreased reaction tendency of the iron itself (accompanied by film formation of the order of monatomic dimensions) may occur independent of the protective oxide layer and this, it is thought, may actually account for the decreased corrosion (definition 1, p. 21). The latter mechanism is made plausible by a measured noble potential for iron (0.1 volt) in normal sodium hydroxide, indicative of passivity, compared to an active potential (-0.3 to -0.4 volt) in more dilute sodium hydroxide. (See O. Bauer, O. Kröhnke, and G. Masing, *Die Korrosion metallischer Werkstoffe*, Vol. 1, p. 151, S. Hirzel, Leipzig, 1936. Consult also *Passivity*, p. 20.)

† In absence of air, sodium ferroate (hypoferrite) Na_2FeO_2, is said to form.(R. Scholder, *Angew. Chem.*, **49**, 255 [1936]).

[12] O. Bauer, O. Kröhnke, and G. Masing, *Die Korrosion metallischer Werkstoffe*, Vol. 1, p. 151, S. Hirzel, Leipzig, 1936.

conductivity of the solution permits operation of anode and cathode areas farther removed from each other. (2) The anode and cathode reaction products tend to combine at a place removed from the iron surface and hence precipitate a less protective corrosion product layer.

As the salt concentration is increased the corrosion rate goes through a maximum and thereafter decreases. (See Fig. 3. See also Fig. 2, p. 168, showing corrosion of mild steel in NaCl solutions at 90° C.) This decrease is caused largely by decrease in oxygen solubility by dissolved salts.

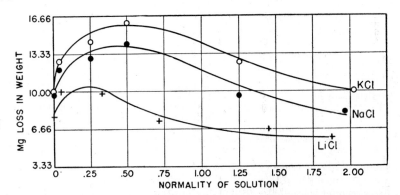

FIG. 3. Effect of Salt Concentration on Corrosion of 0.06% Carbon Cold-Rolled Steel at 35° C (95° F). (48-hr beaker tests, partial immersion, 17.5 sq cm of spec. submerged.)

In 48-hour partial immersion tests at 35° C (95° F) Borgmann[13] showed that the order of increasing corrosiveness of alkali metal ions was Li^+, Na^+, K^+ (Fig. 3). The alkaline earth salts were less corrosive than the alkali metal salts, whereas aluminum and ammonium salts (acid) and chromic (Cr^{+++}) and ferric salts (oxidizing) were more corrosive. Similarly corrosion in 0.1 to 0.25 N potassium or sodium chloride, sulfate, and iodide were reasonably comparable whereas the nitrates were definitely less corrosive at these concentrations, but not necessarily at higher concentrations.[14]

Ammonium salt solutions are exceptional in their active attack on iron and in their continued high rate of attack at relatively high salt concentrations. Dilute solutions of ammonium chloride become markedly more corrosive than similar concentrations of the alkali metal salts above 0.05 N.[13] Corrosion occurs partially with hydrogen evolution since ammonium salt solutions are acid. Ammonium nitrate solutions are considerably more corrosive than either ammonium chloride or sulfate at concentrations above 0.01 N. The corrosion rate in 6 N NH_4NO_3 (the most corrosive concentration) is six times the maximum corrosion rate in sodium chloride or sulfate.[15] High concentrations of ammonium nitrate react with iron to form a yellow solution that, heated to 100° C (212° F), liberates nitrogen and N_2O and precipitates a magnetic oxide, Fe_8O_{11}.[15]

Corrosion rates of steel and cast iron in calcium chloride and some ammonium salt solutions are given in Table 1.

[13] C. W. Borgmann, *Ind. Eng. Chem.*, **29**, 814 (1937).
[14] O. Bauer, O. Kröhnke, and G. Masing, *Die Korrosion metallischer Werkstoffe*, Vol. 1, p. 240, S. Hirzel, Leipzig, 1936.
[15] Bauer, Kröhnke, and Masing, *op. cit.*, p. 241.

Solutions of *acid* salts such as $AlCl_3$ behave much as do dilute solutions of the corresponding acids. Corrosion is both by hydrogen evolution and oxygen depolarization.

TABLE 1*

Salt Solution	Temp.		Duration of Test, days	Corrosion Rate			
				Steel		Cast Iron	
	°C	°F		mdd	ipy	mdd	ipy
5% Calcium chloride	16	60	...	35	0.006	48	0.009
30% " "	−12	10	355	...	...	34	0.006
30% " " (with $K_2Cr_2O_7$ inhibitor)	−12	10	372	...	...	2	0.0004
46% Calcium chloride	180	360	11	28	0.005	...	...
6.6% Ammonium thiocyanate	room	room	13½	30	0.006	15.0	0.003
30% " "	68	154	17	442	0.08	520	0.10
5% Ammonium sulfate	16	60	...	49	0.009	138	0.03
10% " "	16	60	...	50	0.009	151	0.03
25% " "	16	60	...	17	0.003	57	0.012

* Contributed by H. O. Teeple, International Nickel Co., Inc., New York, N. Y.

Solutions of *alkaline* salts (e.g., alkali phosphates, carbonates, borates, aluminates, and silicates) inhibit corrosion of iron. The rate of corrosion is only slight and is much less than for neutral salts. (See "Effect of pH," p. 129.) In alkali sulfide solutions of 0.1 gram/liter at room temperature iron sulfide is formed, but in concentrated solutions the corrosion rate is practically nil.[16]

Oxidizing Salts. Ferric salts are extremely corrosive. They are efficient depolarizing agents, being reduced to ferrous salts in the process of reaction. Mild steel in 3.95% ferric sulfate corroded at a rate of 1.3 ipy at 17° C (63° F) and 2.3 ipy at 45° C (115° F).[17] Likewise cupric and mercuric salt solutions and hypochlorites are very corrosive. *Mine waters* often contain ferric sulfate and owe much of their corrosiveness to this salt rather than to their acid content.

Sodium nitrite is oxidizing, but like *sodium chromate* acts as a corrosion inhibitor (passivator) for iron and steel. Corrosion rates in solutions of these salts at room temperature are practically nil. The rates are measurably higher in presence of chloride or other halide ions. (See *Inhibitors and Passivators*, p. 905. Also refer to Table 1 of this chapter.)

Potassium permanganate also passivates iron. No appreciable corrosion occurs in concentrations above 0.1 gram/liter.[18] Concentrated calcium permanganate solutions are stable in contact with iron up to at least 50° C (120° F), and no visual corrosion occurs in a one-month test.[19]

ACIDS

Dissolved oxygen appreciably accelerates attack of iron by strong or weak non-oxidizing acids.

Hydrochloric Acid. The corrosion rate of electrolytic iron or low-metalloid iron in $1N$ hydrochloric acid is about 0.2 ipy (15° C [59° F]) unagitated acid, open

[16] Bauer, Kröhnke, and Masing, *op. cit.*, p. 243.
[17] F. Ritter, *Korrosionstabellen metallischer Werkstoffe*, p. 63, J. Springer, 1937.
[18] Bauer, Kröhnke, and Masing, *op. cit.*, p. 242.
[19] F. Bellinger, H. Friedman, W. Bauer, J. Eastes, and W. Bull, *Ind. Eng. Chem.*, **38**, 310 (1946).

vessel, 24 hours) and 1.5 to 2.5 ipy in concentrated acid.[20] The corrosion rate in other than dilute solutions increases logarithmically with HCl concentration.

Sulfuric Acid. In dilute *sulfuric acid* the rates are approximately the same as for hydrochloric acid of the same equivalent concentration. The corrosion rate increases with acid concentration, reaching a maximum at about 13 N acid (47 wt. %), after which it decreases to 0.005 to 0.05 ipy for 65 to 100% acid.[21,22] (See "Effect of Velocity on Corrosion in Sulfuric Acid," p. 136.)

Interstate Commerce Commission regulations permit shipment of sulfuric acid of 1.81 specific gravity (65° Baumé, 89%) or greater in steel drums.[23]

The corrosion rate for iron increases again with sulfuric acid concentrations above 100% (fuming sulfuric acid). The rate increases with SO_3 content above 81.5% SO_3 (99.8% H_2SO_4) to a maximum of 0.1 ipy at 85% SO_3, but falls to 0.001 ipy for acid of 90% SO_3 (18° to 20° C, 65° to 68° F, absence of air, 72-hour tests).[22] Cast iron may be embrittled by fuming sulfuric acid, by the reaction, it is thought, of SO_3 with silicon in the metal.[22]

Rise in temperature to 50° C (120° F) increases the corrosion rate in 96.5% sulfuric acid to about 0.2 ipy.[24]

Hydrofluoric Acid. Carbon steel is usefully resistant to hydrofluoric acid in concentrations above 65 to 70% at atmospheric temperatures, but is subject to rapid attack at lower concentrations. Corrosion rates were 0.52 ipy in 48% hydrofluoric acid (duration of test unknown) but only 0.035 ipy in 93% acid at 21° C (70° F), 8-day test.[25] A thoroughly deoxidized or "killed" steel, in which non-metallic inclusions are a minimum, is reported to be more resistant than semi-killed steel to strong hydrofluoric acid concentrations.[26]

Cast iron is generally not recommended for use with hydrofluoric acid solutions.[26]

Interstate Commerce Commission Regulations permit transportation of acid containing not less than 60% HF in steel drums previously "passivated."[27] The passivation treatment consists of filling the drums to 90% of capacity with 58% hydrofluoric acid and allowing them to stand 48 hours at 30° C (80° F) and then 7 hours at 60° C (140° F), the internal pressure being maintained at atmospheric pressure. If containers are washed out with water this treatment must be repeated before filling them again with acid.

Chromic Acid. A chromium plating solution containing 274 grams/liter of Cr_2O_3 and 8.7 grams/liter of sulfuric acid at 50° C (120° F) corroded steel at a rate of 43 mdd (0.008 ipy) in a 19-day test.[28]

Nitric Acid. The corrosion rate of electrolytic iron or low metalloid iron in 1 N nitric acid is about 2 ipy and in 6 N acid (32%) about 13 ipy.[29] Iron becomes passive in acid of specific gravity 1.4 (65%, 15 N) or higher, accompanied by a low corrosion rate (0.005 to 0.02 ipy[30]) and a noble galvanic potential equal to that of platinum.

[20] F. N. Speller, *Corrosion, Causes and Prevention*, p. 510, McGraw-Hill Book Co., New York, 1935.
[21] G. H. Damon, *Ind. Eng. Chem.*, **33**, 67–69 (1941).
[22] R. McKay and R. Worthington, *Corrosion Resistance of Metals and Alloys*, p. 251, Reinhold Publishing Corp., New York, 1936.
[23] Freight Tariff No. 4, "Transportation of Explosives and Other Dangerous Articles by Freight," Section 272, issued by W. S. Topping, 30 Vesey St., New York, 1941.
[24] W. G. Whitman and R. P. Russell, *Ind. Eng. Chem.*, **17**, 348–354 (1925).
[25] W. Z. Friend and H. O. Teeple, *Oil Gas J.*, **44**, 87 (March 16, 1946).
[26] C. M. Fehr, *Oil Gas J.*, **42**, 39–42 (July 22, 1943).
[27] Freight Tariff No. 4, "Transportation of Explosives and Other Dangerous Articles by Freight," Section 264, issued by W. S. Topping, 30 Vesey St., New York, 1941.
[28] Private communication, International Nickel Co., New York, N. Y.
[29] F. N. Speller, *Corrosion, Causes and Prevention*, p. 510, McGraw-Hill Book Co., New York, 1935.
[30] F. Ritter, *Korrisionstabellen metallischer Werkstoffe*, pp. 142–143, J. Springer, Berlin, 1937.

On removing iron from nitric acid of this concentration, passivity is retained for a brief time. Abrasion, heating, or contact with solutions of halide salts accelerates return to the active state. It is stated that iron cannot be made passive in 90% nitric acid above 75° C (170° F) or in 100% acid above 87° C (190° F).[31]

Both steel and cast iron satisfactorily resist concentrated mixtures of nitric and sulfuric acids. Mixtures containing 70 to 95% H_2SO_4, and remainder HNO_3, produced corrosion rates of 0.003 to 0.009 ipy (18° to 22° C [65° to 72° F], 50- to 65-hour test period). Mixtures containing 45 to 65% H_2SO_4, 25% water, and remainder nitric acid corroded cast iron and steel at rates of 0.01 to 0.03 ipy.[32]

Interstate Commerce Commission regulations permit transportation of mixtures of sulfuric acid and nitric acid in tank cars, cargo tanks, tank trucks, or steel drums when the composition of the acid mixture falls within the following ranges.[33]

1. Up to 10% water with not less than 10% H_2SO_4.
2. Up to 15% water with not less than 15% H_2SO_4.
3. Up to 20% water with not less than 20% H_2SO_4.
4. Up to 38% water with not less than 62% H_2SO_4.

Organic Acids. A solution of 6% *acetic acid* at room temperature agitated with oxygen corroded mild steel at a rate of 0.55 ipy, but agitated with hydrogen the rate was only 0.006 ipy (5-hour test).[34] Glacial acetic acid similarly caused attack at a rate of 0.65 ipy under oxygen and 0.01 ipy under hydrogen. Aerated acetic acid of 5, 10, or 25% concentration at room temperature corroded steel at a rate of about 0.1 ipy.[35]

A solution of 5% *citric acid* at room temperature attacked electrolytic iron at a rate of 0.005 ipy.[36] As with other non-oxidizing acids, the rate is increased by higher oxygen concentrations and is also greater for iron of higher carbon content.

Alkalies

At ordinary temperatures iron and steel are satisfactorily resistant to alkalies. Corrosion is practically stopped above a concentration of 1 gram of *sodium hydroxide* per liter or 0.7 gram per liter *potassium hydroxide*.[37] Corrosion rates in 0.1 to 40% sodium hydroxide (open containers, 22 days) are of the order of 0.0001 ipy.[38] Aeration, elevated temperature, higher caustic concentration, presence of dissolved carbon dioxide, and chlorides increase rate of attack.

At 65° C (150° F) mild steel in 50% caustic soda corroded at a rate of 0.008 ipy (30-day test in storage tank).[39] At 105° C (220° F) steel exposed to 70% caustic soda for 90 days corroded at a rate of 0.06 ipy.[40] Moderate (approx. 5%) or high concentrations of sodium hydroxide at elevated temperatures under pressure are severely corrosive to iron and steel. (See *Boiler Corrosion*, p. 520, especially Fig. 3, p. 525.)

[31] E. S. Hedges, *J. Chem. Soc.*, **1928**, 972.
[32] R. McKay and R. Worthington, *Corrosion Resistance of Metals and Alloys*, p. 254, Reinhold Publishing Corp., New York, 1936.
[33] Freight Tariff No. 4, "Transportation of Explosives and Other Dangerous Articles by Freight," Section 267, issued by W. S. Topping, 30 Vesey St., New York, 1941.
[34] W. G. Whitman and R. P. Russell, *Ind. Eng. Chem.*, **17**, 348–354 (1925).
[35] Private communication, International Nickel Co., Inc., New York, N. Y.
[36] F. Ritter, *Korrisionstabellen metallischer Werkstoffe*, p. 187, J. Springer, 1937.
[37] O. Bauer, O. Kröhnke, and G. Masing, *Die Korrosion metallischer Werkstoffe*, Vol. 1, p. 250, S. Hirzel, Leipzig, 1936.
[38] E. Heyn and O. Bauer, *Mitt. kgl. Materialprüfungsamt*, **26**, 1–104 (1908).
[39] G. L. Cox, *Trans. Am. Inst. Chem. Engrs.*, **34**, 657–679 (1938).
[40] Private communication, International Nickel Co., Inc., New York, N. Y.

Fused caustic is also very corrosive. Cast iron immersed for 4 hours in caustic soda carrying 5 to 10% water at 355° C (675° F) corroded at a rate of 1.1 ipy.[41]

Steel, under stress approaching the yield strength, in contact with hot caustic solutions is subject to stress corrosion cracking (caustic embrittlement). (See *Boiler Corrosion*, p. 531.)

Ammonium hydroxide is satisfactorily contained in iron or steel. The corrosion rates at room temperature are stated to be 0.0001 ipy or less for concentrations ranging from 5 to 100% by volume of the commercial concentrated hydroxide.[42] A corrosion rate of 0.01 ipy or less is reported for hot concentrated ammonium hydroxide (8-day tests).[43]

Effect of Velocity

The usual initial effect of velocity in natural waters is to increase corrosion. Acceleration of corrosion has been noted in an investigation involving steel pipes of various sizes with water velocities reaching 8 ft (2.4 meters) per sec.[44] The corrosion rates at the higher velocities were four to thirty times or more those for low velocities, the higher rates applying to the higher water temperatures and smaller diameter pipe (¼ in.).

Other investigators have reported that, above a critical velocity, the corrosion rate of iron in water diminishes with increase in velocity.[45,46,47,48,49] For these tests it is probable that high velocity brought sufficient oxygen to the iron surface to cause passivity in the same manner as high concentration of dissolved oxygen in water passivates iron. (See "Effect of High Oxygen Concentration," p. 127.) This viewpoint is made plausible by a less active potential for iron in moving water (approx. -0.02 volt, hydrogen scale) compared to an active potential in stagnant water (-0.4 volt).[50]

In the work of Speller and Kendall a thin layer of rust may have prevented concentration of oxygen at the metal surface, or the water composition and temperatures used in their experiments may have forestalled passivity. In salt water, where high concentration of chloride ions retards passivation of iron by oxygen, water velocities as high as 25 ft (7.6 meters) per sec continue to increase corrosion. (See Fig. 1, p. 391, *Behavior of Metals and Alloys in Sea Water*.)

In fresh waters, as the velocity approaches very high values, it is expected that corrosion at first increasing, then decreasing, would again increase.[48] This may occur because erosive action serves to break down the passive state.

Whenever high velocities give rise to extremely low pressure areas, as in a jet or rotary pump, vapor bubbles form which on collapse at higher pressure areas destroy protective films or disrupt the metal itself. The effect is observed in all aqueous solutions and in other media as well. The phenomenon of formation and collapse of

---

[41] R. McKay and R. Worthington, *Corrosion Resistance of Metals and Alloys*, p. 259, Reinhold Publishing Corp., New York, 1936.

[42] Private communication, International Nickel Co., Inc., New York, N. Y.

[43] D. M. Strickland, *Ind. Eng. Chem.*, **15**, 566–569 (1923).

[44] F. N. Speller and V. V. Kendall, *Ind. Eng. Chem.*, **15**, 134–139 (1923).

[45] E. Heyn and O. Bauer, *Mitt. kgl. Materialprüfungsamt*, **28**, 62 (1910).

[46] J. N. Friend, "Carnegie Scholarship Memoirs," *J. Iron Steel Inst.*, **11**, 62 (1922).

[47] R. Russell, E. Chappell, and A. White, *Ind. Eng. Chem.*, **19**, 65–68 (1927).

[48] B. Roetheli and R. H. Brown, *Ind. Eng. Chem.*, **23**, 1010 (1931).

[49] U. R. Evans, *Metallic Corrosion, Passivity and Protection*, pp. 228–230, Edward Arnold and Co., London, 1937.

[50] O. Bauer, O. Kröhnke, and G. Masing, *Die Korrosion metallischer Werkstoffe*, p. 140, S. Hirzel, Leipzig, 1936.

cavities attending high velocities is called *cavitation,* and damage to the metal which results is called *cavitation erosion.* (See *Cavitation Erosion,* p. 597.)

Effect of Velocity on Corrosion in Sulfuric Acid. In dilute sulfuric acid exposed to the air, the corrosion rate of steel diminishes at first as velocity is increased (Fig. 4) and then rapidly increases.[51] At the highest velocities (12 ft/sec), various concentra-

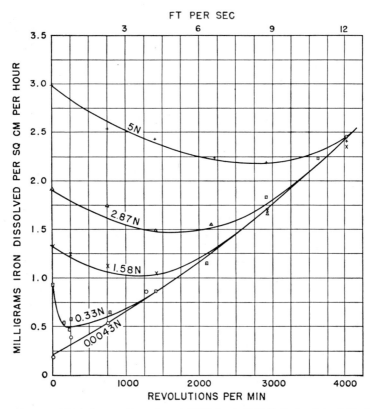

FIG. 4. Effect of Velocity on Corrosion of Mild Steel (0.12% Carbon) in Sulfuric Acid under Air. (23 ± 2° C, 45-min test, rotating spec. — 0.7 in. diam., 2.19 in. long.)

tions of sulfuric acid corrode iron at the *same* rate. In the absence of oxygen the effect of low velocities is to diminish the rate of attack, with higher velocities causing a very gradual increase in corrosion rate (Fig. 5).

For concentrated acid, velocities are not important below the critical velocity needed to erode the protective layer of ferrous sulfate.

GALVANIC COUPLING

The corrosion of iron is retarded in most common aqueous media when iron is coupled with zinc or magnesium. A notable exception for zinc may occur in fresh waters at elevated temperatures, the zinc being either less protective or actually cathodic to iron. (See *Zinc Coatings on Steel,* p. 814.) In sea water, iron is galvanically

[51] W. Whitman, R. Russell, C. Welling, and J. Cochrane, *Ind. Eng. Chem.,* **15**, 672 (1923).

protected when coupled with zinc, magnesium, cadmium, or aluminum. Corrosion of iron is accelerated when iron is coupled with copper, chromium, nickel and other metals of more noble galvanic potential. (See Table 18, p. 416, for galvanic potentials of metals and alloys in sea water. See also *Fundamental Behavior of Galvanic Couples*, p. 481.)

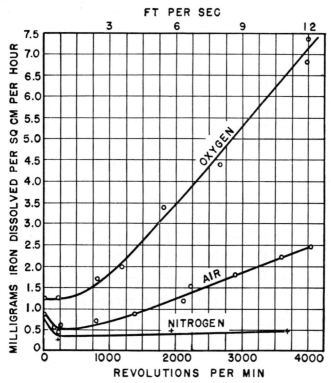

Fig. 5. Effect of Velocity on Corrosion of Mild Steel (0.12% Carbon) in 0.33 N Sulfuric Acid under Air, O_2 or N_2. (Same conditions as for Fig. 4.)

Mechanical Factors

Stress. Iron stressed to near or beyond the yield point is subject to stress corrosion cracking (caustic embrittlement) in hot alkaline solutions, particularly if silicates are present. Although a concentration of 0.13% $NaNO_3$ in 33% $NaOH$ at the boiling point was found to accelerate stress corrosion cracking of steel,[52] nitrates added to waters of high-pressure boilers, to the extent of approximately 40% of the NaOH alkalinity, inhibit cracking.[53]

Iron and steel are subject to rapid stress corrosion cracking in nitrate solutions. A concentrated mixture of calcium and ammonium nitrates and a 60 to 95% NH_4NO_3 at the boiling point are especially active.[54] Concentrated HNO_3 at 20° C can also

---

[52] W. C. Schroeder, A. A. Berk, and R. O'Brien, *Metals and Alloys*, **8**, 320 (1937).

[53] W. C. Schroeder and A. A. Berk, "Intercrystalline Cracking of Boiler Steel and Its Prevention," *Bureau of Mines Bulletin* 443 (1941).

[54] E. Houdremont, H. Bennek and H. Wentrup, *Stahl u. Eisen*, **60**, 757, 791 (1940).

cause cracking of stressed specimens[54] as can dilute nitric acid containing a small amount of $MnCl_2$ as accelerator, held at 60° to 80° C (140° to 175° F). A stressed bridge cable wire was found to crack when immersed in $0.01\ N\ NH_4NO_3$ or $0.01\ N\ NaNO_3$ within 3½ to 9½ months.[55] However, 46% $MgCl_2$ at 125° C (257° F) is ineffective despite its specific ability to crack stressed stainless steels.

A solution of 2.6 to 3.5 grams HCN per liter causes failure within 1 or 2 weeks of iron or low-alloy steels stressed below the elastic limit.[56] It is claimed that HCN

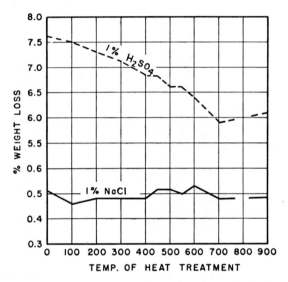

FIG. 6. Effect of Tempering on Corrosion of Cold-Drawn 0.29% Carbon Steel in NaCl and H_2SO_4 at Room Temperature.

contained in illuminating gas can cause stress corrosion failures of steel tanks used to hold the compressed gas. The failures are transgranular in contrast to intergranular failures when alkali or nitrate solutions are the cause.

Stress corrosion cracking can be avoided by altering the environment or sometimes by stress relief treatment. (See *Boiler Corrosion,* p. 531. Stress relief temperatures are given in Table 1, *Stress Corrosion,* p. 574.)

Cold Work. Cold work has no appreciable effect when oxygen diffusion controls the corrosion rate. This has been proved in sea water immersion tests[57] and is also shown by data of Fig. 6.[58] When corrosion is by hydrogen evolution, a definite effect of heat treatment, employed to relieve internal stresses, is apparent (Fig. 6). Corrosion of cold-rolled iron in non-oxidizing acids is usually higher, therefore, than corrosion of annealed iron.[59]

[55] R. E. Pollard, *Symposium on Stress-Corrosion Cracking of Metals (1944),* p. 437, The American Society for Testing Materials and The American Institute of Mining and Metallurgical Engineers, Philadelphia and New York (1945).

[56] H. Bucholtz and R. Pusch, *Stahl u. Eisen,* **62,** 21 (1942).

[57] J. N. Friend, *J. Iron Steel Inst.,* **117,** 639–659 (1928). *Iron Coal Trade Rev.,* **116,** 902 (1928).

[58] O. Bauer, O. Kröhnke, and G. Masing, *Die Korrosion metallischer Werkstoffe,* Vol. 1, p. 314, S. Hirzel, Leipzig, 1936.

[59] W. Eilender, W. Geller, and W. Ausel, *Korrosion u. Metallschutz,* **17,** 314 (1941).

IRON AND STEEL

Vibration. Factors affecting corrosion fatigue of iron and steel are discussed under *Corrosion Fatigue*, p. 578.

Corrosion produced by relative vibration of two contacting surfaces of amplitude high enough to produce slip is called fretting corrosion. Discussion of this type of corrosion begins on p. 590.

Metallurgical Factors

Open Hearth vs. Bessemer Steel. In natural waters where corrosion is controlled by diffusion of dissolved oxygen, no marked difference is found in behavior of open hearth compared with Bessemer steel. Bureau of Standards soil tests of buried specimens showed no material difference in corrosion rate (Fig. 3, p. 452.)

Bessemer Steel is more susceptible to intergranular stress corrosion cracking.[60]

Cast Iron, Wrought Iron, and Steel. Although controversy once existed as to the relative corrosion resistance of wrought iron and steel, it is now recognized that in natural waters the inherent corrosion rates are essentially identical. In acids the rates may differ greatly, depending upon the metal composition.

Cast irons contain, in addition to ferrite, either cementite (white cast iron) or graphite and cementite (gray cast iron). The cementite (Fe_3C) and graphite afford some mechanical protection of ferrite against corrosion, but both act to accelerate corrosion in many media because of their more cathodic galvanic potential.

Gray cast iron is subject to *graphitization* when immersed in salt waters, mine waters, or very dilute acids, or buried underground in some soils, particularly those containing sulfates. It occurs over a period of time as a result of ferrite corroding, leaving the graphite intact. This condition results in porosity of the structure and loss of density and some mechanical strength, but without outward appearance of any damage. White cast iron is immune.

The cast irons corrode more rapidly than mild steel when corrosion is by hydrogen evolution, largely because of their higher carbon content. Table 4, p. 197, contains corrosion rates of cast iron in various media.

Effect of Heat Treatment. In sea water or distilled water there is evidence that a carbon steel quenched from high temperatures (above the critical temperature) corrodes at a slightly higher rate than steel quenched and subsequently tempered. A 0.38% carbon steel quenched from 850° C (1560° F) and immersed in quiet distilled water for 9 months corroded at a rate of 0.0043 ipy. Specimens of the same steel quenched and tempered at temperatures ranging from 300° C (570° F) to 800° C (1470° F) and immersed under identical conditions corroded at an average rate of 0.0033 ipy.[61]

Carbon steels quenched from above the critical temperature and immersed in sea water were also found to corrode somewhat more than tempered or slowly cooled steels.[62] In general, however, it appears that heat treatment is a minor factor with respect to corrosion of carbon steels in natural waters.

The effects differ and are more pronounced in acids. Quenched steel (martensite) in 1% sulfuric acid corroded least, and 400° C (750° F) tempered steel corroded most. The 600° C (1110° F) tempered steel, and steel slowly cooled from above the critical temperature (pearlite) corroded at intermediate rates.[63]

---

[60] E. Houdremont, H. Bennek, and H. Wentrup, *Stahl u. Eisen*, **60**, 793 (1940).
[61] F. N. Speller, *Corrosion, Causes and Prevention*, p. 81, McGraw-Hill Book Co., New York, 1935.
[62] C. Chappell, *J. Iron Steel Inst.*, **85**, 270–293 (1912).
[63] F. N. Speller, *op. cit.*, p. 84.

Effect of Alloying Elements.[64]

CARBON. In fresh waters the carbon content of iron is not of practical importance in determining the corrosion rate. In sea water there is a slight increase in corrosion rate with higher carbon steels,[65] but again not of great significance.

In acids the effect of carbon is complicated by the factor of heat treatment and by the appreciable effect of minor elements often present, such as copper. In general the corrosion rate increases with increasing carbon content.[66,67] High-carbon steels, however, are more readily passivated by nitric acid than are low-carbon steels. High carbon was also found beneficial in steels used to resist mixtures of sulfuric and nitric acid.[68]

Increasing the carbon content of a steel from 0.09 to 0.28% increased the resistance to intergranular stress corrosion cracking.[69]

NITROGEN. Nitrogen normally present in iron or steel appears to increase susceptibility to stress corrosion cracking in alkaline or nitrate solutions.[69,70,71] Steels containing aluminum, in which the dissolved nitrogen presumably has been eliminated by reaction with aluminum, are correspondingly more resistant to stress corrosion failure.

MANGANESE, SULFUR, PHOSPHORUS, SILICON. These elements of ordinary constructional steels, present in usual amounts, have no effect of any commercial significance on corrosion in fresh or salt waters. In acids, silicon in normally small percentages has no appreciable effect on the corrosion rate of iron. The 14 to 15% Si-Fe alloy, however, is acid resistant. (See *Silicon-Iron Alloys*, p. 201.)

Presence of manganese, sulfur, and phosphorus in steels tends to increase corrosion in acids. Manganese has less effect in this respect than sulfur or phosphorus.

COPPER, CHROMIUM, NICKEL. Copper or nickel alloyed in small concentrations has no certain effect one way or another on the corrosion rate of steel in fresh or salt waters. In the atmosphere, the presence of these elements in steel is beneficial. (See *Atmospheric Corrosion of Iron and Steel*, p. 120.)

Nickel appears to be an effective alloying element for increasing corrosion fatigue resistance of steel exposed to oil well brines containing hydrogen sulfide. In the absence of hydrogen sulfide, chromium appears to be an effective alloying element. (See Table 2, p. 585.)

In strongly alkaline solutions, iron or steels containing nickel are more corrosion resistant than nickel-free iron or steels, the resistance increasing with nickel content.[72]

Chromium added to steels in amounts approximating 3% decreases the maximum pit depth in alloys exposed to fresh water (Pittsburgh water, 60° C [140° F], 2-year tests).[73] In flowing sea water, steels containing approximately 0.07% carbon and 2.4% chromium, in combination with less than 1% titanium or molybdenum, showed

---

[64] Refer also to H. J. French and F. L. LaQue, *Alloy Constructional Steels*, p. 164, American Society for Metals, Cleveland, 1942.
[65] F. N. Speller, *op. cit.*, p. 95.
[66] F. N. Speller, *op. cit.*, p. 94.
[67] H. Cleaves and J. Thompson, *The Metal-Iron*, p. 326, McGraw-Hill Book Co., New York, 1935.
[68] J. Eddy and F. Rohrman, *Ind. Eng. Chem.*, **28**, 30–31 (1936).
[69] E. Houdremont, H. Bennek, and H. Wentrup, *Stahl u. Eisen*, **60**, 791 (1940).
[70] J. Waber, H. McDonald, and B. Longtin, *Trans. Electrochem. Soc.*, **87**, 209 (1945).
[71] J. Waber, H. McDonald (and B. Longtin), *Corrosion and Material Protection*, **2** (November, December, 1945); **3** (January, February, March, April, May, 1946).
[72] R. McKay and R. Worthington, *Corrosion Resistance of Metals and Alloys*, p. 259, Reinhold Publishing Corp., New York, 1936.
[73] F. N. Speller, *Trans. Am. Inst. Mining Met. Engrs.*, **113**, 13–32 (1934)

improved behavior compared to 0.25% carbon steel with respect to weight loss and pitting.[74] Cr-Fe alloys containing more than approximately 12% chromium show improved general corrosion resistance an order of magnitude better than the low-chromium or carbon steels. They are the so-called stainless irons or steels (pp. 143 and 150).

Copper addition to iron containing high phosphorus or sulfur appreciably reduces the rate of corrosion in sulfuric, hydrochloric, or citric acids, but affects only slightly the corrosion rates of low-phosphorus or low-sulfur steels. With pure iron, copper additions can increase the corrosion rate in these acids. The corrosion rates of steels in nitric acid are independent of copper content.[75]

Protective Measures

Protective measures applying to iron and steel are discussed in other chapters, particularly in the section, *Corrosion Protection*, beginning p. 801. Additional chapters discussing protective measures in special environments are found under *Special Topics in Corrosion*, beginning p. 382.

NON–AQUEOUS MEDIA

Non-Aqueous Liquids

Iron is used to handle commercial liquid *methyl chloride*. If 0.03% moisture is present, corrosion of iron occurs with slight scale formation, and at 0.05% moisture, corrosion is serious. (Galvanized iron should not be used with anhydrous or moist methyl chloride because zinc reacts.)[76]

Iron is serviceable for dry *carbon tetrachloride* and other chlorinated solvents up to the boiling point. In the presence of moisture corrosion occurs.

Liquid sulfur chloride (98.3%) at 140° C (280° F) in tests of 133 days duration corroded steel at a rate of 0.22 ipy.[77] The vapor at the same temperature was still more corrosive.

Corrosion of 0.06% carbon steel at 310° C (590° F) by *phenol* is greatest for the anhydrous material (0.05 ipy), is a minimum (0.006 ipy) when moisture is 0.2 to 0.6%, and increases to 0.024 ipy at 10% moisture.[78]

The corrosion rates of iron and various steels in *fatty acids* are given in Table 2.

TABLE 2. CORROSION OF IRON AND STEEL IN FATTY ACIDS AT 55° TO 60° C (130° to 140° F)*

16-Day Exposure

Material	Composition						Stearic	Oleic	Soya
	C	Si	Mn	P	S	Cu	ipy		
Mild Steel	0.09	0.01	0.37	0.02	0.046	...	0.0003	0.003	0.006
Tech. Pure Iron	0.03	0.01	0.05	0.01	0.032	0.06	0.0002	0.003	0.005
0.5% C Steel	0.50	0.02	0.68	0.03	0.046	...	0.0001	0.002	0.005
Copper Steel	0.09	0.01	0.48	0.067	0.038	0.78	0.0005	0.002	0.006

* O. Bauer, O. Kröhnke, and G. Masing, *Die Korrosion metallischer Werkstoffe*, Vol. 1, p. 345, S. Hirzel, Leipzig, 1936.

[74] C. P. Larrabee, *Trans. Electrochem. Soc.*, **87**, 161–178 (1945).
[75] P. Bardenheuer and G. Thanheiser, *Mitt. Kaiser-Wilhelm-Inst. Eisenforsch. Düsseldorf*, **14**, 1 (1932).
[76] K. Willson, W. Walker, W. Rinelli, and C. Mars, *Chem. Eng. News*, **21**, 1254 (1943).
[77] Private communication, International Nickel Co., Inc., New York, N. Y.
[78] A. Wachter and N. Stillman, *Trans. Electrochem. Soc.*, **87**, 183–190 (1945).

At higher temperatures the rates are increased. For example, cotton seed fatty acids in the vapor phase at 280° C (530° F) in 51-day tests corroded steel at a rate greater than 0.12 ipy. A mixture of crude oleic and stearic acids as vapor or liquid at 225° C (440° F) in 21-day tests corroded steel at rates of 0.10 and 0.14 ipy, respectively.[79] Cast iron is attacked still more rapidly.

Gases

Iron can be used in continuous contact with dry chlorine gas up to about 200° C (400° F) (Table 6, p. 682). If the ratio of surface to volume of iron is high, the corrosion rate may be excessive at lower temperatures. For example, number 00 steel wool was found to ignite in commercial dry chlorine (0.005% H_2O) at a temperature of 184° C (364° F), but 16-gage sheet mild steel ignited only after holding at 251° C (484° F) for 30 min.[80] The final corrosion rates of 16-gage sheet at temperatures less than 248° C were under 65 mdd (0.012 ipy) in tests of 480-min duration (Fig. 7).

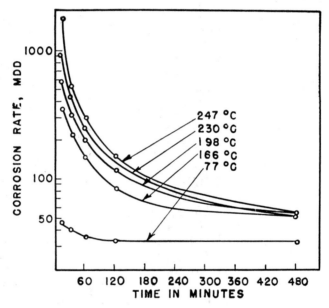

Fig. 7. Rate of Corrosion (Weight Loss) of Mild Steel in Contact with Commercial Chlorine (0.005% Moisture) as a Function of Time.

Moist chlorine is very corrosive at temperatures below the dew point. At 90° C (200° F) liquid chlorine rapidly attacks iron.[81] Iron withstands continuous contact with hydrogen chloride up to about 225° C (450° F) (Table 6, p. 682).

Average corrosion rates of carbon or low-alloy steels in essentially anhydrous hydrofluoric acid (6- to 40-day tests), used in the alkylation process of oil refining, are given in Table 3.

[79] Private communication, International Nickel Co., Inc., New York, N. Y.
[80] G. Heinemann, F. Garrison, and P. Haber, *Ind. Eng. Chem.*, **38**, 497 (1946).
[81] F. Ritter, *Korrosionstabellen metallischer Werkstoffe*, p. 57, J. Springer, 1937.

The resistance of steels to *fluorine* is a sensitive function of their silicon content.[82] Steels containing low silicon satisfactorily resist fluorine up to about 370° C (700° F). However, a steel containing 0.22% silicon is badly attacked at 205° C (400° F). At 500° C (930° F) all steels are severely attacked regardless of silicon content.

TABLE 3. CORROSION OF STEEL IN ESSENTIALLY ANHYDROUS HF*

Temperature		ipy
° C	° F	
15 to 25	60 to 80	0.003
25 to 40	80 to 100	0.006
40 to 50	100 to 120	0.014
55	130	0.014
65	150	0.025
80	175	0.048 (max.)
80 to 90	180 to 190	0.089
120 to 150	250 to 300	0.105

* M. Holmberg and F. Prange, *Ind. Eng. Chem.*, **37**, 1030 (1945).

ACKNOWLEDGMENT

The above chapter was reviewed by F. N. Speller, R. M. Burns, F. L. LaQue, and C. P. Larrabee, to whom the author is indebted for several useful suggestions. F. L. LaQue and H. O. Teeple generously contributed many data contained in the corrosion files of The International Nickel Company.

CHROMIUM-IRON ALLOYS

H. L. MAXWELL*

The addition of chromium to iron imparts an appreciable degree of corrosion resistance to the resulting alloy. Up to 2% of chromium, the mechanical properties are markedly improved, and the resistance to scaling at furnace temperatures is slightly enhanced. However, such small additions of chromium have little or no effect on the general corrosion resistance. Not until the percentage of chromium reaches about 12 does the resistance appear sufficiently improved to impart a useful degree of corrosion resistance, with the exception of certain oil refinery and related high-temperature services.

The principal application of the Cr-Fe alloys in chemical service is under oxidizing conditions, particularly in the handling of nitric acid. In Table 1 typical data, based on a large number of 240-hour tests (five 48-hour periods), may be taken as illustrative of the resistance of different Cr-Fe alloy compositions to boiling 65% nitric acid.

It is worthy of emphasis that although the corrosion resistance is markedly improved as chromium is increased, this addition is accompanied by a gradual decrease in mechanical properties, particularly impact strength. Difficulties in welding are increased irregularly as the chromium content is raised above 16 or 18%. Decisions in the selection of Cr-Fe alloys, while influenced by corrosion data, can, therefore, best be made by giving full consideration to physical properties as well.

* Engineering Department, E. I. du Pont de Nemours and Co., Wilmington, Del.
[82] Private communication, M. H. Brown, E. I. du Pont de Nemours and Company, Wilmington, Del.

TABLE 1. TYPICAL CORROSION RATES OF Cr-Fe ALLOYS IN BOILING 65% HNO_3

Area of specimen — approx. 35 sq cm.
Volume of solution — 600 ml.

Chromium Content, %	Corrosion Rate, *ipy*
10	0.42
12	0.156
16	0.042
18	0.026
20	0.018
24	0.012
30	0.008

Physical Properties

A detailed résumé of the chemical and physical properties and working characteristics of the Cr-Fe alloys is given in Table 2, which includes data for seven compositions ranging between 11.5 and 30% chromium. The rates of cooling during heat treatment as well as the time at maximum temperature share in determining physical properties of the heat-treated alloys. In general, increasing carbon improves response to heat treatment, i.e., it widens the range of physical properties procurable by heat treatment. It must be remembered, however, that corrosion resistance is lowered at the same time. Seldom, except with tool steel applications in services requiring resistance to abrasion, is it found desirable in the selection of a material of construction of the Cr-Fe type to obtain the maximum in mechanical values without regard for corrosion resistance. Most frequently the material, to give maximum service for a specific application, must possess neither the ultimate in corrosion resistance nor the maximum in mechanical properties but rather the optimum combination of properties. A considerable amount of research, correlated with surveys of many plant service applications, has established that the best mechanical properties of Cr-Fe alloys consistent with satisfactory corrosion resistance in nitric acid, to which service these alloys are most regularly applied, lie in the range of 15 to 18% chromium.

Fabrication and Heat Treatment. All the metal compositions described in Table 2 are capable of being formed and welded, within the limitations given. Difficulty in welding and impairment of physical properties due to welding increase with increasing carbon content. Cold-worked material that has been severely deformed by pressing or shaping requires an annealing heat treatment, designated in the table, to restore normal corrosion resistance. Brittleness in the weld areas is characteristic of welded Cr-Fe alloys, and hence an annealing heat treatment following completion of the welding is recommended. Heat treatment procedures are given under the appropriate heading in the table. The indicated heat treatment is essential in order to develop the full potential of corrosion resistance and favorable mechanical properties.

High-Carbon Grades of Chromium-Iron Alloys

Although the high-carbon Cr-Fe alloys, in the strict sense of the term, are not very resistant to a wide variety of corrosive media, they exhibit a combination of properties worthy of mention. These alloys are of two grades: cutlery and tool steels. Both exhibit limited resistance to general corrosion and find use principally where hardness and high strength are required in service.

CHROMIUM-IRON ALLOYS

Cutlery. The Cr-Fe compositions included in this general range have two classifications: 12 to 14% chromium with carbon from 0.30 to 0.40%,* and 16 to 18% chromium with carbon from 0.65 to 0.70%. Both these compositions respond particularly well to heat treatment and are used in services where corrosion resistance is secondary to high physical strength. Yield points in excess of 200,000 psi are obtainable with proper heat treatment.

Tool Steels. When still higher carbon, i.e., 1.40 to 2.25%, is added to the 12.0 to 18% Cr-Fe base compositions, a series of alloys results possessing excellent wear resistance and physical strength, although the resistance to shock is low. These alloys, properly heat-treated, give satisfactory service as drawing and sizing dies and also in gears and valve parts requiring high hardness and wear resistance with actual corrosion resistance a secondary requirement.

GENERAL CORROSION RESISTANCE

The Cr-Fe alloys of greater than 12% chromium, in common with the stainless Cr-Fe-Ni alloys, are outstanding in their resistance to media of an oxidizing character. They are much more likely to be attacked in service involving corrosive solutions under reducing conditions. This effect is well known from operating experience and may be readily illustrated in the laboratory by comparing results of corrosion tests in air-saturated vs. hydrogen-saturated solutions. This behavior follows from their passive properties. (See *Passivity*, p. 21.)†

Deaerated or reducing conditions must be considered unfavorable for corrosion resistance of the Cr-Fe alloys in the same way that aerated or oxidizing conditions are considered unfavorable for iron or steel and most of the non-ferrous metals and alloys. Likewise the presence of halogen ions in appreciable amounts, particularly chlorides, must be looked upon with suspicion even in neutral or alkaline solutions, since the Cr-Fe alloys are subject to a pitting type of attack under such conditions. (See *Pitting in Stainless Steels and Other Passive Metals*, p. 165.) High chloride concentrations in a few specific media also introduce the possibility of stress corrosion cracking. The rather limited evidence available to date indicates that the Cr-Fe alloys are less susceptible to stress corrosion cracking than the austenitic Cr-Ni-Fe alloys. (See discussion of stress corrosion cracking, p. 174 and p. 571.)

The Cr-Fe alloys are much more resistant than ordinary steel to scaling and oxidizing during heat treatment. The resistance to *impingement* is likewise better than steel, and often a cast Cr-Fe pump impeller will give several-fold longer service, particularly where high velocity accompanies bubble formation.

Acids

Comparative results obtained from exposure of two Cr-Fe stainless alloys to acetic, nitric, hydrochloric, and sulfuric acids are listed in Table 3. In general, it is expedient to consider Cr-Fe alloys for processing nitric acid, nitrates, and mixtures of these that are held in an oxidizing condition. The austenitic alloys, i.e., those containing proper amounts of nickel in addition to the chromium, will serve equally well (p. 150). However, in handling most of the other inorganic and many of the

---

* This alloy (A. I. S. I. type 420) and the alloys containing about 0.10% carbon (types 403 and 416) have excellent resistance to high-temperature steam. See *Corrosion by High-Temperature Steam*, p. 511. EDITOR.

† Salt-spray data for Cr-Fe alloys are given in Table 3, p. 184. Sea-water corrosion rates are found in Table 16, p. 413, and Table 2, p. 172. Water-spray data are presented in Fig. 2, p. 28.

CORROSION IN LIQUID MEDIA, ATMOSPHERE, GASES

TABLE 2. COMMERCIAL Cr-Fe ALLOY ANALYSES, PROPERTIES, AND CHARACTERISTICS

	403	416	430	444	446	420	440
American Iron and Steel Institute Type Numbers							
American Society for Testing Materials: Designation Grade			A176 4 / A240 A		A176 6 / A240 B		
Approximate Chemical Composition, %	C 0.12 (max.) Cr 11.50–13.00 Si 0.50 (max.) Mn 0.50 (max.) S 0.03 (max.) P 0.03 (max.)	C 0.12 (max.) Cr 12.00–15.00 Si 0.50 (max.) Mo 0.60 (max.) Mn 0.50 (max.) S 0.07 (min.) P 0.03 (max.)	C 0.12 (max.) Cr 14.00–18.00 Si 0.50 (max.) Mn 0.50 (max.) S 0.03 (max.) P 0.03 (max.)	C 0.08–0.20 Cr 18.00–23.00 Si 0.50–3.50 Mn 0.50 (max.) S 0.03 (max.) P 0.03 (max.)	C 0.35 (max.) Cr 23.00–30.00 Si 0.50 (max.) Mn 0.50 (max.) S 0.03 (max.) P 0.03 (max.)	C 0.30–0.40 Cr 12.0–14.0 Si 0.50 (max.) Mn 0.50 (max.) S 0.03 (max.) P 0.03 (max.)	C 0.65–0.70 Cr 16.0–18.0 Si 0.50 (max.) Mn 0.50 (max.) S 0.03 (max.) P 0.03 (max.)
Physical Properties							
Density grams/cc lb/cu in.	7.78 0.281	7.76 0.280	7.70 0.278	7.60 0.274	7.54 0.272	7.7 0.28	7.7 0.28
Modulus of Elasticity 10^6 psi	29	29	29	29	29	29	29
Structure	Martensitic	Ferritic Martensitic	Ferritic	Ferritic	Ferritic	Martensitic	Martensitic
Specific Heat Btu/°F/lb (32°–212°F)	0.11	0.11	0.11	0.11	0.11	0.11	0.11
Thermal Conductivity Btu/(sq ft) (hr) (°F/ft) 200° 1000°	14.6 16.8	13.3 15.2	14.1 15.1		12.1 14.1	14.4	14.0
Linear Coefficient of Thermal Expansion per °F × 10^6 0–200° F 0–600° F 0–1000° F 0–1500° F	5.7 6.1 6.7 7.0	5.7 6.5 6.7 7.0	5.4 5.6 6.1 6.3	5.4 5.5 6.1 6.4	5.3 5.5 6.0 6.2	5.7 6.6	5.6
Melting Point, °F	2720–2790	2700–2750	2710–2730	2700	2650–2700	2700	2680
Nominal Values of Mechanical Properties							
Hardness Brinell Number, annealed ", heat-treated ", cold-rolled	135–170 180–365	145–185 190–350	145–190 185–270	150–196 190–270	150–190	190 200–490	217 260–530
Rockwell Number, annealed ", heat-treated ", cold-rolled	75–85 B 30–40 C	80–95 B 8–35 C	80–95 B 90–105 B	80–95 B 94–104 B	80–90 B	92 B 51 C	
Ultimate Tensile Strength psi, annealed ", heat-treated ", cold-rolled	60–80,000 85–190,000	70–85,000 80–160,000	65–90,000 80–130,000	70–90,000 90–145,000	75–95,000	90,000 100–210,000	106,000 125–259,000

CHROMIUM-IRON ALLOYS

American Iron and Steel Institute Type Numbers	403	416	430	444	445	420	440
Yield Point psi, annealed ", heat-treated ", cold-rolled	35–45,000 60–150,000	40–50,000 60–130,000	40–55,000 65–120,000	45–55,000 65–130,000	45–60,000	57,000 80–220,000	65,000 90–228,000 ....
% Elongation in 2 in. Annealed Heat-treated Cold-rolled	40–30 25–15	35–25 25–10	35–20 20–2	35–25 20–2	32–20	28 27–5	20 20–2 ....
% Reduction in Area Annealed Heat-treated	75–55 65–50	65–60 60–40	60–45	60–50	55–45	65 65–7	50 40–7
Impact Strength Izod Value, ft-lb Annealed Heat-treated	110–80 100–20	100–60 100–10	60–5	40–5	5–1	30 70	30 14–9
Ductility Olsen, in., annealed Erichsen, mm, annealed	0.40–0.30 8–7	....	0.43–0.30 9–7	 9–7	....	....	....
Heat Treatment Annealing	Furnace-cool from 1500° F to 1100° F or air-cool from 1350°–1400° F.	Furnace-cool from 1550° F to 1100° F or air-cool from 1400°–1450° F.	Furnace-cool from 1500° F to 1100° F or air-cool from 1400°–1450° F.	Air-cool or water-quench from 1450°–1600° F.	Air-cool from 1450°–1650° F. If over 3½-in. water-quench.	Slow-cool from 1550° to 1650° F.	Slow-cool from 1550° to 1650° F.
Hardening	Oil-quench from 1750°–1825° F.	Oil-quench from 1750°–1825° F.	May be hardened slightly if quenched from above 1550° F.*	May be hardened slightly if carbon approaches 0.20%.	May be hardened slightly if carbon is near the maximum.	Oil-quench from 1850°–1900° F.	Oil-quench from 1850°–1950° F.
Tempering	Draw 400°–1300° F.†	Draw 400°–1300° F.				Draw 500°–1400° F.	Draw 600°–1300° F.
Forging	Start 2100° F. Finish at 1500° F.	Start 2100° F. Finish below 1500° F. Cold finish very important.	Start 2000°–2050° F. Finish 1400° F. Cold finish very important.	Start 1900–2050° F. Finish 1300°–1600° F.	Start 2050° F. Finish 1500° F. Cold finish very important.	Start 2000°–2150° F. Finish at 1750° to 1800° F.	Start 1900°–2050° F. Finish at 1750° to 1800° F.
Welding Properties	Fair. Generally should be annealed.‡	Poor-brittle welds. Likely to crack. Usually not recommended.	Fair-brittle welds. Generally should be annealed.	Fair-brittle welds. Generally should be annealed.	Fair-brittle welds. Generally should be annealed.	Fair. Should be preheated and annealed.	Poor-fair. Should be preheated and annealed.
Machining Properties	Fair	Good. About 80–85% compared to screw stock.	Fair. Tough.	Fair	Fair. Tough.	Good. Annealed condition.	Fair. Annealed condition.
Drawing or Stamping	Fair	Poor	Good	Good	Poor	Fair. Annealed condition.	Fair. Annealed condition.

* Hardenability decreases as (a) chromium content increases, (b) carbon content decreases.
† Embrittlement and loss of corrosion resistance occur if drawn in range of 900°–1100° F.
‡ Welding characteristics improved by presence of certain elements, such as Ti and Cb, which render the material nonair-hardening.

organic acids not of an oxidizing character, the rate of corrosion of Cr-Fe alloys may be expected to be excessive and the use of a higher alloy material (addition of nickel and molybdenum, for example) is indicated.

TABLE 3. CORROSION OF Cr-Fe ALLOYS IN ACIDS

Corrosion Rate, ipy

Area of specimen — approx. 35 sq cm.
Volume of solution — 600 ml. (Renewed after each 48-hr period.)
Duration of test — five 48-hr periods.
Velocity — nat. convect.

Alloy	Cr	C	Ni	Mn	Si	S	P
Analysis: 28% Cr:	27.90	0.12	0.47	0.72	0.36	0.014	0.017
16% Cr:	15.70	0.10	0.10	0.47	0.47	0.015	0.012

Alloy	Conc. of Acid: 95%	65%	50%	40%	20%	10%	5%	1%
			Acetic Acid (Boiling)					
28% Cr	Glacial 0.0084*		0.0002†			0.0001		
16% Cr	0.016*		0.0068			0.0005		
			Nitric Acid (Boiling)‡					
28% Cr		0.0084						
16% Cr		0.048		0.012	0.0078		0.0055	0.0010
			Hydrochloric Acid, 25° C (77° F)					
28% Cr					3.9		3.1	
16% Cr					6.0			
			Sulfuric Acid, 50° C (122° F)					
16% Cr	0.057	1.8		2.6		0.72		

* Pitted.
† Specimen was "passivated" after first period of exposure.
‡ Further data for nitric acid are included in Table 2, p. 183. Passivation of stainless steels is described on p. 158.

ALKALIES*

EFFECT OF ALLOY ADDITIONS

Additions of small amounts of nitrogen, molybdenum, silicon, copper, nickel, aluminum, columbium, and titanium have been made to some of the Cr-Fe alloys for special purposes, usually the improvement of mechanical properties. For instance, increase in nitrogen content of the 23 to 30% chromium alloy has proved quite helpful in this respect without deleterious effect on corrosion resistance. Also incorporation of about 4.5% nickel and 1.5% molybdenum in this high chromium range results in an alloy which can be hardened by heat treatment and has shown notable corrosion resistance in certain special applications. In the intermediate chromium ranges resistance to nitric acid is somewhat decreased by nitrogen additions, and this is also true for the other elements mentioned, with the possible exception

* The Cr-Fe alloys satisfactorily resist strongly alkaline solutions at ordinary temperatures, but corrode somewhat more than iron at elevated temperatures. (See McKay and Worthington, *Corrosion Resistance of Metals and Alloys*, p. 309, Reinhold Publishing Corp., New York, 1936). EDITOR.

of copper. The presence of about 1% copper appears to enhance resistance to hot dilute nitric-sulfuric acid solutions and incorporation of about 1 to 2% molybdenum has been reported to be beneficial under non-oxidizing conditions. In general, the degree of resistance of the original alloys and their modifications can be considered of the same order of magnitude, but considerable differences may exist under certain special conditions.

ATMOSPHERIC CORROSION

The effect on corrosion, in an industrial atmosphere, of increasing chromium alloyed with iron is shown in Fig. 1.[1]

TABLE 4. RESULTS OF ATMOSPHERIC CORROSION TESTS ON LOW-CARBON STEELS OF DIFFERENT CHROMIUM CONTENTS*

Surface — annealed and descaled.
Size sample — 7.62 × 15.24 × 0.152 cm.
Corrosion products removed cathodically in 5% H_2SO_4 + 10 ml Rodine inhibitor/liter, 1 amp/sq in., 70°–80° C, 30–60 min.

SEA COAST ATMOSPHERE AT WILMINGTON, N. C.

	Composition, %			Corrosion Rate, mdd	
C	Cr	Mn	Si	540 days	1320 days
0.074	....	0.39	0.30	7.74	6.46
0.060	1.02	0.35	0.23	6.78	5.43
0.046	2.18	0.56	0.30	5.50	3.60
0.044	3.15	0.50	0.32	4.44	2.89
0.066	5.34	0.56	0.32	5.25	2.23
0.056	7.28	0.61	0.33	2.78	1.62
0.060	9.32	0.67	0.31	2.49	0.88
0.056	12.53	0.73	0.31	0.42	0.12
0.042	18.57	0.70	0.40	0.08	0.02

INDUSTRIAL ATMOSPHERE AT NIAGARA FALLS, N. Y

	Composition, %			Corrosion Rate, mdd
C	Cr	Mn	Si	600 days
0.074	....	0.39	0.30	5.36
0.060	1.02	0.35	0.23	4.45
0.046	2.18	0.56	0.30	3.54
0.044	3.15	0.50	0.32	2.93
0.066	5.34	0.56	0.32	2.70
0.056	7.28	0.61	0.33	2.40
0.060	9.32	0.67	0.31	0.97
0.056	12.53	0.73	0.31	0.42
0.042	18.57	0.70	0.40	0.05

METROPOLITAN ATMOSPHERE AT NEW YORK, N. Y.

	Composition, %			Corrosion Rate, mdd
C	Cr	Mn	Si	540 days
0.068	....	0.30	0.10	12.90
0.062	1.01	0.29	0.10	16.20
0.052	2.09	0.26	0.15	11.55
0.042	3.07	0.35	0.19	4.83
0.070	5.50	0.35	0.17	2.60
0.060	7.24	0.33	0.20	3.33
0.060	9.13	0.33	0.18	2.61
0.080	12.58	0.76	0.46	0.005
0.080	17.26	0.45	0.32	0.0008

* Union Carbide and Carbon Research Laboratories.

[1] R. Franks, *Trans. Am. Soc. Metals*, **35**, 616 (1945).

Data on shorter time tests in marine, industrial, and New York City atmospheres are presented in Table 4, through the courtesy of W. O. Binder of Union Carbide

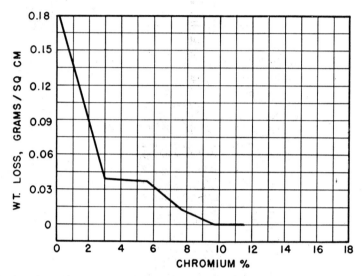

FIG. 1. Results of Atmospheric Corrosion Tests on Low-Carbon (0.1%) Steels of Different Chromium Contents after 10-Year Exposure to an Industrial Atmosphere. (Cleaned after test by brushing off loose oxide.) (R. Franks.)

and Carbon Research Laboratories, Niagara Falls, N. Y. The corrosion rates reach low values characteristic of the stainless steels at compositions above 10 to 12% chromium.

CHROMIUM-NICKEL AUSTENITIC STAINLESS STEELS

RUSSELL FRANKS*

The most widely used austenitic stainless steel is the type generally referred to as 18-8, meaning an iron alloy containing approximately 18% Cr and 8% Ni. There are two other common types of austenitic stainless steels. They contain approximately 25% Cr and 12% Ni (25-12) and 25% Cr and 20% Ni (25-20). All three of these alloys are non-magnetic and are produced in the form of wrought products and castings extensively employed to resist corrosion. Carbon content has a marked influence on corrosion resistance, particularly in fabricated articles, and 18-8 steel has been further divided into types that contain, respectively, a maximum of 0.08% C (type 304), and a maximum of 0.20% C (type 302). The lower the carbon content, the less will be the danger of loss of corrosion resistance associated with heating during fabrication and in use. The maximum carbon content of the 25-12 and 25-20 steels is generally put at 0.25%. These steels have optimum resistance to corrosion when they are fully *annealed* by rapidly cooling from 1050° to 1100° C (1900° to 2000° F).

* Union Carbide and Carbon Research Laboratories, Inc., Niagara Falls, N. Y.

CHROMIUM–NICKEL AUSTENITIC STAINLESS STEELS 151

TABLE 1. MECHANICAL PROPERTIES OF ANNEALED AUSTENITIC Cr-Ni STEELS*

A. I. S. I.† Type No. for Wrought Steels	304		309		310		316		347		321
A. C. I.‡ Type No. for Cast Steels	CF-7		CH-20		CK-25		CF-7M		CF-7C		
Chromium, %	18.40		24.55		25.06		18.70		18.69		18.58
Nickel, %	9.13		12.12		20.22		10.60		9.98		9.68
Carbon, %	0.06		0.15		0.14		0.06		0.06		0.08
Molybdenum, %	……		……		……		2.46		……		……
Columbium, %	……		……		……		……		0.70		……
Titanium, %	……		……		……		……		……		0.42
	Wrought	Cast	Wrought	Cast§	Wrought	Cast‖	Wrought	Cast	Wrought	Cast	Wrought
Elastic Modulus, $E \times 10^6$ psi	28		29		30		28		28		28
Yield Strength, 0.2% Offset, psi	33,400	28,000	45,000	43,800	45,800	45,500	35,000	31,000	36,000	34,740	38,000
Tensile Strength, psi	84,750	70,750	100,250	89,000	95,500	89,400	88,300	74,750	89,500	76,400	86,500
% Elong. in 2 in.	63	55	58	55	47	43	59	50	59	50	59
% Reduction in Area	75	78	70	47	63	55	73	69	67	55	64
Izod Impact, ft-lb	91	86	90	……	90	……	96	98	100	66	100
Brinell Hardness	137	143	149	153	163	179	149	137	143	149	143

* All samples air-cooled from 1040°–1090° C (1900°–2000° F), except as noted below. Samples 0.505 in. in diameter.
† American Iron and Steel Institute.
‡ Alloy Casting Institute.
§ As-cast condition. This metal contained 0.22% carbon.
‖ As-cast condition. This metal contained 0.24% carbon.

MECHANICAL PROPERTIES

Typical mechanical properties of wrought and cast austenitic Cr-Ni steels are given in Table 1. Both the modified (containing Cb or Ti) and the unmodified austenitic Cr-Ni steels have high ductility, toughness, and strength. The strength of all the steels can be materially increased by application of cold work.

AQUEOUS MEDIA

The stainless steels derive their corrosion resistance from the fact that they are passive. The mechanism of passivity in these and similar alloys is discussed in *Passivity*, p. 26. This property accounts for excellent corrosion resistance of the alloys in aerated and oxidizing media, but lower resistance to reducing media. Passivity also accounts for their susceptibility to corrosion when chloride or other halide ions are present in solution.

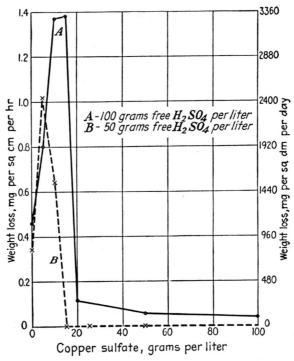

FIG. 1. Effect of Copper Sulfate on the Corrosion of 18-8 in Sulfuric Acid at 15° to 20° C (60° to 70° F). (Monypenny.)

The pH of an aqueous solution is not a sole criterion of whether the annealed and quenched austenitic Cr-Ni steels will become subject to or be free of attack, as the chemical nature of the solution is the major controlling factor. Basic solutions do not rapidly attack the steels and only those acids (reducing) that evolve hydrogen act to destroy passivity. For example, hydrochloric and sulfuric acid solutions readily attack stainless steels, whereas nitric acid in all concentrations, even up to the boiling point, attacks the steel only slowly. One of the important uses of the steels, therefore, is in resisting nitric acid, which is a strong oxidizing agent. Solutions containing dissolved salts vary widely in their action. In general the stainless steels are resistant to neutral and basic salt solutions. They are resistant to many weakly acid salt solutions but are not resistant to salt solutions containing chlorides (e.g., sea water), fluorides, and bromides. These types of salt solutions in time pit stainless steels, and precaution must be used when they are in contact with such media. The heavy metal halide solutions (e.g., $FeCl_3$) cause excessive pitting within

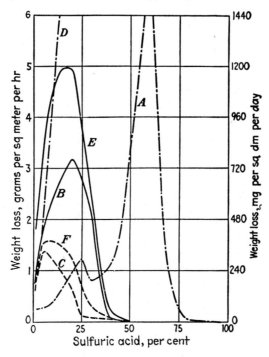

Fig. 2. Effect of Small Additions of Nitric Acid on the Corrosion of 18-8 in Sulfuric Acid. (Concentration of HNO_3 and temperatures of test were: [A] nil, 15° C [60° F]; [B] 1%, 15° C; [C] 3%, 15° C; [D] nil, 40° C [105° F]; [E] 1%, 40° C; [F] 3%, 40° C.) (Monypenny.)

a short time. (Refer to *Pitting in Stainless Steels and Other Passive Metals*, p. 165.)

For all practical purposes the annealed austenitic Cr-Ni steels are satisfactorily resistant to natural fresh waters, mine waters, boiler waters, steam, and condensates from boiler waters. The steels have been exposed to such media for long periods of time, and experience has shown that if they are initially in the passive state they will remain so almost indefinitely.

ACIDS, SALTS, BASES

Corrosion rates in various media are given in Table 2. The 25-12 (type 309) and the 25-20 (type 310) steels are similarly resistant to the same chemicals but usually to a greater degree. The molybdenum 18-8 (type 316) steels are more resistant in general than the 18-8 steels to the same types of corrodents,* whereas the resistance

* But not to nitric acid. In this medium the corrosion resistance of molybdenum 18-8 is lower.

TABLE 2. CORROSION OF 18–8 STEEL (TYPE 304) IN VARIOUS MEDIA
(Air-cooled from 1050° C [1920° F])

Corrosive Medium	Temperature	Duration of Test, hours	Wt. Loss mdd	ipy
20% Nitric acid	Room	1	Nil*	.....
20% Nitric acid	Boiling	18	Nil	.....
3% Nitric acid	Boiling	18	Nil	.....
1% Nitric acid	Boiling	18	Nil	.....
Nitric acid fumes	110° C (230° F)	13	100	0.018
10% Hydrochloric acid	Room	1	360	0.065
10% Sulfuric acid	Room	1	432	0.079
1% H_2SO_4 + 2% HNO_3	Room	17	Nil	.....
0.25% H_2SO_4 + 0.25% HNO_3	Room	17	Nil	.....
10% Acetic acid, C. P.	Room	3	Nil	.....
10% Acetic acid, C. P.	Boiling	12	Nil	.....
Glacial acetic acid, U. S. P.	Room	276	0.1	0.000
Glacial acetic acid, U. S. P.	Boiling	167	130	0.024
Crude acetic acid	Boiling	166	375.5	0.068
10% Phosphoric acid, C. P.	Boiling	17	Nil	.....
10% Carbolic acid, C. P.	Boiling	16	Nil	.....
10% Chromic acid (tech.)	Boiling	41	204	0.037
Concentrated sulfurous acid	Room	22	Nil	.....
0.5% Lactic acid	Boiling	16	4.1	0.001
1.0% Lactic acid	65° C (150° F)	16	Nil	.....
2.0% Lactic acid	Boiling	16	5.1	0.001
50% Lactic acid	Boiling	16	12,240	2.23
85% Lactic acid	Boiling	16	1,560	0.284
10% Tartaric acid	Boiling	39	Nil	.....
1% Oxalic acid	Boiling	39	177.6	0.032
10% Oxalic acid	Room	17	139.2	0.025
10% Formic acid	Boiling	1	3,240	0.590
10% Formic acid	Room	17	2.4	0.000
10% Malic acid	Room	17	Nil	.....
10% Sodium sulfite	Boiling	16	Nil	.....
10% Sodium bisulfate	Boiling	16	Nil	.....
10% Ammonium sulfate	Boiling	16	Nil	.....
10% Ammonium chloride	Boiling	16	Pitted	.....
Lemon juice	Room	89	Nil	.....
Orange juice	Room	91	Nil	.....
Sweet cider	Room	23	Nil	.....
Canned rhubarb	Boiling	16	Nil	.....
Canned tomatoes	Boiling	16	Nil	.....
10% Sodium hydroxide	Boiling	41	Nil	.....

* "Nil" refers to a weight loss of the specimen not detectable within time of the test.

of columbium-bearing 18-8 steel (type 347) is approximately equal to that of the 18-8 steels.

Although sulfuric acid solutions badly attack 18-8, there are some conditions under which the steel can be used in contact with this acid. When oxidizing chemicals such as copper sulfate and nitric acid are present they act as inhibitors. This effect is illustrated by the data of Figs. 1 and 2.

Molybdenum in 18-8 improves the resistance of the alloys to acids that evolve hydrogen. Sulfuric acid is one of this type, and the effect of molybdenum in retarding the action of this acid on 18-8 is shown in Fig. 3. Similar data are given in Fig. 4 for acetic acid.[1] It is pertinent, when economics are considered, that in both acids optimum resistance is obtained with about 2% molybdenum.

In addition, the molybdenum 18-8 (type 316) is generally superior to 18-8 (types 302 and 304) in the following environments:

1. Organic acid vapors, especially when the acid is concentrated.
2. Solutions or vapors containing small percentages of chlorine, hydrochloric acid, or other halogen compounds.
3. Sulfurous acid solutions at high temperatures and pressures, or where small amounts of sulfuric acid may be present as in sulfite pulp processing liquors.
4. Phosphoric acid when hot or concentrated.
5. Fatty acids at high temperatures.

NON-AQUEOUS MEDIA

LIQUIDS

The austenitic Cr-Ni steels are resistant to a number of non-aqueous corrosive liquids but are badly attacked by others. Table 3 summarizes data on the effects of many corrosive media encountered in industrial work.

GASES

In Table 4, data are given for the resistance of the austenitic stainless steels to dry and moist gases at room temperature and at slightly elevated temperatures. Stainless steels resist chlorine and hydrogen chloride when dry, and when moist provided the temperature is above the dew point. The critical upper temperature limits for resistance to these gases are approximately 315° C (600° F) for chlorine and 480° C (900° F) for hydrogen chloride.

The 18-8 types of steel are attacked by moist sulfur dioxide at room temperature, but the 25-12, 25-20, and the molybdenum-bearing 18-8 steel offer good resistance, showing the advantage of the higher alloy content.

CONTACT CORROSION*

Contact or crevice corrosion occurs when surfaces of stainless steels are used in contact with each other and the surfaces are wetted by the corrosive medium, or when a crack or crevice is permitted to exist in a stainless steel part exposed to such media. This type of corrosion has been particularly noticed in stainless steel joints made using gaskets to obtain a tight joint. So long as the joint is tight and does

---

* See *Pitting in Stainless Steels and Other Passive Alloys*, p. 172, for further remarks.

[1] R. Franks, W. O. Binder, and C. R. Bishop, "The Effect of Molybdenum and Columbium on the Structure, Physical Properties, and Corrosion Resistance of Austenitic Stainless Steels," *Trans. Am. Soc. Metals*, **29**, 35 (1941).

156 CORROSION IN LIQUID MEDIA, ATMOSPHERE, GASES

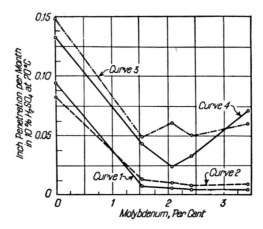

Fig. 3. Curves Showing Effect of Molybdenum and Heat Treatment on Corrosion Resistance of 18-8 Steels in 10% Sulfuric Acid at 70° C (158° F).

Curve 1: Heated 5 min at 1100° to 1150° C (2010° to 2100° F) and air-cooled.

Curve 2: Heated 5 min at 1100° to 1150° C (2010° to 2100° F) and air-cooled; reheated 4 hr at 870° C (1600° F) and air-cooled.

Curve 3: Heated 5 min at 1100° to 1150° C (2010° to 2100° F) and air-cooled; reheated 4 hr at 870° C (1600° F) and air-cooled; reheated 4 hr at 650° C (1200°F) and air-cooled.

Curve 4: Heated 5 min at 1100° to 1150° C (2010° to 2100° F) and air-cooled; reheated 4 hr at 650° C (1200° F) and air-cooled.

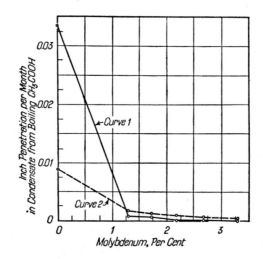

Fig. 4. Curves Showing Effect of Molybdenum on Corrosion Resistance of Cr-Ni Steels in Condensate from Boiling Acetic Acid.

Curve 1: Heated 5 min at 1100° to 1150° C (2010° to 2100° F) and air-cooled. Tests conducted in condensate from boiling 80% acetic acid.

Curve 2: Heated 5 min at 1100° to 1150° C (2010° to 2100° F) and air-cooled. Tests conducted in condensate from boiling 96% acetic acid.

not become wetted, contact corrosion will not occur, but otherwise failure may take place.

It has been observed in experimental work using duplex samples of the stainless steels in a 4% sodium chloride solution, that gaskets of the well-known rubber

TABLE 3. EFFECTS OF NON-AQUEOUS CORROSIVE MEDIA ON ANNEALED AUSTENITIC Cr-Ni STEELS

Corrodent	Temp.		18-8, Type 304	Cb 18-8, Type 347	Mo 18-8, Type 316	Ti 18-8, Type 321	25-12, Type 309	25-20, Type 310
	°C	°F	\multicolumn{6}{l}{Inch Penetration per Year}					
Acetone	21	70	<0.003	<0.003	<0.003	<0.003	<0.003	<0.003
Acetone	Boiling		....	....	<0.0042	....	....	....
Aniline crude			<0.004	<0.004	<0.004	<0.004	<0.004	<0.004
Benzol, tech.	21	70	<0.004	<0.004	<0.004	<0.004	<0.004	<0.004
Benzol, pure	80	176	<0.0044	<0.0044	<0.004	<0.0044	<0.0044	<0.0044
Bromine	21	70	Strong attack	Strong attack	Strong attack	Strong attack	Strong attack	Strong attack
Butyl acetate			No attack	No attack	No attack	No attack	No attack	No attack
Carbon tetrachloride	21	70	<0.004	<0.004	<0.004	<0.004	<0.004	<0.004
Carbon tetrachloride, C. P. + 1% H_2O	Boiling		....	....	0.004-0.042	....	....	....
Carbon tetrachloride, C. P. + 1% HCl	Boiling		....	....	0.004-0.042	....	....	....
Ethyl alcohol	21	70	0.003	0.003	0.003	0.003	0.003	0.003
Fatty acids	21	70	<0.004	<0.004	<0.004	<0.004	<0.004	<0.004
Gelatin			No attack	No attack	No attack	No attack	No attack	No attack
Iodine	21	70	0.42	0.42	0.12-0.42	0.42	0.42	0.42
Phenol, crude	100	212	<0.004	<0.004	<0.004	<0.004	<0.004	<0.004
Sodium hydroxide, fused	315	600	0.004-0.04	0.004-0.04	0.004-0.04	0.004-0.04	0.004-0.04	0.004-0.04

TABLE 4. EFFECTS OF GASES ON ANNEALED AUSTENITIC Cr-Ni STEELS

Gas	Temp.		18-8, Type 304	Cb 18-8, Type 347	Mo 18-8, Type 316	Ti 18-8, Type 321	25-12, Type 309	25-20, Type 310
	°C	°F	\multicolumn{6}{l}{Inch Penetration per Year}					
Cl_2, moist	20	70	0.12-0.42	0.12-0.42	0.04-0.12	0.12-0.42	0.12-0.42	0.12-0.42
SO_2, moist	20	70	<0.04	<0.04	<0.004	<0.004	<0.004	<0.004
SO_2, dry	300	575	<0.004	<0.004	<0.004	<0.004	<0.004	<0.004
NH_3, dry	20	70	<0.004	<0.004	<0.004	<0.004	<0.004	<0.004
HCl, dry	20	70	Attacked	Attacked	Attacked	Attacked	Attacked	Attacked
HCl, moist	20	70	Badly attacked	Badly attacked	Badly attacked	Badly attacked	Badly attacked	Badly attacked
HF, dry			Attacked	Attacked	Attacked	Attacked	Attacked	Attacked

materials cause excessive contact corrosion at the joint. Contact corrosion develops when thin sheets of 18-8 are employed as the gasket material under similar conditions. Powdered metals mixed with rubber cement have been tested as gasket material and surfaces of 18-8 steel have also been painted with the rubber cement, after which different metal powders have been sprinkled on top. After the cement dries, the metal powder not adhering to the rubber cement is removed. Under these conditions the following metal powders do not cause 18-8 steel to become contact corroded: zinc, tin, nickel, lead, and low-carbon ferrochromium (chromium, 69.52%; carbon, 0.05%; silicon, 0.57%; remainder substantially iron.) The best of these metals for use in a gasket for 18-8 is the powdered low-carbon ferrochromium because it does not easily corrode.

The following powdered metals and oxides applied similarly cause severe pitting and contact corrosion of 18-8: magnesium, aluminum, antimony, molybdenum, tungsten, copper, silicon, ferrosilicon, high-carbon ferrochromium (chromium, 69.87%; carbon, 4.71%; silicon, 1.06%), silica, iron oxide, manganese oxide, chromium oxide.

Rapidly moving well-aerated solutions have less tendency to cause contact corrosion at crevices such as threaded joints or in blind pockets than do stagnant solutions. Solutions containing caustic soda, sodium carbonate, sodium bicarbonate, sodium sulfate, manganese sulfate, potassium bicarbonate, and potassium nitrate seem to prevent this type of attack; but solutions containing sodium chloride, sodium sulfide, potassium chloride, potassium sulfate, and magnesium chloride promote it.

Vaseline has been found to be a suitable material for sealing crevices to prevent contact corrosion of stainless steels.* The molybdenum-bearing 18-8 (type 316) is less susceptible to this type of corrosion than are other 18-8's in chloride solutions and is more readily protected by the means suggested above.

"PASSIVATION" TREATMENT

Manufacturers of stainless steels "passivate"† the alloys as a final treatment, using for example nitric acid solutions. This treatment, based on experience, is used to insure best performance in borderline applications and where critical chemicals require that even the initial corrosion rates be at a minimum. It cleans the surface of the stainless steel of contamination, such as free iron, that promotes corrosive attack; and otherwise equipotentializes the surface by removing anodic surfaces. Passivation is particularly effective when the steels are exposed to solutions in which there is a tendency toward corrosion pitting.

A.S.T.M. Subcommittee IV of Committee A-10 recommended (1946) the following conditions to achieve the desired results:

Acid Concentration — 20 to 40% by volume of concentrated nitric acid for all alloy compositions.

Temperature — 55° to 70° C (130° to 160° F) for the austenitic and straight chromium steels containing 17% or more chromium.

— 45° to 60° C (110° to 140° F) for the free machining straight chromium steels and all grades containing 12 to 14% chromium.

Time — 30 to 60 min for all concentrations. The time might possibly be reduced to facilitate continuous operations.

Material should be immersed completely.

* Private Communication, F. L. LaQue, International Nickel Co., Inc., New York, N. Y.

† So-called passivation of stainless steels by immersion in nitric acid is probably more properly called a mild pickle since its major function is to clean the surface chemically. EDITOR.

CHROMIUM-NICKEL AUSTENITIC STAINLESS STEELS

GALVANIC EFFECTS

In the passive state the stainless steels have a noble galvanic potential; hence, when connected under corrosive conditions with more active metals like carbon steel, cast iron, magnesium alloys, and aluminum alloys, accelerated corrosion will occur on the latter metals. This is true particularly if the areas of the parts are equal to each other or if the area of the stainless steel is greater. Stainless steel may be coupled with these metals in some environments if its area is much less than that of the active metal. A good example is the use of an 18-8 steel rivet in a carbon steel plate immersed in sea water, in which instance the 18-8 steel does not appreciably increase the corrosion rate of the carbon steel plate, nor does the carbon steel increase the rate of corrosion of the 18-8 rivet. This subject is discussed further in *Behavior of Metals and Alloys in Sea Water*, pp. 391 and 415.

EFFECT OF MECHANICAL FACTORS

COLD WORK

The effect of cold-rolling various types of 18-8's on corrosion by nitric acid is indicated by data of Table 5. All the steels listed possess good resistance to boiling

TABLE 5. RESULTS OF NITRIC ACID TESTS ON ANNEALED AND COLD-ROLLED AUSTENITIC STAINLESS STEELS

Size of specimen — 2.5 × 3.8 × 0.15 cm (1 × 1.5 × 0.06 in.).
Surface preparation — 2B.
Velocity of acid — still except for boiling action.
Temperature — 166° C (330° F).
Duration — three 48-hr periods. Acid changed after each period.

Material	Composition, %					Condition of Metal*	Inch Penetration per Year in Boiling 65% HNO_3		
	Cr	Ni	C	Mo	Cb		1st 48-hr Period	2nd 48-hr Period	3rd 48-hr Period
304	18.86	8.26	0.06	....	....	1	0.0096	0.0072	0.0060
						2	0.0120	0.0096	0.0096
316	18.38	12.08	0.07	2.00	....	1	0.0084	0.0072	0.0072
						2	0.0084	0.0072	0.0072
316	18.83	14.31	0.06	3.12	....	1	0.0096	0.0084	0.0084
						2	0.0110	0.0110	0.0110
347	18.05	9.52	0.07	....	0.82	1	0.0110	0.0110	0.0110
						2	0.0120	0.0120	0.0110
347	18.40	13.39	0.05	2.02	0.70	1	0.0084	0.0060	0.0060
						2	0.0084	0.0084	0.0072

* 1 = Annealed by air-cooling from 1050°–2000° C (1900°–2000° F).
2 = Cold-reduced 30% by rolling.

65% nitric acid; cold-rolling them uniformly to obtain high strength does not apparently impair their resistance to oxidizing media.* Similar tests have been made on annealed sections of 18-8 in which local areas have been cold-worked by bending,

* However, for media in which corrosion pitting may occur, e.g., $FeCl_3$, cold-working the alloys appears to increase pit nuclei. See *Pitting in Stainless Steels and Other Passive Metals*, p. 171. EDITOR.

but no increase in corrosive attack has been observed. There is evidence, however, from service applications in which flanges are involved that exposure to chemical solutions containing small percentages of sulfuric and nitric acids has resulted in more severe attack in the cold-worked areas than in the annealed sections of the same metal. There is also evidence that under similar conditions the columbium- and molybdenum-bearing 18-8 steels are not affected in the same way by cold work.

STRESS

Refer to *Stress Corrosion Cracking in Stainless Alloys*, p. 174.

EFFECT OF METALLURGICAL FACTORS

Carbon raises the strength of the 18-8 steels and increases susceptibility to intergranular attack. (Discussed opposite page.) Nitrogen raises the strength of the steels but unlike carbon does not detrimentally affect their corrosion resistance.[2]

Manganese and silicon are used in deoxidizing the steels during manufacture to obtain high-quality alloys. The addition of several per cent manganese does not reduce corrosion resistance, but as shown by the data of Table 6, excessive percentages of silicon over those required for deoxidation reduce resistance to boiling concentrated nitric acid.

TABLE 6. EFFECT OF CARBON, SILICON, MANGANESE, AND NITROGEN ON THE CORROSION RESISTANCE OF 18-8 STEEL IN BOILING 65% HNO_3*

Size specimen — 2.5 × 3.8 × 0.6 cm (1 × 1½ × ¼ in.).
Surface preparation — finish-polished with 120 emery cloth.
Velocity of acid — boiling action.
Temperature — 165° C (330° F).
Duration of test — three 48-hr periods. Acid changed after each period.

Main Variable	Composition, %						Corrosion Rate, ipy		
	Cr	Ni	Mn	Si	C	N	1st 48-hr Period	2nd 48-hr Period	3rd 48-hr Period
C	18.12	8.20	1.25	0.21	0.02	0.03	0.0120	0.0096	0.0096
	19.16	8.04	1.44	0.32	0.05	0.04	0.0110	0.0072	0.0072
	18.69	8.72	1.10	0.20	0.07	0.035	0.0110	0.0084	0.0084
	18.05	8.79	0.70	0.35	0.10	0.03	0.0132	0.0120	0.0120
	17.76	8.25	1.19	0.24	0.14	0.03	0.0144	0.0156	0.0156
N	17.96	8.69	1.49	0.36	0.06	0.04	0.0096	0.0072	0.0072
	18.15	8.80	1.50	0.38	0.06	0.07	0.0096	0.0072	0.0072
	18.36	8.72	0.70	0.38	0.07	0.12	0.0110	0.0084	0.0072
	18.55	8.96	1.55	0.37	0.05	0.15	0.0096	0.0084	0.0072
Mn	18.34	9.00	0.59	0.37	0.05	0.04	0.0096	0.0072	0.0072
	18.69	8.72	1.10	0.20	0.07	0.04	0.0110	0.0084	0.0084
	17.96	8.69	1.49	0.36	0.06	0.04	0.0110	0.0072	0.0072
	17.80	8.91	2.25	0.46	0.05	0.03	0.0120	0.0096	0.0084
	18.46	8.34	2.46	0.49	0.07	0.03	0.0110	0.0096	0.0084
Si	17.96	8.69	1.49	0.36	0.06	0.04	0.0096	0.0072	0.0072
	17.87	8.60	0.70	0.70	0.06	0.035	0.0132	0.0132	0.0132
	18.40	8.78	0.70	0.96	0.10	0.03	0.0250	0.0228	0.0240

* All samples annealed by air cooling from 1050° to 1100° C (1900° to 2000° F).

[2] *Nitrogen in Chromium Alloy Steels*, bulletin published by the Electro Metallurgical Co., 30 East 42nd St., New York, N. Y. Copyrighted 1941.

The data of Table 7 show the effect of increasing the chromium and nickel contents of the 18-8 steel on the resistance to hot concentrated nitric acid. It is well demonstrated by these data that the higher chromium and nickel content steels are definitely superior to 18-8 steel for resisting hot nitric acid.

TABLE 7. EFFECT OF INCREASING THE CHROMIUM AND NICKEL CONTENTS ON THE CORROSION RESISTANCE OF 18-8 STEEL IN BOILING 65% HNO_3*

Size of specimen — 2.5 × 3.8 × 0.6 cm (1 × 1½ × ¼ in.).
Surface preparation — finish-polished with 120 emery cloth.
Velocity of acid — boiling action.
Temperature — about 165° C (330° F).
Duration of test — three 48-hr periods. Acid changed after each period.

Type	Composition, %					Corrosion Rate, ipy		
	Cr	Ni	Mn	Si	C	First 48-hr Period	Second 48-hr Period	Third 48-hr Period
304	18.69	8.72	1.10	0.20	0.07	0.0110	0.0084	0.0084
309	24.64	12.17	1.44	0.31	0.08	0.0060	0.0036	0.0036
310	25.70	20.34	1.54	0.42	0.09	0.0048	0.0036	0.0036

* All samples annealed by air cooling from 1050° to 1100° C (1900° to 2000° F).

INTERGRANULAR CORROSION*

The remarks so far have related to the corrosion resistance of the austenitic Cr-Ni steels in the annealed condition. It is now well known that the austenitic steels are metallurgically unstable when heated in the temperature range of 350° to 800° C (650° to 1500° F). After heating in this temperature range they become subject to severe attack at the grain boundaries by even relatively mild corrosive media. This attack is referred to as intergranular corrosion and is so severe that the steel literally disintegrates into separate grains, losing substantially all its recognizable properties. In some instances, the attack may not progress far enough to cause complete disintegration but is sufficiently severe to practically destroy mechanical properties. This condition cannot always be foretold by visual examination, but the steel will usually lack metallic ring when struck with a metal object.

Dr. Benno Strauss[3] and coworkers believed that carbon was responsible for the difficulty, and that if carbon content was as low as 0.07% the 18-8 steels would become free of intergranular attack. The subsequent work of many investigators has shown that lowering the carbon content to this extent does not eliminate the trouble. While carbon content has some influence, the more important factors are time, temperature, and the corrodent. In experiments using a boiling solution containing copper sulfate and sulfuric acid,† Bain et al.[4] obtained the data given in Fig. 5 to illustrate the effect of the time and temperature factors on maximum sensitivity to intergranular attack.

* Refer also to *Intergranular Corrosion Tests* (for Stainless Steels), p. 1019.
† 47 ml H_2SO_4 (sp. gr. 1.84), 13 grams $CuSO_4 \cdot 5H_2O$ per liter.

[3] B. Strauss, H. Schottky, and J. Hinnüber, "Die Carbidausscheidung beim Glühen von nichtrostenden unmagnetischen Chromnickelstahl" (Carbide Precipitation upon Annealing of Stainless Non-magnetic Chrome-Nickel Steel), *Z. anorg. allgem. Chemie*, **188**, 309–324 (1930).
[4] E. C. Bain, R. H. Aborn, and J. J. B. Rutherford, "The Nature and Prevention of Intergranular Corrosion in Austenitic Stainless Steels," *Trans. Am. Soc. Steel Treating*, **21**, 481–509 (1933).

The severity of this type of attack is well illustrated by the fact that even the metal adjacent to welds made in the annealed 18-8 steels is quite subject to intergranular deterioration on exposure to corrosive media, whereas the metal proper

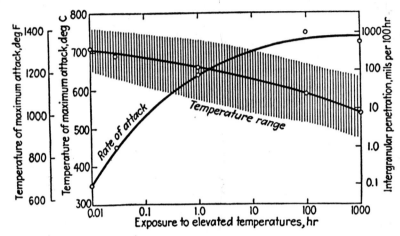

Fig. 5. Effect of Time and Temperature on Maximum Sensitivity of Low-Carbon 18-8 Steel to Intergranular Attack. (Bain, Aborn, and Rutherford.)

Fig. 6. Appearance of an 0.08% Carbon 18-8 Steel Sample After Heating 1 Hour at 650° C (1200° F) and Treating with an Acidified Copper Sulfate Solution. (The metal cracked badly on bending.)

and the weld metal are relatively free of the trouble. (See Fig. 3, p. 1104.) The severity of intergranular attack is also illustrated by the photograph of Fig. 6.

The cause of intergranular corrosion has been the subject of much study. It is stated by some[5,6,7,8] that the decreased corrosion resistance of the steels is due to

[5] V. N. Krivobok and M. A. Grossmann, "Influence of Nickel on the Chromium-Iron-Carbon Constitutional Diagram," *Trans. Am. Soc. Steel Treating*, **18**, 808–836 (1930).

[6] R. H. Aborn and E. C. Bain, "Nature of the Nickel-Chromium Rustless Steels," *Trans. Am. Soc. Steel Treating*, **18**, 837–893 (1930).

depletion of chromium in the area near the grain boundaries, caused by the precipitation of chromium carbide. The reasoning is that under the influence of heat, carbon migrates to the grain boundaries at a more rapid rate than chromium, and a large portion of the chromium at the grain boundaries and in the adjacent zones combines with the carbon and leaves these regions deficient. Miller[9] reasons that the carbon migrates to slip planes and grain boundaries and precipitates as an iron carbide, which is dissolved by weak corrosive media. Kinzel[10] states that internal stress may be responsible, but as yet definite proof as to the cause of intergranular corrosion is lacking.

The chromium impoverishment theory does not seem to be entirely acceptable for several reasons. It has been impossible to increase the chromium content of an austenitic stainless steel sufficiently to obtain even relative freedom from susceptibility to intergranular attack. The 25-12 and 25-20 steels even of extremely low carbon content are as susceptible to intergranular attack as the lower chromium 18-8 steel. The carbon content of 18-8 steel has been decreased below 0.01%, and the metal is still subject to this type of attack, although to a less degree.

Elimination of Intergranular Corrosion. Susceptibility to intergranular corrosion caused by heating in the range of 350° to 800° C (650° to 1500° F) can be eliminated by reannealing the steels at 1050° to 1100° C (1900° to 2000° F), followed by rapid cooling, provided the annealing treatment is applied before the steels are exposed to corrosive environments. This treatment redissolves precipitated carbides so that carbon is in solid solution and the chromium is supposedly redistributed uniformly throughout the steel. In many cases it is impractical to employ the annealing treatment after fabrication, and it is necessary to use an addition agent to the alloy to eliminate the tendency of carbide precipitation.

The announcement by Houdremont and Schafmeister[11] that titanium would inhibit intergranular corrosion was followed by a number of investigations which broadly confirmed their results. The practice at present is to include from four to six times as much titanium as carbon in the 18-8 steels to retard intergranular attack. The work of Becket and Franks[12] showed that columbium also offers a satisfactory solution to the problem. Complete immunity to intergranular attack is obtained by having present at least eight to ten times as much columbium as carbon by weight. The 18-8 steels containing columbium are suitable for welding rods, as the columbium in the rod can be transferred to the weld deposit without excessive loss. The accepted practice is therefore to employ rods of the columbium-bearing 18-8 steel for welding both the titanium- and columbium-bearing 18-8 steels.

The columbium addition does not impair the general overall corrosion resistance of the 18-8 steels, so that these steels are used extensively in chemical equipment that cannot be annealed after welding. The titanium-bearing 18-8 steels are used at

---

[7] B. Strauss, H. Schottky, and J. Hinnüber, "Die Carbidausscheidung beim Glühen von nichtrostenden unmagnetischen Chromnickelstahl" (Carbide Precipitation upon Annealing of Stainless Non-magnetic Chrome-Nickel Steel), *Z. anorg. allgem. Chemie*, **188**, 309–324 (1930).

[8] P. Schafmeister, "Korngrenzenkorrosion und Gefügeätzungen bei Stahl mit 18% Cr und 8% Ni" (Intergranular Corrosion and Etching of 18–8), *Arch. Eisenhüttenw.*, **10**, 405–413 (1936–1937).

[9] J. L. Miller, "An Investigation of Certain Corrosion-Resistant Steels," Iron and Steel Institute, *Carnegie Scholarship Memoirs*, **21**, 111–151 (1932).

[10] A. B. Kinzel, *J. Applied Phys.*, **8**, 341 (1937).

[11] E. Houdremont and P. Schafmeister, "Verhütung der Korngrenzenkorrosion beim Stählen mit 18% Cr und 8% Ni" (Prevention of Intergranular Corrosion in 18–8), *Arch. Eisenhüttenw.*, **7**, 187–191 (1933–1934).

[12] F. M. Becket and R. Franks, "Effects of Columbium in Chromium-Nickel Steels," *Trans. Am. Inst. Mining Met. Engrs.*, **113**, 143–162 (1934).

high temperatures to resist intergranular effects, but are not so widely employed in chemical service. When welded vessels of either steel are constructed in accordance with A.S.M.E. Boiler Code requirements they must be stress-relieved at 845° to 900° C (1550° to 1650° F). It should be emphasized that molybdenum, tungsten, and some other carbide-forming elements do not render the steels free of intergranular attack, although molybdenum additions decrease sensitivity to and severity of attack.

ATMOSPHERIC CORROSION

The austenitic Cr-Ni steels of the 18-8 type and the 25-12 and 25-20 types are all resistant to atmospheric conditions both for interior and for exterior exposures and under conditions of high or low humidity. The steels are resistant to the action of water and in most environments show no signs of permanent staining even after long exposure. In connection with the resistance of these steels to atmospheric corrosion it is important that they be free of surface scale, oxide, and iron contamination. If the surfaces are not properly cleaned there is a good opportunity for the alloys to rust by contact corrosion. However, if the steels are properly cleaned before exposure to the atmosphere, and periodically afterward, they will remain bright almost indefinitely. The resistance of these steels to atmospheric corrosion is shown by data of Table 8.

TABLE 8. RESULTS OF ATMOSPHERIC CORROSION TESTS ON 18-8 STEELS AT NEW YORK CITY AND NIAGARA FALLS, N. Y.

Size of specimens — 17.8 × 25.4 × 0.10 cm (7 × 10 × 0.04 in.).
Surface preparation — 2B finish.
Temperature — not observed.
Samples weighed three times.

Material	% Cr	% Ni	% Mn	% Si	% C	% Mo	% Cb	Duration, months	Location	Corrosion Rate	Pitting
302	18.7	8.8	0.47	0.27	0.12	...	...	96	N.Y.C.	Nil	None
304	18.2	8.7	0.46	0.47	0.07	...	...	96	N.Y.C.	Nil	None
316	18.7	10.50	1.20	0.30	0.08	2.4	...	66	N.Y.C.	Nil	None
304	18.2	8.7	0.46	0.47	0.07	...	...	63	N.F.	Nil	None
347	18.4	9.95	1.25	0.20	0.06	...	0.7	63	N.F.	Nil	None

Note: None of the steels permanently stained.

After periods of 5 to 8 years the surfaces of the 18-8 steels have remained substantially unaffected. The question of cleaning the 18-8 surface cannot be overemphasized and depends to a large extent on the type of atmosphere in which the steels are exposed. It is not so necessary to clean the surface as frequently in rural and urban atmospheres as it is in industrial atmospheres such as in Niagara Falls, N. Y., and Pittsburgh, Pa. It may be necessary to wash the surfaces of the steels frequently when they are exposed to the atmosphere near the sea coast in order to remove deposited salt and dirt particles. The question therefore of keeping the surfaces of the steel bright depends primarily on the type of atmosphere.

Samples of annealed and cold-rolled 18-8 stainless steels were exposed to the industrial atmosphere at Niagara Falls, and to an artificially prepared humid atmosphere for a period of approximately 2 years. Likewise F. L. LaQue* exposed panels

* Private communication.

of 18-8 for one year 800 ft from the ocean at Kure Beach, N. C. The tensile strength of the annealed and cold-rolled 18-8 steels remained unchanged, showing that even under humid atmospheric or marine conditions, the material did not deteriorate any faster within the time of test than under normally dry atmospheric conditions. These results have been amply supported by longer-time application of 18-8 for decorative purposes on store fronts, and on buildings like the Empire State and Chrysler Buildings in New York City.

GENERAL REFERENCES

FRANKS, R., "Effects of Special Alloy Additions to Stainless Steels," *Trans. Am. Soc. Metals*, **27**, 505–529 (1939).
KINZEL, A. B., and R. FRANKS, *The Alloys of Iron and Chromium*, Vol. II, *Alloys of Iron Monographs*, Engineering Foundation, McGraw-Hill Book Co., New York, 1940.
KRIVOBOK, V. N., "A Digest of Stainless Steel," *Year Book of the American Iron and Steel Institute*, pp. 129–167, 1937.
MONYPENNY, J. H. G., *Stainless Iron and Steel*, 2nd Ed., 575 pp., Chapman and Hall, London, 1931.
THUM, E. E., *The Book of Stainless Steels*, 2nd Ed., American Society for Metals, 1935.

PITTING IN STAINLESS STEELS AND OTHER PASSIVE METALS

H. H. UHLIG*

Passive metals such as the stainless steels resist many corrosive media and perform excellently over long periods of time. In some solutions the surface remains literally unattacked, but if corrosion does eventually occur, rapid penetration takes place at several small areas. These are so-called pits. They occur at random within or at the grain boundaries of the alloy. Pitting may be a serious type of corrosion because of the rapidity with which heavy sections of metal are perforated.

Corrosion pitting may occur with any metal, but for reasons discussed below it is the prevalent form of corrosion experienced with the passive alloys and the passive metals such as aluminum, nickel, and chromium. The mechanism of pitting in stainless steels is characteristic of the mechanism in many if not all of the passive metals.

Frequently preventive measures can be taken. However, in specific applications of stainless steels, such as immersion in sea water, pitting will eventually occur in an interval of exposure ranging from months to several years, depending on the alloy and factors of exposure.

THEORY

Pits begin by breakdown of passivity at favored nuclei on the metal surface. The breakdown is followed by formation of an electrolytic cell, the anode of which is a minute area of active metal and the cathode of which is a considerable area of passive metal. The large potential difference characteristic of this "passive-active cell" (0.5 to 0.6 volt for 18-8) accounts for considerable flow of current with attendant rapid corrosion at the small anode. The corrosion-resistant passive metal surrounding the anode and the activating (passivity-destroying) property of the corrosion products within the pit account for the tendency of corrosion to penetrate the metal rather than spread along the surface. The large potential of the cell accounts for a large

* Corrosion Laboratory, Department of Metallurgy, Massachusetts Institute of Technology, Cambridge, Mass.

effective cathode area, thereby drawing upon a considerable volume of depolarizer (dissolved oxygen or oxidizing salt) for maintenance of corrosion currents.
Pit growth is controlled by rate of depolarization at the cathode areas. In sea water, control is exercised by the amount and availability of dissolved oxygen. In NaCl solutions, free of dissolved oxygen, no pitting has occurred in laboratory tests. Ferric chloride, a more effective depolarizer than dissolved oxygen, causes greater numbers of pits and more rapid penetration. Nitrates do not readily depolarize and hence do not take part in or accelerate pitting. On the contrary, they help to stabilize passivity. 18-8 immersed in 10% $FeCl_3 \cdot 6H_2O$ solution containing 3% $NaNO_3$ did not pit and lost only 0.002 mdd in 2¼ years,[1] whereas, without nitrate, the corrosion rate with pitting was a million times greater.

THE ELONGATED PIT

When the corrosive agent contains chloride ions, as is commonly the case, anodic products within the pit consist largely of concentrated ferrous chloride and other chlorides of metals composing the alloy. These solutions are acidic (pH = approx. 1.0) and have been proved effective in breaking down passivity. Hence it is not surprising that as these high-density solutions are exuded by the pit, they make contact with metal beneath it, and destroy passivity in the direction of gravity. Elongated electrolytic cells are established, and pits of this shape are typical of stainless steel specimens immersed in relatively stagnant sea water or in ferric chloride.

CONDITIONS AFFECTING PIT FORMATION

Two heats of stainless steel of the same overall analysis and in the same environment will show different tendencies toward pitting. Here factors are operating obviously characteristic of the alloy. On the other hand, the same alloy will pit profusely or not at all, depending upon conditions of the environment. Hence both factors must be considered.

FACTORS OF THE ENVIRONMENT

Pitting is most likely to occur in the presence of chloride ions, combined with such depolarizers as oxygen or oxidizing salts. An oxidizing environment is usually necessary for preservation of passivity with accompanying high corrosion resistance, but, unfortunately, it is also a condition for occurrence of pitting. The oxidizer can often act as depolarizer for passive-active cells established by breakdown of passivity at a specific point or area. The chloride ion in particular can accomplish this breakdown.

In solutions of chlorides pitting occurs if the standard oxidation-reduction potential is less than approximately -0.15 volt. This is illustrated by data of Table 1. For example, 18-8 (type 304) immersed in 10% nickel chloride in open beakers for 24 hours corroded at an average rate of 0.026 ipy with no evidence of pitting. This is in accord with the fact that no lower valence nickel salts form and the oxidation-reduction potential is zero. If, however, hydrogen peroxide is added to nickel chloride solutions, as in electroplating baths operating at about 60° C (140° F), the stainless steels, including type 316 containing molybdenum, pit severely.* In this instance it is expected that the oxidation-reduction potential is definitely more negative than -0.15 volt.

* Private communication from F. L. LaQue, International Nickel Co., New York, N. Y.

[1] H. H. Uhlig, *Trans. Am. Inst. Mining Met. Engrs.*, **140**, 422 (1940). After 9 years and 8½ months, this same specimen was still free of pits and had corroded at an overall rate of 0.002 ipy.

TABLE 1. RELATION OF OXIDATION–REDUCTION POTENTIAL TO
PITTING OF 18-8 IN IMMERSION TESTS

Solution	Oxidation-Reduction Reaction	Std. Potential, Volts*	Presence of Pits After Test†
Hypochlorous acid	$H^+ + HClO \rightarrow \frac{1}{2}Cl_2 + H_2O - e^-$	-1.63	+
Thallic chloride	$Tl^{+++} + Cl^- \rightarrow TlCl - 2e^-$	-1.36	+
Mercuric chloride	$2Hg^{++} \rightarrow Hg_2^{++} - 2e^-$	-0.91	+
Ferric chloride	$Fe^{+++} \rightarrow Fe^{++} - e^-$	-0.77	+
Cupric chloride	$Cu^{++} + Cl^- \rightarrow CuCl - e^-$	-0.57	+
Stannic chloride	$Sn^{++++} \rightarrow Sn^{++} - 2e^-$	-0.15	–
Nickel chloride (NiCl$_2$)	No lower valence salt	0	–
Manganese chloride (MnCl$_2$)	" " " "	0	–
Chromic chloride	$Cr^{+++} \rightarrow Cr^{++} - e^-$	$+0.41$	–

* These potentials are largely from Latimer and Hildebrand, *Reference Book of Inorganic Chemistry*, The Macmillan Co., 1940, but with signs reversed to conform with convention of sign adopted by The Electrochemical Society.

† Duration of test usually 24 hours or longer.

Aerated neutral or near-neutral chlorides can pit stainless steels, a fact which might be expected because their oxidation-reduction potentials are within the critical range. Pitting is less pronounced in rapidly moving aerated solutions as compared with partially aerated stagnant solutions. The flow of liquid carries away corrosion products which would otherwise accumulate at crevices or cracks. It also insures uniform passivity through free access of dissolved oxygen. Data illustrating the effect of velocity to reduce rate of pitting of stainless steel in sea water are given in Table 17, p. 414.

Pitting rate increases with temperature. In 4 to 10% NaCl solutions, a maximum weight loss produced by pitting is reached at 90° C (195° F), and for more dilute solutions at still higher temperatures. At the boiling point, however, no pitting is known to occur regardless of NaCl concentration. Oxygen is expelled and conditions no longer exist for passive-active cells. In practice it should be emphasized that conditions of complete oxygen exclusion by boiling can be depended upon only with exercise of special precautions.

Pitting diminishes with increase in alkalinity. In aerated 20% NaOH at 90° C (195° F), 18-8 and Mo 18-8 (type 316) did not pit in 24-hour tests varying the NaCl concentration from 0 to 10%. Some weight loss occurred (2 to 18 mdd).[2] It is also reported[3] that addition of 1% Na$_2$CO$_3$ to refrigerating brines avoided pitting of 18-8 in service 2 to 5 years. In solutions of concentrated alkali, it is questionable whether serious pitting would ever occur, because potential measurements prove that passivity is destroyed.*[4]

The effects of temperature, concentration, and pH on weight losses and pitting of 18-8 (type 304) in aerated NaCl solutions above room temperature are shown in Figs. 1 to 4. The number of pits increases with temperature although weight losses go through a maximum, indicating that pits formed in the higher temperature range become smaller. The number of pits also rapidly increases with NaCl concentration, the tendency being toward more uniform corrosion at the highest concentrations. Pit depth in 4% NaCl is greatest at pH = approx. 7.

* Passivity as defined by galvanic potential. See *Passivity*, p. 21.

[2] H. L. Fober, *Thesis*, Massachusetts Institute of Technology, Cambridge, Mass., 1939.
[3] P. Schafmeister and W. Tofaute, *Tech. Mitt. Krupp*, **3**, 223 (1935).
[4] H. H. Uhlig, *Trans. Am. Inst. Mining Met. Engrs.*, **140**, 387–404 (1940).

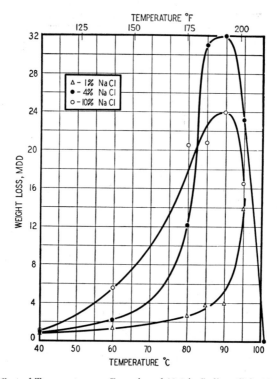

Fig. 1. Effect of Temperature on Corrosion of 18-8 in Sodium Chloride Solutions.

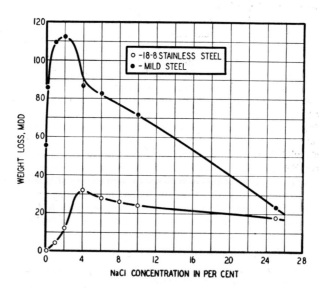

Fig. 2. Effect of Salt Concentration on Corrosion at 90° C (194° F).

Factors within the Metal

Pit nuclei have numerous sources. They occur through minor alloy components or impurities, or through heat treatment or mechanical work. Whether or not they develop into pits depends mainly on the environment. Their exact nature is largely unknown.

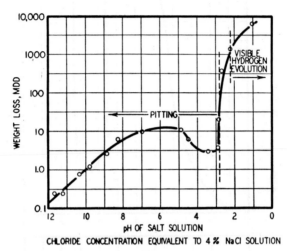

Fig. 3. Dependence of Corrosion of 18-8 on pH of 4% Salt Solution at 90° C (194° F).

Segregation accompanying cooling of the melt introduces pit nuclei, which are eliminated in part by long-time annealing followed by rapid quenching. This is illustrated in Fig. 5, showing weight losses in ferric chloride of low-carbon 18-8 alloys with or without nitrogen in alloy solution.[5] The number of pits approximately parallels weight loss.

Heat treatment of a homogeneous alloy at intermediate temperatures also increases the number of pit nuclei. These temperatures appear within the region that induces carbide precipitation resulting in intergranular corrosion. However, pit nuclei can form apparently independent of carbon. A titanium-stabilized 18-8 heated at 600° C (1110° F) for 112 hours corroded in ferric chloride by pitting, approximately ten times more than the same alloy quenched from 1100° C (2010° F),[6] and the same behavior was noted for a pure 18-8 (0.001% carbon) composed of electrolytic metals and melted in vacuum, also heat-treated at 600° C (1110° F).[7] The nuclei, therefore, are not carbides *per se*. Addition of 0.24% nitrogen[5] to a pure 18-8 indicated that nitrogen, likewise, was not responsible.

α-Phase (body-centered cubic) or ferritic 18-8 corrodes at the same rate as does γ-phase or austenitic 18-8 (face-centered cubic) according to weight losses in ferric chloride, sodium chloride, and the salt spray, the former structure, however, having somewhat greater tendency to form pits in ferric chloride. Some α-phase in γ-phase 18-8 appears to confer a slightly greater tendency toward pitting[5] in ferric chloride.

Titanium and columbium in 18-8 increase pit nuclei in ferric chloride but not as

[5] H. H. Uhlig, *Trans. Am. Soc. Metals*, **30**, 947–980 (1942).
[6] H. H. Uhlig, unpublished observations.
[7] H. H. Uhlig, *Trans. Am. Soc. Metals*, **30**, 963 (1942).

CORROSION IN LIQUID MEDIA, ATMOSPHERE, GASES

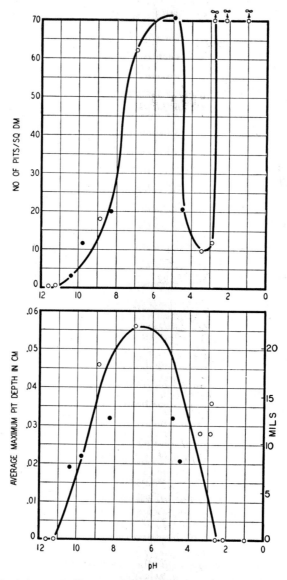

Fig. 4. Relation between the Number of Pits and the Pit Penetration in 18-8, and the pH of 4% Sodium Chloride at 90° C (194° F).

definitely in natural sea water or in aerated NaCl at 90° C (195° F). Comparative data for commercial columbium and titanium 18-8 (types 321 and 347) rolled sheet are given in Tables 2 and 3.

Selenium in 18-8 considerably increases pitting in ferric chloride. Its similar effect in salt solutions is probably less.

Oxide and silicate inclusions have seemingly little effect on pitting compared with the elements above.

Polishing, either mechanical or electrolytic, somewhat reduces the number of pit nuclei but the pits which form, because they are fewer, tend to be larger.

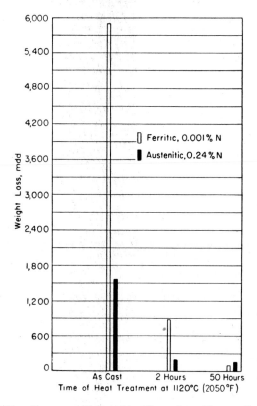

FIG. 5. Effect of Heat Treatment, Followed by Water Quenching, on Corrosion of Ferritic and Austenitic 18-8 in 10% Ferric Chloride.

Effect of Mechanical Work. Severe cold work increases pitting of 18-8 in ferric chloride.[8] Data of Table 2 show that Steckel Mill 18-8, a severely cold-worked strip, also corrodes more than annealed 18-8 in 10% $FeCl_3 \cdot 6H_2O$ and in 4% NaCl at 90° C, but not by the same factor in sea water.

Hot rolling serves to concentrate most of the nuclei at edges of the sheet. This may be caused by exposed impurities, but more likely is due to orientation of grains in such a manner that crystal faces normal to the edge corrode more than those parallel to it. A perfect cylinder cut from a rolled sheet still pitted predominantly at the original edge. Annealing, on the other hand, tended to reduce the ratio of edge pits to those on the sheet surface.[9]

---

[8] John Wulff, *Progress Report* 3 to Chemical Foundation, Corrosion Committee of Massachusetts Institute of Technology, 1937.

[9] H. H. Uhlig and M. C. Morrill, *Ind. Eng. Chem.*, **33**, 876 (1941).

TABLE 2. CORROSION RATES OF VARIOUS STAINLESS STEELS IN
LABORATORY TESTS AND SEA-WATER EXPOSURE*

MDD
(Mill Surface. Pitting Occurs in All Specimens Except as Noted)

Stainless Steel	Sea-Water Immersion, 1 Year, Boston Harbor. Spec.: 7.7 × 12.8 × 0.35 cm (3 × 5 × ⅛ in.). Badly fouled at end of test. (Av. of 5 Spec.)	4% NaCl, 90° C ± 0.1. 24-hour Immersion. Aerated: 40-75 ml air per min, 3½ L. soln. Spec.: 2.5 × 12.8 × 0.35 cm (1 × 5 × ⅛ in.). (Av. of 2 or 3 Spec.)	10% $FeCl_3 \cdot 6H_2O$.† 4-Hour Contact, Room Temp. Area of Spec. Exposed: 5.74 sq cm (0.89 sq in.). (Av. of 2 or more spec.)
18-8, 3% Mo	0.013 (no pits)	2.54 (no pits)	2.1 (no pits)
18-8:			
Manufacturer A	2.08	4.83	570
" B	1.84	....	820
" C (Steckel mill, cold-rolled)	2.76	28.2	5,620
18% Cr	3.34	38.3 (1 spec.)	6,120
18-8, 0.2% Ti	0.78	4.4	8,540
18-8, 0.4% Ti		3.55	16,000
18-8, 0.9% Cb		7.45	4,200
Mild steel (abraded #180 emery)	18.6 (no pits)	86.0 (no pits)	....

* Based on *Progress Report* 6 to Chemical Foundation, Corrosion Committee of Massachusetts Institute of Technology, 1938, and also unpublished data of the author.
† Using "Drop Tester," H. A. Smith, *Progress Report* 2 to Chemical Foundation, Corrosion Committee of Massachusetts Institute of Technology, 1936.

TABLE 3. ANALYSES OF STAINLESS STEELS IN TABLE 2

Stainless Steel	Cr	Ni	C	Mn	Si	S	P	
18-8, 3% Mo	20.92	9.75	0.06	0.40	0.36	0.02	0.014	2.90% Mo
18-8								
A	18.75	10.55	0.06	0.41	0.41	0.01	0.009	
B	18.00	9.66	0.07	0.38	0.25	0.01	0.008	
C	18.18	8.20	0.095	0.45	0.40	0.02	0.018	
(Steckel Mill)								
18% Cr	17.90	....	0.095	0.45	0.40	0.02	0.018	
18-8, 0.2% Ti	17.92	8.97	0.09	0.48	0.45	0.01	0.016	0.2% Ti
18-8, 0.4% Ti	18.6	8.8	0.05	0.36	0.41	0.002	0.02	0.36% Ti
18-8, 0.9% Cb	7.98	9.4	0.07	0.60	0.65	0.006	0.02	0.86% Cb

CONTACT CORROSION

This is corrosion produced at the region of contact of usually non-metallic materials with passive metals. It is also called crevice corrosion.[10] It may occur at washers, under barnacles, sand grains or applied protective films, and at pockets formed by threaded joints. Whether or not stainless steels are free of pit nuclei, they are always susceptible to this kind of corrosion, because a nucleus is not ncessary.

Contact corrosion may begin through action of an oxygen concentration cell. The accumulation of passivity-destroying corrosion products at stagnant areas soon convert the cell to a more rapidly acting passive-active cell. Corrosion then proceeds by a mechanism identical to that described above for pit growth. Contact corrosion of stainless steels is discussed further, beginning p. 155.

[10] E. H. Wyche, L. R. Voigt, and F. L. LaQue, *Trans. Electrochem. Soc.*, **89**, 149 (1946).

METHODS TO AVOID PITTING IN STAINLESS STEELS

According to theory and confirmed in large part by experience, the following expedients are recommended to reduce pitting. The approach is either from the environment or the metal.

Environment

1. Avoid concentration of halogen ions.
2. Insure uniform oxygen or oxidizing solutions. Agitate solutions. Avoid pockets of stagnant liquid.
3. Either increase oxygen concentration or eliminate it. Increasing oxidizing capacity of the solution augments passivity and resistance to attack. 18-8 in 3% H_2SO_4, through which hydrogen bubbled (18-8 not passive), corroded at a rate of 400 mdd, but at only 1/100 this rate when saturated with oxygen (18-8 passive).[11] On the other hand, elimination of oxygen avoids passive-active cells, for example in salt solutions.
4. Increase pH. Appreciably alkaline chloride solutions pit less or not at all compared with neutral or acid chlorides.
5. Operate at lowest temperature possible.
6. Add passivators to corrosive medium. A small concentration of nitrate or chromate is effective in many media. Ferric sulfate and copper sulfate are examples of passivators for stainless steels in sulfuric acid, but when used halogen ions should be absent.
7. Apply cathodic protection. There is evidence that stainless steels protected cathodically by galvanic coupling to steel, aluminum, or zinc do not pit in sea water. (See *Behavior of Metals and Alloys in Sea Water*, p. 416.)

The Alloy

1. Homogenize. Rapidly quench from 1050° to 1100° C (1900° to 2000° F).
2. Treat final article in 10 to 20 vol. % conc. HNO_3 at 55° to 60° C (130° to 140° F) for 15 to 30 min. This process, called passivation, helps remove latent pit nuclei. A more active pickle for removing light oxide scale may precede this treatment. It consists of 10 vol. % conc. HNO_3, 1 to 2 vol. % HF, at 50° to 65° C (130° to 150° F). The latter mixture is especially corrosive and the fumes are irritating. Ventilation is imperative.
3. Cleanse and burnish surfaces periodically. This will help avoid contact corrosion. Alkaline cleaners and stainless steel wool are satisfactory.
4. Use alloy containing 2 to 4% molybdenum (type 316 or 317). Molybdenum added to nickel-stainless steels markedly reduces tendency toward pitting in chloride solutions (see Table 2), although in sea water, for example, pitting eventually may occur (within 2 to 2½ years).[12] This steel (type 316 or 317) is also appreciably more resistant than 18-8 (type 304) to reducing acids and to contact or crevice corrosion.*

---

* In a study of crevice corrosion in sea water this sort of attack occurred on type 304 specimens in 95 cases out of 200 (48%), whereas with the type 316 alloy exposed under the same conditions it occurred in only 11 cases out of 48 (23%). (Private communication from F. L. LaQue, International Nickel Co., New York, N. Y.)

[11] R. McKay and R. Worthington, *Corrosion Resistance of Metals and Alloys*, p. 297, Reinhold Publishing Corp., New York, 1936.
[12] F. L. LaQue, (Discussion) *Trans. Am. Soc. Metals*, **27**, 524 (1939).

STRESS CORROSION CRACKING IN STAINLESS ALLOYS

M. A. Scheil*

References covering observations of the behavior of stainless alloys in several corrosive environments record that stress can cause propagation of cracks: (1) intergranularly, (2) intergranularly and transgranularly, and (3) wholly transgranularly. This type of attack is now receiving widespread recognition because of its possible industrial implications wherever stainless alloys are applied, but in general the recorded cases of failure are surprisingly few in number.

Intergranular corrosion has been successfully and adequately avoided in practice by high-temperature anneal, followed by rapid quench to control harmful carbide precipitation, as well as by use of stabilized types of austenitic alloys such as those containing columbium (type 347) and titanium (type 321). (Refer to discussion of intergranular corrosion, p. 161.) Comparable theory and understanding of transgranular cracking do not exist, however, nor do methods for controlling it other than by stress relief annealing and limitation of applied stresses where the nature of the corrosive medium makes this necessary. Study of stress corrosion cracking of the transgranular type is handicapped because of our inability to define completely the characteristics of corrosive media which can cause it. This situation is not peculiar to stainless steels, but extends to all materials susceptible to stress corrosion cracking.

Figures 1A and 1B illustrate the path of rupture in two molybdenum-bearing 18% Cr-8% Ni alloys that were subject to stress and corrosive conditions in an oil refinery. The active corrosive component of the gasoline vapor in this instance was suspected to be a small concentration of HCl or hydrolyzable chlorides. Figure 1A shows the intergranular separation and Fig. 1B the transgranular separation of the alloys exposed similarly. In both cases failure was localized in the area being stressed.

Aqueous acid chloride solutions appear to be the most active media in producing transgranular cracking of stressed austenitic stainless steels. Also ethyl chloride containing water is reported to be effective above room temperature.[1] However, neither chloride ion nor an acid environment is a necessary condition since transgranular failures have been recorded in acid sulfite cooking liquors where there very likely were no appreciable concentrations of chlorides, and in a caustic soda solution at 350° C (660° F) under pressure.

ACCELERATED TESTS

Many accelerated tests have been devised to study the effect of stress in producing transgranular cracking of the austenitic alloys. A solution containing approximately 42% $MgCl_2$ (based on anhydrous salt), boiling at 154° C (309° F), is an effective accelerant. This solution causes transgranular cracking in almost all the austenitic alloys, sufficiently stressed, regardless of whether they are stabilized, or previously heat-treated to resist intergranular attack.

Testing Solution

A test which has received some study consists of total immersion of the stressed specimen in a boiling refluxed solution of 42% $MgCl_2$. Pyrex flasks are used as containers with 100 ml of solution per sq in. of stainless steel. The same solution has been used for a continuous period of 300 hours. When the extent of contamination

* A. O. Smith Corporation, Milwaukee, Wis.

[1] J. C. Hodge and J. L. Miller, "Stress Corrosion Cracking of the Austenitic Chromium-Nickel Steels and Its Industrial Implications," *Trans. Am. Soc. Metals*, **28**, 25–82 (1940).

by corrosion products reaches 20 mg of stainless steel for each 100 ml of magnesium chloride, the rate of attack increases to about four times normal, and the contaminated solution should be replaced with a fresh solution.

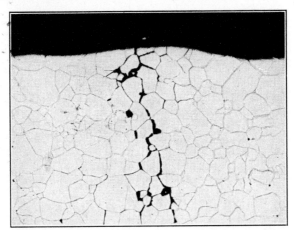

FIG. 1A. Intergranular Stress Corrosion Cracking. (×200.)

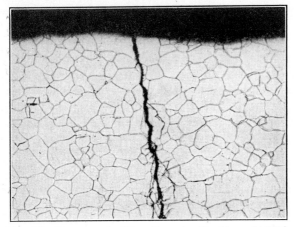

FIG. 1B. Transgranular Stress Corrosion Cracking. (×200.)

FIG. 1. Types of Failure in Two Stressed Mo 18-8 Specimens Exposed to Vapor of Straight-Run Gasoline Containing Corrosive Constituents at 190° C (375° F) and 95 Psi Pressure. Type of specimen: Horseshoe (similar to Fig. 7, p. 1013).

ANALYSIS OF ALLOYS

	A.I.S.I. Type	Cr	Ni	C	Mo	Mn	Si
Fig. 1A	317	18.77	13.35	0.08	3.11	1.67	0.41
Fig. 1B	316	17.43	12.76	0.07	2.79	1.39	0.33

PREPARATION OF TEST SPECIMENS

The test sample assembly should incorporate a simple means of applying stress to a specimen. The horseshoe bend (see Fig. 7, p. 1013) or the beam or cantilever (see Fig. 2) shaped specimens offer a convenient size to test, and stress can be indicated by means of a wire resistance strain gage. Such specimens can be readily heat-treated, surface polished, or pickled prior to stressing.

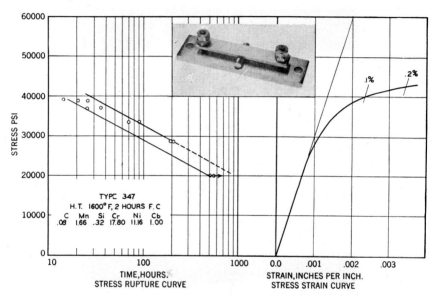

FIG. 2. Stress Rupture Curve for Cb Stabilized 18-8. (Insert: Beam type test specimen.)

EXAMINATION OF SPECIMENS

At frequent intervals, depending upon the stress intensity, the specimens are removed from the boiling $MgCl_2$, washed in tap water, dried, and examined on the stressed surface with a hand lens or under the binocular microscope for superficial cracks. The time to produce the first appearance of a surface crack is recorded and the test is continued until the specimen fails, or until it is decided that the crack is superficial. This is important in the study of surface finish as it has been found that with polished surfaces, cold work produces a typical stress crack in short exposure which does not propagate as the test is continued.

THRESHOLD STRESS FOR STRESS CORROSION CRACKING

In order to arrive at some concept of the effect of stress on stress corrosion cracking in the 42% $MgCl_2$ solution, duplicate sets of stressed beam specimens were fabricated from type 347 stainless steel sheet 0.28 cm (7/64 in.) thick and after machining were heat-treated at 870° C (1600° F) for 2 hours. Each specimen was pickled and duplicate sets were strained to the stress values shown by the stress-strain curve. The highest stress was below the yield strength for 0.2% offset. Figure 2 shows the results in the form of a stress rupture curve. It is obvious that the limiting stress is below 20,000 psi. Figure 3 shows the fractured specimens of each test superimposed upon a diagram showing the value of stress at the point of rupture. It is noted that failure does not always occur at the point of maximum stress but always occurs on the tension surface.

Stress Corrosion Cracking in Tubing

The MgCl₂ test has been used to determine the harmful effects of residual stresses causing stress corrosion cracking in tubing. Figure 4 shows examples of cold-drawn

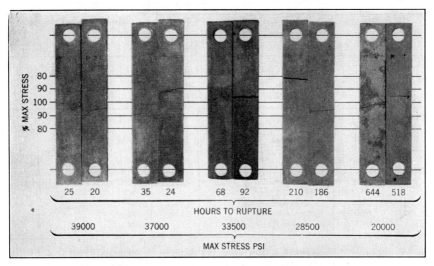

Fig. 3. Stress Rupture Specimens of Columbium-Stabilized 18-8.

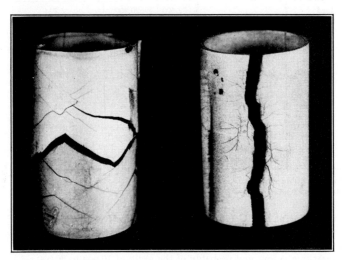

Fig. 4. Stress Cracks Produced by High Longitudinal and Hoop Stresses in Cold-Drawn Types 347 and 316 Stainless Steel Tubing. (×3.)

tubing after immersion for 7½ hours in boiling 42% MgCl₂. The tube exhibiting the spiral cracking was found to have a higher longitudinal than hoop stress. The tube showing longitudinal crack was found to have a higher hoop than longitudinal stress.

TABLE 1. EFFECT OF HEAT TREATMENT ON STRESS CORROSION CRACKING OF STAINLESS STEEL TUBES IN BOILING 42% MgCl$_2$

Heat Treatment of Tube, Specimen			Calculated Residual Stress		Hours to Show Cracking
°C	°F	Time in Hours	Hoop Stress, psi	Longitudinal Stress, psi	

Type 347, Full Hard Tubing, 165,600 psi tensile strength
C   Mn   Cr   Ni   Mo   Cb
0.05%   1.50%   18.86%   10.74%   0.08%   0.88%

°C	°F	Time in Hours	Hoop Stress, psi	Longitudinal Stress, psi	Hours to Show Cracking	
None, full hard			46,200	69,000	7½	↑
None, half hard*			....	....	6	
540°	1000°	24, air-cooled	....	....	7½	
650°	1200°	½, " "	....	....	22	
650°	1200°	8, " "	....	....	14½	
740°	1375°	½, " "	1,850	8,450	245 (1 small crack)	
740°	1375°	½, furnace-cooled	....	....	292 (1 small crack)	Failure
870°	1600°	½, air-cooled	....	....	>292	No cracks
870°	1600°	½, furnace-cooled	....	....	>292	
870°	1600°	24, air-cooled	....	....	>292	↓

Type 316, ¼ hard tubing, 114,600 psi tensile strength
C   Mn   Si   Cr   Ni   Mo
0.05%   1.42%   0.48%   17.60%   10.55%   2.76%

°C	°F	Time in Hours	Hoop Stress, psi	Longitudinal Stress, psi	Hours to Show Cracking	
None, ¼ Hard			52,500	20,200	7½	↑
None, Mill annealed and straightened†			17,000	....	7½	
540°	1000°	24, air-cooled	44,500	....	7½	
650°	1200°	½, " "	39,100	....	7½	
650°	1200°	8, " "	....	....	14½	
740°	1375°	½, " "	26,700	....	22	
740°	1375°	½, furnace-cooled	23,100	....	22	
740°	1375°	8, air-cooled	....	....	22	
790°	1450°	½, " "	10,400	....	24	Failure
845°	1550°	½, " "	3,600	....	>240	No cracks
870°	1600°	½, " "	3,500	8,300	>292	
870°	1600°	½, furnace-cooled	....	....	>292	
870°	1600°	24, air-cooled	....	....	>292	↓

* 133,100 psi tensile strength.
† 91,800 psi tensile strength.

Heat treatment at 870° C (1600° F) was very effective in removing residual stress causing stress corrosion cracking in this solution, as stress calculations and data of Table 1 show. It appears from this information that the threshold stress for transgranular cracking is low, and probably of the order of 10,000 to 20,000 psi hoop stress.

ACCELERATED TESTS OF VARIOUS ALLOYS

Several compositions of stainless alloys were fabricated into horseshoe specimens and tested in the boiling MgCl$_2$ solution. The data from several of these tests are shown in Table 2. They do not represent a systematic exploration of each alloy tested, but serve to show that the MgCl$_2$ solution cracks some of the austenitic alloys more than others. Some ferritic stainless alloys are included in the series. The table shows several alloys that were tested with pickled and polished surfaces.

Stainless steels of types 309, 310, 316, 321, 329, 347, the 20% Cr–24% Ni–3% Mo–3% Si–1.75% Cu alloy,* the 23 Cr–13% Ni–1.5% Mo–1% Cu alloy, and the 16% Cr–27% Ni–3.7% Mo–1% Cu alloy† with polished surface showed cracks. Of these, types 309 and the 20% Cr–24% Ni–3% Mo–3% Si–1.75% Cu* alloy did not fail in the period of the test. The 15% Cr–35% Ni alloy, the 13% Cr–6.5% Fe–79.5% Ni‡ alloy, and the 14% Cr–17% Mo–6% Fe–56% Ni§ alloy did not crack or fail in the polished condition. In Fig. 5 are shown some of the polished specimens that failed and the duplicate specimens with pickled surfaces that did not show stress cracks or failures.

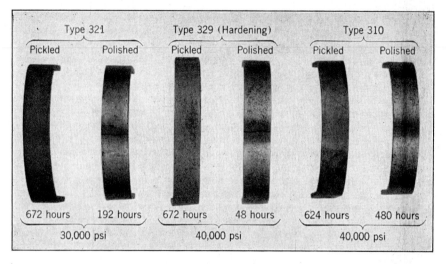

FIG. 5. Stress Corrosion Cracking of Stainless Steels in MgCl$_2$. (Pickled surfaces did not crack.)

These brief remarks are offered as observations of the experiments that were conducted and are being continued to learn more about the mechanism of stress corrosion cracking. It is difficult in our present state of knowledge to offer any definite safe stress values for purposes of engineering design. The safest procedure to follow is to study individual cases with a series of stressed specimens and determine the acceptability of the fabricated alloys to the service conditions. When the contemplated application involves a corrosive environment similar to magnesium chloride, the boiling MgCl$_2$ test may be a valuable tool to determine susceptibility to transgranular failure.

* Worthite. † Durimet T. ‡ Inconel § Hastelloy C.

TABLE 2. THE BEHAVIOR OF VARIOUS STRESSED SPECIMENS OF STAINLESS STEELS TESTED IN BOILING MgCl$_2$ SOLUTION AT 154° C (309° F)*

Alloy	A.I.S.I. Type No.	Heat Treatment After Cold Forming to Horse Shoe			Surface Finish Before Stressing	Stress Applied, psi	Time in Test, hr			Path of Rupture or Cracks—Surface Appearance
		°C	°F	Time in Hours			To first Cracks Formed	To Fracture	No Cracks Formed	
18-8	304	650	1200	1	Original cold finish	Beyond Y.P.		8		Transgranular path
"	304	none	(cold-pressed cup)		Original cold finish	Beyond Y.P.		17		Transgranular path
"	302	none	(cold-formed)		Original 2B finish	Beyond Y.P.		21¾		Transgranular path
Cb 18-8	347	870	1600	½	Polished[1]	40,000	14	48–72		Transgranular path
Mo 18-8	316	870	1600	½	Polished[1]	40,000	7	48–72		Transgranular path
Mo 18-8	317	870	1600	½	Polished[1]	40,000	7			Transgranular path
24 Cr, 15 Ni	309[3]	1065	1950	½	Polished[1]	40,000	24 (fine)		720†	Light surface pitting
24 Cr, 15 Ni	309[3]	1065	1950	½	Pickled[2]	40,000			624	No pitting
25 Cr, 20 Ni	310	1065	1950	½	Polished[1]	40,000	24 (fine)	480		Light surface pitting
25 Cr, 20 Ni	310	1065	1950	½	Pickled[2]	40,000			624	No pitting
Ti 18-8	321	870	1600	½	Polished[1]	30,000	72	192		Transgranular path
"	321	870	1600	½	Pickled[2]	30,000			672	No pitting
15 Cr, 35 Ni[4]			none		Polished[1]	30,000			624	No pitting
15 Cr, 35 Ni[4]			none		Pickled[2]	30,000			624	No pitting
23 Cr, 13 Ni, 1.5 Mo, 1 Cu		595	1100	2	Polished[1]	30,000	48 (severe)	48		No pitting
16 Cr, 27 Ni, 3.7 Mo, 1 Cu‡		595	1100	2	Polished[1]	Beyond Y.P.	120 (severe)	120		No pitting
"		1065	1950	½	Polished[1]	None				No pitting
"		1065	1950	½	Polished[1]	10,000	528 (1 small)		648	No pitting
"		1065	1950	½	Polished[1]	40,000	369			No pitting
"		1065	1950	½	Polished[1]	Beyond Y.P.	72 (1 small)	168		No pitting
20 Cr, 24 Ni, 5 Mo, 3 Si, 1.75 Cu§[5]			none		Polished[1]	40,000	72 (fine)		816†	Very light pitting
"			none		Pickled[2]	40,000			768	
13 Cr, 6.5 Fe, 79.5 Ni‖[6]		595	1100	2	Polished[1]	20,000			456	No surface pitting
"		595	1100	2	Polished[1]	Beyond Y.P.			408	No surface pitting
19 Mo, 20 Fe, 57 Ni¶		none	(as received)		Pickled[7]	40,000			720	
"		1150	2100	⅓	Pickled[7]	40,000			720	
"		650	1200	2	Pickled[7]	40,000			720	
"		1065	1950	3	Pickled[7]	40,000			720	
"		Welded with Hastelloy B (No heat treatment.)			Pickled[7]	40,000			720	

STRESS CORROSION CRACKING IN STAINLESS ALLOYS

Composition	Alloy	Heat treatment	Surface	Stress, psi	Hr. to crack	Total hr.	Remarks
30 Mo, 5 Fe, 64 Ni**		none (as received)	Pickled⁷	40,000		720	No surface pitting
"		1180 2150 ⅓	Pickled⁷	40,000		720	No surface pitting
"		650 1200 2	Pickled⁷	40,000		720	No surface pitting
"		1065 1950 3	Pickled⁷	40,000		720	No surface pitting
"		(Welded with self. No heat treatment.)	Pickled⁷	40,000		720	No surface pitting
14 Cr, 17 Mo, 6 Fe, 56 Ni⁸††		595 1100 2	Polished¹	30,000		456	No surface pitting
"		595 1100 2	Polished¹	Beyond Y.P.		120	No surface pitting
12 Cr, 0.05 C, 0.2 Al	405	none	Polished¹	40,000		504	Surfaces deeply pitted
"	405	{carburized 980 1800 8 w.q.	Polished¹	40,000		504	Very general shallow pitting
"	405	980 1800 ½ w.q.	Polished¹	40,000		504	Surfaces deeply pitted
19 Cr, 0.04 C	430	845 1550 ½ f.c.	Pickled²	40,000		624	Surface deeply pitted
"	430	1065 1950 ½ w.q.	Pickled²	40,000		624	Surface pitted
12 Cr, 2 Ni, 0.10 C	421⁹	none (as received)¹⁰	Polished¹	40,000	258 (fine)		
"	421⁹	1065 1950 ¼ w.q.	Polished¹	40,000	18		Intergranular path
28 Cr, 5 Ni, 1.7 Mo, 0.08 C	329¹¹	1010 1850 ½	Polished¹	40,000	24	672	No surface pitting
"	329¹¹	1010 1850 ½	Polished²	40,000	48	624†	No surface pitting
25 Cr, 2.5 Ni, 1.3 Mo, 0.23C	329¹²	1010 1850 ½	Polished¹	40,000		672	No surface pitting
"	329¹²	1010 1850 ½	Pickled²	40,000			

*Symposium on Stress-Corrosion Cracking of Metals (1944), American Society for Testing Materials and American Institute of Mining and Metallurgical Engineers, Philadelphia and New York, 1945.
† No fracture although cracks formed.   ‡ Durimet T.   § Worthite.   ‖ Inconel.   ¶ Hastelloy A.   ** Hastelloy B.   †† Hastelloy C.
[1] Polished with No. 120 Aloxite belt after heat treatment. All polishing parallel to long direction of specimen.
[2] Pickled in 40% nitric acid and 4% hydrofluoric acid at 80° C (180° F).
[3] Same specimen used — first pickled and tested 624 hr, then ground and tested 720 hr.
[4] Same specimen used — shaped from 1065° C (1950° F) water-quenched annealed bar stock, pickled and tested 624 hr, then severely ground and tested 624 hr.
[5] Same specimen used — first pickled and tested 768 hr, then surface ground, first cracks at 72 hr, no failure after 816 hr.
[6] Same specimen used — first polished and stressed to 20,000 psi, tested 456 hr, then stressed beyond yield point and tested 408 hr.
[7] Pickled — 10% H₂SO₄, 5% by weight NaNO₃ and 5% NaCl used hot at 95° C (200° F). This pickle for ½ hr did not intergranularly attack any Hastelloy B sheet or weld deposit. (2100° F) and 1065° C (1950° F) heat treatments. The pickle for ½ hr heated to 730° C (1350° F). After 1010° C (1850° F) anneal: Rockwell B 96.
[8] Same specimen used — first polished and stressed to 30,000 psi, tested 456 hr, then stressed beyond yield point and tested 120 hr.
[9] This composition was Rockwell B 100 as received, and after 1065° C (1950° F) water quench was Rockwell C 44.
[10] Beam type specimen used.
[11] This composition is of hardening type when heated to 730° C (1350° F). After 1010° C (1850° F) anneal: Rockwell B 96.
[12] This composition is non-hardening type when heated to 730° C (1350° F). After 1010° C (1850° F) anneal: Rockwell B 101.
Note: Specimens when heat-treated were air-cooled except as noted: w.q., water-quenched; f.c., furnace-cooled.

GENERAL REFERENCES (STRESS CORROSION CRACKING IN STAINLESS ALLOYS)

HODGE, J. C., and J. L. MILLER, "Stress Corrosion Cracking of the Austenitic Chromium-Nickel Steels and Its Industrial Implications," *Trans. Am. Soc. Metals*, **28**, 25–82 (1940).
HOYT, S. L., and M. A. SCHEIL, "Stress Corrosion Cracking in Austenitic Stainless Steels," *Trans. Am. Soc. Metals*, **27**, 191–226 (1939).
SCHEIL, M. A., and R. A. HUSEBY, "Stress Corrosion Cracking of Austenitic Types 347 and 316," Supplement, *J. Am. Welding Soc.* (August, 1944)
SCHEIL, M. A., O. ZMESKAL, J. WABER, and F. STOCKHAUSEN, "First Report on Stress Corrosion Cracking of Stainless Steel in Chloride Solutions," Supplement, *J. Am. Welding Soc.* (October, 1943).
Symposium on Stress-Corrosion Cracking of Metals (*1944*), American Society for Testing Materials and American Institute of Mining and Metallurgical Engineers, Philadelphia and New York, 1945.
ELLIS, O. B., "Some Examples of Stress-Corrosion Cracking of Austenitic Stainless Steel," p. 421.
FRANKS, R., W. O. BINDER, and C. M. BROWN, "The Susceptibility of Austenitic Stainless Steels to Stress-Corrosion Cracking," p. 411.
FRASER, O. B. J., "Stress-Corrosion Cracking of Nickel and Some Nickel Alloys," p. 458.
SCHEIL, M. A., "Some Observations of Stress Corrosion Cracking in Austenitic Stainless Alloys," p. 395.

16% CHROMIUM–2% NICKEL STAINLESS STEEL

W. O. BINDER*

The martensitic stainless steels containing 11.5 to 13% chromium and 0.08 to 0.35% carbon respond to heat treatment much the same as plain carbon and low-alloy steels, and are hardened to a high degree by quenching from elevated temperatures. Although the mechanical properties of these steels are excellent, under certain conditions of exposure they are deficient in corrosion resistance. Increasing the chromium content 16 to 18% improves their resistance to corrosion but at the same time decreases their capacity to harden appreciably on quenching. The hardening characteristics of the 16 to 18% chromium steels may be improved by raising the carbon content but not without lowering the corrosion resistance.

A better solution is to use 1 to 2% nickel, since small amounts of nickel increase the hardening capacity of the low-carbon 16 to 18% chromium steels without detrimentally affecting their corrosion resistance. This alloy, which is generally known as the 16-2 steel and has the A.I.S.I designation of type 431 stainless steel, is made in this country to conform with Army-Navy-Aeronautical specification AN–QQ–S–770. The requirements of this specification are listed in Table 1. A similar steel is made in Great Britain under the trade name Twoscore. The British Air Ministry has designated this alloy as S–80, and the steel is made in accordance with the requirements of the British Standards Institution, which are also listed in Table 1. In Great Britain the steel has been used extensively for structural parts in seaplanes and flying boats because of its high strength and toughness coupled with good corrosion resistance.

The addition of 2% nickel to steels containing 15.5 to 17.5% chromium increases the amount of austenite formed in these steels when heated to elevated temperatures.[1] It is necessary, however, to heat the 16% Cr–2% Ni steel considerably above its upper critical range to obtain full transformation and complete homogenization. These steels do not transform completely into austenite at elevated temperatures because of their high chromium content, and a certain amount of ferrite remains in the hardened structure. In wrought metal these stringers of ferrite have a laminated

* Union Carbide and Carbon Research Laboratories, Inc., Niagara Falls, N. Y.

[1] M. E. Tatman, *The Iron Age*, **152** (No. 19), p. 70 (May 11, 1944).

16% CHROMIUM–2% NICKEL STAINLESS STEEL

TABLE 1. SPECIFICATIONS FOR 16% Cr–2% Ni STAINLESS STEELS

Chemical Composition, %	AN–QQ–S–770 (U. S. Government)	S–80 (British Standards Institution)
Carbon	< 0.15	<0.25
Phosphorus	< 0.04	....
Sulfur	< 0.04	....
Silicon	0.2–0.6	<0.50
Manganese	0.3–0.8	<1.0
Chromium	15.5–17.5	16–20
Nickel	1.5–2.5	>1.0

Mechanical Properties	Hardened and Tempered	Annealed	
Tensile strength, psi	>175,000	>115,000	>123,200
Yield strength (0.2% offset), psi	>135,000	> 90,000	....
Elongation in 2 in., %	> 13	> 15	> 15
Izod impact, ft-lb.	> 35	> 60	> 25
Brinell hardness	.....	.....	> 241

appearance and are generally elongated in the direction of final working. There is sufficient transformation, however, to increase the hardness, strength, and toughness of the steel beyond that of a similar steel without nickel.

GENERAL CORROSION PROPERTIES

Although specific corrosion data for the 16% Cr–2% Ni steel are meager, sufficient information has been procured to show that under reducing or weakly oxidizing conditions, its corrosion resistance is only slightly superior to that of plain chromium steels of similar chromium content, whereas under strongly oxidizing conditions, such as in nitric acid (Table 2),[2] its corrosion resistance is slightly inferior. Because the two steels are similar with respect to corrosion, both service test and laboratory

TABLE 2. RESULTS OF TESTS IN BOILING 65% NITRIC ACID ON 16% CHROMIUM STEELS

	% Cr	% Ni	% C	% Mn	% Si
Steel A	16.64	....	0.12	0.63	0.52
Steel B	16.60	2.05	0.11	0.56	0.43

Condition of Metal	Period, 48 hr each	Corrosion Rate, ipy	
		16% Cr Steel A	16% Cr, 2% Ni Steel B
Oil-quenched from 950° C (1740° F), reheated 2 hr at 700° C (1290° F), and air-cooled.	1st	0.040	0.062
	2nd	0.042	0.069
	3rd	0.039	0.065
Oil-quenched from 950° C (1740° F), reheated 3 hr at 300° C (570° F), and air-cooled.	1st	0.047	0.052
	2nd	0.055	0.055
	3rd	0.060	0.058

[2] Unpublished data from the Union Carbide and Carbon Research Laboratories, Inc., Niagara Falls, N. Y.

184 CORROSION IN LIQUID MEDIA, ATMOSPHERE, GASES

corrosion data for plain 16 to 18% chromium steels may be safely used as a guide for applying the 16% Cr–2% Ni steel in various corrosive environments.

According to Watkins,[3] 16% Cr–2% Ni steel shows perfect resistance to tap water after 500 hours of exposure, and is only slightly corroded by salt-spray atmosphere in 100 hours of exposure. In these tests, hardened 16% Cr–2% Ni steel was more resistant than the annealed steel. The appearance of the samples after these tests is shown in Figs. 1 and 2. In salt-spray atmosphere 16% Cr–2% Ni steel has a greater

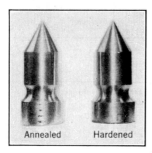

FIG. 1. 16% Cr–2% Ni Steel after 500 Hours, Running Tap Water. (Watkins.)

FIG. 2. 16% Cr–2% Ni Steel after 100 Hours, 4% Salt-Spray Atmosphere. (Watkins.)

tendency to pit than 18% Cr–8% Ni steel when the exposure time is extended beyond 500 or more hours. Monypenny[4] conducted sea-water spray tests on a series of chromium steels that included the 16% Cr–2% Ni type. The tests lasted 88 days, during which time the samples were exposed to spray for 8 hours in each day and for 5 days in the week. The remainder of the time the spray was cut off and the samples allowed to stand in the moisture-laden atmosphere. These tests yielded the results given in Table 3.

TABLE 3. SEA–WATER SPRAY TESTS ON Cr AND Cr-Ni STAINLESS STEELS

Area of sample — 25 sq cm.
Weight of sample — 65 grams.
Condition of surface — not stated.
Duration of test — 88 days, 400-hr spraying time.

Composition, %			Loss in Weight, grams	Remarks
C	Cr	Ni		
0.32	12.7	0.40	0.179	Badly pitted
0.07	13.8	0.38	0.045	Pitted
0.09	17.8	0.12	0.031	Pitted less than low-carbon steel containing 13.8% chromium
0.08	20.7	0.23	0.006	Slightly pitted
0.08	17.0	2.15	0.002	Very slightly pitted
0.16	17.4	1.62	0.019	Slightly pitted
0.12	18.0	8.2	Nil	Minute pits, just visible

Monypenny[4] also completely immersed samples of 16% Cr–2% Ni steels and high-carbon 12% chromium steel in sea water for 52 weeks and found weight losses as indicated by Fig. 3. It will be noted that for the 16% Cr–2% Ni steels, these tests

[3] S. P. Watkins, *Metal Progress*, **44**, 99 (1943).
[4] J. H. G. Monypenny, *Stainless Iron and Steel*, 2nd Ed., Chapman and Hall, London, 1931.

do not show any significant differences between the corrosion resistance of samples in the hardened and in the annealed condition. Owing to the extra resistance to salt water, combined with good hardness, the 16% Cr–2% Ni steel has been used successfully for valve parts and pump rods, in handling hot and cold brines.

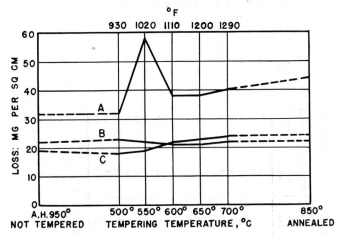

FIG. 3. Effect of Heat Treatment on Resistance of 13% Cr and 17.5% Cr–2% Ni Steels to Sea Water. (Monypenny.)

Curve	% Carbon	% Chromium	% Nickel
A	0.30	12.6	0.78
B	0.20	17.4	2.19
C	0.09	17.9	2.08

While the 12% chromium steel has sufficient resistance to both corrosion and erosion in steam, some pitting is encountered when the steel is required to operate in contact with graphite and asbestos packing materials. Under such circumstances pitting is less likely to develop with the 16% Cr–2% Ni steel; hence it is especially useful for valve seats and spindles of steam valves. Galvanic corrosion caused by contact with copper-base alloys while in the presence of an electrolyte has also been troublesome in the case of the 12% chromium steel.[5] Increasing the chromium content of the steel gives very much greater resistance to this form of attack, and, according to Monypenny,[6] 16% Cr–2% Ni steels have shown no signs of corrosion after 6 months of contact with bronze while immersed in sea water. Under similar conditions the 12% chromium steels were distinctly attacked within a few days.

EFFECT OF TEMPERING

The influence of tempering on the corrosion resistance of the martensitic chromium steels has been studied by Monypenny.[6] His tests showed that tempering in the range of 500° to 600° C (930° to 1110° F) after hardening reduces the corrosion resistance of the cutlery grade of stainless steel (Fig. 3) with less effect on the lower carbon steels of higher chromium content. Tempering the 16% Cr–2% Ni steel in the range of 400° to 500° C (750° to 930° F) should be avoided, however, as this range of temperatures reduces the toughness of the steels and decreases corrosion resistance, as shown in Table 4.

[5] R. Barnaby, *The Book of Stainless Steels*, 2nd Ed., Chapter 12F, p. 347, edited by E. Thum, The American Society for Metals, Cleveland, Ohio, 1935.

[6] J. H. G. Monypenny, *Stainless Iron and Steel*, 2nd Ed., Chapman and Hall, London, 1931.

TABLE 4. TENSILE AND CORROSION PROPERTIES OF 15 TO 17% Cr–2% Ni STEELS*

Composition, %					Condition of Metal	Yield Point, psi	Tensile Strength, psi	% Elong. in 2 in.	% Reduction of Area	Izod Impact, ft-lb.	Brinell Hardness Number	800 Hours in 20% NaCl Spray
Cr	Ni	Mn	Si	C	N							
14.90	1.85	0.54	0.35	0.12	0.04	1	......	......	..	..	20	340	No pitting
						2	167,000	195,000	15	53	22	387	One pit
						3	181,000	203,000	16	51	9	418	No pitting
						4	143,500	152,750	13	57	11	321	Numerous fine pits
						5	98,750	117,600	21	61	53	241	No pitting
15.78	1.84	0.58	0.31	0.11	0.04	1	......	......	..	..	22	321	No pitting
						2	162,000	189,250	16	51	16	387	No pitting
						3	178,000	196,000	16	51	7	418	No pitting
						4	135,000	148,250	14	54	15	321	Several pits
						5	99,500	117,250	21	61	53	241	One pit
16.82	1.85	0.58	0.30	0.11	0.04	1	......	......	..	..	43	302	No pitting
						2	167,900	182,000	16	49	24	387	No pitting
						3	155,000	187,500	17	50	14	387	No pitting
						4	132,500	146,500	16	54	24	311	Few small pits
						5	100,000	119,500	21	59	48	241	No pitting

* Unpublished data from the Union Carbide and Carbon Research Laboratories, Inc., Niagara Falls, N. Y.
1 = Heated 4 hr at 750° C (1380° F) and air-cooled.
2 = Oil-quenched from 950° C (1740° F), heated 5 hr at 275° C (525° F), and air-cooled.
3 = "   "   "   "   "   "   "   "   at 400° C (750° F), "   "
4 = "   "   "   "   "   3 hr at 500° C (930° F), "   "
5 = "   "   "   "   "   "   "   at 600° C (1110° F), "   "

TABLE 5. TENSILE AND CORROSION PROPERTIES OF HIGH-NITROGEN 15 TO 17% Cr–2% Ni STEELS

Composition, %						Condition of Metal	Yield Point, psi	Tensile Strength, psi	% Elong. in 2 in.	% Reduction of Area	Izod Impact, ft-lb	Brinell Hardness Number	800 Hours in 20% NaCl Spray
Cr	Ni	Mn	Si	C	N								
14.61	1.86	0.57	0.29	0.07	0.08	1	......	......	..	..	14	351	No pitting
						2	169,000	196,500	14	46	17	402	No pitting
						3	194,000	211,000	16	52	5	444	No pitting
						4	142,000	157,000	14	53	2	321	Several fine pits
						5	109,000	120,000	19	57	27	241	No pitting
15.91	1.83	0.58	0.30	0.07	0.08	1	......	......	..	..	15	340	No pitting
						2	165,000	195,250	17	57	20	418	No pitting
						3	189,500	208,500	17	56	6	444	No pitting
						4	139,250	152,100	17	58	10	321	Several fine pits
						5	102,500	121,000	20	59	54	241	No pitting
16.96	1.95	0.54	0.30	0.07	0.09	1	......	......	..	..	..	......	No pitting
						2	169,000	199,500	17	54	29	402	No pitting
						3	196,000	210,000	17	55	8	444	No pitting
						4	137,000	164,300	14	55	17	321	No pitting
						5	107,000	126,750	19	58	49	262	No pitting

1 = Heated 4 hr at 750° C (1380° F), and air-cooled.
2 = Oil-quenched from 950° C (1740° F), heated 5 hr at 275° C (525° F), and air-cooled.
3 = "   "   "   "   "   "   "   "   " 400° C (750° F), "   "
4 = "   "   "   "   "   3 hr " 500° C (930° F), "   "
5 = "   "   "   "   "   "   " 600° C (1110° F), "   "

EFFECT OF ADDED NITROGEN

Exceedingly high strength can be obtained in 16% Cr–2% Ni steels with added nitrogen, which would otherwise be obtainable only by an increase in carbon content. Nitrogen[7] effectively reduces the grain size and improves the forgeability of these steels so that they can be fabricated more readily. The good mechanical properties and corrosion resistance obtainable in the 16% Cr–2% Ni type of steel relatively low in carbon and high in nitrogen are given in Table 5. As with the high-carbon low-nitrogen steels, tempering in the range of 400° to 500° C (750° to 930° F) should be avoided, but the low-carbon high-nitrogen steels are somewhat less affected by salt spray than the high-carbon steels when tempered in this range.

CHROMIUM-MANGANESE AND CHROMIUM-MANGANESE-NICKEL STAINLESS STEELS

RUSSELL FRANKS*

Other than nickel, manganese represents the only common metal addition to stainless steels permitting retention of large proportions of austenite on rapidly cooling the alloy from elevated temperatures. Substantially austenitic steels have been obtained in alloys of about 18% chromium, 5% manganese, and 5% nickel, with and without small copper additions. Work in Germany, and also in the United States, has shown that fully austenitic low-carbon content steels are obtained with chromium in the range of 12 to 15%, and manganese in the range of 16 to 20%, a typical steel being one containing about 13% chromium and 16% manganese. As with the austenitic Cr-Ni steels, the corrosion resistance of the Cr-Mn and Cr-Mn-Ni steels reaches a maximum when the steels are annealed by rapidly cooling them from the range of 1050° to 1100° C (1900° to 2000° F).

MECHANICAL PROPERTIES

Typical mechanical properties are submitted in Table 1. These data are representative of the steels in the annealed condition, and reveal that they are comparable in strength, ductility, and toughness to the austenitic Cr-Ni steels.

Inasmuch as Cr-Mn and Cr-Mn-Ni steels are largely austenitic in character, their strength can be greatly increased by application of cold work.

GENERAL CORROSION RESISTANCE OF THE ANNEALED STEELS

The corrosion resistance of the Cr-Mn and Cr-Mn-Ni steels is believed to be due to the presence of a passive film that forms on their surfaces during exposure to oxidizing conditions. (See *Passivity*, p. 21.) The Cr-Mn steels are not so corrosion resistant as the 18% Cr–8% Ni steel, but in general they are more corrosion resistant than plain chromium steels of the same chromium content.

The same precautions to be exercised in applying the austenitic Cr-Ni steel to resist corrosion must also be observed with the Cr-Mn and Cr-Mn-Ni steels. Both steels are subject to pitting when exposed to certain chloride solutions, and are subject to contact or crevice corrosion.

* Union Carbide and Carbon Research Laboratories, Inc., Niagara Falls, N. Y.

[7] *Nitrogen in Chromium Alloy Steels*, Electro Metallurgical Company, New York, N. Y., 1941.

CHROMIUM-MANGANESE STAINLESS STEELS

TABLE 1. MECHANICAL PROPERTIES OF ANNEALED Cr-Mn AND Cr-Mn-Ni STEELS

Chromium	12.90%	12.98%	17.86%	18.04%	18.54%
Manganese	16.37	17.06	9.50	7.90	5.18
Nickel	Nil	2.13	Nil	2.06	4.47
Silicon	0.34	0.30	0.41	0.43	0.41
Carbon	0.09	0.08	0.08	0.07	0.08
Copper	....	....	0.70	....	....
Elastic modulus, E $\times$ 10^6, psi	28	28	28	28	28
Yield strength at 0.2% offset, psi	28,000	30,000	50,000	51,500	45,000
Tensile strength, psi	120,500	110,000	103,500	122,500	104,000
% Elongation in 2 in.	58	56	46	49	60
% Reduction in area	55	58	59	63	70
Izod impact, ft-lb	100	100	100	100	100
Brinell Hardness Number	145	148	160	179	159

Note. All 1-in. round bar samples annealed by air cooling from 1040° C (1900° F); 0.505-in. diameter tensile samples tested.

AQUEOUS MEDIA

Like other stainless steels, the Cr-Mn and Cr-Mn-Ni steels possess optimum resistance to oxidizing media, and their resistance to corrosion is primarily influenced by the percentage of chromium present. Corrosion rates of Cr-Mn and Cr-Mn-Ni steels in boiling 65% HNO_3 are given in Table 2.

TABLE 2. RESULTS OF NITRIC ACID TESTS ON THE Cr-Mn AND Cr-Mn-Ni STEELS*

Size of specimen — 2.5 $\times$ 3.8 $\times$ 0.6 cm (1 $\times$ 1½ $\times$ ¼ in.).
Surface preparation — finish-polished with 120 emery cloth.
Velocity of acid — convection by boiling.
Temperature — about 165° C (330° F).
Duration of test — three 48-hr periods. Acid changed after each period.

Composition, %					Corrosion Rate, ipy, in Boiling 65% HNO_3		
Cr	Mn	Ni	Si	C	Cu	First 48-hr Period	Second 48-hr Period	Third 48-hr Period
12.90	16.37	Nil	0.34	0.09	....	0.30	0.28	0.29
17.86	9.50	Nil	0.41	0.08	0.70	0.036	0.04	0.045
18.04	7.90	2.06	0.43	0.07	....	0.024	0.023	0.026
18.54	5.18	4.47	0.41	0.08	....	0.022	0.018	0.013

*All samples annealed by air cooling from 1040° C (1900° F).

These data illustrate that the steel containing approximately 13% chromium with 16% manganese is not so resistant to oxidizing media as the steel containing about 18% chromium and 9% manganese. Experiments in the spray atmosphere of a 20% sodium chloride solution have shown that steel of higher chromium and lower manganese content is far more resistant to staining. Other experiments under wet atmospheric conditions have revealed that the 13% Cr-16% Mn will rust, whereas the 18% Cr-9% Mn steel will remain bright for a long time. In considering corrosion resistance, the high-manganese steels containing approximately 18% or more chromium are preferred.

Effect of Carbon and Nitrogen Content

The carbon content of the Cr-Mn and Cr-Mn-Ni steels greatly influences their corrosion resistance. When polished samples are exposed to the spray of a 20% salt solution there is a definite tendency for the steels to become stained during a 100-hour exposure test when the carbon content is much above 0.10%, whereas if the carbon content is below 0.10%, the steels will resist the salt-spray atmosphere for longer periods without staining. Continued exposure to the salt-spray atmosphere shows that there is a definite tendency for the steels to become badly pitted when the carbon content is 0.15% or higher. The higher-carbon steels are much more subject to the development of intergranular corrosion than the low-carbon steels. From the standpoint of corrosion resistance it is advisable to maintain a maximum carbon content of 0.12% and preferably 0.10%. The influence of carbon content on the resistance of the alloys to nitric acid is shown by the data of Table 3.

Table 3. EFFECT OF CARBON ON THE CORROSION RESISTANCE OF THE Cr-Mn AND Cr-Mn-Ni STEELS IN NITRIC ACID

Size of specimen — 2.5 × 3.8 × 0.6 cm (1 × 1½ × ¼ in.).
Surface preparation — finish-polished with 120 emery cloth.
Velocity of acid — convection by aeration.
Temperature — about 20° C (70° F).
Duration of test — 12 hr.

Cr	Mn	Ni	C	Corrosion Rate, ipy, in Aerated 20% HNO_3
17.88	8.26	....	0.07	0.013
17.19	8.35	....	0.11	0.029
17.45	8.55	....	0.22	0.037
17.47	8.44	....	0.47	0.048
18.32	5.83	4.50	0.06	Nil*
18.54	5.74	3.84	0.07	0.008
18.37	4.90	4.13	0.09	0.013
18.47	6.02	5.17	0.12	0.027

*No determinable loss in weight using a chemical balance.

When the nitrogen content of the steels is increased, the general resistance to corrosion is not impaired; in fact, the data of Table 4 reveal that the increase in nitrogen content actually improves resistance to boiling nitric acid.

Fresh and Salt Waters

The annealed Cr-Mn and Cr-Mn-Ni steels are resistant to natural fresh water, boiler waters, steam, and condensates from boiler waters. This information has been obtained not from application of the steels under commercial conditions, but from laboratory tests on annealed samples of the steels.

Exposure of the Cr-Mn and Cr-Mn-Ni steels to sea water causes them to become pitted in the same manner as are the Cr and Cr-Ni stainless steels. Samples 0.16 cm (1/16 in.) thick exposed for 700 days to sea water at Fort Tilden, Long Island, N. Y., were pitted and perforated at the end of the exposure period. Sheet specimens of the same thickness as the Cr-Mn-Ni steels have also been tested in flowing sea water at Wilmington, N. C. Inspection after 529 days showed that the specimens were perforated by pitting.

TABLE 4. EFFECT OF NITROGEN ON THE CORROSION RESISTANCE OF THE Cr-Mn AND Cr-Mn-Ni STEELS* IN NITRIC ACID

Size of specimen — 2.5 × 3.8 × 0.6 cm (1 × 1½ × ¼ in.).
Surface preparation — finish-polished with 120 emery cloth.
Velocity of acid — convection by boiling.
Temperature — about 165° C (330° F).
Duration — three 48-hr periods. Acid changed after each period.

Composition, %						Corrosion Rate, ipy, in Boiling 65% HNO₃		
Cr	Mn	Ni	Cu	C	N	First 48-hr Period	Second 48-hr Period	Third 48-hr Period
17.86	9.50	....	0.70	0.08	0.04	0.036	0.040	0.045
18.45	9.60	....	0.69	0.08	0.10	0.026	0.027	0.031
18.04	7.90	2.06	....	0.07	0.04	0.024	0.023	0.026
18.66	8.44	2.09	....	0.07	0.04	0.030	0.024	0.025
18.54	5.18	4.47	....	0.08	0.04	0.022	0.018	0.013
18.63	5.21	4.51	....	0.07	0.12	0.021	0.014	0.010

*All samples air-cooled from 1040° C (1900° F).

ACIDS, ALKALIES, AND SALTS

Corrosion data obtained for the 1100° C (2000° F) annealed and quenched Cr-Mn-Ni steels are summarized in Table 5.

The data show that the Cr-Mn-Ni steels are resistant to a number of important chemicals but, like other stainless steels, are badly attacked by 85% phosphoric (hot), hydrochloric, and sulfuric acid solutions. Dilute solutions of nitric, lactic, and acetic acids have but little effect on the alloys. In like manner, lemon juice, orange juice, sweet cider, etc., leave the alloy substantially unaffected. Chloride solutions except when boiling tend to cause pitting.

SUSCEPTIBILITY TO INTERGRANULAR CORROSION

The Cr-Mn and Cr-Mn-Ni steels are subject to the development of intergranular corrosion after heating in the range 400° to 815° C (750° to 1500° F). After heating at these temperatures the steels will disintegrate if exposed to a variety of corrosive media. The susceptibility to intergranular attack can be eliminated after exposure in the sensitizing temperature range, and prior to exposure to corrosive media by reannealing the steels at 1000° to 1100° C (1800° to 2000° F), followed by rapid cooling.

Susceptibility to intergranular attack can be eliminated by the addition of columbium or titanium. To gain complete immunity to intergranular attack, the columbium should be present to the extent of at least eight to ten times the carbon content, and the titanium four to six times the carbon content. The physical properties of the Cr-Mn steels, however, are detrimentally affected by the presence of these amounts of columbium and titanium. The steels become more ferritic in nature, and *lose ductility and toughness to a marked degree;* but it appears that five to six times as much columbium as carbon can be added to give a marked improvement without seriously affecting physical properties. On the other hand, the titanium requirement of four to six times the carbon content and the columbium requirement of eight times the carbon content can be employed in the Cr-Mn-Ni steel without seriously affecting either the ductility or toughness. It is preferable that the carbon content of both steels treated with these additions be limited to a maximum of 0.08%.

TABLE 5. EFFECT OF VARIOUS MEDIA ON THE Cr-Mn-Ni STEELS

Size of specimen — 2.5 × 3.8 × 0.6 cm (1 × 1½ × ¼ in.).
Surface preparation — finish-polished with 120 emery cloth.
Velocity of acid — nat. convection.

Chromium, 17.76%
Manganese, 6.09
Nickel, 3.83
Copper, 0.85
Silicon, 0.47
Carbon, 0.07
Nitrogen, 0.04

Corrodent	Temperature	Duration of Test, hr	ipy
2% Nitric acid	Boiling	45	0.0002
10% Nitric acid	Boiling	41	0.0008
10% Hydrochloric acid	Room temp.	18	0.12
10% Sulfuric acid	" "	19	0.066
10% Sulfuric and 2% nitric acids	" "	45	0.0003
10% Acetic acid	Boiling	17	Nil
99% " " (glacial)	"	17	0.16
10% Formic acid	Room temp.	26	Nil
10% " "	Boiling	26	0.36
85% Formic acid	Room temp.	20	0.0008
10% Oxalic acid (1st part of test)	Boiling	19	0.075
10% " " (2nd part of test)	" 2nd run 42 hr	42	0.068
10% Phosphoric acid	Boiling	24	0.0007
85% (syrupy) phosphoric acid	140° C	16	Dissolved
Salt spray,* 20% NaCl	38° C	2320	0.020 (by pitting)
10% Potassium chloride	Boiling	47	Nil
10% Calcium "	"	16	"
10% Ammonium chloride	"	20	Pitted
Bleach (calcium hypochlorite, 24 grams Cl per liter)	Room temp.	65	Pitted
2% Lactic acid	Boiling	46	Nil
Lemon juice	Room temp.	89	0.0002
Orange juice	" "	91	Nil
Sweet cider	" "	42	"
Canned tomatoes	100° C	23	"
Canned rhubarb	100° C	23	"
50% NaOH*	Boiling	24	0.20

* These data, contributed by F. L. LaQue, are for 18% Cr-6% Mn-4% Ni-1% Cu-0.12% C alloy water-quenched from 1075° C (1950° F) and surface prepared with No. 150 emery. Similar data for the 18% Cr-8% Mn-0.06% C alloy show somewhat higher corrosion rates.

ATMOSPHERIC CORROSION

Corrosion experiments extending over a period of 5 to 6 years have been completed on sheet samples of the Cr-Mn and Cr-Mn-Ni steels exposed to the atmosphere on top of an office building in New York City. The samples have been cleaned and inspected on two different occasions. Other tests have been made in a highly humid atmosphere in Cleveland, Ohio. Results of both sets of experiments are given in Table 6.

TABLE 6. RESULTS OF ATMOSPHERIC TESTS, AND HIGHLY HUMID ATMOSPHERIC TESTS ON Cr-Mn AND Cr-Mn-Ni STEELS

Size samples — 10.2 × 20.3 × 0.16 cm (4 × 8 × 1/16 in.).
Surface preparation — finish-polished with 120 emery cloth.

Composition, %					Corrodent	Duration, Months	Location	Corrosion Rate, in./yr	Remarks
Cr	Mn	Ni	Cu	C					
17.86	9.48	....	0.80	0.07	Normal atmosphere	61	New York City	Nil	No permanent staining or pitting
"	"	....	"	"	90–95% Humidity 40° C (108° F)	8*	Cleveland, Ohio	Nil	No permanent staining or pitting
17.76	6.09	3.82	0.85	0.07	Normal atmosphere	61	New York City	Nil	No permanent staining or pitting
"	"	"	"	"	90–95% Humidity 40° C (108° F)	8*	Cleveland, Ohio	Nil	No permanent staining or pitting

* Size of samples in this test was 1½ by 6 by 1/16 in. (3.8 by 15.2 by 0.16 cm).

These tests demonstrate that the Cr-Mn and Cr-Mn-Ni steels have good resistance to atmospheric deterioration. The steels retain a bright, pleasing color, which is not dulled appreciably through years of exposure to the atmosphere.*

GENERAL REFERENCES

ANON., "Germans Develop New Stainless Steel," *Iron Age,* **138**, 61 (Dec. 3, 1936).
BECKET, F. M., "Chromium-Manganese Steels," *Year Book American Iron Steel Institute,* pp. 173–194, 1930.
CLARK, C. L., and A. E. WHITE, "Comparative Physical Properties of Chromium-Nickel, Chromium-Manganese, and Manganese Steels," *Trans. Am. Soc. Mech. Eng.,* FSP-53-14, **53**, 177–182 (1931).
FRANKS, R., W. O. BINDER, and C. M. BROWN, "High Manganese Austenitic Steels," *Iron Age,* **150** (No. 14), 51–57 (1942).
KINZEL, A. B., and R. FRANKS, *Alloys of Iron and Chromium,* Vol. II, *Alloys of Iron Monographs,* Engineering Foundation, McGraw-Hill Book Co., New York, 1940.
KLUKE, R., "Einfluss von Manganzusätzen bis 20% auf Gefüge und Eigenschaften von Gusslegierungen mit 30% Cr" (The Effect of Manganese Additions up to 20% on the Structure and Properties of Casting. Alloys with 30% Chromium), *Arch. Eisenhüttenwesen,* **11**, 615–618 (1937–1938).
LEGAT, H., "Beitrag zur Kenntnis der austenitischen Chrom-Mangan-Stähle" (Austenitic Chromium-Manganese Steels), *Arch. Eisenhüttenwesen,* **11**, 337–341 (1937–1938).
RAPATZ, FRANZ, "The Application Possibilities of Stainless and Heat-Resisting Steels Containing Nitrogen Additions," Report 564 of the Werkstoffausschuss des Vereins deutscher Eisenhüttenleute, *Stahl u. Eisen,* **61**, 1073–1078 (Nov. 27, 1941).
THUM, E. E., *The Book of Stainless Steels,* 2nd Ed., American Society for Metals, 1935.

* Cautions relating to periodic cleaning of the Cr-Ni stainless steels to prevent contact corrosion and rusting also apply to these alloys. See pp. 164, 173.

NICKEL-IRON ALLOYS

F. L. LaQue*

Although the addition of nickel effects a considerable improvement in the resistance of iron to corrosion, the principal applications of these alloys are not dependent on their corrosion-resisting characteristics. They are designed primarily to secure some special physical property related to thermal expansion, magnetic or elastic characteristics. As a supplement to such properties, the extra corrosion resistance afforded by the nickel content is frequently of practical advantage.

Exceptions to the foregoing are the austenitic nickel-cast irons† which were developed primarily for corrosion-resisting applications. They will be discussed separately. Another exception is provided by the recent discovery[1] that Ni-Fe alloys containing 50 to 80% nickel are exceptionally resistant to attack by hydrofluoric acid as it is encountered in the alkylation of petroleum hydrocarbons. A useful property of the Ni-Fe alloys in this application is their avoidance of tar formation.

GENERAL CORROSION CHARACTERISTICS

ACIDS AND OTHER MEDIA

Results of several investigations of the resistance of Ni-Fe alloys to attack by various acids were summarized by Marsh,[2] to whose work the reader is referred for a more detailed discussion of this subject. Although investigators used different concentrations of acids under different conditions of test, the combined results demonstrated a fairly consistent effect of nickel on resistance to attack. There is a gradual decrease in corrosion as the nickel content increases to about 30 to 40%, with relatively little further decrease in corrosion with larger amounts of nickel. The same effect of nickel content was demonstrated by tests in 5% sulfuric acid at 25° C (77° F) as shown in Fig. 1 and at the boiling point.[3] (See also Fig. 5, p. 30.)

This effect of nickel content on corrosion should be considered in relation to the Electron Configuration Theory of Passivity (p. 24). This theory suggests that there should be a considerable improvement in corrosion resistance when the nickel content of the alloy is increased to 34%.

In Table 1 are additional data[4] on corrosion of some common Ni-Fe alloys by sulfuric acid. The resistance of Ni-Fe alloys to attack by sulfuric acid may be improved further by additions of copper and silicon.[5]

Nickel improves the resistance of iron to corrosion by *salt solutions* as indicated by the data in Table 2.[6] Corrosion tests in 16% *calcium chloride* solution are summarized in Table 3. Data on the behavior of the 36% nickel alloy‡ in *sea water* are found in Table 13, p. 409.

---

* Development and Research Division, The International Nickel Co., New York, N. Y.
† Ni-Resists.
‡ Invar.

[1] M. P. Matuszak, U. S. Patent 2,360,436 (Oct. 17, 1944).
[2] J. S. Marsh, *The Alloys of Iron and Nickel*, p. 503, Vol. 1, McGraw-Hill Book Co., Inc., New York, 1938.
[3] R. Landau and C. M. Oldach, "Corrosion of Binary Alloys," *Trans. Electrochem. Soc.*, **81**, 521 (1942).
[4] C. G. Fink and C. M. DeCroly, "The Corrosion Rate of Ferro-Nickel Alloys," *Trans. Electrochem. Soc.*, **56**, 239–277 (1929).
[5] J. L. Miller, *An Investigation of Certain Corrosion-Resistant Steels*. Part I: *Some Alloys for Use in Contact with Sulfuric Acid*, Vol. 21, pp. 111–127, Iron and Steel Institute, Carnegie Scholarship Memoirs, 1932.
[6] C. G. Fink and C. M. DeCroly, *loc. cit.*

TABLE 1. BEHAVIOR OF Ni-Fe ALLOYS IN SULFURIC ACID

Alternate Immersion at Atmospheric Temperature for 168 Hours

Acid Concentration, % by Weight	Corrosion Rates, ipy		
	35–38% Nickel*	48% Nickel	79% Nickel†
5	0.25	0.16	0.066
25	0.16	0.15	0.20
75	0.03	0.016	0.03
96	0.09	0.009	0.07

* Invar.   † Permalloy.

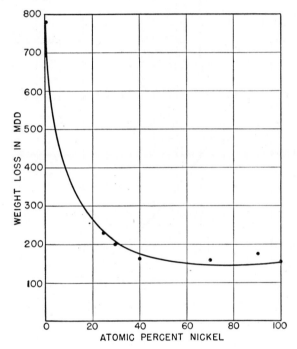

FIG. 1. Effect of Nickel Content on Corrosion of Fe-Ni Alloys by Aerated 5% Sulfuric Acid at 25° C. (Landau and Oldach.)

TABLE 2. RESISTANCE OF Ni-Fe ALLOYS TO CORROSION BY 5% SODIUM CHLORIDE

(Under Conditions of Alternate Immersion at Atmospheric Temperature)

Nickel Content of Alloy, %	Corrosion Rate, ipy
0	0.078
38*	0.005
48	0.0000
80†	0.0004
100	0.0008

* Invar.   † Permalloy.

TABLE 3. RESULTS OF TESTS OF HIGH–NICKEL CONTENT IRON ALLOYS IN AERATED AND AGITATED 16% CaCl$_2$ SOLUTION

At Atmospheric Temperature for 100 Days

Nickel Content of Alloy, %	Corrosion Rate, ipy	Maximum Depth of Pitting, inch
80	0.0006	0.003
90	0.0004	0.004
95	0.0004	0.004
100	0.0007	0.002

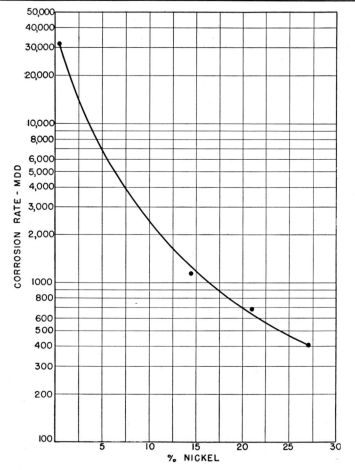

FIG. 2. Corrosion of Nickel Cast Irons in Aerated 5% Sulfuric Acid at Room Temperature.

Nickel-Cast Irons. There has been a considerable practical application of austenitic cast irons containing 20 to 30% nickel, or about 14% nickel and 6% copper plus a small percentage of chromium in each case, along with the usual constituents of cast iron. The general effect of nickel on the resistance of such irons to attack by acids is

TABLE 4. COMPARATIVE DATA ON PLAIN CAST IRON AND AUSTENITIC NICKEL–CAST IRON* IN SEVERAL MEDIA

Corrosive Medium	Concentration	Temperature °C	Temperature °F	Aeration	Corrosion Rate, ipy Cast Iron	Corrosion Rate, ipy Austenitic Nickel-Cast Iron
Acetic acid	5%	16	60	Some	0.68	0.04
Acetic acid	50%	..	Boiling	Some	2.66	0.12
Acetic acid	Concentrated	16	60	Some	0.08	0.02
Ammonium hydroxide	25%	16	60	...	0.0000	0.00002
Ammonium chloride	5%	16	60	Some	0.05	0.003
Ammonium chloride	5%	93	200	Some	0.19	0.006
Ammonium citrate	30%	..	Atmos.	...	22	0.06
Ammonium sulfate	10%	16	60	Some	0.03	0.004
Calcium chloride	5%	16	60	Some	0.009	0.005
Carbon dioxide	Saturated	16	60	Some	0.03	0.001
Citric acid	5%	16	60	Some	0.59	0.09
Fatty acids	....	227	440	...	0.79	0.02
Grapefruit juice	....	30	86	Yes	0.35	0.06
Hydrochloric acid	5% (by vol.)	16	60	Some	1.07	0.01
Hydrochloric acid	50% (by vol.)	16	60	Some	1.24	0.04
Hydrochloric acid	Concentrated	16	60	Some	1.11	0.37
Hydrochloric acid	5%	54	130	Yes	5.33	0.099
Hydrofluoric acid	10%	16	60	...	0.22	0.001
Hydrofluosilicic acid	22%	63	145	None	5.5	0.005
Hydrogen sulfide (moist)	....	93	200	...	0.07	0.01
Lactic acid	10 to 25%	54	130	...	1.77	0.67
Monoethanolamine	....	80–105	180–220	...	0.008	0.004
Nitric acid	5% (by vol.)	16	60	Some	1.04	0.83
Parkerizing solution	....	100	210	None	0.18	0.05
Petroleum vapors	....	370	700	...	0.67	0.13
Phosphoric acid	5%	30	86	None	4.85	0.08
Phosphoric acid	25%	30	86	None	3.79	0.04
Phosphoric acid	5%	88	190	None	34.7	0.28
Phosphoric acid	25%	88	190	None	65.6	0.03
Pineapple juice	....	87	188	Alt. Immer.	0.79	0.07
Rosin vapors	....	370	700	Some	0.58	0.01
Sodium bromide	22°–47° Bé	..	Boiling	...	0.03	0.003
Sodium chloride	Saturated	93	200	...	0.07	0.005
Sodium hydroxide	50–70%	120	250	...	0.29	0.09
Sodium hydroxide	75%	135	275	...	0.07	0.004
Sodium hydroxide	100%	670	1240	...	0.84	0.26
Sulfate black liquor	....	175	344	...	0.34	0.03
Sulfur (molten)	....	127	260	...	0.02	0.02
Sulfur (molten)	....	260	500	...	0.03	0.03
Sulfur (molten)	....	446	835	...	0.40	0.59
Sulfur chloride	98%	125	257	...	0.16	0.001

198 CORROSION IN LIQUID MEDIA, ATMOSPHERE, GASES

TABLE 4 (Continued)

Corrosive Medium	Concentration	Temperature		Aeration	Corrosion Rate, ipy	
		° C	° F		Cast Iron	Austenitic Nickel-Cast Iron
Sulfuric acid	5%	..	Atmos.	None	4.42	0.02
Sulfuric acid	5%	90	194	None	...	0.38
Sulfuric acid	30%	27	80	None	...	0.05
Sulfuric acid	30%	90	194	None	...	1.29
Sulfuric acid	80%	30	86	None	...	0.02
Sulfuric acid	80%	90	194	None	...	0.25
Sulfuric acid sludge	25%	60	140	...	Destroyed	0.008
Sulfuric acid and	⎰10%	32	90	Some	1.54	0.62
copper sulfate	⎱ 2%	32	90	Some	...	...
Sulfite liquor	....	21	69	None	0.91	0.87
Tomato juice	....	52	125	...	0.11	0.05
Whiskey slop	....	..	Boiling	None	0.05	0.01
Whiskey mash	....	..	Hot	...	0.11	0.02
Zinc (molten)	....	510	950	...	42.7	23.9
Zinc chloride	30-70%	..	Boiling	...	0.64	0.08

* Ni-Resist.

shown by the data of Fig. 2 for sulfuric acid.[7] The irons containing 3 to 3.5% total carbon and about 2% silicon.

The improvement due to the presence of nickel results principally from its effect on the corrodibility of the austenitic matrix. In addition, the high-nickel content cast irons possess an advantage over unalloyed irons based on the greater nobility of the alloyed matrix and the resultant decrease in galvanic action between the graphite and the matrix[8,9] which tends to reduce or avoid "graphitic" corrosion. The greater nobility of the austenitic-cast irons is also of value in reducing galvanic corrosion in combinations with copper and nickel alloys and with stainless steels as in pumps, valves, and other chemical processing equipment.

In Table 4 have been assembled some data[10] illustrating the relative performance of plain cast iron and high-nickel content cast irons in several corrosive media.

ALKALIES

Ni-Fe alloys are highly resistant to attack by alkalies. Improvement in corrosion resistance is roughly in proportion to nickel content up to about 30%, at which level corrosion becomes practically negligible under many conditions of exposure. The commercial applications of Ni-Fe alloys have been principally in the form of nickel-cast irons, the performance of which is illustrated by the data in Table 5.[10]

BEHAVIOR OF WELDS IN NICKEL-CLAD STEEL

In welding the nickel side of nickel-clad steel with nickel welding rod there is some dilution of the weld metal with iron. Depending on the thickness of cladding and the

[7] Unpublished data, International Nickel Co., Inc.
[8] W. A. Wesley, H. R. Copson, and F. L. LaQue, "Some Consequences of Graphitic Corrosion of Cast Iron," *Metals and Alloys*, **7**, No. 12, p. 325 (December, 1936).
[9] H. L. Maxwell, "Cast Iron in Chemical Equipment," *Mech. Eng.*, **58**, No. 12, p. 803 (December, 1936).
[10] Unpublished data, International Nickel Co., Inc.

NICKEL–IRON ALLOYS

TABLE 5. RESULTS OF TESTS OF NICKEL–CAST IRONS IN CAUSTIC SODA

While Being Concentrated 50 to 65% under 26-in. Vacuum

Nickel Content of Cast Iron, %	Corrosion Rate, ipy
0	0.091
5	0.049
15	0.030
20	0.006
30	0.0004

welding technique, the iron content of the welds will be between 5 and 20%. Consequently, there has been some interest in examining the corrosion resistance of Ni-Fe alloys of high nickel content in media in which nickel-clad steel is likely to be used.

Practical experience during the past 15 years has demonstrated that, with properly made welds, the resulting Ni-Fe alloys resist corrosion as well as the nickel cladding in the many established applications of nickel-clad steel. The results of tests covered by Tables 1, 2, 3, 6, and 7 bear on this subject.

TABLE 6. RESULTS OF TESTS OF Ni-Fe ALLOYS IN CORROSIVE WELL WATER CONTAINING CONSIDERABLE FREE CARBON DIOXIDE

At 80° C (180° F) for 204 Days

Nickel Content of Alloy, %	Corrosion Rate, ipy	Maximum Depth of Pitting, inch
70	0.006	Perforated (0.050)
80	0.004	0.022
90	0.004	0.031
100	0.001	0.000

TABLE 7. RESULTS OF TESTS OF Ni-Fe ALLOYS IN SODIUM HYDROXIDE SOLUTIONS

Iron Alloys Were Coupled with Pure Nickel in Each Case.
Area of Nickel Ten Times Area of Alloy.

Alloy	Corrosion Rates, ipy		
	75% NaOH at 128° C (262° F)	50% NaOH at 75° C (167° F)	23% NaOH at 105° C (221° F)
Nickel	0.0015	0.0008	0.0004
95% Ni, 5% Fe	0.0010	0.0024	0.0016
90% Ni, 10% Fe	0.0016	0.0019	0.0027
80% Ni, 20% Fe	0.0019	0.0016	0.0036
70% Ni, 30% Fe	0.012	0.008	0.0016
60% Ni, 40% Fe	0.013	0.008	0.0016

STRESS CORROSION CRACKING

There is evidence that some of the low Ni-Fe alloys are subject to stress corrosion cracking in some media. In tests by Schmidt and Wetternick,[11] where stressed specimens were subjected to bend tests after immersion for 2 months at room temperature in 0.4% hydrochloric or 16.4% sulfuric acid, it was shown that alloys with less than

[11] M. Schmidt and L. Wetternick, *Korrosion u. Metallschutz*, **13**, 184–189(1937).

35% nickel would crack. Resistance was excellent above 35% nickel. Houdremont[12] observed cracking of 25% nickel steel when tested for use as a turbine blade material. Thorough investigation did not disclose the cause of the cracks, but they appeared, as in the case of brass, with simultaneous action of corrosion and stress. The cracks resembled stress cracks in that they were predominantly intracrystalline.

ATMOSPHERIC CORROSION

Although the Ni-Fe alloys in the atmosphere do not retain their original appearance as well as the stainless steels, they are not likely to be destroyed by corrosion in even the most corrosive atmospheres. They acquire rusty coatings which vary in thickness, color, and adherence and vary with respect to the extent of attack beneath the rust. The relationship between nickel content and the character of the rust is indicated by Table 8.[13]

TABLE 8. CHARACTER OF RUSTS FORMED ON Ni-Fe ALLOYS EXPOSED TO THE ATMOSPHERE

Nickel Content, %	Character of Rust
0	Adherent, thick, very coarse
20	Adherent, thick, granular
30	Loose, thick, granular
40	Loose, thick, granular
50	Loose, rather thin, finely granular
60	Loose, smooth, thin

The loose rust that forms on alloys that contain 50 and 60% nickel can be brushed off readily to reveal a bright, metallic surface.

Surveyor's tapes made from low-expansion alloys that contain 35 to 42% nickel are much more resistant to corrosion than ordinary steel tapes and are, therefore, easier to keep in serviceable condition.

Some quantitative data from atmospheric corrosion tests of the 36% nickel composition are given in Table 9.

TABLE 9. RESULTS OF ATMOSPHERIC CORROSION TESTS ON WROUGHT 36% Ni-Fe ALLOY*

Location of Test	Duration of Exposure, years	Corrosion Rate, ipy	Maximum Depth of Pitting, in.
Bayonne, N. J.	3	0.0007†	.....
Eastport, Me.	0.75	0.00009	.....
Kure Beach, N. C.	0.5	0.0007	.....
Halifax, N. S.‡ (Just above high tide)	15	0.00005	0.004
Aukland, N. Z.‡ (Just above high tide)	15	0.00005	0.000
Plymouth, England‡ (Just above high tide)	15	0.00014	0.007
Colombo, Ceylon‡ (Just above high tide)	15	0.00014	0.000

*Invar.
†Specimens 0.031 in. thick remained in good condition with no significant pitting after exposure to the industrial atmosphere at Bayonne, N. J., for over 10 years.
‡*Deterioration of Structures in Sea Water*, 18th Report of the Committee of the Institution of Civil Engineers, London, 1938.

[12] E. Houdremont, *Sonderstahlkunde*, p. 170, Berlin, 1935.
[13] J. S. Marsh, *The Alloys of Iron and Nickel*, p. 503, Vol. 1, McGraw-Hill Book Co., Inc., New York, 1938

SILICON-IRON ALLOYS

W. T. BRYAN*

The high Si–Fe alloys, of high intrinsic corrosion resistance, can be produced only as castings. Prior to World War I the use of these alloys in corrosion-resistant equipment was confined to simple shapes as anodes, pipe, and towers; but in recent years applications have been made in more complicated designs. High hardness and brittleness, which are barriers to machining and fabrication by ordinary means, have been overcome by improvement in grinding equipment and novelty of engineering design.

COMPOSITION

Experience and research have demonstrated that a composition of 14 to 15% silicon results in maximum corrosion resistance, and no significant benefits are derived by increasing the silicon beyond this amount. Investigations have been directed toward the improvement of mechanical properties by lowering the silicon content, but the marked decrease of corrosion resistance, as illustrated in Figs. 1 and 2, without improvement of mechanical properties, has discouraged further effort. Therefore, the nominal composition of this alloy as 14.5% silicon, 0.85% carbon, 0.65% manganese, and remainder iron,† is one of long standing. A modification of the standard 14.5% Si–Fe alloy is made by the addition of 3% molybdenum,‡ the latter being specifically resistant to hot hydrochloric acid.

TABLE 1. PHYSICAL CHARACTERISTICS OF 14.5% Si-Fe ALLOY

Density {lb/cu in. / grams/cc}	{0.252 / 7.0}
Melting point	1260° C (2300° F)
Electrical resistivity, Microhm-cm, 0° C	63
Hardness, Rockwell C	52
Relative thermal conductivity (Ag = 1), 0° C	0.125
Tensile strength	17,000
Coefficient of linear expansion:	

Degrees C	
20–100	12.22×10^{-6}
20–200	12.95×10^{-6}
20–300	13.75×10^{-6}
20–400	14.60×10^{-6}
20–500	15.45×10^{-6}
20–600	16.22×10^{-6}
20–700	16.75×10^{-6}

CORROSION RESISTANCE OF 14.5% SILICON-IRON ALLOY

The 14.5% Si–Fe alloy satisfactorily resists sulfuric, nitric, acetic, formic, lactic, and many other commercial acids at any concentration or temperature. The silica film believed to be responsible for corrosion resistance does not depend upon oxidizing, neutral, or reducing conditions; therefore, the type of corrosive agent, from this standpoint, need not be considered. Sea water, mine waters, organic, and most inorganic acids are handled with no difficulty. The alloy, at present, is used for anodes in the electroplating of nickel, copper, chromium, and cadmium; in electrowinning of

* The Duriron Co., Inc., Dayton, Ohio.
† Duriron. (Some other trade names are Tantiron and Corrosiron. EDITOR.)
‡ Durichlor, Patent 1,972,103.

metals such as copper and cadmium; and in the electrolytic pickling of metals employing 10% H_2SO_4 in combination with tin or lead salts.*

However, the 14.5% Si–Fe alloy is not resistant to the halogens, fused alkalies, hydrofluoric acid, crude phosphoric acid containing HF, concentrated hydrofluosilicic acid, sulfurous acid, boiling hydrochloric acid, and aqua regia.

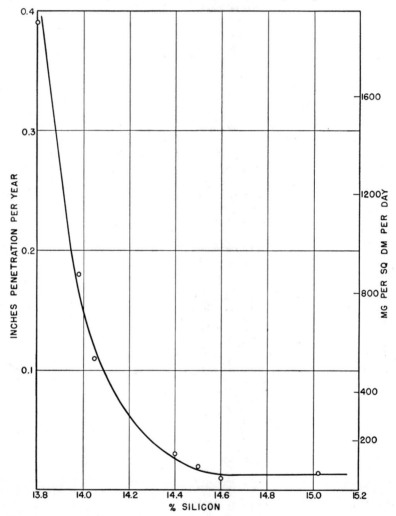

Fig. 1. Effect of Alloyed Silicon on Corrosion Resistance of Iron in 10% H_2SO_4 at 80° C (175° F).

TEMPERATURE

The maximum temperature in many cases is limited only by design of the equipment. Successful installations of 20-in. diameter pipe have been made for acid

* Bullard-Dunn Process.

concentrators handling boiling 95% H_2SO_4 at approximately 285° C (545° F), where products of combustion attain temperatures of the order of 590° C (1100° F). The falling-film and Mantius types of H_2SO_4 concentrators handling 60 to 95% H_2SO_4 at approximately 200° C (400° F) are made of the 14.5% Si-Fe alloy.

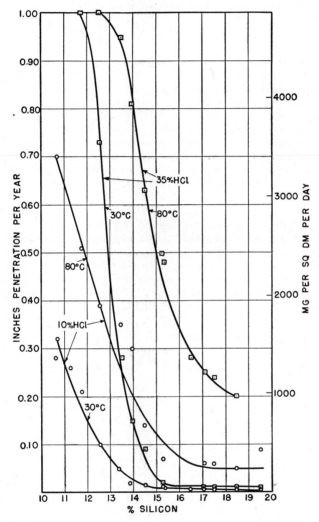

FIG. 2. Effect of Silicon in High-Silicon Iron vs. Hydrochloric Acid (100-Hour Test) 30° C (85° F) and 80° C (175° F).

The limitation of high Si-Fe alloys to applications at elevated temperatures is the rate at which heat is applied or removed. They have relatively low thermal conductivity and are susceptible to cracking from thermal shock. However, satisfactory performance of process equipment is obtained by exercising proper precautions.

pH AND ALKALIES

The 14.5% Si–Fe alloy is not restricted in use to a narrow pH range. Concentrated caustics are in general more corrosive than acids. The alloy will withstand 20% NaOH at boiling and 50% NaOH at 80° C (175° F) with a corrosion rate not exceeding 240 mdd (0.05 ipy). For more severe operating conditions, other alloys have proved more suitable.

DISSOLVED SALTS

The 14.5% Si–Fe alloy resists the action of numerous salts under a variety of conditions. For example, the alloy satisfactorily resists bright nickel plating solution (pH 4.3) under conditions of aeration and mechanical agitation. The corrosion rate at 100° C (212° F) for a test of 250 hours duration was 7 mdd (0.0015 ipy) without occurrence of pitting.

Corrosion, however, occurs in contact with those salts which in specific acid media tend to break down into chemical forms previously outlined as being corrosive. These salts are ammonium fluoride, chlorhydrin, cupric chloride, ferric chloride, mercuric chloride, sodium hypochlorite, sodium sulfite, stannic chloride — or, in general, salts which readily form acids of the halogen family or sulfurous acid.

GASES

The halogen gases are generally corrosive to the 14.5% Si–Fe alloy and this is particularly true of bromine and iodine. In unit operations involving these chemicals, care is usually taken to convert to non-corrosive salts until the final stage. Dry chlorine, hydrogen chloride, fluorine, and sulfur dioxide are not corrosive, but the presence of moisture, and particularly steam, results in rapid attack. Hydrogen sulfide and ammonia have no effect.

VELOCITY

The 14.5% Si–Fe alloy satisfactorily resists erosion. Brinell hardness of approximately 500 makes it suitable for handling corrosive liquids with suspended solids where abrasion resistance is the primary consideration.

GALVANIC COUPLING*

W. D. Richardson[1] (1921) pointed out that 15% Si–Fe alloy is generally anodic to nickel, copper, and carbon, and cathodic to zinc and aluminum. However, in the case of aluminum, the Si–Fe alloy in sulfuric acid is at first cathodic, with the potential gradually dropping to zero and remaining there. The same holds in dilute nitric acid, except that the steady state is reached more quickly. In dilute hydrochloric acid or sodium chloride solutions, the alloy is cathodic to aluminum.

Coupled with iron in sodium chloride or sodium sulfate solutions, the Si–Fe alloy, he stated, acts as the cathode, and the potential decreases with time. In dilute sulfuric acid, the alloy is first strongly cathodic but in time becomes anodic to iron.

METALLURGICAL FACTORS

Numerous efforts have been made to alloy high silicon-iron by the addition of almost all the more common elements, with particular emphasis on nickel, copper,

---

* This summary was prepared by the Editor.

[1] *Trans. Electrochem. Soc.*, **39**, 61 (1921).

molybdenum, tungsten, vanadium, and columbium, but none has developed any marked improvement in mechanical properties. It is reported that the present British practice is to use an alloying addition of 1 to 2% copper, and in Russia about the same amount of nickel is used. Neither of these elements contributes significantly to the corrosion resistance of high silicon-iron, and they are believed to be added primarily to control soundness of the casting. In American practice, this end is accomplished by chemical and metallurgical control.

The most significant discovery relating to the addition of other elements was the profound effect of 3% molybdenum on the resistance to hot hydrochloric acid and various chlorides.

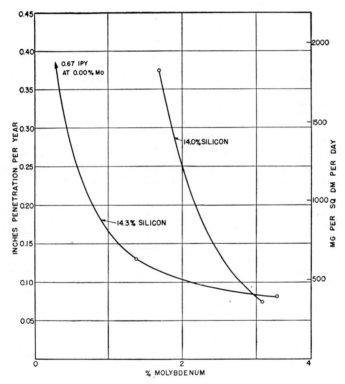

FIG. 3. Effect of Molybdenum in High-Silicon Iron vs. 30% HCl at 65° C (150° F), 64 Hours — Air-Agitated — Four Intermittent Weighings. (Data taken from final 16-hour period.)

CORROSION RESISTANCE OF 14.5% SILICON PLUS 3% MOLYBDENUM-IRON ALLOY

The characteristic resistance of the 3% molybdenum-bearing alloy to chlorides and hot hydrochloric acid is developed after formation on the surface of a molybdenum oxychloride passive film. Therefore, in corrosion evaluation work, it is usually desirable to discard test results from the first two periods consisting of 24 to 48 hours, until passivation has been reached.

The effect of molybdenum in reinforcing resistance of high silicon-iron to hot concentrated hydrochloric acid is shown in Fig. 3. Both silicon and molybdenum contents are critical, and for maximum corrosion resistance should total not less than 17.5%. This is particularly true in 30% HCl at temperatures of the order of 80° C (175° F), as demonstrated in Table 2.

TABLE 2. VARIABLE SILICON — 3% Mo–Fe ALLOY VS. 30% HCl

Volume of solution — 600 ml.
Size of specimen — 3.8 × 1.3 × 1.3 cm (1.5 × 0.5 × 0.5 in.).
Surface preparation — ground.
Aeration —static test.
Duration of tests — 64 hours (four intermittent weighings).
Data taken from final 16-hour period.

Composition		30° C		65° C		80° C	
% Si	% Mo	mdd	ipy	mdd	ipy	mdd	ipy
14.6	2.91	63	0.013	107	0.022	269	0.055
14.2	2.99	83	0.017	127	0.026	531	0.109
13.7	3.01	73	0.015	156	0.032	565	0.116
13.6	2.94	83	0.017	273	0.056	638	0.131
13.5	2.96	141	0.029	229	0.047	716	0.147
13.4	2.93	131	0.027	463	0.095	1102	0.226
12.6	2.83	112	0.023	...	.....	1315	0.270

The order of resistance to hydrochloric acid at various concentrations at 80° C (175° F), air-agitated, in six 16-hour test periods, is shown in Fig. 4.

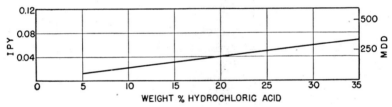

FIG. 4. Corrosion of 14.5% Si–Fe Plus 3% Mo in Hydrochloric Acid. 80° C (175° F)— Air-Agitated — 64 Hours — Four Intermittent Weighings. (Data are from final 16-hour period.)

A few chemicals handled satisfactorily either as dissolved salts or in the presence of concentrated acids by the 14.5% Si plus 3% Mo–Fe alloy up to 80° C (175° F) are the chlorides of aluminum, ammonium, calcium, lead, magnesium, nickel, potassium, sodium, tin, and zinc.

Data of Table 3 illustrate the accelerating effect of ferric chloride on corrosion in a solution of 35% HCl plus 1% H_2SO_4 at various temperatures.

The molybdenum oxychloride film, thought to be responsible for corrosion resistance, is apparently soluble in hot dilute HNO_3, hot concentrated H_2SO_4 and NH_4OH, and is volatile at high temperatures in the order of 640° C (1200° F). The molybdenum alloy is *not* recommended for use with acid salts, such as ferric or cupric chlorides, in the presence of concentrated acids, with mixtures of chlorine gas and steam; or applications where strong H_2SO_4 or weak HNO_3 are alternated with HCl, as these conditions break down the acid-resistant film resulting in a high rate of corrosion.

TABLE 3. 14.5% Si PLUS 3% Mo–Fe ALLOY VS. FeCl₃ IN HCl

Solution — 35% HCl + 1% H₂SO₄ + Variable FeCl₃·6H₂O.
Volume of solution — 400 ml.
Size of specimen — 5 × 5 × 0.64 cm (2 × 2 × 0.25 in.).
Surface preparation — ground.
Aeration — air-agitated.
Duration of test — 100 hours (six intermittent weighings).
Data taken from final test period.

% FeCl₃·6H₂O	21° C (70° F)		38° C (100° F)		60° C (140° F)	
	mdd	ipy	mdd	ipy	mdd	ipy
0.00	70	0.015	173	0.037	458	0.098
0.03	159	0.034	214	0.046	485	0.104
0.09	210	0.045	505	0.108	1100	0.235
0.14	294	0.063	822	0.176	903	0.193
0.50	586	0.125	1430	0.307	3630	0.780

PRACTICAL APPLICATIONS

Centrifugal pumps and valves for handling heavy chemicals are the most common applications of the Si-Fe alloys. Specific applications, in addition to those already mentioned, are sulfuric acid concentrators, dehydrating and denitrating towers and equipment handling nitrating solutions for explosives at all stages of manufacture; steam jets and heating coils in the pickling of steel; reflux condenser tubes in the manufacture of nitric acid; ejectors for mixing chlorine, alum, and bleach in paper mills; absorption towers and packing in the manufacture of acetic acid; mixing nozzles handling sulfuric acid in the petroleum industry; pumps and valves handling dyes and pigments preparatory to filter press operations where alloys must withstand abrasive as well as corrosive solutions; and numerous items such as exhaust fans, sinks, tank outlets, spray nozzles, agitators, thermometer wells, and evaporating dishes.

LEAD AND LEAD ALLOYS*

G. O. HIERS[†]

Lead or lead alloys in commercially available pipe sizes are strong enough to use without reinforcement for 100 to 200 psi internal pressure at 25° C (75° F), and up to 50 psi steam pressure with associated temperatures of 150° C (300° F). Commercial lead-covered copper pipe may be used with steam up to 150 psi.

Small evaporators and stills may be constructed entirely of lead, but, for most chemical equipment, sheet lead is supported or reinforced. Figure 1 shows permissible fiber stresses that may be used with commercial lead piping.[1] Similar long-time tensile stresses may be used in designing equipment employing commercial sheet or other forms of lead. When greater stresses are required, "homogeneous" lead-lined (defined below) or lead-covered equipment is employed. It is a good practice to provide for maximum dynamic stresses not to exceed half the maximum permissible static stresses.

* The reader should also consult *Corrosion of Lead and Lead-Alloy Cable Sheathing*, p. 609.
† Research Laboratories, National Lead Co., Brooklyn, N. Y.

[1] *Chem. and Met. Eng.*, **45** (No. 1), p. 635 (November, 1938).

Frequently sheet lead linings are placed inside steel tanks and strapped thereto with lead-covered quarter round steel straps. Straps should be spaced sufficiently closely so as to minimize buckling of the sheet during heating and cooling since lead has a higher coefficient of expansion than iron.[2,3] Acid brick linings are frequently used inside lead linings to minimize buckling as well as velocity and erosion troubles.

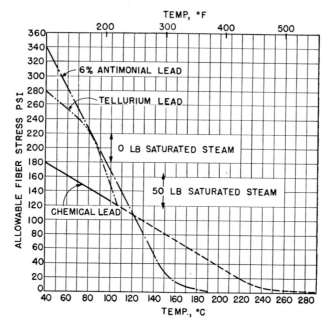

FIG. 1. Maximum Allowable Fiber Stress in Lead at Various Temperatures.

These values are applicable to the use of lead pipe and sheet, adherence to which avoids creep and probably fatigue failures. At 120 psi saturated steam (110° C) and beyond, values for both chemical and tellurium lead coincide.

Homogeneous products are substantial layers of lead, about 0.18 to 0.25 in. (0.45 to 0.65 cm) thick, bonded by fusion to steel or copper. Homogeneous linings or coverings are recommended when good heat transfer through the walls is desired at elevated temperatures.

VARIETIES OF COMMERCIAL LEAD

The specifications of several commercial varieties of lead are given in Table 1.

Corroding Lead

This lead is markedly resistant to general corrosive environments approaching chemical and chemical-tellurium lead in such respect. However, its principal use is the making of white lead (an important white paint pigment) by corrosion. In the Dutch manufacturing process the corrosive agents of lead are humid air (O_2 and H_2O) at about 80° C (180° F), together with small amounts of acetic acid vapor and carbon

[2] E. Mantius and H. F. Freiherr, *Ind. End. Chem.*, **29**, 373 (April, 1937).
[3] G. O. Hiers, *Mech. Eng.* (A. S. M. E.), **58**, 793-798 (December, 1936).

dioxide. With necessary control of temperature and ratio of the corroding agents a permeable corrosion product of white lead, $2PbCO_3 \cdot Pb(OH)_2$, is formed continuously. In this process corrosion takes place for about 90 days at the average rate of about

TABLE 1. COMMERCIAL LEAD SPECIFICATIONS
Per Cent

	Corroding Lead*	Chemical Lead*	Tellurium-Chemical Lead	Common Desilverized Lead A*
Silver, max.	0.0015	0.020	0.020	0.002
Silver, min.	....	0.002	0.002	......
Copper, max.	0.0015	0.08	0.08	0.0025
Copper, min.	....	0.04	0.04	......
Ag + Cu, max.	0.0025	......	......	......
Arsenic, max.	0.0015	......	......	......
Sb + Sn, max.	0.0095	......	......	......
As + Sb + Sn, max.	......	0.002	0.002	0.015
Zinc, max.	0.0015	0.001	0.001	0.002
Iron, max.	0.002	0.002	0.002	0.002
Bismuth, max.	0.05	0.005	0.005	0.15
Tellurium	......	......	0.035–0.050	......
Lead (by difference)	99.94	99.90	99.85	99.85

* A. S. T. M. Designation: B29–43 for pig lead.

0.3 ipy. The mechanism of the corrosive action is explained in *Corrosion of Lead and Lead-Alloy Cable Sheathing*, p. 612.

LEAD FOR CORROSION-RESISTANT CHEMICAL EQUIPMENT

The principal varieties of lead used for corrosion-resistant chemical equipment in this country are chemical lead, chemical-tellurium lead, and hard lead (6% Sb, 94% Pb).

In England, *chemical lead* is a term referring to the purest lead. In this country, it is a special variety, containing copper, that has been found by long experience to possess superior mechanical strength, a high degree of corrosion resistance, and superiority in grain growth inhibition compared with pure lead (approx. 99.995% Pb). The addition of tellurium to chemical lead further inhibits grain growth and also imparts a greater resistance to fatigue.

The general utility of lead for corrosion resistance in a wide variety of environments appears to be the result of formation of self-healing protective coatings of insoluble lead compounds.

COPPER LEAD AND COMMON LEAD

These varieties are made for water pipe, waste pipe, cable sheathing, sheet lead, and for various applications outside of the chemical industry.

GENERAL CORROSION CHARACTERISTICS

NATURAL WATERS — EFFECT OF OXYGEN

In pure distilled water free from dissolved gases lead is not corroded. When oxygen and carbon dioxide are present, the attack is pronounced; hence lead should not be considered a practical material for handling distilled water.

Many potable waters are safely conveyed in lead pipe because they contain sufficient silicate and carbonate to provide a protective coating on the lead. Drinking water containing more than 0.10 ppm of lead is generally considered contaminated and unsafe for consumption. When water is naturally very soft, lime or sodium silicate is often added to reduce the solution of lead. Under the following conditions, drinking water may sufficiently corrode lead so as to require caution in the use of lead equipment:

1. Absence of the above-mentioned protective agents and the presence of aggressive carbon dioxide, or organic acids from peaty soils.
2. Water treatment (softening) for removal of dissolved calcium and magnesium.

Lead pipe heating coils are frequently used to handle steam up to 50-psi gage pressure. Failures due to internal corrosion, despite condensation, have not been encoun-

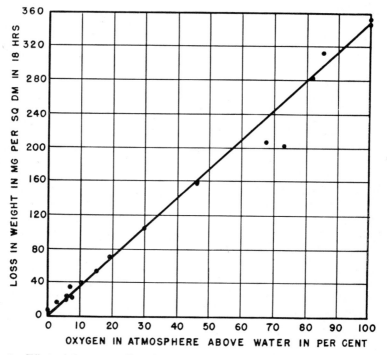

FIG. 2. Effect of Oxygen on Corrosion of Lead Submerged in Distilled Water at 25° C. (R. M. Burns.)

tered probably because of the relative absence of oxygen. Pure water seriously corrodes lead in the presence of oxygen (Fig. 2).[4]

Natural fresh waters and sea water have only slight action on lead. Lead is used therefore for aquaria and artificial ponds. It is used also for roofing, gutters, and down spouts, and in great quantities for sail boat keels and fish-line and net sinkers. It is employed to convey fresh or salt water, either hot, 80° C (180° F), or cold,

[4] R. M. Burns, *Bell System Tech. J.*, **15**, 616 (October, 1936).

4° C (40° F). The corrosion rate of lead in sea water is approximately ipy (see Table 19, p. 417).

EFFECT OF pH

Lead is amphoteric; hence its corrosion is accelerated by both acids and alkalies. The effect of pH on the corrosion rate is illustrated in Fig. 3.[5]

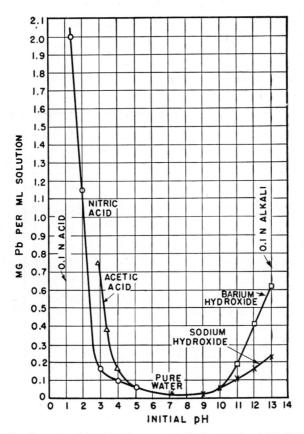

FIG. 3. The Corrosion of Lead in Contact with Various Acid and Alkali Solutions. (Anderegg and Achatz.)

ACIDS

Mineral. The principal mineral acids to which lead is corrosion resistant are *sulfuric* (up to 96%), *sulfurous, phosphoric, chromic,* and *hydrofluoric* (cold). Lead of ⅛ in. or greater thickness usually gives years of service in equipment handling these chemicals. Data on the corrosion rate of lead in sulfuric acid of various concentrations are given in Fig. 4.[6]

[5] F. O. Anderegg and R. V. Achatz, *Purdue University Engineering Experiment Station Bull.*, **18** (1924).
[6] *Chem. and Met. Eng.*, **45** (No. 1), 635 (November, 1938).

High velocities may increase the rate of corrosion. With 20% sulfuric acid at 25° C (75° F) Calcott and Whetzel[7] found in laboratory tests that lead corroded as follows:

Velocity of Solution Across Lead Surface		Rate of Corrosion	
meters/min	ft/min	ipy	in./month
2.6	8.4	0.0066	0.00055
30	97	0.0020	0.00017
47	155	0.0019	0.00016
92	300	0.0103	0.00086

The rates appear to fall with initial increase of velocity, but increase markedly at a velocity of 300 ft per min.

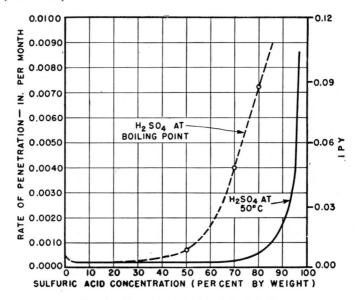

FIG. 4. Corrosion of Lead in Sulfuric Acid.

The specimens were removed daily for 14 days and the protective sulfate coating dissolved in acidified ammonium acetate solution (5%, hot). The corrosion rates are consequently maximum for total immersion in sulfuric acid. Although these values are excessively high, they serve as a guide for commercial practice.

Lead is not satisfactorily resistant to hydrochloric acid and is rapidly attacked by nitric acid, especially in concentrations below 70% by weight.

Mine waters may be used in lead pipes when velocity or erosion effects do not prevent the formation of protective coatings.

Organic. The principal organic acids to which lead is corrosion resistant are: acetic (concentrated), chloroacetic (limited use), fatty acids (but only in absence of oxygen), hydrocyanic (not in pure acid but with sulfuric acid or in aqueous solution with

[7] W. S. Calcott and J. C. Whetzel, *Trans. Am. Inst. Chem. Engrs.*, **15**, 53 (1923).

cyanide), and oxalic and tartaric acids (in absence of oxygen). Lead is rapidly corroded by dilute acetic or by formic acid in the presence of oxygen.

ALKALIES

The action of strong alkalies on lead is not so rapid as by acids like hydrochloric or acetic, but is greater than the action of alkalies on iron. For certain purposes, however, corrosion of lead in contact with NaOH or KOH up to 30% concentration at 25° C (75° F), and up to 10% concentration at higher temperatures, 90° C (190° F), is tolerable. For example, lead has proved useful in the refining of petroleum oil where a sulfuric acid treatment is followed by an alkaline solution treatment in the same lead-lined tank.

Although *lime* ($Ca[OH]_2$) solutions are saturated at about 0.1% $Ca(OH)_2$ at 25° C (75° F) they have been known (as seepage waters from "green" concrete) to corrode lead severely when trickling over the surface. In such cases dissolved oxygen appears to be necessary for corrosion.

SALTS

Lead is generally resistant to corresponding salts of the acids to which it is resistant. Nitrates and alkaline salts tend to be actively corrosive. (See *Corrosion of Lead and Lead-Alloy Cable Sheaths*, p. 609.)

GASES

Lead resists the action of *chlorine,* wet or dry, up to 100° C (210° F) but that of *bromine* only when dry and at lower temperatures. *Sulfur dioxide* and *trioxide,* wet or dry, are frequently handled in lead pipes. *Hydrogen sulfide,* with or without moisture, can be handled with lead provided erosion at high velocities is avoided. *Hydrogen fluoride* is actively corrosive; hence lead is not recommended as a container for this gas.

GALVANIC COUPLING

In acid solutions, iron is anodic to lead; hence the corrosion of iron is accelerated when coupled to lead. In alkaline solutions, however, the reverse situation applies and the corrosion of lead is accelerated by coupling with iron. Copper is anodic to lead in strong acid solutions but cathodic in alkaline solutions.

In the handling of sulfuric acid and sulfates in the chemical industries it is common practice to use antimonial lead (up to 8% antimony) in pumps and valves in electrical contact with sheets and pipes of tellurium- and chemical lead. Galvanic action is inconsequential because of the formation of an insoluble insulating film of lead sulfate.

Serious galvanic action does not occur under sea water with the use of calking lead in cast iron pipe joined by bell and spigot types of joints.

MECHANICAL AND METALLURGICAL FACTORS

Vibration may cause failures in lead equipment and should be avoided as much as possible. In a Bell Telephone Laboratories type of flexure machine[8] for commercially rolled sheet metal the following values of endurance limit for ten million

[8] J. R. Townsend, *Proc. Am. Soc. Testing Materials,* **27**, Part 2, 153 (1927).

cycles at a frequency of one-half million per day and temperature of 26° C (78° F) were found:

Pure lead	440 psi
Chemical lead	540 psi
Tellurium-chemical lead	860 psi
6% Antimonial lead	1040 psi

The advantageous higher endurance limit of tellurium lead compared with pure or chemical lead is probably related to the work-hardening capacity of Te-Pb alloy.

Very pure lead (99.99%) may be subject to grain growth at 25° C (75° F). One or more of the following elements are added to certain varieties of commercial leads as stabilizers or grain growth inhibitors:

Copper	0.04–0.08%
Tellurium	0.035–0.05%
Calcium	0.02–0.04%

In such leads grain growth may occur only at a temperature of 120° C (250° F) or higher. Although fine and uniform grain sizes in fabricated lead are frequently desired, coarser grain sizes seldom give trouble.

In general, internal stresses or work hardness may dissipate with time at ordinary temperature. In the corrosive environments for which lead is suited these metallurgical or mechanical factors are usually of minor importance.

SOIL CORROSION

Refer to *Corrosion by Soils*, p. 454.

PROTECTIVE MEASURES

To prolong the useful life of a lead installation or to use it to best advantage, the following measures have been employed:

(a) Lime or sodium silicate treatment (pH 8 to 9) of potable waters which promotes formation of protective coatings on lead and generally prevents dangerous plumbo solvency.[9]

(b) Coating of sheet lead and pipe with tar, asphaltum, or bituminous paint[10] or waterproof membrane to prevent corrosion due to contact with aerated seepage waters through fresh concrete. After aging one year, the free lime in concrete is usually sufficiently carbonated to eliminate the danger. However, continued seepage of water through some concrete structures may be a source of corrosion for a longer time.

(c) Use of acid brick linings for chemical equipment to prevent erosion effects. They also may lower the lead temperature and reduce buckling of the sheet lead.

(d) Avoidance of large soil particles around bare lead pipe buried underground in certain soils.[11]

(e) Use of automatic steam pressure regulation for use with lead heating coils. The steam should be turned on gradually. A needle valve may often be used.

---

[9] E. A. G. Liddiard, P. E. Bankes and L. de Brauckere, *J. Soc. Chem. Ind.*, **63**, 39–48 (February, 1944).

[10] Master Specifications for Plumbing Fixtures (for land use), Department of Commerce, Bureau of Standards, *Circ.* 310, p. 56 (1926).

[11] R. M. Burns and W. E. Campbell, *Trans. Electrochem. Soc.*, **55**, 271–286 (1929).

LEAD AND LEAD ALLOYS

(*f*) Avoidance of quick-shutting valves in lead pipe handling liquids prevents failure from water hammer.

(*g*) Adoption of lead welding, commonly called *burning*, is the preferred method of making joints in lead chemical equipment. For such purposes, the burning or welding bars should be of the same composition as the material being joined.

ATMOSPHERIC CORROSION

TYPES OF EXPOSURE

Indoor. Sheet lead, either soft or antimonial (hard), is frequently used indoors particularly in chemical plants and laboratories. It is used for ducts (for fumes), hoods, sinks, bench tops, and for floors.

Outdoor. Lead-sheathed cables exposed to the atmosphere are discussed elsewhere (p. 609). Other important outdoor uses of lead are as roofing, flashing, gravel stops, spandrels, gutters, downspouts, garden ornaments, statuary, and wedges between stone building blocks. Sheltering or coating of lead with other materials is not required for atmospheric exposures. Lead-covered nails are available.

Lead resists humid air remarkably well because of the development of protective films. Installed piping and roofing have lasted for hundreds of years.

Atmospheric pollution does not tend to increase the corrosion of lead. Smoke-laden air containing H_2SO_3, H_2CO_3, H_2SO_4, and H_2S is satisfactorily resisted. Lead is also used successfully as ducts for sewer gases.

On account of the electrical-insulating nature of protective films on lead exposed to the atmosphere, lead usually does not produce galvanic corrosion of troublesome magnitude when in contact with other common metals.

TYPES OF ATMOSPHERES

A. S. T. M. Committee B–3, Subcommittee VI, exposed sheet leads (0.035 in., 0.09 cm thick) to various atmospheres for 10 years.[12] The following corrosion rates were observed:

Location	Average Corrosion Rate, ipy	
	Chemical Lead	1% Antimonial Lead
Altoona, Pa. (industrial)	0.000027	0.000021
New York, N. Y. (industrial)	0.000017	0.000012
Key West, Fla. (sea coast)	0.000022	0.000020
La Jolla, Calif. (sea coast)	0.000016	0.000024
State College, Pa. (rural)	0.000019	0.000013
Phoenix, Ariz. (rural)	0.000009	0.000011

Upon careful examination of these sheets, no cracking was discovered and the pitting was found to be insignificant. In most cases the antimonial lead showed less corrosion. Judging from actual service performance of 6% antimonial lead, it would probably corrode at rates comparable with those for chemical and 1% antimonial lead.

J. N. Friend's average data[13] for exposure of lead to city air for 7 years at

[12] A. S. T. M., Committee B–3, Subcommittee VI, *Report*, 1944.
[13] J. N. Friend, *J. Inst. Metals*, **42**, 149–152 (1929).

Birmingham, England, are similar and show by recalculation for comparison with the above:

Soft lead	(99.96%)	0.000032 ipy
Antimonial	(1.6% Sb)	0.000005 ipy

Effect of Wood Atmospheres. Moist air containing acetic acid or other organic acid-type vapors from certain woods such as oak and Douglas fir has been known to cause serious corrosion of lead. Trouble has been ascribed to wooden ducts for lead-sheathed cables, wood used under lead roofing, and to wood in museum cases.[14,15,16] In some applications the lead may be protected by tar, asphaltum, paint, or lacquer.

Permissible woods for contact with lead are cedar or hemlock.* Seasoned wood is preferred over green wood.

MECHANICAL FACTORS

For proper design and installation of hard roofing, recommendations have been made by the Lead Industries Association. Such recommendations are consistent with strength data as given in Fig. 1 and provide for creep resistance and for free movement of metal where necessitated by thermal expansion and contraction.

A large Federal Court House in New York City, the dome on the State Capitol, Harrisburg, Pa., and a large metal dome on the State Reformatory, Rahway, N. J., represent modern installations of lead roofing, gutters, etc.[17,18]

LEAD ALLOYS

LEAD-TIN ALLOYS (TERNE ALLOY AND SOLDER)

Terne plate refers to an alloy-coated (10 to 25% tin-balance lead) sheet iron or steel which is applied by hot dipping. (See also *Lead Coatings*, p. 845.) The alloy has excellent resistance to atmospheric corrosion. Other uses have been for contact with petroleum product containers and paint containers.

During World War I, Pb-Sn alloy coatings of substantial thicknesses were developed and applied by electroplating on steel. They were used successfully for containers of liquid poison gases.

A 93% Pb-7% Sn alloy has been used to some extent as sheet metal for chromium plating solution tank linings and insoluble anodes.

Solders composed of lead and tin are very generally used for joining various sheet metals assembled as roofs, gutters, and downspouts. Solder (50% Pb, 50% Sn) in stream line or sweat fittings for copper pipe is practically unattacked by potable waters, but galvanic action may be serious in sea water. With automobile radiator cores of copper soldered with Pb-Sn alloys in contact with *fresh* water, galvanic corrosion has not been consequential in millions of cases. However, if one were to use brine in the radiators, trouble from this source would follow.

A noble Cu-Sn compound formed where thin sheet copper is soldered may cause serious galvanic action of the copper adjacent to the soldered joint as, for example,

---

* In constructing lead-lined wooden tanks, the wood should be inspected for insect larvae. If these are present, the wood should be rejected. Insects have been known to bore through lead.

[14] S. Beckinsale and H. Waterhouse, *J. Inst. Metals*, Part I, **39**, 375–379 (1928).
[15] R. M. Burns and B. A. Freed, *J. Am. Inst. Elec. Engrs.*, **47**, 576–579 (August, 1928).
[16] R. M. Burns and B. L. Clarke, *Ind. and Eng. Chem.* (Anal. Ed.), **2**, 86–8 (January 15, 1930).
[17] *Lead* (publication of Lead Industries Association), **5** (No. 5), 3–6 (September, 1935).
[18] *Lead* (publication of Lead Industries Association), **5** (No. 6), 5 (November, 1935).

LEAD AND LEAD ALLOYS

on copper roofing.[19] The compound probably results from excessive heat used in soldering and is not frequently encountered.

Solders generally are not recommended for joints of corrosion-resistant chemical equipment and have been known to give trouble in seams in lead tanks used for photographic developers. Lead burning or welding is preferred.

Sea-water corrosion data for Pb-Sn alloys are given in Table 19, p. 417.

LEAD-ANTIMONY ALLOYS

For sheet and pipe, a 6% antimonial lead alloy is frequently employed on account of its superior strength compared to soft lead. It is popular in the construction of equipment for electrolytic refining of copper. In general, the corrosion resistance of antimonial lead is similar to chemical and tellurium leads at 25° to 65° C (75° to 150° F). Its resistance to fatigue is higher. (See p. 214.) Sea-water corrosion data for antimonial lead are given in Table 19, p. 417.

At higher temperatures the corrosion resistance and strength decrease (Fig. 1). Antimonial lead is therefore not usually employed above 95° to 120° C (200° to 250° F), although, as previously noted, chemical or tellurium leads may be employed up to 230° C (450° F). In either case the maximum temperature for use approaches the solidus temperature.

Desirable physical and metallurgical properties account for a large consumption of Pb-Sb alloys as follows:

% By Weight		Use
Sb	Pb	
1	99	Electric cable sheathing
2–4	Bal.	Storage battery connectors and parts
6	94	Roofing and chemical industry
9	91	Storage battery grids
6–8	Bal.	Lead pumps, valves, coatings
15	80 (5 Sn)	Type metal, bearing metal
15–17	Bal. (1 Sn, 1 As)	Bearing metal

LEAD-SILVER ALLOYS

The 99% Pb–1% Ag alloy has been satisfactorily used as an insoluble anode in the production of electrolytic zinc from strongly acid sulfate solutions. Recently a 1% Ag–1% As alloy has been found advantageous as an insoluble anode for electrowinning of manganese from an acid sulfate solution.

The eutectic alloy (2.3% Ag, 97.7% Pb—m.p. 304° C [579° F]) has been used to considerable extent as a soft solder. Unsheltered atmospheric corrosion has proved damaging to the alloy. An alloy in the range of 0 to 1.6% Ag, 1 to 5% Sn, remainder Pb, appears to be more promising as a soft solder. It also has merit as a coating metal on iron or steel for atmospheric exposure.

LEAD-CALCIUM ALLOYS

Alloys containing 0.10% or less of calcium, remainder mostly lead, have been used for cable sheathing, lead pipe, bearing metal, and other purposes. The corrosion resistance is good, permitting use of the alloy satisfactorily in environments of air

---

[19] P. D. Merica, *Chem. Met. Eng.*, **16**, 657 (1917).

and water and in storage batteries. The alloy, so far as is known, is not yet employed in industrial corrosion-resistant chemical equipment.

The 99.9% Pb–0.10% Ca alloy was proposed by the Bell Telephone Laboratories for storage battery grids.[20] The alloy has been shown to cause less self-discharge of a battery than the usual 91% Pb–9% Sb grid metal alloy. An additional advantage derives from absence of stibine formation on overcharge.

MAGNESIUM AND MAGNESIUM ALLOYS

W. S. LOOSE*

COMPOSITIONS

Compositions, mechanical properties, and designations of common magnesium alloys are listed in Table 1.

AQUEOUS MEDIA

The corrosion resistance of magnesium and its alloys is dependent on film formation in the medium to which the alloys are exposed. The rate of formation, solution, or chemical change of the film varies with the medium, and also with the metallic alloying agents or impurities present in the magnesium. Because of these factors, *time* is an important element in evaluating corrosion data. Figure 1 shows the corrosion rate of pure magnesium and controlled purity alloys in a 3% NaCl medium as a log function of the time. Alloys containing harmful cathodic impurities corrode at approximately the same rate as controlled purity alloys during the first few hours, and then, depending apparently on the alloy content and the amount of impurity present, exhibit an increasing rate of attack. Severe localized pitting may develop in some media when critical amounts of certain active cathodic impurities are exceeded.

EFFECT OF DISSOLVED GASES, TEMPERATURE, AND pH

Dissolved Oxygen and Other Gases. Dissolved oxygen does not appear to play a major role in the corrosion of magnesium and its alloys in chloride solutions. Figure 2 shows the effect of aeration on the corrosion rate of pure magnesium and a Mg–6% Al–3% Zn–0.2% Mn alloy in a 3% NaCl solution. Controlled purity metal was used in the test to eliminate the effect of impurities in the metal on the corrosion rate.

Salt solutions as well as distilled water saturated with CO_2 are much more corrosive than similar solutions containing amounts of the gas normally present.

Temperature. The effect of increasing temperature on the corrosion rate of controlled purity alloys in 3% NaCl is also shown in Fig. 2. The increasing rate of corrosion with increase in temperature of the ternary alloy, compared to the relatively constant rate for pure magnesium, may be due to the presence of small amounts of impurities in the former that become active at higher temperature. The curve for this alloy may not be typical for all ternary alloys; variations in alloy and heat treatment have produced differences in the shape of the curve.

* The Dow Chemical Co., Midland, Mich.
[20] E. E. Schumacher and G. S. Phipps, *Trans. Electrochem. Soc.*, **68**, 309–319 (1935).

TABLE 1. MAGNESIUM ALLOYS
Composition, Properties, and Designations

Designations			Form	Composition, % Mg Remainder			Condition	Typical Properties					
A.S.T.M.*	Dow	Am. Mag. Corp.		Al	Mn	Zn		Tensile Strength, 1000 psi	Yield Strength, 1000 psi	Elong. in 2 in., %	Brinell Hardness	Shear Strength, 1000 psi	Endurance limit, 1000 psi (500 million cycles)
AZ63	H-AC	265-C	Sand-cast	6.0	0.2	3.0	C†	29	14	6	50	18	11
..	H-HT	265-T4	"	..	..	..	HT	40	14	12	55	19	14
..	H-HTA	265-T6	"	..	..	..	HTA	40	19	5	73	20	13
AZ92	C-AC	260-C	"	9.0	0.1	2.0	C	24	14	1	65	18	11
..	C-HT	260-T4	"	..	..	..	HT	40	16	10	63	20	14
..	C-HTA	260-T6	"	..	..	..	HTA	40	23	3	84	21	13
AZ90	R	263-C	Die-cast shapes	9.0	0.2	0.6		33	20	3	66	..	..
AZ31X	FS-1	C-52S	"	3.0	0.3	1.0	As extruded	38	26	14	50	..	..
AZ61X	J-1	C-57S	"	6.5	0.2	1.0	"	44	28	16	64	..	..
AZ80X	O-1	C-58S	"	8.5	0.2	0.7	"	46	29	9	67	..	..
AZ80X	O-1 HTA	C-58S-T5	"	"	"	"	Aged	51	35	10	81	..	..
M1	M	3S	"	..	1.5	..	As extruded	34	20	9	46	..	..
AZ61X	J-1	C-57S	Press forged	6.5	0.2	1.0	As forged	43	26	12	55	21	16
AZ80X	O-1	C-58S	"	8.5	0.2	0.7	"	46	31	8	69	22	18
M1	M	3S	Sheet	..	1.5	..	Hard-rolled	37	29	8	56	17	10
AZ31X	FS-1	C-52S	"	3.0	0.3	1.0	"	43	33	11	73	23	14
Z61X	J-1	C-57S	"	6.5	0.2	1.0	"	47	34	9	73	24	14

* Where the designation is followed by X the compositions contain 0.005% Fe max, and 0.005% Ni max, for improved resistance to salt water.
† C is as cast; HT is solution heat-treated; HTA is solution heat-treated and aged to cause precipitation.

As shown by Fig. 3, less concentrated media, such as Midland tap water containing 70 ppm chloride, not only cause little corrosion but also less increase in corrosion rate with rise in temperature.

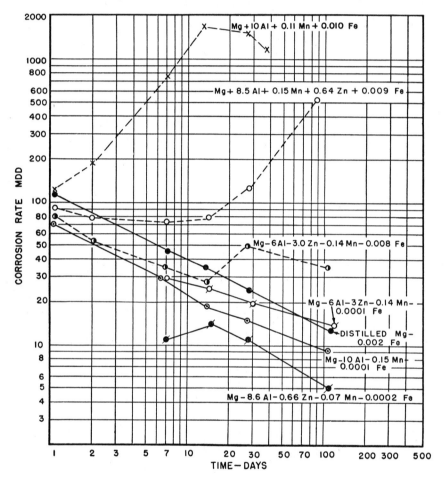

Fig. 1. Effect of Duration of Test on the Corrosion Rate of Magnesium and Magnesium Alloys in a 3% NaCl Medium.

Specimen size — 3.8 cm (1.5 in.) long by 1.9 cm (0.75 in.) diam.
Surface preparation — Aloxite ground. (Aloxite is a trade-mark for an aluminum oxide polishing cloth.)
Temperature — 35° C (95° F).
Aeration — probably air-saturated.
Volume of testing solution — 100 ml.
Type of test — alternate immersion, 30 sec in medium, 2 min in air.

Because of localized impurities the Mg–1.5% Mn alloy shows a greater tendency toward severe pitting with increasing temperature than controlled purity alloys containing aluminum.

When cathodic impurities such as iron, nickel, and copper are present in amounts greater than their tolerance limit,* the corrosion rate and tendency to pit severely increase rapidly, even for small increases in temperature. With impure alloys a thirtyfold increase in rate has been obtained by increasing the temperature of the 3% NaCl medium from 17° C (62° F) to 55° C (131° F).

pH. In general, magnesium alloys are resistant to alkalies and are strongly attacked by acids. Sodium chloride solutions containing appreciable quantities of alkali attack magnesium and its alloys at a very low rate. With increasing pH above the point

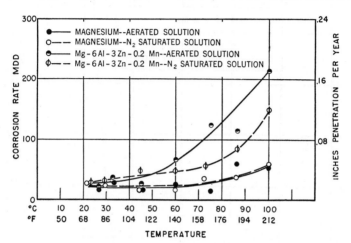

Fig. 2. Effect of Temperature and Aeration on the Corrosion Rate of Cast Magnesium in 3% NaCl.

Specimen size — Mg: 7.5 × 2.5 × 1 cm (3 × 1 × 0.4 in.). Mg–6% Al–3% Zn–0.2% Mn: 3.6 × 1.9 cm diam. (1.4 × 0.75 in. diam.).
Surface preparation — Aloxite ground.
Duration of test — 7 days.
Volume of testing solution — 355 ml /sq cm/(55 ml /sq in.)
Velocity — stirred by gas bubbling through porous alundum crucible.
Aeration — O_2- or N_2-saturated

at which $Mg(OH)_2$ is formed (10.2), the effect of both impurities in the metal and in the corroding media is apparently overshadowed by the greater tendency for film formation. In a medium composed of 4% NaCl, a corrosion rate of 15 to 30 mdd is obtained at 35° C (95° F), whereas in 48% caustic liquor solution containing the same amount of NaCl, the rate is only 1 to 2 mdd.

Salt Solutions

Of the halides, chloride solutions are normally more corrosive, apparently because of the relative ease with which the small ion penetrates the hydroxide film. Neutral or alkaline fluorides form insoluble MgF_2 and are not appreciably corrosive. Ammonium salts usually show more attack than alkali metal salts. Oxidizing salts, especially those containing chlorine or sulfur atoms, are more corrosive than non-

---

* The *tolerance limit* is defined as the critical concentration of an alloyed component of magnesium at which a discontinuity in the corrosion rate occurs.

oxidizing salts, but chromates, vanadates, phosphates, and many others are film forming and tend to retard corrosion at other than elevated temperatures.

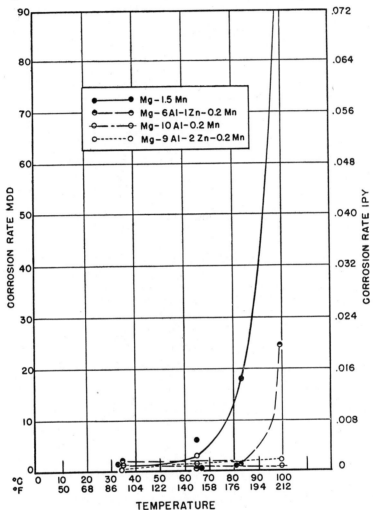

Fig. 3. Corrosion of Magnesium Alloys in Tap Water. (Note expanded vertical scale compared to Fig. 2.)

Specimen size — 5 × 2.3 × 1 cm (2 × 0.9 × 0.4 in.)
Surface preparation — Aloxite ground.
Aeration and velocity — nat. convect.
Volume of testing solution — 200 ml.
Duration of test — 14 days.
Test — continuous immersion.

Table 2 shows the effect of various oxidizing and non-oxidizing salt solutions on the corrosion of a controlled purity commercial magnesium alloy. When impurities are present in the alloys in amounts greater than their tolerance limits, the corrosion rate will increase rapidly in most salt solutions. Chromates and fluorides and the alkaline non-oxidizing salts are exceptions.

TABLE 2. CORROSION OF MAGNESIUM ALLOYS IN SALT SOLUTIONS

Alloy — Mg–6% Al–1% Zn–0.2% Mn.
Specimen size — 7.5 × 2.5 × 0.15 cm (3 × 1 × 0.06 in.).
Surface preparation — HNO_3 pickled.
Temperature — 35° C (95° F).
Volume of testing solution — 100 ml.
Duration of test — 7 days.
Test — alternate immersion — 30 sec in solution, 2 min in air.
Concentration of solution — 3% by weight.

Acid Salts	Corrosion Rate, mdd	Neutral Salts	Corrosion Rate, mdd	Alkaline Salts	Corrosion Rate, mdd
		Non-Oxidizing			
Aluminum sulfate	112	Sodium chloride	14	Sodium silicate, meta	0.7
Zinc chloride	770	Sodium bromide	6	Sodium sulfite	2
Sodium acid tartrate	155	Sodium iodide	29	Sodium borate, meta	5
Sodium dihydrogen phosphate	85	Sodium fluoride	3	Sodium phosphate	3
		Sodium sulfate	8		
		Sodium nitrate	5		
		Oxidizing			
Ammonium persulfate	465	Sodium chromate	2	Calcium hypochlorite	155
Sodium dichromate	4	Sodium chlorate	59	Sodium hypochlorite	5
Ferric sulfate	297	Sodium pyrophosphate	51	Sodium iodate	40

The effect of NaCl concentration on the corrosion of a Mg–0.5% Mn and a controlled purity Mg–6% Al–1% Zn–0.2% Mn alloy is shown in Fig. 4. Tests of bare magnesium alloys in a 20% NaCl spray box at 35° C (95° F) also gave rates approximately double those obtained in 3% NaCl alternate immersion tests at the same temperature, the rate apparently correlating with the concentration.

Sea Water (Refer also to Table 15, p. 412). The corrosion resistance of magnesium alloys subjected to continuous immersion or tidewater immersion in sea water shows good correlation with results obtained in 3% NaCl. Iron, nickel, and copper impurities in the alloy show a deleterious effect on the corrosion resistance.

The effect of increasing iron content is clearly indicated in Table 3. The significant improvement provided by painting is shown, but, even when painted, the effect of purity is apparent.

Similar results were obtained on cast and wrought 35 cm × 10 cm × 0.7 cm (14 in. × 4 in. × 0.25 in.) panels exposed to tidewater immersion at Hampton Roads, Virginia.[1] When painted with two coats of P–27 zinc chromate primer* and two coats of aluminum pigmented V–10 varnish* alloys having an impurity content below the tolerance limit showed good corrosion resistance, whereas alloys above the tolerance limit showed poor resistance. Alloys containing high nickel suffered most.

* P–27 is a Navy specification inhibitive primer containing 85% of the pigment as zinc chromate in a synthetic vehicle. V–10 is a phenol formaldehyde spar varnish containing 680 grams (1.5 lb) aluminum powder per gal.

[1] W. H. Mutchler, unpublished work for NACA, the Army Air Forces, the War Department and the Bureau of Aeronautics, Navy Department.

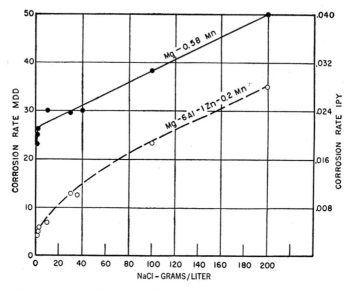

Fig. 4. Effect of Sodium Chloride Concentration on Corrosion Rate.

Specimen size — 7.5 × 2.5 × 0.16 cm (3 × 1 × 0.064 in.).
Surface preparation — Aloxite ground.
Temperature — 35° C (95° F).
Aeration — nat. convect.
Volume of testing solution — 100 ml.
Duration of test — 14 days.
Test — alternate immersion — 30 sec in solution, 2 min in air.

TABLE 3. EFFECT OF IRON CONTENT, HEAT TREATMENT, AND PAINTING ON THE CORROSION RATE OF Mg–6% Al–3% Zn–0.2% Mn ALLOY IN FLOWING TIDEWATER, KURE BEACH, N. C.

Specimen — Mg–6% Al–3% Zn–0.2% Mn casting.
Specimen size — Casting area 360 sq cm (56 sq in.).
Surface preparation — HNO_3-H_2SO_4 pickle.
Temperature — 25° C (77° F) summer; 10° C (50° F) winter.
Aeration — probably saturated with air.
Velocity of sea water — about 60 meters (200 ft) per min.

Composition					Corrosion Rate, mdd					
					Effect of Heat Treatment				As-Cast — Painted	
% Al	% Zn	% Mn	% Fe	% Ni	Duration of Test, days	As Cast	Homogenized	Aged	Duration of Test, days	1 Coat P–27 + 2 Coats V–10AL
6.0	3.1	0.23	<0.001	<0.001	138	14	33	43	320	Nil
6.0	2.8	0.32	0.006	<0.001	138	12	60	82	320	...
6.0	3.0	0.24	0.009	<0.001	138	125	256	195	320	20

Acids

Inorganic. Magnesium is attacked rapidly by *all* mineral acids except *hydrofluoric* and *chromic*. The corrosion rate increases with increase in cathodic impurity content in the metal. Hydrofluoric acid, because of the formation of an insoluble MgF_2 film, does not attack magnesium to an appreciable extent at concentrations above approximately 2%, except at the solution line in open containers. At the solution line, especially at low concentrations of acid, pitting develops. With increasing temperature, the rate of waterline attack increases, but to a negligible extent elsewhere. Table 4 shows the effect of concentration at room temperature on the rate of attack. The increasing rate at lower concentrations represents the pitting developed at and just above the solution line.

TABLE 4. EFFECT OF HYDROFLUORIC ACID CONCENTRATION ON THE CORROSION OF MAGNESIUM

Specimen size — 10 × 1.5 × 0.1 cm (4 × 0.6 × 0.04 in.).
Surface preparation — ground.
Temperature — 25° C (77° F).
Aeration — nat. convect.
Volume of testing solution — 100 ml.
Velocity — quiescent.
Duration of test — 1 week to 4 months.

Composition	Type of Test	Corrosion Rate, mdd					
		HF Concentration, %					
		60	48	30	20	10	5	1
Mg	Total immersion	2	...	...	...	...	...	...
Mg–8% Al–0.2% Mn	Half immersion	...	0.5	2	32	30	23	91

Chromic acid, when pure, attacks magnesium and its alloys at a very low rate, but traces of chloride or sulfate ions greatly increase this rate. There is some tendency for intercrystalline attack. A boiling solution of 20% chromic acid is widely used to remove corrosion product from magnesium alloys because of the solubility of $Mg(OH)_2$ and resistance of magnesium to this medium. A small amount of silver nitrate is normally added to precipitate chlorides.

Organic. Most organic acids appreciably attack magnesium and its alloys. The stronger acids such as *acetic* and *tartaric* are often used at about 10% concentration to pickle cast and wrought products.

The corrosion rates in a variety of acids are shown in Table 5.

Alkalies

Magnesium and its alloys exhibit good resistance to attack by alkalies. Dilute solutions show negligible attack at temperatures up to the boiling point. Sodium hydroxide or caustic liquor solutions containing 50% solids produce little attack at temperatures up to about 60° C (140° F). Temperature is critical, for at 60° C (140° F) the corrosion rate increases appreciably, and at 70° C (160° F) very rapid corrosion ensues. The chloride content of caustic liquor apparently causes an increased rate of attack at 60° C (140° F). These data are shown in Table 6. For temperatures in excess of 100° C (210° F) the rate of attack increases rapidly with temperature.

226 CORROSION IN LIQUID MEDIA, ATMOSPHERE, GASES

TABLE 5. CORROSION OF MAGNESIUM ALLOYS IN ORGANIC ACIDS

Specimen size — 5 × 3 × 0.2 cm (2 × 1.2 × 0.08 in.).
Surface preparation — ground.
Temperature — room.
Aeration — nat. convect.
Volume of testing solution — 100 ml.
Velocity — quiescent.
Duration of test — 30 days.

Acid	% Concentration	Corrosion Rate			
		Mg–1.5% Mn		Mg–6% Al–0.2% Mn	
		mdd	ipy	mdd	ipy
Acetic	1	900	0.720	1000	0.800
Lactic	85	1300	1.04	172	0.138
Tannic	0.5	42	0.034	24	0.019
Tomato juice	Pure	213	0.170	236	0.189

TABLE 6. CORROSION OF MAGNESIUM ALLOYS IN SODIUM HYDROXIDE

Alloy — Mg–10% Al–0.2% Mn.
Specimen size — 7.5 × 2.5 × 0.5 cm (3 × 1 × 0.2 in.).
Surface preparation — acid pickled.
Volume of testing solution — 1 liter (except those in plant settling tank).
Test — continuous immersion in flask fitted with reflux condenser or in caustic plant settling tank.

Concentration, Exposure Conditions	Test Duration, days	Temperature		Aeration	Corrosion Rate	
		°C	°F		mdd	ipy
50% C.P. — refluxing	21	60	140	By convection	1	0.001
48% Liquor — refluxing	7	60	140	By convection	14	0.011
48% Liquor — settling tank	60	30	86	Aerated	3	0.002
48% Liquor — refluxing	7	60	140	Aerated	15	0.012
48% Liquor — settling tank	10	70	160	Aerated	423	0.336

AQUEOUS SOLUTIONS OF ORGANIC COMPOUNDS

Magnesium and its alloys are resistant to aqueous solutions of many organic compounds at room temperature, but at elevated temperatures for extended periods of time inhibitors may be necessary to prevent excessive attack. Alcohols, with the exception of methanol, produce little or at most a moderate attack.

Solutions of ethylene glycol and water at normal temperatures produce negligible attack of magnesium or magnesium galvanically coupled to steel. At temperatures such as 115° C (240° F) the normal corrosion rate increases to some extent, and galvanic corrosion will occur unless the solutions are inhibited. The presence of hydrolyzable chlorides as impurities will greatly increase the rate of attack. Ethylene glycol containing triethanolamine and H_3PO_4 such as that specified for aircraft engines produces negligible attack even at 125° C (260° F) at couples of magnesium and steel. 0.5% NaF is a good inhibitor. Table 7 shows the effect of normal and galvanic corrosion on magnesium alloys in this medium at room and elevated temperatures.

The Mg–1.5% Mn alloy has been used extensively, especially in England, for aircraft gasoline tanks. Gasoline shows no tendency to attack magnesium, and gasoline-water mixtures produce negligible corrosion; but the addition of tetraethyl

MAGNESIUM AND MAGNESIUM ALLOYS

TABLE 7. GALVANIC CORROSION OF MAGNESIUM–STEEL COUPLES IN ETHYLENE GLYCOL AND WATER MIXTURES AT ROOM AND ELEVATED TEMPERATURES

Magnesium alloy — (1) Mg–6% Al–0.2% Mn; (2) Mg–8.5% Al–0.5% Zn–0.2% Mn.
Ratio areas, Mg:Fe — 3 : 1.
Specimen size, Mg — 7.5 × 3.0 × 0.5 cm (3 × 1.2 × 0.2 in.).
Surface preparation — Mg, ground; steel, as rolled.
Aeration — nat. convect.
Volume of testing solution — 100 ml.
Velocity — quiescent.

Solution Composition	Temperature		Time, hours	Mg Alloy	Galvanic Couple	Corrosion Rate		Type Attack
	°C	°F				mdd	ipy	
75% Ethylene glycol	35	95	100	(1)	Mg-Fe	1	0.001	Slight galvanic
95% Ethylene glycol	127	261	100	(2)	Mg only	15	0.012	General
60% Prestone	115	239	100	(1)	Mg-Fe	60	0.048	Severe galvanic
60% Prestone + 0.5% NaF	115	239	100	(1)	Mg-Fe	23	0.019	Some galvanic
95% Air corps Prestone	127	261	100	(2)	Mg only	7	0.006	General
95% Air corps Prestone	127	261	616	(2)	Mg-Fe	..	.....	No galvanic

lead and ethylene dibromide to raise the anti-knock rating causes a greatly increased tendency to attack, especially at the water line. With aviation-type gasoline, tetraethyl-lead attack may best be prevented by the use of slushing compounds containing zinc chromate. When the use of slushing compounds is not feasible, the addition of bags or special capsules containing zinc chromate and an alkaline salt such as $CaCO_3$ or $MgCO_3$ to the sump of the tank will provide satisfactory inhibition. Sodium and chromium glucosates are also effective.

TABLE 8. CORROSION OF MAGNESIUM ALLOYS IN AQUEOUS SOLUTIONS OF, OR MIXTURES WITH ORGANIC COMPOUNDS

Specimen size — 7.5 × 3 × 0.5 cm (3 × 1.2 × 0.2 in.).
Surface preparation — ground.
Aeration — nat. convect.
Volume of testing solution — 100 ml.
Velocity — quiescent.
Test — continuous immersion.

Media	Temperature		Test Duration, days	Composition of Mg Alloy			Corrosion Rate	
	°C	°F		Al	Zn	Mn	mdd	ipy
18% Formaldehyde	Room	...	120	4.0	...	0.3	6	0.004
50% Glycerol	100	212	14	6.0	...	0.2	15	0.012
Buttermilk	8	46	30	...	...	1.5	65	0.052
95% Carbon tetrachloride*	Boil	...	10	6.0	3.0	0.2	6	0.005
95% Perchloroethylene*	Boil	...	10	6.0	3.0	0.2	1420	1.136
50% Soaps	25	77	120	6.0	3.0	0.2	4	0.003

* Half immersed in liquid and half in vapor phase.

Table 8 shows the corrosion rates obtained with a variety of organic compounds dissolved in or mixed with water. In the case of soaps, an average rate for a large number of types is given. The high rate of corrosion in perchloroethylene is due to steam corrosion in the vapor phase below the reflux condenser.

Velocity

Few quantitative data showing the effect of velocity on the corrosion rate are available. Inasmuch as the protection of magnesium from corrosion depends upon film formation, it would be expected that velocity would be critical when it is sufficiently high to affect the protective hydroxide film. In a 15% hydrofluoric acid solution, however, an impeller turning at rates up to 46 meters (150 ft) per min showed no effect of velocity.

Galvanic Coupling

Magnesium and its alloys are normally strongly electronegative (anodic) in character and have a pronounced tendency to go into solution when in electrical contact with other metals in an electrolyte. The corrosion of magnesium being largely cathodically controlled, the polarization characteristics of the coupled cathode will largely control the galvanic corrosion. In a highly conducting medium, such as 3% NaCl, most metals will not polarize to the magnesium potential until a relatively high current density is reached. In contact with metals such as steel or nickel very high corrosion currents are obtained in most highly conducting media. An exception to this is the Al-5% Mg rivet alloy which normally polarizes at a very low current density.

The conductivity and composition of the medium in which a couple is immersed are controlling factors in the rate of galvanic corrosion. Equal areas of various cathodic materials and a magnesium alloy were tested by continuous immersion in

TABLE 9. CORROSION OF Mg-6% Al-3% Zn-0.2% Mn ALLOY GALVANICALLY CONNECTED TO OTHER METALS IN VARIOUS MEDIA

Size of specimens — 4 × 1.3 × 0.2 cm (1.5 × 0.5 × 0.079 in.).
Relative areas — 1:1 (mounted face to face).
Surface preparation — Aloxite 150 ground.
Temperature — room.
Aeration — nat. convect.
Volume of testing solution — 100 ml.
Velocity — quiescent.
Duration of test — 3% NaCl, 3 hours; Midland tap water, 24 hours; Distilled water, 4 days.

Dissimilar Metal	Corrosion Rate, mdd					
	3% NaCl				Midland Tap Water	Distilled Water
	Separation					
	Close Contact	0.35 cm	2.0 cm	10 cm	0.35 cm	0.35 cm
Steel	23,400	25,500	8300	3900	300	18
24ST Aluminum	12,800	25,700	6800	3200	90	6
Nickel	18,800	22,400	6600	....	210	19
2S Aluminum	14,500	15,600	4100	....	40	4
Copper	8,500	8,200	3700	....	90	15
Brass	7,100	4,000	2500	1700	60	14
56S Aluminum	....	1,900	....	....	10	3
Cd-Plated Steel	5,200	2,200	1000	....	40	14
Zinc	6,200	1,300	900	700	30	8
Mg-1.5% Mn	....	50	....	....	2	3
Mg-6% Al-3% Zn-0.2% Mn	....	200	....	....	7	3

3% NaCl, Midland tap water containing approximately 70 ppm chloride, and in distilled water. The effect of anode to cathode distance was also investigated (Table 9).

All commonly used metals will cause galvanic corrosion of magnesium in a strong chloride electrolyte. Cadmium or zinc plating of the more cathodic metals such as iron or steel will reduce the galvanic corrosion to one tenth the rate; a reduction in the conductivity, e.g., a change from 3 NaCl to tap water, will effect an even greater reduction.

Under conditions where the corrosion product is not continuously removed or under conditions of high cathode current density where the surroundings may become strongly alkaline, both the magnesium and an amphoteric contacting metal such as aluminum may suffer severe attack. Aluminum alloys containing appreciable magnesium, such as 52S, 53S, or 56S,* are least severely attacked in chloride media when galvanically coupled. This fact was observed in exposed galvanic couples of magnesium and aluminum alloys to tidewater, and in the atmosphere at Hampton Roads, Virginia.[2]

After one year of tidewater exposure of unpainted magnesium alloy riveted panels, the heads had corroded off 53ST rivets, and A17ST rivets had completely disintegrated; but AM55S (Al–5% Mg) rivets were unattacked. Similar but much less pronounced results were obtained in atmospheric racks located at the ocean shoreline.

The exposure conditions, whether in aqueous media or in the atmosphere, completely dominate the galvanic effects produced. Figure 5 shows the relative amount of corrosion at a contact between a cadmium-plated steel insert and magnesium alloy plate after exposure unpainted to 3% NaCl solution, Midland industrial atmosphere, and to marine atmosphere at Kure Beach, North Carolina. At the latter station, exposures were conducted at two locations — 24 meters (80 ft) and 242 meters (800 ft) from the ocean. The magnesium alloy plate was given a dichromate treatment after the insert was pressed into place. In 3% NaCl more galvanic corrosion had occurred after 18 hours of continuous immersion than in 9 months at the station located 24 meters (80 ft) from the ocean.

The use of ternary Mg-Al-Zn alloys rather than the Mn-Mg alloy may materially affect the galvanic corrosion of the couple. The Mg–6% Al–3% Zn–0.2% Mn was much less severely attacked than the Mg–1.5% Mn alloy in tidewater and sea coast atmospheres when exposed while coupled to aluminum alloys.[3] The use of aluminum alloys containing magnesium, such as 52S or 53S, will, in most cases, satisfactorily reduce galvanic corrosion of the magnesium alloy, and, at the same time, eliminate simultaneous corrosion of the aluminum. Cadmium or zinc plating of the cathode material will diminish galvanic corrosion. Protective treatments applied to the magnesium or to the coupled metal, e.g., anodized aluminum, provide some protection, but in severe media are of value only when assemblies are painted. Good paint adhesion and freedom from porosity are especially important on the cathode, serving thereby to decrease its effective area.

Under severe conditions of exposure where contact, for example, with cadmium-plated steel bolts will cause pronounced galvanic attack, the use of 52S aluminum alloy washers may prevent it. Table 10 shows the results of 20% NaCl spray tests on couples consisting of magnesium plate through which cadmium-plated steel bolts

---
* Compositions of these alloys are given in Table 1, p. 39.

[2] W. H. Mutchler and W. Galvin, *NACA Tech. Notes* 736 and 842.
[3] W. H. Mutchler and W. Galvin, *NACA Tech. Notes* 736 and 842.

were inserted. Area relationships were varied by using different sizes of bolts with cadmium-plated steel, 52S or 56S aluminum alloy washers. The 56S alloy usually produces no galvanic corrosion of magnesium but, when used as a washer in a couple involving a cadmium-plated steel bolt and magnesium, it not only suffers severe attack but also causes galvanic attack of the magnesium. On the other hand,

3% NaCl — Continuous Midland, Mich. — Industrial
Immersion, 18 Hours Atmosphere, 17 Months

Kure Beach, N. C. — 800 Ft Kure Beach, N. C. — 80 Ft
from Ocean, 9 Months from Ocean, 9 Months

FIG. 5. Effect of Atmospheric Exposure and 3% NaCl on Galvanic Corrosion of a Mg–6% Al–1% Zn–0.2% Mn Alloy Containing Cadmium-Plated Steel Inserts.

Specimen size (area) — Steel insert: 12 sq cm (1.9 sq in.); Mg–6% Al–1% Zn–0.2% Mn: 99 sq cm (15 sq in.).
Surface preparation — dichromate treatment.
Temperature — 3% NaCl solution 35° C (95° F), Midland, Mich. Winter: 6° C (21° F), average temp. Summer: 22° C (71° F), average temp.
Aeration of 3% NaCl solution — nat. convect.
Volume of 3% NaCl solution — 300 ml.

in severe media 52S causes attack of magnesium alloys, but in contacts such as these, where the cadmium-plated steel bolt and the magnesium are in the same galvanic couple, the 52S protects the magnesium. When painted with one coat of primer and two coats of lacquer after assembly, no corrosion occurs until after 250 hours in the salt spray, and no serious breakdown takes place after 2000 hours. With cadmium-plated steel washers corrosion takes place after 50 hours, and the attack is serious after 100 hours, even though the couple is painted.

MAGNESIUM AND MAGNESIUM ALLOYS

TABLE 10. USE OF ALUMINUM ALLOY WASHERS TO REDUCE GALVANIC CORROSION BETWEEN CADMIUM–PLATED STEEL BOLTS AND A Mg–6% Al–1% Zn–0.2% Mn ALLOY SHEET IN 20% NaCl SPRAY

Size of magnesium sheet — 7.5 × 15 × 0.15 cm (3 × 6 × 0.06 in.).
Surface preparation — Mg, dichromate treatment; Al, anodized CrO_3; Fe, Cd-plated.
Temperature — 35° C (95° F).
Air pressure — 12 to 15 psi.
Volume of testing solution — 2.63 ml/min.
Duration of test — 200 hours.
Angle of specimen — 10° from vertical.

Corrosion, % Loss in Weight of Sheet and Washer

Washer Alloy	Area of Sheet — 104 sq cm Area of Washer — 1.3 sq cm Area of Bolt Head — 1.2 sq cm		Area of Sheet — 101 sq cm Area of Washer — 1.7 sq cm Area of Bolt Head — 1.6 sq cm	
	Sheet	Washer	Sheet	Washer
Cd-Plated Steel	9.83	0.0	9.75	0.17
52ST Aluminum	0.91	0.67	2.14	1.68
56S Aluminum	1.35	2.57	7.05	3.84
Blank Sheet	0.53	....	0.53	....

Another means of materially decreasing galvanic attack, especially in the case of inserts, is to increase the distance between dissimilar metals by use of inset calking compounds applied in a groove, machined either in the insert or in the magnesium alloy immediately adjacent. Non-conducting washers between dissimilar metals will produce the same result.

Inhibitors in aqueous systems may radically decrease galvanic corrosion. Alkaline chromates, fluorides, and sulfides are especially effective. The addition of magnesium or calcium nitrate to chloride-containing solutions also provides inhibition apparently due to the precipitation of a magnesium or calcium salt on the cathode. Table 11 shows the effect of chromates in diminishing galvanic corrosion.

TABLE 11. EFFECT OF CHROMATES ON GALVANIC CORROSION OF MAGNESIUM COUPLED TO STEEL

Galvanic couple — Mg–6% Al–3% Zn–0.2% Mn : Mild steel.
Relative area Mg : Fe — 1 : 1 (mounted face to face).
Distance apart — 0.35 cm (0.138 in.).
Surface preparation — Aloxite 150 ground.
Temperature — room.
Aeration — nat. convect.
Volume of testing solution — 100 ml.
Velocity — quiescent.
Duration of test — 20 hours.

Solution	Corrosion Rate, mdd
Midland Tap Water	320
Midland Tap Water + 1% Na_2CrO_4	4
Flint City Water	1000
Flint City Water + $ZnCrO_4$	530

MECHANICAL FACTORS

Stress. Stress does not appear to have any effect on the basic general corrosion rate of magnesium and its alloys. Even under a stress approaching the ultimate

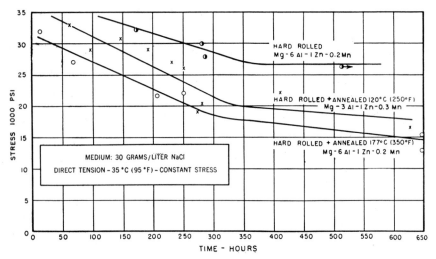

FIG. 6. Effect of Uniform Direct Tension Stress on the Time to Fracture of Magnesium Sheet Alloys in 3% NaCl.

Specimen size — 1.2 cm (0.5 in.). A. S. T. M. sheet tensile test bars, 0.16 cm (0.064 in.) thick.
Surface preparation — chrome-pickled.
Aeration — nat. convect.
Volume of testing solution — 160 ml.
Note: An arrow at a point indicates that the specimen had not failed at this time in the testing cycle.

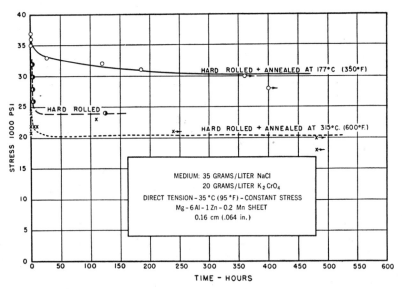

FIG. 7. Effect of Uniform Direct Tension Stress on the Time to Fracture of Mg–6% Al–1% Zn–0.2% Mn sheet in a 3.5% NaCl + 2.0% K_2CrO_4 solution.

Specimen size — 1.2 cm (0.5 in.) A. S. T. M. sheet tensile test bars, 0.16 cm (0.064 in.) thick.
Surface preparation — Aloxite ground.
Aeration — nat. convect.
Volume of testing solution — 160 ml.

strength, the steady state potential of an alloy does not vary from the original potential by more than approximately one millivolt.

However, in some media, at stresses depending on the medium, alloy content, and metallurgical condition of the alloy, stress corrosion cracking may occur. In alkaline media above a pH of 10.2, magnesium alloys appear to be very resistant. In neutral solutions containing chlorides and even in distilled water, sensitivity to cracking exists. Especially rapid cracking occurs in electrolytes containing both NaCl and K_2CrO_4. In chromic acid solutions failures do not occur, although the addition of a small amount of chloride will provide a medium that can induce cracking. Magnesium alloys are resistant to this failure in fluoride or fluoride-containing electrolytes.

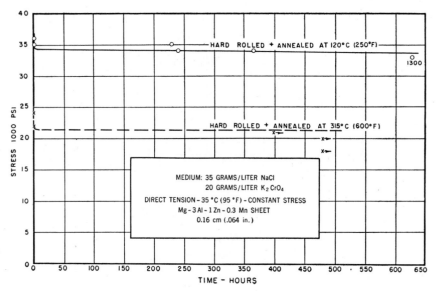

FIG. 8. Effect of Uniform Direct Tension Stress on the Time to Fracture of Mg–3% Al–1% Zn–0.3% Mn sheet in a 3.5% NaCl + 2.0% K_2CrO_4 solution.

Specimen size — 1.2 cm (0.5 in.) A. S T. M. sheet tensile bars, 0.16 cm (0.064 in.) thick.
Surface preparation — Aloxite ground.
Aeration — nat. convect.
Volume of testing solution — 160 ml.

Magnesium, Mg–1.5% Mn alloy, and the casting alloys do not appear to be sensitive to stress corrosion cracking at stresses up to their yield point in chloride-containing media. Sheet alloys containing aluminum, zinc, and manganese are sensitive to a chloride medium at stresses well under their yield point, the sensitivity being dependent on the alloy content and the metallurgical condition of the metal.

Figures 6, 7, and 8 show stress-time curves obtained in 3% NaCl and in the salt-chromate medium for a Mg–6% Al–1% Zn–0.2% Mn and a Mg–3% Al–1% Zn–0.3% Mn sheet in the hard-rolled, low-temperature annealed, and fully recrystallized states. From a comparison of these graphs and Figs. 21 and 22, it is apparent that the precipitated state achieved by cold rolling, followed by low-temperature precipitation of a second phase, provides maximum resistance to stress corrosion in the salt-chromate and atmospheric media. In a 3% NaCl medium, maximum resistance is obtained by cold rolling without subsequent precipitation. In a salt-chromate solution, the stress to produce failure is critical, inasmuch as cracking either occurs within a very few minutes or the material will stand the stress for a prolonged period of time. In the 3% NaCl medium, a much more gradual slope is obtained, cracking apparently being more a function of time.

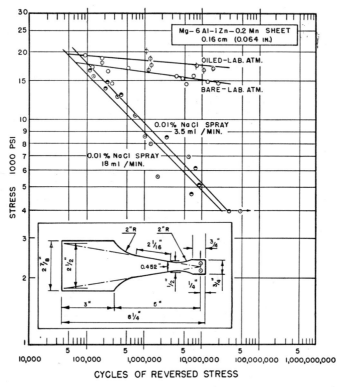

FIG. 9. Effect of Spray Intensity of 0.01% Sodium Chloride on the Resistance to Fatigue of Precipitated Mg–6% Al–1% Zn–0.2% Mn sheet.

Specimen size — Plate-type specimen 0.16 cm (0.064 in.) thick.
Surface preparation — Aloxite ground.
Temperature — about 30° C (90° F).

Stress corrosion cracks are normally transcrystalline in both aqueous and atmospheric media, the cracks being characterized by their many transcrystalline side cracks.

Vibration. Although there is no true endurance limit for magnesium and its alloys tested in fatigue under corrosive conditions, the steepness of the fatigue curve varies with the corrosive environment and the alloy content. Data cited[4] on the effect of both alloys and media show that the Mg–1.5% Mn and Mg–2% Mn–½% Ce alloys are more resistant to corrosion fatigue than alloys containing aluminum and zinc;

[4] A. Beck, *The Technology of Magnesium and Its Alloys*, p. 236, F. A. Hughes and Co., Ltd., London, 1940.

also media such as 3% NaCl or sea water produce a much more rapid drop in the fatigue curve than tap water.

Figures 9 and 10 show data obtained on 0.16 cm (0.064 in.) sheet alloys Mg–6% Al–1% Zn–0.2% Mn and Mg–3% Al–1% Zn–0.3% Mn tested with plate-type bending fatigue equipment in a chloride-containing spray of 0.01% NaCl. Figure 9 also shows data obtained on protected and unprotected sheet to determine the effect of normal laboratory exposure. The two rates of spray shown in Fig. 9 produced the same

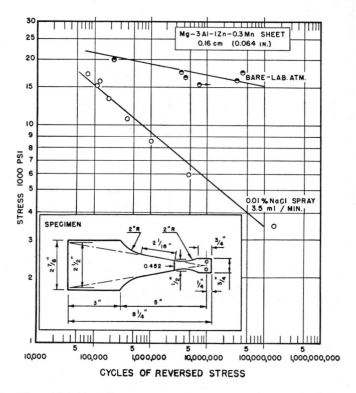

FIG. 10. Effect of 0.01% NaCl Spray on the Resistance to Fatigue of Precipitated Mg–3% Al–1% Zn–0.3% Mn sheet.

Specimen size — plate-type specimen 0.16 cm (0.064 in.) thick.
Surface preparation — Aloxite ground.
Temperature — about 30° C (90° F).

decrease in fatigue strength. Both alloys show approximately an equal susceptibility to fatigue under corrosive conditions. Unprotected metal in the laboratory atmosphere had a slightly lower fatigue strength than when protected.

METALLURGICAL FACTORS

Addition Elements. Impurities such as iron, nickel, and copper have definite tolerance limits, and the quantity of these impurities in magnesium alloys determines the corrosion resistance of the resulting product in aqueous media.[5] This was shown by starting with very pure magnesium produced by multiple distillation and adding definite amounts of purified alloying agents to determine their effect on the corrosion resistance in 3% NaCl. Below certain specified amounts of iron, nickel, and copper, the corrosion rate of magnesium alloys was low. When the impurities exceeded certain

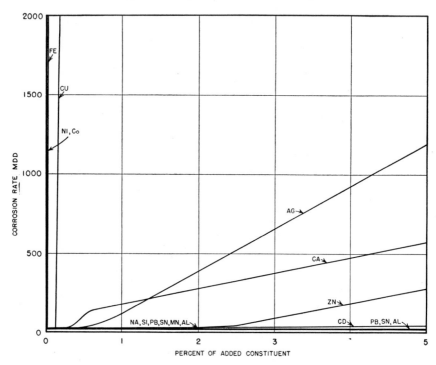

FIG. 11. Corrosion of Binary Alloys.

Corrosive medium — 3% NaCl.
Size of specimen — 3.8 × 2.5 × 0.6 cm (1.5 × 1 × 0.25 in.).
Surface preparation — Aloxite ground.
Temperature — room.
Aeration — nat. convect.
Volume of testing solution — 100 ml.
Duration of test — 16 weeks.
Type of test — alternate immersion, 30 sec in medium, 2 min in air.

small amounts known as the *tolerance limits,* the corrosion rate was greatly accelerated. More recent work has shown that these same impurities largely control the rate of attack in severe exposures and to a lesser extent in atmospheric exposure.

Commercial magnesium has widely varying corrosion rates within the range of 500 to 10,000 mdd, whereas high-purity magnesium has a rate of 15 to 30 mdd in 3% NaCl. Figure 11 gives a broad view of the corrosion of magnesium binary alloys with various common elements added. Some elements are not harmful in large

[5] J. D. Hanawalt, C. E. Nelson, and J. A. Peloubet, *Trans. Am. Inst. Mining Met. Engrs.,* Institute Metals Division, **147**, 273 (1942).

MAGNESIUM AND MAGNESIUM ALLOYS

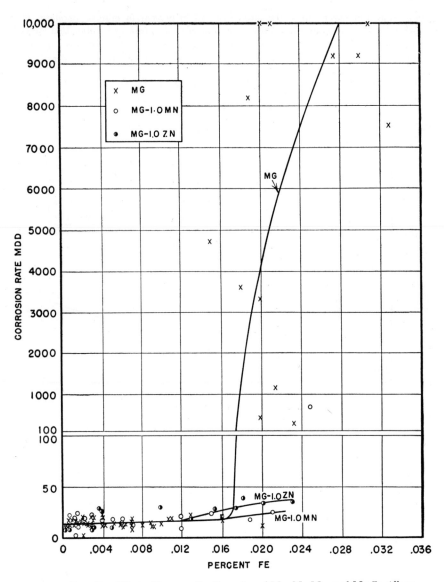

FIG. 12. Effect of Iron on the Corrosion of Mg, Mg-Mn, and Mg-Zn Alloys.

Corrosive medium — 3% NaCl.
Specimen size — 3.8 × 2.5 × 0.6 cm (1.5 × 1 × 0.25 in.).
Surface preparation — Aloxite ground.
Temperature — room.
Aeration — nat. convect.
Volume of testing solution — 100 ml.
Duration of test — 16 weeks.
Type of test — alternate immersion, 30 sec in medium, 2 min in air.

proportion, but others are detrimental even when present in minute amounts. The present important commercial alloy compositions come from a combination of the elements that are not harmful to corrosion. Inasmuch as little magnesium is commercially used in the unalloyed state, the effect of these impurities on alloys is of maximum importance.

Figure 12 shows in detail the effect of iron on the Mg-1% Mn and the Mg-1% Zn alloys in comparison to its marked effect on pure magnesium. For the Mg-1% Mn alloys the tolerance limit of 0.017% is unchanged from that of the magnesium alone. Iron in amounts greater than this tolerance limit causes a smaller increase in the

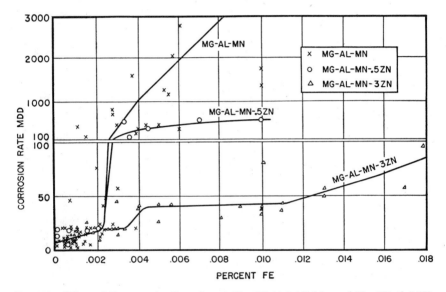

FIG. 13. Effect of Iron on the Corrosion of Mg-6% Al-0.2% Mn and Mg-6% Al-0.2% Mn-0.5% to 3% Zn.

Corrosive medium — 3% NaCl.
Specimen size — 3.8 × 2.5 × 0.6 cm (1.5 × 1 × 0.25 in.).
Surface preparation — Aloxite ground.
Temperature — room.
Aeration — nat. convect.
Volume of testing solution — 100 ml.
Duration of test — 16 weeks.
Type of test — alternate immersion, 30 sec in solution, 2 min in air.

corrosion rate of the alloys than when present in pure magnesium. Similarly, the tolerance limit for copper is unchanged from that of pure magnesium when manganese or zinc is added in small amounts. The addition of 0.2% manganese slightly increases and 1% manganese very greatly increases the tolerance limit for nickel. The addition of 1% Zn to pure magnesium also raises the tolerance limit for nickel, but to a lesser extent than 1% manganese.

With the Mg-Al binary alloys the tolerance limit for iron is strikingly changed. With even as little as a few hundredths per cent aluminum the tolerance limit for iron decreases from 0.017% to a few thousandths per cent. With 7.0% aluminum, it is about 0.0005% iron, whereas with 10.0% aluminum the limit is too low to be determined. When 0.2% manganese is added to the Mg-Al binary alloys, the tolerance limit for iron does not drop below 0.002%, but instead holds very constant over a wide range of aluminum content. Even a few hundredths per cent manganese is sufficient to produce this effect.

Of maximum practical importance is the effect of iron, nickel, and copper on the Mg-Al-Mn and Mg-Al-Zn-Mn alloys. Figures 13, 14, and 15 show the results of adding 0.5% and 3% zinc to the Mg-Al-0.2% Mn alloys. Figure 13 shows that 0.5% zinc does not shift the position of the iron tolerance limit at 0.002%, but somewhat reduces the magnitude of the corrosion rates for higher percentage of iron. The addition of 3% zinc raises the tolerance limit to 0.003% iron and greatly reduces the rate for concentrations of iron up to 0.018%.

Figure 14 shows that 0.5% zinc has little effect on the corrosion behavior of Mg-Al-Mn-Ni alloys; 3% zinc shifts the tolerance limit from 0.001 to 0.002% nickel and reduces the corrosion rate at higher percentages of nickel.

Figure 15 shows that 0.5% zinc raises the copper tolerance limit only slightly, whereas 3% raises it to at least 0.5% copper.

These data serve to show that magnesium alloys are characterized by tolerance limits for iron, nickel, and copper when individually present in alloys. When iron is present in combination with increasing amounts of copper or lead, the tolerance limit for iron in the Mg-Al-Mn alloys decreases with increasing copper or lead. When present with increasing amounts of silicon up to approximately 0.1%, there is no apparent change in the tolerance limit for iron; but above this percentage its tolerance limit appears to be about 0.0005%. With the Mg-Al-Mn-Zn alloys, no evidence for combined effects of iron and copper or lead, as noted with the Mg-Al-Mn alloys, is detected.

Heat Treatment and Cold Work. The effect of heat treatment on the corrosion rate of magnesium alloys varies with the rate of cooling, the alloying agents, and the impurity content of the alloys. With the Mg-1.5% Mn alloy, heat treatment of sheet in the range from 260° C (500° F) to about 485° C (900° F) increases the corrosion rate and tendency to pit. Heat treatment at 540° C (1000° F) to 630° C (1165° F), followed by rapid cooling, completely eliminates pitting and provides increased corrosion resistance.

With controlled-purity Mg-Al-Mn alloys containing 0 to 1% zinc, slight, if any, difference in corrosion rate is noted between the as-fabricated, solution heat-treated, or solution heat-treated and aged states, if air-cooled from the solution heat-treating temperature. With similar cooling, controlled-purity alloys containing 2 to 3% zinc show a slight increase in the corrosion rate in the solution heat-treated or solution-treated and aged states over that of the as-cast state. With alloys containing iron above the tolerance limit, solution heat treatment will increase the corrosion rate by a factor of two to five times, and solution heat treating and aging to a somewhat lesser extent. High-purity Mg-Al-Mn alloys containing no zinc corrode uniformly, whereas the same alloys containing increasing amounts of zinc show an increasing tendency toward pitting. The heat treatment and especially the cooling rate from the solution heat-treating temperature appear to affect the tendency toward pitting. Quenching from the solution heat-treating temperature greatly increases both the tendency to pit and the corrosion rate, even of controlled purity alloys. With increasing zinc content up to 3% the pitting caused by quenching appears to be decreased. This effect of heat treatment and quenching on the corrosion rate is particularly noticeable adjacent to welds corroded in a 3% NaCl solution. Severe attack may take place in the rapidly cooled heat-affected zone. Figure 16 shows the effect of heat treating and

240   CORROSION IN LIQUID MEDIA, ATMOSPHERE, GASES

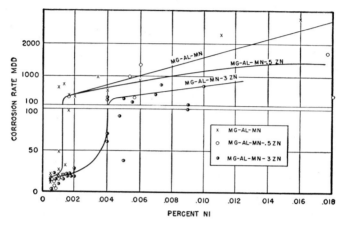

FIG. 14. Effect of Nickel on the Corrosion of Mg–6% Al–0.2% Mn and Mg–6% Al–0.2% Mn–0.5 to 3% Zn.

Corrosive medium — 3% NaCl.
Specimen size — 3.8 × 2.5 × 0.6 cm (1.5 × 1 × 0.25 in.).
Surface preparation — Aloxite ground.
Temperature — room.
Aeration — nat. convect.
Volume of testing solution — 100 ml
Duration of test — 16 weeks.
Type of test — alternate immersion, 30 sec in solution, 2 min in air.

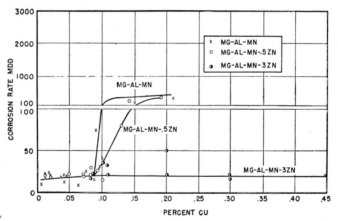

FIG. 15. Effect of Copper on the Corrosion of Mg–6% Al–0.2% Mn and Mg–6% Al–0.2% Mn–0.5 to 3% Zn.

Corrosive medium — 3% NaCl.
Specimen size — 3.8 × 2.5 × 0.6 cm (1.5 × 1 × 0.25 in.).
Surface preparation — Aloxite ground.
Temperature — room.
Aeration — nat. convect.
Volume of testing solution — 100 ml.
Duration of test — 16 weeks.
Type of test — alternate immersion, 30 sec in medium, 2 min in air.

aging (air-cooled) on the physical property loss sustained on corrosion in 3% NaCl by high-purity and commercial Mg–6% Al–3% Zn–0.2% Mn alloys.

Cold working magnesium alloys, as by stretching or bending, does not appear to have any effect on the normal corrosion rate. Surface cold work such as that obtained

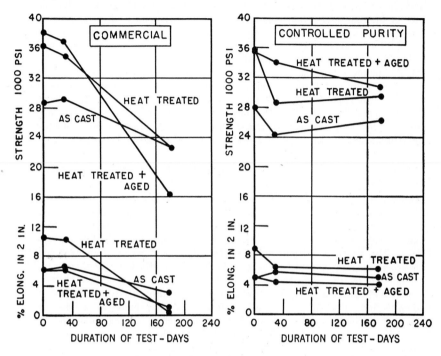

FIG. 16. Effect of Heat Treatment of Controlled Purity and Commercial Mg–6% Al–3% Zn–0.2% Mn Casting Alloy on Tensile Properties after Corrosion.

Corrosive medium — 3% NaCl solution.
Specimen size — A. S. T. M. 1.2 cm (0.5 in.) round test bar.
Surface preparation — machined.
Temperature — 35° C (95° F).
Aeration — nat. convect.
Volume of testing solution — 100 ml per specimen.
Testing method — alternate immersion — 30 sec in medium, 2 min in air.
Heat treatment — 16 hours, 388° C (730° F) + air cool.
Aging treatment — 12 hours, 177° C (350° F) + air cool.

by shot or sandblasting increases the corrosion rate in 3% NaCl. The basic rate may again be obtained by acid pickling about 0.001 in. from the worked surface. Shotblasting or surface cold rolling of residually stressed magnesium alloys greatly increases the stress corrosion resistance, apparently by substituting compressive for tensile surface stress.

PROTECTIVE MEASURES

Magnesium and its alloys may be protected from corrosion in saline solutions by the use of alkali metal or ammonium chromate, sulfide, or fluoride inhibitors. The effectiveness of inhibitors is usually specific for the conditions of exposure. Chromates, especially under alkaline conditions, normally provide maximum inhibition in chloride media. Concentrations equivalent to about 10% by weight of the NaCl content are required. Siebol[6] states that the addition of a few tenths per cent alkali chromate or bichromate to water and non-reducing solutions such as the cooling water of internal combustion engines is sufficient to suppress all attack on water-cooled parts of magnesium alloys as well as that caused by coupling with other metals. The addition of alkaline salts such as calcium or magnesium carbonates or nitrates to chloride-containing corrosive solutions appears to facilitate inhibition.

Maximum protection may be obtained through the use of chemical treatments followed by painting.* Protective chemical treatments are normally of the inorganic chromate type applied either by immersion or anodic treatment. Zinc chromate primers applied over chemical treatments and followed by lacquers, enamels, or varnishes provide maximum protection. The effect of painting on the protection of magnesium against severe exposures is shown in Fig. 24 and in Table 2.

NON-AQUEOUS MEDIA

LIQUIDS

Aliphatic compounds and their halide derivatives cause little or no corrosion of magnesium alloys. Anhydrous alcohols (with the exception of methanol which attacks magnesium vigorously), denatured alcohol, and the higher polyhydric alcohols produce little or no attack. In anhydrous methanol, magnesium methylate is formed. The attack is partially inhibited by addition of water, even 0.1% greatly decreasing the rate of attack. Effective inhibition of attack by methanol can be accomplished by the addition of 1% beet pulp distillate, ammonium sulfide, or chromic acid. Turpentine or dimethyl glyoxime are also effective inhibitors. Magnesium is not attacked by aromatic heterocyclic compounds, by ethers, or by ketones, but aliphatic acids and many of the nitrogen-containing compounds cause appreciable corrosion. Acid-type inks cause pronounced corrosion of magnesium alloys, but alkaline inks show only a moderate rate of attack.

Oils and fats at 195° C (380° F) produce negligible corrosive attack. Fuel and lubricating oils likewise produce negligible corrosion, and the presence of water and galvanic couples does not appreciably increase the rate. Organic acids in lubricating oils will cause some increase in the rate of attack. Neither ordinary gasoline nor high-octane gasoline containing relatively large amounts of tetraethyl lead and ethylene dibromide cause corrosion. Phenol has been stored in magnesium alloy drums for as long as 5 years without showing attack. Methyl bromide causes negligible corrosion.

In Table 12, typical data are given on the attack of magnesium by organic materials.

GASES

Based on limited evidence which unknown conditions may greatly modify, dry Cl_2, I_2, Br_2, and F_2 cause only slight attack of magnesium at room temperature. Dry Br_2 (0.02% water) at its boiling point of 58° C (136° F) apparently does not cause more attack than at room temperature. The presence of a small amount of

* The protection given by chemical treatments is described on pp. 864 to 867. The painting and protection of magnesium alloys is described in Navy Specification SR-15e.

[6] A. Beck, *The Technology of Magnesium and Its Alloys*, p. 303, F. A. Hughes and Co., Ltd., London, 1940.

TABLE 12. CORROSION OF MAGNESIUM ALLOYS IN ORGANIC SOLUTIONS

Magnesium alloy — (1) Mg–6% Al–3% Zn–0.2% Mn; (2) Mg–1.5% Mn.
Specimen size — 7 × 3 × 0.5 cm (2.75 × 1.2 × 0.2 in.).
Surface preparation — ground.
Aeration — nat. convect.
Volume of testing solution — 100 ml.
Velocity — quiescent.
Test — continuous immersion.

Medium	Temperature		Test Duration, days	Mg Alloy	Corrosion Rate	
	°C	°F			mdd	ipy
Ethanol (95%)	Room	Room	14	(1)	0.3	0.0002
Methanol (anhydrous)	Room	Room	1	(1)	3870	3.096
Butanol	Room	Room	120	(1)	2	0.002
Beeswax	50	120	14	(1)	18	0.014
Lard	60	140	30	(1)	3	0.002
Aniline	Room	Room	30	(2)	2	0.002
Fuel oil	Room	Room	50	(1)	0.0	.....
Motor oil	120	250	11	(1)	1.3	0.001

water causes pronounced attack of Cl_2, some attack by Br_2, and negligible attack by F_2.

Sulfur dioxide in either the vapor or liquid phase causes no corrosion. After 7½ years of exposure to NH_3 liquid, and gas, no noticeable corrosion occurred. Natural gas does not attack magnesium, but the presence of water vapor may lead to very slight corrosion.

Liquid or gaseous CCl_2F_2 does not produce corrosion; however, the addition of water creates appreciable attack.

The data for some of these substances are shown in Table 13.

TABLE 13. CORROSION OF MAGNESIUM ALLOYS IN GASES

Magnesium alloy — Mg–4% Al–0.3% Mn.
Specimen size — 3 × 1 × 0.5 cm (1.2 × 0.4 × 0.2 in.).
Surface preparation — ground.
Temperature — room.
Test — continuous exposure.

Medium	Duration Test	Corrosion Rate	
		mdd	ipy
NH_3	7.5 years	0.0	0.0
SO_2	14 days	0.0	0.0
CH_3Cl	14 days	0.7	0.0006
Natural gas	180 days	0.0	0.0
CCl_2F_2 ("Freon")	30 days	0.0	0.0

ATMOSPHERIC CORROSION

INDOOR EXPOSURE

Crystals of distilled magnesium will remain mirror bright for years when covered with a glass jar that prevents deposition of dust and moisture, but does not seal against normal air pressure fluctuations. These same crystals, if left exposed in an open room, will show tarnish in a few weeks, and in a matter of months will have become a dark gray.

OUTDOOR EXPOSURE

Magnesium on exterior exposure gradually assumes a light gray, then dark gray color, exactly the same as occurs on indoor exposure. Bengough and Whitby[7] found that this corrosion product, after 217 days of exposure at Teddington (urban), analyzed as follows:

% $MgCO_3 \cdot 3H_2O$	61.5
% $MgSO_4 \cdot 7H_2O$	26.7
% $Mg(OH)_2$	6.4
% Carbonaceous matter	2.5

Depending on the environment, variable analyses may be obtained. Carbon dioxide absorption by the water layer on the metal with subsequent reaction with $Mg(OH)_2$

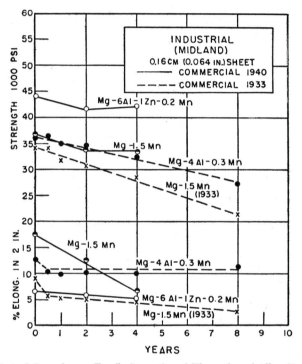

FIG. 17. Effect of Corrosion on Tensile Strength and Elongation of Alloy Sheets Exposed to Industrial (Midland) Atmosphere.

Specimen size — *1933:* 0.16 cm (0.062 in.) with 1.2 cm (0.5 in.) reduced section test bars. *1940:* 0.16 cm (0.062 in.) × 2.5 cm (1 in.) × 17.5 cm (7 in.) sheets.
Surface preparation — Commercial 1933, as rolled. Commercial 1940, chrome-pickled.
Temperature — Winter: 6° C (21° F). Summer: 22° C (71° F).

to form the double salt is believed to play an especially important part in the corrosion and tarnishing of magnesium. As with other metals, the corrosion product on the top of an exposed panel appears to be much thinner and more compact than that on the bottom.

[7] G. D. Bengough and L. Whitby, *Trans. Inst. Chem. Eng.*, **11**, 176 (1933).

Effect of Humidity. Humidity plays a major part in the corrosion or tarnishing of magnesium and its alloys. Thin films of magnesium (400 Å) were enclosed in containers and exposed to atmospheres of humidities of 9.5, 35, 50, 67, 82, and 93% for 18 months at room temperature (no temperature control) and then examined by electron diffraction. Detectable crystalline $Mg(OH)_2$ formed only on specimens exposed to 93% humidity. At this humidity the process consisted of the very slow formation of an amorphous phase, which subsequently was converted to crystalline

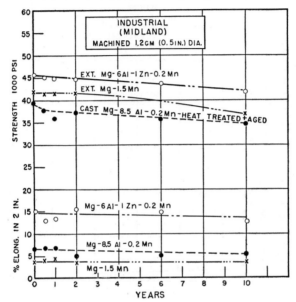

FIG. 18. Effect of Corrosion on Tensile Strength and Elongation of Extruded and Cast Magnesium Alloy Test Bars in Industrial (Midland) Atmosphere. (Temperature — Winter: 6° C [21° F]. Summer: 22° C [71° F].)

$Mg(OH)_2$. With 80% humidity, crystalline magnesium with 30% amorphous material was seen after 18 months; with lower humidities still less of the amorphous phase was present. At 9.5% humidity, no appreciable amorphous phase was formed in this time.

Magnesium alloys react similarly with respect to time and humidity, except that phase mixtures of magnesium and the secondary constituent hydroxides are obtained. When magnesium alloys containing aluminum, manganese, or zinc are exposed to the atmosphere, an analysis of the film formed shows a higher proportion of the secondary constituent than is present in the alloy.

Effect of Atmospheric Pollution. The slow rate of corrosion at Midland, Mich., where the atmosphere contains varying detectable amounts of Cl_2, Br_2, NaOH, HCl, HBr, SO_2, $MgCl_2$ and other organic and inorganic compounds, would indicate that these do not cause rapid breakdown of the protective film on magnesium.

Extensive tensile corrosion data have been obtained over a 10-year period at this location on sheet, extrusions, and casting alloys as commercially produced in 1933. These alloys contained uncontrolled amounts of impurities known to affect the corrosion resistance of the alloys in chloride solutions. More recently data have been obtained over a 4-year period on the tensile corrosion losses of the present commercial controlled purity alloys. These data are shown in Figs. 17 and 18. Although variation in test bar shape and in surface treatment may influence the results to some extent, there appears to be a detrimental effect of impurities even in an industrial atmosphere.

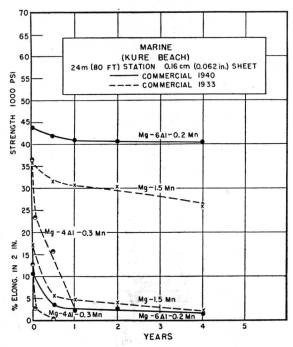

FIG. 19. Effect of Controlled Purity on Corrosion of Magnesium Alloy Sheet Exposed under Marine Conditions.

Specimen size — *1933:* 1.2 cm (0.5 in.) tensile test bars 0.16 cm (0.062 in.) thick. *1940:* 0.16 cm (0.062 in.) × 2.5 cm (1 in.) × 17.5 cm (7 in.) sheets.
Surface preparation — Mg–4% Al–0.3% Mn, as rolled. Mg–6% Al–0.2% Mn and Mg–1.5% Mn, acetic pickle + chrome pickle.

Marine Atmospheres. Extensive data have been compiled on the effect of the marine atmosphere at Kure Beach, N. C., on the tensile corrosion properties of magnesium alloys. Both test bars and sheet or extruded plates were exposed at locations 24 meters (80 ft) and 242 meters (800 ft) from the ocean on racks facing South at an angle of 30° to the horizontal. The effect of corrosion on the tensile properties of both round and flat specimens exposed at the 24-meter station is shown on Figs. 19 and 20. The present commercial controlled-purity alloys are very resistant to severe sea coast atmospheres. At the 242-meter station, corrosion losses of the controlled-purity wrought or even uncontrolled-purity cast alloys are not great enough to be significant after 2 years of exposure.

GALVANIC COUPLING

In contrast to the severe contact corrosion observed in salt solutions, atmospheric exposure causes negligible attack within a year or more, and the corroded area is much more concentrated near the cathode. Figure 5 shows that cadmium-plated steel inserts in a Mg–6% Al–1% Zn–0.2% Mn alloy plate caused more severe attack of the magnesium in 18 hours of continuous immersion in 3% NaCl than was caused in 9 months of exposure at Kure Beach, N. C., 24 meters (80 ft) from the ocean. Exposure 242 meters (800 ft) from the ocean at this location causes very much less severe

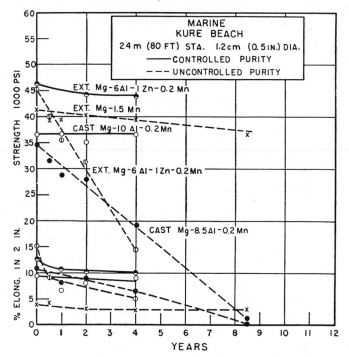

FIG. 20. Effect of Controlled Purity on Corrosion of Magnesium Alloy Test Bars Exposed under Marine Conditions.

Surface preparation — Mg–8.5% Al–0.2% Mn, acid-pickled.
All the others are machined.

attack. In spite of the severe industrial atmosphere at Midland, after 1½ years of exposure only a slight etch had developed. The relative galvanic effect of various metals in contact with magnesium will vary, depending on corrosive conditions, but is approximately the same (although much less severe) as was shown in the 3% NaCl test (Fig. 5).

MECHANICAL FACTORS

Stress. Some magnesium alloys may stress-corrode in the atmosphere.[8] All cases of stress corrosion cracking in magnesium alloy articles noted to date have resulted from residual stresses. The stresses involved have resulted from welding, cast-in inserts, or poorly formed structures subsequently drawn to correct contour. The

---

[8] W. S. Loose and H. A. Barbian, *Symposium on Stress-Corrosion Cracking of Metals (1944)*, p. 273, The American Society for Testing Materials and The American Institute of Mining and Metallurgical Engineers, Philadelphia and New York, 1945.

magnitude of residual stresses in magnesium alloys varies with time, alloy, and method of fabrication.

To determine the effect of residual stresses, magnesium sheet alloys in the hard-rolled, precipitated, and annealed states were subjected to deformation in bending at stresses up to approximately the yield point. Table 14 shows the days to failure for varying stresses at Midland and at Kure Beach, N. C. At Kure Beach, the Mg–6% Al–1% Zn–0.2% Mn alloy in the hard-rolled or fully annealed condition has shown more sensitivity to stress corrosion cracking than when in the precipitated state. In the marine atmosphere this alloy in the precipitated state has not cracked in 15 months at stresses that were originally 80% of the yield point. When exposed in the same manner at 95 and 85% of its yield point at Midland, failures took place

TABLE 14. ATMOSPHERIC STRESS CORROSION CRACKING OF MAGNESIUM SHEET ALLOYS STRESSED BY BENDING TO A PERMANENT DEFORMATION

Specimen size — 36.2–43.2 × 7.5 × 0.1 cm (14.25–17 × 3 × 0.64 in.).
Atmosphere — industrial (Midland); marine (Kure Beach 242 meter [800 ft] Station).
Exposure — tension side up.
Surface finish — A.P. = Acetic Acid Pickle; C.P. = Chrome-Pickle.

Composition			Metallurgical State	Surface Finish	Tensile Yield Strength, 1000 psi	Days to Fracture % Tensile Yield Strength			
Al	Zn	Mn				95	85	80	60
					Midland				
..	...	1.5	Hard-rolled	Ground	22.3	.....	.....	>2200	>2200
3	1	0.3	Annealed 121° C	C.P.	34.2	>450	>450	....	....
6	1	0.2	Annealed 177° C	C.P.	34.0	75	210	....	>1825
					Kure Beach				
...	...	1.5	Annealed 331° C	C.P.	18.2	...	...	>400	>400
3	1	0.3	Annealed 121° C	C.P.	34.2	...	...	>400	>400
3	1	0.3	Annealed 331° C	C.P.	22.3	...	...	>400	>400
6	1	0.2	Hard-rolled	C.P.	40.0	...	...	58	>400
6	1	0.2	Annealed 177° C	C.P .	34.0	...	...	>400	>400
6	1	0.2	Annealed 177° C	A.P.	34.0	...	...	>400	>400
6	1	0.2	Annealed 331° C	C.P.	25.0	...	...	>400	>400
6	1	0.2	Annealed 331° C	A.P.	25.0	...	...	130	165
6	1	0.2	Annealed 331° C	Alox. Abraded	25.0	...	...	>160	>400

in a relatively short time. In either location stress corrosion has not been obtained for the Mg–1.5% Mn alloy or for the Mg–3% Al–1% Zn–0.3% Mn alloy, indicating that these alloys are not particularly sensitive to stress corrosion caused by residual stresses. The commercial casting alloys also show no evidence of marked sensitivity at stresses up to their yield strength.[9]

Figures 21 and 22 show stress-time curves for these sheet alloys based on data obtained from constant stress direct-tension units exposed at Kure Beach, N. C., and at a rural location (Greendale) 17 miles from Midland. Standard A.S.T.M.-type machined test bars were given a chrome-pickle treatment, then axially loaded to a given tensile stress by means of lead weights or by means of large springs. At these two locations, so widely varying in corrosive atmosphere, approximately the same time to failure at a given stress on identical alloy batches was obtained. The

[9] A. Beck, *The Technology of Magnesium and Its Alloys*, p. 294, F. A. Hughes and Co., Ltd., London, 1940.

time of year at which exposure starts has some effect on the shape of the stress-time curve, factors such as rainfall and dew apparently affecting the time to fracture.
Contrary to results obtained using constant deformation tests, the Mg–3% Al–1% Zn–0.3% Mn alloy will fail by stress corrosion, but the time to failure at a given stress is very much longer than that required to crack the higher-aluminum-containing alloy. The stress corrosion limit in the atmosphere for the Mg–3% Al–1% Zn–0.3%

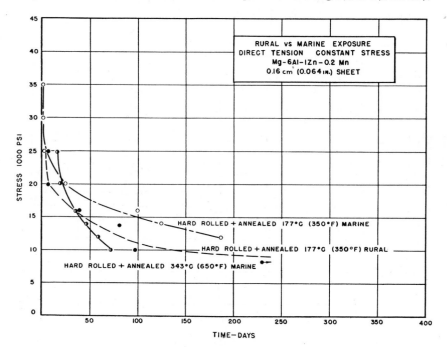

FIG. 21. Stress Corrosion under Rural and Marine Conditions of Mg–6% Al–1% Zn–0.2% Mn Sheet.

Specimen size — standard A. S. T. M. test bars 0.16 cm (0.064 in.) thick
Surface preparation — chrome pickle.

Mn sheet alloy appears to be about 16,000 to 20,000 psi after more than a year of exposure at constant stress. The Mg–6% Al–1% Zn–0.2% Mn alloy is very much more sensitive to stress corrosion, having failed at 10,000 psi in approximately 100 days.

Protection from stress corrosion may be accomplished by a stress relieving heat treatment or, with sheet, by cladding with the more anodic Mg–1.5% Mn alloy. Clad sheet constantly stressed at 90% of its yield point has not failed after 6 months of exposure. As-welded clad structures have not stress corroded after more than two years of exposure. Painting or shot peening are other means of delaying or eliminating the possibility of stress cracking of residually stressed structures.

Vibration. No data are available on the effect of various atmospheres on the fatigue strength of magnesium alloys except those shown in Fig. 9. From these data, the effect of alternating stresses would be expected to vary widely, depending on the corrosive environment.

Metallurgical Factors

As discussed previously, impurities in magnesium and its alloys, such as iron, nickel, and copper, very largely control the corrosion characteristics in aqueous media. While less significant, these impurities also cause some increase in corrosion in industrial atmospheric exposure, and their effect becomes more prominent in atmospheres containing chlorides. The effect on tensile strength losses in Midland and sea coast atmospheres is shown in Figs. 17, 19, and 20.

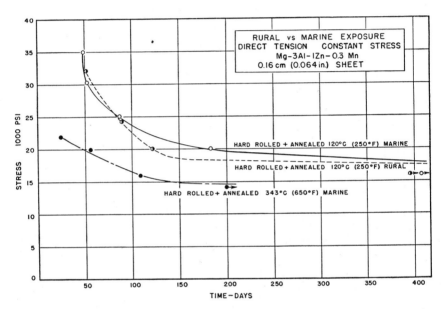

Fig. 22. Stress Corrosion under Rural and Marine Conditions of Mg–3% Al–1% Zn–0.3% Mn Sheet.

Specimen size — Standard A. S. T. M. test bars 0.16 cm (0.064 in.) thick.
Surface preparation — chrome pickle.

In the past it was felt that the Mg–1.5% Mn alloy was superior to the ternary alloys containing aluminum and zinc both in severe atmospheric and aqueous media. A survey of a large number of batches of the Mg–1.5% Mn alloy compared to the high-purity ternary alloys exposed at Kure Beach 24 meters (80 ft) from the ocean shows that there is no marked difference in property losses for these alloys, although the Mg–1.5% Mn alloy has a tendency to pit, owing to localized impurities, and has a poorer appearance.

The addition of zinc to the ternary Mg-Al-Mn alloys causes the same tendency toward slight pitting action in atmospheric corrosion as in salt solutions. Controlled-purity Mg-Al-Mn alloys corrode without pitting.

In ordinary exterior exposure, solution heat treatment of commercial casting alloys has no practical significance; but for severe exposures, as at the 24-meter (80-ft) station at Kure Beach, the "heat-treated" state shows a much greater loss in tensile strength than the "as-cast." Data obtained on 1.2 cm (0.5 in.) round-machined test bars are shown in Fig. 23. Based on data obtained over a 4-year exposure period, high-purity casting alloys do not show appreciable effect of heat treatment.

PROTECTIVE MEASURES

For storage under highly humid conditions, magnesium may be satisfactorily protected by a film of oil or grease.* For structures exposed to severe conditions, zinc chromate primers, followed by lacquers, enamels, or varnishes should be applied over recommended chemical treatments of the metal surface (p. 864). Without these

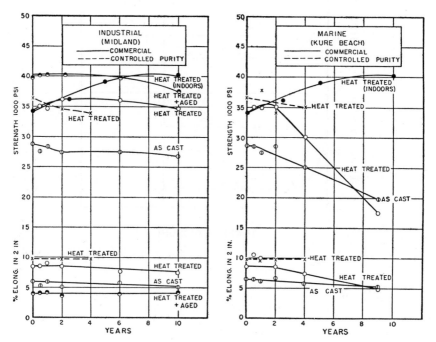

FIG. 23. Effect of Heat Treatment on Property Losses of a Magnesium Casting Alloy Exposed to Industrial and Marine Atmospheres.

Composition — Mg–6% Al–3% Zn–0.2% Mn.
Specimen size — 1.2 cm (0.5 in.) round tensile test bars.
Surface treatment — commercial, sand-blasted. Controlled purity, machined.

Note: "Heat Treated (Indoors)" are data from blanks stored indoors in sealed containers to determine the effect of precipitation on the properties at room temperature.

surface treaments, paint adhesion is normally poor, although special primers showing indications of good adhesion to bare metal have recently been developed. The effectiveness of chemical treatments, followed by painting, on the reduction of tensile corrosion losses of thin sheet is shown by Fig. 24. Initial film formation apparently accounts largely for the losses occurring during the first two years on the thin bare sheet. For thicker materials, as shown in Fig. 19, longer times are necessary to show significant effects.

* See *Temporary Corrosion Preventive Coatings*, p. 916.

Although protection from galvanic corrosion is seldom necessary with magnesium exposed to inland atmospheres, dissimilar metal assemblies exposed to marine condi-

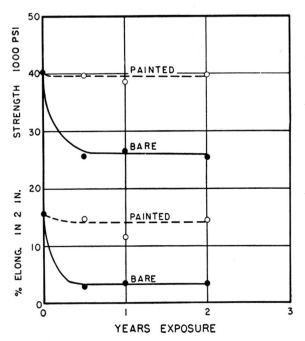

FIG. 24. Effect of Paint on Tensile Loss by Corrosion of Thin Sheet Exposed to Marine Atmospheres.

Composition — Mg–6% Al–1% Zn–0.2% Mn.
Specimen size — 15 × 10 × 0.08 cm (6 × 4 × 0.032 in.).
Surface treatment — Bare: acid-pickled. Painted: dichromate-coated. Paint schedule: 1 coat P-27 + 3 coats V-10e Al.
Atmosphere — Kure Beach, N. C., 24 meters (80 ft) from ocean.

tions, or so designed that electrolytes become trapped, may be protected as suggested under aqueous corrosion (p. 228). Painting of the faying surfaces is usually sufficient.

MOLYBDENUM

W. J. KROLL*

Molybdenum is characterized by the formation of soluble alkali molybdates and resistance to hydrofluoric and hydrochloric acids. The property of embrittlement when overheated and the difficulties in joining restrict use of the metal.

The following information is for molybdenum in wrought form. Cold and hot water do not attack the metal, but water vapor reacts at 700° C (1300° F). Alkali solutions are corrosive only in the presence of oxygen or oxidizing agents like H_2O_2, nitrates, and chlorates.

* Consulting Metallurgist, Albany, Oregon.

Molybdenum rapidly dissolves as anode in caustic solutions. In 20% sulfuric acid the anodic corrosion rate (room temperature) is 610 mg per amp-hour at a current density of 40 amp per sq dm (370 amp per sq ft).[1]

Hydrofluoric acid, cold or hot, is without effect on molybdenum. Concentrated hydrochloric acid at 110° C (230° F) reacts only slowly.[2] Nitric acid and aqua regia dissolve the metal readily, as does concentrated boiling sulfuric acid.

Corrosion rates in dilute acids are reported by Rohn[3] as follows:

MDD

Duration of Test, hr		10% HNO_3	10% H_2SO_4	10% HCl	10% Acetic Acid	10% Phosphoric Acid
24	Room temp.	5,200	6	10	20	30
1	Hot	43,000	36	36	72	216
24	Hot	.....	30	80	..	...

Chromic acid solutions or mixtures with sulfuric acid do not corrode molybdenum readily.

Solutions of $CuCl_2$, $HgCl_2$, and $FeCl_3$ are corrosive.[2] A good etching reagent is a solution of sodium hydroxide and $K_3Fe(CN)_6$. The corrosion characteristics of alloys of molybdenum containing 10 to 20% tungsten have been described.[4]

Fluorine reacts with molybdenum at room temperature, chlorine and bromine at red heat, iodine not at all. The elements sulfur, phosphorus, boron, and silicon and the gases NO, H_2S, CO_2, and $COCl_2$ react with the metal at elevated temperatures.

Molybdenum does not oxidize rapidly in air below 600° C (1100° F) or below 500° C (950° F) in oxygen although oxide films showing interference colors may form. The metal is unaffected by dry or wet hydrogen at low or high temperatures.

Molten caustic alkalies dissolve molybdenum in absence of oxygen, beginning at 660° C, and at lower temperatures in presence of oxygen.

GENERAL REFERENCE

GMELIN, L., *Handbuch der anorganischen Chemie*, Vol. 53, p. 69, Verlag Chemie, 1935.

NICKEL

W. Z. FRIEND*

The principal varieties of nickel used for corrosion-resisting purposes, together with their nominal chemical compositions and mechanical properties, are shown in Table 1.

Nickel in the electromotive series (p. 1134) is more noble than iron and less so than copper. It does not readily discharge hydrogen from any of the common non-oxidizing acids with which it may come in contact so that a supply of some oxidizing agent, such as dissolved air, is necessary for appreciable rates of corrosion. As a

* Development and Research Division, The International Nickel Co., Inc., New York, N. Y.

[1] F. R. Morral and J. L. Bray, *Trans. Am. Electrochem. Soc.*, **75**, 427–440 (1939).
[2] H. Endo and A. Itagaki, *Met. Abstracts*, Institute of Metals, **7**, 20–21 (1940).
[3] W. Rohn, *Z. Metallkunde*, **18**, 387 (1926).
[4] J. A. M. van Liempt, *Rec. trav. chim.*, **45**, 508 (1926) and **46**, 11 (1927).

general rule, oxidizing conditions favor corrosion of nickel, while reducing conditions retard corrosion. However, nickel also has the ability to protect itself against certain forms of attack by the development of a corrosion-resistant or passive film. As a result, oxidizing conditions do not invariably accelerate corrosion.

The galvanic relation of nickel to other metals and alloys is, in general, the same in most corrosive solutions as its relation in sea water as shown by the galvanic series in Table 18, p. 416.

CORROSION CHARACTERISTICS IN LIQUIDS

The following remarks concerning the resistance of nickel in specific corrosive media apply, in general, to all the products shown in Table 1, including nickel plating where the plating is heavy enough to be free from porosity.

FRESH WATER

Nickel possesses a high degree of resistance to corrosion by natural waters and by distilled water. Analysis of distilled water from a nickel storage tank indicated a rate of corrosion under 0.006 mdd (0.000001 ipy). Similarly, tests in natural waters have shown corrosion rates always less than 6 mdd (0.001 ipy), and usually less than 0.6 mdd (0.0001 ipy). Nickel is resistant to corrosion by waters containing hydrogen sulfide or free carbon dioxide, although it will be tarnished by the former.

Carbonated water under pressure is only slightly corrosive toward nickel. Analysis of water from a nickel-lined carbonator operated at a pressure of 200 psi indicated a rate of corrosion of 1.2 mdd (0.0002 ipy). It has been noted that in the presence of a high concentration of chlorides, in one case 2000 ppm, carbonated water may cause pitting of nickel.

STEAM CONDENSATE

The condensate from mixtures of steam, air, and carbon dioxide may be corrosive to nickel if the ratio of carbon dioxide to air in the dissolved gases is within certain limits. The test results given in Table 2 indicate that, at 35 psi pressure and 70° C (158° F), corrosion occurs if the condensate is saturated with a carbon dioxide and air mixture containing between 50 and 90% carbon dioxide. In the presence of corrosion products of iron, the lower limit of the corrosive range may be reduced to approximately 30% carbon dioxide.

MINE WATER

Nickel may be significantly corroded by acid mine waters containing appreciable amounts of ferric or cupric salts, although, in some cases, it appears to be able to protect itself by the formation of a protective oxide film. The results of tests made in three Pennsylvania coal mines, under the auspices of Carnegie Institute of Technology, the U. S. Bureau of Mines, and an Advisory Board of Coal Mine Operators and Engineers[1] are shown in Table 3.

SEA WATER*

Nickel is resistant to corrosion by sea water when it is in motion, rates of attack being usually less than 30 mdd (0.005 ipy). It may suffer local attack or pitting under

---
* Refer also to *Behavior of Metals and Alloys in Sea Water*, p. 402.

[1] W. A. Selvig and G. M. Enos, "Carnegie Institute for Coal Mining Investigations," *Bull.* 4, 69 pp. (1922).

TABLE 1. NOMINAL COMPOSITION AND MECHANICAL PROPERTIES OF COMMERCIALLY PURE NICKEL PRODUCTS

Material	Nominal Composition, % by Wt.							Form of Material	Mechanical Properties				Description
	Ni*	Cu	Fe	Mn	Si	C	S		Tensile Strength, 1000 psi	Yield Strength 0.2% Offset, 1000 psi	Elong. in 2 in., %	Brinell Hardness, 3000 kg	
Nickel†	99.4	0.10	0.15	0.20	0.05	0.10	0.005	Hot-rolled rod	65–80	20–30	45–35	100–140	Commercially pure wrought nickel.
"D" Nickel‡	95.2	0.05	0.15	4.50	0.05	0.10	0.005	Hot-rolled rod	80–95	30–55	50–35	120–170	Wrought alloy with improved resistance to sulfur compounds at elevated temperatures.
"L" Nickel§‡	99.4	0.10	0.15	0.20	0.05	0.02 max.	0.005	Annealed sheet and strip	55–75	15–25	55–35	25–55‖	Low-carbon commercially pure wrought nickel. Sometimes called "carbon-free" nickel.
"Z" Nickel‡	98.0							Hot-rolled rod, heat-treated	160–180	120–140	20–10	300–350	Wrought age-hardening high-nickel alloy applicable where higher hardness or strength are required than are available with nickel.
Cast Nickel	96.7	0.30	0.50	0.50	1.50	0.50	0.01		55–70	20–30	30–15	100–130	

* Includes a small amount of cobalt.
† Also available as nickel-clad steel.
‡ Registered U. S. Patent Office.
§ Also available as "L" Nickel-clad steel.
‖ Rockwell B.

256 CORROSION IN LIQUID MEDIA, ATMOSPHERE, GASES

Table 2. CORROSION OF NICKEL BY DISTILLED WATER SATURATED WITH CARBON DIOXIDE AND AIR MIXTURES

Pressure — 35 psi gage.
Velocity of specimens — 8.4 to 14.7 ft per min (2.6 to 4.5 meters per min).
Duration of tests — 22 hr.
Size of specimens — 3.18 × 3.81 × 0.079 cm (1.25 × 1.5 × 0.031 in.).

% by Volume of Carbon Dioxide and Air in Entering Gas	Corrosion Rate							
	at 70° C (158° F)				at 120° C (248° F)			
	Nickel		Mild Steel		Nickel		Mild Steel	
	mdd	ipy	mdd	ipy	mdd	ipy	mdd	ipy
100% CO_2	11	0.0018	605	0.121	7	0.0011	179	0.035
90% CO_2–10% Air	179	0.029	...	.....	..	......	399	0.080
80% CO_2–20% Air	410	0.066	745	0.149	51	0.0083	252	0.050
70% CO_2–30% Air	568	0.092	...	.....	55	0.0089	295	0.059
60% CO_2–40% Air	129	0.021	655	0.131	32	0.0052	318	0.064
50% CO_2–50% Air	7	0.0011	590	0.118	5	0.0008	323	0.065
40% CO_2–60% Air	8	0.0013	839	0.168	..	......	598	0.120
30% CO_2–70% Air	6	0.0010	...	.....	..	......	316	0.063
20% CO_2–80% Air	...	......	...	.....	6	0.0010	190	0.038
10% CO_2–90% Air	0	0	...	.....	..	......	...	.....
100% Air	1	0.0002	153	0.031	1	0.0002	1354	0.271

Table 3. CORROSION OF NICKEL BY COAL MINE WATERS*

Temperature — atmospheric.
Velocity — 1.8 ft per min (0.55 meter per min).

Coal Mine	Free H_2SO_4, ppm	Ferric Ion, ppm	Duration of Test, days	Corrosion Rate	
				mdd	ipy
Montour Mine No. 1	1430	58	119	146	0.024
Calumet Mine	430	135	98	118	0.019
Edna Mine No. 2	2160	860	135	0.3	0.00005

* W. A. Selvig and G. M. Enos, "Carnegie Institute for Coal Mining Investigations," *Bull.* 4, 69 pp. (1922).

conditions of stagnant exposure, especially where barnacles or other solids collect on the metal surface. Nickel is non-toxic toward marine organisms and will not suppress the growth of barnacles.

Non-oxidizing Inorganic Acids

Sulfuric Acid. Nickel can frequently be used with sulfuric acid solutions although it is usually less resistant than Ni-Cu alloys, Pb, or Si-Fe alloys. The results of a number of laboratory corrosion tests of nickel are shown in Table 4. Rates of corrosion in cold air-free sulfuric acid solutions are usually less than 30 mdd (0.005 ipy) in all concentrations under about 80% by weight. In cold air-saturated solutions in this concentration range, corrosion is usually a maximum at about 5% acid, and decreases uniformly with increasing concentration up to about 80% acid. Above this concentration, behavior is likely to be erratic, with rates of corrosion too high to make the use of nickel economical.

NICKEL

TABLE 4. CORROSION OF NICKEL BY SULFURIC ACID SOLUTIONS

Acid Concentration, % H_2SO_4 by Wt.	Temperature		Duration of Test, hr	Velocity, ft per min	Aeration	Corrosion Rate	
	°C	°F				mdd	ipy
2	20	68	5	No stirring	H_2-saturated	12.3	0.0020
5	30	86	20	15.7	H_2-saturated	19	0.0031
5	30	86	168	Convection	Air-saturated	272	0.044
5	30	86	96	16.0	Air-saturated	356	0.058
5	60	140	100	No stirring	By nat. convect.	58.3	0.0095
5	77	170	120	No stirring	By nat. convect.	127	0.0206
5	70	158	18	16.5	Air-saturated	633	0.103
5	102	216	23	Convection by boiling	Unaerated	201	0.033
10	Room	Room	96	No stirring	By nat. convect.	10.2	0.0017
10	77	170	120	No stirring	By nat. convect.	75	0.0121
10	82	180	20	26.5	Air-saturated	984	0.160
10	103	217	23	Convection by boiling	Unaerated	722	0.117
20	20	68	5	No stirring	H_2-saturated	24.6	0.0040
19	105	221	23	Convection by boiling	Unaerated	685	0.111
25	82	180	20	26.5	Air-saturated	509	0.083
70	38	100	24	15.5	Unaerated	180	0.029
95	Room	Room	20	Convection	N_2-saturated	437	0.071
95	Room	Room	20	Convection	Air-saturated	290	0.047

At temperatures between atmospheric and close to boiling, corrosion rates tend to vary with concentration, temperature, and aeration between 0.05 and 0.10 ipy at acid concentrations up to 25%. Nickel would ordinarily not be used for concentrations above about 25% acid at temperatures much above atmospheric.

In boiling sufuric acid, the use of nickel is usually confined to acid concentrations below 10%. This limiting concentration can sometimes be raised where boiling temperatures are lowered by the use of vacuum. For example, in a 4-month test in a rayon hardening bath evaporator operating at 54° C (130° F) under 27 in. (Hg) vacuum and where the final concentrated solution contained approximately 20% by weight sulfuric acid plus sodium sulfate and other chemicals including hydrogen sulfide, the corrosion rate of nickel was 10 mdd (0.0016 ipy).

Hydrochloric Acid. Corrosion rates of nickel in hydrochloric acid solutions in laboratory tests at 30° C (86° F) are shown in Fig. 1.[2] The upper curve for nickel refers to air-saturated acid at a velocity of 15.7 ft per min (4.8 meters per min), and the lower nickel curve refers to nitrogen-saturated acid at a velocity of 21.6 ft per min (6.6 meters per min). In air-saturated tests, air was bubbled through the solutions in an open container at a rate of 1500 ml per min. In air-free tests, nitrogen was bubbled through in a closed container at 200 ml per min. Applications of nickel at room temperature are in general limited to concentrations of acid under about 20%, although in some cases it may usefully be applied to higher concentrations.

The corrosion rates of nickel in 5% hydrochloric acid, both air-free and air-saturated at temperatures up to 90° C (194° F), are shown in Fig. 2. In unaerated 5% acid, it is usefully resistant up to about 55° C (130° F). In aerated solutions above atmospheric temperatures, its use is normally limited to concentrations of about 2 or 3%.

[2] W. Z. Friend and B. B. Knapp, *Trans. Am. Inst. Chem. Engrs.*, **39**, 731–753 (1943).

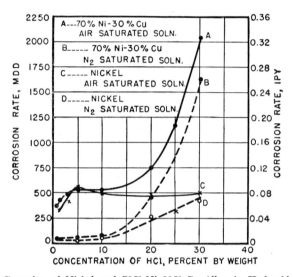

Fig. 1. Corrosion of Nickel and 70% Ni–30% Cu Alloy in Hydrochloric Acid.
Temperature — 30° C (86° F).
Duration of test — Air-saturated: 24 hours. N₂-saturated: 48 hours.

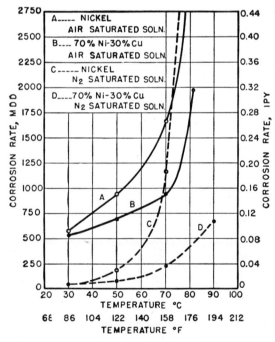

Fig. 2. Effect of Temperature on Corrosion of Nickel and 70% Ni–30% Cu Alloy in 5% Hydrochloric Acid. (Duration of test, 24 hours.)

NICKEL

The results of laboratory corrosion tests of nickel in several boiling hydrochloric acid solutions are shown in Table 5. In most of the processes in which hydrochloric acid is formed as a result of hydrolysis of non-oxidizing chlorides, or of chlorinated solvents, the acid content is less than 0.5% and such solutions are handled satisfactorily.

TABLE 5. CORROSION OF NICKEL BY BOILING HYDROCHLORIC ACID SOLUTIONS

Duration of tests — 10 days.
Velocity — convection by boiling.
Aeration — none.

Acid Concentration, % HCl by Wt.	Corrosion Rate of Nickel	
	mdd	ipy
0.5	1,875	0.304
1.0	4,200	0.680
5.0	35,400	5.74

Phosphoric Acid. Nickel cannot ordinarily be used with crude phosphoric acid solutions containing ferric salts as normally encountered in the processing of acid from phosphate rock. In a 24-hour test[3] in a dilute solution of crude phosphoric containing 0.40% iron at 80° C (176° F) the corrosion rate of nickel was 4960 mdd (0.80 ipy). Pure phosphoric acid solutions containing no oxidizing compounds have slight to moderate action on nickel at atmospheric temperatures, rates being of the order of 20 to 80 mdd (0.003 to 0.012 ipy) for quiet exposure, and 50 to 200 mdd (0.008 to 0.032 ipy) for agitated aerated acid. Hot concentrated solutions of pure acid are very corrosive, rates as high as 3380 mdd (0.55 ipy) having been reported[3] in a 24-hour test in 85% acid at 95° C (203° F).

Sulfurous Acid. Strong sulfurous acid solutions are usually very corrosive to nickel, especially when hot. The low concentrations of sulfur dioxide used in the preservation of food products are not destructive, although they may develop a dark tarnish on the metal. In 20-hour laboratory tests in an unaerated, unagitated water solution containing 1% SO_2 at 20° C (68° F) the corrosion rate of nickel was 331 mdd (0.054 ipy). In a solution containing 1500 ppm the corrosion rate was 42 mdd (0.0068 ipy).

Solutions of Hydrogen Sulfide. Hydrogen sulfide accelerates corrosion of nickel but resistance of the metal to attack is usually adequate in solutions which may contain the gas, such as brines, natural waters, and alkaline solutions. Specimens of nickel suspended over water through which hydrogen sulfide was bubbled at 65° C (150° F) were corroded at a rate of 160 mdd (0.026 ipy). In a 24-hour test in distilled water saturated with hydrogen sulfide at 25° C (78° F) the corrosion rate was 12 mdd (0.0019 ipy).

OXIDIZING ACIDS

Nickel does not possess useful resistance to corrosion by oxidizing acids, such as nitric or nitrous, except in concentrations under 0.5% at atmospheric temperatures. Similarly, high corrosion rates occur in other acids containing oxidizing chemicals such as ferric and cupric salts, peroxides, chromates, and the like.

[3] P. R. Kosting and C. Heins, Jr., *J. Ind. Eng. Chem.*, **23**, 140–150 (1931).

ORGANIC ACIDS AND COMPOUNDS

Organic acids, under the most frequently encountered conditions, are only moderately corrosive toward nickel. However, at elevated temperatures and when highly aerated, such acids as *acetic* and *formic* may cause considerable corrosion. The results of laboratory corrosion tests in several acetic acid solutions are given in Table 6. In cold, air-free solutions, maximum corrosion usually occurs at about 50% concentration and decreases at a somewhat uniform rate at higher and lower concentrations. The combined effects of velocity and aeration are indicated by the results of 20-hour laboratory tests in air-saturated acetic acid solutions at 30° C (86° F) and with specimens moved at a velocity of 18.5 ft per min (5.7 meters per min), where the corrosion rate of nickel in 6% acid was 290 mdd (0.047 ipy), and in 75% acid was as high as 2101 mdd (0.34 ipy).

TABLE 6. CORROSION OF NICKEL BY ACETIC ACID SOLUTIONS
No Stirring

Acid Concentration, % CH$_3$COOH by Wt.	Temperature		Duration of Test, hr	Aeration	Corrosion Rate		Ref.
	° C	° F			mdd	ipy	
6	30	86	96	Immersed-unaerated	21	0.0035	(1)
6	30	86	96	Immersed-N$_2$ saturated	6	0.0009	(1)
6	30	86	96	Immersed-air saturated	67	0.011	(1)
6	30	86	24	Alternate immersion-continuous	141	0.023	(1)
6	30	86	24	Alternate immersion-intermittent	111	0.018	(1)
6	Room	Room	720	Spray	40	0.0064	(1)
6	20	68	5	Immersed-H$_2$ saturated	12	0.0020	(2)
50	20	68	5	Immersed-H$_2$ saturated	62	0.010	(2)
Glacial	20	68	5	Immersed-H$_2$ saturated	25	0.0041	(2)

(1) H. S. Rawdon and E. C. Groesbeck, Bureau of Standards, *Tech. Bull.* 367, 409–446 (1928).
(2) W. G. Whitman and R. P. Russell, *J. Ind. Eng. Chem.*, **17**, 348–354 (1925).

Nickel is corroded only moderately by unaerated boiling acetic acid solutions as shown in Table 7, where specimens were immersed in the boiling acid and also exposed to the vapors therefrom.

TABLE 7. CORROSION OF NICKEL IN ACETIC ACID, BOILING UNDER REFLUX

Duration of tests — 20 hours.

Acetic Acid Concentration, % by Wt.	Corrosion Rate			
	Liquid		Vapor	
	mdd	ipy	mdd	ipy
5	66	0.011	46	0.0075
50	114	0.019	89	0.015
98	73	0.012	24	0.0038
99.9	84	0.014	14	0.0023

Nickel is usefully resistant to fatty acids, such as *stearic* and *oleic*, even at elevated temperatures. In tests for 21 days in a vacuum fatty-acid still at 227° C (440° F) it corroded at a rate of 25 mdd (0.004 ipy).

In laboratory tests with specimens immersed in commercially pure *phenol* with no agitation, the corrosion rate of nickel was 0.41 mdd (0.00007 ipy) in 14-day tests at 53° C (127° F) and 0.17 mdd (0.00003 ipy) in 6-day tests at 184° C (363° F). Samples of phenol taken from a 10,000-gal nickel-clad steel storage tank after 6 months of storage at atmospheric temperature had a nickel content under 0.001%.

Food Products. Common applications of nickel in contact with dilute organic acids and other organic compounds are made in connection with the handling of food products such as fruits and vegetables and their juices and alcoholic beverages.[4,5,6] Corrosion data in some typical *fruit juices* are given in Table 8. Nickel is suitable

TABLE 8. RESISTANCE OF NICKEL TO CORROSION BY FRUIT JUICES

Product	Duration of test, hr	Temperature		Conditions	Corrosion Rate	
		° C	° F		mdd	ipy
Tomato juice	...	Room	Room	Aerated, agitated	75	0.012
Tomato juice	...	Room	Room	Unaerated, agitated	49	0.0080
Tomato juice	9	77	170	Air-saturated, mildly agitated	148	0.024
Tomato juice	7	90	195	Aerated, mildly agitated	122	0.020
Lemon juice	...	Room	Room	Aerated, agitated	124	0.020
Lemon juice	...	Room	Room	Unaerated, agitated	3	0.0005
Lemon juice	24	Boiling	Boiling	Under reflux, unaerated	84	0.014
Pineapple juice	...	Room	Room	Aerated, agitated	110	0.018
Pineapple juice	...	Room	Room	Unaerated, agitated	22	0.0036
Pineapple juice	106	82	180	Air-saturated	224	0.036
Pineapple juice	24	74–80	165–175	Boiling in evaporator under 16–18 in. vacuum	28	0.0046
Grape juice	...	Room	Room	Aerated, agitated	154	0.025
Grape juice	...	Room	Room	Unaerated, agitated	38	0.0062
Grape juice	24	Boiling	Boiling	Under reflux, unaerated	42	0.0068
Orange juice	72	Boiling	Boiling	Under reflux, unaerated	49	0.0080

for use in contact with milk at atmospheric temperature and for heating of milk, but is not recommended for cooling of milk from pasturizing temperatures because of a difference in the type of protective film formed.[7]

Nickel is non-toxic, and a number of investigations[8,9,10] have demonstrated the safety of nickel equipment for the preparation and handling of foods. It has also been demonstrated that nickel is not destructive to vitamins.[11]

[4] International Nickel Co., Inc., "Corrosion Resistance of Materials for Pea, Corn and Tomato Processing," *Tech. Bull.* TS-8 (1939).
[5] International Nickel Co., Inc., "Metals and Wines, Distilled Liquors and Beers," *Tech. Bull.* TS-6.
[6] E. Mrak and W. V. Cruess, *Food Industries*, **1**, 559–563 (1929).
[7] International Nickel Co., Inc., "The Resistance of Pure Nickel and Inconel to Corrosion by Milk," *Tech. Bull.* TS-1.
[8] K. R. Drinker, L. T. Fairhall, G. B. Ray, and C. K. Drinker, *J. Ind. Hygiene*, **6**, 307–356 (1924).
[9] F. B. Flinn and J. M. Inouye, *J. Am. Med. Assoc.*, **90**, 1010–1013 (1928).
[10] R. J. McKay, O. B. J. Fraser, and H. E. Searle, American Institute of Mining and Metallurgical Engineers, *Tech. Publication* 192, 47 pp. (1929).
[11] A. D. Pratt, *J. Nutrition*, **3**, 141–155 (1930).

ALKALIES

Nickel has excellent resistance to corrosion by alkalies such as caustic soda and caustic potash. Tests in evaporators concentrating caustic soda to 50% by weight NaOH show that the rate of corrosion is negligible, being of the order of 0.6 mdd (0.0001 ipy). The average nickel content of 50% caustic processed in nickel equipment is about 0.00005%. In the evaporation of caustic soda to 75% by weight NaOH, corrosion is slightly higher, tests[12] showing corrosion rates normally less than 18 mdd (0.003 ipy). Nickel is free from caustic embrittlement under the above conditions.

In caustic concentrations above 75% by weight, nickel is second only to silver in resisting attack. What actually determines the behavior of nickel in highly concentrated caustic soda or fused caustic is the nature of an oxide film that forms. This film is usually thin and black. In its presence the rate of corrosion by fused caustic is of the order of 20 mdd (0.003 ipy). Without the film, rates up to 150 mdd (0.025 ipy) may be expected. Temperatures above 315° C (600° F) favor development of a protective oxide, whereas at lower temperatures, around 200° C (400° F), a less protective, green form of the oxide may develop.

Nickel-clad steel vessels have been used to hold fused, sulfur-free caustic at 700° C (1300° F) and have shown no appreciable attack after 8 months, where the useful life of steel was 3 weeks. However, if sulfur is added to molten caustic for purification, nickel may suffer intergranular attack by the sulfur.

Under conditions of combined highly stressed metal and concentrated caustic (75 to 98% NaOH) at high temperatures, 300° to 500° C (600° to 900° F), nickel is subject to a form of intergranular attack. Consequently, nickel equipment for use in this service should be annealed before exposure. (See p. 264.) The use of a low-carbon wrought nickel* is preferred over regular wrought nickel for fused caustic applications, because of indications of better resistance to this form of attack, particularly at welds.

Nickel is not attacked by *anhydrous ammonia* and is resistant to *aqueous ammonia* or ammonium hydroxide in concentrations under 1% NH_3 by weight. Aeration may sometimes induce passivity in concentrations under 10%, but in the presence of air more highly concentrated solutions are appreciably corrosive.

SALT SOLUTIONS

Neutral and Alkaline Salts. Neutral and alkaline salt solutions, such as chlorides, carbonates, sulfates, nitrates, and acetates, are resisted well by nickel. Rates of corrosion, even in hot, aerated solutions are rarely more than 30 mdd (0.005 ipy). Applications in the evaporation of such salts as sodium chloride and sodium sulfate are common. The results of a number of plant corrosion tests in neutral and alkaline salts are shown in Table 9.

Acid Salts. Nickel possesses useful resistance to corrosion by acid salts, especially to acid chlorides such as those of ammonium and zinc. Rates of corrosion in boiling concentrated solutions are usually less than 250 mdd (0.04 ipy). The results of a number of plant corrosion tests in solutions of acid salts are shown in Table 10.

Oxidizing Acid Salts. Oxidizing acid salts, as a class, are usually corrosive toward nickel. Nickel is not recommended for use with more than extremely dilute solutions of such salts as ferric chloride, mercuric chloride, and cupric chloride. Likewise, the addition of such oxidizing salts as chromates, dichromates, nitrates, and peroxides to mineral acids may make them highly corrosive to nickel. A possible exception is

---
* "L" nickel.

[12] International Nickel Co., Inc., "Resistance of Nickel and Its Alloys to Corrosion by Caustic Alkalies," *Tech. Bull* T-6.

NICKEL

TABLE 9. CORROSION OF NICKEL BY SOLUTIONS OF NEUTRAL AND ALKALINE SALTS

Corrosive Medium	Exposure Conditions (Concentration of Salt in % by Weight)	Corrosion Rate	
		mdd	ipy
Calcium chloride	In evaporator concentrating solution to 35% at 70°–160° C (160°–320° F) for 225 days	2	0.0003
Cobalt acetate	In evaporator at 105° C (225° F) for 950 hours	25	0.004
Sodium chloride	Saturated solution mixed with steam and air at 95° C (200° F)	13	0.0021
Sodium metasilicate	In evaporator concentrating to 50% solution at 110° C (230° F) for 42 days	0.15	0.00002

TABLE 10. CORROSION OF NICKEL BY SOLUTIONS OF ACID SALTS

Corrosive Medium	Exposure Conditions (Concentration of Salt in % by Weight)	Corrosion Rate	
		mdd	ipy
Aluminum sulfate	Quiet immersion in 25% solution in storage tank at 35° C (95° F) for 112 days	4	0.0006
Aluminum sulfate	In evaporator concentrating to 57% solution at 115° C (240° F) for 44 days	366	0.059
Ammonium chloride	In evaporator concentrating 28% to 40% at 102° C (216° F) for 32 days	52	0.0084
Ammonium sulfate plus sulfuric acid	In saturated solution containing 5% sulfuric acid in suspension tank during crystallization at 41° C (106° F) for 33 days	18	0.0030
Manganese chloride plus some free hydrochloric acid	Immersed in boiling 11.5% solution in flask equipped with reflux condenser at 101° C (214° F) for 48 hours	54	0.0087
Manganese sulfate	In evaporator concentrating 1.250 to 1.350 sp. gr. at 113° C (235° F) for 11 days	18	0.0029
Zinc chloride	In evaporator concentrating 7.9% to 21% at 38° C (100° F) under 26–28 in. vacuum for 210 days	29	0.0046
Zinc chloride	In evaporator concentrating 21% to 69% at 115° C (240° F) under 15–18 in. vacuum for 90 days	245	0.040
Zinc sulfate	Saturated solution containing trace of sulfuric acid in evaporating pan at 105° C (225° F) for 35 days with vigorous stirring	152	0.025

stannic chloride, dilute solutions of which nickel resists to a useful degree. Tests in an aerated 27.5° Bé solution at 210° C (70° F) for a period of 3 weeks showed a corrosion rate of 109 mdd (0.018 ipy).

Oxidizing Alkaline Salts. Nickel is attacked by strong hypochlorite solutions. For continuous exposure its use is limited to the very dilute solutions containing less than 500 ppm available chlorine, frequently used for sterilizing purposes. For discontinuous exposure, as in cyclic textile bleaching operations where bleaching is followed by rinsing and souring in the same vessel, hypochlorite concentrations up to about 3 grams per liter available chlorine can be handled safely.[13] When corrosion occurs, it is likely to be localized in the form of pits.

The resistance of nickel to corrosion by hypochlorite solutions is increased somewhat by a high degree of surface polish.

Attack is often inhibited by as little as 0.5 ml per liter of 1.4 sp. gr. sodium silicate.

[13] International Nickel Co., Inc., "Nickel, Monel and Inconel in Textile Bleaching Operations," *Tech. Bull.* T–22.

The effect of inhibitors on the corrosion of nickel in sodium hypochlorite solutions is shown in Table 11.

TABLE 11. INHIBITING EFFECT OF SODIUM SILICATE AND TRISODIUM PHOSPHATE ON CORROSION OF NICKEL BY SODIUM HYPOCHLORITE SOLUTIONS

Temperature — 40° C (105° F).
Duration of tests — 16 hr.
Beaker tests — no agitation.

Solution Composition Grams per Liter			Corrosion Rate		Max. Depth Pitting	
Available Chlorine	Sodium Silicate	Trisodium Phosphate	mdd	ipy	mm	in.
6.5	...	...	321	0.052	0.56	0.022
6.5	0.5	...	64	0.010	0.36	0.014
6.5	...	0.5	122	0.020	0.61	0.024
6.5	2.0	...	9	0.001	None	None
6.5	...	2.0	57	0.009	None	None
3.3	...	...	183	0.030	0.36	0.014
3.3	0.5	...	22	0.004	None	None
3.3	...	0.5	40	0.006	None	None
0.1	...	...	26	0.004	None	None
0.1	0.5	...	3	0.0005	None	None
0.1	...	0.5	4	0.0006	None	None

Nickel is resistant to alkaline peroxide solutions and does not catalyze the decomposition of these solutions.[13]

CHLORINATED SOLVENTS

A study has been made[14] of the resistance of several metals and alloys to corrosion by chlorinated solvents and mixtures of these solvents with water, both at room temperature and boiling. The corrosion rates obtained for nickel, 70% Ni–30% Cu alloy, and mild steel are shown in Table 12.

EFFECT OF MERCURY

Nickel resists amalgamation by mercury at moderate temperatures. However, in 15-day tests at 400° C (750° F) amalgamation was observed.

EFFECT OF STRESS

Stress corrosion cracking of wrought nickel has been observed[15] only in high-concentration, high-temperature caustic soda and caustic potash. Stress corrosion cracking in this service can be prevented by annealing the equipment briefly (3 to 5 min) at 875° C (1600° F) before placing in use.

High stresses, so long as they are uniform and are not accompanied by physical transformation, do not significantly increase the corrosion rate of nickel as indicated by the corrosion test data presented in Fig. 7 on p. 577.

[14] P. S. Brallier, Private communication.
[15] O. B. J. Fraser, "Stress-Corrosion Cracking of Nickel and Some Nickel Alloys," *Symposium on Stress Corrosion Cracking of Metals (1944)*, The American Society for Testing Materials and the American Institute of Mining and Metallurgical Engineers, Philadelpia and New York, 1945.

NICKEL

TABLE 12. RESULTS OF TESTS OF NICKEL, 70% Ni–30% Cu ALLOY, AND MILD STEEL IN CHLORINATED SOLVENTS AND THEIR VAPORS

In tests at 25°–30° C, specimens were exposed to solvent and air with water absent, and to solvent and air with water present. In both cases, 90% of area of specimen was immersed in solvent.

In tests at boiling point, specimens were exposed to solvent and vapor with water absent, 25% of the area of each specimen being in contact with solvent and 75% with vapor. In tests with water present, about 25% of area of each specimen exposed to solvent, 25% to water layer, and remaining 50% to vapor. Weight loss was averaged over total area of each specimen. No agitation.

Solvent	Corrosion Rate							
	Tests at 25°–30° C (75°–85° F)				Tests at Boiling Point			
	Water Layer Present		Water Layer Absent		Water Layer Present		Water Layer Absent	
	mdd	ipy	mdd	ipy	mdd	ipy	mdd	ipy
(a) Nickel								
Carbon tetrachloride	0.12	0.00002	0.02	0.000003	11.1	0.0018	0.16	0.00003
Chloroform	0.36	0.00006	0.17	0.00003	0.73	0.0001	1.3	0.0002
Ethylene dichloride	0.08	0.00001	0.04	0.000007	2.2	0.0004	0.20	0.00003
Trichloroethylene	2.3	0.0004	0.94	0.00015	5.9	0.001	0.14	0.00002
Carbon tetrachloride * Ethylene dichloride	0.02	0.000003	0.02	0.000003	1.09	0.0002	0.34	0.00006
(b) 70% Ni–30% Cu Alloy								
Carbon tetrachloride	0.66	0.0001	0.06	0.00001	27.0	0.0044	0.23	0.00004
Chloroform	0.10	0.00002	0.08	0.00001	27.3	0.0043	0.93	0.0002
Ethylene dichloride	0.15	0.00002	0.06	0.00001	16.5	0.0027	0.17	0.00003
Trichloroethylene	4.1	0.0007	0.45	0.00007	67.1	0.011	0.36	0.00006
Carbon tetrachloride * Ethylene dichloride	0.12	0.00002	0.06	0.00001	6.1	0.0010	0.59	0.0001
(c) Mild Steel								
Carbon tetrachloride	44.3	0.0082	0.35	0.00006	874.0	0.16	0.30	0.00006
Chloroform	12.9	0.0024	3.7	0.0007	65.9	0.012	45.8	0.0086
Ethylene dichloride	7.6	0.0014	0.30	0.00006	73.4	0.0135	9.0	0.0016
Trichloroethylene	7.0	0.0013	3.3	0.0006	37.8	0.007	....	....
Carbon tetrachloride * Ethylene dichloride	48.4	0.0089	0.04	0.000007	1079.0	0.20	2.5	0.0005

* Mixture contained 90% carbon tetrachloride and 10% ethylene dichloride by volume.

WET AND DRY GASES

Dry gases are not actively corrosive to nickel at or near atmospheric temperatures. Such gases as nitric oxides, chlorine and other halogens, sulfur dioxide, and ammonia are appreciably corrosive only when they contain condensed moisture.

ATMOSPHERIC CORROSION

An important application of nickel is in the form of plating to protect other and more vulnerable materials from atmospheric corrosion and tarnishing. The resistance of nickel coatings to atmospheric attack is discussed beginning p. 817.

Nickel and nickel plating will remain reasonably bright and free from tarnish indoors, being superior to silver, copper, and brass in this respect. Nickel becomes dull when exposed outdoors and tends to acquire a very thin, adherent corrosion product coating which usually is a basic sulfate.

Actual continued corrosion of nickel in the atmosphere is practically nil for indoor exposure, while outdoors the rate of attack is extremely slow and varies with atmospheric conditions. Sulfurous atmospheres encountered in industrial and urban communities are naturally most corrosive. Marine atmospheres are scarcely more corrosive than suburban or rural atmospheres. An indication of the extent of corrosion is provided by the results of 9 to 10 years of exposure of specimens of nickel in several typical atmospheres (Table 13).[16]

TABLE 13. CORROSION OF NICKEL IN VARIOUS ATMOSPHERES*

(Specimens were exposed for 9 to 10 years, followed by chemical cleaning, except where noted.)

Test Location	Type of Atmosphere	Corrosion Rate mdd	Corrosion Rate ipy	Description of Corrosion Product Films on Panels as Removed from Racks
Pittsburgh, Pa.	Industrial†	0.31	0.00013	Uniform dark film ‡
Altoona, Pa.	Industrial	1.02	0.00016	Film slightly rough, mottled dark gray-green and black, adherent
New York, N. Y.	Industrial	0.85	0.00014	Film smooth, mottled gray and green, adherent
Rochester, N. Y.	Industrial†	0.22	0.00004	Smooth mottled film ‡
Sandy Hook, N. J.	Sea coast	0.19	0.00003	Film smooth, dark gray, adherent
Key West, Fla.	Sea coast	0.03	0.000005	Tarnish smooth, gray, adherent
La Jolla, Calif.	Sea coast	0.03	0.000005	Tarnish smooth, gray spotted with many gray-green scale spots
State College, Pa.	Rural	0.04	0.000007	Tarnish smooth and bright, gray
Phoenix, Ariz.	Rural	0.008	0.000001	Tarnish bright, smooth, light gray

* From 1943 Report of Subcommittee VI, Committee B-3, American Society for Testing Materials.
† Specimens weighed after 5 to 6 years of exposure and without chemical cleaning.
‡ Description of films after 5 to 6 years of exposure.

Experience has shown nickel to be free from season cracking and other forms of stress corrosion in atmospheric exposure.

GENERAL REFERENCES

LaQue, F. L., *Mech. Eng.*, **58**, 827–843 (1936).
McKay, R. J., and R. Worthington, *Corrosion Resistance of Metals and Alloys*, 1st Ed., Reinhold Publishing Corp., New York, 1936.

NICKEL-COPPER ALLOYS

W. Z. Friend*

The alloys of nickel and copper containing more than 50% nickel are referred to as Ni-Cu alloys to distinguish them from those containing more than 50% copper, referred to as Cu-Ni alloys.† The nominal compositions and mechanical properties of the principal Ni-Cu alloys used for corrosion-resisting purposes are shown in Table 1.

The Ni-Cu alloys are more resistant than nickel under reducing conditions and more resistant than copper under oxidizing conditions. As a net result, they are, in general, more resistant to corrosion than either of their principal constituents. Some

* Development and Research Division, The International Nickel Co., Inc., New York, N. Y.
† The Cu-Ni alloys are discussed beginning p. 85.
[16] *Proc. Am. Soc. Testing Materials*, **43**, 137–154 (1943).

TABLE 1. NOMINAL COMPOSITION AND MECHANICAL PROPERTIES OF Ni-Cu ALLOYS

Alloy	Nominal Composition, % by Wt.							Mechanical Properties				Description
	Ni*	Cu	Fe	Mn	Si	C	S	Al	Tensile Strength, 1000 psi	Yield Strength, 0.2% Offset, 1000 psi	Elong. in 2 in., %	Brinell Hardness, 3000 kg	
Monel†,‡	67	30	1.4	1	0.1	0.15	0.01	...	80–95§	40–65	45–30	130–170	Commercial wrought alloy
"R"† Monel	67	30	1.7	1.1	0.05	0.10	0.035	...	75–90§	35–60	45–25	130–170	Free-machining alloy particularly adapted for automatic screw machines
"K"† Monel	66	29	0.9	0.4	0.5	0.15	0.005	2.75	135 minimum ‖	100 minimum	20 minimum	260–300	Non-magnetic, age hardening alloy applicable where higher strength or hardness are required than available with Monel
Cast Monel	67	29	1.5	0.9	1.25	0.3	0.015	...	65–80	32.5–40	45–25	125–150	....
"H"† Monel	65	29.5	1.5	0.9	3.0	0.1	0.015	...	70–90	45–65	20–10	175–250	Cast alloy with increased but not maximum hardness and with good ductility
"S"† Monel	63	30	2.0	0.9	4.0	0.1	0.015	...	90–115	70–90	3–1	275–350	Cast alloy with maximum hardness; good gall and erosion resistance

* Includes a small amount of cobalt.
† Registered U. S. Patent Office.
‡ Also available in the form of Monel-clad steel
§ Mechanical properties given for hot-rolled rod.
‖ Mechanical properties given for hot-rolled rod, heat-treated.

CORROSION IN LIQUID MEDIA, ATMOSPHERE, GASES

study[1,2] of the series of Ni-Cu and Cu-Ni alloys has indicated that the alloy containing roughly two thirds nickel and one third copper possesses, in general, the optimum corrosion resistance.

The commercial 70% Ni–30% Cu alloy (70-30 Ni-Cu)* is a simple solid-solution alloy, and is free from such types of corrosion as sometimes result from local galvanic effects between phases of multi-phase alloys. Its galvanic relationship to other metals and alloys is, in general, the same in most corrosive solutions as its relationship in sea water as shown by the galvanic series in Table 18, page 416.

CORROSION CHARACTERISTICS IN LIQUIDS

Experience has shown that all of the Ni-Cu alloys listed in Table 1 are substantially equal in resistance to corrosion. As a general rule, selection of a material from this group is based upon mechanical or physical properties. Similarity in the resistance of two of these alloys to several corrosives is shown in Table 2.

TABLE 2. CORROSION OF Ni-Cu ALLOYS BY SEVERAL CORROSIVE MEDIA

Material*	Corrosion Rate, mdd					
	In unaerated 15% Ammonia at 30°C (86°F), 2-Day Test	In 16% Aerated Calcium Chloride at 21°C (70°F), 28-Day Test	In Aerated Sodium Chloride at 21°C (70°F), 28-Day Test		In unaerated 50% Caustic Soda at 24°C (76°F), 4-Day Test	In Aerated 5% Sulfuric Acid at 24°C (76°F), 3-Day Test
			3% Solution	25% Solution		
"K" Monel, age-hardened	655	33	12	3	3	240
Monel	730	49	13	4	3	233

* See Table 1 for alloy composition.

FRESH WATER

The Ni-Cu alloys are highly resistant to corrosion by distilled water and by natural waters, both hard and soft. Corrosion rates are usually negligible, being less than 6 mdd (0.001 ipy) and often less than 0.6 mdd (0.0001 ipy) under the most severe conditions of temperature, flow, and aeration.

Carbonated water under pressure is only slightly corrosive to Ni-Cu alloys. The corrosion rate in a nickel-lined carbonator under a pressure of 450 psi was only 1.03 mdd (0.0002 ipy).

BOILER WATER AND STEAM CONDENSATE

Tests for 64 days in a boiler feed water heater at 198°C (388°F) under a pressure of 1000 psi at a pH of 8 to 8.5, oxygen content of 0.03 to 0.06 ml per liter, NH_3 content of 0.02 ppm, and carbon dioxide practically zero, showed 70-30 Ni-Cu to have a corrosion rate of 0.54 mdd (0.00009 ipy).

The condensate from mixtures of steam, air, and carbon dioxide may be corrosive to the Ni-Cu alloys if the ratio of carbon dioxide and air in the dissolved gases is

---

* Monel.

[1] H. S. Rawdon and E. C. Groesbeck, Bureau of Standards, *Tech. Bull.* 367, 409–446 (1928).
[2] F. L. LaQue, *J. Am. Soc. Naval Engrs.*, **53**, 29–64 (1941).

within certain limits. The test results in Table 3 indicate that at 35 psi pressure, corrosion occurs if the condensate is saturated with a carbon dioxide-air mixture containing between 30 and about 90% carbon dioxide by volume.

TABLE 3. CORROSION OF 70-30 Ni-Cu BY DISTILLED WATER SATURATED WITH CARBON DIOXIDE-AIR MIXTURES

Pressure — 35 psi.
Velocity of specimens — 8.4 to 14.7 ft per min.
Duration of tests — 22 hours.
Size of specimens — 3.18 × 3.81 × 0.079 cm (1.25 × 1.5 × 0.031 in.).

% by Volume of CO_2 and Air in Entering Gas	Corrosion Rate					
	At 70° C (158° F)		At 100° C (212° F)		At 135° C (275° F)	
	mdd	ipy	mdd	ipy	mdd	ipy
30% CO_2, 70% Air	55	0.009	...	...	...	...
70% CO_2, 30% Air	370	0.060	...	...	...	...
80% CO_2, 20% Air	...	...	197	0.032	32	0.005

MINE WATER

Acid mine waters containing appreciable amounts of ferric or cupric salts may be corrosive to the Ni-Cu alloys as shown by the rates given in Table 4. These tests

TABLE 4. CORROSION OF 70-30 Ni-Cu BY COAL MINE WATERS

Temperature — atmospheric.
Velocity — 1.8 ft per min (0.55 meter per min).

Coal Mine	Free H_2SO_4, ppm	Ferric Ion, ppm	Duration of Tests, days	Corrosion Rate	
				mdd	ipy
Montour Mine No. 1	1430	58	119	136	0.022
Calumet Mine	430	135	98	86	0.014
Edna Mine No. 2	2160	860	79	672	0.110

in three Pennsylvania coal mines were made under the auspices of Carnegie Institute of Technology, U. S. Bureau of Mines, and the Advisory Board of Coal Mine Operators and Engineers.[3]

SEA WATER

The Ni-Cu alloys find their greatest usefulness in sea water under conditions involving high velocity, as in the case of propeller shafts, propellers, pump impellers, pump shafts, and condenser tubes where resistance to the effects of cavitation and impingement is significant. Corrosion rates in strongly agitated and aerated sea water usually do not exceed 6 mdd (0.001 ipy).[4] (See Table 12, p. 408.)

Conditions of stagnant exposure to sea water are less favorable to Ni-Cu alloys, since marine organisms may accumulate and induce local oxygen concentration cell action followed by pitting. Under such conditions, non-fouling Cu-Ni (lower nickel)

[3] W. A. Selvig and G. M. Enos, Carnegie Institute for Coal Mining Investigations, *Bull.* 4, 69 pp. (1922).
[4] F. L. LaQue, *J. Am. Soc. Naval Engrs.*, **53**, 29–64 (1941).

alloys may be more serviceable in spite of their somewhat lower resistance to general corrosion. (See Table 8, p. 399.)

NON-OXIDIZING INORGANIC ACIDS

Sulfuric Acid. The Ni-Cu alloys have useful resistance within certain limits of concentration and temperature to corrosion by all the mineral acids except those of a highly oxidizing nature.

The effects of aeration and acid concentration upon the resistance of 70-30 Ni-Cu to corrosion by sulfuric acid solutions at 30° C (86° F) are shown in Fig. 1.[5] In air-free acid solutions, corrosion rates are less than 50 mdd (0.008 ipy) in all concentra-

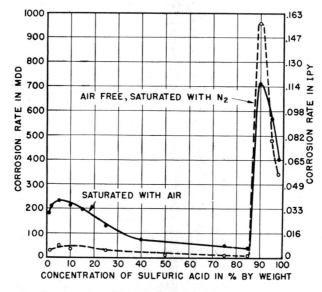

FIG. 1. Corrosion of 70% Ni–30% Cu Alloy in Sulfuric Acid.

Temperature — 30° C (86° F).
Velocity — 17 ft per min.
Duration of test — 24 hr.

tions up to about 80%. In air-saturated acid solutions below 80% concentration, maximum corrosion occurs at about 5% acid concentration where the corrosion rate is 240 mdd (0.04 ipy). In sulfuric acid concentrations above about 80% at atmospheric temperature corrosion of the Ni-Cu alloys is usually too high for economical use. In particular cases, such as in the handling of oil refinery acid sludges and in oil sulfonation processes where mixtures of oil and concentrated sulfuric acid are encountered, corrosion rates are often low because of the inhibiting effect of the oils. Tests in operating sulfonators have shown corrosion rates for 70-30 Ni-Cu of 3 mdd (0.0005 ipy) to 9 mdd (0.0015 ipy) at temperatures up to 60° C (140° F).

The effect of increasing temperature upon the corrosion rates of 70-30 Ni-Cu in 5% sulfuric acid, both air-free and air-saturated, is shown in Fig. 2. It will be noted that in air-free acid, increasing temperature has very little effect upon corrosion. In

[5] W. Z. Friend, *Chem. Met. Eng.*, **46**, 260–262 (1939).

NICKEL–COPPER ALLOYS

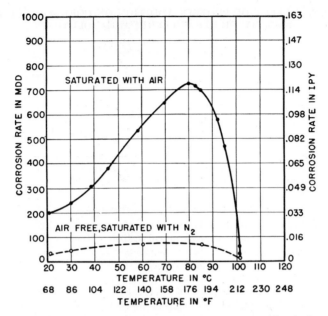

FIG. 2. Effect of Temperature on Corrosion of 70% Ni–30% Cu Alloy in 5 to 6% Sulfuric Acid.

Velocity — 15.5 to 16.5 ft per min.
Duration of test — 24 hr.

air-saturated acid, the most active temperature level is about 80° C (176° F). The lower solubility of oxygen at higher temperatures probably accounts for the lower corrosion rates at higher temperatures.

The Ni-Cu alloys may be used for handling boiling sulfuric acid solutions in concentrations under 20% by weight (Table 5).

TABLE 5. CORROSION OF 70-30 Ni-Cu BY BOILING SULFURIC ACID SOLUTIONS

Acid Concentration, % H_2SO_4 by Wt.	Boiling Temperature		Duration of Test, hour	Corrosion Rate	
	° C	° F		mdd	ipy
5	101	214	23	16	0.0026
10	102	216	23	15	0.0024
19	104	219	23	46	0.0075
50	123	253	23	3,180	0.518
75	182	360	23	10,400	1.70
96	290–295	554–563	3	20,150	3.28

The addition of protein materials such as milk albumin to dilute aerated acid solutions is often effective in reducing corrosion as shown by the results of laboratory tests in Table 6. It is believed that the inhibiting effect is due to the film-forming characteristics of the protein, together with its reaction with some of the oxygen present.

TABLE 6. EFFECT OF PROTEIN ADDITION ON CORROSION OF 70–30 Ni-Cu BY DILUTE SULFURIC ACID SOLUTIONS

Temperature — 30° C (86° F).
Duration of tests — 7 days.

Solution	Corrosion Rate*	
	mdd	ipy
Air-saturated, 2% H₂SO₄	255	0.042
Air-saturated, 2% H₂SO₄ plus 0.5% milk albumin	121	0.020
Air-saturated, 5% H₂SO₄	304	0.049
Air-saturated, 5% H₂SO₄ plus 0.5% milk albumin	144	0.023

* Corrosion rates are high because of the accumulation of cupric sulfate after exposure of the alloy to air-saturated solutions for 7-day periods. For normal corrosion rates in 24-hour tests in air-saturated solutions, see Fig. 1.

Hydrochloric Acid. Corrosion rates of 70-30 Ni-Cu[6] in air-free and air-saturated hydrochloric acid solutions at 30° C (86° F) are shown in Fig. 1, p. 258. The upper curve for 70-30 Ni-Cu refers to air-saturated acid at a velocity of 21.6 ft per min (6.6 meters per min). For most purposes, the practical application of the Ni-Cu alloys is confined to relatively air-free acid in concentrations under about 20% by weight.

The corrosion rates of 70-30 Ni-Cu in 5% hydrochloric acid, both air-free and air-saturated at temperatures up to 90° C (194° F), are shown in Fig. 2, p. 258. In aerated solutions above atmospheric temperatures, the use of Ni-Cu alloys is usually limited to concentrations of about 2% by weight. They are being satisfactorily applied to the handling of aerated hydrochloric acid in 2% solutions at 50° C (122° F) and in 1% solutions at 80° C (180° F).

In unaerated 5% hydrochloric acid, the Ni-Cu alloys are usefully resistant up to about 75° C (167° F). Applications in lower concentrations at higher temperatures are common. The results of laboratory corrosion tests of 70-30 Ni-Cu in boiling hydrochloric acid solutions of several concentrations are shown in Table 7. In most of the processes in which hydrochloric acid is formed as a result of hydrolysis of non-oxidizing chlorides, or of chlorinated solvents, acid concentration is less than 0.5%, and temperatures up to 150° to 200° C (300° to 400° F) are being handled satisfactorily.

TABLE 7. CORROSION OF 70-30 Ni-Cu BY BOILING HYDROCHLORIC ACID SOLUTIONS

Duration of tests — 10 days.
Velocity — none.
Aeration — none.

Acid Concentration, % HCl by Wt.	Corrosion Rate	
	mdd	ipy
0.5	178	0.029
1.0	258	0.042
5.0	1500	0.244

Hydrofluoric Acid. The Ni-Cu alloys are resistant to unaerated hydrofluoric acid solutions of all concentrations, including anhydrous acid, and at all temperatures up

[6] W. Z. Friend, and B. B. Knapp, *Trans. Am. Inst. Chem. Engrs.*, **39**, 731–753 (1943).

to and including boiling. Corrosion rates of 70-30 Ni-Cu are shown in Table 8. Rates may be increased by a high degree of aeration, and by the presence of oxidizing salts in solution.

TABLE 8. RESISTANCE OF 70-30 Ni-Cu TO CORROSION BY HYDROFLUORIC ACID

Acid Concentration, % HF by Wt.	Temperature		Duration of Test, hr	Velocity, ft per min	Corrosion Rate			
					Air-Free		Air-Saturated*	
	°C	°F			mdd	ipy	mdd	ipy
6	76	170	672	None	4.9	0.0008†	...	...
25	30	86	144	42	1.0	0.0002‡	230	0.037
25	80	176	144	42	14.5	0.0024‡	67	0.011
35	120	248	144	None	6.8	0.0011‡	897	0.145
50	30	86	144	42	0.4	0.0001‡	49	0.008
50	80	176	144	42	3.6	0.0006‡	241	0.039
70	50	122	94	None	26.0	0.0042§	...	...
98	115	240	187	None	12.2	0.0020‖	...	...
100	50	122	94	None	3.1	0.0005§	...	...

* Duration of all air-saturated tests, 24 hr.
† In pickling solution. Hydrogen generated in solution during pickling of steel.
‡ Saturated with N_2.
§ In closed steel bomb.
‖ In closed 70-30 Ni-Cu bomb.

Phosphoric Acid. The Ni-Cu alloys have useful resistance to cold pure phosphoric acid solutions without aeration at all concentrations. They are also resistant to pure unaerated concentrated solutions at temperatures up to about 105° C (220° F). Corrosion rates are usually less than 60 mdd (0.10 ipy) in hot aerated solutions, and in concentrated unaerated solutions above 105° C (220° F) corrosion rates are likely to be appreciable, although not invariably so. The results of corrosion tests abstracted from several sources[7,8,9,10] are summarized in Table 9.

Corrosion rates are likely to be high in *crude phosphoric acid* solutions containing ferric salts as encountered in the production of the acid by treatment of phosphate rock with sulfuric acid. Thus, in a 24-hour test[8] in a dilute unaerated crude phosphoric acid solution containing 0.40% iron at 80° C (176° F) the corrosion rate of 70-30 Ni-Cu was 4270 mdd (0.70 ipy).

Corrosion is insignificant in the low concentrations of phosphoric acid encountered in certain beverage syrups.

OXIDIZING ACIDS

The Ni-Cu alloys are severely attacked by oxidizing acids, such as *nitric* or *nitrous*, except in concentrations under 0.5% at atmospheric temperature.

Concentrated *sulfurous acid* solutions are usually very corrosive, especially when hot. The very dilute solutions encountered in the preservation of certain food products are not destructive although they may develop a dark tarnish. In a 20-hour test in

[7] C. E. Hartford and R. L. Copson, *Ind. Eng. Chem.*, **31**, 1123–1128 (1939).
[8] P. R. Kosting and C. Heins, Jr., *Ind. Eng. Chem.*, **23**, 140–150 (1931).
[9] Unpublished data.
[10] C. B. Durgin, J. H. Lum, and J. E. Malowan, *Trans. Am. Inst. Chem. Engrs.*, **33**, 643–668 (1937).

TABLE 9. CORROSION OF 70-30 Ni-Cu IN PURE PHOSPHORIC ACID SOLUTIONS

Acid Concentration, % H₃PO₄ by Wt.	Temperature		Duration of Test, hr	Corrosion Rate				Reference (See footnotes on p. 273.)
	°C	°F		Unaerated		Aerated		
				mdd	ipy	mdd	ipy	
10.0	60	140	60 days	31	0.005	...	...	(7)
10.3	80	176	24	...	...	828*	0.135	(8)
25.0	80	176	24	...	...	425*	0.069	(8)
25.5	95	203	24	25	0.004	292*	0.048	(8)
57.0	105	220	48	19	0.003	...	...	(9)
84.0‡	60	140	48	2	0.0003	9†	0.0014	(10)
85.0	95	203	24	22	0.0037	...	...	(8)
84.0‡	120	248	48	225	0.037	221†	0.036	(10)
84.0‡	180	356	48	...	...	539†	0.088	(10)
90.4	105	220	48	21	0.003	...	...	(9)

* Aerated with 95 ml air per min. Agitated.
† Aerated with 10 liters air per hr.
‡ Per cent P₂O₅ by weight.

an unaerated, unagitated water solution containing 1% sulfur dioxide at 20° C (68° F) the corrosion rate of 70-30 Ni-Cu was 384 mdd (0.062 ipy), whereas in a solution containing 0.15% sulfur dioxide at room temperature the corrosion rate was 3 mdd (0.0005 ipy).

Organic Acids and Compounds

The Ni-Cu alloys are usefully resistant to all the common organic acids, except when highly aerated, and are practically free from corrosion by neutral and alkaline organic compounds.

Corrosion rates in unaerated *acetic acid* at room temperature usually do not exceed 25 mdd (0.004 ipy). In air-saturated solutions at room temperature, rates of attack are between 50 mdd (0.008 ipy) and 150 mdd (0.024 ipy) with maximum corrosion occurring at about 50% concentration, as indicated by Fig. 3.

Corrosion rates are increased somewhat by increasing temperatures, particularly in aerated solutions. In unaerated boiling solutions, the Ni-Cu alloys are resistant at all concentrations as indicated by the performance of 70-30 Ni-Cu in Table 10. In *glacial acetic acid* in a condenser at 110° C (230° F) the corrosion rate was 80 mdd (0.013 ipy).

TABLE 10. CORROSION OF 70-30 Ni-Cu BY BOILING UNAERATED ACETIC ACID

Boiling under reflux.
Duration of tests — 20 hr.

Acid Concentration, % CH₃COOH by Wt.	In Vapors		In Liquid	
	mdd	ipy	mdd	ipy
5	8.0	0.0013	8.0	0.0013
50	27.0	0.0044	13.0	0.0021
98	10.0	0.0016	11.5	0.0019
99.9	12.0	0.0020	38.0	0.0062

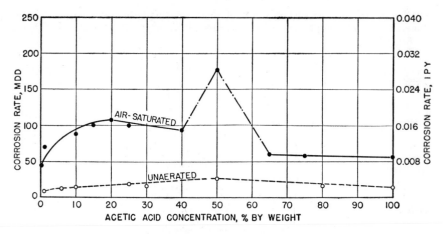

FIG. 3. Corrosion of 70% Ni–30% Cu Alloy in Acetic Acid.

TEMPERATURE — 30° C (86° F)
Air-saturated tests: Duration of tests, 20 hr. Velocity, 16 ft per min.
Unaerated tests: Duration of tests, 72 hr. No agitation.

The performance of Ni-Cu alloys in other organic acids such as *tartaric, malic, citric, formic, lactic,* and *oxalic* acids is somewhat similar to their performance in acetic acid. Corrosion rates in 30% concentrations of several of these acids are shown in Table 11. In a 168-hour test in unaerated boiling 50% *malic acid,* the corrosion rate

TABLE 11. CORROSION OF 70-30 Ni-Cu BY UNAERATED ORGANIC ACID SOLUTIONS OF 30% CONCENTRATION

Duration of tests — 11 days
Beaker tests — no agitation.

Acid	Corrosion Rate			
	Atmospheric Temperature		60° C (140° F)	
	mdd	ipy	mdd	ipy
Tartaric	7.2	0.0012	11	0.0018
Oxalic	3.9	0.0006	49	0.0080
Citric	8.9	0.0015	45	0.0074
Formic	21.0	0.0034	142	0.023

was 11 mdd (0.0018 ipy). Resistance to the dilute acid concentrations encountered in a number of fruit juices is shown in Table 12.

The Ni-Cu alloys are resistant to fatty acids such as *stearic* and *oleic* even at elevated temperatures. Corrosion tests for 84 days in a vacuum fatty acid still at 260° C (500° F) showed a corrosion rate of 26 mdd (0.0042 ipy).[11]

A study[12] has been made of the resistance of several metals and alloys to corrosion

[11] G. L. Cox, *Trans. Am. Inst. Chem. Engrs.,* **34**, 657–679 (1938).
[12] P. S. Brallier, Private communication (1934).

by *chlorinated solvents* and mixtures of these solvents with water, both at room temperature and boiling. The corrosion rates obtained for 70-30 Ni-Cu are shown in Table 12, p. 265.

TABLE 12. RESISTANCE OF 70-30 Ni-Cu TO CORROSION BY FRUIT JUICES

Product	Duration of test, hr	Temperature		Conditions	Corrosion Rate	
		°C	°F		mdd	ipy
Tomato juice	...	Room	Room	Aerated, agitated	18.0	0.0029
Tomato juice	...	Room	Room	Unaerated, agitated	0.2	0.00003
Tomato juice	9	77	170	Air-saturated, mildly agitated	45.0	0.0073
Tomato juice	7	90	195	Aerated, mildly agitated	67.0	0.011
Lemon juice	...	Room	Room	Aerated, agitated	62.0	0.010
Lemon juice	...	Room	Room	Unaerated, agitated	3.0	0.0005
Lemon juice	24	Boiling	Boiling	Under reflux, unaerated	4.0	0.0007
Pineapple juice	...	Room	Room	Aerated, agitated	31.0	0.0051
Pineapple juice	...	Room	Room	Unaerated, agitated	4.0	0.0007
Pineapple juice	106	82	180	Air-saturated	183.0	0.030
Pineapple juice	24	74–80	165–175	Boiling in evaporator under 16–18 in. vacuum	15.0	0.0025
Grape juice	...	Room	Room	Aerated, agitated	28.0	0.0046
Grape juice	...	Room	Room	Unaerated, agitated	13.0	0.0021
Grape juice	24	Boiling	Boiling	Under reflux, unaerated	2.0	0.0003
Orange juice	72	Boiling	Boiling	Under reflux, unaerated	6.0	0.0010

ALKALIES

The Ni-Cu alloys are practically completely resistant to most alkaline solutions, except highly concentrated caustic soda and caustic potash solutions at high temperatures.[13] The results of a number of corrosion tests of 70-30 Ni-Cu in caustic soda solutions under various conditions of concentration and temperature are given in Table 13.

These alloys are resistant to *anhydrous ammonia*. In aqua ammonia or *ammonium hydroxide* their resistance is good up to an NH_3 concentration of about 3% by weight. In higher concentrations, or in the presence of ammonium salts, corrosion is usually appreciable, being accelerated by aeration and temperature.

SALT SOLUTIONS

Neutral and Alkaline Salts. Solutions of neutral and alkaline salts such as chlorides, carbonates, sulfates, nitrates, and acetates have little corrosive action on the Ni-Cu alloys. Even in hot, concentrated solutions, corrosion rates are usually less than 6 mdd (0.001 ipy). The results of corrosion tests of 70-30 Ni-Cu in a number of such solutions are shown in Table 14.

Acid Salts. These alloys have useful resistance to solutions of acid salts such as aluminum sulfate, ammonium sulfate, ammonium chloride, and zinc chloride. Corrosion rates in boiling concentrated solutions are usually less than 125 mdd

---

[13] International Nickel Co., Inc., "Resistance of Nickel and Its Alloys to Corrosion by Caustic Alkalies," *Tech. Bull.* T-6.

NICKEL-COPPER ALLOYS

TABLE 13. RESISTANCE OF 70-30 Ni-Cu TO CORROSION BY CAUSTIC SODA SOLUTIONS

Caustic Concentration, % NaOH by Wt.	Type of Test	Temperature		Duration of Test, hr	Aeration	Velocity, ft/min	Corrosion Rate	
		°C	°F				mdd	ipy
4*	Lab. immersion	30	86	96	Air-saturated	..	1.47	0.0002
4*	Lab. immersion	30	86	96	Nat. convect.	..	1.16	0.0002
10	Lab. immersion	30	86	336	" "	16	0.1	0.00002
10	Lab. immersion	Boiling	Boiling	300	" "	..	0.32	0.00005
14	Plant evaporator	88	190	2160	" "	..	0.33	0.00005
23	Plant receiving tank	104	220	720	" "	..	1.2	0.0002
50	Lab. immersion	30	86	525	" "	16	1.8	0.0003
50	Lab. immersion	Boiling	Boiling	318	" "	..	3.7	0.0006
70	Plant receiving tank	90–115	194–239	2160	" "	..	7.14	0.0011
60–75	Plant evaporator	150–175	302–347	14	" "	..	28.8	0.0047
60–98	Plant evaporator	150–260	302–500	45	" "	..	81.6	0.0134
Anhydrous	Plant evaporator	400–410	752–770	31	" "	..	324.0	0.0532

* H. A. Rawdon and E. C. Groesbeck, Bureau of Standards, *Tech. Bull.* 367, 409–446 (1928).

TABLE 14. CORROSION OF 70-30 Ni-Cu BY SOLUTIONS OF NEUTRAL AND ALKALINE SALTS

Corrosive	Test Conditions	Corrosion Rate	
		mdd	ipy
Calcium chloride	In evaporator concentrating solution to 35% at 70°–160° C (160°–320° F) for 225 days	2	0.0003
Cobalt acetate	In evaporator at 107° C (225° F) for 950 hours	21	0.003
Sodium chloride	Saturated solution mixed with steam and air at 93° C (200° F)	16	0.0026
Sodium metasilicate	In evaporator concentrating to 50% concentration at 110° C (230° F) for 42 days	0.2	0.00003
Sodium nitrate	In 27% solution at 50° C (122° F) for 136 hours	1.1	0.0002

(0.02 ipy). The results of a number of tests in acid salt solutions under plant operating conditions are shown in Table 15.

Oxidizing Acid Salts. The Ni-Cu alloys are not resistant to oxidizing acid salts such as ferric chloride, ferric sulfate, cupric chloride, stannic chloride, mercuric chloride, and silver nitrate, except in extremely dilute solutions. This also applies to acids containing chromates, dichromates, nitrates, peroxides, and other oxidizing compounds, except in a few specific cases, such as acid tanning and textile solutions containing chromates, where the presence of glucose or other organic materials may have an inhibiting effect.[14]

The corrosive effect upon 70-30 Ni-Cu of additions of ferric sulfate to dilute sulfuric acid solutions is shown by the test results given in Table 16.

Oxidizing Alkaline Salts. Concentrated *hypochlorite solutions* are definitely corrosive to Ni-Cu alloys. For continuous exposure, their use is generally limited to the very dilute solutions, containing less than about 500 ppm available chlorine, frequently used for sterilizing purposes. For discontinuous exposure, as in cyclic

[14] W. Z. Friend, *Chem. Met. Eng.*, **46**, 260–262 (1939).

TABLE 15. CORROSION OF 70-30 Ni-Cu BY SOLUTIONS OF VARIOUS
TYPICAL ACID SALTS

Corrosive Medium	Exposure Conditions (Concentration of Salt in % by Wt.)	Corrosion Rate	
		mdd	ipy
Aluminum sulfate	Quiet immersion in 25% solution in storage tank at 35° C (95° F) for 112 days	10	0.0016
Aluminum sulfate	In evaporator concentrating to 57% solution at 116° C (240° F) for 44 days	100	0.0163
Ammonium chloride	In evaporator concentrating 28 to 40% at 102° C (216° F) for 32 days	73	0.012
Ammonium sulfate plus free sulfuric acid	Saturated solution containing 5% sulfuric acid in suspension tank during crystallization at 41° C (106° F) for 33 days	18	0.0030
Magnesium chloride	Specimens immersed in boiling 42% solution in flask fitted with reflux condenser at 135° C (275° F) for 20 days	2.0	0.0003
Manganous chloride	Specimens half submerged in open evaporating pan concentrating to 37 % at 105° C (220° F) for 19 days	116	0.019
Manganese sulfate	In evaporator concentrating 1.250 to 1.350 sp. gr. at 113° C (235° F) for 11 days	29	0.0048
Zinc chloride	In evaporator concentrating 7.9 to 21% at 38° C (100° F) under 26–28 in. vacuum for 210 days	28	0.0045
Zinc chloride	In evaporator concentrating 21 to 69% at 116° C (240° F) under 15 in. vacuum for 90 days	98	0.016
Zinc sulfate	In saturated solution containing trace of sulfuric acid in evaporating pan at 107° C (225° F) for 35 days with vigorous stirring	123	0.020

TABLE 16. CORROSION OF 70-30 Ni-Cu BY AIR–SATURATED SULFURIC
ACID CONTAINING FERRIC SULFATE

Temperature — atmospheric.
Continuous movement.

Acid Concentration, % H₂SO₄ by Wt.	Iron,* %	Corrosion Rate		Ratio of Corrosion to That in Pure Acid
		mdd	ipy	
2.02	Nil	170	0.028	...
2.44	1.0	6990	1.14	41.1
1.60	Nil	140	0.023	...
1.66	0.05	1215	0.20	8.7
0.532	Nil	115	0.019	...
0.604	0.05	1493	0.24	13.0
0.710	0.10	2600	0.42	23.1
0.0635	Nil	76	0.012	...
0.0650	0.005	243	0.040	3.2
0.0857	0.010	421	0.069	5.5

* Added as ferric sulfate.

textile bleaching operations where bleaching is followed by rinsing and (acid) souring in the same vessel, hypochlorite concentrations up to 3 grams per liter available chlorine can be handled safely.[15] When corrosion occurs, it is likely to be localized in the form of pits.

[15] International Nickel Co., Inc., "Nickel Monel and Inconel in Textile Bleaching Operations," *Tech Bull.* T–22.

NICKEL–COPPER ALLOYS

The resistance of Ni-Cu alloys to corrosion by hypochlorite solutions is increased somewhat by a high degree of surface polish. Small concentrations of sodium silicate have been found effective in inhibiting attack by hypochlorites. As little as 0.5 ml per liter of 1.4 sp. gr. sodium silicate will often suffice. Phosphates are also effective (Table 17).

TABLE 17. EFFECT OF INHIBITORS ON CORROSION OF 70-30 Ni-Cu BY SODIUM HYPOCHLORITE SOLUTIONS

Temperature — 40° C (104° F).
Duration of tests — 16 hr.
Beaker tests — no agitation.

Solution Composition, grams per liter			Corrosion Rate		Max. Depth Pitting	
Available Chlorine	Sodium Silicate	Trisodium Phosphate	mdd	ipy	mm	in.
6.5	...	...	692	0.113	0.28	0.011
6.5	0.5	...	107	0.018	0.13	0.005
6.5	...	0.5	51	0.0083	0.15	0.006
6.5	2.0	...	13	0.0021	0.18	0.007
6.5	...	2.0	21	0.0034	0.13	0.005
3.3	...	...	243	0.040	0.18	0.007
3.3	0.5	...	6	0.0010	None	None
3.3	...	0.5	26	0.0042	0.08	0.003
0.1	...	...	23	0.0038	0.08	0.003
0.1	0.5	...	2	0.0003	None	None
0.1	...	0.5	8	0.0013	None	None

These alloys are resistant to alkaline peroxide solutions.

EFFECT OF MERCURY

The 70-30 Ni-Cu alloy resists amalgamation by mercury at moderate temperatures. In 15-day tests at 400° C (750° F) amalgamation was noted. It may be subject to stress cracking and should be stress-relief annealed without subsequent removal of oxide as a safeguard against this.

EFFECT OF STRESS

Stress corrosion cracking of 70-30 Ni-Cu has been observed[16] only in a few specific corrosives such as mercury and solutions of its salts, fluosilicic acid and its salts, and high-concentration, high-temperature caustic soda and caustic potash. Unless very high stresses are imposed on the alloy in service, stress corrosion cracking can be prevented in all the above media by suitable stress relieving heat treatment of the equipment before placing in use. In the mercury and fluosilicate environments, a stress relief anneal at 540° C (1000° F) for about one hour is suitable, and in the high-temperature strong caustic solutions, a brief (3 to 5 minutes) anneal at 875° C (1600° F) is adequate.

[16] O. B. J. Fraser, "Stress Corrosion Cracking of Nickel and Some Nickel Alloys," *Symposium on Stress-Corrosion Cracking of Metals (1944)*, American Society for Testing Materials and American Institute of Mining and Metallurgical Engineers, Philadelphia and New York, 1945.

High stresses, so long as they are uniform and not accompanied by physical transformation, do not significantly increase the corrosion rate in any of the corrosives in which the alloy would ordinarily be used. This is indicated by the corrosion test data in Fig. 7 on p. 577.

WET AND DRY GASES

The Ni-Cu alloys are resistant to corrosion by all the common dry gases at or near atmospheric temperatures. However, they are not resistant to such gases as chlorine, bromine, nitric oxides, ammonia, and sulfur dioxide in the presence of appreciable amounts of water.

The Ni-Cu alloys are corroded at a moderate rate by moist hydrogen sulfide, and corrosion is accompanied by the formation of a sulfide tarnish. The 70-30 Ni-Cu alloy suspended over water, through which hydrogen sulfide was bubbled at 65° C (150° F), was corroded at a rate of 180 mdd (0.030 ipy).

ATMOSPHERIC CORROSION

The Ni-Cu alloys will remain reasonably bright and free from tarnish indoors. In outdoor atmospheres they remain bright only in rural atmospheres that are essentially free from sulfur gases. In sulfurous atmospheres, they will gradually acquire a brownish or greenish film at a rate depending upon the sulfur content of the atmosphere. The effects of weathering are not unpleasing in rural, marine, and mild industrial atmospheres, and the Ni-Cu alloys are used to a considerable extent for outdoor architectural decoration. Even where the rate of tarnishing is greatest the effect on physical and mechanical properties is negligible.

Some quantitative data[17] on the resistance of 70-30 Ni-Cu to corrosion by various atmospheres is provided by 9- to 10-year exposure tests made by Subcommittee VI of Committee B-3 of the American Society for Testing Materials. These results are given in Table 18.

TABLE 18. CORROSION OF 70-30 Ni-Cu IN VARIOUS ATMOSPHERES*

Specimens were exposed 9 to 10 years, followed by chemical cleaning, except where noted.

Test Location	Type of Atmosphere	Corrosion Rate		Description of Corrosion-Product Films at Completion of Test
		mdd	ipy	
Pittsburgh, Pa.	Industrial†	0.23	0.00004	Smooth black film‡
Altoona, Pa.	Industrial	0.37	0.00006	Film fairly smooth, black adherent, sooty
New York, N. Y.	Industrial	0.35	0.00006	Film smooth, yellowish brown, adherent, with loose powdery surface deposit
Rochester, N. Y.	Industrial†	0.06	0.00001	Iridescent dark smooth film‡
Sandy Hook, N. J.	Sea coast	0.16	0.00003	Film smooth, brownish gray, adherent
Key West, Fla.	Sea coast	0.04	0.000007	Tarnish smooth, gray green
La Jolla, Calif.	Sea coast	0.05	0.000008	Film smooth, gray green with spots of loose green scale, adherent
State College, Pa.	Rural	0.03	0.000005	Tarnish smooth, gray
Phoenix, Ariz.	Rural	0.01	0.000002	Tarnish smooth, gray

* From 1943 Report of Subcommittee VI, Committee B-3, American Society for Testing Materials.
† Specimens weighed after 5 to 6 years of exposure and without chemical cleaning.
‡ Description of films after 5 to 6 years of exposure.

---

[17] *Proc. Am. Soc. Testing Materials*, **43**, 137-154 (1943).

NICKEL-CHROMIUM ALLOYS

Experience has shown the 70-30 Ni-Cu alloy to be free from season cracking and other forms of stress corrosion in atmospheric exposure.

GENERAL REFERENCES

LaQue, F. L., *Mech. Eng.*, **58**, 827–843 (1936).
McKay, R. J., and R. Worthington, *Corrosion Resistance of Metals and Alloys*, 1st Ed., Reinhold Publishing Corp., New York, 1936.

NICKEL-CHROMIUM ALLOYS

W. Z. Friend*

GENERAL PROPERTIES

There are available a number of alloys containing 60 to 80% nickel and 13 to 20% chromium, with the remainder mostly iron. These, having a nickel base, are referred to as Ni-Cr alloys to distinguish them from the stainless steels or Cr-Ni-Fe alloys.

TABLE 1. NOMINAL COMPOSITION AND MECHANICAL PROPERTIES OF TYPICAL Ni-Cr ALLOYS AND HARD SURFACING MATERIALS

Material	Nominal Composition, % by Wt.				Range of Mechanical Properties			
	Nickel	Chromium	Iron	Manganese	Tensile Strength, psi	Yield Strength, psi	Elong. in 2 in., %	Hardness, Brinell, 3000 kg
Wrought Alloys								
Inconel‡	80.0	20	....	....	80,000–120,000*	50,000–65,000	40–35	180–200
	79.5	13	6.5	0.25	85,000–120,000†	35,000–90,000	45–30	120–240
	65.0	13	21.0	1	70,000–110,000*	50,000–65,000	35–25	180–200
	60.0	16	23.0	1	70,000–120,000*	50,000–65,000	35–25	180–200
Cast Alloys								
Inconel	77.8	13.5	6	....	70,000–95,000	30,000–45,000	30–10	160–190
	65–68	15–19	....	....	60,000–75,000	30,000–45,000	25–5	180–200
	59–62	10–14	....	....	60,000–75,000	30,000–45,000	25–5	180–200
Hard Surfacing Materials								
	Nickel	Chromium	Boron	Fe, Si, and Cu				Hardness, Rockwell C
Colmonoy	80	11	2	8.0 max.	....	....	....	35–40
Colmonoy	76	13	3	9.0 max.	....	....	....	45–50
Colmonoy	68	18	4	10.0 max.	....	....	....	55–60
Colmonoy	77	19	4	......	....	....	....	.....

* Annealed wire.
† Hot-rolled rod.
‡ Also available in the form of Inconel-clad steel.

The nominal chemical compositions and mechanical properties of several typical alloys of this group are given in Table 1. Of these, the 79.5% Ni–13% Cr–6.5% Fe alloy† is the material which has been used most widely for corrosion-resistant purposes and for which the greatest amount of corrosion test data are available. The

* Development and Research Division, The International Nickel Co., Inc., New York, N. Y.
† Inconel.

other alloys have been used chiefly for heat-resisting purposes. (See *High-Nickel-Chromium (Iron) Alloys*, p. 683.) In general, their corrosion-resistant characteristics are similar to those of the 80% Ni–13% Cr–7% Fe alloy, the data for which may be used as a guide to their probable usefulness in corrosive environments.

In addition, there are several grades of the hard surfacing material known as *Colmonoy* which have a nickel base (68 to 85% nickel, depending upon the grade). Another essential constituent is chromium boride crystals. The nickel and chromium contents of the four grades of Colmonoy (4, 5, 6, and 300 C) are such as to indicate that they would be somewhat similar to the 80% Ni–13% Cr–7% Fe alloy in corrosion resistance.

The chromium content of the Ni-Cr alloys makes them superior to pure nickel under oxidizing conditions, while the high-nickel content enables them to retain considerable resistance under reducing conditions and in strongly alkaline solutions. In very strongly oxidizing solutions, except those of the oxidizing halogens, they are generally resistant to attack. However, the oxidizing effect of dissolved air alone is not sufficient to insure complete passivity and freedom from attack by air-saturated mineral acids or some concentrated organic acids.

It would be expected that the most significant differences in performance of the alloys of this group would occur, on the one hand, in oxidizing solutions such as nitric acid, where higher chromium contents are usually most effective, and, on the other hand, in reducing acids such as hydrochloric, where increased nickel content is usually beneficial. Such limited comparative test data as are available would appear to confirm this. (See Tables 4 and 6.)

The galvanic relationship of the Ni-Cr alloys to other metals and alloys will depend to some extent upon whether the environment is one requiring the maintenance of passivity for corrosion resistance. If so, the alloy when passive will occupy a position in the galvanic series of greater nobility than when active, as indicated in the galvanic series shown in Table 18, p. 416.

CORROSION CHARACTERISTICS IN LIQUIDS

Fresh Water

The Ni-Cr alloys are practically free from corrosion by distilled water and by fresh water, including the most corrosive of natural water which contain free carbon dioxide, iron compounds, chlorides, and dissolved air.

Boiler Water and Steam Condensate

These alloys are also highly resistant to hot boiler waters and to steam condensate containing carbon dioxide and air. In 22-hour laboratory tests in distilled water saturated with various mixtures of carbon dioxide and air under 35 psi pressure at temperatures of 70° C (158° F) to 135° C (275° F) and with velocities of 2.6 to 4.5 meters (8.4 to 14.7 ft) per min, the corrosion rate of the 80% Ni–13% Cr–7% Fe alloy did not exceed 1 mdd (0.0002 ipy).

Mine Water

These materials are highly resistant to mine waters, including acid waters containing considerable amounts of ferric or cupric sulfate. In a coal mine water having a pH of 1.5 and a total iron content of 363 ppm, the corrosion rate of the 80% Ni–13% Cr–7% Fe alloy was 1 mdd (0.0002 ipy). The results of tests with a 56 to 60% Ni, 12 to 14% Cr alloy, in three Pennsylvania coal mines, under the auspices of Carnegie

Institute of Technology, the U. S. Bureau of Mines, and an Advisory Board of Coal Mine Operators and Engineers[1] are shown in Table 2.

TABLE 2. CORROSION OF A 58 TO 60% NICKEL, 12 TO 14% CHROMIUM, IRON ALLOY IN COAL MINE WATERS

Temperature — atmospheric.
Velocity — 0.55 meter (1.8 ft) per min.

Coal Mine	Free H_2SO_4, ppm	Ferric Ion, ppm	Duration of Tests, days	Corrosion Rate	
				mdd	ipy
Montour Mine No. 1	1430	58	119	2.45	0.00041
Calumet Mine	430	135	98	5.8	0.00098
Edna Mine No. 2	2160	860	135	1.81	0.0003

SEA WATER

The Ni-Cr alloys are moderately resistant to rapidly flowing sea water, but in quiet or stagnant exposure, pitting may occur. (Refer to Table 13, p. 409.) In salt-spray tests, these alloys are among the most resistant of commonly available materials. They are useful for applications involving exposure to marine atmospheres.

NON-OXIDIZING INORGANIC ACIDS

Sulfuric Acid. The Ni-Cr alloys have fair resistance to sulfuric acid solutions at atmospheric temperature. As a general rule, the oxidizing effect of aeration alone is not sufficient to provide passivity, and corrosion in air-saturated solutions is greater than in unaerated solutions. Corrosion rates are appreciably increased at elevated temperatures. The addition of a small amount of an oxidizing salt such as ferric or cupric sulfate to acid solutions will increase the passivity range and decrease corrosion (inhibition). The results of several corrosion tests in pure sulfuric acid solutions are given in Table 3.

Hydrochloric Acid. These alloys are useful in hydrochloric acid only in dilute solutions below about 5% concentration by weight at atmospheric temperature. Corrosion is usually severe at elevated temperatures. Performance in hydrochloric acid solutions is indicated by the test results given in Table 4. In the very dilute solutions associated with the hydrolysis of chlorinated solvents in the presence of water, and dilute solutions of some acid chlorides, corrosion resistance is usually adequate even at boiling temperatures. The alloys with nickel content around 80% appear to be somewhat more resistant than those with lower nickel content.

Hydrofluoric Acid. The Ni-Cr alloys are usually resistant to all concentrations of hydrofluoric acid, including anhydrous acid at atmospheric temperature, but may be severely attacked at elevated temperatures. Thus, in 30-day tests, in hydrofluoric acid pickling solutions the corrosion rate of the 80% Ni–13% Cr–7% Fe alloy was less than 1 mdd (0.0002 ipy) in air-free 10% acid at 20° C (68° F), but was 403 mdd (0.068 ipy) in 6% acid at 77° C (170° F).

Hydrogen Sulfide Solutions. Resistance to hydrogen sulfide solutions is good. Laboratory tests of the 80% Ni–13% Cr–7% Fe alloy gave corrosion rates of 0.4 mdd (0.00007 ipy) in tap water saturated with hydrogen sulfide at 25° C (77° F), and 58 mdd (0.010 ipy) in aerated hydrogen sulfide gas saturated with water at 66° C (150° F). Duration of tests was 24 hours.

[1] W. A. Selvig and G. M. Enos, Carnegie Institute for Coal Mining Investigations, *Bull.* 4, 69 pp. (1922).

TABLE 3. CORROSION OF Ni-Cr ALLOYS BY SULFURIC ACID SOLUTIONS

Velocity — 4.7 meters (15.5 ft) per min.

Composition of Alloy, % by Wt.				Acid Concentration, % H_2SO_4 by Wt.	Temperature		Duration of Test, hr	Aeration	Corrosion Rate		Reference
Ni	Cr	Fe	Mn		°C	°F			mdd	ipy	
79.5	13	6.5	0.25	1	30	86	120	Air-sat'd	291	0.049	(1)
(Inconel)				1	78	172	20	Air-sat'd	664	0.11	(1)
				5	30	86	20	H_2-sat'd	53	0.0089	(1)
				5	30	86	120	Air-sat'd	342	0.058	(1)
				5	78	172	20	H_2-sat'd	182	0.030	(1)
				5	78	172	20	Air-sat'd	1200	0.20	(1)
				70	30	86	20	Air-sat'd	271	0.045	(1)
				93	30	86	20	N_2-sat'd	1615	0.0271	(1)
				93	30	86	20	Air-sat'd	62	0.010	(1)
88	11	..	1	10	Room	Room	24	Not stirred. Aeration by nat. convect.	25	0.0042	(1)
64	15	20	1	10	Room	Room	24	"	75	0.013	(1)
64	15	20	1	10	Hot	Hot	24	"	4900	0.82	(1)
65	15	20	..	10	Room	Room	24	"	4	0.0007	(1)
65	15	20	..	10	Hot	Hot	24	"	490	0.082	(1)
84	15	..	1	10	Room	Room	24	"	40	0.0067	(1)
79	20	..	1	10	Room	Room	24	"	10	0.0017	(1)
69	20	10	1	10	Room	Room	24	"	80	0.013	(1)

(1) W. Rohn, Z. Metallkunde, **18**, 387 (1926).

TABLE 4. CORROSION OF Ni-Cr ALLOYS BY HYDROCHLORIC ACID SOLUTIONS

Composition of Alloy, % by Wt.				Acid Concentration, % HCl by Wt.	Temperature		Duration of Test, hr	Velocity, ft per min	Aeration	Corrosion Rate		Reference
Ni	Cr	Fe	Mn		°C	°F				mdd	ipy	
79.5	13	6.5 (Inconel)	0.25	5.0	30	86	20	16.5	H₂-sat'd	78	0.013	(1)
				5.0	30	86	20	16.5	Air-sat'd	581	0.097	(1)
				5.9	30	86	120	Convect.	Air-sat'd	267	0.045	(1)
				5.9	80	176	120	Convect.	Air-sat'd	4,223	0.71	(1)
				5.0	85	185	20	16.5	H₂-sat'd	9,550	1.59	(1)
				5.0	85	185	20	16.5	Air-sat'd	11,630	1.95	(1)
88	11		1	10	Room	Room	24	Not stirred	By nat. convect.	140	0.023	(1)
84	15		1	10	Room	Room	24	"	"	60	0.010	(1)
65	15	20		10	Room	Room	24	"	"	23	0.0039	(1)
65	15	20		10	Hot	Hot	24	"	"	9,050	1.52	(1)
64	15	20	1	10	Room	Room	24	"	"	130	0.022	(1)
64	15	20	1	10	Hot	Hot	24	"	"	39,200	6.66	(1)
79	20		1	10	Room	Room	24	"	"	30	0.0050	(1)
80	20			15	25	77	...	"	"	84	0.014	(2)
80	20			15	96	205	...	"	"	27,500	4.62	(2)
80	20			37	25	77	...	"	"	64	0.011	(2)
80	20			37	96	205	...	"	"	76,500	12.8	(2)

(1) W. Rohn, *Z. Metallkunde*, **18**, 387 (1926).
(2) F. Ritter, *Korrosionstabellen Metallischer Werkstoffe*, p. 148, J. Springer, Vienna, 1937.

TABLE 5. CORROSION OF Ni-Cr ALLOYS BY PHOSPHORIC ACID SOLUTIONS. NO STIRRING

Aeration by Natural Convection Except Where Otherwise Stated.

Composition of Alloy, % by Wt.				Acid Concentration, % H₃PO₄ by Wt.	Temperature		Duration of Test, hr	Corrosion Rate		Reference
Ni	Cr	Fe	Mn		°C	°F		mdd	ipy	
88	11	..	1	10	Room	Room	24	4	0.0007	(1)
84	15	..	1	10	Room	Room	24	3	0.0005	(1)
64	15	20	1	10	Room	Room	24	5	0.0008	(1)
79	20	..	1	10	Room	Room	24	3	0.0005	(1)
69	20	10	1	10	Room	Room	24	23	0.0039	(1)
(Inconel)										
79.5	13	6.5	0.25	57	Room	Room	648	4	0.0007	(2)
				57	105	221	48	3800	0.64	(2)
				84‡	60	140	48	1	0.0002	(3)
				84‡	60	140	48	0	0	(3)
				84*‡	120	248	48	0	0	(3)
				84*‡	120	248	48	2	0.0004	(3)
				84‡	180	356	48	0	0	(3)
				84*‡	180	356	48	1477	0.25	(3)
				90.4	Room	Room	648	0	0	(2)
				90.4	105	221	48	3230	0.54	(2)
80	20	..	..	25.5†	95	203	24	4160	0.69	(4)

* Aerated with 10 liters air per hr by bubbling.
† Aerated with 95 ml air per min by bubbling.
‡ Per cent P₂O₅ by weight.
(1) W. Rohn, Z. Metallkunde, **18**, 387 (1926).
(2) Unpublished data.
(3) C. B. Durgin, J. H. Lum, and J. E. Malowan, Trans. Am. Inst. Chem. Engrs., **33**, 643–668 (1937).
(4) P. R. Kosting and C. Heins, Jr., Ind. Eng. Chem., **23**, 140–150 (1931).

Phosphoric Acid. The Ni-Cr alloys are apparently resistant to phosphoric acid solutions of all concentrations at atmospheric temperature. This applies both to pure solutions and to crude impure solutions containing oxidizing salts such as ferric sulfate. In fact, the range of usefulness is somewhat greater in impure than in pure solutions. However, this resistance does not extend to hot concentrated solutions in which very high corrosion rates have been obtained. The results of a number of corrosion tests of these alloys in pure phosphoric acid solutions are given in Table 5. A corrosion rate of 4780 mdd (0.80 ipy) was reported[2] for an 80% Ni-20% Cr alloy in a 24-hour test at 97° C (207° F) in concentrated crude phosphoric acid from Florida pebble phosphate rock.

OXIDIZING ACIDS

Nitric Acid. The Ni-Cr alloys as a group are highly resistant to strongly oxidizing acids at atmospheric temperature. Apparently the higher the chromium content and the lower the nickel content, the wider the range of concentrations and temperatures in which passivity exists. Thus from Table 6 it will be noted that the 80% Ni-13% Cr-7% Fe alloy has good resistance to nitric acid solutions above about 20% concentration at atmospheric temperature, while at lower concentrations corrosion rates are considerably higher. On the other hand, 79% Ni-20% Cr and 69% Ni-20% Cr-10% Fe alloys are resistant to 10% nitric acid solutions. These alloys are generally inferior to the Cr-Ni-Fe and Cr-Fe stainless steels in hot nitric acid solutions.[3] (See Table 1, p. 144; Table 3, p. 148; and Table 5, p. 159.)

In 48-hour tests in red fuming nitric acid at atmospheric temperature containing 13 to 18% dissolved NO_2 and 2 to 5% water (sp. gr. 1.55 to 159), the corrosion rate of the 80% Ni-13% Cr-7% Fe alloy was 6 mdd (0.001 ipy). In a 6-hour test at 108° to 170° C (227° to 338° F) the rate was 210,000 mdd (35.4 ipy).

Sulfurous Acid. Sulfurous acid solutions are usually corrosive to the Ni-Cr alloys, especially when hot. In 24-hour tests at 20° C (68° F) the 80% Ni-13% Cr-7% Fe alloy corroded at a rate of 72 mdd (0.011 ipy) in a solution containing 1% SO_2, and at a rate of 156 mdd (0.024 ipy) in a solution containing 5% SO_2. However, in a 24-hour test in moist sulfur dioxide gas which had been bubbled through water at room temperature, an 80% Ni-20% Cr alloy showed no weight loss. These alloys are resistant to the very low sulfur dioxide concentrations used in the preservation of some foods.

ORGANIC ACIDS AND COMPOUNDS

The Ni-Cr alloys have generally good resistance to organic acids although this resistance does not extend to boiling concentrated solutions of such acids as *acetic* and *formic*. Corrosion rates in acetic acid solutions are shown in Table 7. The results of corrosion tests in a number of other organic acids are given in Table 8. In the fatty acids, such as *oleic* and *stearic*, resistance is excellent even at distillation temperatures of 200° to 300° C (400° to 600° F).

These alloys have excellent resistance to corrosion by dilute organic acids and other organic compounds as these occur in food products, including dairy products, fruit juices, and alcoholic beverages.[4,5,6,7]

[2] P. R. Kosting and C. Heins, Jr., *Ind. Eng. Chem.*, **23**, 140-150 (1931).
[3] F. Ritter, *Korrosionstabellen metallischer Werkstoffe*, p. 144, J. Springer, Vienna, 1937.
[4] R. J. McKay, *Metals and Alloys*, **4**, 177-180 (1933).
[5] H. A. Trebler, *Metals and Alloys*, **12**, 735-743 (1940).
[6] International Nickel Co., Inc., "The Resistance of Pure Nickel and Inconel to Corrosion by Milk," *Tech. Bull.* TS-1.
[7] International Nickel Co., Inc., "Metals and Wines, Distilled Liquors and Beers," *Tech. Bull.* TS-6.

TABLE 6. CORROSION OF Ni-Cr ALLOYS BY NITRIC ACID SOLUTIONS

Composition of Alloy, % by Wt.				Acid Concentration, %HNO₃ by Wt.	Temperature		Duration of Test, hr	Velocity, ft per min	Aeration	Corrosion Rate		Reference
Ni	Cr	Fe	Mn		°C	°F				mdd	ipy	
79.5	13 (Inconel)	6.5	0.25	5	30	86	24	16	Air-sat'd	36	0.006	...
				5	30	86	24	16	H₂-sat'd	384	0.065	...
				15	30	86	20	16	Air-sat'd	146	0.024	...
				15	30	86	20	16	H₂-sat'd	652	0.110	...
				25	30	86	24	16	Air-sat'd	11	0.0018	...
				25	30	86	24	16	H₂-sat'd	7	0.0012	...
				35	30	86	24	16	Air-sat'd	14	0.0024	...
				45	30	86	48	16	Unaerated*	5	0.0008	...
				55	30	86	48	16	Unaerated*	5	0.0008	...
				65	30	86	24	16	Unaerated*	17	0.0029	...
88	11		1	10	Room	Room	24	Not stirred	By nat. convect.	2565	0.43	(1)
64	15	20	1	10	Room	Room	24	"	"	395	0.066	(1)
65	15	20	...	10	Room	Room	24	"	"	53	0.0089	(1)
65	15	20	1	10	Hot	Hot	24	"	"	260	0.044	(1)
84	15	...	1	10	Room	Room	24	"	"	40	0.0067	(1)
79	20	...	1	10	Room	Room	24	"	"	15	0.0025	(1)
69	20	10	1	10	Room	Room	24	"	"	10	0.0017	(1)
69	20	10	1	10	Hot	Hot	24	"	"	145	0.024	(1)

* Solution exposed to atmosphere. Air was *not* bubbled through.
(1) W. Rohn, *Z. Metallkunde*, **18**, 387 (1926).

TABLE 7. CORROSION OF Ni-Cr ALLOYS BY ACETIC ACID SOLUTIONS

Composition of Alloy, % by Wt.				Acid Concentration, % CH$_3$COOH by Wt.	Temperature		Duration of Test, hr	Other Test Conditions	Corrosion Rate		Reference
Ni	Cr	Fe	Mn		°C	°F			mdd	ipy	
79.5	13	6.5	0.25	0.5	Room	Room	70	Unaerated* — not stirred	13	0.0022	...
(Inconel)				0.5	70	158	100	Unaerated* — not stirred	25	0.0041	...
				2.0	Room	Room	70	Unaerated* — not stirred	13	0.0022	...
				2.0	70	158	100	Unaerated* — not stirred	26	0.0044	...
				2.0	Boiling	Boiling	68	Flask — Reflux condenser	1	0.0002	...
				10.0	30	86	120	Air-sat'd, vel: 16 ft/min	5	0.0008	...
				10.0	Boiling	Boiling	168	Flask — reflux condenser	73	0.013	...
				80 (crude)	Room	Room	912	Immersed in storage tank	1	0.0002	...
				80–100	Boiling	Boiling	8	In a still — in liquid	714	0.12	...
				80–100	Boiling	Boiling	8	In a still — in vapor	48	0.008	...
88	11	...	1	10	Room	Room	24	Not stirred. Aeration by nat. convect.	2	0.0003	(1)
84	15	...	1	10	Room	Room	24	" " " " "	5	0.0008	(1)
64	15	20	1	10	Room	Room	24	" " " " "	14	0.0024	(1)
79	20	...	1	10	Room	Room	24	" " " " "	2	0.0003	(1)
69	20	10	1	10	Room	Room	24	" " " " "	5	0.0008	(1)

* Aeration by natural convection.
(1) W. Rohn, Z. Metallkunde, **18**, 387 (1926).

290 CORROSION IN LIQUID MEDIA, ATMOSPHERE, GASES

TABLE 8. CORROSION OF 80% Ni–13% Cr–7% Fe ALLOY* BY ORGANIC ACIDS

Corrosive	Exposure Conditions (Concentrations in % by Wt. in Water Solutions)	Corrosion Rate	
		mdd	ipy
Acetic acid	See Table 7.		
Citric acid	Immersed in evaporator during concentration from 60 to 78%. Boiling under vacuum at 42° to 64° C (108° to 147° F) for 37.5 days.	14.0	0.0024
Cresylic acid	Lab. test immersed in concentrated acid boiling at atm. pressure for 7 days. No aeration or stirring.	0.9	0.0001
Formic acid	Immersed in commercially pure 90% solution in storage tank at atmospheric temp. for 14 days.	23.8	0.004
Formic acid	Immersed in 90% solution in purification still at 100° C (212° F) for 8 days.	119.0	0.020
Fatty acids	Immersed in mixed oleic and stearic acids in preheater to vacuum still at 246° C (475° F) for 147 days.	2.0	0.0003
Gluconic acid	Lab. test immersed in 4 liters of 1% solution. Boiling at atm. pressure for 43 hr.	1.1	0.0002
Lactic acid	Lab. test immersed in 45% solution at atm. temp. for 14 days. Air-saturated by bubbling. Velocity 4.9 meters (16 ft) per min.	0.8	0.0001
Lactic acid	Immersed in vacuum evaporator during concentration from 10 to 50% at 54° C (130° F) for 360 hr.	47.0	0.008
Levulinic acid	Immersed in pure 98% levulinic acid in storage tank at 32° to 43° C (90° to 110° F) for 57 days.	5.0	0.0008
Maleic acid	Lab. test immersed in 10% solution at 30° C (86° F) for 7 days. Air-saturated by bubbling 100 ml air per min. No stirring.	1.2	0.0002
Maleic acid	Lab. test immersed in 10% solution boiling at 104° C (219° F) for 7 days. No aeration or stirring.	110.0	0.020
Malic acid	Lab. test immersed in 10% solution at 30° C (86° F) for 7 days. Air-saturated by bubbling 100 ml air per min. No stirring.	0	0
Malic acid	Lab. test immersed in 10% solution at 100° C (212° F) for 7 days. No aeration or stirring.	85.0	0.014
Phthalic acid	Lab. test immersed in 4.75% solution boiling at 80° C (176° F) for 7 days. No aeration or stirring.	0.02	0.000002
Tartaric acid	Immersed in 57% solution in vacuum evaporating pan. Boiling under 27 in. Hg vacuum at 54° C (130° F) for 240 hr.	14.0	0.0024

* Inconel.

In laboratory tests with specimens immersed in commercially pure *phenol* with no agitation, the corrosion rate of the 80% Ni–13% Cr–7% Fe alloy was 0.44 mdd (0.00007 ipy) in 14-day tests at 53° C (127° F) and 0.33 mdd (0.00006 ipy) in 6-day tests at 184° C (363° F).

ALKALIES

The Ni-Cr alloys are resistant to corrosion by caustic alkalies and alkaline solutions in most concentrations and temperatures except in highly concentrated sodium or potassium hydroxide at high temperature, such as 90 to 100% caustic at 375° to 475° C (700° to 900° F). In plant and laboratory tests in various concentrations of hot sodium hydroxide up to 70%, the corrosion rate of the 80% Ni–13% Cr–7% Fe alloy did not exceed 2 mdd (0.0003 ipy). In a 35-day test in 75% caustic soda at 135° C (275° F) its corrosion rate was 8 mdd (0.0013 ipy). In laboratory tests on rotating rods of the 80% Ni–20% Cr alloy at high velocity in caustic soda being concentrated from 75% to anhydrous at temperatures up to 480° C (900° F), the corrosion rate was 550 mdd (0.092 ipy).[8]

These alloys are usefully resistant to solutions of alkaline sulfur compounds. For example, in 328-hour laboratory tests with sodium sulfide solution being evaporated

[8] International Nickel Co., Inc., "The Resistance of Nickel and Its Alloys to Corrosion by Caustic Alkalies," *Tech. Bull.* T-6.

to 50% concentration, the corrosion rate of the 80% Ni–13% Cr–7% Fe alloy was 23 mdd (0.004 ipy).
Resistance to *anhydrous ammonia* and to all concentrations of *ammonium hydroxide* is virtually complete. Corrosion rates at atmospheric temperature are less than 1 mdd (0.0002 ipy).

Salt Solutions

Neutral and Alkaline Salts. Neutral and alkaline salt solutions such as chlorides, carbonates, sulfates, nitrates, and acetates are resisted well. Rates of corrosion, even in hot solutions, are usually less than 6 mdd (0.001 ipy). Average corrosion rates in calcium and sodium chloride refrigerating brines are frequently under 1 mdd (0.0002 ipy). Although the Ni-Cr alloys as a group are not entirely free from local attack or pitting in these chloride solutions, serious pitting is not common.

Acid Salts. The resistance of Ni-Cr alloys to solutions of such acid salts as the sulfates and chlorides of zinc, aluminum, and ammonia is usually satisfactory at atmospheric or moderate temperatures. In strong, hot solutions of these salts, however, corrosion is likely to be appreciable because of the formation of the corresponding acids by hydrolysis of the salts. In the acid chloride solutions, corrosion may be accompanied by pitting, although this tendency is considerably less than is the case with Cr-Fe or Cr-Ni-Fe alloys. Pits, if they do occur, are likely to be broad and shallow, and often not progressive.[9]

The results of corrosion tests of the 80% Ni–13% Cr–7% Fe alloy in a number of acid salt solutions are given in Table 9.

Experience with this alloy indicates that it is not subject to *stress corrosion cracking* in acid chloride solutions.

Oxidizing Acid Salts. The Ni-Cr alloys generally have good resistance to solutions of oxidizing acid salts except the heavy metal halides such as ferric chloride, cupric chloride, and mercuric chloride, which are severely corrosive in all but very dilute solutions. The passivating effect of such salts as ferric and cupric sulfates, persulfates, chromates, dichromates, permanganates, nitrates, and nitrites is such that their solutions are usually well resisted except at elevated temperatures. The addition of even small amounts of these salts to acid solutions (other than acid chlorides) decreases the corrosion rate. Table 10 presents the results of corrosion tests in a number of solutions containing oxidizing acid salts.

Oxidizing Alkaline Salts. The Ni-Cr alloys are attacked by concentrated sodium and calcium hypochlorite solutions and are likely to suffer severe pitting, so that they are rarely used except in the very dilute solutions containing 100 to 200 ppm of available chlorine employed for sterilizing purposes.[10] Additions of sodium silicate and trisodium phosphate have been found effective in reducing (inhibiting) attack by these solutions as indicated by the test results in Table 11. Resistance of these alloys is also increased somewhat by a high degree of surface polish.

These materials are highly resistant to alkaline peroxide solutions.

[9] R. J. McKay and R. Worthington, *Corrosion Resistance of Metals and Alloys*, pp. 290, 312, Reinhold Publishing Corp., New York, 1936.
[10] H. S. Haller, F. M. Grant, and C. J. Babcock, U. S. Department of Agriculture, *Tech. Bull.* 756 (1941).

292 CORROSION IN LIQUID MEDIA, ATMOSPHERE, GASES

TABLE 9. CORROSION OF 80% Ni–13% Cr–7% Fe ALLOY* BY SOLUTIONS OF ACID SALTS

Corrosive	Test Conditions (Concentrations of Chemicals in % by Wt.)	Corrosion Rate		Max. Depth of Pitting	
		mdd	ipy	mm	in.
Aluminum chloride	Immersed in solution of 26% AlCl$_3$·6H$_2$O in water in dissolving tank at 50°C (122°F) for 499 hr. Agitated.	9.0	0.0015	0.08	0.003
Ammomium chloride	In evaporator concentrating from 28 to 40% at 102°C (216°F) for 32 days.	3.0	0.0005	0.15	0.006
Ammonium sulfate plus sulfuric acid	In saturated solution of ammonium sulfate with some (NH$_4$)$_2$SO$_4$ crystals in 5% sulfuric acid in crystallizer at 38° to 47°C (100° to 116°F) for 33 days.	5.6	0.0009	None	None
Manganese sulfate	In evaporator concentrating from 1.25 to 1.35 sp. gr. at 113°C (235°F) for 11 days.	5.3	0.0010	0.08	0.003
Manganous chloride	Half submerged in open evaporating pan concentrating to 37% at 104°C (220°F) for 19 days.	165.0	0.028	None	None
Magnesium chloride	Immersed in boiling 42% solution in flask fitted with reflux condenser at 141°C (285°F) for 751 hr.	0.8	0.0001	None	None
Zinc chloride	In evaporator concentrating from 30 to 70% at boiling temperature for 30 days.	145.0	0.024	None	None
Zinc sulfate	Saturated solution containing trace of sulfuric acid in evaporating pan at 107°C (225°F) for 35 days with vigorous stirring.	132.0	0.022	None	None

* Inconel

TABLE 10. CORROSION OF 80% Ni–13% Cr–7% Fe ALLOY* BY SOLUTIONS OF OXIDIZING ACID SALTS

Corrosive	Test Conditions (Concentration of Chemicals in % by Wt.)	Corrosion Rate	
		mdd	ipy
Ammonium persulfate	Immersed in 10% solution at 30°C (86°F) for 45 hr.	5.0	0.0008
Cupric sulfate plus sulfuric acid	Exposed to spray of 10% sulfuric acid plus 2% copper sulfate in brass pickling machine at 32°C (90°F) for 176 hr.	0.46	0.00007
Ferric chloride	Immersed in aerated 10% solution at 30°C (86°F) for 48 hr.	7200.0	1.2
Ferric sulfate plus sulfuric acid	Immersed in 3% sulfuric acid containing 2% Fe$_2$O$_3$, 2.5% Al$_2$O$_3$, and 1.8% V$_2$O$_3$ at room temperature for 96 hr.	1.0	0.0002
Potassium persulfate	Immersed in 4% solution in storage tank at atmospheric temperature for 42 days.	0.2	0.00003
Silver nitrate plus nitric acid	Immersed in silver nitrate solution plus nitric acid to pH 2 at room temperature for 18.5 hr.	2.0	0.0003
Sodium bichromate plus sulfuric acid	Immersed in pickling solution of 5% sulfuric acid plus 0.9% sodium bichromate at 21° to 30°C (70° to 85°F) for 30 days.	2.3	0.0004

* Inconel.

EFFECT OF STRESS

The few cases of stress corrosion cracking which have been reported[11] for Ni-Cr alloys were apparently isolated cases where stresses were exceptionally high and do not point to any particular susceptibility to this type of attack.

[11] O. B. J. Fraser, "Stress-Corrosion Cracking of Nickel and Some Nickel Alloys," *Symposium on Stress-Corrosion Cracking of Metals (1944)*, American Society for Testing Materials and American Institute of Mining and Metallurgical Engineers, Philadelphia and New York, 1945.

TABLE 11. EFFECT OF INHIBITORS ON CORROSION OF 80% Ni-13% Cr-7% Fe ALLOY* BY SODIUM HYPOCHLORITE SOLUTIONS

Temperature — 40° C (104° F).
Duration of tests — 16 hr.
Beaker tests — no agitation.

Solution Composition, grams per liter			Corrosion Rate		Max. Depth of Pitting	
Available Chlorine	Sodium Silicate	Trisodium Phosphate	mdd	ipy	mm	in.
6.5	...	...	69	0.0116	0.69	0.027
6.5	0.5	...	15	0.0025	0.56	0.022
6.5	...	0.5	15	0.0025	0.43	0.017
6.5	2.0	...	6	0.0010	0.20	0.008
6.5	...	2.0	7	0.0012	0.30	0.012
3.3	...	...	29	0.0049	0.81	0.032
3.3	0.5	...	7	0.0012	0.20	0.008
3.3	...	0.5	6	0.0010	0.18	0.007
0.1	...	...	12	0.0020	0.15	0.006
0.1	0.5	...	4	0.0007	0.08	0.003
0.1	...	0.5	4	0.0007	None	None

* Inconel.

High stresses, so long as they are uniform and are not accompanied by physical transformation, do not significantly increase the corrosion rate of the 80% Ni-13% Cr-7% Fe alloy as indicated by the corrosion test data presented in Fig. 7, p. 577.

WET AND DRY GASES

These alloys are completely resistant to all the common gases, when dry, at or near atmospheric temperature. In the presence of appreciable moisture, they are not resistant to chlorine, bromine, hydrogen chloride, or hydrogen bromide, but are resistant to such gases as ammonia, nitric oxides, and hydrogen sulfide.

ATMOSPHERIC CORROSION

The Ni-Cr alloys are resistant to all types of atmospheres, including rural, urban, industrial, and marine. In indoor atmospheres they will often remain bright indefinitely. They are not completely immune to corrosion and tarnishing in sulfurous industrial atmospheres but compare favorably with other alloys that are considered suitable. In clean country air they will remain bright for many years. Free exposure to the atmosphere is more favorable than partially sheltered exposure.

GENERAL REFERENCES

LaQue, F. L., *Mech. Eng.*, **58**, 827–843 (1936).
McKay, R. J. and R. Worthington, *Corrosion Resistance of Metals and Alloys*, Reinhold Publishing Corp., New York, 1936.
Thum, E. E., *et al.*, *The Book of Stainless Steels*, American Society for Metals, Cleveland, Ohio, 2nd Ed., pp. 461–475, 502–512, 1935.

NICKEL-MOLYBDENUM AND NICKEL-MOLYBDENUM-IRON-(CHROMIUM) ALLOYS

BURNHAM E. FIELD*

Although interest in the Ni-Mo alloys for resistance to corrosion began in the United States and in Germany about 1920, or perhaps a year or two earlier, little use has been made of the binary Ni-Mo alloys because of the difficulty of preparing them free from other metallic elements. Figure 1 gives the corrosion curve for binary Ni-Mo alloys in 10% hydrochloric acid at 70° C (158° F), with molybdenum varying from 5 to 47%. When the molybdenum content is above 15% the resistance to attack by acid under these conditions is very good. Within the last few years some alloys have been made by powder metallurgy methods and, although they showed excellent acid resistance, the cost of preparation was too great to make them commercial.

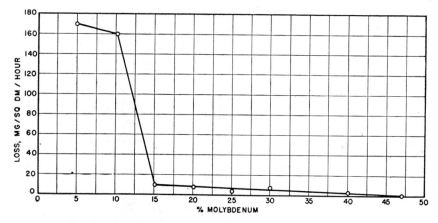

FIG. 1. Corrosion of Ni-Mo Alloys in 10% HCl at 70° C (158° F).

The development of the Ni-Mo alloys for industrial use was started in this country about 1927.[1] A series of these alloys has been marketed under the trade name Hastelloy, in three principal grades, A, B, and C, the approximate compositions of which are given in Table 1. All three grades have been available as castings, but at first only grades A and B were available in rolled products, sheet, bars, etc. More recently, grade C has also been made available in sheet form.

TABLE 1. APPROXIMATE COMPOSITION OF COMMERCIAL
Ni-Mo-Fe-(Cr) CORROSION-RESISTANT ALLOYS (HASTELLOYS)

	% Ni	% Mo	% Fe	% Cr	% W	% C
Grade A	60	20	20	..	..	0.1
Grade B	65	30	5	..	..	0.1
Grade C	59	17	5	14	5	0.1

So far as is known, no use is ordinarily made of the Ni-Mo or Ni-Mo-Fe-(Cr) alloys for resistance to aqueous corrosion except where dissolved corrosive salts, acids, or

* Union Carbide and Carbon Research Laboratories, Inc., Niagara Falls, N. Y.

[1] B. E. Field, "Some New Developments in Acid-Resistant Alloys," *Trans. Am. Inst. Mining Met. Engrs.*, **83**, 149 (1929).

bases are present. The cost of the Ni-Mo alloys is too great to warrant their use in competition with other materials which are sufficiently resistant to water and available at much lower cost.

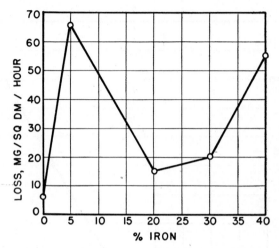

FIG. 2. Weight Loss of Ni-Mo-Fe Alloys Containing 20% Molybdenum in 10% HCl at 70° C (158° F).

ACIDS

The commercial alloys were developed primarily as materials to resist corrosion by mineral acids. Figure 2 shows the weight loss of Ni-Mo-Fe alloys in 10% hydrochloric acid at 70° C (158° F) when the iron content varies from 0 to 40%. Figure 3, in which corrosion loss in 10% sulfuric acid is plotted against iron content, shows that the rate of corrosion of this alloy between 15 and 25% iron changes only slowly.* These curves were prepared during a study of the composition of the Ni-Mo-Fe alloys in an attempt to compromise corrosion resistance, cost, and workability. As will be seen from the compositions of Table 1, Grade A was designed to contain 20% molybdenum and 20% iron and Grade B approximately 30% molybdenum and 5% iron.

Grades A and B are particularly resistant to *hydrochloric acid* in all concentrations, the resistance dropping off as the temperature is increased from room temperature to the boiling point of the solution. Grade A is suitable for use at temperatures to 70° C (158° F), but Grade B is more resistant at temperatures between 70° C (158° F) and the boiling point of the solution. Both grades are resistant to *sulfuric acid* in concentrations up to 50% and also to the concentrated acid up to 70° C (158° F). The chromium present in Grade C confers on this alloy resistance to oxidizing agents such as *nitric acid*, and to acid mixtures such as nitric and sulfuric, and *chromic* and sulfuric, but lowers the resistance to hydrochloric acid, particularly above 50° C (120° F). Laboratory tests showing the loss in weight of the various alloys in acids under conditions of varying concentration and temperature are shown in Table 2. McCurdy[2] gives additional information of physical properties and corrosion resistance.

* The corrosion rates of various Ni-Mo-Fe alloys in boiling 30% sulfuric acid are plotted as a function of molybdenum to iron ratio in Fig. 4, p. 29. For rates in other media, see original reference.

[2] F. T. McCurdy, "Nickel-Molybdenum-Iron and Related Alloys — Their Physical and Corrosion-Resistant Properties," *Proc. Am. Soc. Testing Materials*, **39**, 698 (1939).

The commercial alloys show good resistance to acetic and formic, and to many other organic acids as well which are not so common. Table 3 gives the results of tests in these acids for Grades A and C.

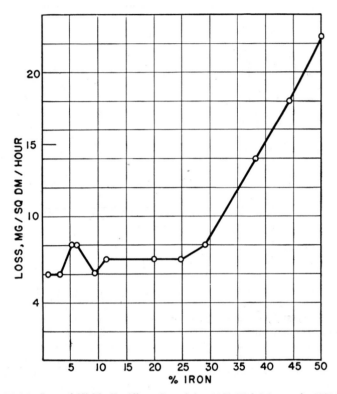

FIG. 3. Weight Loss of Ni-Mo-Fe Alloys Containing 20% Molybdenum in 10% H$_2$SO$_4$ at 70° C (158° F).

ALKALIES AND SALTS

Alkalies and neutral or alkaline salt solutions have very little effect on the alloys. Grades A and B are resistant to salts of the non-oxidizing type, and Grade C shows resistance to oxidizing salts such as ferric and cupric, although pitting is likely to occur unless the solutions are acid.

Grade C shows excellent resistance to pitting and weight loss in sea water. Data for Grades A, B, and C are given in Table 13, p. 409.

METALLURGICAL AND MECHANICAL FACTORS

Additions of small amounts of antimony to the alloys are effective in increasing the resistance to hydrochloric acid. Amounts of 0.3 to 0.5% are sufficient to accomplish the desired result, but any appreciable increase beyond 0.5% is undesirable, particularly in alloys that are to be hot-worked because it tends to cause cracking. Copper has been used as an addition agent to alloys of this type to increase the corrosion

Table 2. TOTAL IMMERSION TESTS IN MINERAL ACIDS

Tests Carried Out in Glassware Using Laboratory Reagents

Grade A specimens — 3.2 × 2.5 × 0.8 cm, weighing approximately 51 grams.
Grade B specimens — 3.2 × 2.5 × 0.7 cm, weighing approximately 45 grams.
Grade C specimens — 3.4 × 2.5 × 0.9 cm, weighing approximately 60 grams.

Acid concentrations are percentage by weight.
Volume — 800 ml.
Duration of test — 48 hr.

MDD

Hydrochloric Acid

Alloy	1%			10%			37% (Conc.)		
	Room Temp.	70° C	Boiling Point	Room Temp.	70° C	Boiling Point	Room Temp.	70° C	Boiling Point
A	242 *	544*	507	88*	515*	1,620	24*	287*	2,280
B	23 *	147*	56	40*	216*	77	12*	123*	108
C	1.1*	10*	635	164*	2095*	16,450	112*	6725*	20,090

Sulfuric Acid

Alloy	10%			60%			96%		
	Room Temp.	70° C	Boiling Point	Room Temp.	70° C	Boiling Point	Room Temp.	70° C	Boiling Point
A	17	170*	220	2.4	74*	23,500	1.9 *	45 *	2200
B	12	139*	15	2.3	14*	239	0.62*	7.7*	310
C	Nil	24*	314	Nil	72*	22,800	Nil	16 *	1940

Nitric Acid

Alloy	10%			70%		
	Room Temp.	70° C	Boiling Point	Room Temp.	70° C	Boiling Point
C	72	239*	269	29	747*	12,320

A and B not recommended for resistance to nitric acid.

Phosphoric Acid, C. P.

Alloy	10%		85%	
	Room Temp.	Boiling Point	Room Temp.	Boiling Point
A	13	162	1.8	5000
B	10	12	1.2	16
C	0.50	9.0	0.48	2990

* Aerated tests.

resistance, but it has not always been found advantageous. For resistance to hydrochloric acid at temperatures approaching the boiling point, copper is definitely detrimental.

Cold work has relatively little effect on the corrosion resistance even when the reductions are fairly heavy. Heating in the range of 500° to 700° C (930° to 1300° F),

TABLE 3. TOTAL IMMERSION TESTS IN ORGANIC ACIDS
Tests Carried Out in Glassware Using Laboratory Reagents.

Grade A specimens — 3.2 × 2.5 × 0.8 cm, weighing approximately 51 grams.
Grade B specimens — 3.2 × 2.5 × 0.7 cm, weighing approximately 45 grams.
Grade C specimens — 3.4 × 2.5 × 0.9 cm, weighing approximately 60 grams.

Acid concentrations are percentage by weight.
Volume — 800 ml.
Duration of test — 48 hr.

MDD

Acetic Acid

Alloy	10%		99% (Glacial)	
	Room Temp.	Boiling Point	Room Temp.	Boiling Point
A	13	15	2.2	52
C	Nil	Nil	Nil	Nil

Formic Acid

Alloy	10% 70° C	85% (Conc.) 70° C
A	169 *	220*
C	4.8*	75*

* Aerated tests.

on the other hand, does tend to make the alloys susceptible to intergranular corrosion. Small amounts of columbium have been helpful in stabilizing the alloys with respect to carbide precipitation or intergranular corrosion. Columbium, however, is not nearly so effective in these alloys as it is in the 18-8 Cr-Ni steels, and, in addition, titanium and tantalum are quite ineffective. Stabilizing heat treatments in the vicinity of 1000° C (1830° F) for one or two hours are quite effective in preventing corrosion of Grade A heated in the 500° to 700° C (930° to 1300° F) range, but are somewhat less effective for grades B and C.

GASES

Alloys A, B, and C are resistant to both oxidizing and reducing flue gases, to carbon monoxide, carbon dioxide, hydrocarbons, hydrochloric acid, and to ammonia, wet or dry. Alloy C is highly resistant to wet chlorine gas and to wet hydrogen sulfide or sulfur dioxide gases up to about 70° C (160° F). Indicated maximum temperatures for use of these alloys in contact with hydrogen chloride and chlorine are included in Table 6, p. 682. For data on accelerated stress corrosion cracking of these alloys in $MgCl_2$ solution, see Table 2, p. 180.

GENERAL REFERENCES (NI-MO-FEALLOYS)

GUERTLER, W., "Molybdän als Legierungsbestandteil," *Z. Metallkunde*, **15**, 151, 251 (1923).
GUERTLER, W., "Das Problem der Säuerfesten Metallischen Werkstoffe," *Z. Metallkunde*, **18**, 365 (1926).
ROHN, W., "Säuerfeste Legierungen mit Nickel als Basis," *Z. Metallkunde*, **18**, 387 (1926).
SCHULZ, E. H., and W. JENGE, "Chemisch Beständige Legierungen und Ihre Eigenschaften," *Z. Metallkunde*, **18**, 377 (1926).

PLATINUM GROUP METALS AND ALLOYS

E. M. WISE*

PLATINUM

Pure platinum is a noble metal and therefore has low chemical affinity. Its standard electrode potential (Pt $\rightleftarrows$ Pt^{++} + 2e^-) is approximately +1.2 volts at 25° C. Its melting temperature is high, 1773° C (3223° F), and mechanical properties are good. (The hardness of annealed pure platinum is 37 VHN†.[1] Platinum can be used, therefore, for a wide range of chemical laboratory and production scale equipment.

Where higher mechanical properties are required, other metals of the platinum group are added to platinum. Of these, rhodium, iridium, and ruthenium are most generally used.

The corrosion resistance of platinum is such that it can be used in very thin frangible disks for protecting pressure vessels from overpressure.[2] Platinum employed for laboratory ware may be hardened with about 0.6% of iridium or about 3.5% of rhodium, whereas platinum used for jewelry normally contains either 5% or 10% of iridium or 5% of ruthenium. Various amounts of palladium can be alloyed with platinum, yielding very ductile materials. These alloys were employed for jewelry in Europe, but have not been used much in this country, except for electrical contacts. Platinum remains bright on exposure to outdoor as well as indoor atmospheres, which is important for jewelry, leaf, and other decorative uses, and for various electrical applications.

Electrical contacts subject to oxidizing atmospheres, including those containing ozone, and also sometimes to atmospheres containing sulfur and halogens, are often made of pure platinum or platinum alloys containing ruthenium or iridium. In this application resistance to chemical corrosion and spark erosion is important.

Platinum can be welded to a wide variety of metals. Clad material is made for chemical plant equipment and allied purposes, generally to reduce the amount of platinum required, and in some instances to secure other special properties, such as high electrical conductivity. For instance, it is sometimes desirable to use a silver- or copper-core platinum-clad rod for insoluble anodes.

Platinum can be electrodeposited, although the production of heavy plates has not been too satisfactory. Generally speaking, it is preferable to apply the platinum by a welded cladding with not less than 5 mils of platinum when substantial corrosion resistance is required.

The corrosion behavior of platinum is important to many industries and has been

---

* Development and Research Division, The International Nickel Co., Inc., New York, N. Y.
† VHN is Vickers Hardness Number.

[1] R. F. Vines, *The Platinum Metals and Their Alloys* (E. M. Wise, editor), The International Nickel Co., New York, 1941.
[2] M. E. Bonyun, *Chem. Met Eng.*, **42**, 260–263 (1935).

TABLE 1. THE EFFECT OF VARIOUS

Weight

Vol. of reagent, 25 ml; area of specimens, 12.9 sq cm; solution

	Temp.	Pt	Ir	Os
98% H₂SO₄	R. T.	0.0 (5 mo)	0.0 (24 hr)	
	100° C	0.0 (7 hr)	0.0 (7 hr)	0.0 (7 hr)
70% HNO₃, sp. gr. 1.42	R. T.	0.05 (5 mo)		
95% HNO₃, sp. gr. 1.5	R. T.	0.0 (5 mo)		
	100° C	0.0 (5 hr)	0.0 (5 hr)	Rapid attack
Aqua regia‡	22° C	4800 (1 hr)	0.0	
	100° C		0.0 (5 hr)	15,000
36% HCl, sp. gr. 1.18	R. T.	0.0 (5 mo)		
	100° C	4.8 (7 hr)	0.0 (7 hr)	96 (7 hr)
5% HCl, intermittent exposure	Boiling	0.7 (5 hr)		
HBr (fuming)	R. T.	2.4 (24 hr)		
	100° C	118 (7 hr)	0.0	46 (7 hr)
HI, sp. gr. 1.75	R. T.	0.003 (5 mo)		
	100° C	336 (4 hr)	0.0	22 (4 hr)
40% HF	R. T.	0.0	0.0	
Phosphoric acid, 100 grams/liter H₃PO₄	100° C	0.0 (5 hr)	0.0 (5 hr)	46 (5 hr)
Perchloric acid, sp. gr. 1.6	R. T.	0.0 (5 hr)		
	100° C	0.0 (2 hr)		
Acetic acid (glacial)	R. T.	0.002 (5 mo)		
	100° C	0.0 (4 hr)	0.0 (4 hr)	
FeCl₃, 100 grams/liter	R. T.	0.005 (5 mo)		
	100° C	408 (4 hr)	0.0	**77.0 (6 hr)**
KCN, 50 grams/liter	R. T.	0.02 (5 mo)		
	100° C	34 (5 hr)		
HgCl₂ solution	100° C	0.0 (2 hr)	0.0 (2 hr)	**5760 (2 hr)**
NaClO+NaCl (100 grams/liter	R. T.	0.002 (5 mo)		
NaOH, saturated with Cl₂)	100° C	0.0 (7 hr)	2.4 (7 hr)	
CuCl₂, 100 grams/liter	R. T.	0.0 (5 mo)		
CuSO₄, 100 grams/liter	R. T.	0.0 (5 mo)	0.0 (5 hr)	
Al₂(SO₄)₃, 100 grams/liter	R. T.	0.0 (5 mo)		
	100° C	0.0 (6 hr)	0.0 (6 hr)	
Cl₂, dry§	R. T.	4.7 (23 hr)	0.0 (23 hr)	1.0 (23 hr)
Cl₂, moist§	R. T.	0.5 (24 hr)	0.7 (24 hr)	
Saturated Cl₂ water	R. T.	0.0 (5 mo)	0.0 (18 hr)	
	100° C	0.0 (2 hr)		
Br, dry (liquid)§	R. T.	86 (21 hr)	0.0	106 (21 hr)
Br, moist§	R. T.	48 (22 hr)	0.0 (22 hr)	2.3 (22 hr)
Saturated Br water§	R. T.	0.0 (5 mo)	0.0 (23 hr)	
I, dry§	R. T.	Trace (23 hr)	0.0 (23 hr)	2.3 (23 hr)
I, moist§	R. T.	0.0 (22 hr)	0.0 (22 hr)	0.0 (23 hr)
I in alcohol, 50 grams/liter§	R. T.	0.0 (5 mo)	0.0 (24 hr)	
H₂S, moist	R. T.	0.0 (45 days)	0.0 (45 days)	Stained (45 days)

* R. Atkinson, A. Raper, and A. Middleton, unpublished research, Mond Nickel Co., Acton, London, England.
† 0.1 N HNO₃ — 0.0 (5 mo). 1 N HNO₃ — 7.2 (5 mo). 2 N HNO₃ — 31.0 (5 mo).
‡ Three volumes 36% HCL plus one volume 70% HNO₃.
§ Reaction film removed before weighing.
R. T. = Room temperature.

PLATINUM GROUP METALS AND ALLOYS

CORROSIVES ON PLATINUM METALS*

Loss, MDD

not aerated and not stirred; duration of test in parentheses.

Pd	Rh	Ru	Alloys			
{0.7 (24 hr) {0.05 (5 mo) 22.0 (7 hr)	7.2 (7 hr)	0.0 (7 hr)				
840.0†						
Rapid attack	0.0 (5 hr)	0.0 (5 hr)				
Rapid			22°C	5%Ru–Pt 700 (1 hr)	5%Rh–Pt 1130.0 (1 hr)	5%Ir–Pt 500 (1 hr)
	0.0 (5 hr)	0.0 (5 hr)				
{0.5 (5 mo) {17.0 (6 hr) 6.0 (5 hr) 2210 (24 hr)	0.0 (7 hr)	0.0 (7 hr)	10%Rh–Pt nil (5 hr)	20%Au–Pd 1.9 (5 hr)	20%Pd–Au 2.2 (5 hr)	30%Pt–Au 2.2 (5 hr)
900 (4 hr)	31 (7 hr)	0.0				
0.0	0.0 0.0	0.0 0.0				
2.4 (5 hr)	0.0 (5 hr)	0.0 (5 hr)				
0.0 (5 mo) 33.6 (2 hr)						
0.005 (5 mo) 0.0 (4 hr)	0.0 (4 hr)	0.0 (4 hr)				
163 (28 days)			10%Pt–Pd 118 (28 days)	30%Pt–Pd 24 (28 days)	10%Au–Pd 8.1 (28 days)	20%Au–Pd 0.8 (28 days)
Rapid attack	0.0	0.0	5%Ir–Pt 19 (4 hr)	5%Ru–Pt 5.5 (4 hr)	5%Rh–Pt 4.8 (4 hr)	
22 (24 hr)			5%Ir–Pt	5%Ru–Pt	5%Rh–Pt	
860 (2 hr)		84 (2 hr)	50 (5 hr)	26 (5 hr)	77 (5 hr)	
0.0 (2 hr)	0.0 (2 hr)	624				
24 (19 days)	4.8 (24 hr)					
204 (7 hr)	4.8 (7 hr)					
4.8 (5 mo)						
0.0 (5 hr)	0.0 (5 hr)	0.0 (5 hr)				
0.0 (5 mo)						
0.0 (6 hr)	0.0 (6 hr)	0.0 (6 hr)				
15.3 (23 hr)	0.0 (23 hr)	0.0 (23 hr)				
192 (24 hr)	0.0 (24 hr)	0.0 (24 hr)				
0.14 (5 mo)	0.0 (18 hr)	19 (18 hr)				
336 (2 hr)						
342 (21 hr)	0.0 (21 hr)	0.0 (21 hr)				
380 (22 hr)	0.0 (22 hr)	0.0 (22 hr)				
10 (23 hr)	0.0 (23 hr)	2.4 (23 hr)				
0.6 (23 hr)	0.0 (23 hr)	0.0 (23 hr)				
7.6 (22 hr)	0.7 (22 hr)	0.0 (22 hr)				
4.8 (24 hr)	0.7 (24 hr)	3.8 (24 hr)				
0.0 (45 days)	0.0 (45 days)	0.0 (45 days)				

studied by individual users in connection with their own problems, but relatively few recent, broadly applicable data are found in the literature. Furthermore, some published data are based on experiments with platinum black, generally containing oxygen, which is known to be much more reactive than normal wrought material. In the laboratory, platinum is frequently employed under very drastic conditions where nothing else will function, and sometimes under conditions where the corrosion rates are appreciable but, nevertheless, acceptable. In plant equipment with large areas of material, the permissible corrosion rates are very low, and laboratory data are not entirely adequate for predicting what these rates will be. It frequently happens that rates decrease with time, so that short-time tests may produce values for the maximum or initial corrosion rates which are in excess of the long-time values.

GENERAL CORROSION BEHAVIOR

Platinum is resistant to reducing or oxidizing acids, but is attacked by a mixture of nitric and hydrochloric acids (aqua regia) and somewhat more slowly by HCl plus other oxidizing agents. Alloying with iridium or ruthenium reduces the rate of attack by aqua regia, often to a point where it is difficult to dissolve these alloys for analysis. The use of aqua regia at high temperature to effect solution of iridium[3] and other metals of this group is discussed in *Noble Metals and Their Alloys*, p. 713.

For the corrosion information presented in Table 1, massive material was used for all the platinum metals except osmium. For the latter, material sintered at a high temperature was employed from necessity, and it is possible that this contributed slightly to its relatively poor behavior. Samples having an area of 12.9 sq cm (2 sq in.) were employed, with few exceptions, and were placed in 25 ml of solution in a stoppered bottle and were not stirred. It is recognized that the absence of aeration may reduce the attack on long exposure to non-oxidizing media, and that where very high corrosion rates are involved, the limited volume of the solution reduces the rate. However, the platinum metals would not be useful under conditions where substantial corrosion is found, so that this effect is unimportant.

Table 2 summarizes information on the probable utility of platinum for use in contact with a number of corrosive agents. In addition to the media listed, platinum is unattacked by solutions of alkalies, nitrates, chlorides, and many organic acids. Generally speaking, this is unimportant, as in most cases common metals are adequate for these compounds, unless exceptional purity is required or some special requirement is involved such as electrolysis.

As will be noted from these tables, hydrofluoric and hydrochloric acids do not attack platinum at room temperature, but hydrobromic acid containing a little bromine attacks it slightly at room temperature, and hydrochloric, hydrobromic, and hydriodic acids attack the metal more or less vigorously at 100° C (212° F). Although commercial chlorine and bromine attack platinum at moderate temperatures, very dry chlorine can be handled at moderate temperatures. Iodine does not attack it appreciably at room temperature. All the free halogens react with platinum over certain ranges of elevated temperature (discussed on p. 710), but HCl gas does not. The addition of iridium or ruthenium to platinum further increases resistance to halogens.

Platinum is attacked by selenic acid at 100° C (212° F), and by sulfuric acid at considerably higher temperatures. In both cases oxidation appears to be involved.

[3] E. Wichers, W. Schlecht, and C. Gordon, *J. Research Natl. Bur. Standards*, **33**, 363–381 (1944); see also **33**, 457–470 (1944).

TABLE 2. CORROSION RESISTANCE OF PLATINUM

	Temperature	Rating
H_2SO_4, conc.	R. T.	A
	100° C	A
H_2SeO_4, sp. gr. 1.4	R. T.	A
	100° C	C
H_3PO_4	100° C	A
$HClO_4$	R. T.	A
	100° C	A
HNO_3, 70%	R. T.	A
HNO_3, 95%	R. T.	A
	100° C	A
Aqua regia	R. T.	D
	Boiling	D
HF, 40%	R. T.	A
HCl, 36%	R. T.	A
	100° C	B
HBr, sp. gr. 1.7	R. T.	B
	100° C	D
HI, sp. gr. 1.75	R. T.	A
	100° C	D
Acetic acid (glacial)	100° C	A
F_2	R. T.	B
Cl_2, dry	R. T.	B
Cl_2, moist	R. T.	B
Br (liq.), dry	R. T.	C
Br (liq.), moist	R. T.	C
Br water	R. T.	A
I, dry	R. T.	A
I, moist	R. T.	A
I, in alcohol	R. T.	A
H_2S, moist	R. T.	A
NaClO solution	R. T.	A
	100° C	A
KCN solution	R. T.	A
	100° C	C
$HgCl_2$ solution	100° C	A
$CuCl_2$ solution	100° C	A
$CuSO_4$ solution	100° C	A
$Al_2(SO_4)_3$ solution	100° C	A

A — No appreciable corrosion.
B — Some attack, but not enough to preclude use.
C — Attacked enough to preclude use.
D — Rapid attack.
R. T. — Room temperature.

Ferric chloride in solution does not corrode platinum appreciably at room temperature but attacks it at somewhat higher temperatures.

Behavior as an Anode

Platinum is resistant as an anode in chloride and sulfate solutions and is passive at usual current densities. At very low current densities in acid-chloride solutions, platinum anodes can be active[4] and in the presence of alternating current at 5-20

[4] G. Grube, F. Oettel, and H. Reinhardt, *Festschrift zum 50-jährigen Bestehen der Platinschmelze*, pp. 108-120, G. Siebert G.m.b.H. Hanau, 1931.

304   CORROSION IN LIQUID MEDIA, ATMOSPHERE, GASES

amp/sq dm are dissolved at an appreciable rate in acid-chloride solutions and in 50 to 60% sulfuric acid. Use is made of this phenomenon in metallographic etching.[5] The nature of the passivity has received considerable study.[6]

Platinum is useful as an insoluble anode under a variety of conditions, including laboratory devices for electroanalysis, as well as very large industrial equipment. In some cases bright platinum is used to secure maximum overvoltage; in others, surfacing with platinum black may be useful when a low overvoltage is sought.

Of the major electrochemical applications of platinum, its use as anode in producing persulfates and persulfuric acid, most of which are converted to hydrogen peroxide, is very important. Also the production of perchlorates utilizing platinum anodes for the final oxidation of chlorate to perchlorate is now a large industry. The use of platinum in these and other processes has the added advantage of avoiding traces of carbon compounds, which may be damaging to the product or hazardous in manufacture from an explosion standpoint.

Platinum-clad anodes have been used for producing chlorine, although graphite or carbon is generally used. However, consideration should be given to platinum in cells so designed to utilize the metal effectively.

The use of platinum-clad anodes for nickel as well as other plating baths is practical, and large installations are in successful operation. Recent data on the corrosion behavior of platinum and certain platinum alloys in typical nickel plating baths are presented in Table 3.[7]

TABLE 3. BEHAVIOR OF PLATINUM ANODES IN NICKEL PLATING BATHS

Anode C. D.		Duration of Test	WEIGHT LOSS IN MDD					
			Pure Pt		10% Ir-Pt		10% Rh-Pt	
amp/sq ft	amp/ sq cm		Watts* Bath	Sulfate† Bath	Watts* Bath	Sulfate† Bath	Watts* Bath	Sulfate† Bath
50	0.05	7 hr	4.1	1.4	5.3	5.3	4.1	7.9
100	0.11	7 hr	4.1	2.9	4.3	6.7	5.8	13.0
100	0.11	39 hr	...	....	...	....	1.7	11.3
200	0.22	7 hr	5.3	10.5	5.3	13.2	6.7	14.6

	*Watts Bath grams per liter	†Sulfate Bath grams per liter
NiSO$_4$·7H$_2$O	300	359
NiCl$_2$·6H$_2$O	45	...
H$_3$BO$_3$	30	30
pH	2.0 approx.	2.0 approx.
Temperature	60° C (140° F)	60° C (140° F)

PALLADIUM

Palladium and platinum are the only two readily workable metals of the platinum group and have about the same mechanical properties. The standard electrode potential of palladium (Pd $\rightleftarrows$ Pd^{++} + 2e$^-$) has been indicated as +0.83 volt at 25° C.

For electrical contacts, the commercially pure metal or the 5% and 10% Ru–Pd alloys are used, whereas in jewelry the 4% and 5½% ruthenium alloys are generally suitable and are extremely resistant to tarnishing.

[5] E. Raub and G. Buss, Z. Elektrochem., **46**, 195–202 (1940).
[6] B. Ershler, Acta Physicochem. (U. R. S. S.), **19** (No. 2–3), 139–147 (1944).
[7] G. P. Gladis, unpublished report, Research Laboratory, International Nickel Co., New York, N. Y.

Palladium is somewhat more reactive than platinum; therefore, the palladium-rich alloys, although having some extremely useful applications, are not generally employed in chemical plant equipment. Various gold-base alloys containing palladium have been used for corrosion-resisting applications, and the ternary Au-Pt-Pd alloy is useful for spinnerets. Palladium is added to many of the gold-base dental alloys, not only to improve their tarnish resistance but also to increase their melting temperature and hardenability.[8] Alloys containing substantial amounts of palladium (15%) are essentially white in color.

Pd-Ag alloys containing 50% or more of palladium have a high resistance to tarnish by indoor atmospheres, and are suitable for electrical contacts and jewelry. Recent quantitative data on the resistance of Pd-Ag alloys to indoor atmospheric attack and to air saturated with sulfur have been obtained and are shown in Fig. 1.[9] Nitric acid will attack these alloys, but the ternary alloy containing 40 to 50% palladium, 10% platinum, and remainder silver[10] is not attacked by a drop of nitric acid, which is one of the tests frequently imposed by jewelers.

The Pd-Cu alloys resemble in many ways the Au-Cu alloys and undergo an ordering reaction corresponding to the composition PdCu and also $PdCu_3$. The Pd-Cu phase is involved in the age hardening of the high-strength dental alloys.[11] The fact that the ordering reaction at PdCu results in a sharp increase in electrical conductivity has led to application of the ordered alloy for electrical contacts.

As a hydrogenation and dehydrogenation catalyst, palladium is extremely effective, and is sometimes preferred to platinum because of its lower cost, or to nickel because of its higher resistance to corrosion, greater catalytic activity, and ease of recovery and reprocessing. Catalysts have specific characteristics which render each one outstanding for an industrial reaction, a fact which is being increasingly recognized, particularly in the vitamin field where the platinum metals are being extensively used. Recent industrial use has been made of palladium catalysts, acting at room temperature, to effect removal of traces of residual oxygen in hydrogen. The catalyst causes reaction of oxygen to form water which is subsequently removed by a drier.

Palladium shows less tendency to remain passive than platinum.[12,13] In its reaction with oxidizing media, such as hot concentrated sulfuric acid and concentrated nitric acid, it is somewhat akin to silver, to which it is related in the Periodic Table. The addition of 10 or 20% of gold to palladium approximately doubles the rate of attack by concentrated sulfuric acid plus chromic acid at 100° C (210° F),[14] but alloys containing 30% or more of gold have the same corrosion rate as gold. Platinum, rhodium, or iridium additions also are effective, particularly in resisting nitric acid. The 10% platinum alloy withstands the nitric acid spot test.

The rate of attack by nitric acid falls sharply with the concentration of the acid. Nitric acid 1 N in concentration showed a corrosion rate of approximately 7 mdd, whereas 0.1 N acid showed no measurable attack.[14] Hydrochloric acid is not active toward palladium, and probably would not attack it at all if free from dissolved oxygen.

[8] W. S. Crowell, E. M. Wise, and J. T. Eash, *Trans. Am. Inst. Mining Met. Engrs.*, **99**, 383 (1932). E. M. Wise and J. T. Eash, *Trans. Am. Inst. Mining Met. Engrs.*, **104**, 276 (1933).
[9] W. E. Campbell, unpublished data, Bell Telephone Laboratories, Murray Hill, N. J.
[10] E. M. Wise, U. S. Patent 2,129,721 (1938).
[11] R. F. Vines and E. M. Wise, *Age-Hardening Precious Metal Alloys*, pp. 190–228 Symposium on Age-Hardening of Metals, American Society for Metals, 1940.
[12] F. Müller, *Z. Elektrochem.*, **34**, 744–752 (1928).
[13] F. Müller and A. Riefkohl, *Z. Elektrochem.*, **36**, 181–183 (1930).
[14] R. Atkinson, A. Raper, and A. Middleton, unpublished research, Mond Nickel Co., Acton, London, England.

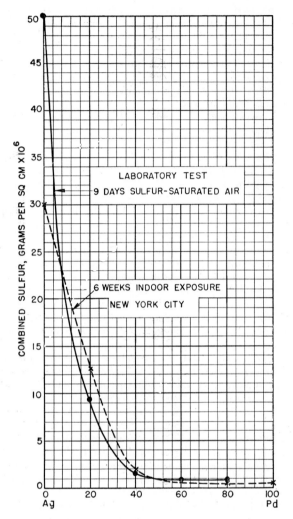

Fig. 1. Tarnish of Pd-Ag Alloys in Sulfur-Saturated Air and Indoor Atmosphere.
(W. E. Campbell.)

BEHAVIOR AS ANODE

As anode in acid chloride solutions[15] and in complex alkali nitrite solutions containing sufficient chloride or bromide ions,[16] high-purity palladium can be quantitatively dissolved. Use is made of this fact in several relatively new palladium plating baths, which, unlike baths for platinum and rhodium, are self-maintaining and operate with high cathode, as well as high anode efficiency.[15,16] For this, and other reasons, palladium can be electrodeposited very satisfactorily in thick ductile coatings.

[15] E. M. Wise and R. F. Vines, U. S. Patent 2,335,821 (1943).
[16] A. Raper, U. S. Patent 1,993,623 (1935).

Palladium anodes may become passive under some conditions, as in 0.01 N HCl, but are active in 1 N and 5 N HCl.[17] Palladium is also passive as anode in chloride-free $NaNO_3$, Na_2SO_4, and HNO_3 solutions.[17]

The corrosion behavior of palladium in a number of corrosive media is presented in Table 1 and is also summarized in Table 4.

TABLE 4. CORROSION RESISTANCE OF PALLADIUM

	Temperature	Rating
H_2SO_4, conc.	R. T.	A
	100° C	C
H_2SeO_4, sp. gr. 1.4	R. T.	C
	100° C	D
H_3PO_4	100° C	B
$HClO_4$	R. T.	A
	100° C	C
HNO_3, 70%	R. T.	D
2 N	R. T.	C
1 N	R. T.	B
0.1 N	R. T.	A
HNO_3, 95%	R. T.	D
	100° C	D
Aqua regia	R. T.	D
	Boiling	D
HF, 40%	R. T.	A
HCl, 36%	R. T.	A
	100° C	B
HBr, sp. gr. 1.7	R. T.	D
	100° C	D
HI, sp. gr. 1.75	R. T.	D
Acetic acid (glacial)	100° C	A
Cl_2, dry	R. T.	C
Cl_2, moist	R. T.	D
Br (liq.), dry	R. T.	D
Br (liq.), moist	R. T.	D
Br water	R. T.	B
I, dry	R. T.	A
I, moist	R. T.	B
I, in alcohol	R. T.	B
H_2S, moist	R. T.	A
NaClO solution	R. T.	C
	100° C	D
KCN solution	R. T.	C
	100° C	D
$FeCl_3$, 10% solution	R. T.	C
	100° C	D
$HgCl_2$ solution	100° C	A
$CuCl_2$ solution	100° C	B
$CuSO_4$ solution	100° C	A
$Al_2(SO_4)_3$ solution	100° C	A

A — No appreciable corrosion.
B — Some attack, but not enough to preclude use.
C — Attacked enough to preclude use.
D — Rapid attack.
R. T. — Room temperature.

[17] F. Müller, Z. Elektrochem., **34**, 744–752 (1928).

ATMOSPHERIC EXPOSURES

Palladium remains untarnished at ordinary temperatures on indoor exposure, and it is, therefore, often used where this is important, as in electrical contacts, decorative leaf, and particularly in jewelry alloys.

Under outdoor exposures to industrial atmospheres the catalytic effect of palladium in oxidizing SO_2 to SO_3 may be involved in the formation of a basic sulfate on the surface of the metal, which causes it to discolor slowly in contrast to its excellent behavior indoors and in jewelry.

Resistance to Tarnish in Household Atmospheres. A number of palladium alloys representing commercial materials employed in dentistry, electrical contacts, or jewelry, plus several experimental palladium alloys and a few comparison metals, have been subjected to lengthy exposure tests to determine their resistance to tarnish by perspiration and prolonged exposure to household atmospheres.[18] The samples, in sheet form, were finished with fine emery paper and scrubbed with pumice. They were repeatedly handled with perspiring hands and then exposed in the kitchen of a home heated with coke. The samples were visually examined at the end of 66 days, at which time samples in group 3 showed tarnish, but the remainder showed little change. All samples were then washed with soap and water. This removed the slight film of dust and grease but did not remove any tarnish. The samples were then exposed for one year in the dining room of the same house. At the end of this exposure one half of each specimen was cleaned with dry alumina to permit the visual detection of even slight tarnish. The samples were arranged in three groups, as shown in Table 5. The order of samples within each group is not significant.

TABLE 5. TARNISH RATING OF ALLOYS AFTER ONE YEAR OF EXPOSURE TO HOUSEHOLD ATMOSPHERES*

	Alloy Composition, %
Group I	100 Pt
	100 Pd
Not sensibly tarnished	95 Pd–3 Rh–2 Ru
	95 Pd–3 Ni–2 Ru
	95 Pd–5 Ir
	90 Pd–10 Pt
	90 Pd–10 Au
	70 Pd–30 Ag
	70 Pd–5 Pt–25 Ag
Group II	50 Pd–50 Ag
	50 Pd–40 Ag–10 Cu
Slightly tarnished	45 Pd–5 Pt–50 Ag
(Acceptable for jewelry)	40 Pd–10 Pt–50 Ag
Group III	40 Pd–60 Ag
	27 Pd–5 Au–8 Cu–2 Zn
Obviously tarnished	34 Pd–16 Cu–50 Ag
	Nickel
	Silver

* R. F. Vines, unpublished report, Research Laboratory, International Nickel Co., Bayonne, N. J.

It is evident that slightly more than 50% of palladium is required for substantially complete resistance to tarnish. This is slightly above 50 atomic per cent palladium, a value in agreement with the findings of W. E. Campbell[19] (Fig. 1). About 70 wt.%

[18] R. F. Vines, unpublished report, Research Laboratory, International Nickel Co., Bayonne, N. J.
[19] W. E. Campbell, unpublished data, Bell Telephone Laboratories, Murray Hill, N. J.

of gold is required for resistance similar to that of 50 to 55% palladium. This corresponds to about the same atomic percentage of tarnish-resistant metal in each case. However, if part of the gold in a complex Au-Ag-Cu alloy, such as is used in dentistry, is replaced with palladium (the atomic percentage of gold plus palladium being held constant), the tarnish resistance is substantially increased. This result is utilized in various commercial dental alloys.

IRIDIUM

Although very pure iridium is forgeable at very high temperatures, it cannot be considered suitable for general fabrication. Barring a few pure iridium crucibles made for very special high-temperature studies, this metal is utilized almost exclusively as an alloying element with platinum. Its standard electrode potential $Ir \rightleftarrows Ir^{+++} +3e^-$ is about 1.0 volt at 25° C (77° F).

TABLE 6. CORROSION RESISTANCE OF IRIDIUM

	Temperature	Rating
H_2SO_4, conc.	R. T.	A
	100° C	A
H_3PO_4	100° C	A
HNO_3, 95%	R. T.	A
	100° C	A
Aqua regia	R. T.	A
	Boiling	A
HF, 40%	R. T.	A
HCl, 36%	R. T.	A
	100° C	A
HBr, sp. gr. 1.7	R. T.	A
	100° C	A
HI, sp. gr. 1.75	R. T.	A
	100° C	A
Acetic acid (glacial)	100° C	A
Cl_2, dry	R. T.	A
Cl_2, moist	R. T.	A
Br (liq.), dry	R. T.	A
Br (liq.), moist	R. T.	A
Br water	R. T.	A
I, dry	R. T.	A
I, moist	R. T.	A
I, in alcohol	R. T.	A
H_2S, moist	R. T.	A
NaClO solution	100° C	B
$FeCl_3$ solution	R. T.	A
	100° C	A
$HgCl_2$ solution	100° C	A
$CuSO_4$ solution	100° C	A
$Al_2(SO_4)_3$ solution	100° C	A

A — No appreciable corrosion.
B — Some attack, but not enough to preclude use.
C — Attacked enough to preclude use.
D — Good solvent for Ir.
R. T. — Room temperature.

Iridium is moderately hard — about 170 Vickers for annealed pure material — and when alloyed with platinum, it hardens and raises the annealing temperature and melting point. It also reduces the rate of attack by aqua regia and other solutions containing active chlorine; hence Ir-Pt alloy is particularly suitable for halogen-evolving anodes. The jewelry industry normally uses large quantities of platinum hardened with iridium or ruthenium. These alloys also are largely used for platinum-base electrical contacts. Platinum alloys containing less than 1% of iridium are used for crucibles and sometimes for glass fiber equipment, but oxidation and volatilization of iridium on heating limit the practical iridium content.

Observations on the corrosion behavior of iridium are summarized in Tables 1 and 6. Of the corrosive media studied, only the NaClO solution showed measureable attack.

Although iridium is resistant to aqua regia at 100° C (212° F), it can be dissolved at a useful rate by heating to 250° to 300° C (480° to 570° F) with a species of aqua regia under pressure.[20]

Iridium is not amalgamated by mercury nor is it soluble in molten lead.

RHODIUM

Rhodium can be produced in a reasonably ductile form by methods somewhat analogous to those used for processing tungsten and molybdenum, but it finds very limited use in fabricated form. The hardness of the pure annealed metal is about 122 Vickers. The standard electrode potential of rhodium (Rh $\rightleftarrows$ Rh^{++} + 2e^-) is about 0.6 volt at 25° C (77° F). Rhodium can be electrodeposited as a bright lustrous plate, and such electrodedeposits are commercially used for surfacing very large (60 in. diameter) metallic mirrors and for finishing jewelry and other decorative items, as it does not tarnish in the atmosphere. The electrodeposited metal is hard (about 600 Vickers), possesses good electrical conductivity, and has found numerous uses for corrosion-resistant contacts which in service are subject to some sliding, as it resists galling. Rhodium sheet can be slowly sublimed by heating in vacuo to within a few degrees of its melting point, to produce front-surfacing glass mirrors.

The major use is as an alloying element with platinum, principally as an ammonia oxidation catalyst. Smaller amounts of the Rh-Pt alloys are used for thermocouples, resistance furnaces, crucibles, and glass-working equipment. These high-temperature uses are considered in *Noble Metals and Their Alloys*, p. 699. The 10% Rh–Pt alloy is also used for low-temperature corrosion-resisting applications, such as insoluble anodes, although pure platinum or 10% Ir–Pt alloy is more frequently used. The 10% Rh-Pt alloy has been found to be most satisfactory for rayon spinnerets, particularly those employed in producing the finer fibers.[21] The merit of this alloy over the older 30% Pt–70% Au alloy appears to be due to its resistance to plugging, coupled with resistance to cleaning solutions, such as boiling dilute hydrochloric acid and hot sulfuric-chromic acid mixtures.

Rhodium is resistant in itself to a wide range of corrosive media and is extremely effective in producing corrosion-resistant alloys. For example, the 37% Rh, remainder Ni, alloy possesses better resistance to corrosion than 14-kt. yellow gold. Rhodium also increases the corrosion resistance of platinum in media such as a boiling 10% ferric chloride solution, which corrodes platinum at the rate of 400 mdd, whereas 5% Rh-Pt alloy corrodes at a rate of only 5 mdd.

[20] E. Wichers, W. Schlecht, and C. Gordon, *J. Research Natl. Bur. Standards*, **33**, 363–381 (1944); see also **33**, 457–470 (1944).
[21] R. V. Williams and E. R. McKee, U. S. Patent 2,135,611 (1938).

Reference to Tables 1[22] and 7 shows that only fuming hydrobromic acid attacks rhodium at a substantial rate. NaClO solution attacks the metal slightly, the rate being about 5 mdd at room temperature and 100° C (212° F), whereas concentrated H_2SO_4 at 100° C (212° F) causes a loss of 7 mdd.

TABLE 7. CORROSION RESISTANCE OF RHODIUM

	Temperature	Rating
H_2SO_4, conc.	R. T.	A
	100° C	B
H_3PO_4	100° C	A
HNO_3, 70%	R. T.	A
HNO_3, 95%	R. T.	A
	100° C	A
Aqua regia	R. T.	A
	Boiling	A
HF, 40%	R. T.	A
HCl, 36%	R. T.	A
	100° C	A
HBr, sp. gr. 1.7	R. T.	B
	100° C	C
HI, sp. gr. 1.75	R. T.	A
	100° C	A
Acetic acid (glacial)	100° C	A
Cl_2, dry	R. T.	A
Cl_2, moist	R. T.	A
Br (liq.), dry	R. T.	A
Br (liq.), moist	R. T.	A
Br water	R. T.	A
I, dry	R. T.	A
I, moist	R. T.	B
I, in alcohol	R. T.	B
H_2S, moist	R. T.	A
NaClO solution	R. T.	B
	100° C	B
$FeCl_3$ solution	R. T.	A
	100° C	A
$HgCl_2$ solution	100° C	A
$CuSO_4$ solution	100° C	A
$Al_2(SO_4)_3$ solution	100° C	A

A — No appreciable corrosion.
B — Some attack, but not enough to preclude use.
C — Attacked enough to preclude use.
D — Rapid attack.
R. T. — Room temperature.

OSMIUM

Osmium is the hardest (approx. 400 Vickers) and the highest melting of the platinum metals, and has the highest modulus of elasticity of all the elements, but is only moderately resistant to corrosion, as is noted from data of Tables 1 and 8.

[22] R. Atkinson, A. Raper, and A. Middleton, unpublished research, Mond Nickel Co., Acton, London, England.

TABLE 8. CORROSION RESISTANCE OF OSMIUM

	Temperature	Rating
H_2SO_4, conc.	R. T.	A
	100° C	A
H_3PO_4	100° C	D
HNO_3, 70%	R. T.	C
HNO_3, 95%	R. T.	D
	100° C	D
Aqua regia	R. T.	D
	Boiling	D
HF, 40%	R. T.	A
HCl, 36%	R. T.	A
	100° C	C
HBr, sp. gr. 1.7	R. T.	A
	100° C	C
HI, sp. gr. 1.75	R. T.	B
	100° C	C
Cl_2, dry	R. T.	A
Cl_2, moist	R. T.	C
Br (liq.), dry	R. T.	D
Br (liq.), moist	R. T.	B
I, dry	R. T.	B
I, moist	R. T.	A
H_2S, moist	R. T.	A
NaClO solution	R. T.	D
	100° C	D
$FeCl_3$ solution	100° C	D

A — No appreciable corrosion.
B — Some attack, but not enough to preclude use.
C — Attacked enough to preclude use.
D — Rapid attack.
R. T. — Room temperature.

It is readily oxidized, and forms a tetroxide boiling at 130° C (265° F) which is very toxic. Osmium appears to be completely non-ductile, and although it can be used for hardening wrought platinum, other hardeners are preferred.

The principal use of osmium is in the production of very hard alloys used for nonrusting pivots for instruments, phonograph needles, tipping the nibs of fountain pens, and for certain types of electrical contacts.

For pen tipping and pivots, alloys containing not less than 60% osmium, or osmium plus ruthenium, are considered desirable for the best-grade material. The other platinum group metals constitute the principal remainder in this type of alloy.

Osmium is the rarest of the platinum group metals, and, although its special uses are important, they are on a rather small scale and are approximately equal to the amount of metal available.

RUTHENIUM

Ruthenium has characteristics somewhat intermediate between those of osmium and iridium, and is perhaps even more difficult to forge than iridium. Annealed pure metal has a hardness of 360 Vickers. Aside from some specialized uses as a catalyst, the metal is employed entirely as a hardening alloying element, principally as an addition to palladium and platinum for electrical contacts and jewelry, where it behaves in a manner similar to iridium, but with a hardening effect about twice as great per unit weight. It also finds some use analogous to osmium in very hard contact and pen-tipping alloys.

As will be seen from Tables 1 and 9, ruthenium is outstanding in its resistance to corrosion and is even resistant to hot aqua regia. Bromine and iodine in solution attack ruthenium slightly, boiling $HgCl_2$ solution is a little more aggresive, while a solution of NaClO attacks it vigorously.

TABLE 9. CORROSION RESISTANCE OF RUTHENIUM

	Temperature	Rating
H_2SO_4, conc.	R. T.	A
	100° C	A
H_3PO_4	100° C	A
HNO_3, 70%	[R. T.	A
HNO_3, 95%	R. T.	A
	100° C	A
Aqua regia	R. T.	A
	Boiling	A
HF, 40%	R. T.	A
HCl, 36%	R. T.	A
	100° C	A
HBr, sp. gr. 1.7	R. T.	A
	100° C	A
HI, sp. gr. 1.75	R. T.	A
	100° C	A
Acetic acid (glacial)	100° C	A
Cl_2, dry	R. T.	A
Cl_2, moist	R. T.	A
Br (liq.), dry	R. T.	A
Br (liq.), moist	R. T.	A
Br water	R. T.	B
I, dry	R. T.	A
I, moist	R. T.	A
I, in alcohol	R. T.	B
H_2S, moist	R. T.	A
NaClO solution	R. T.	D
	100° C	D
$FeCl_3$ solution	R. T.	A
	100° C	A
$HgCl_2$ solution	100° C	C
$CuSO_4$ solution	100° C	A
$Al_2(SO_4)_3$ solution	100° C	A

A — No appreciable corrosion.
B — Some attack, but not enough to preclude use.
C — Attacked enough to preclude use.
D — Rapid attack.
R. T. — Room temperature.

RHENIUM

W. J. KROLL*

Rhenium has some properties in common with manganese, tungsten, and osmium. The metal looks like platinum. Its electrochemical potential indicates that it is more noble than tungsten.[1]

The metal readily dissolves in nitric acid and slowly in sulfuric acid, but is not attacked by either hydrochloric or hydrofluoric acids. Electrolytic deposits of rhenium on brass resist the action of hydrochloric acid.[2]

Alloys of rhenium with various metals have been suggested for fountain pen tips because of good corrosion and wear resistance.

The metal is oxidized when heated in air at moderately high temperatures, being somewhat more resistant to oxidation than tungsten. The higher oxides of rhenium are volatile. It resists oxidation by wet hydrogen at elevated temperatures better than tungsten.

Rhenium does not react with nitrogen at any temperature, nor does it react with carbon, carbon monoxide, or hydrocarbon gases. Mercury does not amalgamate with it up to a temperature of 300° C (570° F).

Chlorine and bromine attack the metal above 500° C (930° F).

Rhenium is extracted from Mansfield furnace iron-pigs, which seem to constitute the main source of supply. Because of its high cost (approximately twice the price of platinum) its use for protective coatings or in alloys is not attractive at present.

GENERAL REFERENCES

GMELIN, L., *Handbuch der anorganischen Chemie*, Number 70, Verlag Chemie, 1940.
MELLOR, J. W., *Comprehensive Treatise on Inorganic and Theoretical Chemistry*, Vol. 12, pp. 471–472, Longmans, Green and Co., London, 1932.
VAN ARKEL, A. E., *Reine Metalle*, pp. 295–300, J. Springer, Berlin, 1939.

SILVER AND SILVER ALLOYS

ROBERT H. LEACH†

Silver is one of the noble metals. It stands high in the Emf Series, having a standard potential of 0.80 volt. This nobility is the principal factor responsible for its high resistance to corrosion. The production of a protective film upon which the corrosion resistance of chromium, stainless steel, and aluminum depends is of secondary importance with silver. It may, however, be a controlling factor in some instances. For example, when silver is immersed in hydrochloric acid or in chloride solutions, a film of silver chloride is formed which inhibits further corrosion.

Silver is a relatively expensive metal with low tensile strength compared to nickel or stainless steel. When high pressures are involved, the thickness of silver required to give the necessary strength on large pieces of equipment makes the cost prohibitive, but silver linings for steel equipment have been used to an appreciable extent.

---

* Consulting Metallurgist, Albany, Ore.
† Handy and Harman, Bridgeport, Conn.

[1] C. Agte, H. Alterthum, K. Becker, G. Heyne, and K. Moers, *Z. anorg. Chem.*, **196**, 129–159 (1931).
[2] C. G. Fink and P. Deren, *Trans. Electrochem. Soc.*, **66**, 471 (1934).

Silver-clad steel is a satisfactory material for providing the corrosion-resistant properties of silver and the necessary strength. Silver plating of steel or other metals is sometimes satisfactory, but it must be remembered that, because of its nobility, silver is cathodic to all metals except gold, the platinum metals, and mercury. Any porosity in the plate which exposes the underlying metal will set up galvanic corrosion with attack of the base metal. This same galvanic corrosion must be considered when silver is brought in contact with a base metal in any structural design.

ACIDS

Silver offers high resistance to most dilute mineral acids, either hot or cold. Although silver fluoride is soluble in water, the metal is not attacked appreciably by *hydrofluoric acid* and is used to handle this acid in any concentration (Table 1). Dissolved oxygen accelerates the attack.

Silver is rapidly attacked by *nitric acid* and by hot *sulfuric acid*, the latter in concentrations above 85%. The attack by nitric acid in all concentrations is greatly accelerated by heat. Data on corrosion of silver in sulfuric acid are given in Table 1.

The degree of attack by *hydrochloric acid* varies with concentration and temperature. Silver has been found satisfactory in many chemical plant operations involving the use of this acid. Nowack and Spanner[1] found that normal hydrochloric acid had no appreciable effect on fine silver after 3 months of exposure at room temperature. In the presence of oxygen this resistance is reduced. McDonald[2] found that corrosion by hydrochloric acid in the presence of an oxidizing agent occurs at elevated temperature, but recommends the use of silver for handling dilute hydrochloric acid if conditions are such that the film produced by the initial attack is not removed. Many investigators have studied the effect of hydrochloric acid on silver, and the variable results reported are without question due to the degree with which the initial chloride film is removed.

Data for 30% and concentrated hydrochloric acid are presented in Table 1.

Silver resists the corrosive action of either hot or cold organic acids in all concentrations (Table 1). Its use for handling phenol (Table 2), fruit juices, essential oils, and many pharmaceuticals is warranted by high corrosion resistance and the resulting minimum contamination and discoloration of these products.

ALKALIES AND SALTS

Silver is highly resistant to alkalies at ordinary or elevated temperatures (Table 3). The presence of oxygen accelerates the corrosion, but, even in the case of fused alkalies, the resistance is high, and its use for production of high-quality alkali is indicated. Silver is corroded by ammonium and substituted ammonium hydroxides.

Silver is attacked by potassium and sodium cyanide solutions, dissolved oxygen accelerating the attack.

Silver resists the attack of urea, and is used in apparatus for the synthesis of urea from ammonia and carbon dioxide.

Sodium and potassium sulfide solutions corrode silver, forming silver sulfide. Silver has been found satisfactory for equipment used in the production of sodium phosphate salts.[3] Corrosion rates in some salt solutions are given in Table 4.

[1] L. Nowack and J. Spanner, "Die Korrosion der Edelmetalle," pp. 765–827, *Die Korrosion Metallischer Werkstoffe* (O. Bauer, Editor), Bd. II, S. Hirzel, Leipzig, 1938.

[2] D. McDonald, "Silver and Its Application to Chemical Plant," *J. Soc. Chem. Ind. (Chem. and Industry)*, **50**, 168–178 (1931).

[3] B. A. Rogers, I. C. Schoonover, and L. Jordan, "Silver: Its Properties and Industrial Uses," National Bureau of Standards, *Circ.* C-412 (1936).

TABLE 1. CORROSION IN ACIDS AND ACID SOLUTIONS
Weight Losses of Silver, MDD

(+ Signifies gain in weight.)

Concentration	Temperature °C	Temperature °F	Test Period, hr	Previous Exposure, hr	Atmosphere	Type of Test* A	B	C	M	N	O
Acetic Acid											
50% by vol.	102	216	24	0	Air	Negligible	+3.3	+1.0			
			24	24		0.0	6.0	0.0			
99.5% by vol.	Boiling	Boiling	24	0		5.0	6.0	2.5			
			24	24		2.0	10.0	20.0			
			72	48		2.5	2.4	0.0			
			24	120		6.0	6.0	...			
			186	24		3.0	4.8	3.0			
Citric Acid											
15% by wt.	102	216	24	0	Air	+0.5	0.0	0.0			
			48	24		0.3	0.0	0.7			
30% by wt.	67	153	48	0						0.4	
Hydrochloric Acid											
30% by vol.	Boiling	Boiling	24	0	Air	0.03	0.50	0.35			
			72	24	Air	0.10	1.00	0.80			
Concentrated	Boiling	Boiling	24	0	Air	+0.10	2.50	1.00			
					Without O₂	0.02	0.85	0.50			
					Excess O₂	0.75	5.00	1.70			
					Excess N₂	0.50	4.75	0.40			
			24	24	Air	0.50	5.00	+2.00			
					Without O₂	0.00	0.50	+0.50			
					Excess N₂	0.10	5.00	+0.80			
Hydrofluoric Acid†											
40%	110	230	48.0							29.2	
100%	38	100	87.5							13.14	
Lactic Acid‡											
50%	Boiling	Boiling	26	0	Air				140		
Phosphoric Acid (Commercial)											
50%	100	212	24	0	Air	2.5	+20.0	+21.0			
						0.2	+ 1.0	+ 4.0			

* Description of tests is given on p. 318.

† *Clad News*, **4**, No. 1, p. 6. (40% Solution: constant boiling solution — otherwise unspecified. 100% Solution: method of test unspecified.)

‡ B. A. Rogers, I. C. Schoonover, and L. Jordan, "Silver: Its Properties and Industrial Uses," National Bureau of Standards, *Circ.* C-412 (1936). (Specimen totally immersed in reflux condenser.)

§ Samples cut into two parts at junction.

TABLE 1. CORROSION IN ACIDS AND ACID SOLUTIONS—*Continued*

Weight Losses of Silver, MDD

Concentration	Temperature		Test Period, hr	Previous Exposure, hr	Atmosphere	Type of Test*					
	°C	°F				A	B	C	M	N	O
Sulfuric Acid											
10%	Boiling	Boiling	24	0	Air	....	2.00	2.40			
			72	24		....	0.25	0.70			
20%	Boiling	Boiling	24	0		....	0.0	4.00			
			72	24		....	+3.2	4.50			
			126	90		....	+2.3	1.00			
60%	Boiling	Boiling	24	0		....	14.40§	26.0			
			72	24		....		252.0			

TABLE 2. CORROSION IN ORGANIC REAGENTS

Weight Losses of Silver, MDD

(+ Signifies gain in weight.)

Reagent	Concentration, %	Temperature		Test Period, hr	Previous Exposure, hr	Type of Test*		
		°C	°F			A	B	C
Carbon tetrachloride	100	185	365	24	0	+4.0	+14.0	+9.0
				24	24	0.0	0.0	0.0
				90	48	+2.0	1.3	0.0
Phenol	90	185	365	24	0	5.0	4.0	1.5
				72	24	1.7	2.2	13.0
	100	185	365	24	0	0.0	+3.3	+8.2
				72	24	+0.66	0.22	+0.80

* Description of tests is given on p. 318.

TABLE 3. CORROSION IN CAUSTIC SODA

% Weight Loss of Silver Crucibles

Reagent	Concentration, %	Temperature		Test Period, hr	Type of Test* O
		°C	°F		
Caustic soda	70	75	167	24	0.0042
				60	0.0041

Crucibles containing approximately 40 ml of molten 70% caustic were successively heated to 75° C (167° F) for 10 hr and then allowed to cool to solidification for 2 hr. This heating cycle was carried on continuously for a period of 6 days. Similar tests were made in which the temperature of the caustic was 110° C (230° F).

The tests were divided into two periods of 1-day and 5-day duration of the heating and cooling cycle. The crucibles used were commercially available 999.7 fine silver.

* Description of test is given on p. 318.

318 CORROSION IN LIQUID MEDIA, ATMOSPHERE, GASES

TABLE 4. CORROSION IN SALT SOLUTIONS

Weight Losses of Silver, MDD

(+ Signifies gain in weight.)

Salt	Temperature °C	°F	Test Period, hr	Previous Exposure, hr	Type of Test* M	N	O
30% solution of aluminum chloride (AlCl$_3$·6H$_2$O)	R. T.†		24	0	1.0		
			48	24	1.2		
			48	72	0.0		
	100	212	24	0		12.	
			48	24		5.0	
			48	72		5.2	
30% solution of aluminum sulfate (Al$_2$[SO$_4$]$_3$·18H$_2$O)	67	153	42	0		2.1	
			48	42		0.0	
30% solution of cadmium sulfate (3CdSO$_4$·8H$_2$O)	R. T.		24	0	0.0		
			72	24	0.0		
	100	212	24	0	0.4		
			72	24	0.7		
Saturated solution of calcium fluoride	100	212	96	0		0.0	
			116	96		0.0	
20% solution of cupric nitrate (Cu[NO$_3$]$_2$·3H$_2$O)	R. T.		24	0	0.0		
			72	24	0.0		
			144	96	0.2		
20% solution of cupric sulfate (CuSO$_4$·5H$_2$O)	R. T.		24	0	+3.2		
			91	24	+0.5		
Saturated solution of sodium fluoride	100	212	24	0		0.0	
			96	24		0.0	
3% solution of sodium chloride‡			168				0.058
			504				0.058

* Description of tests is given below.
† R. T. — Room temperature.
‡ V. de Marchi and C. G. Fink, A. S. P. R. P., Sixth Report, National Bureau of Standards Library (1938). (Alternate Immersion Test: 30-min cycle; 15 min immersion followed by 15 min exposure to air.)

Specific Corrosion Data. The use of silver in industry is of comparatively recent growth. Although some information is available, precise corrosion data, particularly with reference to actual performance under service conditions, are not common. General information, including an extensive bibliography, has been published.[4,5,6]

The quantitative data presented in this chapter (Tables 1 to 4), with few exceptions, were obtained from the work done under the auspices of the American Silver Producers' Research Project. Refluxing, immersion, and evaporation tests on silver were made as follows:

Specimen A: Suspended inside the inner tube of the water-cooled condenser entirely in the vapors, with condensate forming on it. Size 10 × 1 cm.

Specimen B: Suspended at the junction of the liquid and vapor phase. Size 10 × 1.5 cm, about two thirds in the liquid phase.

[4] Lawrence Addicks (Editor), *Silver in Industry*, Reinhold Publishing Corp., New York, 1940.

[5] A. Butts and J. M. Thomas, "Corrosion Resistance of Silver and Silver Alloys," Chapter 15, *Silver in Industry* (Lawrence Addicks, Editor), Reinhold Publishing Corp., New York, 1940.

[6] A. Butts and J. Giacobbe, "Silver Offers Resistance to Many Chemicals," *Chem. Met. Eng.*, **48**, 76–79 (December, 1941).

SILVER AND SILVER ALLOYS

Specimen C: Completely immersed in the liquid. Size 10 × 2.5 cm.
Volume of liquid was 500 ml; vessels were of glass; specimens were suspended with glass rods or hooks. Access of air was allowed to the condenser tubes where the vapor condensed, but no actual aeration was provided.

Specimen M: Beaker immersion test. Specimens, 10 × 2.5 cm, four fifths immersed, were placed in separate 200-ml beakers.

Specimen N: Evaporation test. Same as specimen M test, except that beaker was placed in an oven and the liquid evaporated to dryness. Water was then added and the process repeated several times during the run.

Specimen O: Different methods or different sources are included in the tables under the heating O. Numbered footnotes describing these particular tests follow each table.

Before the initial weighing, each specimen was cleaned by scrubbing both sides with pumice, using a wad of cotton, until a uniform surface was obtained. It was then rinsed with distilled water, immersed in alcohol, dried in an oven at a temperature not above 100° C (210° F), and allowed to cool in a desiccator. After the test, and before subsequent weighing, the specimen was again cleaned in the same manner as before except that the scrubbing was done without an abrasive. With this procedure any non-adherent or soluble corrosion product was removed before weighing. A tightly adherent film was not removed. A degree of uncertainty in the results occurred when the film was adherent to an intermediate degree, but this question was involved in few instances.

The silver used was high-purity, 999.7 to 999.9 fine,* with a thickness of 0.05 mm (0.002 in.). It is probable that in most cases the difference between this high-purity silver and standard electrolytic silver running 999.0 to 999.5 fine would not give materially different results. Except as noted, the corroding liquids were of standard laboratory reagent grade.

GASES

Silver is immune to the attack of most dry and moist atmospheres. Ozone attacks silver, but it is highly resistant to the attack of oxygen even at elevated temperatures. Halogen gases attack silver, but the initial film inhibits the attack. Bromine is more active than chlorine. Dry ammonia has no effect; moist ammonia is corrosive. Carbon monoxide, hydrogen, fluorine, and nitrogen have no effect at ordinary temperatures. There is slight, if any, attack by sulfur dioxide at ordinary temperatures, although some investigators have reported a tarnish, which is probably due to the formation of the sulfide or sulfate.

Tarnishing. Hydrogen sulfide tarnishes silver at room temperature, a reaction which is accelerated by moisture and moderate increases in temperature. The black Ag_2S which forms, even when only traces of H_2S are present, is the cause of blackening commonly observed when silver is exposed to the air. Moisture is necessary for the reaction because H_2S, when thoroughly dried, does not react with silver.†

Ag_2S can be reduced to silver in situ by cathodic hydrogen, a process made use of practically to eliminate tarnish on silver. It is stated that a 1% aqueous solution of morpholine $(O:[CH_2CH_2]_2NH)$ exposed within a cabinet or display case will protect silver against tarnish.[7]

---

* The fineness of silver is the silver content in parts per thousand.
† Oxygen apparently is also necessary. See footnote, p. 15.

[7] H. B. McClure, *Chem. Eng. News*, **22**, 420 (March 25, 1944).

Large sums of money have been spent in the attempt to produce alloys highly resistant to sulfur attack, but so far as sterling silver or coin silver is concerned, nothing has been developed of any considerable commercial importance. Additions of cadmium, zinc, tin, and antimony show added resistance under some conditions sufficient to justify their use. For example, an alloy containing 75% silver and 25% zinc has had some commercial application. If cost is not a factor, Pd-Ag alloys containing up to 40% palladium show greatly improved resistance. (Tarnish resistance of the Pd-Ag alloys is discussed on p. 305. See Fig. 1, p. 306.)

SILVER ALLOYS

Silver alloys are usually less resistant than fine silver to various corrosive substances, with the resistance increasing with the percentage of silver. Fink and de Marchi[8] used a 3% brine solution for testing a series of both low and high fineness alloys. They reported that the addition of a small amount of silver to a few non-ferrous alloys improved the corrosion resistance, but none of the silver-rich alloys were more resistant than fine silver.

Ag-Si alloys containing less than 2.% silicon have higher yield strength and hardness than silver, and A. J. Dornblatt's[9] investigations showed no appreciable loss of corrosion resistance.

Silver brazing alloys (silver solders) are used extensively in the manufacture of equipment for chemical plants and other industries where resistance to corrosion is important.

ACKNOWLEDGMENT

Acknowledgment is made to Mr. John L. Christie, of Handy and Harman, for his help in preparing this chapter.

TANTALUM

CLARENCE W. BALKE*

Tantalum is highly resistant to chemical corrosion. In this respect it resembles glass more nearly than gold or platinum. It is not attacked by most acids at ordinary temperatures. Hydrofluoric acid, as the outstanding exception, attacks it readily.

Tantalum is much less resistant to alkaline solutions, and in boiling solutions of the strong alkalies it is rapidly destroyed. In many cases the rate of corrosion is largely dependent on the temperature of the corrosive agent.

The data presented below have been obtained with the metal in the form of sheet, usually in the annealed state, which is the form in which the metal is used commercially for the construction of chemical equipment. The metal is very pure, being not less than 99.9% tantalum. The main impurities consist of small amounts of iron, carbon, and possibly dissolved oxygen.

* Fansteel Metallurgical Corporation, North Chicago, Ill.

[8] V. S. de Marchi and C. G. Fink, "Corrosion Resistance of Alloys Containing Silver," A. S. P. R. P., *Sixth Report*, National Bureau of Standards Library (1938).

[9] A. J. Dornblatt, "Silver for Plant Use," *Chem. Met. Eng.*, **45**, 601 (November, 1938).

TANTALUM

EMBRITTLEMENT OF TANTALUM

Tantalum can absorb as much as 740 times its own volume if heated to a red heat and cooled in hydrogen. This can also take place at ordinary temperatures if the hydrogen is in the nascent condition. Although molecular hydrogen or oxygen in solution has no action on tantalum, hydrogen absorption and consequent embrittlement of the metal can occur in practically all electrolytes, especially at elevated temperatures, because of electrolytic action resulting from a galvanic coupling or because of stray electrical currents. To prevent this action, the tantalum should be electrically insulated from other metals in the equipment and stray currents should be located and eliminated.

Hydrofluoric acid and alkalies can also produce this type of embrittlement since nascent hydrogen is one product of their reaction with the metal.

Metal once embrittled can be restored only by heating in a vacuum.

AQUEOUS MEDIA

WATER

Natural fresh waters, sea water, and mine waters (usually acidic) have no action on tantalum. In boiler waters and condensates the alkalinity must be controlled. The pH should be less than 9 as a maximum and preferably not more than 8.

ACIDS

Quantitative data for the corrosion of tantalum in acids are given in Table 1. Hydrofluoric acid corrodes and embrittles tantalum very rapidly, and the data indicate that fuming sulfuric acid also has a rapid corrosive action. A mixture of hydrofluoric and nitric acids dissolves the metal rapidly and completely. A number of the laboratory tests shown in Table 1 have been confirmed by the performance of chemical equipment constructed of tantalum. Typical examples are given in Table 2.

ALKALIES

Alkalies attack tantalum. The rate of corrosion and embrittlement depends on the nature of the alkali, its concentration, and temperature. No change in weight resulted when a sample was immersed in a saturated $Ba(OH)_2$ solution for 135 days at atmospheric temperatures. In 5% NaOH solution at 100° C (212° F) there was a loss of 4 mdd (0.0003 ipy) in 60 days. A similar test with KOH resulted in a loss of 0.03 mdd (0.0000 ipy). In both cases there was surface embrittlement, but the center of the sheet remained soft. A 40% solution of either alkali at 110° C (230° F) destroyed and embrittled the specimens in 2 days.

SALT SOLUTIONS

Tantalum is inert to salts or salt solutions, excepting those which contain or hydrolyze to strong alkalies. Fluorides, however, will attack the metal. Unless strongly alkaline, hypochlorites do not affect it, nor is it corroded by a solution of potassium iodine saturated with iodine, or by ferric, mercuric, or stannic chloride solutions at ordinary temperatures. Tantalum also resists the action of ferric chloride at 170° C (365° F) and that of molten crystalline stannic chloride.

Tantalum in the Human Body. Tantalum is not attacked by body fluids under any circumstances and hence certain cells grow on the metal and adhere to it. This behavior has resulted in a rapid development of its use for bone and nerve repair.

TABLE 1. CORROSION OF TANTALUM BY ACIDS

Size of specimen — annealed sheet 0.15 × 11 × 105 mm (0.006 × 7/16 × 4⅛ in.).
Preparation of specimen — cleaned in dichromate cleaning solution.

Corrosive Liquid	Temperature		Duration of Test	Corrosion Rate	
	°C	°F		mdd	ipy
HCl, conc.	19–26	66–79	135 days	0.0	0.0000
Muriatic acid (commercial HCl)	19–26	66–79	135 days	0.0	0.0000
HCl, conc.	110	230	5 days	0.0	0.0000
HNO₃, conc.	19–26	66–79	135 days	0.0	0.0000
	86	187	6 days	0.0	0.0000
H₂SO₄, conc.	19–26	66–79	135 days	0.0	0.0000
	147	297	90 days	0.04	0.0000
	175	347	30 days	1.4	0.0001
	200	392	30 days	17.7	0.0015
	250	482	6 hr	332.0	0.029
	300	572	3 hr	3,960.0	0.342
H₂SO₄ conc. + K₂Cr₂O₇	19–26	66–79	135 days	0.0	0.0000
H₂SO₄ conc. + K₂Cr₂O₇	96	205	90 days	0.04	0.0000
CrO₃ — plating solution	98	208	90 days	0.02	0.0000
H₃PO₄ — 85%	145	293	90 days	0.05	0.0000
	180–210	356–410	31 days	2.7	0.0002
H₂SO₄ — fuming, 15% SO₃	23	73	2 days	3.2	0.0003
	70	158	6 hr	1,060.0	0.092
	130	266	2 hr	45,600.0	3.9
Oxalic acid — saturated solution	96	205	90 days	0.1	0.0000

TABLE 2. PERFORMANCE OF CHEMICAL EQUIPMENT CONSTRUCTED OF TANTALUM

Material Processed	Operation	Duration of Operation	Corrosion or Embrittlement
HCl, conc.	Evaporation	9 yr	None
HCl	Absorption in H₂O	7 yr	None
HNO₃, conc.	Evaporating and condensing	9 yr	None
HBr	Evaporating and condensing; production of 45% HBr	8 yr	None
H₃PO₄, 80%	Heating to 145° C (293° F)	5 yr	Slow progressive embrittlement fluorine less than 5 ppm
Br, liquid	Evaporating and condensing	5 yr	None Chlorine may be present Fluorine should be absent
NH₄Cl, solution	Evaporating	8 yr	None
H₂O₂	Manufacture and concentration	5 yr	None

SOLUTIONS OF ORGANIC COMPOUNDS.

Test data under this heading are limited to lactic, oxalic, and acetic acids and a saturated solution of phenol. They had no effect on tantalum when tested for 135 days at atmospheric temperatures. Boiling tartaric acid is not corrosive to tantalum, but in hot oxalic acid there is a measurable slight corrosion rate. (See Table 1.)

GASES AND NON-AQUEOUS LIQUIDS

At temperatures not exceeding 150° C (300° F) and in the absence of fluorine, free SO_3, or strong alkalies most inorganic and organic liquids have no effect on tantalum. The same is true of nearly all corrosive gases including wet or dry chlorine or bromine. In the case of bromine a temperature of 175° C (350° F) is a safe limit.

ATMOSPHERIC CORROSION

A tantalum plate exposed for 13 years on the outside of a factory building at North Chicago, Illinois, near a chemical plant where the air frequently contains Cl_2, HCl, and sometimes HF, showed no tarnish or pitting.

TIN

BRUCE W. GONSER* AND JAMES E. STRADER*

Since tin is a low-melting metal, it finds no application where high-temperature corrosion is a factor. It is used largely as a protective coating over steel for food cans, and in alloys, such as bronze, pewter, solders, and terne plate. The pure metal, block tin, has a number of applications, however, as in tubing, collapsible tubes, and foil, because of its resistance to corrosion, its ease of fabrication, and its non-toxicity.

AQUEOUS MEDIA

Tin is amphoteric, reacting with both strong acids and strong bases, but is relatively resistant to nearly neutral solutions. Oxygen greatly accelerates corrosion by aqueous solutions. The metal is normally covered with a thin invisible film of stannic oxide, which may be completely removed by acids or alkalies, or penetrated at isolated points to produce pits. The pitting type of attack when corrosion occurs is more common in nearly neutral solutions.

FRESH WATERS

Distilled and soft tap waters are not corrosive to tin. Double distilled water has been shown to have practically no solvent effect on tin as determined by conductivity measurements.[1]

In tap water of pH 7.2 at 25° C (80° F), exposed to air, 3.5 × 2.5 cm (1.4 × 1.0 in.) specimens of 99.99% cold-rolled tin showed an approximate weight gain of 0.2 mg in 50 days (0.023 mdd) and formation of an insoluble film.[2] Hot distilled water in 20 days gave a similar but thinner film. With harder tap waters of pH 7.4 and 8.6, weight losses in 50 days were of the order of 0.4 and 0.1 mg (0.046 and 0.01 mdd), respectively, and slight localized attack near the water line occurred. Practically the same results, except for greater localized attack, were obtained with 99.68% tin. Precipitated carbonate was mainly responsible for localized water line attack with hot and cold hard waters, since no attack occurred without the precipitate. Addition of 5% antimony to the tin prevented localized attack by hard water.

* Battelle Memorial Institute, Columbus, Ohio.

[1] C. L. Mantell, *Tin*, pp. 300–317, Reinhold Publishing Corp., New York, 1929.
[2] T. P. Hoar, "The Corrosion of Tin and Its Alloys. Part 1. The Tin Rich Sn-Sb-Cu Alloys," *J. Inst Metals*, **55**, 135–145 (1934).

Sea Water

Cast tin corroded to the extent of 0.00003 to 0.00009 ipy in 4-year tests. Deepest pits formed in this period extended to about 0.50 mm (0.02 in.) for high-grade tin but were negligible for common tin. (See Table 19, p. 417.)

Acids

Tin is only moderately resistant to corrosion by acids in the presence of air. In the absence of air, the corrosion rate decreases because of the high hydrogen overvoltage of tin.[3] Oxidizers in acids, consisting of either dissolved oxygen or oxidizing salts, will depolarize the surface and accelerate corrosion. It has been shown that the corrosion of tin by weak non-oxidizing acids is determined primarily by the dissolved oxygen in the acid.[4] Table 1 shows that this factor also operates with the stronger non-oxidizing acids.

TABLE 1. EFFECT OF OXYGEN ON ACID CORROSION OF TIN*

Size of specimen — 3.2 × 7.6 × 0.079 cm (1.3 × 3.0 × 0.031 in.).
Surface preparation — slightly roughened, No. 1 emery paper
Aeration — gas rate 250 ml/min in fine bubbles.
Temp. — 20° C (68° F).
Duration — 5 hr.
Analysis of tin — unknown.

Acid	Conc., % by Wt.	Corrosion Rate			
		Under H_2		Under O_2	
		mdd	ipy	mdd	ipy
Sulfuric	6	35	0.007	4,400	0.865
Hydrochloric	6	60	0.012	11,300	2.240
Nitric	3	640	0.126	650	0.128
Acetic	6	15	0.003	2,300	0.465

*W. G. Whitman and R. P. Russell, "The Acid Corrosion of Metals; Effect of Oxygen and Velocity," Ind. Eng. Chem., 17, 348 (1925).

Inorganic Acids. The corrosion rates of tin in dilute sulfuric, hydrochloric, and nitric acids at room temperature are given in Tables 1 and 2. Nitric acid attacks tin slowly when cold and dilute, more rapidly with rising temperature and concentration. When the attack is rapid, as with the hot acid, the tin is partially converted to metastannic acid.[5]

To illustrate the effect of oxidizers other than oxygen, the addition of 55 grams per liter of sodium chlorate to 1 N sulfuric acid increased the rate of solution of tin (specimen area, 64 sq cm) in a 19-hour test at 38° C (100° F) from 74 to 16,500 mdd.[6] Under similar conditions, the addition of 55 grams per liter of potassium permanganate increased the rate of solution of cast tin from 187 mdd to 7135 mdd. Addition of potassium dichromate to sulfuric acid acted as an inhibitor. Tin salts in solution

[3] R. J. McKay and R. Worthington, *Corrosion Resistance of Metals and Alloys*, pp. 180–199, Reinhold Publishing Corp., New York, 1936.
[4] W. G. Whitman and R. P. Russell, "The Acid Corrosion of Metals; Effect of Oxygen and Velocity," Ind. Eng. Chem., **17**, 348 (1925).
[5] C. L. Mantell, *Tin*, pp. 300–317, Reinhold Publishing Corp., New York, 1929.
[6] O. P. Watts and N. D. Whipple, "Corrosion of Metals by Acids," Trans. Electrochem. Soc., **32**, 268, (1917).

also tend to reduce the corrosion rate, as has been demonstrated by a marked reduction in tin dissolved by concentrations of 1% or less of hydrochloric acid in absence of air when as little as 10 ppm of stannous chloride are added to the solution.

Halogen acids attack tin, particularly when hot and concentrated. The rate of attack by hydrofluoric acid is relatively slow, however.

Chloric and hypochlorous acids attack tin, and the rate is accelerated by presence of other acids. Pure hydrocyanic acid has no action on tin, but tin causes decomposition of the acid. Phosphoric acid, 0.1 N, has little action on tin in 7 days at room temperature.[7]

Organic Acids. A comparison of the action of various organic acids on 2.5 × 4.0 cm (1 × 1.6 in.) sheet tin specimens, exposed for 210 hours at room temperature to 150 ml of air-free solutions of concentration equivalent to 0.75% malic acid, showed a rate of attack by *citric, succinic, malic, malonic,* and *acetic acids* ranging from 0.5 to 0.£7 mdd, in the order given.[8] Solutions of 1% acetic and 1% *lactic acid* produced about the same rate of corrosion as sulfuric acid, and a third that of hydrochloric acid, when run under comparable conditions in sealed, but not air-free, containers at 20° C (70° F).[9] Under the same conditions, butyric acid produced no measurable attack, but slight attack (4 mdd in a 5-hour test) was noted at 63° C (145° F).

Acetic acid attacks tin only very slowly, even in hot, fairly concentrated solutions, and scarcely at all in cold dilute solutions. The rate of corrosion in 20, 60, and 100% acetic acid in closed, but not air-free, containers has been found to be 25, 31, and 97 mdd at 25° C (77° F). In boiling acetic acid of the same concentrations, the rate of attack increased to 55, 80, and 854 mdd.[10]

Citric acid gave weight losses of 15 and 12 mdd, respectively, on partly immersed 6.5 × 2.5 cm (2.5 × 1.0 in.), 99.99 and 99.68% tin specimens, when using a 0.1 N solution in contact with air for 6 days at 25° C (77° F).[11] This was 20 to 30% less than with 0.1 N hydrochloric acid under similar conditions. In contact with air, the attack by citric acid was largely at the water line, whereas tin fully immersed was only lightly etched.

Stearic and *oleic* acids attack tin readily at high temperatures. *Oxalic acid* is probably the most corrosive of the common organic acids toward tin.[12]

ALKALIES

As shown in Table 2, dilute solutions of weak alkalies, such as ammonium hydroxide and sodium carbonate, have little effect on tin; but a strong alkali, like sodium hydroxide, is corrosive even when cold and in dilute solution. Tin is dissolved by strong alkalies with formation of stannates. As with acids, the rate of attack is greatly enhanced by aeration. Tests made by immersing tin of 0.6 sq dm (9.3 sq in.) area in 190 ml of 0.1 N sodium hydroxide for 45 hours at 37.5° C (100 °F) gave a weight loss of 27 mdd. This rate was raised to 600 mdd by addition of 10 grams of potassium

---

[7] J. M. Bryan, *Corrosion of Tin*, pp. 185–193, Department of Scientific Industrial Research (British), Food Investigation Board, Special Report 44, 1937.

[8] E. F. Kohman and N. H. Sanborn, "Factors Affecting the Relative Potentials of Tin and Iron," *Ind. Eng. Chem.*, **20**, 1373–1377 (1928).

[9] O. F. Hunziker, W. A. Cordes, and B. H. Nissen, "Metals in Dairy Equipment," *J. Dairy Sci.*, **12**, 140–181 (1929).

[10] W. S. Calcott and J. C. Whetzel, "Laboratory Corrosion Tests," *Trans. Am. Inst. Chem. Engrs.*, **15** (1), 75 (1923).

[11] T. P. Hoar, "The Corrosion of Tin and Its Alloys. Part 1. The Tin Rich Sn-Sb-Cu Alloys," *J. Inst. Metals*, **55**, 135–145 (1934).

[12] J. M. Bryan, *loc. cit.*

permanganate, and the rate became 3310 mdd by raising the concentration to 20% and adding 5 grams of picric acid.[13]

The rate of attack of tin by sodium carbonate has been shown to be negligible with a solution of pH 9.5, but weight losses of 9 and 15 mdd were obtained with a solution of pH 10 and 11.2, respectively, in a 45-day test at 25.8° C (78° F), using 10 × 2.5 cm (4.0 × 1.0 in.) specimens in 100 ml sealed, but not air-free, containers.[14] No measur-

TABLE 2. THE EFFECT OF DILUTE ACIDS, ALKALIES, AND SALTS ON THE CORROSION OF TIN*

99.2% commercial tin (0.61% Pb, 0.15% Zn).
Area of specimens — 1 sq dm (15.5 sq in.).
Surface preparation — rolled and polished.
Aeration — nat. convect.
Temperature — 17° to 20° C (63° to 68° F).

Solution, 0.2 N	Wt. Loss, grams			Wt. Loss, mdd		
	Vol. Soln.: 1 liter† 4 hr	Vol. Soln.: ½ liter‡ 7 days	Vol. Soln.: ½ liter† 28 days	Vol. Soln.: 1 liter 4 hr	Vol. Soln.: ½ liter 7 days	Vol. Soln.: ½ liter 28 days
HNO₃	1.62	4.0	7.2	570		257
HCl	0.02	0.42	0.9	120	60	32
H₂SO₄	0.02	0.22	0.25	120	31	9
NaOH		0.30	0.50		42	18
Na₂CO₃		0.01	0.00		1	0
NH₄OH		0.00	0.00		0	0
NaCl		0.00	0.08		0	3
CaCl₂		0.13	0.08		18	3
MgCl₂		0.16	0.10		23	4

* A. J. Hale and H. S. Foster, "The Action of Dilute Solutions of Acids, Alkalies, and Salts upon Certain Metals," *J. Soc. Chem. Ind.*, **34**, 464 (1915).
† Same solution throughout.
‡ Solution renewed daily.

able effect was noted with a bicarbonate solution of pH 8.4. Addition of various anions, such as phosphate, perborate, and chromate, decreased corrosion. Likewise, addition of sodium sulfite to alkaline detergents containing 0.5% sodium carbonate has been shown to reduce the corrosive action on tin effectively.[15]

SALT SOLUTIONS

Salts with acid reaction in solution attack tin in the presence of oxidizers or air. Thus a solution of aluminum chloride is particularly corrosive when aerated. Oxidizing salts, such as ferric chloride, are markedly corrosive.[16] Table 2 compares the action of some salts with acids and alkalies.

The oxide film on tin is not entirely removed by neutral solutions of the chloride or sulfate type, and when corrosion occurs, if at all, it is by thickening the oxide film

[13] O. P. Watts and N. D. Whipple, "Corrosion of Metals by Acids," *Trans. Electrochem. Soc.*, **32**, 268 (1917).

[14] G. Derge and H. Markus, "Studies upon the Corrosion of Tin," *Trans. Am. Inst. Mining Met. Engrs.*, **133**, 295–301 (1939); **143**, 198–208 (1941).

[15] R. Kerr, "The Use of Sodium Sulphite as an Addition to Alkaline Detergents for Tinned Ware," *J. Soc. Chem. Ind.*, **54**, 217–221T (1935).

[16] R. J. McKay and R. Worthington, *Corrosion Resistance of Metals and Alloys*, pp. 180–199, Reinhold Publishing Corp., New York, 1936.

to give a visible tarnish, or by localized attack to give pitting and black spots.[17] Solutions which form no precipitate with stannous ions (chloride, bromide, chlorate, perchlorate, sulfate, nitrate) form such black spots, chlorides producing the greatest attack. No such action is obtained with solutions forming stable precipitates, such as iodate, borate, phosphate, chromate, thiocyanate, iodide, nitrite, sulfite, bicarbonate, ferricyanide, and ferrocyanide.[18]

AQUEOUS SOLUTIONS OF ORGANIC COMPOUNDS

Milk has only a slight action on tin even when strongly aerated, or when sour.[19] Tests with sweet milk, sour milk, and condensed milk have shown weight losses between 1.5 and 3.8 mdd in the 6° to 62° C temperature range. Sweet and sour cream caused insignificant attack at 20° C (70° F) but showed losses of 11 and 6.7 mdd, respectively, in 5 hour tests at 63° C (145° F). Fruit juices corroded tin at a rate of 1 to 25 mdd under mildly aerated conditions at room temperature, but weight losses of 128 to 350 mdd have been noted for boiling lemon, tomato, red grape, and apple juices (listed in order of increasing corrosion).

Tin is considered to have only limited application in wineries because of relatively poor resistance to corrosion,[20] but extensive use in handling beer has indicated reasonably satisfactory performance, even though cloudiness is imparted after many hours contact in pipe lines with beer which has not been partially neutralized to diminish acidity.

NON-AQUEOUS MEDIA

Tin is practically unattacked by common petroleum products, such as *gasoline, kerosene,* and *lubricating oils. Benzol,* a mixture of 192 proof *alcohol* plus 5% benzine, and *absolute alcohol* gave respective weight losses of 0.038, 0.109, and 0.127 mdd using tin disks in 150 to 200 day tests in a closed, but not air-free, container at room temperature.[21] By constant shaking, the rate was increased to 0.76, 1.15, and 2.23 mdd, respectively, in 12-day tests.

In a 6-month test, tin was found to corrode in *carbon tetrachloride* at the rate of 0.738 mdd at room temperature. The rate in a 100-hour test in carbon tetrachloride and water vapor at the mutual boiling point was 5.62 mdd.[22]

Chlorine, bromine, and *iodine* rapidly attack tin, even at low temperatures; *fluorine* reacts at about 100° C (210° F).

ATMOSPHERIC CORROSION

Figure 1 shows the linear relationship between time and weight gain from tarnish on exposure of tin indoors. The weight gain amounts to 0.004 mdd. This compares with weight gains of 0.007, 0.014, and 0.019 mdd for copper, zinc, and cadmium tested

[17] S. Brennert, "Black Spots on Tin and Tinned Ware," *International Tin Research and Development Council, Tech. Publ.* D-2, 1–27 (1935).
[18] T. P. Hoar, "The Corrosion of Tin in Nearly Neutral Solutions," *Trans. Faraday Soc.*, **33**, 1152–1167 (1937).
[19] R. J. McKay and R. Worthington, *Corrosion Resistance of Metals and Alloys,* pp. 180–199, Reinhold Publishing Corp., New York, 1936.
[20] H. E. Searle, F. L. Laque, and R. H. Dohrow, "Corrosion Resistance of Metals in Wine Making and Tolerance of Wine for Metals," *Ind. Eng. Chem.*, **26**, 617–627 (1934).
[21] W. Wawrzinick, "New Method for Determining the Corrosion of Sheet Metals in Motor Fuels," *Automobiltech. Z.*, **33**, 28–30 (1930).
[22] F. H. Rhodes and J. T. McCarty, "The Corrosion of Certain Metals by Carbon Tetrachloride," *Ind. Eng. Chem.*, **17**, 909–911 (1925).

for 220 days under the same conditions as tin. Table 3 gives comparable data for outdoor exposure. In a Stevenson screen* test of exposure to vigorous air currents and

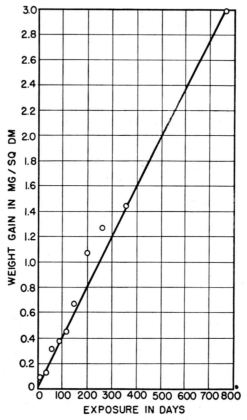

Fig. 1. Weight Gain of Tin in an Indoor Atmosphere. (Kenworthy.)

TABLE 3. EFFECT OF OUTDOOR EXPOSURE ON TIN*

Size of specimen — 10 × 5.0 × 0.2 cm (4 × 2 × 0.08 in.).
Surface preparation — rinsed in acetone.
Duration — 1 year (in England).
Corrosion product removed largely by mechanical means.

Material	Average Wt. Gain, mdd	Wt. of Metal Corroded, mdd (By difference)
Pure tin	+0.178	0.30
Tin + 2.2% antimony	+0.126	0.31
Tin + 0.2% copper	+0.189	0.36

* L. Kenworthy, "The Atmospheric Corrosion and Tarnishing of Tin," *Trans. Faraday Soc.*, **31**, 1331–1345 (1935); International Tin Research and Development Council, A-24 *Tech. Publication* (1935).

* Stevenson screens are wooden boxes with double roofs and double louvered sides. They are used by meteorologists to shelter their instruments from rain and direct solar radiation. For complete description, see J. C. Hudson, *Trans. Faraday Soc.*, **25**, 186 (1929).

fluctuations of temperature and humidity, but sheltered from direct rainfall, the rate of attack on tin in a year was 0.27 mdd, compared to 0.24 for copper and 0.47 for nickel.[23]

A. S. T. M. exposure tests of 22.8 × 30.5 cm (9 × 12 in.) commercial 99.85% tin panels in rural, industrial, and marine locations for 10 years gave average weight gains of 0.007, 0.067, and 0.11 mdd, respectively.[24] After chemically cleaning, the weight losses were 0.09, 0.27, and 0.45 mdd, respectively, and respective losses in tensile strength were 19, 23.5, and 30%.

TITANIUM

W. J. Kroll*

In the group of elements comprising silicon, titanium, zirconium, and hafnium, titanium is the least corrosion resistant. The metal is very malleable, hot or cold, when free of oxygen and nitrogen, but the smallest traces of these impurities embrittle it. It is made in small quantities by sodium reduction of titanium tetrachloride in a bomb, or by reduction of the chloride with magnesium under argon. Both reduction processes yield malleable metal. It may be shaped either by sintering, or by melting in special furnaces followed by rolling.[1] The ductile metal embrittles when overheated because of grain growth. Difficulties in joining are considerable.

Corrosion data available in the literature concern mostly impure brittle metal. Pure titanium is said to behave in general like 18-8 stainless steel.[2]

Titanium is not attacked by cold or by boiling water. Steam reacts with it at 800° C (1475° F).

The metal is attacked by cold dilute hydrochloric or sulfuric acids. Concentrated nitric acid "passivates" the metal, after which it is more resistant to hydrochloric and sulfuric acids.[2] Hydrofluoric acid is corrosive to titanium. Mixtures of hydrofluoric and nitric acids attack the metal more rapidly than hydrofluoric acid alone.

Titanium can be heated to 600° C (1100° F) without impairing its properties by oxygen penetration.[2] It reacts with fluorine, chlorine, and iodine at elevated temperatures. The iodine reaction is reversible, permitting the production of high-purity metal by dissociation of titanium iodide on a hot filament.[3] Hydrogen chloride reacts at high temperatures to form titanium tetrachloride.

Sulfur, H_2S, and CS_2, when heated in contact with titanium, form the metal sulfide. Hydrogen and nitrogen react at high temperatures to produce the hydride and nitride respectively. Ammonia forms the nitride above 800° C (1475° F).

Molten KOH reacts with titanium with evolution of hydrogen.

At elevated temperatures, titanium reduces all known oxides.

* Consulting Metallurgist, Albany, Ore.

[23] L. Kenworthy, "The Atmospheric Corrosion and Tarnishing of Tin," *Trans. Faraday Soc.*, **31**, 1331–1345 (1935); International Tin Research and Development Council, *Tech. Publ.* A–24 (1935).

[24] W. H. Finkeldey, "Report of Subcommittee VI on Atmospheric Corrosion of Nonferrous Metals and Alloys," *Proc. Am. Soc. Testing Materials*, **43** (1), 137–154 (1943)

[1] W. J. Kroll, *Trans. Electrochem. Soc.*, **78**, 35 (1940).

[2] W. J. Kroll, *Metallwirtschaft*, **18**, 77 (1939).

[3] J. D. Fast, *Z. anorg. Chem.*, **241**, 42 (1939).

TUNGSTEN

W. J. KROLL*

The general behavior of tungsten is analogous to that of molybdenum. The formation of soluble alkali tungstates and good resistance to hydrofluoric and hydrochloric acids are the main characteristics of this metal. Its use is limited by its brittleness, except in very thin sections and by the difficulties of joining. Powder is much more readily attacked than wrought metal. The data given below refer to wrought metal only.

Cold and hot water do not attack tungsten, but water vapor reacts at red heat. Alkali solutions and ammonia are not corrosive in absence of oxygen.[1] Hydrofluoric acid either hot or cold has no corrosive action. Concentrated nitric acid dissolves the metal slowly at 110° C (230° F), but concentrated hydrochloric acid has very little action at the same temperature.[2] Concentrated sulfuric acid reacts only slightly.[3] The corrosion rates in dilute acids are reported by Rohn[2] as follows:

Duration of Test		10% HNO_3, mdd	10% H_2SO_4, mdd	10% HCl, mdd
24 hr	Cold	0.0	0.0	0.0
1 hr	Hot	1440	240	240

Chromic acid solutions or mixtures with sulfuric acid do not appreciably attack tungsten.

Alloys of tungsten with iron have been made by sintering the powders, and they were tested for corrosion resistance in various media.[4] The 76% W–24% Fe alloy in normal hydrochloric or sulfuric acid is stated to corrode at a rate of about 1150 mdd at 20° C.

Aqueous solutions of $CuCl_2$ or $FeCl_3$ are corrosive to tungsten.[5] A good etching reagent is a solution of NaOH and $K_3Fe(CN)_6$.

As anode in 20% sulfuric acid at room temperature and current density of 0.18 amp per sq dm (1.7 amp per sq ft) the metal loss was 40 mg per amp-hour.[6] Tungsten as anode in caustic solutions is rapidly dissolved.

Fluorine reacts with tungsten at room temperature; bromine and iodine at red heat. Chlorine reacts in presence of oxygen at 600° C (1100° F).[7] The gases CO_2, CS_2, S_2Cl_2, CO, NO, and NO_2 react at elevated temperatures. The elements S, B, C, and Si form compounds with tungsten at elevated temperatures.

Tungsten does not rapidly oxidize below 600° C (1100° F) in air or below 500° C (950° F) in oxygen, although oxide films showing interference colors may form. At high temperatures, tungsten reacts with MgO, Al_2O_3, ThO_2, and ZrO_2.[8]

Molten alkalies are corrosive to the metal in presence of oxygen or oxidizing agents. Molten nitrates, nitrites, and peroxides react violently. Tungsten can be used as anode in the electrolysis of fused $AlCl_3$ and $BeCl_2$.[9]

---

* Consulting Metallurgist, Albany, Ore.
[1] S. L. Malowan, Z. Metallkunde, 23, 69 (1931).
[2] W. Rohn, Z. Metallkunde, 18, 387 (1926).
[3] C. G. Fink, Trans. Electrochem. Soc., 17, 229 (1910).
[4] G. Grube and K. Schneider, Z. anorg. Chem., 168, 28 (1928).
[5] Hikozo Endo and Akira Itagaki, Institute of Metals, Met. Abstracts, 7, 20–21 (1940).
[6] F. R. Morral and J. L. Bray, Trans. Electrochem. Soc., 75, 427–440 (1939).
[7] R. Spitzin and L. Kaschtanow, Z. anorg. Chem., 157, 154 (1926).
[8] H. von Wartenberg and H. Moehl, Z. physik. Chem., 128, 439 (1927).
[9] I. G. Farbenindustrie, German Patent 514,125 (1930).

ZINC

GENERAL REFERENCES (TUNGSTEN)

GMELIN, L., *Handbuch der anorganischen Chemie (Tungsten)*, Vol. 54, pp. 86–90, Verlag Chemie, 1933.
LI, K. C., and CHUNG YU WANG, *Tungsten*, p. 205, Reinhold Publishing Corp., 1943.

URANIUM

W. J. KROLL*

The element uranium (m.p. approx. 1150° C [2100° F]) resembles the rare earths in a great many respects. The reduction of the chloride with sodium in a bomb, the electrolysis of a fused fluoride bath, and the reduction of the oxide with calcium in presence of $CaCl_2$ under argon yield malleable pure metal. It is fairly soft and can readily be rolled. Only a few data are available for the rolled metal.† The powder is probably more readily attacked than compact metal.

Uranium tarnishes slowly in air, becoming yellow, brown and finally black. It starts oxidizing in air at 150° to 170° C (300° to 340° F). Hydrogen at moderately high temperatures readily forms a hydride. In presence of nitrogen at elevated temperatures reaction occurs with formation of a nitride. Chlorine, bromine, and iodine react at 180°, 210°, and 260° C (355°, 410°, 500° F), respectively; sulfur at 500° C (930° F). Fluorine reacts at room temperature.

Water does not react with uranium at room temperature, but at boiling temperature attack occurs. Dilute hydrochloric and sulfuric acids dissolve the metal slowly when cold, rapidly when hot. Concentrated sulfuric acid also rapidly attacks the metal. Acetic acid is slightly corrosive. Dilute nitric acid is without effect, but the concentrated acid reacts violently. Solutions of ammonium or sodium persulfate rapidly corrode the metal.

Alkali solutions in general have no action, but ammonium hydroxide reacts when hot.

GENERAL REFERENCE

GMELIN, L., *Handbuch der anorganischen Chemie*, System No. 55, Verlag Chemie, 1936.

ZINC‡

E. A. ANDERSON§ AND C. E. REINHARD§

Information on the compositions and forms of zinc available commercially will be useful in interpreting the discussion of corrosion.

The natural impurities in zinc as ordinarily produced are lead, iron, and cadmium. Rolled strip and sheet, ingot slabs, and coatings on steel are available in a range of compositions. The American Society for Testing Materials recognizes six grades of slab zinc, as follows:[1]

---

* Consulting Metallurgist, Albany, Ore.
† F. H. Driggs and W. C. Lilliendahl, *Ind. Eng. Chem.*, **22**, 516 (1930).
‡ Refer also to *Zinc Coatings on Steel*, p. 803.
§ Research Division, The New Jersey Zinc Co. (of Pa.), Palmerton, Pa.

[1] A. S. T. M. Standards, Part 1, 815 (1942).

Grade	% Maximum			Sum of Lead, Cadmium, and Iron
	Lead	Iron	Cadmium	% Maximum
Special High Grade	0.007	0.005	0.005	0.010
High Grade	0.07	0.02	0.07	0.10
Intermediate	0.20	0.03	0.50	0.50
Brass Special	0.60	0.03	0.50	1.0
Selected	0.80	0.04	0.75	1.25
Prime Western	1.60	0.08	....	....

Zinc is also available in rolled form as alloys containing about 1% of copper with and without supplementary additions of magnesium, chromium, etc. The hot-rolled form is most common for both rolled zinc and zinc alloys, although for certain purposes various degrees of cold rolling are required.

Die castings are made from zinc alloys, the most frequently used being those designated by A.S.T.M. as numbers XXIII and XXV. The compositions specified for these alloys are:[2]

	Alloy XXIII	Alloy XXV
Copper, %	0.10 maximum	0.75 to 1.25
Aluminum, %	3.5 to 4.3	3.5 to 4.3
Magnesium, %	0.03 to 0.08	0.03 to 0.08
Iron, maximum, %	0.100	0.100
Lead, maximum, %	0.007	0.007
Cadmium, maximum, %	0.005	0.005
Tin, maximum, %	0.005	0.005
Zinc, %	Remainder	Remainder

Zinc coatings range in composition through the entire series of A.S.T.M. grades. Electrodeposited zinc is generally very pure. The hot-dip coatings sometimes contain added tin or aluminum and always contain iron. The corrosion of zinc coatings is considered in another section of this book (p. 803).

The density of rolled zinc is 7.14 grams/cc; that of the die casting alloys is 6.8.

AQUEOUS MEDIA

Zinc in the form of coatings on steel is widely used in the handling of water supplies (galvanized pipe, range boilers, etc.) with generally satisfactory results. Uses of rolled zinc and zinc die castings in contact with water are less common.

The rate of corrosion of zinc in water is, in general, much lower than that of iron (making its use as a protective coating for iron economical) and somewhat higher than that of copper. Like other common metals, the rate of corrosion will vary greatly with variations in the exposure conditions such as temperature, pH, and concentration of dissolved oxygen. Some of these factors have fairly consistent effects, and data obtained under laboratory conditions have some value in the analysis of corrosion problems.

EFFECT OF OXYGEN, CARBON DIOXIDE, AND OTHER GASES IN WATER

Increase in oxygen concentration results in increased corrosion rate (Table 1). With high oxygen availability, the corrosion tends to be reasonably uniform. However,

[2] A.S.T.M. Standards, 1943 Supplement, Part 1, Specification B86-43.

when the oxygen concentration falls below that required for uniform saturation of the water, concentration cells develop between areas rich and areas poor in oxygen, which result in pitting and marked acceleration of the corrosion rate, with the formation of bulky zinc corrosion products. Typical practical cases are the attack of zinc alloy carburetors by stagnant pools of water trapped under the gasoline and the corrosion of stacked zinc or zinc-coated steel sheets by entrapped moisture.

TABLE 1. EFFECT OF DISSOLVED OXYGEN CONCENTRATION AND TEMPERATURE ON THE CORROSION RATE OF ZINC IN DISTILLED WATER*

Material — hot-rolled high-grade zinc.
Specimen size — 3.8 × 10 × 0.05 cm (1.5 × 4 × 0.021 in.).
Surface preparation — degreased, cathodically electrocleaned, alkaline bath; brief dip in 25 vol. % H_2SO_4.
Water — distilled laboratory supply.
Volume — 500 ml in wide-mouthed Erlenmeyer flasks; duplicate specimens in each flask.
Temperatures — Room, 40° C (104° F) and 65° C (149° F).
Number of runs — two with duplicate specimens in each.
Duration of test — 7 days.
Specimen motion — none.

Aeration	Temperature, °C	Corrosion Rate†					
		Run 1		Run 2		Average	
		mdd	ipy	mdd	ipy	mdd	ipy
1. Water boiled — specimens inserted — flask sealed with latex casings	Room	6	0.0012	4	0.0008	5	0.0010
2. No aeration except through air-water interface	Room	10.5	0.0021	10.5	0.0021	10.5	0.0021
3. Slow bubbling of oxygen	Room	48	0.0096	39	0.0076	43	0.0086
1. Water boiled — specimens inserted — flask sealed with latex casings	40	12	0.0024	7	0.0014	9.4	0.0019
2. No aeration except through air-water interface	40	20	0.0040	17	0.0034	18.4	0.0037
3. Slow bubbling of oxygen	40	91.5	0.0184	45	0.0091	68.6	0.0138
1. Water boiled — specimens inserted — flask sealed with latex casings	65	16	0.0032	16.5	0.0033	16.5	0.0033
2. No aeration except through air-water interface	65	27	0.0054	41	0.0083	34	0.0069
3. Slow bubbling of oxygen	65	61	0.0123	62.5	0.0126	62	0.0125

* Unpublished data, The New Jersey Zinc Co.
† Averages of two in each run. Corrosion rate determined after chemical removal of corrosion products. All surfaces etched and finely pitted.

Rapid replenishment of the oxygen supply at the corroding surface, which takes place when thin films of moisture condense on a zinc surface, has the usual accelerating effect on the corrosion rate. Both this and the stagnant water type of attack can be minimized by the use of chromate films (see p. 862).

Aerated water usually contains carbon dioxide, as well as air, and may contain sulfur dioxide or even hydrogen sulfide. Carbon dioxide in the amounts normally encountered alters the corrosion product from zinc oxide or hydroxide to a basic carbonate of the formula $4ZnO \cdot CO_2 \cdot 4H_2O$.[3] The reaction takes place in two stages,

[3] E. A. Anderson and M. L. Fuller, *Metals and Alloys*, **10**, 282–287 (1939).

the zinc oxide forming first and later reacting with the dissolved CO_2 to form the basic salt. Carbon dioxide increases the rate of corrosion of zinc by aerated water (Table 2).

Dissolved sulfur dioxide also increases the corrosion rate. It is probable that a basic sulfate forms by a mechanism similar to that reported for carbon dioxide.

TABLE 2. THE EFFECT OF CARBON DIOXIDE ON THE CORROSION RATE OF ZINC IN WATER*

Zinc composition — 0.008% Sn, 0.007% Fe, 0.01% Cd, Pb – nil; rolled.
Surface preparation — polished on emery paper (00), washed in 15% acetic acid, rinsed in water, rinsed in alcohol.
Water — distilled.
Volume — 50 ml per sq cm of Zn.
Temperature — room.
Suspension — glass hooks.
Duration of test — 24 hr (a few tests — 72 hr)
Removal of corrosion products — 15% acetic acid.

Test Condition	Approximate Specimen Area, sq cm	Corrosion Rate†	
		mdd	ipy
Distilled water	41	62	0.0124
Distilled water with air bubbled through water	40	183	0.0368
Distilled water with air washed in KOH bubbled through water	22	33	0.0066
Conductivity water in atmosphere free from CO_2	23.8	19	0.0038

* G. D. Bengough and O. F. Hudson, *J. Inst. Metals*, **21**, 37–210 (1919).
† Averages of values reported by Bengough and Hudson.

TABLE 3. EFFECT OF TEMPERATURE ON THE CORROSION OF ZINC*

Material — rolled high-grade zinc.
Specimen size — 8 × 12 × 0.076 cm (3.2 × 4.7 × 0.030 in.).
Surface preparation — not known.
Water — distilled.
Volume — nearly 15 liters (4 gal); duplicate specimens immersed.
Duration of test — 15 days.
Aeration — unwashed air; other details not given.
Specimen motion — 56 rpm on about 2-in. radius; specimens horizontal.

Temperature, °C	Corrosion Rate†		Appearance of Corrosion Product
	mdd	ipy	
20	3.9	0.00078	Definitely gelatinous — very adherent
50	13.7	0.0028	Slightly less gelatinous — adherent
55	76.2	0.015	Mostly granular — non-adherent
65	577	0.12	Decidedly granular, becoming flaky, and compact — non-adherent
75	460	0.092	Decidedly granular, flaky, and compact — non-adherent
95	58.7	0.012	Compact, dense, and flaky — adherent
100	23.5	0.0047	Varies from grayish white to black, very dense, resembling enamel — very adherent

* G. L. Cox, *Ind. Eng. Chem.*, **23**, 902–904 (1931).
† Determined after removal of corrosion products. Average of two specimens. Values chosen from Fig. 2 of original reference.

ZINC

EFFECT OF TEMPERATURE AND pH

The corrosion rate of zinc increases with increasing water temperature (Table 1). Sharp changes in rate are related to changes in the nature of the corrosion product film as the temperature rises (Table 3). At the higher temperatures, decreased oxygen solubility in the water becomes an important factor in decreasing the rate.

The corrosion rate of zinc is high in acid and strongly alkaline solutions. The lowest corrosion rates are observed in weakly alkaline solutions (Table 4) in the pH range 7 to 12.5. The usefulness of zinc in handling water or aqueous solutions is greatest in this pH range.

TABLE 4. EFFECT OF pH ON THE CORROSION OF ZINC BY WATER*

Material — hot-rolled special high-grade zinc.
Specimen size — not given.
Surface preparation — pickled in 10% HCl; washed; polished lightly with very fine steel wool; rinsed in alcohol; dried.
Water — distilled; pH nearly 7.
Volume — 4-liter beaker nearly full; quadruplicate specimens in each beaker.
Temperature — 30° ± 1° C.
Duration of test — 5 to 30 days to give approximately same weight change in each test.
Specimen motion — none, solution agitated with stirrer operated at 120 rpm.
pH Control — by additions of HCl and NaOH.
Aeration — CO_2 free air to saturate separate solution supply; new aerated solution supplied to corrosion vessel at 2 liters per hr.

	Corrosion Rate†		Corrosion Film
pH	mdd	ipy	
2.8	958	0.19	
3.5	352	0.071	Acid film dissolving
5	137	0.028	
7	49	0.0098	
9	20	0.0040	
10.5	10	0.0020	Stable film
12	4	0.0008	
13	68	0.014	Dilute alkaline film dissolving
13.5	352	0.071	
14	342	0.069	Strong alkaline film dissolving
14.5	313	0.063	

* B. E. Roetheli, G. L. Cox, and W. B. Littreal, *Metals and Alloys*, **3**, 73–76 (1932).
† Average of four specimens; points chosen from Fig. 2 original reference; not all points taken; corrosion products removed mechanically.

NATURAL FRESH WATERS

Natural fresh waters vary widely in composition and in corrosiveness to zinc. Oxygen and other gas content, hardness, pH, temperature, and other factors already discussed have definite influences on the corrosion of zinc. Hard waters which tend to deposit protective scale films on metal surfaces generally are less corrosive than soft waters. Typical data for a hard natural water and soft (distilled) water are listed in Table 5.

In closed hot water systems involving soft water, a highly localized pitting type of attack frequently takes place. This is much less likely to occur with hard waters.

TABLE 5. THE CORROSION OF VARIOUS GRADES OF ZINC IN WATER*

Series I

Specimen size — 2.5 × 5.0 × 0.13 cm (1.0 × 2.0 × 0.050 in.) = 27.7 sq cm.
Preparation of specimens — cleaned with ether and alcohol.
Check specimens — three tested in individual containers.
Solution volume — 600 ml in Erlenmeyer flask.
Test duration — 30 days.
Temperature — room.
Aeration — none.
Corrosion criterion — change in weight after chemical removal of corrosion products.
Type of test — simple immersion in open container.

Grade of Zinc†	Distilled Water	
	mdd	ipy
Spectroscopically pure; rolled	12	0.0025
High-grade; rolled	11	0.0022
Brass special; rolled	10	0.0020
Selected; rolled	10	0.0020

Series II

Specimen size — 19.35 sq cm surface area (3.0 sq in.).
Preparation of specimens — degreased in trichloroethylene.
Check specimens — one.
Solution volume — 100 ml.
All other test conditions same as in Series I.

Grade of Zinc†	Distilled Water		Tap Water‡	
	mdd	ipy	mdd	ipy
High-grade; rolled	27	0.0054	3	0.0007
Die cast Alloy XXV	28	0.0057	11	0.0023

* Unpublished data, the New Jersey Zinc Co.
† A. S. T. M. designation.
‡ Temporary hardness of 41 ppm. Permanent hardness of 86 ppm.

The widespread use of galvanized pipe in the handling of water supplies furnishes a background of service data from which the probable service behavior of other zinc materials may be estimated.

Data have been published[4] which indicate that the chlorine additions usually made to water supplies for health protection do not increase the rate of corrosion of zinc in water.

Sea Water

Refer to *Behavior of Metals and Alloys in Sea Water*, p. 418, particularly Table 19, p. 417.

Dissolved Salts, Acids, and Bases

Solutions of acid or strongly alkaline salts are not normally used in contact with zinc because of rapid corrosive attack. Zinc-coated steel is used in handling refrigerating brines, which may contain calcium chloride. In such cases the corrosion rate is

[4] E. A. Anderson, C. E. Reinhard, and W. D. Hammel, *J. Am. Water Works Assoc.*, **26**, 49–60 (1934).

TABLE 6. EFFECT OF COMMON NEUTRAL SALTS ON THE CORROSION OF ZINC BY WATER*

Material — rolled pure zinc.
Specimen size — 3 × 6 cm (1.2 × 2.4 in.); thickness not given.
Surface preparation — not given.
Water — not stated but presumed to be distilled.
Volume — 750 ml per test; one specimen per container.
Temperature — NaCl, 12.4° C; KCl, 12.6° C; KNO$_3$, 8.0° C; Na$_2$SO$_4$, 12.6° C; K$_2$SO$_4$, 12.6° C.
Duration of test — NaCl, 64 days; KCl, 65 days; KNO$_3$, 189 days; Na$_2$SO$_4$, 60 days; K$_2$SO$_4$, 57 days
Specimen motion — none, still solution; specimen 6 cm below surface of solution.
Aeration — none. Corrosion products removed by scraping.

Salt	Conc., grams/liter	Corrosion Rate — Average of Two Specimens													
		0	5	10	15	30	50	60	100	120	150	200	300	400
NaCl	mdd	10.2	18	15.6	13.9	10	†	7.2	†	3.8	†	†	†	2.5‡
	ipy	0.0021	0.0036	0.0031	0.0028	0.0020	†	0.0014	†	0.0008	†	†	†	0.0005
KCl	mdd	10.7	18.2	†	11.7	†	3.4	†	†	†	1	†	2.0‡	†
	ipy	0.0022	0.0037	†	0.0024	†	0.0007	†	†	†	0.0002	†	0.0004	†
KNO$_3$	mdd	7.6	3.2	1.9	1.7	†	1.5	†	1.1‡	†	†	†	†	†
	ipy	0.0015	0.0007	0.0004	0.0004	†	0.0003	†	0.0002	†	†	†	†	†
Na$_2$SO$_4$·10H$_2$O	mdd	9.8	12.8	†	10.2	†	8.1	†	6.0	†	†	3.1	1.7‡	†
	ipy	0.0020	0.0026	†	0.0021	†	0.0016	†	0.0012	†	†	0.0006	0.0004	†
K$_2$SO$_4$	mdd	9.4	10.5	9.6	9.5	†	6.8	†	0.5‡	†	†	†	†	†
	ipy	0.0019	0.0021	0.0019	0.0019	†	0.0014	†	0.0001	†	†	†	†	†

* J. Newton Friend and J. S. Tidmus, *J. Inst. Metals*, **31**, 177–185 (1924).
† No test.
‡ Test at saturation.

kept under control by adding sufficient alkali to bring the pH into the mildly alkaline range and by the addition of suitable inhibitors such as sodium chromate.

The effect of neutral salts on the aqueous corrosion of zinc is dependent upon the concentration (Table 6). The rates given in the table are minima since the test was made under non-aerated, quiescent conditions, and the mechanical removal of the corrosion products prior to reweighing was probably incomplete.

Certain salts, such as the dichromates, borates, and silicates, act as inhibitors to the aqueous corrosion of zinc and frequently are added to closed water systems. Inhibitive chromate films serve the same purpose where the water contact is occasional. The use of inhibitors is further discussed in *Inhibitors and Passivators*, p. 905.

Aqueous Solutions of Organic Compounds

Liquid and moist foodstuffs represent a substantial branch of the general field of water solutions of organic materials. It is recommended that the use of zinc in contact with such foodstuffs be avoided. While authorities agree that zinc is not essentially toxic to the human system,[5] the salts formed with strong acids are irritating. Any long contact of acid foodstuff with zinc will produce sufficient of such salts to cause nausea and general discomfort.

Effect of Velocity

Zinc is not ordinarily used under conditions involving impingement attack. However, it should be noted that, like the corrosion of many other metals, the corrosion of zinc is stimulated by relative motion between the metal and the solution (Table 7).

Table 7. EFFECTS OF AERATION AND MOTION ON THE CORROSION OF ZINC IN WATER*

Material — hot-rolled high-grade zinc.
Specimen size — 2.54 × 5 × 0.127 cm (1 × 2 × 0.050 in.) in quadruplicate.
Specimen preparation — unknown; probably washed with ether and alcohol.
Volume — 25 liters; eight specimens for 15 days; four specimens last 15 days.
Temperature — 20°–25° C (68°–77° F).
Duration of test — 15 and 30 days.
Specimen motion — 6 rpm around center axis at 4 in. radius plus 42 rpm independent rotation about specimen axis.
Aeration — 5–8 ml air per sec when used.
Corrosion products removed chemically at end of test.

Type Test	Corrosion Rate†							
	Distilled Water				3.5% NaCl Solution			
	15 Days		30 Days		15 Days		30 Days	
	mdd	ipy	mdd	ipy	mdd	ipy	mdd	ipy
No aeration; no rotation	24.8	0.0050	24.0	0.0048	20.2	0.0041	25.9	0.0052
No aeration; rotation	46.4	0.0093	32.1	0.0064	70.2	0.0141	47.0	0.0094
Aeration; no rotation	37.5	0.0075	24.9	0.0060	24.7	0.0050	22.9	0.0046
Aeration; rotation	37.6	0.0075	23.7	0.0048	37.6	0.0075	43.8	0.0088

* Unpublished data, the New Jersey Zinc Co.
† Average of four specimens.

[5] P. Drinker, *J. Ind. Hygiene*, **4**, 177–197 (1922).

EFFECTS OF GALVANIC COUPLING

Zinc is anodic to practically all commercial metals and would be expected, in theory, to protect these metals at the expense of increased zinc corrosion. Few data on the practical realization of this possibility are available. Die-cast zinc alloy carburetors are fitted with brass inserts, which cause no practical trouble even when water is present. In this case the cathode surface is small and the anode surface large. The use of small zinc or zinc-coated parts in contact with large areas of strongly electropositive (noble) metals is undesirable in most cases.

Zinc is widely used in the galvanic protection of steel in the form of zinc coatings and of zinc plates bolted to ship hulls. The protection is generally quite good and is obtained with relatively little expenditure of zinc. The amount of zinc corroded is usually equal to or greater than the sum of the losses which would occur from the zinc and the iron in the uncoupled state, depending upon the environment and the galvanic current density on the zinc.

EFFECT OF STRESS AND VIBRATION

Stress and vibration effects may be classified as stress corrosion cracking, corrosion fatigue, and the acceleration of general corrosion by stress. None of these appears to function in zinc materials under normal conditions. It is possible that the intercrystalline corrosion noted in the impure zinc die casting alloys used many years ago was of a stress corrosion cracking type, although this was never demonstrated. The question is of academic importance only since the modern zinc die casting alloys[6] are free from this type of failure.[7]

METALLURGICAL FACTORS

The corrosion rate of zinc is not largely altered by variations in composition (Table 5). As stated earlier, the known behavior of one type of zinc may be used as a guide to the probable service behavior of other forms of zinc. Mechanical working has relatively little effect, largely because zinc materials are not capable of absorbing large amounts of cold work. In the case of the A.S.T.M. die casting alloys, the presence of more than the limiting lead, tin, and cadmium contents or the omission of the specified magnesium content[8] causes serious increases in the corrosion of these alloys, particularly in steam and hot water.

NON-AQUEOUS MEDIA

LIQUIDS

Zinc may safely be used in contact with many organic liquids provided they are nearly neutral in pH and are substantially free from water. The uses of glycerin in die-cast zinc alloy door checks and inhibited trichloroethylene in steel-coated tanks are common. Zinc-coated steel and zinc alloy die castings are used in the handling of gasoline.

Anhydrous methyl chloride is reported to cause appreciable corrosion of zinc.[9]

When free water is present, rapid local corrosion frequently results. In addition, zinc tends to catalyze the decomposition of materials such as trichloroethylene in the

---

[6] For compositions, see A. S. T. M. Standards, 1943 Supplement, Part 1, Specification B86–43.
[7] A. C. Street, *Metal Industry* (London), **50**, 600 (1937).
[8] A. S. T. M. Standards, 1943 Supplement, Part 1, Specification B86–43.
[9] K. Willson, W. Walker, W. Rinelli, and C. Mars, *Ind. Eng. Chem.*, News Edition, **21**, 1254 (1943).

presence of water, with resultant acid attack. Some organic substances, such as low-grade glycerin, tend to attack zinc because they contain acid impurities.

GASES

In the absence of water, zinc may be used freely in contact with most common gases at normal temperatures. Elevated temperatures are not normally encountered because of the low structural strength of zinc at high temperatures. Moisture content tends to stimulate attack unless, as in the case of illuminating gas, a tar film forms, which is mechanically protective.

PROTECTIVE MEASURES

The protective measures employed with zinc are comparable to those commonly used with other metals and consist of mechanical shielding with protective coatings, the adjustment of the solution to a composition and pH most suitable for zinc, and the use of soluble inhibitors or inhibitive films.

Protection may be obtained with *paint* films and *electrodeposited metal coatings*. Satisfactory adhesion of paint films is secured by the use of phosphate pretreatments of the zinc surface or by the use of zinc dust primers. The finish coating should, of course, be capable of withstanding attack by the solutions to be encountered. Metallic coatings generally will be cathodic to zinc and hence must be applied in sufficient thickness (at least 1 mil) to prevent severe galvanic attack on the zinc at pores in the coating. Copper and nickel are most commonly used, but chromium coatings, if at least 0.2 mil in thickness, have a limited field.

It is far more common practice to control the corrosion of zinc in water by adjusting the water composition and pH, and by adding *inhibitors*. The practice selected must take the conditions of use into consideration. In closed water systems, soluble inhibitors may be employed to advantage. In open systems, the regulation of conditions is more difficult.

A variety of inorganic inhibitors is available for use with zinc, including sodium dichromate, borax, sodium silicate, and sodium hexametaphosphate. Also available are a number of adsorptive-type organic compounds, such as lanolin, which largely function as mechanical exclusion films. (Refer to *Temporary Corrosion-Preventive Coatings*, p. 916.) For most purposes the adjustment of the pH to the mildly alkaline range and the addition of sodium dichromate is preferred. The dichromate is usually neutralized to the chromate state. The concentrations for most effective use have been studied (Table 8). Many authors warn against the danger of intensified pitting when an insufficient amount of inhibitor of this type is added.

In automobile radiator systems, zinc may come in contact with hot water containing alcohol, ethylene glycol, glycerin, etc., together, in most cases, with quantities of iron rust. Satisfactory inhibition of the resultant corrosion has been achieved with the use of borax and lanolin in patented proprietary anti-freeze mixtures.

In open systems, such as community water supplies, there is only limited opportunity to alter conditions to suit a given metal. These opportunities are still further restricted by the fact that conditions most suitable for one metal may be unsatisfactory for another metal used in the same system. However, by proper control of water composition, pH, and hardness, the formation of protective scales on metal surfaces may be facilitated, with the result that all metals in the circuit are protected.

One interesting procedure involves the addition of very small quantities of sodium hexametaphosphate. These additions are made primarily to control scale formation,[10]

[10] G. B. Hatch and O. Rice, *Ind. Eng. Chem.*, **32**, 1572–1579 (1940).

TABLE 8. EFFECT OF SODIUM DICHROMATE ON THE CORROSION OF ZINC IN WATER AND IN NaCl SOLUTIONS*

Material — rolled zinc; 99.87% pure. Specimen size — rectangular plates about 190 sq cm area. Thickness not given.
Surface preparation — pickled in 10% HCl; rinsed; scrubbed with steel wool with soap and water; rinsed in water then in alcohol; dried.
Water volume — about 12 liters; duplicate specimens immersed. Duration — 30 days at 20° C (68° F); 14 days at 75° C (167° F); 8 days at 95° C (203° F).
Water — distilled and distilled plus NaCl and neutralized $Na_2Cr_2O_7 \cdot 2H_2O$†. Aeration — solutions were aerated but conditions not given.
Motion — specimens rotated at velocity of 36.6 cm (14.5 in.) per sec. Corrosion products removed with steel wool.

NaCl, Weight %	Temperature, °C	\\	0	0.1		0.5		$Na_2Cr_2O_7 \cdot 2H_2O$, grams/liter 1		2		3		4	
		Average Corrosion Rate, mdd	Maximum Pit Depth, ipy	Average Corrosion Rate, mdd	Maximum Pit Depth, ipy	Average Corrosion Rate, mdd	Maximum Pit Depth, ipy	Average Corrosion Rate, mdd	Maximum Pit Depth, ipy	Average Corrosion Rate, mdd	Maximum Pit Depth, ipy	Average Corrosion Rate, mdd	Maximum Pit Depth, ipy	Average Corrosion Rate, mdd	Maximum Pit Depth, ipy
0	20	4.5	0.00090	1	0.00019	0.53	0.00010	0.27	0.00006	0.33	0.00007	**	**	0.00	0.00000
	75	455	‡	7.2	0.028§	1.8	0.00037	1	0.00020	**	**	0.33	0.00007	**	**
	95	58	0.49	30	0.41§	24	0.34§	4.7	0.00095	**	**	0.00	0.00000	**	**
0.002	20	1.4	0.00028	0.25	0.00005	0.16	0.00003	0.18	0.00004	0.12	0.00002	**	**	0.18	0.00004
	75	547	0.47§	4.2	0.015§	2.1	0.032§	2.6	0.00052	**	**	1.4	0.046	**	**
	95	65	0.47§	26	0.55§	0.00	0.00000	4.5	0.00091	**	**	0.82	0.00016	**	**
0.05	20	8.2	0.0052§	1	††	0.90	0.0039§	0.63	0.0039§	0.59	0.0039§	**	**	0.47	0.0026§
	75	499	‡	12	0.20§	12	0.059§	17	0.047§	**	**	8.6	0.12§	**	**
	95	211	0.83§	22	0.22§	11	0.22§	17	0.11§	**	**	3.5	0.054§	**	**
3.5	20	5.0	0.0065§	3.4	0.0039§	1.8	0.0039§	1.5	0.0039§¶	1.2	0.0039§	**	**	0.92	0.0039§
	75	450	‡	31	0.0061§	30	0.0061§	23	0.0045§	**	**	18	0.0037§	**	**
	95	207	‡	196	‡	59	0.16§	114	0.22§	**	**	55	0.011¶	**	**
22	20	15	0.20§	13	0.13§	8.2	0.13§	6.6	0.13§	5.4	0.091§	**	**	5.3	0.077§
	75	175	‡	30	‡	25	0.44§	50	0.21§	50	0.19§	**	**	103	0.31§
	95‖	450	‡	27	‡	44	0.24§	51	0.15§	36	‡	**	**	131	0.48§

* B. E. Roetheli and G. L. Cox, *Ind. Eng. Chem.*, **23**, 1084–1090 (1931).
† Neutralized to sodium chromate; weight % as $Na_2Cr_2O_7 \cdot 2H_2O$. ‡ Failed by complete perforation before end of test period. § Severe pitting.
‖ 14-day tests. ¶ Negligible pitting. ** No test. †† No data.

but seem capable of inhibiting the corrosion of zinc (Table 9), when used above a threshold concentration.

Chromate films produced on zinc surfaces (p. 862) are used to a considerable extent where accidental or limited contact with water is expected. They have been used successfully in preventing white corrosion product formation in water passed through zinc-coated steel cooler heads.[11]

TABLE 9. EFFECT OF SODIUM HEXAMETAPHOSPHATE ON THE CORROSION OF ZINC COATINGS BY TAP WATER*

Material — steel; hand hot-dip zinc-coated with high-grade zinc after cutting specimens.
Size — 7.6 × 12.7 × 0.09 cm (3 × 5 × 0.035 in.).
Water volume — 500 ml.
Water temperature — 65°–68° C (149°–154° F) (cooled to room temperature over weekends).
Water changes — twice weekly; evaporation losses made up with distilled water as needed.
Solutions — tap water (soft) with and without sodium hexametaphosphate additions of 5 and 10 ppm.
Failure criterion — rusting of steel base.

Sodium Hexameta-phosphate, ppm	Time in Days to				
	First Rust	5–10% Rust	10–20% Rust	20–50% Rust	Over 50% Rust
0	19–29	37–61	47–68	54–75	61–96
5	9	33–83	61–89	103	152
10	61–75	292	355	446	>621†

* Unpublished data, The New Jersey Zinc Co.
† Test discontinued.

SOIL CORROSION

Refer to *Corrosion by Soils*, p. 454, especially Table 10, p. 460. Data on corrosion rates of zinc coatings are given in Tables 11 and 12, pp. 461 and 462.

ATMOSPHERIC CORROSION

Many data are available from a variety of sources on the corrosion of zinc and its alloys in the atmosphere. Only those data are included here which are based on weight changes after complete removal of adhering corrosion products.

The overall data support the conclusion that zinc corrodes slowly and uniformly in normal atmospheres at a rate determined by the factors to be discussed below. Deep pitting is not observed in a significant number of cases.

INDOOR EXPOSURE

Indoor exposures may be classified into three groups: (1) simple exposure in normal room atmospheres at moderate relative humidities; (2) exposure in normal room atmospheres under conditions conducive to condensation and slow drying; and (3) exposure to indoor industrial atmospheres which may at times be very corrosive.

In normal indoor atmospheres, zinc corrodes very little. Generally a visible tarnish film forms slowly, starting at spots where dust particles have fallen on the surface. Over a period of time, such films grow slowly until the surface has lost much of its original luster. The appearance of the reaction film and the degree of corrosive attack are related to the relative humidity. Increasing humidity in sheltered exposures has

[11] Unpublished data, The New Jersey Zinc Company.

little influence on the corrosion rate of zinc up to about 70% relative humidity.[12] At and above this point, the zinc corrosion products appear to absorb sufficient moisture to stimulate the corrosion reaction, and the attack increases with increasing relative humidity. It is possible that the initiation of attack at dust particles is related to the relative humidity at which such particles absorb moisture from the air. Visible moisture films are not normally present even at high humidities.

A special case under the foregoing is that of metal parts cooled in use below the dew point of the ambient atmosphere. In this case, visible moisture films form which,

TABLE 10. CORROSION OF ROLLED ZINC IN INDOOR INDUSTRIAL ATMOSPHERES*

Specimen size — 10 × 15 × 0.05 cm (4 × 6 × 0.020 in.) = 300 sq cm.
Specimen position — two specimens horizontal and two vertical for each metal. Held in porcelain insulators on painted wooden racks.
Surface preparation — washed with ether and alcohol.
Test duration — 10 years.
Corrosion criterion — loss in weight after chemical removal of corrosion products.
Removal of corrosion products — 3-min immersion in 200 grams/liter CrO_3 at 80° C. Small correction applied based on control specimens

Metal A — hot-rolled high-grade zinc. Metal B — hot-rolled brass special zinc. Metal C — hot-rolled selected zinc. Metal D — hot-rolled selected zinc plus 1% Cu and 0.01% Mg.

Test Location	Corrosion Rate							
	Metal A		Metal B		Metal C		Metal D	
	mdd	ipy	mdd	ipy	mdd	ipy	mdd	ipy
Pyrometer shed — ZnO furnace building — typical general atmosphere	0.15	0.00003	0.15	0.00003	0.15	0.00003	0.15	0.00003
ZnO furnace building over furnaces — heat, moisture, SO_2, etc.	0.4	0.00008	0.45	0.00009	0.5	0.0001	0.5	0.0001
Cement crusher building — industrial gases and cement dust	0.05	0.00001	0.05	0.00001	0.05	0.00001	0.10	0.00002
Cement kiln furnace building — heat, gases, moisture, cement dust	0.10	0.00002	0.10	0.00002	0.10	0.00002	0.05	0.00001
Coal breaker building — industrial gases,† moisture, coal dust	0.10	0.00002	0.10	0.00002	0.15	0.00003	0.15	0.00003

* Unpublished data, the New Jersey Zinc Co.
† Locomotive entered building frequently.

because of the small volume of water and large water-air interface area, are well supplied with oxygen. Considerably accelerated corrosion takes place with the formation of a sufficient thickness of film to be objectionable in parts with close tolerances. Means for minimizing such attack with inhibitive films are discussed, beginning p. 862.

Many industrial processes produce atmospheres containing moisture, SO_2, and other corrodents which may condense on cooler parts of the building structure. The data in Table 10 illustrate the range of corrosion rates observed in one set of tests where the racks were suspended near but not in contact with the building roof.

[12] J. C. Hudson, *Trans. Faraday Soc.*, **25**, 177–252 (1929).

Outdoor Exposure

Outdoor exposures may be divided into two classes: (1) completely exposed, and (2) exposed to the atmosphere but sheltered from direct contact with rain. The sheltered exposures are free from the leaching-washing action of rain but are likely to dry more slowly after they have been wet with condensed moisture.

The rate at which zinc corrodes outdoors is controlled primarily by three factors: (1) the frequency and duration of moisture contact; (2) the extent of industrial pollution of the atmosphere; and (3) the rate of drying. The worst conditions involve frequent wetting of the metal with dews and fogs in industrial atmospheres where the condensed moisture is distinctly acid. In such cases, the metal is attacked by the acid moisture and the formation of the partially protective basic corrosion product film is hindered.

The effect of rainfall alone is less pronounced, since such water is not usually acid. Rainfall may at times be somewhat beneficial, because it washes away corrosion accelerants, such as chlorides at seacoast locations, and acid dust particles at industrial locations.

The rate of drying is important because thin, highly aerated moisture films promote corrosion. Sheltered locations where drying may be slow might be expected to show such effects. Hudson[13] has reported data for sheltered and unsheltered exposures in the same locations. Although he used two different criteria for determining the corrosion rate, the data indicate that the corrosion rate in the sheltered exposures is roughly half that in the unsheltered.

TABLE 11. THE CORROSION RATE OF ZINC IN VARIOUS ATMOSPHERES*

Specimen size — rolled metal, 10.2 × 15.2 × 0.061 cm (4 × 6 × 0.024 in.).
  Die cast metal, 7.6 × 15.2 × 0.32 cm (3 × 6 × 0.125 in.).
Mounting of specimens — vertically 1½ in. apart on Bakelite rods through three ⅜ in. diameter holes in each specimen in the form of a triangle.
Specimen preparation — numbered, washed with ether and alcohol before weighing.
Check specimens — duplicates exposed; values given are averages.
Test racks — wooden with wood supports for rods spaced 12 in.; for two parallel rows of specimens; 36 in. long. Painted.
Test duration — 10 years.
Corrosion criterion — weight change after chemical removal of corrosion products.
Removal of corrosion products — 10 min in 10% NH_4Cl solution at 88° ± 1° C; 30 min on specimens from seacoast locations. Control specimens were used and correction, which was small, was made.

Grade of Zinc†	Corrosion Rate, mdd (ipy)					
	Palmerton, Pa.	New York City	Pittsburgh, Pa.	Key West, Fla.	Montauk Point, L. I., N. Y.	Hanover, N. M.
Rolled high-grade	0.27 (0.000055)	1.4 (0.00028)	0.9 (0.00018)	0.7 (0.00014)	0.6 (0.00012)	0.04 (0.000008)
Rolled brass special	0.29 (0.000058)	1.3 (0.00027)	0.8 (0.00017)	0.7 (0.00014)	0.6 (0.00013)	0.04 (0.000009)
Rolled high-grade + 1.0% Cu and 0.01% Mg	0.26 (0.000052)	1.1 (0.00023)	0.7 (0.00015)	1.3 (0.00026)	0.8 (0.00016)	0.07 (0.000014)
Die cast alloy XXIII	0.4 (0.00008)	1.1 (0.00022)	0.5 (0.00011)	1.5 (0.00030)	2.1 (0.00043)	Not Tested

* Unpublished data, the New Jersey Zinc Co.
† A. S. T. M. designation.

[13] J. C. Hudson, *Trans. Faraday Soc.*, **25**, 177–252 (1929).

TABLE 12. CORROSION OF ROLLED ZINC IN TYPICAL OUTDOOR ATMOSPHERES*

Specimen size — 22.9 × 30.5 × 0.09 cm (9 × 12 × 0.035 in.).
Specimen position — vertical at 3 in. spacing in porcelain insulators; duplicate specimens.
Test duration — 10 years (approx.).
Corrosion criterion — change in weight after chemical cleaning.

Type Zinc	Corrosion Rate													
	Altoona, Pa.†		New York City†		Sandy Hook, N. J.‡		Key West, Fla.‡		La Jolla, Calif.‡		State College, Pa.§		Phoenix, Ariz.∥	
	mdd	ipy	mdd	ipy	mdd	ipy	mdd	ipy	mdd	ipy	mdd	ipy	mdd	ipy
Spec. high-grade	0.94	0.00019	0.99	0.00020	0.29	0.000058	0.080	0.000016	0.26	0.000053	0.19	0.000039	0.025	0.0000051
High-grade	1.0	0.00021	1.1	0.00023	0.29	0.000059	0.094	0.000019	0.34	0.000069	0.18	0.000037	0.038	0.0000077
Prime western	1.0	0.00021	1.0	0.00021	0.29	0.000059	0.11	0.000022	0.37	0.000075	0.21	0.000043	0.055	0.000011

* Report of Sub Committee VI on Atmospheric Corrosion Tests of Non-Ferrous Metals and Alloys, *Proc. Am. Soc. Testing Materials*, **44** (1944).
† Industrial.
‡ Seacoast.
§ Rural.
∥ Desert.

Types of Atmospheres. Some data on a number of typical locations have been compiled in Table 11.

Very useful information has been obtained by Sub-committee VI of Committee B-3, A. S. T. M., at seven exposure locations selected to cover definite types of atmospheres. Table 12 compiles the corrosion rates for three grades of rolled zinc obtained by this committee.

By far the most rapid corrosion has taken place at the industrial locations, Altoona, Pennsylvania, and New York City. The seacoast locations show a much smaller rate of attack, while in the pure dry atmosphere of Phoenix, the attack over the entire 10-year exposure period was almost negligible.

An important factor involved in seacoast exposures appears in the data for the Key West exposures in Tables 11 and 12. It will be noted that rolled high-grade zinc has a corrosion rate of 0.094 mdd in Table 12, and 0.7 mdd in Table 11, an eightfold difference. The specimens for the tests in Table 11 were exposed near the Sand Key lighthouse some miles southwest of Key West and showed definite evidence of contact with wind-borne salt water. The windward edges of the specimens were definitely more severely corroded than the leeward edges. In other tests definite indications were noted of abrasion by wind-borne sand. It seems probable that sharp differences exist in seacoast exposures between those immediately adjacent to the shore line and those which are a reasonable distance inland.

GALVANIC COUPLING

The most complete published work on the atmospheric corrosion of zinc in contact with other metals is that of Subcommittee VIII of Committee B-3, A. S. T. M.[14] The following conclusions may be drawn from this work:

1. Contact with aluminum, copper, iron, lead, nickel, and tin results in increased corrosion of zinc.

2. The extent of increase in attack is surprisingly small (usually less than two and not over three times the normal rate in the more corrosive atmospheres) and does not parallel the order of the metals in the electromotive series.

3. With the exception of lead, all the metals received substantial to complete protection when in contact with zinc.

In the A. S. T. M. tests, at least partially successful efforts were made to insure metallic contact in the couples. Frequently such perfection is not obtained in commercial parts and anticipated possible galvanic attack does not occur. As in the case of aqueous corrosion, small zinc areas in contact with large areas of electropositive (more cathodic) metals should be avoided. Small areas of other metals in contact with large areas of zinc are relatively safe, as shown by the use of aluminum and of tin-coated steel rivets in the fabrication of corrugated sheet zinc roofs.

EFFECT OF STRESS AND VIBRATION

No cases are known where stress and vibration have clearly influenced the atmospheric corrosion of zinc. However, zinc is not normally used under conditions involving heavy stresses. The degree of cold working in rolled zinc has only a very minor effect on the corrosion rate.

METALLURGICAL FACTORS

The rate of attack on zinc in the atmosphere is not markedly influenced by composition (Table 11). Hudson[15] and others have offered similar observations.

[14] *Proc. Am. Soc. Testing Materials*, **39**, 247–255 (1939).
[15] J. C. Hudson, *Trans. Faraday Soc.*, **25**, 177–252 (1929).

Protective Measures for Atmospheric Exposure

Most zinc exposed to the atmosphere is used without protection. Paint films and electrodeposited coatings, used at times for decorative purposes, exert definite protection if properly applied. The procedures and limitations discussed under "Protective Measures" (p. 340) apply in the case of atmospheric exposure. Inhibitive chromate films have not shown large advantages for the prevention of atmospheric corrosion over any substantial period of time.

ZIRCONIUM*

Ductile zirconium was first produced in 1926 by J. H. deBoer and J. D. Fast, who discovered that the iodides of zirconium could be decomposed in vacuum on an incandescent filament. Metal deposited in this manner was quite soft and sufficiently

TABLE 1. EFFECT OF COMMON AGENTS ON DUCTILE ZIRCONIUM

Size of specimen — 6.2 × 1.3 × 0.13 mm (2.5 × 0.5 × 0.005 in.).
Surface preparation — degreased.

Temperature — Boiling Water Bath, 100° C (212° F)

Agent	Concentration	Time, days	Corrosion rate, ipy
H_2SO_4	10%	14	0.0007
HCl	Conc.	14	0.0002
HCl	5%	14	No attack*
HNO_3	Conc.	14	0.00005
HNO_3	10%	14	0.00003
H_3PO_4	10%	14	0.00005
Citric acid	10%	14	0.00002
Oxalic acid	10%	14	0.00004
NaOH	10%	14	0.00002
NaOH	50%	4	0.00017
KOH	10%	14	No attack *
NH_4OH	28%	14	Gained weight
$HgCl_2$	Sat. soln.	14	No attack *
NaCl	20%	5	Slight tarnish

Temperature — Environment

HF	All dilutions	...	Rapid attack
H_2SiF_6	Conc.	...	Rapid attack
$AgNO_3$	0.8%	3	No attack *
H_2SO_4	10%	14	0.0002
HCl	Conc.	14	0.0001
HCl	5%	14	No attack *
HNO_3	Conc.	14	0.00001
HNO_3	10%	14	0.00001
H_3PO_4	Conc.	14	0.00004
H_3PO_4	10%	14	0.00002
Citric acid	10%	14	No attack *
Oxalic acid	10%	14	No attack *
NaOH	10%	14	No attack *
KOH	10%	14	0.00002
NH_4OH	28%	14	0.00003
NaF	10%	14	0.0003
I_2 in KI	0.1%	14	No attack *
Atmosphere	...	Years	No attack *

* No attack = loss less than 0.1 mg.

* Research Laboratory, Foote Mineral Co., Philadelphia, Pa.

ductile to withstand severe cold deformation. The ductile metal has only recently become available in this country in quantities sufficient for the precise determination of its corrosion-resistant properties.*

It is considered that an invisible film of oxide is formed on the surface of zirconium which governs its behavior toward corrosion. Although the metal is practically inert at temperatures up to 200° C (390° F), very thin filaments will burn in air and the metal combines with nitrogen and chlorine at a dull red heat. Nevertheless, except for HF and hot concentrated H_2SO_4 and H_3PO_4, its acid resistance is very good. Zirconium is also resistant to strong alkalies and molten caustic.

The effects of some of the common corrosive media are listed in the table on p. 347.

NON-METALLIC MATERIALS AND SEMI-METALS

Introduction

Non-metallic materials may deteriorate in various ways and manners, but not by corrosion. Only metals corrode according to the accepted definition of corrosion used in this book. (See the Glossary.) The mechanism of deterioration of the two classes of materials serves to distinguish the corrosion reaction of metals, largely electrochemical in character, from the physical and sometimes chemical processes that attend degradation of non-metals. However, the practical usefulness of many non-metals and the potential uses of semi-metals (e.g., silicon and boron) in corrosive environments have dictated inclusion of this section discussing their properties.

In considering the service of non-metallic materials in corrosive environments a somewhat different concept from that applied to metals must be adopted. Weight loss or penetration per unit of time can be used as a criterion to a limited extent in judging carbon and graphite, ceramics and glass, but to even a lesser degree in considering plastics. Pitting, likewise, is not a good measure. In general, the best criterion is a measure of the change in physical or mechanical properties resulting from contact with a given environment.

The failure of a non-metallic material may occur by swelling, by the leaching out of binder or matrix, by failure of cements that are used to bond parts together, by absorption of liquids, by softening in the case of elevated temperatures, by devitrification in the case of glass or ceramics, and in other ways. Indication of attack in any of the ways mentioned within a prescribed time limit is usually evidence that the material is not suited to the application in question.

Carbon and Graphite

F. J. Vosburgh[†]

Commercial carbon and graphite are produced from discrete particles of carbon bonded with materials which upon subsequent processing carbonize so that the final article is a homogeneous structure essentially carbon. The final properties of such a bonded carbon article are dependent on the processing temperature. Articles produced

---

* Zirconium absorbs hydrogen under certain conditions, thereby becoming embrittled. In this respect it behaves like columbium and tantalum. The hydrogen can be removed and ductility restored by heating in vacuum. Private communication, C. W. Balke.

† National Carbon Co., Inc., New York, N. Y.

at temperatures under 1400° C (2550 F°) are defined as carbon; those produced at temperatures sufficiently elevated (over 2000° C [3600° F]) to develop a crystalline structure are defined as graphite. Karbate materials are carbon or graphite impregnated with resin which is then polymerized in the pores so that natural permeability is essentially eliminated.

Carbon and electric furnace graphite differ to some extent in their resistance to some media. Concentrated sulfuric acid, bromine, and fluorine, for example, have little or no effect on carbon but disintegrate graphite quite rapidly. Neither material is affected by ordinary atmospheric exposure.

Carbon, graphite, and the impregnated materials satisfactorily resist most mineral and organic acids, aqueous solutions of alkalies, salts and organic compounds, and organic solvents at temperatures up to the boiling points.

The types of corrosive agents which both carbon and graphite do not resist are the strong oxidizing agents. Even in some cases of this kind, however, they may be found to be more serviceable than other materials. For example, in nitric acid pickling baths, carbon brick has given satisfactory service as a lining material for periods of more than ten years.

AVAILABLE FORMS

Carbon, graphite, and the impregnated materials* are produced in a variety of forms — cylindrical, rectangular, and tubular. In size they vary from small rods $\frac{1}{16}$ in. in diameter to cylinders 40 in. in diameter, rectangles $\frac{1}{2}$ in. square to blocks 24 × 30 × 180 in. and cylindrical tubes of $\frac{1}{2}$ to 10 in. inside diameter. Graphite, being relatively soft, can be shaped on woodworking machines, whereas carbon, being somewhat abrasive, is worked with ordinary machine shop equipment.

POROSITY

The porosity of carbon and graphite normally varies between 18 and 32%, depending on the type and grade of material and size. The pores in the standard materials vary in size and are not regularly connected, so they are seldom usable for filtration and aeration purposes.

Special porous carbon and graphite are available in cylinders and plates in which the size of the pores is well controlled and the pores interconnected, so that the materials are satisfactory for filtration, diffusion, and aeration. The porosity of the special materials is about 48%, with a wide range of controlled permeabilities many times those of the regular carbon and graphite. (See Table 1.)

PURITY

The purity of electric furnace graphite is high, usually better than 99% carbon. Specially processed graphite for spectroscopic electrodes is made containing less than 0.000075% ash. Carbon products are less pure, dependent on the prospective use. Carbon blocks for furnace liners run about 90% carbon, electrodes for electric furnaces 93% carbon, and carbon plates, electrodes, tubes, etc., primarily designed to be graphitized but frequently used for a variety of other purposes, are 98% carbon or higher. SiC, SiO_2, Fe_2O_3, and Al_2O_3 make up 75 to 80% of the ash for the latter materials.

* Karbate.

TABLE 1. PHYSICAL PROPERTIES OF POROUS CARBON AND GRAPHITE (AVERAGE VALUES)

Grade	Material	Weight, lb/cu ft	Effective Porosity, %	K^*	Electrical Resistivity, Room Temp		Minimum Strengths			Average Pore Diameter		Filter Action Minimum; Diam. Particle Retained, in.	Average Water Permeability,‡ gal/sq ft/ min at 5 psi pressure	Average Air Permeability,‡ cu ft/sq ft/ min at 2 in. H_2O pressure
					Ohm-in.	Ohm-cm	Tensile, psi	Transverse, psi	Compressive, psi	Inches	Microns			
C	Carbon	84	36	6	0.0020	0.0051	600	1530	2700	0.0002	5	...	0.30	...
C	Graphite	84	36	5	0.00045	0.0114	500	1080	1680	0.0002	5	...	0.30	...
60	Carbon	69	48	27	0.0070	0.0178	190	600	850	0.0013	33	0.00047	14.0	...
60	Graphite	69	48	21	0.0012	0.0030	110	250	500	0.0013	33	0.00047	14.0	...
50	Carbon	69	48	27	0.0070	0.0178	180	500	830	0.0019	48	0.00079	30.0	...
50	Graphite	69	48	21	0.0012	0.0030	110	250	500	0.0019	48	0.00079	30.0	...
40	Carbon	69	48	27	0.0057	0.0145	120	320	900	0.0027	69	0.00098	45.0	4.0
40	Graphite	69	48	21	0.0013	0.0033	100	190	500	0.0027	69	0.00098	45.0	4.0
30	Carbon	69	48	27	0.0070	0.0178	100	250	770	0.0039	99	0.00173	80.0	8.5
30	Graphite	69	48	21	0.0017	0.0043	80	200	520	0.0039	99	0.00173	80.0	8.5
20	Carbon	68	48	27	0.0070	0.0178	90	240	700	0.0055	140	0.0030	120.0	17.0
20	Graphite	68	48	21	0.0020	0.0051	60	140	310	0.0055	140	0.0030	120.0	17.0
10	Carbon	68	48	26	0.0080	0.0204	80	160	300	0.0075	190	0.0059	175.0	30.0
10	Graphite	68	48	21	0.0020	0.0051	50	140	270	0.0075	190	0.0059	175.0	30.0

*Coefficient of thermal expansion /°F to temperature $t(°F) = [K + 0.0039t(°F)] \times 10^{-7}$. Coefficient of thermal expansion /°C to temperature $t(°C) = [1.8K + 0.0077t(°C)] \times 10^{-7}$.
† Water at 21° C (70° F); 1-in. thick sample.
‡ 70° F; 760-mm Hg pressure; 15% relative humidity; 1-in. thick sample.

RESISTANCE TO THERMAL SHOCK

Both carbon and graphite are so highly resistant to thermal shock that most forms may be heated repeatedly to incandescence and plunged into ice water without causing fracture or spalling. They are unaffected physically by temperatures normally encountered in commercial operations and may be used safely so far as temperature is concerned up to 2000° C (3600° F).

MISCELLANEOUS PROPERTIES

The thermal conductivity of graphite is high — 0.31 to 0.41 gram-cal/(sec) (sq cm) (°C/cm); (75–99 Btu/(hr) (sq ft) (°F/ft) — at normal temperatures, several times that of corrosion-resistant metals. Carbon, on the other hand, has relatively low thermal conductivity, namely, 0.012–0.025 gram-cal/(sec) (sq cm) (°C/cm); (3–6 Btu/(hr) (sq ft) (°F/ft). The thermal conductivity of carbon increases with increasing temperature, whereas that of graphite decreases.

The behavior of carbon and graphite is different from that of metals as their elastic and ultimate limits are closer together than for most metals, they are not as strong, and they are more susceptible to damage by impact. With proper design they can give satisfactory service in the chemical industries. Neither material exhibits any evidence of cold flow. The impregnated materials have up to twice the strength of regular carbon and graphite. (See Table 2.)

Both carbon and graphite are electrical conductors, although their specific resistances are high compared to the metals. The specific resistance of graphite at room temperature is approximately 0.0009 ohm-cm (0.00035 ohm-in.) or 900 microhm-cm (350 microhm-in.). For carbon the corresponding figures are approximately 0.0038 ohm-cm (0.0015 ohm-in.) and 3800 microhm-cm (1500 microhm-in.). The specific resistances of carbon and graphite increase with the size of the stock considered and are affected differently by temperature changes. (See Table 2.)

Carbon and graphite are used in innumerable ways. In the concentration of sulfuric acid by the Cottrell Process complete tower structures as large as $15 \times 15 \times 27$ ft are made of carbon. The floor, walls, roof, inlet and outlet tubes, and tube supports and even the power lead-ins are of carbon in one form or another.

IMPREGNATED CARBON AND GRAPHITE

The impregnated materials have all the normal characteristics of the carbon or graphite from which they are made, and are stronger and more impervious to gases and liquids under pressure. They are used in the form of tubes for conveying corrosive liquids and gases and are used in heat interchange operations where high thermal conductivity is an important consideration. Plates, blocks, and tubes are often machined and cemented together for special equipment. Because of the impregnants used to make the carbon or graphite impervious, these materials should not be used at temperatures above 170° C (340° F).

These materials are not attacked by hydrochloric acid of any concentration or temperature. They are unaffected by sulfuric acid up to 96%, and by phosphoric acid or by hydrofluoric acid up to 60%.

TABLE 2. PHYSICAL CHARACTERISTICS OF CARBON AND GRAPHITE PRODUCTS

Product	Apparent Density, grams/cc	Weight, lb/cu ft	Strength, psi			Elastic Modulus. Multiply by 10^5	Specific Resistance, Room Temp.		K^* (Thermal Expansion)	Thermal Conductivity, Btu/(hr)(sq ft)(°F/ft)
			Tensile	Compressive	Transverse		ohm-in.	ohm-cm		
CARBON										
Cylinders										
8 in. diameter	1.54	96.0	660	2920	1320	5.5	0.0013	0.0033	13	6.0
10–14 in. diam.	1.53	95.0	470	2120	950	5.4	0.0013	0.0033	12	6.0
17–24 in. diam.	1.54	96.0	400	2200	790	5.4	0.0014	0.0036	13	6.0
30–40 in. diam.	1.54	96.0	400	1910	810	4.3	0.0026	0.0066	12	6.0
Rectangulars										
4 × 4 in. to 6 × 6 in.	1.51	97.8	840	4100	1670	9.4	0.0018	0.0046	14	4.0
8 × 8 in. to 20 × 20 in.	1.55	96.7	500	2140	990	7.1	0.0016	0.0041	15	4.0
15 × 30 in. to 24 × 30 in.	1.54	96.0	400	1910	810	4.3	0.0026	0.0066	12	4.0
GRAPHITE										
Cylinders										
To 5½ in. diam.	1.56	97.3	760	3050	1750	8.8	0.00036	0.00092	5–12	84.0
6–12 in. diam.	1.55	96.7	610	3420	1810	8.0	0.00037	0.00094	6–12	79.0
14–20 in. diam.	1.53	95.3	500	3180	1490	6.7	0.00040	0.00102	8–12	70.0
Rectangulars										
To 5 in. thick	1.56	97.3	700	3050	1750	8.8	0.00036	0.00092	5–12	94.0
6 in. thick to 1.44 sq in.	1.55	96.7	700	3420	1810	8.0	0.00037	0.00094	6–12	84.0
Over 1.44 sq in.	1.53	95.3	570	3180	1490	6.7	0.00039	0.00094	8–12	79.0
KARBATE										
Carbon	1.76	110.0	2000	10500	4650	25.0	0.0016	0.0041	14	2.8
Graphite	1.91	120.0	2350	10500	5000	22.0	0.00033	0.00084	24	75.0

* Coefficient of thermal expansion per degree. To temperature t °F = $[K + 0.0039t(°F)]10^{-7}$. To temperature t °C = $[1.8K + 0.007t(°C)]10^{-7}$.

Chemical Stoneware

JAMES A. LEE*

Chemical stoneware is a generic term for ceramic equipment produced from carefully selected and processed clays that differs from earthenware in its comparatively higher physical strength, resistance to thermal shock, and resistance to attack by chemicals in general, except hydrofluoric acid and strong caustic alkalies. It is manufactured from mixtures of clays and other crystalline or amorphous materials, either natural or artificially prepared in the electric furnace. These materials are carefully beneficiated to remove undesirable impurities and blended with water to form a plastic, workable mass.

Jiggering, casting, and extruding are the methods most commonly employed in its manufacture, although all of the forming methods common to the ceramic industry may be used. After forming, the ware is carefully dried to eliminate mechanically held water and then subjected to a cycle of heat treatment which is rigidly controlled to develop the desired pyrochemical reactions. During this firing operation, salt is introduced into the kiln to form the glaze on the body. As chemical stoneware has a very dense body, the glaze is unnecessary except to present an attractive appearance. In many instances unglazed ware is used.

Because of high shrinkage during the drying and firing operations, dimensional accuracy greater than plus or minus 2% cannot be readily obtained except by machining the fired material. This is accomplished by utilizing precision grinding methods, using silicon carbide or diamond abrasives.

TABLE 1. PHYSICAL PROPERTIES OF CHEMICAL STONEWARE*

This table, prepared by the General Ceramics Co., gives the physical properties of an average grade of chemical stoneware. It should be emphasized here that "chemical stoneware" is not the name of a definite material, such as an alloy, but a generic term applied to a wide variety of ceramic compositions, and hence that in any particular composition designed to give optimum properties in one respect, it will ordinarily be impossible to secure optimum properties in all other respects.

Specific gravity	2.2	Modulus of elasticity, psi	8,000,000
Hardness, scleroscope	100	Specific heat	0.2
Ultimate tensile strength, psi	2,000–3,000	Thermal cond., Btu/(hr)(sq ft)(°F/ft)	0.8–2.9
Ultimate compressive strength, psi	80,000	Linear thermal expansion, per °F	0.0000020
Modulus of rupture, psi	5,000–13,000	Water absorption, %	0–2

* Copyrighted, *Chemical and Metallurgical Engineering*.

Since most of the inorganic non-metallic minerals can be incorporated in chemical stoneware compositions, an endless variation of bodies is possible so that the physical and chemical properties of the product can be varied within wide limits. Table 1 lists physical properties of a typical stoneware body. It must be remembered that while these values give an indication of what can be expected from the average body, they will vary for bodies manufactured for a specific purpose. Although low resistance to thermal shock is considered an inherent weakness of chemical stoneware, special bodies containing up to 40% of carborundum have a high resistance to thermal shock, with accompanying variations in the other physical properties. The latter is an example of a body fabricated for a particular purpose.

Larger shapes are more prone to fail under elevated temperature conditions than

* *Chemical and Metallurgical Engineering*, McGraw-Hill Publishing Co., Inc., New York, N. Y.

smaller ones and for that reason a multiplicity of units should be used whenever possible. Maximum operating temperatures for standard chemical stoneware bodies range from 80° C (170° F) for small shapes not larger than 18 in. diameter to 55° C (130° F) for large shapes such as tower sections and vessels up to 60 in. diameter. The above limitations of temperature will depend upon such design considerations as gradual changes of direction, reinforcing fillets, and avoidance of thick sections.

Mechanical equipment, such as pumps and blowers, are encased in cast iron armor to provide resistance to mechanical stresses and shock. Piping and valves, used under stringent conditions, are also armored. Installations of this type provide the chemically inert properties of chemical stoneware with the strength of metals and are ideal for the most drastic corrosive conditions.

Chemical stoneware, by the nature of its physical properties and methods of manufacture, lends itself to the fabrication of almost all types of chemical equipment such as processing and storage vessels, piping and heating coils, pumps and blowers, and reaction and absorption towers. Installations are limited usually by the size of the individual piece and the thermal conditions at which it is to operate.

Glass and Vitreous Silica

JAMES A. LEE*

GLASS

Resistance to practically all corrosive agents except hydrofluoric acid (and the caustic alkalies under some circumstances), combined with ease of fabrication into almost any desired form, makes glass suitable for many applications. It is used for stills, absorbers and columns, valves, tanks, pumps, industrial piping and fittings, heat exchangers, filter blankets, and a variety of other industrial equipment.

TABLE 1. COMPOSITION OF COMMERCIAL GLASSES*

Type of Glass	Lime Glass for Plate	Soft Lead Glass	Borosilicate Glass	96% Silica Glass
Constituent	Amount of Constituent, Wt. %			
SiO_2	70 to 75	55 to 65	75 to 82	96.0
Na_2O	12 to 14	5	3 to 10	....
K_2O	....	7 to 8	0.4	....
CaO	13 to 14	0 to 6	....	....
PbO	....	25 to 30	0 to 10	....
B_2O_3	....	....	5 to 20	3.0
Al_2O_3	0.5 to 1.5	....	2.2	....

* Courtesy of Corning Glass Works.

It is general practice in the glass industry to control the physical constants of the product rather than the chemical composition. Although control of physical constants results in fairly close control of chemical composition, no great amount of data are available relating the two. In Table 1 the range of composition of four common types of commercial glasses are shown.

Borosilicate glass resists solution by aqueous media in general with the outstanding

* *Chemical and Metallurgical Engineering*, McGraw-Hill Publishing Co., Inc., New York, N. Y.

exception of hydrofluoric acid. Data of Tables 2 and 3 indicate very low rates of solution in boiling hydrochloric acid (0.45 mdd) and approximately 2000 times this

TABLE 2. COMPARATIVE RESISTANCE OF CERTAIN GLASSES TO AQUEOUS SOLUTIONS AND STEAM*

Reagent	Time (Hours)	Borosilicate Glass†	96% Silica Glass†	Alkali-Resistant Glass
		\multicolumn{3}{c}{Wt. Loss, mg/sq cm}		
Steam (100 psi, gage press.)	96	0.5	0.1	0.5
1% NaOH, 100° C (212° F)	6	....	....	0.02
5% NaOH, boiling	6	2.0	1.5	0.04
5% NaOH, 100° C (212° F)	6	1.4	0.9	....
50% NaOH, 100° C (212° F)	6	2.7	....	0.49
$N/50$ Na_2CO_3, 100° C (212° F)	6	0.12	0.07	....
$N/50$ Na_2CO_3, boiling	24	....	....	0.01
5% HCl, 100° C (212° F)	24	0.0045	0.0005	0.01

* Courtesy of Corning Glass Works.
† Very resistant to all acids except hydrofluoric and hot glacial phosphoric. Attacked at appreciable rates by steam and alkalies. However, the borosilicate glass is widely used as gage glasses for steam, and as sight glasses for hot caustic solutions.

TABLE 3. RESISTANCE OF A BOROSILICATE GLASS TO SODIUM HYDROXIDE SOLUTIONS. EFFECT OF CONCENTRATION AND TEMPERATURE*

Concentration of NaOH, Wt. %		5	10	15	20	25	30	35	40	45	50
\multicolumn{2}{c}{Temperature}	\multicolumn{10}{c}{Loss in Wt., mg/sq cm/hr}
°C	°F										
40	104	0.0017	0.0025	0.0031	0.0030	0.0025	0.0022	0.0019	0.0017	0.0016	0.0015
51	124	0.0038	0.0052	0.0062	0.0060	0.0051	0.0044	0.0035	0.0031	0.0028	0.0026
65	149	0.014	0.016	0.018	0.018	0.018	0.017	0.016	0.014	0.013	0.013
75	167	0.032	0.039	0.045	0.048	0.050	0.048	0.042	0.038	0.036	0.035
100	212	0.22	0.30	0.36	0.40	0.44	0.47	0.48	0.48	0.46	0.45

* Courtesy of Corning Glass Works.

rate in boiling 5% NaOH. Steam at 100 psi measurably attacks the glass, but at a very low rate (12 mdd). The rate of attack by alkali appears to fall off with time (Table 4).

TABLE 4. RESISTANCE OF A BOROSILICATE GLASS TO BOILING $N/50$ Na_2CO_3 SOLUTION*

Effect of Time on Rate of Attack

Time of Exposure, hr	Loss in Weight, mg/sq cm
3	0.07
6	0.12
24	0.16

* Courtesy of Corning Glass Works.

Physical properties of borosilicate glass are given in Table 5. The weight losses when held at various temperatures are given in Table 6.

TABLE 5. PHYSICAL PROPERTIES OF LOW-EXPANSION GLASSES AND VITREOUS SILICA*

Material	Specific Gravity	Specific Volume, cu in./lb	Tensile Strength, psi	Modulus of Elasticity, psi (Multiply by 10^5)	Hardness†	Thermal Expansion per °C (Multiply by 10^{-5})	Thermal Cond., cal/(sec) (sq cm) (°C/cm) (Multiply by 10^{-4})	Specific Heat, gram cal/°C	Softening Point °C	Softening Point °F	Dielectric Constant, 60 cycles	Refractive Index, n_D	Forms Available‡
Borosilicate glass	2.23	12.4	10,000	98	....	0.32	28	0.20	818	1505	4.6	1.474	S, R, T, other
96% silica glass	2.18	12.7	....	....	....	0.080	....	....	1500	2732	3.8 Approx.	1.458	R, T, other
Vitreous silica, transparent	2.20	12.6	4,000	105–126	4.9	0.054	33.2	0.25	1425	2600	3.8	1.459	S, R, T, other
Vitreous silica, non-transparent	2.07	13.4	400–800	94–114	....	0.054	33.1	....	1425	2600	3.7	....	S, R, T, other

* Copyrighted, *Chemical and Metallurgical Engineering*.
† Hardness: 2.5-mm ball, 25-kg load, depth in 1/200 mm.
‡ S — sheets; R — rods; T — tubes.

NON-METALLIC MATERIALS AND SEMI-METALS

TABLE 6. WEIGHT LOSS OF A BOROSILICATE GLASS HELD AT VARIOUS TEMPERATURES*

Temperature		Wt. Loss in 6 hr,
°C	°F	mg/sq cm
125	257	0.002
150	302	0.010
175	347	0.045
200	392	0.12
225	437	0.20
250	482	0.27

* Courtesy of Corning Glass Works.

Borosilicate glass tubing is available in sizes from ¼ to 4½ in. O.D. and in lengths of 48 to 60 in. Flanged piping and fittings are available in sizes between 1 in. and 4 in. I.D. suitable for working pressures up to 100 psi.

A glass which has been given a special heat treatment is available in the form of tanks in two types: (1) all-glass or transparent and (2) steel or wood shells lined with glass. Both types are shipped ready for service. In the case of tanks that are too large to be shipped complete, linings can be installed in tank shells at the job site. The heat treatment permits the glass to withstand instantaneous temperature shock of 225° C (400° F) and continuous working temperatures of 250° to 315° C (500° to 600° F).

Fastenings are of non-corrosive metal, such as nickel alloys or stainless steel. Side walls are grooved to take gaskets which are impregnated glass cloth. Impregnated materials vary with the solution to be contained.

PHOSPHORUS GLASS

A new glass is available which offers major resistance to hydrofluoric acid. It contains no sand, and the principal ingredient is said to be phosphorus pentoxide. It is believed that this glass will perform satisfactorily in most applications, but users are cautioned that it is slowly attacked. The rate of attack by hydrofluoric acid is much slower however than with silicate glasses and the glass retains its transparency. In one test a piece was immersed in a bath of acid for 500 hours. At the end of that time the glass was substantially transparent and to the naked eye showed no obvious attack.

96% SILICA GLASS

A 96% silica glass (remainder is mostly boric oxide) has been developed primarily for high-temperature applications. It has resistance to extremes of thermal shock and a high-temperature softening point, necessary for elevated-temperature applications. It may be used continuously at temperatures as high as 900° C (1650° F), and for very short periods it may be used at temperatures approaching 1100° C (2000° F). Physical properties are listed in Table 5.

The 96% silica glass is extremely stable toward all acids except hydrofluoric, and even in this acid it is attacked considerably more slowly than are the conventional glasses. The rate in cold 60% hydrofluoric acid is about one fifth as great as for borosilicate glass and still less when compared with lime glasses. It is also very stable toward water and all other neutral or acid solutions. Its durability is less marked in alkaline solutions, but even in such solutions it is generally superior to conventional chemical-resistant glasses. Values for rates of solution in alkalies, hydrochloric acid, and steam are listed in Table 2. In this connection, it can be pointed out that because the glass contains only traces of elements other than silicon, boron, oxygen, and aluminum, the contamination of solutions, notably alkalies, in contact with it will be extremely small. For this reason, the glass is particularly well adapted to certain high-precision analytical work. An example of this is its use in glass durability measurements that involve the determination of alkali released by the sample.

The rate of devitrification of 96% silica glass is very low at temperatures below 900° C (1650° F) but becomes appreciable in the neighborhood of 1000° C (1830° F). Up to 1200° C (2200° F), this devitrification rate is greater than that of fused silica, owing to the lower viscosity of the 96% silica glass. Above 1200° C (2200° F) the rate falls progressively below that of fused silica, and at no point does it reach the relatively rapid rates observed for the latter in the range 1300° to 1500° C (2375° to 2730° F). It is accelerated by reducing gases, alkalies, alkaline earths, and certain other metallic oxides.

A number of standard shapes, such as beakers, flasks, tubing, rod, containers, trays, and flat glass, are available. Special shapes made by conventional pressing, blowing, or drawing can be obtained, but, in general, wider dimensional tolerances are required than for products made from softer glasses.

VITREOUS SILICA

All types of vitreous silica are equal in chemical resistance, but are distinguished by their degree of transparency, by the materials from which they are made, and by their mechanical strength. Except where transparency is needed, the cheaper, opaque fused silica, which is made from a high grade of glass sand, is generally satisfactory. For special purposes demanding transparency, such as sight glasses and tubes, or for the application of ultraviolet light, the more expensive material produced by the fusion of rock quartz is required. Both products are generally similar in physical properties apart from the exceptions noted. Both consist of equally pure silica (about 99.8%) but the fused quartz is stronger than the fused silica in the tensile strength ratio of 1700 to 6000. Vitreous silica is suitable for operating temperatures to 1000° to 1100° C (1830° to 2010° F) in the absence of certain mineral salts and reducing gases which may accelerate breakdown due to devitrification. Both qualities of quartz are characterized by low thermal expansion and high resistance to thermal shock. Physical properties are listed in Table 5.

Vitreous silica is so little affected by boiling water under normal pressures that it is used for the construction of stills for highest purity or "conductivity" water. However, solubility of silica is appreciable at elevated temperatures and pressures so that silica gage glasses have been found unsuitable for use with high-pressure steam. Volatility of silica in superheated steam has been reported.

The mineral acids do not react with vitreous silica and are suitably contained in silica equipment. Hydrofluoric acid is the outstanding exception. Phosphoric acid reacts with silica at elevated temperatures, but chemical action is only slight below 250° C (480° F).

Alkalies are more reactive than acids. Aqueous ammonia, nevertheless, can be distilled in silica equipment without fear of contamination.

Vitreous silica is not attacked by the usual gases. The exceptions are those containing fluorine and phosphorus which may be quite reactive. Vitreous silica, though impermeable to most gases, is appreciably permeable to helium and hydrogen at elevated temperatures.

NON-METALLIC MATERIALS AND SEMI-METALS

Among the important uses for fused silica are burners, combustion furnaces, S-bend coolers, and absorption vessels for HCl gas. The properties of impermeability, freedom from iron, resistance to thermal shock, and inertness to HCl have made fused silica of practical importance in the production of hydrochloric acid.

Opaque equipment has been produced in pieces up to 150-gal capacity, but a 50-gal vessel may be considered the largest practical size at the present time. Drawn tubes are obtainable in all sizes up to 4½ in. internal diameter and lengths to 10 ft, and molded pipes are made in sizes of 2 to 30 in. internal diameter with lengths to 10 ft for not over 6-in. bore, and lesser lengths for larger bores. S-bends, elbows, and connecting pieces are made to fit such tubes and pipes. Tubes are produced by drawing from a larger diameter; pipes are made by expanding from a smaller diameter in a temperature-resisting mold. Although the drawn material is usually considered more resistant to severe temperature changes, improvements in the molded product have made it equally suitable for severe service in many cases.

Plastics

JAMES A. LEE*

Plastics are organic materials which possess certain useful characteristics and advantages for structural design. They are derived from synthetic resins, natural resins, cellulose derivatives, coal tar, petroleum or protein substances, and are blended with fillers to produce molding compounds.

The data in Table 1 are indications of the chemical resistance of some of the important plastics. Different evaluation may result if the test period is greater than the seven days specified by A. S. T. M. and used in the table. Important evaluation of the chemical resistance of plastics in structural applications is obtained by measuring the influence of contact by chemicals on the tensile strength.

A phenol-formaldehyde-asbestos composition† is finding considerable use for handling chemicals. It is a finished material available in the form of complete units of equipment or parts molded to shape. Three different grades of the material are available, depending upon the chemicals to be handled, and, if properly selected, are resistant to strong alkalies, salt solutions, to practically all acids, to halogen gases and sulfur dioxide and to organic solvents. They are not resistant to strong oxidizing agents. The material is unaffected by temperatures up to 130° C (265° F) and is not subject to damage caused by rapid temperature changes. Physical properties are given in Table 2.

Acrylic resins are used often for their transparency and are said to be useful in the handling of many chemicals, for example, in construction of pumps. Acrylic resins are being used in the lining of tanks holding chromic and sulfuric acid, for anodic treatment of aluminum alloys. This resin is said not to be affected by sudden temperature changes or by the 15% chromic acid solution used. Chemical properties are listed in Table 3 and physical properties in Table 4.

Vinylidene chloride, used to a large extent as piping and tubing, is tough, durable, and resistant to abrasion and corrosion. It is heat resistant to about 80° C (170° F). At present piping is available in ½-, ¾-, 1-, 1¼-, and 2-in. sizes. Tubing is fabricated in a wide range of wall thicknesses from ⅛ to as large as ¾ in. in diameter. Physical properties are listed in Table 4 and chemical resistance properties in Table 5.

* *Chemical and Metallurgical Engineering*, McGraw-Hill Publishing Co., Inc., New York, N. Y.
† Haveg.

TABLE 1. EFFECT ON PLASTICS OF IMMERSION FOR

	Phenol-Formaldehyde Molded	Phenol-Formaldehyde Cast	Phenol-Formaldehyde Laminated	Urea-Formaldehyde Molded	Urea-Formaldehyde Laminated	Vinyl Chloride-Acetate Resin
30% Sulfuric acid	Surface roughened	None	Edges swollen	Surface roughened	Surface attacked	None
3% Sulfuric acid	Surface roughened	None	Edges swollen	Surface roughened	Surface attacked	None
10% Nitric acid	Surface roughened	None	Edges swollen	Surface roughened	Delaminated	None
10% Hydrochloric acid	Surface roughened	None	Edges swollen	Surface roughened	Delaminated	None
5% Acetic acid	None	None	Edges swollen	None	None	None
Oleic acid	None	None	None	None	None	None
10% Sodium hydroxide	Decomposed	Decomposed	Delaminated	None	Surface attacked	None
1% Sodium hydroxide	Surface roughened	Decomposed	Edges swollen	None	None	None
10% Ammonium hydroxide	Surface dulled	Discolored	Discolored edges swollen	None	None	None
2% Sodium carbonate	None	Discolored	Discolored	None	None	None
10% Sodium chloride	None	None	Edges swollen	None	None	None
3% Hydrogen peroxide	None	Discolored	None	Surface dulled	Delaminated	None
Distilled water	None	None	None	None	None	None
50% Ethyl alcohol	None	None	None	None	None	None
95% Ethyl alcohol	None	None	None	None	None	None
Acetone	None	Softened	Blistered	None	None	Dissolved
Ethyl acetate	None	None	None	None	None	Decomposed
Ethylene chloride	None	None	None	None	None	Dissolved
Carbon tetrachloride	None	None	None	None	None	None
Toluene	None	None	None	None	None	Soft, rubbery
Gasoline	None	None	None	None	None	None

* From a paper, *Resistance of Plastics to Chemical Reagents*, by G. M. Kline, R. C. Rinker and H. F.

7 DAYS IN CHEMICAL REAGENTS AT 25° C (77° F)*

Vinyl Butyral Resin	Methyl Methacrylate Resin	Styrene Resin Molded	Cellulose Nitrate	Cellulose Acetate	Ethylcellulose No. 1	Cold-Molded Phenolic	Casein Plastic
None	None	None	None	Crazed; softened	None	None	Rubbery
Cloudy	None	None	None	Swollen	None	None	Swollen; rubbery
Cloudy	None	None	None	Decomposed	None	None	Swollen; cracked
Cloudy	None	None	None	Decomposed	None	Cracked on drying	Swollen; cracked
Cloudy	None	None	None	Swollen	None	None	Rubbery; split
Tacky	None	None	None	None	Decomposed	None	None
None	None	None	Crazed	Decomposed	None	Decomposed	Decomposed
Slightly cloudy	None	None	Crazed	Surface attacked	None	Decomposed	Broken up
Opaque	None	Discolored	Crazed; discolored	Opaque; soft	None	None	Swollen; split
Slightly cloudy	None	None	None	Swollen	None	None	Swollen; rubbery
None	None	None	None	None	None	None	None
Cloudy	None	None	None	None	None	None	Swollen; rubbery
Cloudy	None	None	None	None	None	None	Swollen; rubbery
Swollen; rubbery	Slightly swollen	None	None	Partly dissolved	Swollen; cracked	None	Swollen; rubbery
Dissolved	Surface attacked	None	Dissolved	Partly dissolved	Dissolved	None	None
Swollen; opaque	Dissolved	Dissolved	Dissolved	Dissolved	Dissolved	None	None
Decomposed	Dissolved	Dissolved	Dissolved	Dissolved	Dissolved	None	None
Decomposed	Dissolved	Dissolved	Partly dissolved	Soft; swollen	Dissolved	None	None
Swollen; rubbery	Surface attacked	Dissolved	None	None	Dissolved	None	None
Swollen; rubbery	Dissolved	Dissolved	Partly dissolved	None	Dissolved	None	None
None	None	Partly dissolved	None	None	Swollen; cracked	None	None

Meindl, presented before Chicago meeting of The American Society for Testing Materials, June 24, 1941

Because of its acid and alkali resistance, the use of vinyl resin filter cloth in processes involving strong acid or alkali solutions eliminates the expense of frequent replacements. These filter cloths are being used in connection with dyestuffs, bleaches, pigments, pharmaceuticals, concentrated caustics, and phosphoric acid.

TABLE 2. PROPERTIES OF PHENOL–FORMALDEHYDE–ASBESTOS PLASTICS*

Properties†	Haveg 41	Haveg 43	Haveg 50
Specific gravity	1.6	1.6	1.6
Tensile strength, psi	2,500	2,500	1,800
Compressive strength, psi	10,500	8,000	6,000
Flexural strength, psi	5,600	4,500	3,500
Shearing strength, psi	3,500	3,000	2,500
Modulus of elasticity	1,000,000	850,000	....
Thermal conductivity, Btu/(hr) (sq ft) (°F/ft)	0.203	0.607	0.203
Coefficient of expansion, per °F 18 $\times$ 10^{-6}; per °C 33 $\times$ 10^{-6}			

* Copyrighted, *Chemical and Metallurgical Engineering.*
† 41 is acid-resistant type; 43 is primarily for hydrofluoric acid and related materials; 50 is alkali-resistant type.

TABLE 3. EFFECT OF TOTAL IMMERSION ON ACRYLIC PLASTICS*

Solution†	% Wt. Gain‡
30% Sulfuric acid	0.6
3% Sulfuric acid	1.0
10% Hydrochloric acid	0.7
10% Sodium hydroxide	0.8
1% Sodium hydroxide	1.0
10% Nitric acid	0.9
5% Acetic acid	1.0
2% Sodium carbonate	1.0
10% Sodium chloride	0.9
10% Ammonium hydroxide	0.9
3% Hydrogen peroxide	1.0
100% Distilled water	0.9

* Approved tests of the Committee on Plastics of the American Society for Testing Materials using pieces of material 1 $\times$ 3 $\times$ 0.125 in. totally immersed in the various chemical solutions for 192 hr at 25° C. Data supplied for Plexiglas by Rohm & Haas Co., Philadelphia, Pa.
† All concentrations given in percentage by weight in distilled water.
‡ A gain in weight of 1% or less is considered negligible except in unusual applications. None of the above solutions appreciably affects the appearance or strength characteristics.

The use of plastic-sprayed coatings as linings for car tanks, drums, and process equipment has shown many advantages in the field of corrosion protection. These coatings are usually combinations of several inert resins. The vinyl and phenolic resins are those most used for this purpose at present. Plastic coatings find their greatest use at atmospheric temperatures, although the phenolic resins have some applications for above-room temperature service.

TABLE 4. AVERAGE PROPERTIES OF SOME PLASTIC MATERIALS

Material	Physical Constants			Mechanical Properties							Chemical Properties — Resistance to $P = poor, F = fair, G = good$						
	Specific Gravity	Thermal Conductivity,* gram-cal/(sec)(sq cm)(°C/cm)	Thermal Expansion,† cm/cm/°C × 10^{-5}	Tensile Strength, 10^{-3} psi	Tensile Modulus of Elasticity, 10^6 psi	Impact Strength, ft-lb/notch-in.	Hardness, Brinell, 2.5-mm ball, 25-kg load	Strength-Weight Ratio, Tensile Strength, 1000 sp. gr.	Maximum Operating Temperature, °F, Sulfur-Free Atmospheres	Water Absorption, % in 24 hr	Weak Mineral Acids	Strong Mineral Acids	Oxidizing Acids	Weak Alkalies	Strong Alkalies	Organic Solvents	Weathering
Acrylic acid resins	1.18	6.0	8.5	8.0	0.5	0.3	19.0	5.9	160	0.4	G	G	P	G	G-F	P	F
Casein	1.35	Very low	5.5	10.0	0.5	1.0	23.0	7.4	110	10.5	F	P	P	P	P	P	P
Cellulose acetate	1.32	6.2	12.0	7.5	0.2	2.8	10.0‡	5.2	180	3.4	F	P	P	F	P	P	F
Cellulose acetate butyrate	1.18	6.2	13.5	5.0	0.2	3.2	9.0	4.2	180	1.9	F	P	P	F	P	P	F
Cellulose nitrate (pyroxylin)	1.38	4.1	14.0	8.5	0.3	5.0	9.5‡	6.2	140	1.3	F	P	P	G	F	P	P
Ethylcellulose	1.13	5.1	12.0	5.5	0.3	3.6		4.9	170	1.5	F	P	P	G	F	P	F
Phenolics, cast, electrical, and mechanical grades	1.31	4.0	9.5	9.3	0.4	0.6		7.1	160	0.5	F	F	F-P	P	P	P	F
Phenolics, laminated, paper-base	1.33	6.5	2.1	12.5	1.7	4.1	32.0	9.4	250	4.7	G	F	F	F	P	F-G	F
Phenolics, molded, fabric-filled	1.39	4.0	4.0	6.8	1.0	2.8	36.0	4.9	230	1.5	G	G	P	F	P	G	F
Phenolics, molded, mineral-filled	1.84	14.0	3.3	6.0	2.8	0.5		3.3	390	0.2	F	G	P	F	P	G	F
Phenolics, molded, woodflour-filled	1.39	8.0	5.6	6.5	1.3	0.3	38.0	4.7	250	0.4	F	F	P	F	P	G	F
Styrene resins	1.06	1.9	7.0	7.0	0.3	0.4	25.0	6.6	170	0	G	F	P	G	G	F	F
Urea resins	1.48	7.1	2.8	6.3	1.4	0.3	51.0§	4.3	170	2.0	F	P	P	F	P	F-P	F
Vinyl resins	1.45	4.0	16.0	9.0	0.4	0.8	13.5	3.6	140	0.3	G	G-F	F	G	G	F-P	F-P
Vinylidene chloride resins	1.72	2.2	15.8	5.5	0.2	5.0		3.2	180	0	G	F	F	G	F	F-P	G-F

* To convert to Btu/(hr)(sq ft)(°F/ft), multiply by 242.
† To convert to in./in. °F, multiply by 0.555.
‡ 10-kg load.
§ 500-kg load.

TABLE 5. CHEMICAL RESISTANCE OF VINYLIDENE CHLORIDE PIPE

At Room Temperature After 3 Months Exposure

Reagent	Wt. Change*	Stability Rating†
	(Calc. to % change per yr)	
98% (conc.) H_2SO_4	+0.3	Good
60% H_2SO_4	+0.2	Excellent
10% H_2SO_4	+0.3	Excellent
35% (conc.) HCl	+0.5	Excellent
30% H_2SO_4	+0.2	Excellent
10% HCl	+0.3	Excellent
65% (conc.) HNO_3	+0.5	Excellent
10% HNO_3	+0.2	Excellent
Glacial acetic	+0.7	Excellent
10% Acetic	+0.4	Excellent
5% H_2SO_3	+0.5	Excellent
Conc. oleic	.....	Excellent
50% NaOH	+0.5	Fair
10% NaOH	+0.3	Good
28% NH_3	Unsuitable	Unsuitable
10% NH_3	Poor	Poor
Ethyl alcohol	−0.3	Excellent
Ethyl acetate	+0.5	Fair
Acetone	+2	Fair
Methyl Iso-butyl Ketone	+3	Fair
Carbon tetrachloride	+10	Good
Ethylene dichloride	+10	Poor
Diethyl ether	−3	Poor
Dioxane	+13	Unsuitable
Benzene	+4	Fair
o-Dichlorobenzene	+4	Poor
Ethyl gasoline	.....	Excellent
Turpentine	.....	Excellent
Triethanolamine	−0.5	Excellent
Lubricating oil	.....	Excellent
Linseed oil	.....	Excellent
Bromine water	+3	Unsuitable
Chlorine water	+0.5	Unsuitable
Bleaching solution	+0.4	Excellent
10% Duponol	.....	Excellent
10% Zinc hydrosulfite	+0.5	Excellent
15% $CaCl_2$	+0.4	Excellent
15% $FeSO_4$	+0.7	Good
Water-air	−0.02 to +0.11 (6 months)	Excellent
Air	0.01 to −0.10 (6 months)	Excellent

* Such information should not be taken too literally and, more often than not, a general rating, as the third column, will give a better picture of what the true comparison is. This is so since the number of properties that must be considered in connection with chemical resistance may affect not only weight change or volume change, but also tensile and impact strength, electrical properties, color, flexibility, etc. The general rating is more or less a combination of all those factors.

† The stability rating given is based upon observed changes in color, weight, dimension, tensile strength, and hardness of the samples tested.

Porcelain

JAMES A. LEE*

Porcelain is an extremely dense and homogeneous body which permits the fabrication of such pieces as rolls, ground and lapped down to a mirror finish for such jobs as coating photographic papers and carrying the tender filaments from rayon baths. A typical analysis is silica, 69%; alumina, 26%, and alkalies, 5%. Physical data are given in Table 1.

TABLE 1. CHEMICAL PORCELAINWARE*

Specific gravity	2.41
Ultimate tensile strength, psi	5,000–8,000
Ultimate compressive strength, psi	100,000
Modulus of rupture, psi	12,000–15,000
Modulus of elasticity, psi	10,400,000
Specific heat	0.2
Thermal cond., Btu/(hr) (sq ft) (°F/ft)	0.7
Linear thermal expansion, per °F	0.0000023
Water absorption, %	0

* Copyrighted, *Chemical and Metallurgical Engineering*. Data supplied by Lapp Insulator Co.

Porcelain is unaffected by all acids except hydrofluoric, but is only slightly resistant to alkali solutions.

This material is now available in numerous types and sizes of plant equipment. There is a complete line of process valves, including Y- and angle-type for shutoff service, as well as pop-type safety valves, check valves, flush bottom valves for tanks, and motor control valves. Pressure piping with a complete line of fittings up to 8 in. are available. In addition there is the usual range of kettles, tanks, etc., that are made in many other ceramics. However, there is some limit in size. At present 63 gal in a single vessel is the maximum.

Porcelain is limited in mechanical strength and, therefore, is somewhat vulnerable to both mechanical abuse and thermal strains. About 150° C (300° F) is a reasonable limiting temperature for application of this material with a sudden temperature change not to exceed about 50° C (90° F). As temperature increases above these values, the problem of avoiding harmful thermal shock increases.

Natural and Synthetic Rubber†

JAMES A. LEE*

Natural rubber has been used for the construction of chemical equipment for many years. More recently several synthetic rubber products have been developed and are now available for services where their special characteristics make them preferable. Among the rubber-like materials are Buna S, Buna N, Butyl, Neoprene, Thiokol. The physical properties of these and other rubber-like substances are given in Table 1.

Natural rubber may be molded into any form and compounded for special purposes. No one type is applicable to all conditions for maximum performance. For example,

* *Chemical and Metallurgical Engineering*, McGraw-Hill Publishing Co., Inc., New York, N. Y.
† Refer also to *Rubber and Rubber-like Coatings*, p. 902.

TABLE 1. PHYSICAL PROPERTIES OF

Material	Specific Gravity of Base Material	Tensile Strength, psi	Hardness, Shore Durometer	Maximum Temp. for Use, °F†
Chemigum, oil-resistant	1.0–1.5	800–4,000	30–90	300
Chemigum, tire	1.0–1.15	1,000–4,000	50–65	450
GR–I (Butyl)	0.91	500–3,000	15–90	250–300
GR–M (Neoprene)	1.25–1.30	1,000–4,500	10–95	300
GR–N (Perbunan)	0.96	500–5,000	30–90	300
GR–P (Thiokol FA)	1.34	1,400	25–90	200
GR–P (Thiokol ST)	1.27	500–2,000	30–90	250–300
GR–S (Buna S), hard	0.94	4,000–11,000	70–95§	220
GR–S (Buna S), soft	0.94	500–3,000	25–95	300
Hycar OR–15, soft	1.00	500–4,000	20–95	300
Hycar OR–25, soft	0.99	500–3,000	20–95	300
Hycar OR–15, hard	1.00	4,000–11,000	70–95	275
Hycar OS–10, soft	0.98	500–3,500	20–95	300
Koroseal, soft	1.40	500–2,500	30–80	190
Koroseal, hard	1.40	2,000–9,000	80–100	212
Pliolite, No. 40	1.06	4,000–5,000	....	160–248
Resistoflex	1.26	2,000–5,000	55–95	250
Tygon T	1.33–1.36	9,000	....	175
Vistanex, medium	0.9	200	....	....
Vistanex, high	0.9	550	....	....
Natural rubber, hard	0.93	4,000–11,000	70–95§	220
Natural rubber, soft	0.93	500–5,000	20–95	300

SYNTHETIC AND NATURAL RUBBERS*

Dielectric Strength, volts/mm	Effect of Heat	Abrasion Resistance	Effect of Sunlight‡	Effect of Aging	Machining Qualities
....	Stiffens	Excellent	Equal to rubber	Stiffens	Can be ground
....	Stiffens	Good	None	Better than rubber	....
25,000	Stiffens slightly	Excellent	None	Highly resistant	Can be ground
....	Stiffens slightly	Excellent	None	Highly resistant	Can be ground
....	Stiffens	Excellent	Slight	Highly resistant	Can be ground
....	Hardens slightly	Fairly good	None	None	Excellent
....	Hardens slightly	Good	None	None	Excellent
....	....	....	....	Highly resistant	Excellent
....	Stiffens	Excellent	Deteriorates	Highly resistant	Can be ground
....	Stiffens	Excellent	Slightly better than natural rubber	Highly resistant	Can be ground
....	Stiffens	Excellent	Slightly better than natural rubber	Highly resistant	Can be ground
....	....	....	....	Highly resistant	Excellent
....	Stiffens	Excellent	Deteriorates	Highly resistant	Can be ground
15,000–30,000	Softens	Good	None	Highly resistant	Can be ground
30,000–50,000	Softens	Excellent	None	Highly resistant	Good
....	Softens	....	None	None	....
6,000–10,000	Softens	Good	None	None	....
35,000–50,000	Softens	Good	None	....	Excellent
....	....	....	None	Better than rubber	Cannot be machined
....	....	....	None	Better than rubber	Cannot be machined
....	....	....	....	Highly resistant	Excellent
....	Softens	Excellent	Deteriorates	Moderately resistant	Can be ground

* Copyrighted, *Chemical and Metallurgical Engineering*.
† Maximum temperature suitable for service depends greatly upon the exact service conditions. Maximum temperature for use as a packing can be much higher than the maximum temperature suitable for tank lining. Individual cases should be referred to the supplier for recommendations.
‡ Effect of exposure to sunlight under tension.
§ Type "D".

linings range in characteristics from a soft, pure gum, non-contaminating material to flexible hard rubber. In the case of metallic equipment of irregular shapes, adherent protective coatings of soft or hard rubber are readily applied by the electrolytic anode process. Remarkably high tear resistance is characteristic of anode rubber. Physical properties are given in Table 1 and chemical resistance in Table 2.

Buna S rubber is a copolymer of butadiene and styrene. It is not oil-resistant, but does have good resistance to abrasion and chemicals, and good electrical properties. Hard rubber products can be made from Buna S. It can be bonded to metal satisfactorily and hence is useful as a tank or pipe lining. It is useful in handling the following solutions: (1) All inorganic acids except strong oxidizing acids such as nitric and chromic. (2) All inorganic salt solutions except those which are strongly oxidizing. (3) Plating solutions. (4) Inorganic bases, such as sodium hydroxide. (5) Many organic acids, such as acetic. (6) Pickling solutions. (7) Chlorine water and hypochlorite solutions.

Buna N or *Perbunan* is a copolymer of butadiene and acrylonitrile. It will vulcanize to a hard product. Vulcanized Buna N is very resistant to aliphatic oils, quite resistant to heat, and has low permanent set or good cold flow resistance. Methods have been devised for bonding this rubber to metal. The heat conductivity is about 20% higher than that of natural rubber.

Perbunan has excellent aging properties and possesses good heat and fatigue resistance. With proper compounding it can be used at temperatures up to 150° C (300° F) and will remain flexible at temperatures as low as $-45°$ C ($-45°$ F). Perbunan compounds of the ebonite or hard rubber type can be used for much more severe temperature conditions than natural rubber ebonite compounds.

Perbunan is unaffected by dilute acids or alkalies, or salt solutions of any concentration. It swells very slightly in aliphatic hydrocarbons, vegetable and animal oils, and fats. The reduction in physical properties as a result of swelling is small, making Perbunan especially suitable for gasoline and oil-resistant applications. Crude Perbunan or uncured Perbunan compounds are soluble in aromatic hydrocarbons. However, cured compounds swell to a lesser extent than cured natural rubber compounds. It is superior to natural rubber for mixtures of gasoline with aromatic hydrocarbons.

Chlorinated hydrocarbons and organic bases have a strong swelling action on Perbunan. Ketones, organic acids, alcohols, and esters have a greater swelling effect than on natural rubber.

Butyl rubber is a hydrocarbon polymer obtained by the copolymerization of an olefin and a diolefin. It undergoes thermal curing in the presence of sulfur.

The general physical properties of vulcanized Butyl rubber compounds can be compared with natural rubber as follows. Elasticity and extensibility is in the range of natural rubber; resilience is lower than in natural rubber at room temperatures but increases remarkably with increasing temperatures. Abrasion resistance is equal to that of natural rubber for some services and slightly inferior for other services. Compression set is somewhat higher than in natural rubber. Tear resistance is equal to that of natural rubber as is flexibility at low temperatures (Table 1).

Butyl rubber is far superior to natural rubber for resistance to: (1) water absorption; (2) aging in storage and sunlight; (3) the action of air and ozone; (4) deterioration by heat; (5) the action of nitrogen-containing solvents such as nitrobenzene and aniline; (6) oxygenated solvents such as ethers, alcohols, and esters; (7) swelling in vegetable and animal oils; (8) many corrosive chemicals; (9) permeability by gases.

Several *Neoprene* rubbers (polymers of chloroprene) are in use. Neoprene is better

NON-METALLIC MATERIALS AND SEMI-METALS

than natural rubber with respect to oils; deterioration by heat, oxygen, and sunlight; flexing life; and gas diffusion. Animal, vegetable, and petroleum-base products cause slight swelling but have little effect on the physical properties. Neoprene does not dissolve, become gummy, or slough off. In general, it may be used in contact with inorganic chemicals, and even strong alkalies may be successfully handled. Mineral acids cause deterioration when used in concentrations above 50%. Strong oxidizing acids such as concentrated sulfuric, nitric, and chromic should be avoided. Halogens in liquid form cause embrittlement; however, dilute gas and aqueous solutions may be handled satisfactorily.

Neoprene resists the attack of most organic compounds, the more highly saturated compounds having less effect than the less saturated. It is not recommended for use with the chlorinated hydrocarbons or with creosote. Certain aromatic solvents such as benzol also cause rapid swelling.

Neoprene linings are satisfactory under the following conditions:

Chemical	Concentration	Probable Safe Max. Temp.	
		°C	°F
Chlorine water	Up to 3%	28	82
Hydrochloric acid	Up to 20%	100	212
Hydrochloric acid	Over 20%	28	82
Phosphoric acid	Up to 85%	105	220
Sulfuric acid	Up to 50%	105	220
Aluminum chloride	Up to 25%	70	160
Ammonium nitrate	Up to 50%	105	220
Calcium chloride	Up to sat.	105	220
Ferric chloride	Up to 25%	93	200
Silver plating solution	Standard	105	220
Sea water		105	220
Sodium chloride	Up to sat.	105	220
Sodium hydroxide	Up to 70%	105	220
Sodium sulfide	Any	93	200
Acetone		Boiling point	
Formaldehyde	Up to 40% aqueous sol.	28	82

There are a number of polysulfide rubbers known as *Thiokols*. They are made from an organic dihalide and sodium polysulfide. This group is best known for its exceptional resistance to oils — both alphatic and aromatic. They are quite resilient but have poor resistance to high temperatures and severe mechanical stresses.

However, unlike natural rubber, the Thiokols are practically unaffected by petroleum hydrocarbons and most commercial solvents. They are resistant to alcohol, esters and ketones. The swelling of these products in aromatic hydrocarbons varies from practically none to around 100% increase in volume, depending on the type. Even when Thiokol swells in aromatic solvents, it still retains good physical characteristics. It is not materially affected by carbon tetrachloride but is attacked by ethylene dichloride in varying degrees. Thiokol is not affected by sunlight, air, ozone, or ultraviolet light.

It can be compounded to five various hardnesses and characteristics suitable for each particular application. Although not possessing original tensile strength as great as natural rubber, it exceeds natural rubber (it is said) in both strength and abrasion resistance when in contact with gasoline and oils.

Silicone rubber is an organo-silicone oxide polymer characterized by its exceptional heat resistance. This material remains resilient at temperatures up to $260°$ C ($500°$ F) and retains flexibility at temperatures down to $-60°$ C ($-70°$ F).

TABLE 2. CHEMICAL RESISTANCE OF NATURAL RUBBER COMPOUNDS*

	Concentration by Weight	Maximum Temperature, °F	Degree of Vulcanization	Design of Compound
1. Solutions of Inorganic Acids				
Arsenic acid	Any concentration	150	Soft	Specific
			Hard	General
Carbonic acid	Up to saturation at atmospheric pressure	150	Soft or hard	General
Chlorine water (hypochlorous acid)	Up to saturation at atmospheric pressure	100	Soft	Specific
		150	Hard	Specific
Fluoboric acid	Any concentration	150	Soft or hard	General
Fluosilicic acid	Any concentration	150	Soft or hard	General
Hydrobromic acid	Any concentration	100	Soft	Specific
		150	Hard	Specific
Hydrofluoric acid	Up to 50 %	150	Soft or hard	General
Hydrogen sulfide water	Up to saturation at atmospheric pressure	150	Hard	General
Hydrochloric acid	Any concentration	150	Soft or hard	General
Phosphoric acid	Up to 85%	150	Soft	Specific
			Hard	General
Sulfuric acid	Up to 50%	150	Soft or hard	General
Sulfurous acid	Up to saturation at atmospheric pressure	150	Hard	Specific
2. Solutions of Inorganic Salts and Alkalies				
Aluminum chloride	Up to saturation	150	Soft or hard	General
Aluminum sulfate	Up to saturation	150	Soft or hard	General
Alums	Up to saturation	150	Soft or hard	General
Ammonium chloride	Up to saturation	150	Soft or hard	General
Ammonium hydroxide	Up to saturation	100	Hard	General
Ammonium persulfate	Up to saturation	100	Soft	General
		150	Hard	General
Ammonium sulfate	Up to saturation	150	Soft or hard	General
Barium sulfide	Up to saturation	150	Soft or hard	General
Calcium bisulfite	Up to saturation	150	Hard	Specific
Calcium chloride	Up to saturation	150	Soft or hard	General
Calcium hypochlorite	Up to saturation	150	Soft	Specific
			Hard	General
Sodium hydroxide	Up to saturation	150	Soft or hard	General
Potassium hydroxide	Up to saturation	150	Soft or hard	General
Copper chloride (cupric)	Up to saturation	150	Hard	General
Copper cyanide (in solution with alkali cyanides)	Up to saturation	150	Soft or hard	General
Copper sulfate (cupric)	Up to saturation	150	Soft or hard	General
Ferric chloride	Up to saturation	150	Soft	Specific
			Hard	General
Ferrous sulfate ("copperas")	Up to saturation	150	Soft or hard	General
Nickel acetate	Up to saturation	150	Hard	Specific
Plating solutions: Brass, Cadmium, Copper, Gold, Lead, Nickel, Silver, Tin, Zinc	....	150	Soft or hard	General
Potassium cuprocyanide	Up to saturation	150	Soft or hard	General
Potassium dichromate	Up to saturation	150	Hard	General
Sodium (or potassium) antimonate	Up to saturation	150	Soft or hard	General

TABLE 2. CHEMICAL RESISTANCE OF NATURAL RUBBER COMPOUNDS*
(Continued)

	Concentration by Weight	Maximum Temperature, °F	Degree of Vulcanization	Design of Compound
Sodium (or potassium) bisulfite	Up to saturation	150	Hard	General
Sodium (or potassium) acid sulfate	Up to saturation	150	Soft or hard	General
Sodium (or potassium) chloride	Up to saturation	150	Soft or hard	General
Sodium (or potassium) cyanide	Up to saturation	150	Soft or hard	General
Sodium (or potassium) hypochlorite	Up to saturation	150	{ Soft / Hard	Specific / General
Sodium (or potassium) sulfide	Up to saturation	150	Soft or hard	General
Sodium (or potassium) sulfite	Up to saturation	150	Soft or hard	General
Sodium (or potassium) thiosulfate	Up to saturation	150	Soft or hard	General
Silver nitrate	Up to saturation	150	Soft	Specific†
Tin chloride (either stannous or stannic)	Any aqueous solution	150	{ Hard / Soft or hard	General / General
Zinc chloride	Up to saturation	150	Soft or hard	General
Zinc sulfate	Up to saturation	150	Soft or hard	General
3. Organic Materials				
Acetic acid	Any concentration	150	Hard	Specific
Acetic anhydride	....	150	Hard	Specific
Acetone	Any concentration	150	{ Soft / Hard	Specific / General
Amyl alcohol	Any concentration	150	{ Soft / Hard	Specific / General
Aniline hydrochloride	Any concentration	150	Soft or hard	General
Butyl alcohol	Any concentration	150	{ Soft / Hard	Specific / General
Casein	Any concentration	150	Soft or hard	General
Castor oil	....	150	Hard	Specific
Citric acid	Up to saturation	150	Soft	Specific
Coconut oil	....	150	Hard	Specific
Cottonseed oil	....	150	Hard	Specific
Dyestuffs	....	150	Hard	Specific
Ethyl alcohol	Any concentration	150	{ Soft / Hard	Specific / General
Ethylene glycol	Any concentration	150	Soft or hard	General
Formaldehyde (formalin)	40% aqueous solution	100	Hard	Specific
Formic acid	Any concentration	100	Hard	Specific
Furfural	....	100	Hard	Specific
Gallic acid	Up to saturation	150	Soft or hard	General
Glucose	Any concentration	150	Soft or hard	General
Glue	Any concentration	150	Soft or hard	General
Glycerin	Any concentration	150	Soft or hard	General
Lactic acid	Any concentration	150	Hard	Specific
Malic acid	Up to saturation	150	Soft or hard	Specific
Methyl alcohol	Any concentration	150	{ Soft / Hard	Specific / General
Mineral oils	....	100	Hard	Specific
Propyl alcohol	Any concentration	150	{ Soft / Hard	Specific / General
Soaps	Any concentration	150	Soft or hard	General
Tannic acid	Up to saturation	150	Soft or hard	General
Tartaric acid	Up to saturation	150	{ Soft / Hard	Specific / General
Triethanolamine	Any concentration	150	Soft or hard	General
Vinegar	....	150	Hard	Specific

* Copyrighted, *Chemical and Metallurgical Engineering*.
† If discoloration is to be avoided.

Wood

JAMES A. LEE*

Wood is largely used in the chemical process industries for tanks, pipe, and cooling towers, and to some extent for filter presses, agitators, and other equipment. In general it may be said that a solution which destroys the fiber of a wooden vessel dissolves either the cellulose or the lignin or both. Hardwoods contain more lignin, softwoods contain more cellulose; hence the nature of the solutions determines the type of wood to be used in a particular installation. Wood is usually more resistant to dilute solutions than to concentrated solutions, all other things being equal.

Sulfuric and hydrochloric acids appear to affect all woods adversely, although dilute solutions are not injurious to any serious extent. Acetic acid at any concentration seems to be without destructive action on any of the woods. The wood selected for use with this acid should be that which yields the least coloring matter to the solution.

In general, all woods appear to withstand the action of acids and acid solutions better than they do strongly alkaline and caustic solutions. The latter attack and disintegrate the fiber, although dilute or weak alkaline solutions may not affect the wood to any great extent. Lime and magnesia solutions, however, are most detrimental to wood, whereas potassium and sodium cyanide solutions under normal conditions are practically without action.

Strong oxidizing agents should not be brought into contact with wood as they rapidly destroy it by structural decomposition. The cellulose is easily converted to oxycellulose, a brittle substance with practically no coherent properties. Nitric acid, chromic acid, potassium permanganate, and free chlorine are particularly active in this respect.

At ordinary temperatures a solution may have no appreciable action on wood for an indefinite period, but an increase in temperature may cause immediate and rapid destruction. The effect of temperature of various media in contact with wood is illustrated by information given in Table 1.

Plywood offers many opportunities for engineering design. Because of the wide variety and many types of construction, it is preferable to give the characteristics of wood and adhesives rather than to detail properties for special methods of construction. (See Tables 2 and 3.)

* *Chemical and Metallurgical Engineering*, McGraw-Hill Publishing Co., Inc., New York, N. Y.

TABLE 1. CONDITION OF WOODS AFTER 31 DAYS IMMERSION IN COLD SOLUTIONS*,† (Examined after 7 days drying)

	Fir	Oak	Oregon Pine	Yellow Pine	Spruce	Redwood	Maple	Cypress
Hydrochloric acid, 5%	NAC	NAC	NAC	SS	SS	SS	NAC	NAC
10%	NAC	NAC	NAC	SS	SS	SS	NAC	NAC
50%	SS,SB,SWF	SS,WF	S,WF	S,WF	S,WF	S,WF	S,WF	S,WF
Sulfuric acid, 1%	NAC	NAC	NAC	SS	SS	NAC	NAC	SS,SB
5%	SS	SS	SS	SS,SB	SS,SB	SS,SB	SS,SB	SS,SB
10%	S,FSD	S,FSD	S,FSD	S,FSD	S,FSD	S,FSD	S,FSD	S,FSD
25%	SSp,FSD	SSp,FSD	SSp,FSD	SSp,FSD	SSp,FSD	SSp,FSD	SSp,FSD	SSp,FSD
Caustic soda, 5%	S,NAC	MSh,SWp	SS	SS,FSD	SSp,FSD	SSp,FSD	MSh	SSp,FSD
10%	S,FSD	MSh,WF,Horny	SS	SS,SB,FSD	SS,SB,FSD	SS,SB,FSD	MSh	S,SB,FSD
Alum, 13%	NAC	NAC	NAC	NAC	NAC	NAC	NAC	NAC
Sodium carbonate, 10%	SB, GC	NAC	GC	SB,GC	SB,GC	SB,GC	GC	SB,GC
Calcium chloride, 25%	NAC	NAC	NAC	NAC	NAC	NAC	NAC	NAC
Common salt, 25%	NAC	NAC	SS,GC	SS,GC	SS,GC	SB,GC	NAC	NAC
Water	NAC	NAC	NAC	NAC	NAC	NAC	NAC	NAC
Sodium sulfide	SS,SB	MSh,WF	SB	SB	SB	SB	MSh,FSD	FSD

CONDITION OF WOODS AFTER 8 HOURS IN BOILING SOLUTIONS† (Examined after 7 days drying)

	Fir	Oak	Oregon Pine	Yellow Pine	Spruce	Redwood	Maple	Cypress
Hydrochloric acid, 10%	SB,S	FSD	FSD	FSD	FSD	FSD	FSD	FSD
50%	FD,Ch,B,S,NG	FD,Ch,B,S,NG	FD,Ch,B,S,NG	FD,Ch,B,S,NG	FD,Ch,B,S,NG	FD,Ch,B,S,NG	FD,Ch,B,S,NG	FD,Ch,B,S,NG
Sulfuric acid, 4%	SB,GC	SB,GC	SB,GC	SB,GC	SB,GC	SB,GC	SB,GC	SB,GC
5%	SS,GC	SB,GC	SB,GC	SB,GC	SB,FSD	SB,FSD	SB,GC	SB,FSD
10%	SS,GC	B,FD,Wpd,NG	Sp,FD,NG	B,Sp,FD,NG	B,Sp,FD,NG	SB,FSD	SB,FSD	B,FD
Caustic soda, 5%	SS	MSh	S	GC	S,GC	S,GC	Sh	SSp
Alum, 13%	SB,GC	NAC	NAC	SB,GC	SB,GC	SB,GC	NAC	SB,GC
Sodium carbonate, 10%	SB,GC	GC	GC	GC	GC	GC	GC	SB,GC
Calcium chloride, 25%	SB,GC	SB,SS,GC	NAC	SB,GC	SB,GC	NAC	NAC	SB,GC
Common salt, 25%	NAC	NAC	NAC	SB,GC	NAC	SB,GC	NAC	NAC
Water	NAC	NAC	NAC	SB,GC	NAC	NAC	NAC	NAC

* Copyrighted, *Chemical and Metallurgical Engineering*.

† The condition of eight varieties of woods used for tanks and other chemical-resistant uses are based on a report of James K. Stewart, consulting chemist to the Mountain Copper Co., Martinez, Calif. Tests were conducted on samples 1 × 4 × ¼ in. in size, seasoned and chosen so as to be as nearly as possible in the same physical condition as the woods would be when used for equipment construction. Results of the tests are described by terms explained in the following key:

Abbreviation Key

FD	— Fiber Disintegrated	NAC	— No Apparent Change	Sh	— Shrunk	SWF	— Slightly Weakened Fiber
FSD	— Fiber Slightly Disintegrated	NG	— No Good	Sp	— Spongy	SWp	— Slightly Warped
GC	— Good Condition	S	— Softer	SS	— Slightly Softer	WF	— Weakened Fiber
MSh	— Much Shrunk	SB	— Slightly Brittle	SSp	— Slightly Spongy	Wpd	— Warped
B	— Brittle						
Ch	— Charred						

TABLE 2. PROPERTIES

Wood (Common Name)	Specific Gravity, Oven-Dry, 12% Moisture	Weight per cu ft, Oven-Dry, 12% Moisture	Hardness, Load to Embed a 0.444-in. Diam. Ball to ½ Its Diam., lb		Static Bending			Impact Bending, Height of Drop of 50-lb Hammer to Cause Complete Failure, in.‡
			End	Side	Proportional Limit, psi	Modulus of Rupture†, psi	Modulus of Elasticity, 10^{-6} psi	
SOFTWOODS								
Bald cypress	0.46	32	660	510	7,200	10,600	1.44	24
Douglas fir	0.48	34	760	670	8,100	11,700	1.92	30
Hemlock, western	0.42	29	940	580	6,800	10,100	1.49	26
Larch, western	0.52	36	1,110	760	7,900	11,900	1.71	32
Pine, ponderosa	0.4	28	550	450	6,300	9,200	1.26	17
Redwood, virgin	0.4	28	790	480	6,900	10,000	1.34	19
Redwood, second growth	0.34	24	710	400	5,500	8,300	1.12	16
Spruce, Sitka	0.4	28	760	510	6,700	10,200	1.57	25
White-Cedar, Port Orford	0.42	29	730	560	7,700	11,300	1.73	28
HARDWOODS								
Ash, commercial white‖	0.6	42	1,720	1,320	8,900	15,400	1.77	43
Aspen ¶	0.38	26	510	350	5,600	8,400	1.18	21
Balsa ¶	0.15	11	.....	.....	.....	2,500	.....	.....
Basswood, ¶ American	0.37	26	520	410	5,900	8,700	1.46	16
Beech, American	0.64	45	1,590	1,300	8,700	14,900	1.72	41
Birch, yellow, sweet	0.62	43	1,480	1,260	10,100	16,600	2.01	55
Cherry, black	0.5	35	1,470	950	9,000	12,300	1.49	29
Chestnut, American	0.43	30	720	540	6,100	8,600	1.23	19
Cottonwood, ¶ eastern	0.4	28	580	430	5,700	8,500	1.37	20
Elm, American	0.5	35	1,110	830	7,600	11,800	1.34	39
Elm, rock	0.63	44	1,510	1,320	8,000	14,800	1.54	56
Hickory, shagbark	0.72	50	.....	.....	10,700	20,200	2.16	67
Khaya (African mahogany)	0.45	32	1,080	790	7,890	10,700	1.48	22
Magnolia, southern	0.5	35	1,280	1,020	6,800	11,200	1.4	29
Mahogany	0.7	49	1,350	1,330	7,100	9,600	1.2	.....
Maple, sugar (hard)	0.63	44	1,840	1,450	9,500	15,800	1.83	39
Maple, silver (soft)	0.47	33	1,140	700	6,200	8,900	1.14	25
Oak, commercial red	0.63	44	1,580	1,290	8,500	14,300	1.82	43
Oak, commercial white	0.68	48	1,520	1,360	8,200	15,200	1.78	37
Pecan	0.66	46	1,930	1,820	9,100	13,700	1.73	44
Sycamore	0.49	34	920	770	6,400	10,000	1.42	26
Sweetgum	0.49	34	950	690	8,100	11,900	1.49	32
Tupelo, water (Tupelo Gum)	0.5	35	1,200	880	7,200	9,600	1.26	23
Tupelo, black (Black Gum)	0.5	35	1,240	810	7,300	9,600	1.2	22
Walnut, black	0.55	38	1,050	1,010	10,500	14,600	1.68	34
Yellowpoplar ¶	0.4	28	560	450	6,100	9,200	1.5	20

* "Strength and Related Properties of Woods Grown in the United States," L. J. Markwardt and T. R. C. Wilson. Tech. Bull. 479, U. S. Dept. of Agriculture (September, 1935).

† Assumed to be equal to maximum tensile strength parallel to the grain, which is developed seldom in actual use; although tensile strength is actually higher than modulus of rupture for straight-grained material, it is reduced more seriously by steeper slopes of grain.

OF WOODS*

Compression			Ultimate Tensile Strength perpendicular to Grain, psi	Maximum Shearing Strength Parallel to Grain, psi	Cleavage, Load to Cause Splitting, lb/in. of Width	Workability§	Paint Holding§	Nailing§	Resistance to Decay§
Proportional Limit Parallel to Grain, psi	Ultimate Strength Parallel to Grain, psi	Proportional Limit Perpendicular to Grain, psi							
4,470	6,360	900	270	1,000	170	Fair	Very good	Splits rarely	Very good
6,450	7,420	910	300	1,140	180	Poor	Poor	Splits easily	Fair to Good
5,340	6,210	680	310	1,170	200	Fair	Fair	....	Poor
5,950	7,490	1,080	310	1,360	160	Poor	Poor	Splits, use blunt nails	Fair
4,060	5,270	740	400	1,160	220	Good	Fair	Does not split readily	Fair
4,560	6,150	860	240	940	150	Fair	Very good	....	Very good
3,750	5,240	640	280	930	160	Fair	Very good	....	Very good
4,780	5,610	710	370	1,150	210	Good	Fair	Little tendency to split	Fair
5,890	6,470	760	400	1,080	220	Fair	Excellent	Splits rarely	Very good
5,790	7,410	1,410	940	1,950	480	Poor	Good	Tendency to split	Poor
3,040	4,250	460	260	850	210	Good	Fair	Does not split	Very poor
......	......	......	......	......	......	....	....	....	Poor
3,800	4,730	450	350	990	230	Good	Very good	Does not split	Very poor
4,880	7,300	1,250	1,010	2,010	490	Poor	Good	Tendency to split	Poor
6,130	8,170	1,190	920	1,880	520	Poor	Good	Tendency to split	Poor
5,960	7,110	850	560	1,700	350	Fair	Good	Tendency to split	Good
3,780	5,320	760	460	1,080	250	Good	Good	Splits frequently	Very good
3,490	4,910	470	580	930	270	Good	Fair	Does not split	Very poor
4,030	5,520	850	600	1,510	......	Fair	Good	Difficult to split	Fair
4,700	7,050	1,520	......	1,920	......	Fair	Good	Difficult to split	Fair
......	9,210	2,170	......	2,430	......	Very poor	Good	Splits readily	Poor
......	5,680	980	......	1,340	......	Good	Good	Tendency to split	Good
3,420	5,460	1,060	740	1,530	430	Fair	Good	Tendency to split	Poor
......	6,240	1,770	......	1,480	......	Good	Good	Tendency to split	Good
5,390	7,830	1,810	......	2,330	......	Fair to Poor	Good to Fair	Tendency to split	Poor
4,360	5,220	910	500	1,480	340	Fair to Poor	Good to Fair	Tendency to split	Poor
4,580	6,760	1,250	800	1,780	410	Poor	Good	Tendency to split	Good
4,760	7,440	1,320	800	2,000	450	Poor	Good	Tendency to split	Good
5,180	7,850	2,130	......	2,080	660	....	....	....	....
3,710	5,380	860	720	1,470	400	Poor	Good	Fair	Poor
4,700	5,800	860	800	1,610	380	Poor	Good	Tendency to split	Fair
4,280	5,920	1,070	700	1,590	360	Poor	Good	Tendency to split	Fair
3,470	5,520	1,150	500	1,340	340	Poor	Good	Tendency to split	Fair
5,780	7,580	1,250	690	1,370	320	Fair	Stains well	....	Very good
3,550	5,290	580	520	1,100	280	Good	....	Does not split readily	Poor

‡ Standard test specimen 2 × 2 × 30 in. (A. S. T. M. D143–27).
§ These properties of plywoods will be different from those of normal wood, depending upon the adhesive used and the type of construction.
‖ Includes white, blue, and green.
¶ Soft species of hardwoods.

TABLE 3. PROPERTIES

Adhesive	Source	Form Supplied	Cold or Hot Press	Added Ingredients of Mix	Pressing Temp., °F	Pressing Time	Moisture Content (M.C.) Allowed in Veneers, %	Conditioning of Panels After Pressing	Dry Bond Strength
Animal	Hides and bones of animals cooked in water, evaporated, jellied, dried	Powder, pearl flake, shreds	Cold, and cold with warm cauls	Water, some preservatives	Cold with warm cauls	2–18 hr	2–3 for faces; 5 or less for cores and crossbands	Unnecessary for hot press; necessary for cold press. Dry to 7% M.C. for indoors, 12% M.C. for outdoors	Very high
Blood albumen	Dried beef blood	Powder, flake	Hot press	Water, alkalies, paraformaldehyde, corn syrup, invert sugar, hydrated lime, albuminous adhesives	180–250	3 min for hot press; varies with thickness of panel. Overnight cold press	Same as above	Same as above	Medium
Blood and phenol-formaldehyde combined	Phenol resin plus dried blood	Resin and blood supplied separately or mixed	Hot press	Water and reagents	220–250	3–11 min	2–5	Dead piled for 6–8 hr	High
Blood and urea-formaldehyde combined	Urea resin plus dried blood	Resin and blood separately	Hot press	Water, accelerators, wheat and rye flours	220–260	3–8 min	5–15	None usually	High
Casein	Precipitated from milk	Powder	Mainly cold; some hot	Water, formaldehyde, preservatives and alkalies (lime and sodium salt)	Ordinary room temp. for cold press, 180–250 hot press	2–18 hr for cold press; 3 min for thin panels, hot press	Same as Animal	Same as Animal	Medium
Phenol-formaldehyde resins	Phenol-formaldehyde condensation	Lumps, powder, liquid and film	Mainly hot; some cold	Alcohol and water, flour fillers	Room temp. –320	3–6 min for thin panels, hot press; overnight for cold press	2–14 for liquid resin. For film, cores and crossbands 5–10; faces 2–6	Generally none, sometimes dipped or sprayed with water; dead piled	Medium to high
Soybean and vegetable protein	Mainly soya bean meal; also peanut and cottonseed meal	Flour	Mainly cold; some hot	Water, lime, silicate of soda, carbon bisulfide	Cold at room temp. Hot press 180–250	2–5 hr cold; 3–8 min, hot	Same as Animal	Same as Animal	Medium

OF ADHESIVES*

Wet Bond Strength	Water Resistance	Mold Resistance	Staining Power	Heat and Fire Resistance	Exterior Durability	Relative Cost	Principal Uses	Special Features
None	Poor	Poor	Some staining	Poor	None	High	Lumber cores and joints	Relatively expensive. Easy to use
Medium	Medium	Poor to good	Low to high	Poor	Fair to good	Medium to high	Refrigerator, truck and car panels; special panels; packaging	Especially suitable for low-temperature hot pressing; most durable of adhesives except resins. Also mixed with casein and soybean flours
Medium	Good	Uncertain	No stain May diffuse	Moderate	Good	Medium	Douglas fir	Popular for Douglas fir because good durability is attained with fast low-temperature cycle
Medium	Good	Uncertain	No stain May diffuse	Moderate	Good	Medium	Panel stock, furniture	Rather critical to handle. Partially resistant to boiling water
Poor to medium	Poor to good	Generally poor	Generally marked	Poor	Poor	Medium to high	Furniture and interior panels, concrete forms	Superior to vegetable glue for wet strength
Excellent in both hot and cold water	Excellent	Excellent	Some staining with liquids; none with film	High	Excellent	High	2-ply facing; exterior panels; aircraft skins; boats	By far the most durable bond; resists all forms of deterioration
Poor to medium	Fair	Poor	Marked	Poor	Poor	Very low	Douglas fir panels; building boards; concrete forms	Mostly soybean meal used on Douglas fir

378 CORROSION IN LIQUID MEDIA, ATMOSPHERE, GASES

TABLE 3. PROPERTIES OF

Adhesive	Source	Form Supplied	Cold or Hot Press	Added Ingredients of Mix	Pressing Temp., °F	Time	Moisture Content (M.C.) Allowed in Veneers, %	Condition-ing of Panels After Pressing	Dry Bond Strength
Thermo-plastic resins	Chiefly plasticized polyvinyl butyralde-hyde	Resin solution in organic solvent	Hot press	None, thin with solvent if necessary	230–300	Few min to several hr—must be cooled in press or mold	3–12	Generally none, sometimes sprayed	Low to medium
Urea-formal-dehyde resins	Urea-formalde-hyde con-densation	Powder and liquid	Mainly hot; some cold	Water, accelerators, wheat and rye flours	210–240	3 min for thin panels, hot press; 8 hr for cold press	Wide range — 2–100 under special conditions	None	High
Vegetable starch	Vegetable carbo-hydrate. Base mainly cassava (tapioca); also potato flour	Flour	Cold	Water and alkalies (sodium hy-droxide)	Ordinary room temp.	4–18 hr	Same as Animal	Same as Animal	Medium

* Copyright 1942 by Bakelite Corporation, New York, N. Y.

Boron*

W. J. KROLL†

This metalloid resembles carbon and silicon in many respects.[1] Despite the fact that boron is widely available in nature, no suitable methods have been found up to now for producing the element in a state of high purity at low cost. Weintraub[2] made high-grade material by reducing boron trichloride with hydrogen in an arc. Recently, pure crystals of the element have been prepared by reduction on a heated filament.[3]

Commercial boron is prepared by reduction of boric oxide with magnesium and usually contains large amounts of impurities such as magnesium oxide and boron suboxide. Many samples of so-called pure boron contain less than 75% of the element.[4] Naturally the chemical properties of this material will be different from those of the massive crystalline form.

* Much information in this chapter was contributed by D. T. Hurd, Research Laboratory, General Electric Co., Schenectady. N. Y.
† Consulting Metallurgist, Albany, Ore.

[1] J. W. Mellor, *Comprehensive Treatise on Inorganic and Theoretical Chemistry* (*Boron*), Vol 5, Longmans, Green and Co., London, 1924.
[2] E. Weintraub, *Ind. Eng. Chem.*, **5**, 106–115 (1913).
[3] A. Laubengayer, D. Hurd, A. Newkirk, and J. Hoard, *J. Am. Chem. Soc.*, **65**, 1924 (1943).
[4] E. Winslow and H. Liebhafsky, *J. Am. Chem. Soc.*, **64**, 2725 (1942).

ADHESIVES* — Continued

Wet Bond Strength	Water Resistance	Mold Resistance	Staining Power	Heat and Fire Resistance	Exterior Durability	Relative Cost	Principal Uses	Special Features
Medium	Medium	High — not affected	None	Poor	Poor to good, depending on conditions	High	Molded plywood	Flows at elevated temperature. Must be cooled in press. Will reseal. Plies retain flexibility
Medium to high	Cold water, good; hot water, poor	High	None	Moderate	Good	Resin alone, high; with fillers, low	Wall panels; radio cabinets; furniture; boats and aircraft	Very economical when mixed with flour. Provides a highly water-resistant bond set hot or cold
None	Poor	Poor	Generally marked	Poor	None	Very low	Furniture panels; games; production work; packages	Inexpensive, popular adhesive. Most eastern plywood made with cassava flour

The corrosion properties of the element may be summarized as follows. Ignition of powdered commercial boron in air takes place at 700° C (1300° F), although the massive element in crystalline form is little affected at temperatures under 1200° C (2200° F). Boron is inert to the action of boiling hydrochloric acid and to hydrofluoric acid. It is slowly oxidized by hot concentrated nitric or sulfuric acid, but unless the material is finely powdered the reaction proceeds only at a very slow rate.

The massive element is inert to boiling concentrated alkali and reacts only slowly with fused sodium hydroxide below 500° C (930° F). It reacts vigorously with fused sodium peroxide and with a fused mixture of sodium carbonate and potassium nitrate.

The halogen and halogen hydried gases attack boron at elevated temperatures as do CO, CO_2, N_2, NO_2, SO_2, H_2S, and NH_3 although attack is much less rapid when the element is in other than finely divided form.

Boron forms metallic borides readily with most of the metals in the last four groups of the Periodic System.[5] These show corrosion properties in general similar to those of the element itself.

GENERAL REFERENCE

GMELIN, L., *Handbuch der anorganischen Chemie (Boron)*, 8th Ed., System 13, Verlag Chemie, Berlin, 1926.

[5] K. Becker, *Hochschmelzende Hartstoffe und ihre technische Anwendung*, Verlag Chemie, Berlin, 1935.

Silicon

W. J. KROLL*

The similarity of this metalloid to both boron and carbon appears in its ability to form hydrides, soluble alkali silicates, and volatile halides. Silicon can be produced on a large scale at fairly low cost and the commercial product is reasonably pure. It can be further purified to 99.8% silicon by extraction with concentrated hydrochloric and hydrofluoric acids. Impurities remaining are iron, aluminum, calcium, and manganese in addition to carbon and oxygen. The purest substance is made by dissociation of $SiCl_4$ in presence of hydrogen on a hot filament.

Silicon, being brittle, is formed only by casting. Recently, however, it has been successfully sintered by adding titanium hydride to the powder.

Most available literature on the chemical properties of silicon concerns the crystalline element in powder form. The resistance of silicon to corrosive media is excellent except in alkalies. Hot or cold water has no effect nor has hot or cold, dilute or concentrated hydrochloric, nitric, and sulfuric acids. Concentrated sulfuric acid reacts at high temperatures. Hydrofluoric acid does not react, but a mixture with nitric acid attacks silicon readily. Slight attack is observed for the fine colloidal material in presence of either oxidizing agents or of metals cathodic to silicon.[1] Copper sulfate, CrO_3, aqua regia, mixtures of sulfuric with hydrofluoric acid, and solutions of heavy metal chlorides, such as $FeCl_3$, do not react with silicon.

Hot potassium and sodium hydroxide solutions rapidly dissolve silicon, and molten nitrates, peroxides, chlorates, and carbonates react violently. Fluorine reacts at room temperature, chlorine at 300° C (570° F), bromine at 500° C (930° F), and iodine at red heat. The hydrogen halide gases react at elevated temperatures.

Lumps of silicon may be heated up to 1100° C (2000° F) without visible change, either in air or in oxygen. This is believed due to a protective coating of SiO_2. Nitrogen has no action up to 1300° C (2400° F).

Phosphorus, arsenic, and antimony do not react up to the boiling points of these elements. Sulfur and sulfur chloride form a sulfide at high temperatures. Carbon dioxide reacts to produce carbon monoxide above 1200° C (2200° F). Carbon combines to form SiC at 1400° C (2550° F).

The oxides of Mg, Ca, Ba, and Sr at high temperatures are reduced to metal, with silicates as reaction products, whereas the oxides of Th, Ti, and Zr are more stable. Silicon is said to react with SiO_2 at 1450° C (2640° F) with formation of a volatile monoxide.[2]

The numerous known silicides display in part the chemical resistance of the element. Many metals can be surface alloyed by heating in presence of the element, master alloys, or $SiCl_4$.[†3]

GENERAL REFERENCE

MELLOR, J. W., *Comprehensive Treatise on Inorganic and Theoretical Chemistry*, Vol. 5, Longmans, Green and Co., London, 1924.

* Consulting Metallurgist, Albany, Ore.
† Ihrigizing.

[1] H. Funk, *Ber.*, **67B**, 464 (1934). Charles Bedel, *Compt. rend.*, **189**, 643 (1929). Charles Bedel, *Compt. rend.*, **188**, 1255, 1294 (1929).
[2] E. Zintl, *Z. anorg. Chemie*, **245**, 1 (1940). U. S. Patents 2,286,663; 2,242,497.
[3] H. K. Ihrig, *Metal Progress*, **36**, 380–381 (1939). A. Fry, *Stahl u. Eisen*, **43**, 1039 (1923).

SPECIAL TOPICS IN CORROSION

CORROSION BY SEA WATER

Behavior of Metals and Alloys in Sea Water

F. L. LAQUE*

This section contains data on materials in actual contact with natural sea water either continuously or intermittently as at half-tide level.

The information in the chapter on the physical and chemical nature of sea water and in the chapter on its biological characteristics with reference to corrosion (p. 433) will serve to explain why sea water differs from simple salt solutions with regard to the kind and intensity of corrosion. Exposure of specimens to simple salt solutions or synthetic sea water in the laboratory usually fails to measure properly the resistance of a material to attack by sea water as it is encountered naturally.

In spite of wide variations in temperature, salinity, and marine organism growth from place to place, there are surprisingly small differences in the corrosion of common metals and alloys when they are exposed to corrosion by sea water at different points throughout the world. This is illustrated by the data referring to steel in Table 1. Evidently the controlling factors change in compensating ways, e.g., a high water temperature which tends to promote higher reaction rates also serves to encourage the development of protective calcareous deposits and marine growths which stifle attack. Of course, in localities where there is considerable pollution of the sea water, or some special kind of contamination, ordinary sea water corrosion data will not be applicable and the results of special tests or local experience must be applied.

The data to be provided here should be considered as illustrative rather than definitive. However, where rates of attack are observed to differ greatly from those recorded, it may be assumed that some unusual factor is involved and that excessively high rates of corrosion may be due to correctible causes.

STEEL AND IRON

CONTINUOUS IMMERSION

Illustrative data for steel and iron, largely from long-time tests, collected in Table 1 show a surprising uniformity of rates of attack (as expressed in inches penetration per year) for specimens exposed under conditions of continuous immersion at several points throughout the world. The spread is between 0.001 and 0.0077 ipy, with an average of 0.0043. Evidently the compensating factors mentioned in the introductory paragraphs tend to hold normal corrosion rates within rather narrow practical limits irrespective of water temperature, salinity, etc. Consequently, for a rough estimate in the absence of data applying to a particular locality or condition of exposure, it would be reasonable to use a figure of about 0.005 ipy, or about 25 mdd, for the expected average rate of corrosion of steel or iron continuously immersed in sea water under natural conditions. Observed average rates of attack much higher than this may be assumed to be the result of some peculiar corrosion accelerating factor which may be discovered by investigation and subject to control. The weight loss of steel continu-

* Development and Research Division, The International Nickel Co., Inc., New York, N. Y.

384 SPECIAL TOPICS IN CORROSION

TABLE 1. RESULTS OF EXPOSURE OF STEEL AND IRON TO CORROSION BY SEA WATER UNDER NATURAL CONDITIONS

Steel Specimens—Continuously Immersed

Line	Composition of Steel, %								Original Condition of Surface	Duration of Exposure, years	Weight Loss		Maximum Depth of Pitting, inches	Geographical Location	Reference (see p. 429)
	C	Mn	Si	S	P	Cu	Ni	Cr	Other			mdd	ipy			
1	.07	.31	.02	.06	.11					Pickled	2	31	.0058	...	Sandy Hook, N. J.	1
2	.20	.69	.005	.055	.06	.04				Mill scale	4	23	.0042	.009	Bristol Channel, Weston-super-Mare	2
3	.08	.40	.05	.02	.024	.09				Pickled	1	21	.0038	.018	Dunkerque, France	3
4	.08	.40	.05	.02	.024	.09				Pickled	1	18	.0033	.017	Concarneau, France	3
5	.08	.40	.05	.02	.024	.09				Pickled	1	23	.0042	.031	Le Havre, France	3
6	.08	.40	.05	.02	.024	.09				Pickled	1	42	.0077	.031	Brest, France	3
7	Average of carbon steels included in tests															
8	"	"	"	"	"	"				Pickled	15	26	.0048	.075	Halifax, N. S.	4
9	"	"	"	"	"	"				Mill scale	15	23	.0043	.068	"	4
10	"	"	"	"	"	"				Pickled	15	17	.0030	.043	Aukland, N. Z.	4
11	"	"	"	"	"	"				Mill scale	15	18	.0033	.147	"	4
12	"	"	"	"	"	"				Pickled	15	13	.0024	.065	Plymouth, Eng.	4
13	"	"	"	"	"	"				Mill scale	15	13	.0024	.156	"	4
14	"	"	"	"	"	"				Pickled	15	20	.0036	.064	Colombo, Ceylon	4
15	Average of 9 mild steels submitted									Mill scale	15	21	.0039	.240	"	4
16	.20	.59	.02		.04	.02				Machined	3	27	.005	...	Eastport, Me.	5
17	.21	.61	.03		.04	.48				Sandblasted	1.2	27	.0049	...	Gosport, Eng.	6
18	.21	.61	.03		.04	.48				Sandblasted	1.2	28	.0051	...	"	6
19										Mill scale	0.5	33	.0060	.050	"	6
20										Pickled	1	23	.0041	...	Kill-van-Kull, N. J.	7
21				.032	.025					Pickled	0.5	36	.0066	.030	San Francisco, Calif.	7
22	.16	.60	.017	.032	.025					Pickled	1	20	.0038	.028	Cape May, N. J.	7
23	.16	.60	.017							Mill scale	1	27	.005	.080	"	7
24	.13	.51		.021	.013	.05				Pickled	5	22	.004	.115	Freeport, Texas	7
25	.13	.51		.021	.013	.05				Mill scale	7.5	33	.0060	.25 (Perf.)	"	7
26	.10	.52	.10	.04	.05	.34				Pickled	6.5	23	.004	.21	Kure Beach, N. C.	7
27	.08	.61	.27	.04	.05		.11	2.92		Pickled	6.5	30	.005	.035	"	7
28	.08	.40	.19	.04	.05		.03	6.00		Pickled	6.5	13	.002	.091	"	7

TABLE 1. RESULTS OF EXPOSURE OF STEEL AND IRON TO CORROSION BY SEA WATER UNDER NATURAL CONDITIONS — *Continued*

Steel Specimens — Continuously Immersed

Line	Composition of Steel, %								Original Condition of Surface	Duration of Exposure, years	Weight Loss		Maximum Depth of Pitting, inches	Geographical Location	Reference (see p. 429)
	C	Mn	Si	S	P	Cu	Ni	Cr	Other			mdd	ipy			
29	.06	.39	...	.03	.016	...	2.15	...	...	Mill scale	7.5	26	.005	.094	Kure Beach, N. C.	7
30	.18	.66	.27	.016	.016	...	1.80	.26	...	Mill scale	7.5	23	.004	.063	"	7
31	1.36	11.92	.23	...	...	...	...	...	...	Mill scale	4	19	.0035	.029	Bristol Channel	2
32	Carbon steel pipe, 6 ft diam.									Mill scale	10	...	.0045	.084	Kure Beach, N. C.	7
33	Steel piling									Mill scale	31	...	.001	...	Dry Tortegas Island	8
34	Steel piling									Mill scale	7.5	...	.0037	.098	Boston, Mass.	9

Wrought Iron Specimens — Continuously Immersed

35	.04	.10	.10	.028	.018	.06	...	...	...	Mill scale	4	19	.0035	.031	Bristol Channel, Weston-super-Mare	2
36	.015	.02	.15	.013	.11	...	...	...	...	Pickled	15	25	.0046	.060	Halifax, N. S.	4
37	.015	.02	.15	.013	.11	...	...	...	...	Pickled	15	12	.0022	.040	Aukland, N. Z.	4
38	.015	.02	.15	.013	.11	...	...	...	...	Pickled	15	11	.0020	.061	Plymouth, Eng.	4
39	.015	.02	.15	.013	.11	...	...	...	...	Pickled	15	17	.0031	.108	Colombo, Ceylon	4
40	.03	.02	.12	.017	.08	...	...	...	...	Mill scale	6.5	28	.0051	.139	Kure Beach, N. C.	7
41	.025	.05	.15	.014	.031	.01	...	...	...	Mill scale	4	28	.0052	.022	Bristol Channel, Weston-super-Mare	2
42	...	...	...	...	...	...	...	...	...	Pickled	0.5	33	.0061	...	San Francisco, Calif.	7
43	...	...	...	...	...	...	...	...	...	Sandblasted	6	35	.0065	.191	Kure Beach, N. C.	7
44	...	...	...	...	...	...	...	...	...	Mill scale	7.5	37	.0069	.164	"	7

Cast Iron Specimens — Continuously Immersed

45	3.52	.88	1.5	.09	.17	...	...	...	...	As cast	4	21	.0040	.25	Bristol Channel, Weston-super-Mare	2
46	3.41	.74	1.73	.07	.48	...	...	...	...	As cast	15	32	.0059	G	Halifax, N. S.	4
47	3.41	.74	1.73	.07	.48	...	...	...	...	As cast	15	14	.0026	.058	Plymouth, Eng.	4
48	3.41	.74	1.73	.07	.48	...	...	...	...	As cast	15	46	.0085	.198	Colombo, Ceylon	4
49	Average of 8 cast irons in test									...	3	24	.0048	G	Eastport, Me.	5
50	...	...	...	...	...	...	...	...	...	Machined	0.5	34	.0068	...	San Francisco, Calif.	7
51	...	...	...	...	...	...	...	...	...	Machined	0.2	32	.0065	...	East River, New York, N. Y.	7

G = Severe graphitic corrosion.

TABLE 1. RESULTS OF EXPOSURE OF STEEL AND IRON TO CORROSION BY SEA WATER UNDER NATURAL CONDITIONS — *Continued*

Cast Iron Specimens — Continuously Immersed

Line	Composition of Steel, %									Original Condition of Surface	Duration of Exposure, years	Weight Loss		Maximum Depth of Pitting, inches	Geographical Location	Reference (see p. 429)
	C	Mn	Si	S	P	Cu	Ni	Cr	Other			mdd	ipy			
52	...	...	...	...	...	...	...	...	...	As cast	6	103	.021	G	Kure Beach, N. C.	7
53	3.0	1.2	1.7	...	...	6.5	15.0	2.0	...	As cast	6	4	.0008	0	"	7
54	2.8	1.0	1.7	...	...	...	20.0	2.5	...	As cast	6	8	.0016	0	"	7

Steel Specimens — Exposed at Half-Tide Level

55	.13	.55	.12	.06	.04	...	...	...	...	Mill scale	3	9.4	.0017	...	Southampton, Eng.	10
56	Average of carbon steels included in test									Pickled	15	15	.0028	.124	Halifax, N. S.	4
57	Average of carbon steels included in test									Mill scale	15	16	.0029	.126	"	4
58	Average of carbon steels included in test									Pickled	15	5	.0009	.023	Aukland, N. Z.	4
59	Average of carbon steels included in test									Mill scale	15	3	.0006	.022	"	4
60	Average of carbon steels included in test									Pickled	15	23	.0042	.049	Plymouth, Eng.	4
61	Average of carbon steels included in test									Mill scale	15	14	.0026	.082	"	4
62	Average of carbon steels included in test									Machined	3	17	.0031	...	Eastport, Me.	5
63	.17	.52	...	.03	.02	...	...	...	...	Sandblasted	1	71	.013	...	Pacific Ocean, Panama	7
64	Steel piling									Mill scale	21	...	.0044	...	Glenwood Landing, N. Y.	8
65	Steel piling									Mill scale	7.5	...	.0055	...	Boston, Mass.	7
66	Carbon steel									Pickled and with mill scale	1	52	.0096	...	Seabright, N. S.	14
67	.09	.63	.14	.028	.098	.58	.36	.50	...	Pickled and with mill scale	1	42	.0077	...	"	14
68	.09	.37	.68	.028	.141	.39	...	1.00	...	Pickled and with mill scale	1	48	.0086	...	"	14
69	.09	.47	.14	.022	.008	1.02	1.89	...	...	Pickled and with mill scale	1	39	.0072	...	"	14
70	.09	.78	...	.020	.010	1.10	1.32	...	.10 Mo	Pickled and with mill scale	1	44	.0081	...	"	14
71	.10	.40	.03	.025	.08	.60	.70	...	.08 "	"	1	38	.0070	...	"	14
72	Copper steel									"	1	53	.0098	...	"	14
73	Carbon steel										1	29	.0053	...	Esquimalt, B. C.	14

G = Severe graphitic corrosion.

TABLE 1. RESULTS OF EXPOSURE OF STEEL AND IRON TO CORROSION BY SEA WATER UNDER NATURAL CONDITIONS — *Continued*

Steel Specimens — Exposed at Half-Tide Level

Line	Composition of Steel, %								Original Condition of Surface	Duration of Exposure, years	Weight Loss		Maximum Depth of Pitting, inches	Geographical Location	Reference (see p. 429)
	C	Mn	Si	S	P	Cu	Ni	Cr	Other			mdd	ipy			
74	.09	.63	.14	.028	.098	.58	.36	.50	...	Pickled and with mill scale	1	25	.0046	...	Esquimalt, B. C.	14
75	.09	.37	.68	.028	.141	.39	...	1.00	...	"	1	20	.0037	...	"	14
76	.09	.47	.14	.022	.008	1.02	1.89	...	...	"	1	26	.0048	...	"	14
77	.09	.78	...	.020	.010	1.10	1.32	...	.10 Mo	"	1	28	.0051	...	"	14
78	.10	.40	.03	.025	.08	.60	.70	...	.08 "	"	1	28	.0051	...	"	14
79	Copper steel									"	1	28	.0051	...	"	14

Wrought Iron Specimens — Exposed at Half-Tide Level

80	.015	.02	.15	.013	.11	...	...	...	...	Pickled	15	13	.0024	.059	Halifax, N. S.	4
81	.015	.02	.15	.013	.11	...	...	...	...	Pickled	15	4	.0007	.026	Aukland, N. Z.	4
82	.015	.02	.15	.013	.11	...	...	...	...	Pickled	15	19	.0035	.043	Plymouth, Eng.	4
83	.06	.10	.12	.03	.27	.06	...	...	...	Mill scale	3	7	.0014	...	Southampton, Eng.	10

Cast Iron Specimens — Exposed at Half-Tide Level

84	3.41	.74	1.73	.07	.48	...	...	...	...	As cast	15	8	.0015	.052	Halifax, N. S.	4
85	3.41	.74	1.73	.07	.48	...	...	...	...	As cast	15	11	.0020	.180	Plymouth, Eng.	4
86	Average of 9 cast irons in test									Machined	3	8	.0015	...	Eastport, Me.	5

ously immersed in sea water is for all practical purposes a linear function of time. Although the accumulation of corrosion products and marine growths seems to set up conditions at the metal : water interface, which is less corrosive than if the metal were in free contact with the sea water, the rate of corrosion under these layers does not continue to change with time. In other words, the accumulations serve to establish a limiting rate of corrosion without having much further effect on the rate once it has been established. The limiting rates seem to become established during the first year or so of exposure provided there are no gross changes in the corrosive nature of the water in which the steel is immersed.

In connection with recent observations related to the role of bacteria in promoting corrosion (see *Corrosion by Micro-organisms in Aqueous and Soil Environments*, p. 466), it may be stated that specimens of steel have been exposed in sea water where sulfate-reducing bacteria were known to be present, and, in fact, were found in the corrosion products which contained appreciable percentages of iron sulfide. The observed rates of corrosion and pitting of such steel specimens fell within the normal range previously defined. From this it may be concluded that the action of such bacteria is common to the exposure of steel in many localities. It is possible, also, that if it were not for these organisms the corrosion of steel would be reduced appreciably by the accumulation of corrosion products. The action of the bacteria that grow beneath the original coating of corrosion products may simply be to take the place of the excluded oxygen in enabling the corrosion reactions to continue at the rates commonly observed.

PITTING AND THE EFFECT OF MILL SCALE

The average rates suggested do not take into account non-uniformity of corrosion in the form of pitting. The depth of pitting is influenced considerably, especially during the early stages of exposure, by the presence of mill scale. So long as mill scale remains in place, it acts to accelerate corrosion at bare spots or breaks in the scale through the action of galvanic cells set up between the scale-covered steel and the bare steel. The extent of this acceleration is determined roughly by the relative areas of bare steel and scale-covered steel, since the effect is to concentrate all corrosion on the bare spots with consequent increase in the depth of attack in such areas. It is observed from Table 1 that the weight losses of scaled steels are about the same as descaled steels under similar exposure conditions. The galvanic action tends to become stifled in time through the development of insulating calcareous deposits as a result of the cathodic reaction at the scaled surfaces. These cathodic deposits function also to retard the removal of scale and, therefore, serve to prolong the galvanic action, though at a reduced intensity. The extent of the acceleration of corrosion by mill scale and the relative area effect are illustrated by the data in Table 2 obtained by immersion of scaled steel specimens from which definite areas of the scale had been removed prior to exposure.

It may be stated as a general rule that steel exposed with mill scale present will be pitted about three times as deeply as descaled steel for short periods of exposure, such as a few months. This ratio decreases as the exposure is prolonged and would be about 1.5 to 1 for a 10-year exposure period. It follows, then, that the desirability of removing mill scale prior to exposure is greatest when the thickness of the steel is small and when pitting would result in failure of the metal to accomplish its purpose, e.g., in a pipe, tank, or other vessel required to hold sea water. Removal is less important when the sections are relatively large, for pitting is then less important and the period of exposure is likely to be long as with steel piling.

BEHAVIOR OF METALS AND ALLOYS IN SEA WATER 389

TABLE 2. RELATIONSHIP BETWEEN AREA OF BARE METAL AND DEPTH OF ATTACK OF PARTIALLY DESCALED STEEL*

Specimens Immersed in Salt Water for 4 Months at Fore River, Mass.

Bare Area as Percentage of Total	Ratio of Scaled to Bare Area	Observed Depth of Attack in Bare Area, ipy
5	19 : 1	.045
10	9 : 1	.035
25	3 : 1	.015

* Data courtesy Malcolm Mosher, Bethlehem Steel Co., Shipbuilding Division, Quincy, Mass.

The pitting factor varies between scaled and descaled steel. It also varies with the incidental conditions and with the duration of exposure, that is, the longer the exposure, the lower the pitting factor. A normal pitting factor for exposures of about 10 years would be about 2.5 for descaled steel, and 3.5 for steel exposed with mill scale. Applying these factors to the normal corrosion rate as indicated by weight loss, the deepest attack on descaled steel would be about 0.015 ipy, and on steel exposed with mill scale 0.020 ipy over about a 10-year period.

However, for short periods of exposure, e.g., less than a year, the rate of pitting might be as much as 0.04 ipy for descaled steel, and 0.1 ipy for steel exposed with mill scale. In view of the variability of the several factors that influence pitting, these suggestions as to what may be considered normal rates should be used primarily as a basis for deciding whether there may be correctible causes for much higher observed rates of pitting, rather than for estimating probable life of steel under conditions of exposure that have not been investigated. However, they are probably better than a guess for the latter purpose.

CORROSION AT HALF-TIDE LEVEL

Frequently corrosion of steel is most severe in the between-tides region and just above high tide where the metal is wet with salt water and salt spray in the presence of air. However, corrosion is not invariably greater in these regions, since in many harbors there is enough pollution of the water by silt and oils and greases that float on the surface to keep the metal between tides coated with mildly protective films. A typical example is provided by the data from tests in Southampton Harbor at half-tide level — line 55, Table 1 — as compared with data from tests on continuously immersed specimens of a similar steel in the clean water of the Bristol Channel — line 2, Table 1. The specimens at Southampton acquired protective deposits from the water which served to stifle attack.

In the absence of such extraneous protection the ratio of corrosion between tides to corrosion below low tide will vary with the temperature. In warm climates, the ratio will be high, since the higher temperature not only promotes the development of marine growths and calcareous deposits which tend to reduce underwater corrosion, but also accelerates attack in the absence of such protection in the tidal range and just above it.

In view of the many disturbing factors, such as water and atmospheric pollution, it is not possible to assign any precise figures to show the normal ratio of tidal to submerged corrosion. In temperate climates this ratio is not likely to exceed 2 to 1, whereas in humid tropical climates it may reach 4 to 1. Pitting factors will not be quite so high as in submerged exposure, a reasonable value for about 10 years exposure being 1.5 for descaled steel, and 2.5 for steel exposed with mill scale present. This leads to estimates of normal maximum rates of penetration of 0.015 ipy for descaled steel, and 0.025 ipy for scaled steel in temperate climates, and double these rates in tropical climates.

In the case of steel piling extending well above the water line, the most severe corrosion often occurs in a zone that extends a foot or so from the top of the piling where attack may be about four times as great as on the rest of the piling. It is suggested that this may be connected with the fact that the rust takes the form of loosely bonded vertical strata open at the top so that the space between the strata and the steel becomes in effect a reservoir for the corrosive medium. Further down, the accumulated rust will have a much better chance to dry out and act to some extent as a barrier between the underlying steel and the salt spray. For this reason it is good practice to apply caps over the tops of steel piling.

Attack at Mud Line

Although quantitative data are not available to the author, there have been several reports of accelerated corrosion of steel piling in the region of the mud line. This may be the result of oxygen concentration cell action, the scouring effect of shifting sand, or some peculiar acceleration of corrosion by sulfate-reducing bacteria. In any event, it would be in order where practicable to provide for extra corrosion in a region extending about 2 ft above and below the expected mud line. This might be accomplished by attaching pads by welding in the critical region. The extent of corrosion to be provided for is not known definitely, but is not likely to be more than two to three times the normal rate of attack for continuous immersion in sea water.

Effect of Velocity

The corrosion of steel by sea water increases as the velocity increases. The trend of the velocity effect is illustrated by Fig. 1, which shows that the rate of attack is proportional to the velocity until some critical velocity is reached, beyond which there is little further increase in corrosion. The tests were made using weighed 6-in. lengths of mild steel pipe varying in size from ½ to 2 in. I.D. connected in series. Sea water of 2.8% average salt content at about 23° C (73° F) was pumped through a recirculatory system, with renewal of water twice a week. The duration of the test was 36 days.

This effect of velocity in increasing corrosion applies only when the steel is in free contact with the moving water. Accumulations of fouling organisms will reduce the velocity at the metal : water interface so that corrosion will not be affected much by variations in the rate of flow past the attached organisms. Macro-organisms are not likely to become attached at steady flow rates much above about 3 ft per sec (1 meter per sec), but once attached during periods of lower velocity flow, they may be able to retain a foothold and grow in contact with water moving at much higher velocities. Furthermore, films of bacterial slime are able to develop in contact with water moving at velocities too high for fouling by macro-organisms. These slime films also function to prevent an increase in corrosion with increase in velocity. Consequently, it is only when conditions are such that neither macro-organisms nor slime films are present that an increase in velocity will have the effect indicated by Fig. 1. It may also be expected that beyond some critical velocity or degree of turbulence where macro-organisms and slime films cannot persist, there will be an abrupt increase in corrosion rates.

From corrosion test data and practical experience, it may be said that the normal corrosion of steel or iron piping handling sea water at the usual velocities and temperatures encountered in ships' piping systems should not exceed 0.05 ipy. If rates of attack much higher than this should be encountered, a search for correctible causes is warranted.

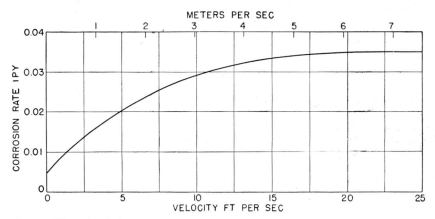

FIG. 1. Effect of Velocity on Corrosion of Steel by Sea Water at Atmospheric Temperature.

Effect of Temperature

Since there is little occasion to use steel in contact with sea water at elevated temperatures, reliable data on the temperature effect are scarce. It would be expected that rates of attack would increase with temperature to a maximum extent of doubling for about each 10° C (20° F) rise in temperature, whereas in the higher temperature ranges the rate of attack might be lessened by the deposition of protective calcareous deposits and by loss of oxygen from solution. In a particular test in a sea water evaporator at 80° C (170° F) steel was corroded at a rate of about 0.1 ipy in a short time test interrupted by occasional exposure to cold water. The indicated rate of attack was calculated only for the time at operating temperature.

Effect of Welding

Properly made welds resist corrosion by sea water as well as the parent metal; stress relief annealing after welding is not necessary from the corrosion standpoint. The potential difference between unalloyed weld metal and adjoining unalloyed plate is negligible. Weld deposits slightly more noble than unalloyed plate may be obtained by incorporating small amounts, e.g., 2% of nickel, in the welding rod. Such low-alloy content weld deposits are given substantial galvanic protection by the surrounding unalloyed steel without detriment to the steel. Similarly, more highly alloyed welds, such as austenitic stainless steels, are given complete protection without serious acceleration of corrosion of carbon or low-alloy steel plates. Alloy steel rivets or bolts are protected in much the same way as low-alloy steel weld deposits.

Effect of Composition

Within ranges normally encountered the common constituents of steel do not seem to have any significant effect on corrosion by sea water. None of the common alloying elements (with the possible exception of chromium) in amounts under 5% has any commercially significant effect on the underwater corrosion of steel as measured by weight loss. The corrosion reaction appears to be controlled by external factors rather than by the composition of the steel.

Chromium has the greatest effect in reducing weight loss with the optimum amount being about 3%, since with more chromium pitting is increased.

With respect to pitting, the situation is different, since most of the common alloy steels resist pitting significantly better than unalloyed steel or iron. The most useful alloy additions are nickel, chromium, and molybdenum, either with or without copper. Copper itself has no appreciable effect on either weight loss or pitting. Illustrative data are provided in lines 26 to 31 and 66 to 79 of Table 1.

In submerged exposure, also, the alloy steels benefit from galvanic protection when combined with larger areas of unalloyed steel, as in the case of welds, rivets, or other fastenings as described previously.

Under conditions of partial or intermittent immersion, the nickel steels and other alloy steels which perform well in marine atmospheres, as discussed in *Atmospheric Corrosion of Iron*, p. 120, demonstrate commercially significant superiority over carbon steel or copper steel.

CAST IRON

The iron phase in ordinary cast iron is vulnerable to attack by sea water to the same extent as low-alloy steel. However, the graphite phase in cast iron exerts a controlling effect on the extent and distribution of corrosion after the first stages of attack.

In most instances, the graphite is left mixed with iron corrosion products as a more or less compact layer over the unattacked iron. To the extent that this layer is impervious to water, corrosion will cease or slow down. If the layer is porous, corrosion will be accelerated by galvanic action between the graphite and the iron. The usual result is continuous corrosion at a rate not much different from the maximum penetration of steel by pitting. For example, rolled steel condenser water boxes 1 in. thick failed by pitting in 6 years, where cast iron water boxes 2 in. thick required replacement in 10 years when graphitic corrosion (graphitization) had penetrated to an average of half way through the 2-in. section.

Even after considerable graphitic corrosion has occurred, cast iron may be able to function satisfactorily as a container for salt water provided the stresses are not too high as, for example, in pipes, valves, and pump casings handling water at low pressure. The layer of graphite may also be effective in reducing the galvanic action between cast iron and more noble alloys, such as for pumps fitted with bronze impellers.

Low-alloy content cast irons frequently demonstrate superior resistance to graphitic corrosion, evidently as a result of their more dense structure and the development of more compact and more protective graphitic coatings. Highly alloyed, austenitic cast irons* show considerable superiority over cast iron due in large part to the more noble potential of the austenitic matrix plus more protective graphitic coatings. Illustrative data are provided by lines 45 to 54 in Table 1.

The casting skin appears to have no long-time effect on corrosion of cast iron by sea water. A comparison of "as cast" and machined pipe specimens through which sea water was passed at low velocity for about a year showed a rate of attack of 0.013 ipy for the "as cast" specimens and 0.011 ipy for the machined specimens.

* Ni-Resist.

COPPER AND COPPER ALLOYS

In Tables 3 to 9 are assembled results of tests made by exposing specimens of copper and copper alloys to corrosion by sea water under natural conditions at several locations. It will be observed that, with the exception of those alloys susceptible to dezincification, there is little practical difference in apparent durability under such conditions of exposure. Likewise, there is little practical difference in the extent of attack at different test sites, with the exception of polluted harbor waters in which attack is likely to be increased by sulfur compounds. Under the latter conditions, the high-tin bronzes, aluminum bronzes, and Cu-Ni alloys demonstrate an appreciable advantage over other compositions.

All the copper alloys are given complete protection from corrosion when in low-resistance electrical contact with uncoated steel having an exposed area at least equal to that of the copper alloy.

Of the various types of copper, *arsenical copper* (0.5% As) is most resistant to attack by sea water under conditions of alternate immersion, as at half-tide level.[1] Cu-Si alloys behave much like copper.

Among the *brasses*, the best all-round performance under conditions of quiet immersion is likely to be given by alloys containing 65 to 85% copper. Alloys of higher copper content corrode at higher rates and are more susceptible to pitting and water line attack. Alloys of higher zinc content are more susceptible to dezincification. Dezincification of yellow brass and Admiralty brass is suppressed effectively by the presence of small amounts of arsenic, antimony, or phosphorus. (See *Copper-Zinc Alloys*, p. 69.) Larger amounts of these inhibiting elements are also helpful in the higher zinc alloys — e.g. Muntz metal and Naval Rolled brass. The presence of aluminum in aluminum brass induces a degree of passivity which suppresses weight loss appreciably, and, under simple immersion conditions, results in localization of attack (usually shallow pits) in well-defined areas.

The *Cu-Sn* alloys (*bronzes*) show good resistance to attack, with alloys of higher tin content (8%) being definitely superior to those of lower tin content (5%).

The *Cu-Al* alloys generally are attacked less than other high-copper compositions, but in some instances have been found to be more susceptible to pitting.

With the *Cu-Ni* alloys, resistance to attack increases roughly in proportion to the nickel content.[2] Attack is reduced also by the presence of small amounts of iron or aluminum. Although the effects of these elements are favorable under conditions of quiet immersion, they are most important where erosion is a factor. The Cu-Ni alloys demonstrate a special advantage over copper and other high-copper alloys in resisting local attack in the water line region.

There is not much difference in the extent of attack on the various common bronze casting compositions under conditions of quiet immersion in sea water. However, the high-zinc alloys, commonly called manganese bronzes, are susceptible to dezincification under such conditions of exposure.

Effect of Velocity

Where erosion and other effects of high velocity must be withstood, the tin bronzes are most reliable — the tin content should be between 5 and 10% in cast

---

[1] A. W. Tracy, D. H. Thompson, and J. R. Freeman, Jr., "The Atmospheric Corrosion of Copper," *Proc. Am. Soc. Testing Materials*, **43**, 615 (1943).

[2] F. L. LaQue, "The Behavior of Nickel-Copper Alloys in Sea Water," *J. Am. Soc. Naval Engrs.*, **53** (No. 1), 29 (February, 1941).

TABLE 3. RESULTS OF EXPOSURE OF COPPER TO CORROSION BY SEA WATER

Continuous Immersion

Composition, %		Surface Condition	Duration of Exposure, years	Weight Loss		Maximum Depth of Pitting, inches	Loss in Tensile Strength, %	Geographical Location	Reference (see p. 429)
Cu	As			mdd	ipy				
99.8	...	Cold-rolled bar	4	17.5	.0028	...	4.37	Bristol Channel, Weston-super-Mare	2
99.3	.45	Cold-rolled bar	4	16.4	.0027	...	4.98	"	2
99.98	...	Cold rolled sheet, tough pitch	.67	12.2	.0020	...	11.7	Bridgeport, Conn.	12
99.97	...	Cold-rolled sheet P copper	.67	14.7	.0024	...	14.4	"	12
99.9	...	Machined	3	2.4	.0004	...	...	Eastport, Me.	5
99.9	...	Cold-rolled sheet	.5	24.4	.004	...	...	San Francisco Harbor	7
99.9	...	Cold-rolled sheet	1	4.1	.0007	...	...	Kure Beach, N. C.	7
99.9	...	Hot-rolled plate, pickled	2.1	7.4	.0012	...	...	"	7
99.9	...	Hot-rolled plate, pickled	3.5	8.4	.0014	.015	...	"	7
99.9	...	Cold-rolled strip	1	9.7	.0016	.003	...	"	11
99.6	.33	Cold-rolled strip	1	10.7	.0017	.0015	...	"	11
99.9	(.01 P)	Cold-rolled strip	1	9.5	.015	.022	...	"	11
Exposed at Half-Tide Level									
99.9	...	Cold-rolled bar	3	.6	.0001	...	...	Southampton Harbor	10
99.9	...	Machined	1	7.3	.0011	...	...	Eastport, Me.	5
99.9	...	Cold-rolled sheet	1	.9	.0002	...	...	Pacific Ocean, Panama	7
99.9	...	Cold-rolled sheet	.5	28.2	.005	...	...	San Francisco Harbor	7
99.97	...	Cold-rolled sheet	2	4.6	.0008	...	8.3	Long Island Sound	13
99.35	.56	Cold-rolled sheet	2	3.0	.0005	...	5.2	"	13
99.9	...	Hot-rolled plate, pickled	4.1	5.0*	.0008	.007	...	Kure Beach, N. C.	7

*Specimens partially immersed.

TABLE 4. RESULTS OF EXPOSURE OF Cu-Si ALLOYS TO CORROSION BY SEA WATER

Continuous Immersion

Composition, %						Surface Condition	Duration of Exposure, years	Weight Loss		Maximum Depth of Pitting, inches	Loss in Tensile Strength, %	Geographical Location	Reference (see p. 429)
Cu	Zn	Sn	Si	Fe	Other			mdd	ipy				
96	...	...	2.9	.09	...	Cold-rolled sheet	.67	11.8	.0019	...	15.4	Bridgeport, Conn.	12
94	...	...	4	...	...	Machined	3	2.5	.0004	...	...	Eastport, Me.	5
96	1	...	3	...	...	Pickled plate	1	2.3	.0004	...	...	Kure Beach, N. C.	7
91	2.5	1.0	3.5	1.25	...	Machined casting	1	0.5	.0001	...	...	"	7
98	...	...	2	...	...	Cold-rolled strip	1	9.7	.0017	.009	...	"	11
97	...	...	3	.1	...	Cold-rolled strip	1	10.3	.0017	.004	...	"	11
97.5	...	1.5	1	...	...	Cold-rolled strip	1	9.4	.0016	.005	...	"	11

Exposed at Half-Tide Level

94	...	...	4	...	...	Machined	5	3.1	.0005	...	...	Eastport, Me.	5
Bal.	...	...	3	.15	...	Hot-rolled plate, pickled	4.1	7.0*	.0012	...	...	Kure Beach, N. C.	7

* Specimens partially immersed.

TABLE 5. RESULTS OF EXPOSURE OF Cu-Zn ALLOYS (BRASSES) TO CORROSION BY SEA WATER

Composition, %							Surface Condition	Duration of Exposure, years	Weight Loss		Maximum Depth of Pitting, inches	Loss in Tensile Strength, %	Geographical Location	Reference (see p. 429)
Cu	Zn	Sn	Pb	As	Al	P			mdd	ipy				

Red Brass — Continuous Immersion

Cu	Zn	Sn	Pb	As	Al	P	Surface Condition	Duration	mdd	ipy	Max Pit	Loss TS	Location	Ref
86	14	...	...	...	...	...	Cold-rolled sheet	.67	15.2	.0025	...	14.7	Bridgeport, Conn.	12
85	15	...	...	...	...	...	Machined	3	2.6	.0004	...	...	Eastport, Me.	5
85	15	...	...	...	...	...	Cold-rolled sheet	1	2.3	.0004	...	...	Kure Beach, N. C.	7
90	10	...	...	...	...	...	Cold-rolled strip	1	10.7	.0018	.002	...	"	11
85	15	...	...	...	...	...	Cold-rolled strip	1	10.3	.0018	.003	...	"	11
80	20	...	...	...	...	...	Cold-rolled strip	1	11.0	.0018	...	...	"	11
75	25	...	...	...	...	...	Cold-rolled strip	1	10.8	.0018	...	...	"	11

Yellow Brass — Continuous Immersion

Cu	Zn	Sn	Pb	As	Al	P	Surface Condition	Duration	mdd	ipy	Max Pit	Loss TS	Location	Ref
70	28	...	2	...	...	...	Cold-rolled bar	4	23	.004	...	6.1	Bristol Channel, Weston-super-Mare	2
69.7	30	...	...	...	...	...	Cold-rolled sheet	.67	13.5	.0023	Dezincified	15.3	Bridgeport, Conn.	12
65	35	...	...	...	...	...	Cold-rolled sheet	.5	44	.0076		...	San Francisco Harbor	7
65	35	...	...	...	...	...	Cold-rolled strip	1	12.1	.0021	...	...	Kure Beach, N. C.	11

Aluminum Brass — Continuous Immersion

Cu	Zn	Sn	Pb	As	Al	P	Surface Condition	Duration	mdd	ipy	Max Pit	Loss TS	Location	Ref
76	21.4	...	.07	.05	2	...	Cold-rolled strip	1	1.7	.0003	...	...	Kure Beach, N. C.	7
76	22	...	...	.02	2	...	Cold-rolled strip	1	4.5	.0008	.007	...	"	11

Admiralty Brass — Continuous Immersion

Cu	Zn	Sn	Pb	As	Al	P	Surface Condition	Duration	mdd	ipy	Max Pit	Loss TS	Location	Ref
69.8	29	1.22	...	...	...	...	Cold-rolled sheet	.67	11.9	.0020	...	12.3	Bridgeport, Conn.	12
72	26.8	1.1	.01	...	...	...	Cold-rolled sheet	1	4.2	.0007	...	...	Kure Beach, N. C.	7
71	27	1.1	.03	...	...	.014	Cold-rolled sheet	1	2.8	.0005	...	...	"	7
70	29	1	.04	...	...	...	Cold-rolled strip	1	9.0	.0015	...	...	"	11
70	29	1	...	...	...	...	Cold-rolled strip	1	10.0	.0018	...	...	"	11

TABLE 5. RESULTS OF EXPOSURE OF Cu-Zn ALLOYS (BRASSES) TO CORROSION BY SEA WATER — Continued

Muntz Metal — Continuous Immersion

Composition, %							Surface Condition	Duration of Exposure, years	Weight Loss		Maximum Depth of Pitting, inches	Loss in Tensile Strength, %	Geographical Location	Reference (see p. 429)
Cu	Zn	Sn	Pb	As	Al	Fe			mdd	ipy				
60.8	38.8	.10	.35	...	...	...	Cold-rolled bar	4	13	.0022	...	9.4	Bristol Channel, Weston-super-Mare	2
60	38.6	...	1.37	...	...	...	Cold-rolled bar	4	6.5	.0011	...	3.8	"	2
60	40	...	...	...	...	...	Machined	3	3.9	.0007	Dezincified	...	Eastport, Me.	5
60	40	...	...	...	...	...	Cold-rolled strip	1	12.7	.0022	Dezincified	...	Kure Beach, N. C.	11
63	37	...	...	.1	...	...	Cold-rolled strip	1	12.8	.0020	.003	...	"	11
61	38	...	.3	.2	...	...	Cold-rolled strip	1	9.2	.0016	.002	...	"	11
59	40	.6	...	...	...	.8	Cold-rolled strip	1	11.4	.0020	.005	...	"	11

Naval Rolled Brass — Continuous Immersion

62	36.7	1.01	.23	...	...	...	Cold-rolled bar	4	25	.0043	Slight dezincification	4.8	Bristol Channel, Weston-super-Mare	2
60	39	1	...	...	...	...	Machined	3	3.5	.0006	...	...	Eastport, Me.	5
62	37	.7	...	...	...	...	Cold-rolled strip	1	10.0	.0017	.001	...	Kure Beach, N. C.	11
60	39	.75	...	.07	...	...	Cold-rolled strip	1	10.7	.0018	...	...	"	11

Cu-Zn Alloys — Exposed at Half-Tide Level

85	15	...	...	...	...	...	Machined	1	7.3	.0012	Dezincified	...	Eastport, Me.	5
65	35	...	...	...	...	...	Cold-rolled sheet	.5	31	.0053	...	...	San Francisco, Calif.	7
60	40	...	...	...	...	...	Machined	5	5.2	.0009	Dezincified	...	Eastport, Me.	5
60	39	1	...	...	...	...	Machined	1	3.8	.0007	...	...	"	5
60.6	Bal.	.72	.04	...	...	...	Hot-rolled plate, pickled	4.1	5.0	.0008	...	...	Kure Beach, N. C.	7

TABLE 6. RESULTS OF EXPOSURE OF PHOSPHOR BRONZE TO CORROSION BY SEA WATER

Continuous Immersion

Composition, %			Surface Condition	Duration of Exposure, years	Weight Loss		Maximum Depth of Pitting, inches	Loss in Tensile Strength, %	Geographical Location	Reference (see p. 429)
Cu	Sn	P			mdd	ipy				
92.2	7.9	.03	Cold-rolled sheet	.67	6.5	.0011	...	6.9	Bridgeport, Conn.	12
92	7.6	...	Cold-rolled sheet	4	3.5	.0006	.006	...	Kure Beach, N. C.	7
94.8	5.1	...	Cold-rolled sheet	4	4.9	.0008	.015	...	"	7
95	5	.1	Cold-rolled sheet	1	7.6	.0012	.004	...	"	11

TABLE 7. RESULTS OF EXPOSURE OF ROLLED ALUMINUM BRONZE TO CORROSION BY SEA WATER

Continuous Immersion

Composition, %					Surface Condition	Duration of Exposure, years	Weight Loss		Maximum Depth of Pitting, inches	Loss in Tensile Strength, %	Geographical Location	Reference (see p. 429)
Cu	Al	Ni	Fe	As	Sn			mdd	ipy			
94.8	5.1	...	.12	...	...	Cold-rolled sheet	.67	8.2	.0015	...	5.4	Bridgeport, Conn.	12
92.1	7.8	...	.12	...	...	Cold-rolled sheet	.67	2.8	.0005	...	3.6	"	12
89.9	8.3	.53	.86	...	.37	Cold-rolled sheet	.67	5.9	.0011	...	30.3	"	12
92	4	4 Si	...	...	...	Pickled plate	1.8	2.8	.0005	...	...	Kure Beach, N. C.	7
91	7	2	...	...	...	Cold-rolled strip	1	6.7	.0013	.003	...	"	11
95	5	...	...	.25	...	Cold-rolled strip	1	10.6	.0019	...	...	"	11
95	5	...	...	...	...	Cold-rolled strip	1	6.5	.0012	.001	...	"	11

Partial Immersion

| 95.2 | 4.7 | ... | ... | ... | ... | Pickled plate | 4.1 | 2.0 | .0003 | .007 | ... | Kure Beach, N. C. | 7 |

TABLE 8. RESULTS OF EXPOSURE OF ROLLED Cu-Ni ALLOYS TO CORROSION BY SEA WATER

Continuous Immersion

Composition, %						Surface Condition	Duration of Exposure, years	Weight Loss		Maximum Depth of Pitting, inches	Loss in Tensile Strength, %	Geographical Location	Reference (see p. 429)
Cu	Ni	Fe	Mn	Sn	Zn			mdd	ipy				
98.2	1.8	...	...	...	...	Cold-rolled bar	4	14.5	.0024	...	1.5	Bristol Channel, Weston-super-Mare	2
82.6	17.3	.07	...	...	...	Cold-rolled sheet	.67	8.8	.0014	...	10.4	Bridgeport, Conn.	12
69.1	28.6	.019	.4	1.04	.6	Cold-rolled sheet	.67	5.5	.0009	...	2.5	"	12
74.5	20.1	...	.5	...	4.9	Cold-rolled sheet	.67	8.6	.0014	...	8.9	"	12
75	20	...	...	...	5	Machined	3	3.0	.0005	...	...	Eastport, Me.	5
70	30	...	...	...	...	Machined	3	3.2	.0005	...	...	"	5
70	30	...	...	...	...	Cold-rolled sheet	1	9.1	.0015	.000	...	Kill-van-Kull, N. J.	7
70	30	...	...	...	...	Cold-rolled sheet	1	3.2	.0005	.000	...	Cape May, N. J.	7
70	30	...	...	...	...	Cold-rolled sheet	1	1.7	.0003	.001	...	Kure Beach, N. C.	7
69.4	29.4	.26	.80	...	...	Cold-rolled sheet	3	.9	.00015	.005	...	"	7
69.5	29.5	1.81	...	...	...	Pickled plate	4.4	4.9	.0008	.026	...	"	7
Bal.	12.6	...	.43	...	...	Cold-rolled strip	1	.7	.0001	.001	...	"	7
69.1	30.2	.42	...	...	...	Pickled plate	3.9	4.2	.0007	.021	...	"	7
70	30	.42	...	...	...	Cold-rolled strip	1	2.1	.0003	...	...	"	11
70	30	.02	...	...	...	Cold-rolled strip	1	13.4	.002	.003	...	"	11

Exposed at Half-Tide Level

Cu	Ni	Fe	Mn	Sn	Zn	Surface Condition	Duration	mdd	ipy	Pitting	Tensile	Location	Ref
98.0	1.9	.04	...	...	...	Cold-rolled bar	3	.7	.0001	...	...	Southampton	10
90.1	9.6	.26	.10	...	...	Cold-rolled bar	3	.4	.00007	...	...	"	10
79.7	20.1	.11	.07	...	...	Cold-rolled bar	3	.2	.00003	...	...	"	10
54	44	.90	.33	...	...	Cold-rolled bar	3	.04	.000007	...	...	"	10
75	20	...	...	...	5	Machined	1	1.5	.0002	...	...	Eastport, Me.	5
70	30	...	...	...	...	Machined	1	1.9	.0003	...	...	"	5

Partial Immersion

Cu	Ni	Fe	Mn	Sn	Zn	Surface Condition	Duration	mdd	ipy	Pitting	Tensile	Location	Ref
69.5	29.5	...	...	...	...	Pickled plate	4.1	3.6	.0006	.012	...	Kure Beach, N. C.	7
54.3	44.8	.31	...	...	...	Pickled plate	4.1	2.4	.0004	.010	...	"	7
64.9	17.9	.08	...	...	17	Pickled plate	4.1	3.0	.0005	...	...	"	7

TABLE 9. RESULTS OF EXPOSURE OF CAST BRONZES TO CORROSION BY SEA WATER

Continuous Immersion

Composition, %									Duration of Exposure, years	Weight Loss		Maximum Depth of Pitting, inches	Geographical Location	Reference (see p. 429)
Cu	Sn	Zn	Pb	Ni	Mn	Al	Fe	Si		mdd	ipy			
75.2	9.9	...	14.7	...	...	...	...	...	.07	22.3	.0037	...	Bridgeport, Conn.	12
88	10	2	...	...	...	...	...	...	5	3.1	.0007	...	Eastport, Me.	5
88	9	3	...	...	...	...	...	...	1	1.1	.0002	...	Kure Beach, N. C.	7
85	6	3	...	6	...	...	...	...	1	1.1	.0002	...	"	7
87	5	3	...	4	...	...	...	...	1	1.1	.0002	...	"	7
88	6.7	3.9	1.5	...	...	...	...	...	1	3.2	.0005	...	"	7
88	3.8	3.8	1.6	3.0	...	...	...	...	1	3.2	.0005	...	"	7
85	5	5	1.6	...	...	...	...	...	1	2.5	.0004	...	Eastport, Me.	5
85	5	5	5	...	...	...	...	...	1	2.6	.0004	...	Kure Beach, N. C.	7
85	2.5	5	5	2.5	...	...	...	...	1	1.5	.0002	...	"	7
92	...	...	...	...	...	8	...	...	3	1.4	.0002	...	Eastport, Me.	5
89	...	...	...	...	...	10	1	...	1	1.1	.0002	...	Kure Beach, N. C.	7
82	...	...	...	...	...	13	4.4	...	1	2.0	.0003	...	"	7
91	1	...	...	...	...	...	1.25	3.5	1	.5	.0008	...	Eastport, Me.	5
65	...	25	...	...	...	...	...	...	5	15.0	.0026	Dezincified	"	5
60	...	40	...	...	4	...	2	...	1	9.6	.0017	Dezincified	Bridgeport, Conn.	12
65	...	23	...	...	1	6	1	...	.67	20.5	.0036	Dezincified	Kure Beach, N. C.	7
58	...	39	...	...	1	1	1	...	.5	.9	.0001	...	"	7
55.5	...	39	...	2.5	...	...	...	...	.5	1.3	.0002	...	"	7

Exposed at Half-Tide Level

88	10	2	...	...	...	...	...	...	1	1.9	.0003	...	Eastport, Me.	5
85	5	5	5	...	...	...	...	...	5	1.5	.0002	"	5
92	...	...	...	...	...	8	...	...	3	1.1	.00015	"	5
88.2	...	...	...	...	...	8.1	3.1	...	1	2.2	.0004	Pacific Ocean, Panama	7
65	...	25	...	...	...	...	...	...	5	2.0	.0004	Sl. dezincified	Eastport, Me.	5
60	...	40	...	...	.5	.5	1.3	...	5	4.6	.0008	"	"	5
57	...	40	...	...	5.5	6.55	2.8	...	1	.4	.00007	...	Pacific Ocean, Panama	7
65	...	20.2	...	...	...	...	...	...	1	1.6	.0003	...	"	7

alloys.[3] A 12% tin alloy has been proposed for condenser tubes.[4] *Silicon bronzes* and other compositions containing less than 5% tin appear to be unable to form adequately protective films in sea water at high velocity, especially when the water is warm. The extent of acceleration of corrosion by velocities high enough to erode protective corrosion product films is illustrated by tests on a low-tin (2.5%) bronze which was corroded at a rate of 0.0002 ipy in water flowing at 1 to 2 ft per sec (0.3 to 0.6 meter per sec) and at a rate of 0.037 ipy when moved through the same water at 20 ft per sec (6 meters per sec).

The high-zinc manganese bronze compositions resist erosion very well, but suffer dezincification during contact with sea water at low velocity. This dezincification may be reduced by the presence of tin, by such inhibitors as arsenic, antimony, or phosphorus, and by galvanic contact with substantial areas of steel.

The ultimate performance of copper-base alloys is determined by the nature of the corrosion product films that they acquire, the adherence and continuity of these films, and their ability to form, be maintained, and survive under the action of the erosive forces associated with contact with sea water at high velocity. Effects of velocity are complicated by the presence of entrained air bubbles which generally aggravate film erosion and by temperature which also accelerates film breakdown and corrosion. However, there is also some evidence that certain compositions, such as high-tin bronzes, aluminum brass, and Cu-Ni alloys, form and maintain protective films more readily when the water contains dissolved and entrained air, so that with these compositions the presence of substantial amounts of air in the water may actually be helpful. The presence of iron corrosion products in sea water also helps these alloys in the formation of protective films.[5]

Among the brasses, resistance to *erosion* or *impingement attack* increases with the zinc content of the alloy. Yellow brass is better than red brass. The presence of 1% tin in Admiralty brass and Naval Rolled brass results in a slight improvement in resistance to dezincification. Much greater resistance to impingement attack is provided by aluminum brass[6] (22% Zn, 2% Al) which usually contains arsenic, antimony, or phosphorus to avoid dezincification. Aluminum brass has been found to perform very well in contact with polluted harbor waters, as in the case of steam power plants using polluted tidewater in condensers.[7]

Anti-Fouling Characteristics

As discussed elsewhere, copper and high-copper alloys frequently are *anti-fouling*. This characteristic is associated with their corrosion, since it has been established that in order to suppress fouling through either a toxic effect or sloughing off of organisms along with non-adherent corrosion products, it is necessary that a certain amount of corrosion take place. There is evidence that the required corrosion to prevent fouling is that which will yield copper in corrosion products at a rate of about 5 mg of

---

[3] A. H. Hesse and J. L. Basil, "Nickel in Bronze. The Effect of Replacing Tin by Nickel on the Porosity, Mechanical Properties and Corrosion Resistance," *J. Am. Soc. Naval Engrs.*, **55** (No. 1), 44 (February, 1943).

[4] J. Chapman and J. W. Cuthbertson, "Corrosion-Resisting Properties of Bronze Condenser Tubes," Part I. Resistance to Sediment and Impingement Attack," *J. Soc. Chem. Ind.*, **58** (No. 3), 100 (March, 1939).

[5] R. May, "Condenser Tube Corrosion. Some Trends of Recent Research," *Trans. Inst. Marine Engrs.*, **49**, Part 8 (1937).

[6] R. May, "Eighth Report to the Corrosion Research Committee of the Institute of Metals," *J. Inst. Metals*, **40** (No. 2), 141 (1928).

[7] C. W. E. Clarke, A. E. White, and C. Upthegrove, "Condenser Tubes and Their Corrosion," *Trans. Am. Soc. Mech. Engrs.*, **63**, 513 (1941).

copper per sq dm of surface per 24 hours.[8] With rates of copper solution less than this value, fouling is likely to occur to an increasing extent as the rate of solution of copper decreases. It has been observed also that copper alloys that contain aluminum are especially likely to foul either because of lower solution rates or because of some specific effect of aluminum in the corrosion products.[9]

Since limitations of space do not permit a sufficiently detailed discussion of all the factors that may influence the choice among copper alloys for particular services, an effort has been made to provide some condensed qualitative information in Table 10, referring particularly to the use of alloys where erosion and dezincification are important factors, as in condenser service and salt water piping.

With particular reference to condenser tubes and salt water piping, especially on board ship, a word of caution is warranted. It is prudent to assume that velocities well above those figured from average requirements may be encountered frequently on occasions of concentrated demands for water, e.g., in sanitary lines and in fire lines, or as a result of imperfect distribution of flow through tubes in a condenser. For this reason, it is desirable to provide materials able to withstand erosion at velocities well above the expected average water velocity.

Under conditions of intermittent exposure to sea water, as between tides, copper and high-copper alloys are likely to be corroded slightly more than when continuously immersed. When there is a well-defined water line, these alloys are likely to suffer accelerated attack in this region. The high-zinc brasses, aluminum bronzes, and Cu-Ni alloys are corroded less under conditions of intermittent immersion than when continuously immersed. These alloys also are less susceptible to water line attack.[10]

NICKEL AND NICKEL ALLOYS

Pure nickel is passive in sea water and is subject to rather severe local attack under fouling organisms or other deposits that may set up oxygen concentration cells in sea water at low velocity. The heat-hardenable nickel* behaves much like ordinary nickel and its performance is not altered by the heat treatment used for hardening the alloy. Both ordinary nickel and heat-hardenable nickel are highly resistant to impingement attack and erosion and perform well in sea water at high velocity.

The 13.5% Cr–77.5% Ni alloy† is also subject to severe local attack under marine organisms and other deposits, and, although it resists erosion and impingement attack very well, it is not particularly well suited for use under conditions of quiet immersion in sea water. At the same time, the alloy demonstrates outstanding resistance to attack and discoloration by marine atmospheres.

The general effect of nickel content on the corrosion of Ni-Cu alloys is to reduce total corrosion as the nickel content is increased. With nickel contents up to about 40%, the effect of nickel is supposedly to favor the formation of protective corrosion product films which develop quickly and repair themselves rapidly after damage. The formation of these films is aided considerably by the presence of small percentages

---

* "Z" Nickel.
† Inconel.

[8] F. L. LaQue and W. F. Clapp, "Relationships between Corrosion and Fouling of Copper-Nickel Alloys in Sea Water," *Trans. Electrochem. Soc.*, **87**, 103 (1945).

[9] C. L. Bulow, "Corrosion and Biofouling of Copper-Base Alloys in Sea Water," *Trans. Electrochem. Soc.*, **87**, 127 (1945).

[10] F. L. LaQue, "The Behavior of Nickel-Copper Alloys in Sea Water," *J. Am. Soc. Naval Engrs.*, **53** (No. 1), 29 (February, 1941).

of iron or aluminum. The alloys containing less than 40% nickel are discussed along with the other copper-base alloys (p. 393).

With nickel contents higher than 40%, the alloys demonstrate passivity which also is favored by the presence of iron or aluminum. Under conditions of quiet exposure where marine organisms may become attached, this passivity may be destroyed at areas of contact with the organisms so that local attack (pitting) occurs. In the case of 70% Ni–30% Cu alloy and other alloys containing less than about 70% nickel, such local attack slows down with time. For example, the depth of pitting observed after exposure for 5 years was only slightly greater than after two years. For Ni-Cu alloys containing more than 70% nickel, local attack is likely to progress so that these materials are not desirable for use in sea water at low velocity. In the early stages of exposure, local attack on 70% Ni–30% Cu alloy and other high-nickel alloys is least in the case of hot-rolled material exposed either with the mill scale in place or after removal by pickling. Local attack develops most quickly on cold-rolled or other highly finished surfaces, but, as mentioned above, slows down soon after the surface layers have been penetrated.

The 66% Ni–29% Cu–2.75% Al alloy* performs about like 70% Ni–30% Cu alloy in resisting corrosion by sea water. Its resistance to attack is not changed by the heat treatment used to harden it and it is not necessary to remove the thin scale resulting from heat treatment in order to maintain corrosion resistance.

Resistance to erosion and impingement attack increases with the nickel content of the alloys and reaches a very high level in the 70% Ni-Cu alloys. Addition of aluminum improves the erosion resistance of all Cu-Ni alloys. Cast 70% Ni–30% Cu alloy and cast 63% Ni–30% Cu–2% Fe–4% Si alloy specimens moved through sea water at a velocity of 20 ft per sec (6 meters per sec) were corroded at a maximum rate of 0.0016 ipy. Such high resistance to erosion accounts for the good performance of these alloys as used for propeller shafts, propellers, pump impellers, pump shafts, valve trim, hull fastenings, eductor-nozzles, and the like.

The 70% Ni–30% Cu alloys show a relatively noble potential in sea water and are given substantial protection by practically all the materials with which they may be combined in service. This, along with their inherent resistance to impingement attack and erosion, accounts for their choice as valve seat materials in salt water valves, especially when these are used for throttling and the velocity of flow over the valve seat surfaces may be very high.

Welds of all types on Ni-Cu alloys resist corrosion as well as the parent metal. Neither the heat of welding nor internal stresses that may be associated with welding operations have any detrimental effect on corrosion resistance. The 70% Ni–30% Cu alloy is a satisfactory material for welding 70% Cu–30% Ni alloy for use in sea water.

When nickel is alloyed with substantial amounts of chromium and molybdenum† exceptional resistance to attack by sea water results. This includes freedom from local attack under marine organisms or other deposits and extends to exposure at high velocity. Consequently, these alloys may be considered as the most reliable materials for exposure to sea water over the broadest range of exposure conditions.

The Ni-Mo-Fe alloys which do not contain chromium‡ are corroded somewhat more than nickel, but are less subject to local attack. They are not outstandingly resistant to sea water corrosion, and, consequently, have not been used extensively for this particular purpose.

* "K" Monel.
† As in Hastelloy C and Illium.
‡ Hastelloys A and B.

TABLE 10. RELATIVE QUALITIES OF COPPER ALLOYS IN RESISTING DIFFERENT TYPES OF ATTACK BY SEA WATER

Alloy	Resistance to Dezincification or Analogous Corrosion	Ability to Maintain Corrosion Resistance with Increase in Temp. in Absence of Impingement Effects	Resistance to Corrosion and Impingement Attack in Absence of Entrained Air at Velocities*				Resistance to Corrosion and Impingement Attack in Presence of Entrained Air at Velocities*				Freedom from Fouling under Severe Fouling Conditions at Low Velocity
			Under 3 fps	4 to 7 fps	8 to 15 fps	Over 15 fps	Under 3 fps	4 to 7 fps	8 to 15 fps	Over 15 fps	
Copper	Immune	Good	Fair	Poor	Bad	Bad	Fair	Bad	Bad	Bad	Good[1]
Arsenical Cu	Immune	Good	Fair	Poor	Bad	Bad	Fair	Bad	Bad	Bad	Good[1]
Si Bronze	Immune	Good	Fair	Poor	Bad	Bad	Fair	Bad	Bad	Bad	Good[1]
Phosphor Bronze[9] (8% Sn)	Immune	Good	Excel.	Good to excel.	Good	Fair	Excel.	Good to excel.	Good	Fair	Fair[2]
Admiralty Brass	Fair	Fair	Good[5]	Fair to good[5]	Fair to poor	Poor to bad	Good[5]	Fair to poor	Poor to bad	Bad	Fair[2] to good[1]
Inhibited Adm. Brass	Excel.	Fair to good	Good	Fair to good	Fair to poor	Poor to bad	Good	Fair to poor	Poor to bad	Bad	Fair[2] to good[1]
Aluminum Brass	Poor	Poor	Excel.[5]	Excel.[5]	Good[5]	Fair	Excel.[5]	Excel.[5]	Fair to good[5]	Poor	Poor[3]
Inhibited Al Brass	Excel.	Fair to good	Excel.	Excel.	Good	Fair	Excel.	Excel.	Fair to good	Poor	Poor[3]
Red Brass	Good	Fair to good	Good	Poor	Bad	Bad	Fair	Poor	Bad	Bad	Good[1]
Yellow Brass	Poor to bad	Bad	Good[5]	Fair	Poor	Bad	Fair	Poor	Bad	Bad	Fair[2] to good[1]
Muntz Metal	Bad	Bad	Good[5]	Good[5]	Good[5]	Fair[5]	Good[5]	Good[5]	Fair[5]	...	Fair[2] to poor[3]
Naval Rolled Brass	Fair to poor	Poor to bad	Good[5]	Good[5]	Good[5]	Fair[5]	Good[5]	Good[5]	Fair[5]	...	Fair[2] to poor[3]
70:30 Cu-Ni Alloy with Fe under 0.15%	Immune	Good	Excel.	Fair to good	Fair to poor	Poor to bad	Excel.	Fair to good	Poor	Bad	Fair[2] to good[1]
70:30 Cu-Ni Alloy with Fe over 0.15%	Immune	Good	Excel.	Excel.	Excel.	Good	Excel.	Excel.	Excel.	Good	Poor[3]
Cu-Ni Alloys containing over 40% Ni	Immune	Good	Fair[6]	Good to excel.[6]	Good to excel.	Excel.	Fair[6]	Good to excel.[6]	Excel.	Excel.	Poor[3] to bad[4]
Tin (Solder) Coated Cu[7]	Immune	Fair	Excel.	Excel.	Good to excel.	Fair to good	Excel.	Excel.	Good to excel.	Fair to good	Bad[4]

TABLE 10.—Continued

RELATIVE QUALITIES OF COPPER ALLOYS IN RESISTING DIFFERENT TYPES OF ATTACK BY SEA WATER

Alloy	Resistance to Dezincification or Analogous Corrosion	Ability to Maintain Corrosion Resistance with Increase in Temp. in Absence of Impingement Effects	Resistance to Corrosion and Impingement Attack in Absence of Entrained Air at Velocities*				Resistance to Corrosion and Impingement Attack in Presence of Entrained Air at Velocities*				Freedom from Fouling under Severe Fouling Conditions at Low Velocity
			Under 3 fps	4 to 7 fps	8 to 15 fps	Over 15 fps	Under 3 fps	4 to 7 fps	8 to 15 fps	Over 15 fps	
Cast Tin Bronze Containing over 5% Sn	Immune	Good	Excel.	Good to excel.	Fair to good	Fair	Excel.	Good to excel.	Fair to good	Fair	Fair[2]
Cast Tin Bronze Containing under 5% Sn	Immune	Good	Good	Fair to good	Poor	Poor	Good	Fair	Poor	Poor	Good[1]
Cast Silicon Bronze	Immune	Good	Good	Fair to poor	Bad	Bad	Fair	Poor to bad	Bad	Bad	Good[1]
Cast Mn Bronze	Poor to bad	Poor to bad	Good[5-10]	Good[5-10]	Good[5-10]	Good[5-10]	Good[5-10]	Good[5-10]	Good[5-10]	Fair[5-10]	Fair[2] to poor[3]
Cast Al Bronze	Good[8]	Fair to good	Good	Good	Good	Good	Good	Good	Good	Good	Bad[4]

* Ratings take into account the probability of aggravated attack in regions of excessive disturbance of stream line flow or turbulence.
[1] Good. Generally free from fouling.
[2] Fair. Occasionally subject to fouling.
[3] Poor. Occasionally free from fouling.
[4] Bad. Generally subject to fouling.
[5] Alloys subject to dezincification will suffer such damage if exposed to sea water at low or no velocity. Consequently, their relatively good resistance to impingement attack cannot be used to advantage where conditions favoring dezincification may be encountered also.
[6] Given less than excellent rating because of possibility of local attack under deposits or organisms that may remain attached at low velocity. Excellent rating even at low velocity would apply to 70% Ni-30% Cu alloy (Monel) in all applications involving contact with less noble materials, such as iron and bronze, as in valves and pumps.
[7] Such tin (solder) coatings should be applied in substantial thickness, e.g., 1/8 in. by wiping. Thin coatings, such as may be applied by dipping, are unreliable, and may cause serious acceleration of corrosion of copper exposed at bare spots.
[8] Not rated as immune because of occasional susceptibility to a dezincification type of attack.
[9] Data do not refer to use as pipe or tubing, but for such purposes as staybolts.
[10] Data refer particularly to use as propellers where good performance is aided by galvanic protection from steel hulls.
fps — feet per sec. Multiply by 0.305 to obtain meters per sec.

The 36% Ni–Fe alloy* is superior to binary alloys of iron with other elements with respect to overall corrosion and pitting but demonstrates no advantage over Ni–Cu alloys which are used much more extensively to resist sea water attack.

Some illustrative data are assembled in Tables 11, 12, and 13.

ALUMINUM AND ALUMINUM ALLOYS

Pure aluminum and stronger alloys containing magnesium, or magnesium and silicon, demonstrate good resistance to corrosion by sea water and marine atmospheres. The high-strength alloys that contain copper are definitely less corrosion resistant but may be protected adequately by cladding with pure aluminum, e.g., Alclad 24ST,† which is used widely in the construction of flying boats.

Typical results of exposure of specimens of wrought alloys immersed either continuously or alternately as at half-tide level are shown in Table 14.

The clad alloys are most resistant, 52S is next, 53S and 61S follow.† The copper alloys 17ST and 24ST suffer considerable attack and ordinarily should not be used in salt water except when protected in special ways.

As is the case with the wrought alloys, castings that contain magnesium, manganese, or silicon, singly or together, are preferred over those which contain copper as the principal alloying element.

Preferred methods of protecting aluminum and its alloys from attack by sea water are described elsewhere in this book (see p. 49). Aluminum applied as a sprayed metal coating provides excellent protection to steel immersed in sea water. Coatings 10 mils thick have prevented corrosion under conditions of immersion for over five years.

MAGNESIUM AND MAGNESIUM ALLOYS‡

As in other corrosive media, the behavior of magnesium and its alloys in sea water is determined principally by the purity of the alloy, particularly with respect to iron, nickel, and copper, which should not be present in more than trace amounts. For this reason, previously available data from tests on impure compositions must be discarded in appraising the ability of magnesium to resist sea water corrosion.

Although it is unlikely that magnesium would be exposed to sea water without some protective coating, some illustrative data obtained from half-tide exposure of bare specimens are given in Table 15. The Dow 7 and 8 chemical treatments§ afford some improvement in corrosion resistance. More detailed information on surface treatments and painting of magnesium alloys will be found on pp. 241, 251 and 864.

With the cast alloys, maximum resistance to corrosion is provided in the "as cast" condition. Solution heat-treated metal is definitely inferior, whereas aging restores part of the corrosion resistance lost by the solution heat treatment.

* Invar.
† Code numbers and corresponding compositions are given in Table 1, p. 39.
‡ Additional data appear in Table 3, p. 224, and on p. 223
§ No. 7 treatment (same as type III treatment, p. 864) involves contact with 15–20% HF or 5% sodium, potassium or ammonium acid fluoride, followed by immersion in boiling 10–15% $Na_2Cr_2O_7$ and final rinsing.

No. 8 treatment involves immersion in 15–20% HF, followed by 45-min immersion in a boiling solution containing $(NH_4)_2SO_4$, $Na_2Cr_2O_7$, and NH_4OH, and finally 15-min treatment in a boiling 1% solution of arsenius oxide.

TABLE 11. RESULTS OF EXPOSURE OF NICKEL TO CORROSION BY SEA WATER

Continuous Immersion

Composition, %			Form in Which Exposed	Duration of Exposure, years	Weight Loss		Maximum Depth of Pitting, inches	Geographical Location	Reference (see p. 429)
Ni	Cu	Fe			mdd	ipy			
99.8	.08	.31	Cold-rolled strip	.67	5.0	.0008	.035 (Perf.)	Bridgeport, Conn.	12
99.8	.04	...	Hot-rolled rod with mill scale	4	2.0	.0003	...	Bristol Channel, Weston-super-Mare	2
99+	...	...	Cold-rolled strip	.5	7.4	.0012	...	San Francisco Harbor	7
99+	...	...	Hot-rolled plate	3	6.1	.001	.092	Kure Beach, N. C.	7
Heat-Hardenable Nickel			Sheet, heat-treated	4	4.4	.0007	.073	"	7

Exposed at Half-Tide Level

99.6	.06	...	Hot-rolled rod with mill scale	3	.05	.000008	...	Southampton Harbor	10
99.6	.02	.07	Hot-rolled plate	1	.4	.00007	...	Pacific Ocean, Panama	7
99+	...	...	Cold-rolled strip	.5	10.1	.0016	...	San Francisco Harbor	7

TABLE 12. RESULTS OF EXPOSURE OF Ni-Cu ALLOYS (MONELS) TO CORROSION BY SEA WATER

Composition, %				Form in Which Exposed	Duration of Exposure, years	Weight Loss		Maximum Depth of Pitting, inches	Geographical Location	Reference (see p. 429)
Ni	Cu	Fe	Al			mdd	ipy			

Continuous Immersion

Ni	Cu	Fe	Al	Form	Dur.	mdd	ipy	Pit	Location	Ref
65.6	31.6	1.7	...	Cold-rolled strip	.67	6*	.001	...	Bridgeport, Conn.	12
68	29	1	...	Cold-drawn rod	3	7.3	.0012	...	Eastport, Me.	5
68	29	1	...	Cold-rolled strip	.5	15	.0024	...	San Francisco Harbor	7
68	29	1	...	Hot-rolled plate	1	7	.0012	.008	Kill-van-Kull, N. J.	7
68	29	1	...	Hot-rolled plate	1	10	.0017	.014	Cape May, N. J.	7
68	29	1	...	Hot-rolled plate	3	2	.0003	.00	Kure Beach, N. C.	7
68	29	1	...	Hot-rolled plate	2.8	1.8	.0003	.016	"	7
66	29	.9	2.8	Cold-rolled, heat-treated	4	1.3	.0002	.005	"	7

Exposed at Half-Tide Level

Ni	Cu	Fe	Al	Form	Dur.	mdd	ipy	Pit	Location	Ref
68	29	1	...	Cold-drawn rod	3	.06	.00001	...	Eastport, Me.	5
70.3	28	.67	...	Cold-drawn rod	3	.0	.0000	...	Southampton Harbor	10
68	29	1	...	Cold-rolled strip	.5	15	.0024	...	San Francisco Harbor	7
67.7	29.9	1.1	...	Hot-rolled plate	1	.6	.0001	.010	Pacific Ocean, Panama	7
68.8	28.7	1.3	...	Hot-rolled plate	4.1	1.2†	.0002	...	Kure Beach, N. C.	7
64	31	.01	2.4	Hot-rolled plate	4.1	1.0†	.0002	.018	"	7

* Loss in tensile strength 5.0%.
† Partial immersion-fluctuating level.

TABLE 13. RESULTS OF EXPOSURE OF SOME NICKEL ALLOYS TO CORROSION BY SEA WATER

Continuous Immersion

Alloy	Composition, %						Form in Which Exposed	Duration of Exposure, years	Weight Loss		Maximum Depth of Pitting, inches	Geographical Location	Reference (see p. 429)
	Ni	Cr	Mo	Cu	Fe	W			mdd	ipy			
Inconel	80	13.5	...	...	6.5	...	Cold-rolled strip	.5	1.4	.0002	.026	San Francisco Harbor	7
Inconel	79.5	13.6	...	...	6.4	...	Cold-rolled sheet	2.7	4.5	.0008	.063	Kure Beach, N. C.	7
Hastelloy A	56	...	22	...	22	...	Hot-rolled plate	6	7.1	.0012	.023	"	7
Hastelloy B	62	...	32	...	6	...	Hot-rolled plate	6	8.5	.0013	.034	"	7
Hastelloy C	53	17	19	...	6	5	Hot-rolled plate	6	.3	.00005	.000	"	7
Illium G	61	21	6	4	6	2	Cast	3	1.6	.0003	.000		7
Illium R	61	21	5	3	8	2	Hot-rolled plate	3	.3	.00005	.000		7
Invar	36.55	...	...	...	Bal.	...	Hot-rolled bar with mill scale	15	13.2	.0023	.138	Halifax, N. S.	4
Invar	36.55	...	...	...	Bal.	...	Hot-rolled bar with mill scale	15	5.0	.0009	.042	Aukland, N. Z.	4
Invar	36.55	...	...	...	Bal.	...	Hot-rolled bar with mill scale	15	7.8	.0014	.072	Plymouth, Eng.	4
Invar	36.55	...	...	...	Bal.	...	Hot-rolled bar with mill scale	15	7.7	.0013	.098	Colombo, Ceylon	4
Invar	36	...	...	...	Bal.	...	Hot-rolled plate	1	9.0	.0015	.000	Kure Beach, N. C.	7
Exposed at Half-Tide Level													
Invar	36.55	...	...	...	Bal.	...	Hot-rolled bar with mill scale	15	2.2	.0004	.102	Halifax, N. S.	4
Invar	36.55	...	...	...	Bal.	...	Hot-rolled bar with mill scale	15	.9	.0002	.010	Aukland, N. Z.	4
Invar	36.55	...	...	...	Bal.	...	Hot-rolled bar with mill scale	15	3.3	.0006	.010	Plymouth, Eng.	4

TABLE 14. RESULTS OF EXPOSURE OF ALUMINUM AND ALUMINUM ALLOYS TO CORROSION BY SALT WATER

Sheet Specimens — 7.6 × 22.8 × 0.16 cm (3 × 9 × 0.064 in.)

Material	Composition, %							Type of Immersion	Duration of Exposure, years	Loss in Tensile Strength, %	Geographical Location	Reference (see p. 429)
	Al	Mg	Si	Fe	Cu	Mn	Cr					
52S½H	Bal.	2.5	...	...	...	...	.25	Cont.	1	0	Montreal, Canada	23
"	Bal.	2.5	...	...	...	...	.25	Alt.	1	4	"	23
"	Bal.	2.5	...	...	...	...	.25	Cont.	1	0	Black Rock Harbor, Conn.	23
"	Bal.	2.5	...	...	...	...	.25	Alt.	1	0	"	23
"	Bal.	2.5	...	...	...	...	.25	Cont.	2	5	East River, New York, N.Y.	23
"	Bal.	2.5	...	...	...	...	.25	Alt.	2	2	"	23
"	Bal.	2.5	...	...	...	...	.25	Cont.	1	2	Norfolk, Va.	23
"	Bal.	2.5	...	...	...	...	.25	Alt.	2	4	"	23
"	Bal.	2.5	...	...	...	...	.25	Cont.	4	2	Miami, Fla.	23
"	Bal.	2.5	...	...	...	...	.25	Alt.	4	1	"	23
"	Bal.	2.5	...	...	...	...	.25	Cont.	2	4	Mobile, Ala.	23
"	Bal.	2.5	...	...	...	...	.25	Alt.	2	3	"	23
"	Bal.	2.5	...	...	...	...	.25	Cont.	2	2	New Orleans, La.	23
"	Bal.	2.5	...	...	...	...	.25	Alt.	2	1	"	23
53ST	Bal.	1.3	.7	...	...	...	.25	Cont.	1	3	Montreal, Canada	23
"	Bal.	1.3	.7	...	...	...	.25	Alt.	1	4	"	23
"	Bal.	1.3	.7	...	...	...	.25	Cont.	1	5	Black Rock Harbor, Conn.	23
"	Bal.	1.3	.7	...	...	...	.25	Alt.	1	2	"	23
"	Bal.	1.3	.7	...	...	...	.25	Cont.	2	39	East River, New York, N.Y.	23
"	Bal.	1.3	.7	...	...	...	.25	Alt.	2	6	"	23
"	Bal.	1.3	.7	...	...	...	.25	Cont.	1	0	Norfolk, Va.	23
"	Bal.	1.3	.7	...	...	...	.25	Alt.	2	11	"	23
"	Bal.	1.3	.7	...	...	...	.25	Cont.	2	4	Mobile, Ala.	23
"	Bal.	1.3	.7	...	...	...	.25	Alt.	2	3	"	23
"	Bal.	1.3	.7	...	...	...	.25	Cont.	2	2	New Orleans, La.	23
"	Bal.	1.3	.7	...	...	...	.25	Alt.	2	1	"	23
61ST	Bal.	1	.6	...	.25	...	.25	Cont.	1	11	Black Rock Harbor, Conn.	23
"	Bal.	1	.6	...	.25	...	.25	Alt.	1	1	"	23

TABLE 14. RESULTS OF EXPOSURE OF ALUMINUM AND ALUMINUM ALLOYS TO CORROSION BY SALT WATER—Continued

Sheet Specimens — 7.6 × 22.8 × 0.16 cm (3 × 9 × 0.064 in.)

Material	Composition, %							Type of Immersion	Duration of Exposure, years	Loss in Tensile Strength, %	Geographical Location	Reference (see p. 429)
	Al	Mg	Si	Fe	Cu	Mn	Cr					
Alclad 24ST	...	...	...	...	...	...	...	Cont.	1	0	Montreal, Canada	23
"	...	...	...	...	...	...	...	Alt.	1	1	"	23
"	...	...	...	...	...	...	...	Cont.	1	0	Black Rock Harbor, Conn.	23
"	...	...	...	...	...	...	...	Alt.	1	0	"	23
"	...	...	...	...	...	...	...	Cont.	2	1	East River, New York, N.Y.	23
"	...	...	...	...	...	...	...	Alt.	2	0	"	23
"	...	...	...	...	...	...	...	Cont.	1	0	Norfolk, Va.	23
"	...	...	...	...	...	...	...	Alt.	2	1	"	23
"	...	...	...	...	...	...	...	Cont.	4	0	Miami, Fla.	23
"	...	...	...	...	...	...	...	Alt.	4	0	"	23
"	...	...	...	...	...	...	...	Cont.	2	1	Mobile, Ala.	23
"	...	...	...	...	...	...	...	Alt.	2	1	"	23
"	99.4	...	...	.32	...	...	...	Cont.	2	0	New Orleans, La.	23
"	99.4	...	...	.32	...	...	...	Cont.	4	15.9	Bristol Channel, Weston-super-Mare*	2
	...	...	...	...	...	...	...	Alt.	3	...	Southampton, England†	10
Alclad 3S	...	...	...	...	...	...	...	Cont.	1	1	Black Rock Harbor, Conn.	...
"	...	...	...	...	...	...	...	Alt.	1	0	"	...
"	...	...	...	...	...	...	...	Cont.	1	0	Norfolk, Va.	...
"	...	...	...	...	...	...	...	Alt.	2	...	"	...
"	...	...	...	...	...	...	...	Cont.	2	5	New York Harbor	...
"	...	...	...	...	...	...	...	Alt.	2	0	"	...
"	...	...	...	...	...	...	...	Alt.	2	7	Norfolk, Va.	...

*Exposed as polished bar 1⅛ in. diam., 22 in. long.
Weight loss — 47.3 grams.
Corrosion rate — 6.5 mdd.
Equivalent penetration — .0034 ipy.

† Exposed as polished bar 1⅛ in. diam., 22 in. long.
Weight loss — .15 grams.
Corrosion rate — .03 mdd.
Equivalent penetration — .00002 ipy.

TABLE 15. RESULTS OF EXPOSURE OF MAGNESIUM AND ITS ALLOYS TO CORROSION BY SALT WATER

Alternate Exposure, Kure Beach, N. C.*

A.S.T.M. Designation	Composition, %							Form in Which Exposed	Duration of Exposure, years	Loss in Tensile Strength, %
	Mg	Al	Zn	Mn	Fe	Ni	Cu	Si			
AZ63	Bal.	6.8	2.7	.12	.013	.002	.026	.13	Casting with Dow No. 8 treatment	.86	35.4†
AZ63X	Bal.	5.5	3.1	.37	.001	.001	.01	.01	"	.86	2.5†
AZ92X	Bal.	9.2	2.3	.35	<.001	.001	<.01	...	Cast tension bar, ½ in. diam., machined	.67	31
A10X	Bal.	10.3	...	.46	.001	.001	<.01	...	"	.67	31
AZ63X-HT	Bal.	6.2	2.8	.57	<.001	<.001	.016	...	"	.67	51
AS90X	Bal.	9.4	0.0	.24	...	.002	.01	.29	Die cast tension bar, ¼ in. diam., as cast	.29	33
AZ90X	Bal.	8.6	0.60	.21	.003	<.01	...	.01	"	.29	19
A6X	Bal.	6.3	...	.46	.001	<.001	<.01	...	Sheet — 1 in. × 8 in. × .064 in. — no treatment	.4	...
AZ61X	Bal.	6.0	1.5	.58	.0004	.001	<.01	...	Extruded bar, ½ in. diam., machined	.67	59
M1	Bal.	.004	0.0	2.3	.024	.000	<.01	...	"	.67	56

* Unpublished data from Dow Chemical Co., Midland, Mich.
† % loss in weight.

Of the wrought alloys, the 6.5% Al–1% Zn (AZ61X) composition (J-1) is considered best for sea water exposure. Its advantage over the 1.5% Mn alloy (M1), although not disclosed by the data in Table 15, is less susceptibility to pitting. The AZ61X alloy also can be protected by paint more readily than can the M1 alloy.

STAINLESS STEELS

When it occurs, the corrosion of stainless steels by sea water is always localized either as pits distributed at random or at cut edges, or as attack in crevices (contact corrosion) such as may be formed by fouling organisms, deposits, packing, and at lap joints or other faying surfaces.

EFFECT OF COMPOSITION

Differences between types of stainless steel are related principally to the probability of pitting or the number of pits in a given area, although the best compositions tend to be attacked to a lesser depth, as well as less frequently.

From the results of a large number of tests of a variety of stainless steel compositions, the data shown in Table 16 have been selected to provide what is believed

TABLE 16. TYPICAL RESULTS OF TESTS ON STAINLESS STEELS IN SEA WATER AT LOW VELOCITY UNDER CONDITIONS PERMITTING ATTACHMENT AND GROWTH OF MARINE ORGANISMS

Alloy	AISI Type No.	Composition				Duration of Exposure, days	Weight Loss in Grams per Day 12 in. × 12 in. Specimen	Maximum Depth of Pitting, ipy
		Cr	Ni	Mo	C			
13% Cr	410	13.75	.22	...	.08	388	.30	.070
17% Cr	430	17.28	...	...	.10	568	.18	.069
18% Cr–9% Ni	304	18.5	9.01	...	.05	685	.01	.069
18% Cr–10% Ni with Mo	316	18.8	10.2	2.74	.05	1923	.007	.007
25% Cr–20% Ni	310	24.0	21.0	...	.12	988	.013	.007

to be a suitable basis of comparison among some of the standard alloys. These data refer to tests made at one site (Kure Beach, N. C.), and, although the duration of exposure was not the same for each composition, those exposed for the longest periods were in test with the others so that the results are truly comparable.

The data in Table 16 refer to adverse conditions of exposure, that is, continuous, complete, or partial immersion in stagnant or slowly moving warm sea water under conditions where marine organisms may become attached and accelerate corrosion either directly — by establishing crevices — or indirectly through the corrosion accelerating effects of decomposition products after the organisms die.

It will be noted that the best performance under the adverse conditions described may be expected from stainless steel which contains molybdenum in addition to chromium and nickel. Such compositions also represent the best choice of stainless steel for sea water service under more favorable conditions — as at high velocity — since there may occasionally be exposure at low velocity or other unfavorable conditions where the benefits of superior corrosion resistance will be needed.

The weight loss and pitting data of Table 16 do not suffice to define the resistance of the alloys to attack by sea water. Depending on such factors as water temperature, velocity, and the types of fouling organism encountered, rates of attack either lower or higher than those indicated may be experienced. Nevertheless, the relative behavior of the different compositions is not likely to be altered. The alloys indicated as best by these data are most dependable.

In view of the tendency of the alloys to corrode by pitting, and for rates of pitting, with the better compositions, to decrease with time, it is desirable to use as heavy a gage as practical circumstances permit. For example, if other considerations should require the use of stainless steel for tubes in heat exchangers employing salt water as a cooling medium, the preferred composition of tubes would be AISI Type 316 Cr-Ni-Mo alloy and the preferred minimum gage would be 0.078 in.

Effect of Velocity

The stainless steels perform much better when in contact with sea water at high velocities, e.g., over 5 ft per sec (1.5 meters per sec). In a particular case, duplicate specimens of Type 316 stainless steel which contained about 17% chromium, 12% nickel, and 2.4% molybdenum were exposed simultaneously to relatively quiet sea water and to the same water as it flowed past the specimens at 4 to 5 ft per sec in a channel. At the first test location the specimens become covered with marine growths, whereas in the faster water only a few organisms became attached. After exposure for 13 months, the two sets of specimens were attacked as shown in Table 17.

TABLE 17. BEHAVIOR OF Mo 18–8 (TYPE 316) STAINLESS STEEL IN SEA WATER AT LOW VELOCITY WITH FOULING, AND AT HIGH VELOCITY WITH PRACTICALLY NO FOULING

Duration of Exposure — 13 Months

Condition of Exposure	Weight Loss of 8 × 12 × ¼ in. Specimens, grams			Maximum Depth of Pitting, inches		
	1	2	Mean	1	2	Mean
In quiet water where heavy fouling occurred	9	17	13	.105	.040	.073
In water flowing at 4 to 5 ft per sec (1.2 to 1.5 meters per sec) with little fouling	0	0	0	.004*	.000	.002

* Single shallow pit.

It is evident that the corrosive conditions in the faster-flowing water were much more favorable to the stainless steel. This was probably due to the absence of adherent organisms or other deposits. It is also evident that results of exposure tests of stainless steel under conditions of contact with salt water at low velocity, such as are covered by Table 16, provide no measure of ability to resist attack at higher velocity and vice versa. When using stainless steel in salt water, an effort should be made to provide conditions similar to those covered by the tests in sea water at high velocity.

Examples of applications under favorable velocity conditions are pump impellers in fairly continuous operation, and tubes and piping through which sea water flows continuously at high velocity. Stainless steel impellers have been found to be particularly satisfactory in pumps handling polluted harbor waters, especially when hydrogen sulfide, which accelerates attack of copper-base alloys, is present.

Referring again to the example of the stainless steel heat exchanger, it would be preferable to have the water flow through the tubes at a velocity of at least 5 ft per sec rather than through a shell around the tubes, since, under the former conditions of operation, the desired velocity of flow over the alloy surface could be assured most readily. Likewise, it would be desirable to maintain the flow of water as continuous as possible and avoid long shutdown periods with the alloy tubes in contact with stagnant salt water.

Effect of Finish

Under the unfavorable conditions associated with exposure to sea water at low velocity, the performance of austenitic stainless steels is not improved by providing a very smooth finish or high polish. There is some evidence to indicate that where marine organisms and other deposits cannot be avoided, a relatively rough pickled or sandblasted finish is to be preferred over a smooth, bright finish, since, with the rougher finish, although the number of points of attack may be greater, their depth is likely to be less than for the relatively few pits that may develop from a polished surface. Furthermore, protective calcareous deposits adhere more firmly to the rougher surfaces.

A very smooth finish is of no help in preventing attachment of marine organisms. However, a smooth surface is desirable for exposure to sea water at high velocity, since it will be less likely to hold deposits that might initiate local attack.

Effect of Welding

The corrosion resistance of properly made welds has been found to equal that of rolled plate of the same chemical composition. The alloy used for welding rod should always be of a composition equal to, or superior to, that of the base metal with respect to resistance to sea water attack. Welds should be made and finished carefully so as not to leave craters or fissures that might become the seats of local attack.

The alloy in the heat-affected zone adjacent to weld deposits or at cross welds may be attacked at an accelerated rate unless the stainless steel contains a stabilizing element, such as columbium or titanium, or has been heat-treated by rapid cooling from about 1075° C (1950° F) after welding, or after other heat treatment operations in the critical temperature range between 475° and 750° C (900° and 1400° F). Neither columbium nor titanium used as stabilizing elements appears to have any effect on the overall susceptibility of the stainless steels to corrosion or pitting by sea water.

Tests of high-speed spot (shot) or seam welds in thin gage high-tensile stainless steel have shown no preferential or accelerated attack of such welds or the alloy adjacent to them. In these cases, of course, there was no heat treatment after welding.

Galvanic Relationships

Stainless steels are normally passive in sea water and exhibit a relatively noble potential. However, any pits that develop on stainless steel surfaces will be in the active condition and behave galvanically as indicated by the positions of the "active" stainless steels in Table 18. These active areas are subject to acceleration of their corrosion by metals and alloys below the "active" positions of stainless steels in Table 18, including the stainless steels of the same composition in the passive state.

Table 18. GALVANIC SERIES IN SEA WATER

Magnesium
Magnesium alloys

Zinc
Galvanized steel or galvanized wrought iron

Aluminum 52SH*
Aluminum 4S
Aluminum 3S
Aluminum 2S
Aluminum 53S–T
Alclad

Cadmium

Aluminum A17S–T
Aluminum 17S–T
Aluminum 24S–T

Mild steel
Wrought iron
Cast iron

Ni-Resist

13% chromium stainless steel, type 410 (active)

50-50 lead tin solder

18-8 stainless steel, type 304 (active)
18-8, 3% Mo stainless steel, type 316 (active)

Lead
Tin

Muntz metal
Manganese bronze
Naval brass

Nickel (active)
78% Ni -13.5% Cr–6% Fe (Inconel) (active)
Yellow brass
Admiralty brass
Aluminum bronze
Red brass
Copper
Silicon bronze
5% Zn–20% Ni, Bal. Cu (Ambrac)
70% Cu–30% Ni
88% Cu–2% Zn–10% Sn (Composition G-bronze)
88% Cu–3% Zn–6.5% Sn–1.5% Pb (Comp. M-bronze)

Nickel (passive)
78% Ni-13.5% Cr–6% Fe (Inconel) (passive)

70% Ni–30% Cu (Monel)

18-8 stainless steel, type 304 (passive)
18-8, 3% Mo stainless steel, type 316 (passive)

---

* Aluminum code numbers and corresponding compositions are listed in Table 1, p. 39.

Steel, zinc, and aluminum are less noble than active stainless steel, so that when stainless steel is in metallic contact with them, it may be expected to receive galvanic protection to an extent influenced by the relative areas of the metals involved. Even under the unfavorable conditions of exposure to sea water at low velocity, 18-8 stainless steel (Type 304) has been found to remain unattacked and free from pitting when in contact with an equal area of carbon steel, aluminum, or galvanized steel. Similarly, 13% chromium steel impellers have performed well in pumps with cast iron casings handling brackish harbor water.

An area of carbon steel equal to that of the stainless steel is ordinarily sufficient to protect the latter from corrosion by sea water. The protective effect of smaller areas of steel is uncertain and probably should not be depended upon, especially since galvanic corrosion of the steel will be accelerated to an intolerable extent as its relative area is decreased. Similarly, relatively small areas of copper-base alloys, especially brasses, will suffer severe acceleration of corrosion when coupled with large areas of stainless steel. Brass fastenings should not be used to join large stainless steel sections. The reverse relationship, that is, of a small area of stainless steel with a large area of carbon steel, is quite satisfactory, as for example, stainless steel fastenings in steel assemblies, or stainless steel welds on steel plates. In these the relatively little galvanic corrosion of the carbon steel has not been found to concentrate in the immediate vicinity of the contact, but tends to become dissipated over the relatively large area of exposed carbon steel. Austenitic stainless steel welds, e.g., 25% Cr–20% Ni, in

TABLE 19. RESULTS OF EXPOSURE OF MISCELLANEOUS MATERIALS TO CORROSION BY SEA WATER
Continuous Immersion

Material and Composition	Form in Which Exposed	Duration of Exposure, years	Weight Loss mdd	Weight Loss ipy	Maximum Depth of Pitting, inches	Geographical Location	Reference (see p. 429)
Babbitt-Lead (80% Pb, 15% Sb, 5% Sn)	Cast plate	1.4	3.5	.0006	....	Kure Beach, N. C.	7
Babbitt-Tin (86.9% Sn, 7.4% Sb, 3.7% Cu)	Cast plate	1.4	12	.002	....	" " "	7
Cadmium Plate (0.0022 in. on steel)	Plated panel	1	3.7	.0007	.004	" " "	7
Galvanized Steel	Plate	1	5.1	.001	.0037	Southampton	15
Galvanized Steel	Bar	4	12.6	.0025	....	Bristol Channel, Weston-super-Mare	2
Galvanized Steel — 2.5 oz per sq ft	Plate	1	5.7	.0011	....	Kure Beach, N. C.	7
Galvanized Steel — 3.7 oz per sq ft	Plate	.5	10.4	.0021	....	" " "	7
Lead (99.96% Pb)	Bar	4	4.0	.0005	....	Bristol Channel, Weston-super-Mare	2
Lead-Antimonial (98.4% Pb, 1.6% Sb)	Bar	4	3.0	.0004	....	" " "	2
Lead (Chemical)	Sheet	.5	3.1	.0004	.007	San Francisco Harbor	7
Lead	Plate	1.7	4.7	.0006	....	Kure Beach, N. C.	7
Lead-Tin Solder (50% Pb, 50% Sn)	Sheet	.5	15	.0021	....	Bogue Inlet, N. C.	7
Lead-Tin Solder (60% Pb, 40% Sn on Cu)	Plate	2.1	2.1	.0003	....	Kure Beach, N. C.	7
Stellite 6 (28-31% Cr, 4-6 W, 1-1.25% C, Bal. Co)	Plate	2	.7	.0001	.007	" " "	7
Tin (Common) (99.2% Sn)	Cast bar	4	.15	.00003	....	Bristol Channel, Weston-super-Mare	2
Tin (High Grade) (99.75% Sn)	Cast bar	4	.44	.00009	....	" " "	2
Zinc (99.1% Zn)	Sheet	1	5.6	.0011	....	Eastport, Me.	5
Zinc (Commercial) (1.12% Pb)	Bar	4	18	.0036	.034	Bristol Channel, Weston-super-Mare	2
Zinc (0.5% Pb, 0.15% Cd, 0.015% Fe)	Sheet	.5	9.5	.0019	....	Kure Beach, N. C.	7
Zinc (0.01% Pb, 0.003% Cd, 0.01% Fe)	Sheet	.5	9.5	.0019	....	" " "	7
Galvanized Steel	Plate	1	5.4	.0011	.0037	Southampton	15
Lead (Chemical)	Sheet	.5	3.6	.0005	.006	San Francisco Harbor	7
Zinc (99.1% Zn)	Sheet	3	5.1	.0010	....	Eastport, Me.	5
Zinc (Prime Western) (1.0% Pb, 0.30% Cd, 0.04% Fe)	Cast plate	1	5.3	.001	....	Pacific Ocean — Panama	7
Zinc (Special) (0.002% Pb, 0.0017% Cd, 0.001% Fe)	Cast plate	1	5.1	.001	....	" "	7
Zinc (0.06% Pb, 0.003% Cd, 0.007% Fe)	Sheet	1	7.1	.0014	....	" "	7

carbon steel or low-alloy steel plates have remained completely free from attack after exposure in slowly moving sea water for several years.

MISCELLANEOUS MATERIALS

In Table 19 data have been collected on the performance of several additional metals and alloys exposed to natural sea water. *Zinc* coatings, such as are produced in galvanizing, provide considerable protection to steel. This effect includes galvanic protection which persists until most of the zinc is gone, and also serves to prevent pitting of the steel while zinc remains. It may be expected that under ordinary conditions of exposure to sea water at low velocity, protection of steel by zinc is effective over a period equivalent to about one year for each one thousandth of an inch (1 mil) of zinc thickness, or a little less than one year per ounce when coatings are measured in ounces per square foot of sheet.

Zinc also provides effective protection when applied as a sprayed coating. In a particular case, a sprayed coating 10 mils thick prevented corrosion of steel immersed in sea water for about 5 years.

The usual impurities in rolled or cast zinc have no significant effect on normal corrosion by sea water. Their effect appears of greatest practical significance in the use of "zincs" for galvanic protection in sea water. In this service, the purer the zinc, the greater will be the output of protective galvanic currents over an extended period of exposure.

Both *lead* and *tin,* as well as *Pb-Sn alloys,* are highly resistant to attack by sea water at low velocity. Lead is subject to erosion by sea water at high velocity, but tin resists such action very well and contributes this property to the Pb-Sn alloys. As a result, the 60% Pb–40% Sn alloy is commonly applied to copper to protect it from attack by sea water at high velocity. Such "tin" coatings should be applied in substantial thickness by "wiping" rather than by dipping, since the very thin-dipped coatings may be imperfect or are readily damaged so as to expose bare copper. Such exposed copper suffers severely accelerated attack, especially when in contact with sea water at high velocity. Consequently, thin "tin" or solder coatings on copper are likely to be worse than none at all.

The Co-Cr-W* alloys are highly resistant to attack by sea water although they are susceptible to appreciable local corrosion in cracks or crevices.

GALVANIC EFFECTS

Since sea water is a good electrolytic conductor, and since it is common practice to combine dissimilar metals and alloys in structures exposed to sea-water attack, galvanic corrosion is encountered quite frequently.

Galvanic action in sea water follows the general laws of galvanic corrosion, as discussed in detail beginning p. 481. However, the calcium, magnesium, and strontium present in sea water tend to precipitate as carbonates on cathodic surfaces. The effect of such precipitated deposits, plus heavy growths of marine organisms, is to stifle the galvanic effect and to distribute the galvanic protection over larger areas of cathodic surfaces than would be the case in their absence. Marine growths also tend to distribute galvanic action over the anodic surfaces by interposing a common resistance which reduces the relative importance of the initial resistance of the electrolyte.

On the basis of practical experience and experimental observations, a chart has

* Stellites.

been constructed as a qualitative guide to what may be expected when different metals and alloys are combined with different area relationships in sea water exposure. Both in the chart and in Table 18 the common materials have been arranged in a galvanic series with respect to sea water. In the case of active-passive materials, like the stainless steels, it has been assumed in the chart that these alloys may suffer accelerated corrosion when in contact with all materials more noble than their active state and that they may accelerate corrosion of all materials less noble than their passive state. In other words, the chart tends to err in the safe direction.

BIOFOULING CHARACTERISTICS OF SOME COMMON METALS AND ALLOYS

In Tables 20 to 22 (pp. 427–429) a number of common metals and alloys are listed with respect to their tendency to become fouled when exposed in natural sea water under what may be considered severe fouling conditions. Where the environment is less favorable to the fouling organisms, it may be expected that those alloys which show variable fouling tendencies will be shifted to the non-fouling group, whereas exceptionally bad fouling conditions will shift them to the group most likely to foul. All listings are based on actual observations of the alloys, or ones like them, after exposure to natural sea water under conditions where the vulnerable materials were fouled severely by most of the common fouling organisms, including *Balanus* (barnacles), annelids, Bryozoa, hydroids (*Tubularia*), oysters, mussels, and algae.

All the alloys that are listed as being not likely to foul contain copper as an essential constituent. Observations have shown also that in order to remain free from adherent fouling organisms, such materials must be corroded sufficiently to release copper as a corrosion product at a rate of more than 5 mdd of exposed surface. This is equivalent to thinning at a rate of about 0.001 ipy.

Whenever conditions of exposure are such as to prevent release of copper above this critical rate, fouling will then occur. Factors likely to cause this suppression of corrosion and subsequent fouling include:

1. Galvanic protection as a result of metallic contact with a substantial area of some less noble metal or alloy, e.g., iron or zinc.
2. The development of calcareous deposits which stifle attack by sea water.
3. The formation of adherent corrosion product films which may lower the copper solution rate below the critical value upon prolonged exposure. Occasional exposure to the atmosphere and drying out of the corrosion product films may increase their protective effect and thus promote fouling.
4. Dezincification type of corrosion reaction in which copper released does not persist in the ionic form and may be kept from redissolving by galvanic action of the alloy corroding under the layer of copper.
5. The accidental or deliberate coating of the surface with films, such as oil or grease, which stifle corrosion and allow fouling to occur during the early stages of exposure.

In addition to copper and its alloys, there are some other metals that remain free from fouling because of the continual sloughing off of marine organisms along with corrosion products to which they are attached. Zinc acts in this way when its corrosion is accelerated by contact with substantial areas of steel or other more noble metals. Cadmium, aluminum, and magnesium probably would behave in the same way.

While it must be remembered that the following classifications are qualitative, it is

SEA WATER CORROSION OF GALVANIC COUPLES

Legend

☐ The corrosion of the metal under consideration will be reduced considerably in the vicinity of the contact.
○ The corrosion of the metal under consideration will be reduced slightly.
△ The galvanic effect will be slight with the direction uncertain.
▲ The corrosion of the metal under consideration will be increased slightly.
● The corrosion of the metal under consideration will be increased moderately.
◀ The corrosion of the metal under consideration will be increased considerably.

S Exposed area of the metal under consideration is small compared with the area of the metal with which it is coupled.
E Exposed area of the metal under consideration is approximately equal to that of the metal with which it is coupled.
L Exposed area of the metal under consideration is large compared to that of the metal with which it is coupled.

BEHAVIOR OF METALS AND ALLOYS IN SEA WATER 421

SEA WATER CORROSION OF GALVANIC COUPLES — Continued

Legend

- □ The corrosion of the metal under consideration will be reduced considerably in the vicinity of the contact.
- ○ The corrosion of the metal under consideration will be reduced slightly.
- △ The galvanic effect will be slight with the direction uncertain.
- ▲ The corrosion of the metal under consideration will be increased slightly.
- ◆ The corrosion of the metal under consideration will be increased moderately.
- ● The corrosion of the metal under consideration will be increased considerably.

- S — Exposed area of the metal under consideration is small compared with the area of the metal with which it is coupled.
- E — Exposed area of the metal under consideration is approximately equal to that of the metal with which it is coupled.
- L — Exposed area of the metal under consideration is large compared to that of the metal with which it is coupled.

BEHAVIOR OF METALS AND ALLOYS IN SEA WATER 423

SEA WATER CORROSION OF GALVANIC COUPLES — Continued

Legend

- ☐ The corrosion of the metal under consideration will be reduced considerably in the vicinity of the contact.
- ○ The corrosion of the metal under consideration will be reduced slightly.
- △ The galvanic effect will be slight with the direction uncertain.
- ▲ The corrosion of the metal under consideration will be increased slightly.
- ● The corrosion of the metal under consideration will be increased moderately.
- ◀ The corrosion of the metal under consideration will be increased considerably.

- S Exposed area of the metal under consideration is small compared with the area of the metal with which it is coupled.
- E Exposed area of the metal under consideration is approximately equal to that of the metal with which it is coupled.
- L Exposed area of the metal under consideration is large compared to that of the metal with which it is coupled.

METAL CONSIDERED	COUPLED WITH →	Muntz Metal			Manganese Bronze			Naval Brass			Nickel			Inconel (13% Cr, 6.5% Fe, Bal. Ni)		
		S	E	L	S	E	L	S	E	L	S	E	L	S	E	L
Magnesium		☐	☐	○	☐	☐	○	☐	☐	○	☐	☐	○	☐	☐	○
Magnesium Alloys		☐	☐	○	☐	☐	○	☐	☐	○	☐	☐	○	☐	☐	○
Zinc		☐	☐	○	☐	☐	○	☐	☐	○	☐	☐	○	☐	☐	○
Galvanized Steel		☐	☐	○	☐	☐	○	☐	☐	○	☐	☐	○	☐	☐	○
Aluminum 52S		☐	☐	○	☐	☐	○	☐	☐	○	☐	☐	○	☐	☐	○
Aluminum 4S		☐	☐	○	☐	☐	○	☐	☐	○	☐	☐	○	☐	☐	○
Aluminum 2S		☐	☐	○	☐	☐	○	☐	☐	○	☐	☐	○	☐	☐	○
Alclad		☐	☐	○	☐	☐	○	☐	☐	○	☐	☐	○	☐	☐	○
Aluminum 3S		☐	☐	○	☐	☐	○	☐	☐	○	☐	☐	○	☐	☐	○
Aluminum 53S-T		☐	☐	○	☐	☐	○	☐	☐	○	☐	☐	○	☐	☐	○
Aluminum 61S-T		☐	☐	○	☐	☐	○	☐	☐	○	☐	☐	○	☐	☐	○
Cadmium		☐	☐	○	☐	☐	○	☐	☐	○	☐	☐	○	☐	☐	○
Aluminum A17S-T		☐	☐	○	☐	☐	○	☐	☐	○	☐	☐	○	☐	☐	○
Aluminum 17S-T		☐	☐	○	☐	☐	○	☐	☐	○	☐	☐	○	☐	☐	○
Aluminum 24S-T		☐	☐	○	☐	☐	○	☐	☐	○	☐	☐	○	☐	☐	○
Mild Steel		☐	☐	○	☐	☐	○	☐	☐	○	☐	☐	○	☐	☐	○
Wrought Iron		☐	☐	○	☐	☐	○	☐	☐	○	☐	☐	○	☐	☐	○
Low-Alloy Steels		☐	☐	○	☐	☐	○	☐	☐	○	☐	☐	○	☐	☐	○
Cast Iron		☐	☐	○	☐	☐	○	☐	☐	○	☐	☐	○	☐	☐	○
Low-Alloy Cast Irons		☐	☐	○	☐	☐	○	☐	☐	○	☐	☐	○	☐	☐	○
4–6% Chromium Steel		☐	☐	○	☐	☐	○	☐	☐	○	☐	☐	○	☐	☐	○
Ni Cast Iron		☐	☐	○	☐	☐	○	☐	☐	○	☐	☐	○	☐	☐	○
12–14% Chromium Steel		◀	△	△	◀	△	△	◀	△	△	☐	○	△	☐	○	△
Lead-Tin Solders		◀	△	△	◀	△	△	◀	△	△	△	△	△	△	△	△
16–18% Chromium Steel		●	△	△	●	△	△	●	△	△	☐	○	△	☐	○	△
Lead		◀	△	△	◀	△	△	◀	△	△	△	△	△	△	△	△
Tin		◀	△	△	◀	△	△	◀	△	△	△	△	△	△	△	△
Muntz Metal		△	△	△	☐	○	△	△	△	△	☐	○	△	☐	○	△
Manganese Bronze		●	△	△	△	△	△	●	△	△	☐	○	△	☐	○	△
Naval Brass		△	△	△	☐	○	△	△	△	△	☐	○	△	☐	○	△
Nickel		●	△	△	●	△	△	●	△	△	△	△	△	△	△	△
Yellow Brass		●	△	△	●	△	△	●	△	△	●	△	△	●	△	△
Admiralty Brass		●	△	△	●	△	△	●	△	△	●	△	△	●	△	△
Aluminum Bronze		●	△	△	●	△	△	●	△	△	●	△	△	●	△	△
Red Brass		●	△	△	●	△	△	●	△	△	●	△	△	●	△	△
Copper		●	△	△	●	△	△	●	△	△	●	△	△	●	△	△
Silicon Bronze		●	△	△	●	△	△	●	△	△	●	△	△	●	△	△
Nickel Silver		●	△	△	●	△	△	●	△	△	●	△	△	●	△	△
70–30 Copper Nickel		●	△	△	●	△	△	●	△	△	●	△	△	●	△	△
Composition G Bronze		●	△	△	●	△	△	●	△	△	●	△	△	●	△	△
Composition M Bronze		●	△	△	●	△	△	●	△	△	●	△	△	●	△	△
Inconel		●	△	△	●	△	△	●	△	△	●	△	△	△	△	△
Silver Solder		●	△	△	●	△	△	●	△	△	●	△	△	△	△	△
70–30 Nickel Copper		●	△	△	●	△	△	●	△	△	●	△	△	●	△	△
25–30% Chromium Steel		●	△	△	●	△	△	●	△	△	●	△	△	●	△	△
Cr-Ni Stainless Steel		●	△	△	●	△	△	●	△	△	●	△	△	●	△	△
Cr-Ni-Mo Stainless Steel		●	△	△	●	△	△	●	△	△	●	△	△	●	△	△
Graphite		●	●	△	●	●	△	●	◀	△	●	◀	△	●	◀	△

BEHAVIOR OF METALS AND ALLOYS IN SEA WATER

SEA WATER CORROSION OF GALVANIC COUPLES — *Continued*

Legend

□ The corrosion of the metal under consideration will be reduced considerably in the vicinity of the contact.
○ The corrosion of the metal under consideration will be reduced slightly.
△ The galvanic effect will be slight with the direction uncertain.
▲ The corrosion of the metal under consideration will be increased slightly.
◆ The corrosion of the metal under consideration will be increased moderately.
● The corrosion of the metal under consideration will be increased considerably.

S Exposed area of the metal under consideration is small compared with the area of the metal with which it is coupled.
E Exposed area of the metal under consideration is approximately equal to that of the metal with which it is coupled.
L Exposed area of the metal under consideration is large compared to that of the metal with which it is coupled.

METAL CONSIDERED ↓ / COUPLED WITH →	S/E/L	Magnesium	Magnesium Alloys	Zinc	Galvanized Steel	Aluminum 52S	Aluminum 4S	Aluminum 2S	Alclad	Aluminum 3S	Aluminum 53S-T	Aluminum 61S-T	Cadmium	Aluminum A17S-T	Aluminum 17S-T	Aluminum 24S-T	Mild Steel	Wrought Iron	Low-Alloy Steels	Cast Iron	Low-Alloy Cast Irons	4-6% Chromium Steel	Ni Cast Iron	12-14% Chromium Steel	Lead-Tin Solders	16-18% Chromium Steel	Lead	Tin	Muntz Metal	Manganese Bronze	Naval Brass	Nickel	Yellow Brass	Admiralty Brass	Aluminum Bronze	Red Brass	Copper	Silicon Bronze	Nickel Silver	70-30 Copper Nickel	Composition G Bronze	Composition M Bronze	Inconel	Silver Solder	70-30 Nickel Copper	25-30% Chromium Steel	Cr-Ni Stainless Steel	Cr-Ni-Mo Stainless Steel	Graphite
Silver Solder	S	□	□	□	□	□	□	□	□	□	□	□	□	□	□	□	□	□	□	□	□	□	□	□	□	○	□	□	□	□	□	□	□	□	□	□	□	□	□	△	△	△	△	△	△	△	△	△	●
	E	□	□	□	□	□	□	□	□	□	□	□	□	□	□	□	□	□	□	□	□	□	□	○	○	○	○	○	○	○	○	△	○	○	○	○	○	○	○	△	△	△	△	△	△	△	△	△	△
	L	○	○	○	○	○	○	○	○	○	○	○	○	○	○	○	○	○	○	○	○	○	□	△	△	△	△	△	△	△	△	△	△	△	△	△	△	△	△	△	△	△	△	△	△	△	△	△	△
70-30 Nickel Copper	S	□	□	□	□	□	□	□	□	□	□	□	□	□	□	□	□	□	□	□	□	□	□	□	□	□	□	□	□	□	□	□	□	□	□	□	□	□	□	□	□	□	□	□	□	△	△	△	●
	E	□	□	□	□	□	□	□	□	□	□	□	□	□	□	□	□	□	□	□	□	□	□	○	○	○	○	○	○	○	○	○	○	○	○	○	○	○	○	○	○	○	○	○	○	△	△	△	△
	L	○	○	○	○	○	○	○	○	○	○	○	○	○	○	○	○	○	○	○	○	○	○	○	○	○	○	○	○	○	○	○	○	○	○	○	○	○	○	○	○	○	○	○	○	△	△	△	△

TABLE 20. METALS AND ALLOYS LEAST LIKELY TO FOUL

Material	Nominal Composition, %						
	Cu	Zn	Sn	Ni	Pb	Fe	Other
Admiralty brass	70	29	1	....	....	....	....
Admiralty brass with As	70	29	1	....	....	....	As 0.05
Admiralty brass with Sb	70	29	1	....	....	....	Sb 0.05
Admiralty brass with P	70	29	1	....	....	....	P 0.02
"Adnic"	70	....	1	29	....	....	....
"Ambrac"	75	5	....	20	....	....	....
Arsenical copper	99.5	....	....	....	....	....	As 0.3
Beryllium copper	97.4	....	....	0.25	....	....	Be 2.3
Brasses containing more than 65% Cu	>65	<35	....	....	....	....	....
Brasses containing Sn and more than 80% Cu	>80	Bal.	1–2.5	....	....	....	....
Bronzes — tin	>80	....	<10	....	....	....	....
Bronze — Comp. G	88	2	10	....	....	....	....
Bronze — Comp. M	88	3	6.5	....	....	1.5	....
Bronze — nickel	>80	....	....	1–10	....	....	....
Cartridge brass	70	30	....	....	....	....	....
Chain bronze	95	....	5	....	....	....	....
Commercial bronze	90	10	....	....	....	....	....
Copper	99+	....	....	....	....	....	....
Cu-Ni Alloys containing less than 30% Ni and less than 0.15% Fe	>70	....	....	<30	....	<0.15	....
"Duronze"	97	....	2	....	....	....	Si 1
"Everdur"	96	....	....	....	....	....	Si 3
Free-cutting leaded brass	>65	Bal.	....	....	1.5	....	....
German silver	64	18	....	18	....	....	....
Gilding metal	96	4	....	....	....	....	....
Government bronze	88	2	10	....	....	....	....
Gun metal	88	2	10	....	....	....	....
Hardware bronze	89	9	....	....	2	....	....
"Herculoy"	96	....	0.5	....	....	....	Si 3.25
Low brass	80	20	....	....	....	....	....
Ni-Cu alloys containing less than 30% Ni and 0.15% Fe	>70	....	....	<30	....	<0.15	....
Nickel silver	64	18	....	18	....	....	....
Olympic bronze	96.5	1	....	....	....	....	Si 2.75
Ounce metal	85	5	5	....	5	....	....
Phosphor bronze	Bal.	....	4–10	....	....	....	....
P. M. G. bronze	95.5	....	....	....	....	1.5	Si 3
Red brass — cast	85	5	5	....	5	....	....
Red brass — wrought	85	15	....	....	....	....	....
Rich low brass	85	15	....	....	....	....	....
Silicon bronze	97	....	....	....	....	....	Si 3

428 SPECIAL TOPICS IN CORROSION

TABLE 21. METALS AND ALLOYS MOST LIKELY TO FOUL

Material	Nominal Composition, %								
	Al	Cu	Ni	Zn	Sn	Cr	Fe	Mo	Other
Aluminum	....	....	....	....	....	....	....	....	....
Aluminum alloys	....	....	....	....	....	....	....	....	....
Aluminum brass with or without dezincification inhibitors	2	76	....	22	....	....	....	....	....
Aluminum bronze	10	10	....	....	....	....	....	....	....
Antimonial lead	....	....	....	....	....	....	....	....	....
Antimony	....	....	....	....	....	....	....	....	....
Armco iron	....	....	....	....	....	....	99+	....	....
Babbitt	....	35	....	....	89	....	....	....	Sb 7.5
Bronze — aluminum	10	90	....	....	....	....	....	....	....
Cadmium plate	....	....	....	....	....	....	....	....	....
Cast iron	....	....	....	....	....	....	....	....	....
Cast steel	....	....	....	....	....	....	....	....	....
Chemical lead	....	....	....	....	....	....	....	....	....
Chromium plate	....	....	....	....	....	....	....	....	....
Cobalt	....	....	....	....	....	....	....	....	....
Constantan	....	55	45	....	....	....	....	....	....
Copper steel	....	0.25	....	....	....	....	99+	....	....
Cu-Ni alloys containing more than 40% Ni	....	....	....	....	....	....	....	....	....
"Dowmetal" (Magnesium alloys)	....	....	....	....	....	....	....	....	....
Duralumin	....	....	....	....	....	....	....	....	....
"Duriron"	....	....	....	....	....	....	82	....	Si 14.5
"Hastelloy A"	....	....	Bal.	....	....	....	22	22	....
"Hastelloy B"	....	....	Bal.	....	....	....	6	32	....
"Hastelloy C"	....	....	Bal.	....	....	17	6	19	W 5
"Hastelloy D"	....	3	Bal.	....	....	....	....	....	Si 10
"Inconel"	....	....	Bal.	....	....	13	6.5	....	....
Ingot iron	....	....	....	....	....	....	....	....	....
"Invar"	....	....	36	....	....	....	64	....	....
Iron	....	....	....	....	....	....	....	....	....
"K-Monel"	2.75	29	66	....	....	....	0.9	....	....
Lead	....	....	....	....	....	....	....	....	....
Pb-Sn alloys	....	....	....	....	....	....	....	....	....
Magnesium	....	....	....	....	....	....	....	....	....
Magnesium alloys	....	....	....	....	....	....	....	....	....
Manganese steel	....	....	....	....	....	....	Bal.	....	Mn 11–15
"Monel"	....	30	67	....	....	....	1.4	....	....
"Nichrome"	....	....	62	....	....	15	Bal.	....	....
Nickel	....	....	....	....	....	....	....	....	....
Ni-Al bronze	4	92	4	....	....	....	....	....	....
Ni-Cr alloys	....	....	....	....	....	....	....	....	....
Ni-Cr-Fe alloys	....	....	....	....	....	....	....	....	....
Ni-Cu alloys containing more than 40% Ni	....	....	....	....	....	....	....	....	....
"Ni-Resist"	....	5–7	12–15	....	....	....	Bal.	....	....
Nitrided steel	....	....	....	....	....	....	....	....	....
Pewter	....	....	....	....	85–90	....	....	....	Pb-Bal.
"S-Monel"	....	30	63	....	....	....	2	....	Si 4
Soft solder	....	....	....	....	50	....	....	....	Pb 50
Stainless steels	....	....	....	....	....	....	....	....	....
"Stellite"	....	....	....	....	....	....	....	....	....
Tin	....	....	....	....	....	....	....	....	....
Sn-Pb alloys	....	....	....	....	....	....	....	....	....
Wiping solder	....	....	....	....	40	....	....	....	Pb 60
"Worthite"	....	....	24	....	....	19	Bal.	3	....
Wrought iron	....	....	....	....	....	....	....	....	....

felt that they will be useful to those who wish to choose from a group of otherwise suitable materials those most likely to demonstrate the fouling or anti-fouling characteristics desired in a particular application.

TABLE 22. METALS AND ALLOYS SHOWING CONSIDERABLE VARIATION IN FOULING TENDENCY

Material	Nominal Composition, %						
	Cu	Zn	Sn	Ni	Pb	Fe	Other
Brasses containing less than 65% Cu	<65	Bal.	....	....	....	....	....
Bronze — Manganese	58	40	....	....	....	....	Mn 2
Commercial brass	65	35	....	....	....	....	....
Common brass	65	35	....	....	....	....	....
Cu-Ni alloys containing less than 30% Ni and more than 0.15% Fe	>70	....	....	<30	....	>0.15	....
Cu-Ni alloys containing 30 to 40% Ni	60–70	....	....	30–40	....	....	....
High brass	65	35	....	....	....	....	....
Leaded high brass	65	34	....	....	1	....	....
Manganese bronze	58	40	....	....	....	....	Mn 2
Muntz metal	60	40	....	....	....	....	....
Naval brass	60	39	1	....	....	....	....
Ni-Cu alloys containing less than 30% Ni and more than 0.15% Fe	>70	....	....	<30	....	>0.15	....
Ni-Cu alloys containing 30 to 40% Ni	60–70	....	....	30–40	....	....	....
Silver Tobin bronze	60	39	1	....	....	....	....
Yellow brass	65	35	....	....	....	....	....
Zinc	....	99+	....	....	....	....	....

LITERATURE REFERENCES FOR TABLES 1, 3–9, 11–14, 19

[1] H. M. Howe, *Effets relatifs de la corrosion sur le fer, l'acier doux et l'acier au nickel, dans l'air, l'eau de mer, et l'eau ordinaire*, Congrès International de Methodes d'Essais des Matériaux de Construction, Vol 2, Part 1, pp. 229–266, 1901.

[2] J. N. Friend, "The Relative Corrodibilities of Ferrous and Non-Ferrous Metals and Alloys. Part 1," *J. Inst. Metals*, **39** (No. 1), 111 (1928).

[3] E. Herzog, "Les aciers semi-inoxydables et la resistance aux eaux naturelles," *Journées de la lutte contre la corrosion*, Paris, November 19–24, 1938. *Chimie et industrie*, **41** (No. 4), pp. 453C–459C.

[4] *Deterioration of Structures in Sea Water*, Eighteenth Report of the Committee of the Institution of Civil Engineers, London, 1938.

[5] *Materials Corrosion Investigation at Eastport, Me.*, Fifth Interim Report, U.S. Engineer Office, Boston, Mass., June, 1946.

[6] J. C. Hudson, *The Corrosion of Iron and Steel*, p. 154, Chapman & Hall, Ltd., London, 1940.

[7] Unpublished data from International Nickel Co., Inc., New York, N.Y.

[8] G. G. Greulich, "Steel Piles for Subaqueous Work," *Civil Eng.*, **12** (No. 10), 566 (October, 1942).

[9] Unpublished data from Wm. F. Clapp, William F. Clapp Laboratories, Duxbury, Mass.

[10] J. N. Friend, "The Relative Corrodibilities of Ferrous and Non-Ferrous Metals and Alloys, Part III," *Jour. Inst. Metals*, **48** (No. 1), 109 (1932).

[11] C. L. Bulow, "Corrosion and Biofouling of Copper-Base Alloys in Sea Water," *Trans. Electrochem. Soc.*, **87**, 127 (1945).

[12] Report of Subcommittee VII, Committee B-3, *Proc. Am. Soc. Testing Materials*, **35**, 160 (1935).

[13] A. W. Tracy, D. H. Thompson, and J. R Freeman, Jr., "The Atmospheric Corrosion of Copper," *Proc. Am. Soc. Testing Materials*, **43**, 615 (1943).

[14] J. F. J. Thomas and A. C. Halferdahl, "Comparative Corrosion Resistance to Sea Water of Low-Alloy, High Strength Steels," *Canadian Chem. Process Indus.*, **43** (January, 1945).

[15] *Deterioration of Structures in Timber, Metal and Concrete Exposed to the Action of Sea Water*, Fifteenth Report of the Committee of the Institution of Civil Engineers, London, 1935.

[16] R. B. Mears and R. H. Brown, "Resistance of Aluminum-Base Alloys to Marine Exposures," *Trans. Soc. Naval Architects and Marine Engrs.*, **52**, 91 (1944).

GENERAL REFERENCES (BEHAVIOR OF METALS AND ALLOYS IN SEA WATER)

BASSETT, W. H., and C. H. DAVIS, "Corrosion of Copper Alloys in Sea Water," *Trans. Am. Inst. Mining Met. Eng.*, **71**, 745 (1925).
BENGOUGH, G. D., and R. MAY, "Report to the Corrosion Research Committee of the Institute of Metals," various volumes of *J. Inst. Metals*.
BROWNSDON, H. W., and L. C. BANNISTER, "A Modified Impingment Corrosion Apparatus," *J. Inst. Metals*, **XLIX** (No. 2), 123 (1932).
COOK, MAURICE, and I. BOODSON, "The Effect of Crystal Size on the Resistance of Copper Alloys to Corrosion by Sea Water," *Trans. Faraday Soc.*, **38**, 391 (1942).
FFIELD, P., "Recommendations for Using Steel Piping in Salt Water Systems," *J. Am. Soc. Naval Engrs.*, **57** (No. 1) (1945).
MORRIS, A., "Seasonal Variation in Rate of Impingement Corrosion," *Trans. Institute of Metals Division, Am. Inst. Mining Metal Engrs.*, **99**, 274 (1932).
NOYES, M. S., "Copper-Nickel (70:30) Alloy," *J. Am. Soc. Naval Engrs.*, **48** (No. 1), (1936).

Paints and Organic Coatings for Sea-Water Exposure

G. W. SEAGREN*

The problem of providing protection in sea water is aggravated by the high conductivity of sea water which favors local cell action, a high content of sodium ions which promotes the development of alkalinity at cathodic areas, a high total ion content which facilitates the establishment of ion concentration gradients and resultant osmotic action through coating films, and a high content of chloride ions which are active in the breaking down of passivity. The ideal protective coating is one which would completely and permanently bar the metal surface from contact with the corroding environment. This condition is never realized in practice, and particularly where the exposure is to sea water. The kind of structure to be painted, the facilities available for coating application, etc., usually necessitate marked departures from the ideal. Even if the organic coating could be applied and maintained free of mechanical imperfections, such as pinholes, cracks, or thin spots, it is usually sufficiently permeable to permit water to diffuse through it at a rate sufficient for corrosion to proceed beneath the film. For this reason, pigments are usually incorporated with the object of inhibiting corrosion in the presence of any water which diffuses through the coating. Some of the most effective pigments for this purpose are the zinc and lead chromates, basic lead sulfate, white lead, and red lead.

Those metal areas not covered by the organic coating, owing either to imperfection or to injury after application, will not be protected by inhibitive pigments in the adjacent coating. The extent of corrosion of bare areas is a function of the area of metal exposed. Tiny scratches or pinholes usually cease to corrode when the area is small enough to retain the corrosion products which are formed, and thereby stifle further corrosion. Larger areas will continue to corrode and will damage the adjacent coating by undercutting or by causing blistering.

Blistering adjacent to corroding areas probably results from a combination of electrolysis and osmosis. The corroding areas are anodic and the adjacent coated areas cathodic. There is formation of alkali at the cathode areas under the paint film and water diffuses through the coating at an accelerated rate by virtue of the ion concentration gradient. The resulting internal osmotic pressure can thereby produce blisters. Thus the formation of alkali at the metal coating interface degrades the coating both by blister formation and by actual chemical attack if the coating

---

* Stoner-Mudge, Inc., Multiple Industrial Fellowship, Mellon Institute, Pittsburgh, Pa.

vehicle is not alkali-resistant. Hydrogen evolution at cathodic areas can also cause blistering.

The Vehicle

Among the most generally employed vehicles are those containing a relatively high percentage of phenol-formaldehyde modifying resins, "oil lengths" of no more than 25 to 35 gal of oil per 100 lb of added resin being preferred. The oils most frequently encountered are China wood and linseed, usually in roughly equal proportions. Vehicles of this type are not particularly resistant to alkali, however, and for this reason the newer and perhaps more efficient chlorinated rubber and vinyl type vehicles such as polyvinyl chloride acetate and the polyacrylics are becoming increasingly important. Probably the greatest volume of orthodox marine coatings continue to employ rosin or rosin derivatives (ester gum, limed rosin, zinc soaps, etc.) and coal tar as vehicle.

Problems of Dry Docking

Use of the newer types of organic coatings for metal protection in sea water has been retarded by the usual shipbuilding and maintenance practice. Dry-docking costs make it imperative that a ship be painted and water-borne as quickly as possible. The usual procedure is to scrape the loose paint from the hull of the ship as soon as it is dry-docked and apply the new paint immediately, even though high humidity or even wet conditions exist. The ship is then undocked as soon as the final paint coat is applied, almost invariably before it dries completely. Such methods cannot permit the coating to acquire its maximum properties. Recognition of the fact that dry docking costs and the time factor play important roles in the maritime industry has resulted in attempts to provide marine paints which will develop maximum protective ability under these adverse conditions of application. These efforts have not been entirely successful, though progress is being made.

Surface Preparation

Shipbuilders have long recognized that metal surface preparation is one of the prime factors in obtaining optimum paint life. It has been general practice to remove mill scale because even the original tight scale will become loosened during the life of the structure, because of corrosion reactions or differences in thermal expansion, thereby resulting in loss of both scale and paint coat at those areas. The earlier methods of descaling were to "weather" the ship plates before assembly or to launch the hull without painting so that the scale was loosened and later scraped off. These methods have not been satisfactory in that *complete* scale removal is never insured. The recent trend is either to pickle the plates, or to sandblast after fabrication. The latter method has the double effect of removing mill scale or other preformed oxide together with non-metallic contaminants such as oils, soaps or greases, as well as roughening the surface to present a greater specific area for adhesion forces. Another method of descaling attracting much attention is *flame cleaning* in which an intensely hot brush-like flame is rapidly passed over the surface of the metal to be descaled. The thermal gradient through metal and mill scale and the difference in expansion coefficient act to cause the scale to split off; it may then be wire-brushed away. Although this method does not remove all scale, it has been particularly effective in loosening thick scale and in preparing weathered surfaces for painting or repainting. It has the unique advantage that the local heat reduces the moisture film on the metal surface, which, if immediately painted, adds greatly to the life of the paint. In experienced hands this technique offers real promise.

There is still considerable difference of opinion as to whether painting over old paint coats is a bad practice. It can be fairly stated that new application over intact paint (the British practice) probably gives good results. The difficulty lies in determining whether the existing paint coat is in sufficiently good condition to serve as a secure foundation for the new paint. For this reason, there is a pronounced trend toward complete removal of old paint where time and facilities permit.

Much emphasis has been placed on chemical pretreatment of the metal surface to prevent corrosion by the permeated water. The wide variety of proprietary inhibitive-wash systems currently offered to the metal fabricator testifies to the increasing importance of corrosion-prevention efforts along this line. These systems, primarily designed for application to steel, are, in their simplest form, chemical methods for producing dense insoluble surface films of oxides, chromates, phosphates, oxalates, or silicates. These films — if developed properly and if subsequently cleaned of all reactant chemicals — form ideal bases for coating application. The difficulty lies, however, in the development of surface layers having all these desired characteristics. Films of significant thickness, to be effective, are formed only under ideal and controlled conditions. If the chemical which develops the film is not completely reacted and the excess is not washed off before applying the paint system, adverse results may be encountered.

GALVANIC EFFECTS

Marine structures generally, and hulls of ships in particular, are normally constructed of several metals in electrical contact and, being in an invironment of high conductivity, galvanic corrosion can be serious. If, for example, a metal is corroding under the accelerating influence of an adjacent cathode, painting only the anodic portion is dangerous practice. Complete coverage of the corroding area is usually impractical since discontinuities in the paint film are almost certain to be present initially, or will be produced later in service. Because galvanic corrosion in sea water is largely under cathodic control,* incomplete reduction of the anode area will not appreciably affect the total amount of corrosion, but will serve only to concentrate the attack on a smaller area; serious pitting and even perforation can thus occur. Conversely, if only the cathode element of the galvanic couple is painted, the amount of corrosion of the anode is reduced to that of freely corroding uncoupled metal. In this case, however, the alkali resistance of the paint system over the cathode area is of great importance in maintaining the effect of cathode insulation. A paint system which is degraded by alkali soon peels and washes off. It has been shown that best corrosion protection is afforded if both anode and cathode areas are coated.

It is at present impractical to paint propellers; the erosion at the high operating velocities rapidly scours off the paint. As a result, if bare bronze propellers are in electrical contact with painted steel hulls, severe localized pitting is occasionally experienced in the hulls. Some beneficial results have been obtained by applying insulating coatings of diverse kinds to the shaft or in other manner breaking the couple between propeller and shaft or hull.

NEWER TRENDS

Recent shipbuilding practice, necessitated by the war, has resulted in the introduction of large prefabricated sections. The application of primer coats to large areas of

* The term *cathodic control* is discussed on page 486.

ship plating in fabricating shops is thus done under better conditions than obtain in dry docks or on ship ways. Such practice should materially add to the life of a paint system and also make it practical to apply some of the newer synthetic resin coatings which in themselves can offer inherently better protection and service. The old method of applying paint in dry dock or on ships under adverse conditions of cold, dampness, and high humidity, with little or no surface preparation, has retarded the art in that the difference between the best and worst paint formulations applied under these conditions is usually insignificant.

For details on typical useful marine paint formulations, see *Paints and Varnishes*, p. 889.

Macro-organisms in Sea Water and Their Effect on Corrosion
WILLIAM F. CLAPP*

FOULING ORGANISMS

Many thousands of species of invertebrate animals and marine plants have been described and named. In some locations, even far from shore, pelagic marine life is so prolific and forms such dense mats on the surface of the ocean that the speed of a vessel may be appreciably reduced

Only a few kinds of these animals and plants should be considered as fouling organisms. The vast majority are free swimming and do not possess the necessary organs or have the power to become attached to any material. Others must burrow in mud, sand, or some other substance in the ocean for protection in order to survive. None of the vertebrates, such as the fish or mammals which are so plentiful in salt water, are capable of maintaining a permanent foothold on any material and none, therefore, should be considered as being true fouling organisms.

Organisms which might affect metals or other submerged materials in salt water can be roughly divided into three groups: sessile organisms, semi-motile fouling organisms, and motile organisms.

SESSILE ORGANISMS

This group cannot survive without becoming firmly attached to a suitable base.

All organisms found firmly attached to a material in salt water secured this attachment while still in a very minute embryonic form. No organisms have the ability to become firmly attached to any base after having passed from the minute embryonic form to the mature form. No true fouling organism can change its position once it has become attached. Very few can survive if they become separated from the base upon which they have grown, and none which has become attached has the power to secure a new foothold.

The following list includes most of the common forms of true sessile fouling organisms:
I. Organisms which build hard calcareous or chitinous shells:
   1. Annelids, which form coiled or twisted tubes.
   2. Barnacles, which build cone-shaped shells built up of laminated plates.
   3. Encrusting Bryozoa, colonial animals which form flat, spreading, multi-cellular, coral-like patches.
   4. Mollusks, including several species, such as oysters and mussels.
   5. Corals.

* William F. Clapp Laboratories, Duxbury, Mass.

II. Organisms without hard shells:
   1. Marine algae: green, brown, or red filament-like growths which occur usually near the water line and include such forms as *Ceranium, Fucus, Polysiphonia,* and *Ulva.*
   2. Filamentous Bryozoa: fern-like or tree-like growths.
   3. Coelenterates (hydroids) such as *Tubularia* with stalk-like or branching growths, each branch terminating in an expanded tip; also *Bougainvillia* and *Campanularia.*
   4. Tunicates (sea squirts): soft, spongy masses.
   5. Calcareous and siliceous sponges.

All the above are completely sessile fouling organisms. There are, however, other animals which may occasionally contribute to some extent, since they are semi or optionally motile.

Semi-Motile Fouling Organisms

Some of these organisms may become attached early in life and remain on the same spot until they die. Other organisms may grow on all sides, thus preventing any change of position. Unless completely covered and thus deprived of food and water, they can survive and thrive, but they have the power of locomotion and frequently do move from one location to another. They include:
   1. Sea anemones and allied forms, flower-like animals which may become firmly fastened to metals or other materials, but they also have the ability to move very slowly by sliding along over a mucilaginous slime which they excrete. However, even when in motion, these animals are in firm contact with the base upon which they are living.
   2. Some of the worms which construct more or less temporary, loosely adherent tubes of mud and sand for protection. These organisms can readily, and frequently do, abandon these tubes (which may be 8 or more in. in length) and move to another location. The old tubes are left behind and contribute to fouling.
   3. Certain Crustacea, such as *Corophium,* build small temporary sand and mud tubes which they cement to materials submerged in salt water. The tubes are quite adherent, but the builders frequently abandon them and move to other locations.
   4. Various mollusks, such as many of the numerous species of mussels, become firmly attached to any convenient base by means of a mat of very strong chitinous hairs. The tip of each hair becomes cemented to any suitable base, thus forming a dense mat on the surface, but these organisms have the power to loosen their holdfasts and migrate to new locations. When they die, the mat of chitinous hairs remains firmly attached to the material last occupied.

Motile Organisms

This third group of organisms may influence the corrosion of metals in salt water, even though very indirectly. They constitute:
   1. Some of the worms, particularly the scavenger varieties which leave mucilaginous excretions.
   2. Mollusks, such as the sea slugs and snails.

The slimy film execreted by these animals may have little or no effect on a metal, but because of its consistency frequently smothers and kills the more minute fouling organisms covered by it.

FILM-FORMING ORGANISMS

In addition to marine bacteria. large numbers of other unicellular marine organisms are usually among the first to appear on any material submerged in salt water. This film of microscopic organisms is of importance, principally because it may provide a favorable foothold for macro-organisms. It has been thought that if these microorganisms could be prevented from becoming permanently attached, or the firmness of the attachment could be decreased, later fouling by the larger multicellular plants and animals would be greatly lessened or eliminated. Experiments show, however, that barnacles, at least, can become directly attached to the surface of a material without the aid of the microbiological film, although it is possible that such attachment of

Fig. 1. Mud Tubes and Other Marine Organisms on Steel Specimen 12 In. Square After Immersion in Sea Water. (Courtesy F. L. LaQue.)

the larger fouling organisms is not so permanent as if it occurs on the organic film.

The appearance of a steel panel covered with various fouling organisms after immersion in sea water is shown in Fig. 1.

DISTRIBUTION

Fouling organisms are plentiful from the Arctic Circle to the Antarctic. Although there are many more kinds of organisms in the tropics, the bulk of growth in a given period of time may be as great in Labrador as at the Equator.

Most of the fouling organisms have very definite depths at which they survive or thrive. Practically all true sessile fouling organisms live in comparatively shallow water. Certain species of barnacles live at depths up to 200 ft, but these species thrive only at such depths and are rarely found attached to any objects near the surface of the water. Other species of barnacles exist only in the vicinity of the high-water mark and are very rarely found at a depth of more than 2 or 3 ft below mean high water. Other organisms which inhabit this zone are the green algae, such as *Enteromorpha*, which are commonly seen in the vicinity of the water line on vessels and wharves.

Fouling does not occur in the deep water off the coast because there are no objects upon which the organisms may secure the foothold necessary for their survival.

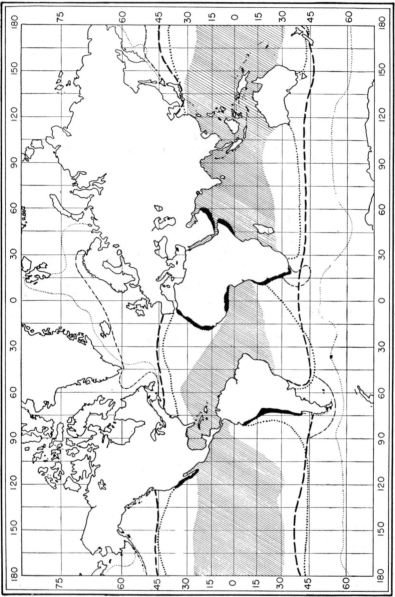

Fig. 2. Map of Marine Zoogeographic Regions as Related to Temperature. Shaded tropical water is always above 20° C (68° F). Boundaries of the temperate zones after Ortmann: dotted lines; and, after Meisenheimer: dashed lines. Heavy black coastal areas are regions of upwelling cold bottom water. After Hesse, Allee, and Schmidt, *Ecological Animal Geography*, John Wiley and Sons (1937). Prepared by L. W. Hutchins, Woods Hole Oceanographic Institute, Woods Hole, Mass.

FACTORS AFFECTING FOULING

TEMPERATURE

Temperature is a very important factor in the life of marine organisms. In northern waters the breeding season of most fouling organisms is restricted to the warmer summer months. Since fouling continues only to the end of the breeding season, it follows that most fouling occurs in northern waters only during the summer. However, the organisms which do become attached during this period continue to thrive and increase in size indefinitely, dependent upon the length of life of the particular species involved. In the warmer waters farther south, the breeding season is longer, until in the tropics it is almost continuous. Therefore, while a vessel in the north may be completely free from fouling from November to June, in the tropics it might become badly fouled during any month of the year.

Effect of Ocean Currents and the Seasons.* A map referring only to the distribution of fouling has not yet been published. For certain sections of the world, as for the coasts of Asia, the data are inadequate. However, since the most important factor governing fouling distribution is temperature, distribution of marine organisms will probably agree closely with maps based on temperature. Several such maps are combined in Fig. 2. The demarcations depend on the limits of empirically known critical maximum or minimum temperatures, localities in which the range of temperatures is exceptionally great, or places where there is a rapid transition from one to another set of conditions. In no case will there be a complete change in the population, but at any such point a sufficient number of organisms will find their limits of tolerance to give a distinctive character to the faunas of the two sides of the boundary. The map shows a shaded tropical zone in which the water is always above 20° C (68° F). Outside of this is a subtropical zone with water still generally warm. It is bounded by the temperate regions, differently limited by Ortmann (1896) and Meisenheimer (1906), outside of which in turn are the circumpolar cold waters. The greatest difficulties lie in localizing the boundaries in coastal waters, which are the most important in fouling considerations. There is justification from either the general or fouling point of view both for Meisenheimer's emphasis on the faunal and temperature break at Cape Cod and Ortmann's stress on the similar situation at Cape Hatteras. As a first approximation of a fouling map, Ortmann's divisions seem somewhat preferable. His north temperate zone corresponds well, for example, with the known distribution of the common fouling form, the mussel *Mytilus edulis*. In practical terms these zones indicate roughly the regions in which similar types of fouling may be expected.

The composition of the fouling population is not the only feature dependent on temperature. Figure 3 shows the seasons of attachment of various fouling types at six different locations with varied temperature conditions. Most organisms in regions with cold waters part of the year do not breed continuously, but only at times when certain critical temperatures are attained. Differential requirements in this respect result in the successions shown. Such differences are often found for separate species of the same group, as illustrated by the barnacles at Woods Hole. Other cases are concealed in the graphs by temporal overlaps. A further point is the fact that for each species there are one or more peak periods of attachment, when it becomes especially abundant. These peaks characterize even the tropical fouling, in which, as at Madras and Hawaii, given species may be attaching to some extent throughout the year. Because they are contingent in large part on critical temperatures for breed-

* This topic is discussed by L. W. Hutchins, Woods Hole Oceanographic Institution, Woods Hole, Mass.

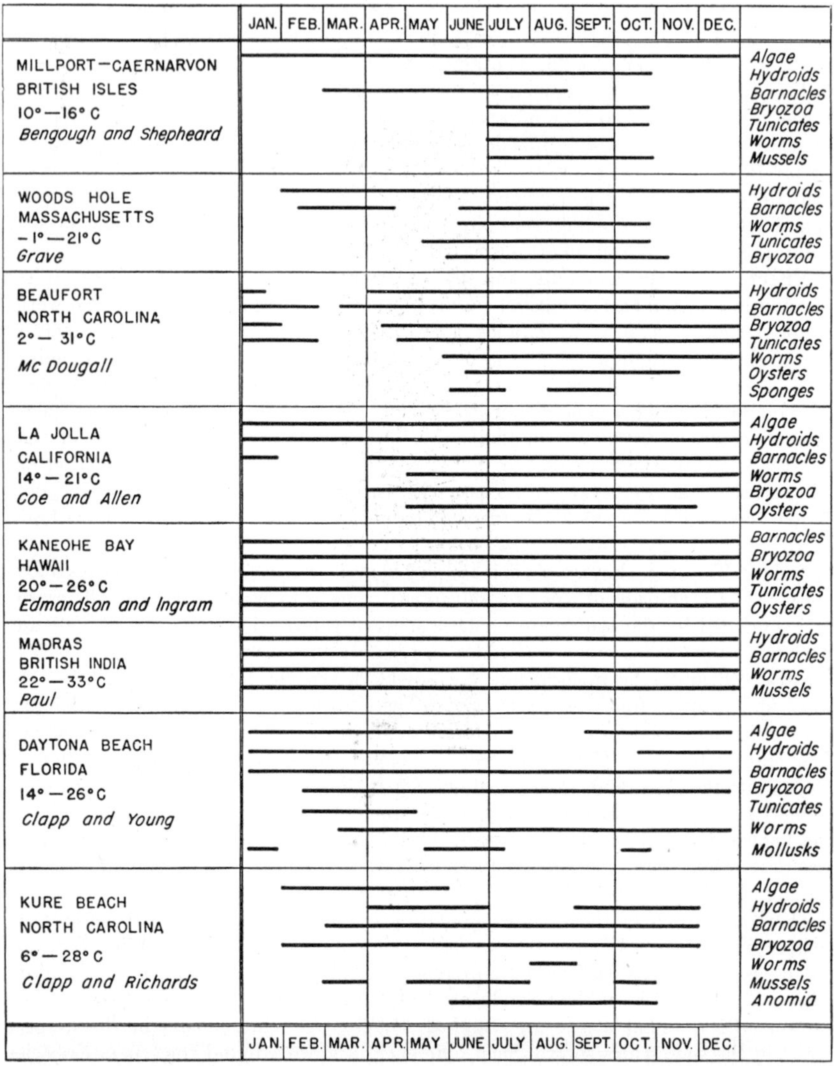

	JAN.	FEB.	MAR.	APR.	MAY	JUNE	JULY	AUG.	SEPT.	OCT.	NOV.	DEC.	
MILLPORT–CAERNARVON BRITISH ISLES 10°–16° C *Bengough and Shepheard*													Algae Hydroids Barnacles Bryozoa Tunicates Worms Mussels
WOODS HOLE MASSACHUSETTS –1°–21° C *Grave*													Hydroids Barnacles Worms Tunicates Bryozoa
BEAUFORT NORTH CAROLINA 2°–31° C *Mc Dougall*													Hydroids Barnacles Bryozoa Tunicates Worms Oysters Sponges
LA JOLLA CALIFORNIA 14°–21° C *Coe and Allen*													Algae Hydroids Barnacles Worms Bryozoa Oysters
KANEOHE BAY HAWAII 20°–26° C *Edmandson and Ingram*													Barnacles Bryozoa Worms Tunicates Oysters
MADRAS BRITISH INDIA 22°–33° C *Paul*													Hydroids Barnacles Worms Mussels
DAYTONA BEACH FLORIDA 14°–26° C *Clapp and Young*													Algae Hydroids Barnacles Bryozoa Tunicates Worms Mollusks
KURE BEACH NORTH CAROLINA 6°–28° C *Clapp and Richards*													Algae Hydroids Barnacles Bryozoa Worms Mussels Anomia
	JAN.	FEB.	MAR.	APR.	MAY	JUNE	JULY	AUG.	SEPT.	OCT.	NOV.	DEC.	

Fig. 3. Seasons of Attachment of Various Types of Fouling at Eight Localities.

Prepared by L. W. Hutchins, Woods Hole Oceanographic Institute, Woods Hole, Mass. Data for Daytona Beach were contributed by G. H. Young, Mellon Institute, Pittsburgh, Pa., and data for Kure Beach by Beatrice Richards, Wm. F. Clapp Laboratories, Duxbury, Mass.

ing, the peaks for different organisms do not necessarily coincide; but the great aggregate of them tends to occur in the warm months.

Motion

Since fouling organisms must become attached to some base while still very minute, it is very difficult for them to secure a firm foothold on objects which are in motion. On floating materials, in motion because of tide or current, it is conceivable that the larvae in the vicinity may be drifting at the same approximate speed and direction and hence have no difficulty in becoming attached.

The exact speed which would be too rapid for the various species of marine organisms to become attached has not been determined. It appears to lie in the range of 2 to 4 miles per hour (3 to 6.5 kilometers per hour) and is influenced by the roughness of the surface involved.

Color

Many of the fouling organisms are affected by light. The majority occur most plentifully on shaded, or dark, surfaces. The heaviest fouling is generally found on the northern side of stationary objects in the northern hemispheres. This is true of mussels, hydroids, and many of the algae. The green algae, such as *Enteromorpha*, thrive best in the light. In general, the lighter shades of protective coatings on submerged materials, the whites and light grays and yellows, are somewhat less subject to fouling than the darker shades, all other factors being equal.

Attachment of Fouling Organisms

When in the minute larval or undeveloped juvenile form, the various fouling organisms secure the necessary permanent foothold in a number of different ways, each characteristic of a particular group. Many species of the algae, shortly after landing on a favorable surface, secure a firm attachment by exuding a mucilaginous material which hardens to a glue-like consistency. As the plant grows, root-like growths spread over the surface of the base, greatly increasing the firmness of attachment. A very similar method is used by the plant-like animals known as hydroids.

Other groups of fouling organisms, such as the barnacles and some of the mollusks, excrete a calcareous material which is deposited on any suitable submerged base to serve as a point of attachment. With still others, the cementing material is of a siliceous nature.

Effect of Surface. A hard, smooth surface generally provides a more secure footing for fouling organisms than a soft material. The adherence on hard vulcanized rubber is much stronger than it is on soft rubber. On a smooth polished surface of stainless steel a barnacle (*Balanus eburneus*) can obtain an exceptionally firm foothold. It is very difficult to remove such organisms even with metal scraping tools. This is also true to a somewhat lesser degree when these organisms land on glass. On ordinary steel the firmness of attachment is much less. Similar effects are observed with protective coatings on metals. Non anti-fouling paint coats with a hard glossy surface provide a firmer foothold for the organisms than softer paint films.

Effect of Corrosion Products. The physical condition of a firm hard surface which provides an excellent base for fouling organisms may change after being submerged in salt water. The initial growth on ordinary steel may be as great as on stainless steel, but corrosion products form very quickly on ordinary steel and the character of the surface is changed. The organisms become fastened not to the metal, but to the surface of the film of corrosion products. The firmness of foothold of the organism is then equal only to the adhesion of the corrosion film. With a material such as stainless steel, where the corrosion-product film formed on the surface of the metal is negligible, a fouling organism may be expected to maintain its foothold throughout its normal life.

With copper and certain high-copper alloys, the relationship between corrosion and fouling is very apparent. As compared with steel, these materials form corrosion products that are less voluminous, and a larger proportion is carried away either in solution or in suspension. In addition to any toxic effect, this continual loss or sloughing of corrosion products serves to prevent any prolonged attachment of foul-in organisms. If the usual corrosion of copper or its alloys is arrested, as by galvanic action, the normal formation of corrosion products does not occur and fouling can then proceed as on other inert surfaces.

EFFECTS OF ENCRUSTING MATERIALS ON CORROSION

Physical Factors

The first spot of cementing material enlarges as the animal grows. With the oyster, the organic lime deposit may form a tightly adherent base which covers an area of several square inches. If the base is equally adherent to a metal over the entire area, this base may be expected to protect against corrosion. If the surface is uneven, and water penetrates, there will be different oxygen concentrations at points under the lime base and elsewhere on the surface of the metal. Oxygen concentration cells may then accelerate corrosion. This occurs with many of the fouling organisms, particularly with many of the numerous species of barnacles. It also may be caused by one or more micro-organisms, such as diatoms or Foraminifera, having become attached to a portion of the area over which the macro-organism grows. Any uneven growth of micro-organisms on a portion of the metallic surface over which a macro-organism extends its base may result in unequal adherence of the latter. Because of this unequal firmness of attachment beneath some portions of the base, oxygen concentration cells may develop, resulting in corrosion.

Chemical Factors

Another entirely different and very important type of corrosion is encountered when, in the process of expansion of the base by growth, an organism completely surrounds and covers a later arrival, or smaller organism. The covered organism quickly dies. Degeneration sets in, followed by the probable production of hydrogen sulfide. An acid condition results which causes accelerated corrosion.

Destruction of Protective Coatings

Some organisms, particularly certain species of barnacles, such as *Balanus eburneus*, are capable of penetrating or otherwise damaging protective coatings on metals. In the process of growth and the spreading out of the base of the organisms over the surface of any material there is also exerted a very strong downward pressure. As a result of this pressure, the outer edge of the base may penetrate the film of protective coating. As the diameter of the base increases, the paint film is pushed up on the sides of the growing organism. In some of the thick bituminous enamels and asphalt coatings, the organisms may penetrate through one quarter of an inch of coating to the base metal, which subsequently becomes exposed to corrosion aggravated by its localization.

On a hard, dense, protective coating, penetration by these organisms is prevented, but in many cases the growing organisms becomes more firmly bonded to the protective coat than is the latter to the metal. When any action dislodges the organism, the complete paint system remains attached to the organism rather than to the metal. Accelerated corrosion of the exposed metal follows.

It has been found, at least with some paint coats, that the difference in the firmness of adhesion of the paint coat beneath the base of an organism and that of the paint coat outside of, but adjacent to, the base of the organism, results in accelerated corrosion in the form of pits directly beneath the organism even while the paint system and the fouling organisms are apparently undisturbed. This may be the result of oxygen concentration cells having been formed beneath two adjacent areas, over one of which the paint coat is firmly adherent and at least somewhat impervious, whereas in the other the paint coat, because of the action of the fouling organism, is not adherent.

GENERAL REFERENCES

BAERENFAENGER, C., Verein deutscher Ingenieure, Fachausschuss für Anstrichtechnik, *Bücher der Anstrichtechnik*, **3**, 21–24 (1938).

Bureau of Ships, Navy Dept., Washington, D. C., "Guide to Fouling Organisms and Instructions Regarding the Docking Report," *Docking Report Manual*, 1942.

GAWN, R. W. L., *Roughened Hull Surface*, North East Coast Institution of Enineers and Shipbuilders, Nolbec Hall, London, 1942.

MCDOUGALL, K. D., "Sessile Marine Invertebrates at Beaufort, North Carolina," *Bull.* 1, Duke University Marine Station.

POMERAT, C. L., and E. R. REINER, "The Influence of Surface Angle and of Light on the Attachment of Barnacles and Other Sedentary Organisms," *Biol. Bull.*, **82** (No. 1), 14–25 (February, 1942).

PRATT, H. S., *A Manual of the Common Invertebrate Animals*, The Blakiston Co., Philadelphia, Pa., 1935.

VISSCHER, J. P., "Reactions of the Cyprid Larvae of Barnacles at the Time of Attachment," *Biol. Bull.*, **54**, 327, 350 (1928).

VISSCHER, J. P., "Nature and Extent of Fouling of Ships' Bottoms," *Bull.*, U. S. Bureau of Fisheries, **43**, 193–252 (1928).

ZOBELL, C. E., and E. C. ALLEN, "The Significance of Marine Bacteria in the Fouling of Submerged Surfaces," *J. Bacteriology*, **29** (No. 3), 239–251 (1935).

ZOBELL, C. E., and C. B. FELTHAM, "The Occurrence and Activity of Urea-Splitting Bacteria in the Sea," *Science*, **81**, 2096, 234–236 (1935).

ZOBELL, C. E., and C. B. FELTHAM, "Preliminary Studies on the Distribution and Characteristics of Marine Bacteria," *Bull.* 3, pp. 279–296, Scripps Institute of Oceanography, Technical Survey.

Anti-fouling Measures

G. H. YOUNG*

The problem of fouling of surfaces subjected to immersion in sea water has long been serious. The major alleviative efforts to date have involved the use of an *anti-fouling paint*, applied in the too-often vain hope that, by its nature or content of *toxic* agents, fouling growths would be prevented or at least reasonably inhibited. To this end an almost unlimited number of specific compositions, for which antifouling properties are claimed, appear in the patent and technical literature. In many preparations the beneficial effect is attributed to the composition of the vehicle; in others, to the peculiar fashion in which the several ingredients are compounded together. In by far the greatest number, the virtue is ascribed to the presence of one or more *poisons*.

The phototropicity of certain types of fouling organisms has frequently been demonstrated by biological studies. The suggestion has therefore been made that color

* Mellon Institute of Industrial Research, Pittsburgh, Pa.

alone is the most significant factor in anti-fouling paint performance; the argument of light versus dark paints is still unsettled.

It has more recently been asserted that the only effective approach from the paint standpoint is to strive for extreme exfoliation or *underwater chalking* properties; but the relative merit of chalking coatings versus toxic coatings is still undetermined.

THE FOULING PROCESS

The fouling of a ship's bottom, which begins as soon as the ship is water-borne, can be divided into three phases taking place more or less simultaneously: the formation of the slimy microbiological film; the attachment of macroscopic fouling organisms (almost always in larval form); and their growth into the mature forms, visible as barnacles, mollusks, annelids, bryozoans, algae, and other fouling growths.[1] The probable role in the fouling process of the microbiological film has been described by ZoBell, Clapp, and others.[2,3] (See also the previous chapter.) Experiments indicate that slime formation on anti-fouling paints may eventually vitiate effects of the paint by a simple blanketing action preventing lethal concentrations of the toxic agents from building up at the slime film-water interface. Where sufficiently high toxic concentrations can be realized, in spite of the slime film, slime-covered surfaces may be maintained free from macro-growths for many months.

As discussed in the previous chapter, all the important macroscopic forms mature from tiny microscopic larvae. For a short time these larvae swim freely in the water. Then, after a period varying from a few minutes to several weeks, depending upon the specific organism, they must become attached to a surface in order to survive. After attachment, metamorphosis into the adult form and growth are very rapid. The effectiveness of an anti-fouling paint depends upon its ability to hinder or prevent the primary attachment of the larvae. Once attached, they grow regardless of the toxicity of the paint since the poisons have little or no effect on the adult organisms at concentrations attainable in the paint-water interface zone.

USE OF PAINTS

ANTI-FOULING PAINT FORMULATION

With a few exceptions, every commercial anti-fouling paint carries copper in some form or other as the active toxicant. By far the major number of published formulas employ red cuprous oxide, alone or fortified with minor amounts of other toxic agents — usually mercury and arsenic compounds. For the yachting trade the so-called racing bronzes, which employ metallic copper powders as the active pigment, have come into prominence.

It is now well established that the anti-fouling efficiency of a bottom paint is substantially a linear function of the percentage of dissolvable toxic pigment carried in the dried paint film. (The amount of toxic in a gallon of paint is a meaningless figure so far as predicting anti-fouling life is concerned, unless the total non-volatile content also is reported.) The copper content of a representative group of commercial bottom paints selected at random[4] averaged 20 to 25% calculated as Cu_2O. The mercury content of a random group of fortified bottom paints averaged 3.5% cal-

[1] U. S. Navy Department (BuShips), *Docking Report Manual*, p. 5, 1942.
[2] W. F. Clapp, American Association for the Advancement of Science, *Symposium on Corrosion*, 1941.
[3] C. E. ZoBell, National Paint, Varnish, Lacquer Association, Scientific Section, *Circ.* 588 (1939); Federation of Paint & Varnish Production Clubs, *Official Digest*, **17**, 379–385 (1945).
[4] A. J. Cox, State of California Department of Agriculture, *Special Publication* 200, 1942–1943.

culated as HgO. (Both figures are on the dry-film basis.) Such paints seldom last longer than 6 months without fouling. Researches carried out over a period of years (many of them unpublished for security reasons) have indicated that anti-fouling performance in excess of 6 to 7 months demands copper concentrations approximating twice the concentration found in most commercial paints. At this toxic level, the advantageous effect of adding mercury compounds is no longer so apparent.

Toxic Properties of Metallic Copper versus Cuprous Oxide. The effective soluble toxic complex (probably $[CuCl_2]^-$) can be built up equally in the surface water interface zone from (1) sheet copper, (2) metallic copper powder, (3) cuprous oxide, or (4) other soluble and dissociable copper compounds. That is, the toxic sources differ only in their ability to furnish the active toxic ions; it is the soluble complex which has biocidal properties. Thus the primary toxicity of such a pigment is best measured by its *available* copper content. Cuprous compounds are effective because they dissolve to liberate cuprous ions; cupric compounds are ineffective because they are insoluble in sea water; metallic copper is effective because under most conditions it *corrodes* to yield the necessary cuprous ions.

Inhibition of Anti-fouling Efficiency. It is thus apparent that any factors influencing the primary rate of solution of the toxic agents will profoundly affect the performance of the anti-fouling paint. Typical of such factors is the nature of the vehicle or binder employed in the paint. If this contains substances capable of tight compound formation with cuprous ions, the life of the paint may be very short; the blanketing effect of *slime*, with subsequent overgrowth of fouling organisms is an illustrative example. The slime film *concentrates* and holds the copper ions; precipitation takes place and fouling ensues.

One of the most serious causes of anti-fouling paint malfunctioning is attributable to accidental contact of this paint with the bare steel hull. The mechanism probably involves the local deposition of copper by interaction with iron, resulting in great depletion of the copper *reservoir* in the interface zone. The phenomenon is demonstrable with both metallic copper and cuprous oxide paints. For this reason the importance of having adequate primer coats on steel hulls before applying the anti-fouling paint cannot be exaggerated.

Alternative Anti-fouling Pigments. The technical literature recites numerous compounds to which anti-fouling activity has been assigned. Typical are copper resinates, copper arsenite, calomel, copper oxychloride, basic copper carbonate, selenium oxide, copper and zinc selenides, cuprous cyanide, and numerous arsenic compounds. In addition, a wide variety of organic compounds have been described and patented as anti-fouling toxicants. The use of certain metallo-organic complexes has been widely explored in recent years. In general, it can be stated of these alternative pigments that few if any are as effective on a pound-for-pound basis as a simple copper scale, copper, or cuprous oxide, and none can presently compete on a cost-per-pound per month of effective paint life.

Vehicle Considerations. It is important to understand clearly a marked and fundamental difference between anti-fouling paints as a type and all other paints. The anti-fouling paint is not, and dare not, be a *permanent* coating. Its very efficacy depends upon a constant sacrifice of one or more of its ingredients to the adjacent water, thus guaranteeing that this immediate contacting layer of water is lethal to marine micro-organisms. It is not the *paint* which is lethal; it serves merely as a reservoir for undissolved toxic substance to replace that which is rapidly lost from the neighboring water owing to current effects, precipitation, diffusion, and other dissipating influences.

Therefore vehicles which are resistant to permeation by water, and are inert to hydrolysis and similar chemical breakdown, are very inefficient *anti-fouling* vehicles. The toxic agent is permanently locked up within such a film; only a very small percentage is ever made available in the interface zone and its effective anti-fouling life, if any, is short.

On the other hand, a vehicle which is too rapidly permeated by water, and is readily broken down by hydrolysis, swelling stresses, etc., may give an equally short-lived paint, by wasting the toxic agent at an excessive rate and thus becoming rapidly depleted.

The art of formulating an anti-fouling paint thus revolves around the choice of an optimum vehicle.[5,6] A vehicle optimum for metallic copper is not necessarily right for cuprous oxide, and may be very wrong for calomel. In practice, it has been found that rosin as a major ingredient of the vehicle affords an excellent carrier for cuprous oxide; this component must be used judiciously where it is desired to employ metallic copper, else too rapid sacrifice takes place. A completely inert vehicle, such as chlorinated rubber unmodified with other resinoids, will not allow functioning except at excessively high pigmentations.

Accelerated Corrosion under Anti-fouling Paints

Considerable attention has been accorded the question as to whether bottom paints accelerate pitting of the hull, but few published data are available. It has been shown[7] by quantitative measurement of the comparative rates at which prescribed painted steel panels corrode when immersed in sea water that there can be accelerated attack on the exposed steel. The intensity of attack appears to vary directly with the concentration of copper pigment in the anti-fouling coat, and may be as severe with cuprous oxide paints as with metallic copper paints. The rate of such accelerated attack decreases materially if even a single barrier coat is interposed between the steel surface and the anti-fouling paint.

The practical effect on the corrosion at damaged areas, "holidays," and exposed rivet heads of having relatively large concentrations of copper going into solution near by is still an open question. It seems likely that as more efficient undercoat systems are developed, and particularly as paint application is more carefully supervised, difficulties attributable primarily to the anti-fouling paint ingredients will diminish.

The Evaluation of Anti-fouling Paints

The fouling process is so complicated by localized environment, eddy currents, seasonal fluctuations, transient contamination, and the like as to make strict laboratory duplication prohibitively costly. The closest approach is by panel exposure in fouling waters, and even here conditions are often far different from those actually encountered on a ship's bottom. The necessity for multiple exposures, preferably at a variety of locations and under varying application conditions, is thus understandable.

A number of attempts have been made to evaluate toxic efficiency artificially under standardized conditions; correlation with panel exposure is usually fair but with actual ship's bottom service disappointing. Attention has been given to leaching

[5] J. D. Ferry and B. H. Ketchum, *Ind. Eng. Chem.*, **38**, 806–810 (1946).
[6] B. H. Ketchum, J. D. Ferry, and A. E. Burns, Jr., *Ind. Eng. Chem.*, **38**, 931–936 (1946).
[7] G. H. Young and coworkers, *Ind. Eng. Chem.*, **36**, 341–344 (1944).

ANTI-FOULING MEASURES

determinations as a means for prognosticating paint life;[8] the U. S. Navy has been reportedly quite active in exploring this approach. Most investigators agree that no sure criterion exists except actual service tests after lengthy panel exposure. This factor alone is a major reason for the apparent slowness with which anti-fouling techniques change in comparison with other protective techniques.

OTHER MEASURES

Fresh Water Anchorage

The traditional advantage of periodic anchorages in fresh water to kill and remove accumulated fouling has perhaps been overemphasized.[9] Although it is true that 24 to 72 hours exposure of mature marine organisms to fresh water usually results in death, the shells are seldom detached by such exposure. Where shell-forming organisms are sloughed, it generally will be found that this has been accompanied by gross loss of paint. The action is probably the result of osmotic pressure stresses and may be severely damaging to the anti-corrosive coating on the ship's bottom. Sudden changes in water temperature can and will have the same disruptive action on most paints. Thus the benefit derived from shedding accumulated fouling may well be nullified by the accompanying damage to the bottom paint.

Use of Electric Currents

Numerous investigators have suggested the possibility of applying either direct or alternating current to the hull at intermittent intervals, to prevent primary attachment. Thus far, such experiments have been disappointing. The larval forms of most fouling organisms are remarkably insensitive to d-c electric currents as such, unlike fish and higher animal life. While application of sufficient potential to plate out hydrogen should result in generation of enough alkali at the (cathodic) hull surface to be lethal, the blistering effect on any paint would be severe. Furthermore, the simultaneous deposition of calcium, magnesium, and strontium basic salts as an adherent film appears to be a limiting factor. Power consumption increases proportionately, and eventually fouling occurs on the built-up insulating calcareous deposit.

High-frequency a-c current has also been employed on an experimental scale, with little success to date. Nevertheless, this type of approach would appear to offer greater likelihood of an ultimate solution.

Incorporation of Toxic Agents in the Hull Plate Metal

Controlled tests by many investigators confirm that the amounts of toxic metals which can be incorporated in steel, of which cadmium, copper, arsenic, and manganese are typical, are far too low to impart any anti-fouling properties to the bare plates. The chances for development of a ferrous alloy for ships' plates having the anti-fouling properties of copper and certain of the high-copper alloys appear to be remote.

Introduction of Biocidal Agents

The continuous sterilization of the water adjacent to the hull would be a prohibitively costly operation, even were techniques for efficient distribution of the sterilizing reagent available. However, this approach becomes feasible in salt water piping aboard ship. Either continuous or intermittent treatment of the water with chlorine or other

[8] A. J. Cox, State of California Department of Agriculture, *Special Publication* 200, 1942–1943.
[9] U. S. Navy Department (BuShips), *Docking Report Manual*, p. 5, 1942.

toxic reagent is practicable, and the continuous treatment is being widely adopted. It offers a simple and effective solution to an annoying problem aboard ship, or wherever sea water is circulated in semi-closed systems.

CORROSION BY SOILS

K. H. LOGAN*

The prevention of damage to underground structures arising from corrosion has long been an engineering problem of major importance. There are in the United States about 500,000 miles of pipe lines used for transporting water, gas, oil, and gasoline. The value of these lines is in the order of 6 billion dollars.

Accurate figures for the corrosion losses of these lines are not available. However, the average life of steel mains has been taken as 33⅓ years by the Interstate Commerce Commission.[1] On this basis the annual cost for replacement of pipe lines would be 200 million dollars. Most of the loss is probably due to corrosion.

DEFINITION

The deterioration of the exterior of metals exposed to soils, but not to stray electric currents from street railways or similar sources of electric current, is usually called soil corrosion and attributed to soil characteristics. It can be shown, however, that factors other than soil properties influence and sometimes control the corrosion attributed to soils. Nevertheless, soil properties are responsible for much corrosion, and the ability to recognize soil properties enables the pipe line engineer to avoid many corrosion troubles.

It is generally assumed that the corrosiveness of a soil controls the service life of a metal in that soil, but this service life depends also upon the thickness of the metal, the area exposed, and the repairs or maintenance applied. Rates of corrosion in soils are functions of the lapsed time over which they are computed and usually are inverse functions of this time. They also depend upon whether loss of metal, loss of strength, or depth of pits is the criterion of corrosion. It has been proposed that the corrosivity of a soil be expressed by an equation[2] which takes account of the initial rate of corrosion, the change in the rate of pitting or loss of weight, and the relation of the exposed area to the maximum corrosion. Although these factors are recognized, their values and relationships have not been established definitely and the proposed equations are of uncertain value.

THE CORROSIVE PROPERTIES OF SOILS

The Department of Agriculture[3] recognized eleven great soil groups as shown in Fig. 1, and divided the groups into some three thousand soil series, most of which have at least three horizons or layers (A, B, C), differing in physical and chemical properties. The name of a soil is divided into two parts, the first indicating the series to which the soil belongs and the second the texture of the A horizon or uppermost soil layer.

* National Bureau of Standards, Washington, D. C.

[1] *Valuation Docket* 1203, Atlantic Pipe Line Co., Interstate Commerce Commission, 1937.

[2] K. H. Logan, S. P. Ewing, and I. A. Denison, *Soil Corrosion Testing*, Symposium on Corrosion Testing Procedures, American Society for Testing Materials, Philadelphia, Pa., 1937.

[3] *Soils and Men*, 1938, Yearbook of Agriculture, Part V, *Soils of the United States*, U. S. Government Printing Office, 1938.

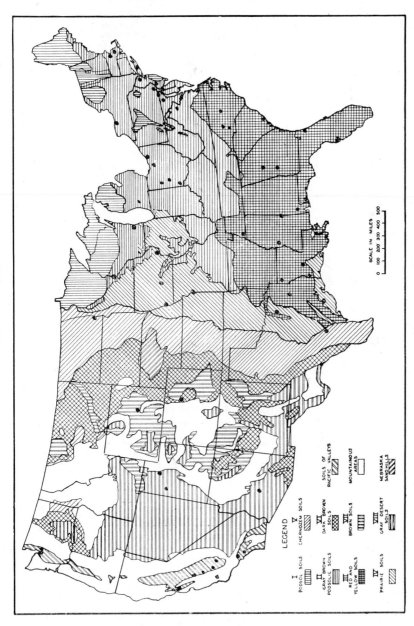

Fig. 1. Soil Groups of the United States. (Reproduced from Marbut.) Solid circles indicate present or former soil corrosion test sites of the National Bureau of Standards.

The texture of a soil horizon is determined by the percentages of the particles of various size groups. As to particle size, there are two grand subdivisions — those materials having diameters of 2 or more mm (0.079 in.), which include gravel, cobbles, and larger stone, and a group of materials of smaller diameter subdivided as in Table 1.[4]

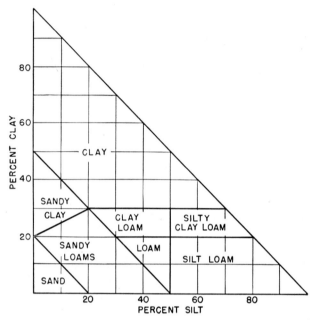

FIG. 2. The Whitney Diagram Showing Classification of Soils According to Silt and Clay Content.

TABLE 1. CLASSIFICATION OF SOIL PARTICLES AS TO SIZE*

Class	Diameter	
	millimeters	inch
Gravel and stones	>2	>0.08
Fine gravel	1 to 2	0.04 to 0.08
Sand	0.05 to 1	0.002 to 0.04
Silt	0.002 to 0.05	0.00008 to 0.002
Clay	<0.002	<0.00008

* C. F. Marbut, *Atlas of American Agriculture*, Part III, *Soils of the United States*, U. S. Goverment Printing Office, 1935.

The percentages of silt and clay in the principal soil classes with respect to texture are shown in Fig. 2.[5] The texture indicated by the second part of the soil name is

[4] C. F. Marbut, *Atlas of American Agriculture*, Part III, *Soils of the United States*, U. S. Government Printing Office, 1935.

[5] M. Whitney, "The Use of Soils East of the Great Plains Region," U. S. Department of Agriculture Bureau of Soils, *Bull.* 78 (1911).

TABLE 2. COMPOSITION OF WATER EXTRACTS OF SOME CORROSIVE AND NON-CORROSIVE SOILS*

No.	Soil Series	Corrosivity Max. Pit Depth in 12 Years, mils	Electrical Resistivity, ohm-cm	Acidity pH	Acidity Total, mg equivalents per 100 grams soil	Composition of Water Extract, mg equivalents/100 grams soil Na + K as Na	Ca	Mg	CO$_3$	HCO$_3$	Cl	SO$_4$
\multicolumn{13}{l}{Least corrosive with respect to max. pit on 126 sq in. of Bessemer steel at 12 years}
47	Unidentified silt loam	29	1,770	7.6	3	0.67	0.72	0.39	0.00	0.88	0.06	0.48
24	Merrimae gravelly sandy loam	31	11,400	4.5	12.6	....	....	....	....	....	....	....
26	Miami silt loam	39	2,980	7.3	2.6	0.27	0.50	0.31	0.00	0.70	0.03	0.12
6	Everett gravelly sandy loam	40	45,000	5.9	12.8	....	....	....	....	....	....	....
31	Norfolk fine sand	45	20,500	4.7	1.8	....	....	....	....	....	....	....
\multicolumn{13}{l}{Most corrosive with respect to max. pit on 126 sq in. of Bessemer steel at 12 years}
23	Merced (alkali) silt loam	173	278	9.4	A	8.38	0.38	0.22	0.02	1.87	1.12	5.57
28	Montezuma (alkali) clay adobe	153	408	6.8	†	1.50	0.06	0.18	0.00	0.12	0.99	0.89
29	Muck	146	1,270	4.2	28.1	2.15	1.92	1.55	0.00	0.00	1.69	2.30
45	Alkali soil, unidentified	137	263	7.4	A	8.15	3.70	0.70	0.00	0.24	0.18	11.98
8	Fargo clay loam	119	350	7.6	A	1.42	1.72	2.55	0.00	0.71	0.01	4.43

*"Underground Corrosion," *National Bureau of Standards, Circular C450* (1945). † Not determined. A = alkaline.

not necessarily the texture of the B or C soil horizon in which pipes are usually laid. Usually the subsoils are similar throughout a soil series.

Generally poor aeration and high values for acidity, electrical conductivity, salt content, and moisture content are characteristics of corrosive soils. Poorly aerated neutral soils are favorable to the development of sulfate-reducing bacteria which accelerate corrosion. Well-aerated soils, especially those derived from limestone, usually are not corrosive. High-resistance soils are usually non-corrosive unless they are poorly aerated.

Soils frequently have properties or characteristics, some of which indicate that the soil is corrosive and others of which indicate the opposite. On this account it is sometimes difficult to tell from soil analysis whether or not a soil is corrosive. This fact is illustrated by Table 2, which shows some of the properties of five corrosive and five non-corrosive soils in the Bureau of Standards soil corrosion tests.[6]

It will be noted that, with one exception, the soils in the corrosive group are low in electrical resistivity and contain considerable quantities of salts. The non-corrosive group have high resistivities. There is no apparent relation between the total acidity and the corrosivity of the soils in Table 2. However, when total acidity is the chief variable, as in Table 3, the relation between acidity and corrosion is more evident.

TABLE 3. RELATION OF ACIDITY TO REPAIRS IN SOIL TYPES*
Resistivity 4000 to 5000 ohm-cm

Soil Type	Total Acidity, milligram equivalents	Pipe-Line† Repairs, %
Wauseon fine sandy loam ‡	7.5	6.3
Caneadea silt loam	12.7	13.3
Miami silt loam	16.8	22.8
Mahoning silt loam	18.1	20.9
Trumbull clay loam	21.1	20.0
Crosby silt loam	22.0	30.8

* I. A. Denison and A. P. Ewing, "Corrosiveness of Certain Ohio Soils," *Soil Sci.*, **40** (No. 4), 287 (1935).

† Includes spot welds, half soles, and replacements necessitated by leaks or corrosion. Expressed as percentage of pipe line requiring repairs. Average age of 5 pipe lines, 33 years.

‡ One sample.

Table 4 shows the physical properties of twenty-four soils arranged in order of decreasing corrosiveness. In general, the corrosive soils in this table are poorly aerated. That poor aeration does not always make a soil corrosive is indicated by the twenty-second and twenty-fourth soils. The table also shows that some high-resistance soils are corrosive. The reason for lack of close relationship between physical properties and corrosion is that chemical properties and bacteria also influence corrosion. In many cases the conflicting indications of the chemical and physical properties of a soil cannot be evaluated with sufficient accuracy to justify a statement as to which is the controlling factor, or as to the corrosiveness of the soil, except as experience indicates.

[6] "Corrosion in Soils" *National Bureau of Standards Letter Circ.* LC 689 (1942).

TABLE 4. CORRELATION OF SOIL CORROSIVITY WITH THE
PHYSICAL PROPERTIES OF SOILS*

Relative Corrosiveness	Soil No.	Moisture Equivalent, %	Aeration	Air-pore space, %	Apparent Specific Gravity	Volume shrinkage, %	Resistivity at 15.6° C (60° F), ohm-cm
1†	28	19.6	VP	2.5	‡	5.9	408
2	43	‡	VP	‡	‡	‡	60
3	14	12.2	F	14.4	1.76	1.0	3,520
4	4	22.2	P	7.0	1.78	2.2	6,670
5	40	31.0	P	2.3	1.78	16.4	970
6	42	24.8	F	14.9	1.79	4.7	13,700
7	37	7.0	F	‡	‡	0	11,200
8	8	34.8	P	8.7	1.56	21.0	350
9	11	31.3	G	15.5	1.49	8.6	11,000
10	16	16.5	F	12.0	1.65	0.6	8,290
11	39	18.3	P	7.5	1.72	3.8	7,440
12	41	28.1	F	6.9	1.61	14.6	1,320
13	44	25.3	G	7.2	1.55	6.0	1,000
14	3	29.9	G	18.2	1.60	7.0	30,000
15	30	24.0	P	7.2	1.81	7.5	1,500
16	32	11.8	G	11.7	1.85	0.1	5,700
17	19	26.3	F	3.9	1.76	11.8	1,970
18	2	35.2	P	2.0	1.95	23.0	684
19	22	28.4	G	9.6	1.67	3.0	5,150
20	18	22.0	G	16.6	1.26	1.3	1,410
21	25	18.6	F	9.5	1.95	7.6	1,780
22	7	36.4	P	3.7	2.02	34.5	2,120
23	36	14.9	G	16.0	1.62	0	11,200
24	17	27.7	P	4.4	1.72	5.4	5,980

* "Underground Corrosion," *National Bureau of Standards Circular* C450 (1945).
† Most corrosive soil.
‡ Not determined.
VP = very poor; P = poor; F = fair; G = good.

THE BEHAVIOR OF METALS AND ALLOYS IN SOILS

Low-Alloy Steels

If representative samples of the commonly used ferrous materials are buried side by side, they will corrode at somewhat different rates. If the experiment is repeated under the same or different soil conditions, the relative merits of the materials may appear somewhat different. Most exposure tests of the past have not included a sufficient number of identical specimens exposed to identical conditions to justify a conclusion as to whether the apparent differences in merit are accidental or real.

If the depths of the maximum pits on specimens of commonly used ferrous materials are averaged for a number of soils, the differences in the rates of corrosion of most of them lie within the sum of their standard errors. This is illustrated in Fig. 3. It will be noted that the pit depths on the 1½-in. diam. specimens are less than those on the 3-in. diam. specimens, i.e., the pit depth is a function of the area of the specimen.[7]* Figures 4 and 5 permit a comparison of the pit depths on cast iron, wrought iron, and steel.

* The relation of pit depth to area of the specimen and time of exposure is discussed on p. 1051.
[7] K. H. Logan, "Engineering Significance of National Bureau of Standards Soil Corrosion Data," *J. Research Natl. Bur. Standards*, **22**, 109 (1939). R.P. 1171.

Table 5 shows the pit depths for low-alloy steels after 4 years exposure to fourteen soils. With one exception, each material appears best for at least one soil condition and worse for some other soil conditions. This, if not accidental, may mean that

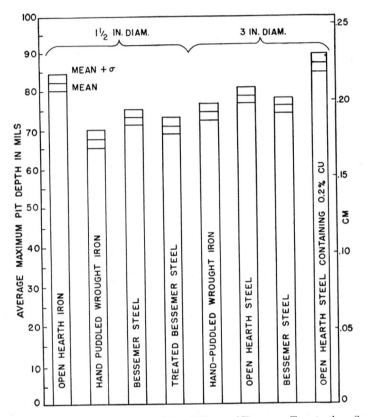

FIG. 3. Average Maximum Pit Depth for 12 Years of Exposure, Twenty-three Soils.

the choice of pipe material to be exposed to a definite soil condition should not be based on its average performance in many soils.

HIGH-ALLOY STEELS

Table 6 shows the performance of some high-alloy steels. None was free from pits in all soils. The weight losses indicate that the diameters of the pits were small.

COPPER AND COPPER ALLOYS

Copper and copper alloys corrode much more slowly than steel under most soil conditions, as is shown in Tables 7 and 8. Table 7 is too favorable to the copper alloys subject to dezincification since their appreciable loss of strength is not measured by their loss of weight. For example, Muntz metal is unsatisfactory for use in many soils because of dezincification.

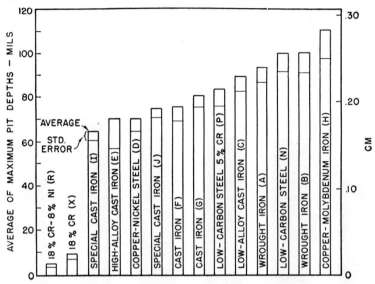

Fig. 4. Averages of Maximum Pit Depths on Ferrous Pipes Removed from Merced Clay Adobe After an Exposure of 5 Years.

ANALYSES OF MATERIALS IN FIGURE 4
Per Cent

Material	C	Si	Mn	S	P	Cr	Ni	Cu	Mo	Oxide + Slag
A	0.016	0.10	0.29	0.018	0.160	....	....	....	....	2.56
B	0.017	0.125	0.41	0.018	0.106	....	....	....	....	2.68
C	3.50	2.50	0.70	0.050	0.400	0.30	0.15	....	....	....
D	0.14	0.19	0.21	....	....	....	2.47	1.08	....	....
E	2.98	2.13	1.00	....	....	2.61	15.0	6.58	....	....
F	3.58	1.64	0.48	0.074	0.70	....	....	....	....	....
G	3.58	1.64	0.48	0.074	0.79	....	....	....	....	....
H	0.04	0.05	0.32	0.027	0.016	....	....	0.52	0.15	....
I	2.53	1.43	0.28	0.077	0.128	....	....	0.51	....	....
J	2.90	2.04	0.83	0.060	0.248	....	....	0.62	....	....
N	0.15	....	0.49	0.030	0.013	....	....	....	....	....
P	0.13	....	0.46	0.025	0.012	5.05	....	....	....	....
R	0.05	0.28	0.46	0.011	0.015	17.52	8.85	....	....	....
X	0.12	0.277	0.42	0.017	0.016	17.72	....	0.287	....	....

454 SPECIAL TOPICS IN CORROSION

OTHER MATERIALS

Lead is more corrodible than copper and high-copper alloys, but less corrodible than steel in most soils, as is indicated in Table 9. Lead corrodes rapidly in some soils which contain organic acids and very little in soils high in sulfates. Behavior of lead buried in soils is also discussed on p. 612.

Zinc is not used extensively underground except as a protective coating in which probably the protection of the base metal is largely at the expense of the zinc. Table 10 indicates that zinc corrodes rapidly in some soils.

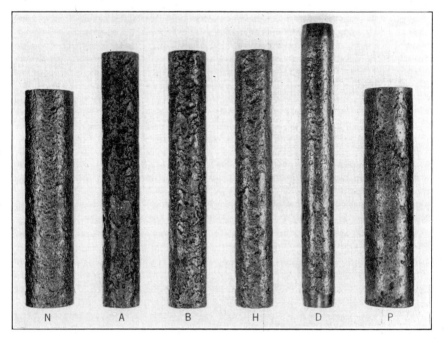

FIG. 5. Wrought-Iron and Steel Pipe Buried 9 Years in Susquehanna Clay at Meridian, Miss. (Soil 62).

N, low-carbon tube; A, hand-puddled wrought iron; B, machined-puddled wrought iron; H, copper-molybdenum open-hearth iron; D, nickel-copper steel; P, 5% chromium steel.

Data on aluminum and aluminum alloys buried for five years are presented in Table 6, p. 50.

Asbestos-cement pipe exposed to fourteen soils for 4 years showed no signs of deterioration other than a softening of the surface layer which was not applied under pressure.[8] Although the tests do not cover a sufficient period of exposure to justify final conclusions, they indicate that probably asbestos-cement pipe formed under pressure is not affected by soil acids or salts.

[8] K. H. Logan and Melvin Romanoff, "Soil-Corrosion Studies, 1941," *J. Research Natl. Bur. Standards*, **33**, 145 (1944). R.P. 1602.

TABLE 5. MAXIMUM PENETRATION OF ALLOY IRON AND STEEL PLATES EXPOSED FOR FOUR YEARS*
In Mils

No.	Soil Type	Open Hearth Steel	Open Hearth Iron		Low-Alloy Steel			4 to 6% Chromium Steel				High-Alloy Steel
			0.45% Cu, 0.07% Mo	0.54% Cu, 0.13% Mo	0.95% Cu, 0.52% Ni	1.01% Cu, 1.96% Ni	Cr-Si-Cu-P Steel, 1.02% Cr	2.01% Cr, 0.57% Mo	5.02% Cr	4.67% Cr, 0.51% Mo, 0.030% Al, 0.022% Ti	5.76% Cr, 0.43% Mo, 0.027% Ti	18% Cr
53	Cecil clay loam	76	74	72†	57	56	64	52	57	50	57	....
55	Hagerstown loam	54	44	51	50	52	51	52	48	47	39	....
56	Lake Charles clay	100‡	116	100	96	139‡	77‡	60‡	95	80‡	90‡	....
58	Muck	61‡	48‡	44	64‡	52‡	52‡	42‡	44‡	46‡	44‡,§	42§
60	Rifle peat	40	28	26†	40‡	28	67‡	26‡	51‡	36	32	....
61	Sharkey clay	50‡	66	54	63	56	41	35	36	36	32	....
62	Susquehanna clay	47	38	49	60	69	44	56	52	46	58	....
63	Tidal marsh	26§	48†	47	28	25	41	24	70	73	72	....
64	Docas clay	78	75	76	84	88	70	70	60	66‡	72	....
65	Chino silt loam	51	65	57†	60	84†	44	55	46	48	56	....
66	Mohave fine gravelly loam	188+§,‖	84	98	85	73	80‡	130+	99‡	88	117	....
67	Cinders	132+†,‡,¶	55¶,**,††	74‡,¶	90‡,**,††	84‡,**,††	47‡	68‡,¶	57‡	52‡,¶	44‡,**,††	....
69	Houghton muck	22	20	16	15	12	42	27	39	26	33	....
70	Merced silt loam	77‡	97‡	122‡	82	78	94‡	94	121	106‡,‡‡	94	....

* K. H. Logan and Melvin Romanoff, "Soil-Corrosion Studies, 1941," *J. Research Natl. Bur. Standards*, **33**, 145 (1944). R.P. 1602.
† Data for the individual specimens differed from each other by more than 50%.
‡ No original surface; impossible to measure true penetration.
§ Data on one specimen only
‖ The + mark in all cases indicates one or more specimens punctured.
¶ Severe corrosion at one end.
** Plate entirely destroyed at one end.
†† Data for one specimen — impossible to obtain data from other specimen because of loss of original surface caused by corrosion.
‡‡ Hole from both sides.

TABLE 6. AVERAGE LOSS OF WEIGHT AND MAXIMUM PENETRATION OF HIGH-ALLOY SHEETS EXPOSED FOR NINE* YEARS†

M, shallow metal attack, roughening of surface, but no definite pitting.
P, definite pitting, but no pits greater than 6 mils.
U, apparently unaffected by corrosion.
$+$, one or more specimens contained holes because of corrosion, rendering the computation of the exact penetration impossible. The thickness of the specimen has been used as the maximum pit in this case.

Soil No.	11.95% Cr, 0.48% Ni, 0.38% Mn (5)‡		17.08% Cr, 0.09% Ni, 0.36% Mn (5)		17.72% Cr, 9.44% Mn (2)		17.76% Cr, 3.83% Ni, 6.09% Mn (1)		17.2% Cr, 8.95% Ni, 0.44% Mn (2)§		18.69% Cr, 9.18% Ni, 0.36% Mn (5)		22.68% Cr, 12.94% Ni, 1.50% Mn (5)	
	Loss, Average	Maximum Penetration, Average	Loss, Average	Maximum Penetration, Average	Loss, Average	Maximum Penetration, Average	Loss	Maximum Penetration	Loss, Average	Maximum Penetration, Average	Loss, Average	Maximum Penetration, Average	Loss, Average	Maximum Penetration, Average
	mdd	mils	mdd	mils	mdd	mils	mdd	mils	mdd	mils	mdd	mils	mdd	mils
53	0.00039	$\ddot{M}$	0.0013	$\ddot{U}$	...	...	...	...	0.00093	U	0.00049	$\ddot{P}$	0.0014	$\ddot{P}$
55	...	...	...	...	...	...	...	...	0.012‖	25+(5,7)**	...	...	...	...
56	0.00078	$\ddot{P}$	0.0015	$\ddot{P}$	...	...	...	...	0.0010	U	0.00069	$\ddot{P}$	0.0020	$\ddot{P}$
58	...	...	...	...	...	...	...	...	0.013	M	...	...	...	...
59	0.0018¶	$P(5,7)$	0.14††	$P(5)$††	...	...	...	...	0.0011	$M(5)$	0.0011	P	0.0022	P
60	...	...	...	...	...	...	...	...	0.0015	P	...	...	...	...
61	...	...	...	...	...	...	...	...	0.0010	M	...	...	...	...
62	...	...	...	...	...	...	...	...	...	...	...	...	...	...
63	2.95	63+(2,5,7)	1.04	63+(2,5,7)	0.45	63+(2,5,7)	0.41	$P(2,7)$¶	0.72††	14+††	0.0024	$\ddot{P}$	0.0058	$\ddot{P}$
64	0.037	53+(2,7)	0.22††	43+(2,5,7)¶,††	...	...	...	...	...	...	...	...	...	...
65	...	...	...	...	...	...	...	...	0.0013	$P(2,7)$¶	0.0014	$P(2)$	0.0019	$P(2)$
66	0.52	55+(2,5,7)	0.65	63+(7)	0.00061	$\dot{M}(7)$¶	...	...	0.23††	14+(5,7)††	0.00049	M	0.11††	14+††
67	...	...	...	...	...	...	0.0039	$\ddot{M}$	0.0014	$\ddot{P}$	...	...	...	...

* In this and subsequent tables of this chapter the weight losses are expressed as average *rates* for the given time of exposure. It should be remembered that these rates combine in most instances, but not always, a higher initial rate with a lower final rate and hence in any extrapolation represent the maximum expected corrosion. Values of maximum penetration are for the total period of exposure.

† K. H. Logan and Melvin Romanoff, "Soil-Corrosion Studies, 1941," *J. Research Natl. Bur. Standards*, **33**, 145 (1944). R.P. 1602.

‡ The number in parentheses indicates the number of specimens removed from each test site.

§ Polished surface.

‖ Data for one specimen only.

¶ Average loss of weight or pit depth of 1939 removal is greater.

** The number in parentheses after the pit depth indicates that at least one specimen of a previous removal was punctured, e.g., (2) indicates a puncture after two years.

†† Data for the individual specimens differed from each other more than 50%.

TABLE 7. LOSS OF WEIGHT OF COPPER AND COPPER ALLOYS EXPOSED FOR NINE YEARS*

mdd

Soil			Alloy										
No.	Type	Exposure, years	Tough Pitch Copper	Deoxidized Copper	Red Brass	Admiralty Metal	2-and-1-Leaded Brass	Brass: 66% Cu, 33% Zn	Muntz Metal	Bronze: 97% Cu, 1% Si, 1.8% Sn	98% Cu, 1.5% Si, 0.2%Mn	98% Cu, 1.5% Si, 0.2% Mn†	95% Cu, 3% Si, 1% Mn	75% Cu, 20% Ni, 5% Zn	Low-Carbon Steel
53	Cecil clay loam	9.47	0.21	0.19	0.22	0.26	0.42	0.36	0.46	0.33	0.26	....	0.32	0.24	3.62
55	Hagerstown loam	9.11	0.18	0.16	0.23	0.23	0.33	0.28	0.71	0.32	0.28	....	0.26	0.18	3.50
56	Lake Charles clay	9.42	0.64	0.69	0.62	0.51	0.79	0.83	1.07	0.67	0.55	....	0.55	0.69	25.5
58	Muck	9.51	1.72	1.85	1.81	1.84	1.46‡	3.93	4.62	1.85	1.73	....	1.87	1.23	14.3
59	Carlisle muck	9.12	0.11	0.10	0.11	0.08§	0.18	0.03	0.03	0.16	0.22‖	....	0.14	0.08	4.32
60	Rifle peat	9.24	6.57	4.54	3.22	2.58‖	2.92‖	3.80‖	4.12‖	2.92‖	3.58‖	1.65	4.03	3.26	15.1
61	Sharkey clay	9.53	0.29	0.31	0.36	0.45	0.76	1.07	2.26	0.54	0.42	....	0.46	0.38	5.06
62	Susquehanna clay	9.47	0.32	0.42	0.38	0.47	0.63	0.83	1.58	0.61	0.50	....	0.53	0.42	5.87
63	Tidal marsh	9.55	3.90	3.70	0.66	0.16	0.46‖	0.07	0.09	3.83‖	6.09	....	4.65	3.12	7.90
64	Docas clay	9.21	2.54	4.82	1.02‖	0.89	0.97‖	1.76	10.50	2.08	4.42	....	2.62	0.67‖	D
65	Chino silt loam	9.25	0.23	0.22	0.25	0.52	1.25	1.45	1.31	0.87	0.56	....	0.43	0.40	11.6
66	Mohave fine gravelly loam	9.23	0.41	0.56‖	0.70	0.72	0.86	1.05	1.28	1.39	0.46‖	0.31	0.50	0.54‖	16.8
67	Cinders	9.24	8.90	10.40	7.65	7.48	Z	20.9¶	Z	7.92	20.4	20.9	12.2	5.68	52.9**

* Kirk H. Logan and Melvin Romanoff, "Soil-Corrosion Studies, 1941," *J. Research Natl. Bur. Standards*, **33**, 145 (1944). R.P. 1602.
† These specimens had brazed joints; data for one specimen only.
‡ Data for one specimen only.
§ Average for three specimens.
‖ Data for the individual specimens differ from each other by more than 50%.
¶ Data for one specimen; the other specimen was destroyed by dezincification.
** Data for one specimen; the other specimen was destroyed by corrosion.
Z indicates specimens destroyed by dezincification.
D indicates specimens destroyed by corrosion.

TABLE 8. MAXIMUM PENETRATION OF COPPER AND COPPER ALLOYS EXPOSED FOR NINE YEARS*

In Mils

M, shallow metal attack, roughening of the surface but no definite pitting.
P, definite pitting, no pits greater than 6 mils.
s, uniform corrosion, impossible to measure true penetration.
D, selective corrosion, such as dezincification over large areas.
d, selective corrosion over small areas.
Z, specimens destroyed by corrosion (dezincification).
+, one or both specimens punctured.

Soil No.	Tough-Pitch Copper	Deoxidized Copper	Copper with Soldered Joints†	Red Brass	Admiralty Brass	Two-and-One-Leaded Brass	Brass: 66% Cu, 33% Zn	Muntz Metal	Bronze: 97% Cu, 1% Si, 1.8% Sn	Alloy 98% Cu, 1.5% Si, 0.2% Mn	Alloy 98% Cu, 1.5% Si, 0.2% Mn‡	Alloy 95% Cu, 3% Si, 1% Mn	Alloy 75% Cu, 20% Ni, 5% Zn	Low-Carbon Steel
53	6	P	13	10, D	20, D	P, D	6, D	6, D	12	P	...	8	8, D	59
55	8	P	8‖	7, D	20, D	7, D	10, D	6, D	20§	P	...	8	6, D	59
56	P	P	15‖	P, D	P, D	P, D	P, D	P, D	12	P	...	M	P, D	154+ (7)¶
58	10	14§	18	10, D	26, D	P, D**	8, D	6, D	53	12, s	...	12, s	P, D	110, s
59	6	P	9	M	P‖	P	M, D	P, D	20	M	...	P	P	40§
60	40, s	38, s	17, s§	34, D	16, s, d§	12, D§	33, D§	27, s, D§	18, s	10, s	8	21, s	32, D	27, s§
61	8	8	10	7, D	35, D§	6, D	6, D	12, D	37§	10	...	P	P, D	96§
62	8	8	10	14, D	24, D	P, D	13, D	P, d	22	12, s	...	6	12, D	87
63	6	10, s	6, s	6	P, d	P	M, D	42+, D§, ††	10, s	21, s	...	9	P	54§
64	14§	16§	13§	26, d	43, D§	15, D§	20, D		34			16§	18, D§	154+ (5,7)
65	10	P	8	P, D	13, D	P, D	18, D¶	P, D	22	20	...	11	P, D	112
66	13	10	13	P, D	P, D	P, D	8, D	P, D	22	13	P, ...	9	P, D	154 (5,7)
67	51, s	58, s	145+ (5)	54, D§	68, s, D	Z‡(5,7)	132+ (2,5,7)‡‡	Z(2,5,7)	102	145+	90, s	80, s	36, D	154+ (2,5,7)

* Kirk H. Logan and Melvin Romanoff, "Soil-Corrosion Studies, 1941," *J. Research Natl. Bur. Standards*, **33**, 145 (1944). R.P. 1602.
† These specimens had streamlined caps and couplings soldered in place.
‡ These specimens had brazed joints; data for only one specimen.
§ Maximum pits for individual specimens differed from each other by more than 50%.
‖ Data for three specimens.
¶ A number in parentheses after the pit depth indicates that the specimens of a previous removal were punctured, e.g., (5) indicates that the specimen was punctured after five years.
** Data for one specimen only.
†† Hole in one specimen due to dezincification. The mate, although it had no measurable pits greater than 8 mils, was badly dezincified, as indicated by the flat sound when the pipe was struck with another piece of metal.
‡‡ One specimen destroyed by dezincification.

TABLE 9. LOSS OF WEIGHT AND MAXIMUM PENETRATION OF LEAD PIPE EXPOSED TWO AND FOUR YEARS*

Soil		Chemical Lead†				Tellurium Lead‡				Antimonial Lead§			
		Loss of Weight		Maximum Penetration		Loss of Weight		Maximum Penetration		Loss of Weight		Maximum Penetration	
No.	Type	2 years	4 years	2 years	4 years	2 years	4 years	2 years	4 years	2 years	4 years	2 years	4 years
		mdd	mdd	mils	mils	mdd	mdd	mils	mils	mdd	mdd	mils	mils
53	Cecil clay loam	0.92	0.44	18	12	1.04	0.65	12	20	1.04	0.46	10	10
55	Hagerstown loam	1.55	0.42	24	26	1.42	0.59	26	26	0.80	0.31	26	18
56	Lake Charles clay	0.88	0.94	38	37	1.59	1.71	30	48	1.30	1.04	39	52
58	Muck	6.53	5.03	34	28	7.03	5.85	55	56	6.06	4.43	50	58
60	Rifle Peat	0.75	0.59	18	15	0.63	0.42	29	10	0.42	0.46	6	P
61	Sharkey clay	6.10	4.62	35	39	5.06	3.66	33	30	3.93	3.66	31	42
62	Susquehanna clay	1.26	1.94	32	29	1.51	1.34	19	31	1.13	2.16	12	30
63	Tidal marsh	0.23	0.031	14	18	0.23	0.031¶	10	12¶	0.16	0.03	P	16
64	Docas clay	0.84	0.40	24	16	1.04	0.38	21	11	0.50	0.40	12	12
65	Chino silt loam	0.59	0.27	40	24	0.71	0.33	22	16	0.71	0.44	6	15
66	Mohave fine gravelly loam	0.42	0.21	44	34	1.04	0.25	23	41	0.27	0.25	12	15
67	Cinders	15.35	25.5	79	104	14.0	27.6	71	94	13.3	8.80	56	90
69	Houghton muck	1.51	1.69	21	15	0.96	2.26	8	12	0.84	2.18	9	7
70	Merced silt loam	0.14	0.25	48	14	0.40	0.31	16	27	0.42	0.29	11	12

*Kirk H. Logan and Melvin Romanoff, "Soil-Corrosion Studies, 1941," *J. Research Natl. Bur. Standards*, **33**, 145 (1944). R.P. 1602. See asterisk footnote Table 6.
† Cu, 0.056%; Bi, 0.002%; Sb, 0.0011%.
‡ Cu, 0.082%; Te, 0.043%; Sb, 0.0011%.
§ Cu, 0.036%; Bi, 0.016%; Sb, 5.31%.
P indicates definite pitting but no pits greater than 6 mils.
¶ Data for one specimen only.

460 SPECIAL TOPICS IN CORROSION

TABLE 10. LOSS OF WEIGHT AND MAXIMUM PENETRATION OF ZINC PLATES EXPOSED FOR FOUR YEARS*

Metals — rolled zinc (0.095 Pb-0.0038 Cd-0.009 Fe). Die cast A. S. T. M. Alloy XXV.
Specimen size — rolled zinc — 30.5 × 5.9 × 0.38 cm (12 × 2.3 × 0.15 in.); die castings — 17.3 × 11.3 × 0.318 cm (6.81 × 4.44 × 0.125 in.).
Number — duplicate.
Exposure period — 4 years.
Temperature — natural for soil.

	Soil	Rolled Zinc		Die-Cast Zinc	
No.	Type	Loss of Weight	Maximum Penetration	Loss of Weight	Maximum Penetration
		mdd	mils	mdd	mils
53	Cecil clay loam	1.30	10	1.13	22
55	Hagerstown loam	1.25	8†	1.27	20†
56	Lake Charles clay	7.15	26‡	10.35	30
58	Muck	10.6	66	13.2	25+(2)§
60	Rifle peat	21.6	100†	31.2	125+
61	Sharkey clay	2.00	8	2.34	28
62	Susquehanna clay	2.59	9	1.25	16
63	Tidal marsh	4.80‡	34	2.98	24
64	Docas clay	1.19	18	5.28	20
65	Chino silt loam	1.59	36	1.59	16
66	Mohave fine gravelly loam	5.45‡	28‡	9.90	124+
67	Cinders	25.4‖	118+(2)†	27.3	125+
69	Houghton muck	3.55	10	3.42	36
70	Merced silt loam	3.38‖	102+‡	4.57‖	80+‡

* K. H. Logan and Melvin Romanoff, "Soil-Corrosion Studies, 1941," *J. Researh Natl. Bur. Standards*, **33**, 145 (1944) R.P. 1602. See asterisk footnote Table 6.
† Uniform corrosion; no reference surface left.
‡ Data for individual specimens differed from the average by more than 50%.
§ + Indicates that one or both specimens punctured by corrosion from one side of the plate. (2) Indicates that one specimen from the previous removal was punctured after 2 years.
‖ Data for one specimen only; the other specimen was destroyed by corrosion.

PROTECTIVE MEASURES

PROTECTIVE COATING CONSIDERATIONS

Protective coatings on pipe lines serve under adverse conditions. Because they are usually applied before the pipe is in final position, pipe coatings are subject to injury while the pipe is being transported and laid. A large part of the failures of coatings is attributable to such injuries. Stones and hard clods frequently injure coatings while the trench is being back-filled. Later, movements of the pipe because of changes in temperature and movement of soil, as the coating shrinks on drying or swells on absorbing moisture, work stones into or through it. The weight of the pipe may cause the coating beneath it to flow with similar effects produced by the weight and settling of the soil.

Some pipe lines are coated and laid in weather frequently unfavorable for satisfactory field application. Moisture, dust, the necessity for speed, inadequate equipment, and insufficient skilled labor are the causes of many coating failures.

METALLIC COATINGS — LEAD, CADMIUM, AND ZINC

Lead coatings usually contain pinholes or are too thin to resist corrosion except in soils containing sulfates. Lead is cathodic to iron; hence the lead coating tends to accelerate the corrosion of iron as soon as the coating is punctured. (Soil-exposure tests on lead-coated steel are discussed further beginning p. 849.)

All the specimens of cadmium-coated pipes in the A. P. I. tests were rusted or pitted within 4 years.[9]

Zinc is the only metallic coating used extensively for underground service, and most of this use is limited to small-diameter pipe. In some soils the formation of zinc carbonate provides a partly protective coating; in others after corrosion has punctured the zinc the iron beneath is cathodically protected. The distance from the zinc

TABLE 11. LOSS OF WEIGHT AND DEPTH OF MAXIMUM PENETRATION OF GALVANIZED AND BLACK IRON PIPE EXPOSED FOUR YEARS*

Soil		Galvanized Pipe, 3.08 oz Zn/sq ft		Black Iron Pipe		Condition of Coat†
No.	Type	Loss of Weight	Penetration	Loss of Weight	Penetration	
		mdd	mils	mdd	mils	
53	Cecil clay loam	2.9	6	6.0	98	2
55	Hagerstown loam	2.55‡	8	5.4	50	2
56	Lake Charles clay	8.1	7	33.5	104	2
58	Muck	11.3	21‡	18.2	46	3
60	Rifle peat	15.0	12	16.8	38‡	3
61	Sharkey clay	3.1	12	10.4	45	2
62	Susquehanna clay	4.8	9	9.0	56	2
63	Tidal marsh	4.5	10	19.2	38	1
64	Docas clay	3.3	9	12.4	67	2
65	Chino silt loam	4.7‡	6	9.5	59	3
66	Mohave fine gravelly loam	7.0	8	25.8	145+(2)§	2
67	Cinders	11.3	45	77.4‖	145+(2)§	3
69	Houghton muck	7.0	11	6.9	20	3
70	Merced silt loam	9.5	12	20.3‡	118+	3

* K. H. Logan and Melvin Romanoff, "Soil-Corrosion Studies, 1941," *J. Research Natl. Bur. Standards*, **33**, 145 (1944). R.P. 1602. See asterisk footnote Table 6.

† 1, Coating on more than 50% of surface; 2, coating on less than 50% of surface; 3, little or no coating remaining.

‡ Data for individual specimens differed from each other by more than 50%.

§ + indicates hole in one or both specimens caused by corrosion; (2) indicates that one or both specimens were punctured after 2 years exposure.

‖ Data for one specimen; the other specimen was destroyed.

at which the iron is protected ranges from a fraction of an inch to several feet, depending upon the character of the soil. The duration of the protection of iron by zinc is usually proportional to the thickness of the zinc. Table 11 indicates that in corrosive soils a zinc coating of 3 oz per sq ft of exposed surface will not prevent the rusting of galvanized iron for 4 years.[10] Earlier tests by the Bureau of Standards[11] in some of the same soils as those of Table 12 indicated that a 3 oz per sq ft zinc coating would prevent pitting in most soils for 10 years. No explanation has been found for the lack of agreement between the two experiments except that the specimens in the earlier tests (Table 12) were passed through the zinc bath twice. (Zinc coatings in soils are discussed further beginning p. 814.)

[9] K. H. Logan, "A. P. I. Coating Tests, Final Report," *Proc. Am. Petroleum Inst.*, (IV), **32** (1941).

[10] K. H. Logan and Melvin Romanoff, "Soil-Corrosion Studies, 1941," *J. Research Natl. Bur. Standards*, **33**, 145 (1944). R.P. 1602.

[11] K. H. Logan and S. P. Ewing, "Soil-Corrosion Studies, 1934," *J. Research Natl. Bur. Standards*, **18**, 361 (1937). R.P. 982.

TABLE 12. CORROSION OF GALVANIZED PIPE*

Average wt. of coating — 2.8 oz Zn/square ft.
Average time buried — 10 years.
Z — zinc continuous over specimen.
A — blue or black alloy layer exposed over at least a portion of specimen.
R — rusted or bare steel exposed.
M — shallow metal attack; no pit as great as 10 mils total depth.

Soil No.	Soil Type	mdd	Condition of Coating
1	Allis silt loam	2.3	R
2	Bell clay	0.29	Z
3	Cecil clay loam	0.34	Z
4	Chester loam	1.5	R
5	Dublin clay adobe	1.5	R
6	Everett gravelly, sandy loam	0.10	Z
7	Maddox silt loam	2.1	R
8	Fargo clay loam	0.61	Z
9	Genesee silt loam	0.96	A
10	Gloucester sandy loam	1.0	R
11	Hagerstown loam	0.71	A
12	Hanford fine sandy loam	...	...
13	Hanford very fine sandy loam	0.72†	R
14	Hempstead silt loam	0.20	Z
15	Houston black clay	0.29	Z
16	Kalmia fine sandy loam	0.83	M
17	Keyport loam	2.9	R
19	Lindley silt loam	0.54	Z
20	Mahoning silt loam	0.95	R
22	Memphis silt loam	1.0	R
23	Merced silt loam	7.9	6.7 mils per year
24	Merrimac gravelly, sandy loam	0.20†	Z
25	Miami clay loam	0.28	Z‡
26	Miami silt loam	0.57	A
27	Miller clay	0.76	A
28	Montezuma clay adobe	1.7†	R
29	Muck	5.0†	R
30	Muscatine silt loam	0.38	A
31	Norfolk fine sand	0.13†	Z
32	Ontario loam	0.47†	R
33	Peat	1.4	R
35	Ramona loam	0.25	A
36	Ruston sandy loam	0.19	Z
37	St. Johns fine sand	1.7	R
38	Sassafras gravelly, sandy loam	0.17†	Z
40	Sharkey clay	0.78	R
41	Summit silt loam	0.43	A
42	Susquehanna clay	0.59	R
43	Tidal marsh	1.1†	A
44	Wabash silt loam	...	...
45	Unidentified alkali soil	1.5†	R
46	Unidentified sandy loam	0.13	Z
47	Unidentified silt loam	0.84	A

* Kirk H. Logan and S. P. Ewing, "Soil-Corrosion Studies, 1934," *J. Research Natl. Bur. Standards*, **18**, 361 (1937). R.P. 982.
† There were two specimens of this material. The condition is for the worst of these specimens.
‡ No specimens 10 years old were removed from soil 25. The data given are for specimens exposed 8 years.

Cement Mortar

Cement and concrete coatings having thicknesses of an inch or more have proved very satisfactory, especially if the coating is evenly applied, dense, and made of sulfate-resisting cement. The coating retains its alkalinity for 8 years at least. However, rust and pits have been observed on pipes where cement coatings have been punctured. Some porous cement coatings have not afforded complete protection. A handy metal form for applying concrete coatings has been described.[12]

[12] J. H. Peper, *Oil and Gas J.*, **32** (No. 33), 9; (No. 37), 34 (1934).

Grease Coatings

Heavy petrolatum base coatings, with or without a chromate inhibitor, and usually reinforced or shielded, or reinforced by a metal fabric or wax, are the most effective of the cold applied coatings. Their makers claim that it is not essential to have the

TABLE 13. SUMMARY OF CONDITIONS OF LINE PIPE UNDER A. P. I. COATINGS AFTER TEN YEARS*

Thickness, mils	Coating Character	Total Feet Inspected	Unaffected Per Cent	Rusted Per Cent	Metal Attack Per Cent	Pitted Per Cent	Depth Deepest Pit (millimeters)	(mils)
		1. Cold Applications						
21	Cutback coal tar	166	1.2	5.4	9.6	84.5	8.18	322†
65	Asphalt emulsion	178	0	0	1.7	98.3	8.18	322†
		2. Enamels						
60	Coal-tar-asphalt enamel	152	13.8	13.8	10.7	62.0	6.71	264
69	Coal-tar-asphalt enamel	183	7.2	8.2	29.6	55.0	4.95	195
58	Coal-tar enamel	151	0	0	2.0	98.0	5.87	231
80	Coal-tar enamel	157	19.8	3.8	6.4	70.1	8.18	322†
		3. Mastic						
519	Asphalt mastic	213	72.5	18.4	6.6	2.8	0.97	38
		4. Shielded Coatings						
29	Cutback asphalt	164	1.2	10.4	18.3	70.0	5.08	200
419	Asphalt emulsion	170	3.5	39.0	21.8	35.9	2.72	107
63	Coal-tar-asphalt enamel	202	29.3	17.9	25.3	27.8	8.18	322†
81	Coal-tar enamel	177	43.0	4.0	5.1	48.0	6.10	240
		5. Reinforced Coatings						
107	Grease	208	0	4.3	29.4	66.3	2.26	89
150	Asphalt	166	0	1.7	23.5	75.3	7.37	290
151	Asphalt	192	4.7	8.9	36.5	50.0	5.00	197
201	Asphalt	228	0.4	9.7	46.1	44.1	4.06	160
143	Asphalt enamel	208	20.2	9.2	20.2	50.5	5.00	197
171	Coal-tar-asphalt enamel	218	30.8	6.0	23.5	39.9	3.20	126
351	Coal-tar enamel	175	14.9	14.3	33.1	37.7	1.47	58
230	Asphalt	229	5.2	13.5	61.1	20.1	3.15	124
	Total	3,537	14.7	10.0	22.9	52.2	....	....

* Kirk H. Logan,"A. P. I. Pipe Coating Tests, Final Report," *Proc. Am. Petroleum Inst.* (IV), **21**, 32 (1940).
† Through pipe.

pipe free from moisture or rust. The coatings seem to serve best in wet soils. The perfomance of one such coating (Table 13) is included below under bituminous coatings. The pits formed beneath this type coating were frequently very small in diameter.

Bituminous Coatings*

Kelly[13] has stated that 50% of the steel pipe lines in the United States have protective coatings and that 80% of the coatings are coal-tar base materials. Most of

* See also *Bituminous Coatings*, p. 898.

[13] T. P. F. Kelly, *Performance of Coal Tar Protective Coatings in Commercial Use*, Fifth National Bureau of Standards Underground Corrosion Conference, 1943. Unpublished.

the other coatings are derived from petroleum asphalt. Thin bituminous and paint coatings have only temporary value as preventives of corrosion, although recent data suggest that they may be effective in reducing the current required for cathodic protection. There is inconclusive evidence that under some conditions corrosion is accelerated by a defective coating.[14]

Thick bituminous coatings, especially coal-tar pitches, have a tendency to cold flow, and on this account practically all thick coal-tar pitch coatings and some asphalt coatings contain 10 to 30% of finely divided fillers, such as ground slate, silica flour, or mica.[15] Some coating manufacturers use slaked lime (calcium hydroxide) with the intention that if moisture penetrates the coating the alkalinity of the lime will retard corrosion. Some coating specifications prohibit the use of lime because of the belief that it makes the coating less moistureproof. There are few data to support either view. Bituminous-base coatings containing up to 50% of mineral filler are commonly referred to as enamels. They vary greatly in physical properties because of differences in the characteristics of the bitumen and the kind and amount of filler. The early coal-tar enamels had high-temperature coefficients. They tended to sag in hot weather and to crack if the weather turned suddenly cold. To overcome these features, some coal-tar enamels are now modified (plasticized) and show temperature susceptibilities similar to those of asphalts. Plasticized enamels are said to be somewhat more expensive than the older enamels and to require more careful application.

In order to obtain a thicker coating and one more resistant to rough handling and soil stress, many coating manufacturers apply alternate layers of enamel and rag or asbestos felt to the pipes to be protected. This insures greater freedom from pinholes and holidays, more moisture resistance because of the greater thickness of the bitumen, and more resistance to shock, pressure, and abrasion. Some of the rag felts rot or absorb moisture when the coating is exposed to wet soils and thus act as wicks which conduct moisture into the coating and to the pipe. Because of this, asbestos felt is preferred by many, although it does not distribute external pressures as well because it is usually thinner and not so stiff as rag felt.

One of the most detailed specifications for a coal-tar enamel coating is that of the American Water Works Association.[16] The specification includes the amount of filler, softening point, shock, flow, and low temperature tests, electrical inspection and method of application, as well as method of handling the coated pipe.

In lieu of reinforcements, some coating manufacturers use a graded aggregate with only enough bitumen to fill the voids between the solid particles. The material, which has a structure similar to that of concrete, is known as a mastic.[17] The coating containing about 12% of asphalt is usually half an inch to an inch thick. It is extruded over the pipe either at a coating plant or in the field. The field joints are applied under pressure by means of a metal form.

The Usefulness of Bituminous and Other Coatings. The value of a protective coating has been variously estimated. The Interstate Commerce Commission[18] and a committee of the American Petroleum Institute agreed upon the following additions to the life of a pipe line on account of protective coatings: paint coatings, 1 year;

---

[14] K. H. Logan, "The Effect of Protective Coatings on the Rate of Pitting of Pipe Lines," *Proc. Am. Petroleum Inst.*, **22**, 34 (1941).

[15] O. G. Strieter, "The Effect of Mineral Fillers on the Serviceability of Coating Asphalts," *Proc. Am. Soc. Testing Materials*, **36** (No. 2), 486 (1936).

[16] AWWA Standard Specifications for Coal Tar Enamel for Steel Pipe, 7A.5 (1940), 7A.6 (1940), American Water Works Association.

[17] A. J. Swank, "Moulded Coating Halts Oil Pipe Corrosion," *Elec. World*, **101**, 134 (1933).

[18] Valuation Docket 1203, Atlantic Pipe Line Co., Interstate Commerce Commission, 1937.

bituminous coatings, single ply, 5 years; double ply, 7 years; cement coating, 20 years. Probably the values represent the average performance of coatings at the time of the agreement rather than the maximum usefulness of these types of coatings.

It is claimed that present-day coatings are better than those applied a number of years ago. Unfortunately, there are few unbiased data on the usefulness of the newer coatings. A most extensive and carefully planned investigation was made jointly by the more important coating makers, the American Petroleum Institute, and the National Bureau of Standards.[19] The coatings were applied to working pipe lines in various soils. Table 13 shows their effectiveness at the close of 10 years exposure in terms of the condition of the pipe to which the coatings were applied. None of the coatings completely protected all the pipes to which it was applied. No coating failed to reduce the total amount of corrosion, but in some cases the maximum pit depths on coated pipe were greater than those on adjacent pipe. It is impossible to establish the reason for this condition, although several explanations have been offered.[20] In general the effectiveness of the coating was proportional to its thickness. Since the results were published, several of the coatings have disappeared from the market. Several of the tested coatings, however, are still being sold with but little change except more careful application. More coatings are now machine applied, and this has eliminated some of the causes of failure. Electrical testing of coatings has insured more perfect coatings.

Service Tests for Coating Failures. Pearson[21] has developed an apparatus making use of a system of electrical measurements for locating pinholes and injuries in coatings on buried pipe lines without uncovering the line. A crew of three to four men can electrically inspect about 3 miles of line per day. Such inspections enable the pipe line owner to determine whether the coating is protecting his line and where the coating has failed.

An inspection of the coating by means of this apparatus a year or so after the pipe is laid will show whether the coating has been properly applied and handled, and subsequent inspections by the same method will indicate the progress of coating deterioration.

SOIL TREATMENT

Soil treatment may take three forms: (1) the addition of chemicals to neutralize corrosive soil properties or accelerate the formation of protective films; (2) the replacement of corrosive soil next to the pipe by less corrosive soil; (3) water drainage.

The Dutch Corrosion Committee[22] reported the examination of a 9-year-old pipe laid in a bog, surrounded by a layer of sand to which lime had been added. Analysis of the sand showed that it contained 5.1% $CaCO_3$ and had a pH of 7.9. The pipe was not corroded except at a point where the sand was no longer in contact with the trench.

Wichers,[23] who made extensive studies of underground corrosion in Groningen, Holland, suggested three ways of improving soil conditions: (1) using impermeable

---

[19] K. H. Logan, "A. P. I. Pipe Coating Tests, Final Report," *Proc. Am. Petroleum Inst.* (IV), **21**, 32 (1940).

[20] K. H. Logan, "The Effect of Protective Coatings on the Rate of Pitting of Pipe Lines," *Proc. Am. Petroleum Inst.*, **22**, 34 (1941).

[21] J. M. Pearson, "Electrical Examination of Coatings on Buried Pipe Lines," *Petroleum Engr.*, **12** (No. 10), 82 (1941).

[22] *Second Report of the Dutch Corrosion Committee for the Study of the Corrosive Effect of Soils on Pipes*, 1935. Unpublished.

[23] C. M. Wichers, *The Corrosion of Pipe Lines Due to the Earth Contact*, Fourth National Bureau of Standards Underground Corrosion Conference, 1937. Unpublished.

earth for the upper layers of the trench and water drainage where necessary; (2) kneading plastic clay or loam around the pipe; (3) surrounding the pipe with sand or earth neutralized with lime.

Clay and sand have been used to surround short stretches of pipes in the United States, especially when the pipe passed through "made" ground containing cinders. The practice is not general, and no definite results have been reported; but no doubt these materials are less corrosive than cinders and rubbish. More frequently a heavy bituminous protective coating or one of concrete is used for such conditions.

CORROSION BY MICRO-ORGANISMS IN AQUEOUS AND SOIL ENVIRONMENTS

RAYMOND F. HADLEY*

INTRODUCTORY CONSIDERATIONS

Microbiological corrosion may be defined as the deterioration of a metal by corrosion processes which occur, either directly or indirectly, as a result of the metabolic activity of micro-organisms.

It is essential in studying the microbiological aspect of corrosion that such corrosion be clearly differentiated from the microbiological fouling so frequently observed in water pipe lines and on metallic surfaces exposed to miscellaneous aqueous environments. Microbiological corrosion and microbiological fouling are not synonymous, and the occurrence of one of these processes does not necessarily involve occurrence of the other, although they may frequently coexist.

THE INFLUENCE OF MICRO-ORGANISMS ON CORROSION PROCESSES

Micro-organisms contribute to corrosion by one or more of the following factors, each of which is dependent upon the physiological characteristics of the micro-organism:

1. Direct influence on the rate of anodic or cathodic reaction.
2. Change of surface metal film resistance by their metabolism or products of metabolism.
3. Creation of corrosive environment.
4. Establishment of a barrier by growth and multiplication so as to create electrolytic concentration cells on the metal surface.

To develop and grow, most micro-organisms must have available certain inorganic and organic chemical compounds to supply the oxygen, carbon, nitrogen, hydrogen, or sulfur necessary to their metabolic processes. These compounds are subject to considerable variation in regard to both chemical composition and the relative quantities of each required by a specific bacterium. Microbiological literature reports a number of organisms with which we are concerned in microbiological corrosion that are actually autotrophic and thus able to develop and sustain growth in environments devoid of organic nutrients.

The micro-organisms associated with corrosion should not be considered in any sense as being restricted to a soil or soil type as compared with an aqueous environment, or vice versa. Availability of various inorganic and organic nutrients in a given environment together with other factors, such as pH, oxygen concentration, and temperature, will determine whether or not microbiological development can take place.

* Susquehanna Pipe Line Co., Philadelphia, Pa.

It is possible, so far as corrosion is concerned, to select any non-corrosive washed silica sand and make this sand severely corrosive by the simple expedient of suitably adjusting both the chemical and associated environmental conditions. Little, if any, fundamental difference exists between microbiological corrosion in aqueous media and in soils, except so far as the physical characteristics of the environment itself may limit optimum concentration of chemical nutrients required to insure active microbiological growth. Once the knowledge has been gained as to how each microbiological corrosion process functions, it is necessary only to evaluate the environmental conditions at any location to determine whether or not any of the several microbiological corrosion processes can occur.

THE MICRO-ORGANISMS ASSOCIATED WITH CORROSION

Micro-organisms are frequently grouped according to their ability or inability to grow in an environment containing atmospheric oxygen. Aerobic micro-organisms readily grow in such environments, whereas anaerobic micro-organisms develop most favorably in environments in which the concentration of dissolved oxygen approaches zero.

The nature of the corrosion reactions promoted by the several different orders of micro-organisms in the aerobic group is considerably different from that fostered by the anaerobic group. All micro-organisms which appreciably influence the corrosion of metals can, therefore, be classified in either one or the other of two groups. From the standpoint of ferrous corrosion particularly, they may be listed in the order of importance given in this outline.

I. ANAEROBIC MICRO-ORGANISMS

A. Sulfate-Reducing Vibrios. Type species *Sporovibrio desulfuricans* (*Beijerinck*) *Starkey*. Curved rods, 0.5 micron by 1 to 5 microns in size. These bacteria are motile and frequently occur in chains having a spiral shape.

The role of these micro-organisms in contributing to corrosion has been extensively investigated.

Unsatisfactorily classified by Bergey[1] under Schizomycetes; order I Eubacteriales; family III Pseudomonadaceae; tribe I Spirilleae; genus not mentioned; species referred to after genus IV *Spirillum* in Appendix as No. 6. See classification of Prevot.[2]

Included in this genus are several species, namely, *Sporovibrio desulfuricans* (*Beijerinck*[3] 1895, *Rev. Baars*, 1931) *Starkey*, which includes the halophilic strain [Salt (NaCl) tolerant] of Van Delden[4] and the thermophilic strain of Elion[5]; *Sporovibrio rubentschikii* (*Baars*)[6] *Starkey*,[7] and *Sporovibrio Rubentschikii var. anomalous*[1](*Baars*) *Starkey*. The strains are differentiated by their salt and temperature characteristics; the species are differentiated one from the other by their dehydrogenating characteristics.

[1] D. H. Bergey, *et al.*, *Bergey's Manual of Determinative Bacteriology*, 5th Ed., Williams and Wilkins Co., Baltimore, Md., 1939.
[2] A. R. Prevot, "Etudes de systematique bacterienne. V. Essai de classification des vibrions anaerobies," *Ann. l'inst. Pasteur*, **64**, 117–125 (1940).
[3] M. W. Beijerinck, "Uber Spirillum desulfuricans als Ursache von Sulfate Reduktion," *Centr. Bakteriologie*, Series II, **1**, 1 (1895).
[4] A. van Delden, "Beitrag zur Kenntnis der Sulfatreduktion durch Bakterien," *Centr. Bakteriologie*, Series II, **11**, 81 (1904).
[5] L. Elion, "A Thermophilic Sulfate-Reducing Bacterium," *Centr. Bakteriologie*, Series II, **63**, 58 (1924).
[6] J. K. Baars, *Over Sulfaatreductie door Bacterien*, Dissertation, Delft, 1930.
[7] R. L. Starkey, "A Study of Spore Formation and Other Morphological Characteristics of *Vibrio desulfuricans*," *Archiv Mikrobiologie*, **IX**, (No. 3), 268 (1938).

B. Nitrate-Reducing Micro-organisms.

Nitrate reduction is accomplished by a considerable number of the more common bacteria, including *Escherichia coli, Pseudomonas fluorescens, Pseudomonas aeruginosa, Micrococcus denitrificans, Proteus vulgaris,* and *Serratia marcescens.* It is difficult at the present time to classify satisfactorily the nitrate-reducing micro-organisms supposedly associated with corrosion.

The degree to which the nitrate-reducing bacteria influence the corrosion of ferrous metals in the soil is still largely a matter of conjecture. No proof has yet been offered to establish definitely their part in microbiological corrosion, either in the laboratory or in the field.

C. The Methane-Producing Bacteria.

At least one species is classified by Bergey under Schizomycetes; order I Eubacteriales; family VI Micrococcaceae; genus IV *Sarcina;* species 14 *Methano sarcina methanica.* Microbiological methane fermentation, however, may result from the metabolic processes of a number of micro-organisms, including *Methanobacterium omelianskii* (rod-shaped), *Methanococcus mazei* (spherical), and *Methanobacterium söhngenii* (rod-shaped).

The part played by the methane bacteria in the corrosion of ferrous metals is still hypothetical and remains to be established.

II. AEROBIC MICRO-ORGANISMS

A. Sulfur Bacteria.
1. The Thiobacilli. Type species *Thiobacillus thioparus Beijerinck.*[8] Small rod-shaped cells.

Classified by Bergey under Schizomycetes; order I Eubacteriales; family I Nitrobacteriaceae; tribe III Thiobacilleae; genus VII *Thiobacillus.*

The microbiological corrosion of iron or steel under environmental conditions favorable for the growth of the thiobacilli, *Thiobacillus thiooxidans* in particular, is well established. The only question is concerned with the prevalence and economic importance of this corrosion process.

2. The Sulfur Bacteria.

Some of the most widely studied of these micro-organisms are classified by Bergey under Schizomycetes; order V Thiobacteriales; family II Beggiatoaceae; genera I–III.

These sulfur bacteria are so called because of their ability to store sulfur temporarily in their cells before oxidizing it to sulfate. They include the non-pigmented genera *Thiothrix, Beggiatoa.* and *Thioploca,* with numerous other genera of the red or purple pigmented organisms containing bacteriopurpurin to be found in the extensive literature referring to the subject.

B. The Iron Bacteria.

Classified by Bergey under Schizomycetes; order III Chlamydobacteriales; family I Chlamydobacteriaceae; genera I–IV.

Included in this group of higher micro-organisms, commonly referred to as the iron bacteria, are the genera *Crenothrix, Leptothrix, Gallionella, Siderocapsa, Sideromonas,* etc.

These higher micro-organisms, being typical water forms, have been extensively studied by numerous investigators interested in water microbiology.

[8] M. W. Beijerinck, "Ueber die Baketerien welche sich im Dunkeln mit Kohlensäure als Kohlenstoffquelle ernähren können," *Centr. Bakteriologie,* Series 2, **11**, 592–599 (1904).

C. Miscellaneous Organisms.
Included under this subheading are the numerous slime-forming bacteria, fungi, algae, protozoa, diatoms, bryozoa, etc. The role of these organisms in contributing to corrosion consists primarily in the establishment and maintenance of differential effects (concentration cells). The corrosion which occurs following their growth results more from other known corrosion reactions than from any effect attributable to their specific physiological characteristics.

ANAEROBIC MICROBIOLOGICAL CORROSION PROCESSES

SULFATE REDUCTION

Of the many micro-organisms associated with microbiological corrosion listed in the outline, the sulfate-reducing bacteria of the type species *Sporovibrio desulfuricans* (*Beijerinck*) *Starkey* have the most prominent role, both as to prevalence and as to economic importance.

Since this corrosion process results from the metabolic processes of anaerobic micro-organisms, it occurs only under environmental conditions where a high degree of anaerobiosis can be established and maintained. Such locations most frequently occur in aqueous environments, bogs, and water-saturated soils. However, even in waters and soils of high oxygen content, a sufficient degree of anaerobiosis may frequently occur at isolated points and permit the active development of these bacteria. A decrease in the oxygen concentration at such a point, due to biochemical reducing reactions, particularly cellulose fermentation, or even the presence of reducing metal ions, is often sufficient to permit the development of the sulfate-reducing bacteria

In studying the serious corrosion encountered in Holland, von Wolzogen Kühr[9] concluded that the microbiological anaerobic corrosion of iron could be represented as:

Anodic solution of iron:

$$8H_2O \rightleftarrows 8H^+ + 8(OH)^-$$

$$4Fe + 8H^+ \rightleftarrows 4Fe^{++} + 8H$$

Cathodic depolarization:

$$H_2SO_4 + 8H \longrightarrow H_2S + 4H_2O$$

Corrosion products:

$$Fe^{++} + H_2S \rightleftarrows FeS + 2H^+$$

$$3Fe^{++} + 6(OH)^- \rightleftarrows 3Fe(OH)_2$$

The net overall reaction for the process was written:*

$$4Fe + H_2SO_4 + 2H_2O \longrightarrow 3Fe(OH)_2 + FeS$$

---

[9] C. A. H. von Wolzogen Kühr, "De sulfaatreductie als corzaak der aantasting vani jzeren buisleddingen" !("Sulfate Reduction as the Cause of the Corrosion of Iron Pipe Lines"), *Water and Gas*, **VII** (No. 26), 277 (1923). C. A. H. von Wolzogen Kühr and L. S. Van der Vlugt, "The Graphitization of Cast Iron as an Electro-biochemical Process in Anaerobic Soils," *Water* (No. 16), 147 (August, 1934).

* From the standpoint of the electrochemical theory of corrosion, the above reactions can be analyzed somewhat differently, as follows:

Anodic reaction: $\quad 4Fe \longrightarrow 4Fe^{++} + 8e^-$

Cathodic reaction: $\quad 8H_2O \longrightarrow 8H + 8OH^- - 8e^-$

(The hydrogen atoms are adsorbed on the iron surface or dissolved in the lattice, in either instance impeding the reaction by polarizing the cathodic elements.) Continued on next page.

The above equation indicates a theoretical ratio of the total iron in the corrosion products to the iron existing as sulfide in the same corrosion products of 4:1. Experimental determinations by von Wolzogen Kühr of this ratio varied, in ten analyses, between 2.4:1 and 3.4:1, whereas a saturated aqueous solution of hydrogen sulfide in contact with cast iron corrosion specimens resulted in a ratio equal to 1. Since differential electrolytic cell effects may also exist in anaerobic environments the agreement between the experimentally determined ratio and the theoretical ratio is considered satisfactory.

Subsequent investigations of the microbiological anaerobic corrosion process have been reported.[10,11,12,13,14,15,16] Starkey[17] and Wight[18] have studied ferrous corrosion in connection with the development and practical application of oxidation-reduction potential measurements for locating active microbiological sulfate-reducing areas in the field.

The laboratory investigations of Berkalof and Cabasso[19] show that the emf of a ferrous electrode during the period of bacterial growth is 52 mv negative (anodic) to the emf of a similar electrode in sterile culture medium. Employing corrosion cells containing two parallel ferrous metal electrodes, they found, upon inoculation of one

Bacterial depolarization:

$$8H + CaSO_4 \longrightarrow 4H_2O + CaS$$

and in the presence of carbonic acid:

$$CaS + 2H_2CO_3 \longrightarrow 2Ca(HCO_3)_2 + H_2S$$

Summary:

$$4Fe + 4H_2O + CaSO_4 + 2H_2CO_3 \longrightarrow 4Fe(OH)_2 + H_2S + 2Ca(HCO_3)_2$$

or

$$4Fe + 2H_2O + CaSO_4 + 2H_2CO_3 \longrightarrow 3Fe(OH)_2 + FeS + 2Ca(HCO_3)_2$$

EDITOR.

---

[10] H. J. Bunker, "The Role of the Sulfate-Reducing Bacteria in Metallic Corrosion," *Proc. Brit. Assoc. Refrig.*, **35** (No. 1), 152 (1936–1937). "Anaerobic Soil Corrosion. The Function of the Sulfate-Reducing Bacteria," *Fifth Report*, Corrosion Committee, Iron and Steel Institute, Section F, Part 3, p. 431 (1938). "Microbiological Experiments in Anaerobic Corrosion," *J. Soc. Chem. Ind.*, **58**, 93 (March, 1939). "Microbiological Anaerobic Corrosion," *Chemistry and Industry*, **59**, 412 (1940). "Activities of the Sulfate-Reducing Bacteria with Particular Reference to Underground Metallic Corrosion," *Proc. Soc. Agr. Bacteriologists*, pp. 8–9 (1941).

[11] T. D. Beckwith and J. R. Moser, "The Reduction of Sulfur-Containing Compounds in Wood Pulp and Paper Manufacture," *J. Bact.*, **24** (No. 1), 43 (July, 1932).

[12] C. L. Clark and W. J. Nungester, "Corrosion of Water Pipes in a Steel Mill," *Trans. American Society for Metals*, **31**, 304 (1943).

[13] Richard Pomeroy, "Corrosion of Iron by Sulfides," *Water Works and Sewerage*, **92**, 133 (1945).

[14] Edwin G. Pont, "Association of Sulfate Reduction in the Soil with Anaerobic Iron Corrosion," *J. Australian Inst. Agr. Sci.*, **5**, 170–171 (1939).

[15] K. W. H. Leeflang, *Laboratory Experiments with Regard to the Corrosion of Cast Iron in the Soil*, Bureau of Standards Corrosion Conference, 1937.

[16] P. Ganser, "Pipe Line Corrosion Caused by Anaerobic Bacteria," *Pipeline Gas J.* **231**, 271-3 (1940).

[17] R. I. Starkey and K. M. Wight, "Correlations between Oxidation-Reduction Potentials and the Development of Sulfate-Reducing Bacteria," paper presented to the meeting of the combined New York and New Jersey sections of the Society of American Bacteriologists, Dec. 29, 1942. "Soil Areas Corrosive to Metallic Iron Through Activity of Anaerobic Sulfate-Reducing Bacteria," *Am. Gas Assoc. Monthly*, pp. 223–228 (May, 1943), and also contained in *Report* of Distribution Committee Technical Section, American Gas Association, pp. 19–29, 1943. *Detection of Bacterial Activity in Soils*, National Bureau of Standards Corrosion Conference, March, 1943.

[18] K. M. Wight, *Corrosion of Iron by Sulfate-Reducing Bacteria and Detection of Corrosive Locations in Soil*, M. Sc. Thesis, Rutgers University, New Brunswick, N. J., May, 1943.

[19] E. Berkalof and V. Cabasso, "Sur une souche de vibrion reducteur des sulfates isolée de l'eau d'un puits artesien, en Tunisie," *Arch. Inst. Pasteur (Tunis)*, **29**, 262–284 (1940). P. Menard and E. Berkalof, "Corrosion biologique dans les puits artesiens en Tunisie," *Arch. Inst. Pasteur (Tunis)*, **29**, 455–462 (1940)

of the cells with cultures of the sulfate-reducing bacteria, an overall internal resistance decrease from the original value of 15 ohms to approximately 1 ohm, whereas the sterile cell actually showed an increase in resistance.

Development of the sulfate-reducing bacteria in truly inorganic media where bicarbonate is the only carbon source has been recently demonstrated,* thus indicating that sulfate-reducing bacteria are adaptive in regard to their ability to utilize gaseous hydrogen in an environment devoid of organic nutrients.

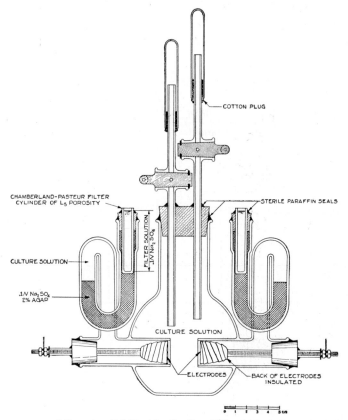

FIG. 1. One Type of Corrosion Cell Used for Studies of Microbiological Anaerobic Corrosion.

Electrode Behavior in Contact with Sulfate-Reducing Bacteria. The current and potential data[20] reported herein were collected from experiments carried out in the corrosion cells illustrated in Fig. 1. Two electrodes were machined from the same piece of steel. Each electrode was identically cleaned and sterilized which resulted in the formation of a relatively thick layer of iron oxide over the entire surface. Later studies were made using zinc electrodes of similar size.

The potential behavior of all-steel electrodes studied, before, during, and after

---

* Private communication from R. L. Starkey and K. M. Wight.

[20] R. F. Hadley, *The Influence of Sporovibrio desulfuricans on the Current and Potential Behaviour of Corroding Iron*, National Bureau of Standards Corrosion Conference, March, 1943.

bacterial inoculation and growth, was practically identical and can be represented by the average curve shown in Fig. 2.

The potential between A and B prior to inoculation appears to be determined by the concentration of hydrogen ions in the culture solution ($pH = 7.1$). Hydrogen depolarization occurs during the time interval of active bacterial growth between B and C. Upon removal of polarizing hydrogen by the bacteria, ferrous ions immediately go into solution, and the potential of the steel electrode rapidly changes 0.05 volt in the anodic direction. The time interval during which this anodic change of potential occurs in the lactate culture solution is 350 hours or less.

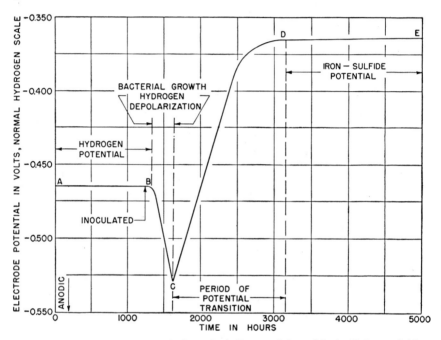

Fig. 2. Typical Potentials of Steel Electrodes in Lactate Culture Solution Before and After Inoculation.

The third period, occurring between C and D, may be denoted as the potential transition stage which always follows the anodic potential change B–C of a steel electrode. The activity of the sulfate-reducing bacteria during this period is probably negligible; otherwise the potential of the steel electrode should have remained at the relatively anodic value reached at C. This potential transition period generally occurs during a time interval of approximately 1000 hours.

The final phase of the steel electrode potential between D and E appears to be the potential as determined by the iron-sulfide reaction, the constancy of which indicates a reaction of uniform rate. It is doubtful that the sulfate-reducing bacteria contribute to the corrosion of the metal electrode during this final phase. However, if under the conditions encountered in nature, the sulfate-reducing bacteria are capable of growing in one location on a ferrous metal surface adjacent to an area in which earlier growth has resulted in the establishment of this iron-sulfide potential, a local corrosion cell potential difference of at least 150 mv may be established and maintained during the period of bacterial activity.

When dissimilar iron-zinc couples are placed in sterile corrosion cells and later inoculated, the potential of the steel electrode behaves as shown in Fig. 3. This potential change of the steel electrode in the cathodic direction during bacterial activity is accompanied by another change not heretofore reported which may be fully as important as the reduction of the polarizing cathodic hydrogen by the sulfate-reducing bacteria. This characteristic is the reduction in film resistance of steel electrodes when exposed to an active microbiological anaerobic environment as

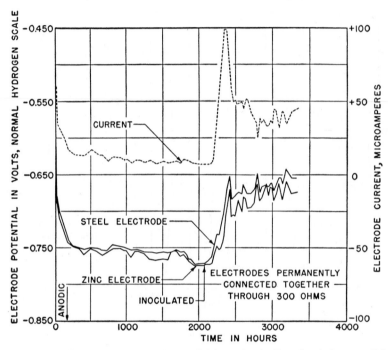

Fig. 3. Potential and Current of Zinc-Steel Electrodes in Inoculated Lactate Culture Solution.

illustrated in Fig. 4. Although the local corrosion-cell resistances are in parallel so far as the external circuit is concerned, they are in series for the local corrosion cell circuit. It is therefore apparent that any pronounced reduction in either, or both, the anode and cathode resistances which are in parallel for the corrosion cell currents must likewise represent a similar reduction in the total series resistance of the local corrosion cell circuit.

These results substantiate the observations of severe corrosion in the field,[21] as such corrosion is frequently found in locations where electrical measurements at the soil surface indicate that a pipe line is not discharging current (positive) to the earth.

[21] R. F. Hadley, "Microbiological Anaerobic Corrosion of Steel Pipe Lines," *Oil and Gas J.*, **38** (No. 19), 92 (September, 1939). "Methods of Studying Microbiological Anaerobic Corrosion of Pipe Lines," Part I, *Petroleum Eng.*, **XI** (No. 6), 171 (March, 1940); Part II, **XI** (No. 7), 112 (April, 1940). "Studies in Microbiological Anaerobic Corrosion," *Proc. Am. Gas Assoc.*, Tech. Sect. 764–788 (1940).

It is now believed that numerous interspaced anodes and cathodes simultaneously exist in a small area, the local corrosion cell resistances of which have decreased to a low value. The electrical effects of the local corrosion cell currents do not extend more than a few inches from the pipe under these conditions.

Although the results of these experimental investigations definitely indicate the removal or reduction of the cathodic polarizing hydrogen by the sulfate-reducing bacteria, the mechanism by which this process is accomplished still remains to be determined. The conflicting reports in the literature[22,23] regarding the bacterial utilization of hydrogen are such as to warrant additional research on the subject.

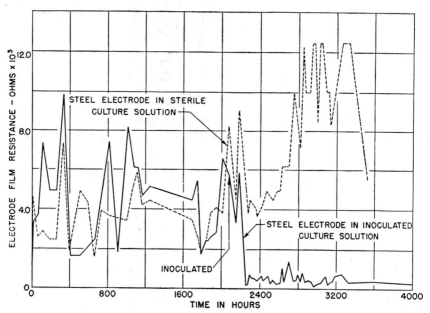

Fig. 4. Film Resistance of Steel Electrodes (Area 2.55 Sq Cm) in Sterile and Inoculated Lactate Culture Solution.

Bacterial Action throughout a Soil Area Not Necessarily Corrosive. C. M. Wichers[24] in discussing anaerobic corrosion mentions that when microbiological sulfate reduction occurs in the vicinity of a pipe line serious corrosion results. However, where the microbiological sulfate reduction process occurs throughout the entire soil environment, corrosion does not necessarily occur. An examination of the surface of ferrous metals in the latter locations reveals not only lack of appreciable corrosion, but also, what is equally important, little ferrous sulfide although active cultures of the sulfate-reducing bacteria may be isolated from almost any part of the soil as well as from the soil in the immediate vicinity of the metal. The most active cultures of the sulfate-reducing bacteria ever isolated from the soil by the writer were col-

[22] Edwin G. Pont, "Association of Sulfate Reduction in the Soil with Anaerobic Iron Corrosion," *J. Australian Inst. Agr. Sci.*, **5**, 170–171 (1939).

[23] R. F. Hadley, *The Influence of Sporovibrio desulfuricans on the Current and Potential Behaviour of Corroding Iron*, National Bureau of Standards Corrosion Conference, March, 1943.

[24] C. M. Wichers, "Bodenkorrosion guBeiserner Wasserleitungen," *Korrosion u. Metallschutz*, **12**, 81–85 (April, 1936).

lected from such a location. The apparent explanation for the occasional absence of serious corrosion in areas where intense microbiological sulfate reduction occurs prior to the burial of the ferrous metal is that the metal superficially reacts with sulfide and becomes cathodic (*D–E* portion of Fig. 2).

Action of Sulfate-Reducing Bacteria in Pitting. In combination with the differential aeration circuits set up by clod action, a more satisfactory explanation of the randomness of pitting of steel pipe lines, so long sought by many corrosion investigators, is suggested by the microbiological anaerobic corrosion process, for even in the very aerobic soils, the sulfate-reducing micro-organisms are readily cultured from the ferrous corrosion products contained in the pits. Similar attempts to obtain cultures of the same micro-organisms from the large adjacent cathodic surfaces meet with little success. Local colonies of the sulfate-reducing bacteria may develop at small isolated points on the surface of the pipe where organic debris (Fig. 5) has been included in the backfill. Organic nutrients, upon the decomposition of this cellulose, are made available for the sulfate-reducing bacteria which, in turn, suppress the oxygen concentration at this contact. Microbiological cathodic depolarization under

Fig. 5. Microbiological Anaerobic Corrosion Accompanying the Decomposition of Hemp Rope (Cellulose) in the Soil.

such conditions is of little consequence, since the microbiological contribution to corrosion in any aerobic environment results more from the maintenance of anaerobic conditions and low film resistances in the anodic areas.

It is a common occurrence in clay and boggy soils to find these isolated colonies of the sulfate-reducing bacteria growing in the pits and in the organic material contacting the external surface of the steel or cast iron pipe. Upon removal of the organic material, which is generally saturated with the ferrous corrosion products of the metal, severe pits are observed in the pipe under this organic debris. The sulfate-reducing bacteria with proper care can nearly always be cultured from these corrosion products. Corrosion of this nature is more the "deposit type," and is typical of the microbiological anaerobic corrosion encountered in locations where the oxygen concentration or other factors prevent the mass growth of the sulfate-reducing bacteria throughout the entire environment.

Detection of Corrosion by Sulfate-Reducing Bacteria. The detection of microbiological anaerobic corrosive areas is difficult since such areas are not readily located by the ordinary electrical measurements. Locations where such electrical measurements indicate that a metal structure is not discharging current to the surrounding environment and yet is severely corroded should be studied for evidence of microbiological anaerobic corrosion. A chemical analysis of the corrosion products associated with corroded metallic surfaces is particularly helpful. Appreciable quantities of ferrous sulfide indicate microbiological anaerobic corrosion, particularly if sulfides are not present in the soil or available from some other source.

Great caution must be used in handling, transporting, and storing the corrosion products formed under anaerobic conditions[25] since the ferrous sulfide in these corrosion products is usually precipitated in an extremely porous and finely divided form. The oxidation of this ferrous sulfide by atmospheric oxygen is so rapid that erroneous results are obtained unless the sample is prepared for analysis under vacuum or an inert gas. The transportation and storage of these anaerobic corrosion products are best accomplished under water collected from the same environment in which they were formed.

Rapid qualitative determination for ferrous sulfide is made in the field by adding dilute hydrochloric acid to the corrosion products and observing evolution of hydrogen sulfide. Use of microbiological techniques for culturing the sulfate-reducing micro-organisms from the soil or the corrosion products is also of considerable value; however, positive results do not necessarily constitute conclusive evidence that microbiological anaerobic corrosion is taking place.

The measurement of oxidation-reduction potentials in areas where microbiological anaerobic sulfate reduction is suspected is also of assistance in evaluating the degree and extent of the process; however, the apparatus and techniques used for these measurements are still in the development stage. The Redox potential method, based on field measurements and correlative studies of the severity of the corrosion actually observed, displays considerable promise as an early solution to the problem whereby such areas may be readily located in the field.[26,27]

Preventive Measures. The mitigation of microbiological anaerobic corrosion has been satisfactorily accomplished by the use of carefully selected asphalt, bituminous enamel, and concrete coatings, the first two coatings being supplemented by cathodic protection.

The relocation of pipe lines to avoid soils known to be corrosively aggressive due to microbiological anaerobic sulfate reduction is also occasionally possible.

Soil drainage of certain areas has also been reported; however, care must be exercised to prevent corrosion by oxidized products of microbiological anaerobic corrosion. The oxidation of ferrous sulfide to ferric sulfate can produce an environment which is very corrosive.

Use of a mixture of sand and lime[28,29] around pipe lines to maintain an alkaline environment and to permit an influx of oxygen has also been suggested. However, in water-saturated soils where optimum environmental conditions permit the mass

---

[25] Melvin Romanoff, *Effect of Aeration on the Hydrogen-Ion Concentration of an Anaerobic Soil*, National Bureau of Standards Corrosion Conference, March, 1943.

[26] R. L. Starkey and K. M. Wight, *Correlations between Oxidation-Reduction Potentials and the Development of Sulfate-Reducing Bacteria*, paper presented to the meeting of the combined New York and New Jersey sections of the Society of American Bacteriologists, Dec. 29, 1942. "Soil Areas Corrosive to Metallic Iron Through Activity of Anaerobic Sulfate-Reducing Bacteria," *Am. Gas Assoc. Monthly*, pp. 223–228 (May, 1943), and also contained in *Report* of Distribution Committee Technical Section, American Gas Association, pp. 19–29, 1943. *Detection of Bacterial Activity in Soils*, National Bureau of Standards Corrosion Conference, March, 1943.

[27] K. M. Wight, *Corrosion of Iron by Sulfate-Reducing Bacteria and Detection of Corrosive Locations in Soil*, M. Sc. Thesis, Rutgers University, New Brunswick, N. J., May, 1943.

[28] C. M. Wichers, *The Corrosion of Pipe Lines Due to Earth Contact*, U. S. Department of Commerce, National Bureau of Standards Conference on Underground Corrosion, 1937.

[29] *Report of Dutch Committee for the Study of Corrosion of Pipe Lines*, Groningen, Holland, Feb. 18, 1937.

development of the sulfate-reducing bacteria, the effectiveness of mitigation measures of this kind is doubtful.[30]

Heavy clay or loam packed tightly around the pipe has likewise been recommended,[28] particularly if the clay is sufficiently impervious to limit the movement of water and at the same time is devoid of various organic nutrients required by the sulfate-reducing bacteria.

Although some clays may be satisfactory for this purpose, sealing the pipe in an oxygen-deficient environment may result in the anodic solution of iron because of differential aeration cells, the cathodes of which are far removed to an electrically remote aerobic area.

The application of cathodic protection has been successfully demonstrated, although the current densities required to achieve complete protection may be excessive, particularly if the areas to be protected are very extensive. In practice, a minimum protective current density of 15×10^{-6} to 50×10^{-6} ampere per sq cm has actually been measured in an active microbiological anaerobic corrosive environment.

The use of sacrificial anodes for the cathodic protection of ferrous metal surfaces under environmental conditions favoring microbiological anaerobic corrosion is a subject still requiring investigation. Laboratory experiments in aqueous environments tend to substantiate the reported successful utilization of zinc; however, field measurements have been made from which contrary conclusions are drawn. Data from installations where zinc fails to provide adequate protection in areas of anaerobic bacterial activity appear to indicate that an insoluble anodic film forms accompanied by anodic polarization of the zinc. The use of zinc in locations where this polarizing anodic film does not form, or where the film can be removed either periodically or as it forms, may provide adequate protection.

In aqueous systems the successful use of chlorine or chlorine plus ammonia for the mitigation of microbiological anaerobic corrosion has been reported.[31,32,33] The residual chlorine content of the water which must be maintained to inhibit the development and growth of the sulfate-reducing bacteria, however, is apparently subject to considerable variation since the values in the several successful installations reported in the literature have varied between 0.1 and 1.0 ppm. Organic materials and inorganic salts, together with the degree of microbiological infection, are all instrumental in determining the chlorine residual necessary.

Certain organic corrosion inhibitors as well as a number of oxidizing agents have likewise been successfully employed in reducing or eliminating corrosion of this nature. The use of organic dyes has been studied by Rogers,[34] who reports success in mitigating microbiological anaerobic corrosion with acriflavine and proflavine dye concentrations of 1:750,000.

NITRATE REDUCTION

The possibility that the microbiological nitrate reduction process may contribute to corrosion has been mentioned by von Wolzogen Kühr more as a possibility than

[30] H. J. Bunker, "The Control of Bacterial Sulfate Reduction by Regulation of Hydrogen-Ion Concentration," *Proc. Soc. Agr. Bacteriologists*, **13**, 8–10 (1942).

[31] H. F. Simons, "Combat Corrosion in West Texas," *Oil and Gas J.*, **41**, 61 (February, 1943).

[32] L. H. Dial, "Design, Construction, and Operation of Salt Water Disposal System," *The Petroleum Engr.*, **15**, 51–56 (August, 1944).

[33] A. H. Thomas, "Role of Bacteria in the Corrosion of Iron and Steel," *Proc. Second Annual Water Conference*, Pittsburgh, Pa. (November, 1941).

[34] T. H. Rogers, "The Inhibition of Sulfate-Reducing Bacteria by Dyestuffs," *J. Soc. Chem. Ind. (Trans.)*, **59**, 34 (February, 1940).

as an established fact. The overall equation for the cathodic process is given as

$$HNO_3 + 8H \longrightarrow NH_3 + 3H_2O$$

The degree of reduction of the nitrate will depend upon the physiological characteristics of the organisms involved in the reduction process, for in some instances nitrite is produced, whereas in others gaseous nitrogen or ammonia is formed. There are many organisms which reduce nitrate to nitrite, but not nitrite to ammonia.

It is doubtful that this corrosion process, even if it does occur, is of any great economic importance.

METHANE FERMENTATION

von Wolzogen Kühr[35] has suggested that the microbiological anaerobic methane fermentation may be of sufficient importance in regard to corrosion to warrant additional study. The carbonic acid reduction reaction is given as

$$CO_2 + 8H \longrightarrow CH_4 + 2H_2O$$

and the overall reaction for the iron corrosion process as

$$4Fe + 6H_2O + CO_2 \longrightarrow 4Fe(OH)_2 + CH_4$$

Studies[36] of the factors which influence microbiological methane fermentation indicate that this process occurs between the pH limits of 5.2 and 8.7, the reaction being most active between pH 6.0 and 8.0. It is of interest to note that these pH limits are approximately the same as those reported by Starkey for the most favorable development of *Sporovibrio desulfuricans*.

The universal distribution of the methane fermentation in the water-logged soils, bogs, river and lake bottoms, etc., has been established by numerous studies. The micro-organisms responsible for this fermentation, however, generally exist in association with the sulfate-reducing bacteria, these two microbiological anaerobic processes co-existing in nature to such an extent that they are often considered to some degree interdependent.

The occurrence of severe corrosion in soils near manure heaps and in environments contaminated by offal and sewage is frequently observed. It is significant, however, that the corrosion observed in these locations is always associated with appreciable quantities of ferrous sulfide in the corrosion products. Active cultures of the sulfate-reducing bacteria can nearly always be isolated from these soil locations. The relationship which exists between methane fermentation micro-organisms and corrosion requires additional investigation in an effort to ascertain the degree to which these micro-organisms influence cathodic depolarization of the metal.

AEROBIC MICROBIOLOGICAL CORROSION PROCESSES

OXIDATION OF SULFUR BY THIOBACILLUS THIOOXIDANS

The oxidation of sulfur or reduced inorganic sulfur compounds, under aerobic conditions, is a biochemical process promoted by specific micro-organisms of the genus *Thiobacillus*, principally by *Thiobacillus thiooxidans*. While most of these organisms are strictly aerobic, one species, *Thiobacillus denitrificans*, is actually

---

[35] C. A. H. von Wolzogen Kühr, *The Unity of the Anaerobic and Aerobic Iron-Corrosion Process in the Soil*, National Bureau of Standards Fourth Corrosion Conference (confidential), 1937.

[36] H. A. Barker, "Studies on Methane-Producing Bacteria," *Archiv Mikrobiologie*, **7**, 420–438 (1936).

anaerobic and thus able to oxidize sulfur and sulfides in the absence of atmospheric oxygen. *Thiobacillus thiooxidans* has been extensively investigated by numerous microbiologists.[37,38]

The primary source of energy for the development and growth of *Thiobacillus thiooxidans* is the oxidation of elementary sulfur, and the principal end product is sulfuric acid. This bacterium develops best in an acid environment with a terminal pH between 0 and 1.0. Waksman and Starkey obtained growth of the bacterium in a medium which initially contained 5% sulfuric acid, and Starkey reported that the organisms produced as much as 10% sulfuric acid (the equivalent of 2 N H_2SO_4).

Beckwith[39] has reported the results of some quantitative laboratory experiments in which weighed iron wires were placed in separate flasks of culture medium. One of the flasks was then inoculated with a small amount of soil known to contain *Thiobacillus thiooxidans*. At the end of the incubation period the weight of the wire in the control flask was found to be practically unchanged; however, the wire in the flask in which the micro-organisms developed lost 66% of its weight.

To those engaged in the operation of the pipe lines in which sulfur compounds for sealing joints[40] have been used, the hazard is ever present that this microbiological sulfur oxidation process may occur and cause serious corrosion in the immediate vicinity of these joints. Likewise, in the construction of welded or screw joint pipe lines through soil areas which contain elemental sulfur or incompletely oxidized inorganic sulfur compounds, the potential danger of this same microbiological process must be considered.

The remedial measures employed in the mitigation of microbiological corrosion of this nature have not been extensively publicized. In areas where the mass growth of the *Thiobacillus thiooxidans* takes place throughout the entire environment the use of asphalt or bituminous enamel pipe coatings supplemented by cathodic protection, as required, will unquestionably prove effective. The application of cathodic protection to bare structures is possible, but probably economically unattractive because of its high operating cost. The use of concrete for the mitigation of this corrosion is not considered entirely satisfactory since numerous reports are to be found in the literature where *Thiobacillus thiooxidans* has caused the rapid disintegration of concrete.[41]

Investigations have not as yet been reported in sufficient number to determine the prevalence and economic importance of corrosion of this nature.

[37] S. A. Waksman, "Micro-organisms Concerned in the Oxidation of Sulfur in the Soil. I. Introductory," *J. Bact.*, **7**, 231–238 (1922). "Micro-organisms Concerned in the Oxidation of Sulfur in the Soil. III. Media Used for the Isolation of Sulfur Bacteria from the Soil," *Soil Sci.*, **13**, 329–336 (1922). "Micro-organisms Concerned in the Oxidation of Sulfur in the Soil. IV. A Solid Medium for the Isolation and Cultivation of *Thiobacillus thiooxidans*," *J. Bact.*, **7**, 605–608 (1922). "Micro-organisms Concerned in the Oxidation of Sulfur in the Soil. V. Bacteria Oxidizing Sulfur under Acid and Alkaline Conditions," *J. Bact.*, **7**, 609–616 (1922).

[38] R. L. Starkey, "Concerning the Physiology of *Thiobacillus thiooxidans*, an Autotrophic Bacterium Oxidizing Sulfur under Acid Conditions," *J. Bact.*, **10** (No. 2), 135 (March, 1925). "Concerning the Carbon and Nitrogen Nutrition of *Thiobacillus thiooxidans*, an Autotrophic Bacterium Oxidizing Sulfur under Acid Conditions," *J. Bact.*, **10** (No. 2), 165 (March, 1925). "Products of Oxidation of Thiosulfate by Bacteria in Mineral Media," *J. Gen. Physiol.*, **18** (No. 3), 325–349 (January, 1935).

[39] T. D. Beckwith, "The Bacterial Corrosion of Iron and Steel," *J. Am. Water Works Assoc.*, **33**, 147 (No. 1) (January, 1941). *Corrosion of Iron by Biological Oxidation of Sulfur and Sulfides*, National Bureau of Standards Corrosion Conference, March, 1943.

[40] T. D. Beckwith and P. F. Bovard, "Bacterial Disintegration of Sulfur-Containing Sealing Compound in Pipe Joints," *Univ. Calif. (at Los Angeles) Pub. Biol. Sci.*, **1** (No. 7), 121–132 (February, 1937).

[41] W. M. Barr and R. E. Buchanan, "The Production of Excessive Hydrogen Sulfide in Sewage Disposal Plants and Consequent Disintegration of the Concrete," *Ind. Eng. Chem.*, **4**, 564–567 (August, 1912).

The Sulfur Bacteria

The published references to corrosion specifically attributable to the sulfur bacteria are few in number. Some investigators[42] have inferred that the sulfur bacteria, upon death and decomposition, may be responsible for the occurrence of hydrogen sulfide in estuarine waters although their reference to this possibility is more suggestive than the report of an ascertained fact. It is now known that hydrogen sulfide may be readily produced in estuarine waters by the action of the anaerobic halophilic (salt-tolerant) sulfate-reducing bacteria.

The extent to which the so-called sulfur bacteria[43,44] influence metallic corrosion in environments favorable for their growth is not certain at the present time. The fact that these micro-organisms contain sulfur, however, would suggest that they represent, at least potentially, a corrosive influence which requires consideration in any study of microbiological corrosion.

The Iron Bacteria

The role of the iron bacteria in microbiological corrosion[45,46,47] is still rather obscure despite the many studies and investigations which have been reported in the literature. Although their physiological and morphological characteristics have been extensively studied by microbiologists, the exact part which these micro-organisms play in the corrosion of ferrous metals has not been generally agreed upon.

The most important contribution of the iron bacteria to corrosion probably results when their growth develops to such an extent that they create a barrier capable of maintaining oxygen concentration gradients between the metal and the solution, the end result of such a process being tuberculation of the metal. (See also discussion of tuberculation, p. 497.)

The iron gathered by the bacteria may be obtained from water flowing in the pipe.[48]

It is particularly significant in the reported studies of tuberculation that evidence of microbiological anaerobic corrosion is observed in the interior of these tubercles. Ellis,[47] referring to tuberculation, says that each incrustation consists of a cone-like structure, which grows by the addition of concentric layers. The central portion is black when fresh and soft. Often it contains a little sulfide of iron, and it becomes red on exposure to the air.

It is unlikely that the corrosion accompanying tuberculation will be severe unless anaerobic conditions are maintained on the inside of the tubercle. Additional data are admittedly required to clarify the relationship which exists between the iron bacteria, the sulfate-reducing bacteria, and tuberculation.

Mitigation of the corrosion caused by the iron bacteria has been successfully accomplished by the use of internal pipe coatings. Cement-lined pipe has also been recommended as satisfactory for this purpose, particularly if the surface of the cement is treated to prevent leaching of the more soluble constituents. Chlorine, maintained

---

[42] Ulick R. Evans, *Metallic Corrosion, Passivity and Protection*, Edward Arnold and Co., London, 1937.

[43] D. Ellis, *Sulfur Bacteria*, Longmans, Green and Co., New York, 1932.

[44] H. J. Bunker, "A Review of the Physiology and Biochemistry of the Sulfur Bacteria," His Majesty's Stationery Office, Chemistry Research, *Special Report* 3 (1936).

[45] N. Cholodny, *Die Eisenbakterien — Beitrage zu einer Monographie*, Pflanzenforschung, herausgegeben von Prof. Dr. R. Kolkwitz, Berlin-Dahlem, 1926.

[46] P. Dorff, *Die Eisenorganismen — Systematik und Morphologie*, Pflanzenforschung, herausgegeben von Prof. Dr. R. Kolkwitz, Berlin-Dahlem, 1934.

[47] D. Ellis, *Iron Bacteria*, Frederick A. Stokes Co., New York, 1919.

[48] H. G. Reddick and S. E. Linderman, "Tuberculation of Mains as Affected by Bacteria," *J. New England Water Works Assoc.*, **46** (No. 2), 146 (June, 1932).

as a residual of approximately 0.5 ppm, has likewise been successfully used[49,50] for preventing growth of these micro-organisms.

MISCELLANEOUS ORGANISMS

A numerous variety of organisms, including the slime-forming bacteria, fungi, algae, protozoa, diatoms, and bryozoa, contribute to the corrosion of metals by establishing a microbiological film capable of maintaining concentration gradients of the dissolved salts and gases of the electrolyte in contact with the metal. The maintenance of such a microbiological barrier results in the establishment of differential effects which contribute to the corrosion of the underlying metal. Frequently beneath these barriers the concentration of oxygen is reduced to a value which permits development of anaerobic sulfate-reducing micro-organisms.

O'Connell[51] has published a comprehensive report of the corrosion and bio-fouling of metallic surfaces exposed to a variety of aqueous environments. Unfortunately, little attempt has been made, thus far, to study and evaluate the nature of the corrosion reactions observed in connection with the growth of these various organisms. The fact, however, may be significant that even in many aerobic waters sulfide films are frequently observed beneath the miscellaneous microbiological population which develops on such surfaces.

The mitigation practices successfully employed to eliminate corrosion resulting from these miscellaneous organisms are numerous, depending to some extent upon the nature of the organisms involved and the degree of infection of the system. Coatings of paint, asphalt, bituminous and concrete, as well as cathodic protection are being successfully used. Chemical water treatment, chlorination, pH adjustment, bactericidal and bacteriostatic agents, organic and inorganic corrosion inhibitors, etc., have also had considerable application in aqueous systems. The elimination of bio-fouling and corrosion on the water side of heat-exchange condensers of petroleum refineries and the steam generation stations of the electrical industry is receiving increased attention,[51,52] and numerous reports are to be found in which this bio-fouling and corrosion have been successfully mitigated by the use of the proper water treatment.

FUNDAMENTAL BEHAVIOR OF GALVANIC COUPLES
W. A. WESLEY* AND R. H. BROWN†

DEFINITIONS

When dissimilar metals in electrical contact with each other are exposed to an electrolyte, a current flows from one to the other and is called a galvanic current. Galvanic corrosion is that part of the corrosion of the anodic member of such a couple directly related to the galvanic current by Faraday's Law.

Simultaneous additional corrosion taking place on the anode will be called *local*

---

* Research Laboratory, The International Nickel Co., Inc., Bayonne, N. J.
† Chemical Metallurgy Division, Aluminum Research Laboratories, Aluminum Company of America, New Kensington, Pa.

[49] A. H. Thomas, "Role of Bacteria in the Corrosion of Iron and Steel," *Proc. Second Annual Water Conference*, Pittsburgh, Pa. (November, 1941).

[50] K. W. Brown, "The Occurrence and Control of Iron Bacteria in Water Supplies," *J. American Water Works Assoc.*, **26** (No. 11), 1684 (November, 1934).

[51] W. J. O'Connell, Jr., "Characteristics of Microbiological Deposits in Water Circuits," *Proc. Am. Petroleum Inst.*, Eleventh Mid-year Meeting, **22**, (III), 66–83 (May, 1941).

[52] A. E. Griffin, "Anaerobic Corrosion," *Proc. Fourth Annual Water Conference*, Pittsburgh, Pa. (1943).

corrosion; corrosion taking place when there is no contact with a dissimilar metal will be called *normal corrosion.* When galvanic corrosion takes place, the local corrosion of the anode may be equal to the normal corrosion, or it may be altered. Such a change is often called the *difference effect,* and it may be either positive, if the local corrosion increases when the galvanic current flows, or negative.

A galvanic current generally causes a reduction in the total rate of corrosion of the cathodic member of the couple.* This is called *galvanic* or *cathodic protection.* Under certain conditions the rate of corrosion of some metals as cathode may actually be increased, in which case the term *cathodic corrosion* is applied.

The corrosion potential of a metal or alloy is the steady state irreversible potential it assumes under fixed corrosive condition.† The open circuit potential is the potential which a metal attains when no current is flowing in the external circuit. The polarized potential is the potential which a metal attains when current is flowing in the external circuit to or from the electrode.

FUNDAMENTAL CONSIDERATIONS

Potentials of Corroding Metals

It is important to recall the differences between steady state potentials of corroding metals and the thermodynamically reversible potentials of pure metals listed in the electromotive force series. The potentials measured in studying galvanic corrosion are not equilibrium potentials because the electrode reactions going on are continuously dissipating energy. Metal ions go into solution and hydrogen is liberated or hydroxyl ions formed. This does not mean that galvanic potentials are not reproducible. They usually *are* reproducible and can be held constant for long periods of time by control of the corrosive conditions, but this constancy is simply evidence that the electrode reactions are taking place at a constant rate.

In presenting corrosion potential data, as much information as possible should be given concerning the corrosive conditions maintained. Many of the data in the literature are of limited usefulness owing to failure to include enough such information. A general discussion of the potentials of corroding metals is available[1] as is also consideration of the potentials of metals and alloys in sea water.[2]

Kirchhoff's Law

The fundamental relationship involved in galvanic corrosion is described by Kirchhoff's Second Law:

$$\Sigma EMF = \Sigma IR \tag{1}$$

* However, conditions can exist where the anode member of a couple may corrode less than the same metal exposed uncoupled. In ammonium chloride, aluminum is anodic to copper. If coupled to copper, current in the electrolyte (positive electricity) flows from the aluminum to the copper, with the result that the corrosion of the copper is less than if the aluminum were not coupled to it. If both the aluminum and copper are exposed uncoupled in the same container to the ammonium chloride solution, both the metals corrode at a higher rate than if coupled. The copper corrosion products which are relatively soluble in ammonium chloride can come in intimate contact with the aluminum and are reduced to a lower state of valence, with the result that there is relatively rapid corrosion. However, when coupled, the current flow prevents the formation of the copper compounds which accelerate corrosion of the uncoupled aluminum.

† Strictly true only if the resistance of the electrolyte-metal circuit of the local-action cells is negligible. The corrosion potential is the potential of anode or cathode polarized to the same value by current flowing through the cell. Refer to Fig. 2, p. 484 and also to discussion of corrosion potential beginning p. 1040. EDITOR.

[1] O. Gatty and E. C. Spooner, *Electrode Potential Behaviour of Corroding Metals in Aqueous Solutions,* Oxford University Press, London, 1938.
[2] F. L. LaQue and G. L. Cox, *Proc. Am. Soc. Testing Materials,* **40,** 670–687 (1940).

FUNDAMENTAL BEHAVIOR OF GALVANIC COUPLES

Hence,
$$E_c - E_a = IR_e + IR_m \qquad (2)$$

where R_e = resistance of the electrolytic portion of the galvanic circuit,

R_m = resistance of the metallic portion,

E_c = effective (polarized) potential of the *cathodic* member of the couple,

E_a = effective (polarized) potential of the *anodic* member.

Polarization

In many cases E_a and E_c are functions of the galvanic current I; hence the potential difference accounting for the current through the electrolyte is not equal to the open circuit cell potential.

In Eq. 2 the terms E_c and E_a can be written as functions of the open circuit potentials E_c' and E_a' in the following way:

$$E_c = E_c' - f_c \frac{I}{A_c} \qquad (3)$$

$$E_a = E_a' + f_a \frac{I}{A_a} \qquad (4)$$

where f_c and f_a = the polarizing functions of cathode and anode, respectively,* A_c and A_a = areas of cathode and anode, respectively, and I = total current of the cell.

From Eqs. 3 and 4 it can be seen that polarization, defined as the change of potential of an electrode resulting from current flow, alters the potential of the anode in the cathodic direction and the potential of the cathode in the anodic direction.

Equations 3 and 4 can be considered graphically. Assume that an electrode in some electrolyte has an open circuit potential E'. If current (positive electricity) leaves the electrode and enters the electrolyte, the electrode is functioning as anode. Generally speaking, the electrode will polarize anodically, and its potential will be altered in the cathodic direction (curves 1a, 2a, or 4a, Fig. 1) unless $f_a \frac{I}{A_a}$ is zero (that is, the electrode does not anodically polarize). In the latter event, the potential will not change with increased current flow and is represented in Fig. 1 by the horizontal line AC. It is very important to note that if the electrode does polarize anodically, the area of the electrode as well as the amount of current flowing influences the amount of the polarization.

If current leaves the electrolyte and enters the electrode, the electrode is functioning as a cathode. Generally speaking, the electrode will polarize cathodically, and its potential will be altered in the anodic direction (curves 1c, 2c, or 4c, Fig. 1) unless $f_c \frac{I}{A_c}$ is zero. In the latter instance, the potential again will not change with increased current and the potential curve will be the horizontal line AC.

Graphic Estimation of Galvanic Currents

The maximum galvanic current which can flow in a given couple, known as the *limiting galvanic current*, is that at the intersection of the polarization curves of the

---

* In some cases the polarization is linear with current. The terms $f_c \frac{I}{A_c}$ and $f_a \frac{I}{A_a}$ can then be simplified to $k_c \frac{I}{A_c}$ and $k_a \frac{I}{A_a}$, respectively.

484　　　　　　SPECIAL TOPICS IN CORROSION

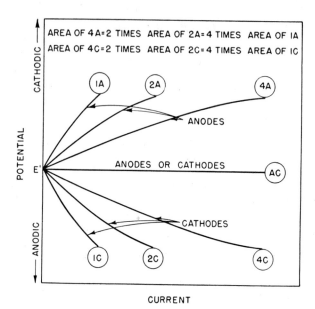

Fig. 1. Change of Potential of Anodes or Cathodes with Flow of Current.

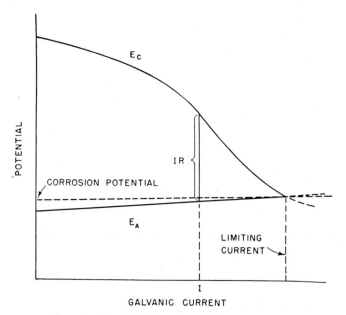

Fig. 2. Graphic Estimation of Galvanic Current.

anode and cathode. The reason for this is evident from Fig. 2. When R is infinite no current flows, and $E_c - E_a$ is the open circuit value of the cell potential. As R is made smaller, I increases and $E_c - E_a$ grows smaller because of polarization. Finally I approaches its maximum value (the limiting current) as R approaches zero, and $E_c - E_a$ simultaneously approaches zero.

Suppose that the value of $R = R_e + R_m$ for a particular zinc and copper couple has been measured and the polarization curves for the electrodes are known. It is desired to calculate what the steady state galvanic current will be for the given value of R. This is done quite simply graphically by locating the value of I in Fig. 2, at which the E_c and E_a curves are exactly IR volts apart.[3] This value of I must be less than the limiting current.

Determination of Resistance R. If it is possible to break the circuit in a practical galvanic couple, it becomes a simple matter to measure R_e, the resistance of the electrolytic portion of the galvanic circuit. This is done with sufficient precision by means of an inexpensive 60-cycle a-c bridge testing set with an a-c galvanometer. In lieu of this, an estimate can sometimes be made with a d-c bridge, in which case the resistance should be measured with the testing current applied in both directions and an average of the two readings taken.

When it is inconvenient to measure directly the unpolarized resistance R_e of a galvanic circuit, that quantity may often be roughly estimated by formulas based on the general theory of continuity of flow.[4] The resistance between two electrodes immersed in an electrolyte is related to the capacity of the condenser which would be formed if the electrolyte were replaced by air in the following way:

$$R = \frac{r}{4\pi C} \times \frac{1}{9 \times 10^5} \quad (5)$$

where r is the specific resistivity of the electrolyte in ohm-cm and C is the capacity of the condenser in microfarads. Formulas for the capacities of many types of metal electrodes forming a condenser are available in handbooks of physics and of electrical engineering.[5]

The internal resistance in a system of two electrodes placed in an infinite medium, such as the ocean, does not increase indefinitely as the electrodes are separated by increasing distance. On the contrary, the maximum resistance between them is probably approached soon after they have been separated by a distance of a higher order than the dimensions of the electrodes. It appears that the total resistance may be regarded as the sum of two electrode surface resistances, one at the crossing of the current from the first electrode to the medium, the other at the return of the current from the medium to the second electrode. A useful formula for reference is that the resistance in absence of polarization at the junction of a disk electrode of radius b cm in a solution of specific resistance r is $r/8b$. The resistance of one face of such a disk is $r/4b$. Formulas can be worked out for other simple electrode shapes.[4]

RELATIONSHIP OF GALVANIC CURRENT TO GALVANIC CORROSION

The source of the energy manifested as galvanic current is primarily the dissolution of the anodic metal. For every coulomb of galvanic current which flows, a definite

[3] U. R. Evans, *Metallic Corrosion, Passivity and Protection*, p. 533, Edward Arnold and Co., London, 1937.
[4] J. H. Jeans, *The Mathematical Theory of Electricity and Magnetism*, pp. 350–363, Cambridge University Press, Cambridge, 1925.
[5] C. D. Hodgman, *Handbook of Chemistry and Physics*, 6th Ed., p. 1866, Chemical Rubber Publishing Co., Cleveland, 1936.

SPECIAL TOPICS IN CORROSION

weight of metal must dissolve. The relation between the galvanic corrosion of the anode and the galvanic current is Faraday's Law:

$$W = \frac{tM}{Fn} I \tag{6}$$

where W is the weight of metal dissolved to produce the galvanic current (grams), t is the time of flow of current (seconds), F the Faraday constant (96,501 coulombs), M the atomic weight of the anode metal, n the charge of the metal ions formed, and I the galvanic current in amperes.

The loss in weight of the anode will usually be greater than W because some local action corrosion accompanies the galvanic corrosion. The total weight loss of the anode will be equivalent to the sum of the galvanic current and the local action currents on the anode.

Table 3, on p. 1133 of the General Tables lists the electrochemical equivalents for common metals based upon Faraday's Law.

TYPES OF CONTROL

Because it is possible for the degree of polarization to vary with the metal and solution, it is helpful to classify couples by the type of polarization produced on each

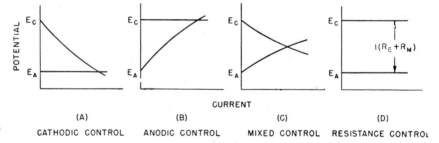

FIG. 3. Types of Controlling Factors in Galvanic Corrosion.

electrode. If the anode does not polarize and the cathode does, then in solutions of low resistivity the current flow will be controlled entirely by the cathodic electrode. If the area of the anode is either doubled or reduced by one-half, the galvanic weight loss of the anode will remain unchanged although the intensity of attack will increase with decreasing area. However, if the cathode area is reduced by half, the total galvanic corrosion of the anode will be reduced by half; or, if the area of the cathode is doubled, the galvanic corrosion of the anode will be doubled. The polarization curves for such a system are shown in Fig. 3A and the system is said to be under *cathodic control*.

If the anode polarizes and the cathode does not, as in Fig. 3B, the status is reversed and the system is said to be under *anodic control*.

If both the anode and cathode polarize, then in solutions of low resistivity the current flow will be controlled by the polarization of both electrodes. Hence, the areas of both electrodes will affect the galvanic corrosion of the anode. The polarization curves for such a system are represented by Fig. 3C, and the system is said to be under *mixed control*.

The above discussion has been confined to those systems in which the IR drops

FUNDAMENTAL BEHAVIOR OF GALVANIC COUPLES

can be neglected. If neither electrode polarizes, the current flowing and attendant corrosion will be controlled by the resistance of the liquid and metallic path (*resistance control*). Since the metallic path usually has a resistance small compared with that of the liquid path, the resistance of the latter will usually be the factor controlling corrosion of the anode. Figure 3D shows the current-potential diagram for complete resistance control.

In general, polarization as well as resistance will control the amount of galvanic current flowing.

Diffusion Control

The fundamental factor which is probably of greater importance than any other in determining corrosion rates is diffusion. Because the rate of diffusion of a substance which causes either polarization or depolarization can control the shape of the polarization curves, some couples can be said to be corroding under diffusion control.

For example, consider a metal such as iron totally immersed in a stagnant neutral salt solution in a container so that the surface of the liquid in contact with the air is relatively small. The corrosion of iron in such solutions free of dissolved oxygen is practically nil, whereas if dissolved oxygen is present, the amount of iron corroded is practically equivalent to the amount of oxygen consumed. If the area of the iron is sufficiently large, the rate of corrosion is controlled by the rate at which oxygen diffuses through the liquid-air interface. As long as these conditions exist, the total corrosion loss will not be increased even if the size of the specimen is increased. Totally immersing and connecting a cathodic metal such as copper to the iron will not increase the rate of oxygen diffusion through the liquid-air interface; hence the corrosion of iron will not increase over that occurring on the uncoupled iron. If service conditions are not under diffusion control, it is essential that any studies of galvanic corrosion are made under conditions where diffusion is not the controlling factor.

Effect of Relative Areas

The effect of increasing cathode area on the galvanic current flowing from anodes of the same area for three types of polarization control (neglecting resistance) is shown in Fig. 4. Curve 1 is for ideal cathodic control, and the galvanic current is proportional to the cathodic area. Curve 2 is the ideal mixed control.* Although the galvanic current increases with increased area, the increase is less than proportional to the cathode area. Curve 3 shows how the galvanic current is influenced by increasing cathode area for a couple corroding under conditions close to ideal anodic control.

The effect of increasing the anodic area on the galvanic current per unit area of anode (galvanic current density on the anode) for the three types of polarization control is shown in Fig. 5. This is drawn on the basis that the areas and the slopes of polarization curves for the cathodes are the same regardless of the type of control. It will be observed that for ideal cathodic control, decreasing area of the anode greatly increases the galvanic current density on the anode (curve 1). If the couple is under ideal mixed control, decreasing the anode area does not greatly increase the galvanic current density on the anode (curve 2). The more closely the conditions approach those of ideal anodic control, the less the galvanic current density on the

---

* Ideal mixed control is where the polarization of the anode is equal to the polarization of the cathode for equal areas of both electrodes.

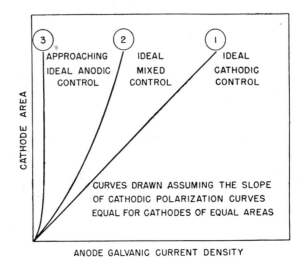

Fig. 4. Effect of Increasing Cathode Area on Galvanic Current per Unit Anode Area.

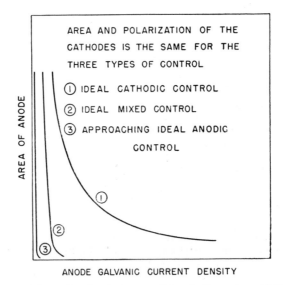

Fig. 5. Effect of Increasing Anode Area on Galvanic Current per Unit Anode Area

FUNDAMENTAL BEHAVIOR OF GALVANIC COUPLES 489

anode is increased by decreasing the anodic area (curve 3). The situations are summarized in Table 1.

TABLE 1. EFFECT OF POLARIZATION AND RESISTANCE ON THE GALVANIC CORROSION OF COUPLES

Type of Control	Polarized Electrode	Effect of Increasing Anode Area*		Effect of Increasing Cathode Area†		Example
		Total Weight Loss of Anode	Intensity of Anodic Corrosion	Total Weight Loss of Anode	Intensity of Anodic Corrosion	
Cathodic	Cathode	No effect	Decreased	Increased	Increased	Zn-Fe and Zn-Cu in neutral chloride solutions
O₂-Starvation	Cathode	No effect	Decreased	No effect	No effect	Total immersion of Fe-Cu in solutions of restricted liquid-air interface
Mixed	Both	Increased	Decreased	Increased	Increased	Fe-Cu in neutral salt solutions
Anodic	Anode	Increased	No effect	No effect	No effect	Al-Cu in neutral chromate solutions
Resistance	Neither appreciably	Depends upon effect of area of electrode on ohmic resistance. Decrease in resistance will increase total weight loss.				Most metal combinations in distilled water.

* Cathode area constant.   † Anode area constant.

INFLUENCE OF CORROSION VARIABLES

The influence upon galvanic currents of each of the many variables which affect corrosion is predictable from its influence upon anodic and cathodic polarization curves or upon the resistivity of the electrolyte. Thus a change in any factor which increases anodic polarization tends to cause a decrease in the galvanic current. The formation of insoluble films upon anodic areas is an example of this. Changes which decrease cathodic polarization, like increase in degree of aeration, in velocity of motion, or in temperature, tend to increase galvanic corrosion when the process is under cathodic or mixed control, just as they do in normal corrosion.

EFFECT OF NATURE OF CATHODE MATERIAL

With Oxygen Depolarization Control. Whitman and Russell[6] exposed steel plates partly coated with copper to corrosion by water and found that, regardless of the fraction of surface coated, the total corrosion remained the same as that of the bare steel. They suggested that the velocity of attack was controlled by the amount of oxygen reaching the metal surface. The cathodic reaction took place over all the metal area, and the anodic reaction was confined to the portion of steel not covered. Evans coined the term *Catchment Area Principle* for this case, since the copper acted as a catchment area for oxygen. He has discussed this with other cathode area data[7] and pointed out that the principle should be applicable when the oxygen supply is sufficiently restricted that diffusion of oxygen is the sole controlling process. Under such conditions cathode polarization curves for different metals lie close together, and the corrosion of the anode is accelerated equally by contact with more noble metals irrespective of their degree of nobility.[7]

[6] W. G. Whitman and R. P. Russell, *Ind. Eng. Chem.*, **16**, 276–279 (1924).
[7] U. R. Evans, *Metallic Corrosion, Passivity and Protection*, pp. 513–516, Edward Arnold and Co., London, 1937.

With a liberal supply of oxygen the cathode polarization curves of the common metals and alloys diverge considerably. Although equal galvanic currents are not to be expected when different cathode materials are coupled with the same anode material under such conditions, the several currents are likely to be at least of the same order of magnitude, as shown in Table 2.

TABLE 2. GALVANIC CURRENTS CAUSED BY CONTACT OF ALUMINUM ALLOY 52S–0* WITH VARIOUS DISSIMILAR METALS†

Electrolyte: 6% NaCl solution.
Anode size: 15 sq cm.
Cathode size: 15 sq cm.
Agitation: Solution stirred.
Aeration: Air bubbled through solution.
Temperature: Room.

Cathode of Couple	Galvanic Current, ma
Copper	9.0
Aluminum bronze	8.25
Low-carbon steel	6.3
Stainless steel (18–8)	4.35

* Al, 2.5% Mg, 0.25% Cr.
† R. B. Mears and R. H. Brown, *Ind. Eng. Chem.*, **33**, 1009 (1941).

In Corrosion with Hydrogen Evolution. The nature of the cathode material, specifically, its hydrogen overvoltage, plays an all-important role in galvanic corrosion when the cathode process is hydrogen evolution. The cathode polarization curve represents a logarithmic relationship between the potential and the current density so

TABLE 3. GALVANIC CORROSION OF MAGNESIUM*

Electrolyte: 3% NaCl solution.
Duration of test: 3 hours.
Anodes: Technically pure magnesium (0.03% Al, 0.04% Fe, and 0.02% Si).
Anode and cathode areas were equal.

	Mg Alone	Hg	Zn	Mn	Pb	Cu	Ni	Fe	Al	Pt
H_2 evolved on the cathode (ml)	...	0.5	6	13	22	27.5	45	58	126	142
H_2 evolved on the anode (ml)	13.0	14	15	22	34	43	54.5	70	80	110
Galvanic current (ma)	...	0.1	4.5	..	13	19	30	32	12 to 40	67

* W. Kroenig and G. Kostyler, *Z. Metallkunde*, **25**, 144–145 (1933).

that galvanic currents measured with a given anodic material vary tremendously with different cathodes. Thus in Table 3 the galvanic currents observed when magnesium anodes were coupled with various other metals bore no relationship to the corrosion potentials of those metals nor to their positions in the standard EMF Series, but was governed entirely by the relative hydrogen overvoltages of the cathodes, the metals of lowest overvoltage permitting the greatest galvanic currents.

Corrosion of Cathode. The corrosion of most alloys and metals is usually less when functioning as a cathode of a couple than the normal rate of corrosion. The quantitative aspects of this behavior are dealt with in the chapters on cathodic protection, pp. 923 and 935. It is true also that, as a result of current flow to the cathode, reaction products are released which can cause corrosion of certain cathodes to be greater than if the material were corroding under normal conditions. Since the solution around the cathode becomes more basic, amphoteric metals such as aluminum, zinc, and lead may corrode rapidly as a cathode if the current density is too high.*

DISTRIBUTION OF CORROSION

The theory of distribution of current over the surface of electrodes is prohibitively complex for any but the simplest symmetrical shapes.[8] In practical galvanic corrosion problems, interest usually lies in the distribution of the current as related to the

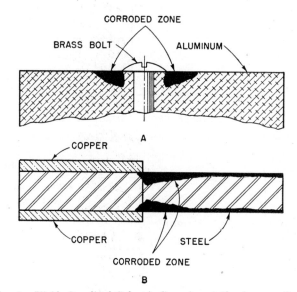

FIG. 6. Highly Localized Galvanic Corrosion of Aluminum or Steel.

distance from the line of contact or the junction of two coplanar surfaces. The factors are obvious enough. If the anode polarization curve is very steep and the conductivity of the electrolyte is good, the galvanic corrosion should be well spread out over the anode surface and vary little with distance from the junction. At the other extreme, corrosion will be highly localized along the boundary, as indicated in Fig. 6, if the solution is of low conductivity and the anode polarization is feeble. Other combinations of these factors yield intermediate types of distribution.

Figure 7 shows the results of a careful study of the distribution of galvanic currents about a junction between nickel and steel surfaces exposed to tapwater.[9] In this cell the electrolyte was of low conductivity and the anode was one showing but little polarization. The observed galvanic corrosion of the steel adjacent to its line of contact with nickel was about 3.5 times greater than the average galvanic corrosion

---

* The alkaline products of electrolysis at the cathode decrease the normal corrosion rate of iron.
[8] C. Kasper, *Trans. Electrochem. Soc.*, **77**, 353, 365 (1940); **78**, 131, 147 (1940); **82**, 153 (1942).
[9] H. R. Copson, *Trans. Electrochem. Soc.*, **84**, 71–81 (1943).

rate. On the other hand, exposure of a steel plate in contact with a copper plate in natural sea water (good conductivity) showed no appreciable concentration of galvanic corrosion at or near the junction.[10] A complicating factor was the deposition of a calcareous adherent scale on the cathode surface which probably tended to spread out the cathode current and reduce concentration of the anode current near the junction.

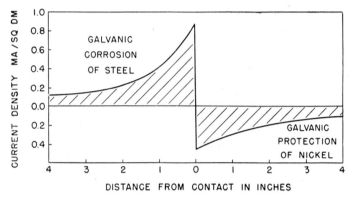

Fig. 7. Galvanic Current Distribution of Steel-Nickel Couple in Tap Water.

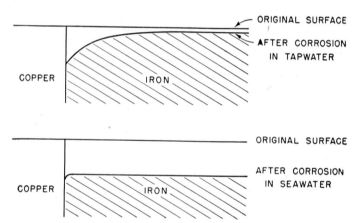

Fig. 8. Relative Distribution of Galvanic Corrosion in Electrolytes of High and Low Conductivities (Schematic).

It is fortunate that the total galvanic current is usually less in a poorly conducting solution than in a highly conducting one. This makes the localized corrosion of the former case less dangerous than would appear at first sight. The relative contours of the iron surfaces in the two examples are shown schematically in Fig. 8. The distribution resulting from a combination of good electrolyte conductivity with very low anode polarization is illustrated in experiments on aluminum alloys in sodium chloride solution.[11]

[10] F. L. LaQue, Unpublished data, 1943.
[11] R. H. Brown, *Am. Soc. Testing Materials Bull.* 21–26 (January, 1940).

USEFUL LIMITING RULES

At times it is necessary to make an estimate of possible galvanic corrosion when it is difficult to measure galvanic effects under the particular corrosive conditions to be faced, or with electrodes even approximating in size or shape those in the practical application. The following relationships can then be of some assistance.

1. The difference effect can often be neglected. Then the total rate of corrosion of the anode will be approximately equal to its normal rate of corrosion plus the rate equivalent to the galvanic current.[12]

$$cA = cN + cG \qquad (7)$$

where cA is the total corrosion of the anode, cN the normal rate, and cG the galvanic corrosion.

2. The galvanic current is proportional to the cathode area

$$I = kA_c \qquad (8)$$

where k is a constant and A_c is the cathode area. This principle applies when both anode polarization and resistance drop can be neglected, that is, when the cell is under cathodic control.[13]

3. If the conditions described under 2 are obtained, the average galvanic current density on the anode will be given by

$$i_a = \frac{I}{A_a} = \frac{kA_c}{A_a} \qquad (9)$$

where i_a is the average current density and A_a is the total area of the anode member of the couple.

4. An absolute upper limit of the galvanic current can be set, if the open circuit (unpolarized) potential and the resistance of the electrical circuit of the couple are known, by the following equation:

$$I = \frac{E_c' - E_a'}{R} \qquad (10)$$

From the previous discussion $E_c' - E_a'$ is the maximum potential difference that the couple can attain, and which will normally be diminished by polarization.

There are two precautions to be observed in using this principle. First, with metals and alloys which can shift from passive to active condition and vice versa, there is uncertainty as to what value of E_c' or E_a' should be used. Second, even though the estimated maximum value of I may represent a safe anode current density, the possibility of cathodic corrosion must also be considered before it is concluded that the selected pair of metals can be safely employed.

COMPLEX CELLS*

Another type of differential metal corrosion may also occur in some cases. This results from complex cell action,[14] the nature of which can be best understood by

---

* This subject is discussed by R. B. Mears, Aluminum Company of America, New Kensington, Pa.

[12] W. A. Wesley, *Trans. Electrochem. Soc.*, **73**, 539–549 (1938).

[13] W. A. Wesley, *Proc. Am. Soc. Testing Materials*, **40**, 690–702 (1940).

[14] See also R. B. Mears and R. H. Brown, *Ind. Eng. Chem.*, **33**, 1010 (1941); R. B. Mears and J. R. Akers, *Proc. Am. Soc. Brewing Chemists*, p. 9 (1942).

the aid of a few simple diagrams. Comparisons between complex cells and simple cells are illustrated in Fig. 9. A practical case of corrosion resulting from complex cell action has been described by Zurbrügg.[15] Three aluminum fermenting tanks which were in metallic contact with each other were equipped with copper attemperator coils (Fig. 10). These coils were insulated from the tanks. This combination was

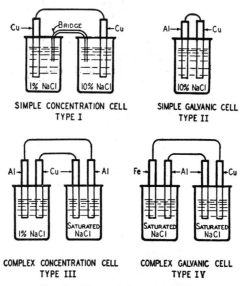

Fig. 9. Types of Corrosion Cells.

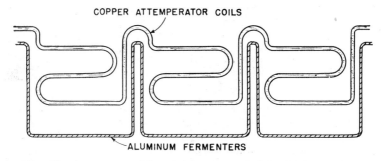

Fig. 10. Arrangement of Copper Attemperator Coils in Aluminum Tanks.

operated successfully for some time. However, eventually one of the copper coils was replaced by an aluminum coil. After this replacement, attack of the aluminum coil and of the two fermenters containing the copper coils occurred. This is an example of complex galvanic cell action.

As can be seen, corrosion resulting from complex cell action may be rather difficult to recognize. It is, therefore, fortunate that it is generally not so severe as corrosion resulting from direct galvanic action.

[15] E. Zurbrügg, *Schweizer Archiv angew. Wiss. u. Technik*, **6**, 40 (1940).

ALLEVIATION OF GALVANIC CORROSION

Designers of structures and equipment may be required because of both certain physical and certain chemical characteristics to employ coupled dissimilar metals. Unless proper precautions are taken, severe galvanic corrosion of one or more of the parts may occur. The galvanic current can be reduced by decreasing the open circuit potentials, increasing the polarization characteristics of one or both members of the couple, and increasing the resistance of either the metallic or electrolytic portions of the circuit or both. Because the relative areas of the anode and cathode members of the couple will also determine the intensity of attack, it is important that these be controlled within the limits that a given design will permit. Below are listed some of the precautions which may be employed to minimize galvanic corrosion.

Effect of Relative Areas

It is good practice to avoid combinations where the areas of a more anodic alloy are relatively small as compared with the area of the more cathodic alloy. Unless this is done, corrosion on the anode will be very intense. It has been found good engineering procedure to use alloys which are cathodic to the main structure for fastenings of relatively small area such as rivets.

Insulation

Wherever it is practical it is desirable to insulate the members of a couple electrically. In many cases it will not be found practical to employ insulation, but this method of minimizing galvanic corrosion should not be disregarded without adequate consideration.*

Painting

Paint coatings serve to increase the resistance of the metal-liquid interface and, therefore, will reduce galvanic current. However, the more anodic member of the couple should not, in general, be painted unless the more cathodic member of the couple is also painted; otherwise, the galvanic current will be concentrated at breakdowns in the coatings on the anode, with highly intensified corrosion as a result. Generally, the coatings on cathodic members should be resistant to alkalies unless the couples are exposed in acid media. The coatings on both the anodes and cathodes should be kept in good repair.

Avoidance of Depolarizing Conditions

In some cases it may be possible to prevent or limit the amount of aeration reaching the cathode surfaces. On most metals this will tend to increase the cathodic polarization and hence reduce galvanic current.

Inhibitors

The galvanic current can be reduced in many cases by the addition of inhibitors. Dangerous inhibitors, if not employed at sufficiently high concentrations, will cause localization of corrosion on the anode member of the couple, and intensity of the attack may be greater than if no inhibitor were employed. (See *Inhibitors and Passivators*, p. 907.) Therefore periodic checks must be made to assure that the inhibitor concentration is above the required minimum level. Inhibitors which

---

* Insulating washers have been used between magnesium and dissimilar metal bolts to increase the electrolytic resistance and thereby decrease galvanic corrosion. See p. 231. EDITOR.

increase polarization of the cathodic member of the couple, for instance those which increase hydrogen overvoltage, will generally reduce galvanic current without increasing the intensity of attack on the anode.

CATHODIC PROTECTION

If both electrodes when coupled can be made cathodic by a current from some outside source, galvanic corrosion in many cases can be eliminated. The source of the current may be from an inert electrode with an externally applied potential or from a metal more anodic than the two comprising the couple. In the latter case there will be sacrificial corrosion of the auxilliary anodic electrode; hence it will have to be replaced from time to time. (Refer to chapters on cathodic protection, pp. 923 and 935.)

POLARIZATION

The coupling of highly anodic metals to more cathodic metals which do not polarize should be avoided. One of the most commonly encountered cases of metals which do not polarize rapidly are those which have low hydrogen overvoltage.

HOT AND COLD WATER SYSTEMS

F. N. SPELLER*

This chapter refers mainly to the action of water as commonly used in metal storage tanks and conduits, and to protective measures. Therefore, the controlling factors in these problems are those discussed in previous chapters on total or alternate immersion in distinction to atmospheric corrosion.

With respect to the water, the rate of attack is accelerated as the dissolved oxygen, carbonic acid, temperature, and acidity are increased, and is retarded by the reverse of these conditions. The rate of motion generally accelerates corrosion up to a certain point mainly by bringing more dissolved oxygen to the metal surface. Above a certain velocity, in the case of iron, however, there is a retarding effect due to the more rapid formation of protective layers of corrosion products on the metal. Metal surface films and deposits formed from substances in solution in the water, or by corrosion products, often determine the rate of underwater corrosion.

The distribution of damage due to corrosion usually determines the useful life of metal containers. For example, water pipe often fails by perforation through pitting before more than 5% of its weight has been lost. Pitting is caused by contact of dissimilar materials, or solutions of dissimilar composition or concentration in contact with a metal in an electrolyte.

It is well known that metals differ materially in their resistance to the action of water. In ferrous metals, such as malleable iron castings, wrought iron, soft steel, ingot iron and low-alloy steels, and copper-bearing steel or iron used for water pipes and boiler tubes, no material difference is found in the rate of distribution of corrosion. Higher percentages of alloying elements such as chromium and nickel may increase the resistance to corrosion. Stainless steel such as 18-8 is a notable example.†
The resistance of various metals to the action of water is taken up under the discussion of these metals in previous chapters and need not be referred to here except

* Consultant, Pittsburgh, Pa.
† However, this and similar alloys are susceptible to pitting when halogen ions (e.g., Cl^-) are present in the water. See *Pitting in Stainless Steels and Other Passive Metals*, p. 165.

where specific metals have proved most economical for certain specific types of service. It can be definitely said that *the environment*, i.e., the water in contact with a particular metal such as iron, for example, is *usually much more influential in controlling the rate of attack than minor changes in the composition of the metal.*

Corrosion can be brought under control in nearly every case so that the system will give an uninterrupted period of service by (1) selecting a suitable metal or alloy or (2) by changing the chemical or physical properties of the environment (for example, by removing dissolved oxygen, or otherwise treating the water so that it is less active toward the metal; or by the application of cathodic protection [discussed beginning p. 923]), or by (3) applying a suitable protective coating on the metal. The problem, like most engineering problems, is to find the remedy that is the most economical in the long run.

TUBERCULATION

Tuberculation is closely associated with pitting of the interior of iron and steel pipes. Tubercles consist of mounds of corrosion products covering anodic areas where pits are developing. Tubercles at first increase the rate of penetration underneath, but later on, when they become more impermeable, the rate of attack may be greatly reduced. Quite often the main objection to tuberculation is the increase in friction and reduction in flow which for smaller water pipes sometimes result in complete clogging of the piping system.

J. R. Baylis[1] has made a detailed study of tubercles from Baltimore water mains. He found that the pH of the water in an active tubercle was about 6 and slightly higher in those that were dormant (6.4 to 6.8). The pH of the treated city water was 7.9. He also observed that negative ions such as chlorides and sulfates were materially concentrated in tubercles and that the black ferrous-ferric hydroxide (hydrous Fe_3O_4) in these tubercles, being magnetically attracted, tended to become attached to the iron in the form of porous columnar fibers.

The process of formation of tubercles is not clearly understood. A thin shell or membrane of ferric oxides forms over the corrosion products after they attain certain thickness, so that diffusion to and from the external water is impeded. However, free oxygen is greatly reduced inside the tubercles, the concentration of negative ions is increased, and the pH lowered. These conditions tend to cause corrosion with the evolution of hydrogen. A differential oxygen cell is set up so that corrosion can proceed in a tubercle that seems impervious. Examination of tubercles indicates active and dormant periods. Frequently Baylis found that the surface membrane had been broken and the ferrous salts had diffused out only to be oxidized and covered with another shell.

Bacterial activity in water mains sometimes seriously increases the formation of corrosion products. (Refer to *Corrosion by Micro-organisms in Aqueous and Soil Environments*, p. 466.) The anaerobic sulfate-reducing type that forms hydrogen sulfide may flourish in tubercles where the oxygen concentration is low and may help to maintain a high oxygen gradient which increases localized attack in the pit.[2]

Iron bacteria, such as crenothrix and leptothrix, do not attack iron but require ferrous iron for their sustenance which they exude as ferric iron compounds through their outer skin. Tubercles may be formed in this way. Such accumulations have

[1] *Ind. Eng. Chem.*, **18** (No. 4), 370 (1926). Baylis' observations were confirmed in experiments, "Synthetic Corrosion Pits and the Analysis of Their Contents," by E. D. Parsons, H. H. Cudd, and H. L. Lochte, *J. Phys. Chem.*, **45**, 1339–1345 (1941).

[2] H. J. Bunker, *J. Soc. Chem. Ind.*, **59**, 412 (1940).

caused much trouble from "red water" and obstructions in water systems. Bacterial action can be eradicated by water treatment such as chlorination.

Water treatment with an alkali or sodium silicate (discussed on p. 502) tends to retard both corrosion and tuberculation.

TREATMENT OF WATERS TO REDUCE CORROSION

Classification of Natural Waters

At present no general system of classification of natural waters has been devised. A. S. T. M. Committee D-19 has given preliminary consideration to the problem, and in the symposium on this subject in June, 1944, W. D. Collins gave the composition of several typical waters used for industrial purposes. They may be divided roughly into soft, medium, hard, and salty. Of natural waters, soft waters are inherently the most corrosive.

Hot Water Heating Systems

Here the water is usually obtained from domestic supplies and is either soft or of medium-hard quality. These systems are closed except for a small air vent so that the free oxygen is soon exhausted, and further corrosion is therefore negligible. The water in such systems should not be removed when the system is not in use, and the system should be maintained free from leaks so that make-up is reduced to a minimum. Under these conditions, hot water heating systems constructed of ungalvanized iron or steel pipe have shown no serious deterioration from corrosion after 70 years of service. A small amount of corrosion with the evolution of hydrogen does occur, however, in water which has been deactivated in this way. This gas may be drawn off through pet cocks provided near the tops of radiators. By adding sufficient caustic soda to bring the pH of the water up to about 8.5, the generation of hydrogen can be prevented and the gas nuisance eliminated.

Fire Protection

Sprinkler systems of the *wet type* are kept filled with water under pressure at room temperature. Under these conditions, as in the hot water heating systems, corrosion is not a serious matter. Such systems may be constructed of black steel pipe and malleable iron fittings. In the absence of oxygen the galvanic effect between nonferrous sprinkler heads and steel pipe is reduced to a minimum and has not been found to be so serious as to require water treatment.

In *dry* sprinkler systems, corrosion is also negligible as the pipes are kept filled with air under pressure.

Brine Refrigerating Systems

Calcium and sodium chloride brines are commonly used in refrigerating systems at a concentration of about 20%. When oxygen is present, these brines are decidedly corrosive, especially where dissimilar metals are in contact. Damage may be considerably reduced on black and galvanized surfaces in refrigerating brines by treating with caustic soda to a pH of 8.5

Better protection may be obtained by treating the brine with sodium bichromate ($Na_2Cr_2O_7 \cdot 2H_2O$) and adding sufficient caustic soda to adjust the alkalinity to pH 8 to 8.5 or by using sodium chromate ($Na_2CrO_4 \cdot 10H_2O$). When zinc and other metals are in contact with iron, the general recommendations of the A. S. R. E. Committee prescribe 100 lb (1600 ppm) sodium bichromate per 1000 cu ft for calcium brines and

200 lb (3200 ppm) for sodium brines,[3,4] preferably with the necessary amount of caustic soda to form the normal chromate. About 27 lb per 100 lb of bichromate are required. This treatment has been in use for a number of years in ice plants and has proved satisfactory in maintaining the galvanized ice cans and steel work in the brine system in good condition. The dezincification of brass in brines contaminated with ammonia is reported to be prevented by this treatment. Chromates in proper amount also reduce galvanic contact effects materially. Sodium chromate with a small amount of sodium metasilicate has been found effective in aluminum and aluminum-iron systems.[5]

COOLING WATER

Cooling water that is recirculated in contact with the atmosphere, as in spray ponds, can generally be inhibited to prevent serious corrosion of steel by the addition of about 300 ppm of sodium chromate. Aeration does not impair the efficiency of chromates. The optimum amount required depends on the composition of the water and may be determined approximately by laboratory tests. The manufacturers who specialize in the production of chromate for such service are usually able to give useful information as to the amount required.[6]

The tendency to pit when insufficient chromate is used may be considerably reduced by occasional removal of rust tubercles, and by cleaning or rubbing the metal where this is possible. In old systems not previously inhibited it is desirable to start with about 500 ppm. The concentration may be roughly estimated by comparing the yellow color of the treated water with that of solutions of known strength in test tubes. The pH should be kept about 8.5, which gives a faint pink color with phenolphthalein. In starting treatment, tests should be made once a week.

Water-soluble organic inhibitors (such as the glucosates or some tannin compounds) may also be used as inhibitors in cooling systems although they are not so effective as sodium chromate. They are usually somewhat more expensive than the chromates, but have no toxic effect in the concentrations employed. Advice on concentrations required for specific conditions may be obtained from manufacturers.[6]

Automobile Radiators. Corrosion may be inhibited in the bimetallic systems of radiators (using fresh water) by the addition of about 2 oz of sodium chromate to the radiator. Calcium chloride anti-free mixtures, even when treated with inhibitors, are not advisable, for, should the concentrations of inhibitor be reduced below a certain minimum, corrosion proceeds rapidly. Organic anti-freeze mixtures are very much safer because they may be used with less danger of corrosion. Organic inhibitors are now available for use with organic anti-freeze compounds. Where anti-freeze mixtures are used more than one season, more inhibitor is usually required to prevent corrosion, and the pH should be on the alkaline side.

Diesel and Gas Engines. The recirculating cooling systems of gas engines may also be protected by chromates. When the volume of water is small it is better to use a higher concentration (1500 to 2000 ppm). When scale from hard water is encountered, a combination of the polyphosphates with suitable organic corrosion inhibitors or chromate may be used.

Power Rectifiers and Power Transformers. Close temperature control is essential in these systems. Efficient inhibitors properly maintained will help materially in pre-

[3] *Refrig. Eng.*, **14**, 173–182 (1927).
[4] F. N. Speller, *Corrosion, Causes and Prevention*, pp. 615–619, McGraw-Hill Book Co., New York, 1935.
[5] Marc Darrin, *Ind. Eng. Chem.* **37**, 747 (1945).
[6] "A. C. S. Symposium on Inhibitors for Aqueous Corrosion, September, 1944," *Ind. Eng., Chem.*, **37**, 702–752 (1945).

SPECIAL TOPICS IN CORROSION

venting impairment of heat transfer by rust scale. Hexametaphosphate and other polyphosphates are very effective in preventing the formation of lime scale, and when used in sufficient concentration they act as corrosion inhibitors. When the volume of water is not great and perfect protection against both scale and corrosion is essential, it seems advisable to use 5 to 10 ppm of the phosphate with about 500 ppm of chromate, or sufficient to more than prevent bimetallic corrosion.[7]

Boilers. Chromates have not been found generally applicable to protection of operating steam boilers but are applicable when boilers are out of service. At room temperatures, where it is necessary only to protect iron, and the loss of water is small and the volume large, sufficient protection may often be obtained with 100 ppm chromate at pH 8.5.

Limitations of Chromate Treatment. Where the chromate concentration against the metal drops below a certain minimum (which depends on many conditions), the corrosion may be localized in the form of pits. It is therefore advisable to maintain a much higher concentration than the minimum necessary to prevent localized corrosion. The depletion of chromate is less at the higher concentrations, assuming that there is no leakage or other losses.

The chromates are sometimes reduced to the chromic salts, the latter having no value as inhibitors; therefore the chromates cannot be used economically where there is more than a very small amount of certain organic reducing matter in the water. The reducing effect of organic substances or contaminated waters can be easily determined in the laboratory.*

TOXICITY OF CHROMATES. Wherever chromates are used in solution they may produce an irritation of the skin, if exposure is prolonged, though washing after handling chromate solution is sufficient to prevent such trouble in most cases. Where cooling towers are used they should be well baffled to reduce windage losses. High concentrations of chromate are usually inadvisable when the temperature of cooling water is reduced by spray ponds or cooling towers.

NON-RECIRCULATING COOLING SYSTEMS

These may be treated with lime or caustic soda to raise the pH to about 11 unless aluminum or other metals are present that are attacked by the high alkalinity water. Polyphosphates, to be effective, require circulation or agitation. Organic inhibitors, sodium silicate, or chromates may also be used in systems of this type. With due regard to limitations stated above, chromates probably will give the best results in this type of system. Mechanical deaeration has been applied recently to such systems. (See p. 506.) When the volume of water is very large the cost of treatment is often the limiting factor.

In moving or surging water systems such as hydraulic elevators and gas holders 100 to 200 ppm chromate usually give satisfactory results.[8]

CATHODIC PROTECTION†

This method, which has proved satisfactory and economical for the exterior protection of gas and oil pipe lines underground, has also been applied successfully to

---

* It appears that chromates are reduced by ethyl alcohol but not as readily by methyl alcohol anti freeze solutions.

† Discussed further in chapters beginning pp. 923 and 935.

[7] "A. C. S. Symposium on Inhibitors for Aqueous Corrosion, September, 1944," *Ind. Eng. Chem.*, **37**, 702-752 (1945).

[8] R. C. Ulmer and J. M. Decker, *Combustion*, **10**, 31-32 (1928).

the protection of the inside of water-storage tanks and other structures that are in contact continuously with water. A protective paint or bituminous coating that insulates a large portion of the surface will very materially reduce the total amount of protective current that is required to arrest corrosion on bare anodic areas.

For each structure it is necessary to determine or estimate the minimum cathode current density required and design the anode or anodes so that the necessary protection can be obtained most economically. In waters having relatively high electrical conductivity, such as sea water, this requirement is comparatively easy compared with fresh water. In fresh water the composition of the water is a major factor. It is therefore desirable to obtain experienced engineering advice in the analysis of these problems in order to obtain an accurate estimate of the minimum current density required. The current is then controlled by the potential between the anode and the structure to be protected.

Rectifiers and sacrificial anodes have generally proved to be practical for supplying the necessary direct current for protection of surfaces in contact with neutral waters.[9]

Oil Product Lines*

Gasoline and other oil product lines underground sometimes suffer serious loss of capacity by corrosion tubercles that form as in water mains. Water containing dissolved oxygen is condensed on the pipe walls when the temperature of the oil or gasoline drops. There is usually enough oxygen dissolved in the fluid to keep the corrosion process going on. Practical remedies have been found in the addition of suitable water-soluble inhibitors, which include sodium nitrite or chromate, and, in a few cases, in dehydration of the oil or gasoline before it enters the transportation pipe lines. An organic inhibitor soluble in gasoline and slightly soluble in water has been developed.[10]

Domestic Water Systems

These waters are selected and often treated primarily for domestic use, but are often used for boiler feed and cooling, with or without local treatment.

Treatment of Domestic Water. Water treatment for very large systems by reason of cost involved is usually limited to pH control and intentional deposition of $CaCO_3$ scale. For small systems, more expensive treatment is warranted as a rule. It is occasionally more economical to treat water used in individual buildings where hot water piping, for example, is subject to corrosion from soft water than to treat the entire city supply.

The treatment of potable water is also limited to that which will not render it objectionable or injurious for drinking or cooking even though much of the supply may be used for cooling or fire extinction. As corrosion prevention is ultimately an economic problem, it is always important to consider cost in selecting a system of treatment for any specific purpose.

Sometimes only the hot water supply of individual buildings requires treatment, because metal parts of hot water systems generally last less than one third as long as those of cold water systems, unless the dissolved oxygen has been reduced or other protection provided. The maximum temperature in hot water supply systems should be limited to 60° C (140° F) unless the water is well treated or deaerated.

---

* Further discussion appears in *Inhibitors and Passivators*, p. 914.
[9] L. P. Sudrabin, *Proc. Eng. Soc. W. Pa.*, Fifth Water Conference (October, 1944).
[10] "Symposium on Internal Corrosion of Oil Product Lines," *Proc. A.P.I. Production Div.*, 1943.

The Carbonate Balance System (the Saturation Index). The carbonate balance system of treatment aims at having an excess of $CaCO_3$ present so that a continuous protective layer of $CaCO_3$ is deposited on the inside of the pipe or container. This is accomplished by adjusting the pH with respect to the carbonate, bicarbonate, and free CO_2.[11]

Where the scale deposited is mainly $CaCO_3$, the tendency to form this scale (and the resulting retardation of overall corrosion that may be expected at normal temperatures) is indicated roughly by the *Saturation Index*.[12,13,14,15,16] This index is a mathematical evaluation of the ability of a water to precipitate calcium carbonate. Compounds like calcium carbonate are more likely to deposit on cathodic areas, by reason of the greater alkalinity in these regions; and by decreasing the rate of diffusion of oxygen to the metal, they tend to slow down corrosion.

The Saturation Index, pH (actual) $- p$H$_s$ (also called Langelier Index), is based on total solids, total alkalinity, pH, and calcium content of the water. It is calculated using the expression as derived by Langelier:[12]

$$p\text{H}_s = (p\text{K}_2' - p\text{K}_s') + p\text{Ca}^{++} + p\text{Alk}.$$

The last two terms are negative logarithms of the molal and equivalent concentrations of calcium and titratable base (methyl orange indicator), respectively, $p\text{K}_2'$ is the negative logarithm of the second dissociation constant for carbonic acid and $p\text{K}_s'$ is the negative logarithm of the activity product of $CaCO_3$. The difference $(p\text{K}_2' - p\text{K}_s')$ varies with ionic strength (salinity) and temperature. Its value for soft water at 20° C is 2.1.

The Saturation Index is equal to the difference between the actual measured pH of a water and the value of pH$_s$ calculated by the above expression. A positive index indicates that calcium carbonate is being deposited; a negative index indicates a tendency of calcium carbonate to dissolve. Where the index is over $+0.5$ at normal temperature (oxygen and other factors being the same), corrosion is decreased, whereas with an index of -0.5, or less, corrosion may increase. As the index is essentially an indication of the tendency for a calcium carbonate scale to be deposited, it must be used with considerable judgment and with due regard to other important factors. It is not claimed that the index predicts the rate or distribution of corrosion (pitting) at elevated temperatures, particularly since calcium carbonate and other scales deposited from water often break down locally at these temperatures.

Silicate of Soda. Silicate of soda has been used since 1920 to protect iron, lead and brass water pipe. For most waters a solution of $Na_2O : 3SiO_2$ is recommended, which may be fed by any convenient type of proportional feed apparatus.

In all cases where inhibitors are added, the metal surface should preferably be free from thick corrosion products or scale. Most of the corrosion is found in the hot water supply pipes, but it is convenient and often desirable in large buildings to feed the silicate near the cold water inlet. Sodium silicate, equivalent to about 12 or 16 ppm silica, should be fed to the water for the first month, after which it may be reduced to give 8 ppm added silica or even less in some cases. It is sometimes desirable to double this minimum dose every few months. The water is not injured

---

[11] F. N. Speller, *Corrosion, Causes and Prevention*, p. 362, McGraw-Hill Book Co., New York, 1935.
[12] W. F. Langelier, *J. Am. Water Works Assoc.*, **28**, 1500 (1936).
[13] V. V. Kendall, *Proc. Am. Soc. Testing Materials*, **40**, 1318 (1940).
[14] C. P. Hoover, *J. Am. Water Works Assoc.*, **30**, 1802 (1938). (This is a convenient graphical method for determining the index.)
[15] F. E. deMartini, *J. Am. Water Works Assoc.*, **30**, 85 (1938).
[16] S. T. Powell, H. E. Bacon, and J. R. Lill, *Ind. Eng. Chem.*, **37**, 842 (1945).

for domestic use by this treatment. The rate of corrosion of iron pipe has been reduced by 70% and dezincification of brass pipe practically stopped by this simple treatment. The same treatment may be used to retard solution of copper by regulating the silicate of soda so as to give the water a pH of about 8. However, the amount required and the effect are not the same in all waters, and important applications may require the services of a water-treating chemist to secure the best results.

Hexametaphosphate. Hexametaphosphate and other polyphosphates of soda have proved highly useful in preventing the formation of scale deposits due to excess of calcium and magnesium salts in water; 2 to 5 ppm are usually sufficient for this purpose. These compounds also prevent discoloration from iron when added in the proportion of 2 ppm for each 1 ppm of iron in solution. For this purpose the phosphate must be added before the water is chlorinated or exposed to air.

When the water at normal temperature is in turbulent flow (about 1 ft per sec or higher in pipe under 1 in. diameter) 8 to 10 ppm hexametaphosphate have been shown to retard corrosion materially, apparently by formation of an insoluble film on the metal.[17]

A 30% solution of hexametaphosphate can easily be mixed with a similarly concentrated solution of sodium silicate and fed through the same apparatus. This mixture appears to give better protection to water systems, hot or cold, than by using either of these compounds separately. The protective deposit from two different inhibitors used together is sometimes more effective than either one by itself.

Organic Substances. Organic substances such as glucosates, dextrine, or the tannins compounded with other inhibitors have been found promising, especially in water at high temperatures, 80° to 90° C (180° to 200° F), where the cost is not prohibitive.

The materials now available as corrosion inhibitors for potable water are limited. More exploratory work by chemists is needed in this field as the damage that may be caused by a relatively small amount of raw hot water in the intricate piping system of a large building generally warrants treatment at a cost that would be prohibitive for a large municipal supply.[18]

THE SELECTION OF METALS OR COATINGS

PIPE MATERIALS

For fresh waters the choice lies mainly between galvanized ferrous pipe, copper, and two or three grades of brass. Each is available in three thicknesses and standard inside nominal diameters with threads and couplings of about the same metal composition. Copper is available in thinner wall pipe provided with solderable streamline copper fittings. For all but very temporary installations, the thickest grade of this type of copper pipe should be used.

Soft waters (nearly neutral) are much more reactive with iron than those of higher primary alkalinity or where supersaturated with calcium carbonate. Galvanized pipe, at least in cold water lines, has given good results in many localities, especially where the waters are of moderate carbonate hardness.

In water, salt or fresh, there is no material difference in rate of pitting of wrought iron, steel, low-metalloid steels, or copper-bearing steels; this is contrary to the relative performance of these metals in the atmosphere. Low-alloy steels containing

---

[17] G. Hatch and O. Rice, *Ind. Eng. Chem.*, **37**, 752–759 (1945).
[18] "A. C. S. Symposium on Inhibitors for Aqueous Corrosion, September, 1944," *Ind. Eng. Chem.*, **37**, 702–842, (1945).

nickel and chromium are not enough better in this respect to warrant their extra cost for water supply services.

For monumental buildings, copper or 85% Cu–15% Zn brass of the standard pipe gage is advisable unless the water is deactivated by removal of the dissolved oxygen or by other adequate treatment. Brasses with 60 to 67% copper are dezincified in some corrosive waters (see Fig. 10, p. 1107, and Fig. 11, p. 1108), and in certain localities are not much more serviceable than galvanized iron or steel pipe. The zinc may be leached out locally, leaving a plug of porous copper, or the dezincification may be spread out, resulting in a general weakening of the pipe, which is especially noticeable under threads.[19]

Corrosion of copper and brass piping varies considerably with the water composition. The solubility of copper increases mainly with the free CO_2 content and to a lesser extent by half-bound CO_2, or waters high in bicarbonates. Copper in solution may produce a greenish tinge on vitreous enamel fixtures. This condition may be remedied by raising the pH to about 8.5 with lime or sodium silicate. As a rule, galvanized pipe gives its best service in waters containing a moderate amount of bicarbonate of lime (as in Lake Erie), whereas copper shows up better in softer waters when the free CO_2 is not in excess of 2 ppm.

A low concentration (0.5 ppm) of copper in water seems to accelerate the corrosion of galvanized iron or zinc by deposition of finely divided copper on the zinc.[20] Copper heating coils attached to galvanized hot water storage tanks cause the latter to corrode more rapidly because the high temperature, free oxygen of the water, and rapid circulation actively dissolve copper. Cement or vitreous enamel lining in tanks or suitable water treatment will remedy this condition. The amount of copper taken up by the water does more harm to the galvanized equipment than to the pipe that supplies the copper. In water used for cooking, copper tends to destroy vitamin C (ascorbic acid).[21]

Cupro-nickel (70-30) pipe and tubing give the best service in sea water systems, although red brass and extra heavily galvanized pipe are used with certain precautions for this type of service. Galvanic contact effects must be guarded in sea water systems. They are of less importance in dealing with fresh water.

Aluminum pipe may be used for distilled water lines but it is attacked seriously by some domestic and industrial waters.

Obviously there is no one piping material that is suitable for all water systems, so that water treatment and precautionary measures are often necessary to obtain the best service with any of the pipe materials commonly used.

Protection of Pipe by Coatings

Both cast iron and steel water pipe corrode more or less, and lose carrying capacity rapidly by tuberculation unless protected from corrosive waters. Interior protection is now usually obtained by the hot application of bituminous enamel about 1/16 in. thick by the centrifugal process usually under A. W. W. A. Standard Specifications 7A.5 and 7A.6–1940.

Air-drying paints have been materially improved for service in water, but except perhaps as a priming coat for bituminous enamel they lack the desired durability for protection of water lines. Certain baking-type phenolic-resin formulations are,

---

[19] E. P. Polushkin and H. L. Shuldener, *Trans. Am. Inst. Mining Met. Engrs.*, **161**, 214 (1945).
[20] L. Kenworthy, *J. Inst. Metals*, **69**, 67 (1943).
[21] L. Tronstad and R. Veimo, *Water and Water Eng.*, **42**, 225 (1940).

however, much more durable than air-drying paints and promise to prolong considerably the life of steel pipe and tanks in hot and cold water systems and similar service.

PAINTS FOR UNDERWATER SERVICE

Paints for underwater service at normal temperatures have been greatly improved during recent years by the use of phenolic, chlorinated rubber, and other synthetic resin bases. Substitution of tung oil for all or 60% of the linseed oil previously used and the addition of inhibitive pigments have also played an important part in prolonging the useful life of paints in water.[22] (See chapters on paints, beginning p. 878, and chapter on paints for sea water exposure, p. 430.) Several of these paints are now being made by leading manufacturers.

Portland cement coatings have given long service as interior coatings for large water mains and can be readily applied to old tuberculated mains after cleaning.[23] The exteriors of water mains are usually protected from soil corrosion by reinforced bituminous coatings (see p. 898), preferably with cathodic protection. (See pp. 923 and 935.)

Preparation of Metal Surface for Painting. Sand or steel grit blasting or acid pickling is now recognized as the best method of preparing the metal for painting. All mill scale and rust should be completely removed and the paint applied when the surface is dry and at a temperature above 10° C (50° F). Mill scale is cathodic to iron and therefore promotes pitting.

Aqueous inhibitive solutions containing phosphoric acid applied to the clean metal before painting have been found to reduce but not entirely prevent blistering of paints in water, but their effect is markedly different with different paints. These inhibitors are usually most effective with the better-class paints. Under atmospheric exposure the beneficial influence of these inhibitive washes is much less noticeable or entirely lacking. (Report of Subcommittee XXIX of Committee D–1, A.S.T.M., 1944.)

STORAGE TANKS

Copper, 70% Ni–30% Cu alloy (Monel), and other rust-resisting non-ferrous alloys are satisfactory for hot water storage in residences. Galvanized tanks are generally used, but frequent replacement is often necessary in soft water districts. A.S.T.M. Specification A120–43 requires a minimum of 2 oz of zinc per sq ft of covered surface for pipe. This requirement should be met on the interior of galvanized pipe and tanks as the protection obtained is about proportional to the thickness of zinc.

The interior of cold water storage tanks of steel may be protected with three or four coats of selected air-drying synthetic resin-base paint preferably supplemented by cathodic protection. Steel tanks lined with certain new baked synthetic resin base coatings are promising for hot water equipment such as for low-pressure heating boiler tubes, heat-exchanger tubes, and piping.

Steel tanks lined with neat Portland cement have given satisfactory service with hot water for over 12 years. A coating of this type can be applied so as to cover all exposed metal. Cement-lined standard steel pipe is listed in sizes ½ to 12 in. These products have been subjected to severe heating and cooling tests in water with satisfactory results. Small pipe lined with hydraulic cement by hand has given over 50 years of satisfactory service with New England soft water. Until better materials

---

[22] F. N. Speller, Third Water Conference, *Proc. Eng. Soc. W. Pa.* (1942)
[23] F. N. Speller, *Corrosion, Causes and Prevention*, pp. 349–355, McGraw-Hill Book Co., New York, 1935.

are available, this well-known means of protecting the inside of steel tanks and pipe may be used with confidence when properly applied. The cement should preferably carry a small amount of iron and manganese and the coating in tanks must be at least 5/16 in. thick. Tanks provided with manholes may be lined in situ by hand with neat cement mortar about ½ in. thick. The cement after initial set should be cured 24 hours with steam or hot water. Before coating, old loose rust should be removed, but the metal does not have to be perfectly clean and dry. Special attention should be given to protection of exposed metal in joints of cement-lined pipe as prescribed by the manufacturer.

DEACTIVATION AND DEAERATION OF WATER

F. N. SPELLER*

Dissolved oxygen is usually the controlling factor in corrosion of iron in water. In the return condensate lines of heating systems, free CO_2 is also of prime importance. (See *Corrosion of Steam-Condensate Lines*, p. 538.) These facts have been established by laboratory experiments and by experience in service.

The degree of removal of free oxygen required to prevent serious corrosion depends mainly upon the service temperature and, to a lesser extent, upon the amount of water passing through the system. In cold water systems it is desirable to reduce the oxygen to not over 0.30 ppm (0.2 ml/liter). The maximum dissolved oxygen that can be tolerated in cold water systems depends on the type of service and flow capacity of the system. When lower oxygen is desired than can be attained with single-stage deaerators, it may be obtained by deactivating the effluent from deaerators with sodium sulfite or other chemical treatment, or by using a multiple-stage deaerator. At 70° C (160° F), as in hot water supply systems, it is usually not necessary to reduce free oxygen below 0.10 ppm (0.07 ml/liter); for low-pressure steam boilers operating under 250 psi without economizers the desirable maximum should not exceed about 0.03 ppm (0.02 ml/liter); for high-pressure boilers or when economizers are used practically zero oxygen is required.† Standard methods and discussion on precision of determination of dissolved O_2 and CO_2 will be found in A.S.T.M. *Proceedings*, Reports of Committee D-19 on Water for Industrial Uses. (See *Boiler Corrosion*, footnote, p. 525.)

DEACTIVATION

Removal of corrosive gases by chemical means has been accomplished by allowing the hot water at about 70° C (160° F) to remain in contact with a large area of perforated steel sheet or scrap for half an hour or more until the oxygen has been sufficiently used up in corrosion of the metal. Apparatus has been designed for hot water supply systems and used for that purpose with sand filters, which demonstrated that the principle is sound, but the apparatus was often too bulky and required regular attention. This method has, therefore, been largely superceded by mechanical deaeration.

Sodium sulfite is used for removal of residual dissolved oxygen and is reasonable in cost when 95% of the free oxygen is removed previously by mechanical deaeration. About 8 lb of sodium sulfite are required to remove 1 lb of oxygen. About 30 ppm of

---

* Consultant, Pittsburgh, Pa.
† Zero oxygen, commercially speaking, usually means oxygen below 0.005 ppm (0.0035 ml/liter).

excess sulfite should be maintained in boilers to insure complete removal of oxygen.* Ferrous sulfate, neutralized with caustic soda, has been used to a lesser extent for this same purpose.

DEAERATION

Dissolved gases can be removed from water if the temperature and pressure conditions are regulated so that the gases become insoluble and conditions are favorable for their complete separation from the water. During the years just past, the

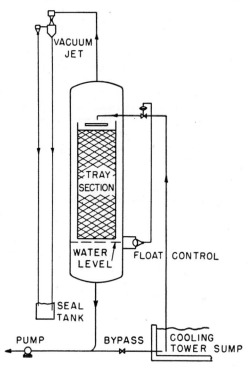

FIG. 1. General Layout of Cold Water Deaerator. (The Texas Company.)

design of mechanical apparatus for this purpose has been greatly improved, and there are now available several types of deaerators, each of which is adapted for a specific service. This equipment is also designed to remove free CO_2, H_2S, and NH_3 from water.

Cold Water Deaeration

Apparatus is now available for deaeration of water without heating. One large installation handling 4,000,000 gal per day reduces the oxygen to 0.33 ppm, which was found sufficient to prevent objectionable corrosion and tuberculation in a long steel pipe line.[1] The water is sprayed over trays in a chamber under high vacuum

---

* The rate of reaction of sodium sulfite with dissolved oxygen has been studied by R. M. Hitchens and R. W. Towne, *Proc. Am. Soc. Testing Materials*, **36**, Pt. II, 687 (1936).

[1] S. T. Powell and H. S. Burns, *Chem. Met. Eng.*, **43**, 180–184 (1936).

(about 28.5 in. of Hg). The non-condensable gases may be removed by steam ejectors with condensers or by vacuum pumps. A diagram of a plant which embodies this principle is shown in Fig. 1.[2] Some very large installations have been constructed recently.*

HOT WATER DEAERATION

The main condition in any hot deaerating device is to maintain the finely divided water for a sufficient time at the boiling point corresponding to the pressure at which dissolved gases separate freely. The simple type of open feed-water heater which precedes the deaerating heater reduces the dissolved oxygen to about 0.5 ppm (0.3 ml/liter) when operated between 88° and 93° C (190° and 200° F) and freely vented to the atmosphere. This has given considerable relief from corrosion in low-pressure steam boilers. However, in economizers or high-pressure boilers, corrosion increases so rapidly with the temperature that complete removal of dissolved oxygen is necessary. This can be practically accomplished in deaerators of modern design.

Deaerators for Hot Water Supply Systems. This type is designed for use mainly in large buildings like hospitals and hotels. The water is heated to the boiling point under a vacuum so that the water temperature will not exceed 60° to 80° C (140° to 175° F). The steam passes through heating coils so that there is no danger of oil contaminating the water. The water is sprayed downward over trays and is heated by two sets of steam coils. The steam admitted to the lower coils is above the temperature of the water and generates free steam which sweeps the non-condensable gases out through a vent condenser cooled by the incoming cold water. The condensate from the vent condenser is drained back into the tray chamber while the non-condensable gases are drawn off by a vacuum pump or steam ejector. This type of deaerator may be located in the basement of the building, in which case a hot water circulating pump is required, or it may be placed at a sufficiently high level to permit gravity flow. For such service a minimum of 0.06 ppm (0.04 ml/liter) of residual oxygen will give good protection at temperatures below about 70° C (160° F).

Deaerators for Boiler Feed Water. These are designed for direct contact between the water and steam. The deaerators most generally in use at present are of the *tray* type operated under pressure or vacuum. The *spray* type operated under light pressure is coming into more general use in steam boiler plants. In the tray-type deaerating heater, the cold feed water passes first through a vent condenser, then into a steam-heated chamber, where it is cascaded over metal trays and thence to the storage space. Steam fills the entire space, its course being so directed as to heat the water and eliminate non-condensable gases. Practically zero oxygen can be obtained in this way (Fig. 2).

A later type of deaerator has been designed for complete removal of gases by mechanically atomizing the water into an atmosphere of steam at about 1½ psi pressure. This type of deaeration has been developed satisfactorily for marine boiler service, and its wider use for stationary boiler service is generally predicted (Fig. 3).†
It consists essentially of a vent condenser, a steam-heated section, a deaerating section surrounding the steam inlet, and a storage section below. The cold feed water passes through the vent condenser, then through the atomizing nozzles into the steam-heated

---

* A multiple-jet deaerating apparatus intended to effect greater removal of non-condensable gases has been patented by S. T. Powell (U. S. P. 2080151, May 11, 1937).

† Deaerators designed on the principles illustrated in Figs. 2 and 3 are also available from other manufacturers.

[2] W. H. Attwill, *Proc. Am. Petroleum Inst.*, 11th Mid-year meeting, Sect. III, **22**, 90–93 (May, 1941).

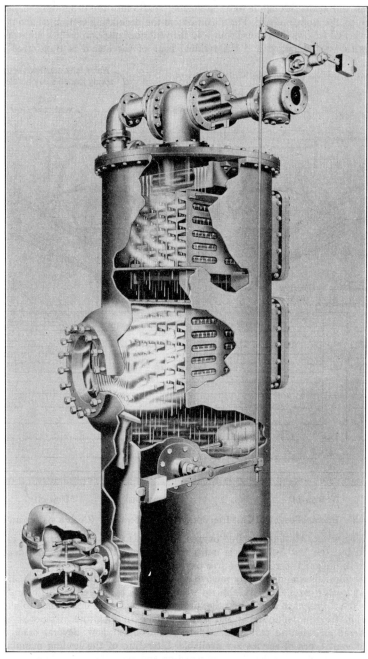

Fig. 2. Stationary Tray-Type Deaerator. (Elliott Company.)

section, and again through a water atomizing nozzle into the deaerating section, and thence to the storage space. The steam enters the deaerating section at about 1½ lb pressure and follows the reverse course to the vent condenser, where the non-condensable gases are discharged and the residual heat of the steam is transferred to the

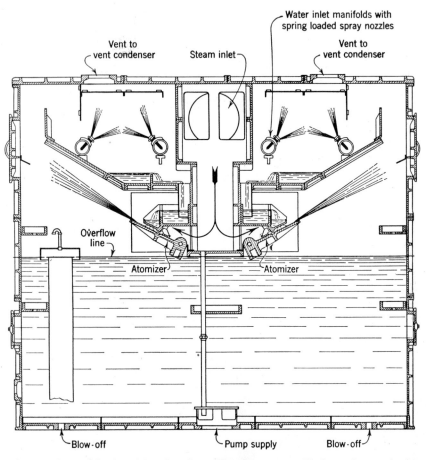

FIG. 3. Internal Section of Cast Iron Spray-Type Deaerator. (Cochrane Corporation.)

incoming water. Most of the free oxygen is removed from the water when it is first heated but the last 5% or so is much more difficult to eliminate. The deaerating section is designed to reduce the free oxygen to practically zero.

The more efficient deaerators remove all the free CO_2 and part of the half-bound CO_2 and other gases. The pH is raised when the CO_2 is lowered in this operation.

The development of these and other deaerating devices has done much to eliminate corrosion from water systems and steam boilers. In fact, this apparatus may be considered an essential part of a modern steam boiler plant. Several other designs for different types of service are available. Literature of the leading manufacturers should be consulted for details of construction and as to the design best adapted for any specific service.

CORROSION BY HIGH-TEMPERATURE STEAM

G. A. Hawkins* and H. L. Solberg*

The introduction of high temperatures in modern central station steam-generating equipment and industrial processes has resulted in an intensive study of the corrosion produced by steam in contact with various alloy steels. The materials investigated are used for superheater tubes, turbine blades and buckets, valve parts and piping, and also in other apparatus for various industrial processes.

TESTS UNDER LABORATORY CONDITIONS

Effect of Pressure

Corrosion tests[1] conducted on S. A. E. 1010 steel (0.1% C steel) at 593° C (1100° F) for a period of 36 hours at pressures ranging from 400 to 1200 psi indicated that, within the limits of the test conditions, pressure did not influence the type or rate of corrosion. Although the tests were of short duration, they covered the period of time during which corrosion is most rapid.

Influence of Test Duration

The effect of duration of exposure time on the corrosion of S. A. E. 1010; 1.25 to 9% Cr-Mo, 12% Cr, and 18-8 columbium stabilized stainless steels (analyses in Table 1)

TABLE 1. CHEMICAL ANALYSIS (LADLE) OF STEELS IN FIGS. 1 TO 6
Percentage

Fig.	Steel	C	Mn	P	S	Si	Cr	Ni	Mo
1, 2, 3, 4, 6	S. A. E. 1010	0.08	0.30	0.017	0.034	...	...	...	...
6	2% Cr-Mo	0.11	0.47	0.017	0.016	0.40	1.98	...	0 51
1, 3, 4, 5, 6	3% Cr-Mo	0.11	0.51	0.014	0.016	0.36	2.95	...	0.98
1, 2, 6	4-6% Cr-Mo	0.12	0.41	0.017	0.016	0.28	4.60	...	0.54
3, 4, 5	4-6% Cr-Mo	0.10	0.33	0.020	0.027	0.28	5.66	0.22	0 50
2	7% Cr-Mo	0.11	0.43	0.012	0.011	0.92	7.33	...	0.59
1, 2, 3, 5, 6	9% Cr-Mo	0.11	0.38	0.010	0.016	0.27	9.00	...	1.22
1	12% Cr	0.05	1.19	0.021	0.025	0.38	11.92	...	...
2	12% Cr	0.10	0.51	0.022	0.029	0.40	12.70	...	...
3, 4	12% Cr	0.10	0.52	0.014	0.015	0.32	12.92	0.12	...
1, 2, 3, 4, 5	18-8-Cb stabilized	0.07	0.36	0.015	0.012	0.39	18.62	9.9	1.11 Cb
2	18-8	0.08	1.13	0.014	0.078	0.39	18.79	9.08	0.27
5	25-20	0.07	1.62	...	...	0.34	24.45	20.30	...
5	25-15-2% W	0.10	1.75	...	...	0.56	24.18	14.34	...	2.06 W

has been investigated at 595° C (1100° F). Reported tests[2] cover time intervals of 200, 500, 1000, and 2000 hr. The data averaged for over 200 test samples are plotted in Fig. 1. It is apparent that the corrosion decreases as the chromium content of the steel increases. The high-chromium steels are particularly corrosion-resistant for the time intervals indicated. The corrosion rate is very rapid during the first 100 hours, but levels off under constant temperature conditions. Sixty-five to more than 80% of the corrosion occurring in 2000 hours took place during the first 500 hours.

* School of Mechanical Engineering, Purdue University, Lafayette, Ind.
[1] A. A. Potter, H. L. Solberg, and G. A. Hawkins, *Trans. Am. Soc. Mech. Engrs.*, **59**, 725–732 (1937).
[2] H. L. Solberg, G. A. Hawkins, and A. A. Potter, *Trans. Am. Soc. Mech. Engrs.*, **64**, 303–316 (1942).

Influence of Chromium Content

The amount of chromium in steel, as Fig. 1 shows, has a marked influence on the resistance to corrosion by high-temperature steam. All reported tests verify this conclusion. A representative set of tests data[2] for approximately 745° C (1370° F) and 300-hour exposure is shown in Fig. 2. The steels having 7% chromium or more showed very little corrosion. The 18-8 stainless steels, both stabilized and unstabilized,

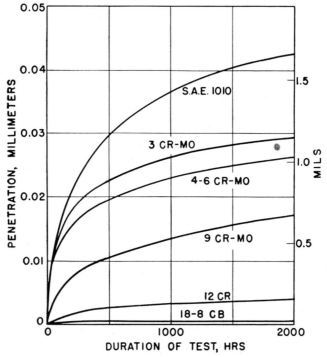

FIG. 1. Corrosion of Steel Bars in Contact with Steam at 595° C (1100° F) for Various Time Intervals.

were practically unattacked at the test temperature. The addition of chromium appears to be the most effective means for obtaining corrosion protection in high-temperature steam.

Effect of Temperature

Steam temperatures materially influence the corrosion of steels, as shown in Fig. 3.[3] All except 18-8 stainless steel begin to corrode rapidly at some temperature below 910° C (1670° F). The temperature at which rapid corrosion begins increases with the chromium content. The 18-8 stainless steel shows the same tendency toward rapid corrosion above its limiting temperature of 910° C (1670° F) as does the S. A. E. 1010 steel at a much lower temperature (580° C, 1075° F).

[3] G. A. Hawkins, J. T. Agnew, H. L. Solberg, *Trans. Am. Soc. Mech. Engrs.*, **66**, 291–295 (1944).

Figures 4 and 5 show the results obtained in two tests[4] conducted at steam temperatures of 815° C (1500° F) and 940° C (1725° F) for a period of 500 hours. The marked influence of chromium is evident. The 25% Cr–20% Ni and the 25% Cr–15% Ni–2% W steels which were tested at a steam temperature of 954° C (1750° F) showed no corrosion at the end of 500 hours. Additional tests showed that these two steels

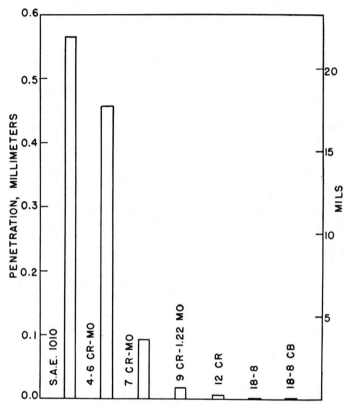

FIG. 2. Corrosion of Steels in Contact with Steam at Approximately 745° C (1370° F) for 300 Hours.

were also extremely resistant to steam corrosion for 1300-hour test periods. At 980° C (1800° F) the 25% Cr–20% Ni steel lost only 0.11% of its original weight in 1300 hours.

EFFECT OF EXTREME TEMPERATURE FLUCTUATIONS

In order to determine the influence of rapid temperature changes on the corrosion rate, specimens of S. A. E. 1010 and Cr-Mo steels were subjected to superheated steam at 650° C (1200° F) for 500 hours. At the end of this period three specimens were removed from the apparatus. One half of the remaining samples was held at 650° C (1200° F) with sudden cooling to room temperature in nitrogen, and heating

[4] G. A. Hawkins, H. L. Solberg, J. T. Agnew, and A. A. Potter, *Trans. Am. Soc. Mech. Engrs.*, **65**, 301–308 (1943).

again to 650° C (1200° F) every 100 hours until the total time in contact with high-temperature steam was 1200 hours. The remainder of the specimens was treated similarly except for a 50-hour cooling cycle. Specimens were removed after 700 hours and at the end of 1200 hours in order to measure the extent of corrosion.

Visual observation indicated that the layer of corrosion products is thinner and more brittle as the chromium content increases, up to 5%. The scale on the low-carbon-steel specimens was very thick, porous, and, in spite of the severe temperature

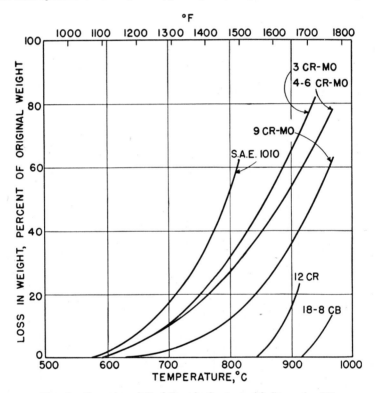

Fig. 3. Corrosion of Steel Bars in Contact with Steam for 500 Hours at Various Temperatures.

shocks, had spalled only slightly as compared to the 4 to 6% Cr steel. The 2% Cr steel spalled less than the 3% Cr steel, and the third layer of scale (discussed below) was cracking in the 4 to 6% steel at the end of 1200 hours. Little scale formed on the 9% Cr steel, and a microscopic examination of the surface at the end of the test failed to show evidence that the scale had cracked or checked.

Figure 6 is a chart of the results.[5] It is apparent that the difference in results employing the 50- or 100-hour-cycle tests is not significant. The thickness of the scale or, in other words, the time of exposure, is the factor which controls cracking and spalling of any given steel subject to temperature fluctuations. Here, too, the corrosion decreases as the amount of chromium in the steel increases.

[5] H. L. Solberg, G. A. Hawkins, and A. A. Potter, *Trans. Am. Soc. Mech. Engrs.*, **64**, 303-316 (1942).

Types of Scale Formations

In general, the steels which have been tested[5] at 650° C (1200° F) under rapid temperature fluctuations may be grouped into three classes according to the type of scale formed. The first group consists of low-carbon steel, carbon-0.5% Mo, and the low-chromium steels, which are covered with a thick, porous, tightly adhering scale.

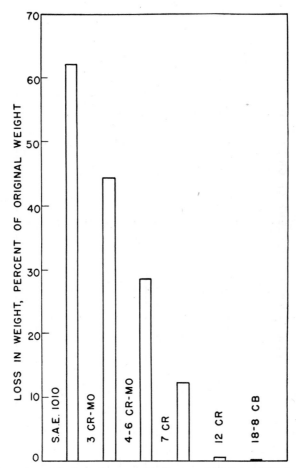

FIG. 4. Corrosion of Steels in Contact with Steam at 815° C (1500° F) for 500 hours.

The scale which forms on the steels of the second group, that is, the 4 to 6% chromium steels and the 2% Cr–0.6% Mo–0.7% Al–1.3% Si steel is very brittle and flakes off under impact or fluctuating temperatures. The third group consists of steels having a chromium content of 7% or more, upon which a very thin non-porous tightly adhering scale is formed.

The scales which formed on 3% Cr–Mo, 4 to 6% Cr–Mo, 7% Cr–Mo, and 9% Cr–Mo steels during 500-hour tests at temperatures of 815° C (1500° F) and 943° C

(1730° F) were composed of three layers.[6] In general the inner layer was too thin to remove for laboratory analysis. In all cases a concentration of chromium, silicon, and molybdenum occurred in the middle-scale layer. In every case the alloy content in

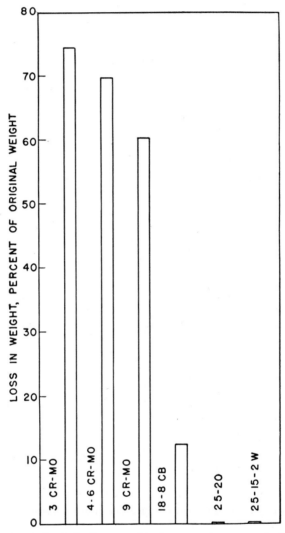

FIG. 5. Corrosion of Steels in Contact with Steam at Approximately 940° C (1725°F) for 500 Hours.

the outer layer was lower than in the middle layer. This agrees with many of the oxidation theories which have been advanced, in that since the outer layer is composed chiefly of iron, the iron must diffuse through the alloy-rich middle layer.

[6] G. A. Hawkins, J. T. Agnew, H. L. Solberg, *Trans. Am. Soc. Mech. Engrs.*, **66**, 291–295 (1944).

CORROSION BY HIGH-TEMPERATURE STEAM

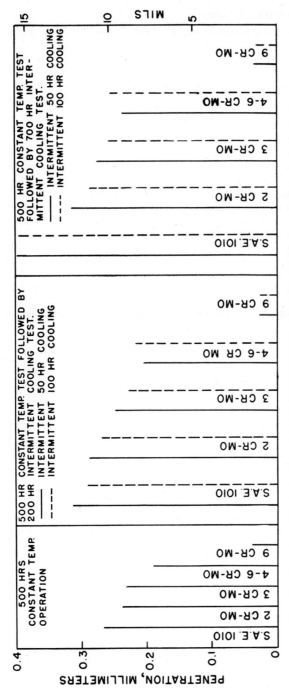

FIG. 6. Corrosion of Steels by Steam at 650° C (1200° F) When Subjected to Extreme Temperature Fluctuations.

Influence of Shape of Surface

Tests[7] have been conducted at 650° C (1200° F) for 1300 hours to determine the effect of specimen surface shape. Each specimen was machined so as to have a convex, concave, and flat surface. The scale formed on the inner concave surface did not flake off as readily as the scale which formed on the convex or flat surfaces. It is possible, therefore, that alloy tubes of small internal diameter may resist corrosive attack better than would be indicated from an analysis of corrosion data obtained from small diameter solid bar stock.

Effect of Tensile Stresses

In order to determine the influence of tensile stress, tests[8] have been conducted for 1030 and 2000 hours at steam temperatures ranging between 593° C (1100° F) and 650° C (1200° F). The range in stresses used for the various steels were: carbon–0.5% Mo, 5000 to 1000 psi; 1.25% Cr–Mo, 11,300 to 810 psi; 2% Cr–Mo, 9500 to 1300 psi; 7% Cr–Mo, 7500 to 1000 psi; 9% Cr–Mo, 4030 to 810 psi; 18-8, 10,500 to 2160 psi. Within the range of the test conditions, tensile stress did not influence the amount of corrosion produced on these steels.

Cast Steels

Figure 7 shows the total corrosion in 570 hours at 650° C (1200° F) for several cast steels,[9] the approximate analyses of which are given in Table 2. The corrosion

TABLE 2. CHEMICAL ANALYSIS (LADLE) OF CAST STEELS IN FIG. 7
Percentage

Steel	C	Mn	P	S	Si	Cr	Ni	Mo
Carbon	0.24	0.85	0.021	0.014	0.44	....	....	....
Carbon-Mo	0.20	0.67	0.02	0.013	0.40	....	....	0.49
Ni-Cr-Mo	0.31	0.61	0.019	0.010	0.45	0.68	2.19	0.26
5% Cr–Mo	0.24	0.64	0.029	0.010	0.82	5.28	....	0.45
9% Cr–Mo	0.23	1.05	0.032	0.015	0.84	9.09	....	1.56

decreases as the chromium increases, as is the case with other steels having similar compositions. In general, the corrosion of cast steel specimens is not materially different in amount from the corrosion of rolled steel specimens.

TESTS UNDER SERVICE CONDITIONS

Tests[10] on the corrosion of steels by steam have been conducted under service conditions using average steam power plant conditions of fluctuating temperature and intermittent operation over very long periods of time (16,000 hours). Some of the data obtained at 595° C (1100° F) are shown in Fig. 8. In many instances only one specimen was used at any one temperature and exposure period.

The rate of corrosion in these tests (Detroit Edison, Fig. 8) is higher than that obtained in the previously described tests (Purdue University, Fig. 1). The difference

[7] H. L. Solberg, G. A. Hawkins, and A. A. Potter, *Trans. Am. Soc. Mech. Engrs.*, **64**, 303–316 (1942).

[8] H. L. Solberg, A. A. Potter, G. A. Hawkins. and J. T. Agnew, *Trans. Am. Soc. Mech. Engrs.*, **65**, 47–52 (1943).

[9] H. L. Solberg, G. A. Hawkins, and A. A. Potter, *Trans. Am. Soc. Mech. Engrs.*, **64**, 303–316 (1942).

[10] I. A. Rohrig, R. M. Van Duzer, Jr., and C. H. Fellows, *Trans. Am. Soc. Mech. Engrs.*, **66**, 277–290 (1944).

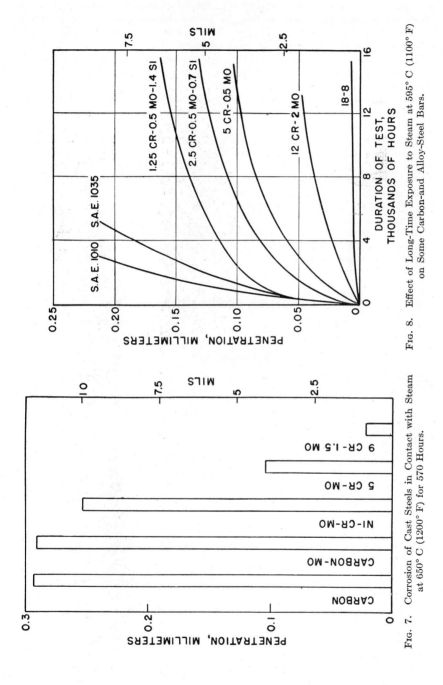

Fig. 8. Effect of Long-Time Exposure to Steam at 595° C (1100° F) on Some Carbon-and Alloy-Steel Bars.

Fig. 7. Corrosion of Cast Steels in Contact with Steam at 650° C (1200° F) for 570 Hours.

is probably accounted for by the fact that the latter tests were conducted under steady temperature conditions, whereas the Detroit Edison tests simulated service conditions, which included dropping the temperature to room temperature for several days at a time on several different occasions.

The tests conclusively show again that chromium enhances the corrosion resistance in steam and that ferrous alloys containing high percentages of chromium are adequately corrosion-resistant at 595° C (1100° F).

Non-ferrous alloys containing high percentages of nickel and copper were found to corrode more at steam temperatures of 500° C (925° F) and 595° C (1100° F) than the best of the ferrous materials.

The differences found in corrosion rates of carbon steels and medium- or low-alloy steels at 595° C (1100° F) become insignificant at a lower temperature of 500° C (925° F).

The corrosion resistance of low-alloy content (0.5% Mo or 1% Cr steels), on the other hand, compares favorably with the 4 to 6% Cr steels at test temperatures of both 500° C (925° F) and 595° C (1100° F). The high Cr-Ni and 12% Cr stainless steels were the most corrosion-resistant of the metals and alloys tested.

The corrosion rates at a steam temperature of 595° C (1100° F) do not increase with time, but tend to flatten out. In many of the low-alloy steels the rates were materially reduced after 15,000 to 16,000 hours.

BOILER CORROSION*

Max Hecht† W. C. Schroeder§
E. P. Partridge‡ S. F. Whirl∥

INTRODUCTION

In the broadest sense, corrosion in a steam boiler is always the same process, represented by the overall equation:

$$3Fe(s) + 4H_2O(l \text{ or } g) \longrightarrow Fe_3O_4(s) + 4H_2(g) \tag{1}$$

Fundamentally, from the standpoint of corrosion, a boiler is but a thin film of magnetic iron oxide supported by steel; this oxide film is continually being damaged and repaired during boiler operation, with corresponding production of hydrogen. Since the surface film of magnetic iron oxide is the major protection for the steel, it must be maintained in such a condition that it is least permeable to water.

The bulk of the material used for the construction of boilers, boiler auxiliaries, and water and steam piping is plain carbon or low-alloy iron and steel.[1] The corroding medium in all cases is water (both liquid and vapor) of purity ranging from distilled water having a conductivity of less than one reciprocal microhm-cm to complex solutions of highly soluble salts. The corrosive behavior of the water is often influenced by dissolved gases.

* The contents of this chapter have been contributed by the authors named, all of them members of the A. S. T. M. Committee D-19 on Water for Industrial Uses.
† Consultant, Pittsburgh, Pa.
§ Office of Synthetic Liquid Fuels, U. S. Bureau of Mines, Washington, D. C.
‡ Hall Laboratories, Inc., Pittsburgh, Pa.
∥ Power Stations Dept., Duquesne Light Co., Pittsburgh, Pa.

[1] The compositions of the steels are those specified by the *Am. Soc. Mech. Eng. Boiler Code*, published by the American Society of Mechanical Engineers, 29 West 39th St., New York 18, N. Y., or as given in A. S. T. M. Standards, American Society for Testing Materials, Philadelphia, Pa.

The temperature at which corrosion is encountered varies from that of an idle boiler to the boiling point of saturated solutions in an operating boiler, which may be as high as 700° C (1300° F). A solution can exist at such a temperature, considerably above the critical point of pure water, 374° C (705° F). High concentrations of dissolved salts, however, occur only locally in boilers during operation.[2]

The mechanisms by which physical or chemical agents penetrate or destroy the film on boiler surfaces are essentially the same as those experienced at lower temperatures on less critical equipment.* However, the rate of attack in boilers is greatly accelerated by the higher temperatures and pressures. In addition, the high heat transfer rates, locally attaining 200,000 Btu (sq ft)$^{-1}$ (hr)$^{-1}$, tend to accentuate conditions for corrosion.

PITTING–TYPE CORROSION

Pitting-type corrosion is characterized by local attack of the metal in such a manner that a small cavity, called a pit, is produced in the surface. The shape of the pits and their distribution, both of which may vary widely, determine the appearance of the corroded area. Pits may occur within a pit, and very frequently they are so close together that the surface is merely irregular.

Identification of Pitting

The identification of any particular type of corrosion is often difficult because several mechanisms may be acting simultaneously, and the changes which occur when a high-temperature boiler is being cooled down and emptied may mask what happened during service. Experience assists materially in recognizing and identifying pitting attack in boilers. For example, it has been noted that the presence of black magnetic iron oxide in the pit or in the "barnacle" or "tubercle" built up over it is evidence of active corrosion.[3] This observation is often used in studying the effectiveness of corrosion-preventive treatments.

The oxide in these locations of active corrosion should not be confused with the black magnetic iron oxide carried in suspension by many boiler waters. It must be remembered that neither the total quantity of finely divided black iron oxide nor the quantity of hydrogen evolved in the boiler are reliable indices of the amount of corrosion taking place. Ferrous hydroxide from sources external to the boiler, such as the condensate storage tanks and surge lines, may account in part for both the oxide and the hydrogen. The ferrous hydroxide entering with the feed water is converted in the boiler according to the following reaction:

$$3Fe(OH)_2 \longrightarrow Fe_3O_4 + 2H_2O + H_2 \qquad (2)$$

---

* Although the same factors, e.g., galvanic corrosion, may continue to operate at high as well as low temperatures, the overall corrosion by steam involves two distinct electrochemical mechanisms. Oxidation of iron, particularly at the higher temperatures, involves diffusion of iron ions through the continuous oxide layers covering the metal surface. Reaction with steam or oxygen then occurs at or near the outer oxide surface, which is the cathode of the operating cell. The entire iron surface in contact with oxide is the anode.

This mechanism is distinct from that occurring largely at lower temperatures where molecular diffusion of water and oxygen takes place through a film of hydrated iron oxides, and where the anodes and cathodes, unlike the situation above, are located on the metal surface. See *Fundamentals of Oxidation and Tarnish*, p. 11. EDITOR.

[2] R. E. Hall, *Trans. Am. Soc. Mech. Engrs.*, **66**, 457–488 (1944).
[3] U. R. Evans, *Metallic Corrosion, Passivity and Protection*, p. 346, Edward Arnold and Co., London, 1937. R. Stumper, *Korrosion u. Metallschutz*, **3**, 169 (1927).

FACTORS THAT AFFECT PITTING

Impurities and Stress. Segregated impurities or their solid solutions in the steel, as well as stress, either applied or residual, are notable for their ability to form anodic areas. In general, pits are of various sizes and randomly scattered. When stress is present, the pits tend to align themselves with the stress. Typical examples of this condition are found in fin tubes at the point where the fin has cracked and in the rolled-in area (expanded portion) of boiler tubes (Fig. 1).

Dissolved Oxygen. Probably the greatest promoter of pitting-type corrosion is dissolved oxygen in the water. At all temperatures, and even in an alkaline solution, it is an active depolarizer. In addition, the conditions encountered in the boiler favor

Fig. 1. A Pit Formed in a Boiler Tube at End of Expanded Portion of Tube.

the formation of oxygen concentration cells, particularly under scale or sludge deposits where relatively stagnant conditions exist. The usual remedies involve removal of oxygen through mechanical deaeration of the feed water and chemical scavenging with sulfites or ferrous hydroxide or various organic reducing chemicals. Pits similar to the one shown in Fig. 2 may be produced by oxygen attack.

Dissolved Carbon Dioxide. (Refer also to *Corrosion of Steam Condensate Lines*, p. 538.) Carbon dioxide, because of the weakly acidic character of its solutions, also tends to promote pitting in boilers. An alkaline boiler water combats the aggressiveness of dissolved carbon dioxide, but the benefits do not extend into the steam space or the condensate lines. Removal, along with dissolved oxygen by mechanical deaeration, is the usual remedy, although cyclohexylamine has been recently tried for control of corrosion in the steam and condensate lines of heating systems.[4]

Deposits. Very frequently pits are found under or along the periphery of deposits such as mill scale, boiler sludge, boiler scale, corrosion products, and oil films. It is

[4] A. A. Berk, U. S. Bureau of Mines, Department of the Interior, R. I. 3754.

common experience that once pitting-type corrosion has started, unless the corrosion products are removed, the process will continue. This localized corrosion is promoted by the cathodic character of the deposits with respect to the boiler steel or by depletion of oxygen under the deposits.

Copper in Boiler Waters. When one considers the large amount of copper alloys used in the construction of auxiliary equipment such as condensers, heaters, coolers, and pumps, it is not surprising that copper is present in most boiler deposits. (See *Copper*, p. 64.) Although generally found in the metallic state, it is sometimes present as the oxide. The amount of copper in a deposit may vary from a fraction of one per cent to almost pure copper.

The significance of these copper deposits and their relation to the corrosion of steel in boilers is a controversial subject among water-treating authorities. Some claim

FIG. 2. Pitting Due to Oxygen Attack.

that copper is but an innocent bystander in the corrosion process; others hold it to be a serious offender in promoting pitting-type attack by acting as cathode to the steel of the boiler. Neither viewpoint has been substantiated by direct experimental proof.

There are numerous cases where little or no corrosion is encountered although the deposits throughout the boiler contain appreciable metallic copper. There is also some evidence that when copper and low-carbon steel are in contact with an alkaline boiler water at elevated temperatures the copper is attacked rather than the steel. For example, copper ferrules around the ends of re-rolled tubes, and copper rivets and screens in auxiliary equipment through which boiler water flowed have been almost completely eaten away, even at low boiler temperatures. In view of this experience, it has been reasoned that metallic copper does not promote the corrosion of boiler steel. Instead, it is regarded as merely the end product of the reduction of cuprous oxide by nascent hydrogen resulting from local attack of the steel by the boiler water.

Conversely, severe pitting of boiler metal has often been noted in the vicinity of deposits abnormally rich in copper. This observation has led to the theory that copper is functional in the corrosion process because copper is cathodic to the steel, and thereby promotes the pitting action. In boilers one is seldom dealing with metallic iron alone, but rather with a complex protective coating consisting chiefly of iron oxide. Where breaks occur in this coating, it is quite possible that areas exist anodic with respect to copper which enhance pitting. This condition may in some cases account for accelerated pitting once a small pit is started, as well as the severe pitting sometimes encountered following the acid cleaning of boilers.

Improper Care of Idle Boilers. One of the most frequent causes of pitting-type corrosion in boilers is failure to care properly for idle boilers. An idle boiler should either be kept absolutely dry or filled with water treated in such manner that corrosion will not occur. Water on the surface of a non-operating boiler will absorb oxygen from the air and initiate pitting-type corrosion. These pits are then focal points for further attack when the boiler is placed in service. The usual instructions for protecting idle boilers against corrosion are one of the following:

1. Drain the boiler while hot (approximately 90° C [200° F]). Circulate air to insure complete drying, and maintain boiler in dry condition.
2. Fill the boiler completely with alkaline water (pH approximately 11) containing an excess of sulfite ion (approximately 100 ppm), and place under water or steam seal.
3. Fill the boiler completely with an alkaline chromate solution (approximately 200 to 300 ppm CrO_4).[5,6]

In the chemical cleaning of boilers, it is obvious that the protective iron oxide coating on the surface will be removed in many locations. During subsequent operation, these areas may not heal properly, and even in the absence of copper they may become focal points for pitting-type corrosion. Therefore, immediately after the cleaning, it appears advisable to recondition the surfaces by an alkaline boiling-out procedure similar to that used for new boilers.

ECONOMIZER CORROSION

The statements made concerning boiler corrosion are equally applicable to the economizer. However, the economizer being a once-through heater and being located ahead and independently of the boiler proper is especially susceptible to pitting-type attack. It is the first high-temperature surface to encounter the destructive action of dissolved oxygen in the feed water. Furthermore, the water passing through it has normally a relatively low pH value and is devoid of inhibiting chemicals.

The combating of economizer corrosion is based on the same principles as are applied in protecting boiler surfaces; namely, mechanical deaeration of the feed water and treatment with alkaline and reducing chemicals. In some cases, treatment is accomplished by recirculation of a portion of boiler water through the economizer. When this is done, care must be exercised to avoid sludge deposits in the economizer. Consideration must also be given to the effect of such recirculation on steam quality in those instances where the boilers are equipped with scrubbing-type steam purifiers.

[5] R. C. Ulmer and J. M. Decker, *Combustion*, **10**, 3, 31–33 (1938).
[6] Marc Darrin, "Chromate Corrosion Inhibitors in Bimetallic Systems," *Ind. Eng. Chem.*, **37**, 741 (1945).

BOILER WATER TREATMENT*

The formation and preservation of a protective film on metal surfaces is the primary objective of all corrosion-inhibiting boiler water treatments. The combination of chemicals employed will depend on operating conditions, particularly the pressure, temperature, heat absorption rates, and quality of feed water; but in all

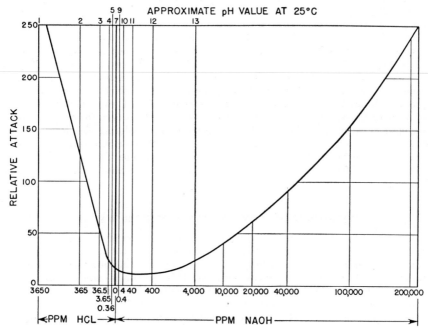

FIG. 3. Attack on Steel at 310° C (590° F) by Water of Varying Degrees of Acidity and Alkalinity. (Curve by Partridge and Hall, based on data of Berl and van Taack.)

cases treatment is founded on three rules: namely, keeping the boiler water (1) alkaline, (2) free of dissolved oxygen, and (3) maintaining clean heating surfaces.

The effect of acidity from hydrochloric acid and of alkalinity from caustic soda on the corrosion rate of steel at 310° C (590° F) is shown in Fig. 3. It is evident that caustic soda confers maximum protection at pH values between 11 and 12. In practice, with complex boiler waters, a pH value in the vicinity of 11 has proved most satis-

---

* Papers on this subject may be found in the *Transactions* of the American Society of Mechanical Engineers since 1926. For recent trends in modes of chemical treatment, and in apparatus, the reader is referred to the *Proceedings* of the Engineers' Society of Western Pennsylvania, Pittsburgh, Pa., special publications of the Annual Water Conferences, published annually since 1941.

Committee D-19, on Water for Industrial Uses, has prepared the following methods for Water Analysis, and additional methods are in preparation. These methods are published in the current issue of the *Book of Standards*, issued by the American Society for Testing Materials, 1916 Race Street, Philadelphia, Pa.

"Total Carbon Dioxide, and Calculation of the Carbonate and Bicarbonate Ions in Industrial Waters, Determination of." D513, *Book of Standards*, Part 3-B.

"Tendency of Boiler Water to Cause Embrittlement Cracking of Steel, Field Test for." D807, *Book of Standards*, Part 3-B.

"Hydroxide Ion in Industrial Waters, Determination of." D514. *Book of Standards*, Part 3-B.

factory. For boilers operating below 250 psi the pH is usually maintained between 11.0 and 11.5. For higher pressures, because of the increased possibility of metal attack as a result of faulty circulation and local concentration of solution, the operating range is usually between pH 10.5 and 11.0.

For scavenging residual oxygen, the chemicals previously mentioned (sulfites, ferrous hydroxide, and organic reducing materials) are widely used. Although iron compounds are effective scavengers, they yield a sludge that may be objectionable with high heat transfer rates. The organic reducing materials, because of their instability at high temperatures, are generally not recommended for boilers operating above 500 psi. Decomposition of sulfites has also been reported at the higher temperatures,[7] but their use in low concentrations is quite common in boilers operating up to 1400 psi. Many high-pressure plants operate without scavenging chemicals.

While decidedly advantageous, the cost of specially designed deaerating equipment cannot always be justified for small plants operating at relatively low pressures. At pressures below 200 psi partial deaeration by means of feed water heaters capable of reducing the oxygen content to approximately 0.7 ppm has been found to be more practical in many cases. At these lower boiler pressures, reducing chemicals appear to be quite effective in preventing oxygen attack of boiler metal, especially when the pH of the boiler water is above 11, and the scavenging chemicals are added sufficiently far from the boiler so that the oxygen is consumed before the water enters the boilers.[8]

CORROSION BY CONCENTRATED BOILER WATER

Experience has shown that a low concentration of the order of 100 ppm of sodium hydroxide in the boiler water helps to maintain the oxide coating of the steel in a condition which affords maximum protection; higher concentrations developed locally may, however, cause serious corrosion. This increase in rate of reaction is illustrated by Fig. 3,[9,10] based on experiments with iron powder in contact with solutions in a bomb at 310° C (590° F).

The regions in a boiler where the desirable low-caustic content of the normal boiler water may be concentrated to a damaging degree are generally characterized by high input of heat in relation to circulation of water past the heated surface. The concentrating film may be produced in various regions, but the resulting corrosion occurs generally as a band or elongated area of attack, sometimes clean, sometimes filled with hard, compact magnetic iron oxide.

Horizontal or slightly inclined tubes subjected to intense radiation from above tend to be corroded internally along the top, as illustrated by the cross section shown in Fig. 4. Attack of this type has been produced experimentally by Straub.[11] Several

"Dissolved Oxygen in Industrial Waters, Methods of Test for." D888, *Book of Standards*, Part 3-B.
"pH of Aqueous Solutions with the Glass Electrodes, Method for the Determination of." E70, *Book of Standards*, Part 2.
"Total Orthophosphate, and Calculation of the Respective Orthophosphate Ions in Industrial Waters, Determination of." D515, *Book of Standards*, Part 3-B.

[7] R. E. Hall, *Trans. Am. Soc. Mech. Engrs.*, **66**, 457–488 (1944).
[8] R. M. Hitchens and R. W. Towne, *Proc. Am. Soc. Testing Materials*, **36**, Part II, 687–696 (1936).
[9] E. Berl and F. van Taack, "Forschungsarbeiten auf dem Gebiete des Ingenieurwesens," *Heft* 330 (Berlin) (1930).
[10] E. P. Partridge and R. E. Hall, *Trans. Am. Soc. Mech. Engrs.*, **61**, 597–622 (1939); discussion, **62**, 711–717 (1940).
[11] F. G. Straub, *Mech. Eng.*, **61**, 199–202 (1939).

cases of damage of this kind to tubes in high-capacity boilers have been described by Partridge and Hall.[10]

Tubes in which circulation is uncertain or inadequate under high load may suffer general attack in a band along the bottom, sometimes with more pronounced loss of metal along the fluctuating water level at either side. Heavy accumulations of

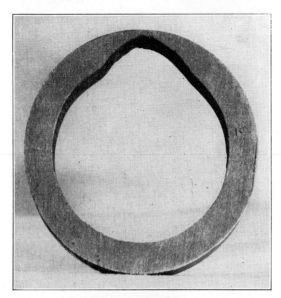

FIG. 4. Grooving of the Ceiling of a Slightly Inclined Slag-Screen Tube.

The development of a more or less continuous band of steam along the strongly heated top of this tube caused overheating of the steel and attack by the concentrated film of boiler water for a distance of several feet.

magnetic iron oxide, sometimes loose, sometimes cemented together into hard masses, are frequently found covering the corroded areas. Figure 5 is an example. A less common type of damage which leads to failure without appreciable deformation of the tube is illustrated in Fig. 6.

The attack is often accelerated by overheating of the steel. This may result from the blanketing of the top of an inclined tube by a band of steam. *Steam blanketing* may also develop in vertical tubes subjected to very high heat input, as has been indicated by the temperature measurements of Davidson and his associates[12] on an experimental tube in an operating boiler. A representative series of their data showing this effect is reproduced in Fig. 7. Limited regions of overheating in clean vertical tubes with normal temperature above and below the "hot spot" probably are caused by *film boiling*. This is a condition analogous to that of a drop of water on a very hot griddle, separated from the metal surface by a continuous insulating film of steam.

Every time that a bubble of steam forms at the surface of a boiler tube there must be some slight local increase in the temperature of the underlying metal. Hall[13] has

[12] W. F. Davidson, P. H. Hardie, C. G. R. Humphreys, A. A. Markson, A. R. Mumford, and T. Ravese. *Trans. Am. Soc. Mech. Engrs.*, **65**, 553–591 (1943).
[13] R. E. Hall, *Trans. Am. Soc. Mech. Engrs.*, **66**, 457–488 (1944).

528   SPECIAL TOPICS IN CORROSION

FIG. 5.   Attack along Bottom of Slightly Inclined Top-Row Generating Tube.

A hard mass of magnetic iron oxide extending for about a foot along the bottom of the tube marked the area of attack.

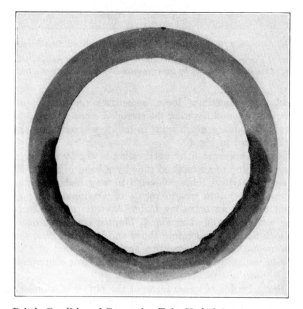

FIG. 6.   Brittle Condition of Generating Tube Underlying Area of Corrosion.

The portion of the tube wall showing dark after macro-etching exhibits little, if any, ductility and shows intergranular damage under the microscope.

pointed out that concentration of boiler water must result at the interface created by the bubble forming in contact with the water and the heat transfer surface. Figure 8 indicates that even an increase of only a few degrees in the temperature of the film of water in contact with the metal and with the expanding bubble will result in a caustic content measurable in percentage rather than in parts per million. The

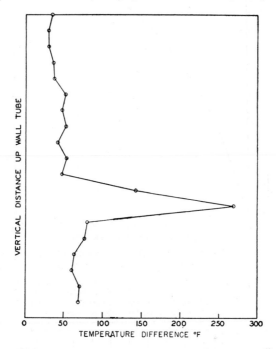

FIG. 7. Evidence of a Hot Spot in a 50-Foot Vertical Water-Wall Tube.

This figure represents the elevation in temperature of the outer surface of the boiler tube above the temperature of the main body of boiler water. The thermocouple measurements show a short region of severe overheating, presumably because film boiling produced a nearly continuous blanket of steam on the tube surface. Conditions both above and below the hot spot are normal.

concentrating film resulting from the formation of each bubble affects only a very small area of metal for a very brief period of time. Nevertheless, the cumulative effect of a high rate of steam generation on a heat transfer surface may be equivalent to continuous exposure of the hot steel to a corrosive caustic liquor, even though the main body of boiler water contains but a few parts per million of hydroxide.

Several ways of meeting the problem posed by local concentration of sodium hydroxide at heat transfer surfaces have been tried. A means of reducing the tendency of the concentrating film to attack the boiler steel even when some hydroxide is present in the boiler water has been suggested by Hall.[13] In brief, this involves maintaining a neutral salt, such as chloride, in the boiler water in an amount that is high relative to the amount of hydroxide.* The most obvious method is not to have

---

* It has been reported by Berl and Van Taack and by Straub (see footnotes 9 and 11 on p. 526) that attack of iron by caustic soda solutions at elevated temperatures can be inhibited by various neutral salts. Consult *Iron and Steel*, p. 636.

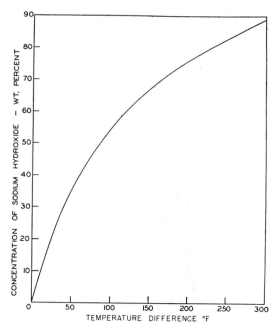

FIG. 8. Caustic Content Attainable in Concentrating Film of Boiler Water. (Based on data from International Critical Tables, **3**, 370 [1928].)

When the temperature of the film of water at the heat-transfer surface exceeds the temperature of the main body of the boiler water by only a small value of ΔT, a relatively high concentration of sodium hydroxide can develop in the film. This curve approximates the limiting equilibrium in a solution containing only sodium hydroxide; the precise values depend somewhat on boiler pressure.

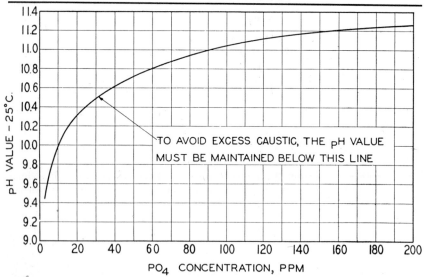

FIG. 9. Relation of pH Values to Trisodium Phosphate Concentration.

any sodium hydroxide in the boiler water. This can be accomplished by the coordinated phosphate-pH control method of Purcell and Whirl,[14] using the curve illustrated in Fig. 9. This provides what has been termed *captive alkalinity*,[13] that is, alkalinity at a level corresponding to that of alkali phosphate with no excess caustic present. Although a boiler water conditioned in this manner will exhibit a high pH, it may be concentrated without increasing the concentration of hydroxyl ion appreciably.

It must be remembered that elimination of sodium hydroxide means only that one accelerating factor in the corrosion process has been removed. With a boiler water substantially free of sodium hydroxide, tubes subject to steam blanketing may still corrode, although at a slower rate than if caustic soda were concentrated to a high degree on the metal surface.[15]

A mechanical solution to the problem is currently being sought by boiler designers. At the same time, there has been a continuing tendency to crowd more and more heat through each square foot of tube surface, which tends to intensify this type of attack.

EMBRITTLEMENT IN BOILERS*

Embrittlement can be defined as a type of cracking occurring at riveted joints or other areas in a boiler where contact between metal surfaces permits the accumulation of concentrated solution and where the metal stresses are high. Trouble of serious proportions in steam boilers has been almost completely confined to the riveted joints. This has occasionally led to an explosion, and frequently to heavy replacement costs even on relatively new boilers. One American railroad has reported cracking in as many as forty locomotives a year, with estimated annual repair costs running up to $60,000.

Embrittlement has been found also in rolled tube ends, tube ligaments, headers, and threaded pipe connections. Many of these cases have been troublesome, but they are not generally so dangerous or so costly as failures in riveted seams.

Stress Necessary for Embrittlement

The experimental evidence shows that there is little likelihood of embrittlement in ordinary boiler steel unless stresses above the yield point have existed. The stress created by the steam pressure or uniformly distributed structural load has only slight bearing on the cracking. On the other hand, stresses created by rolling the plate to form the drum, distortion during riveting, or any cold work which causes permanent deformation can create the necessary stress conditions for cracking.

The existence of external applied stress is not necessary to cause cracking. A bar of boiler steel given a permanent set by bending with a stress applied and then released can be cracked in an alkaline solution such as might result from concentration of a boiler water.

Concentration of the Boiler Water

Caustic concentrations existing in the boiler drum will not cause embrittlement cracking because under normal conditions there would be less than 1000 ppm of

---

* This section is published by permission of the Director, Bureau of Mines.

[14] T. E. Purcell and S. F. Whirl, "Embrittlement Symposium," *Trans. Am. Soc. Mech. Engrs.*, **64**, 397–402 (1942); *Proc. Third Annual Water Conference*, Engineers' Society of Western Pennsylvania, 45–60 (1942); *Trans. Electrochem. Soc.*, **83**, 279 (1943).

[15] M. K. Drewry, *Combustion*, **9**, 8: 18–24 (1938); anon., **11**, 7: 33–34 (1940).

NaOH. The minimum concentrations necessary to cause embrittlement are approximately 100 times this value.

These very high concentrations may be attained by extremely slow diffusion of boiler water out of a riveted seam or other joint in the manner represented in Fig. 10. As the boiler water passes point A the pressure drops rapidly toward atmospheric,[16] and the heat in the metal and liquid will cause evaporation to produce a concentrated solution. Ultimately, salts such as sodium sulfate and sodium chloride crystallize from the solution. It is this mechanism that causes the formation of solids outside most riveted boiler seams. The most dangerous leakage is probably one which cannot be easily observed, but which leaves the deposit of solid within the riveted joint where the residual stresses are high. The combined action of stress and concentrated solution can result in embrittlement cracking.

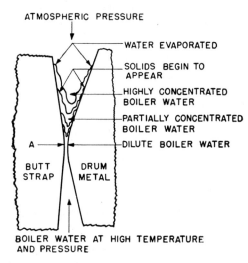

FIG. 10. Greatly Magnified Diagram of a Leak Showing Concentration Mechanism.

EMBRITTLEMENT DETECTOR

The embrittlement detector has been developed to permit the concentration of boiler water on a stressed steel specimen in the same manner in which it may occur in a riveted seam. If the specimen cracks, the boiler water is one that can cause embrittlement, and treatment of the water to eliminate this characteristic may be necessary. Cracking of the specimen does not mean, however, that the boiler is embrittled or will be embrittled, since in the riveted seams or other joints proper conditions of leakage, boiler water concentration, and stress may not exist simultaneously as they do in the embrittlement detector.

The embrittlement detector, as illustrated in Fig. 11, is suitable for attachment directly to the boiler[17] and permits differentiation between boiler waters that are capable of producing embrittlement cracking and those that are not, as well as testing methods of chemical control to eliminate embrittlement tendency. The test

---

[16] F. G. Straub and T. A. Bradbury have also described a method of concentration that seems to depend on evaporation against the boiler pressure. *Mech. Eng.*, **60**, 371-376 (1938).

[17] See Tentative Standard D807, *A. S. T. M. Book of Standards*, Part IIIB

is carried out over a period of 30 or more days, with the boiler water constantly circulating through the detector.

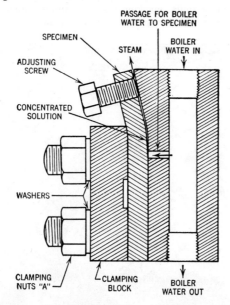

FIG. 11. Embrittlement Detector.

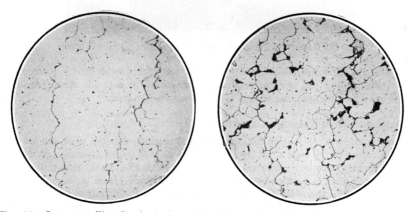

FIG. 12. Intercrystalline Cracks in Stressed Boiler Steel Subject to Corrosion by Alkaline Solution. ×200.

IDENTIFYING EMBRITTLEMENT CRACKS

Microscopic examination reveals that embrittlement cracking in ordinary boiler steel has unique characteristics that distinguish it from stress, fatigue, or corrosion-fatigue cracking.[18,19] This is illustrated in the photomicrographs in Fig. 12, which

[18] For a more complete discussion on this subject see "Intercrystalline Cracking of Boiler Steel and Its Prevention," Bureau of Mines, *Bull.* 443 (1941).
[19] H. N. Boetcher, *Mech. Eng.*, **66**, 593–601 (1944).

show that the embrittled steel contains a honeycomb of fine cracks which do not cut across the grains of the metal but follow the grain boundaries. The difference between this type of intergranular cracking and a corrosion fatigue crack in boiler metal can be seen readily by comparing Fig. 12 with Fig. 13.

Fig. 13. Corrosion Fatigue Crack in Boiler Steel. Etched, ×100.

In alloy steels, such as the low-nickel or silicon-manganese steels used in locomotive boiler construction, the cracking retains its honeycomb or network appearance but is distinctly less intergranular than it is in ordinary boiler steel.

Theory of Embrittlement Cracking

The present best explanation for embrittlement cracking depends upon the heterogeneous nature of the steel structure. The current view of physical metallurgists indicates that grain boundary atoms are attached to crystals of different orientation, and may be maintained in position by atomic force lines distorted from their normal position. It is possible, therefore, to remove such atoms from their strained position more easily than to remove atoms from the body of a crystal. Careful adjustment of the corrosive activity of many solutions would be expected to produce this selective removal of grain boundary atoms, and it has been found experimentally that acid solutions, neutral solutions (aided by a small electric current to create corrosive conditions), and concentrated alkaline solutions can cause intercrystalline cracks. If a solution that produces general overall corrosion is modified by adding a chemical that tends to protect the crystal faces, the general activity at these surfaces may be reduced and corrosion focused on the boundary atoms.

The corrosive agent, so far as embrittlement of boilers is concerned, is sodium hydroxide. Sodium silicate appears to be the major chemical that can act to protect the crystal faces and leave the grain boundaries exposed. The balance between these corrosive and protective effects depends upon many factors, such as concentration, temperature, metal stress, and the presence of other chemicals.

A colloidal explanation of embrittlement has been proposed by Tajc.[20] Zapffe has attributed embrittlement to the action of hydrogen at high temperatures.[21] His paper includes a comprehensive bibliography.

PROTECTION AGAINST EMBRITTLEMENT

Protection from embrittlement may be obtained by using welded drums to eliminate leakage (this has also been done in some cases with tube ends), by using a steel resistant to intercrystalline cracking, or by chemically treating the boiler water so that it will not cause cracking.[22] In riveted boilers now in service, the latter method is the only one that can be used conveniently. It is believed that tests under controlled conditions, such as those in the embrittlement detector, are the best way to establish the effectiveness of protective agents.

Sodium sulfate does not prevent cracking of embrittlement detector specimens. Sodium nitrate has been used successfully to prevent cracking in detector specimens at pressures up to 750 psi.* It is now in extensive use in stationary boilers with concentrations maintained at about 20 or 30% of the sodium hydroxide alkalinity. Figure 14 shows how effective this salt has been in eliminating cracking in locomotive boilers on one American railroad.

Crude quebracho extract, obtained from South American trees of the same name, is effective in preventing cracking of detector specimens. The protection afforded by this material does not seem to be quite as certain as that offered by sodium nitrate, which may be due in part to difficulties in analysis and control. Quebracho is now in extensive use to prevent embrittlement as well as to help retard corrosion and formation of scale.

Since caustic soda is the primary chemical responsible for embrittlement, its elimination from the boiler water is the most obvious method of controlling cracking. This can be done by the method of Purcell and Whirl.[23] This, as well as methods for reduction of concentration, are discussed in this chapter, page 526, under the heading "Corrosion by Concentrated Boiler Water."

SUPERHEATER TUBE CORROSION†

The corrosive attack on the internal surfaces of superheater tubes results primarily from the reaction between the metal and steam at high temperatures and to a lesser degree from carryover of boiler water salts into the steam. In the latter case, high concentrations of sodium hydroxide may develop on the metal wall, permitting direct chemical attack, or deposits may bake on the tube wall that give rise to blistering.

* Concentrated nitrate solutions boiling at atmospheric pressure may cause stress corrosion cracking of mild steel. Refer to effect of stress on corrosion of iron, p. 137, and effect of nitrogen, p. 140. EDITOR.
† Acknowledgment is made to C. H. Fellows, Research Department, Detroit Edison Co., Detroit Mich., for assistance in preparing this section.

[20] J. A. Tajc, *Proc. Am. Soc. for Testing Materials*, **37**, Part II, 588–599 (1937).
[21] C. A. Zapffe, *Trans. Am. Soc. Mech. Engrs.*, **66**, 81–117 (1944).
[22] Symposium on Caustic Embrittlement, *Trans. Am. Soc. Mech. Engrs.*, **64**, 393–444 (1942).
[23] T. E. Purcell and A. F. Whirl, "Embrittlement Symposium," *Trans. Am. Soc. Mech. Engrs.*, **64**, 393–444 (1942); *Proc. Third Annual Water Conference*, Engineers' Society of Western Pennsylvania, 45–60 (1942); *Trans. Electrochem. Soc.*, **83**, 279 (1943).

With the boiler banked or out of service, and with steam condensed in relatively cold superheaters, pitting may develop because of dissolved oxygen and carbon dioxide.

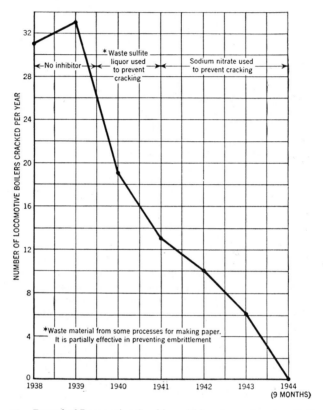

FIG. 14. Record of Locomotive Cracking within a Large American Railroad.

CORROSION BY HIGH-TEMPERATURE STEAM

Refer to the chapter on this subject, p. 511.

HYDROGEN AS A MEASURE OF RATE OF ATTACK

Steam temperatures in modern boilers are approaching those once used in the commercial production of hydrogen by the direct reaction between steam and iron. (See Eq. 1.)

Fellows in his initial investigation to determine rate of attack by steam on carbon and alloy steel tubes heated to 650° C (1200° F) collected the hydrogen evolved by the reaction.[24] Several operating companies determine the rate of hydrogen evolution from boiler water as an approximation of overall corrosion, specifically that resulting from steam blanketing in the generating tubes. In recent years, three miniature

[24] C. H. Fellows, *J. Am. Water Works Assoc.*, **21**, 10: 1373–1387 (1929).

degassing or deaerating units have been used in power plants in this country.* Virtually complete extraction of all gases is secured with these units and the gas-freed condensate is available to determine the salts dissolved therein as an indication of salt impurities carried in the steam.

An approximation of overall corrosion in the superheater during operating runs of the boiler may be secured by observing the difference in hydrogen concentration obtained by analysis of samples extracted from steam entering and leaving the superheater.

Corrosion Due to Impurities in Steam

The saturated steam entering the superheater elements carries in it very small but measurable amounts of gases and boiler water salts. The more generally detected gases include oxygen, ammonia, and carbon dioxide. These gases pass through the superheater without discernible change in concentration. Little, if any, corrosion of superheater metal has been attributed to them.

It has not yet been demonstrated that the boiler water salts, when dried and deposited on the superheater element, contribute directly to corrosion of the metal. However, sodium hydroxide, when a major constituent of the entrained boiler water salts, may contribute to corrosion of the highly heated element, especially if this substance adheres to the wall of the metal. (See Fig. 4 and discussion under "Corrosion by Concentrated Boiler Water," p. 526.)

The purification of saturated steam by removal of gases requires effective deaeration of the boiler feed water.† Reduction in the amount of boiler water salts carried with the saturated steam requires effective baffling in the steam off-take drum, the use of mechanical separators, the washing of the saturated steam with feed water, or adjustments in the chemical treatment of the water. Various systems are in operation.

The estimation of the quantity and character of the gases carried in the saturated steam may be secured by utilizing the apparatus mentioned above for the sampling of the steam and analysis by appropriate methods. The estimation of the amount of boiler water salts carried in the saturated steam may be most conveniently made by utilizing electrical conductivity procedure or, if desired, by the evaporation of a large volume of condensed steam. Hecht and McKinney[25] suggested refinements of the conductivity procedure, applying corrections for some of the dissolved gases. The steam condensate secured in the operation of the above-mentioned miniature degasifying units offers a convenient supply for measuring the electrical conductivity.

Steam Condensation in Superheaters

The superheater functions as a condenser when the boiler is banked or out of service. Typical underwater pitting may occur if oxygen or carbon dioxide has access to the steam condensate and is dissolved therein. This reaction has been discussed in the preceding paragraphs under "Dissolved Oxygen" and "Dissolved Carbon Dioxide."

---

* They include the Elliott Company Tray-Type Unit, the Cochrane Corporation Atomizing Degasifier Unit, the Straub and Nelson Degasifier. P. B. Place, in his paper, "The Degasification of Steam Samples for Conductivity Tests" [Symposium on Problems and Practices in Determining Steam Purity by Conductivity Methods, *Proc. Am. Soc. for Testing Materials* **41**, 1302–1313 (1941); includes extensive bibliography] shows illustrations in Figs. 3, 4, 5, respectively.

† Deaeration and deactivation are discussed beginning p. 506.

[25] Max Hecht and D. S. McKinney, *Trans. Am. Soc. Mech. Engrs.*, **53**, 139–159 (1931).

CORROSION OF STEAM CONDENSATE LINES

LEO F. COLLINS*

This chapter deals with corrosion of the *internal* surfaces of steam condensate lines. Except in special cases, such lines are of low-carbon steel and operate at temperatures below 120° C (about 250° F). They are found in steam power plants, certain process industries, and in the steam-heating systems of buildings.

NATURE OF THE PROBLEMS PRESENTED

Corrosion of the internal surfaces of steam condensate lines presents two general types of problems: (1) those originated by plugging of small lines with insoluble deposits and (2) pitting or grooving of the pipe walls.

The insoluble materials originated by corrosion, which plug condensate lines,† are not always found at the spot where the metal was attacked but, frequently downstream in the system. Thus, although they are an indication of corrosion upstream, they do not indicate the rate of attack, since such deposits may or may not be an accumulation. Invariably, they consist of the oxides and hydrates of iron, and an analysis will, therefore, serve to differentiate between a deposit resulting from corrosion and extraneous insoluble matter which, at times, also causes plugging.

BASIC CAUSES FOR CORROSION

Dissolved oxygen and carbonic acid account for practically all the corrosion in condensate lines. Hydrogen sulfide and sulfur dioxide are known to have caused such trouble, but these gases are rarely encountered.

OXYGEN ATTACK

Corrosion accelerated by oxygen is typified by the presence of reaction products which, when removed, disclose a roughened surface. In some cases pits are formed which are partly or completely filled with reaction products. In other cases the entire surface may be uniformly covered with corrosion products. When an examination of a deposit shows it to be gelantinous in texture and of a tan-to-red color, pitting of the supporting metal is unlikely. When only the outer skin of a deposit is of this texture, and the remainder is green-to-black and more or less crystalline, active corrosion of the supporting metal may be expected.

Measurements made in the condensate lines of the steam heating systems of about twenty-five large office buildings[1] indicate that when the temperature is below about 70° C (160° F) and the pH of the condensate is 6 or higher, oxygen concentrations below 0.5 ppm cause negligible corrosion. In the pH range 6 to 8 and at oxygen concentrations within the range 0.5 to 4 ppm (approximately), the rate of attack is given by the equation:

$$R = 24(C - 0.4)^{0.9}$$

where R = average rate of penetration in mdd, C = oxygen concentration of the condensate in ppm.

---
* Chemical Engineer, Detroit, Mich.
† Occasionally, condensate lines are plugged with extraneous materials left in the system during construction.

[1] L. F. Collins and E. L. Henderson, *Heating, Piping, Air Conditioning*, **11** (September to December, 1939), **12** (January to May, 1940).

This equation does not take into consideration a pitting type of corrosion wherein the rate of penetration is invariably much higher than the average rate of penetration, nor does it take into consideration the accelerating effect of temperature. Laboratory tests[2] indicate that an increase in temperature from 60° C (140° F) to 90° C (194° F) about doubles the rate of corrosion by oxygen.

CARBONIC ACID ATTACK

Corrosion caused by dissolved carbon dioxide is characterized by clean surfaces. The attacked surfaces are uniformly thinned below the water line and, in many cases, exhibit a bright metallic sheen. Measurements made in laboratory experiments[3] showed the rate of attack to follow the equation:

$$R = 5.7W^{0.6}$$

where R = corrosion rate in mdd,
 W = weight of CO_2, contacting the test specimen, in pounds per hour $\times$ 100,000 (equals ppm of CO_2 in condensate $\times$ pounds of condensate flowing per hour $\times$ 0.1).

To what extent this equation holds in actual practice has not yet been definitely established. It does not take into consideration the accelerating effect of temperature. Laboratory tests[2] indicate that raising the temperature of a carbonic acid solution from 60° C (140° F) to 90° C (194° F) increases the rate of attack on low-carbon steel by a factor of 2.6.

MECHANISM OF SOLUTION OF DELETERIOUS GASES

The amount of oxygen and carbon dioxide which will dissolve in condensate under equilibrium conditions is directly proportional to their partial pressures. The solubility decreases with increase in condensate temperature. It is important to remember, however, that in steam-condensing equipment, it is not always possible to predict accurately when equilibrium has been attained. Experimentally[4,5] it has been demonstrated that in units operating at high condensing rates the ratio of non-condensable gases to steam is much higher in the vapor space of the unit than in the steam supply line. It has been shown, also, that in the vapor space of conventional types of equipment a gradient mixture of steam and non-condensable gases occurs, with the highest concentration of gas being at the vapor condensate interface. It is the partial pressure at the *interface* which determines the amount of gas dissolved in the condensate leaving the equipment.

These observations explain why in actual practice the condensate lines from condensers which operate at high condensing rates for long periods of time on badly contaminated steam are often seriously corroded. Water heaters, unit (space) heaters, blast coils, and condensers in certain process industries are most often so affected.

Condensers which operate at low condensing rates usually produce condensates whose gas concentration is less than in the incoming steam, owing to the escape of undissolved gas into the condensate line. In the condensate line, a progressive increase in the gas content of the condensate may occur as the condensate chills, and air may be sucked into the line, thus increasing the oxygen content of the condensate.

[2] G. T. Skaperdas and H. H. Uhlig, *Ind. Eng. Chem*, **34** (No. 6), 748 (1942).
[3] L. F. Collins, *Power Plant Eng.*, **48**, 88–89 (March, 1944).
[4] L. F. Collins, *Heating, Piping, Air Conditioning, Journal* Section, **17**, 36–41 (January, 1945).
[5] D. S. McKinney, J. J. McGovern, C. W. Young, and L. F. Collins, *Heating, Piping, Air Conditioning, Journal* Section, **17**, 97–104 (February, 1945).

When CO_2 is virtually the only gas present, there is a progressive increase in the iron content of the condensate as well as in the CO_2 content as progress is made downstream in the system. The iron content more than offsets the CO_2 content, as is attested by a progressive increase in the pH of the condensate. Thus, where ferrous lines are involved, the evidences of corrosion (thinning of the pipe walls) becomes less pronounced as progress is made downstream. When CO_2 is virtually the only gas present in the steam, but considerable inleakage of air occurs in the condensate lines, deposits of iron oxide are usually found immediately following the point of air infiltration.

In condensers which operate cyclically or under vacuum, such as in the steam-heating systems of many buildings, the purging action of the incoming air (through leaks or purposely admitted through vacuum breakers) or the suction of the vacuum pump removes considerable CO_2 from the system but simultaneously brings about a higher oxygen content in the condensate than otherwise would occur. As a result, such systems are characterized by oxygen corrosion, but even so corrosion troubles in such systems are not so frequent as in systems operated under positive pressure.

PREVENTION OF CORROSION

Because of an imperfect technology and the prohibitive cost of resistant materials, corrosion in steam condensate lines cannot be entirely prevented, but more can be done than is generally appreciated. At present four different expedients are usually considered when corrosion troubles present themselves:

1. Treatment of the boiler feed water so as to minimize the amount of deleterious gases entrained with the steam.
2. Improved designs of condensing equipment to minimize solution in the condensate of the deleterious gases entrained with the steam.
3. Chemical treatment of the condensate.
4. Use of corrosion-resistant metals or alloys.

BOILER FEED WATER TREATMENT

Almost complete elimination of oxygen from boiler feedwater and, therefore, from the steam developed can be accomplished with the better types of mechanical equipment now available. Where the small residuals are of concern, they are usually dissipated by chemical means. (See *Deactivation and Deaeration of Water*, p. 506.)

Where boilers of industrial proportion are required to use make-up water that is initially carbonate bearing, the production of a CO_2-free steam upon a commercial basis is an undertaking that should be attempted only when better than average talent is available.

Tests have indicated[6] that in low-pressure heating boilers, where the boiler input contains less than about 50 ppm of carbonate hardness, the CO_2 in the steam can be controlled by adding calcium hydroxide to the boiler. In Fig. 1 is shown the equilibrium conditions proposed for boilers operating at pressures up to about 5 psi gage pressure. This expedient is hardly practical in higher pressure boilers because of the possibilities of scale and sludge formations. In these boilers, the method usually used consists in (1) removing the alkaline earth salts, (2) subsequently acidulating to a residual carbonate alkalinity of 0 to 5 ppm, (3) followed by aeration at room temperature, and then (4) deaeration at temperatures near the atmospheric boiling point of water.

[6] F. N. Speller, *Trans. Am. Soc. Heating Ventilating Engrs.*, **34**, 195 (1928).

Design of Condensing Equipment

It has been demonstrated that in the design of water heaters and comparable types of condensing equipment, it is possible to shift the accumulation of non-condensable gases to a location away from the condensate level and, subsequently, vent these gases to the atmosphere. Both theory[7] and experimentation[8] indicate that venting

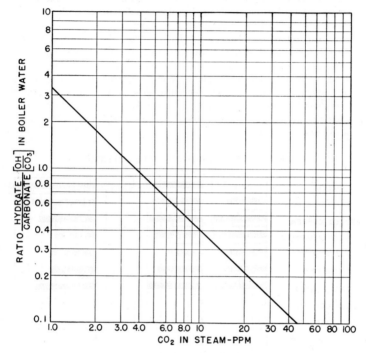

Fig. 1. Relation of Hydrate/Carbonate Content in "Hard" Boiler Water and CO_2 in Steam at Operating Pressure of about 5 Pounds Gage. All Analytical Values Are Ppm by Weight.

an amount of steam equal to about ½% of the total steam entering the condenser is the optimum vent rate for minimizing CO_2 in the condensate. How to dispose of the vented steam is a problem which must be solved upon the basis of local conditions.

Apparently, venting as a corrosion control expedient is of little practical value when the CO_2 content of the incoming steam is below about 5 ppm. When the steam contains more than 5 ppm, venting provides a means of producing a condensate containing a minimum of about 3 ppm. However, even as little as 3 ppm of dissolved CO_2 may produce active corrosion if large amounts of condensate are flowing.

Chemical Treatment of Condensate

Condensates containing comparatively large amounts of oil, such as are present in the exhausts from reciprocating engines, are practically non-corrosive. This is

[7] E. W. Guernsey, *Trans. Am. Soc. Heating Ventilating Engrs.*, **51** (1945).

[8] D. S. McKinney, J. J. McGovern, C. W. Young, and L. F. Collins, *Heating, Piping, Air Conditioning, Journal* Section, **17**, 97–104 (February, 1945).

ascribed to the protective film provided by the oil. Tests[9] made with oil intentionally added to condensate show that inadequate quantities may accelerate rather than decelerate corrosion on those surfaces not covered by the oil.

Sodium silicate added to CO_2-bearing condensate has been shown to decrease but not entirely prevent corrosive action. The protection afforded by silicate solutions may be due to the establishment of a protective film on the metal surface or to neutralization of CO_2 by the alkali in the silicate solution, or both.

It has been postulated that ammonia[10] cyclohexylamine,[11] ethylene diamine, and morpholine[12] will retard corrosion of condensate lines, but no direct measurements of their effectiveness have, as yet, been published. Where copper and its alloys are involved, the use of alkaline inhibitors is believed inadvisable. Where ferrous metals are involved, sodium hexametaphosphate is not regarded as an effective inhibitor where the condensate contains high concentrations of CO_2 and oxygen. Whether chemical treatment of steam or condensate is feasible must be determined not only upon the basis of the acuteness of corrosion troubles but also upon the uses to which the steam or condensate are put.

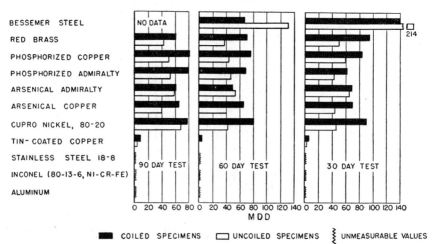

Fig. 2. Average Test Conditions. (Measured with N.D.H.A. corrosion tester.)

Average O_2 concentration of condensate	= 4.6 ppm (3.2 ml/l)
Average CO_2 concentration of condensate	= 14 ppm (9.8 ml/l)
Average pH value of condensate	= 5.5 ppm
Average temperature of condensate	= 57° C (135° F)
Average flow of condensate over test specimens	= 50 lb/hr

Use of Resistant Metals

For economic reasons the metals known to resist corrosion can seldom be used exclusively for condensate lines in any sizeable installation. Nevertheless, there are

---

[9] L. F. Collins and E. L. Henderson, *Heating, Piping, Air Conditioning*, **11** (September to December, 1939), **12** (January to May, 1940).

[10] F. N. Speller, *Proc. Natl. District Heating Assoc.*, **24**, 203 (1933).

[11] M. E. Dreyfus, *Heating and Ventilating*, **39**, 31–33 (June, 1942).

[12] H. L. Cox, U. S. Patent 1,903,287 (1933).

instances where limited use of the more costly corrosion-resistant materials is justified. The data in Fig. 2 are the results of tests[13] designed to compare the corrosion resistance of the more commonly used metals to attack by condensate containing oxygen and CO_2. The results shown for the copper alloys compare favorably with the findings of Harrison[14] for condensates containing ammonia.

According to the writer's experience, galvanic corrosion resulting from contact of dissimilar metals in a condensate line is not commonly serious. In contemplating the use of a resistant metal in a section of the condensate line, however, it should be remembered that corrosive attack may merely be transferred downstream in the system.

Paints or similar protective coatings have not so far proved satisfactory. Tests of cement-lined and vitreous-lined pipe have shown the linings to be readily dissolved by hot condensates.

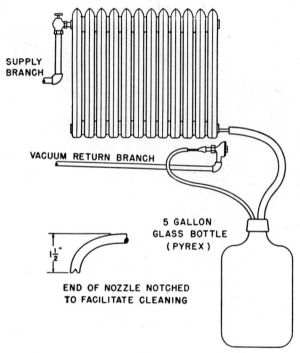

FIG. 3. Arrangement of Radiator-Cleaning Equipment.

REMOVAL OF INSOLUBLE DEPOSITS

When plugged lines are accessible, mechanical cleaning can usually be accomplished without special equipment. Where they are not accessible, it is sometimes possible to adopt the principle suggested by Winans[15] for cleaning radiators (Fig. 3).

In this scheme, a strong glass bottle, equipped with a two-hole rubber stopper,

---

[13] L. F. Collins, *Proc. Fourth Annual Water Conference*, Engineers' Society of Western Pennsylvania pp. 33–46 (1943).
[14] E. B. Harrison, *Power Plant Eng.*, **48**, 84–87 (June, 1944).
[15] G. D. Winans, *Heating, Piping, Air Conditioning*, **5**, 607–608 (December, 1933).

serves as a catch-all. One of the holes is linked by means of thick-walled rubber tubing to a vacuum. The other hole is linked, by means of thick-walled rubber tubing, to a short length of steel tubing (or ¼-in. pipe), one end of which is bent and notched.

In cleaning, water is introduced into the vessel and the nozzle is raked back and forth to loosen the deposit mechanically. The material so freed is sucked into the catch-all along with the water.

Chemical cleaning with an inhibited acid solution is also possible.[16,17] It should be attempted only by those experienced in the use of inhibitors and acids. There are commercial firms which specialize in this type of work.

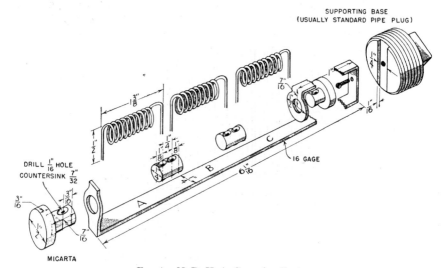

Fig. 4. N. D. H. A. Corrosion Tester.

MEASURING CORROSION

The most satisfactory means known to the writer for gaging the rate of corrosion in a condensate line is the tester shown in Fig. 4. It has been used in laboratory tests for gaging the potential corrosivity of steams and the effectiveness of inhibitors. Detailed instructions for its use may be obtained through the National District Heating Association.* To calculate corrosion rates, as average penetration in inches per year, two factors must be known:

1. The loss in weight of the specimen (i.e., helical coils).
2. The length of time the specimen was exposed to the corroding medium (normally, the number of days it was in test).

The formula used is:

$$P = \frac{D}{2T} \frac{W_1 - W_2}{W_1 + \sqrt{W_1 W_2}} \qquad (1)$$

* Office of the Secretary, 827 North Euclid Ave., Pittsburgh, Pa.

[16] F. N. Alquist, C. H. Groom, and G. F. Williams, *Trans. Am. Soc. Mech. Engrs.*, **65**, 719 (1943).
[17] F. N. Speller, E. L. Chappell, and R. P. Russell, *Trans. Am. Inst. Chem. Engrs.*, **19**, 165 (1927).

where P = ipy,
   D = initial diameter of wire in inches (usually 0.05 ± 0.001 in.),
   T = time of exposure in years,
   W_1 = initial weight of wire in grams,
   W_2 = final weight of wire in grams.

In the hands of a trained investigator who is aware of its limitations, the tester facilitates a more rapid measure of corrosion than otherwise is possible. Conceivably, there are times when the corrosion rates measured with this device are absolute values, but it is not always possible to determine when this is true. Until more correlating is done, attempts to translate quantitatively the rate of corrosion, as shown by the tester, into actual life of equipment are not justified.

CONDENSER CORROSION

C. A. Gleason*

Any complete discussion of corrosion as it affects condensers would of necessity require consideration of corrosive action on the water boxes, tube sheets, the shell, support plates, baffles, and the condenser tubes. Practically, however, severe or serious corrosion of condensers is usually limited to that which occurs at the inside or outside surface of the condenser tubes.

The large majority of condenser tubes in use at the present time is manufactured of copper or copper-base alloys, primarily because these materials exhibit excellent corrosion-resistant properties under the various conditions of service to which they are exposed. The high thermal conductivity and favorable physical properties of copper and copper alloys are also of importance in this application.

Table 1 gives the chemical requirements of the A. S. T. M. Specification B–111 for the common condenser tube alloys.

During their initial stages of service, condenser tubes usually undergo slight surface corrosion. The corrosion products and, in some cases, solids from the circulating media, may precipitate on the tube walls. The resulting coating or film will protect the metal against further corrosion provided that it is insoluble, impervious, continuous, and adherent. If such a film fails to form, or, if initially formed, it is dissolved, mechanically broken, or destroyed during subsequent service, corrosion of the metal will again proceed. This corrosion may be slight or severe, depending upon conditions of service and the inherent film-forming characteristics of the tube alloy.

Effect of Velocity, Impingement Attack

The impingement of rapidly moving water, particularly where air bubbles are present or where intermittent cavitation occurs, may result in breakdown of protective films and subsequent severe localized or general attack of tube surfaces. (See Fig. 9, p. 1107.) This action commonly occurs near tube inlet ends but may extend along the entire length. Abrasive solid matter (sand, etc.) carried in suspension in circulating water results in a similar action, although erosion by abrasive solids usually results in a uniform thinning, particularly on the bottom side of the tube. Many tube failures can be averted if the causes of impingement attack are understood and attention is given to changes in design or operation of condensers to minimize this source of trouble.

* Scovill Manufacturing Co., Waterbury, Conn.

TABLE 1. CHEMICAL REQUIREMENTS

Alloy	Cu,* %	Sn, %	Al, %	Ni† %	Pb max., %	Fe max., %	Zn, %	Mn, max., %	As, %	Sb, max., %	P, max., %
Muntz metal	59.0 to 63.0	.....	.....	.....	0.3	0.07	remainder	.....	.....	.....	.....
Admiralty metal:											
Type A	70.0 to 73.0	0.90 to 1.20	.....	.....	0.075	0.06	remainder	.....	.....	.....	.....
Type B	70.0 to 73.0	0.90 to 1.20	.....	.....	0.075	0.06	remainder	.....	0.10 max.	.....	.....
Type C	70.0 to 73.0	0.90 to 1.20	.....	.....	0.075	0.06	remainder	.....	.....	0.10	.....
Type D	70.0 to 73.0	0.90 to 1.20	.....	.....	0.075	0.06	remainder	.....	.....	.....	0.10
Red brass	84.0 min.	.....	.....	.....	0.075	0.06	remainder	.....	.....	.....	.....
Aluminum brass:‡											
Type A	76.0 min.	.....	1.75 to 2.50	.....	0.075	0.06	remainder	.....	.....	.....	.....
Type B	76.0 min.	.....	1.75 to 2.50	.....	0.075	0.06	remainder	.....	0.10 max.	.....	.....
Type C	76.0 min.	.....	1.75 to 2.50	.....	0.075	0.06	remainder	.....	.....	0.10	.....
Type D	76.0 min.	.....	1.75 to 2.50	.....	0.075	0.06	remainder	.....	.....	.....	0.10
Aluminum bronze	93.5 min.	.....	5.00 min.	.....	.....	.....	.....	.....	.....	.....	.....
70–30 copper-nickel	remainder	1.50 max.	.....	29.0 to 33.0	0.05	0.5	1.0 max.	1.0	.....	.....	.....
80–20 copper-nickel:											
Type A	remainder	1.0 max.	.....	19.0 to 23.0	0.05	0.5	1.0 max.	1.0	.....	.....	.....
Type B	remainder	1.0 max.	.....	19.0 to 23.0	0.05	0.5	3.0 to 6.0	1.0	.....	.....	.....
Copper	99.90 min.	.....	.....	.....	.....	.....	.....	.....	.....	.....	0.035
Arsenical copper	99.45 min.	.....	.....	.....	.....	.....	.....	.....	0.15 to 0.50	.....	.....

* Silver counting as copper.
† Cobalt counting as nickel.
‡ Total other impurities, 0.30 %, max.

CONDENSER CORROSION

Inlet-end erosion may be defeated temporarily by applying an appropriate paint or cement, and more permanently by the use of properly designed metallic inserts. The most effective preventive, however, is streamlined design of water boxes, injection nozzles and piping, avoiding abrupt, angular changes in direction, low-pressure pockets, obstructions to smooth flow, and any other feature which can cause local

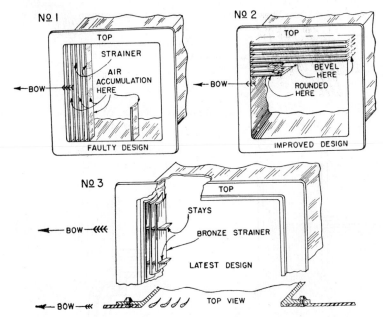

FIG. 1. Faulty and Improved Design of Injection Strainers for Use on Ships.

high velocities or turbulence of the circulating water.[1,2] The injection nozzle (the terminal portion of pump discharge pipe) should be gradually increased in area in smooth curves, avoiding sharp angles as it approaches the water box. The water box should be ample in depth and its roof should be vented to by-pass the air to some portion of the system where pressure is less than that of the water box (e.g., the discharge pipe).

Propeller-type pumps are preferable for circulating the water. Where suction types are used, they should be profiled to reduce the suction lift to a minimum in order to avoid cavitation; also the sump should be ample to prevent funneling of air. Funneling often can be eliminated by barriers in the vortices to interrupt the circular motion of the water. Also care must be exercised to seal pump glands with water of higher pressure than that in the pump chamber to insure that air cannot be drawn into the circulating water at this location.

Suggestions of design to avoid impingement attack are given in Figs. 1 to 9.

The fact that very slow rates of flow favor dezincification of brasses and that high

---

[1] A. J. German, "Condenser-Tube Life as Affected by Design and Mechanical Features of Operation," *Trans. Am. Soc. Mech. Engrs.*, **61**, 125–132 (1939).

[2] R. E. Dillon, G. C. Eaton, and H. Peters, "The Prevention of Failures of Surface-Condenser Tubes," *Trans. Am. Soc. Mech. Engrs.*, **59**, 147–150 (1937).

548　SPECIAL TOPICS IN CORROSION

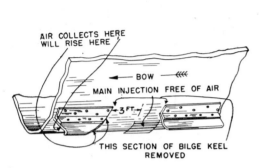

Fig. 2. Section of Bilge Keel Cut Away to Prevent Air Entrapment.

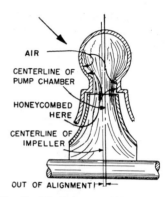

Fig. 3. Misalignment of Pump Impellers Increases Turbulence.

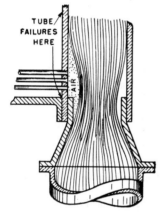

a. *Objectionable.* Restricted entrance due to shallow water box.

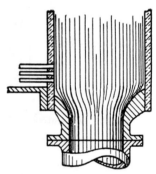

b. *Improved.* Air pocket eliminated by streamlined entrance and deepened water box.

Fig. 4. Designs of Injection Nozzle and Water Box.

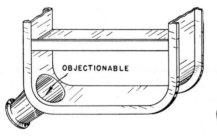

a. *Objectionable.* Sharp angles at junction of nozzle and water box.

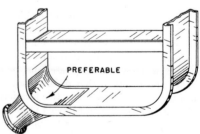

b. *Improved.* Gradually increased nozzle area at approach to water box.

Fig. 5. Designs of Entrance to Water Box.

rates of flow add to the aggressiveness of inlet-end corrosion-erosion seem to justify limitation of the velocity of water through the tubes of steam condensers to 5 to 7 ft per sec (1.5 to 2.1 meters per sec). Modifications of these limits are permissible, depending upon selection of material and appropriate stream lining of the water flow.

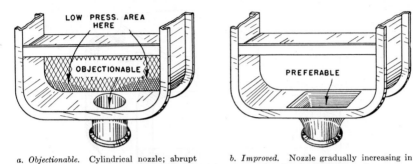

a. *Objectionable.* Cylindrical nozzle; abrupt increase in area.

b. *Improved.* Nozzle gradually increasing in area at approach to water box.

Fig. 6. Designs of Entrance to Water Box.

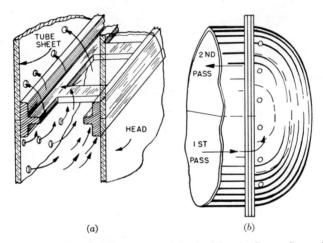

Fig. 7. (a) Reverse-Flow Water Box with Reinforcing Plate in Center Causes Turbulence and Air Separation in Water Entering Second Pass Tubes. (b) Well-Designed Reverse-Flow Water Box with Test Holes Drilled as Shown.

CONCENTRATION CELL CORROSION

Many cases of tube corrosion are experienced when an electrochemical action, generally termed *concentration cell corrosion* is set up.* Corrosion of this type usually results in intense local pitting of the tube surface. (See Fig. 8, p. 1106.) It is occasionally serious at or adjacent to baffles and tube sheets, either as a result of crevices which exist between tubes and tube sheet or baffles, or because of accumulation of corrosion products or foreign material at these locations. This type of attack may

* The action is probably caused largely by concentration differences of oxygen at one part compared with other parts of the tube. EDITOR.

550 SPECIAL TOPICS IN CORROSION

also be severe beneath non-protective film coatings such as corrosion products of sulfides.

Deposit attack or contact corrosion is a form of concentration cell action initiated by lodgment of sand, mud, cinders, stones, wood, shells, iron scale, etc., on the tube

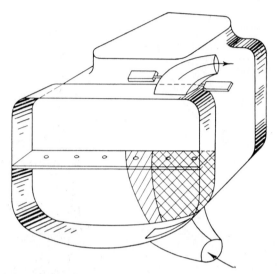

FIG. 8. Injection Nozzle Too Far from Tube Sheet. Air Vents in Division Plate Reduce the Harmful Effect.

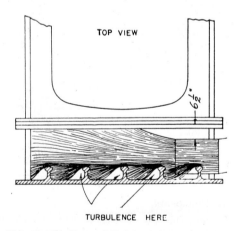

FIG. 9. Objectionable Internal Reinforcing Ribs in Inlet Water Box.

surface. Intense pitting of the tube wall may take place at or adjacent to such deposits. If the deposit is of a size and shape to form an obstruction, turbulence and air separation will set up erosion and corrosion at the obstruction and also downstream from it. Some tube alloys have better resistance to corrosion of the concentration cells type than others. There is no assurance, however, that any particular alloy will serve economically under conditions where tube failures of this type are experienced. On the other hand, such failures can be prevented or minimized by taking the following precautionary measures: (1) Avoid all cracks or crevices at the sheets, baffles, or other areas where seepage and trapping of liquid can occur. (2) Clean tubes frequently and thoroughly, avoiding any cleaning methods which may injure the tube surface or destroy protective films.* (3) Provide adequate screening of the circulating water to prevent entrance of harmful solids into the tubes.

Dezincification

Brasses containing less than 85% copper, in certain service conditions, are subject to a type of corrosion which results in the replacement of the brass by a porous, brittle copper deposit. This may occur locally in very small areas, in which case it is termed *plug-type dezincification* (Fig. 11B, p. 1108) or more uniformly over larger areas when it is described as *layer-type dezincification* (Fig. 11A, p. 1108). Conditions favoring dezincification of condenser tubes are:

1. Contact with slightly acid or alkaline waters not highly aerated.
2. Low rates of flow of the circulating media.
3. Relatively high tube wall temperature.
4. Permeable deposits or coatings over the tube surfaces.

In the past, dezincification has been a common cause of failure of brass condenser tubes, but more recently the addition of very small percentages of alloying elements such as antimony, arsenic, or phosphorus has been shown to retard or inhibit such action, particularly with Admiralty metal and aluminum brass. Cu-Zn alloys containing 85% or more of copper do not dezincify under most conditions of service. Results of laboratory dezincification tests on an inhibited and an uninhibited Admiralty metal are shown graphically in Figs. 10 and 11.

Effect of Temperature†

Figure 1, p. 72, shows the effect of temperature on the rate of pitting of Admiralty metal in 3% NaCl, cotton being used to simulate substances found in condenser tubes that are active in forming differential aeration cells. The rapid rise of corrosion with temperature as found in the laboratory is confirmed by the tendency of condensers to fail prematurely when operated at too high a temperature. Actually, seasonal variations in temperature of cooling waters are adequate to account for higher pitting rates of condenser tubes in summer. (See Fig. 2, p. 73.)

Hot Wall Effect.[3] The *hot wall effect* is usually characterized by deep local pitting on the water side of heater tubes, the rate of perforation increasing with an increase in tube wall temperature. Water heaters and evaporators with steam or flame on one side of the tube illustrate the most common equipment where this type of pitting occurs. Laboratory tests have shown that both ferrous and non-ferrous metals fail quite rapidly when exposed to 3% sodium chloride solutions when the hot wall effect is a factor.

The pitting takes place where bubbles of gas repeatedly separate from the water (usually at scratches or sharp indentations). Hot spots form on the heated metal surface at those points where the bubbles of gas cling to the metal, preventing cooling

---

* Soft rubber plugs forced through the tubes often suffice to remove organic slime deposits. Proper chlorination of cooling water will prevent formation of these slimes without harming the tube material.
† This discussion is by C. L. Bulow, Bridgeport Brass Co., Bridgeport, Conn.

[3] C. Benedicks, *Trans. Am. Inst. Mining Met. Engrs.*, **71**, 597–626 (1925).

of the metal surface. In non-ferrous tubes, deep pits form on the underside of horizontally held tubes, where the bubbles make contact. Frequently, localized corrosion is virtually absent on the topside of the tubes. The movement of gas bubbles up around the side of tubes sometimes produces a noticeable upward streaming in the

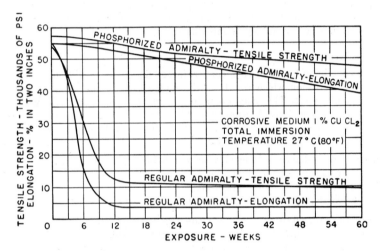

FIG. 10. Accelerated Corrosion Tests of Inhibited and Regular Admiralty Metal at Room Temperature.

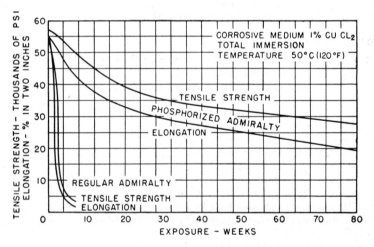

FIG. 11. Accelerated Corrosion Tests of Inhibited and Regular Admiralty Metal at 50° C (120° F).

corrosion products. As corrosion progresses, nodules of black or greenish-black corrosion products (essentially cupric oxide, CuO) form over deep pits, which are filled with bright, red, crystalline cuprous oxide (Cu_2O). This localized corrosive action proceeds at rates ranging from 0.025 ipy up to ten or more times this value, depending upon the alloy, temperature, and composition of the corrosive solution.

The rate of penetration may be ten to two hundred times the normal uniform corrosion rate.

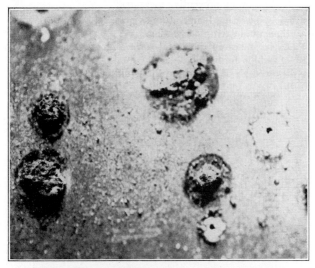

FIG. 12. Hot Wall Effect. ×8. (C. L. Bulow.)

Nodules of corrosion product covering pits formed on the water side of a copper water heater tube. Soft non-scale forming water.

FIG. 13. Side View of Same Nodules of Corrosion Product Associated with the Hot Wall Effect shown in Fig. 12. (C. L. Bulow.)

There are several important factors which together accelerate corrosion at the point of bubble contact:

1. Local high temperature.
2. High concentration of corrosive agents separating out where the bubbles form.
3. Available depolarizing gases.

In addition, the gas composition of the bubbles also probably plays a part.

In Figs. 12, 13, and 14 are shown several views of the nodules which formed over pits produced by the hot wall effect in copper tubing. The mechanism of this corrosive action suggests the following preventive measures:

1. Employ as low operating temperature as possible.
2. Avoid continued uninterrupted heating.
3. Produce new scratches on the metal surface or wipe the surface so as to shift the point of bubble formation.
4. Remove dissolved air or gases from the liquid being heated.

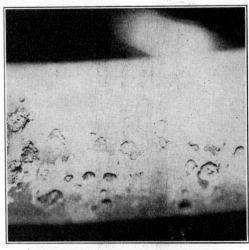

FIG. 14. Nodules and Pits Formed on Copper Tubing in 3% NaCl by Hot Wall Effect. (C. L. Bulow.)

Cold Wall Effect. Occasionally, condensed water droplets absorb small amounts of corrosive materials such as hydrochloric acid, ammonium hydroxide, or sulfur

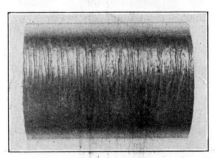

FIG. 15. Grooves Produced by Corrosive Condensate Flowing Down the Sides of a Cold Metal Tube Held in a Horizontal Position. (C. L. Bulow.)

dioxide, which may lead to the formation of deep corrosion grooves in the side of the tube as illustrated in Fig. 15. This occurs where the droplets of water form repeatedly at the same spot and roll down the same area on the cold tube surface. This may also take place where two immiscible liquids condense side by side. Sometimes, considerable longitudinal grooving or pitting may occur on the bottom side of the tube where the droplets of corrosive liquid may hang for some time before dropping off. Usually considerably more thinning of the tube wall occurs on the bottom side of the tube as compared with the top side.

Effect of Stresses

Stresses which may contribute to tube failure may be residual in the metal (and not removed by proper heat treatment), or applied tensile stress, or cyclic stresses. Stress cracking of brass tubes is the result of action by specific reagents, particularly by mercury or ammonia and their compounds on statically stressed material. The term *season cracking** sometimes is applied to failures of this type when the stresses are residual in the metal. Cracking predominantly follows the grain boundaries.

Tube specifications usually include a mercurous nitrate test (p. 1013). This test establishes that internal stresses from manufacturing operations sufficient to cause season cracking have been adequately removed by annealing. Tubes shipped from the mill by and large are free from internal stress, but care must be exercised not to induce excessive strains during installation. The tube alloys which are least susceptible to stress corrosion cracking are copper-nickel and copper.

Corrosion fatigue of tubes may occur from the simultaneous action of corrosion and *cyclic stresses* acting on the metal. In this connection, it should be noted that cyclic stress tends to accelerate corrosion of the pitting type, the latter in turn decreasing the fatigue limit very markedly. Corrosion fatigue cracks are transcrystalline.

The cyclic stresses contributing to fatigue failures usually result from vibration of the tubes which can be minimized satisfactorily in many cases (1) by introducing a bias stress, accomplished by elevating the intermediate support plates, (2) by forcing slats between tube rows, and (3) by increasing the number of support plates.

Galvanic Action

The ordinary combinations of condenser materials, i.e., iron or steel shells and tube sheets with copper alloy tubes, usually do not cause serious trouble of a galvanic nature. Nevertheless, where cost is not a consideration, copper alloy tube sheets are preferred because of their better corrosion resistance. Perhaps one of the most important controlling factors in corrosion of this type is the ratio of the cathode area to the anode area. In a good conducting solution the larger the cathode area (i.e. copper alloy), the greater is the corrosion at the anode (iron). (Discussed in *Fundamental Behavior of Galvanic Couples,* p. 481.) Since the area of tube sheets is large relative to tube ends, it is permissible to use steel tube sheets with nonferrous tubes.

SELECTION OF MATERIAL

The nominal composition and physical properties of the generally used condenser tube materials are listed in Table 2. A selection involves a survey of all the known service conditions, consideration of the alloy previously used, and the type or form of corrosion previously experienced in the unit in question or in similar units.

---

* The term *stress corrosion cracking* has recently been directed to situations where the stress is either residual or applied, and hence includes the type of failure represented by *season cracking*. The term does *not* apply to cracking by action of molten metals such as mercury, since it is thought that corrosion (chemical reaction) is not involved. Thus mercury causes stress cracking of brass, but ammonia produces stress corrosion cracking. Editor.

SPECIAL TOPICS IN CORROSION

TABLE 2. PHYSICAL AND MECHANICAL PROPERTIES

Material	Composition, %	Tensile Strength, psi		Yield Point Soft, psi	% Elong. 2 in., Soft	Endurance Limit,† psi	Modulus of Elasticity	Melting Point	
		Soft	Hard					°C	°F
Deoxidized Copper	Cu 99.9+	32,000	55,000	10,000	50	10,000	15 × 10⁶	1082	1981
Arsenical Copper	0.15 to 0.50 As	35,000	60,000	12,000	40	10,000	15 × 10⁶	1082	1981
Red Brass	Cu 85 Zn 15	40,000	75,000	18,000	55	15,000	15 × 10⁶	1021	1868
Admiralty	Cu 70 Sn 1 Zn 29	50,000	90,000	20,000	60	15,000	15 × 10⁶	935	1715
Inhibited Admiralty	Cu 70 Sn 1 Zn 29 As or P or Sb–0.10 max.								
Inhibited Aluminum Brass	Cu 76 Al 2 Zn 22 As or P or Sb 0.10 max.	52,000	100,000	15,000	50	15,000	15 × 10⁶	971	1780
Special Muntz	Cu 62 Zn 38	45,000	85,000	18,000	50	20,000	15 × 10⁶	921	1688
Naval Brass	Cu 60 Sn ¾ Zn 39¼	55,000	95,000	23,000	40	18,000	15 × 10⁶	Approx. 885	Approx. 1625
Cupro Nickel	Cu 80 Ni 20	50,000	90,000	23,500	50	20,000	21 × 10⁶	1200	2192
Cupro Nickel	Cu 70 Ni 30	55,000	95,000	23,500	50	30,000	26 × 10⁶	1225	2237

* Variations from these values must be expected in practice.
† From "The Resistance of Copper and Its Alloys to Repeated Stress," a correlated abstract by H. W. Gillett, *Metals and Alloys* **3**, 200, 236, 257, 275 (1932). (Ideal condition determinations.)

MUNTZ METAL

Use of this alloy at the present time is confined chiefly to steam condensers, operating at low temperature, and at inland stations with fresh circulating water from rivers, lakes, or wells. Muntz metal has good resistance to hydrogen sulfide and other active sulfur compounds, but is seldom used in applications where corrosion by such compounds is severe, primarily because other factors such as high temperature, high velocities, and badly contaminated waters are usually active at the same time. Failure of Muntz metal tubes occurs usually as a result of gradual dezincification of the "layer" type.

ADMIRALTY METAL

This alloy has excellent corrosion-resistant properties and is used widely in tube form in such diverse services as steam condensers using fresh, salt, brackish, or acid

OF CONDENSER-TUBE MATERIALS*

Density, 20° C‡			(Range, 77°–572° F) Linear Expansion, ° F‡	Electrical Conductivity (Copper, Soft = 100%)	Thermal Conductivity, ‖ Btu/(hr)(sqft)(° F/ft)	Gram cal/ (sec)(sq cm) (° C/cm)	No Appreciable Change in Physical Properties¶ up to	
Lb per cu in.	Lb per cu ft	Grams per cc					° C	° F
0.323	556.4	8.91	0.0000098	80 to 90	186 to 20	0.77 to 0.84	200	390
0.323	556.4	8.91	0.0000098	50 to 30	121 to 72	0.50 to 0.30	250	480
0.316	544.3	8.73	0.0000104	37	89	0.37	....	....
0.308	534.0	8.55	0.0000112	25	65	0.27	250	480
0.301	520.1	8.33	0.0000102	23	55	0.23	150	300
0.304	525.3	8.41	0.0000115	26	70	0.29	200	390
0.303	523.6	8.39	0.0000119	25	68	0.28	200	390
0.323	558.1	8.94	0.0000091	6	22	0.089	300	570
0.323	558.1	8.94	0.0000090	5	17	0.069	300	570

‡ From *Scientific Paper* 410, U. S. Bureau of Standards.
‖ From "Thermal Conductivity of Copper Alloys. I. Cu-Zn Alloys," C. S. Smith, *Trans. Am. Inst. Mining Met. Engrs.*, Institute of Metals Div., **89**, 84, (1930).
¶ Paper by W. B. Price, discussion W. H. Bassett, 1931 Vol., A. S. T. M. and A. S. M. E., Joint Symposium, Short Time Test Data.

mine water for circulating purposes; as heat exchangers and coolers in oil refineries where corrosion from sulfur compounds, acids, and contaminated waters may be very severe; and as feed water heaters and heat exchanger equipment in industrial processes. Admiralty metal tubes are often used in apparatus operating at temperatures of 200° C (400° F) or higher, although severe dezincification of the alloy may be experienced under certain service conditions at elevated temperatures.

Inhibited Admiralty Metal. The addition of a very small percentage of phosphorus, antimony, or arsenic to Admiralty metal increases the dezincification resistance of the alloy to a marked degree. This so-called inhibited Admiralty metal was developed, therefore, particularly for installations where severe dezincification of Admiralty metal has been experienced or is anticipated.

Red Brass

Red brass is sufficiently high in copper so that it does not dezincify under normal service conditions; it is also not so susceptible to stress corrosion cracking as brasses of lower copper content. It is a very serviceable alloy for many heat exchanger installations. Red brass tubes have been used chiefly for service in contact with fresh waters. This alloy is not generally recommended for use where corrosion by hydrogen sulfide or other active sulfur compounds is liable to be severe, nor in brakish or salt waters.

Deoxidized Copper

Tubes of deoxidized copper are limited to applications where the high purity of the material is essential, or where good thermal and electrical conductivity are necessary. It has excellent resistance to practically all types of fresh water and to many chemicals. It is not usually serviceable in contact with salt water or active sulfur compounds. In the hard-drawn temper, copper has a tensile strength of approximately 60,000 psi but it does not retain this strength at temperatures of 200° C (390° F) or higher.

Arsenical Copper

The alloying of a small percentage (0.15 to 0.50) of arsenic with copper reduces the electrical and thermal conductivity of the copper 25 to 50%. Compared with deoxidized copper, arsenical copper has (1) a higher softening or annealing temperature (300° C vs. 200°C), (2) increased strength, hardness, and stiffness of cold-drawn material with no impairment of ductility, and (3) higher fatigue limit. Arsenical copper finds its principal use in condensers and heat exchangers employing fresh circulating water. It is not recommended for service in contact with (1) hydrogen sulfide or other active sulfur compounds, (2) media carrying an appreciable percentage of organic acids, particularly in the presence of air (oxygen), (3) acid solutions of an oxidizing nature, such as acid mine waters, and (4) salt or brackish waters.

Aluminum Brass

The addition of approximately 2% of aluminum and an inhibitor to brass of 76% copper content improves the resistance of the resulting alloy very materially, particularly to corrosion by salt and brackish water* and to combined erosion and corrosion in salt water service. The excellent corrosion resistance of this alloy probably depends upon the resistant, self-healing nature of the film which develops over the surface of the tube. Tubes of this alloy are very serviceable in installations where tubes of other brass alloys might suffer rapid impingement attack. Aluminum brass tubes have been used quite extensively during the past 10 years in marine condensers and in steam condensers of tidewater stations where relatively high velocities of the circulating water, together with turbulence and air bubble impingement, have a severe local wearing action on Admiralty metal and other brass tubes.

70-30 Copper-Nickel

Tubes of this alloy are used in heat exchanger equipment under conditions where severe wear and corrosion have been experienced, particularly at elevated temperatures and at high velocity of the circulating media. Cu-Ni tubes have given excellent service, for example, in ship condensers of the U. S. and British Navy, in steam

---

* However, aluminum brass is somewhat susceptible to pitting in sea water. (See p. 393.) EDITOR.

condensers of tidewater stations where brackish or salt circulating waters are very corrosive, and in various oil-refining units operating at relatively high temperatures and in contact with corrosive materials. Cu-Ni of 70-30 composition combines excellent corrosion-resistant properties with very favorable physical characteristics, such as high strength and good ductility. The strength of the alloy is retained at relatively high temperatures. The thermal conductivity is approximately one fourth that of Admiralty metal; actually, however, the overall heat transfer coefficient of a new unit is reduced only slightly when Cu-Ni tubes are used in place of Admiralty metal. Some designers of heat transfer equipment allow 5% more tube surface area when Cu-Ni replaces Admiralty or other brass tubes. Cu-Ni tubes are not subject to stress corrosion cracking, so they may be used with safety either in the finish annealed or hard-drawn condition. The preference at the present time is for finish annealed tubes.

80-20 COPPER-NICKEL

The Cu-Ni alloy containing 20% nickel has similar physical and chemical properties to the 70-30 Cu-Ni alloy, and is, therefore, used in the same type of service. The corrosion-resistant properties of the alloy are not equal to those of the higher nickel alloy, so that the 70-30 alloy is still generally used under the most severe conditions.

The author gratefully acknowledges his indebtedness for metallurgical and chemical data to Messrs. W. B. Price and F. M. Barry, and for Figs. 1 to 9 to Mr. A. J. German.

CORROSION BY LUBRICANTS

C. M. LOANE*

INTRODUCTION

Metals are not corroded by the hydrocarbon components of lubricants, although corrosion does occur under certain conditions as the result of the presence of impurities or additives in lubricants, and as a result of the development of oil oxidation products. Certain bearing metals are more susceptible to corrosive attack than others. This is particularly true of Cu-Pb and cadmium-base bearing metals which are fairly readily attacked by oil oxidation products. Copper and copper alloys may be attacked, not only by oxidized oil, but also by the active sulfur compounds which may be present. Steel and iron parts are seldom corroded by lubricants. When such corrosion does occur, it is usually under very limited conditions involving the use of additive-containing oils.

Although lubricants may tend to corrode working parts under some conditions, they afford considerable protection against corrosion caused by the presence of moisture, and, where necessary, a high degree of anti-rust protection can be achieved by the use of additives.

CORROSION OF BEARINGS

GENERAL ASPECTS

Bearing corrosion became an important problem to the petroleum and automotive industries during the period 1934 to 1935, at which time the use of Cu-Pb, Cd-Ag,

* Standard Oil Co. (Ind.), Whiting, Ind.

and Cd-Ni bearings instead of the previously used babbitt bearings became fairly widespread. The critical aspect of the situation was overcome by changes in engine design and improved lubricants. The problem of bearing corrosion has persisted, however, and today is one of the important factors in the evaluation of lubricants.

Cu-Pb and cadmium-base bearings have greater mechanical strength than babbitt bearings, particularly at elevated temperatures, but unfortunately they are much more subject to chemical attack by oxidation products of the oil. The various types of bearings are classified, on their resistance to corrosion,[1] as:

1. High resistance — tin-base babbitt.
2. High-to-moderate resistance — lead-base babbitt; alkali-hardened lead; As-Pb base babbitt; Ag-Pb base babbitt.
3. Low resistance — Cd-Ag; Cd-Ni; Cu-Pb.

Cu-Pb bearings have been very widely used, both because of their cheapness and their relatively great mechanical strength. A typical analysis of Cu-Pb bearing metal shows 65% copper and 35% lead. It is not a true alloy, but consists of lead distributed through a dendritic copper matrix. Cu-Pb bearings are readily corroded by oil oxidation products, which preferentially attack the lead, leaving a porous copper structure. A bearing weakened by lead corrosion may lose whole sections of the bearing metal, which may tear away from the steel support. A typical example of lead corrosion is shown by the photomicrograph of Fig. 1.

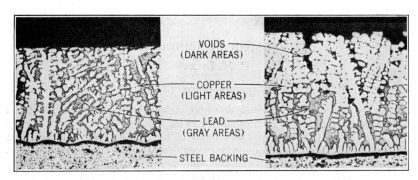

Fig. 1. Photomicrographs of Cross Section of Copper-Lead Bearings. Bearings from G. M. Diesel Tests after 500 Hours Operation. ×100. (Courtesy of Research Laboratories Division, General Motors Corporation.)

Left. Non-corroded bearing. Note lead present right up to the surface of the bearing. *Right. Corroded bearing.* Note voids where lead has been removed to a depth approximating half the thickness of the bearing.

Under exceptional conditions, corrosion of copper may occur, generally through formation and tearing off of CuS. However, this is the exception, and a darkened or even a black bearing ordinarily indicates a certain measure of protection against corrosion.

The phenomenon of bearing corrosion in service is so complicated that simple beaker tests have limited significance in relation to engine performance, and even full-scale engine tests must be interpreted with caution in relation to service performance.

[1] L. M. Tichvinsky, *Trans. Soc. Automotive Engrs.*, **51**, 69 (1943).

FACTORS AFFECTING THE CORROSION OF BEARINGS

Bearing corrosion is closely related to oil oxidation, since Cu-Pb and cadmium-base bearings are readily attacked by oil oxidation products. The major factors which influence oil oxidation, and consequently corrosion, can be listed as follows:

1. Temperature.
2. Catalysts.
3. Oxidation characteristics of the lubricant.

Other variables which also have some effect are, briefly:

4. The availability of oxygen in the crankcase and efficiency of crankcase ventilation — only a limited amount of information is available.
5. The structure of the bearing — discussed in a later section.
6. The load carried by the bearing — in general, bearing corrosion is more extensive when higher loads are carried. Even limiting consideration to one bearing, the load is by no means distributed evenly. And the observation has been made that, under ideal conditions, "the shape and depth of the corrosion pattern may approximately parallel the hydrodynamic film pressure diagram."[2]
7. Mechanical irregularities, such as misfit bearings, tend to accelerate corrosion because of local high temperatures. Abnormal conditions resulting in decreased oil flow have a similar effect.
8. Time of operation — obviously the longer the engine is run between oil drains, the more likely corrosion will occur. However, the rate of corrosion, after corrosion has started, may increase, decrease, or remain constant with time, depending on both the nature of the oil and the type of service.
9. Characteristics of the fuel — fuels and fuel oxidation products which find their way into the crankcase may affect bearing corrosion. Thus the decomposition products of lead tetraethyl may either accelerate or inhibit corrosion, depending on the type of lubricating oil used. High-sulfur fuels appear to afford some protection against bearing corrosion, and even the characteristics of the fuel hydrocarbons themselves appear to affect corrosion in accelerated engine tests.

TABLE 1. EFFECT OF CRANKCASE TEMPERATURE ON BEARING CORROSION

Chevrolet 36-Hour Tests*

Oil	Cu-Pb Bearing (Connecting Rod) Corrosion Loss in Grams per Half Bearing†		
	130° C (266° F) Oil Sump	138° C (280° F) Oil Sump	143° C (289° F) Oil Sump
No. 1. Experimental, detergent-inhibitor-type, S. A. E. 30 grade.	0.07	0.68	....
No. 2. Experimental, detergent-inhibitor-type, S. A. E. 30 grade.	....	0.08	0.55

* See *Tests for Corrosion by Lubricants*, p. 1034.
† One gram per half bearing = 3600 mg/sq dm.

Temperature. The rate of bearing corrosion is especially sensitive to crankcase temperature changes in the range of 100° to 160° C (210° to 320° F). An increase in

---

[2] A. J. Assessor (Federal-Mogul Corp.), private communication.

oil temperature as small as 5° C (9° F) can cause a change from no corrosion to extensive corrosion, other test conditions remaining constant.

This behavior is illustrated by the following laboratory beaker and engine tests:

Type Test	Temp. Range Covered, ° C	Summarized Observations	For Data, refer to:
Shell thrust bearing corrosion test*	107–160 (225°–320° F)	Six oils change from non-corrosive to corrosive when the temp. is increased 10° C.	Fig. 2
Chevrolet 36-hour test†	130–143 (265°–290° F)	In tests on two oils, an increase of 5°–8° C causes about a tenfold increase in corrosion.	Table 1
G. M. C. Model 71 Diesel test‡	110–127 (230°–260° F)	With five out of five oils, an increase of 17° C causes more corrosion than a fivefold increase in operating time.	....

* S. K. Talley, R. G. Larsen, W. A. Webb, American Chemical Society Petroleum Division, *Preprints*, p. 173, September, 1941.
† Unpublished data, Standard Oil Co. (Ind.).
‡ L. Raymond, *Trans. Soc. Automotive Engrs.*, **50**, 533 (1942).

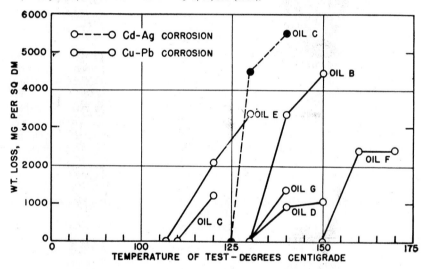

FIG. 2. Shell Thrust Bearing Corrosion Test. Effect of Temperature on Corrosion. (Test conditions: 20-hour run, 125 psi thrust, 2400 rpm.)

Fortunately, oil temperatures encountered in service (for appreciable periods of time) are not so high as those involved in accelerated tests. Consequently, it is only under unusual operating conditions that rapid bearing corrosion occurs.

Catalysts. The enormous effect of catalysts on lube oil oxidation stability, and consequently on bearing corrosion, is indicated by the observation[3] that the stability of an oil (as measured by rate of O_2 absorption) may be decreased to 1% of its original value after less than 1000 miles of service in an engine. The decrease in stability is largely due to the acquisition of a small concentration of suspended solids

[3] R. G. Larsen and F. A. Armfield, *Ind. Eng. Chem.*, **35**, 581 (1943).

or crankcase catalyst. The crankcase catalyst obtained from service where leaded fuels are used is usually more active catalytically than that obtained where clear fuels are used. The most active constituents of this catalyst on "undoped" lube oils are the ferric halides from the reaction of the halides in ethyl fluid with the metals in the engine. The most active constituents, as concerns acceleration of bearing corrosion, for some additive type oils are the inorganic lead compounds, particularly PbO from the decomposition of leaded fuel.[4] The effect of ethyl fluid decomposition products as catalysts of bearing corrosion is highly specific, and actually some oils are less corrosive when leaded fuels are used (Table 2).

TABLE 2. EFFECT OF CATALYSTS ON Cu-Pb BEARING CORROSION
Chevrolet Tests*

Oil	Engine Cleanliness	Fuel	Cu-Pb Bearing (Connecting Rod) Corrosion Loss in Grams per Half Bearing†
	A. Accelerating Effect of Leaded Fuel		
Detergent-type, S. A. E. 20 grade	Chemically clean‡	Unleaded	0.13 gram in 60 hours
Same	" "	3 ml $PbEt_4$/gal	1.68 grams in 60 hours
	B. Inhibiting Effect of Leaded Fuel		
Detergent-inhibitor type, S. A. E. 20 grade	Chemically clean‡	Unleaded	0.80 gram in 60 hours
Same	" "	3 ml $PbEt_4$/gal	0.18 gram in 60 hours
	C. Accelerating Effect of Clean Metal Surfaces		
Detergent-type, S. A. E. 20 grade	Chemically clean‡	3 ml $PbEt_4$/gal	1.15 grams in 15 hours
Same	No cleaning other than flushing with oil	3 ml $PbEt_4$/gal	0.28 gram in 30 hours

* Very similar in test conditions to the standard 36 hour Chevrolet test (see p. 1034).
† One gram per half bearing = 3600 mg/sq dm.
‡ Engine parts boiled in alkaline cleaner.

In addition to crankcase catalyst, the massive metal surfaces in an engine (principally Fe and Cu) exert a pronounced catalytic action on oil oxidation and bearing corrosion. This catalytic effect is more severe when the engine has been thoroughly cleaned before use (Table 2).

Oxidation Characteristics of Lubricating Oil. Lubricants themselves can vary perhaps a hundredfold in their bearing corrosion tendencies. Among uncompounded oils, the highly solvent-extracted paraffinic oils are most likely to corrode bearings. Naphthenic-type oils, especially those high in sulfur content, are the least likely of "undoped" oils to corrode bearings.

The addition of detergents* (excepting some containing divalent sulfur) to oils almost invariably increases the tendency to corrode bearings. However, practically all lubricating oils, either straight or compounded, can be brought to a satisfactory level of bearing corrosion performance by the addition of corrosion inhibitors.

* Detergents are metal salts of organic acids or phenols added to lubricants to prevent deposition of varnish and sludge on engine parts.

[4] C. M. Loane and J. W. Gaynor, *Ind. Eng. Chem.*, Anal. Ed., **17**, 89 (1945).

Detection of Corrosion

It is often difficult to ascertain whether corrosion of bearings has occurred since chemical attack may be difficult to distinguish from fatigue, erosion, or other causes of bearing losses. It is necessary to examine a carefully prepared cross section of the bearing under the microscope. The procedure for preparing Cu-Pb bearings involves alternately polishing and etching with acid to bring the lead and copper to a common plane.[5]

Some short-cut methods involve (1) analysis of the used oil for oxidation products and bearing metals and (2) insertion of a lead-plated strip into the crankcase during operation of the engine and noting the extent of lead removal.[6] Although these methods are of some value in indicating the condition of the oil, and have the practical advantage that they can be applied without dismantling the engine, they are only indications and do not constitute proof that bearing corrosion has occurred or will occur.

Mechanism of Bearing Corrosion and Corrosion Prevention

Only recently has any light been shed on the mechanism of bearing corrosion. It has been generally recognized that the acidity (from oxidation) of an oil, although concerned in some indirect manner, by no means controls the rate of corrosion. Denison[7] demonstrated very convincingly that the rate of corrosion is governed by the concentration of organic peroxides present. His data showed approximately that

$$\frac{d \text{ (corrosion loss)}}{dt} = -k \text{ (conc. of peroxide)}$$

Corrosion is postulated to occur as a result of the following reactions:

$$M + AO_2 \longrightarrow AO + MO$$
$$MO + 2HA \longrightarrow MA_2 + H_2O$$

where M = metal; AO_2 = peroxide; HA = organic acid.

Lead oxide was isolated as an intermediate in tests on the corrosion of lead. The shapes of the peroxide-time and the corrosion rate-time curves are strikingly similar. However, the same relation between peroxide concentration and corrosion rate does not hold for widely different oils, since peroxides from different sources vary considerably in activity.

A study[8] of the corrosion of lead and cadmium in hydrocarbon solvents containing added fatty acids and (1) molecular oxygen and (2) peroxides led to the following:

1. Confirmation of Denison's conclusions that the metal oxide is an intermediate in the corrosion process, that the corrosion rate is largely independent of acid concentration, and that corrosion of bearing metals by acidic oils above 100° C (210° F) is caused principally by organic peroxides.

2. The observation that at temperatures lower than 100° C (210° F) corrosion by acidic oils may be caused by dissolved oxygen and that the limiting rate of corrosion varies directly with oxygen pressure and may be represented as a function of temperature by the Arrhenius equation. (See footnote, p. 16.)

3. The observation that the rate of corrosion is mainly controlled by the rate of diffusion of oxidizing agent into the metal surface and by the chemical rate of oxidation of the metal.

[5] H. L. Grange, *Metal Progress*, **38**, 674 (November, 1940).
[6] R. G. Larsen, F. A. Armfield, and L. D. Grenot, *Ind. Eng. Chem.*, Anal. Ed., **17**, 19 (1945).
[7] G. H. Denison, Jr., *Ind. Eng. Chem.*, **36**, 477 (1944).
[8] C. F. Prutton, D. R. Frey, D. Turnbull, and G. Dlouhy, *Ind. Eng. Chem.*, **37**, 90 (1945).

Later work[9] on the corrosion of lead in hydrocarbon media containing organic acids showed that organic oxidizing agents other than peroxides or oxygen may be effective in causing corrosion. Oxynitrogen compounds, diacetyl, and quinone—compounds that could possibly occur in crankcase oils in service—also promote lead corrosion. Although lead oxide or hydroxide may be formed as an intermediate when hydroperoxides or oxygen are the oxidizing agent, other types of peroxides and non-peroxidic oxidizing agents probably form no metal oxide or hydroxide intermediate in the process of corrosion. In service, hydroperoxides and non-peroxidic oxidizing agents are more likely to be damaging than highly active peroxides since the latter are more rapidly destroyed in contact with heated crankcase oil.

Although the above work, concerned with the mechanism of corrosion in simplified systems, is probably applicable to some extent to corrosion in engines under service conditions, it is again emphasized that generally the rate of oxidation of the oil controls the rate of bearing corrosion in an engine.

Additives used to prevent corrosion can work in a variety of ways. Probably their chief action is to destroy organic peroxides and inhibit oil oxidation. Some inhibitors are primarily metal deactivators (i.e., they neutralize the catalytic effect of metals) in addition to being oxidation inhibitors. For example, some of the aromatic amines are powerful deactivators of iron; phosphites show a specific effect for copper; and sulfur compounds are effective lead deactivators. In addition, some additives, particularly sulfur compounds, form a film on the bearing, protecting it from contact with corrosive oil oxidation products. This protective film formation is also characteristic of the "undoped," lightly refined, naphthenic-type oils previously mentioned as being non-corrosive.

Minimizing Bearing Corrosion

Improvements in Bearings. Modern Cu-Pb bearings are much more resistant to corrosion than those originally used. Changing from coarse to fine structure bearings (see Fig. 3) has a markedly beneficial effect on bearing corrosion. This effect is brought out clearly in the Diesel tests[10] shown in Table 3.

Table 3. COMPARISON OF COARSE AND FINE STRUCTURE Cu-Pb BEARING CORROSION
Diesel Engine Tests*

	Cu-Pb Bearing (Connecting Rod) Corrosion Loss in Grams per Half Bearing†	
	Coarse Structure	Fine Structure
Oil 1	0.837	0.046
Oil 2	0.383	0.088
Oil 3	5.755	2.740
Oil 4	1.650	1.037
Oil 5	3.518	0.268
Oil 6	5.084	2.661
Oil 7	8.596	3.061

* Similar to G. M. C. Diesel, Model 71, 500-hour test (see *Tests for Corrosion by Lubricants*, p. 1034), but run for 100 hours at 127° C (260° F) oil sump temperature.
† One gram per half-bearing = 2400 mg/sq dm.

[9] C. F. Prutton, D. Turnbull, D. R. Frey, *Ind. Eng. Chem.*, **37**, 917 (1945).
[10] L. Raymond, *Trans. Am. Soc. Automotive Engrs.*, **50**, 533 (1942).

Cu-Pb and other type bearings have been rendered more resistant to corrosion by coating with Pb-In,[10,11] Pb-Sn, or Pb-Sb approximately 0.013–0.025 mm (3.3–6.4 mils) thick. The coatings are made by applying indium, tin, or lead electrolytically and subjecting the bearing to a heat-treating process to diffuse the inhibiting metal into the lead.

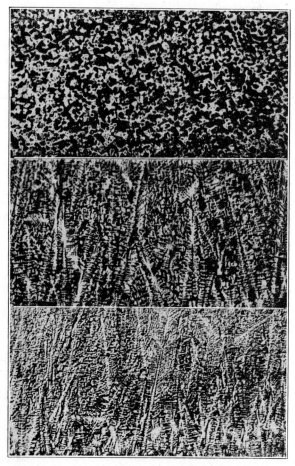

FIG. 3. Photomicrographs of Cross Sections of Copper-Lead Bearings. Typical Bearing Structures. ×50. (Courtesy of Federal-Mogul Corporation.)

Top. Coarse, globular-type structure. Sintered copper-lead S.A.E. No. 480 analysis. *Middle.* Medium-coarse, dendritic-type structure. Cast copper-lead S.A.E. No. 48 analysis. *Bottom.* Fine, dendritic-type structure. Cast copper-lead S.A.E. No. 48 analysis.

Reducing the Severity of Engine Operating Conditions. This is primarily a question of reducing temperatures. Although it has been done in some cases by providing more efficient cooling, changes in this direction are limited since high engine temperatures are necessarily encountered in modern, high-output engines.

[11] J. M. Freund, H. B. Linford, P. W. Schutz, *Trans. Electrochem. Soc.*, **84**, 65 (1943).

Generally more effective crankcase ventilation, resulting in removal of volatile oil oxidation products, tends to lessen corrosion difficulties.

Additives in Lubricants. Although some steps can be taken in improving bearings and in improving engine operating conditions, the burden of the corrective measures for minimizing bearing corrosion has fallen upon the lubricating oil. Oils can be sufficiently stablilized by additives (used in concentrations varying from $\frac{1}{10}\%$ to several per cent, depending upon the specific additive and base oil used) to stand up under all reasonable conditions of service. The more important of these additive types are:

Sulfur-containing additives — sulfurized hydrocarbons, sulfurized esters, metal alkyl thiophosphates.

Phosphorus-containing additives — alkyl and aralkyl phosphites. The phosphite esters are not particularly effective in preventing corrosion of Cu-Pb bearings, and sometimes show adverse effects where leaded fuels are used.

Aromatic amine additives — conventional oxidation inhibitors such as phenyl-α-naphthylamine and tetramethyldiaminodiphenylmethane.

Phenolic-type additives — sulfurized alkyl phenols and metal salts thereof and metal salts of sulfurized alkyl salicyclic acids have been used commercially. Other phenolic compounds such as aminophenols and polyhydroxybenzene derivatives are also effective.

CORROSION OF COPPER

Corrosion by Active Sulfur Compounds

Modern, well-refined lubricants give little trouble with copper. A copper strip test (p. 1037), involving heating the oil in the presence of a copper strip, is extensively used, but often grossly misinterpreted. This test is used to detect the presence of corrosive sulfur or sulfur compounds in lubricants. A black, scaly deposit indicates that the oil may corrode copper lines or copper alloy parts in service. Moderate staining of the strip, however, is no indication that corrosion of copper will occur in service. The presence of water often accelerates the corrosion of copper by active sulfur compounds.

Corrosion by Oil Oxidation Products

Lubricating oils, particularly highly refined oils when exposed to oxygen and extensive surfaces of copper, especially in the presence of small amounts of water, may develop high organic acidity. In such cases very severe corrosion of copper and copper alloy parts takes place to such an extent that the oil may become deep green in color from dissolved copper salts.

A very troublesome type of copper corrosion occurs in compressor systems, particularly refrigeration units. Copper is dissolved by the oil and plates out on steel parts of the system. Faulty valve action and actual freezing of the piston to the cylinder wall have occurred because of such deposits. Although this phenomenon is believed to be basically due to the solution of copper in oxidized oil, it has been observed in units where only minute amounts of air are present. Copper plating troubles are more pronounced where chlorinated hydrocarbons are used as refrigerants and where traces of moisture are present. It can be minimized by (1) using highly refined lubricants, preferably white oils, and (2) using oils containing suitable additives (aromatic amines, naphthols, polyamine-aldehyde condensation products, or other copper deactivators).

CORROSION OF FERROUS METALS

Corrosive Wear Caused by Sulfur-Type Additives

Although most lubricating oils, even when highly oxidized, do not corrode iron or steel, some compounded oils attack ferrous metals by a mechanism called *corrosive wear*. For example, active-sulfur extreme pressure lubricants cause accelerated wear of steel gears, and compounded cutting oils may considerably shorten tool life by forming surface films which are more readily worn off than is the steel surface itself.

Corrosion Caused by Chlorine-Containing Additives

Extreme pressure lubricants containing chlorine compounds have, in some cases, caused serious corrosion of steel gears and housings. This type of corrosion is caused by HCl liberated from the additive and is especially severe when the service includes high-temperature operation (favors HCl liberation) followed by cooling (promotes moisture condensation and corrosion by the aqueous HCl formed).

Chemical Polishing by Anti-Wear Additives[12]

Certain additives in lubricating oil may reduce wear by a very special kind of corrosive action called *chemical polishing*. Even highly polished metal surfaces are microscopically rough. When two such surfaces are pressed together, contact is made by only a few high spots. When the surfaces slide over each other under heavy loads, the extreme pressure at the few points of contact causes excessively high temperatures, followed by welding, tear, and further roughening of the surface. Lube oil additives such as tricresyl phosphate, triphenyl phosphine, or triphenyl arsine react with the iron to form compounds (e.g., Fe_3P) capable of forming eutectics with the iron. The lowering of the melting point (e.g., m.p. Fe_3P-Fe is 515° C lower than m.p. Fe) aids greatly in maintaining a polished surface, and in laboratory test apparatus reduces overall wear manyfold.

PREVENTION OF AQUEOUS CORROSION

One of the functions of almost all lubricants is the prevention of corrosion of the lubrication system by water. There are a number of types of industrial lubricants, such as circulating oils and hydraulic oils, where prevention of aqueous corrosion is especially important. For example, steam turbine oils must possess high anti-rust characteristics to protect adequately the turbine system from rusting during both operation and shutdown periods. Long-chain organic acids, especially dicarboxylic acids, represent one type of additive which may be used in low concentrations in oil (e.g., 0.1%) to impart such properties. Other anti-rust oils, where the anti-rust function is stressed more than the lubricating function, are discussed in *Temporary Corrosion Preventive Coatings*, p. 916.

[12] O. Beeck, J. W. Givens, A. E. Smith, and E. C. Williams, *Proc. Royal Soc.* (A), **177**, 90 (1940).

EFFECT OF MECHANICAL FACTORS ON CORROSION

Stress Corrosion*

H. R. Copson†

Broadly the term stress corrosion includes any combined effect of stress and corrosion on the behavior of metals. Generally stresses are divided into two types, cyclic and static. Phenomena involving cylic stresses are dealt with under *Corrosion Fatigue*, p. 578.

Static stresses may effect the general corrosion behavior or they may lead to cracking. The term stress corrosion cracking is used to indicate combined action of static stress and corrosion which leads to cracking or embrittlement of metal.

FACTORS INVOLVED IN STRESS CORROSION CRACKING

The principal factors are stress, environment, time, and internal structure of the alloy.[1] The importance of these varies, and they may interract, one accelerating the action of another.

STRESS

If stress corrosion cracking is to occur there must be tensile stress at the surface. The stress may be internal or applied. Internal stresses are produced by non-uniform deformation during cold working, by unequal cooling from high temperature, and by internal structural rearrangements involving volume changes. Stresses induced when a piece is deformed, those induced by press and shrink fits, and those in rivets and bolts may be classed as internal stresses. These concealed stresses are of greater importance than applied stress, especially in view of the factor of safety used in design. The magnitude of the stress varies from point to point within the metal. Generally tensile stresses in the neighborhood of the yield strength are necessary to promote stress corrosion cracking, but failures have occurred at lower stresses.

ENVIRONMENT

Stress corrosion cracking has been observed in almost all metal systems. Yet for each metal, stress cracking is associated with specific environments. The environment that induces cracking frequently attacks the metal only superficially if stresses are absent.

TIME

The time required for failure may vary from minutes to years. With some alloys there is an incubation period during which structural changes such as precipitation are occurring. Thus Al-Mg alloys (over 6% magnesium) become more susceptible to stress corrosion cracking as aging occurs.

INTERNAL STRUCTURE

The internal structure of a metal depends upon composition, fabrication, thermal treatments, and often the extent of aging. It is of considerable importance in stress corrosion.

---

* See also *Tests for Stress Corrosion Cracking*, p. 1009.
† Research Laboratory, The International Nickel Co., Inc., Bayonne, N. J.

[1] A. Beerwald, *Metallwirtschaft*, **23**, 174 (1944).

EXAMPLES OF STRESS CORROSION CRACKING

Some examples are discussed in other chapters. Mention is made here of most common examples.

Aluminum Alloys

Pure aluminum is quite resistant to stress corrosion cracking, but Al-Zn alloys containing more than 12% zinc and Al-Mg alloys containing more than 6% magnesium are very susceptible to cracking under mild corrosive conditions.[2] The cracking is intergranular. Grain boundary constituents have been observed microscopically, and potential measurements have shown the grain boundaries to be anodic to the grains in solutions known to produce cracking.

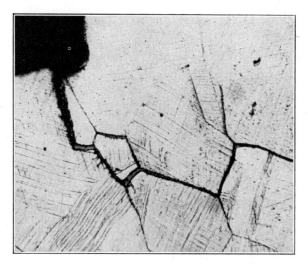

FIG. 1. Section through Silicon Bronze Showing Intergranular Stress Corrosion Cracking. ×500.

Brass*

Perhaps the best-known example of stress corrosion cracking is the *season cracking* of brass,[3,4,5] so called because the cracks may resemble those in seasoned wood. Exposure to moist atmospheres containing traces of ammonia is particularly drastic in producing cracking. Ammonia and compounds of ammonia are believed to be the potent agencies, but moisture and oxygen are necessary, and carbon dioxide may have a contributing effect. Susceptibility to cracking increases with zinc content.

Susceptibility also increases with tensile stress. Cracking rarely occurs with internal stress less than about 840 kg/sq cm (12,000 psi), but stresses of 840 to 1400 kg/sq cm (12,000 to 20,000 psi) produce cracking. The cracking is usually intergranular, but transcrystalline cracking has been reported. Special brasses which contain other

* Refer also to *Copper-Zinc Alloys*, p. 78.

[2] E. H. Dix, Jr., *Trans. Am. Inst. Mining Met. Engrs.*, **137**, 11 (1940).
[3] Bibliography on Season Cracking, *Proc. Am. Soc. Testing Materials*, **41**, 918–926 (1941).
[4] H. P. Croft and G. Sachs, *Iron Age*, **151** (No. 10), 47, and (No. 11), 62 (1943).
[5] G. Sachs, *Iron Age*, **146** (No. 14), 21 (1940).

elements, such as iron, tin, manganese, silicon, aluminum, cadmium, and lead behave similarly to straight zinc brasses.[5]

OTHER COPPER ALLOYS

Compared with the brasses, other copper alloys, such as tin bronze, aluminum bronze, silicon bronze, and arsenical copper, show comparatively little tendency to season crack.[5] Nevertheless, they will crack if conditions are right. Figure 1 shows an intergranular stress corrosion crack in silicon bronze. The cracking was confined to regions of high internal tensile stress.

IRON

A well-known example of stress corrosion cracking is the *caustic embrittlement* of mild steel in steam boilers.[6] (Refer also to *Boiler Corrosion*, p. 531.) Sodium hydroxide must be present in the water; hence the name caustic embrittlement. The cracking is predominantly intercrystalline, but transcrystalline cracks may be present. Oxides are usually present in the cracks. High tensile stresses must be present, and usually the cracking takes place along rows of rivets. Salt deposits have been observed in some cracked rivet seams. It has been established that small leaks or capillary spaces must be present so that caustic concentrations can build up to high values. The presence of a little silicate greatly accelerates cracking. Other constituents in the water also influence cracking.

Stress corrosion cracking of iron has been observed in media other than alkalies. (Refer to "Stress," *Iron and Steel*, p. 137.)

STAINLESS STEEL*

In certain acid chlorides and other corrosive media, transcrystalline cracking has been observed in austenitic Cr-Ni steels.[7] Figure 2 shows an instance of transcrystalline cracking of Type 316 stainless steel exposed in a spool-type corrosion test (see p. 1053). The cracks radiate from the stencil marks. Away from the cracks the metal was ductile.

Figure 3 shows another instance of cracking in boiling calcium magnesium chloride brine. As-received specimens cracked through the stencils. Specimens water-quenched from 1070° C (1950° F) cracked even more profusely, but the cracks were no longer located at the stencils. Heating relieved internal stresses at these points, but quenching stresses were high enough to produce cracking. A stress relief anneal consisting of heating at 730° C (1350° F) for one hour, followed by furnace cooling overnight, eliminated the cracking. The transcrystalline nature of the cracks is shown in Fig. 4.

LEAD

Lead cable sheaths simultaneously exposed to stress and corrosion fracture in intergranular manner.[8]

---

* Refer also to *Stress Corrosion Cracking in Stainless Alloys*, p. 174.

[6] W. C. Schroeder and A. A. Berk, U. S. Department of the Interior, Bureau of Mines, *Bull.* 433 (1941).
[7] S. L. Hoyt and M. A. Scheil, *Trans. Am. Soc. Metals*, **27**, 191 (1939). J. C. Hodge and J. L. Miller, *Trans. Am. Soc. Metals*, **28**, 25 (1940). H. J. Rocha, *Stahl u. Eisen*, **62**, 1091 (1942). M. A. Scheil, O. Zmeskal, J. Waber, and F. Stockhausen, *Welding J.* (Supplement), **22**, 493–s (1943).
[8] H. S. Rawdon, *Ind. Eng. Chem.*, **19**, 613 (1927).

Magnesium Alloys*

Some magnesium alloys are quite sensitive to stress corrosion cracking in the atmosphere and under a variety of corrosive conditions.[9] Pure magnesium, Mg-Mn alloys, and the magnesium casting alloys are relatively resistant to cracking at stresses up to or exceeding their tensile yield strength. Wrought alloys show increasing

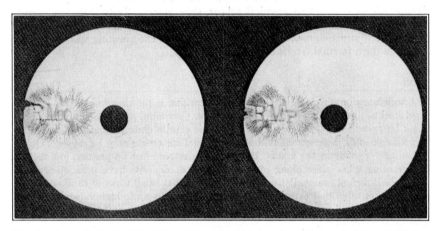

FIG. 2. Stress Corrosion Cracking of Type 316 Stainless Steel at Punched Stencil Marks. (2¼ in. diam., .031 in. thick).

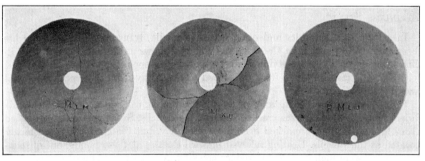

	Water-Quenched	Furnace-Cooled
As	from	from
Received	1950° F	1350° F

FIG. 3. Stress Corrosion Cracking of Type 316 Stainless Steel (2¼ in. diam., 0.031 in. thick) in a Boiling Concentrated Calcium Magnesium Chloride Brine. (Duration of test: 17 days.)

sensitivity with increasing aluminum and zinc content. Additions of lead, tin, or cadmium seem to have little effect. Cold-rolled and fully annealed materials are more sensitive than fully precipitated materials. Welded seams on the more sensitive materials increase the possibility of cracking because of the high stress and the effect

* Refer also to *Magnesium and Magnesium Alloys*, pp. 231–234, 247–249.

[9] A. Beck, *The Technology of Magnesium and Its Alloys*, English Translation from the German by F. A. Hughes and Co., London, 2nd Ed., 1940.

of the annealed area adjacent to the weld. Riveted seams are not prone to cracking. The cracks are predominantly transcrystalline. With wrought magnesium alloys, especially those low in aluminum and zinc, residual stresses are gradually relieved by creep.

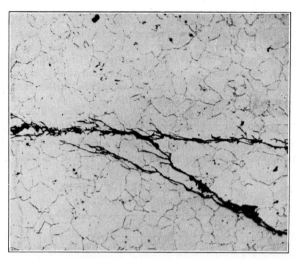

FIG. 4. Section through Type 316 Stainless Steel Showing Transcrystalline Stress Corrosion Cracking. ×100. (Specimen similar to as-received specimen of Fig. 3.)

FIG. 5. Section through Cold-Worked 70% Ni–30% Cu Alloy Showing Intergranular Stress Corrosion Cracking after Exposure to Hydrofluosilicic Acid. ×100.

NICKEL ALLOYS

Environments known to cause stress corrosion cracking of nickel and nickel alloys are few and failures have been rare. Figure 5 shows an intergranular stress corrosion crack in a severely cold-worked 70% Ni–30% Cu alloy* tube exposed to hydrofluosilicic acid. Another medium which sometimes produces stress corrosion cracking is fused caustic soda. Other examples have been listed by O. B. J. Fraser.[10] (See also *Nickel-Iron Alloys*, p. 199.)

ZINC

Cracking of older type zinc die castings is thought not to be an example of stress corrosion cracking because stress did not appear to be necessary and the attack was more definitely in the form of intergranular corrosion. The offending impurities were found to be lead, tin, and cadmium, which produce this type of corrosion only when aluminum is present in the alloy. (See pp. 339 and 346 for further remarks concerning zinc. Refer also to *Steam Test for Zinc*, p. 1023.)

REMEDIES FOR STRESS CORROSION CRACKING

The most common method of preventing stress corrosion cracking is to relieve the internal stress by heating and cooling slowly (Table 1).[11] With suitable temperature

TABLE 1. STRESS RELIEF ANNEALS
To Be Followed by Slow Cooling

Material	Temperature		Time, Hours
	° C	° F	
Aluminum alloys	230–260	450–500	1
Brass	200–300	390–570	½–1
Iron	500–600	930–1100	½–1
Stainless steel	730–870	1350–1600	1–2
Magnesium alloys	130–300	265–570	¼–1
Nickel alloys	600–650	1100–1200	½–1

Manufacturers should be consulted to obtain best temperature and time for specific alloy compositions.

and time, loss of strength and hardness can be minimized. This method cannot be used in composite structures with metals having different coefficients of expansion.

In some cases stress corrosion cracking may be prevented by adding appropriate inhibitors to the corrosive medium, for example, in caustic embrittlement of boilers. In other cases the corrosive medium may be prevented from contacting the metal by protective coatings or paints. Thick electrodeposits have worked well. An aluminum cladding on Duralumin prevents stress corrosion cracking of the latter. Painting magnesium alloys susceptible to stress corrosion cracking increases their life four or five times.

Occasionally altering the composition of the alloy may sufficiently increase resistance to cracking. Thus lowering the zinc content of brass or the magnesium content of an aluminum magnesium alloy may work. However, substituting one austenitic stainless steel for another has not prevented chloride stress corrosion cracking, and alloying of steel has been of little help in caustic embrittlement.† Suitable

* Monel.
† Some evidence is reported that aluminum additions to steel effectively improve resistance to intergranular stress corrosion cracking. See "Nitrogen" in *Iron and Steel*, p. 140. EDITOR.

[10] *Symposium on Stress-Corrosion Cracking of Metals (1944)*, American Society for Testing Materials and American Institute of Mining and Metallurgical Engineers, Philadelphia and New York, 1945.
[11] G. Sachs, *Iron Age*, **146** (No. 14), 21 (1940).

heat treatments or other processes may change the structure of the alloy and bring about improvement.

Another preventive is to change the fabrication so as to produce compressive stresses at the surface. Shot peening, tumbling, rolling, swaging, and the like have

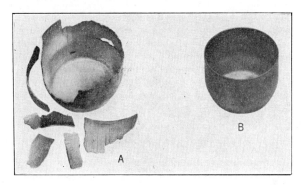

FIG. 6. Effect of Peening on Season Cracking of Brass. (J. O. Almen, General Motors Research Corp. [Classified O.S.R.D. Research Report].)

A. Non-peened specimen failed in 2½ hours in ammonia atmosphere. *B.* Lightly peened specimen is undamaged after 100 hours in ammonia atmosphere.

worked in certain instances. An example of the beneficial effect of peening on season cracking of brass is shown in Fig. 6. In shot peening magnesium, non-metallic shot should be used to prevent metallic contamination and subsequent electrolytic corrosion.

THEORY OF STRESS CORROSION CRACKING

An electrochemical theory has been proposed by E. H. Dix, Jr.,[12] as follows:

First there must exist in the alloy a susceptibility to selective corrosion along more or less continuous paths, as, for instance, the grain boundaries. This susceptibility is present when the internal structure of the alloy is microscopically heterogeneous and the phase forming the continuous paths is anodic in the specific corrosive medium to the areas composing the major part of the structure. Second, there must exist a condition of high stress acting in a direction tending to pull the metal apart along these continuous paths. If these two conditions exist simultaneously and the metal is subjected to a corrosive environment producing the specified potential relations, corrosion will start along the anodic paths. This will produce a concentration of stress at the bottom of the notches so formed. As the stress increases, fissures will begin to develop, destroying any protective film and thus exposing fresh anodic material to the corrosive medium. Corrosion will proceed more rapidly and at the same time the stress will increase at an accelerated rate, especially if the material is subjected to a high external load. These mutually accelerated actions will continue at an increasing rate until the metal fails.

This theory explains why cracking is observed only in specific corrosive media. It has been well substantiated for the intergranular cracking of aluminum alloys, and it probably applies to many alloy systems.[13] There are other continuous paths through metals besides grain boundaries, e.g., slip planes and planes of precipitated constitu-

[12] E. H. Dix, Jr., *Trans. Am. Inst. Mining Met. Engrs.*, **137**, 11 (1940).
[13] R. B. Mears and R. H. Brown, *Ind. Eng. Chem.*, **33**, 1002 (1941).

ents. Thus the theory may apply to transcrystalline stress corrosion cracking. Support for the theory is found in the fact that an applied potential making the system cathode prevented stress corrosion cracking of brass, aluminum, stainless steel, and magnesium.[14]

The above theory has been developed by J. T. Waber and H. J. McDonald[15] into a general "precipitation" theory. The distinctive added feature is that high local stresses are said to accelerate the formation of galvanic cells by accelerating precipitation. Cracks then grow by dissolution of the newly formed anodic material. In the case of mild steel the precipitated material is said to be iron nitride. In the case of austenitic stainless steels a martensitic decomposition product is postulated. The precipitation theory also serves to accommodate strain etching, in that more rapid corrosion may occur at macro areas of heterogeneity produced by strain-accelerated precipitation.

EFFECT OF STATIC STRESS ON GENERAL CORROSION

Where there is no cracking or embrittlement the presence of static stress may still affect corrosion. This subject has been discussed by U. R. Evans[16] and F. N. Speller,[17] and referred to frequently in the literature.[16] Examination of the data gives a confused picture; some authors have found static stress to increase corrosion, others have found no effect, and still others report a decrease in corrosion. The effect seems specific both for the metal and for the environment.

Usually the effects are relatively small.*[18] Thus Edwards, Phillips, and Thomas found that in 6% sulfuric acid the increase in corrosion of steel caused by cold rolling was overshadowed by small variations in carbon, phosphorus, sulfur, and copper content of the steel. Some results on nickel and nickel alloys are given in Fig. 7 where hardness is given as a measure of cold reduction.

It is frequently stated that the energy stored in distorted metal makes such metal anodic to stress-free material, and that as a result galvanic effects may cause rapid corrosion of distorted metal. Attempts have been made to arrive at the potential difference.[19] Results vary, but in general indicate that the difference in emf is only a few millivolts. This difference is too small to have much effect on corrosion and could easily be upset by many factors such as concentration cells. Hence where static stress alters the corrosion rate it is likely to be the result of secondary factors. Fre-

* See Fig. 6, p. 138, showing effect of stress on corrosion of steel. EDITOR.

[14] *Symposium on Stress-Corrosion Cracking of Metals (1944)*, American Society for Testing Materials and American Institute of Mining and Metallurgical Engineers, Philadelphia and New York, 1945.

[15] J. T. Waber, H. J. McDonald, and B. Longtin, *Trans. Electrochem. Soc.*, **87**, 209 (1945). J. T. Waber and H. J. McDonald, *Corrosion and Material Protection*, **2** (No. 8), 13 (1945); (No. 9), 13 (1945); and **3** (No. 1), 13; (No. 2), 13; (No. 3), 13; (No. 4), 13; (No. 5), 13; (No. 8), 13 (1946).

[16] U. R. Evans, *Metallic Corrosion, Passivity and Protection*, Edward Arnold and Co., London, 1937.

[17] F. N. Speller, *Corrosion, Causes and Prevention*, McGraw-Hill Book Co., New York, 1935.

[18] V. P. Severdenko, *Metallurg*, **12** (No. 7), 53 (1937). J. A. Duma, *J. Am. Soc. Naval Engrs.*, **49**, 566 (1937). A. Skapski and E. Chyzewski, *Metaux & Corrosion*, **13**, 21 (1938). C. A. Edwards, D. L. Phillips, and D. F. G. Thomas, *J. Iron Steel Inst.*, **137**, 223 (1938). J. D. Grogan and R. J. Pleasance, *J. Inst. Metals*, **64**, 57 (1939). C. W. E. Clark, A. E. White, and C. Upthegrove, *Trans. Am. Soc. Mech. Engrs.*, **63** (No. 6), 513 (1941). J. A. Addlestone and M. W. Burke, *Virginia J. Science*, **2**, 192 (1941). H. L. Solberg, A. A. Potter, G. A. Hawkins, and J. T. Agnew, *Trans. Am. Soc. Mech. Engrs.*, **65**, 47 (1943).

[19] P. D. Merica, U. S. Bureau of Standards, *Tech. Publication*, 83 (1916). G. Tammann, *Chem. Zentralblatt*, **3**, 90 (1919). T. F. Russel, *J. Iron Steel Inst.*, **107**, 497 (1923). T. Turner and J. Jevons, *J. Iron Steel Inst.*, **111**, 169 (1925). H. Endo and S. Kanazawa, *Sci. Reports Tohoku Univ.*, **20**, 124 (1931). G. I. Taylor and H. Quinney, *Proc. Roy. Soc.*, **143**, 307 (1934). R. W. France, *Trans. Faraday Soc.*, **30**, 450 (1934). L. V. Nikitin, *J. Gen. Chem. (U. S. S. R.)*, **9**, 794 (1939) and **11**, 146 (1941).

quently the presence of stress is only one of several possible explanations of observed corrosion damage.

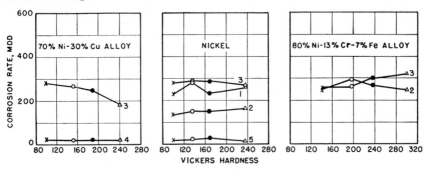

Fig. 7. Effect of Internal Stress, Produced by Cold Working, on Corrosion of Nickel and Nickel Alloys.

1 in 5.9% HCl, 30° C, Aerated
2 in 5.9% HCl, 30° C, Unaerated
3 in 5.0% H_2SO_4, 30° C, Aerated
4 in 3.5% NaCl, 30° C, Aerated
5 in 50% NaOH, 130° C, Unaerated

x Dead soft
o Quarter hard
● Three-quarter hard
△ Full hard

OTHER STRESS CORROSION EFFECTS

The borderlines between corrosion fatigue, stress corrosion cracking, and the effects of stress on general corrosion are not sharp. Thus in steel boilers corrosion has been

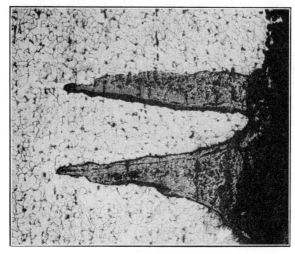

FIG. 8. Section through Locomotive Steel Firebox Sheet Showing Nature of Cracks or Grooves Which Follow Lines of Stress. ×100.

known to produce shallow grooves which follow lines of stress. The grooves may be so narrow as to be called cracks. Figure 8 shows their typical appearance. They have been attributed to corrosion of less noble stress metal[20] and to corrosion fatigue.[21] They could be stress corrosion cracks that had been widened by subsequent crevice corrosion. When several factors are operative, results may not be clear cut.

Corrosion Fatigue

BLAINE B. WESCOTT*

The damaging effect of the simultaneous action of corrosion and cyclic stresses on metals, known as corrosion fatigue, was mentioned first in 1917 by Haigh.[1] The results of numerous investigations of this problem since published have emphasized the influence which the concurrent action of corrosion and cyclic stresses has upon the performance of metals in service. No metal is immune from some reduction of its resistance to cyclic stressing if it is corroded by the environment in which it is stressed.

NATURE OF CORROSION FATIGUE

The serious lowering of the resistance of metals to cyclic stresses while exposed to a corrosive medium is closely related to the type of pitting induced by such conditions. Steel provides a representative example for the purposes of this discussion. The pits produced in steel, unstressed or under static stress, by differential aeration generally are saucer-shaped.[2] This shape is maintained as corrosion proceeds. However, under cyclic stressing, sharp, deep pits are formed and, as corrosion fatigue continues, many of these pits become the origin of cracks which are filled with corrosion products. Failure occurs subsequently from the progression of one of the cracks across the section of steel involved. The characteristic appearances of pits produced by stressless corrosion and cracks resulting from corrosion fatigue are shown in Fig. 1.[3]

The damage from corrosion fatigue is greater than the sum of the damage arising from cyclic stresses and that due to corrosion. The net effect of both factors is to increase the rate of crack propagation principally because of the effect of cyclic stresses on corrosion. Evans[4] believes that the rate of corrosion of metals is controlled by the properties of the primary film (usually oxides) formed upon the metal. On metals resistant to corrosion, this film forms rapidly and is quickly restored when damaged. Usually the films have low strength and low ductility so that cyclic stressing causes rupture at a greater rate than repair can occur, and this allows corrosion to proceed rapidly at points where the normal protection provided by the film is absent. Secondary corrosion products are formed which clog the pits and retard diffusion of oxygen required to repair the primary film; thus the metal in the anodic area at the base of the pit is maintained in an active condition and has, consequently, an increased tendency to corrode.

* Gulf Research and Development Co., Pittsburgh, Pa.

[20] F. G. Straub, *University of Illinois Bull.* 216 (1933).
[21] H. N. Boetcher, *Mech. Eng.*, **66** (No. 9), 593 (1944).

[1] B. P. Haigh, "Experiments on the Fatigue of Brasses," *J. Inst. Metals*, **18** (No. 2), 55 (1917).
[2] U. R. Evans, *Metallic Corrosion, Passivity and Protection*, Edward Arnold and Co., London, 1937.
[3] D. J. McAdam, Jr., and G. W. Geil, "Influence of Cyclic Stress on Corrosion Pitting of Steels in Fresh Water and Influence of Stress Corrosion on Fatigue Limits," *J. Research Natl. Bur. Standards*, **24**, 685 (1940). (R.P. 1307.)
[4] U. R. Evans, *Metallic Corrosion, Passivity and Protection*, Edward Arnold and Co., London, 1937.

It is shown in Fig. 1 that the pits formed as a result of corrosion fatigue are sharp. These chemically formed notches function as stress raisers of unusually high intensity. As crack growth extends away from the surface into the metal, the effect of stress concentration becomes the predominant factor contributing to eventual failure.

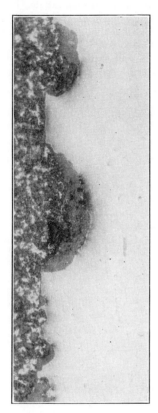 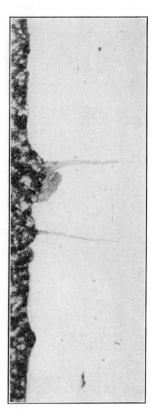

FIG. 1. Comparison of Pits and Corrosion Fatigue Cracks Produced by Well Water in 3.7% Ni, 0.28% C Steel. (Courtesy of D. J. McAdam, Jr.)

Left: Pits after stressless corrosion of 50 days. *Right:* Corrosion fatigue cracks after stressing at 1300 psi, at 1450 cycles per minute for 5 days.

The process of failure by corrosion fatigue is comprised of two stages.[5] During the first stage the combined action of corrosion and cyclic stresses damages the steel by pitting and crack formation to such a degree that fracture by cyclic stressing would occur ultimately even if the corrosive environment were removed altogether. The second stage is essentially a fatigue stage in which failure proceeds by propagation of the crack and is controlled primarily by stress concentration effects and the physical properties of the steel. This conception of the mechanism is identical with that of failure from fatigue if corrosion is regarded simply as the

[5] D. J. McAdam, Jr., "Stress-Strain-Cycle Relationship and Corrosion Fatigue of Metals," *Proc. Am. Soc. Testing Materials,* **26** (II), 224 (1926).

method by which stress concentration effects of particularly damaging magnitude are produced by a chemical rather than a mechanical process.

Cracks resulting from corrosion fatigue progress predominately in a transcrystalline rather than in an intercrystalline manner, and this is further evidence that there is no fundamental difference between fatigue and corrosion fatigue. Ludwick[6] demonstrated conclusively that neither at the origin nor subsequently were corrosion fatigue cracks intercrystalline in either pure aluminum or soft iron. Likewise a corrosion fatigue crack in normalized 3.5% Ni–0.10% C steel (Fig. 2) shows no evidence to indicate intercrystalline attack. Experiments with two-crystal specimens of aluminum[7] and with aluminum specimens having four to six crystals[8] produced no evidence of intercrystalline corrosion or cracking.

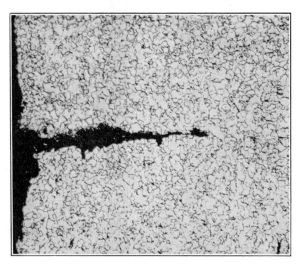

Fig. 2. Origin of Corrosion Fatigue Crack in Normalized 3.5% Ni–0.10% C Steel. ×100.

The normal fatigue fracture in lead is intercrystalline,[9] a behavior unique among metals. Corrosion fatigue failures of lead may also be intercrystalline.[10] It is suggested that fatigue of a metal with a low melting point such as lead at atmospheric temperature is analogous to high temperature tests on alloys of high melting points; therefore, intercrystalline fracture may not be exceptional. Rawdon[11] reported intercrystalline failure of Duralumin fatigued by reversed flexure when immersed in a solution of sodium chloride and hydrogen peroxide for a short time at frequent periodic intervals. However, this was also the method of failure under stressless corrosion and by corrosion under static stress. The normal manner of failure from

[6] P. Ludwik, "Kerb- und Korrosionsdauerfestigkeit," *Metallwirtschaft*, **10**, 705–710 (1931).
[7] H. J. Gough, H. L. Cox, and D. G. Sopwith, *J. Inst. Metals*, **54**, 193 (1934).
[8] H. J. Gough and G. Forrest, *J. Inst. Metals*, **58**, 97 (1936).
[9] B. P. Haigh and B. Jones, "Atmospheric Action in Relation to Fatigue in Lead," *J. Inst. Metals*, **43**, 271 (1930).
[10] J. R. Townsend, "Fatigue Studies of Telephone Cable Sheath Alloys," *Proc. Am. Soc. Testing Materials*, **27** (II), 153 (1927).
[11] H. S. Rawdon, "The Effect of Corrosion, Accompanied by Stress, on the Tensile Properties of Sheet Duralumin," *Proc. Am. Soc. Testing Materials*, **29** (II), 314 (1929).

corrosion fatigue for nearly all commonly used metals and alloys is by intracrystalline (transcrystalline) fracture.

Corrosion fatigue of a single crystal of aluminum[12] under cyclic torsional stresses caused general surface pitting having no relation to stress conditions. Slip bands were formed only in the initial stages of stressing, but preferential corrosion occurring along the slip planes was the essential cause of failure. Such behavior is also characteristic of fatigue. Since slip is a consequence of cold working, this result explains why cold-worked metal is particularly susceptible to corrosion fatigue although it possesses higher fatigue resistance. Gough concluded that failure of ductile metals

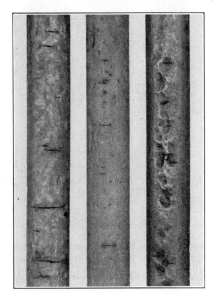

FIG. 3. Various Stages in Development of Corrosion Fatigue Failures in Sucker Rods Used for Pumping Oil Wells.

either by fatigue or corrosion fatigue is inseparably associated with failure of elasticity by slip. X-ray analysis has proved that the structure is fragmented into crystallites at slip planes, and it is probable that fatigue failure originates at local junctions of the crystallites where some lattice bonds rupture under continued stressing. Local strains at such points obviously are high and bear no definite relation to the average strain.

EVIDENCE OF CORROSION FATIGUE

Usually it can be ascertained readily whether failure has been caused by corrosion fatigue. The surface of the metal is pitted, and frequently cracks are visible at the bottoms of the pits. The direction of the cracks is normal to that of the stresses which produced them. If the cracks are not detectable under a low-power magnifier, they can be made so by deep etching of the surface, by plastic deformation, or by magnetic inspection. In this respect, corrosion fatigue differs from fatigue because

[12] H. J. Gough, "Crystalline Structure in Relation to Failure of Metals, Especially by Fatigue," *Proc. Am. Soc. Testing Materials*, **33** (II), 3 (1933).

rarely is there any evidence of more than one crack in a metal part which has failed from fatigue. Characteristic evidence of failure in service because of corrosion fatigue is presented in Figs. 3 and 4. In Fig. 5 the appearance of the fracture of a sucker rod used for pumping an oil well clearly shows the origin in a corrosion pit and the progressive manner of crack propagation in the second stage identical with ordinary fatigue failure.

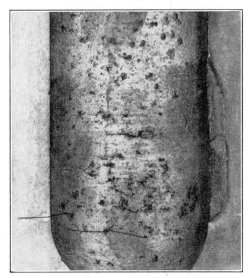

FIG. 4. Section of Hollow Water-Cooled Stainless Steel Plunger of Hot Oil Pump Which Failed in Service by Corrosion Fatigue.

Cracks originated at interior water-cooled surface and progressed outward through wall of plunger.

SIGNIFICANCE OF CORROSION FATIGUE LIMITS

Through common usage the corrosion fatigue limit of a metal is defined as the maximum repeated stress that will not cause failure in a stated number of stress applications under defined conditions of corrosion and stressing. The value is significant only when the conditions are completely defined because, if the metal is attacked by its environment, eventually it will be corroded completely and will possess no load-carrying capacity whatever. In other words, when test results are plotted in the manner normally used for fatigue testing, the S-N* curve never becomes truly asymptotic to the abscissa. Time and rate of stress application have much greater importance than in fatigue. Depending upon the bases chosen, a metal may have numerous corrosion fatigue limits, the principal value of which is to provide a method for ranking metals in the probable order of performance in service. Rarely may the corrosion fatigue limit be used for design purposes because the value is usually determined for complete reversal of stresses, and this condition is not often encountered in service. Other factors which exert a marked influence on the corrosion fatigue limit are the nature of the corroding medium, nature and range of stress, rate of stressing, temperature, degree of aeration, inherent corrosion resistance of the metal,

* For description of S-N curves, refer to *Tests for Corrosion Fatigue*, p. 987.

and physical properties of the metal, particularly its notch sensitivity. Judgment and experience are essential, therefore, in the application of corrosion fatigue data to operational problems.

FIG. 5. Corrosion Fatigue Failure of Oil Well Sucker Rod Showing Origin in Corrosion Pit and Progressive Crack Propagation.

DATA ON CORROSION FATIGUE LIMITS

The corrosion fatigue limits of several steels are given in Table 1. These results were obtained by exposing the specimens to fresh well water and moderately saline river water in air. Certain pertinent generalizations can be drawn from the data: (1) There is no relation between corrosion fatigue limit and tensile strength. (2) For carbon steels there is no marked effect attributable to carbon content. (3) Medium-alloy steels have only slightly higher corrosion fatigue limits than carbon steels. (4) Heat treatment does not improve the corrosion fatigue limits of either carbon or medium-alloy steels and residual stresses remaining have a deleterious effect. (5) Corrosion-resistant steels have higher corrosion fatigue limits than other steels; in this respect chromium is more effective than nickel as an alloying element. (6) Corrosion fatigue limits of all steels are lower in salt water than in fresh water. (7) Resistance to corrosion fatigue depends largely on corrosion resistance.

Results of corrosion fatigue tests of low-alloy steels in oil well water and the same water saturated with hydrogen sulfide, both with natural gas present but air excluded, are given in Table 2.[13] Generalizations from these data are:[14] (1) Damage from corrosion fatigue in salt water in the absence of air is less than in fresh water with

[13] B. B. Wescott and C. N. Bowers, "Economical Selection of Sucker Rods," Trans. Am. Inst. Mining Met. Engrs., **114**, 177–192 (1935).

[14] B. B. Wescott, "Fatigue and Corrosion Fatigue of Steels," Mech. Eng., **60**, 813 (November, 1938).

air present. (2) Some medium-alloy steels are damaged much less by corrosion fatigue than carbon steels. (3) Low-carbon steels are damaged proportionately less by corrosion fatigue than medium-carbon and high-carbon steels. (4) Steels in the heat-treated

TABLE 1. CORROSION FATIGUE LIMITS OF STEELS IN FRESH WELL WATER* AND SALINE RIVER WATER†
Stressed for at least 10^7 cycles at 1450 cycles per min.
(McAdam)

Material	Condition	Tensile Strength, psi	Endurance Limit,‡ psi	Corrosion Fatigue Limit, psi		Damage Ratio§	
				Well Water	Salt Water	Well Water	Salt Water
Carbon Steels							
Ingot iron	Annealed	42,300	21,000	15,000	...	0.71	...
Ingot iron	Quenched, drawn	43,900	24,000	20,000	...	0.83	...
0.11 C steel	Annealed	46,400	25,000	16,000	...	0.64	...
0.16 C steel	Annealed	52,000	25,600	17,000	5,000	0.66	0.20
0.16 C steel	Quenched, drawn	65,700	35,000	20,000	7,000	0.57	0.20
0.24 C steel	Annealed	59,200	27,000	16,000	...	0.59	...
0.26 C steel	Quenched, drawn	84,500	39,000	23,000	...	0.59	...
0.36 C steel	Annealed	79,200	34,000	25,000	...	0.74	...
0.36 C steel	Quenched, drawn	103,600	52,000	19,000	...	0.37	...
0.49 C steel	Annealed	82,900	34,000	23,000	...	0.68	...
0.49 C steel	Quenched, drawn	110,900	53,000	20,000	...	0.38	...
1.09 C steel	Annealed	103,400	42,000	23,000	...	0.55	...
Ni Steels							
3.47 Ni, 0.32 C Steel	Annealed	92,500	49,000	29,000	...	0.59	...
3.47 Ni, 0.32 C steel	Quenched, drawn	124,800	69,000	25,000	...	0.36	...
5.37 Ni, 0.52 C steel	Annealed	121,800	55,000	21,000	...	0.38	...
5.37 Ni, 0.52 C steel	Quenched, drawn	132,700	66,000	19,000	...	0.29	...
3.11 Ni, 1.58 Si, 0.47 C steel	Annealed	137,500	65,000	22,000	...	0.34	...
3.11 Ni, 1.58 Si, 0.47 C steel	Quenched, drawn	251,300	108,000	12,000	...	0.11	...
Cr-V Steels							
0.88 Cr, 0.14 V, 0.46 C steel	Annealed	58,800	42,000	22,000	...	0.52	...
0.88 Cr, 0.14 V, 0.46 C steel	Quenched, drawn	160,600	69,000	17,000	...	0.25	...
Stainless Steels							
14.5 Cr, 0.38 C steel	Annealed	94,400	52,000	36,000	36,000	0.69	0.69
14.5 Cr, 0.38 C steel	Quenched, drawn	117,400	53,000	37,000	27,000	0.70	0.51
17.3 Cr, 8.2 Ni, 0.16 C steel	Hot-rolled	125,300	50,000	50,000	25,000	1.00	0.50
17.7 Cr, 25.3 Ni, 0.39 C steel	Hot-rolled	115,200	54,000	45,000	34,000	0.83	0.63
10.9 Cr, 34.7 Ni, 0.39 C steel	Hot-rolled	112,300	57,000	41,000	22,000	0.72	0.39
11.8 Cr, 0.11 C iron	Annealed	79,800	41,000	34,000	14,000	0.83	0.34
13.8 Cr, 0.09 C iron	Quenched, drawn	84,900	50,000	35,000	18,000	0.70	0.36

* Well water contained 2 ppm of calcium sulfate, 200 ppm of calcium carbonate, 17 ppm of magnesium chloride, 140 ppm of sodium chloride.
† Water from the Severn River having about one-sixth the salinity of sea water.
‡ Fatigue limit in air.
§ Corrosion fatigue limit divided by endurance limit.

condition suffer greater damage than when in the normalized, annealed, or hot-rolled condition. (5) Chromium is the most effective alloying element for increasing the corrosion fatigue strength in the absence of hydrogen sulfide, and nickel is the most

TABLE 2. CORROSION FATIGUE LIMITS OF STEELS IN OIL WELL BRINE*

Based on 10^7 stress cycles at 1750 cycles per min.

Steel	Composition				Tensile Strength, psi	Fatigue Limit, psi			Damage Ratio†	
	C	Ni	Cr	Other		Air	Brine	Brine Containing Hydrogen Sulfide	Brine	Brine Containing Hydrogen Sulfide
S. A. E. 1035, hot-rolled	0.30–0.40	...	...	...	88,500	40,600	24,600	10,600	0.61	0.26
S. A. E. 1050, hot-rolled	0.45–0.55	...	...	...	93,400	31,600	19,900	10,900	0.63	0.34
S. A. E. 1050, water-quenched and drawn	0.45–0.55	...	...	...	130,100	60,100	25,400	13,900	0.42	0.23
S. A. E. 2315, hot-rolled	0.10–0.20	3.25–3.75	...	...	79,700	51,900	31,600	23,900	0.61	0.46
S. A. E. 2335, normalized	0.30–0.40	3.25–3.75	...	...	104,300	53,900	39,900	24,900	0.74	0.46
S. A. E. 3130, normalized	0.25–0.35	1.00–1.50	0.45–0.75	...	100,500	55,100	31,600	15,900	0.57	0.29
S. A. E. 4130, water-quenched and drawn	0.25–0.35	...	0.50–0.80	0.15–0.25 Mo	128,500	70,100	26,900	14,100	0.38	0.20
S. A. E. 4615, normalized	0.10–0.20	1.65–2.00	...	0.20–0.30 Mo	91,300	48,600	33,100	22,400	0.68	0.46
S. A. E. 9260, normalized	0.55–0.65	...	...	1.80–2.20 Si	143,800	72,100	24,900	14,900	0.35	0.21
Pearlitic manganese, normalized, 0.38 C, 1.91 Mn	0.38	...	...	1.91 Mn	118,200	56,400	29,400	12,100	0.52	0.21
5% Chrome, oil-quenched and drawn, 0.20 C, 4.80 Cr	0.20	...	4.80	...	130,600	73,900	52,900	15,500	0.72	0.21
Ni-Mn, normalized, 0.30 C, 1.34 Mn, 1.66 Ni	0.30	1.66	...	1.34 Mn	113,000	58,900	26,100	21,900	0.44	0.37
Cu-Ni, 0.40 C, 1.69 Cu, 0.75 Ni	0.40	0.75	...	1.69 Cu	77,300	54,100	33,600	19,600	0.62	0.36
Wrought iron, hot-rolled	...	...	...	...	47,700	30,400	19,600	16,400	0.64	0.54
3.5% nickel wrought iron, hot-rolled	...	...	...	...	57,400	39,600	26,900	19,100	0.68	0.48
3% nickel-molybdenum iron, hot-rolled	...	...	...	...	63,600	45,100	25,400	21,900	0.56	0.49

* The natural oil well brine contained 67,940 ppm of sodium chloride, 2710 ppm of calcium chloride, 988 ppm of magnesium chloride, and the pH was 7.6. This brine was saturated with natural gas and hydrogen sulfide for the sulfide corrosion fatigue tests.

† Ratio of corrosion fatigue limit to air fatigue limit.

effective in the presence of hydrogen sulfide. Service records over a period of many years have shown that these tests furnished a reliable basis for classifying steels in order of performance ability.

TABLE 3. CORROSION FATIGUE LIMITS OF NON-FERROUS METALS*
Stress for 10^8 Cycles at 1450 cycles per min.

Material	Condition†	Tensile Strength, psi	Endurance Limit,‡ psi	Corrosion Fatigue Limit, psi		Damage Ratio§	
				Well Water	Salt Water	Well Water	Salt Water
Nickel, 98.96 Ni	Annealed at 600° F	131,700	51,000	24,000	21,500	0.47	0.42
Nickel, 98.96 Ni	Annealed at 1400° F	77,600	33,000	23,500	21,500	0.71	0.65
Monel metal, 67.5 Ni, 29.5 Cu	Annealed at 800° F	127,200	51,500	24,500	27,500	0.48	0.53
Monel metal, 67.5 Ni, 29.5 Cu	Annealed at 1400° F	81,900	36,500	26,000	28,000	0.71	0.77
Nickel-copper, 48.3 Ni, 48.4 Cu	Cold-rolled	85,800	36,500	21,000‖	25,000‖	0.58	0.68
Nickel-copper, 48.3 Ni, 48.4 Cu	Annealed at 1400° F	78,000	31,500	21,000‖	25,000‖	0.67	0.79
Nickel-copper, 21.2 Ni, 77.9 Cu	Annealed at 400° F	62,400	25,500	21,500‖	23,500‖	0.84	0.92
Nickel-copper, 21.2 Ni, 77.9 Cu	Annealed at 1400° F	47,300	19,000	18,000	18,000	0.95	0.95
Copper, 99.99 Cu	Annealed at 250° F	46,500	16,500	17,500	17,500	1.06	1.06
Copper, 99.99 Cu	Annealed at 1200° F	31,200	9,800	10,000	10,000	1.02	1.02
Aluminum, 99.4 Al	Hard temper	20,500	8,400	4,600	2,800	0.55	0.33
Aluminum, 99.4 Al	Half-hard temper	16,000	7,300	2,800‖	2,800	0.38	0.38
Aluminum, 99.4 Al	Annealed	12,600	5,900	...	2,100	...	0.36
Aluminum alloy, 98.0 Al, 1.22 Mn	Hard temper	29,600	10,700	5,500	3,800	0.51	0.36
Aluminum alloy, 98.0 Al, 1.22 Mn	Half-hard temper	23,800	10,100	5,500	3,800	0.54	0.38
Aluminum alloy, 98.0 Al, 1.22 Mn	Annealed	16,700	6,800	3,200	...	0.47	...
Duralumin	Tempered	69,100	17,000	7,700	6,500	0.45	0.38
Duralumin	Annealed	33,400	13,500	7,500	6,700	0.56	0.50
Brass, 60.0 Cu, 40.0 Zn	Cold-rolled	84,000	24,000	18,000	...	0.75	...
Brass, 60.0 Cu, 40.0 Zn	Annealed	53,000	21,000	18,000	...	0.86	...
Magnesium and Mg alloys (See *Magnesium and Magnesium Alloys*, p. 234)							

* Waters used were same as those described for Table 1.
† Metals treated by low-temperature anneal were in cold-rolled condition.
‡ Fatigue limit in air.
§ Corrosion fatigue limit divided by endurance limit.
‖ Estimated result.

Corrosion fatigue limits of several non-ferrous metals are listed in Table 3.[15] It is clear from these data that the corrosion fatigue limits of all metals listed except

[15] D. J. McAdam, Jr., "Corrosion Fatigue of Non-Ferrous Metals," *Proc. Am. Soc. Testing Materials*, **27** (II), 102 (1927).

copper and aluminum, whether in the cold-worked or annealed condition, are practically the same. The aluminum alloys are damaged severely by corrosion fatigue.

Composition over a wide range has only a slight effect on the corrosion fatigue limits of Ni-Cu alloys. The results for copper are unique, as the corrosion fatigue limit is higher than the endurance limit. McAdam attributes this to the cooling effect of the water, but Gough[16] suggests that the endurance limit is low because of the corrosive action of the air and calls attention to the fact that when this value is determined in high vacuum it is increased by 14%.[17] Under similar conditions the maximum increase for steels was 5%, but annealed brass (70-30) increased 26%.

NOTCH EFFECT FROM CORROSION

The reduction in the endurance limit because of the notch effect of pits produced by previous stressless corrosion has been investigated by Moore[18] and McAdam.[19] Moore found that the endurance limit of 0.45% carbon steel was reduced 22% after it had been previously corroded for 40 days in a spray of 20% sodium chloride solution. Under similar conditions the endurance limit of Duralumin was lowered 35%. McAdam found that the rate of damage decreased with increase in time of prior stressless corrosion for common steels, but that the reverse was true for stainless steels, aluminum bronze, and 70% Ni–30% Cu alloy. After a prior stressless corrosion period of 10 days in well water, the reduction of endurance limit was 17% for ingot iron, 32% for hardened S. A. E. 2330 steel, 27% for annealed S. A. E. 6145 steel, and 34% for this steel in the heat-treated condition. In no case was the damage sufficient to lower the endurance limit to the corrosion endurance limit. McAdam[20] states that when the rate of damage decreases with corrosion time, as for common steels, the corrosion rate is under cathodic control, and when the rate of damage increases with corrosion time, as for stainless steels, 70% Ni–30% Cu alloy, and aluminum bronze, then the corrosion rate is anodically controlled. The damage from corrosion pitting is accentuated when stress is combined with corrosion in the prefatigue stage and it requires less time for equal damage. A given amount of damage can result from numberless combinations of stress range, cycle frequency, and time. A complete discussion of these factors may be found in the papers by McAdam.

Dolan[21] determined the additive effect of mechanically produced and chemically produced notches by corrosion fatigue tests of specimens with fillets or holes in flowing tap water under conditions of reversed cycles of torsional stress. Corrosion fatigue increased the stress concentration resulting from the mechanical notches. The stress concentration factor due only to corrosion fatigue was negligible for hot-rolled S. A. E. 1020 steel, but was found to be as high as 1.85 for heat-treated S. A. E. 3140 steel.

[16] H. J. Gough, "Corrosion Fatigue of Metals," *J. Inst. Metals*, **49** (No. 2), 17 (1932).

[17] H. J. Gough and D. G. Sopwith, "Atmospheric Action as a Factor in Fatigue of Metals," *J. Inst. Metals*, **49**, 93 (1932).

[18] R. R. Moore, "Effect of Grooves, Threads, and Corrosion upon the Fatigue of Metals," *Proc. Am. Soc. Testing Materials*, **26** (II), 255 (1926). "Effect of Corrosion upon the Fatigue Resistance of Thin Duralumin," *Proc. Am. Soc. Testing Materials*, **27** (II), 128 (1927).

[19] D. J. McAdam, Jr., and R. W. Clyne, "Influence of Chemically and Mechanically Formed Notches on Fatigue of Metals," *J. Research Natl. Bur. Standards*, **13**, 527 (October, 1934). (R.P. 725).

[20] D. J. McAdam, Jr., and G. W. Geil, "Influence of Stress on the Corrosion Pitting of Aluminum Bronze and Monel Metal in Water," *J. Research Natl. Bur. Standards*, **26**, 135 (February, 1941). (R.P. 1366.)

[21] T. J. Dolan, "The Combined Effect of Corrosion and Stress Concentration at Holes and Fillets in Steel Specimens Subjected to Reversed Torsional Stresses," University of Illinois Engineering Experiment Station, *Bull.* 293 (1937).

INFLUENCE OF TYPE OF STRESS

The influence of the mean axial stress for stresses ranging from zero to a maximum, and for stresses fluctuating from a mean stress value when the corroding medium was a spray of 3% salt solution, was found[22] to be similar to the effect on the endurance strength in air for 0.5% carbon, 18-8 stainless, 15.0% chromium steels, and for Duralumin and a magnesium alloy containing 2.5% aluminum.

Torsional corrosion fatigue[23] tests made at 350 cycles per min in well water on S. A. E. 6145 steel in both the annealed and heat-treated conditions showed that the corrosion fatigue limit was the same for both conditions and, compared to the corresponding values for reversed bending, the damage ratios were 0.37 and 0.50, respectively. Surface cracks intersected each other at right angles (X-shaped) and occurred at angles of 45 degrees with the axis of the specimen, i.e., normal to the direction of maximum tension.

CORROSION FATIGUE IN STEAM

Corrosion fatigue in steam has been investigated by Fuller.[24,25] The steels included were S. A. E. 2330, 12.5% chromium, 0.10% carbon; three nitriding steels; and three 18-8 stainless steels of different carbon contents between 0.07 and 0.15%. They were tested in various suitable conditions of heat treatment. In the absence of appreciable liquid water and oxygen at temperatures as high as 370° C (700° F), the S. A. E. 2330 steel (119,000 psi tensile strength) and the 12.5% chromium steel were not affected appreciably. Damage to the other steels increased with temperature but, even at 370° C (700° F), a nitrided steel had the highest corrosion fatigue limit. When exposed to a jet of steam in air the corrosion fatigue limits of all the steels except the nitrided steels were lowered markedly. The damage ratio of one of the nitrided steels was 0.80. The nitrided steels tested in steam at 150° C (300° F) also had damage ratios of 0.74 to 0.83 for notched specimens compared to unnotched specimens.

PREVENTION OF CORROSION FATIGUE

Previous discussion has emphasized the importance of the corrosion resistance of a metal in resisting corrosion fatigue. In some instances the corrodent can be treated to reduce its corrosivity (discussed below). The effectiveness of surface coatings of all kinds has been studied extensively. The essential requirements for successful performance of a coating are:[26] (1) It must adhere to the base metal and withstand deformation which the base metal undergoes without rupture. (2) It should preferably be anodic to the base metal; if cathodic, it must be impermeable. (3) It should not reduce the air endurance strength of the base metal. Chromium plating provided considerable protection to S. A. E. 2335 exposed in air to a steam jet.[27] Platings of

---

[22] H. J. Gough and D. G. Sopwith, "The Influence of the Mean Stress of the Cycle on the Resistance of Metals to Corrosion Fatigue," *J. Iron Steel Inst.*, **135**, 293 p (1937); *Engineering*, **143**, 673 (1937).

[23] D. J. McAdam, Jr., "Some Factors Involved in Corrosion and Corrosion-Fatigue of Metals," *Proc. Am. Soc. Testing Materials*, **28** (2), 117 (1928).

[24] T. S. Fuller, "Endurance Properties of Steel in Steam," *Trans. Am. Inst. Mining Met. Engrs.*, Iron and Steel Division, *Tech. Publ.* 294, pp. 3–13 (1930).

[25] T. S. Fuller, "Endurance Properties of Some Well-Known Steels in Steam," *Trans. Am. Soc. Steel Treating*, **19**, 97–111 (1931).

[26] D. G. Sopwith and H. J. Gough, "The Effect of Protective Coatings on the Corrosion-Fatigue Resistance of Steel," *J. Iron Steel Inst.*, **135**, 315 p (1937).

[27] T. S. Fuller, "Endurance Properties of Some Well-Known Steels in Steam," *Trans. Am. Soc. Steel Treating*, **19**, 97–111 (1931).

copper, nickel, and chromium, 0.2 mm (8 mil) thick, on soft steel afforded perfect protection against corrosion fatigue in fresh water for 100 million cycles.[28] Lowering of the corrosion fatigue limit below the air endurance limit for galvanized and sherardized specimens was only half that of uncoated specimens when 0.47% carbon steel of 100,000 psi tensile strength was tested in a stream of tap water for 10 million cycles.[29] In galvanizing, the brittle Zn-Fe alloy formed at the interface is a probable origin of cracks that cause loss in fatigue strength. There was no reduction of fatigue strength for electroplated zinc specimens which were not pickled in preparation for plating. When such a coating is provided, the steel can be stressed up to its endurance limit in air. Electroplated zinc also has been found effective in improving the resistance of Duralumin to corrosion fatigue.[30] Metal-sprayed coatings of zinc, cadmium, and aluminum on steel have been used.[31] Electroplated coatings which are cathodic to steel, such as tin, nickel, and copper, are not recommended[32] for resistance to corrosion fatigue, unless the coatings are impermeable.

Reference was made above to the exceptional resistance of nitrided steels to corrosion fatigue in a steam jet. Inglis and Lake[33] found the corrosion fatigue limit of nitrided steel to be 68% of the endurance limit when tested in river water for 170 million cycles. Another useful property of nitrided steels is low notch sensitivity arising from the high compressive stresses in the nitrided case. It may be partly for this same reason that surface rolling[34] was found to increase the resistance of certain steels to corrosion fatigue and that shot blasting, currently arousing much interest in this country, may produce some beneficial results.

Thus far paints have not proved to be of much protective value, but some useful coating materials might well result from the intensive development of resins now in progress.

USE OF INHIBITORS

It was found[35] that 200 ppm of sodium dichromate completely prevented corrosion fatigue of normalized 0.35% carbon steel by water containing 25 ppm each of sodium chloride and sodium sulfate. In the presence of the inhibitor at stresses above the endurance limit more stress cycles were required to produce failure than for testing in air. Only partial protection was afforded, however, when localized corrosion was induced by rubber washers or bands of paint around the specimens. Sodium dichromate was ineffective in amounts up to 16,000 ppm in waters containing 2000 ppm (synthetic oil well drilling mud water) and 30,000 ppm (synthetic sea water) of sodium chloride, but sodium chromate prevented damage when added in amounts of 2500 ppm and 8000 ppm, respectively.[36] Hot-rolled 0.42% carbon steel was used

[28] R. Cazaud, "Journées de la lutte contre la corrosion," *Chimie et industrie*, **41**, 381–384 (April, 1939).
[29] W. E. Harvey, "Zinc as a Protective Coating against Corrosion-Fatigue of Steel," *Metals & Alloys*, **1**, 458 (April, 1930).
[30] I. J. Gerard and H. Sutton, "Corrosion-Fatigue Properties of Duralumin with and without Protective Coatings," *J. Inst. Metals*, **56** (1), 29, 1935.
[31] Battelle Memorial Institute, *Prevention of Fatigue of Metals*, John Wiley and Sons, New York, 1941.
[32] Krystof, "Über die Haltbarkeit von Zinn- und Zinküberzügen bei Korrosionsdauerbeanspruchungen," *Metallwirtschaft*, **14**, 305 (1935).
[33] N. P. Inglis and G. F. Lake, *Trans. Faraday Soc.*, **27**, 803 (1931) and **28**, 715 (1932).
[34] A. Thum and H. Ochs, "Die Bekämpfung der Korrosionsermüdung durch Druckvorspannung," *Ver. deut. Ing.*, **76**, 915 (1932).
[35] F. N. Speller, I. B. McCorkle, and P. F. Mumma, "Influence of Corrosion Accelerators and Inhibitors on Fatigue of Ferrous Metals," *Proc. Am. Soc. Testing Materials*, **28** (II), 159 (1928).
[36] F. N. Speller, I. B. McCorkle, and P. F. Mumma, "Influence of Corrosion Accelerators and Inhibitors on Fatigue of Ferrous Metals," *Proc. Am. Soc. Testing Materials*, **29** (II), 238 (1929).

for the latter tests, and the conclusions are based on the number of cycles to failure at a stress of 2000 psi less than the air endurance limit. Again, the inhibitors were ineffective when localized corrosion occurred.

In an investigation of corrosion fatigue of steel wire[37] in 0.01 to 1.0 molal sodium chloride solutions it was found that zinc yellow gave better protection than the equivalent concentration of chromate in the form of potassium chromate. This result was attributed to the ability of the zinc ion to act as a cathodic inhibitor. In these tests it was found that increasing the pH from 6.52 to 7.57 was slightly beneficial at high stresses but not at low stresses.

Fretting Corrosion

J. O. ALMEN*

The rapid corrosion that occurs at the interface between contacting, highly-loaded metal surfaces when subjected to slight relative (vibratory) motions has become known as *fretting corrosion*.[1] This form of corrosion is damaging to a large variety of machine parts because:

1. It increases the susceptibility to fatigue failures of dynamically loaded machine elements.
2. It destroys the dimensional accuracy of closely fitted parts.
3. It ruins bearing surfaces, particularly the surfaces of ball and roller bearings.[2]

CHARACTERISTICS OF FRETTING CORROSION

FATIGUE FAILURES

Fretting corrosion is usually characterized by surface discoloration and deep pits in regions where slight relative movements have occurred between highly loaded surfaces. Since fretting corrosion affects loaded surfaces, it follows that the effects of the damage are frequently catastrophic. The deep pits are points of stress concentration which accentuate stresses due to applied loads and, therefore, fatigue failures often have their origin in fretted areas. Unless better corrosion preventives are found than are now available, increasing trouble from fretting corrosion can be expected as machine parts are operated at higher stresses.

Fatigue failures traceable to fretting corrosion are found in many aircraft engine parts, such as connecting rods, knuckle pins, splined shafts, clamped and bolted flanges, couplings, and many others.[3] Examples of two of these failures in different

* Research Laboratories Division, General Motors Corp., Detroit, Mich.

[37] C. G. Fink, W. D. Turner, and G. T. Paul, "Zinc Yellow in the Inhibition of Corrosion Fatigue of Steel in Sodium Chloride Solution," *Trans. Electrochem. Soc.*, **83**, 377 (1943).

[1] G. A. Tomlinson, P. L. Thorpe, and H J. Gough, "Investigation of the Fretting Corrosion of Closely Fitting Surfaces," and Discussion at London Meeting and Manchester Meeting, *Proc. Inst. Mech. Engrs.*, **141**, 223-249 (1939).
Fretting corrosion has also been known as "chafing," "bleeding," and "cocoa."

[2] J. O. Almen, "Lubricants and False Brinelling of Ball and Roller Bearings," *Mech. Eng.* **59**, 415-422 (June, 1937).

[3] H. C. Gray and R. W. Jenny, "An Investigation of Chafing on Aircraft-Engine Parts," *S. A. E. J.* (Trans. Section), **52**, 511-518 (November, 1944).

FIG. 1a. Section through Link Pinhole of Master Connecting Rod Showing Fatigue Cracks with Nuclei (Shown by Arrows) in Fretting Corrosion Areas.

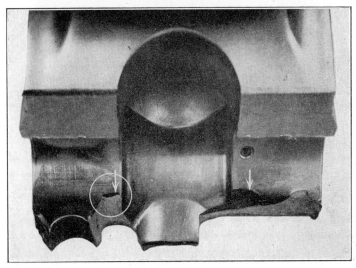

FIG. 1b. Another View of the Same Piece Showing the Fretting Corrosion to Better Advantage. Arrows Indicate Nuclei.

592   SPECIAL TOPICS IN CORROSION

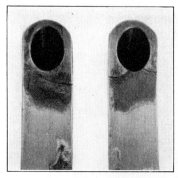
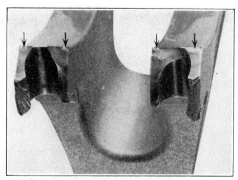

FIG. 2a. Another Example of Fatigue Cracks Originating in Fretting Regions of a Connecting Rod.

FIG. 2b. Fractures of Same Connecting Rod. Arrows Indicate Nuclei of Fatigue Cracks.

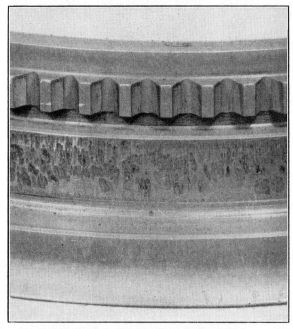

FIG. 3. Section of a Bearing Support Showing Fretted Area Which Caused Loss of Dimensional Accuracy and Resulted in Malfunctioning of the Parts.

types of connecting rods are shown in Figs. 1 and 2. Such failures also occur in railway axle shafts at the wheel seats and in automobile axle shafts, suspension springs, steering knuckles, and other machine parts too numerous to mention.

DIMENSIONAL ACCURACY LOST

The loss of material by fretting corrosion causes malfunctioning of many machine and structural parts such as the bearing support shown in Fig. 3 and the splined shaft in Fig. 4. A long-standing and unsolved service difficulty occurs in the clamp systems joining the adjacent ends of railroad rails. The movements between the rails and their clamp plates as each car wheel passes over the joint cause severe fretting corrosion and loss of material. The loss of material appears as looseness of the clamping bolts

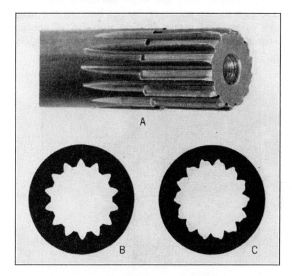

FIG. 4. Views of Splined Shaft Showing Damage Resulting from Fretting.

    A. Shows splined end of a failed pump shaft
    B. Shows original cross section of spline.
    C. Is a cross section showing loss of material due to fretting

which, therefore, require frequent tightening. Since rail joints are not lubricated or otherwise protected, fretting corrosion progresses at a rapid rate and creates a service problem that is as old as the railroad business.[4] Loss of material also results in excessive clearances in splines, taper shafts, and couplings which destroy the desired fit, and the consequent looseness may increase the working load by impact through lost motion or backlash and thereby cause eventual fracture.

BOLT FATIGUE FAILURES

Similarly, fretting corrosion causes loss of material in highly loaded bolted assemblies and consequent looseness of clamping bolts and studs. Looseness of bolts and studs increases their liability to fatigue failure, a hazard that is particularly serious when the bolts or studs are short as in aircraft engine cylinder hold-down studs.[5]

In some instances, such as in press and taper fits, the products of fretting corrosion accumulate in the corroded region to such extent that difficulty is experienced in disassembling the contacting parts.

---

[4] Walter S. Lacher, "Experience of the Railroads Justifies Confidence in Fasteners," *Fasteners*, **1** (No. 4), 4–7, published by the American Institute of Bolt, Nut and Rivet Manufacturers.
[5] J. O. Almen, "On the Strength of Highly Stressed, Dynamically Loaded Bolts and Studs," *S. A. E. J.* (Trans. Section), **52**, 151–158 (April, 1944).

Destruction of Bearing Surfaces

Malfunctioning of ball and roller bearings and sometimes of gears results from the loss of material by fretting corrosion. When such bearings are not in motion except for slight oscillating movements while loaded, as during shipment of electric motors, automobiles, machine tools, and other assembled machines in freight cars, fretting corrosion occurs at the points of contact between the balls or rollers and their raceways. An example of severe fretting corrosion of needle bearings on a universal joint cross is shown in Fig. 5. The shafts on which this joint was mounted

FIG. 5. Universal Joint Cross Showing Fretting Corrosion in Needle Bearing.

were in close but imperfect alignment; thus the needle bearings were subjected to slight oscillating movements as the shafts rotated. The grooves or indentations that are formed by the corrosion ruin the bearings. Such bearing damage is often erroneously attributed to pressure indentation (brinelling) because the products of corrosion are washed away by the bearing lubricant, leaving only reasonably smooth indentations.[6]

Destruction of Protecting Films

The reason for the almost universal susceptibility of metals to fretting corrosion and the circumstances under which fretting corrosion will occur are imperfectly understood. Probably the actual process of corrosion differs little, if at all, from corrosion that occurs under more familiar circumstances, the name *fretting* being merely descriptive of special conditions or environments. Under other circumstances the undisturbed products of corrosion often serve as protective coatings that retard the progress of corrosion to great depths. In fretting corrosion, however, the slipping movements at the interface of the contacting surfaces destroy the continuity of protective films and, therefore, corrosion may advance at a rapid rate and to relatively great depths.[7,8]

[6] J. O. Almen, "Lubricants and False Brinelling of Ball and Roller Bearings," *Mech. Eng.*, **59**, 415–422 (June, 1937).

[7] S. J. Rosenberg and L. Jordan, "The Influence of Oxide Films on the Wear of Steels," *Trans. Am. Soc. Metals*, **23**, 577–613 (1935).

[8] G. A. Tomlinson, "Rusting of Steel Surfaces in Contact," *Proc. Royal Soc.* (A), **115**, 472–483 (1927).

Fretting corrosion may be related to *stress corrosion* or, more accurately, *tension corrosion* since the friction from the relative movements of the contacting surfaces will necessarily stress these surfaces and thereby promote corrosion.

Welding and Corrosion

It is probable that surface damage from another primary cause is sometimes mistaken for fretting corrosion. Highly loaded surfaces subjected to relative vibratory motions may become bonded or welded together in small areas under the influence of high local temperature resulting from friction.[9] The relative motion of the compressively loaded surfaces not only generates sufficient local heat to fuse the contacting surfaces, but also cleans the surfaces by rubbing away contaminating surface films that might prevent welding.[10,11] Subsequent relative motion of the fused areas pulls material out of the surfaces causing pits, scratches, and fragments which may then corrode. Careful inspection is sometimes necessary to determine whether the original trouble was caused by welding or by corrosion.

SUSCEPTIBILITY TO FRETTING CORROSION

When the heat of friction from relative motion between contacting surfaces is not sufficient to cause local welding, damage by corrosion will still occur. Since this corrosion damage will occur at the interface of any two highly loaded contacting surfaces, provided at least one of them is a metal, it seems that the process is corrosion without fusion. Thus fretting corrosion has been observed between such materials as paper and steel,[12] wood and steel,[12] agate and steel,[13] glass and steel,[12,13] and between many combinations of metals and alloys.[12] That fretting corrosion occurs without fusion is also indicated by the reduced rate of corrosion when tests are conducted in nitrogen[14] and in vacuum.[15]

The susceptibility of the different metals and metal combinations to damage by fretting corrosion varies considerably. Among the ordinary metals such as steel, brass, chromium, aluminum, nickel, and glass tested dry, the best resistance to fretting was found in combinations in which one of the metals was brass, and the most susceptible appeared to be stainless steel upon itself or in combination with any other metal. When lubricated, all combinations of metals showed reduced rates of fretting corrosion, the least susceptible being nickel in all combinations except upon itself, and the most susceptible still being combinations that included stainless steel.[16]

[9] H. C. Gray and R. W. Jenny, "An investigation of Chafing on Aircraft-Engine Parts," *S. A. E. J.* (Trans. Section), **52**, 511–518 (November, 1944).

[10] F. P. Bowden and K. W. W. Ridler, "Physical Properties of Surfaces. III. The Surface Temperature of Sliding Metals. The Temperature of Lubricated Surfaces," *Proc. Royal Soc. London* (A), **154**, 640–656 (1936).

[11] H. C. Mougey and J. O. Almen, "Extreme Pressure Lubricants," *Proc. Am. Petroleum Inst.*, Sec. III, 12th Annual Meeting, **12**, 76–81 (1931).

[12] G. A. Tomlinson, P. L. Thorpe, and H. J. Gough, "Investigation of the Fretting Corrosion of Closely Fitting Surfaces," and Discussion at London Meeting and Manchester Meeting, *Proc. Inst. Mech. Engrs.*, **141**, 223–249 (1939).

[13] G. A. Tomlinson, "Rusting of Steel Surfaces in Contact," *Proc. Royal Soc.* (A), **115**, 472–483 (1927).

[14] Max Fink, "Wear Oxidation, a New Component of Wear," *Trans. Am. Soc. Steel Treating*, **18**, 1026–1034 (1930).

[15] J. O. Almen, "Lubricants and False Brinelling of Ball and Roller Bearings," *Mech. Eng.*, **59**, 415–422 (June, 1937).

[16] G. A. Tomlinson, P. L. Thorpe, and H. J. Gough, *loc. cit.*

PREVENTION OF FRETTING

SLIPPING

Since fretting corrosion is caused by slipping movements between loaded surfaces, it follows that prevention of slipping will also prevent corrosion. Thus it is sometimes possible to overcome fretting corrosion by increasing the load at the interface sufficiently to prevent relative motions of the contacting surfaces. In other cases, friction may be increased by roughening the surfaces to stop slipping and corrosion.

HARDNESS

Corrosive attack under fretting conditions progresses at a more rapid rate in soft steels than in hard steels. It is, therefore, common practice to correct service fretting corrosion troubles by increasing the hardness of one or both contacting steel surfaces.

LUBRICATION

Fretting corrosion is greatly retarded when the contacting surfaces are well lubricated so as to exclude direct contact with air.[16,17,18,19] In many practical cases lubrication is an adequate though incomplete remedy. The ability of a lubricant to exclude air from the regions of contact will vary with its characteristics. High-viscosity oils and greases may not be sufficiently mobile to maintain continuous films, particularly when relatively large movements occur. Low-viscosity oils may flow away too easily from the regions of contact and thus admit air. Special lubricants that combine low viscosity and tenacious qualities can be compounded by the use of aluminum soap. Such lubricants will often afford better protection against fretting corrosion than the more common lubricants, provided operating conditions permit their use.

More complete exclusion of air by mercury-wetted steel surfaces prevented the occurrence of fretting corrosion under circumstances that otherwise produced extensive damage when the surfaces were well lubricated.[20] However, no practical use has been found for this mercury treatment.

Lubrication may also be an effective way to reduce damage by welding and subsequent corrosion. Under favorable conditions an oil film can aid in dissipating local high temperatures, can reduce fretting friction, and can supply a contaminating antiwelding film.

INDUCED RESIDUAL COMPRESSION STRESS

It has been observed that both formation of deep pits by corrosion and tension corrosion cracking are greatly reduced in surfaces that are residually stressed in compression. Likewise, fretting corrosion is retarded when the contacting surfaces have been treated to induce residual compression stress.[21] Surface compression stress

---

[17] J. O. Almen, "Lubricants and False Brinelling of Ball and Roller Bearings," *Mech. Eng.*, **59**, 415–422 (June, 1937).
[18] H. C. Mougey and J. O. Almen, "Extreme Pressure Lubricants," *Proc. Am. Petroleum Inst.*, Sec. III, 12th Annual Meeting, **12**, 76–81 (1931).
[19] K. Dies, "Fretting Corrosion, a Chemical-Mechanical Phenomenon," *Z. des. V. D. I.*, **87** (29/30), 475–476 (1943).
[20] J. O. Almen, "Lubricants and False Brinelling of Ball and Roller Bearings," *Mech. Eng.*, **59**, 415–422 (June, 1937).
[21] J. O. Almen, "Shot Blasting to Increase Fatigue Resistance," *S. A. E. J.* (Trans. Section), **51**, 248–268 (July, 1943).

can be induced by mechanical operations such as shot peening or surface rolling or by surface treatments such as nitriding and carburizing.

Deep penetration of corrosion to form pits is resisted by the surface residual compression stress, and any pits that are formed are less hazardous as nuclei for fatigue fractures. Shot peening offers the double advantage of inducing residual surface compression stress to retard corrosion and roughening the surfaces, thereby increasing friction.[22]

Cavitation Erosion*

S. LOGAN KERR†

DEFINITION AND NATURE OF CAVITATION DAMAGE

Cavitation is basically a dynamic action within a fluid with formation and collapse of cavities in regions periodically below the residual absolute barometric pressure. Damage to metallic or other structures in the presence of cavitation has been called *pitting erosion* or *cavitation erosion*.

The mechanical nature of damage by cavitation has been demonstrated by tests on glass with distilled water where no chemical action is normally expected. The surface of polished glass can be roughened so that it has a dull sand-blasted appearance. Stainless steel can likewise be damaged by this purely mechanical action in the absence of any chemical or electrolytic action. A corrosive environment can, however, accelerate damage to a metal by cavitation. Illustrations showing the nature of cavitation damage are given in Figs. 1, 2 and 3.

The cycle of cavitation action is as follows:

1. Extreme low-pressure areas are produced by flow irregularities.
2. Pockets or "cavities" of vapor form.
3. Pressure and flow conditions change abruptly.
4. Pockets or "cavities" collapse with resultant shock pressures reaching several hundred atmospheres in local areas.

The resultant impact tears out sections of porous or brittle material or causes plastic flow or deformation leading to cracking and breaking out of material. Work hardening may occur in the surface layer of some metals with a corresponding change in physical characteristics.

The cavitation cycle repeats many thousands of times. Ductile metals by this action may resist erosion for a period of time, but breakdown of large areas that take on the appearance of hard hammered surfaces eventually occurs. If the metal is brittle and of relatively low strength, such as cast iron, the appearance of pitting will be quite pronounced. If the material is dense and with high fiber strength, there will be only a roughening of the surface.

EFFECT OF CORROSION

Where conditions conducive to corrosion exist in the presence of cavitation, they can be expected to accelerate the damage, as the products of corrosion will be

* Refer also to *Tests for Damage by Cavitation*, p. 993.
† Consulting Engineer, Philadelphia, Pa.

[22] R. L. Mattson and J. O. Almen, "Effect of Shot Blasting on the Mechanical Properties of Steel (NA-115)," Parts I and II, National Defense Research Committee, *Research Project* NRC-40, OEMsr-1123.

removed more rapidly than in static conditions, and new and fresh surfaces are exposed to action of the environment. Such conditions are found in sea water with cast iron or steel, where more rapid loss of weight occurs than would be experienced with the same cavitation conditions in fresh waters.*

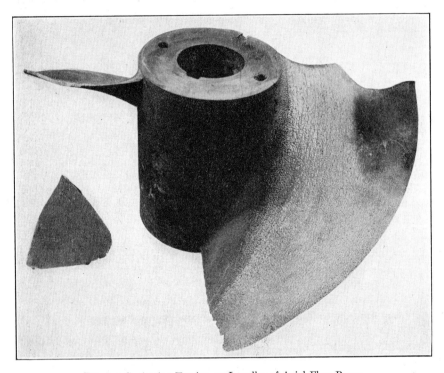

FIG. 1. Cavitation Erosion on Impeller of Axial Flow Pump.

Note "hammered" appearance of Bronze blade and flow lines as well as broken portions of inlet edge. Impeller is reversed in position to show underside of blade. Top surface had little damage.

EFFECT OF TEMPERATURE

Mousson[1] in tests on 18-8 stainless steel found that damage by cavitation was sensitive to temperature (Fig. 4). Damage in service, therefore, is expected to be more severe in summer than in winter.

* Since disintegration produced by cavitation is traced to surface fatigue of a metal, it is logical that a corrosive environment can also accelerate cavitation damage in a manner identical with that operating in corrosion fatigue. (H. Boetcher, *Trans. Am. Soc. Mech. Engrs.*, **58**, 355 [1936].) Private communication, F. N. Speller.

It should be mentioned that creation of superficial high temperatures on moment of impact is another factor that probably accelerates disintegration of the metal through chemical reaction or corrosion. EDITOR.

[1] J. M. Mousson, "Pitting Resistance of Metals under Cavitation Conditions," *Trans. Am. Soc. Mech. Engrs.*, **59** (No. 5), 399 (July, 1937).

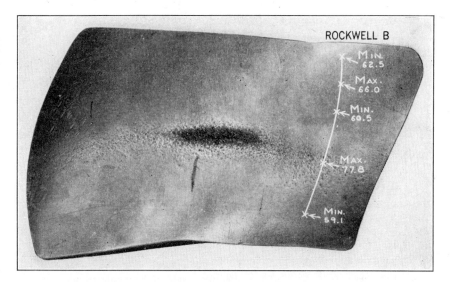

FIG. 2. Cavitation Erosion on Blade of High-Speed Marine Propeller. (Courtesy of Research Laboratory, International Nickel Co.)

Note eroded area in manganese bronze blade in position similar to that on pump impeller in Fig. 1.

FIG. 3. Photomicrograph of Section through Eroded Area of Fig. 2 Showing the Displacement and Rearrangement of Crystals Due to Mechanical Action Resulting from Cavitation. (Courtesy of Research Laboratory, International Nickel Co.)

OCCURRENCE OF DAMAGE

Conditions of cavitation related to the unstable state of fluid flow where the local or general pressures fall below the absolute zero pressure are found in the water passages of hydraulic turbines and pumps and in free discharge valves, throttling valves, propellers, or in fact any place where high velocity flow under medium or low pressure condition is encountered. In at least one high-pressure conduit, a sudden change in direction of flow with high local velocity conditions caused areas of cavitation that damaged the interior of a reinforced concrete tunnel.

The places that are most frequently affected are the discharge side of hydraulic turbine runners, the suction side of pump impellers, the discharge side of regulating valves, and on portions of marine propellers. Local cavitation damage has been produced with a free jet under atmospheric conditions impinging upon an irregular surface. The area beyond the surface irregularity has shown pitting damage identical to that found in pumps or hydraulic turbines.

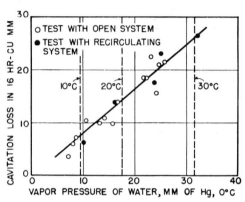

FIG. 4. Effect of Temperature and Vapor Pressure of Water on Cavitation Loss of 18-8 in Venturi-Type Tests.

RELATIVE RESISTANCE OF METALS

Two extensive investigations have been published on the determination of the relative resistance of various metals to damage by cavitation. Although they were made with radically different test procedures, the metals tested show the same relative behavior. Tables 1 and 2 (pp. 995–996) summarize the results for fresh water tests. Vibratory tests reported in Table 1 were also conducted in sea water and showed some differences compared to the fresh water tests, but not enough to warrant a major shift in relative resistance to cavitation erosion. The non-ferrous specimens in sea water and fresh water as a class agreed within the limits of experimental error, whereas alloy cast irons showed lower resistance in sea water.

The austenitic Cr-Ni stainless steels in these tests suffered least in weight loss.

REMEDIES FOR DAMAGE BY CAVITATION

The first and foremost remedy is through design of the turbine, pump, or valve, with proper velocity distribution and sufficiently high residual pressures to eliminate cavitation conditions, either general or local. The correction of existing conditions can be divided into three options: (1) Sufficient air can be admitted to the flowing fluid to relieve the local or general low pressure areas and thus eliminate the cause of cavitation. (2) If this is not practicable, the areas affected can be welded or coated with a dense high-tensile material that will resist damage from cavitation. The chromium stainless steels, including those with nickel components such as 18-8, have proved superior to other surface replacement materials when the composition as deposited on the surface permits grinding to a smooth finish. (3) The reforming by chipping of eroded surfaces may slow down or eliminate the damage. (4) Coatings of resilient materials, such as Thiokol, Neoprene, etc., have been applied and have successfully reduced or eliminated cavitation erosion. A strong bond with the metal is essential. Further experience in varied applications is necessary to establish this method as a complete remedy.

GENERAL REFERENCES

BEECHING, R., "Resistance to Cavitation Erosion," *J. Inst. Engrs. and Shipbuilders in Scotland*, **85**, 210–276 (1942).

Symposium on "Cavitation in Hydraulic Structures," *Proc. Am. Soc. Civil Engrs.*, **71**, 999–1068 (September, 1945).

CORROSION BY STRAY CURRENTS

SCOTT P. EWING*

DEFINITION OF STRAY-CURRENT CORROSION

Stray-current corrosion occurs on buried or submerged metallic structures, and it differs from all other forms of corrosion damage in that the current which causes the corrosion has a source external to the affected structure. This source is usually an electric generator. At some part of the circuit, the current or part of the current passes through the earth. The conductivity of most metals is in the order of one billion times the conductivity of average soils, and therefore any buried or submerged metallic structure acts as a low-resistance path and tends to collect current which may be flowing through the surrounding earth electrolyte. The corrosion of the structure caused by these currents is stray-current corrosion.

SOURCES OF STRAY CURRENTS

Stray currents may be produced by electric railways, grounded electric power systems, electric welding systems, cathodic protection systems, electroplating plants, and other sources. The most important source of stray current is electric railways, because these systems use large currents and in most cases the track rails are used as return conductors. As these rails are not well insulated from the ground, a portion of this current leaks from the rails to the earth.

Most power systems use alternating current, although some d-c systems with grounded neutral may cause occasional local damage. Recent extensive use of welding in ship construction and lack of appreciation of the dangers involved have resulted in occasional damage to ships' hulls during the fitting-out period or while repair work was being done. Damage to ships' hulls from welding currents can be prevented by connecting each welding generator to only one ship.

The tendency toward higher generator voltages in electroplating plants imposes higher voltages between some of the tanks and ground, where it is difficult to maintain adequate insulation. These relatively small leakage currents may produce electrical hazards[1] and may cause damage to tanks and other equipment.

* Bureau of Ships, Navy Department, Washington, D. C. Present address: Research Laboratory, Carter Oil Co., Tulsa, Okla.

The statements, opinions, and conclusions here published are those of the author and in no way express the official opinion of the Navy Department or the Naval Service at large.

[1] V. F. Hanson and F. G. LaViolette, "Location of Ground Faults on Series Electrolytic Cell Systems," *Trans. Electrochem. Soc.*, **82**, 341–352 (1942).

Although instances have been reported[2] where telegraph lines as long as 110 miles have been operated by natural earth currents during magnetic storms, these disturbed conditions occur on the average of only about 16 days per year. The rest of the time the earth currents are so small that their potential gradients seldom exceed one millivolt per kilometer (1.6 millivolts per mile). Such a gradient would produce a current density in a steel conductor of only 0.83 milliampere per square centimeter of its cross section. This is equivalent to 32 milliamperes in a 6-in. steel pipe and is less than the currents usually produced in long steel pipe lines by galvanic potentials.

EFFECTS OF STRAY CURRENTS

Alternating-Current Corrosion

The results of numerous experiments indicate that the corrosive effects of alternating or frequently reversed direct current are usually negligible compared with the natural corrosion rate of the metal in the medium. Damage from alternating current of commercial frequencies is in the order of 1% of the damage produced by direct current of the same magnitude.[3,4,5] In view of this, stray-current corrosion is usually associated only with direct current.

Direct-Current Corrosion

Anodic Areas. Since the anodic and cathodic areas resulting from stray currents are usually widely separated, the anodic and cathodic products of electrolysis are not likely to react with each other. At the anodic areas, metal goes into solution, oxygen may be evolved, and the electrolyte tends to become acid. In most soils and waters, and at the current densities usually encountered in practice, the principal anodic reaction with the commonly used metals is solution of the metal. If the metal dissolves according to Faraday's Law, whereby 96,500 coulombs (26.8 amp-hours) removes one gram-equivalent of the metal, the following amounts of metal would be removed by one ampere in one year:

	Kg	$Pounds$
Iron (Fe^{++})	9.128	20.06
Copper (Cu^+)	20.770	45.60
(Cu^{++})	10.39	22.80
Lead (Pb^{++})	33.866	74.50
Zinc (Zn^{++})	10.665	23.60

The ratio of the actual corrosion loss to the theoretical loss expressed as percentage is called the *corrosion efficiency* or *anode efficiency*. Tests have shown that for current densities varying from 5.0 to 0.05 ma per sq cm (4.6 to 0.046 amp per sq ft) the corrosion efficiency of ferrous metals varies from about 20 to 140%, with the higher efficiency at the lower current density.[6] When the electrical conditions are considered

---

[2] O. H. Gish, "Electrical Messages from the Earth; Their Reception and Interpretation," *J. Wash. Acad. Sci.*, **26** (No. 7), 267–289 (1936).

[3] B. McCollum and G. H. Ahlborn, "Influence of Frequency of Alternating or Infrequently Reversed Current on Electrolytic Corrosion," *Bureau of Standards Tech. Paper* 72 (August, 1916). (Out of print. Summary in *Bureau of Standards Circ.* 401 [1933].)

[4] P. D. Morgan and E. W. W. Double, "The Corrosion of Lead by Alternating Current," British Electrical and Allied Industries Research Association, *Rep.* F/T73 (1934). Published by Communication to 1937 Conference on Underground Corrosion.

[5] L. F. Greve and D. L. Levine, *Prevention of Corrosion of Lead Cable Sheaths*, Bureau of Standards Soil Corrosion Conference, 1937.

[6] B. McCollum and K. H. Logan, "Electrolytic Corrosion of Iron in Soils," *Bureau of Standards Tech. Paper* 25 (June, 1913). (Out of print. Summary in *Bureau of Standards Circ.* 401 [1933].)

moderately severe under average conditions of soil moisture, the corrosion efficiency is usually between 50 and 110%. At low current densities, the loss in weight closely approximates the normal corrosion rate added to the Faraday equivalent of the discharged current. (Refer to *Fundamental Behavior of Galvanic Couples*, p. 481.) At higher current densities, oxygen or chlorine is evolved at the anode, with a reduction in the corrosion efficiency. The usual and recommended practice in calculating metal loss from the ampere-year discharge is to assume the Faraday equivalent given above.[7]

Anodic attack of lead is indicated if the corrosion product contains greater concentration of chlorides than the surrounding environment.[8,9]

Cathodic Areas. At the cathodic areas hydrogen is evolved or it combines with oxygen if oxygen is available, and the solution tends to become alkaline. High cathodic current densities do not injure ferrous metals or copper, because these metals are not appreciably affected by high-alkali concentrations. However, coatings applied to pipes and cable sheaths to prevent soil corrosion may be injured by excessive cathodic voltages. The applied potential tends to drive moisture into the coating (electro-osmosis), which lowers its electrical resistance. Accumulation of alkali on the metal under the coating causes saponification of oil paints. Hydrogen evolution causes excessive blistering and loss of bond of many types of coatings.

Care must be taken not to make lead cable sheaths excessively negative, especially in soils containing appreciable amounts of sodium or potassium salts, because the highly alkaline products which form on cathodic areas are corrosive to lead.[8] The alkali reacts with the lead to form plumbites, and red crystals of lead monoxide or litharge are precipitated from the saturated plumbite solution.

METHODS FOR PREVENTING STRAY-CURRENT CORROSION

It is obvious that stray-current corrosion would not occur if electrical systems were entirely insulated from earth. It is possible to build insulated electric railway systems, but it is more economical to use the rails as return conductors. Proper grounding of electric power circuits is necessary to reduce electric shock hazards. Hence stray-current corrosion will probably never be eliminated by the method of completely insulating electrical systems.

Some structures are more vulnerable to stray-current corrosion than others. A small metal loss from an underground cable sheath may cause penetration and failure of the cable if concentrated in one pit. The relatively thin metal used in these sheaths and the relatively greater volume of lead removed by a given current add to their vulnerability. Penetration of a gas pipe line may result in fire or explosion. On the other hand, railway rails are on or near the surface so that corrosion can be readily observed and repaired. Heavy rails will stand considerable corrosion before they become seriously weakened.

The following procedure helps prevent damage by stray currents: (1) Arrange the electric circuits so as to keep the current within the metallic system, so far as this is economically feasible. (2) If currents must discharge so that metallic structures are anodic, arrange to have the discharge occur from structures that are not readily

---

[7] O. Gatty and E. C. R. Spooner, *Electrode Potential Behaviour of Corroding Metals in Aqueous Solutions*, Oxford University Press, London, 1938.

[8] R. M. Burns, "The Corrosion of Metals. II. Lead and Lead Alloy Cable Sheathing," *Bell System Tech. J.*, **15**, 603–625 (1936).

[9] W. G. Radley and C. E. Richards, *Determination of the Cause of Corrosion Failure of Lead Sheathed Cable in Buried Conduit*, Report, Third Bureau of Standards Soil Corrosion Conference, 1937.

damaged or from purposely built anodes. (3) Make the more vulnerable structures cathodic. (4) Keep the anodic current densities as low as possible, except on purposely built anodes where the current densities should be as high as is economically possible.

There are two diametrically opposed methods for accomplishing the above objectives, namely, the so-called *drainage method* and the *insulation method*. Every system of stray-current control is modeled after one or the other of these methods. In both methods stray currents from railway systems are controlled as much as is feasible by keeping the potential drops in the rail system as low as possible. This is accomplished by keeping the resistance between rails and ground high, by keeping the rail bonds in good repair, and by the use of insulated negative feeders extending for some distance from the substations. The current flow in the feeders is adjusted with series resistances in short feeders, and boosters in long feeders, so as to keep the entire rail system at nearly the same potential by reducing IR drops in the rails.[10,11] In cities where the water table is high, soil resistances are in general low, and track insulation is difficult to maintain. The above drainage method is then used more extensively.

The insulation method is adapted to the opposite condition where the soil resistances are high. Insulated joints are inserted if necessary so that the underground systems are broken up into a large number of relatively short conductors. Railway currents are kept in the rails so far as this is possible, and the current that leaks off does not tend to flow in pipes and cables.

If all underground structures were about equally vulnerable and if the resistances between structures and earth were about the same, most stray-current systems could be bonded so that little damage would occur. Unfortunately cast iron pipe joints are usually of rather high resistance, and the surface resistance of the pipe is low. Insulating joints cannot be easily placed in welded steel lines, and these lines are sometimes heavily surface coated. Telephone and power cables are in ducts so that the surface insulation is comparatively high, and these structures are therefore especially vulnerable to stray currents. Hence, within the same city, it may be advantageous to consider cast iron pipes as part of an insulated system, whereas drainage is feasible and preferable on cable sheaths.

DETECTION OF STRAY-CURRENT CORROSION

Stray-current failures can usually be distinguished from corrosion failures by electrical tests (discussed below). Failures which repeatedly occur in the same place are likely to be stray-current failures. Corrosion failures will probably be more widely scattered and gradually increase in frequency with the age of the pipe line or cable.

The stray-current conditions at cable failures and pipe leaks are usually investigated, but failures and leaks should not be the only means for locating stray-current exposures. Adequate maintenance of underground structures requires more or less continuous electrolysis testing so that serious exposures will be discovered and corrected before too much damage has been done.

Test Methods

Voltage measurements between metallic structures are useful in that a fluctuating voltage indicates the presence of stray currents and, if the time variations in voltage

---

[10] E. B. Rosa and B. McCollum, "Electrolysis and Its Mitigation," *Bureau of Standards Tech. Paper* 52 (November, 1918). (Out of print. Summary in *Bureau of Standards Circ.* 401 [1933].)

[11] *Report of the American Committee on Electrolysis,* American Institute of Electrical Engineering, 1921

can be correlated with the load on some power station, the source of the stray current is fixed. A small steady potential between two structures gives no indication of the extent or location of corrosion, or whether a steady current flows, or what the appropriate remedy may be. If the potential is caused by an outside source, the positive structure is corroding and the remedy is to bond the structures. If the potential is galvanic, the negative structure may be corroding and the remedy is to leave the metallic circuit open.

The potential drop along a metallic conductor of known dimensions may be used to determine the current in the conductor.

The potential difference between a pipe or cable sheath and a driven pipe or a non-polarizing or reference electrode in contact with near-by soil will give only a rough indication of whether the pipe or cable is anodic or cathodic with respect to an external current. The potential differences between two non-polarizing electrodes, along with certain other calibration data, may be used to calculate the current flow to or from an isolated pipe or cable. The electrodes may be placed near the pipe in an excavation[12] or on the surface of the soil over the pipe.[13]

Current measurements by the potential drop or any other convenient method, taken at intervals along a pipe line, may be used to determine the anodic and cathodic areas on the pipe, provided the currents are reasonably steady. When currents are continuously fluctuating, as is usually the case, simultaneous readings of two or more instruments, with appropriate methods of calculation, are necessary in order to determine just what is taking place.

Network methods are based on the fact that the only quantities which remain reasonably constant in stray-current circuits are the resistances. In order to determine these constants, it is necessary to take simultaneous readings on at least two separate instruments. The quotient of two measurements or the slope of a straight line usually obtained by plotting pairs of simultaneous measurements against each other provides a means for obtaining the true constants of the circuit. Methods for quickly obtaining the essential data in the field, and interpreting these data so as to get the essential facts, have been developed to a high state of perfection and usefulness.[13]

COOPERATIVE INVESTIGATIONS

In most cities the various utility systems are owned by several independent organizations and each organization is primarily concerned with its own structures. Each organization would prefer to have its structures cathodic to all other structures, but, of course, this is impossible. The electric railways produce most of the stray currents, whereas the pipes and cables of other utilities are likely to be damaged by these currents. Since the separately owned systems are intimately interwoven with each other, anything that any one system does is likely to affect the structures of all the others. It is obvious that control of stray-current corrosion can best be accomplished by the cooperative efforts of all utilities in the area. Cooperating Electrolysis Committees have been organized in many cities. In some cases these committees have continued to function, the only limit to their effectiveness being the skill of the engineers. In other cases, some utilities have felt their interests could best be protected by legal methods, and this usually forces all the utilities in the area to rely for protection on legal rather than engineering talent. However, stray-current damage

[12] B. McCollum and K. H. Logan, "Practical Application of the Earth Current Meter," *Bureau of Standards Tech. Paper* 351 (1927).

[13] J. M. Pearson, *Electrical Instruments Applied to the Study of Pipe Line Corrosion*, A. P. I. Sixteenth Annual Meeting, November, 1935.

to underground structures is now certainly very much less than it was some twenty-five years ago, and the amount of damage per year from this cause is probably decreasing. There are several reasons for this. Many poorly maintained suburban electric lines have been abandoned, and the loads on the remaining lines are lower because much of the traffic is being carried by automobiles, busses, and trucks. Also the causes of stray-current damage are better understood, and effective mitigation measures have been adopted in most cases.

CORROSION-RESISTANT ANODES

Colin G. Fink*

In the electrolytic refining of copper, as well as in the electrowinning of copper, zinc, manganese, cadmium, and other metals, certain electrolytic cells are equipped with corrosion-resistant or so-called insoluble anodes.

The purpose of these anodes is a simple one in the case of the copper refinery. Here the electrolyte builds up in copper because of higher anode current efficiencies than cathode current efficiencies. This build-up in copper tends to increase the cell voltage, but there is also the danger of salts crystallizing out, interfering with the normal operation of the refinery. Furthermore, impurities such as nickel tend to build up in the electrolyte and so it becomes imperative to keep the concentration, of, say, nickel in the electrolyte below a fixed maximum. Appreciable codeposition of nickel otherwise takes place which will seriously reduce the electrical conductivity of the copper. Cells with insoluble anodes are employed as part of the procedure adopted for the elimination of nickel.[1]

In the electrowinning of metals such as copper, zinc, and manganese from leach liquors, the insoluble anodes serve a double purpose: (1) the prevention of any anode metal going into solution; (2) the regeneration of sulfuric acid or building up of the acid content of the electrolyte so as to become available again for further leaching of fresh quantities of ore. It will be readily appreciated that this second function of the insoluble anode is a most essential one in completing the electrowinning cycle.

In Table 1 are listed the principal insoluble anodes in commercial use today.

MISCELLANEOUS APPLICATIONS OF INSOLUBLE ANODES

In 15 to 20% (by weight) sulfuric acid, lead, of many metals tested, was found to be most resistant to anodic dissolution.[2] Furthermore, the addition of as little as 0.05% tellurium reduced the corrosion loss for lead to one-tenth.

Iron with 13 to 14% silicon is highly resistant to sulfuric acid attack, but the silicious protective surface film is brittle and easily dislodged. Furthermore, some iron enters the solution.

Titanium in amounts of 25%, added to the metal, protects iron, nickel, and copper against anodic corrosion.

Anodes in the electrolysis of molten salts are exclusively carbon or graphite for the fused halide baths and nickel for the fused alkali hydrates. For fused borates and phosphates, graphite anodes are preferred.

In cathodic pickling, that is, making the "work" or article being pickled cathode,

* Division of Electrochemistry, Columbia University, New York, N. Y.

[1] For further details, see C. G. Fink, *Handbook of Nonferrous Metallurgy*, D. M. Liddell, Editor, 2nd Ed., Chapter XVIII, McGraw-Hill Book Co., New York, 1945.
[2] F. R. Morral and J. L. Bray, *Trans. Electrochem. Soc.*, **75**, 427 (1939).

TABLE 1. GROUP A: FOR ACID SULFATE ELECTROLYTES

Composition	Surface Film	Remarks
Magnetite*	Fe_2O_3	Limited use in Europe. Very brittle.
Lead + 4% to 12% Sb†	$PbO_2 + Sb_2O_3$	Universally used in most electrolytic copper refineries.
Duriron (13% Si)	$Fe_2O_3 + SiO_2$	Limited use in copper refineries. Brittle.
Copper silicide‡	PbO_2, MnO_2, SnO_2	Limited use by Chile Copper Co. for making high-purity, silver-free copper. Brittle.
Lead + Ag§	$PbO_2 + Ag_2O(+ Sb_2O_3)$	Used in all electrolytic zinc plants; at Chuquicamata; and in one of the Mn recovery plants

GROUP B: FOR MISCELLANEOUS ELECTROLYTES

Platinum	PtO_2	For electrolytic chlorate production.
Lead and Lead-tin ‖	PbO_2, SnO_2	For chromium plating.
Iron, nickel-plated¶	Ni_2O_3	For electrolytic $H_2 + O_2$ production.
Graphitized carbon	CO_2	In over 90% of the alkali-chlorine cells.
Lead and silver	$PbO_2, Ag_2O, AgCl$	In a few alkali-chlorine plants in Europe.
Lead-tin-antimony-cobalt**	$PbO_2, SnO_2, Sb_2O_3, Co_2O_3$	In the Knoxville plant for electrolytic manganese.

* The magnetite anodes as used in Germany and Italy are hollow and closed at the lower end. A heavy copper plate on the inside wall counteracts the high electric resistance of magnetite. Aside from being decidedly brittle, these anodes do dissolve slowly, introducing objectionable iron into the electrolyte.
† The antimony was originally added to the lead to counteract sagging or "creep." But it also adds to its insolubility. Antimonial lead cannot be used in the electrolytic zinc cells since a trace of antimony in the electrolyte seriously interferes with the electrodeposition of zinc. In these cells lead-silver anodes are used instead. (See below.)
‡ The copper silicide anode has been in use at Chuquicamata for over 20 years. Colin G. Fink, U. S. Patents, 1,441,567; 1,441,568 (1923). Now being partly replaced by a lead-antimony-silver anode.
§ See footnote †. Also U. C. Tainton et al., *Trans. Am. Inst. Mining Met. Engrs.*, **85**, 192 (1929). Colin G. Fink and L. C. Pan, *Trans. Electrochem. Soc.*, **46**, 349 (1924); **49**, 85 (1926). Colin G. Fink and R. W. Low, U. S. Patent 1,740,291 (1929).
‖ Chas. H. Eldridge, U. S. Patent 1,975,227 (1934).
¶ B. P. Sutherland, *Trans. Electrochem. Soc.*, **85**, 183 (1944).
** Colin G. Fink and M. Kolodney, U. S. Patent 2,320,773 (1943); C. L. Mantell, 2,340,400 (1944). With ordinary lead anodes or lead-antimony anodes about half the manganese is deposited on the anode (as MnO_2) and half (as metal) on the cathode. With the above anode, practically all the manganese is deposited on the cathode.

the anode is graphite or carbon in muriatic acid pickles; lead or antimonial lead in sulfate pickles; and lead or tin anodes in Bullard-Dunn pickling tanks.

In electroplating, the anode is usually composed of the same metal or alloy that is being deposited on the article acting as cathode. There are, however, exceptions. In chromium plating, the anode is lead, or preferably lead with 6% tin.[3] In gold plating from cyanide baths, the insoluble anodes used are either hard carbon, stainless steel (18-8), Nichrome, or platinum.

[3] Charles H. Eldridge, U. S. Patent 1,975,227 (1934).

ACTUAL LIFE OF ANODES IN COMMERCIAL SERVICE

The actual life of an insoluble anode depends upon a large number of factors: the thickness of the original anode; the thickness of the protective film or crust; the presence of small amounts of impurities in the electrolyte (for example, small amounts of chlorides in sulfate electrolytes are detrimental); temperature fluctuations; unintentional but effective impurities in the original anode; etc. Accordingly, it has been common practice to determine relative merits of one anode as against those of another by running two or more cells in series and trying out two or more different anodes at a time.

Lead anodes used in copper winning should last 15 years; Pb-Ag anodes in zinc plants, 10 to 15 years; carbon anodes in alkali-chlorine plants, at least 2 to 3 years.

WHAT MAKES AN ANODE INSOLUBLE

Aside from the fact that a metal such as platinum is most frequently used in the laboratory as an insoluble anode, simply because it is practically insoluble in the electrolyte used — with or without current applied — the insoluble anodes used commercially are not only much cheaper than platinum in first cost, but also cheaper as far as electric power consumption is concerned. However, contrary to the usual behavior of platinum, a number of the insoluble anode metals used commercially are relatively soluble without current applied but decidedly insoluble when serving as anode.

The theoretical basis of the insolubility of an anode was propounded some years ago[4] and has been found to apply generally. At the anode surface there are primarily two fundamental reactions: (1) the attack on the metal of the anode by the discharged anion, such as SO_4^{--}, causing the metal to go into solution; and (2) the discharge of OH^- ions (supplied by the water) to form oxygen:

$$SO_4^{--} + Me \longrightarrow MeSO_4 + 2e^- \qquad (1)$$

$$H_2O \longrightarrow H^+ + OH^- \longrightarrow 2H^+ + O + 2e^- \qquad (2)$$

The relative velocity of these two reactions more or less determines the degree of solubility of the anode under the action of the current. It is not difficult to picture the ideal case for a perfect insoluble anode. The velocity of the second reaction is infinitely greater than that of the first reaction. The more we can catalyze the formation and evolution of gaseous oxygen, the more insoluble will be the anode.

One very important result in catalyzing the formation and evolution of gaseous oxygen is the marked drop in cell voltage. Accordingly, not only is the ideal insoluble anode "everlasting," but also, because of the catalytic film, the cell voltage is decidedly lower than with anodes without such catalytic film.

THE TWO COMPONENTS

In general there are two essential components of an insoluble anode: the metal or alloy proper of the body of the anode and the catalytic surface film. Among the catalytic film-forming metals that have given good results — low anode solubility and low anode potential — are Ni_2O_3, Co_2O_3, MnO_2, PbO_2, Ag_2O, SnO_2, and combinations of these. Usually a combination of metal oxides is more effective and efficient than one or the other oxide alone.

[4] C. G. Fink, *Ind. Eng. Chem.*, **16**, 566 (1924).

These oxides are usually "formed" on the surface by anodic action. Thus, lead dissolved in silver, so treated, forms a catalytic film composed of $PbO_2 + Ag_2O$. Similarly, metallic cobalt forms a film of Co_2O_3.

These oxides have thus made it possible to so lower the anode potential below the minimum voltage at which anode metal will go into solution that the anode energy efficiency for oxygen production is well nigh 100%. A pure copper plate can be rendered anodically insoluble in an acid copper sulfate solution by applying an adherent film of a combination of manganese, lead, and tin oxides to the copper surface. But without current such a plate will go into solution as though the film were non-existent. The addition of silver to various anode metals has had a most remarkable effect in reducing the solubility of anode alloys as well as lowering the anode potential.[4]

CORROSION OF LEAD AND LEAD-ALLOY CABLE SHEATHING

R. M. BURNS*

Telephone and power cables are sheathed with extruded coatings of lead or certain lead-rich alloys, the principal purpose of which is to exclude moisture and electrolytes. Although unalloyed commercial lead is generally used for power cables, it is customary practice to employ lead alloys for telephone cables. The first of these alloys to be used contained 3% tin, but for more than 30 years the standard alloy for telephone cables has been lead containing 1% antimony. More recently lead hardened with 0.03% calcium has been used to a limited extent. The most suitable type of lead for the manufacture of Pb-Sb sheathing is one designated as chemical grade lead. This product runs high in copper and low in bismuth.

The selection of lead and its alloys for cable sheathing has been dictated by such physical properties as ease of extrudability, its flexible character over the range of outdoor temperatures, its resistance to fatigue cracking, and finally by its relatively high degree of corrosion resistance. The low incidence of cable failures due to corrosion in the more than 200,000 miles of sheathed cables in this country is the result of the inert character of lead and, in particular, the vigilance of corrosion engineers who, recognizing the environmental conditions likely to give rise to corrosion, contrive to eliminate these from the cable plant. Laboratory studies and long experience in the field have shown that the composition of cable sheath is of less importance than the nature of the environment in determining whether or not corrosion will occur.

THE NATURE AND ACTIVITY OF CORROSION CELLS ON CABLE SHEATHING

Corrosion cells on cable sheathing owe their origin either to sheath composition or to some inhomogeneity of the surrounding environment.[1] Metallic impurities such as copper, bismuth, and zinc or alloying elements such as calcium, cadmium, or antimony may, if present as separate phases, form corrosion cells of galvanic type with lead. A measure of the activity of these cells may be obtained by comparing the relative rates of sulfation of lead specimens containing these constituents with that of spectroscopically pure lead. Taking the rate of sulfation of the high purity

* Bell Telephone Laboratories, Murray Hill, N. J.

[1] R. M. Burns, *Bell System Tech. J.*, **15**, 603–625 (1936).

grade as unity, the rates of reaction of several grades of lead and certain cable sheath alloys have been determined in the laboratory to be as follows:

Spectroscopically pure lead (99.999% Pb)	1.00
Corroding lead (A. S. T. M. — Grade I, 99.94% Pb)	1.25
Chemical lead (A. S. T. M. — Grade II, 99.90% Pb; contains 0.06% Cu)	1.53
Common lead (A. S. T. M — Grade III, 99.85% Pb; contains 0.13% Bi)	1.42
Corroding lead alloyed with 0.15% Sn, and 0.25% Cd	1.18
Chemical lead alloyed with 1.5% Sn, and 0.25% Cd	1.33
Corroding lead alloyed with 3% Sn	1.42
Chemical lead alloyed with 0.04% Ca	1.25
Corroding lead alloyed with 0.04% Ca	1.33
Common lead alloyed with 0.04% Ca	1.80
Corroding lead alloyed with 0.5% Sb and 0.25% Cd	4.00
Chemical lead alloyed with 0.5% Sb and 0.25% Cd	4.00
Chemical lead alloyed with 1.0% Sb	5.00

These results indicate that the surface reactivity of lead is markedly increased by the presence of small proportions of other metals, notably antimony, and in a lesser degree by copper, bismuth, cadmium, and tin. In practical terms this means that when lead so alloyed is used in sulfuric acid it develops corrosion-resistant sulfate films more rapidly than pure lead. Calcium, on the other hand, appears to have little influence.

Inhomogeneity in physical state such as strained structures, or difference in grain size, is capable of giving rise initially to corrosion cells of 2 or 3 millivolts, but it appears that the self-annealing tendency of lead at ordinary temperatures relieves these strains, causing the cells to disappear.

Inhomogeneity of the environment at the surface of cable sheathing is one of the most important sources of corrosion cells. Of these, differential aeration which induces oxygen concentration cells is the most common. Cells of this kind may develop as a result of contact of the sheath with soil particles, making it unsafe practice to bury lead and lead alloy covered cables directly in the soil without the use of protective coatings. The areas of reduced oxygen concentration at the points of contact become the corroding anodes of differential concentration cells.

Cables in conduit are seldom subject to contact with inert objects of the character which lead to the establishment of oxygen concentration cells. The environment of these cables is very complex, however. It consists of surface and soil waters, silt deposition, and subterranean atmospheres which are usually higher in carbon dioxide and lower in oxygen content than the outside air, and which may be contaminated with illuminating gases and sewage vapors. Free lime leached from concrete structures, alkalies derived from the electrolysis of sodium or calcium chlorides used to melt ice on streets, and seepage from sewers or industrial plants are sometimes contaminants. Cables housed in conduit made from acidic wood or in wood conduit which has been improperly processed in the preservative treatment may be subjected to acetic acid vapor.

EFFECT OF ENVIRONMENT ON THE OPERATION OF CORROSION CELLS

From the foregoing it is apparent that a diversity of electrolytic cells may exist at the surface of cable sheathing and that these are subject in the underground plant to environmental conditions of great complexity. There are present constituents which may accelerate cell activity, that is, cause corrosion, and there are other elements of the environment which may retard or stifle cell activity and thereby prevent corrosion. The influence of these constituents depends both upon concen-

CORROSION OF CABLE SHEATHING

tration and upon specific properties such as solubility and depolarizing strength. The occurrence of corrosion and the nature of corrosive attack, if it does take place, may be determined by the ratio of the corrosive agents to the protective ones in the environment. For high values of this ratio the sheathing will be more or less uniformly corroded, for low values it will be protected, but for intermediate values of this ratio the surface of the sheathing will be only partly protected, with the result that corrosion will be localized in the form of destructive pitting. Some of the principal constituents of the environment which affect the corrosion behavior of cable sheathing may be classified as follows:

Corrosive	Protective
Oxygen	Silicates
Nitrates	Carbonates
Alkalies	Sulfates
Chlorides	Certain organic compounds
Organic acids	Colloidal substances
Soils of large texture	Soils of small texture

OXYGEN

Oxygen is listed here as a corrosive element, and in the underground plant where electrolytic corrosive cells exist, of the type previously described, it is one of the most influential corrosive agents. Its action consists in the depolarization of the cathodic areas of the sheath. However, under other circumstances oxygen may have a protective influence in the formation of a continuous oxide film. For example, in the case of aerial cables and even underground cables which are maintained relatively dry, oxygen contributes to the development of a more or less invisible film which may preserve the metal indefinitely if not physically disturbed. Cables on reels or cables recently installed may become superficially coated with a hydrated oxide which may become carbonated. This is loosely adherent and is of little significance.

The protective film formed on cable sheathing in the atmosphere does not prevent cracking of the sheath which sometimes occurs in sections that are subjected to repeated stresses or prolonged vibration. There is, on the contrary, some evidence that corrosion reactions contribute to this process of intergranular fracture which is one of the principal causes of failure in aerial cables.

NITRATES

The presence of nitrates in acid soil waters leads to accelerated corrosion because under these circumstances nitrates are cathodic depolarizers. Furthermore, owing to the high solubility of lead nitrate, there may be some impedance to the formation of protective films which might otherwise develop. The nitrate content of soil waters is very low — a few parts per million ordinarily — but occasionally contamination from sewers and surface drainage leads to a several-fold increase. Nitrate concentrations of 20 to 425 ppm have been found to cause serious sheath corrosion. In these cases the corroded area is usually black in appearance and covered with loosely adherent finely divided lead and antimony, if antimony is a constituent of the sheath.

ALKALIES

Alkalies seriously corrode cable sheathing. The attack is characterized usually by the presence of red tetragonal crystals of lead monoxide that crystallize out of saturated solutions of the unstable alkali plumbites which are the primary corrosion products of alkaline attack on lead. If detected before failure of the cable, the action usually can be stopped by removing the source of the alkali and thoroughly flushing the cable conduit with water.

CHLORIDES

Chlorides in dilute solution may have a mildly accelerating influence on sheath corrosion owing in part to the increase in electrolytic conductance which is produced, and in part to the destructive action of chloride ions on protective films produced by other constituents which are present. However, in solutions in which the concentration of chlorides approaches that of sea water, corrosive action appears to be reduced probably because of chloride film formation and the reduction in oxygen solubility.

ACETIC ACID

The severe action on lead produced by dilute organic acids, particularly acetic acid, has long been known.[2] In the presence of the carbon dioxide of the atmosphere, the product is the pigment, white lead. This forms as a porous white incrustation over the surface of the sheath. The corrosiveness of air laden with acetic acid vapor depends upon the fact that the corrosion cells in this case do not become polarized or suffer reduction in voltage during the course of the action. A constant source of hydrogen ions is provided by the acid, the precipitation of lead as carbonate by the action of carbon dioxide maintains a low concentration of lead ions, and oxygen depolarizes the cathodic areas. The carbonate or basic carbonate is precipitated at an appreciable, although very small, distance from the metal surface, and offers no hindrance to the corrosion.

SOILS

The action of soils in setting up oxygen concentration cells at points of contact of soil particles with sheathing has been mentioned. The activity of these cells depends largely upon the texture of the soils.[3] In coarse soils the corrosive action may cause perforation of the sheath, whereas in soils of fine silty nature there may be little or no attack. In the latter case the relatively smaller cathodic area becomes so highly polarized that the cells are maintained in a polarized state in spite of the depolarizing action of the oxygen present in soil air. It may be observed, incidentally, that as soil particle size decreases, the environment at the metal surface becomes increasingly homogeneous until ultimately differential aeration disappears.

The corrosion resistance of underground cables depends principally upon the presence of silicates, carbonates and to a lesser extent sulfates which induce passivity by a process of anodic polarization, that is, formation of protective films. The silicate ion appears to be the most important single protective constituent although there is evidence that the carbonate ion exerts a marked retarding influence on corrosion. It seems likely that inhibitive effects of these components are additive.

Analyses of water samples from cable manholes and subways have shown silicate contents of 2 to 143 ppm. The rate of corrosion of Pb-Sb cable sheathing in distilled water to which was added about 10 ppm of silicate ion (derived from a suspension of acid-washed silica flour) has been found to be 5 mdd as against 85 mdd in distilled water. Similarly specimens of sheathing are not appreciably attacked in saturated solutions of calcium silicate which contain less than 10 ppm of silicate ion. The

---

[2] R. M. Burns and B. A. Freed, *J. Am. Inst. Elec. Engrs.*, **47**, 576 (1928).
[3] R. M. Burns and D. J. Salley, *Ind. Eng. Chem.*, **22**, 293 (1930).

corrosiveness of certain natural waters has been greatly reduced by the addition of only 10 ppm of sodium silicate (expressed as silicic acid).[4]

Carbonate ions derived either from the soil or soil atmosphere are invariably present in the underground cable plant. The protective quality of the carbonate (or possibly basic carbonate) film may be somewhat reduced by the higher carbon dioxide content of soil air. Laboratory studies of the treatment of natural water supplies with small concentrations of lime have indicated a reduction in the plumbo-solvent character of these waters, probably because of the removal of free carbon dioxide.[5]

The results of laboratory immersion-type corrosion tests* carried out on lead and lead alloy cable sheathing in water samples taken over a period of years from a large number of cable conduit and manholes at locations where observations had been made on the condition of the sheathing may be summarized as follows:

Corrosion Conditions	Loss in Weight in Test
Markedly corrosive	>75 mdd
Mildly corrosive	20 to 50 mdd
Non-corrosive	<10 mdd

This correlation of laboratory tests and field experience is not concerned with corrosion induced by stray electrical currents, which is determined more by electrical than chemical conditions.

STRAY-CURRENT CORROSION[†]

In the era of street railways, the commonest kind of corrosion was that resulting from the flow of electrical currents from the sheathing to earth. Prevention of stray-current corrosion now as in the past depends upon maintenance of the cable system at or near earth potential by bonding to other underground structures and the use of electrical drainage if necessary. (See *Application of Cathodic Protection*, p. 935.)

In stray-current corrosion the cable has the general characteristics of an anode, whereas the cathode is some remote structure. The potentials of this large-scale cell may be and generally are greater than those of electrolytic-type corrosion cells. The electrolytic resistance of the path will depend upon soil moisture conditions, presence of ground waters, and their alkali, acid, or salt contents.

The corroded area of the sheathing may be clean-cut pitting or burrowing or it may be roughly etched. There may be evidence of brown lead peroxide or of white salts, particularly chlorides.

PROTECTIVE COATING OF BURIED CABLES

In sparsely inhabited areas it is sometimes more economical to bury cables directly in the soil. A suitable coating for protection against corrosion in those cases is composed of asphalt-impregnated paper next to the sheathing, followed by one or more layers of jute impregnated with a preservative compound, and, in some cases, steel tape over the final layer of jute. The structure is flooded with asphalt before and after each serving of paper and jute. The steel tape is employed where danger of induction from power lines or depredation by gophers exists.

---

* In this test the metallic specimens are scratch-brushed, weighed, degreased, rinsed, and placed in jars of water samples open to laboratory air. The volume of water is 1 liter for 1 sq dec. of metallic surface. At the conclusion of the test the specimens are rinsed, dried, and weighed. Corrosion products when present in these tests are loosely adherent and readily removable. The period of exposure is usually 18 to 48 hours, although longer periods have been used.

† Refer also to chapter on this subject, p. 601.

[4] J. C. Thresh, *Analyst*, **47**, 459 (1922).
[5] E. A. G. Liddiard and P. E. Bankes, *J. Soc. Chem. Ind.*, **63**, 39 (1944).

HIGH-TEMPERATURE CORROSION

ALUMINUM AND ALUMINUM ALLOYS

R. B. MEARS*

GASES

Aluminum (m.p., 660° C, 1220° F) is highly resistant to corrosion at elevated temperatures when exposed in dry atmospheres containing air, oxygen, nitrogen, carbon dioxide, hydrogen, or most other common gases (other than the halogens or their compounds). A thin film of reaction products forms on the surface of the aluminum when exposed to certain of these gases (especially oxygen), and this film is highly protective.

In the presence of water vapor, there is little increase in action at temperatures up to about 350° C (700° F). However, upon exposures to atmospheres containing both oxygen and water vapor above this temperature, internal oxidation of certain aluminum alloys may occur. The aluminum-base alloys which are most susceptible to such internal oxidation are those which contain appreciable amounts of magnesium. The other aluminum alloys are definitely less susceptible. This phenomenon has been noted most frequently at heat-treating temperatures (above 480° C [900° F]).[1]

The presence of small amounts of *sulfur dioxide* increases the susceptibility of the alloys to internal oxidation, whereas appreciable amounts of *carbon dioxide* inhibit this form of attack. Small amounts of proprietary compounds are also effective and are widely used in conditioning the air in heat-treating furnaces.

Most aluminum alloys are highly resistant to *hydrogen sulfide* either alone or when mixed with air and water vapor at elevated temperatures.

Steam causes a definite protective white film to form on aluminum alloys. This film is highly protective at temperatures up to 180° to 250° C (350° to 500° F). At temperatures above this range, under some conditions at least, the steam reacts with aluminum with the formation of aluminum oxide and hydrogen.

MOLTEN SALTS AND MOLTEN METALS

Aluminum-base alloys resist the action of many molten salts which are nearly neutral in reaction. The Duralumin-type alloys are frequently heat-treated by immersion in molten sodium nitrate (or mixtures of sodium nitrate and potassium nitrite) at temperatures in excess of 500° C (900° F).

Molten metals generally attack aluminum alloys. The melting points of many metals are above those of the aluminum-base alloys so that melting of the aluminum-base alloys occurs if the exposure period is at all prolonged. Molten *zinc* readily alloys with and dissolves aluminum so that, although the aluminum does not melt, its characteristics are completely altered. Molten *tin* is somewhat similar to molten zinc in its behavior, although the action is less rapid. Molten *lead* is quite inert at temperatures below the melting point of the aluminum. Molten lead baths are sometimes used for heating aluminum alloys.

---

* Development Division, Aluminum Company of America, New Kensington, Pa. Present address: Research Laboratory, Carnegie-Illinois Steel Corp., Pittsburgh, Pa.

[1] P. T. Stroup, *Controlled Atmospheres*, pp. 207-220, American Society for Metals, Cleveland, Ohio, 1942.

618   *HIGH-TEMPERATURE CORROSION*

Mercury, which may be considered to be a molten metal, amalgamates readily at room temperature with aluminum alloys if their naturally formed oxide films are removed or scratched. Once a small area of the aluminum has been amalgamated, rapid corrosion will occur if the material is subsequently exposed to moist air or water. In many cases, the amalgamation of stressed aluminum alloy articles will result in cracking, since the mercury penetrates into the aluminum alloy selectively at grain boundaries, thus weakening the material.

CHROMIUM

See *Nickel and Chromium Coatings*, page 825.

COBALT ALLOYS

W. A. WISSLER*

The commercial cobalt-base alloys containing chromium and either tungsten or molybdenum† are most effective when used at elevated temperatures because they lose less of their inherent hardness while hot than other usable alloys. Heating to temperatures of 1000° C (1830° F) or 1100° C (2010° F) for reasonably short periods of time has no permanent effect upon their original hardness and strength. For example, at 1000° C (1830° F) the Grade 6‡ has a hardness of 70 Brinell and a tensile strength of 36,000 psi but returns to normal properties (360 Brinell and 105,000 psi) at room temperature. The alloys are especially resistant to high-temperature creep. This thermal stability has made them useful not only for cutting tools but also for other applications subject to wear and erosion, and their resistance to chemical attack makes them capable of withstanding combined wear and corrosion.

As such alloys are made of expensive materials, with high cost of fabrication, they are most frequently used in the form of small castings or as welded coatings applied to the particular part or surface that needs protection. Some of the alloys can be forged and rolled, but with difficulty.

AIR

When heated in air, the alloys begin to tarnish on short exposure at 400° C (750° F) and lose appreciable weight at temperatures of 750° C (1380° F) or higher. The loss is difficult to determine accurately, since the protective and tightly adhering scale that usually forms after the first initial loss prevents further oxidation. At temperatures above 1000° C (1830° F) scaling may be progressive. Moisture seems to have but little accelerating effect, and wear by impingement, such as is encountered in valves and turbine service, is comparatively moderate.

Table 1 gives typical data on the effect of heating these alloys in air. At the end of each test, the samples were wire-brushed to remove all loose material, and then weighed. It will be seen that the loss at 850° C (1560° F), although fairly high on the first 24-hour exposure, decreases, and in fact some specimens show a gain after 72 hours. In the longer period of exposure the samples become coated with an impervious adherent layer of oxide.

* Union Carbide and Carbon Research Laboratories, Niagara Falls, N. Y.
† Haynes Stellite.
‡ For compositions of various grades, see Table 1, p. 619, and also Table 1, p. 57.

TABLE 1. OXIDATION OF Co-Cr-W ALLOYS

Tests in Air

Specimen size — 1.3 × 2.5 × 5 cm (½ × 1 × 2 in.).
Surface — ground.
Heated in electrical muffle furnace.
Weight loss in mdd.

Composition				750° C (1380° F)	850° C (1560° F)	
Grade	Cr	W	C	Co	24 Hours Exposure	24 Hours Exposure	72 Hours Exposure
No. 6	30	4.5	1.10	Bal.	98	765	196 gain
No. 12	30	8	1.30	Bal.	11	365	350
No. 1	31	13	2.50	Bal.	98	970	76 gain
Star J	32	17	2.40	Bal.	73	875	204

STEAM

High-temperature steam will oxidize the surface of the alloys, but the action is not progressive. Grade 6 is commonly used for valve trim and other accessories operating at high temperatures and pressures. Quantitative data on weight loss are not available, but typical examples include a control valve handling steam at 650 psi and 455° C (850° F), which was still in satisfactory condition after 8 years of continuous service, and another handling steam at 1350 psi at 495° C (925° F), which was found in perfect condition after 2 years of service.[1,2]

Steam turbine blades, especially those in the last row where the erosion is most severe, are usually protected by a sheath of rolled Grade 6 alloy fastened over the leading edge. Comparative tests show the superiority of the Co-Cr-W alloy for such service.[3]

HYDROGEN, AMMONIA, HYDROCARBONS

Hydrogen at high temperatures, and dried to avoid oxidation of the alloys, will decarburize the surface with a resultant slight softening, but with little decrease in the ability to resist erosion. Decarburization is extremely slow below 1000° C (1830° F).

Heated in nitrogen, the alloy may absorb the gas to a slight degree but no scaling or deleterious effects occur. Ammonia may both decarburize and nitride slightly, neither action affecting the ability of the alloy to resist corrosion and erosion. Many hydrocarbon gases will slightly carburize the alloy with no effect except that if the action is excessive the melting point will be slightly reduced.

MOLTEN SALTS

Molten salts, especially those that act as fluxes with respect to chromium oxide, may cause progressive attack. Alkali carbonates and hydroxides are especially active in this way, and valve failures may occur in superheated steam installations if sodium compounds are allowed to collect and remain molten on the seating surfaces. The

---

[1] R. L. Lerch, "Reducing Valve Wear by Hard Facing," *Power Plant Eng.*, **47**, 71–73 (March, 1943).
[2] W. F. Crawford and L. H. Carr, "Welding in the Manufacture of Valves," *Welding J.*, **18**, 713–722 (November, 1939).
[3] C. Richard Soderberg, "Turbine Blade Erosion," *Elec. J.* (December, 1935).

degree of attack under these conditions seems to be more severe than if the samples were completely immersed in the molten salts.

USE FOR VALVES

The Co-Cr-W alloys have been extensively used for the seating surfaces of aircraft and automotive engine valves, in which service they insure a tightly fitting valve for a comparatively long period of time. Laboratory tests for this service have been made by heating a specimen in contact with litharge at temperatures up to and above those at which the engine valve operates. The nature of such tests precludes the collection of exact quantitative data, but shows that at temperatures up to 650° C (1200° F), the Co-Cr-W alloys, especially Grade 6, are very resistant to oxidation.[4] Recent experiments indicate that the results of this test are greatly influenced by the atmosphere of the furnace and that in non-oxidizing gases, such as nitrogen or engine exhaust gas, the attack on this particular alloy is still further reduced.

THE 65% Co–30% Cr–5% Mo ALLOY (VITALLIUM)

This alloy has excellent high-temperature properties similar to those of the Co-Cr-W alloys. Its most recent use is in the form of investment-cast buckets for the turbosuperchargers of aircraft motors. These buckets are subject to erosion by high-velocity exhaust gas from leaded fuels and high-tensile stresses at the elevated temperature at which the turbine operates[5] (approx. 730° C [1350° F]).

COLUMBIUM

CLARENCE W. BALKE*

Heated in air, columbium becomes coated with oxide films which gradually change color with increasing temperature. They become visible before any increase in weight can be detected, as shown in Table 1.

TABLE 1. CORROSION OF COLUMBIUM IN AIR

Size of spec. — approx. 40 sq cm (6.2 sq in.) of surface.
Sheet 0.1 mm (0.004 in.) thick.
Duration of test — 20 hr.

Temperature		Gain in Weight		Color of Specimen
° C	° F	mdd	oz/sq ft/day	
180	356	0.0	0.0000	Very faint yellowish tinge
215	419	0.0	0.0000	Pale yellow
230	446	0.2	0.0001	Distinct yellow
260	500	0.5	0.0002	Bright brass yellow; some purple spots
280	536	1.2	0.0004	Purple, bright blue, and yellow areas
300	572	1.8	0.0006	Bright blue
325	617	4.4	0.0014	Greenish blue
350	662	6.7	0.0022	Darker greenish blue
390	734	7.6	0.0025	Very dark; white oxide forming

* Fansteel Metallurgical Corporation, North Chicago, Ill.
[4] S. D. Heron, O. E. Harder, and M. R. Nestor, *Exhaust Valve Materials for Internal-Combustion Engines*, Symposium on New Materials in Transportation, American Society for Testing Materials, March 6, 1940.
[5] A. W. Merrick, "Precision Castings of Turbosupercharger Buckets," *Iron Age*, **153**, 52–58 (Feb. 10, 1944).

Table 2 gives data covering the absorption of various gases by columbium. The elongation of the test specimens shows the effect of gas absorption on the ductility

TABLE 2. EFFECT OF VARIOUS GASES ON COLUMBIUM AT ELEVATED TEMPERATURES

Size of spec. — strips of sheet 0.10 × 6.3 × 127 mm (0.004 × 0.25 × 5 in.).

Temperature		Duration of Test, hr	Gain in Weight, gram	Gain in Weight, %	Room Temp. Elongation, % in 1 in.	
°C	°F					
Hydrogen						
200	424	5	0.0000	0.000	(Original 16.1)	16.7
250	482	3	0.0000	0.000		17.0
300	572	5	0.0000	0.000		15.6
350	662	5	0.0017	0.025		14.9
400	752	1	0.0034	0.046		14.4
400	752	2	0.0047	0.071		...
400	752	3	0.0065	0.098		4.5
					Partial embrittlement	
Nitrogen						
300	572	5	0.0000	0.000	(16.1)	14.2
400	752	1	0.0030	0.048		7.8
Air						
350	662	5	0.0045	0.068	(20.1)	13
Oxygen						
350	662	5	0.0034	0.052	(20.1)	22.9

of the metal. It is interesting that oxygen has less effect on this property than air. Nitrogen penetrates the metal, whereas oxygen produces only a surface effect at these temperatures.

COPPER AND COPPER ALLOYS

FREDERICK N. RHINES*

GENERAL

Copper and its alloys are attacked by hot oxygen, sulfur vapor, sulfur dioxide, hydrogen sulfide, phosphorus, halides, and some acid vapors; they are generally inert toward hot nitrogen, hydrogen, carbon monoxide, carbon dioxide, and reducing gases except so far as cyclic exposure to these may accelerate oxidation. Oxygen-bearing copper is embrittled by hydrogen, dissociated water vapor, and dissociated ammonia (hydrogen disease). Copper alloys annealed without proper precleaning may develop "red stain."

OXIDATION OF PURE COPPER

TEMPERATURE AND PRESSURE LIMITS

The oxide scale which forms upon copper at elevated temperatures (under 1025° C [1875° F]) in air is composed of a relatively thick inner layer of Cu_2O (ruby red)

* Metals Research Laboratory, Carnegie Institute of Technology, Pittsburgh, Pa.

and a very thin outer layer of CuO (black) (Fig. 1b). Dissociation pressures of Cu_2O and CuO appear in Tables 1[1] and 2[2] respectively. Extrapolated values of the dissociation pressure from 500° to 1000° C, Table 1, although probably inaccurate, will serve to indicate roughly their order of magnitude at usual working temperatures. In

TABLE 1. DISSOCIATION PRESSURE OF Cu_2O

Temperature		Pressure, mm Hg
°C	°F	
500	932	6×10^{-18} extrapolated
600	1112	5×10^{-15} extrapolated
700	1292	6×10^{-12} extrapolated
800	1472	2×10^{-9} extrapolated
900	1652	2×10^{-7} extrapolated
1000	1832	1×10^{-5} extrapolated
1064	1947	1.4×10^{-4}
1090	1994	3.7×10^{-4}
1150	2102	1.4×10^{-3}
1250	2282	3.0×10^{-2}
1350	2462	2.3×10^{-1}

TABLE 2. DISSOCIATION PRESSURE OF CuO

Temperature		Pressure, mm Hg
°C	°F	
900	1652	15.8
950	1742	37.5
1000	1832	99
1010	1850	121
1020	1868	143
1030	1886	171
1040	1904	204
1050	1922	239

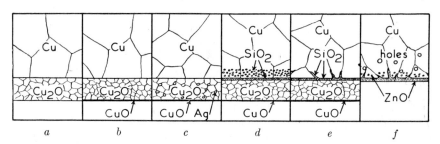

FIG. 1. Types of Scale Forming upon Copper and Its Alloys at High Temperatures.

(a) Pure copper oxidized at low pressure or in air above 1025° C (1875° F). (b) Pure copper oxidized in air below 1025° C (1875° F). (c) Cu-Ag alloy (Class I) oxidized in air. (d) Dilute Cu-Si alloy (Class II) oxidized in air above 700° C (1290° F). (e) Dilute Cu-Si alloy (Class II) oxidized in air at 600° C (1110° F) or below. (f) Yellow brass (Class III) oxidized in air.

practice, oxidation is often inappreciable at pressures well above the theoretical minima. Above 1025° C in the atmosphere, CuO does not appear in the oxide scale on copper (Fig. 1a).

[1] N. P. Allen and T. Hewitt, J. Inst. Metals, **51**, 257-272 (1933).
[2] H. W. Foote and E. K. Smith, J. Am. Chem. Soc., **30**, 1344-1350 (1908).

CRYSTAL HABIT

Complex cubic Cu_2O crystals grow with their cube axes parallel to those of the parent copper crystal.[3] Prolonged growth with recrystallization destroys this relationship; thick Cu_2O scales are usually coarse grained. No crystallographic relationship between CuO and its parent Cu_2O crystal has yet been found; CuO is normally fine grained and of random orientation.

INITIAL RATE OF OXIDATION

At relatively low temperatures (100° C [210° F]) the initial oxide film on pure copper increases in thickness linearly with the logarithm of time,[4] and there is an abrupt increase in the rate at a thickness of about 2 to 4×10^{-8} cm. There is an irregular increase of rate with temperature and a rapid increase of rate with pressure up to 12 mm Hg, after which a decrease with rising pressure to a steady rate above about 150 mm Hg is found. The rate also varies with the crystallographic orientation,[4] accounting for the temper color differences observed among the grains of tarnished polycrystalline copper.

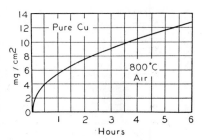

FIG. 2. Weight Gain of Pure Copper Oxidized at 800° C (1470° F) in Air. (From Pilling and Bedworth.)

SUBSEQUENT RATE OF OXIDATION

Beyond temper film thickness the rate of oxide film growth (Fig. 2) is described approximately by the parabolic equation:

$$W^2 = kt \qquad (1)$$

where W is the weight increase (or equivalent thickness of film), t is the time, and k is a constant. There are significant differences among the oxidation rates found by various investigators, but the values reported by N. B. Pilling and R. E. Bedworth[5] appear to be most widely accepted (Table 3). The rate increases with temperature

TABLE 3. VALUES OF THE CONSTANT k FOR PURE COPPER
Use with Eq. 1 expressing W in grams per sq cm and t in seconds

Pure Oxygen		Air	
Temp., ° C	$k \times 10^{10}$	Temp., ° C	$k \times 10^{10}$
400	0.044	700	8.03
500	0.44	800	79.7
600	3.24	900	336
700	16.0	1000	1350
800	86.9		
900	349		
950	730		
1000	1780		

[3] R. F. Mehl, E. L. McCandless, and F. N. Rhines, *Nature*, **134**, 1009 (1934).
[4] B. Lustman and R. F. Mehl, *Trans. Am. Inst. Mining Met. Engrs.*, **143**, 246-267 (1941).
[5] N. B. Pilling and R. E. Bedworth, *J. Inst. Metals*, **29**, 529-582 (1923).

(Fig. 3) according to the expression:

$$\log \frac{W^2}{t} = \frac{a}{T} + b \qquad (2)$$

where T is the absolute temperature, a and b are constants, and W and t have their previous significance. A shaded area in Fig. 3 denotes the range within which rates are variable owing to spalling at intermediate temperatures. At 800° C (1470° F) the rate has been found to increase rapidly with oxygen pressure up to about 0.3 mm Hg, beyond which the effect of pressure is small (Fig. 4). There is little difference between the rates of oxidation in air and in pure oxygen at normal pressures.

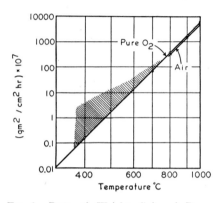
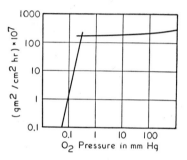

FIG. 3. Rate of Weight Gain of Pure Copper Oxidized in Oxygen or in Air versus Temperature. (From Pilling and Bedworth.) Shaded area represents range of variations resulting from spalling.

FIG. 4. Rate of Weight Gain of Pure Copper at 800° C (1470° F) versus Oxygen Pressure. (From Pilling and Bedworth.)

OXIDATION ABOVE 1065° C

Above 1065° C (1950° F) in air or oxygen, pure copper develops grain boundary and "rosette" liquation (Cu-Cu$_2$O eutectic liquid) as oxygen diffuses inward. The rate, though known to be very high, has not been reported. Upon subsequent cooling a network of solid Cu$_2$O-Cu eutectic remains at the grain boundaries.

OXIDATION OF COPPER ALLOYS

CLASSIFICATION

With respect to their oxidation behavior the alloys of copper may be divided into three more or less distinct classes.

Alloys with the Noble Elements. The scale is similar to that on pure copper but with metallic inclusions of the noble metals (Fig. 1c). There is a subsurface enrichment of the noble metal in the copper and the scale carries a deficiency of the noble metal.[6]

Dilute Alloys with the Baser Elements. (α solid solution up to a maximum of 20% of the alloying agent.) The usual layers of CuO and Cu$_2$O form with an additional layer of an oxide of the baser metal between the Cu$_2$O and the alloy. A

[6] K. W. Frölich, Z. Metallkunde, **28**, 368–375 (1936).

subscale composed of particles of an oxide of the baser element deposited within a matrix of copper occupies a zone just beneath the surface of the metal. When formed at and above 700° C (1290° F), the oxide of the subscale is distributed throughout the copper grains (Fig. 1d); when formed at 600° C (1110° F) and below, the oxide of the subscale is chiefly located at grain boundaries (Fig. 1e).[7] There is depletion of the metal in the alloying component and corresponding enrichment of the external scale. In extreme cases the withdrawal of the alloying component from the metal produces holes within and beyond the zone of oxidation.

Alloys Rich in the Baser Elements. The normal CuO-Cu_2O scale is usually absent. There is a relatively thin external scale composed of an oxide of the baser element and in addition a subscale, commonly of the grain boundary type (Fig. 1f).[8] Depletion of the alloy and enrichment of the scale occur as in the second class, but the first effect is frequently less pronounced than in the second class. Exceptions are not uncommon; for example, Cu-Ni alloys form $CuO + NiO$ instead of simple NiO external scales.[9]

RATES OF OXIDATION OF ALLOYS

In general, the regular oxidation laws (Eqs. 1 and 2) apply to the rates of "total" oxidation of the alloys and also to the rates of growth of individual layers in a complex scale. Irregularly accelerated rates accompany spalling and liquation of the

TABLE 4. VALUES OF THE CONSTANT k FOR COPPER ALLOYS OXIDIZED IN AIR AT 800° C (1470° F)*

Use with Eq. 1 expressing W in grams per sq cm and t in seconds.

Composition	$k \times 10^{10}$	Composition	$k \times 10^{10}$
1.0% Al	23.1	0.05% P	75.0
2.5% Al	0.925	0.54% P	251 (irregular)
3.0% Al	0.231	2.4% Sb	45.0
9.5% As	545 (irregular)	0.5% Si	66.9
0.38% Be	28.0	1.8% Si	14.8
1.0% Be	3.80	3.0% Si	14.8
2.4% Be	0.0578	2.8% Sn	75.0
1.0% Ca	39.2	8.6% Sn	3.7(?)
5.0% Ce	168	20.0% Sn	0.925(?)
2.0% Cr	112	4.0% Ti	33.2
7.75% Fe	52.1	5.0% Zn	45.0
0.67% Li	28.0	20.0% Zn	1.44
0.7% Mg	33.2	40.0% Zn	3.70
2.76% Mg	0.231	5.2% Ni + 0.38% B	20.8
5.0% Mn	59.2	7.0% Ni + 0.5% B	11.3
1.0% Ni	45.4	6.1% Sn + 0.54% P	52.1
20.0% Ni	66.5	Pure Cu	45.4

* K. W. Frölich, Z. Metallkunde, **28**, 368–375 (1936).

oxide (Ce, Cr, P, As). Retarded rates result from the growth of a "protective" oxide film (Al, Be, Sn, Zn). A compilation of "total" oxidation rates in air at 800° C (1470° F) is presented in Table 4. These rates are to be compared with the accompanying oxidation rate for pure copper, which is smaller than that given in Table 3.

[7] F. N. Rhines, Trans. Am. Inst. Mining Met. Engrs., **137**, 246–286 (1940).
[8] F. N. Rhines and B. J. Nelson, Trans. Am. Inst. Mining Met. Engrs., **156**, 171 (1944).
[9] S. Miyake, Scientific Paper, Inst. Phys. Chem. Research (Tokyo), **31**, 161–173 (1937).

TABLE 5. VALUES OF THE CONSTANTS k', a', AND b' FOR COPPER ALLOYS OXIDIZED IN AIR

Use with Eqs. 3 and 4, expressing X in centimeters and t in seconds.

Composition	External Scale		Subscale	
	Temp., °C	$k' \times 10^{10}$	a'	b'
0.03% Al	750	42.5	−11,890	3.614
0.17% Al	750	25.4	−12,020	2.912
	875	390	....	....
0.45% Al	750	63.5	−10,800	1.215
	875	316	....	....
	1000	1690.	....	....
0.72% Al	....	....	−12,140	2.015
0.05% B	1000	1130	−10,420	2.762
0.10% Ba	....	....	−13,480	4.248
0.018% Be	1000	1720	−11,150	3.492
0.054% Be	1000	1610	−14,290	4.487
0.101% Be	....	....	−11,980	2.572
0.01% Ca	....	....	−10,080	1.955
0.04% Cb	....	....	−10,250	2.681
0.01% Ce	1000	1830	−10,870	2.984
0.14% Co	1000	1310	−10,500	2.812
0.08% Co	1000	1790	−11,910	3.824
0.10% Fe	1000	2070	−10,390	2.620
0.56% Fe	1000	1400	−10,730	2.072
1.52% Fe	1000	1110	−10,630	1.533
2.65% Fe	1000	1350	−10,310	0.878
0.03% Ga	....	....	−15,880	7.135
0.02% Ge	1000	1590	−10,210	1.790
0.25% In	....	....	−11,300	3.274
0.02% Li	....	....	−10,330	2.011
0.10% Mg	1000	1660	−11,830	3.362
0.033% Mn	....	....	−10,710	2.765
0.42% Mn	875	446	−10,570	1.648
1.55% Mn	....	....	−10,570	1.186
0.115% Ni	1000	2740	−11,110	3.660
5.0% Ni	1000	1780	−13,710	3.510
0.03% P	....	....	−12,050	3.703
0.07% P	....	....	−11,700	2.955
0.24% P	....	....	−11,050	1.749
0.03% Pb	1000	184	....	....
0.076% Si	750	21.1	−10,680	2.181
0.3% Si	750	28.4	−10,050	0.952
	875	341.	....	....
0.59% Si	750	23.0	−8,950	−0.331
	875	177.	....	....
1.93% Si	....	....	−7,010	−2.667
0.31% Sn	....	....	−11,820	3.536
0.10% Sr	1000	1690	−11,140	3.344
0.04% Ta	....	....	−10,810	3.000
0.05% Ti	1000	2250	....	....
0.09% V	....	....	−13,000	4.492
0.16% Zn	1000	1800	....	....
0.21% Zn	....	....	−12,180	3.861
0.16% Zr	1000	1730	−11,110	2.809

Binary brasses in the range 14 to 20% zinc behave abnormally.[10] Some measurements of the individual rates of external scale and subscale growth upon the second class of alloys in air at temperatures above 700° C (1290° F) are recorded in terms of the increase in thickness in Table 5.[11] These data are to be employed with the oxidation laws written in the forms:

$$X^2 = k't \qquad (3)$$

and

$$\log \frac{X^2}{t} = \frac{a'}{T} + b' \qquad (4)$$

where X is scale thickness (instead of weight increase as in previously mentioned cases). Below 700° C (1290° F) the rates of subscale growth are irregular. As the alloy concentration increases, the rate of growth of the subscale diminishes rapidly. The rate of growth of the external scale usually decreases more gradually.

PROTECTIVE MEASURES

Reducing or completely neutral atmospheres prevent oxidation, but atmospheres containing hydrogen, water vapor, or ammonia should not be used with copper containing oxygen except in very short heat treatments at relatively low temperatures (425° C [800° F] maximum).

It is possible to form a subscale upon a dilute alloy in an atmosphere that is just non-tarnishing with respect to pure copper. This may interfere with buffing and may also lead to the appearance of surface cracks during subsequent working operations. Either strongly oxidizing or fully reducing atmospheres are to be preferred in such cases.

REACTIONS WITH OTHER GASES (AND LIQUIDS)

STEAM

The only important effects of water vapor on record are to induce "hydrogen disease" (see "Hydrogen," p. 629) and to accelerate the attack by sulfur compounds (see "Sulfur," p. 628). However, wet steam at high velocities causes severe attack. (See p. 83.) The ratios of steam to hydrogen that are just oxidizing to copper at various temperatures are given in Table 6.

TABLE 6. EQUILIBRIUM CONDITIONS FOR THE REACTION:*
$Cu_2O + H_2 \leftrightarrows 2Cu + H_2O$

Temperature		Ratio $\dfrac{Partial\ Pressure\ H_2O}{Partial\ Pressure\ H_2}$
°C	°F	
450	842	19 (from Wöhler and Baltz)
900	1652	50,100 (from Maier)
950	1742	34,700 (from Maier)
1000	1832	25,100 (from Maier)
1050	1922	18,600 (from Maier)

* M. Randall, R. F. Nielsen, and G. H. West, *Ind. Eng. Chem.*, **23**, 388–400 (1931).

[10] J. S. Dunn, *J. Inst. Metals*, **46**, 25–48 (1931).
[11] F. N. Rhines, W. A. Johnson, and W. A. Anderson, *Trans. Am. Inst. Mining Met. Engrs.*, **147**, 205–221 (1942).

Carbon Dioxide, Carbon Monoxide, and Hydrocarbon Gases

The gaseous compounds of carbon are generally inert toward copper and its alloys.* Some carbon may be absorbed by alloys high in manganese and nickel. (See also "Cyclic Oxidation and Reduction," p. 630.)

Sulfur, Sulfur Dioxide, and Hydrogen Sulfide

Sulfur and its gaseous compounds attack copper in a manner closely resembling oxidation. Scales composed of CuS and Cu_2S are formed[12] in thickness ratio 3 : 5. Growth generally follows the parabolic relation [13,14,15] (Eq. 1). The rates of growth are greater and increase more rapidly with temperature than those of oxidation. Hydrogen sulfide and sulfur form simple sulfide scales, whereas sulfur dioxide produces mixed oxide and sulfide scales according to the reaction:

$$6Cu + SO_2 \rightleftarrows Cu_2S + 2Cu_2O$$

Equilibrium temperatures and pressures for this reaction appear in Table 7.[16,17]

TABLE 7. EQUILIBRIUM CONDITIONS FOR THE REACTION:
$6Cu + SO_2 \rightleftarrows 2Cu_2O + Cu_2S$

Temperature		Pressure, mm Hg
°C	°F	
500	932	30
600	1112	133
700	1292	502
730	1346	760
800	1472	1480
900	1652	3200
1000	1832	4630

Above 730° C (1350° F) at one atmosphere pressure no sulfide appears in the scale. Small quantities of sulfur dioxide in flue and furnace gases, particularly if water vapor is also present, accelerate oxidation without forming sulfides. Alloying generally decreases the rate of attack by sulfur and its compounds (Tables 8[18] and 9[19]). Aluminum, magnesium, and beryllium are thought to be most effective in this respect.[20] When slightly oxidized, Cu-Al and Cu-Be alloys are notably resistant to attack by sulfur.[21] With high nickel contents, sulfur is absorbed from hot sulfur dioxide or hydrogen sulfide and embrittlement results.

* However, ethane and methane at elevated temperatures (450° to 650° C, 850° to 1200° F) may be decomposed into carbon and hydrogen on the surface of copper, acting as a mild catalyst. The film of carbon or possibly copper carbide which forms on the copper surface peels off from time to time and removes some of the metal. This process continues until the copper is completely disintegrated. (C. Thomas, G. Egloff, and J. Morrell, *Ind. Eng. Chem.*, **31**, 1091 [1939]). Private communication, C. L. Bulow.

[12] S. Satake, *J. Soc. Rubber Ind.* (*Japan*), **8**, 461–471 (1935).
[13] T. Murakami and K. Nagasaki, *Nippon Kinzoku Gakkai-si*, **4**, 201–202 (1940).
[14] G. Tammann and W. Köster, *Z. anorg. u. allgem. Chem.*, **123**, 196–224 (1922).
[15] G. Tammann, *Rec. trav. chim. des Pays-Bas et de la Belgique*, **42**, 547–551 (1923).
[16] C. M. Stubbs, *J. Chem. Soc.* (*Trans.*) **103**, 1445 (1913).
[17] M. Randall, R. F. Nielsen, and G. H. West, *Ind. Eng. Chem.*, **23**, 388–400 (1931).
[18] O. Bauer and H. Arndt, *Z. Metallkunde*, **18**, 85–88 (1926).
[19] G. H. McGregor and J. W. Stephens, *Paper Trade J.*, **97**, 40–41 (1933).
[20] W. Baukloh and W. W. G. Krysko, *Metallwirtschaft*, **19**, 157–160 (1940).
[21] L. E. Price and G. J. Thomas, *J. Inst. Metals*, **63**, 21–38 (1938).

TABLE 8. CORROSION OF Cu-Mn ALLOYS BY FREE SULFUR IN PLASTILIN

Composition, % Mn	Weight Gain, grams/5 sq cm/5 hr		
	30°–40° C (85°–105° F)	130° C (265° F)	400° C (750° F)
0	0.0362	0.0913	0.0994
2	0.0113	0.0145	0.0165
5	0.0086	0.0090	0.0099
10	0.0042	0.0037	0.0045
33.3	0.0008	0.0009	0.0012
80	0.0004	0.0004	0.0005
90	0.0002	0.0002	0.0002
100	0.0001	0.0001	0.0001

TABLE 9. CORROSION OF COPPER ALLOYS IN HOT PAPER MILL VAPORS CONTAINING SULFUR DIOXIDE

Temperature — 200° to 220° C (390° to 430° F).
Atmosphere — 17 to 18% SO_2 + 1 to 2% O_2.
Time of test — mostly 30 days, some longer.

Material	Composition								Weight Loss, mdd
	Cu	Ni	Si	Mn	Al	Zn	Sn	P	
Nickel silver	55	18	...	...	...	27	...	...	637.8
Nickel silver	65	18	...	...	...	17	...	...	674.2
Nickel silver	75	20	...	...	...	5	...	...	355.6
Phosphor bronze	95.5	...	...	...	...	...	4.3	0.2	285.8
Phosphor bronze	91.8	...	...	...	...	...	8.0	0.15	394.3
Bronze	90.0	...	...	...	...	...	10.0	...	220.3
Nickel bronze	88.5	5	1.5	...	...	...	5.0	...	704.9
Aluminum bronze	90	...	...	...	10	...	...	...	264.2
Silicon bronze	94–96	...	3–4	1–1.2	...	...	...	...	502.3

HYDROGEN

Copper and its alloys are not subject to attack by hydrogen except when they contain oxygen in some form. When oxygen is present in solution, as free Cu_2O within the metal, or as another oxide, hydrogen diffusing inward forms H_2O under pressure and ruptures the metal (hydrogen disease).[22] Cast tough-pitch copper (containing free Cu_2O) is most sensitive to hydrogen embrittlement. Cases have been observed at temperatures at least as low as 400° C (750° F), and the severity of the effect increases with temperature. The most common source of the damaging hydrogen is water vapor decomposed by hot iron. (See also "Cyclic Oxidation and Reduction," p. 630.)

So-called oxygen-free coppers are available for use where heating is necessary in the presence of hydrogen. These can become slightly sensitive to hydrogen embrittlement should they be heated in an oxidizing atmosphere.

NITROGEN AND AMMONIA

Nitrogen is generally inert toward copper and its alloys. NO_2 and NO promote oxidation. NH_3 present in furnace atmospheres is said to accelerate oxidation. Under

[22] F. N. Rhines and W. A. Anderson, *Trans. Am. Inst. Mining Met. Engrs.*, **143**, 312–322 (1941).

conditions where the dissociation of ammonia occurs, the hydrogen embrittlement of copper is encountered.[23] (See "Hydrogen," p. 629.)

PHOSPHORUS

Molten phosphorus attacks copper, forming a moderately protective film[24] of Cu_3P. Vaporized phosphorus reacts violently with copper.

HALOGENS AND HALOGEN COMPOUNDS

Below the melting points of their respective salts, chlorine, bromine, and iodine attack copper in accordance with the parabolic relation[25] (Eq. 1). At higher temperatures the reaction with chlorine becomes very rapid.[26,27] Vapors of halogen compounds, particularly HCl, accelerate high-temperature oxidation; the resulting scale may or may not contain halides. Some molten halogen compounds also attack copper.[28]

CYCLIC OXIDATION AND REDUCTION

Alternate exposure to oxygen and one of the reducing gases (H_2, CO, hydrocarbon gases) greatly accelerates oxidation by destroying the protective continuity of the scale. Grain boundary attack is notably increased. Cyclic variations in temperature have a somewhat similar effect. Alloys not ordinarily subject to the hydrogen disease may be embrittled by alternate exposure to oxygen and hydrogen.

RED STAIN

Volatile metals (Zn, Cd, and sometimes Pb) are subject to evaporation from alloys during annealing, particularly when no protective oxide film is present, as in bright annealing. Clean brass surfaces exposed to hot oxidizing atmospheres lose only minor quantities of zinc, but initially tarnished or dirty surfaces lose zinc rapidly in isolated spots. Upon subsequent pickling, the "dezinced" areas remain copper colored (red stain). Sulfur dioxide is thought by some to promote red staining.

IRON AND STEEL

J. B. AUSTIN* AND R. W. GURRY*

The rate of corrosion, or scaling, of iron or steel at elevated temperature is influenced by a number of factors, those best established being composition of the metal and of

* Research Laboratory, U. S Steel Corporation, Kearny, N. J.

[23] J. S Vanick, Proc. Am. Soc. Testing Materials, 24 (II), 348–372 (1924).
[24] N. W. Ryssakov and I. N. Bushmakin, Zhur. Priklad. Khim., 5, 715–721 (1932).
[25] G. Tammann, Rec. trav. chim. des Pays-Bas et de la Belgique, 42, 547–551 (1923).
[26] L. E. Price and G. J. Thomas, J. Inst. Metals, 63, 21–38 (1938).
[27] To use copper in contact with gaseous HCl or Cl_2 the temperature should be below 225° C (450° F) and 260° C (500° F) respectively. (See p. 682.) EDITOR.
[28] P. G. Rudenski, Khim. Mashinostroenie, 5, 42 (1936).

its environment, temperature of the metal, time at temperature, cleanliness or roughness of the metal surface, and contact with other materials. Finally, it is determined largely by the degree to which the scale formed under particular conditions blocks further action between metal and environment. In the determination of the rate of any heterogeneous reaction, it is difficult, indeed often impossible, precisely to control all the significant variables; consequently there are virtually no unexceptionable measurements of an absolute rate of scaling. All observations must therefore be regarded as relative only. Moreover it is difficult to generalize from such data because each steel or alloy behaves more or less as an individual; it forms its own characteristic type of scale whose composition and imperviousness are specific to the given alloy and atmosphere, to the temperature, and to the length of sojourn at temperature. Consequently, even a slight difference in composition of steel or atmosphere — for instance, the presence of sulfur or of steam — may exert a substantial influence upon the type of scale, and hence upon progress of scaling. Furthermore, in an alloy the elements usually differ greatly in tendency to react. Therefore, at the scale-metal interface the proportion of the more active element (e.g., chromium, in the case of oxidation) may be considerably less than in the body of the metal. All these possible differences emphasize the fact that it is not safe to predict even relative scaling resistance over a long period of time from observations extending over a relatively short period, such as a week, nor in one atmosphere from observations in another, nor to extrapolate beyond the temperature range for which direct measurements are available.

INFLUENCE OF THE SURFACE

Cleanliness of the surface is a factor whose significance is too often neglected. The presence of grease frequently causes the scale formed in air to be pourous, hence more pervious than that formed on a clean surface, with a consequent increase in rate of corrosion, particularly in the initial stage of the process. Likewise, the presence of a particle of silica may result in local formation of an iron silicate which adheres so tightly to the metal that it is difficult to remove.

Closely allied to the influence of cleanliness is that of contact with other material. It has long been known that the rate of corrosion of a metal in an aqueous medium may be greatly altered by contact with another metal (galvanic corrosion), but it is not so widely recognized that such contact may also affect the progress of high-temperature corrosion, though usually for a somewhat different reason. Such effects have not been extensively investigated; but, on the basis of present knowledge, they appear to be due either to the formation of some fusible compound of iron, such as the silicate described above, or to the diffusion of some more reactive element, for instance chromium or silicon, from one of the metals to the other.

The influence of roughness of the surface is well illustrated by data reported by Day and Smith,[1] who compared the rate of oxidation in air at 595° C (1100° F) of steel surfaces which had been polished, pickled, milled, or milled and then dipped in dichromate solution. Of these several surfaces, those which had been polished not only yielded the most reproducible results but also were least attacked. Corrosion was most rapid on the milled samples; those which had been pickled behaved much like those which had been milled and treated with dichromate. The influence of roughness is, however, greatest at the beginning, and diminishes as the original surface corrodes away.

[1] M. J. Day and G. V. Smith, *Ind. Eng. Chem.*, **35**, 1098-1103 (1943).

SCALING IN AIR OR OXYGEN

The rate of oxidation of iron or of carbon steel in air or oxygen at constant temperature is initially rapid but falls off markedly as the scale increases in thickness. Quantitatively, the progress of scaling can usually be represented by a simple parabolic relation, that is, the thickness of scale (roughly also the gain in weight) is proportional to the square root of the elapsed time, as would be expected if the scale is continuous and uniform and if the rate of oxidation is controlled by a rate of diffusion. This is illustrated by the data presented in Fig. 1 in which typical measurements of rate of oxidation of commercial iron in air at 700°, 900°, or 1100° C (1290°, 1650° or 2010° F) are plotted on double-logarithmic coordinates. The circles represent observations, the straight lines being drawn with the ideal slope of ½.*

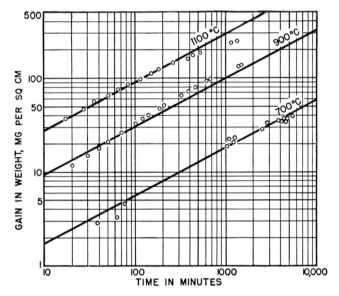

FIG. 1. Rate of Scaling of Commercial Iron (Typical Analysis 0.04% C, 0.06% Mn, 0.01% Si) in Air Plotted on Logarithmic Coordinates to Illustrate Validity of Parabolic Law. (After Heindlhofer and Larsen, *Trans. Am. Soc. Metals*, **21**, 865–895 [1933].)

Circles represent observations; heavy lines are drawn with ideal slope of ½.

The presence of one or more alloying elements may significantly alter this relationship, especially if the free energy of formation of the oxide of the element is considerably greater than that of iron oxide, as is true for chromium, silicon, or aluminum. In this case, although the oxidation-time curves are roughly parabolic in form, they do not fan out regularly from the origin, but often cross. For example, at 815° C (1500° F) a steel containing 2% molybdenum is more resistant after exposure for

---

* Data for oxidation rates at 252° to 385° C (486° to 725° F) are given in Fig. 1, p. 12. Note that within this lower temperature range the rates are represented by a logarithmic function of time. EDITOR.

a day or a week than one containing 2% chromium, yet after 6 weeks exposure the latter is appreciably the better. Marked deviations from the parabolic law occur when the scale cracks or blisters to expose fresh metal, as is illustrated in Fig. 2. The blistering is often caused by dirt or grease on the surface, or differential expansion within the scale resulting from changes in temperature.

Iron and steel commonly scale more rapidly in oxygen than in air, the magnitude of the difference being indicated by the data presented in Fig. 3, which shows bands representing the range of rate of oxidation in air or oxygen at constant temperature as reported in the literature.

The rate of oxidation increases rapidly with increase in temperature, the change for a given change in temperature being somewhat greater for oxygen than for air (see Fig. 3). This difference may be associated with the fact that the temperature

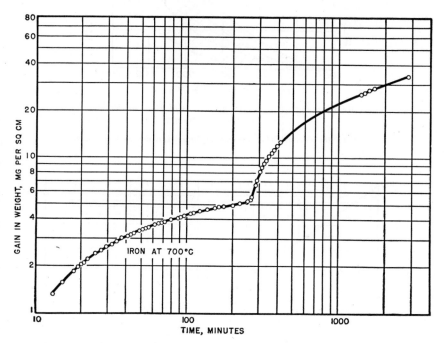

FIG. 2. Typical Example of Irregular Rate of Scaling of Iron (Typical Analysis 0.04% C, 0.06% Mn, 0.01% Si) Caused by Blistering or Cracking of Scale. (After Heindlhofer and Larsen.)

which controls the rate of oxidation is the true temperature at the metal-scale interface, which, because of the rather large heat of oxidation and of the insulating effect of the scale, may be significantly higher than the temperature of the surrounding atmosphere.

When the scale formed is continuous, uniform, and of constant composition over a range of temperature, as is often the case with iron or carbon steel, the logarithm of the gain in weight per unit surface per unit time varies inversely with the absolute

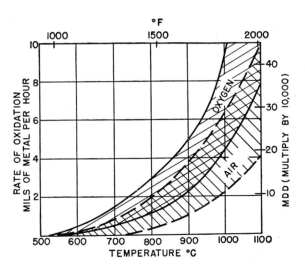

Fig. 3. Bands Representing Temperature Variations of Rate of Scaling of Iron or Mild Steel in Air or in Oxygen.

Data are for initial period of exposure and hence represent maximum rates.

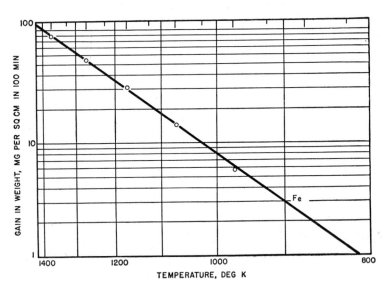

Fig. 4. Diagram Illustrating Linear Variation of the Logarithm of Increase in Weight per sq cm in 100 min with the Reciprocal of the Absolute Temperature. (After Heindlhofer and Larsen.)

temperature (see Fig. 4); but this relation does not hold for many alloy steels for which the type of scale changes with temperature. Moreover, temperature fluctuations often cause the scale to crack or blister with the result illustrated in Fig. 2.

EFFECT OF OTHER GASES WITH OXYGEN OR AIR*

The presence of moisture or of carbon dioxide increases the rate of oxidation in air or oxygen, probably owing to the formation of a somewhat different, more pervious scale. The effect is more marked the higher the temperature.

The presence of sulfur dioxide, as in furnace gases, increases rate of scaling and often results in a deep intergranular penetration of the steel through the formation of a liquid iron oxide–iron sulfide eutectic. This increased penetration and oxidation can be largely avoided by having excess of free oxygen to prevent the formation of this eutectic. For instance, it is reported[2] that the rate of oxidation of low-carbon steel at 1150° C (2100° F) in an atmosphere containing 5% of oxygen is unaffected by the presence of sulfur dioxide.

EFFECT OF VELOCITY

The rate of scaling increases with velocity of gas flow so long as this is small; at higher velocities it increases more slowly and finally remains substantially constant. The velocity at which this happens varies with the conditions of exposure, especially with the turbulence of the atmosphere. In general, however, the rate of oxidation is substantially constant so long as a high concentration of oxygen is maintained at the metal-gas interface.†

EFFECT OF ALLOYING ELEMENTS

Alloying elements present in steel may greatly alter the rate of scaling, among the most effective in this respect being chromium, aluminum, and silicon. These when present in significant concentration oxidize rapidly, yielding a relatively impervious film which retards the rate of further attack on the underlying metal. On the other hand, a high concentration of sulfur in the steel increases the rate of attack just as does sulfur in the atmosphere, and for the same reason, that is, it forms a highly corrosive molten iron oxide-iron sulfide eutectic. The influence of carbon content is still not established, but is in any case relatively small. It has also been reported that rimming steel oxidizes more rapidly than killed steel. This difference, if real, is presumably due to the presence of some excess of the deoxidizing element, aluminum or silicon.

* Data on rate of attack of mild steel at elevated temperatures are listed as follows:

Air containing SO_2, CO_2, and these gases plus water:	Table 5, p. 648. Fig. 4, p. 649.
O_2	Table 3, p. 646.
CO_2	Table 4, p. 648.
SO_2	Table 6, p. 649.
H_2S	Table 9, p. 651. Fig. 6, p. 652.

† It is reported that in a 2-in. diam. furnace, the effect of velocity on rate of oxidation of iron in air, CO_2, or H_2O at 1100° to 1375° C (2000° to 2500° F) is pronounced up to 10 ft per min (3 meters per min) but that increasing gas flow above 50 ft per min (15 meters per min) has practically no effect on the rate. W. E. Jominy and D. W. Murphy, *Trans. Am. Soc. Steel Treating*, **18**, 19 (1930).

D. W. Murphy, W. P. Wood, and W. E. Jominy, *ibid.*, **19**, 193 (1931–1932). EDITOR.

[2] A. Preece and R. V. Riley, Iron and Steel Institute (British), **149**, 253–270 (March, 1944).

SCALING IN OTHER ATMOSPHERES*

Although the rate of corrosion of iron and steel in gases other than air or oxygen has not been widely investigated, it is established beyond question that carbon dioxide, steam, and sulfur dioxide are active scaling agents. Quantitative data are not at all concordant, but qualitatively there is general agreement that, at a given temperature, the rate in carbon dioxide is less than in air, whereas in steam it is significantly greater, presumably because the scale formed is much more pervious; moreover, the rate in moist air is usually greater than in dry air.

Since scaling, like all corrosion, is a reaction between a metal and its environment, it may be lessened, in some cases even prevented, by changing the composition of the atmosphere or of the metal or of both. As an example of a change in atmosphere, the deleterious effect of sulfur dioxide in furnace gases may be offset by providing excess oxygen; or it may be possible to alter the conditions of combustion so that the concentration of carbon monoxide is increased relative to that of carbon dioxide until the equilibrium ratio is exceeded and the atmosphere becomes inert toward the metal. If such expedients are not feasible, the resistance of the metal may be increased by addition of chromium, aluminum, silicon, or other alloying element which tends to form a protective film. Each instance must, however, be considered by itself, since there is no simple or generally valid rule for the prevention of scaling except that of discovering and applying a coating which is protective and permanent under the particular operating conditions.

MOLTEN SALTS

Iron and mild steel are attacked by many molten salts, especially those which are not neutral or are strongly oxidizing. The rate of attack increases rapidly with temperature and in most cases becomes severe at temperatures above about 800° C (1500° F).

CORROSION BY STEAM

See *Corrosion by High-Temperature Steam*, p. 511.

CORROSION BY ALKALINE† AND SALT SOLUTIONS‡

Limited data at 310° C (590° F) for alkaline and salt solutions have been published[3] using mild steel powder (C = 0.11%) in a bomb of the same material in tests of 7½ hours duration. NaOH was found to corrode iron with evolution of hydrogen at a rate which increased with NaOH concentration. Addition of Na_2SO_4 to the alkaline solution acted as an inhibitor. The relative corrosion rates are shown in Fig. 5. Presence of 0.1 mole per liter of NaCl, however, eliminated protection by Na_2SO_4.

---

* High-temperature corrosion data for *cast iron* are given in Table 4, p. 197, for *fatty acids, petroleum vapors* (data for mild steel on p. 654), *rosin vapors, sodium hydroxide, sulfate black liquor, sulfur* (molten), and *sulfur chloride*. See also asterisk footnote, p. 635.

To use carbon steel in contact with *gaseous HCl* or *chlorine*, the temperature should be below 225° C (450° F) and 200° C (400° F) respectively. (See p. 682)

Equilibrium constants at various temperatures for the reaction of iron with H_2O and CO_2 are given in Fig. 1, p. 1123; for the reaction with CO in Fig. 2, p. 1124; and for the reaction with CH_4 in Fig. 3, p. 1125.

† Refer also to *Boiler Corrosion*, p. 520.

‡ This summary was prepared by the Editor.

[3] E. Berl and F. Van Taack, *Forschungsarbeiten auf dem Gebiete des Ingenieurwesens*, Berlin, Heft **330** (1930).

If the iron surface was first conditioned by treating with Na_2SO_4 solutions, or steam at 310° C for 7½ hours, but not allowed to dry, a protective film was formed resistant to the action of NaOH solutions at 310° C at concentrations below but not above 5%. These results are depicted in Fig. 6.

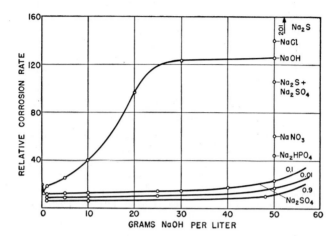

FIG. 5. Attack of Iron by NaOH Solutions at 310° C (590° F) as Affected by Na_2SO_4 Additions (0.01 to 0.9 mole per liter) and Other Salts (0.10 mole per liter).

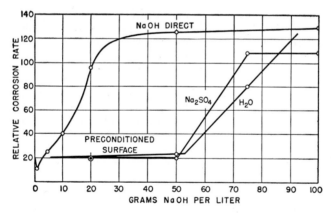

FIG. 6. Attack of Iron at 310° C (590° F) with and without Surface Preconditioning by H_2O or Na_2SO_4 Solution.

Some salt solutions were found to be corrosive to iron independent of the presence of caustic soda (Fig. 7). Na_2SO_4 and NaCl solutions, however, either had little effect, or slightly retarded corrosion at 310° C (590° F). There was no evidence that SO_4 was reduced in presence or absence of alkali at this temperature.

Chromates and bichromates at 310° C react with iron and are themselves reduced. Added to NaOH solutions at this temperature they accelerate corrosion.

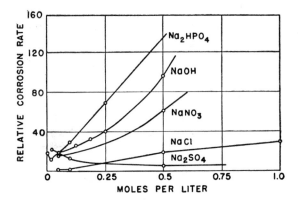

FIG. 7. Attack of Iron by Various Salt Solutions at 310° C (590° F) as a Function of Concentration.

GENERAL REFERENCES

Review of Oxidation and Scaling of Heated Solid Metals, London, H. M. Stationery Office, 1935.
A. PREECE, G. T. RICHARDSON, and J. W. COBB, Second Report of the Alloy Steels Research Committee, Iron and Steel Institute, *Special Report* 24, 1939.

ALUMINUM-IRON ALLOYS

W. E. RUDER*

Aluminum is one of the most effective alloying elements that can be added to other metals (Fe, Cu, Ni, Mo, W) for the purpose of resisting oxidation at high temperatures. Unfortunately, its presence in sufficient quantity to be effective (approx. 10%) develops such a degree of brittleness as to impose serious limitations on its usefulness. For this reason, surface coatings have been used (Calorizing) for most practical applications.

Data on Al-Fe alloys up to 10% aluminum are given in the following table. These alloys were all hot-forged to ¼-in. bars and, although brittle, still exhibited useful ductility up to about 6% aluminum (No. 3). This alloy showed no signs of breakdown after 283 hours at 800° C (1470° F), but scaled badly after a short run at 1000° C (1830° F).

Small amounts of aluminum, as found in Nitralloy, impart improved resistance to corrosion in *high-temperature steam* compared to similar alloys without aluminum.[1,2]

DIFFUSED ALUMINUM COATINGS

Calorizing,[3] a process by which aluminum is diffused into the surface of iron, copper, and other easily oxidizable metals, produces on the outer surface an aluminum-rich layer which adequately protects the surface against scaling at any temperature

* Research Laboratory, General Electric Co., Schenectady, N. Y.

[1] I. A. Rohrig, R. M. Van Duzer, Jr., C. H. Fellows, "High-Temperature-Steam Corrosion," *Trans. Am. Soc. Mech. Engrs.*, **66**, 277 (1944).

[2] E. Houdremont and G. Bandel, "Reactions of Hot Gases with Heat-Resisting Steels," *Arch. Eisenhüttenw.*, **11**, 131 (1937).

[3] W. E. Ruder, "Calorizing Metals," *Trans. Electrochem. Soc.*, **27**, 253 (1915).

ALUMINUM–IRON ALLOYS

TABLE 1. GAIN IN WEIGHT OF Fe-Al ALLOYS IN AIR

Area of specimens — 0.0415 sq dm (0.644 sq in.).
Time of test — 168 hours.
MDD

No.	% Al	600° C (1110° F)	700° C (1290° F)	800° C (1470° F)	1000° C (1830° F)
1	1.96	Loose scale — no	measurement taken		
2	3.98	72.5	Loose scale		
3	6.34	82.5	82.5	24	Loose scale
4	7.85	52	17.3	13.8	87*
5	10.08	24	10.4	3.5	10.4†

* This specimen after 67 hours showed some sign of failure at the ends.
† No sign of failure on this specimen.

short of melting. A number of processes have been successfully used for applying the coating. On heavy iron and steel parts it is usually done by cementation at about

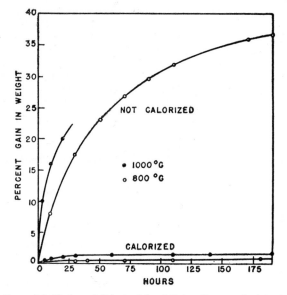

FIG. 1. Rate of Oxidation of Calorized Steel Tubes Compared with Plain Tubes.

1000° C (1830° F) in a protective hydrogen atmosphere, or by dipping into molten aluminum. For lighter parts, spray and cold-dip mixtures are applied followed by a short heating cycle to alloy the surfaces.

This coating is effective until the aluminum is depleted by diffusion to less than 8%. The diffusion depends upon the temperature, amount of aluminum originally deposited, and thickness of the base metal.

Aluminum-coated iron surfaces are particularly resistant to atmospheres containing sulfur dioxide. Tubes so treated have found wide application in the oil industry, and are used in soot blowers, pyrometer tubes, and furnace parts for long-time service at temperatures up to 1000° C (1830° F). (See Fig. 1.) Contact with iron oxide or any of the refractory oxides which form low melting slags with Al_2O_3 must be avoided, as such slags rapidly attack the protective coating.

Diffused-aluminum coatings give no protection against ordinary atmospheric or aqueous corrosion of iron, but on copper the rich aluminum-bronze coating formed has good protective properties and has been extensively used for condenser tubes and the like.

CHROMIUM-IRON, AUSTENITIC CHROMIUM-NICKEL-IRON, AND RELATED HEAT-RESISTANT ALLOYS

W. O. Binder*

INTRODUCTION

Because of their excellent resistance to oxidation and corrosion at elevated temperatures, the Cr-Fe and Cr-Ni-Fe alloys are widely used for industrial heat-resistant purposes. To simplify discussion, the commercially important heat-resistant alloys containing iron, chromium, and nickel may be divided roughly into three groups: (1) chromium steels and cast irons; (2) austenitic Cr-Ni steels; and (3) Ni-Cr alloys. The alloys in group 2 contain at least 50% iron, and more chromium than nickel, whereas the alloys in group 3 contain less than 30% iron and considerably more nickel than chromium. The 15% Cr–35% Ni alloy is the principal exception to these groupings as its composition falls between groups 2 and 3. This discussion will cover the resistance of the first two groups to corrosion at elevated temperatures.† The heat-resistant characteristics of the austenitic Cr-Ni-Si and Cr-Mn alloys will also be covered in this chapter, as these alloys have been employed commercially to some extent for high-temperature service.

PROPERTIES OF CHROMIUM AND AUSTENITIC CHROMIUM-NICKEL STEELS

Chromium is used in heat-resistant alloys in amounts ranging from 2 to 35% because of its utility in imparting oxidation and corrosion resistance at elevated temperatures. Marked differences in the properties of the alloys exist as a result of this wide range in chromium content, and it is apparent from the constitutional diagram for Cr-Fe alloys that within the range of 2 to 35% chromium, the alloys fall within two structural zones; namely, martensitic steels containing 2 to 14% chromium, and ferritic steels containing 14 to 35% chromium. These groupings are only approximate, and the composition limits may be shifted in major fashion by minor alloy additions. For example, the presence of the austenite-forming elements: carbon, nitrogen, manganese, and nickel extends the range of martensitic steels into regions of higher chromium content, whereas the presence of elements such as silicon, tungsten, molybdenum, titanium, columbium, and aluminum, because of their stabilizing effect on ferrite, decreases the range by lowering the upper limit of the chromium content.

Chromium steels in the martensitic range respond to heat treatment much the same as plain carbon steel and pearlitic alloy steels. Because of their air-hardening properties, the martensitic steels must be annealed after hot-forming or welding to restore softness, ductility, and toughness. Inasmuch as carbon is of little benefit in wrought

---

* Union Carbide and Carbon Research Laboratories, Inc., Niagara Falls, New York.
† The Ni-Cr alloys are discussed beginning p. 683.

martensitic steels for heat-resistant purposes, it is generally kept below 0.1% to avoid excessive hardness in the steel when cooled from temperatures above the critical range. In castings, however, the carbon content ranges from 0.25 to 1%, as carbon improves the fluidity of these steels.

Molybdenum is added to the martensitic 2 to 12% chromium steels to eliminate brittleness which develops in these steels at about 480° C (900° F) when cooled to normal temperatures, after having been in service for prolonged periods at elevated temperatures. The addition of 0.5% molybdenum practically eliminates the tendency of these steels to exhibit temper brittleness and imparts a greater degree of stability to the structure of these steels at elevated temperatures.* Small amounts of titanium, columbium, or aluminum are sometimes added to the martensitic steels to reduce air hardening. The use of these elements is resorted to when annealing after fabrication is impractical. To further increase their resistance to oxidation, small amounts of silicon and aluminum are frequently employed in the 2 to 10% chromium steels.

When the carbon content is less than 0.25%, the air hardening characteristics of chromium steels diminish rapidly as the chromium content exceeds 12 to 14%, and steels in the annealed condition containing 16 to 25% chromium are composed primarily of ferrite. Because of their higher chromium content, the ferritic steels are more resistant to oxidation and corrosion at elevated temperatures than the martensitic chromium steels. The ferritic steels containing 20 to 28% chromium are used primarily where oxidation resistance and corrosion resistance at high temperatures are required. The latter steels, however, are subject to grain growth at elevated temperatures and become quite brittle unless the grain size of the steel is controlled by the addition of nitrogen. The steels containing more than about 22% chromium become embrittled when held in the temperature range of 425° to 550° C (800° to 1000° F) so that slow cooling through this range should be avoided. The brittleness is only evident at room temperature, and the toughness can be restored by heating the steel to 650° C (1200° F) or above, and cooling rapidly.

When only resistance to oxidation is required, the carbon content of the ferritic steels may exceed 0.5% and may go as high as 2.5% in the higher chromium compositions; but, when good fabricating characteristics are required, the carbon content is held to a maximum of 0.25%, and in the 16-18% chromium grades it usually does not exceed 0.12%. A large proportion of the steel in the range of 16 to 30% chromium is used in the form of castings with 1 to 2.5% carbon. Such castings possess high resistance to oxidation and abrasion. High-chromium cast irons also fall into this classification and are used for applications in which resistance to oxidation and abrasion is required.

The addition of 8 to 35% nickel to low-carbon 18 to 25% chromium steels produces the so-called austenitic steels. These have greater strength at elevated temperatures than the plain chromium steels. They cannot be hardened except by cold work and are adaptable to cold-forming operations. Because of their ease of fabrication by welding, they have been used more widely than any of the plain chromium steels. Typical austenitic alloys besides the well-known 18% Cr–8% Ni composition (18-8) are the 25% Cr–12% Ni (25-12), the 25% Cr–20% Ni (25-20), and the 15% Cr–35% Ni (15-35) steels. These steels, principally because of their greater oxidation resistance and strength at elevated temperatures, are more frequently used for heat resistance than the 18-8 steel when the temperature exceeds 800° C (1500° F).

Carbon is generally kept low in the austenitic Cr-Ni steels for heat-resistant purposes, and seldom exceeds 0.5%. The low-carbon alloys are more ductile and

* Molybdenum also increases creep strength.

tough, and possess greater resistance to corrosion than the high-carbon alloys. Other additions are made to these steels for specific purposes. For example, manganese is present in all the austenitic steels to improve their soundness and hot workability. The alloys containing manganese as a major alloying element to replace all or part of the nickel content of the austenitic 18-8 steels have been employed to a greater extent in Europe than in the United States.

The oxidation resistance of the Cr-Fe alloys and austenitic steels is further improved by the addition of 2 to 3% silicon.* Silicon in excess of about 1.5%, however, tends to embrittle the steel at normal temperatures, after long exposure of temperatures between 650° and 900° C (1200° and 1650° F). The addition of 2 to 3% molybdenum to 18-8 greatly improves its strength at elevated temperatures, but its resistance to high-temperature oxidation is not correspondingly improved.

In the annealed condition the low-carbon austenitic steels have a homogeneous solid solution structure, but when heated for short periods of time in the temperature range of 425° to 815° C (800° to 1500° F), carbide precipitation occurs at the grain boundaries. Carbide precipitation, known to cause a marked decrease in corrosion resistance at normal temperatures (see p. 161), also decreases resistance at elevated temperatures. The loss of corrosion resistance from this cause may be greatly reduced in low-carbon 18-8 by adding small amounts of titanium or columbium to the steel (producing stabilized 18-8), or by annealing the steel to redissolve the carbides and quenching rapidly. Carbide precipitation does not occur in the stabilized 18-8 steels when they are welded or stress-relieved, and therefore their fabricating characteristics are correspondingly improved.

All these steels derive their heat resistance from a surface oxide layer that forms when they are heated, and serves as protection against further attack. If this layer is non-porous and tightly adherent, it protects the underlying metal. A porous or loose oxide has no similar protective action and will grow in thickness at the expense of the underlying metal. The latter conditions are typical of ordinary steel at temperatures above about 550° C (1000° F). The former characterize a heat-resisting steel.

High-Temperature Strength

Although the question of chemical stability is of paramount importance in the selection of an alloy for resisting corrosion, it would be unwise to choose an alloy for elevated temperature service solely on the basis of its corrosion resistance. If the alloy lacked strength at elevated temperatures, the structure would fail prematurely or would be inefficient because of its size.

The austenitic Cr-Ni steels, as shown in Table 1, are superior to the plain chromium steels in creep resistance. The relatively low creep strength of the ferritic chromium steels at 650° to 730° C (1200° to 1350° F) may be associated with the temper aging that occurs in this range and with the fact that the body-centered cubic lattice arrangement of the ferritic steels is fundamentally less creep-resistant than the face-centered cubic lattice arrangement of the austenitic steels. Small amounts of molybdenum and higher chromium and nickel contents improve the creep strength of the austenitic steels. The 18-8 steels containing molybdenum, the 25-12, and the 25-20 steels are stronger than the 18-8 steels.

The relative strength of various cast austenitic Cr-Ni steels and Ni-Cr alloys may

---

* See *Chromium-Silicon-Iron Alloys*, p. 664. The austenitic steels containing silicon are discussed on p. 662.

CHROMIUM–IRON AND CHROMIUM–NICKEL–IRON ALLOYS

TABLE 1. CREEP STRENGTH OF WROUGHT CHROMIUM AND AUSTENITIC Cr-Ni STEELS
(Collected Data)

Type of Steel	Stress in psi for 1% Creep in 10,000 hours					
	480° C (900° F)	540° C (1000° F)	595° C (1100° F)	650° C (1200° F)	705° C (1300° F)	815° C (1500° F)
Carbon steel with 0.5% Mo	24,500	12,400	...	...	...	...
1.25% Cr, 0.5% Mo	20,500	12,000	4,300	1,600	...	...
2% Cr, 0.5% Mo	19,300	8,900	6,000	3,500	...	...
5% Cr, 0.5% Mo	15,200	10,100	5,800	2,800	...	...
9% Cr, 1% Mo	...	12,000	7,000	2,300	...	...
12% Cr	14,000	11,000	5,000	2,200	...	...
18% Cr	13,000	9,000	4,600	1,600	...	...
25% Cr	13,000	5,500	2,000	1,000	...	...
18% Cr, 8% Ni	24,000	18,300	13,200	8,200	4,000	...
18% Cr, 8% Ni + Cb	...	26,500	23,500	10,000	6,200	1,000
18% Cr, 8% Ni + Ti	...	23,000	21,000	9,000	5,700	...
16% Cr, 13% Ni, 3% Mo	...	24,500	21,000	14,500	9,200	4,200
25% Cr, 12% Ni	...	17,000	13,000	...	4,800	1,000
25% Cr, 20% Ni	...	27,500	...	15,000	8,800	1,200

be obtained from data of Fig. 1.[1] As in the case of the wrought steels, increasing chromium and nickel improves strength at elevated temperatures. Although maximum

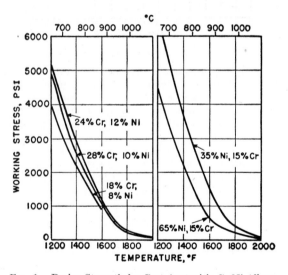

FIG. 1. Design Strength for Cast Austenitic Cr-Ni Alloys.

strength is associated with the 15-35 steel, increasing the nickel content to 65% does not further increase strength. Although comparative data are not shown, castings generally have higher creep strength than wrought metal at temperatures above 650° C (1200° F.).

[1] *Metals Handbook*, American Society for Metals, Cleveland, Ohio., 1939.

HIGH-TEMPERATURE CORROSION OF CHROMIUM AND AUSTENITIC CHROMIUM-NICKEL STEELS

AIR[2, 3, 4, 5, 6, 7, 8, 9]

As shown in Fig. 2, the addition of chromium greatly improves the resistance of steel to oxidation. Steels containing 3% chromium show a marked improvement in oxidation resistance at 650° C (1200° F) over plain carbon steel, and, with higher chromium contents, the steels become practically free of scaling at this temperature.

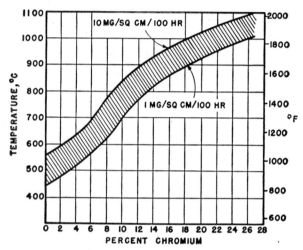

FIG. 2. Effect of Increasing Chromium on Oxidation Resistance of Steel.

Steels containing 11 to 14% chromium are resistant to scaling at temperatures up to about 815° C (1500° F). Increase of chromium content to 16 to 18% raises the range of usefulness to about 870° C (1600° F), and further increase in chromium content to 20 to 35% raises the limiting temperature to about 1150° C (2100° F). With the exception of some of the Fe-Ni-Cr steels and Fe-Cr-Al alloys (see p. 667), the steels containing 20 to 35% chromium are the most oxidation-resistant of the ferrous alloys. The Cr-Ni steels, depending upon their chromium content, are useful at temperatures within the range of 870° to 1150° C (1600° to 2100° F). The limiting temperature for 18-8 is about 870° C (1600° F), and for the 25-12 and 25-20 steels it is about 1150° C (2100° F). The data in Table 2[10] and Fig. 3[11] show the comparative resistance of typical chromium steels to oxidation in air at elevated temperatures.*

* Further data on oxidation of cast alloys in air are given in *High-Nickel Chromium-(Iron)Alloys* beginning p. 684.

[2] Carnegie-Illinois Steel Corp., Pittsburgh, Pa., *U. S. S. Stainless and Heat-Resisting Steels for the Petroleum Industry*, 1940.
[3] R. F. Miller, W. G. Benz, and M. J. Day, *Trans. Am. Soc. Metals*, **32**, 381 (1944).
[4] Republic Steel Corporation, Cleveland, Ohio, *Enduro HCN, NC-3, HC*, 1936.
[5] R. Franks, *Trans. Am. Soc. Metals*, **35**, 616-636 (1945).
[6] K. Heindlhofer and B. M. Larsen, *Trans. Am. Soc. Steel Treating*, **21**, 865-895 (1933).
[7] R. L. Rickett and W. P. Wood, *Trans. Am. Soc. Metals*, **22**, 347-384 (1934).
[8] W. H. Hatfield, *J. Iron Steel Inst.*, **115** (I), 483-508 (1927).
[9] E. Thum, *Book of Stainless Steels*, 2nd Ed., Chapter 13 (C. E. MacQuigg), pp. 351-368, American Society for Metals, Cleveland, Ohio, 1935.
[10] Republic Steel Corporation, Cleveland, Ohio, *Enduro HCN, NC-3, HC*, 1936.
[11] R. F. Miller, W. G. Benz, and M. J. Day, *loc. cit.*

TABLE 2. RESULTS OF INTERMITTENT OXIDATION TESTS ON TYPICAL HEAT-RESISTANT STEELS

Size of specimen — not given.
Surface preparation — scale-free.
Velocity of air — nat. convect.
Duration of test — tests were conducted by heating specimens to temperature in an electric furnace, holding 30 min and air-cooling.
This cycle was repeated 12 times.

Composition			Loss in Weight, mg/sq dm												
Cr	Ni	Other Element	540° C (1000° F)	595° C (1100° F)	650° C (1200° F)	705° C (1300° F)	760° C (1400° F)	815° C (1500° F)	870° C (1600° F)	925° C (1700° F)	980° C (1800° F)	1040° C (1900° F)	1095° C (2000° F)	1150° C (2100° F)	1205° C (2200° F)
0	...	...	543	620	1010	1780	2870	5040	...	...	...	...	...	...	...
5	...	...	388	465	620	1085	1940	3560	...	...	...	...	...	...	...
12	...	...	388	465	465	1775	1160	2320	4960	...	...	...	...	...	...
15	...	...	310	310	388	465	930	1780	3720	...	...	...	...	...	...
17	...	...	...	...	...	...	388	620	1160	2020	3260	4960	...	...	...
18	8	...	...	...	...	...	697	1550	2640	4340	6200	...	...	...	...
18	11	2–4% Mo	...	...	...	...	388	542	1085	2320	4340	6970	...	...	...
18	8	2–3% Si	...	...	...	...	...	465	1010	1550	2170	2790	3720	4950	6200
27	...	...	...	...	...	...	...	...	697	1010	1320	1700	2090	2790	3720
25	12	2% Si (max)	...	...	...	...	...	...	465	620	930	1240	1860	2032	3100
25	20	2% Si (max)	...	...	...	...	...	...	388	542	697	1085	1395	1940	2480

Characteristics of the Oxide Layer. Pfeil[12] has shown that most of the chromium in the oxide scale is concentrated at the inner layer that forms during exposure to air at elevated temperatures. The scale produced on low-chromium steels consists of three layers, but as the chromium content is increased to 25 to 30% a single layer is produced. According to Rickett and Wood,[13] the chromium is present as mixtures of $(FeCr)_2O_3$ and $(FeCr)_3O_4$. The study of Baeyertz[14] showed that the oxidation of chromium steel is due to preferential oxidation of chromium. The oxidation is accompanied by grain boundary attack, and the higher the chromium content the higher the temperature required before this occurs.

OXYGEN

The behavior of chromium and austenitic Cr-Ni steels in pure oxygen at elevated temperatures has been investigated by Hatfield (Table 3). The attack on mild steel in oxygen is severe; however, the addition of chromium produces a substantial increase in resistance, the resistance increasing with chromium content.

TABLE 3. OXIDATION OF WROUGHT CHROMIUM STEELS IN OXYGEN AT ELEVATED TEMPERATURES*

Size of specimen — 1 in. diameter weighing 20 grams.
Surface preparation — polished with 00 emery paper.
Velocity of oxygen — 4 cu ft per hr.
Duration of test — 24-hr continuous exposure.

Composition, %					Weight Increase, mg per sq dm			
C	Mn	Si	Cr	Ni	700° C (1290° F)	800° C (1470° F)	900° C (1650° F)	1000° C (1830° F)
0.17	0.67	0.18	....	....	1,040	2,980	8,340	17,090
0.32	0.25	1.32	13.12	0.29	20	190	280	9,130
0.09	0.39	0.37	18.53	0.26	60	70	190	290
0.11	0.34	0.21	14.84	10.16	60	110	410	4,360
0.12	0.28	0.31	17.74	8.06	110	100	270	570

* W. H. Hatfield, *J. Iron Steel Inst.*, **115** (I), 483–508 (1927).

CARBON DIOXIDE

As shown by Hatfield, chromium increases the resistance of steel to carbon dioxide at elevated temperatures. The results of his tests on mild steel and chromium steels are summarized in Table 4.

AIR CONTAINING VARIOUS ADDITIONS*

The influence of pure air, and also of atmospheres modified with sulfur dioxide, carbon dioxide, and steam at 900° C (1650° F) upon mild steel and 18-8 is given in Table 5. It is evident that the presence of water, carbon dioxide, sulfur dioxide, and

* Additional data on rate of attack of cast alloys in reducing and oxidizing gases containing sulfur are presented in *High-Nickel Chromium-(Iron) Alloys* beginning p. 691.

[12] L. B. Pfeil, *J. Iron Steel Inst.*, **119** (No. I), 501 (1929); **123** (No. 1), 237 (1931).
[13] R. L. Rickett and W. P. Wood, *Trans. Am. Soc. Metals*, **22**, 347–384 (1934).
[14] M. Baeyertz, *Trans. Am. Soc. Metals*, **24**, 420–450 (1936).

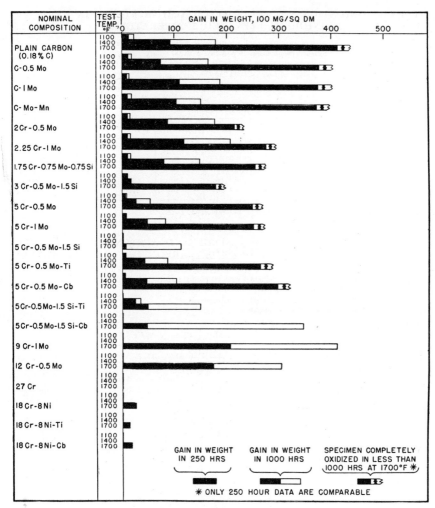

FIG. 3. Oxidation of Steels in 250 and 1000 Hours at 595°, 760°, and 925° C (1100°, 1400°, and 1700° F).

water plus carbon dioxide or sulfur dioxide increases the corrosion of mild steel. The corrosion resistance of 18-8 is much greater throughout than that of mild steel. The ratio of mild steel to 18-8 oxidation rates diminishes, however, in the presence of water, carbon dioxide, or sulfur dioxide.

SULFUR-BEARING GASES*

The presence of sulfur-bearing compounds in the atmosphere to which the Cr and Cr-Ni steels are exposed, as previously mentioned, increases the rate of scaling of

* Data for some high-nickel chromium alloys are presented in Tables 3 and 4, p. 689.

TABLE 4. OXIDATION OF WROUGHT CHROMIUM STEELS IN CARBON DIOXIDE AT ELEVATED TEMPERATURES*

Size of specimen — 1 cm diameter weighing 20 grams.
Surface preparation — polished with 00 emery paper.
Velocity of carbon dioxide — 4 cu ft per hr.
Duration of test — 24-hr continuous exposure.

Composition, %					Weight Increase, mg per sq dm			
C	Mn	Si	Cr	Ni	700° C (1290° F)	800° C (1470° F)	900° C (1650° F)	1000° C (1830° F)
0.17	0.67	0.18	....	....	1000	3670	7490	9350
0.32	0.25	1.32	13.12	0.29	80	90	1600	6230
0.09	0.39	0.37	18.53	0.26	20	80	290	290
0.11	0.34	0.21	14.84	10.16	140	210	1240	3700
0.12	0.28	0.31	17.74	8.06	40	110	330	590

* W. H. Hatfield, *J. Iron Steel Inst.*, **115** (I), 483–508 (1927).

TABLE 5. OXIDATION OF CARBON STEEL AND AUSTENITIC Cr-Ni STEEL IN AIR CONTAINING SULFUR DIOXIDE, CARBON DIOXIDE, OR WATER*

Size of specimen — 1 cm diameter weighing 20 grams (wrought material).
Surface preparation — polished with 00 emery paper.
Velocity of gas — not stated.
Duration of test — 24-hr continuous exposure.
Temperature — 900° C (1650° F).

	% C	% Mn	% Si	% Cr	% Ni
Mild Steel	0.17	0.67	0.18	....	0.25
18-8	0.12	0.28	0.31	17.7	8.06

Condition of Atmosphere	Increase in Weight, mg/sq dm		Ratio $\frac{\text{Mild steel}}{\text{18-8}}$
	Mild steel	18-8	
Pure air	5,524	40	138
Atmosphere	5,717	46	124
Pure air plus 2% sulfur dioxide	6,517	86	76
Atmosphere plus 2% sulfur dioxide	6,576	113	58
Atmosphere plus 5% sulfur dioxide plus 5% water	15,242	358	43
Atmosphere plus 5% carbon dioxide plus 5% water	10,044	458	22
Pure air plus 5% carbon dioxide	7,688	118	65
Pure air plus 5% water	7,421	324	23

* W. H. Hatfield, *J. Iron Steel Inst.*, **115** (I), 483–508 (1927).

these steels. The data in Table 6 describe the behavior of typical steels in sulfur dioxide at temperatures between 700° and 1000° C (1290° and 1830° F). In this connection the investigation of Preece and his co-workers[15] on scaling of carbon and alloy steels, including 12% chromium and 18-8, is also of special interest. They heated samples 1.53 cm (0.6 in.) in diameter by 1.02 cm (0.4 in.) long, for 1.5 hours, in an atmosphere containing 80% nitrogen, 10% steam, and 10% carbon dioxide, to which oxygen, carbon monoxide, or sulfur dioxide was added. As shown in Fig. 4, the increase in sulfur dioxide decreases the scaling rate of 18-8 slightly, whereas the 12%

[15] A. Preece, G. T. Richardson, and J. W. Cobb, British Iron and Steel Institute, *Special Report* **24**, pp. 9–63 (1939).

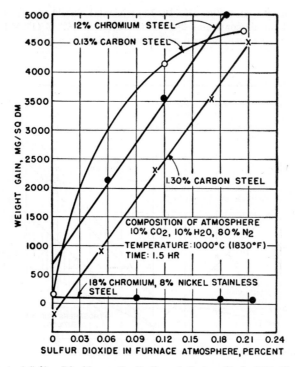

FIG. 4. Effect of Sulfur Dioxide on the Scaling of Carbon Steel, 12% Chromium Steel, and 18% Cr–8% Ni Steel.

TABLE 6. EFFECT OF SULFUR DIOXIDE ON CARBON STEEL, Cr-Fe ALLOYS, AND AUSTENITIC Cr-Ni STEELS AT ELEVATED TEMPERATURES*

Size of specimen — 1 cm diameter weighing 20 grams.
Surface preparation — polished with 00 emery paper.
Velocity of sulfur dioxide — 2 cu ft per hr.
Duration of test — 24-hr continuous exposure.

Composition, %					Weight Increase, mg/sq dm			
					700° C (1290° F)	800° C (1470° F)	900° C (1650° F)	1000° C (1830° F)
C	Mn	Si	Cr	Ni				
0.17	0.67	0.18	....	....	990	4,160	17,700	Converted
0.32	0.25	1.32	13.12	0.29	40	100	990	5,860
0.09	0.39	0.37	18.53	0.26	50	50	80	140
0.11	0.34	0.21	14.84	10.16	110	120	170	370
0.12	0.28	0.31	17.74	8.06	140	160	180	260

* W. H. Hatfield, *J. Iron Steel Inst.*, **115** (I), 483–508 (1927).

chromium steel and the plain carbon steels scale more heavily.* Figure 5 indicates the effect on the scaling rate of carbon monoxide and oxygen added to atmospheres in

* The effect of sulfur dioxide on scaling rate of carbon steels is avoided by maintaining an excess of free oxygen. (See p. 635.)

presence or absence of sulfur dioxide. Experimental data[16] on cast 25-12, 25-20, and 30-18 steels in air containing 2% sulfur dioxide at 980° C (1795° F) are shown in Table 7. It is seen that sulfur dioxide has little influence on the scaling of these steels in air.

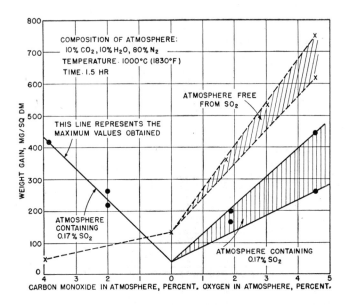

FIG. 5. Effect of Carbon Monoxide, Oxygen, and Sulfur Dioxide on the Scaling of 18-8 Steel at 1000° C (1830° F).

TABLE 7. RESULTS OF OXIDATION TESTS ON Cr-Ni STEEL CASTINGS IN AIR CONTAINING 2% SULFUR DIOXIDE AT 980° C (1795° F)

Size of specimen — 6.35 × 2.54 × 0.475 cm (2½ × 1 × 3⁄16 in.).
Surface preparation — samples polished with 120 emery paper.
Composition of atmosphere — air containing 2% sulfur dioxide by volume, continuous slow stream of gas.
Duration of test — 168-hr continuous exposure.

Composition, %						Condition of Metal	Loss of Weight in mg/sq dm in 168 hr
C	Mn	Si	Cr	Ni	N		
0.26	1.79	0.74	24.52	13.12	0.11	Cast	1240
"	"	"	"	"	"	Annealed*	620
0.23	1.54	0.79	25.10	13.10	0.18	Cast	775
"	"	"	"	"	"	Annealed*	620
0.24	1.66	0.88	25.33	18.87	0.07	Cast	1090
"	"	"	"	"	"	Annealed*	465
0.24	1.24	0.99	30.63	18.38	0.11	Cast	310
"	"	"	"	"	"	Annealed*	465

* Held ½ hr 1050° C (1920° F), air-cooled.

[16] Unpublished data from Union Carbide and Carbon Research Laboratories, Inc., Niagara Falls, N. Y.

The experimental data[17] given in Table 8 obtained on 18-8 show that hydrogen sulfide is more active at elevated temperatures than sulfur dioxide. At 600° C (1110° F), in moist hydrogen sulfide gas, the alloy lost about 5% in weight, and at

TABLE 8. EFFECT OF SULFUR AND ITS COMPOUNDS AT ELEVATED TEMPERATURE ON A STEEL CONTAINING 18.45% CHROMIUM AND 8.56% NICKEL

(Plus 0.10% Carbon, 0.47% Manganese, 0.23% Silicon)

Corrosive Agent	Temperature of Test		Duration of Test, hr	Weight Loss, %
	° C	° F		
Molten sulfur	140	285	24	Nil
Moist SO_2 gas	600	1110	166	0.05
	900	1650	24	2.18
Moist H_2S gas	300	570	158	0.052
	600	1110	65	4.30
	900	1650	21	Converted to sulfide
H_2S gas (1 part) plus nitrogen (10 parts)	600	1110	65	2.08

900° C (1650° F) it was converted completely to sulfide, whereas in moist sulfur dioxide at 900° C (1650° F) it was relatively resistant to attack.

The information published by Scholl,[18] summarized in Table 9, on the action of

TABLE 9. RESISTANCE OF CHROMIUM AND AUSTENITIC Cr-Ni STEELS TO H_2S AT ELEVATED TEMPERATURES

Composition, %			Increase in Weight, Mg per sq dm						
Cr	Ni	Other Elements	480° C (900° F), 28.5 hr	650° C (1200° F), 24 hr.	700° C (1290° F), 1 hr.	800° C (1470° F), 1 hr.	900° C (1650° F), 1 hr	1000° C (1830° F), 1 hr	Investigator
...	...	Low-carbon steel	3,400	...	...	...	...	...	Scholl
8	...	...	3,000	...	...	...	...	...	"
14	...	...	1,900	...	...	...	...	...	"
19	...	...	560	...	...	...	...	...	"
25	...	...	560	...	...	...	...	...	"
26	22	3.2 Si	500	...	...	...	...	...	"
17	26	...	480	...	...	...	...	...	"
13	...	...	...	15,100	...	...	...	...	Sayles
20	...	...	...	11,700	...	...	...	...	"
20	8	...	...	9,400	...	...	...	...	"
12	60	...	...	17,100	...	...	...	...	"
...	...	Iron	...	...	1,320	9,600	17,400	Converted	Gruber*
25	...	...	...	...	1,010	5,500	13,800	23,000	"
24	25	...	...	...	750	5,500	14,100	21,800	"
15	61	...	...	...	1,690	11,800	27,200	48,600	"

* Specimen size — 6 × 3 × 1.2 cm (2.4 × 1.2 × 0.5 in.).
Surface preparation — surface ground to remove scale.
Conditions of test — specimen was heated to test temperature in hydrogen. The hydrogen was then replaced for 1 hr by hydrogen sulfide. Specimen cooled in hydrogen.

[17] A. B. Kinzel and R. Franks, *Alloys of Iron and Chromium*, II, The Engineering Foundation, McGraw-Hill Book Co., New York, 1940.
[18] W. Scholl, Test results quoted by H. J. French, *Trans. Electrochem. Soc.*, **50**, 47–89 (1926).

hydrogen sulfide upon chromium steels, indicates that at 480° C (895° F) the amount of attack in hydrogen sulfide decreases continuously as the chromium content increases up to about 18%, beyond which there is little further change. The action of hydrogen sulfide at temperatures of 400° C (750° F) or higher is given in Table 9 and Figs. 6 and 7.[19,20,21,22] The data published by Rickett and Wood,[22] shown in

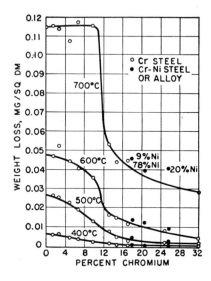
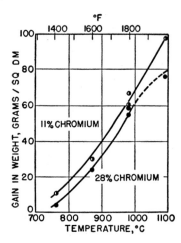

FIG. 6. Effect of Chromium Content and Temperature on Reaction of Fe-Cr, Cr-Ni, and Fe-Cr-Ni Alloys with Pure Hydrogen Sulfide; 120 Hours.

FIG. 7. Effect of Temperature on Reaction of Chromium Steels with Hydrogen Sulfide in 25 Hours.

Fig. 7, were obtained on polished samples 1.27 cm (0.5 in.) in diameter by 3.82 cm (1.5 in.) long, which were exposed to a mixture of hydrogen sulfide and nitrogen containing 20% by volume of hydrogen sulfide, the final polish being with 00 emery paper. At temperatures above 500° C (930° F) it is seen that additions of 23 to 30% chromium somewhat reduce the attack by hydrogen sulfide.

Greenwood and Roennfeldt[23] tested a number of commercial alloys in a zinc-roasting furnace in which the alloys were exposed to gases containing 5% sulfur dioxide, with the possibility that condensed lead sulfide was also present (Table 10). Their tests indicated that the alloy giving the best service consists of 1.5 to 3% carbon and 25 to 30% chromium. The authors expressed the opinion that from a practical point of view the corrosive action in zinc-roasting furnaces is caused by the presence of galena in the concentrates, and that any improvement in ore dressing practice that lowers the lead content will be reflected in longer life of rabbles constructed of these

[19] B. J. Sayles, *Fuels and Furnaces*, **8**, 835 (1930).
[20] H. Gruber, *Z. Metallkunde*, **23**, 151–157 (1931).
[21] P. Schafmeister and F. K. Naumann, *Chem. Fabrik*, **8**, S.83–90 (1935).
[22] R. L. Rickett and W. P. Wood, *Trans. Am. Soc. Metals*, **22**, 347–384 (1934).
[23] J. N. Greenwood and A. J. Roennfeldt, *Eng. Mining J.*, **128**, 851–852 (1929).

CHROMIUM-IRON AND CHROMIUM-NICKEL-IRON ALLOYS

TABLE 10. CORROSION RESISTANCE OF CAST CHROMIUM AND AUSTENITIC Cr-Ni STEELS IN ZINC-ROASTING FURNACE ATMOSPHERE

Composition, %			Rate of Penetration, ipy	
C	Cr	Ni	24 hr, 1050° C (1920° F)	96 hr, 900° C (1650° F)
1.5-3	25-30	...	4.7-8.8	1.1-1.5
1	22	25	11	4.3
0.9	17	34	14	3.3
0.67	19	40	8.8	4.0

alloys. According to Brown,[24] rabble shoes of high chromium steel operate satisfactorily in reducing and oxidizing atmospheres high in sulfur.

Scharschu[25] investigated the corrosion of low-carbon 25 to 30% chromium steel in sulfur vapor. The test was continued for 24 hours and the corrosion rate determined by removing the scale and measuring the reduction in cross-sectional area. The results are given in Table 11. It will be noted that above 445° C (835° F) the corrosion

TABLE 11. RESISTANCE OF HIGH-CHROMIUM STEELS TO SULFUR VAPOR

Steel	Composition, %					
	C	Mn	Si	Cr	Ni	Al
28% Cr	0.12	....	....	27.60	0.25	....
25% Cr, 1.5% Al	0.15	0.81	0.85	25.06	0.72	1.53

28% Cr Steel				25% Cr, 1.5% Al Steel			
Temp. of Test		Corrosion Rate		Temp. of Test		Corrosion Rate	
°C	°F	Penetration, ipy	Weight Loss, mdd	°C	°F	Penetration, ipy	Weight Loss, mdd
295	560	0.0090	48.7	295	560	0.018	97.5
360	680	0.0090	48.7	375	710	0.018	97.5
455	850	0.028	149.5	495	925	0.054	292.5
575	1070	0.046	247	620	1150	0.072	390
700	1290	0.102	553	745	1375	0.24	1300
820	1510	0.31	1690	845	1550	0.61	3355
880	1620	0.58	3120	915	1675	0.86	4680
930	1705	0.72	3900	955	1750	1.04	5655
950	1745	0.90	4875	980	1795	1.04	5655
955	1755	1.00	5395	990	1810	1.04	5655

resistance of the steels is relatively low. The addition of 1.5% aluminum to the 25 to 30% chromium decreases its resistance to sulfur vapors (increases attack).

In summarizing the corrosive action of sulfur on the chromium and austenitic Cr-Ni steels, it may be stated that under certain conditions excessive attack renders the steels unsuitable. The extent of the corrosive attack, however, depends upon

[24] E. Thum, *Book of Stainless Steels*, 2nd Ed., Chapter 19-B (R. E. Brown) pp. 645-655, American Society for Metals, Cleveland, Ohio, 1935.
[25] A. B. Kinzel and R. Franks, *Alloys of Iron and Chromium*, II, The Engineering Foundation, McGraw-Hill Book Co., New York, 1940.

concentration of the sulfur compounds, temperature, and relative oxidizing nature of the atmosphere. The oxidizing properties of the atmosphere are an important factor, and it is necessary to distinguish between scaling in hydrogen sulfide, either pure or mixed with non-oxidizing gases, and scaling in mixtures of sulfur dioxide and oxidizing gases, as, for example, oxygen, carbon dioxide, and water vapor. Although the protective effect of chromium in resisting corrosive attack by sulfur compounds cannot be stated precisely, practical observations show that more severe sulfur conditions can be tolerated the higher the chromium content of the steel. The protective action of chromium in the presence of sulfur is in part explained by the work of Vogel and Reinbach,[26] who made a thorough study of the system Fe-Cr-FeS. They showed that chromium progressively raises the melting point of liquid components containing sulfur from 985° C (1805° F), when there was no chromium, to 1175° C (2147° F) when the chromium content was 50%.

Available data indicate that for high-temperature use, when sulfur compounds are present, the 20 to 35% chromium steels are more satisfactory than most of the austenitic Cr-Ni steels and the Ni-Cr alloys. The greater attack of sulfur compounds on the nickel-bearing alloys is attributed by Kayser[27] to the formation of nickel sulfide, which melts at about 650° C (1200° F). With chromium at 20 to 35%, however, nickel contents of 0 to 35%, under oxidizing conditions, do not decrease resistance to sulfur compounds to any great extent.

PETROLEUM PRODUCTS

The action of sulfur-containing gases and other gases on chromium steels has been investigated in connection with the studies of corrosion of oil refinery equipment. Refinery corrosion is caused chiefly by sulfur, chlorine, and their compounds. The rate of corrosion is often affected by turbulence and velocity of flow, as these factors to a certain extent eliminate by erosion the protective coating of corrosion products.

The increase in life of 4 to 6% chromium steels over ordinary steel in petroleum refinery service is summarized in Table 12.[28] Other data in connection with petroleum refinery corrosion are given in Tables 13[29] and 14[29] and Fig. 8.[30] From a study of these data, it will be noted that the resistance of steel to petroleum products at high temperatures is dependent upon the chromium content of the alloy.

TABLE 12. INCREASE IN LIFE OF 4-6% CHROMIUM STEELS OVER CARBON STEELS IN PETROLEUM REFINERY SERVICE

Service Conditions	Life of Carbon Steel, years	Life of Chromium Steel, years	Ratio Increase in Life of Chromium Steel over Carbon Steel
Mild	8	10	1.25
Moderate	2	8	4
Severe	½	3	6
Extra severe	⅕	2	10

[26] R. Vogel and R. Reinbach, *Arch. Eisenhüttenw.*, **11**, 457 (1938).
[27] J. F. Kayser, *Trans. Faraday Soc.*, **19**, 184 (1923).
[28] *Science of Petroleum*, A. E. Dunstan, Editor, p. 2289, Oxford University Press, New York, 1938.
[29] B. B. Morton, *The Petroleum Engineer*, May, 1944.
[30] Carnegie-Illinois Steel Corp., Pittsburgh, Pa., *U. S. S. Stainless and Heat-Resisting Steels for the Petroleum Industry*, 1940.

TABLE 13. CORROSION RESULTS FROM A BUBBLE TOWER

Location: Temperature:	Bottom of Tower 370° C (700° F), approx.	Top of Tower 85° C (185° F)
Metals	Corrosion Rate, ipy	
Mild steel	0.033	0.091
18% Cr, 8% Ni	0.0006	0.029
80% Ni, 13% Cr	0.00015	0.035
18% Cr, 12% Ni, 3% Mo	0.00003	0.025

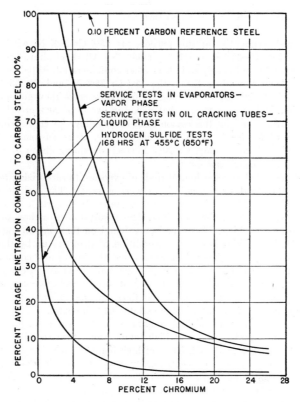

FIG. 8. Beneficial Effect of Chromium on Corrosion Resistance of Steel in Hydrogen Sulfide and Petroleum Refinery Products.

HIGH-PRESSURE STEAM

The resistance of steel to high-temperature steam is greatly increased by the amount of alloyed chromium. See *Corrosion by High-Temperature Steam*, p. 511.

INFLUENCE OF CARBONACEOUS SOLIDS AND HYDROCARBON GASES

The heat-resisting Cr-Ni steels and Ni-Cr alloys have found wide application in equipment for use in the carburization of steel. The chemical compositions of the

HIGH-TEMPERATURE CORROSION

TABLE 14. EFFECTS OF CHROMIUM CONTENT ON RESISTANCE TO ATTACK BY SULFUR IN PETROLEUM PRODUCTS

Temperature 345°–400° C (650°–750° F)

Gas Oil Cracking Operation.	
Metal	Penetration, ipy
9% Cr, 20% Ni	0.00130
14% Cr, 80% Ni	0.00036
18% Cr, 8% Ni	0.00002

Note. Ordinary steel severely attacked in same environment.

alloys used most frequently for this purpose are 25% Cr–20% Ni with and without 1.5 to 2.5% Si; 15% Cr–35 Ni; and 15% Cr–65% Ni. All these alloys are available in the form of castings and sheet, and sheet steel containers fabricated by welding are sometimes used. Of the alloys listed, the 15% Cr–35% Ni and the 15% Cr–65% Ni are used extensively in cast carburizing containers because they are more resistant to carburization. It has been stated that nickel in excess of 20% greatly retards the absorption of carbon, and for this reason most of the alloys for carburizing containers employ 35 to 65% nickel. With high-sulfur fuels, it is customary to use the 15% Cr–35% Ni steel and the 25% Cr–20% Ni steel containing silicon as they are more resistant to sulfur-bearing atmospheres.

The indications are that, after long periods of exposure to carburizing temperatures, the corrosive attack follows the carbide network inwardly. The carbide itself or the chromium-deficient solid solution adjacent to the carbide particle is oxidized and eventually causes failure of the container. The absorption of carbon over a period of time also causes cracking. These cracks are found on the container and according to MacQuigg[31] the cracking is due to a condition of stress set up by the absorption of carbon. It should be emphasized, however, that the life of any container depends upon many other factors besides the composition and carburization of the metal and that, in general, premature failures are caused mainly by defective castings, by high-sulfur fuels, by overloading, by rough handling, by non-uniform heating, and by chemical action of energizers used in the carburizing compound.

Considerable quantities of high-chromium steel tubes are used in processing petroleum and other hydrocarbons at high temperature. Because of the detrimental catalytic effect of nickel on the efficiency of high-temperature hydrocarbon processes, many concerns use seamless or centrifugally cast 28% chromium steels or 28% chromium steels containing 2 to 3% nickel. The addition of 2 to 3% nickel increases toughness of these steels. The tubes operate for prolonged periods at temperatures as high as 1100° C (2000° F) before failing by excessive carburization.

Specific data on the rate of carburization of Cr and Cr-Ni steels have been published (Table 15).[32]

On the basis of these results, it was concluded that at 915° C (1680° F) all the intermediate pearlitic steels containing 5 to 9% chromium are subject to carburization. Silicon appears to be the most effective alloying element for holding carburization to a minimum. All the austenitic steels with the exception of the 25-20 steel carburize,

[31] E. Thum, *Book of Stainless Steels*, 2nd Ed., Chapter 13 (C. E. MacQuigg), pp. 351–368, American Society for Metals, Cleveland, Ohio, 1935.
[32] The Timken Roller Bearing Company, *Résumé of High Temperature Investigations Conducted during 1943–1944*, 1944.

CHROMIUM-IRON AND CHROMIUM-NICKEL-IRON ALLOYS

TABLE 15. CARBURIZING TEST RESULTS ON CHROMIUM STEELS

Size of specimen — not stated.
Surface preparation — not stated.
Test temperature — 915° C (1680° F).
Test conditions — Specimens were carburized in gas-carburizing furnace. Length of each cycle was 36 hr, and specimens were submitted to 20 cycles. For the first 4 hr of each cycle the composition of the gas was 20.7% CO, 38.7% H_2, 0.8% CH_4, 39.8% N_2. After 4 hr, 15 to 20% natural gas was added. Between each 36-hr cycle, carbon was burned out with air, the temperature being held at 915° C (1680° F).

Steel	Carbon Content in Percentages at Indicated Distance from Surface					
	Surface	0.127 mm (0.005 in.)	0.254 mm (0.010 in.)	0.381 mm (0.015 in.)	0.508 mm (0.020 in.)	0.660 mm (0.025 in.)
5 Cr, Mo	4.23	3.71	3.30	3.06	2.87	2.68
5 Cr, Mo, Ti	3.71	3.39	2.81	2.50	2.34	2.24
5 Cr, Mo, Si	2.80	2.73	2.72	2.50	2.30	2.11
7 Cr, Mo, Si	3.16	3.26	3.08	2.84	2.46	2.19
9 Cr, Mo, Si	4.20	3.09	2.59	2.52	2.44	2.36
18 Cr, 8 Ni	0.13	0.18	0.16	0.13	0.11	0.10
18 Cr, 8 Ni, Cb	0.32	0.26	0.19	0.16	0.11	0.10
16 Cr, 13 Ni, 3 Mo	....	0.16	0.18	0.18	0.16	0.15
25 Cr, 20 Ni	0.06	0.07	0.08	0.10	0.08	0.08
27 Cr	0.19	0.18	0.16	0.15	0.13	0.11

but they carburize at a much slower rate than the pearlitic steels. At 915° C (1680° F) the 27% chromium steel does not carburize any more rapidly than the austenitic 18-8 steels.

INFLUENCE OF AMMONIA AND HYDROGEN AT HIGH TEMPERATURES

The 12 to 16% chromium and 18-8 steels have been used for nitriding containers, but they have not been found satisfactory because the surface of the container is nitrided and eventually causes an increase in ammonia consumption and falling off of surface hardness and case depth of the work.[33,34] In Germany[35] nitriding containers of 25-20 steel have been used, but in this country the high-nickel alloys such as those containing 65% nickel and 15% chromium seem to be the most satisfactory.

Vanick[36] states that a 16% Cr-15% Ni steel containing 3% silicon, and 0.4% carbon is unusually resistant to deterioration in a synthesizing ammonia atmosphere at 510° C (950° F) under 15,000 psi pressure. Ammonia under these conditions is reducing, and the highly penetrative hydrogen present under such conditions severely attacks most metals. Experimental data[37] obtained on samples exposed one year to a mixture of nitrogen and hydrogen gases (ratio of 3 to 1) plus 12 to 13% ammonia, 5% methane, and 5% argon at 290 atmospheres and 500° C (930° F), show that 5 to 17% chromium steels are subject to cracking and excessive nitriding. Under similar conditions ingot iron cracked but did not become nitrided. Although the sample of 18-8 steel containing columbium was nitrided, the depth of nitriding was relatively shallow, and the steel retained some ductility in spite of the formation of a hard case. The latter steel was also free of cracking.

[33] R. Sergeson and H. J. Deal, *Trans. Am. Soc. Steel Treating*, **18**, 474 (1930).
[34] J. W. Harsch and J. Muller, *Metal Progress*, **20**, 41–44 (December, 1931).
[35] R. McKay and R. Worthington, *Corrosion Resistance of Metals and Alloys*, Reinhold Publishing Corp., New York, 1936.
[36] J. S. Vanick, *Proc. Am. Soc. Testing Materials*, **24** (II), 348–372 (1924).
[37] Unpublished data from Union Carbide and Carbon Research Laboratories, Inc., Niagara Falls, N. Y.

Cox[38] observed that hydrogen at 500° C (930° F) and up to 3000 psi pressure does not attack 18-8 or 25-20 steels when the carbon content is kept low. Inglis and Andrews[39] have shown, however, that a low-carbon (0.09–0.14% C) 18-8 steel absorbed large quantities of hydrogen at 450° C (840° F) and 250 atmospheres and became embrittled. The ductility of the alloy was restored by heating (900°–1150° C; 1650°–2100° F) to expel the absorbed hydrogen. According to Naumann,[40] chromium greatly retards the penetration of hydrogen at elevated pressures and temperatures (Fig. 9).

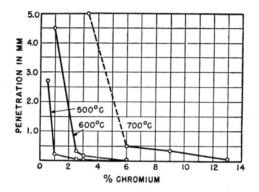

FIG. 9. Influence of Chromium Content on Resistance of Steel to Penetration by Hydrogen at 300 Atmospheres in a Period of 300 Hours.

MOLTEN METALS AND SALTS

Chromium and austenitic Cr-Ni steels are attacked by molten aluminum, zinc, brasses, copper, babbit, and mercury vapors. The corrosive attack is chiefly due to intergranular penetration of the steels, which causes brittleness. Tin and solder do not penetrate the grain boundaries of these steels unless they are excessively superheated. The influence of time and temperature on the rate of attack of 18-8 steels by molten zinc is brought out in Table 16.[41] Because of rapid embrittlement, zinc-coated parts should not be placed in contact with chromium and austenitic Cr-Ni steels that will be exposed to elevated temperatures. Since cadmium has shown no tendency of similar attack, Campbell[42] has suggested that cadmium-plated parts be used if contact with these steels is made at elevated temperatures. Under certain conditions, to be discussed below, molten lead can be satisfactorily handled with chromium and austenitic Cr-Ni steels.

Molten lead as a liquid-heating medium is used in the heat treatment of steel. At temperatures below 700° C (1300° F), cast iron and cast steel are suitable, but, for higher temperatures, alloy steels and castings are required for long life. The 65% Ni–15% Cr alloy and the 35% Ni–15% Cr and the 25% chromium steels are the materials most generally recommended for lead-bath containers. According to

[38] J. L. Cox, *Chem. Met. Eng.*, **40**, 405–409 (1933).
[39] N. P. Inglis and W. Andrews, *J. Iron Steel Inst.*, **128**, 383–408 (1933).
[40] E. Houdremont, *Sonderstahlkunde*, p. 268, Julius Springer, Berlin, 1935.
[41] Unpublished data from Union Carbide and Carbon Research Laboratories, Inc., Niagara Falls, N. Y.
[42] H. A. Campbell, *S. A. E. Journal* (Trans. Section), **52** (No. 2), 67–72 (1944).

CHROMIUM-IRON AND CHROMIUM-NICKEL-IRON ALLOYS 659

TABLE 16. BRITTLENESS TESTS ON 18% Cr-8% Ni TYPE STEELS IN CONTACT WITH ZINC

At Temperatures beween 390° C (735° F) and 800° C (1470° F)

Condition of Sample*	Results of Bend Tests on 18-8, 18-8 Ti and 18-8 Cb Steels
Heated for 140 hr at 390° C (735° F)	No signs of brittleness
Immersed 20 hr at 450° C (840° F)	No signs of brittleness
Immersed 15 hr at 500° C (930° F)	No signs of brittleness
Immersed 44 hr at 500° C (930° F)	Signs of brittleness after bending 180°
Immersed 2 hr at 600° C (1110° F)	Cracked after bending 180°
Immersed 15 hr at 600° C (1110° F)	Cracked after bending 45°
Immersed 2 hr at 700° C (1290° F)	Cracked after bending 180°†
Immersed 15 min at 800° C (1470° F)	Cracked after bending 10°

* Samples annealed by heating at 1075° C (1970° F).
† Dissolving rapidly in solution.

Klouman,[43] corrosion of these materials may be rapid at the surface of the molten lead because of the slagging action of the lead oxide on the protective oxide covering the metal. This type of failure is materially lessened by removing the lead oxide frequently and by maintaining a carbonaceous covering on the lead. It should also be pointed out that oxides of other low melting metals may have a similar corrosive effect on these alloys, unless precautions are taken to keep the surface of the molten metal clean.

For *salt* and *cyanide containers,* steel of 35 to 40% nickel and 15 to 20% chromium has been found very satisfactory. The Cr and austenitic Cr-Ni steels are less generally used for this purpose, although the 25% chromium steel is used for electrodes in immersed electrode-type salt bath heat-treating furnaces. Severe corrosion of Cr and austenitic Cr-Ni steels in salt baths is due to the dissolving action of these salts on the protective scale. *Caustic soda* is particularly corrosive from this standpoint, and salts such as sodium carbonate that may transform in time to caustic soda will cause excessive corrosion. Nickel-bearing chromium alloys are more resistant to molten salts than the plain chromium steels.

CORROSION IN EXHAUST GASES FROM INTERNAL COMBUSTION ENGINES

Conditions in aircraft exhaust manifolds are extremely severe, as the products of combustion from the engine range in temperature from 550° to 950° C (1000° to 1700° F) and contain carbon dioxide, oxygen, carbon monoxide, hydrogen, methane, nitrogen, and traces of lead oxides traveling at relatively high velocities. The metals employed must withstand severe vibration and the cooling action of rain and sea water on the outer surface of the hot manifold, as well as corrosion by condensate on the inner surface during initial periods of operation.

The problem of selecting metals for exhaust systems is complicated by the fact that design has considerable influence on service life. Improved service life with a given material has resulted from better cooling of the system through changes in design. Because of their good fabricating characteristics, austenitic Cr-Ni steels of the 18-8 type are used in exhaust manifolds.[44] The products of combustion, however,

[43] E. Thum, *Book of Stainless Steels,* 2nd Ed., Chapter 19-A (H. Klouman), pp. 637-644, American Society for Metals, Cleveland, Ohio, 1935.
[44] H. A. Campbell, *S. A. E. Journal* (Trans. Section), **52** (No. 2), 67-72 (1944).

have a corrosive effect on the grain boundaries of unstabilized 18-8 steel, as the temperature of these gases is in the carbide precipitation range. For this reason steels stabilized with titanium or columbium must be employed. The titanium-bearing 18-8 alloy has proved adequate for many applications. However, in other cases the titanium-bearing alloy has failed whereas the columbium-bearing alloy has been used successfully. The better performance of the latter is attributed to its higher strength at elevated temperatures.

TABLE 17. HEAT-RESISTING PROPERTIES OF EXHAUST VALVE STEELS*

Specimen size — 5.08 cm (2 in.) × 0.95 cm (.375 in.) diameter.
Surface preparation — not stated.
Conditions of test — samples exposed 200 hr in Ethyl Gasoline Corp. scaling furnace at maximum temperature of 980° C (1800° F) and minimum temperature of 345° C (650° F).
Type of fuel — Special kerosene treated with 10 ml of tetraethyl lead per U. S. gallon in form of Q lead anti-knock fluid containing both bromine and chlorine. The kerosene had about the same ultimate analysis as average automobile gasoline.

Composition, %								Loss in Weight in mg†	
Cr	Ni	C	Mn	Si	W	Mo	Others	Oxidizing Conditions‡	Reducing Conditions§
8.5	...	0.45	...	3.2	...	...	...	50	140
2.9	...	0.40	...	4.0	...	...	...	160	360
8.0	...	0.50	...	1.5	...	0.75	...	10,000	12,000
21.0	1.5	0.72	...	2.0	...	...	...	40	75
13.3	0.65	1.35	...	0.65	...	0.70	2.5 Co	5,000	2,200
13.0	0.5	1.1	...	0.5	...	...	...	19,000‖	19,000‖
19.0	8.0	0.38	...	2.9	...	...	...	90	70
21.5	11.5	0.25	...	2.8	...	...	...	50	90
14.0	14.0	0.45	...	0.55	2.4	0.5	...	2,600	1,600
14.0	14.0	0.45	...	3.0	2.4	0.5	...	100	150
14.0	26.0	0.50	...	1.25	3.5	...	...	840	370
25.0	14.0	0.45	...	0.55	...	2.5	...	50	40
15.5	14.0	1.10	...	2.5	...	...	...	140	160
20.0	32.0	0.50	1.0	1.0	...	...	...	60	40
3.5	12.0	0.55	5.0	0.50	...	...	...	11,000	11,000
18.0¶	8.0	0.07	...	0.75	...	...	...	2,400	1,700
12.8	7.4	0.30	...	2.5	...	...	...	2,800	2,300
23.8	4.8	0.45	...	1.0	...	3.0	...	40	50

* S. D. Heron, O. E. Harder, and M. R. Nestor, American Society for Testing Materials, *Symposium on New Materials in Transportation*, pp. 1–25, 1940.
† Loss expressed in mg for entire test because of uncertainty of area.
‡ 15 to 16 lb of dry air per pound of fuel.
§ 12 to 13 lb of dry air per pound of fuel.
‖ In 170 hr.
¶ Not used as an exhaust valve material but included for comparative purposes.

The action of exhaust gases from internal combustion engines on valve steels has been reported.[45] It is stated that corrosion of the exterior of the hot valve is due to the combustible charge and its products of combustion. The more important components in the charge that may combine with the steel before or after combustion are: oxygen, hydrocarbons, aldehydes, carbon dioxide, carbon monoxide, hydrogen, water vapor, sulfur dioxide and sulfur trioxide, decomposition products of lead anti-knock compounds, halogen acids, nitrogen oxides, nascent nitrogen, and free carbon. Of these

[45] S. D. Heron, O. E. Harder, and M. R. Nestor, *Symposium on New Materials in Transportation*, A. S. T. M., pp. 1–25, 1940.

components, the most important reacting media appear to be lead compounds, oxygen, and reducing agents.

The relative scaling properties of some of the more important valve steels in exhaust gases containing lead are given in Table 17. These data show that the scaling resistance is chiefly a function of the chromium content of the steel and that chromium appears to be the outstanding useful alloying addition. Silicon has a beneficial effect on resistance to scaling when the chromium content is less than 20% but it has negligible effect when the chromium content is greater.

CORROSION BY CONDENSING FLUE GASES

A very comprehensive series of tests on the corrosion behavior of metals in condensing flue gases has been carried out.[46] In one series, 18-8 steel showed practically no corrosion. However, in another series of tests, all but three of twenty-two samples of stainless steel showed severe attack that resulted in perforation of some test strips. The 17% chromium steel was attacked first. Increasing the chromium content to 27% delayed but did not prevent attack. Among the six samples of 18-8 steel a considerable difference in degree of attack occurred. A similar variation in corrosion resistance was noted in the 18-8 columbium and 18-8 titanium steels. Corrosion of the molybdenum-bearing 18-8 steel was found to be practically nil, showing the marked influence of molybdenum in improving the resistance of the Cr-Ni steels to condensing flue gases containing sulfur compounds. The samples of 25-12 steel showed a resistance to corrosion somewhat higher than that of straight 18-8 steel. However, the 25-20 steels were much less resistant than the 25-12 steels and the 25% chromium steels. As a result of these tests, it was estimated that in an atmosphere of condensing flue gas vapors resulting from the combustion of manufactured city gas, 20-gage sheet metal, when exposed from one side only, would fail by pitting within the approximate time limits given in Table 18.

TABLE 18. MAXIMUM LIFE OF 20-GAGE STAINLESS STEEL COMPARED WITH ORDINARY STEEL IN CONDENSING FLUE GASES*

Size specimen — 3.8 × 17.8 × 0.081 cm (1.5 × 7 × 0.032 in.) (20 gage).
Surface preparation — 2B finish, with exception of two steels having No. 1 finish, and carbon steel sample, which was ordinary black iron.
Velocity of gas — not stated, condensing vapors in constant motion.
Temperature — not stated.
Duration of test — The samples were removed every 6 months, and the loss by corrosion determined. Tenacious corrosion product removed with stiff wire brush. Total duration of test 24 months.
Composition, % by volume, of combustible gas producing flue-gas atmosphere (purified manufactured gas) — 2.9% CO_2; 3.5% illuminants; 0.2% O_2; 12.0% CO; 46.8% H_2; 25.7% CH_4; 8.9% N_2; 539 Btu per cu ft; 2 grams per 100 cu ft of NH_3; no HCN; no H_2S; 9 grams per 100 cu ft of organic sulfur as S.

Type of Steel	Estimated Maximum Life in Years
Carbon steel	1†
17% Chromium steel	3
27% Chromium steel	3–4
18% Cr-8% Ni steel	4–8
18% Cr-8% Ni titanium steel	2–5†
18% Cr-8% Ni columbium steel	2–5†
18% Cr-8% Ni molybdenum steel	Completely resistant
25% Cr-12% Ni steel	5–9
25% Cr-20% Ni steel	2–3†

* L. Shnidman and J. S. Yeaw, *Ind. Eng. Chem.*, **34**, 1436–1444 (1942).
† Some of the test samples perforated during exposure.

[46] L. Shnidman and J. S. Yeaw, *Ind. Eng. Chem.*, **34**, 1436–1444 (1942).

AUSTENITIC CHROMIUM-NICKEL-SILICON STEELS

The effect of silicon on the oxidation of 18-8 steel is indicated by data given in Table 19.[47,48] The improvement in resistance to oxidation shown by these data is

TABLE 19. EFFECT OF SILICON ON OXIDATION RESISTANCE OF 18-8

Silicon, %	Held 36 hr at 980° C (1800° F)		Silicon, %	Held 72 hr at 980° C (1800° F)	
	Penetration, ipy	Loss, mdd		Penetration, ipy	Loss, mdd
0.40	0.88	4769	0.52	1.51	8184
1.10	0.49	2656	1.86	0.39	2114
2.14	0.043	233	3.62	0.009	48.8

undoubtedly related to the nature of the scale formed when silicon is present. Silicon is also added to the 25-12 and 25-20 steels to improve oxidation resistance. If more than 2% silicon is present, cracking and checking will occur on hot working. Silicon is added to castings to improve fluidity as well as oxidation resistance, and it is common to add 1 to 1.5% to improve soundness. The influence of silicon on scaling of exhaust valve steels has already been mentioned.

AUSTENITIC CHROMIUM–MANGANESE STEELS

Wrought low-carbon steels containing about 18% chromium, 9% manganese, with and without small copper or nickel additions,[49,50] have oxidation resistance similar to that of the 18% Cr-8% Ni steels. As in the Cr-Ni steels, the limiting temperature for resistance to oxidation is dependent upon the chromium content. At elevated temperatures, however, the Cr-Mn steels have neither the strength nor the creep resistance of the Cr-Ni steels. They have the advantage, however, of superior resistance to attack by gases containing sulfur dioxide and hydrogen sulfide. Experimental data[51] showing the comparative resistance of the Cr-Mn and the 18-8 steel to sulfur-bearing compounds are shown in Table 20.

According to Kinzel and Franks,[52] Cr-Mn steels have been used for cover and carrier sheets in annealing furnaces because of resistance to hot gases containing sulfur compounds. Electric smelting furnace covers have also been made of the Cr-Mn steels.

CHROMIUM CAST IRONS

Chromium-bearing cast irons* are frequently employed for heat-resistant purposes. Their important properties from an industrial standpoint are the following: (1)

* Typical compositions are:
                  Low Cr   High Cr
    C   2.5–3.5%   1–3.5%
    Si 1.5–3.5%   0.25%–2.5%
    Cr 0–3%   12–30%

[47] J. A. Mathews, *Ind. Eng. Chem.*, **21**, 1158–1164 (1929).

[48] E. Thum, *The Book of Stainless Steels*, pp. 431–444 (J. A. Mathews) American Society for Metals, Cleveland, Ohio, 1935.

[49] A. B. Kinzel and R. Franks, *Alloys of Iron and Chromium*, II, The Engineering Foundation, McGraw-Hill Book Co., New York, 1940.

[50] F. M. Becket and R. Franks, *Book of Stainless Steels*, Chapter 16-C, pp. 497–501, American Society for Metals, Cleveland, Ohio, 1935.

[51] Unpublished data from Union Carbide and Carbon Research Laboratories, Inc., Niagara Falls, N. Y.

[52] A. B. Kinzel and R. Franks, *Alloys of Iron and Chromium*, II, The Engineering Foundation, McGraw-Hill Book Co., New York, 1940.

resistance to oxidation and to structural decomposition; (2) resistance to wear; (3) resistance to corrosion; and (4) high fluidity.

TABLE 20. COMPARATIVE RESISTANCE OF WROUGHT Cr-Mn AND 18% Cr-8% Ni STEELS TO SULFUR-BEARING COMPOUNDS

	Composition of Alloys				
	Cr	Ni	Mn	Cu	C
A	17.90	9.04	0.40	....	0.10
B	17.73	....	8.63	0.86	0.08

Corrosive Gas	Conditions of Test	Duration of Test, hr	A % Loss in Weight	B % Loss in Weight
Sulfur dioxide	400° C (750° F)	1500	0.04	0.13
" "	550° C (1020° F)	750	0.38	0.10
" "	900°–950° C (1650°–1740° F)	66	1.51 pitted	0.47 no pitting
Hydrogen sulfide	550° C (1020° F)	750	4.0	2.70
" "	850° C (1560° F)	66	Converted to sulfide	Converted to sulfide

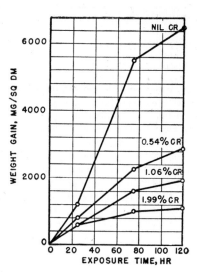

FIG. 10. Effect of Chromium on the Oxidation Resistance of Cast Iron in Air at 815° C (1500° F).

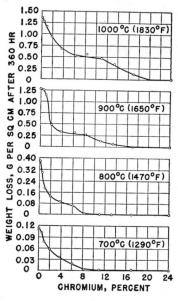

FIG. 11. Effect of Chromium on the Oxidation of Cast Iron Containing Approximately 3% Carbon and 2.25% Silicon, at 700° to 1000° C (1290° to 1830° F).

Cast irons containing more than 1% chromium show substantial improvement in oxidation resistance over ordinary gray iron. In many applications the use of 1% chromium cast iron has been found satisfactory at temperatures up to 760° C

(1400° F). The addition of 1 to 1.25% chromium greatly retards "growth" and, because of less warpage, improves the life of the casting under repeated heating and cooling conditions. For temperatures above 760° C (1400° F) the cast iron should contain 15% or more chromium for high resistance to oxidation. The upper limit of chromium in oxidation-resistant irons is approximately 35%, and such irons have not been harmed by repeated short periods at temperatures as high as 1150° C (2100° F). Table 21 and Figures 10 and 11 show the beneficial effect of chromium on the resistance of cast iron[53,54] to oxidation.

TABLE 21. RESISTANCE TO OXIDATION OF POLISHED 0.5-IN. CUBES OF CHROMIUM CAST IRON AT 815° C (1500° F) AND 1000° C (1830° F)

Chromium, %	Test Temp.		Duration of Test, hr	Change in Weight, %
	° C	° F		
0	815	1500	148	8.0 loss
0.77	"	"	"	5.3 loss
1.29	"	"	"	5.3 loss
15.45	"	"	"	0.03 gain
0	1000	1830	48	24.2 loss
15	"	"	"	0.4 gain to 6.5 loss*
23–25	"	"	"	0.02 gain to 0.10 loss*

* Depending upon carbon and silicon contents.

CHROMIUM-SILICON-IRON ALLOYS

C. L. CLARK*

Increased silicon content is beneficial in improving the high-temperature corrosion resistance of chromium-containing steels in certain types of environments. It is particularly effective when the alloy is exposed to air or to products of combustion normally encountered in commercial furnaces. On the other hand, it exerts little, if any, improvement when the attacking medium is highly superheated steam.

The effectiveness of increased silicon content on oxidation resistance is illustrated in Fig. 1. It will be observed that an increase in the silicon content of a 5% Cr–Mo steel from 0.18 to 0.93% caused a marked improvement in the resistance to oxidation at temperatures up to 870° C (1600° F). In fact, the higher silicon steel is as resistant to scaling at 870° C (1600° F) as the lower silicon steel at 650° C (1200° F). An additional increase to 1.55% silicon results in only slight further improvement.

In hot air and in furnace combustion gases the available data indicate that the beneficial effect of silicon within the above range in the presence of chromium is about seven times as effective as chromium. In other words, if the silicon content of a steel is multiplied by seven and added to the chromium content, the oxidation resistance is about as would be expected for the equivalent chromium content.

Increased silicon content often improves the corrosion resistance of an alloy in hot petroleum products as encountered in thermal cracking processes provided the corrosion rate is not too severe. For example, in one particular installation, a 1%

* The Timken Roller Bearing Co., Canton, Ohio.

[53] Electro Metallurgical Company, New York City, *Chromium in Cast Iron*, 1939.

[54] A. B. Kinzel and R. Franks, *Alloys of Iron and Chromium*, II, The Engineering Foundation, McGraw-Hill Book Co., New York, 1940.

chromium steel gave a life four times that of carbon steel. When the silicon content of this chromium steel was raised from 0.24 to 1.17% the life was further increased to 6.5 times that of the carbon steel. In another application a 3% chromium steel with 1.35% silicon gave a longer service life than a 5% chromium steel with a normal silicon content.

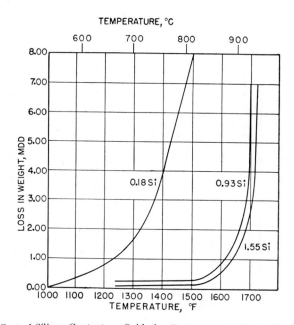

FIG. 1. Effect of Silicon Content on Oxidation Resistance of 5% Cr–Mo Steel in Air.

Analyses

Si	C	Cr	Mo	Mn	P	S
0.18	0.10	5.09	0.55	0.45	0.018	0.010
0.93	0.11	5.24	0.59	0.40	0.009	0.012
1.55	0.10	4.83	0.51	0.38	0.011	0.010

Under conditions in which the corrosion by hot petroleum products is more severe, and in the presence of large amounts of chromium, the beneficial effect of increased silicon content is not so apparent. For example, Fig. 2, presenting the results obtained from three different refineries, shows a fairly definite relation between chromium content and corrosion loss, regardless of the silicon content of the individual steels. Under these particular conditions chromium content appears to be the controlling factor. An increase in the chromium content from 2 to 4% doubles the corrosion resistance. Likewise a 7% chromium steel is twice as resistant as a 5% chromium steel, and a 9% steel is twice as resistant as a 7% steel.

In superheated steam, silicon content has little if any influence on the corrosion resistance and may, in fact, be slightly detrimental. Results obtained at Purdue University on steam corrosion tests at 650° C (1200° F) are presented in Fig. 3. Even though some of the steels contained in excess of 1% silicon, all the results fall on a continuous curve when plotted as chromium content versus the ratio to plain carbon steel. Results for oxidation tests in air at 680° C (1250° F) are included in this same figure, showing that, unlike data for steam, increased silicon content definitely improves resistance to attack.

From the results presented, it follows that general statements cannot be made concerning the influence of increased silicon content on the high-temperature corrosion resistance of chromium-bearing steels without reference to the environment. However, in very few applications is silicon addition detrimental, and even in the

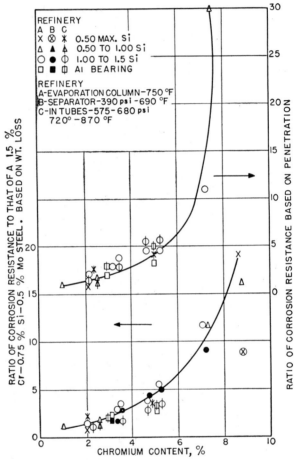

FIG. 2. Effect of Chromium and Silicon Content on the Corrosion Resistance of Steel in Hot Petroleum Products.

A. P. I. strip corrosion specimens: 6 × 1 × ⅛ in., installed in units as designated.

case of steam superheated tubes, increased silicon content is beneficial in combating the oxidation of the outer tube wall in contact with products of combustion. It is believed that proper consideration has not been given to the Cr-Si steels and that when their corrosion resistant characteristics are properly appreciated their use will be greatly broadened.*

* The Cr-Ni-Si and related steels for high-temperature service are discussed on p. 662.

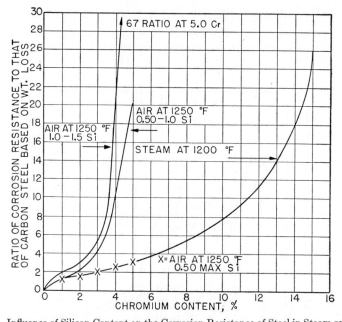

FIG. 3. Influence of Silicon Content on the Corrosion Resistance of Steel in Steam and in Air. Alloys exposed to steam contain variable amounts of silicon.

CHROMIUM-ALUMINUM-IRON ALLOYS

W. E. RUDER*

A variety of compositions[1] of these alloys has been suggested for stability at high temperatures and for electrical resistors suitable for operation up to 1350° C (2460° F). A partial list of trade names, compositions, and properties is given in Table 1.

TABLE 1. COMPOSITION AND OPERATING TEMPERATURE

Name	Approx. Composition			Approx. Resistivity, microhm-cm (room temp.)	Max. Operating Temp. Claimed		Reference
	Cr	Al	Other		° C	° F	
Alcres I	15–35	5–12	min. imp.	200	1300	2375	(1)
Alcres II	12–15	4–5	" "	120	800	1475	...
Chromal	28–31	3–4.5	" "	...	...	...	(2)
Fechral	13–14.5	3.5–4.5	+0.2 Ti	...	...	...	(2)
Kanthal A1	24	6.5	Co 2.0	145	1350	2460	(3)
Kanthal A	24	5.5	" "	139	1300	2375	...
Kanthal D	24	4.5	" "	135	1150	2100	...
Megapyr	30	5	...	140	1300	2375	...
Ohmaloy	12–15	4–5	...	120	800	1475	...
Ohmax	20	8.5	...	166	1300	2375	...
Smith No. 10	37.5	7.5	...	170	1315	2400	(4)

(1) W. E. Ruder, U. S. Patent 1,833,723 (1931).
(2) I. Kornilov and V. Mikheev, Comp. rend. Acad. Sci. (USSR), 24, 904, 907, 911 (1939).
(3) E. Schoene, Elektrowärme, 7, 220–226 (1937).
(4) S. L. Hoyt, Metal Progress, 28, 38 (July, 1935).

* Research Laboratory, General Electric Co., Schenectady, N. Y.
[1] I. Kornilov et al., Comp. rend. Acad. Sci. (USSR), 24, 904, 907, 911 (1939).

Effect of Composition[2]

Figure 1 shows the effect of aluminum and chromium content on loss of weight due to oxidation in air after 240 hours at 1100° C (2010° F). The authors conclude that with 9 to 10% aluminum there is no advantage in adding more than 20% chromium.

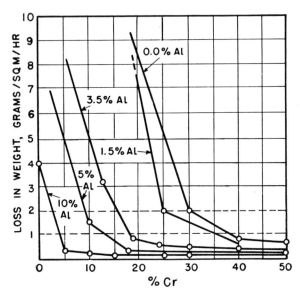

FIG. 1. Oxidation of Al-Fe Alloys at 1100° C (2010° F) in Air as a Function of Chromium Content. (Multiply by 240 to obtain mdd.)

Effect of Atmosphere[3]

Figure 2 illustrates the relative effect of atmosphere on a 30% Cr–5% Al–65% Fe alloy at 1200° C (2190° F) in terms of percentage increases in electrical resistivity and life of 16-mil wires. Comparative data for the Ni-Cr and Ni-Cr-Fe alloys are included. Life test data of these alloy wires in various atmospheres are given in Fig. 17, p. 699. Nitrogen is particularly harmful to the Fe-Cr-Al alloy. Figure 3[4] gives similar data for the 37.5% Cr–7.5% Al–bal. Fe alloy.

These alloys are very resistant to sulfur-bearing gases* and in this respect have a decided advantage over the nickel-rich alloys (Table 2).

Mechanical Properties[5]

The tensile strength, as drawn, of these alloys containing up to about 5% aluminum is comparable to that of the Ni-Cr alloys but the ductility is less than one-half at room temperature.

---

* However, data on exposure of 28% Cr–1.5% Al alloy to *sulfur vapor* appear to indicate that aluminum decreases resistance to attack (Table 11, p. 653). EDITOR.

[2] I. Kornilov and R. Minz, *Comp. rend. Acad. Sci.* (USSR), **34**, 78–82 (1942).
[3] W. Hessenbruch, E. Horst, and K. Schichtel, *Arch. Eisenhüttenw.*, **11**, 225–229 (1937).
[4] S. L. Hoyt, *Metal Progress*, **28**, 38 (July, 1935).
[5] W. Hessenbruch, *Elektrowärme*, **7**, 7–12 (1937).

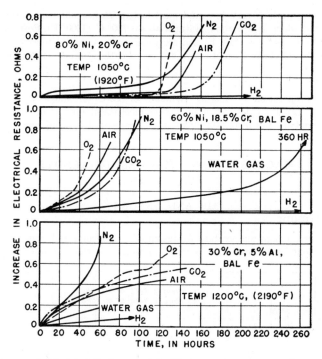

FIG. 2. Increase in Electrical Resistance of Cr-Ni, Cr-Ni-Fe, and Cr-Fe-Al Alloys at 1050° C or 1200° C in Various Gases as a Function of Time.

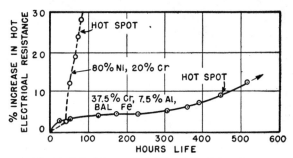

FIG. 3. Increase in Electrical Resistance of Cr-Ni and Cr-Al-Fe Alloys at 1260° C (2300° F) in Air as a Function of Time. ($\frac{1}{8}$ in. wires.)

TABLE 2. COMPARATIVE EFFECT OF VARIOUS GASES ON NICKEL- AND ALUMINUM-BEARING ALLOYS

Alloy	Increase in Wt.: grams/sq meter/hr at 800° C (1470° F)		
	H_2S	$SO_2 + H_2O$	50% CO-50% CO_2 (dry)
24% Cr, 5.5% Al, 2% Co, Bal. Fe	31.5	0.0	0.7
50% Ni, 33% Cr, 13% Fe	244.0	2.0	2.6
80% Ni, 20% Cr	1182.0	2.8	...

The creep strength is very low, and electrical resistance units must therefore be well supported when used at temperatures above 700° C (1300° F). Softening begins at 500° to 600° C (930° to 1110° F) and recrystallization is complete at 700° C. Wires operated at 800° C (1470° F) and above are very brittle at room temperature.

PRECAUTIONS

Satisfactory operation of heating elements of Fe-Cr-Al alloys at temperatures up to 1350° C (2460° F) depends upon proper mechanical support of the element to prevent sag and the use of refractory brick and cements free from silica, iron oxide, or alkalies.

MAGNESIUM AND MAGNESIUM ALLOYS

W. S. LOOSE*

AIR AND OXYGEN

PURE MAGNESIUM

Experiments on high-temperature oxidation of magnesium and magnesium alloys were carried out by following the increase in weight of thin specimens of approxi-

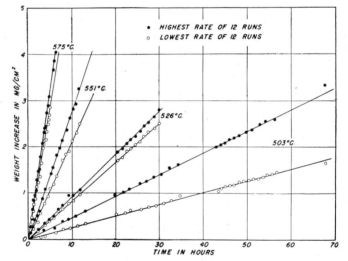

FIG. 1. Oxidation of Pure Magnesium in Pure Oxygen.

Specimen size — 1.3 × 2.54 × 0.1 cm (0.5 × 1.0 × 0.040 in.). Specimen preparation — finished with 9/0 Aloxite paper.

mately 1 sq in. in area as a function of time.[1] This confirmed the earlier prediction[2] that magnesium oxidizes according to a linear law, as illustrated for the temperature range 500° to 575° C (930° to 1065° F) in Fig. 1. Because a wide range of rates at each temperature was obtained, sufficient specimens were oxidized to obtain a repre-

* The Dow Chemical Co., Midland, Mich.

[1] T. E. Leontis and F. N. Rhines, *Trans. Am. Inst. Mining Met. Engrs.*, **166**, 265 (1946).
[2] N. B. Pilling and R. E. Bedworth, *J. Inst. Metals*, **29**, 372 (1923).

sentative average. The results are summarized in Fig. 2, where it is observed that the Arrhenius Equation is closely observed; that is, the temperature dependence of the rate of oxidation can be expressed by an equation of the following type:

$$k = Ae^{-(Q/RT)}$$

where k = rate of oxidation,
 Q = heat of activation for the reaction,
 T = absolute temperature,
 A = reaction constant,
 R = gas constant.

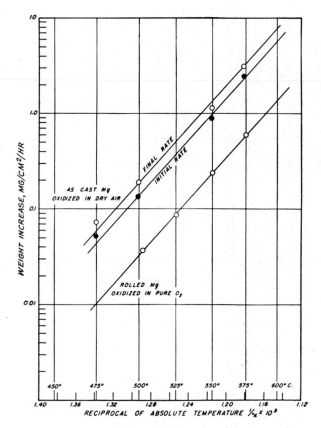

FIG. 2. Effect of Temperature on the Rate of Oxidation of Pure Magnesium in Air and Oxygen.

Specimen size — 1.3 × 2.54 × 0.1 cm (0.5 × 1.0 × 0.040 in.). Surface preparation — finished with 9/0 Aloxite paper. Duration of test — 12 to 72 hours.

In Fig. 2 are also shown the results for the oxidation of cast and rolled magnesium in dry air and in pure oxygen, respectively. In general, it was found that the rate was always higher in air than in pure oxygen.

Another interesting feature is that after a length of time, depending on the temperature, an abrupt increase in the rate of oxidation occurs, but the linear law is still obeyed. It is believed that the slower initial rate is due to the protective action of the normally formed hydroxide film, which, at a certain critical point, breaks down with a resulting higher rate of oxidation.

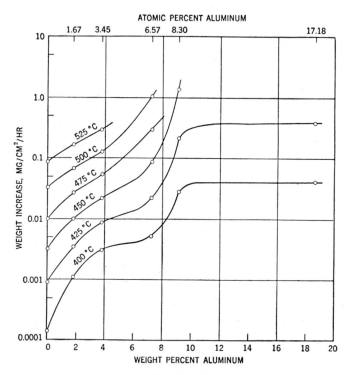

FIG. 3. Effect of Aluminum Content on Rate of Oxidation of Magnesium in Pure Oxygen.

Specimen size — 1.3 × 2.54 × 0.1 cm (0.5 × 1.0 × 0.040 in.). Surface preparation — finished with 9/0 Aloxite paper. Duration of test — 12 to 48 hours.

MAGNESIUM ALLOYS

A study of the effect of various alloying elements on the oxidation of magnesium, with special emphasis on the Mg-Al and the Mg-Zn alloys, showed that the linear law of oxidation is obeyed, with slight deviations in some cases toward slower rates at the beginning of the oxidation curve. Both aluminum and zinc increase the rate of oxidation as is shown in Figs. 3 and 4. The rate of oxidation rapidly increases with increasing aluminum content up to 10%, at which point there is a marked leveling off. It should be noted that this corresponds approximately to the limit of solid solubility and that the 18.66% alloy has a two-phase structure, whereas all the other alloys studied were single phase at the temperature of oxidation. Zinc increases the rate of oxidation by about the same amount as aluminum, but no two-phase alloys were investigated in this system.

Copper and *nickel* produce the most marked increase in the rate of oxidation of any elements investigated, including aluminum and zinc; 0.23% copper increases the rate about the same amount as 9.49% nickel. *Tin* and *gallium* in amounts of 3.78% and 2.38%, respectively, increase the rate of oxidation about the same amount, but decrease the heat of activation of the reaction considerably more than any other elements; hence, their effect is more marked at the lower temperatures of the range investigated, 500° to 550° C (930° to 1020° F). *Silver* and *lead* increase the rate of oxidation moderately, whereas the elements *indium, thallium, cadmium, iron, manganese, silicon,* and *calcium* have hardly any effect.

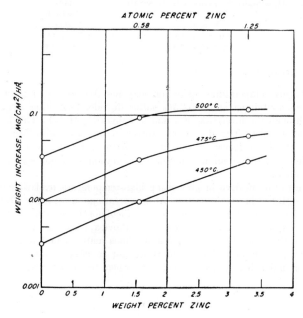

FIG. 4. Effect of Zinc Content on Rate of Oxidation of Magnesium in Pure Oxygen.

Specimen size — 1.3 × 2.54 × 0.1 cm (0.5 × 1.0 × 0.040 in.). Surface preparation — finished with 9/0 Aloxite paper. Duration of test — 12 to 48 hours.

There is an indication that silicon and calcium have a beneficial effect at the higher temperatures. An alloy containing 0.24% *cerium* plus 0.32% *lanthanum* was found to be the most oxidation-resistant of all the alloys investigated and had an oxidation rate considerably below that of pure magnesium. *Beryllium*, even in quantities of less than 0.001%, has the most marked effect of any known alloying element in decreasing high-temperature corrosion. The presence of less than 0.01% in some magnesium alloys has been shown completely to prevent oxidation in air at temperatures up to incipient fusion.

EFFECT OF WATER VAPOR, SO_2, AND CO_2

The effect of various gases on the oxidation rate of pure magnesium at 550° C (1020° F) is shown in Table 1. Both carbon dioxide and sulfur dioxide are very effective in deceasing the rate of oxidation of magnesium, whereas water vapor has a marked tendency to increase it.

TABLE 1. EFFECT OF VARIOUS GASES ON THE OXIDATION RATE OF PURE
MAGNESIUM AT 550° C (1020° F)

Specimen size — 1.3 × 2.54 × 0.1 cm (0.5 × 1.0 × 0.040 in.).
Surface preparation — Finished with 9/0 Aloxite paper.
Duration of test — 12 to 48 hours.

Gas	Rate of Oxidation Weight Gain, mdd
Pure O_2	576
O_2 saturated with H_2O at 28° C (82° F)	1464
Pure CO_2	Negligible
Pure SO_2	Negligible
$O_2 + 5\% SO_2$	Negligible
$O_2 + 11\% CO_2$	Negligible

PROTECTIVE MEASURES

At heat-treating temperatures, oxidation may best be prevented through the use of atmospheres containing at least 0.5% sulfur dioxide. Protection against burning is achieved through the formation of a film of anhydrous $MgSO_4$. This film is of the order of 3000 Å thick. The degree of protection increases with increasing sulfur dioxide content, the velocity of the film reaction increasing with concentration. Sulfur dioxide is effective in minimizing surface oxidation up to and above the melting point of the alloy.

Fluorides are also effective in preventing high-temperature oxidation through the formation of a MgF_2 film on the surface. At heat-treating temperatures (350° to 450° C [660° to 840° F]) ammonium borofluoride and silicofluoride will decompose and form a fluoride protective film. This film is not stable above approximately 375° C (700° F), breaking down to form MgO and some amorphous material in 4 hours. Because of their instability, magnesium fluoride films are not so effective as films formed by sulfur dioxide under heat-treating conditions when relatively long times may be involved.

STEAM

The corrosion of magnesium alloys in steam is shown for typical alloys in Fig. 5. At 100° C (212° F) the corrosion rate in the steam phase is distinctly greater than in the liquid phase. Above this temperature the Mg–1.5% Mn alloy exhibits a greater tendency to corrode in steam than alloys containing aluminum or aluminum and zinc. With the ternary alloys there appears to be a flat portion of the curve between 100° C (210° F) and 120° C (250° F) where increasing temperature does not affect the corrosion rate. Above this temperature hydroxide formation proceeds rapidly. With the Mg–1.5% Mn alloy, this flat portion of the curve does not appear to extend above approximately 110° C (230° F).

OTHER GASES

Magnesium and its alloys are protected in an atmosphere of 100% *hydrogen*. Hydrogen becomes more soluble in magnesium with increasing temperatures, but the possibility of its combining with magnesium at temperatures below the melting point has not yet been studied.

Magnesium may be melted under an atmosphere of *nitrogen*, although some Mg_3N_2 is formed. Whether any magnesium nitride is formed below the melting point has not been determined. Hydrocarbon gases may be used to protect solid magnesium

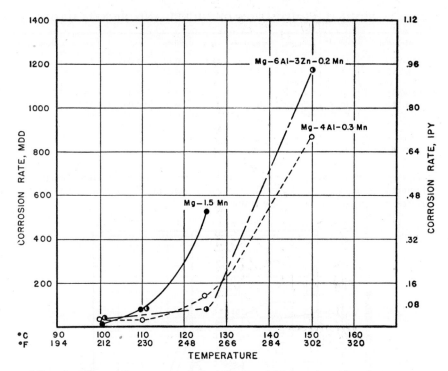

FIG. 5. Effect of Temperature on Corrosion Rate of Magnesium Alloys in Steam. Specimen size — 7.5 × 3 × 0.5 cm (3 × 1.2 × 0.2 in.). Surface preparation — ground. Duration of test — 30 days at temperature up to 126° C (259° F), 7 days at 150° C (300° F).

from high-temperature oxidation and are, in fact, being used in some cases for heat-treating atmospheres.

Although only qualitative data are available, magnesium apparently reacts slowly with dry *chlorine* at temperatures up to the melting point of the metal. Dry *bromine* reacts slowly with magnesium alloys at 150° C (300° F). With *fluorine*, a protective film of MgF_2 is formed. In the presence of oxygen or moisture, this film apparently breaks down at high temperatures.

Magnesium is completely protected in both the solid or liquid state by atmospheres of *helium* or *argon*. These gases are widely used for arc-welding magnesium.

Pure magnesium is not attacked by carbon dioxide or sulfur dioxide at 550° C (1020° F). (See Table 1.)

NICKEL AND NICKEL-COPPER, NICKEL-MANGANESE, AND RELATED HIGH-NICKEL ALLOYS

W. A. MUDGE*

Nickel and the commercial high-nickel alloys considered in this chapter have the nominal chemical compositions summarized in Table 1. At high temperatures the

* Development and Research Division, The International Nickel Co., Inc., New York, N. Y.

HIGH-TEMPERATURE CORROSION

TABLE 1. NOMINAL COMPOSITION OF NICKEL AND NICKEL ALLOYS, %

Material	Nickel*	Copper	Manganese	Iron	Silicon	Sulfur	Carbon	Aluminum
(1) Nickel	99.4	0.1	0.2	0.15	0.05	0.005	0.1	....
(2) "L" Nickel†	99.5	0.02	0.2	0.05	0.15	0.005	0.02 max	....
(3) "D" Nickel†	95.2	0.05	4.5	0.15	0.05	0.005	0.1	....
(4) Monel†	67.0	30.0	1.0	1.4	0.1	0.01	0.15	....
(5) "K" Monel†	66.0	29.0	0.85	0.9	0.5	0.005	0.15	2.75
(6) Constantan	45.0	bal.	0.5-1.85	0.15-0.5	0.05	0.005	0.06	....

* Includes a small amount of cobalt.
† Registered U. S. Patent Office.

usefulness of nickel and high-nickel alloys not containing appreciable amounts of chromium is determined largely by the sulfur content of the atmosphere and is greatest in atmospheres which are sulfur-free.

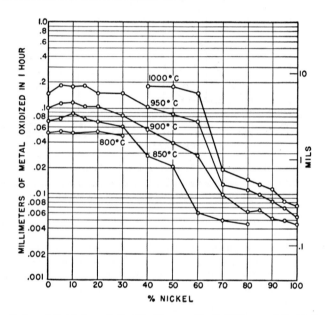

FIG. 1. Rates of Oxidation of Cu–Ni Alloys. (N. B. Pilling and R. E. Bedworth.)

AIR (AND OXYGEN)*

The extent of oxidation of nickel, copper, and their alloys, one hour in air at 800° to 1000° C (1470° to 1830° F) is given in Fig.1.[1] (Some data for the 1% Ni, 5% Ni,

---

* Data on oxidation rates of nickel at 485° to 685° C are given in Fig. 2, p. 12. Note that the rate within this temperature range is represented as a logarithmic function of time. At higher temperatures the rate is stated to be a parabolic function of time. (N. B. Pilling and R. E. Bedworth, *J. Inst. Metals*, **29**, 529–582 [1923].) EDITOR.

[1] N. B. Pilling and R. E. Bedworth, "Oxidation of Copper-Nickel Alloys at High Temperatures," *Ind. Eng. Chem.*, **17**, 372–376 (1925).

NICKEL AND NICKEL–COPPER ALLOYS

and 20% Ni-Cu alloys are given in Tables 4 and 5, p. 625–626.) The addition of copper to nickel, and nickel to copper, causes specific changes in the rate of oxidation but not in the mechanism of reaction which, in the binary alloys of the two metals, resembles that of simple metals.

Specimens, ¾ in. in diameter, of commercial materials heated in moving air for 15 days have given the results summarized in Table 2.[2]

TABLE 2. OXIDATION IN AIR AFTER 15 DAYS AT TEMPERATURES OF 700° C TO 1000° C (1290° TO 1830° F)

Material*	Depth of Metal Penetration							
	700° C (1290° F)		800° C (1470° F)		900° C (1650° F)		1000° C (1830° F)	
	cm	in.	cm	in.	cm	in.	cm	in.
(1) Nickel	...		0.0	0.0	0.0025	0.001	0.023	0.009
(3) Ni, 4.5% Mn†	...		0.0	0.0	0.043	0.017	0.053	0.021
(4) Ni, 30% Cu‡	0.0		0.025	0.01	0.470	0.185	0.820	0.323
(5) Ni, 29% Cu, 3% Al§	...		...	...	0.66	0.26	...	...

* Numbers correspond to analyses listed in Table 1. † "D" Nickel. ‡ Monel. § "K" Monel.

The results of laboratory studies of the effect of small amounts of added elements upon the rate of oxidation of nickel in air at 980° C (1800° F) for periods up to 4 hours are given in Table 3.[2]

TABLE 3. THE EFFECT OF SMALL AMOUNTS OF ADDED ELEMENTS ON THE RATE OF OXIDATION OF NICKEL AT 980° C (1800° F)

No Appreciable Effect		Oxidation Increased	
Element	Max. Amount Present, %	Element	Max. Amount Present, %
Barium	0.05	Aluminum	1.
Boron	0.02	Chromium	0.2
Carbon	0.24	Cobalt	5.
Copper	10.	Columbium*	0.5
Magnesium	0.1	Iron	10.
Silver	0.25	Manganese	5.
Sulfur	0.012	Molybdenum*	1.
Tellurium	0.02	Phosphorus	0.02
		Selenium	0.02
		Silicon	2.
		Tantalum	0.25
		Titanium*	0.4
		Tungsten*	1.
		Vanadium*	0.5

* Contains iron.

Protective measures, such as metallizing, are usually not employed to increase resistance of the high-nickel materials to oxidation.

[2] Unpublished data, The International Nickel Co., Inc., New York, N. Y.

Color of Oxide

The oxide formed on nickel by heating in air or oxygen is dark green in color.[3] The effects on color produced by small amounts of several elements are indicated in Table 4.[4] Iron, copper, and manganese influence the color of nickel oxide markedly.

TABLE 4. THE EFFECT OF SMALL AMOUNTS OF ADDED ELEMENTS ON THE COLOR OF NICKEL OXIDE PRODUCED BY HEATING IN OXYGEN AT 1010° C (1850° F) FOR 15 MINUTES

Oxide Color Unchanged		Oxide Color Changed		
Element	Maximum Content Studied, %	Element	Per cent at Which Effect Was Apparent	Resultant Color
Barium	0.5	Aluminum	0.1	Apple green
Boron	0.02	Cobalt	5.0	Dull green
Carbon	0.24	Columbium*	0.5	Brown green
Chromium	0.2	⎰Copper	0.5	Brown green
Lithium	0.02	⎱Copper	5.0	Dark brown
Magnesium	0.15	Iron	0.2	Dark brown
Selenium	0.02	Manganese	2.0	Black
Silver	0.25	Molybdenum*	1.0	Brown green
Sulfur	0.012	Phosphorus	0.02	Olive green
Tantalum*	0.25	Silicon	2.0	Light brown
Tellurium	0.02	Tungsten*	1.0	Brown green
Titanium	0.1	Vanadium*	0.5	Brown

* Contains iron.

Spalling

The tendency toward spalling depends largely upon the expansion characteristics and composition of the oxide formed. The majority of the elements present in the commercial high-nickel materials, as given in Table 1, do not cause spalling. Of the others, the minimum concentration of alloying element to induce spalling varies with the element. The elements having the greatest spall-producing tendency are, in decreasing order of their effect, chromium, iron, and aluminum. The elements vanadium, tungsten, and molybdenum aggravate spalling in the presence of iron.

Thermocouple Alloys

The 45% Ni–55% Cu alloy,* coupled with iron, and the 10% Cr–90% Ni alloy,† coupled with 2% Al–2% Mn–1% Si–95% Ni alloy,‡ are used for thermocouples. Constantan-iron couples are recommended for long-time use up to 875° C (1600° F) without protection and may be used for short-time work at temperatures up to 1100° C (2000° F) when enclosed in suitable protection tubes to minimize oxidation. Chromel-Alumel couples are recommended for long-time use up to 1100° C (2000° F) without protection and may be used for short-time application at temperatures up to 1300° C (2400° F) when enclosed in suitable protection tubes.

* Constantan.
† Chromel.
‡ Alumel.

[3] N. B. Pilling and R. E. Bedworth, "The Oxidation of Metals at High Temperatures," *J. Inst. Metals*, **29**, 529 (1923).
[4] Unpublished data, The International Nickel Co., Inc., New York, N. Y.

In reducing atmospheres containing carbon monoxide or hydrogen the usual practice is to protect the couple. Neither thermocouple should be used in sulfidizing atmospheres without enclosure in the proper kind of protection tube.

STEAM

The high-nickel materials are resistant to steam below temperature limits which are specific to the particular alloy. Beyond these limiting temperatures, embrittlement may occur.[5,6] Embrittlement results from intercrystalline oxidation, which may be accelerated by the precipitation of graphite in those alloys containing carbon and which also are low in iron and manganese. Table 5 summarizes the maximum temperatures at which uses of the high-nickel materials are recommended.[5]

OXYGEN–CONTAINING GASES, SULFUR–FREE

CARBON DIOXIDE (CO_2)

All the high-nickel materials are oxidized slowly by carbon dioxide at elevated temperatures, but the rate of oxidation is less than that by air at any specific temperature.

CARBON MONOXIDE (CO)

Carbon monoxide is reducing to all the materials considered in this chapter except the Ni–29% Cu–3% Al alloy.* Limiting service temperatures for some of the others are given in Table 5.

Carbon monoxide is a suitable medium for bright annealing nickel and Ni–Cu alloys.

SULFUR AND SULFUR–CONTAINING GASES (SO_2 and H_2S)

Elemental sulfur reacts with nickel to form Ni_2S below 600° C (1110° F) and Ni_3S_2 above 600° C (1110° F). With the solid-solution Ni–Cu alloys, an excess of sulfur produces a double-layer sulfide with Cu_2S on the outside and NiS on the inside.[7]

High-nickel materials, except those containing chromium, are not usefully resistant to sulfur in any form above approximately 315° C (600° F). Sulfur reacts along grain boundaries of nickel to form a low-melting Ni-nickel sulfide (Ni_3S_2) eutectic which causes brittleness. In oxidizing atmospheres, this sulfide is converted to nickel oxide, with retention of a high degree of brittleness. In reducing atmospheres, the formation of sulfide is cumulative, and the net effect is several times greater than that in oxidizing atmospheres due to the absence of the semi-protective oxide surface layer.

Additions to nickel of chromium, iron, and manganese, in order named, produce alloys which are resistant to sulfur. As service temperatures increase, the chromium content should be increased and the nickel content decreased. Among the alloys used for mildly sulfidizing conditions are 24% Cr–12% Ni (Type 309), or 25% Cr–20% Ni (Type 310), or 16% Cr–35% Ni (Type 330) stainless steels. (Refer to *Chromium-Iron, Austenitic Chromium-Nickel-Iron, and Related Heat-Resistant Alloys*, p. 640.)

---

* "K" Monel. Binary alloys of nickel and iron also react with carbon monoxide.

[5] Unpublished data, The International Nickel Co., Inc., New York, N. Y.

[6] J. F. Saffy, "Profound Alteration of an Alloy of Nickel-Copper in Super-saturated Steam at 350° to 400° C," *Compt. rend.*, **186**, 1116–1118 (1928).

[7] I. P. Podolskii and N. M. Zarubin, "The Action of Sulfur on Non-Ferrous Alloys," *Vestnik Metalloprom.*, **11** (No. 5), 57–62 (1931); *C. A.*, **26**, 4015.

TABLE 5. LIMITING SERVICE TEMPERATURES

Material†	Sulfur-Free Atmospheres						Steam		Sulfurous Atmospheres							
	Oxidizing		Reducing H$_2$		Reducing CO				Temp. Below Which These Materials May Be Used*				Temp. Above Which These Materials Cannot Be Used			
									Oxidizing		Reducing		Oxidizing		Reducing	
	°C	°F	°C	°F	°C	°F	°C	°F	°C	°F	°C	°F	°C	°F	°C	°F
(1) Nickel	1050	1900	1250	2300	1250	2300	425–475	800–900	315	600	250	500	540	1000	375	700
(3) Ni; 4.5% Mn‡	900	1650	1150	2100	1150	2100	§	§	315	600	§	§	540	1000	§	§
(4) Ni; 30% Cu¶	540	1000	1100	2000	815	1500**	375–425	700–800	315	600	250	500††	540	1000	340	650
Ni; 13% Cr, 6.5% Fe‡‡	1100	2000	1150	2100	1150	2100	590	1100§§	815	1500	540	1000	815	1500	540	1000

* These materials should not be used above the recommended minimum temperatures without consulting supplier.
† Numbers correspond to analyses listed in Table 1.
‡ "D" Nickel.
§ Insufficient practical data to date. Applications approaching upper temperature limit should be referred to supplier for specific recommendations.
‖ Inconel and "D" Nickel are superior to nickel and to Monel at all temperatures.
¶ Monel.
** Estimated.
†† Not recommended for use in contact with crude oils containing sulfur at temperatures above 500° F.
‡‡ Inconel.
§§ No experience at higher temperatures.

NICKEL AND NICKEL-COPPER ALLOYS

OTHER GASES

HYDROGEN

None of the high-nickel materials is affected adversely by hydrogen, provided no oxides are present from insufficient deoxidation. If deoxidation is not complete, hydrogen embrittlement may occur as with tough-pitch copper. (See p. 629.) Hydrogen atmospheres are used regularly for commercial bright annealing of the alloys.

NITROGEN

Nitrogen has no effect upon nickel. The 4.5% Mn alloy,* the 30% Cu alloy,† and the 29% Cu–3% Al alloy‡ are embrittled slightly because of the formation of brittle manganese nitrides.

HYDROCARBON GASES

The saturated and unsaturated hydrocarbons have no effect upon the high-nickel alloys which are low in iron. Nickel alloys that contain appreciable amounts of iron will be carburized. Under certain conditions, the high-nickel materials cause catalytic cracking of hydrocarbon gases with formation of carbon, although the alloys themselves are not affected adversely.

AMMONIA

Up to the temperature at which anhydrous ammonia is dissociated into nitrogen and hydrogen, it has no effect upon the high-nickel materials. At higher temperatures, used for nitriding steels (480° to 550° C, 900° to 1000° F), nitriding results. The 13% Cr–6.5% Fe-Ni alloy resists nitriding well.

HALOGEN AND HALOGEN COMPOUNDS

Nickel and some high-nickel alloys provide useful resistance to hydrogen chloride and chlorine over a wide range of concentration and temperature.[8] Ultimate performance is determined to a considerable extent by the amount of protection afforded by corrosion products. In the case of nickel and, to some extent, with the 13% Cr–6.5% Fe-Ni alloy§, the corrosion product films (metallic chlorides) are protective and weight losses in 20-hour tests are not much greater than those in 4-hour tests. The protective value of corrosion product films decreases considerably near, and above, temperatures at which sublimation or vaporization may occur at an appreciable rate.

Table 6 represents upper temperature limits for the use of nickel and other common metals and alloys in contact with dry and wet hydrogen chloride and chlorine under conditions where the temperatures of the metallic surfaces are always above the dew points of the moist gases and where oxidizing chemicals are absent. When air is present, there is a possibility of more severe attack, and caution is suggested in applying these limits. Where the major consideration is resistance to corrosion at temperatures below the dew point, but where resistance to high-temperature attack is desired also, the indicated order of preference would be: (1) 30% Mo–5% Fe-Ni

* "D" Nickel
† Monel.
‡ "K" Monel
§ Inconel.

[8] W. Z. Friend and B. B. Knapp, "Behavior of Nickel and High-Nickel Alloys in Hydrochloric Acid and Hydrogen Chloride," *Trans. Am. Inst. Chem. Engrs.*, **39** (No. 6), 731–752 (1943).

alloy; (2) 20% Mo–20% Fe-Ni alloy; (3) 30% Cu-Ni alloy; (4) nickel; (5) 13% Cr–6.5% Fe-Ni alloy; (6) copper; and (7) carbon steel.

TABLE 6. RECOMMENDED MAXIMUM TEMPERATURES FOR USE OF VARIOUS METALS AND NICKEL ALLOYS WITH HCl OR CHLORINE

Material	Indicated Maximum Temperature for Use in Contact with:			
	Hydrogen Chloride		Chlorine	
	°C	°F	°C	°F
Nickel	540	1000	540	1000*
Ni, 13% Cr, 6.5% Fe (Inconel)	540	1000	540	1000
Ni, 30% Mo, 5% Fe (Hastelloy B)	450	850	540	1000
Ni, 20% Mo, 20% Fe (Hastelloy A)	370	700	540	1000
Ni, 17% Mo, 15% Cr, 8% Fe (Hastelloy C)	...	....	510	950
Ni, 30% Cu (Monel)	425	800	450	850
Silver	425	800	425	800
Carbon steel	225	450	200	400
Copper	below 225	450	below 260	500

* In dry chlorine, at 315° C (600° F), nickel is superior to Inconel, Monel, or any of the Hastelloys.

CYCLIC OXIDATION AND REDUCTION

Alternate oxidation and reduction of the grain boundary material occurs rapidly and progressively when nickel or the high-nickel alloys are heated in atmospheres which change repeatedly from reducing to oxidizing even though the atmospheres are sulfur-free. Such an atmosphere may contain either hydrogen or carbon monoxide as reducing gases and either moisture or carbon dioxide as oxidizing agents. This condition may be encountered in gas carburizing, pack carburizing, annealing, and in heating for hot working. Eventually, the entire cross section of the material becomes embrittled. Other than the elimination of the oxidizing gases to as great a degree as possible by maintaining an excess of reducing gases, no remedial measure is known.

The probable resistance of Ni-Cr alloys to this type of attack in decreasing order are: 13% Cr–6.5% Fe-Ni alloy, 15% Cr–25% Fe–60% Ni alloy, 15% Cr–50% Fe–35% Ni alloy, and 20% Cr–80% Ni alloy. Cyclic embrittlement is rare in 24% Cr–12% Ni (Type 309), 25% Cr–20% Ni (Type 310), and 20% Cr–25% Ni (Type 311) stainless steels.

MOLTEN METALS

Nickel is attacked more or less by all molten metals of low-melting point at a rate which increases rapidly with temperature. The Ni-Cr-Fe alloys, the Ni-Fe alloys, and the Ni-Cu alloys, in order named, are more resistant than nickel. None of the materials considered in this chapter gives economical life in contact with molten metals.

FUSED SALTS

Carbon-free nickel* was developed especially for the nitrate and nitrate-nitrite baths use for thermal treatment of aluminum and its alloys and for fused caustic soda.†

* "L" Nickel.
† For corrosion rates, see p. 262.

Nickel containing carbon is embrittled slowly at service temperatures because of the intercrystalline precipitation of graphite. Nickel, the 70% Ni–30% Cu alloy, and the 13% Cr–6.5% Fe–Ni alloy resist the corrosive action of the neutral chloride-carbonate and the high-speed salt baths if they are sulfur-free. If sulfur or sulfate is present, originally or by contamination from operation, a Ni-Cr-Fe alloy is required.

HIGH-NICKEL CHROMIUM-(IRON) ALLOYS

M. A. Hunter*

The Ni-Cr-Fe alloys with high-nickel content are preeminently useful in all operations carried on at high temperatures. In the form of castings they are used for carburizing boxes and fixed and rotary retorts in operations at 900° to 930° C (1650° to 1700° F), for retorts and auxiliary parts in the construction of furnaces operating at 975° to 1150° C (1800° to 2100° F), and for containers of molten liquids such as salt, cyanide, or lead operating at lower temperatures. Furnace containers and auxiliary parts are also made from rolled sheet where the composition of the sheet lends itself to rolling operations. In the form of rod, wire, or strip this group of alloys plays a predominant role in the manufacture of electrical resistance materials.

Ranges in Nickel and Chromium Content

The nickel contents of alloys largely considered in this chapter range from more than 55 to as high as 80%, with a chromium content of 12 to 20%. The alloys in the lower range of both nickel and chromium are used as castings or wrought material in furnace parts, and the higher contents in the more extreme service of electrical resistance materials. All these alloys are outstandingly resistant when they are subjected to oxidizing or reducing atmospheres free from sulfur. In atmospheres which contain large amounts of sulfur, the high-nickel alloys are susceptible to rapid attack. Under these conditions alloys with lower nickel and higher chromium content are more serviceable.

OXIDIZING AND REDUCING ATMOSPHERES

The first application of this class of alloys on an industrial scale in furnace operation was made by J. C. Henderson.[1] He cast an alloy of composition 60% nickel, 12% chromium, remainder iron (employed at that time as an electrical resistance material) in the form of carburizing boxes and used them with eminent success. The use of this type of alloy has been extended since then to many other furnace parts operating at high temperatures. In spite of their extended use in industry, the alloys have been subject to few precise investigations of corrosion behavior under controlled conditions. Their continually expanding use has been based entirely on meritorious performance in industrial practice.

Dickenson[2] made a direct comparison of wrought carbon steel, Cr-Fe, and Ni-Cr-Fe alloys in which specimens were heated in air for 5½ hours at temperature, followed by removal of scale from the surface by scraping (Table 1).

The results indicate evident superiority of the high-nickel chromium alloy.

* Department of Metallurgical Engineering, Rensselaer Polytechnic Institute, Troy, N. Y. Consulting Engineer, Driver-Harris Company, Harrison, N. J.

[1] J. C. Henderson, U. S. Patents 1,190,652 (1916); 1,241,971 (1917); 1,270,519 (1918).
[2] J. H. Dickenson, *J. Iron Steel Inst.*, **106**, 103–140 (1922).

HIGH-TEMPERATURE CORROSION

TABLE 1. OXIDATION OF Fe, Cr-Fe, AND Ni-Cr-Fe ALLOYS IN AIR

Temp.		Rate of Metal Loss, mdd		
		0.30 C Steel	14.7 Cr, 0.26 C	11.7 Cr, 65 Ni
°C	°F			
675	1250	2,900	180	<36
875	1605	34,000	2,900	400
1025	1880	130,000	51,000	730
1100	2010	200,000	100,000	2900

Hatfield[3] compared the action of oxygen, carbon dioxide, and steam in 24-hour continuous tests on Ni-Cr-Fe alloys. (Refer also to Tables 3, p. 646, and 4, p. 648.) The rates for the high-nickel alloys are given in Table 2.

TABLE 2. AVERAGE RATES OF OXIDATION OF Ni-Cr-Fe ALLOYS BY O_2, H_2O, AND CO_2

Alloy		Weight Increase, mdd		
		800° C (1470° F)	900° C (1650° F)	1000° C (1830° F)
11 Cr	35 Ni	200–730	615–1170	1165–1930
12 Cr	60 Ni	23–41	30–49	115–180

OXIDATION OF CASTINGS IN AIR

An important investigation was recently carried out at Battelle Memorial Institute[4] in a series of reports issued to the Alloy Casting Institute. A summary of the more important conclusions is published here through the courtesy of the Alloy Casting Institute.

Machined specimens 0.953 cm (0.375 in.) diameter by 2.54 cm (1.00 in.) long were degreased and exposed at temperature in specially constructed metal boats inserted in an electrically heated muffle furnace. Ordinary air was passed through the furnace at the rate of 200 ml per min, after being saturated with water vapor at 32° C (90° F). After 100 hours exposure, specimens were descaled electrolytically as cathode in a molten salt bath (60% NaOH, 40% Na_2CO_3) at a current density of 400 amp per sq ft for a period of 4 min; then water-quenched and flash-dipped in concentrated HCl saturated with As_2O_3 as an inhibitor. After being washed in distilled water, the specimens were reweighed. In addition to the major constituents, the alloys contained 0.40% carbon, 1.2% silicon, and 0.8% manganese.

Figure 1 shows the effect of chromium on the oxidation rate of Cr-Fe and Cr-Fe-Ni alloys. In Figs. 2, 3, and 4 are placed a variety of metal loss data on the ternary diagram for iron, nickel, and chromium. In these and succeeding figures, HH corresponds to an alloy of 12% nickel and 25% chromium, HT corresponds to an alloy of 36% nickel and 16% chromium, and HW to an alloy of 60% nickel and 12% chromium.

It is significant that the chromium for optium corrosion resistance is relatively low for the high-nickel alloys, being in the range of 12% to 16% even at the maximum

---

[3] W. H. Hatfield, *J. Iron Steel Inst.*, **115**, 483–508 (1927).

[4] J. T. Gow, A. S. Brasunas, and O. E. Harder, *Special Reports* to Alloy Casting Institute, Battelle Memorial Institute, Columbus, Ohio. See also: *Proc. A.S.T.M.*, **46**, 129 (1946).

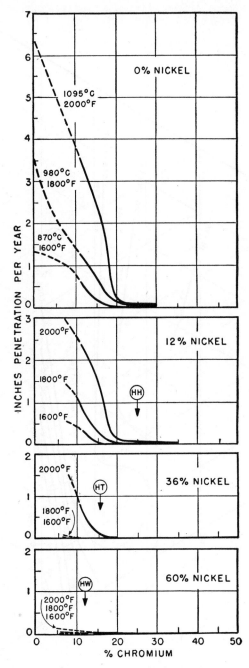

FIG. 1. The Effect of Chromium at Various Nickel Levels on Air Oxidation of Cast Ferrous Alloys. (Courtesy of Alloy Casting Institute.)

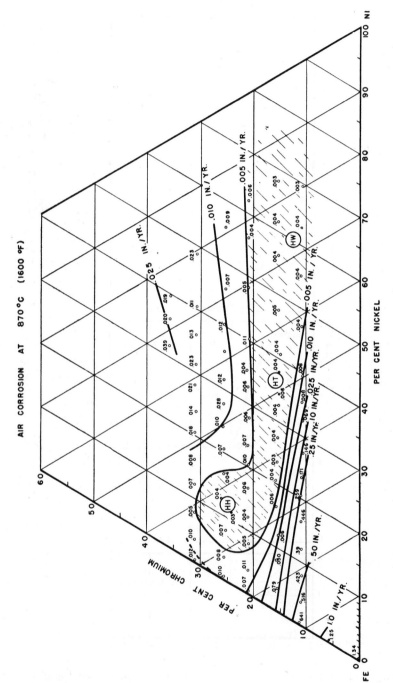

Fig. 2. Metal Loss by Oxidation in Air at 870° C (1600° F). (Courtesy of Alloy Casting Institute.) Corrosion is expressed in ipy. (100-hour tests.)

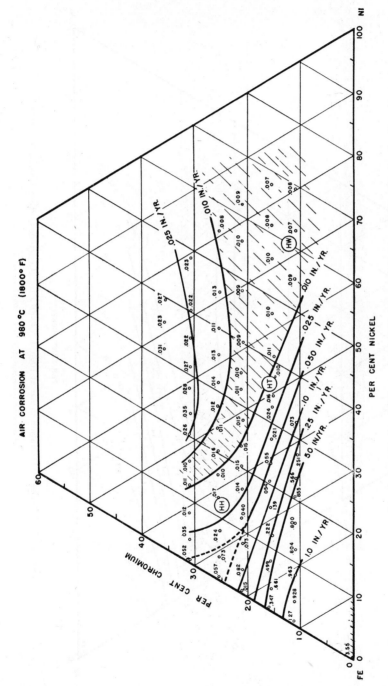

FIG. 3. Metal Loss by Oxidation in Air at 980° C (1800° F). (Courtesy of Alloy Casting Institute.) Corrosion is expressed in ipy. (100-hour tests.)

688 HIGH-TEMPERATURE CORROSION

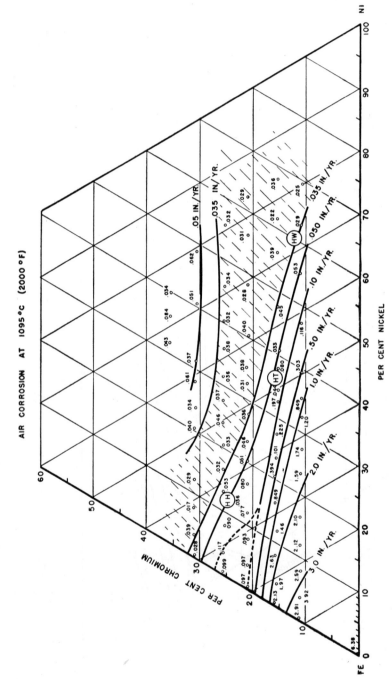

Fig. 4. Metal Loss by Oxidation in Air at 1095° C (2000° F). (Courtesy of Alloy Casting Institute.) Corrosion is expressed in ipy. (100-hour tests.)

temperature studied. As the nickel content decreases, the amount of chromium must be increased for maximum corrosion resistance.

Temperature plays an important part in determining the effectiveness with which a given amount of nickel or chromium is capable of imparting oxidation resistance to these alloys. The effectiveness of increased amounts of nickel is enhanced as the temperature is increased from 870° to 1100° C (1600° to 2000° F), especially for alloys with chromium contents below 20%.

The extent of the corrosion loss in air for alloys of various compositions is summarized in the nomographic charts (Figs. 5, 6, and 7) at temperatures of 870°, 980°, and 1095° C (1600°, 1800°, and 2000° F) respectively. The corrosion loss for any alloy can be read from the chart by drawing suitable lines through points of corresponding composition.

Sulfur Dioxide

Hatfield[5] determined the gain in weight of various Ni-Cr-Fe alloys exposed 24 hours to sulfur dioxide. The results obtained for the high-nickel alloys are given in Table 3. (Refer also to Table 6, p. 649.)

TABLE 3. OXIDATION OF Ni-Cr-Fe ALLOYS BY SO_2

Alloy		Weight Gain, mdd		
		800° C (1470° F)	900° C (1650° F)	1000° C (1830° F)
11 Cr	35 Ni	41	42	790
12 Cr	60 Ni	500	5500	9100

Chromium is the essential element in building up resistance to pure sulfur dioxide or sulfur dioxide in air mixtures. The best behavior is afforded by a 30% chromium alloy to which nickel may be added up to 35%. In excess of this amount, nickel decreases the resistance to sulfur attack. The high-nickel chromium alloys with 20% chromium or less offer little resistance and are rapidly deteriorated.

Hydrogen Sulfide

In test results of Scholl[6] various alloys were exposed to the action of hydrogen sulfide for 28½ hours at 480° C (900° F). Results for the high-nickel alloys are given in Table 4. (See also Tables 8 and 9, p. 651.)

TABLE 4. REACTION OF Ni-Cr-Fe ALLOYS WITH H_2S AT 480° C (900° F)

Alloy		Weight Gain, mdd
14.7 Cr	35 Ni	2400
13.8	62	2900
18	75	5800

In this atmosphere, high-nickel chromium alloys are rapidly attacked. The action is much more damaging than under oxidizing conditions.

[5] W. H. Hatfield, *J. Iron Steel Inst.*, **115**, 483–508 (1927).
[6] Quoted by H. J. French, *Trans. Electrochem. Soc.*, **50**, 47–81 (1926).

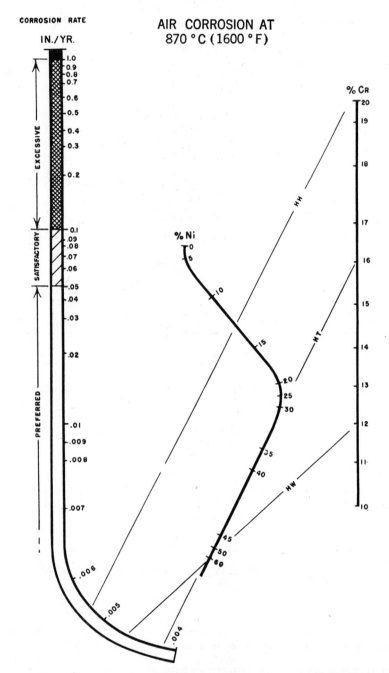

Fig. 5. Nomographic Chart Illustrating Relationship between Nickel-Chromium Content of Ferrous Castings and Air Corrosion. (Courtesy of Alloy Casting Institute.)

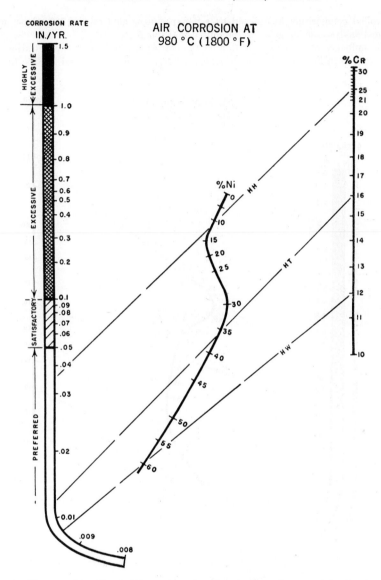

Fig. 6. Nomographic Chart Illustrating the Relationship between Nickel-Chromium Content of Ferrous Castings and Air Corrosion. (Courtesy of Alloy Casting Institute.)

FLUE-GAS ATMOSPHERES CONTAINING SULFUR

Tests of cast Ni-Cr-Fe alloys in gases containing sulfur were conducted at Battelle Memorial Institute[7] in similar manner to those on oxidation discussed above. Gas of

[7] J. T. Gow, A. S. Brasunas, and O. E. Harder, *Special Reports* to Alloy Casting Institute. See also: *Proc. A.S.T.M.*, **46**, 129 (1946).

controlled composition was passed over the specimens at a rate of 100 ml per min. Typical analyses of the gases are presented in Table 5.

The influence of alloy composition on the corrosion rate in inches penetration per year is given in Figs. 8 and 9 for an oxidizing flue-gas atmosphere, and in Figs. 10 and

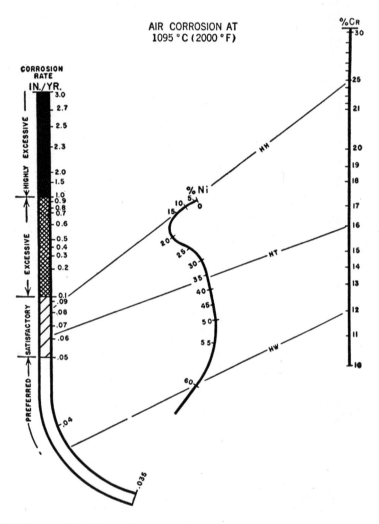

Fig. 7. Nomographic Chart Illustrating the Relationship between Nickel-Chromium Content of Ferrous Castings and Air Corrosion. (Courtesy of Alloy Casting Institute.)

11 for reducing flue-gas atmospheres. A nomographic chart giving corrosion rates at 980° C (1800° F) in a reducing atmosphere containing 100 grains (6.5 grams) of sulfur per 100 cu ft is given in Fig. 12.

TABLE 5. TYPICAL FLUE-GAS ATMOSPHERES

Intake Sulfur Grains/100 cu ft	Intake Composition						Exhaust Composition					
	CO_2	O_2	CO	H_2	H_2O	N_2	CO_2	O_2	CO	H_2	H_2O	N_2
Reducing Flue Gas												
5	7.2	5.5	4.7	1.0	...	bal.	10.8	0	3.2	2.1	7.9	bal.
100	8.4	5.3	5.1	8.2	...	bal.	11.4	0	3.5	1.9	7.1	bal.
Oxidizing Flue Gas												
5	13.7	8.1	0	10.4	...	bal.	16.1	3.4	0	0	10.4	bal.
100	16.4	6.8	0	7.1	...	bal.	18.3	4.0	0	0	7.1	bal.

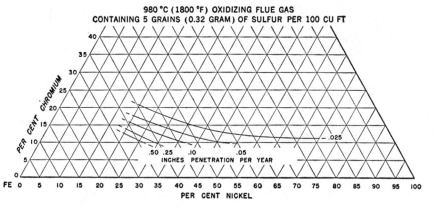

FIG. 8. Influence of Alloy Content on Extent of Corrosion in Flue Gas at 1800° F as Indicated by 100-Hour Tests. (Courtesy of Alloy Casting Institute.)

The following conclusions were drawn by the authors[7] as a result of this investigation.

1. Under oxidizing conditions with sulfur as SO_2, corrosion is analogous to that encountered in oxidation by air. Additions of nickel in the range studied (above 20%) increase corrosion resistance of alloys markedly at the 11 and 16% chromium levels. The influence is less noticeable at 21% chromium, and it disappears at 26%, 31%, and 36% chromium.
2. Under oxidizing conditions variations in sulfur content from 5 grains to 100 grains per 100 cu ft resulted in no increased amount of corrosion.
3. Under reducing conditions the sulfur content of the corrosive atmosphere is more important. With 11% chromium the high-sulfur gas (100 grains sulfur per 100 cu ft) results in corrosion rates varying from 1 to 30 times that noted in a low sulfur gas. At 16% chromium the variations of corrosion rate with sulfur content are less noticeable, while with 21% chromium there are no noticeable effects.
4. In reducing flue gases containing substantial amounts of sulfur, chromium is much more effective than nickel in reducing the corrosion rate of alloys. At 20% nickel, the optimum chromium content appears to be approximately 26% chromium, whereas at 70% nickel the optimum chromium would be 16%. Speaking generally, when nickel is high, chromium should be relatively low, and when nickel is low chromium should be increased for optimum corrosion resistance.

In general, the Ni-Cr-Fe compositions for maximum corrosion resistance at 1100° C (2000° F) in air and oxidizing flue-gas atmospheres appear to be confined to those compositions of intermediate chromium content (12 to 20% chromium) and high-nickel content (more than 50% nickel) approximating the HW type composition (60% Ni–15% Cr). However, under reducing flue-gas conditions when sulfur is present as H_2S, the compositions for maximum corrosion resistance shift towards alloys of higher chromium and lower nickel contents.*

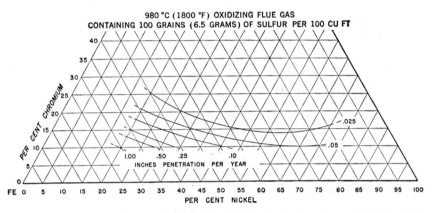

Fig. 9. Influence of Alloy Content on Extent of Corrosion in Flue Gas at 1800° F as Indicated by 100-Hour Tests. (Courtesy of Alloy Casting Institute.)

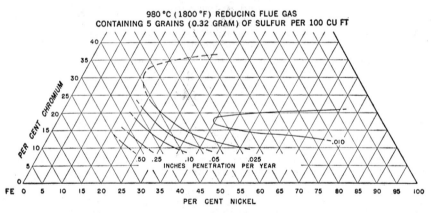

Fig. 10. Influence of Alloy Content on Extent of Corrosion in Flue Gas at 1800° F as Indicated by 100-Hour Tests. (Courtesy of Alloy Casting Institute.)

ELECTRICAL HEATING ALLOYS

THE NICKEL-CHROMIUM ALLOYS

Nickel at high temperatures can carry as much as 40% of chromium in solid solution. The solubility decreases with temperature, but the true solubility, although subject to much investigation, is still indeterminate by reason of the slowness in

---

* Recommended limiting service temperatures for the 80% Ni–13% Cr–6.5% Fe alloy (Inconel) in both sulfur-free and sulfur atmospheres are given in Table 5, page 680.

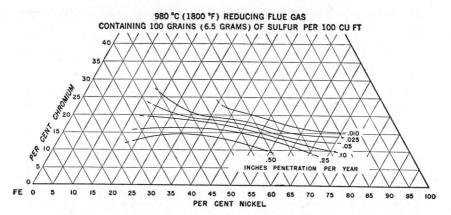

FIG. 11. Influence of Alloy Content on Extent of Corrosion in Flue Gas at 1800° F as Indicated by 100-Hour Tests. (Courtesy of Alloy Casting Institute.)

reaching equilibrium. In practice, alloys with more than 25% chromium after long use at elevated temperatures become heterogeneous with precipitation of a solid solution containing 80 to 90% chromium. Alloys of 25% chromium with manganese and silicon have a still more restricted solubility. This concentration in general marks the limit of chromium additions to the Ni-Cr alloys as electrical resistors because of the mechanical difficulty in working alloys beyond this range.

In actual practice the 80% Ni-20% Cr alloy is the best of the commercial oxidation-resistant alloys. The original patent on its use for an electrical resistor was issued to Marsh[8] in 1905.

In Fig. 13, Matsunaga[9] gives the increase in weight of various Ni-Cr wires oxidized at various temperatures.

Carbon, Manganese, and Silicon Additions. The 80-20 Ni-Cr alloys, as produced industrially, contain carbon, manganese, and silicon, either intentionally added or as impurities arising from the raw materials of their manufacture. Carbon has, in general, a damaging effect on the life of the alloy. The chromium carbides which are thereby formed lower the chromium content of the

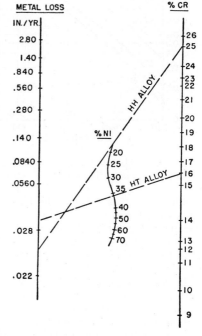

FIG. 12. Nomographic Chart Expressing Corrosion Rate of Ni-Cr-Fe Alloys (1.2% Si, 0.8% Mn, 0.45% C) at 980° C (1800° F) in a Reducing Flue-Gas Atmosphere Containing 100 Grains (6.5 grams) of Sulfur per 100 cu ft. (Courtesy of Alloy Casting Institute.)

[8] A. L. Marsh, U. S. Patent 811,859 (March, 1905).
[9] Y. Matsunaga, *Japan Nickel Rev.*, **1**, 347 (1933).

solid solution and act in this way together with other factors to reduce high-temperature oxidation resistance.

Manganese slightly lowers the high-temperature resistance of Ni-Cr alloys, the first tenths of 1% having the greatest effect. With additions in excess of 1%, the effect is, however, still small.

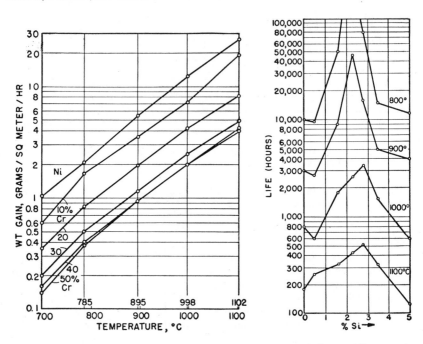

Fig. 13. Oxidation Rates of Nickel and Cr-Ni Alloys at Various Temperatures. (Matsunaga.)

Fig. 14. Effect of Silicon on the Life of Cr-Ni Alloys. (Horioka.)

Silicon additions increase the high-temperature resistance. Horioka[10] has determined the life in hours of Ni-Cr alloys of varying silicon content at various temperatures (Fig. 14). The maximum effect is obtained with 2 to 2.5% silicon. Additions in excess of this amount probably give rise to too high a silicon content in the scale. This lowers the melting point and thereby the protective qualities of the oxide scale.

Aluminum additions improve the life but lower the workability of the material.[11]

In this connection, Smithells, Williams, and Grimwood,[12] found that 1% of silicon increased and 1% of manganese decreased the life of Ni-Cr alloys, whereas 0.5% of aluminum had no effect. They found that 1% of aluminum had a damaging influence.

Iron as one of the important constituents occurs as a common addition to some of the present-day industrial alloys. Its effect on oxidation resistance has been implied in previous discussion and is treated further below.

The Effect of Some Other Alloy Additions. Minor additions of readily oxidizable metals exert an influence on the life of Ni-Cr wires out of all proportion to the small

[10] M. Horioka, *Japan Nickel Rev.*, **1**, 292 (1933).
[11] W. Hessenbruch, *Zunderfeste Legierungen*, p. 85, Berlin, 1940.
[12] C. J. Smithells, S. Williams, and E. Grimwood, *J. Inst. Metals*, **46**, 443 (1931).

amounts present. Small additions of calcium exert a large effect in increasing the life.[13] Additions of both zirconium and calcium have been found very useful.[13]

A summary of the effect on life at elevated temperatures of a number of such additions to 80% Ni–20% Cr alloy and 60% Ni–18.5% Cr–bal. Fe alloy is given in Figs. 15 and 16.

Calcium and zirconium additions in small amounts are widely used commercially in the United States. The use of cerium, prevalent abroad, has not been adopted here.

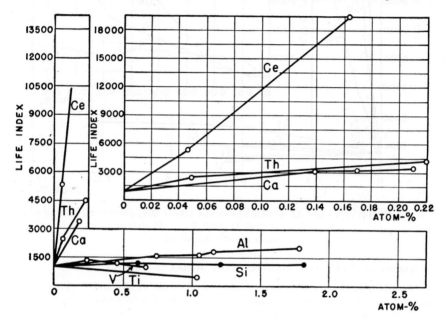

Fig. 15. Effect of Various Alloy Additions on the Life of 60% Ni–18.5% Cr–bal. Fe Alloy, 0.4-mm Diam. Wires at 1050° C (1920° F). (Hessenbruch.)

Life index equals number of cycles for 2 min at temperature, 2 min off, until wire fails.

In seeking for an explanation of the improvement in life at high temperatures brought about by these minor additions, the author of this chapter has maintained that the improvement is due to the high degree of deoxidation which is produced by these additions. In a highly deoxidized alloy, grain boundaries are less susceptible to precipitation of oxides and, in consequence, less vulnerable to attack at high temperatures. The fact that silicon, aluminum, zirconium, and calcium produce good effects in the order named would support this contention. The behavior of thorium and cerium in the matter of deoxidation is not at present known to the writer.

Another theory has been given by Hessenbruch.[14] The small additions of elements which give the best results have atomic volumes that are greater than the molecular volume of nickel oxide. It is suggested that this situation is necessary to impede diffusion of metal ions outward through the oxide and thus retards the oxidation reaction.

[13] M. Hunter, U. S. Patent 2,005,423 (June, 1935). J. M. Lohr, U. S. Patents 2,047,917; 2,047,918 (July, 1936), 2,019, 686; 2,019,687 (November, 1935).
[14] W. Hessenbruch, *Zunderfeste Legierungen*, p. 117, Berlin, 1940.

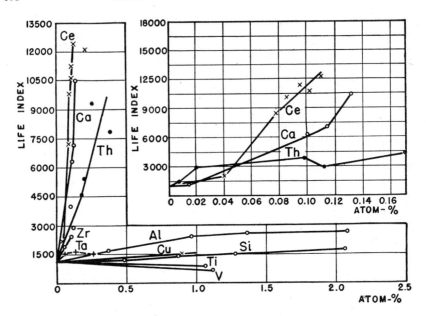

FIG. 16. Effect of Various Alloy Additions on the Life of 80% Ni–20% Cr Alloy, 0.4-mm Diam. Wires at 1050° C (1920° F). (Hessenbruch.)
For definition of life index, see Fig. 15.

BEHAVIOR OF Ni-Cr AND Ni-Cr-Fe ALLOY WIRES IN VARIOUS GASES*

Hessenbruch, Horst, and Schichtel[15] heated wires electrically and determined the life in hours to destruction in various atmospheres. Small spirals 0.040 cm (0.0157 in.) in diameter were heated and cooled in 2 min intervals until the entire section of the wire had scaled.

The compositions of the alloys studied are given in Table 6.

TABLE 6. COMPOSITION OF WIRE ALLOYS IN PERCENTAGES

	Ni	Cr	Fe	Mn	Si	Al	C	Mo
Ni-Cr alloy	77.10	19.80	0.68	1.50	0.81	....	0.02	...
Ni-Cr-Fe alloy	60.4	18.40	18.10	2.80	0.80	....	0.02	...
Ni-Cr-Fe alloy	33.7	19.50	43.10	1.97	0.85	....	0.02	...
Ni-Cr-Fe-Mo alloy	49.85	15.03	15.31	1.90	0.65	....	0.02	7.1
Cr-Al-Fe alloy	0.20	29.90	63.40	0.80	0.53	5.30	0.03	...

The scaling resistance of these alloys was determined in: (1) air, (2) carbon dioxide, (3) nitrogen (2% oxygen), (4) city gas (53% H_2, 1.7% CO_2, 3.85% SO_2, 30.3% CH_4), (5) oxygen, (6) hydrogen, (7) water gas (49% H_2, 40.5% CO_2, 4% CO_2). The life in hours of the alloys in the various gases is given in Figure 17.

The deterioration of the wires with time was determined by measuring increase in resistance during the test. These results are plotted in Fig. 2, p. 669.

The results indicate that the 80% Ni–20% Cr alloy constitutes an excellent heating

---

* Life test data in air for some high-nickel alloy wires at various temperatures are given in Fig. 1, *High Temperature Tests*, p. 999.

[15] W. Hessenbruch, E. Horst, and K. Schichtel, *Archiv Eisenhüttenwesen*, **11**, 225–229 (1937).

wire material. Micrographic investigations showed that oxidation proceeded slowly along the grain boundaries. In water gas, carburization, with formation of chromium carbides, is extensive but in carbon monoxide or carbon dioxide is slight. Nitrogen, ordinarily assumed to be inert, penetrates into the grain interior, forming presumably chromium nitrides.

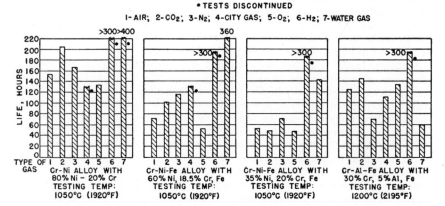

FIG. 17. Scaling Resistance of Electrical Heating Wire Alloys When Annealed in Different Gases. (Hessenbruch, Horst, and Schichtel.)

Fe-Ni-Cr alloys give results similar to those of the Ni-Cr alloys, but increasing contents of iron lower the scaling resistance in oxidizing atmospheres (air or oxygen) to a marked degree. The life of the iron-containing alloys in carbon dioxide or nitrogen is slightly longer than in air. In contrast to the Ni-Cr alloys, the difference between life in hydrogen and in water gas is quite marked. Although the useful life in water gas is many times that in air, the absorption of carbon is so considerable that overheating and failure occur the more readily with increasing iron content. When heated in nitrogen, increased amounts of nitride appear with increasing iron or decreasing nickel content.

With the Ni-Cr alloys containing molybdenum, heated in a carbonaceous atmosphere of nitrogen, the interior of the grain is practically free of nitrides or carbides while the grain boundaries are surrounded by continuous bands of these compounds.

CYCLIC OXIDATION AND REDUCTION

The Ni-Cr and Ni-Cr-Fe alloys are subject to embrittlement when exposed to high-temperature alternate reducing and oxidizing atmospheres. This is discussed on p. 682.

NOBLE METALS AND THEIR ALLOYS

E. M. WISE*

OXIDATION (AIR AND OXYGEN)

GOLD AND GOLD ALLOYS

Pure gold, Au-Ag, and the high Au-Pt, and Au-Pd alloys do not oxidize on heating in air, but some, particularly those containing large amounts of silver or palladium,

* Development and Research Division, The International Nickel Co., Inc., New York, N. Y.

may dissolve enough oxygen to be damaged by subsequent heating in reducing atmospheres containing hydrogen. The latter heat treatment opens up the grain boundaries and may produce minute blisters.

In general, the carat jewelry golds contain sufficient copper or other base metal to be rather susceptible to oxidation on heating in air. The rate of tarnishing on heating in air and oxygen has been determined.[1] Many of the dental alloys also are slightly oxidized on heating, but some are essentially free from base metals and hence are not attacked. Sulfur in high-temperature gases behaves like oxygen, but is more active. Sulfur has caused much trouble where dental gold alloys are heated in contact with plaster of Paris (hydrated $CaSO_4$) plus a little carbon.

Oxidation results in surface darkening and deeper-seated oxidation, comprising a dispersion of cuprous oxide or oxides of other elements in the metal, which may extend to an appreciable depth. The surface darkening can be removed by pickling, but the deep-seated oxidation, generally known as "Fire," does not respond to the usual pickling process. It may be removed by dissolving an appreciable amount of the alloy as, for example, using an aqua regia type pickle or by stripping as anode in a cyanide bath, or it may be reduced by heating at about 700° C (1300° F) for some time in hydrogen. However, it is best to prevent oxidation either through the use of a thoroughly protective coating of boric acid or borax plus ocher or, better still, through the use of reducing atmospheres in the annealing process. Another device, useful particularly in soldering, is to permit some methyl borate to vaporize and be carried along by the incoming gas stream, which results in deposition of a very thin protective film of boric acid where the blowpipe flame impinges on the work.[2] Data on the rate of oxidation of a 5% Cu-Au alloy as reported by Kubaschewski[3] are shown in Fig. 1.

PLATINUM AND PLATINUM ALLOYS

Platinum remains bright on heating in air but platinum, particularly in the form of sponge, oxidizes slowly when heated in oxygen over a narrow range of temperature. The monoxide is produced by heating platinum black in oxygen within the range of 510° to 560° C (950° to 1050° F).[4] Above this temperature the oxide decomposes. The stability of platinum, however, is such that small wires maintain a very constant resistance, as is required in resistance thermometers, minute fuses, and the like.

The rate of volatilization of platinum is greater in the presence of oxygen[5,6] than in vacuo, whereas in a strictly inert gas volatilization might be expected to be not more than one fiftieth of the rate in vacuo. The data of Jones, Langmuir, and Mackay[7] on the rate of evaporation of platinum in vacuo are plotted in Fig. 2, in addition to data from Rideal and Wansbrough-Jones[8] on the rate of loss in oxygen at low pressure, which are slightly above the loss in vacuo. The latter data apparently were reported on a temperature scale used in the earlier experiments of Langmuir and have been adjusted to agree with the revised scale used by Jones, Langmuir, and

[1] E. Raub and M. Engel, Vortrage der Hauptversammlung, *Deut. Ges. Metallkunde*, 83–89 (1938).
[2] Devised by author, 1925, unpublished research. See also N. B. Pilling and T.E. Kihlgren, *J. Am. Welding Soc.*, **8** (No. 4), 20–28 (1929).
[3] O. Kubaschewski, *Z. Elektrochem.*, **49**, 446–454 (1943).
[4] F. Ephraim, *Inorganic Chemistry*, p. 462, Oliver and Boyd, Edinburgh, Scotland, 1943.
[5] L. Holborn and L. W. Austin, *Phil. Mag.*, **7**, 388 (1904).
[6] J. H. T. Roberts, *Phil. Mag.*, **25**, 270 (1913).
[7] H. Jones, I. Langmuir, and G. Mackay, *Phys. Rev.*, **30**, 201–214 (1927).
[8] E. K. Rideal and O. H. Wansbrough-Jones, *Proc. Royal Soc.* (A), **123**, 202 (1929).

Mackay. Oxidation by ionized gases was studied by Güntherschulze and Betz.[9] In all these cases it appears that PtO_2 is involved. At 1200° C (2195° F) in oxygen under

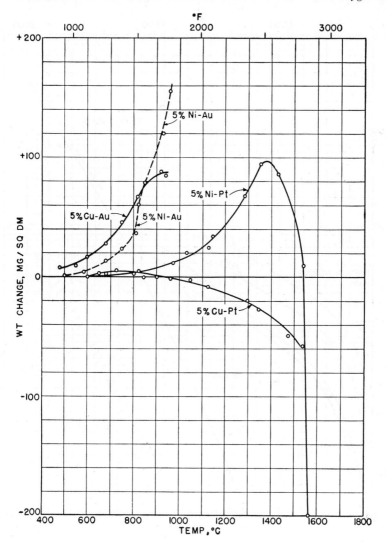

FIG. 1. Oxidation of Platinum and Gold Alloys at Various Temperatures (According to Kubaschewski). Heated 4 Hours in Air.

pressures corresponding to several atmospheres pressure, it is possible that a higher oxide may be found.[10]

The weight loss of platinum in air at 1360° C (2480° F) from a series of exposures

[9] A. Güntherschulze and H. Betz, Z. Elektrochem., **44**, 253–255 (1938).
[10] A. Schneider and U. Esch, Z. Elektrochem., **49**, 55–56 (1943).

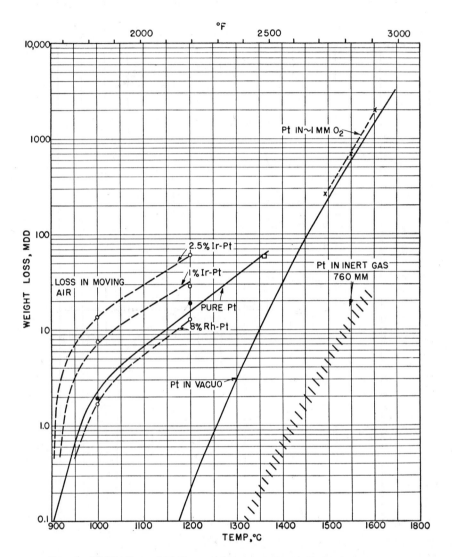

FIG. 2. Weight Loss of Platinum and Some Platinum Alloys Heated in Air, Inert Gas, and Vacuum.

——— Jones, Langmuir, and McKay.
X   Rideal and Wansbrough-Jones.
☐   Kubaschewski.
●   Burgess and Waltenberg.

totalling 7 hours[11] and points at 1000° and 1200° C (1830° and 2195° F) from 4-hour tests[12,13] are also shown in Fig. 2. The latter experiments (on crucibles) indicated that the loss would drop to zero at about 900° C (1650° F) and that a slight gain in weight might be expected at lower temperatures, probably due to oxygen picked up by a

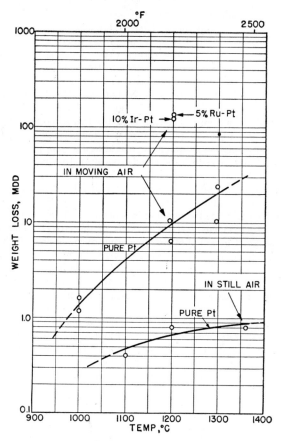

Fig. 3. Effect of Moving and Still Air on Weight Loss of Platinum at Various Temperatures. (Carter and Lincoln.)

trace of iron or other oxidizable impurities. Iridium substantially increased the weight loss at high temperatures, and rhodium decreased it.

Data by Carter and Lincoln[14] on the weight loss of platinum and certain platinum alloys are shown in Fig. 3. These results are based on 24-hour tests of usually duplicate thin 1 by 2 in. samples heated in an electrically heated tube furnace. Great

[11] O. Kubaschewski, Z. Elektrochem., **49**, 446-454 (1943).
[12] G. K. Burgess and R. G. Waltenberg, Bureau of Standards Scientific Paper 280 (1916).
[13] G. K. Burgess and P. D. Sale, Bureau of Standards Scientific Paper 254 (1915).
[14] Unpublished research by Carter and Lincoln of Baker and Co., Newark, N. J.; communicated by F. E. Carter.

care was used to avoid sublimation of impurities onto the platinum, a high-alumina refractory tube being employed, and the samples were rigorously acid-washed and reduced in hydrogen after heating to avoid any spurious weight gains from traces of iron or silica. The latter sources of error are not always given sufficient consideration.

In still air, with the furnace tube closed at one end, the weight loss was minute — only about 0.8 mdd for platinum at 1200° C (2195° F) and 1360° C (2480° F) — but when the furnace was inclined 15° and open at both ends so that a slow stream of air moved through the tube by convection, over the range of 1000° to 1300° C (1830° to 2370° F) the losses increased tenfold to twentyfold.

For the 10% Ir-Pt and 5% Ru-Pt alloys (these compositions were selected to be of approximately equal hardness), the losses at 1200° C (2195° F) are almost identical, and the values for the pair of samples differ so little that they were not separately plotted. Neither of these alloys would be used for long exposure in air at so high a temperature because Rh-Pt or pure platinum is more suitable.

It is evident that in platinum equipment operating for long periods above 1000° C (1830° F), it is desirable to reduce the circulation of air to a minimum. This is safer than sharply reducing the oxygen content of the atmosphere, as limited oxygen is likely to lead to reduction of the refractory and thereby damage the platinum. In 20% Rh-Pt alloy electric tubular furnaces, operating up to 1500° C (2700° F), it has been found desirable to place the winding in grooves in the alumina supporting tube and to smear alumina cement over the opening of the grooves. The wire is free to move within the closed groove, but no circulation of air is possible. With this construction the life is long and most of the sublimed platinum can be recovered from the surrounding refractory. In furnaces where the change in electrical resistance of a platinum winding is used for temperature control, it may be worth while to paint the groove lightly with platinum black to minimize the initial loss of platinum from the winding.

Kubaschewski[15] measured the change in weight of 5 and 10% Ni-Pt and 5 and 10% Cu-Pt alloys as well as the corresponding gold alloys on heating in air. Results for the 5% alloys heated in air for 4 hours at various temperatures are shown in Fig. 1. The gain in weight, due to oxidation of the alloyed base metal, is non-linear with time so that no singular oxidation rate can be assigned.

The Rh-Pt alloys deserve comment in view of their extensive use at high temperatures. Such alloys containing up to 1.5% rhodium, it is reported,[16] do not oxidize visibly on heating in air at any temperature. At 800° C (1475° F) in air, alloys containing up to 2% rhodium do not oxidize, and neither do alloys with less than 10% rhodium at 900° C (1650° F). However, in oxygen at 900° C (1650° F) the 10% rhodium alloy oxidizes visibly. The rates of oxidation are very low, as is attested by the long and successful use of the 10% Rh-Pt and occasionally the 13% Rh-Pt thermocouple over a wide range of temperatures. On the whole, the Rh-Pt alloys, especially with 10% rhodium, are the most generally useful platinum alloys for high-temperature service. The 10% Rh-Pt alloy is in general use as gauze for the catalytic oxidation of ammonia at 900° to 950° C (1650° to 1750° F),[17] for certain types of feeder dies for molten glass,[18] and also for small electric furnace windings, with the 20% Rh-Pt alloy also used for the latter application where maximum temperatures are required.

[15] O. Kubaschewski, Z. Elektrochem., **49**, 446-454 (1943).
[16] Experiments of F. Zimmermann, private communication from F. E. Carter, Baker and Co., Newark, N. J.
[17] S. L. Handforth and J. N. Tilley, Ind. Eng. Chem., **26**, 1287-1292 (1934).
[18] H. K. Richardson, U. S. Patent 2,190,296 (February, 1940).

In the case of gauze catalyst, the loss of material is due not to any usual type of corrosion but probably to a progressive change in the contour of the surface and ultimately to the mechanical removal of minute particles of the metal, some of which can be recovered. The rate of loss is so low that 1,000,000 lb of nitric acid can be secured from one troy ounce of Rh-Pt with the old atmospheric pressure plants, but the loss of catalyst is several times this in modern pressure plants.

PALLADIUM

Palladium oxidizes on heating in air over the range of 350° to about 790° C (660° to 1450° F). At temperatures above 790° C (660° F), the oxide decomposes,[19] resulting in bright metal. Some data on oxidation on heating, and also by anodic treatment, have been published.[20]

Palladium is capable of taking some oxygen into solid solution at high temperatures. If palladium containing dissolved oxygen is subsequently heated in hydrogen, the hydrogen diffuses into the metal rapidly and reacts to form steam.[21] The steam does not diffuse appreciably, so that its formation produces a disruptive effect at the grain boundaries and may also cause small blisters. Silver behaves similarly, whereas platinum is only slightly affected. Alternately oxidizing and reducing atmospheres should be avoided in processing the metal.

IRIDIUM

Iridium oxidizes slowly on heating in air over a considerable range of temperatures, but the oxide is rather volatile. Slight oxidation becomes apparent on heating in air within the range of 600° to 1000° C (1100° to 1850° F). Above the latter temperature the metal remains bright, but the formation and volatilization of an oxide occur, which results in high losses. Values of the order of 3500 mdd[22] have been reported for iridium heated in air at 1210° C (2210° F).

RHODIUM

Rhodium oxidizes slowly on heating in air to about 600° C (1100° F). Finely divided rhodium appears to oxidize most rapidly at about 800° C (1475° F). The oxide, which is slightly volatile, decomposes at about 1000° C (1830° F) when heated in air. Above this temperature the metal remains bright; nevertheless it appears that oxygen increases the rate of volatilization in the high-temperature region. In Crookes' experiments the weight loss of rhodium was the lowest of the platinum metals examined.[23]

OSMIUM

Osmium is easily oxidized. The tetroxide has a boiling point of only 130° C (265° F) and is very toxic. The production of this oxide affords a convenient method for separating osmium from other metals. The metal itself has a very low vapor pressure and high melting point, but in melting alloys containing much osmium precautions

[19] L. Wöhler, *Z. Elektrochem.*, **11**, 839 (1905).
[20] G. Tammann and J. Schneider, *Z. anorg. Chem.*, **171**, 367–371 (1928).
[21] R. F. Vines and E. M. Wise, unpublished research, Research Laboratory, The International Nickel Co., New York, N. Y.
[22] Private communication, F. E. Carter, Baker and Co., Newark, N. J.
[23] W. Crookes, *Proc. Royal Soc.* (A), **86**, 461 (1912).

must be taken to keep OsO_4 out of the eyes and noses of the workers. Osmium alloys are not suitable for use at high temperatures in oxidizing atmospheres.

RUTHENIUM

Ruthenium resembles iridium in high temperature behavior. RuO_2 is not very volatile, a vapor pressure of 40 mm of Hg at 960° C (1760° F) having been reported. The higher oxide, RuO_4, can be produced and has a vapor pressure of 183 mm at 100.8° C (213.4° F). This compound, like the similar osmium compound, is useful in analytical work for separating ruthenium from other metals. Ruthenium oxidizes slowly on heating in air above about 450° C (840° F).

Some care is required to avoid loss of ruthenium on initial melting of alloys, but the usual 5 to 10% Ru-Pt alloys and 5% Ru-Pd alloys respond to the regular manufacturing processes without special attention. However, neither the 10% ruthenium nor the 10% iridium alloys are desirable for exposure to high-temperature oxidizing atmospheres, the rhodium alloy being preferable for this type of service where pure platinum may not be adequate.

EFFECT OF REDUCING ATMOSPHERES AND REDUCIBLE OXIDES

In addition to the effects of alternate oxidation and reduction mentioned under palladium, other factors are to be considered where a choice of atmospheres is possible. Platinum and high-platinum alloys generally perform best under oxidizing conditions because there is then the least possibility of contamination with the base metal impurities, in particular iron, silicon, lead, phosphorus, and manganese. When platinum or platinum-alloy crucibles are heated under reducing conditions, as in a strongly reducing flame, some damage results from the carbon content of the gas; but hydrogen is also present, which diffuses through the crucible and tends to reduce a portion of the charge. If any baleful elements are present in the charge, they will alloy with the inside surface of the crucible, greatly to its detriment. In the case of thermocouples, protection tubes have been developed which effectively exclude reducing gases and prevent contamination of the thermocouple which otherwise would occur. In certain types of experimental work, where fused silica sheaths must be used with thermocouples, great care should be taken to avoid the presence of reducing material within the sheath and to check the thermocouple from time to time to be sure that contamination from reduced silicon has not occurred.

It has been observed that an iron oxide, such as magnetite, when heated in contact with platinum, will yield some iron which alloys with the platinum, and the oxygen liberated increases the state of oxidation of the remainder of the charge.[24] This same general phenomenon applies to some of the other reducible oxides and must be avoided.

It must not be concluded that strong oxidation is desirable under all circumstances for, in the presence of molten salts and alkalies in particular, oxidation sharply increases the corrosion rate, as will be discussed in a later section (p. 716), and, as previously noted, increases the volatilization loss in gases.

SULFUR-GROUP COMPOUNDS

Although gold sulfides exist, *gold* is not attacked by hot sulfidizing gases. Selenium probably does not attack it either, but tellurium vapor does.

[24] Private communication, F. E. Carter, Baker and Co., Newark, N. J.

Platinum, often used in contact with a mixture of sulfur dioxide, air, and sulfur trioxide at about 400° to 450° C (750° to 850° F) as the catalyst in sulfuric acid manufacture, survives this exposure for years with only minute loss.

Although much over 0.02% sulfur alloyed with platinum reduces its high-temperature ductility, sulfur dioxide does not appear to damage platinum.[25,26] One hour exposure at 800 C (1475° F) caused no weight change, and a similar exposure at 1000° C (1830° F) produced a loss of but 0.0013 gram/sq dm.[25] The platinum remained bright and ductile after these treatments. Heating in hydrogen sulfide over the range of 400° to 1000° C (750° to 1830° F) liberated sulfur and produced a slight blue film on the platinum, which may be protective. Twelve hours exposure to hydrogen sulfide at 1000° C (1830° F) caused a weight gain of 0.024 gram/sq dm[25,26] but no loss in ductility. Calculations[27,28] indicate that PtS should not form on platinum exposed to hydrogen sulfide above 900° C (1650° F). Similar calculations for platinum in contact with gaseous sulfur indicate that PtS should not form above 1270° C (2320° F). Runs of 1 hour each at 1100° C (2010° F) in sulfur vapor, followed by metallographic examination, showed a gray corrosion product film of the order of 0.1 mil thickness on pure platinum, with perhaps a slightly thicker film on 0.4% Ir-Pt, 3.5% Rh-Pt, and 5% Au-Pt alloys.[26] No intergranular penetration was observed with any of these materials, but 10% Rh-Pt is intergranularly attacked under similar circumstances.[26] Possibly this behavior of 10% Rh-Pt is responsible for the reputed bad effect of sulfur compounds on platinum thermocouples.

Palladium in sulfur dioxide exposed for one hour at 800° C (1475° F) underwent a weight gain of 1.3 mg/sq dm and formed a purple-blue film.[26] Exposures for one hour at 1000° C (1830° F) produced a weight gain of 5.1 mg/sq dm, and the specimen appeared lightly etched and was coated with a purple film, but the metal was not embrittled. Some of this weight gain may have been due to oxidation on cooling. In contrast to the above, heating in hydrogen sulfide at temperatures above 600° C (1110° F) caused rapid attack through the formation of a sulfide eutectic melting at approximately 600° C (1110° F).

SULFURIC ACID

The behavior of *gold* and *platinum* toward hot sulfuric acid was at one time of great importance, as these metals were used for evaporating pans in concentrating this acid. Data on the behavior of gold and the platinum metals in sulfuric acid at 240° C (464° F) and 300° C (572° F) are presented in Table 1.[29] *Iridium* shows no loss and gold very little loss. The potent effect of adding a reducing agent like sulfur is evident by the reduction in corrosion rate for platinum at 300° C (572° F) to 3% of the weight loss for concentrated pure sulfuric acid and by the reduction in rate of attack on gold to zero in 7 hours exposure.

Osmium showed about one half the corrosion rate of platinum in sulfuric acid at 300° C (572° F), whereas *palladium* showed about twice, and *rhodium* over ten times the rate of platinum. The addition of sulfur to sulfuric acid increased the attack on both rhodium and palladium.[29]

[25] E. M. Wise and J. T. Eash, *Trans. Am. Inst. Mining Met. Engrs.*, **128**, 282 (1938).
[26] E. M. Wise and R. F. Vines, unpublished data, Research Laboratory, The International Nickel Co., New York, N. Y.
[27] W. Biltz and R. Juza, *Z. anorg. Chem.*, **190**, 161 (1930).
[28] *Bulletin* 406, U. S. Bureau of Mines.
[29] R. Atkinson, A. Raper, and A. Middleton, unpublished research, Mond Nickel Co., Acton, London, England.

TABLE 1. HIGH-TEMPERATURE CORROSION OF PLATINUM GROUP METALS IN SULFURIC AND PHOSPHORIC ACIDS

Weight Loss in mdd

Acid	Temp. °C	Time of Exposure, hr	Au	Pt	Ir	Os	Pd	Rh	Ru	Alloys			
										5%Ir-Pt	5%Ru-Pt	5%Rh-Pt	5%Pd-Pt
98% H₂SO₄	240	4	...	65	...	...	...	...	...	12	16.8	14.4	14.4
98% H₂SO₄	300	7	9.6	281	0.0	154	600	3240	...				
98% H₂SO₄ + sulfur	300	7	0.0	9.6	...	...	9860	8530	...				
Phosphoric acid	300	2½	0.0	0.0	0.0	192	0.0	22	0.0				

The effect of oxidation on the behavior of sulfuric acid toward platinum at 400° C (750° F) is reported in Table 2,[30] wherein the rates of corrosion under reduced

TABLE 2. CORROSION OF PLATINUM IN SULFURIC ACID AT 400° C (750° F)

Conc. of Acid	Loss in Weight, mdd
Commercial Pt	
94%, vacuum	24.0
94%, O_2	2976.0
Pure Pt	
94%, vacuum	14.4
94%, O_2	544.8
94% + 2% SO_3, vacuum	648.0
98.6%, vacuum	182.4
96.75%, vacuum	33.6

pressure are compared with those prevailing with oxygen or SO_3 present. With pure platinum the corrosion rate in 94% sulfuric acid under reduced pressure was 14 mdd, as compared to about 550 and 650 mdd with oxygen or SO_3 added.

With changes in production processes, corrosion under these conditions has become less important. Nevertheless, the results are instructive and indicate the means for sharply reducing the corrosion of gold or platinum under very severe conditions. In addition to sulfur, it is known that carbon, SO_2, and ammonium sulfate reduce the rate of attack.

HYDROGEN

Hydrogen has little effect on pure *gold* at elevated temperatures, although *gold alloys* containing large percentages of metals capable of dissolving hydrogen are mildly affected. Gold alloys containing oxygen are more seriously affected, as discussed on p. 699. Gold and *platinum group* metals heated in hydrogen while also in contact with reducible oxides of elements which alloy with the noble metals will be damaged. This is well recognized in the case of platinum, but is frequently overlooked in the case of gold alloys, particularly during melting. Silicon is often picked up, which makes some gold alloys "hot short."

The direct effect of hydrogen on platinum group elements is most apparent in *palladium*, as hydrogen has a very high solubility in this element and diffuses through it so rapidly that practical diffusors can be made for supplying extremely pure hydrogen. These operate preferably at about 450° C (850° F), at which temperature 300 cu ft/sq ft/day of hydrogen at N.T.P. will diffuse through a septum 0.0016 in. (0.04 mm) thick with a difference of one atmosphere in the partial pressure of hydrogen on the two sides of the septum.* Since absorption of hydrogen by palladium results in a volume change sufficient to roughen the surface, it is desirable to avoid cyclic changes in the hydrogen content of diffusor elements to prevent progressive changes in shape.

The effect of hydrogen on the other platinum group metals has not been of much concern, except from the standpoint of the analyst, who should be aware of a slight increase in weight which results from cooling assay samples in hydrogen. To avoid

---

* The rate attained in practice is sensitive to the surface condition of the palladium. (Private communication from E. F. Rosenblatt, Baker and Co., Inc.)

[30] L. Quennessen, *Compt. rend. Acad. Sci.* (Paris), **142**, 1341 (1906).

this error, carbon monoxide is often employed for reducing finely divided material prior to weighing.

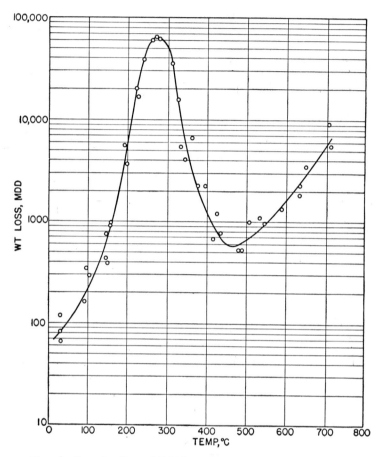

FIG. 4. Corrosion Rate of Gold in Dry Chlorine. Velocity 1.3 ft/min. (Brown, Delong, and Auld.)

CHLORINE (AND OTHER HALOGENS)

Chlorine attacks *gold* even at room temperatures, and the attack is vigorous at 200° C (390° F). The vapor pressure of Au_2Cl_6 reaches a maximum at 270° C (520° F).[31] Above 480° C (895° F) the lower valence compound Au_2Cl_2 becomes important and exerts a partial pressure of about 3.5 mm of Hg at the melting point of gold. If chlorine is bubbled through molten impure gold, such as that resulting from the cyanide process, silver and the base metals are rapidly converted to chlorides and can be recovered. By stopping the process at the right point, it is possible to secure gold **997 fine** with very little loss.

[31] W. Biltz, W. Fischer, and R. Juza, Z. anorg. Chem., **176**, 121-142 (1928).

Quantitative data on the corrosion of gold in flowing chlorine at various temperatures have recently been determined by M. H. Brown et al.[32] In these tests of 2-hour duration the chlorine flowed at a velocity of 1.3 ft/min, measured at room temperature. After exposure the specimens were washed in water and rubbed gently with a piece of rubber to remove any reaction film prior to weighing.

Thin but visible films were apparent on samples exposed at temperatures up to about 400° C (750° F). At higher temperatures no film was apparent. The film is probably Au_2Cl_6, which has quite a high vapor pressure, making it unlikely, therefore, that it offers much protection. It will be noted from Fig. 4 that the rate of corrosion increases sharply with the temperature and reaches a maximum of about 64,000 mdd at 270° C (520° F). On increasing the temperature to 460° C (860° F) the corrosion rate drops to about one hundredth of this value, but the rate is still too high to render gold useful. A further increase in temperature again increases the corrosion rate. This maximum corrosion at 270° C (520° F) corresponds exactly to the maximum vapor pressure of Au_2Cl_6. At higher temperatures the lower valence compound Au_2Cl_2 forms.

Chlorine and the other halogens react with the *platinum metals* at elevated temperatures, but the reactions are complicated experimentally by the appreciable dissociation and vapor pressures of the compounds formed. The behavior of platinum compounds at high temperatures has been studied[33,34] as well as the reaction of platinum with low-pressure (0.005-0.02 mm of Hg) chlorine,[35] bromine,[36] and iodine.[37,38]

Platinum can be used freely in chlorine up to about 100° C (210° F). Above this temperature, and up to about 250° C (480° F), particularly where the gas may not be dry, the nature of the service and the desirability of using *rhodium, iridium*, or *Ru-Pt* should be considered. For example, exposure of pure platinum to flowing chlorine at 400° C (750° F) for one hour produced a weight gain of 0.65 mg/sq dm, but the film, which appeared to be $PtCl_2$, was tenacious and was probably protective. On reducing and removing this film a net loss of 2.5 mg/sq dm occurred. The corrosion rate for longer exposures would be expected to be reduced somewhat by this film and probably would be much less than 60 mdd.

The behavior of platinum exposed to moving chlorine for 2 hours over a wide range of temperatures has recently been examined by Brown et al[39] (Fig. 5). After exposure, the specimens were washed and rubbed slightly to remove any films prior to weighing. Slight films were observed on samples heated to temperatures up to about 600° C, but no films were observed at higher temperatures.

It is difficult to assess the possible protective effect of chloride films but it seems likely that they will offer some protection up to temperatures of the order of 400° C (750° F), and that the corrosion rates for longer exposures at this temperature or below will tend to be lower than those indicated by the short-time experiments. Samples exposed above 600° C (1110° F) showed no evidence of films, and an independent determination by Atkinson[40] for one-hour exposure at 700° C (1295° F)

[32] M. H. Brown, W. B. Delong, and J. R. Auld, *Ind. Eng. Chem.*, **39**, 839–844 (1947).
[33] L. Wohler and S. Streicher, *Ber.*, **46**, 1577–1586, 1591–1597 (1913).
[34] L. Wohler and W. Muller, *Z. anorg. Chem.*, **149**, 125–138 (1925).
[35] C. Nogareda, *Anales soc. españ. fis. quím.*, **32**, 289–344 (1934).
[36] C. Nogareda, *Anales soc. españ. fis. quím.*, **32**, 567–589 (1934).
[37] C. Nogareda, *Anales soc. españ. fis. quím.*, **32**, 658–665 (1934).
[38] G. Van Praagh and E. K. Rideal, *Proc. Royal Soc.* (A), **134**, 385–404 (1931).
[39] M. H. Brown, W. B. Delong, and J. R. Auld, E. I. duPont de Nemours Co., *loc. cit.*
[40] R. H. Atkinson, unpublished research, Research Laboratory, The International Nickel Co., New York, N. Y.

showed a weight loss equivalent to 400 mdd, in agreement with the observations of Brown et al.

The corrosion rate reaches a maximum at 570° C (1060° F) and then drops to about one fiftieth of this value at 650° C (1200° F), undoubtedly due to change in the vapor pressure and nature of the platinum chlorides formed.

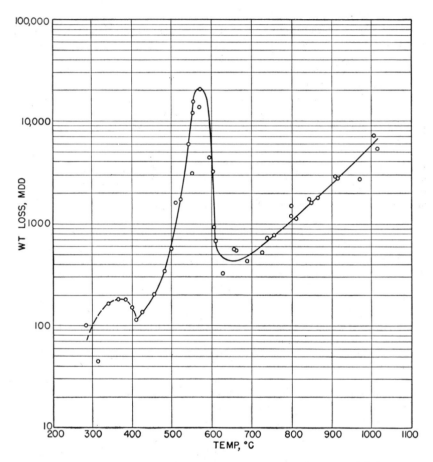

Fig. 5. Corrosion Rate of Platinum in Dry Chlorine. Velocity 1.3 ft/min. (Brown, Delong, and Auld.)

It will be observed that a slight hump occurs in the corrosion rate at about 360° C (680° F) but this corresponds to specimen weight losses in the order of a milligram and therefore is less accurately established than other features of the rate curve.

In the experiments of Nogareda,[35,36,37] a platinum filament was electrically heated in a halogen gas at low pressure contained in a glass flask maintained at 0° C (32° F). The change in pressure with time was followed by Langmuir's technique. The dimensions of the flask and filament differed in the numerous experiments, and in some cases uncertainty exists concerning the dimensions from which losses in mdd are calculated. Despite this, it is believed that the derived values shown in Fig. 6 are at least approximately correct.

With chlorine at 0.02 mm of Hg pressure, negligible attack was observed at 500° C (930° F), but within the range 600° to 850° C (1110° to 1560° F) the rate increased rapidly with the temperature. Within 850° to 1200° C (1560° to 2200° F) the rate was somewhat variable, and at times was below that at lower temperatures. From 1200° C (2200° F) to above 1300° C (2400° F) the rates increased sharply with the temperature as shown. In the case of bromine and iodine at the same pressure, no attack was observed in the 600° to 800° C (1110° to 1475° F) region, but attack did occur above 1200° C (2200° F) at rates of the same order as for chlorine.

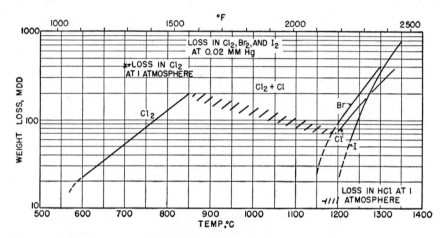

Fig. 6. Rate of Reaction of Platinum with Halogen Gases as a Function of Temperature. (Calculated from data of Nogareda.)

$PtCl_4$ seems to be the primary product at 600° to 850° C (1110° to 1560° F), but $PtCl_2$ is probably first formed in the 1200° to 1300° C (2200° to 2400° F) region and then changes to $PtCl_4$. In the case of bromine, $PtBr_2$ is first formed, but the final product condensed on the walls is a mixture of $PtBr_2$ and $PtBr_4$. In the case of iodine, PtI_2 is the only product formed.

Over the range of low pressure studied, the corrosion rate in chlorine at 600° to 850° C (1110° to 1560° F) varies with the square of the pressure, whereas in the 1200° to 1300° C (2200° to 2400° F) range with chlorine and bromine, it varies essentially with the first power of the pressure. In the case of iodine, it is independent of the pressure according to Nogareda.

Although quantitative data on the behavior of Ir-Pt and Ru-Pt in halogens at high temperature are not available, it is probable that the corrosion rates in the range up to 500° C (930° F) will be lower than for pure platinum. However, nickel behaves so well in reasonably dry chlorine in this region that generally it would be preferred below 500° C (930° F).

CHLORINE IN SOLUTION AT ELEVATED TEMPERATURES

One of the difficult steps in the analysis of platinum metal alloys high in rhodium, ruthenium, or iridium is to get them into solution, for they are relatively unattacked by boiling aqua regia. Sometimes the alloy is fused with lead, and this alloy is attacked successively by nitric acid, hot sulfuric acid, and aqua regia. This treatment dissolves gold and most of the platinum, palladium, and rhodium, leaving an impure Ir-Ru metallic phase.

A study was undertaken[41,42] concerning the possibility of using sealed tubes containing chlorine plus HCl at temperatures up to 300° C (570° F), associated with pressures as high as 4000 psi. In this work, pure iridium sheet was used, as it was considered to be the most difficult to dissolve. Although high concentrations were explored, 36% HCl was adequate for most purposes. Ten to 20 ml of this acid per gram of sample is required and 0.025 gram of chlorine per ml volume of the reaction tube. This chlorine may be introduced either as the element or by any chlorine-producing oxidant. Where no organic matter is present to cause an explosion and the reaction is to be conducted above 250° C (480° F), perchloric acid is considered to be the most convenient oxidant. If the sample is mostly platinum, iridium, or osmium, 0.22 ml of 70% perchloric acid should be used per gram of sample, correspondingly more if the sample contains metals of lower atomic weight. If the reaction is to be conducted at lower temperature (100° to 150° C [210° to 300° F]), which is appropriate for Ir-Pt alloys, sodium chlorate (0.37 gram per gram of high atomic weight sample) or nitric acid should be used as the oxidizing agent. For dissolving 10% Ir-Pt in the form of 0.1 to 0.2 mm chips, a one-gram sample will require 9 ml of 36% HCl and 1 ml of 70% nitric acid. The reactants are sealed in a small-stemmed 15-mm bore heavy-walled Pyrex tube, having a volume of about 20 ml. Heating at 110° C (230° F) effects solution overnight. The tubes are cooled in a mixture of dry ice and 1 to 1 chloroform and carbon tetrachloride before opening. If osmium or ruthenium are present, provision is made to absorb the gaseous products.

The rate of attack on iridium at 300° C (570° F) in 36% HCl containing 0.05 gram of chlorine per ml is of the order of 160,000 mdd. Ruthenium and alloys containing other platinum metals are also rapidly dissolved under these circumstances.

HYDROGEN CHLORIDE

The good behavior of *platinum* in contact with gaseous HCl at elevated temperatures is of considerable industrial importance. In one series of experiments,[43] comprising twenty-five tests of 2 hours each over the range of 425° to 1200° C (800° to 2195° F), specimens of platinum having a surface area of about 0.066 sq dm were exposed to HCl gas flowing at 40 cm/min. The attack did not exceed 90 mdd in any case; hence platinum is considered useful up to 1100° C (2000° F) and probably to 1200° C (2200° F). The weight losses of the specimens at 1165° C (2130° F), 1180° C (2155° F), and 1200° C (2195° F) were 0.1 mg, 0.1 mg, and 0.0 mg, corresponding to an average of 12 mdd.

In another series of tests,[44,45] pure platinum and 10% Ir-Pt samples approximately 0.25 sq dm in area were exposed to HCl at 540° C (1000° F) containing about 0.25% H_2O and moving at 10 cm/min. The duration of the tests was 4 hours, and the weight loss was less than 0.1 mg. These results are confirmed by several years operation of industrial platinum equipment exposed to HCl plus a little carbon at 700° to 750° C (1295° to 1380° F).

[41] E. Wichers, W. Schlecht, and C. Gordon, *J. Research Natl. Bur. Standards*, **33**, 363-381 (1944); **33**, 457-470 (1944).
[42] C. L. Gordon, Research Paper 1521, *J. Research Natl. Bur. Standards*, **30**, 107-111 (1943).
[43] Private communication, E. I. duPont de Nemours and Co., Inc., Wilmington, Del. (Consult ref. 32.)
[44] B. B. Knapp, Report, Research Laboratory, The International Nickel Co., New York, N. Y.
[45] W. Z. Friend and B. B. Knapp, *Trans. Am. Inst. Chem. Engrs.*, **39**, 731-753 (1943).

The presence of air or other oxidizing agents or iron salts may be expected to increase corrosion by HCl.[43] Some indication of the behavior of mixtures containing a low partial pressure of chlorine can be obtained from the work of Nogareda,[46] previously cited. From that work it may be inferred that a low partial pressure of free bromine or free iodine would not be objectionable at temperatures not over 1000° C (1850° F) or perhaps 1100° C (2000° F).

The behavior of *gold* was also examined[43] in HCl flowing at 40 cm/min and generally showed somewhat greater weight loss than did platinum under the same circumstances. Above 930° C (1700° F) definite corrosion was apparent, and at 980° C (1795° F) the loss reached the value of 650 mdd in a 2-hour test.

Below 540° C (1000° F)[45] nickel is often adequate against HCl above the dew point, but in the higher-temperature range platinum should be considered with *palladium* as an alternate possibility. At temperatures below 540° C (1000° F) platinum may be justified if the conditions are such that the dew point is reached.

PHOSPHORIC ACID

The behavior of phosphoric acid toward *gold* and *platinum* is of some technical importance. Tests reported in Table 1 show that the acid at 300° C (570° F) attacks *osmium* at a fair rate and *rhodium* slowly, but it has no effect upon the other metals within the period of test (2½ hours).

CARBONACEOUS GASES

Unstable hydrocarbons which crack in contact with hot *platinum metals* often damage the metals by a mechanism which is not completely understood. The end result is frequently an intergranular precipitation of carbon extending some distance into the metal. Heating platinum crucibles over reducing, and particularly over smoky, flames is damaging for this reason, and also because the hydrogen picked up diffuses through the crucible and may reduce the contents (Si, P, Pb, etc.), which in turn causes damage.

Carbon dioxide is fairly inert toward *gold* and platinum metals, but commercial carbon monoxide reacts in various ways. *Palladium* annealed in the latter is somewhat hardened, whereas ruthenium reacts at 180° C (355° F) and 200 atmospheres, forming carbonyls.[47] The remainder of the platinum metals apparently does not react in this way, but their halides tend to react, forming chlorocarbonyls.[47]

Recent investigation of the effect of various types of flames used in hard-soldering palladium jewelry at temperatures up to about 1400° C (2550° F) has shown that an oxy-acetylene flame, operated slightly oxidizing, is best. The use of this flame almost completely avoids surface and subsurface damage to the palladium. Damage will occur with oxy-city gas flame or oxy-hydrogen flame, being greatest when reducing flames are used. Both oxy-city gas and oxy-hydrogen can, however, be used with platinum.

GLASS

The early work on platinum-lined feeder dies for special glass-working machinery indicated that *10% Rh-Pt* was superior to pure *platinum*,[48] and therefore this alloy

[46] C. Nogareda, *Anales soc. españ. fís. quím.*, **32**, 286–344 (1934).
[47] H. J. Emeleus and J. S. Anderson, *Modern Aspects of Inorganic Chemistry*, pp. 390–431, D. Van Nostrand Co., New York, 1938.
[48] H. K. Richardson, U. S. Patent 2,190,296 (February, 1940).

was extensively used for this purpose and later for fiber-glass-producing equipment. Subsequent work led to the use of other high-platinum alloys for fiber glass, including one containing a small percentage of nickel with or without about 1% of iridium.[49] These developments, plus much attention to the composition of the glass and to the exclusion of materials detrimental to platinum, have led to negligible corrosion and very long life.

The life of glass feeder dies operating on a dark-colored glass containing various metal oxides, including lead, is normally satisfactory. However, any tramp metallic iron will reduce some of the lead, which promptly damages the platinum. Arsenic and phosphorus under reducing conditions also are destructive, as they form low-melting phases with platinum.

In addition to the use of platinum for various types of nozzles, a considerable amount is used in the form of large crucibles for handling special glasses. It is also used for the crucibles employed for producing single crystals of various bromides and fluorides, of optical quality.[50] This last use is perhaps the most searching test that can be made of resistance to molten halides, as the product would be ruined by a minute amount of contamination.

TABLE 3. CORROSION OF GOLD AND THE PLATINUM GROUP METALS IN FUSED SALTS*

One-Hour Exposure, Wt. Loss in mdd

	Temp., °C	Au	Pt	Ir	Pd	Rh	Ru
KHSO₄	440	0.0	72.0	36.0	432.0	1,320	505
KCN	700	3600.0	28,000.0	1800.0	32,000.0	11,000	1,680
NaCN	700	384.0	7450.0	4800.0	14,200.0	25,200	3,600
1 KCN + 2 NaCN	550	480.0	840.0	2160.0	8,160.0	25,400	2,160
KNO₃	350	0.0	0.0	72.0	0.0 slight stain	36	1,200
NaNO₃	350	0.0	0.0	3.6	0.0 slight stain	120	264
NaOH, reducing condition	350	9.6	...	3.6	192	180	720
Na₂O₂	350	2000.0	0.0	72.0	360	336	33,600
Na₂CO₃, reducing condition	920	48.0	72.0	−132.0	108	−48	−890
HCl moist	540	...	0.0 (4 hr)	...	...	...	...
HCl + carbon	750	...	0.0	...	...	...	...

* R. Atkins on, A. Raper, and A. Middleton, unpublished research, Mond Nickel Co., Acton, London, England.

MOLTEN SALTS

Quantitative data on corrosion rates in platinum crucibles containing molten salts involve numerous variables, including the composition of the crucible itself. The loss on simple ignition has been thoroughly studied,[51] and a limited amount of corrosion information on gold and the platinum metals in fused salts is presented in Tables 3 and 4.

Nowack and Spanner[52] present a long discussion of the corrosion behavior of

[49] M. B. Vilensky, U. S. Patent 2,188,636, (January 1940).
[50] D. C. Stockbarger, *Rev. Sci. Instruments*, **7**, 133–136 (1936).
[51] G. K. Burgess and R. G. Waltenberg, *Natl. Bureau of Standards Scientific Paper* 280 (1916).
[52] L. Nowack and J. Spanner, *Korrosion metallischer Werkstoffe*, O. Bauer, Editor, Vol. 2, pp. 782–785 and 809–827, Verlag Hirzel, Leipzig, 1938.

TABLE 4. PLATINUM CRUCIBLE FUSIONS (ONE–HALF HOUR)*

— little or no attack.   .. appreciable attack but usable.   Beyond termination of lines, corrosion rates are excessive.

Reagent	300	400	500	600	700	800	900	1000° C	Remarks
NaOH	—	—							
NaOH (Covered Crucible)	—	—	—						
KOH	—	—							Very active above 600° C (1110° F)
Ba(OH)$_2$	—	—	—	—					
NaNO$_3$	—	—	—	—	—	.....			
KNO$_3$	—	—	—	—	—	.....			
Ba(NO$_3$)$_2$	—	—	—	—					Increases rapidly above 700° C (1300° F)
1NaOH + 2NaNO$_3$	—	—	—	—	—				
2NaOH + NaNO$_3$	—	—	—	—	—	—	.....		Slight attack 800° C (1475° F)
1KOH + 2KNO$_3$	—	—	—	—	.....				Much more active than Na salts
K$_2$CO$_3$	—	—	—	—	—				
Na$_2$CO$_3$	—	—	—	—	—	—			
Na$_2$CO$_3$ (covered)	—	—	—	—	—	—	—		
2Na$_2$CO$_3$ + 1NaNO$_3$	—	—	—	—	—				
1Na$_2$O$_2$ + 2NaOH	—	—	—						Loss 1.6 mg at 500° C (930° F), rapid at higher temp.
NaCN	.....	.....	.....	..					Severe attack at 800° C
2NaOH + 1NaCN	—	—	—	—	.....	.....	.....		Loss 2–4 mg at 800° C
KOH + KCN	—	—	—	—					Serious at 800° C (1475° F)
KOH + K$_2$S	—	—							
KOH + K$_2$S covered or reducing	—	—	—	—	—				
NaOH + Na$_2$S	—	—	—	—	—	.....			
Na$_2$S	—	—	—	—	—	.....			
Alkaline polysulfides	.....	.....	...						Corrosive
K$_2$SO$_4$	—	—	—	—	—	—	—		Corroded if reducing agents are present
Na$_2$SO$_4$	—	—	—	—	—	—	—		Not quite so sensitive to reducing agents
Alkaline sulfates plus reducing agent	.....	.....							Corrosive
KHSO$_4$	—	.....							Useful for cleaning
KCl	—	—	—	—	—				
NaCl	—	—	—	—	—				

* G. Bauer, *Chem.-Zeitung*, **62**, 257–262 (1938).

platinum metals and gold in which there is some information on molten salts. They report that the *carbonates* are not corrosive to platinum, even at fairly high temperatures. Covering the crucible reduces losses of metal and the introduction of carbon dioxide completely prevents attack at 900° C (1650° F). The fused *alkali halides* are not corrosive to either gold or platinum, but if mixed with oxidizing agents, such as nitrates or sulfates, corrosion occurs with all the noble metals. NaCl + Fe$_2$(SO$_4$)$_3$ yields S$_2$O$_5$Cl$_5$, which is a potent corrosive agent. The general destructiveness of the *cyanides* is noteworthy, but using the sodium salt, and diluting with NaOH, greatly reduces the attack. Alkaline *sulfides*, particularly *polysulfides*, are also active, but again the sodium salts are least corrosive, and dilution with NaOH is helpful. K$_2$S + KOH is corrosive in open crucibles above 500° C (930° F), but exclusion of air reduces the attack on platinum below approximately 800° C (1475° F).

With the *caustic alkalies*, the presence of oxygen accelerates attack. The rate also varies with the alkali, sodium hydroxide being rather mild, potassium hydroxide more active, whereas rubidium hydroxide is said to be most active. $Ba(OH)_2$ behaves much like KOH and is fairly destructive above 400° C (750° F). Covering the crucible, or otherwise excluding oxygen, helps to reduce attack, and reduction in temperature is even more beneficial.

The molten *peroxides* are very corrosive to gold and ruthenium at 350° C (660° F) and corrode platinum considerably at still higher temperatures. The *nitrates*, in contrast, are very well tolerated at 350° C (660° F) and are not active toward platinum up to about 700° C (1300° F).

In platinum crucibles used for ignitions, in contrast to those used for fusions, a strongly oxidizing atmosphere within the crucible and the exclusion, as far as possible, of carbon in the charge are desirable. Some of the *magnesium phosphates* and also potassium salts yielding *pyrophosphates* are particularly likely to cause damage on ignition, in some cases apparently even in the absence of a reducing agent. The ashing of *lecithin* is particularly likely to cause trouble.

The use of electrical resistance furnaces operated with the door open for heating crucibles during ignitions undoubtedly increases the life of crucibles, as the temperature can be controlled and kept within a safe range, and no possibility exists for diffusing hydrogen through the walls of the crucible, as may happen in a reducing gas flame.

SILVER

Robert H. Leach*

The increase in the use of silver for equipment in the chemical industries necessitates the consideration of its corrosion resistance at moderately elevated temperatures. The strength of fine silver is low at high temperatures as compared with that of most metals and alloys commonly used for their corrosion-resistant properties. Consequently, instead of all silver it is practical to use silver linings, silver-clad or silver plate, except in the case of small equipment. Care must be taken to provide a sufficient thickness of silver as a coating to reduce the rate of oxygen diffusing through the silver and attacking the base metal.

It is difficult to obtain specific data on results in actual service but the following information briefly reviews what may be expected when silver is exposed to several well-known corrosive substances at elevated temperatures.

AIR AND OXYGEN

Silver is not corroded in the usual acceptance of this term by oxygen or air at elevated temperatures and normal pressures. Heavy silver sheets heated in air for several hours at temperatures of 600° to 800° C (1110° to 1475° F) do not show any appreciable attack or tarnish. High pressures may influence the attack by oxygen, for Le Chatelier[1] reported that he produced a black oxide of silver at 300° C (570° F) and 15 atmospheres pressure.†

There is general agreement that oxygen will diffuse through commercially pure silver (fineness 999.7–999.9) at elevated temperatures, the rate of diffusion increasing

* Handy and Harman, Bridgeport, Conn.
† The dissociation pressure of Ag_2O is 1 atmosphere at 184° C (363° F). EDITOR.
[1] H. Le Chatelier, "On Oxidation of Silver," *Z. physik. Chem.*, **1**, 516–518 (1887).

rapidly above 400° C (750° F). The effect of this diffusion is apparently detrimental only when the silver is subsequently heated in hydrogen. Martin and Parker[2] report that solid silver will dissolve enough oxygen at high temperatures to become embrittled if heated in hydrogen at 500° C (930° F) or above.* Chaston[3] observed this same embrittling effect when 99.98% silver, which had been annealed in air at temperatures from 300° to 800° C (570° to 1475° F) was afterwards heated in hydrogen at temperatures of 400° to 800° C (750° to 1475° F). Similar tests on silver containing less than 0.01% total impurities did not show the same embrittlement, and he advanced the theory that impurities in silver were the cause of hydrogen embrittlement.

CARBON MONOXIDE

This gas shows no effect on silver annealed at elevated temperatures. However, M. Berthelot[4] found that silver was attacked when heated to 300° C (570° F) in a sealed tube filled with carbon monoxide. Forstner[5] also observed this effect and claimed that silver oxide was formed at 300° C (570° F). Information available is not conclusive, and, if any contemplated use of silver involves exposure to carbon monoxide at elevated temperatures and pressures for long periods of time, further tests should be made.

OTHER GASES

STEAM

Silver annealed in an atmosphere of superheated steam at temperatures of 600° to 700° C (1100° to 1300° F) does not show any attack.

GASES CONTAINING SULFUR

Silver is attacked by gases containing sulfur with accelerated attack at elevated temperatures.

AMMONIA

Ammonia gas at high temperatures reacts with silver, probably resulting in nitrides and decomposition of the gas.

SULFUR DIOXIDE

At red heat silver is attacked by this gas, the effect increasing with temperature.

HYDROGEN AND NITROGEN

Hydrogen does not react with silver, but diffuses slowly through the metal at elevated temperatures, reacting with any oxygen which may be present and producing embrittlement and surface blisters. (See "Air and Oxygen.")

---

* This effect parallels that observed in copper containing oxygen. (See p. 629.)

[2] D. Martin and E. Parker, *Trans. Am. Inst. Mining Met. Engrs.*, Institute of Metals Division, **152**, 269 (1943).

[3] J. C. Chaston, "Some Effects of Oxygen in Silver and Silver Alloys", *J. Inst. Metals* (London), **71**, 23 (1945).

[4] M. Berthelot, "Carbon Dixoide and Silver," *Compt. rend.*, **131**, 1167–1169 (1900).

[5] H. M. Forstner, "Silver Coatings and Their Application for Apparatus of Chemical Factories," *Oberflächentech.*, **12**, 91–94, 109, 119, 137, 147 (1935).

Silver is inert to nitrogen at elevated temperatures, and no embrittlement has been observed.

HYDROGEN CHLORIDE AND CHLORINE

To use silver in contact with gaseous HCl or Cl_2, the temperature should be below 425° C (800° F). (See Table 6, p. 682.)

MOLTEN SALTS

CHLORIDES

Molten chlorides attack silver. The effect is more noticeable at the surface of the liquid salt where there is contact with air.

POTASSIUM AND SODIUM HYDROXIDE

The fused caustics have only very slight effect on silver. The corrosion is more pronounced at the junction of the molten salt with air. Tests made by holding molten sodium hydroxide in a silver crucible at 500° C (930° F) for 312 hours showed a loss of 10 mdd.[6] The resistance to corrosion is sufficiently high to warrant use of silver crucibles for fusions of alkalies in laboratory work. Silver or silver-lined vessels are used in chemical plants where these salts are handled at elevated temperatures.

SODIUM CARBONATE

Silver resists corrosion of fused sodium carbonate, but at temperatures above 500 °C (930° F) some corrosive effect was noted.[7]

SULFIDES

Fused alkali sulfides attack silver rapidly.

TANTALUM

CLARENCE W. BALKE*

AIR

Tantalum is not resistant to air at elevated temperatures. When the metal is heated it becomes covered with a film of oxide showing light interference colors. A short exposure to a low red heat produces a black color which on longer exposure changes to white tantalum oxide. These oxide films can be observed before an increase in weight is detected. See Table 1 for quantitative data showing the effect of temperature on the rate of oxidation.

HYDROGEN, OXYGEN AND NITROGEN

Table 2 gives data concerning the effect of hydrogen, nitrogen, oxygen, and air

* Fansteel Metallurgical Corporation, North Chicago, Ill.
[6] Handy and Harman Laboratory.
[7] L. Nowack and J. Spanner, "Corrosion of Noble Metals," O. Bauer, O. Kröhnke, and G. Masing, *Korrosion metallischer Werkstoffe*, Bd. II, S. Hirzel, Leipzig, 1938.

TABLE 1. HIGH-TEMPERATURE CORROSION OF TANTALUM IN AIR

Size of specimen — approx. 40 sq cm (6.2 sq in.) of surface.
Sheet — 0.1 mm (0.004 in.) thick.
Duration of test — 20 hr.

Temperature		Gain in Weight		Color Change
°C	°F	mdd	oz/sq ft/day	
260	500	0.0	0.0000	None
280	536	0.4	0.00013	Very slight
300	572	0.7	0.00024	Very slight
325	617	2.5	0.00082	Showing interference colors
350	662	5.0	0.0016	Blue
390	734	12.1	0.0040	Darker interference colors
435	815	18.7	0.0061	Darker interference colors
480	896	106.8	0.035	Very dark. White oxide forming. Increase = 1% of wt. of sample

TABLE 2. EFFECT OF VARIOUS GASES ON TANTALUM AT ELEVATED TEMPERATURES

Size of specimen — strips of sheet 0.13 × 6.3 × 127 mm (0.005 × 0.25 × 5 in.).

Temperature		Duration of Test, hr	Gain in Wt., gram	Gain in Wt., %	Room Temp. Elongation, % in 1 in.	
°C	°F					
Hydrogen						
100	212	2	0.0000	0.000	Original (31)	31.0
200	424	5	0.0000	0.000		33.0
250	482	3	0.0000	0.000		32.8
300	572	5	0.0000	0.000		32.8
350	662	5	0.0012	0.0073		30.6
400	752	1	0.0017	0.011		25.8
400	752	2	0.0024	0.018		...
400	752	3	0.0041	0.031		16.0 Some embrittlement
Nitrogen						
300	572	5	0.0000	0.000	(31)	28.2
400	752	1	0.0008	0.0053		22.0
Air						
350	662	5	0.0009	0.013	(33.2)	27.1
Oxygen						
350	662	5	0.0008	0.011	(33.2)	26.6

on tantalum at various temperatures. The elongation of the specimens has been used to give an approximate indication of the effect of gas absorption on the ductility of the metal.

STEAM

Tantalum is not corroded or embrittled by steam up to at least 165 psi. However, the pH of any condensate should not be over 8.

OTHER GASES

All gases except the noble gases react with tantalum at high temperatures. The oxide, carbide, nitride, chloride, etc., will be formed, depending upon the gas in contact.

The action of chlorine is resisted below 150° C (300° F) and bromine below 175° C (350° F).

Hydrogen at high temperatures, and particularly if under pressure, may cause embrittlement.

NH_3 at elevated temperatures and pressures pits the metal without embrittlement but with rapid loss in weight.

HIGH TEMPERATURES IN VACUUM

High temperature alone does not embrittle tantalum, as is the case with molybdenum and some other metals. Many substances, including some metals, can be fused in tantalum in a vacuum without destruction of the metal. Copper wets the surface but does not dissolve tantalum.

ZINC

E. A. ANDERSON*

The high-temperature corrosion of zinc is of little practical interest since the mechanical properties at elevated temperatures largely prevent its use under such conditions.

The invisible oxide film formed on hot zinc surfaces, such as those of cast slabs or hot-rolled strip, has been studied.[1] By using electron diffraction methods, zinc oxide films less than 0.00005 cm (0.02 mil) thick were identified on cast slabs of special high-grade zinc (cast in chromium-plated steel, closed mold. Mold temperature: 130° C [265° F], metal temperature: 475° C [885° F]). The layer adjacent to the zinc and the outer layer were randomly oriented, but the central layer of the film was highly oriented.

Pilling and Bedworth[2] state that the formation of zinc oxide at 400° C (750° F) follows a straight line when the weight increase squared is plotted against time. The slope of this line is 0.88×10^{-10} (grams/sq cm)^2 hr^{-1}. Heavy coatings are milky white in appearance. The rates of formation at other temperatures do not seem to have been established.

The attack on zinc by oxygen at elevated temperatures appears to take place at the surface only, there being no evidence of oxygen solution with subsurface effects (internal oxidation).

NON-METALLIC MATERIALS

Introduction

As in applications to liquid or gaseous environments at ordinary temperatures, the non-metals find important uses at elevated temperatures. Their limited physical properties more than their chemical behavior restrict their application. They some-

* The New Jersey Zinc Co. (of Pennsylvania), Palmerton, Pa.
[1] M. L. Fuller, unpublished data, The New Jersey Zinc Company.
[2] N. B. Pilling and R. E. Bedworth, *J. Inst. Metals*, **29**, 529–582 (1923).

times can be used in competition with metals but more often are applied in special situations involving extreme conditions of temperature or environment where metals are not suitable.

The non-metals deteriorate in various ways but not by corrosion, according to the definition of corrosion accepted in this book. (See Glossary.) Strictly speaking, only metals corrode. The importance, however, of non-metals in high-temperature applications has warranted inclusion of the following chapters.

High-temperature properties of glass and vitreous or fused silica are described beginning p. 354.

Carbon and Graphite

F. J. Vosburgh*

Carbon and graphite are extremely useful materials where a non-oxidizing atmosphere can be provided, or where selective oxidation can be arranged so that the carbon will not be attacked. The strength of carbon is not affected by temperatures used commercially, and it is not subject to thermal shock.

Carbon and graphite have been used successfully for commercial applications combining pressure and heat at temperatures in excess of 2000° C (3600° F). Carbon generally, and graphite less frequently, are used as the lining of practically all ferroalloy furnaces, for furnaces producing elemental phosphorus and phosphoric acid, in furnaces producing magnesium by the carbothermic process, in many furnaces making calcium carbide, and in all aluminum pots. In these applications carbon resists attack by corrosive slags.

In Europe carbon blocks are generally used for lining blast furnaces in the hearth section and as far up as the mantel. The carbon is not attacked by slags of iron and consequently no salamander is formed. In almost all such furnace applications it is unnecessary to provide cooling for the carbon, although cooling may be required for other refractories used back of the carbon or to protect the furnace shell.

Carbon and graphite may be used as bricks or blocks similar in size to the usual ceramics or preferably, in the case of carbon, in blocks weighing up to 7200 lb each. The graphite blocks may be easily shaped on any woodworking equipment, but carbon is fabricated using the sturdier metal-working machines or by using diamond or abrasive saws.

Both carbon and graphite, particularly the latter, are being used more widely as molds for casting iron, steel, non-ferrous metals of all sorts, and the various types of metal carbides used in the metal-working industry where resistance to corrosion, high temperatures, and thermal shock is particularly desirable.

Carbon and graphite are especially suitable for many types of resistance heating, such as rods, tubes, rings, and disks, where resistance to chemical attack is added to desirable electric characteristics.

Refractories

James A. Lee†

Of the various factors which determine the kind of refractory to use in a particular case, the most important are working temperature, rapidity and extent of temperature

---

* National Carbon Co., Inc., New York, N. Y.
† *Chemical and Metallurgical Engineering*, McGraw-Hill Publishing Co., Inc., New York, N. Y.

change, strength requirements, and the chemical reactions to be encountered. The last may be one of the most fertile sources of refractory failure in the presence of metallurgical slags, many fuel ashes, and products such as metallic oxides, glasses, and cement clinker, alkalies, and alkaline earth oxides. Various furnace gases have a deleterious effect, such as carbon monoxide within a critical range of 420° to 470° C (788° to 878° F) on some kinds of brick, and sulfur dioxide or steam on others.

While the chemical compositions of the usual commercial refractories classify them in general groups — acid, basic, and neutral — a sharp distinction cannot be made in all cases. Silica bricks are decidedly acid and best suited to resist fluxes of an acidic character at high temperatures. Magnesite brick are strongly basic, and chrome, high-alumina, and fireclay brick belong to the less precise neutral class. Among the fireclay brick are those ranging in silica content from 70% or slightly higher to as low as 50%. The high-alumina brick comprise those containing 50 to 80% alumina, or higher for some special brick. This is a wide range in the alumina-silica series of refractories and it follows that there also are great differences in their chemical reactions with various fluxes. In a given class of refractories, chemical analysis alone may not necessarily serve as good criteria of their chemical behavior or of their physical properties. The proper criteria are chemical analysis, mineral composition, character of the bond, and the permeability. With all other factors the same, the permeability becomes of prime importance since it is a property which the manufacturer can control within limits.

Physical properties of various refractories are listed in Table 1.

It is the rule rather than the exception that several classes of refractory are used in the construction of a furnace because one type of brick is better adapted than another to withstand the widely different conditions that prevail in various parts of some furnaces. In order to be assured of the best results from each kind of refractory used, it is necessary to guard against reaction between them. This can be done by the use of a separating course of brick or cement which has least tendency to react with either of the adjacent kinds of brick, or by designing the construction in such manner that the reacting bricks are in contact with each other in the cooler or in protected portions of the furnace.

Other Metals

Some high-temperature properties of metals and semi-metals not included in this Section are discussed elsewhere in the *Handbook*. The page references are: Beryllium, p. 56; Boron, p. 378; Chromium, p. 825; Molybdenum, p. 252; Rhenium, p. 314; Silicon, p. 380; Titanium, p. 329; Tungsten, p. 330; Uranium, p. 331.

NON-METALLIC MATERIALS

TABLE 1. PHYSICAL PROPERTIES OF REFRACTORY MATERIALS*

Type of Brick	Silica	High Heat Duty (No. 1) Fireclay	High Heat Duty (No. 1) Kaolin	Super Duty Fireclay	Alumina-Diaspore 70% Al₂O₃	Silli-manite (Mullite)	Chrome	Unburned Chrome	Magnesite	Unburned Magnesite	Bonded Silicon Carbide (Grade A)	Bonded Fused Alumina	Insul. Firebrick (2600° F)
Typical composition, per cent													
SiO_2	96	50–57	53	52	22–26	35	6	5	3	5	7–9	8–10	50–57
Fe_2O_3	1	1.5–2.5	2	1	1–1.5	0.5			6	8.5	0.3–1	1–1.5	1.5–2.5
FeO							15	12					
Al_2O_3	1	36–42	42	43	68–72	62	23	18	2	7.5	2–4	85–90	36–42
TiO_2	2	1.5–2.5	2	2	3.5	1.5					1	1.5–2.2	1.5–2.5
CaO										2			
MgO							17	32	3	64			
Cr_2O_3							38	30	86	10			
SiC											85–90		
Flux‡													
P.C.E. (with approx. equivalent temp. ° F)§	31–32 (3056–3092°)	1–3.5 (3056–3173°)	1 (3173–3200°)	2 (3173–3200°)	1–1.5 (3290°)	0.5 37–38 (3308–3335°)	41+ (3578+)	41+ (3578+)	41+ (3578+)	41+ (3578+)	1.5 39 (3389°)	0.8–1.3 39+ (3389°+)	1–3.5 29–30 (2984–3002°)
Deformation under load, %‖ (at psi and temp., ° F, shown)	Shears 25 psi 2900°	2.5–10† 25 psi 2460°	6–7 25 psi 2640°	2–4† 25 psi 2640°	1–4† 25 psi 2640°	0.0–0.5† 25 psi 2640°	Shears 28 psi 2740°	Shears 28 psi 2955°	Shears 28 psi 2765°	Shears 28 psi 2940°	0–1 50 psi 2730°	1 50 psi 2730°	0.3 10 psi 2200°
Resistance to spalling, %¶	5–20	5–20	5–15	0–4	No loss	No loss	Poor	Fair	Poor	Fair	Good	Good	Good
Permanent linear change on reheating‡‡ (after 5 hr at temp., ° F, shown)	(+)0.5–0.8 2640°	(±)0–1.5 2550°	(±)0–1.5 2550°	(±)0–1.5 2910°	(−)2–4 2910°	(−)0–0.8 2910°	(−)0.5–1 3000°	(−)1–2 3000°	(−)1–2 3000°	(−)0.5–1.5 3000°	(+)2** 2910°	(+)0.5 2910°	(−)1.5†† 2550°
Porosity (as open pores), %	20–30	15–25	24–28	12–15	27–36	20–25	20–26	10–12	20–26	10–12	13–28	20–26	75
Weight per brick (std. 9 in. straight), lb	6–6.5	7.5	7–7.5	8.5	7.5–8.5	8.5	11	11.3	10	10.7	8–9.3	9–10.6	2.25–3
Specific heat (60°–1200° F)	0.23	0.23	0.22	0.23	0.23	0.23	0.20	0.21	0.27	0.26	0.20	0.20	0.22
Relative slag resistance§§													
Acid steel slag	Good	Fair	Fair	Fair	Good	Good	Poor	Poor	Poor	Poor	Good	Good	Poor
Basic steel slag	Poor	Poor	Poor	Poor	Fair	Fair	Good	Good	Good	Good	Good	Good	Poor
Mill scale	Fair	Fair	Fair	Good	Fair	Fair	Good	Good	Good	Good	Fair	Fair	Poor
Coal ash slag	Poor	Fair	Fair	Fair	Fair	Fair	Fair	Fair	Good	Good	Good	Good	Poor

* Copyrighted, *Chemical and Metallurgical Engineering*. Compiled by L. J. Trostel, General Refractories Co., Baltimore, Md.
§ Pyrometric Cone Equivalent; terms "fusion," "softening," "deformation," and "melting" points heretofore loosely used.
‖ Data marked (†) are from A. S. T. M. test C 16–36 with High Heat Duty time-temperature schedule; those marked (†) are from same test with Super Duty time-temperature schedule; others determined by other commonly used tests.
¶ Loss in appropriate A. S. T. M. Panel Test.
** Oxidizing atmosphere.
†† After 24 hr.
‡‡ (+) means expansion; (−) means shrinkage.
§§ Ratings affected somewhat by varying temperatures and type of atmosphere prevailing. Resistance to coal ash slag affected by furnace temperature as well as analysis and fusion point of slag.

HIGH-TEMPERATURE RESISTANT MATERIALS

HIGH-TEMPERATURE RESISTANT MATERIALS

G. F. Geiger[*]

FOREWORD

Corrosion resistance is only one of the factors that must be taken into consideration when selecting an alloy for high-temperature service. The other factors are creep resistance, thermal expansion, thermal shock, and thermal stability. If a relatively short life is adequate, one of the lower cost alloys may be used, but if a long life is required it may be more economical to use one of the more highly alloyed materials.

The tables should be used only as a guide. More detailed information on the properties of the alloys should be obtained from the references or from the producers of the alloys before making a final selection.

[*] Development and Research Division, The International Nickel Co., Inc., New York, N. Y.

TABLE 1. CORROSION RATES OF ALLOYS IN FURNACE ATMOSPHERES
AIR

A = annealed. N = normalized.
T = tempered. HR = hot-rolled.
Q = quenched.

Temp.		No.*	Alloy (Bal. Fe)				Cast or Wrought	Heat Treatment	Notes (See p. 732)
°C	°F		Cr	Si	Ni	Mo			
			< 0.005 ipy						
870	1600	8	15		60		C	...	1
"	"	10	15		35		C	...	1
"	"	21	20		10		C	...	1
"	"	19	25		12		C	...	1
"	"	22	29		10		C	...	1
"	"	20	28		15		C	...	1
815	1500	81	9	1		1	W	NT	2
"	"	82	9	1		0.5	W	NT	2
"	"	83	7	1		1	W	NT	2
"	"	84	7	1		0.5	W	NT	2
"	"	85	5	1.5		1	W	A	2
"	"	87	5	2		0.5	W	A	2
"	"	95	5 (Al 0.5)			0.5	W	NT	2
"	"	23	18		8		W	Q	2
675	1250	86	5	1		1	W	A	2
540	1000	90	2	1		0.5	W	A	2
"	"	91	1	1		0.5	W	A	2
"	"	92	1.5	1		0.5	W	A	2
"	"	93		1.5		0.5	W	A	2
"	"	96	C-Mo Steel				W	A	2
"	"	97	1015 Steel				W	A	2
			0.005–0.15 ipy						
1175	2150	6	14		80		W	HR	3
"	"	7	14		80+ Cb		W	HR	3
"	"	10	15		35		C	...	3
"	"	11	15		35		W	HR	3
"	"	12	20		30		W	HR	3
"	"	14	25		20		W	HR	3
"	"	15	25		20+ Si		W	HR	3
"	"	19	25		12		C	...	3
"	"	18	27		11		C	...	3
"	"	17	29		9		W	HR	3
"	"	25	28		3		C	...	3
1095	2000	8	15		60		C	...	1
"	"	13	25		20		C	...	1
"	"	21	20		10		C	...	1
"	"	99	29		5		C	...	1
"	"	22	29		10		C	...	1
"	"	20	28		15		C	...	1
955	1750	23	18		8		W	Q	2
"	"	95	5	(Al 0.5)		0.5	W	NT	2
870	1600	98	20		...		C	...	2
815	1500	86	5	1		1	W	A	1
"	"	88	5	1		0.5	W	A	2

*Numbers in this column refer to alloy compositions listed in Table 7, p. 740.

TABLE 1.—*Continued*
AIR

A = annealed. N = normalized.
T = tempered. HR = hot-rolled.
Q = quenched

Temp.		No.*	Alloy (Bal. Fe)				Cast or Wrought	Heat Treatment	Notes (See p. 732)
°C	°F		Cr	Si	Ni	Mo			
0.005–0.15 ipy									
815	1500	89	3	1		0.5	W	A	2
"	"	94	3	(Al 0.5)		0.5	W	A	2
675	1250	90	2	1		0.5	W	A	2
"	"	91	1	1		0.5	W	A	2
"	"	92	1.5	1		0.5	W	A	2
"	"	93		1.5		0.5	W	A	2
"	"	96	C-Mo Steel				W	A	2
"	"	97	1015 Steel				W	A	2
0.15–0.50 ipy									
980	1800	98	20				C	...	1
955	1750	87	5	2		0.5	W	A	2
815	1500	90	2	1		0.5	W	A	2
"	"	91	1	1		0.5	W	A	2
"	"	92	1.5	1		0.5	W	A	2
"	"	93		1.5		0.5	W	A	2
"	"	96	C-Mo Steel				W	A	2
"	"	97	1015 Steel				W	A	2
0.50–1.0 ipy									
1095	2000	98	20				C	...	1
955	1750	81	9	1		1	W	NT	2
"	"	82	9	1		0.5	W	NT	2
"	"	83	7	1		1	W	NT	2
"	"	84	7	1		0.5	W	NT	2
"	"	85	5	1.5		1	W	A	2
"	"	86	5	1		1	W	A	2
"	"	88	5	1		0.5	W	A	2
"	"	89	3	1		0.5	W	A	2
"	"	90	2	1		0.5	W	A	2
"	"	91	1	1		0.5	W	A	2
"	"	94	3	(Al 0.5)		0.5	W	A	2

OXIDIZING, SULFUROUS
(5 TO 50 GRAINS SULFUR/100 CU FT)

0.010–0.025 ipy									
980	1800	8	15		60		C	...	1
"	"	13	25		20		C	...	1
0.025–0.10 ipy									
980	1800	10	15		35		C	...	1

(100 GRAINS SULFUR/100 CU FT)

0.010–0.025 ipy									
980	1800	8	15		60		C	...	1

*Numbers in this column refer to alloy compositions listed in Table 7, p. 740.

TABLE 1.—Continued

A = annealed. N = normalized.
T = tempered. HR = hot-rolled.
Q = quenched.

Temp.		No.*	Alloy (Bal. Fe)		Cast or Wrought	Heat Treatment	Notes (See below)
°C	°F		Cr	Ni			
			OXIDIZING, SULFUROUS—Continued (100 GRAINS SULFUR/100 CU FT)				
			0.025–0.10 ipy				
980	1800	10	15	35	C	...	1
"	"	13	25	20	C	...	1
			REDUCING, SULFUROUS (5 TO 50 GRAINS SULFUR/100 CU FT)				
			<0.010 ipy				
980	1800	8	15	60	C	...	1
			0.010–0.025 ipy				
980	1800	13	25	20	C	...	1
			0.025–0.10 ipy				
980	1800	10	15	35	C	...	1
			(100 GRAINS SULFUR/100 CU FT)				
			0.010–0.025 ipy				
980	1800	8	15	60	C	...	1
			0.025–0.10 ipy				
980	1800	13	25	20	C	...	1
			0.10–0.50 ipy				
980	1800	10	15	35	C	...	1

*Numbers in this column refer to alloy compositions listed in Table 7, p. 740.

Note 1. *Alloy Casting Bull.* 4 (March, 1945). These data were obtained at Battelle Memorial Institute, Columbus, Ohio, for the Alloy Casting Institute. For details of these tests, consult *High Nickel Chromium-(Iron) Alloys*, pp. 684 and 691.

Note 2. *Digest of Steels for High Temperature Service*, 1939. The Timken Roller Bearing Co., Steel and Tube Division, Canton, Ohio.

Oxidation tests were conducted in the laboratory using specimens from commercial heats. The specimens were heated in air for 1000 hours at temperatures ranging from 540° to 955° C (1000° to 1750° F). After exposure, the scale was removed by an electrolytic method. The specimens were then reweighed to determine the metal loss.

Note 3. From data developed in field tests by The International Nickel Company.

Test bars 5 by 1 by about ¼ in. (12.5 by 2.5 by 0.6 cm) were accurately measured for thickness and then assembled by welding into a rack in the form of a ladder. The test specimens formed the rungs of the ladder.

The original thickness of all specimens was recorded before the test. The racks were exposed for 2000 hours in furnaces used in the production of magnesium by the ferro-silicon process of magnesia reduction. After the test the thickness of sound metal between the oxidized surfaces was measured on a micrometer microscope stage. From this the rate of oxidation was calculated.

HIGH-TEMPERATURE RESISTANT MATERIALS

TABLE 2. CORROSION RATES OF ALLOYS IN STEAM

A = annealed. WQ = water-quenched.
D = drawn. HR = hot-rolled.
OQ = oil-quenched.

Temp. °C	Temp. °F	No.*	Alloy Cr	Alloy Ni	Alloy	Cast or Wrought	Heat Treatment	Note (See Note 4, p. 734)
\<.0001 ipy								
595	1100	10	15	35		C	A	
"	"	13	25	20		C	A	
"	"	19	25	12		C	A	
"	"	24	18	8	Cb 2	C	A	
"	"	55	12	—	C 0.32	W	OQD	
"	"	56	12	—	C 0.10	W	A	
"	"	57	12	—	S 0.3	W	A	
0.0001–0.001 ipy								
595	1100	6	14	80		W	A	
"	"	16	25	20	Cb 2	C	A	
"	"	23	18	8		W	WQ	
"	"	58	12		C 0.13	C	OQD	
510	950	63	1.25		Si 0.8, Mo 0.5	W	A	
"	"	66	1		Mo 0.5	W	A	
"	"	67	1.25		Al 1, Mo 0.2	...	...	
"	"	73			C 0.17, Mo 0.5	W	A	
"	"	74			Steel, C 0.10	W	...	
"	"	76			Steel, C 0.35	W	HR	
0.001–0.01 ipy								
595	1100	2	...	99+,	Low C†	C	A	
"	"	1	...	99+,	High C†	C	A	
"	"	27	...	66	Cu 29, Al 3	W	A	
"	"	28	...	63	Cu 30, Si 4	C	A	
"	"	54	12		Mo 2	W	A	
"	"	59	5		Si 1.5, Mo 0.5	W	A	
"	"	60	5		Ti 0.5, Mo 0.5	W	A	
"	"	61	5		Mo 0.5	W	A	
"	"	62	2.5		Si 0.8, Mo 0.5	W	A	
"	"	63	1.25		Si 0.8, Mo 0.5	W	A	
"	"	64	1.25		Si 1.5, Mo 0.5	W	A	
"	"	65	1		V 0.18	W	A	
"	"	67	1.25		Al 1, Mo 0.2	...	...	
"	"	68	...	3	Mo 0.4	C	A	
"	"	69	0.75	1.25	Mo 0.5	C	A	
"	"	70	...		Si 1.35, Mo. 0.5	C	A	
"	"	71	...		C 0.15, Mo 0.5	C	AD	
"	"	72	...		C 0.15, Mo 0.5	W	A	
"	"	75	...	Steel,	C 0.18	C	A	
"	"	77	4	14	Cu 6	C	...	
"	"	78	2	15	Cu 6	C	...	
"	"	79	2	20		C	...	

TABLE 2. CORROSION RATES OF ALLOYS IN STEAM — *Continued*

A = annealed.
D = drawn.
OQ = oil-quenched.
WQ = water-quenched.
HR = hot-rolled.

Temp.		No*	Alloy		Cast or Wrought	Heat Treatment	Note (See Note 4 below)
°C	°F		Cr	Ni			
0.01 ipy							
595	1100	74	Steel, C 0.10		W	...	
"	"	76	Steel, C 0.35		W	HR	
"	"	80	...	35 C 2	C	...	
"	"	4	...	95 Mn 4.5	W	A	

* Numbers in this column refer to alloy compositions listed in Table 7, p. 740.
† Subject to intergranular corrosion.
Note 4. "High-Temperature Steam Corrosion Studies at Detroit," by I. A. Rohrig, R. M. Van Duzer, Jr., and C. H. Fellows, Detroit Edison Company, Detroit, Michigan, in *Trans. Am. Soc. Mech. Engrs.*, **66** (No. 4), 277–290 (May, 1944).

Unstressed specimens were exposed to superheated steam under plant-operating conditions. The rate of steam flow was 2.4 meters (8 ft)/sec in the 595° C (1100° F) section and 1.8 meters (6 ft)/sec in the 510° C (950° F) section.

The specimens, 1.3 cm (½ in.) in diameter by 15.2 cm (6 in.) in length, with ground finish, were assembled nine to a cage. The cages were exposed in steam lines for periods ranging from 4000 to 16,000 hours. The steam was of high purity, containing no carbon dioxide and less than 0.01 ppm of oxygen.

At the conclusion of the test periods the specimens were removed from the cages for metallographic examination, and hardness and weight loss determinations.

TABLE 3. POTS FOR MOLTEN METALS[1]

Molten Metals[2]	Temp. Range		Pot Alloy						
			1st Choice			2nd Choice			3rd Choice
	°C	°F	No.*	Cr	Ni	No.*	Cr	Ni	No.*
Aluminum	690–775	1275–1425	74	Steel		...	...	...	...
Magnesium	675–900	1250–1650	74	Steel		...	...	...	...
Zinc	450–480	850–900	74	Ste 1		...	...	...	...
Type metal	310–380	600–725	74	Steel		...	...	...	...
Solder	310–510	600–950	74	Steel		...	...	...	...
Lead	590–900	1100–1650	8, 9	15	60	10, 11	15	35	74	Steel

* Numbers in this column refer to alloy compositions listed in Table 7, p. 740.
[1] Only the low melting point metals and alloys may be melted in metal pots. The governing factor is whether or not the pot material will alloy with the charge. Such an alloying action has two undesirable effects: it reduces the thickness of the pot and destroys the purity of the molten metal.
[2] Compositions are the pure metals or alloys as commonly used in commercial practice.

TABLE 4. CARBURIZING AND NITRIDING CONTAINERS

Atmosphere	Temp. Range		Container Alloy								
			1st Choice			2nd Choice			3rd Choice		
	°C	°F	No.*	Cr	Ni	No.*	Cr	Ni	No.*	Cr	Ni
Carburizing†	790–950	1450–1750	6 8, 9	14 15	80 60	10, 11	15	35	13, 14, 15	25	20
Nitriding‡	510–540	950–1000	6 8, 9	14 15	80 60	10, 11	15	35	13, 14, 15	25	20

* Numbers in this column refer to alloy compositions listed in Table 7, p. 740.
† Carbon monoxide, methane, ethane, propane, and butane.
‡ NH₃ plus dissociation products.

TABLE 5. ALLOYS FOR

Fused Salts	Use
"Cyanide" (NaCN + NaCl)	Case-hardening steel
"Carburizer" (NaCN + NaCl + Na₂CO₃ plus activators)	Case-hardening steel
"Nitriding" (NaCN + KCN)	Case-hardening steel
Neutral (KCl + Na₂CO₃ + small amount of NaCN) (NaCl + KCl + small amount of BaCl₂ or Na₂CO₃) (NaCL + KCl + BaCl₂)	Heat-treating steel and non-ferrous alloys
"Tempering" (Combinations of NaNO₃, KNO₃, and NaNO₂)	Tempering hardened steel
"High-speed preheat" (NaCl + KCl + BaCl₂)	Preheating high-speed steel
"High speed" (BaCl₂ with rectifiers)	Hardening high-speed steel
"High-speed quench" (NaCl + KCl + BaCL₂ + NaCN) (KCl + NaCN + Na₂CO₃) (KCl + NaCN + Na₂CO₃ + BaCO₃)	Quenching high-speed steel
"Aluminum heat treating" (NaNO₃ + KNO₃ with or without nitrites)	Heat-treating aluminum

* Numbers in this column refer to alloy compositions listed in Table 7, p. 740.
¹ Salt pot life is dependent upon a number of factors. Furnace conditions are very important. The pot should be heated uniformly, preferably by radiation and never by flame impingement. Salt should never be permitted to enter the combustion chamber for the vapors developed destroy the protective film on the alloy. When starting a cold furnace, ample time should be allowed for the preheat and melt-down periods to minimize the thermal stresses

FUSED SALT POTS [1]

Temp. Range		Pot Alloy								
		1st Choice			2nd Choice			3rd Choice		
°C	°F	No.*	Cr	Ni	No.*	Cr	Ni	No.*	Cr	Ni
840–950	1550–1750	8, 9	15	60	10, 11	15	35	74	Steel	
840–950	1550–1750	8, 9	15	60	10, 11	15	35	74	Steel	
480–590	900–1100	74	Steel		...	..		...	...	
650–870	1200–1600	8, 9	15	60	10, 11	15	35	74	Steel	
700–900	1300–1650	8, 9	15	60	10, 11	15	35	74	Steel	
590–980	1100–1800	8, 9	15	60	10, 11	15	35	74	Steel	
150–590	300–1100	2	... 99.40[2]		6	14	80	74	Steel	
					8, 9	15	60			
					10, 11	15	35			
590–980	1100–1800	8, 9	15	60	10, 11	15	35	74	Steel	
950–1290	1750–2350	...	Ceramic		8, 9	15	60	17, 19	28	12
540–700	1000–1300	8, 9	15	60	10, 11	15	35	74	Steel	
230–590	450–1100	2	... 99.40[2]		6	14	80	74	Steel	
					8, 9	15	60			

set up in the pot and prevent overheating and oxidation. A longer service life may be expected from a pot that is in continuous service than one subjected to frequent heating and cooling cycles.

[2] Nickel (99.40%) containers are of low-carbon nickel ("L" Nickel) or low-carbon nickel-clad steel, usually of welded construction.

738 HIGH-TEMPERATURE RESISTANT MATERIALS

TABLE 6. HEAT-RESISTANT ALLOYS FOR MISCELLANEOUS APPLICATIONS

Use	Temp. Range	No.*	Alloy	Usual Form	Name
Heating elements and resistors	Up to 1310° C (2400° F)	31	37.5 Cr, 7.5 Al, Bal. Fe	Wire	Alloy No. 10
	" " 1310° C (2400° F)	32	24 Cr, 6.5 Al, 2 Co, Bal. Fe	Wire and strip	Kanthal
	" " 1150° C (2100° F)	5	20 Cr, 80 Ni	Wire and strip	80–20
	" " 980° C (1800° F)	9	15 Cr, 60 Ni	Wire and strip	60–15
	" " 760° C (1400° F)	11	15 Cr, 35 Ni	Wire and strip	35–15
	" " 540° C (1000° F)	29	55 Cu, 45 Ni	Wire and strip	Copper-Nickel
	" " 540° C (1000° F)	30	30 Fe, 70 Ni	Wire	Nickel-Iron
Thermocouples	" " 1540° C (2800° F)	35	99.99 Pt	Wire	Platinum
	" " "	36	90 Pt, 10 Rh	Wire	Platinum-Rhodium
	" " 1095° C (2000° F)	33	10 Cr, 90 Ni	Wire	Chromel P
	" " "	34	2 Al, 95 Ni, 2 Mn, 1 Si	Wire	Alumel
	" " 875° C (1600° F)	38	99 + Fe	Wire	Iron
	" " "	29	55 Cu, 45 Ni	Wire	Constantan
	" " 400° C (750° F)	37	99.9 Cu	Wire	Copper
Resistance thermometers	Room to 595° C (1100° F)	35	99.99 Pt	Wire	Platinum
	Room to 300° C (575° F)	3	99.90 Ni	Wire	Nickel
	Sub zero to 95° C (200° F)	37	99.90 Cu	Wire	Copper
	Up to 595° C (1100° F)	30	30 Fe, 70 Ni	Wire	Nickel-Iron
Spark plug electrodes		39	4 W, Bal. Pt	Wire	Platinum-Tungsten
		40	4 Mn, 1 Si, Bal. Ni	Wire	Nickel Manganese Silicon
		41	3–4 Cr, 94 Ni, 2 Mn, 0.4 Ba	Wire	A.C. Spark Plug Alloy
		100	3–4 Cr, 1 Mn, 2 Cb, 0.4 Ba, Bal. Ni	Wire	" " "
		42	96 Ni, 2 Mn, 1 Zr, 1 Si	Wire	Champion Sp. Plug Alloy
		23	18 Cr, 8 Ni	Wire and rod	18–8
		1	99 + Ni	Wire, rod and strip	Nickel

HIGH-TEMPERATURE RESISTANT MATERIALS 739

TABLE 6.—Continued

Valves — automotive and marine	700°–750° C (1300°–1400° F)	43	9 Cr, 3.5 Si	Forgings	Silcrome No. 1
		44	21 Cr, 1.5 Ni, 2 Si	"	XB
		45	21 Cr, 12 Ni	"	2112
		46	0.60 Cr, 1.25 Ni	"	3140
Valves — truck and bus	750°–875° C (1400°–1600° F)	44	21 Cr, 1.5 Ni, 2 Si	Forgings	XB
		47	24 Cr, 5 Ni, 3 Mo	"	X CR
		48	21 Cr, 12 Ni, 3 Si	"	Silcrome No. 8
		49	14 Cr, 14 Ni, 2.5 W, 0.5 Mo	"	TPA
		45	21 Cr, 12 Ni	"	2112
Valves — aircraft	650°–700° C (1200°–1300° F)	49	14 Cr, 14 Ni, 2.5 W, 0.5 Mo	Forgings	TPA
		50	13 Cr, 8 Ni, 2.5 Si	"	CNS
		51	13 Cr, 1 C, 3 Co	"	PRK
		52	20 Cr, Bal Ni[1]	Forgings or facings	80–20
		53	28–31 Cr, 1–2 Si, 1–2 Fe, 4–6 W, 1–1.25 C, Bal. Co[2]	Facings	Stellite No. 6

* Numbers in this column refer to alloy compositions listed in Table 7, p. 740.
[1] Used in the wrought form for valve heads and applied with a torch as facings for valve heads and seats.
[2] Applied with a torch for valve seat facings.

HIGH-TEMPERATURE RESISTANT MATERIALS

TABLE 7. DETAILED COMPOSITION

No.	Type No. or Designation		Cast or Wrought	C	Mn	Si
	A. I. S. I.	A. C. I.				
1	...	...	W	0.10	0.20	0.05
2	...	...	W	0.02 max.	0.20	0.05
3	...	...	W	...	...	...
4	...	...	W	0.10	4.50	0.05
5	...	...	W	0.25 max.	2.5 max.	0.40 max.
6	...	...	W	0.08	0.25	0.25
7	...	...	W	0.08	0.25	0.25
8	...	HW	C	0.40–0.60	1.5 max.	2.0 max.
9	...	...	W	0.25 max.	3 max.	0.75–1.5
10	...	HT	C	0.35–0.55	2 max.	2.5 max.
11	330	...	W	0.05–0.40	0.1–1.2	1.0–1.5
12	...	...	W	0.12	0.71	0.55
13	...	HK	C	...	...	...
14	310	...	W	0.25 max.	2.0 max.	2.0 max.
15	...	...	W	0.20 max.	...	2–3
16	...	...	C	0.39	0.75	0.51
17	...	...	W	0.14	1.48	1.38
18	...	...	C	0.29	0.03	1.11
19	...	HH	C	0.20–0.45	2.5 max.	1.75 max.
20	...	HI	C	...	...	...
21	...	HF	C	...	...	...
22	...	HE	C	...	...	...
23	304	...	W	0.08 max.	2.0 max.	0.75 max.
24	347	...	W	0.10 max.	2.0 max.	0.75 max.
25	...	HC	C	...	...	...
26	329	...	W	0.20 max.	...	...
27	...	...	W	0.15	0.40	0.50
28	...	...	C	...	...	4.00
29	...	...	W	...	...	...
30	...	...	W	...	...	...
31	..	...	W	0.06	0.50	0.50
32	..	...	W	0.06	0.25	0.50
33	...	...	W	...	...	...
34	...	...	W	...	2.5	1.0
35	...	...	W	...	...	...
36	...	...	W	...	...	...
37	...	...	W	...	...	...
38	...	...	W	...	...	...
39	...	...	W	...	...	...
40	...	...	W	...	4.0	1.0
41	...	...	W	...	2.0	...
42	...	...	W	...	2.0	1.0
43	...	...	W	0.40–0.50	...	3.0–3.5
44	...	...	W	0.60–0.86	...	1.25–2.75
45	...	...	W	0.20–0.25	...	0.70–0.90
46	...	...	W	0.40	...	0.20
47	...	...	W	0.40–0.50	...	1 max.
48	...	...	W	0.25	...	3
49	...	...	W	0.45	...	0.55
50	...	...	W	0.25–0.35	...	2–3

OF ALLOYS IN TABLES 1-6

Cr	Ni	Fe	Other Elements	No.
...	99+	0.15	0.10 Cu	1
...	99.40	0.15	0.10 Cu	2
...	99.9	...	...	3
...	95.20	0.15	0.05 Cu	4
19-20	77-79	1.5 max.	...	5
13.00	79.50	6.50	0.20 Cu	6
13.00	77.50	6.50	2.0 Cb	7
10-14	59-62	Bal.	...	8
14-18	57 min.	Bal.	...	9
13-17	34-37	Bal.	...	10
15-20	35-39	Bal.	...	11
20.09	30.37	Bal.	...	12
23-26	19-21	Bal	...	13
24-26	19-22	Bal.	...	14
24-26	19-21	Bal.	...	15
24.00	19.44	Bal.	2.43 Cb	16
28.24	10.16	Bal.	...	17
27.36	11.20	Bal.	...	18
23-28	10-14	Bal.	0.20 N max.	19
26-30	13-16	Bal.	...	20
18-23	8-11	Bal.	...	21
27-30	8-11	Bal.	...	22
18-20	8-10	Bal.	...	23
18 min.	8 min.	Bal.	Cb = 10 × C, min.	24
26-29	3 max.	...	...	25
23-28	2.5-5.0	...	1-2 Mo	26
...	66.00	0.90	29.0 Cu	27
...	63.00	2.00	30.0 Cu	28
...	45	...	55.0 Cu	29
...	70	30	...	30
37.5	...	Bal.	7.5 Al	31
23.0	...	Bal.	6 Al, 2 Co	32
10.0	90	...	...	33
...	94	0.5	2 Al	34
...	...	...	99.99 P	35
...	...	...	10-20 Rh, Bal. Pt	36
...	...	...	99.9 Cu	37
...	...	99+	...	38
...	...	...	4 W, Bal. Pt	39
...	Bal.	...	...	40
3-4	94	...	0.4 Ba	41
...	96	...	1.0 Zr	42
8-9	...	Bal.	...	43
19-23	1-2	Bal.	...	44
20.5-22	10.5-12.0	Bal.	...	45
0.60	1.25	Bal.	...	46
23.25-24.25	4.5-5.0	Bal.	2.5-3.0 Mo	47
21	12	Bal.	...	48
14	14	Bal.	1.75-3.0 W, 0.50 Mo	49
12.0-13.5	7.0-8.5	Bal.	...	50

Table 7. DETAILED COMPOSITION

No.	Type No. or Designation		Cast or Wrought	C	Mn	Si
	A. I. S. I.	A. C. I.				
51	...	...	W	1.3	...	0.65
52	...	...	C, W	0.15–0.30	0.60–1.00	0.30 max.
53	...	...	C	1.25	...	2.7
54	...	...	W	0.05	0.13	0.30
55	420	...	W	0.32	0.36	0.22
56	403	...	W	0.10	0.35	0.20
57	416	...	W	0.11	0.35	0.25
58	...	...	C	0.13	0.29	0.24
59	...	...	W	0.10	0.30	1.55
60	...	...	W	0.06	0.36	0.41
61	...	...	W	0.13	0.45	0.32
62	...	...	W	0.11	0.41	0.78
63	...	...	W	0.09	0.41	0.78
64	...	...	W	0.10	0.41	1.40
65[1]	...	...	W	0.24	0.65	0.28
66	...	...	W	0.10	0.36	0.25
67[2]	...	...	W	0.35	0.40	0.25
68	...	...	C	0.37	0.65	0.17
69	...	...	C	0.41	0.77	0.28
70	...	...	W	0.11	0.19	1.35
71	...	...	C	0.15	0.40	0.34
72	...	...	C	0.18	0.70	0.35
73	...	...	W	0.17	...	0.15
74[3]	...	...	W	0.10	...	...
75[4]	...	...	C	0.18	0.51	0.15
76[5]	...	...	W	0.35	...	...
77[6]	...	...	C	3.13	1.18	1.77
78[6]	...	...	C	2.18	1.54	2.86
79[6]	...	...	C	2.62	1.17	1.83
80[7]	...	...	C	2.04	0.92	1.88
81	...	...	W	0.15 max.	0.50 max.	0.50–1.00
82	...	...	W	0.15 max.	0.50 max.	0.50–1.00
83	...	...	W	0.15 max.	0.50 max.	0.50–1.00
84	...	...	W	0.15 max.	0.50 max.	0.50–1.00
85	...	...	W	0.15 max.	0.50 max.	1.00–1.50
86	...	...	W	0.15 max.	0.50 max.	0.50–1.00
87	...	...	W	0.15 max.	0.50 max.	1.00–2.00
88	...	...	W	0.15 max.	0.50 max.	0.50–1.00
89	...	...	W	0.15 max.	0.50 max.	1.00–1.40
90	...	...	W	0.15 max.	0.50 max.	1.00–1.40
91	...	...	W	0.15 max.	0.50 max.	1.00–1.40
92	...	...	W	0.15 max.	0.30–0.60	0.50–1.00
93	...	...	W	0.15 max.	0.50 max.	1.15–1.65
94	...	...	W	0.15 max.	0.50 max.	0.50–1.00
95	...	...	W	0.15 max.	0.50 max.	0.50–1.00
96	...	...	W	0.10–0.20	0.30–0.60	0.25 max.
97	...	...	W	0.10–0.20	0.30–0.60	0.25 max.
98	...	HB	C	...	...	...
99	...	HD	C	...	...	...
100	...	...	W	...	1.0	...

[1] S. A. E. 6120.   [2] Nitrided Nitralloy.   [3] S. A. E 1010.   [4] Aluminum-killed.

OF ALLOYS IN TABLES 1–6 — *Continued*

Cr	Ni	Fe	Other Elements	No.
13	0.65 max.	Bal.	0.75 Mo, 3.0 Co	51
19–21	Bal.	1 max.	...	52
27	...	...	4 W, 65 Co	53
12.18	...	Bal.	2.0 Mo	54
13.57	...	Bal.	...	55
12.33	...	Bal.	...	56
12.27	..	Bal.	...	57
11.87	...	Bal.	...	58
4.83	...	Bal.	0.51 Mo	59
5.18	...	Bal.	0.58 Mo, 0.46 Ti	60
4.96	...	Bal.	0.52 Mo	61
2.50	...	Bal.	0.50 Mo	62
1.21	...	Bal.	0.58 Mo	63
1.24	...	Bal.	0.58 Mo	64
0.94	...	Bal.	0.18 V	65[1]
0.97	...	Bal.	0.55 Mo	66
1.25	...	Bal.	0.20 Mo, 1.20 Al	67[2]
...	3.03	Bal.	0.49 Mo	68
0.73	1.25	Bal.	0.59 Mo	69
...	...	Bal.	0.50 Mo	70
...	...	Bal.	0.49 Mo	71
...	...	Bal.	0.51 Mo	72
...	...	Bal.	0.56 Mo	73
...	...	Bal.	...	74[3]
0.01	0.05	Bal.	0.01 Mo, 0.04 Cu	75[4]
...	...	Bal.	...	76[5]
3.80	13.29	Bal.	6.00 Cu	77[6]
1.92	15.52	Bal.	6.05 Cu	78[6]
2.12	20.07	Bal.	0.10 Cu	79[6]
3.58	35.58	Bal.	0.02 Cu	80[7]
8.00–10.0	...	Bal.	0.90–1.10 Mo	81
8.00–10.0	...	Bal.	0.45–0.65 Mo	82
6.00–8.00	...	Bal.	0.90–1.10 Mo	83
6.00–8.00	...	Bal.	0.45–0.65 Mo	84
4.00–6.00	...	Bal.	0.90–1.10 Mo	85
4.00–6.00	...	Bal.	0.90–1.10 Mo	86
4.00–6.00	...	Bal.	0.45–0.65 Mo	87
4.00–6.00	...	Bal.	0.45–0.65 Mo	88
2.75–3.25	...	Bal.	0.45–0.65 Mo	89
1.75–2.25	...	Bal.	0.45–0.65 Mo	90
0.75–1.25	...	Bal.	0.45–0.65 Mo	91
1.00–1.50	...	Bal.	0.45–0.65 Mo	92
...	...	Bal.	0.45–0.65 Mo	93
2.75–3.25	...	Bal.	0.45–0.65 Mo, 0.30–0.70 Al	94
4.00–6.00	...	Bal.	0.45–0.65 Mo, 0.30–0.70 Al	95
...	...	Bal.	0.45–0.65 Mo	96
...	...	Bal.	...	97
18–22	...	Bal.	...	98
27–30	3–6	Bal.	...	99
3–4	Bal.	...	2 Cb, 0.4 Ba	100

[5] S. A. E. 1035.   [6] Ni Resist Cast Iron.   [7] Invar Cast Iron.

CHEMICAL RESISTANT MATERIALS

CHEMICAL RESISTANT MATERIALS*

FOREWORD

In evaluating the ability of a material to resist corrosion, consideration must always be given to the fact that the study of corrosion is the study of material-fluid systems. In the majority of cases, the material which is a component of one of these systems will be a commercially pure metal or an alloy, no one of which is completely resistant to corrosion by all fluids. Many of these materials have been discussed in considerable detail in the preceding chapters, and attention has been directed to the apparently extraneous factors that frequently affect the rates at which metals corrode.

When selecting a material or a combination of materials for use in a particular system it must be remembered that the specifications for one of these systems must include not only the characteristics of the fluid and of the material — the two essential components — but also extraneous factors which may be present and may influence corrosion. Since these other factors are peculiar to a particular system, no two of which are exactly alike and the total number of which is infinitely large, it is obviously impracticable to offer a set of hard and fast "corrosion recommendations" that cover all situations.

It is possible, however, to tabulate many of the available corrosion data in such a form that by comparing the known behavior of certain type materials in a large number of different environments, a reasonably good prediction can be made of the probable behavior of: (1) any one of the listed materials in a *supposedly* identical

* Prepared by L. G. Vande Bogart, Chairman, Crane Co., Chicago, Ill.
C. L. Bulow, Bridgeport Brass Co., Bridgeport, Conn.
M. H. Brown, E. I. du Pont de Nemours and Co., Wilmington, Del.
G. O. Hiers, National Lead Co., Brooklyn, N. Y.
F. L. LaQue, The International Nickel Co., New York, N. Y.
H. L. Maxwell, E. I. du Pont de Nemours and Co., Wilmington, Del.
R. B. Mears, Aluminum Company of America, New Kensington, Pa.
W. E. Pratt, Worthington Pump and Machinery Corp., Harrison, N. J.
G. A. Sands, Electrometallurgical Co., New York, N. Y.

In the preparation of this section, information was obtained from many sources. Acknowledgment is especially due:
   Allegheny Ludlum Steel Corp.
   Ampco Metal, Inc.
   Burgess-Parr Co.
   Carbide and Carbon Chemicals Corp.
   Dow Chemical Co.
   The Duriron Co., Inc.
   Fansteel Metallurgical Corp.
   Gates Engineering Co.
   B. F. Goodrich Co.
   Haveg Corp.
   Haynes Stellite Co.
   Monsanto Chemical Co.
   National Carbon Co., Inc.
   Phillips Petroleum Co.
   Revere Copper & Brass, Inc.
   A. O. Smith Corp.

environment; (2) a similar material — not listed — in one of the same environments; or (3) one of the listed or related materials in a somewhat different environment. This section consists largely of such tabulations. In using these tables, it should be remembered that they are based on the behavior of material-fluid systems, the exact specifications for which may never again be exactly reproduced in some other such system.

EVALUATION OF CORROSION RESISTANCE

The method of presenting these data was selected on the basis of several considerations. In the first place, it is not too difficult to differentiate between two materials that are corroded at a widely different rate by the same fluid under approximately similar conditions. It is difficult, however, to measure precisely the differences in rates at which two or more materials corrode when these rates are of the same general order of magnitude. This is particularly true when measurements are made in the field, where conditions can not be controlled as well as in the laboratory. In the second place, the evaluation of corrosion resistance can be based on laboratory corrosion tests, field corrosion tests, or on actual performance in one or more pieces of equipment. Data from any one of these three sources can be translated into units of penetration per unit of time, but too often the data obtainable from actual performance are not available in a form which permits the calculation of corrosion rates. Usually the records will show only the results obtained in length of time of satisfactory service. It is just such data, however, which are the most valuable and carry the most weight with the builders and users of chemical plant equipment. Since performance data usually cannot be translated into corrosion rates, even though they may be supported by laboratory and field test data which can be so presented, it was decided for the purpose of this section to define the three degrees of corrosion resistance mainly on the basis of known performance in the field. The three ratings are defined as follows.

*Class A.** Materials suitable for critical parts where very little dimensional change can be tolerated and whose behavior is not likely to be altered materially by turbulence, changes in oxidizing or reducing power, or other incidental conditions of exposure.

*Class B.** Materials in common use for non-critical parts where some corrosion can be tolerated or which are equivalent in corrosion resistance to those materials in common use.

Class C. Materials not ordinarily considered suitable.

Class A Materials. Examples of parts which normally require Class A materials are valve seats, pump shafts and rings, small springs, etc. The definition is based on the normal requirements for such parts, however, and not on what *can* be used in the absence of something better. With extremely corrosive fluids, e.g., hot, concentrated hydrochloric acid, it may be necessary to make critical parts out of materials which corrode with appreciable dimensional change and to replace these parts as frequently as may be necessary. The fact that a critical part is made of a particular material does not justify a Class A rating for that material unless it also meets the requirement of "very little dimensional change." In other words, a Class A material

---

* Materials conforming to Class A will, in general, be those which corrode at a rate no greater than 0.005 ipy. Materials conforming to Class B will, in many cases, be those which corrode at rates between 0.005 ipy and 0.05 ipy. As indicated in the discussion, the limiting corrosion rates which determine this classification cannot be defined precisely.

will always be among the best materials, but the best material is not necessarily a Class A material.

Class B Materials. Examples of non-critical parts which can tolerate some corrosion, i.e., which can be made of Class B materials, are tanks, piping, valve bodies, many structural parts, etc. Again the definition is based on the normal requirements for such parts; and, since these requirements vary over a much wider range than do those which provide the basis for the definition of Class A materials, the definition of Class B materials is correspondingly less rigid. In this classification, usefulness is the determining factor.

Circumstances frequently convert one of these non-critical pieces of equipment into a critical part which cannot tolerate much dimensional change. Springs and other load-bearing members, for example, can, in the absence of pitting and stress corrosion phenomena, tolerate an appreciable amount of corrosion when the amount of metal removed does not represent too large a percentage of the total cross-sectional area. For parts having an initial small cross section, however, *any* loss of metal may be too large a percentage. Practically, then, there are limiting sizes below which such parts are critical and require Class A materials and above which they are non-critical and can employ Class B materials. Similarly, light-wall tubing must usually be made of a material that is Class A in any environment with which the tubing is used, whereas standard or extra heavy pipe can be, and frequently is, made of Class B material. The corrosion rates for a number of materials with the same environment may vary over a fairly wide range, within which range one material may be better than another, yet all may be useful. It is with this class of materials that the incidental conditions of exposure must be considered with particular care, as reliable and economical service from one of them depends on complete information about the corroding environment *and also* about all other factors which may affect the usefulness of a material.

Class C Materials. Class C materials will in most cases be the materials which corrode at an excessively rapid rate, but they may also be those materials which corrode just enough to produce objectionable contamination of the fluid with corrosion products. In general, materials which will normally be Class B will be Class C with environments which cannot tolerate contamination with corrosion products; and materials which would normally be Class A might be expected to retain their Class A rating. Actually, a material which corrodes appreciably, but yields colorless or tasteless corrosion products, may be less objectionable than another material which is almost, but not quite, completely resistant to corrosion and yields highly colored, poisonous, or otherwise objectionable corrosion products. Some of the high-chromium alloys, for example, might be rated Class A with certain electroplating solutions from the standpoint of any dimensional change resulting from corrosion, yet they actually rate Class C because a small amount of corrosion contaminates the solution with highly objectionable chromium salts. In general, the influence of corrosion on contamination will depend, among other things, upon the ratio between the area of material exposed to the volume of fluid which may become contaminated. A material acceptable for some small critical part may introduce too much contamination when used for piping, tanks, etc. Conversely, a material not acceptable for parts which have a large surface area might find use for some other part having a much smaller area and for which it is especially well suited from the standpoint of physical properties or some other reason.

The probability of stress corrosion may also be cause for assigning a Class C rating to an otherwise useful material. Here again the use to be made of a material may determine the rating. A material in the form of a part which does not retain residual internal stresses resulting from fabrication or is not stressed as a phase of its use may resist corrosion well enough to be entitled to a Class B or even a Class A rating. When resistance to both stress and corrosion are involved the rating may change to Class C.

In considering the above classification, the fact that material-fluid systems are involved must again be emphasized, for frequently a material will be given one rating with one chemical under one set of conditions and a different rating with the same chemical under different conditions. For example, gray cast iron is commonly used with 94 to 98% sulfuric acid, hot or cold; and in such service this material would certainly be assigned a Class B rating and might, by some, be rated Class A. The same material is not recommended at all for use with hot sulfuric acid of concentration over 100%, and in such service it would just as certainly be rated Class C.

Originally, it had been intended to follow the usual practice of rating a material on the basis of *ideal* conditions and explaining the failure of a material to come up to expectations as being due to departures from ideal conditions. In the adopted classification, however, Class A materials are defined as those whose "behavior is not likely to be altered materially" by departures from the ideal. This is still not a rigid definition, but it comes closer to being practical than one based on conditions which seldom obtain in the field. For example, cast iron and carbon steel are commonly used for valves, piping, and other equipment exposed to 94 to 98% sulfuric acid, yet these materials are not rated Class A for any phase of sulfuric acid service because their behavior is altered materially by incidental conditions of exposure. In the absence of exposure to the air, as in the case of a valve in a pipe line which is completely filled with acid on both sides, these materials give excellent results and can even be used for valve seats and other critical parts. In some other location, where a part may be exposed alternately to acid and to air (containing moisture), the acid may become diluted as a result of absorption of moisture and become actively corrosive. Since this possibility is always present as an incidental condition of exposure, these materials are not given a Class A rating with sulfuric acid.

The line of demarcation between Class B and Class C materials is less sharp than the division between Class A and Class B (or between Class A and Class C) because the definitions of the amount of corrosion which can be tolerated and that which is excessive are, again, peculiar to a particular situation. A part which can be easily repaired or replaced can, as a rule, tolerate an amount of corrosion which would be excessive for a similar part used in some less accessible location. Similarly, what may be acceptable for an experimental or temporary installation may not be acceptable for another installation within which conditions cannot be so well controlled and which may be expected to operate for a longer period of time. Materials suitable for use with *anhydrous* hydrogen chloride in pilot plant installations, for example, may be entirely unsuited for use with the same chemical in large-scale manufacturing plants from which moisture cannot be completely excluded. Again, the handling of abnormally corrosive fluids may necessitate the use of a material that corrodes at a rate which would be considered excessive when dealing with other less corrosive chemicals, but is tolerated with a fluid like hydrochloric acid — with which very few materials have Class A resistance. In such situations, extreme caution must be used, as incidental conditions of exposure, e.g., aeration, are quite likely to change the corrosion rate from one which can be tolerated to one which is excessive. In general, any material to which a Class B rating has been assigned might, under very adverse conditions, be entitled to no better than a Class C rating, but is not entitled to a Class A rating on the basis of probable or average behavior, even though it may sometimes exhibit Class A properties under the most favorable conditions. Similarly, materials which have been given a Class C rating might under very favorable or unusual conditions approach a Class B rating and occasionally be used. The fact that they are not in common use or equivalent in corrosion resistance to those in common use would prevent them from attaining a full Class B rating.

An additional cause for a change from Class A or Class B rating to a Class C rating is the change in the corrosion characteristics of a fluid resulting from changes in its composition. When such changes are made intentionally, the result is usually classified as another and different fluid. It is not unusual for a change in composition to occur as a result of normal service conditions, however, and these changes in composition are frequently responsible for the departures from ideal conditions which complicate the evaluation of corrosion resistance. In the tables which follow, the Class A and Class B ratings are — unless otherwise stated — based on the assumption that the fluid is chemically pure, or, at worst, commercially pure. Under Class C ratings, however, attention will frequently be directed to the effect of contaminants — i.e., changes in composition — on the corrosion of various materials by the chemical under consideration.

List of Corrosion-Resistant Materials

The selection of a material for use with a particular fluid is often complicated by the large number of materials commercially available. Between many of these materials, however, there are more outstanding differences in physical properties than in the ability to resist corrosion, a situation which becomes particularly apparent when materials are evaluated in accordance with the aforementioned definitions of Class A, Class B, and Class C materials. In this section, therefore, no attempt will be made to rate *all* materials. Instead, metals and alloys have been classified into "types," with one to several representatives for each type. (See Table A.) With a few exceptions, which will be noted later, it is believed that this list from the standpoint of corrosion resistance covers the total number of available metals and alloys which are industrially important. To these metals and alloys has been added a number of non-metallic materials, without which chemical industry would be seriously handicapped.

Magnesium provides the exception to the general rule that the evaluation of corrosion resistance for the purposes of this section is based on actual performance in chemical plant equipment. Magnesium has not been used extensively as a corrosion-resisting material, and references to it in the tables will, in most cases, be based on laboratory rather than field experience.

Conspicuous by their absence from Table A are the three metals zinc, cadmium, and tin. Their omission does not imply any suggestion that they are not industrially important but is based on the following considerations.

Zinc, like magnesium, is used for many articles which are not ordinarily exposed to corrosive environments and do not require especially good resistance to chemical corrosion. When used in corrosive environments, the serviceability of both magnesium and zinc is likely to depend to a large degree upon the supplementary use of protective coatings. Zinc and cadmium are used as corrosion-resistant materials in the form of protective coatings on iron and steel (occasionally other materials), but only when the exposure is to mild or moderately severe environments.

Tin, like zinc and cadmium, finds considerable use as a protective coating for other metals; and also is used in the form of block tin pipe for handling distilled water and carbonated beverages, and in sheet and foil for collapsible tubes and packaging. From the standpoint of the handling of chemicals in general, however, it is not a corrosion-resistant material.

Also omitted from Table A are the many organic protective coatings of the paint, varnish, or lacquer type. Some of these coatings are of considerable industrial importance, but they have been omitted for two general reasons. In the first place, a coating cannot be evaluated on the basis of its composition alone. Skill of application and control of the conditions under which a coating is applied are often of greater importance than variations in the coating itself. In the second place, most protective coatings are sold under trade names or numbers, any one of which may apply to more than one formula or to a basic formula which has been modified to meet changed conditions. When coatings of this type can be evaluated on the basis of specified compositions rather than on the basis of arbitrary designations, it may then be possible to assign ratings to them just as ratings have been assigned to metals and alloys and a few non-metallic materials. In contrast to protective coatings of this type are protective linings, comprised of materials which are prefabricated into layers of substantial thickness and then attached to a metal surface. Type 25 and some Type 24 materials fall within this category, but all of them are not restricted for use in this manner.

The original plans for this section contemplated ratings for a much longer list of chemicals. Since the method of rating materials used here is somewhat experimental, and since the preparation of additional tables would have delayed publication, it was decided to limit this section in this edition of the *Corrosion Handbook* to the chemicals listed in Tables 1 to 12, inclusive. Additional tables will be compiled and published at some later date.

Effect of Temperature

It should be noted that the headings at the top of the second and third "Temperature" columns use the words "up to" instead of the single word "at," the assumption being that a material which is entitled to a given rating at some elevated temperature is entitled to the same rating at any and *all lower temperatures*. One exception is cited in the case of magnesium with hydrofluoric acid.

CHEMICAL RESISTANT MATERIALS

TABLE A. LIST OF MATERIALS

Type	Nominal Composition
1. Aluminum and Aluminum-Base Alloys	(A) Commercially pure aluminum. Not more than 1% total impurities, mainly iron and silicon. (B) High-purity aluminum, containing up to 99.85% Al. (C) Al-Mn alloy, containing 1.2% Mn. (D) Duralumin-type alloy; containing 4.0% Cu, 0.5% Mn, and 0.5% Mg. (E) Al-Mg silicide alloy; containing 0.7% Si, 1.3% Mg, and 0.25% Cr. (F) Al-Si alloys, containing 5 to 12% Si. (G) Al-Mg alloys, containing 3.8 to 10% Mg.
2. Iron and Steel (See Note 1)*	Includes commercially pure iron, wrought iron, low-alloy irons, mild steel, low-alloy steels, gray cast iron (with and without small alloy additions), and malleable cast iron.
3. High-Silicon Cast Irons (See Note 2)	(A) Not less than 14.25% Si. (B) Not less than 14.25% Si; approximately 3% Mo.
4. High-Nickel Cast Irons	(A) 13.5 to 17.5% Ni; 5.5 to 7.5% Cu; 1.5 to 2.5% Cr. (B) 18.0 to 22.0% Ni; nil to 4.0% Cu; 1.5 to 3.0% Cr. (C) 28.0 to 32.0% Ni; 2.5 to 4.5% Cr.
5. 4 to 10% Chromium Steels	4 to 10% Cr; 0.5 to 1.5% Mo or W; 0.15% C (max.).
6. Stainless Steels (Martensitic) (See Notes 3 and 4)	(A) 11.5 to 14.0% Cr; 0.15% C (max.). (B) 11.5 to 14.0 % Cr; 0.35 % C (max.). (C) 16.0 to 18.0% Cr; 0.65 to 0.75% C.
7. Stainless Steels (Ferritic) (See Notes 3 and 4)	(A) 15.0 to 18.0% Cr; 0.15% C (max.). (B) 23.0 to 28.0% Cr; 0.25% C (max.).
8. Stainless Steels (Austenitic) (Notes 3, 4, 5, and 6)	(A) 18.0 to 20.0% Cr; 8.0 to 10.0% Ni; 0.08% C (max.). (B) 17.5 to 20.0% Cr; 10.0 to 14.0% Ni; 2.0 to 4.0% Mo; 0.10% C (max.). (C) 22.0 to 24.0% Cr; 12.0 to 15.0% Ni; 0.20% C (max.). (D) 24.0 to 26.0% Cr; 19.0 to 22.0% Ni; 0.25% C (max.).
9. Special Fe-Cr-Ni Alloys (Austenitic) (See Note 7)	(A) 19 to 20% Cr; 22 to 24% Ni; 2 to 3% Mo; 1.0 to 1.75% Cu; 1.0 to 3.25% Si; 0.07% C. (B) 20% Cr; 29% Ni; 2% Mo; 4% Cu; 1% Si; 0.07% C. (C) 25% Cr; 35% Ni; 5% Mo; 0.20–0.50% C.
10. Nickel	Commercially pure nickel.
11. Ni-Cr-Fe Alloy	79.5% Ni; 13.0% Cr; 6.5% Fe; 0.2% Cu.
12. Ni-Cu Alloys (more than 50% Ni)	63.0 to 67.0% Ni; 29.0 to 30.0% Cu; 0.9 to 2.0% Fe; 0.10 to 4.0% Si; nil to 2.75% Al.
13. Ni-Mo-Fe (Cr) Alloys	(A) 55.0 to 60.0% Ni; 15.0 to 20.0% Mo; 15.0 to 20.0% Fe. (B) 60.0 to 65.0% Ni; 25.0 to 35.0% Mo; 5.0% Fe. (C) 55.0 to 60.0% Ni; 15.0 to 20.0% Mo; 6.0% Fe; 12.0 to 16.0% Cr; 5.0% W.

* See Notes on pp. 755, 756.

Table A. LIST OF MATERIALS — Continued

Type	Nominal Composition
14. Ni-Cr-Cu-Mo Alloy	(A) 56.0 to 57.0% Ni; 23.0 to 24.0% Cr; 8% Cu; 4% Mo; 2% W; 1.0% Si or Mn.
15. Ni-Si Alloy	85% Ni; 10% Si; 3% Cu; 2% Al.
16. Cu-Ni Alloys (more than 50% Cu)	(A) 80% Cu; 20% Ni. (B) 75% Cu; 20% Ni; 5% Zn. (C) 70% Cu; 30% Ni.
17. Copper and Copper-Base Alloys	(A) Commercially pure copper. Not more than 0.1% total impurities. (B) Cu-Al alloys; containing 8.5 to 11.0% Al and nil to 4% Fe. (C) Cu-Si alloys; containing 1 to 4% Si; nil to 1% Mn; nil to 1.75% Sn; and nil to 1% Zn. (D) Cu-Sn alloys; containing 4 to 12% Sn with not more than 0.25% of either Pb or Zn. (E) Cu-Sn-Zn alloys; containing Sn up to 20%; Zn less than 17%; and Zn less than Sn. With or without Pb. (F) Cu-Sn-Zn and Cu-Zn alloys; containing Zn less than 17% with Sn less than Zn. With or without Pb. (G) Cu-Zn alloys; containing more than 17% of Zn.
18. Lead	(A) 99.9% Pb; 0.08% Cu (max.); 0.005% Bi (max.). (B) 99.85% Pb; 0.08% Cu (max.); 0.005% Bi (max.); 0.05% Te. (C) 4 to 10% Sb; remainder Pb; additions as in (A) or (B) above.
19. Silver (See Note 8)	"Fine" silver; 99+ %
20. Noble Metals and Tantalum (See Note 9)	(A) Tantalum. (B) Platinum and platinum alloys. (C) Gold.
21. Magnesium and Magnesium-Base Alloys (See Note 10)	
22. Ceramic Materials	(A) Glass. (B) Porcelain. (C) Chemical stoneware.
23. Carbon and Graphite (See Notes 11 and 12 below)	(A) Carbon. (B) Graphite. (C) Impervious carbon. (D) Impervious graphite.
24. Plastics (See Notes 12 and 13)	(A) Vinylidene chloride-type resin. (B) Phenolic resin, asbestos-filled. (C) Phenolic resin, graphite-filled. (D) Furane resin, asbestos or graphite-filled. (E) Polyvinyl chloride-type resin.
25. Rubbers and Synthetic Elastomers (See Notes 12, 13, and 14)	(A) Natural rubber: (1) soft; (2) hard. (B) GRA (butadiene-acrylonitrile): (1) soft; (2) hard. (C) GRS (butadiene-styrene): (1) soft; (2) hard. (D) GRI (butadiene-isobutylene). (E) GRM (chloroprene).

CHEMICAL RESISTANT MATERIALS

Note 1. The various modifications of the above materials are so numerous, and the differences in behavior are so slight in most cases that they have not been divided into subgroups. Where convincing evidence of outstanding differences in behavior is available, appropriate footnotes are inserted.

Note 2. These materials resist corrosion by many environments at elevated temperatures, but are not always usable at those temperatures because of danger of breakage resulting from thermal shocks. The ratings assigned do not take this factor into consideration and are based on corrosion resistance only.

Note 3. Types 6, 7, 8, and 9 materials depend for maximum resistance to corrosion upon proper heat treatment and, in some cases, upon proper surface preparation. Such treatment may not be necessary with all environments with which they may be used, but the ratings assigned to these materials are all based on the assumption that the materials have been properly conditioned in accordance with standardized and approved procedures.

Note 4. The machining properties of steels in Groups 6A, 7A, and 8A may be improved by the addition of phosphorus, sulfur, or selenium (in amounts over 0.07%) with or without additions of molybdenum or zirconium (in amounts up to 0.60%). In some environments these free machining grades are inferior to the corresponding grades which do not contain the free machining additions. Prospective users should ascertain whether these additions have any adverse effect on resistance to corrosion by the proposed environment.

Note 5. Unless otherwise stated, the ratings assigned to Type 8 materials in the Corrosion Tables are based on the assumption that the carbon content is not over 0.10%. There are many situations in which the same materials with higher carbon content are entirely satisfactory, but this should be determined by actual test in the proposed environment.

Note 6. Stainless steels in this austenitic group, heated in the range 350° to 850° C (650° to 1550° F) and lacking proper final heat treatment, are subject to intergranular corrosion. They may be stabilized to prevent this type of attack by the addition of columbium or titanium (see discussion of intergranular corrosion, p. 161). The carbon content of alloys so stabilized should not exceed 0.10%.

Note 7. These materials — when not properly heat-treated or containing too much carbon to permit effective heat treatment (9c) — are susceptible to intergranular corrosion by many of the same environments which cause intergranular corrosion of Type 8 materials. The use of 9c with any environment which is known to be capable of causing intergranular corrosion of any of the other austenitic materials should be questioned unless the carbon content is limited to 0.10% (max.), or unless experience has demonstrated that 9c with some higher carbon content is immune to this type of attack. Unless otherwise stated, the ratings for 9c are based on the presumption that the carbon will be over 0.10%.

Note 8. Sterling silver and other silver alloys are sometimes used because of more desirable physical properties. The corrosion resistance of these alloys should not be assumed equivalent to that of fine silver.

Note 9. Alloys of platinum with iridium or rhodium may be more useful than pure platinum because of better physical properties. Unless otherwise noted, the recommendations for platinum apply to these alloys also.

Note 10. There are numerous variations of these materials. Prospective users should ascertain which type or types can be used with the proposed environment.

Note 11. There are several varieties of impervious carbon and impervious graphite, and there are important differences in heat transferability and other physical properties between grades made from the same stock. Prospective users should ascertain which type or types can be used with the proposed environment. Impervious graphite is used more generally than impervious carbon.

Note 12. Characteristic properties of Types 23, 24, and 25 materials may be modified by formulation and processing (e.g., addition of or changes in plasticizers, fillers, etc.) to meet specific conditions. Consequently, some of the letter designations refer to types rather than to rigidly controlled representatives of a type, and prospective users of these materials should consult the suppliers for more specific recommendations. When any of these materials are used as linings, both the bond used to obtain adhesion to the steel or other metal and the thickness of the lining have a decided influence on the limiting factors of service.

Note 13. Most of these materials are likely to break as a result of sudden thermal shocks when formulated into the "hard" forms. The ratings in the Corrosion Tables do not take this factor into consideration, but are based on corrosion resistance only.

Note 14. Grades A, B, and C under Type 25 are available in non-rigid forms commonly designated as "soft" and in rigid or "hard" forms. In general, the latter retain the characteristic chemical properties of the corresponding soft forms (sometimes to a different degree), but they differ in physical properties and, as a consequence, differ in their behavior with respect to such factors as temperatures, abrasion, mechanical stresses, etc. Grades D and E are available only in soft forms.

Table 1. ACETIC ACID GROUP

Types of Material
(See p. 753 for subgroups and further details of composition.)

1. Al and Al-Base Alloys
2. Iron and Steel
3. High-Si Cast Irons
4. High-Ni Cast Irons
5. 4 to 10% Chromium Steels
6. Stainless Steels (Martensitic)
7. Stainless Steels (Ferritic)
8. Stainless Steels (Austenitic)
9. Special Fe-Cr-Ni Alloys (Austenitic)
10. Nickel
11. Ni-Cr-Fe Alloy
12. Ni-Cu Alloys (>50% Ni)
13. Ni-Mo-Fe-(Cr) Alloys
14. Ni-Cr-Cu-Mo Alloy
15. Ni-Si Alloy
16. Cu-Ni Alloys (>50% Cu)
17. Cu and Cu-Base Alloys
18. Lead
19. Silver
20. Noble Metals and Tantalum
21. Mg and Mg-Base Alloys
22. Ceramic Materials
23. Carbon and Graphite
24. Plastics
25. Rubber and Synthetic Elastomers

Acetic Acid and Acetic Anhydride

% By Weight	Class A Materials (See definition of Class A Materials on p. 748.)		
	At Room Temperature Only (up to 30° C, 85° F)	Up to Intermediate Temperature Indicated	Up to Boiling Point
Up to 10%	1a, b, c, e, f; 12; 24a, b		3a,b; 8a,b,c,d; 9a,b,c; 11; 13a, b, c; 14a; 15; 19; 20a, b, c; 22a,b,c; 23a,b,c,d; 24d
10 to 20%	1a, b, c, e, f; 12; 24a, b		3a, b; 8a, b, c, d; 9a, b, c; 13a, b, c; 14a; 15; 19; 20a, b, c; 22a, b, c; 23a, b, c, d; 24d
20 to 50%	1a, b, c, e, f; 12; 24a,b	8a up to 80° C (175° F)	3a, b; 8b, c, d; 9a, b, c; 13a, b, c; 14a; 15; 19; 20a, b, c; 22a, b, c; 23a, b, c, d; 24d
50 to 80%	1a, b, c, e, f; 12; 24a	8a up to 80° C (175° F)	3a, b; 8b; 9a, b, c; 13a, b, c; 14a; 15; 19; 20a, b, c; 22a, b, c; 23a, b, c, d; 24d
80 to 99.9%	1a, b, c, e, f; 12; 24a	8a up to 80° C (175° F)	3a, b; 8b; 9a, b, c; 13a, b, c; 14a; 15; 19; 20a, b, c; 22a, b, c; 23a, b, c, d; 24d
100%	24a		
Hot vapors (See Note 1 below)			8b; 9a, b, c; 12; 14a; 19; 20a, b, c; 22a, b, c; 23a, b, c, d (See Note 1 below)
Acetic anhydride	1a, b, c, e, f; 13a, b, c; 15		8b; 9a, b, c; 13b, c; 14a; 19; 20a, b, c; 22a, b, c; 23a, b, c, d

Note 1. The limiting temperature for most metallic materials will be that at which "High-Temperature Corrosion" begins. Among the non-metallic materials, 23a, b are Class A at all temperatures, and 23c, d are Class A up to 170° C (340° F) and Class C above that temperature.

TABLE 1. ACETIC ACID GROUP — *Continued*

ACETIC ACID AND ACETIC ANHYDRIDE

% By Weight	Class B Materials (See definition of Class B Materials on p. 749. Other materials in common use are listed under Class A Materials.)		
	At Room Temperature Only (up to 30° C, 85° F)	Up to Intermediate Temperature Indicated	Up to Boiling Point
Up to 10%	4A, B, C, D; 6A	1A, B, C, D, E, F up to 70° C (160° F); 24B up to 100° C (210° F); 24C up to 50° C (130° F); 25A–2 and 25D up to 65° C (150° F)	6B, C; 7A, B; 10; 12; 16A, B, C; 17A, B, C, D
10 to 20%	6B, C	1A, B, C, D, E, F up to 70° C (160° F); 24B up to 100° C (210° F); 24C up to 50° C (130° F); 25A–2 and 25D up to 65° C (150° F)	7A, B; 8A; 10; 11; 12; 16A, B, C; 17A, B, C, D
20 to 50%	6B, C	1A, B, C, D, E, F up to 70° C (160° F); 24B up to 100° C (210° F); 24C up to 50° C (130° F); 25A–2 and 25D up to 65° C (150° F)	7A, B; 8A; 10; 11; 12; 16A, B, C; 17A, B, C, D
50 to 80%	6B, C; 24B	1A, B, C, D, E, F and 7A, B up to 80° C (175° F); 24C up to 50° C (130° F); 25A–2 and 25D up to 65° C (150° F)	8A; 10; 11; 12; 16A, B, C; 17A, B, C, D
80 to 99.9%	24B	1A, B, C, D, E, F and 7A, B up to 80° C (175° F); 24C up to 50° C (130°F); 25A–2 and 25D up to 65° C (150° F)	6B, C; 8A, C, D; 10; 11; 12; 16A, B, C; 17A, B, C, D
100%	24B	24C up to 50° C (130° F); 25A–2 and 25D up to 65° C (150° F)	
Hot vapors (See Note 1, p. 757)			Type 2 Materials and 4A, B, C above dew point; 8C, D; 10; 11; 12; 16A, B, C; 17A, B, C, D (See Note 1, p. 757)
Acetic anhydride	4A, B, C; 24A, B, C	1A, B, C, E, F up to 80° C (175° F); 24D up to 50° C (130° F); 25A–2 and 25D up to 65° C (150° F)	3A, B; 6A, B, C; 7A, B; 8A, C, D; 10; 11; 13A, B, C; 14A; 15; 16A, B, C; 17A, B, C, D

CHEMICAL RESISTANT MATERIALS

TABLE 1. ACETIC ACID GROUP — *Continued*

ACETIC ACID AND ACETIC ANHYDRIDE

% By Weight	Class C Materials (Materials not ordinarily considered suitable)	
	At All Temperatures	Above Indicated Temperatures
Up to 99.9%	Types 1G, 2, 4, (above 10%), 5, 17G, 18, and 21 materials. All Type 1 materials with solutions which are contaminated with formic acid or with as much as 1 ppm of copper, tin, or lead. All Types 1, 6, 7, and 8A, C, D materials with solutions contaminated with sulfuric acid, hydrochloric acid, or salt. Types 16 and 17 materials with highly aerated solutions, 17E, F materials which contain lead with edible products. All Type 18 materials, 25A-1, B-1, C-1, D, E	4A, B, C above 30°C (85°F) with concentrations up to 10%. 6A, B, C above 30°C (85°F) with concentrations above 10% and up to 80%. 7A, B above 80°C (175°F) with concentrations above 50%. 24A above 50°C (130°F); 24c above 50°C (130°F); 25A-2 and 25D above 65°C (150°F)
100%	As above	All Type 1 materials above 105°C (120°F) except when acetic anhydride is also present. Otherwise as above.
Hot vapors (See Note 1, p. 757)		(See Note 1, p. 757)
Acetic anhydride	1G; 2; 5; 17G; 21; 25A-1, B-1, C-1, D, E	1A, B, C, E, F above 80°C (175°F); 4A, B, C above 30°C (85°F); 24A, B above 50°C (130°F); 24c above 30°C (85°F); 25A-2 and 25D above 65°C (150°F)

RELATED ACIDS

Butyric acid Formic acid Lactic acid Propionic acid Pyroligneous acid	These acids behave much like acetic acid, data for which apply here also. The cautions regarding the handling of acetic acid contaminated with salt, mineral acids, and heavy metals should be observed. *Exceptions.* Formic acid corrodes aluminum, which is ordinarily Class C, and behaves differently towards Types 24 and 25 materials than acetic acid. Consult suppliers of these materials for more specific recommendations.

Class A, B, or C Materials

FOOD PRODUCTS CONTAINING ACETIC ACID OR RELATED ACIDS

Butter and buttermilk Sauerkraut juice Vinegar and fruit juices	These may be more troublesome than pure acids with materials which tend to pit because of tendency to deposit solid substances on metal surface. Class B materials frequently not acceptable because of adverse effect of dissolved metal salts on color and flavor. Type 18 materials and other materials rich in lead are Class C.

Table 2. CITRIC ACID GROUP

Types of Material
(See p. 753 for sub-groups and further details of composition.)

1. Al and Al-Base Alloys
2. Iron and Steel
3. High-Si Cast Irons
4. High-Ni Cast Irons
5. 4 to 10% Chromium Steels
6. Stainless Steels (Martensitic)
7. Stainless Steels (Ferritic)
8. Stainless Steels (Austenitic)
9. Special Fe-Cr-Ni Alloys (Austenitic)
10. Nickel
11. Ni-Cr-Fe Alloy
12. Ni-Cu Alloys (>50% Ni)
13. Ni-Mo-Fe-(Cr) Alloys
14. Ni-Cr-Cu-Mo Alloy
15. Ni-Si Alloy
16. Cu-Ni Alloys (>50% Cu)
17. Cu and Cu-Base Alloys
18. Lead
19. Silver
20. Noble Metals and Tantalum
21. Mg and Mg-Base Alloys
22. Ceramic Materials
23. Carbon and Graphite
24. Plastics
25. Rubber and Synthetic Elastomers

CITRIC ACID

% By Weight	Class A Materials (See definition of Class A materials on p. 748.)		
	At Room Temperature Only (up to 30° C, 85° F)	Up to Intermediate Temperature Indicated	Up to Boiling Point at Indicated Concentration
All concentrations	1 A, B, C, E, F; 8A, C, D	All Types 24 and 25 materials up to limiting temperature for the material	3A, B; 8B; 9A, B, C; 11 u to 5%; 13c; 14A, B; 15; 19; 20A, B, C; 22A, B, C; 23A, B, C, D

% By Weight	Class B Materials (See definition of Class B materials on p. 749. Other materials in common use are listed under Class A materials.)		
	At Room Temperature Only (up to 30° C, 85° F)	Up to Intermediate Temperature Indicated	Up to Boiling Point at Indicated Concentration
Up to 15%		1A, B, C, E, F up to 100° C (210° F); 4A, B, C up to 50° C (120° F) with 5 to 15%; 8A, C, D up to 65° C (150° F); 18A, B, C up to 65° C (150° F) with non-edible products	4A, B, C up to 5%; 6A, B, C; 7A, B; 10; 11 with 5 to 15%; 12; 13A, B; 16A, B, C; 17A, B, C, D
Over 15%	4A, B, C; 6A, B, C; 7A, B	1A, B, C, E, F up to 100° C (210° F); 8A, C, D up to 65° C (150° F); 18A, B, C up to 65° C (150° F) with non-edible products	10; 11; 12; 13A, B; 16A, B, C; 17A, B, C, D

TABLE 2. CITRIC ACID GROUP — *Continued*

CITRIC ACID

% By Weight	Class C Materials (Materials not ordinarily considered suitable.)	
	At All Temperatures	Above Indicated Temperatures
All concentrations	Types 1G, 2, 5, 17G, 18 (with edible products) and 21 materials. All Type 1 materials with solutions which are contaminated with as much as 1 ppm of copper, tin, or lead. All Types 1, 6, 7, and 8A, C, D materials with solutions which are contaminated with sulfuric acid, hydrochloric acid, or salt.	4A, B, C; 6A, B, C and 7A, B above 15% above 30° C (85° F); 18A, B, C (with non-edible products) above 65° C (150° F)

FOOD PRODUCTS CONTAINING CITRIC ACID OR RELATED ACIDS

Carbonated beverages Fruit juices Vegetable juices	Recommendations for these products may be the same as for the corresponding concentrations of the characteristic acid, but are not necessarily so. Mineral acids or carbon dioxide may also be present, in which case consult the proper table containing recommendations for these media.

CHEMICAL RESISTANT MATERIALS

TABLE 3. FATTY ACID GROUP

Types of Material
(See p. 753 for sub-groups and further details of composition.)

1. Al and Al-Base Alloys
2. Iron and Steel
3. High-Si Cast Irons
4. High-Ni Cast Irons
5. 4 to 10% Chromium Steels
6. Stainless Steels (Martensitic)
7. Stainless Steels (Ferritic)
8. Stainless Steels (Austenitic)
9. Special Fe-Cr-Ni Alloys (Austenitic)
10. Nickel
11. Ni-Cr-Fe Alloy
12. Ni-Cu Alloys (>50% Ni)
13. Ni-Mo-Fe-(Cr) Alloys
14. Ni-Cr-Cu-Mo Alloy
15. Ni-Si Alloy
16. Cu-Ni Alloys (>50% Cu)
17. Cu and Cu-Base Alloys
18. Lead
19. Silver
20. Noble Metals and Tantalum
21. Mg and Mg-Base Alloys
22. Ceramic Materials
23. Carbon and Graphite
24. Plastics
25. Rubber and Synthetic Elastomers

The data for "Fatty Acids" and "Related Products" are based on the handling of substantially anhydrous products in the solid, liquid, and vapor states. Recommendations for "Room Temperature" and "Up to 100° C" include the handling of solutions of these products in otherwise inert organic solvents. Recommendations for higher temperatures include the handling of boiling liquids and vapors, usually at reduced pressures low enough to prevent "cracking."

To the extent that they are water soluble, when aqueous solutions are encountered use the recommendations for "Acetic Acid, up to 10%."

FATTY ACIDS

Fatty Acids	Class A Materials (See definition of Class A materials on p. 748.)			
	At Room Temperature Only (up to 30° C, 85° F)	Up to 100° C (210° F)	Up to 170° C (340° F)	Up to 300° C (570° F)
Abietic Arachidic Capric Caproic Caprylic Cerotic Clupanodonic Eleomargeric Eleostearic Erucic Isovaleric Lauric Linoleic Linolenic Myristic Oleic Palmitic Ricinoleic Stearic	4A, B, C; 6A, B, C; 24A with acids which are liquid at room temperature and with solutions in organic solvents	7A, B; 8A, C, D; 10; 12; 13A, B; 15	23C, D	1A, B, C, E, F*; 3A, B; 8B and 9A, B, C up to 285° C (545° F); 11; 13C; 14A; 19; 23A, B
	Class B Materials (See definition of Class B materials on p. 749. Other materials in common use listed under Class A materials.)			
	At Room Temperature Only (up to 30° C, 85° F)	Up to 100° C (210° F)	Up to 200° C (390° F)	Up to 300° C (570° F)
	2; 18 with non-edible products when sulfuric acid is present	6A, B, C; 7A; 24A up to 75° C (165° F) with liquefied acids and with solutions in organic solvents	2†; 5†; 7B; 8A, C, D	2†; 4A, B, C; 5†; 13A, B; 10; 12; 16A, B, C; 17A, B, C, E, F

TABLE 3. FATTY ACID GROUP — Continued

FATTY ACIDS

	Class C Materials (Materials not ordinarily considered suitable.)	
	At All Temperatures	Above Indicated Temperatures
	18A, B, C with edible products; 21; all Type 25 materials.	All Type 1 materials with hot vapors under vacuum conditions where oxygen and water vapor are absent from the environment, especially where turbulence is great. 6A, B, C; 7A above 100° C (210° F). 7B, 8A, C, D and all Types 16 and 17 materials above 200° C (390° F). 18A, B, C (with non-edible products) above 30° C (85° F). 24A above 75° C (165° F).

General Note. These fatty acids, while related structurally to the acids in the preceding two groups, differ from them by being either almost completely insoluble in water or only slightly soluble. In general, they behave much alike, and the above data should apply equally well to all of them. However, there is some evidence that the fatty acids which are slightly water soluble tend to be more corrosive than the others. When these acids are dealt with, therefore, the above data should be used cautiously.

* The recommndations for Type 1 materials at temperatures over 100° C are based on the provision that there is enough oxygen or water vapor in the environment to maintain the protective oxide film.

† Sometimes used where considerable corrosion can be tolerated.

TABLE 4. HYDROCHLORIC ACID GROUP

Types of Material
(See p. 753 for sub-groups and further details of composition.)

1. Al and Al-Base Alloys
2. Iron and Steel
3. High-Si Cast Irons
4. High-Ni Cast Irons
5. 4 to 10% Chromium Steels
6. Stainless Steels (Martensitic)
7. Stainless Steels (Ferritic)
8. Stainless Steels (Austenitic)
9. Special Fe-Cr-Ni Alloys (Austenitic)
10. Nickel
11. Ni-Cr-Fe Alloy
12. Ni-Cu Alloys (>50% Ni)
13. Ni-Mo-Fe-(Cr) Alloys
14. Ni-Cr-Cu-Mo Alloy
15. Ni-Si Alloy
16. Cu-Ni Alloys (>50% Cu)
17. Cu and Cu-Base Alloys
18. Lead
19. Silver
20. Noble Metals and Tantalum
21. Mg and Mg-Base Alloys
22. Ceramic Materials
23. Carbon and Graphite
24. Plastics
25. Rubber and Synthetic Elastomers

HYDROCHLORIC ACID

% By Weight	Class A Materials (See definition of Class A materials on p. 748.)		
	At Room Temperature Only (up to 30° C, 85° F)	Up to Intermediate Temperature Indicated	Up to Boiling Point at Indicated Concentration
Up to 1%	13A	13C and 24A up to 50° C (120° F); 24E up to 65° C (150° F); 25A, B, C, D, E up to 80° C (180° F)	13B; 19; 20A, B, C; 22A, B, C; 23A, B, C, D; 24B, C, D
1 to 5%	13A, C	24A up to 50° C (120° F); 24E up to 65° C (150° F); 25A, B, C, D, E up to 80° C (175° F)	13B; 19; 20A, B, C; 22A, B, C; 23A, B, C, D; 24B, C, D
5 to 10%	13A, C	13B up to 70° C (160° F); 24A up to 50° C (120° F); 24B, C, D up to 100° C (210° F); 24E up to 65° C (150° F); 25A, B, C, D, E up to 80° C (175° F)	20A, B, C; 22A, B, C; 23A, B, C, D
10 to 20%	13A, C	13B up to 70° C (160° F); 24A up to 50° C (120° F); 24B, C, D up to 100° C (210° F); 24E up to 65° C (150° F); 25A, B, C, D, E up to 80° C (175° F)	20A, B, C; 22A, B, C; 23A, B, C, D
Over 20%	13A, C	13B up to 70° C (160° F); 24A up to 50° C (120° F); 24B, C, D up to 100° C (210° F); 24E up to 65° C (150° F); 25A, B, C, D, E up to 80° C (175° F)	20A, B, C; 22A, B, C; 23A, B, C, D

TABLE 4. HYDROCHLORIC ACID GROUP — *Continued*

HYDROCHLORIC ACID

% By Weight	Class B Materials (See definition of Class B materials on p. 749. Other materials in common use are listed under Class A materials.)		
	At Room Temperature Only (up to 30° C, 85° F)	Up to Intermediate Temperature Indicated	Up to Boiling Point at Indicated Concentration
Up to 1%	1A, B, C*; 2*; 8B; 17A, B, C, D; 18A, B	4A, B, C up to 50° C (120° F); 9A, B, C up to 80° C (175° F); 13C, 14A up to 70° C (160° F); 15 up to 60° C (140° F); 16A, B, C up to 50° C (120° F); 24A above 50° C (120° F) up to 75° C (165° F)	3A, B; 10; 11; 12; 13A; 18C
1 to 5%	8B; 15; 16A, B, C; 17A, B, C, D; 18A, B	3A up to 70° C (160° F); 4A, B, C up to 40° C (105° F); 9A, B, C up to 45° C (115° F); 10 and 11 up to 55° C (130° F); 12 up to 75° C (165° F); 13A, C up to 70° C (160° F); 14A up to 45° C (115° F); 24A above 50° C (120° F) up to 75° C (165° F)	3B; 18C
5 to 10%	9A, B, C; 10; 11; 12; 14A; 15; 16A, B, C; 17A, B, C, D; 18A, B	3A up to 50° C (120° F); 13A up to 70° C (160° F); 13C up to 50° C (120° F); 19 up to 70° C (160° F); 24A above 50° C (120° F) up to 75° C (165° F)	3B; 13B; 18C
10 to 20%	10; 11; 12; 15; 16A, B, C; 17A, B, C, D	3A up to 50° C (120° F); 13A up to 70° C (160° F); 13C up to 50° C (120° F); 18A, B up to 80° C (175° F); 18C up to 100° C (210° F); 19 up to 70° C (160° F); 24A above 50° C (120° F) up to 75° C (165° F)	3B; 13B
Over 20%	10; 13C; 19	13A up to 70° C (160° F); 18A, B up to 30% up to 80° C (175° F); 24A above 50° C (120° F) up to 75° C (165° F)	3B; 13B

* See asterisk footnote on p. 768.

TABLE 4. HYDROCHLORIC ACID GROUP — Continued

HYDROCHLORIC ACID

% By Weight	Class C Materials (Materials not ordinarily considered suitable.)	
	At All Temperatures	Above Indicated Temperatures*
Up to 1%	All Types 1 and 2 materials except as indicated under Class B materials at room temperature; all Types 5, 6, 7, 8A, C, D, 17G, and 21 materials	4A, B, C above 50° C (120° F); 8B above 30° C (85° F); 9A, B above 80° C (175° F); 13C above 70° C (160° F); 14A above 70° C (160° F); 15 above 60° C (140° F); 16A, B, C above 50° C (120° F); 17A, B, C and 18A, B above 30° C (85° F); 24A above 75° C (165° F); 24E above 65° C (150° F); 25A, B, C, D, E above 100° C (210° F)
1 to 5%	All Types 1, 2, 5, 6, 7, 8A, C, D, 15, 17G, and 21 materials	3A above 70° C (160° F); 4A, B, C above 40° C (105° F); 8B above 30° C (85° F); 9A, B above 45° C (115° F); 10 and 11 above 55° C (130° F); 12 above 75° C (165° F); 13A, C above 70° C (160° F); 14A above 45° C (115° F); 16A, B, C, 17A, B, C, D, and 18A, B above 30° C (85° F); 24A above 75° C (165° F); 24E above 65° C (150° F); 25A, B, C, D, E above 100° C (210° F)
5 to 10%	All Types 1, 2, 5, 6, 7, 8, 17G, and 21 materials	3A above 50° C (120° F); 9A, B, 10, 11, and 12 above 30° C (85° F); 13A above 70° C (160° F); 13C above 50° C (120° F); 14A, 15, 16A, B, C, 17A, B, C, D, and 18A, B above 30° C (85° F); 24A above 75° C (165° F); 24E above 65° C (150° F); 25A, B, C, D, E above 100° C (210° F)
10 to 20%	All Types 1, 2, 5, 6, 7, 8, 9, 14, 17G, and 21 materials	3A above 50° C (120° F); 10, 11, and 12 above 30° C (85° F); 13A above 70° C (160° F); 13C above 50° C (120° F); 15, 16A, B, C, and 17A, B, C, D above 30° C (85° F); 18A, B above 80° C (175° F); 18C above 100° C (210° F); 24A above 75° C (165° F); 24E above 65° C (150° F); 25A, B, C, D, E above 100° C (210° F)
Over 20%	All Types 1, 2, 3, 4, 5, 6, 7, 8, 9, 11, 12, 14, 15, 16, 17, 18, and 21 materials	10 above 30° C (85° F); 13A above 70° C (160° F); 13C and 19 above 30° C (85° F); 24A above 75° C (165° F); 24E above 65° C (150° F); 25A, B, C, D, E above 100° C (210° F)

All the above materials, except as indicated below under "Oxidizing Chlorides — Acid Solutions," are Class C with solutions of hydrochloric acid contaminated with free chlorine or with oxidizing chlorides. The presence of such contaminants should always be suspected in solutions which have been or are being handled in iron or copper alloys, as the ferrous chloride and cuprous chloride, formed initially as a result of corrosion of these materials by hydrochloric acid, oxidize readily to ferric chloride and cupric chloride, respectively. Aeration accelerates the corrosion rate manyfold; and *air-saturated* solutions behave much as solutions contaminated with oxidizing chlorides. Same is true of solutions containing free chlorine.

* See asterisk footnote on p. 768.

TABLE 4. HYDROCHLORIC ACID GROUP — *Continued*

HYDROGEN CHLORIDE

The ratings below are based on the assumption that the gas is dry or at temperatures far enough above the dew point to prevent condensation of hydrochloric acid. When water is present and the temperature drops below the dew point, the recommendations for hydrochloric acid apply. There is evidence that in some cases water vapor may increase the corrosion rate above the dew point. At elevated temperatures the presence of oxygen accelerates the corrosion, with resultant lowering of the critical temperature up to which a material may be used.

Class A Materials*,†	Class C Materials*
Type 2 materials up to 200° C (390° F); (See Notes 1 and 2, p. 768); 3A, B up to 150° C (300° F); 4A, B, C up to 200° C (390° F); 10 and 11 up to 500° C (930° F); 12 up to 200° C (390° F); 13A up to 400° C (750° F); 13B, C up to 450° C (840° F); 14 and 15 up to 400° C (750° F); 16A, B, C and 17A, B, C up to 100° C (210° F); 19 up to 200° C (390° F); 20A up to 150° C (300° F); 20B up to 1100° C (2000° F); 20C up to 800° C (1475° F)	All Type 1 materials at all temperatures; cast iron over 315° C (600° F) and carbon steel over 400° C (750°F); (See Notes 1 and 2, p. 768); 3A, B over 150° C (300° F); 4A, B, C over 200° C (390° F); Types 5, 6, 7, 8, and 9 materials; (See Notes 1, 2, and 3, p. 768) 10 and 11 over 500° C (930° F); 12 over 425° C (800° F); 13A over 400° C (750° F); 13B, C over 450° C (840° F); 14 and 15 over 400° C (750° F); 16A, B, C and 17A, B, C over 100° C (210° F); 17D, E, F, G at all temperatures; 18A, B, C over 100° C (210° F); 19 over 425° C (800° F); 20A over 150° C (300° F); 20B over 1100° C (2000° F); 20C over 800° C (1475° F)

NON-OXIDIZING CHLORIDES — ACID SOLUTIONS

Aluminum chloride (Anhydrous) (See below)	May be encountered as molten salt or as anhydrous vapor; or as mixtures with hydrocarbons and other similarly inert fluids. Recommendations are the same as for anhydrous hydrogen chloride.
Aluminum chloride Ammonium chloride Antimony chloride Nickel chloride Stannous chloride Zinc chloride	Solutions of these salts in water and in hydrochloric acid behave as do solutions of hydrochloric acid alone. Recommendations for 1 to 5% hydrochloric acid usually apply. 8B and 9A, B, C, which are Class A or B with some concentrations of hydrochloric acid, are Class C with electroplating solutions which cannot tolerate contamination with chromium. These alloys are subject to stress corrosion cracking with hot, concentrated solutions.
Cuprous chloride Ferrous chloride	Solutions of these salts should be expected to behave like those above. Practically, however, they oxidize so readily to the corresponding cupric and ferric salts that, unless protected by an inert atmosphere or reducing conditions, they should be considered as oxidizing salts. (See "Oxidizing Chlorides.")

* See asterisk footnote on p. 768.
† See dagger footnote on p. 768.

768   CHEMICAL RESISTANT MATERIALS

TABLE 4. HYDROCHLORIC ACID GROUP — Continued

OXIDIZING CHLORIDES — ACID SOLUTIONS

Cupric chloride Ferric chloride Mercuric chloride Stannic chloride	Solutions of these salts in water and in hydrochloric acid do not behave as do solutions of hydrochloric acid alone of equivalent acidity. 13c is rated as Class A with cupric chloride up to 10% at room temperature and with the others up to 10% up to 70° C (160° F). 20A is Class A with all concentrations and temperatures. 3B is Class B up to 50° C (120° F). (Differential aeration promotes corrosion. It is important to drain or flush pumps and valves of this alloy when not in use.) All other metallic materials are Class C. Type 22 and 23 materials are Class A up to boiling point. 23 is Class A up to 80° C (175° F), Class C over 100° C (210° F).

* Some of the materials listed are apparently entitled to a Class B rating at temperatures substantially higher than the limiting temperatures for Class A ratings. In some situations, however, particularly under oxidizing conditions, the rate of attack may increase so rapidly with temperature that no attempt has been made to assign Class B ratings. Instead, the higher temperatures indicate the approximate values above which a material becomes Class C. It should not be assumed that these higher temperatures are safe at all times and with all types of equipment.

† Note that some of these recommended maximum temperatures are slightly lower than values given in Table 6, p. 682. Variations in the order of 25° to 50° C are to be expected. EDITOR.

Note 1. When exposed alternately to anhydrous hydrogen chloride and to air, the ferrous chloride formed initially on the surface of iron alloys oxidizes to ferric chloride. This ferric chloride then absorbs moisture from the air, followed by increased corrosion of the metal beneath.

Note 2. Carbon steels and stainless steels in contact with hydrogen chloride are subject to possible stress corrosion unless stress-relieved after severe cold working or welding.

Note 3. With respect to Notes 1 and 2 above, Types 5, 6, 7, 8, and 9 materials are no better than Type 2 materials and may be inferior; hence they have been given a Class C rating. Under favorable conditions Type 8 materials are probably not Class C at any temperature below 480° C (900° F).

CHEMICAL RESISTANT MATERIALS

Table 5. HYDROFLUORIC ACID GROUP*

Types of Material
(See p. 753 for sub-groups and further details of composition.)

1. Al and Al-Base Alloys
2. Iron and Steel
3. High-Si Cast Irons
4. High-Ni Cast Irons
5. 4 to 10% Chromium Steels
6. Stainless Steels (Martensitic)
7. Stainless Steels (Ferritic)
8. Stainless Steels (Austenitic)
9. Special Fe-Cr-Ni Alloys (Austenitic)
10. Nickel
11. Ni-Cr-Fe Alloy
12. Ni-Cu Alloys (>50% Ni)
13. Ni-Mo-Fe-(Cr) Alloys
14. Ni-Cr-Cu-Mo Alloy
15. Ni-Si Alloy
16. Cu-Ni Alloys (>50% Cu)
17. Cu and Cu-Base Alloys
18. Lead
19. Silver
20. Noble Metals and Tantalum
21. Mg and Mg-Base Alloys
22. Ceramic Materials
23. Carbon and Graphite
24. Plastics
25. Rubber and Synthetic Elastomers

Hydrofluoric Acid

% By Weight	Class A Materials (See definition of Class A materials on p. 748.)		
	At Room Temperature Only (up to 30° C, 85° F)	Up to Intermediate Temperature Indicated	Up to Boiling Point at Indicated Concentration
Up to 40%	10; 13B, C; 24A (See Notes 1 and 2)†	(See Note 1)	12; 19; 20B, C; 23A, B, C, D (See Note 1)
40 to 50% (See Note 1)	10; 24A (See Note 2)	13A up to 55° C (130° F)	12; 19 up to 145° C (295° F); 20B, C; 23A, B, C, D
60%	10; 13A, B (See Note 2)		12; 19; 20B, C; 23A, B, C, D
70%	(See Note 2)		12; 19; 20B, C; 23A, B, C, D
80%	(See Note 2)		19; 20B, C; 23A, B, C, D
80 to 100%	13A, B (See Note 2)		12 up to 115° C (240° F); 19; 20B, C; 23A, B, C, D
Anhyd. HF‡; Plant acid§	(See Note 2)		12 up to 150° C (300° F); 19 (anhydrous only); 20B, C; 21 above 55° C (130° F) and up to 70° C (160° F) (See Note 2); 23A, B, C, D

* Incomplete data indicate that Types 9 and 14 materials can be used with both anhydrous and aqueous solutions of HF. The limiting temperatures and concentrations have not been properly defined; hence these materials are not classified.

† See Notes on p. 771.

‡ In practice, anhydrous hydrogen fluoride is referred to as "A.H.F."

§ "Plant Acid" — A.H.F. plus hydrocarbons. May contain up to several per cent of water.

TABLE 5. HYDROFLUORIC ACID GROUP — *Continued*

HYDROFLUORIC ACID

% By Weight	Class B Materials (See definition of Class B materials on p. 749. Other materials in common use are listed under Class A materials.)		
	At Room Temperature Only (up to 30° C, 85° F)	Up to Intermediate Temperature Indicated	Up to Boiling Point at Indicated Concentration
Up to 40%	4A, B, C (up to 10%); 21; 25A-2 (See Note 1)*	11 (up to 10%) up to 50° C (120° F); 24A up to 55° C (130° F); 25A-1, C-1, D, E up to 65° C (150° F) (See Note 1)	(See Note 1)
40 to 50% (See Note 1)	25A-2	21 up to 85° C (185° F); 24A up to 55° C (130° F); 25A-1, C-1, D, E up to 65° C (150° F)	10; 12; 16A, B, C; 17A, D, E, F (all up to 145° C, 295° F); 18A, B (See Note 3)
60%	2 (See Note 4); 21	17A, D, E, F (See Note 1); 18A, B (See Note 5); 21 up to 85° C (185° F); 25D, E up to 65° C (150° F)	12; 16A, B, C (See Note 3)
70%	2 (See Note 4)	16A, B, C; 17A, D, E, F (See Notes 1 and 3)	12 (See Note 3); 21 up to 85° C (185° F)
80%	2 (See Note 4)	16A, B, C; 17A, D, E, F (See Notes 1 and 3)	12 (See Note 3); 21 up to 85° C (185° F)
80 to 100%	2 (gray cast iron); 4A, B, C (See Note 4)	2 (carbon steel) (See Note 4); 16A, B, C; 17A, D, E, F (See Notes 1 and 3)	12 (See Note 3); 21 up to 85° C (185° F)
Anhyd. HF; Plant acid	1A for special applications (packing, gaskets, etc.)		12; 16A, B, C (See Note 3); 21 up to 55° C (130° F)

% By Weight	Class C Materials (Materials not ordinarily considered suitable.)	
	At All Temperatures	Above Indicated Temperatures
Up to 60%	All Types 1, 2, 4, 5, 6, 7, 8, 17G, 20A, 25A, B, C; 4A, B, C (10 to 60%); Types 3, 15, 17C, 22 (See Note 6); Type 13 (See Note 7)	4A, B, C (up to 10%); 21 above 30° C (85° F); 24A above 55° C (130° F); 25A-2 above 30° C (85° F); 25A-1, C-1, D, E above 65° C (150° F)
60 to 80%	All Types 1, 5, 6, 7, 8, 17G, 20A, 25; Types 3, 15, 17C, 22 (See Note 6)	2, 21 above 30° C (85° F); 17A, D, E, F (See Note 1); 18A, B (See Note 5)

* See Notes on p. 771.

TABLE 5. HYDROFLUORIC ACID GROUP — *Continued*

HYDROFLUORIC ACID

% By Weight	Class C Materials — *Continued* (Materials not ordinarily considered suitable.)	
	At All Temperatures	Above Indicated Temperatures
80 to 100%	All Types 1, 5, 6, 7, 8, 17G, 18, 20A, 25; Types 3, 15, 17C, 22 (See Note 6)	2, 17A, D, E, F; 21 above 30° C (85° F); 17A, D, E, F with high-velocity streams
Anhyd. HF; Plant acid	Types 5, 6, 7, 8 (See Note 8); 1A for all containing equipment; Types 17 and 19 with plant acid; 2 (gray cast iron); 4A, B, C	2 (carbon steel) above 55° C (130° F)

Note 1. Corrosion of materials by hydrofluoric acid has been studied more extensively with 40 to 50% acid (including the constant boiling mixture) than with other concentrations. Scattered results indicate that the data for "40 to 50%" apply equally well to "up to 40%."

Note 2. Laboratory corrosion data supplemented by limited field experience indicate that magnesium is entitled to a Class A rating with A.H.F. at elevated temperatures and with all aqueous solutions at room temperature. In the absence of thorough testing in the field to determine the effect of "incidental conditions," however, the rating for this material is limited to Class B except as indicated.

Note 3. The corrosion rates for Types 10, 12, 16, and 17 materials increase rapidly with aeration, and with air saturated solutions they may change from Class B to Class C. Type 17 materials are sensitive to velocity as the concentration goes over 80%. With anhydrous HF they are definitely Class C with high velocity streams. Types 12 and 16 materials are not similarly affected.

Note 4. Gray cast iron and low to medium carbon steel have been used successfully with cold 60% hydrofluoric acid, but usually are not recommended for use with concentrations below 80%. Over 80% both are acceptable at room temperature, but at elevated temperatures cast iron, including Type 4 materials, is subject to graphitic corrosion. As the concentration approaches 100%, the limiting temperature for steel approaches that reported for A.H.F., and the tendency towards graphitic corrosion of cast iron becomes apparent at room temperature.

Interstate Commerce Commission Regulations permit transportation of greater than 60% HF in steel containers previously passivated. (See p. 133.)

Note 5. Lead has been used with 60% acid up to the boiling point, but the limiting concentration is about 65%, for which the limiting temperature drops. Use cautiously at temperatures above room temperature.

Note 6. High-silicon alloys are not recommended for use in any phase of HF service, even though reported corrosion rates indicate that they can be used. The same is true for ceramic materials. All these materials are therefore rated Class C.

Note 7. Incomplete data indicate that Type 13 materials resist some concentrations of HF well enough to justify a Class A or a Class B rating, but that they are susceptible to intergranular corrosion with other concentrations. The limiting conditions have not been properly defined; hence these materials are not rated for all concentrations. 13A and 13B are definitely Class C with 40 to 50% acid at elevated temperatures.

Note 8. These materials are not superior to carbon steel in any phase of HF service, and may be inferior. In particular, Types 6 and 7 are velocity sensitive; and, when used in combination with carbon steel in contact with aqueous solutions, they may become anodic to carbon steel. They are rated therefore Class C.

TABLE 6. NITRIC ACID GROUP
Types of Material
(See p. 753 for sub-groups and further details of composition.)

1. Al and Al-Base Alloys
2. Iron and Steel
3. High-Si Cast Irons
4. High-Ni Cast Irons
5. 4 to 10% Chromium Steels
6. Stainless Steels (Martensitic)
7. Stainless Steels (Ferritic)
8. Stainless Steels (Austenitic)
9. Special Fe-Cr-Ni Alloys (Austenitic)
10. Nickel
11. Ni-Cr-Fe Alloy
12. Ni-Cu Alloys (>50% Ni)
13. Ni-Mo-Fe-(Cr) Alloys
14. Ni-Cr-Cu-Mo Alloy
15. Ni-Si Alloy
16. Cu-Ni Alloys (>50% Cu)
17. Cu and Cu-Base Alloys
18. Lead
19. Silver
20. Noble Metals and Tantalum
21. Mg and Mg-Base Alloys
22. Ceramic Material
23. Carbon and Graphite
24. Plastics
25. Rubber and Synthetic Elastomers

NITRIC ACID

% By Weight	Class A Materials (See definition of Class A materials on p. 748.)		
	At Room Temperature Only (up to 30° C, 85° F)	Up to Intermediate Temperature Indicated	Up to Boiling Point at Indicated Concentration
Up to 0.5%	6A, B; 8C, D if over 0.10% carbon; 9C; 11	23A, B, C, D up to 85° C (185° F); 24A, B, C, D up to 50° C (120° F)	3A, B; 7A, B; 8A, B, C, D; 9A, B; 13C; 14A; 20A, B, C; 22A, B, C
0.5 to 20%	6A; 11; 24C (5 to 20%); 24D (5 to 10%)	13C up to 65° C (150° F); 23A, B (0.5 to 10%) up to 85° C (185° F); 24A up to 50° C (120° F); 24B, C, D (0.5 to 5%) up to 50° C (120° F)	3A, B; 7A, B; 8A, B, C, D; 9A, B; 14A; 20A, B, C; 22A, B, C
20 to 40%	11	7A up to 70° C (160° F); 13C up to 65° C (150° F); 24A up to 50° C (120° F)	3A, B; 7B; 8A, B, C, D; 9A, B; 14A; 20A, B, C; 22A, B, C
40 to 70%	11; 13C; 24A above 50%	7A up to 60° C (140° F); 7B, 8A, B, 9A, B up to 80° C (175° F); 14A up to 85° C (185° F); 24A (40 to 50%) up to 50° C (120° F)	3A, B; 8C, D; 20A, B, C; 22A, B, C
70 to 80%	7A	7B, 8A, B, C, D and 9A, B up to 50° C (120° F)	3A, B; 20A, B, C; 22A, B, C
80 to 95%	1A, B, C, E, F; 7A	7B, 8A, B, C, D, 9A, B up to 50° C (120° F)	3A, B; 20A, B, C; 22A, B, C
Over 95%	1A, B, C, E, F		3A, B; 20A, B, C; 22A, B, C

CHEMICAL RESISTANT MATERIALS

TABLE 6. NITRIC ACID GROUP — *Continued*

NITRIC ACID

% By Weight	Class B Materials (See definition of Class B materials on p. 749. Other materials in common use are listed under Class A materials.)		
	At Room Temperature Only (up to 30° C, 85° F)	Up to Intermediate Temperature Indicated	Up to Boiling Point at Indicated Concentration
Up to 0.5%	1A, B, C, D, F	6C up to 50° C (120° F); 24B, D up to 100° C (210° F); 24E up to 65° C (150° F); 25A-2 up to 65° C (150° F)	6A, B
0.5 to 20%	8C, D if over 0.10% carbon; 9C; 25A-2	6A, B up to 70° C (160° F); 6C up to 50° C (120° F); 23C, D (0.5 to 10%) up to 85° C (185° F); 24E up to 50° C (120° F)	
20 to 40%	6A, B; 8C, D if over 0.10% carbon; 9C; 24E		
40 to 70%	6A, B	7A up to 90° C (195° F)	
70 to 80%	1A, B, C, E, F	7A up to 50° C (120° F); 7B, 8A, B, C, D and 9A, B up to 70° C (160° F)	
80 to 95%		1A, B, C, E, F (80 to 90%) up to 60° C (140° F); 7A up to 50° C (120° F); 7B, 8A, B, C, D, and 9A, B up to 70° C (160° F)	1A, B, C, E, F (90 to 95%)
Over 95%	7A, B; 8A, B; 9A, B	8C, D up to 50° C (120° F)	1A, B, C, E, F

% By Weight	Class C Materials (Materials not ordinarily considered suitable.)	
	At All Temperatures	Above Indicated Temperatures
Up to 0.5%	All Types 1G; 2; 4; 5; 10; 12; 13A, B; 15; 16; 17; 18; 19	1A, B, C, D, F above 30° C (85° F); 6C above 50° C (120° F); 8C, D if over 0.10% carbon and 9C above 30° C (85° F); 23A, B, C, D above 85° C (185° F)
0.5 to 20%	All Types 1, 2, 4, 5, 10, 12, 13A, B, 15, 16, 17, 18, 19, 21, 23 (above 10%), 25A-1, 25B-1, 25C-1, and 25E, F	6A, B above 70° C (160° F); 6C above 50° C (120° F); 8C, D if over 0.10% carbon and 9C above 30° C (85° F); 23A, B, C, D (0.5 to 10%) above 85° C (185° F); 24A, B, C, D, E above 50° C (120° F)
20 to 40%	All Types 1; 2; 4; 5; 6C; 10; 12; 13A, B; 15; 16; 17; 18; 19; 21; 23; 24B, C, D and 25 materials	6A, B above 70° C (160° F); 8C, D if over 0.10% carbon and 9C above 30° C (85° F); 24A above 50° C (120° F); 24E above 30° C (85° F)

CHEMICAL RESISTANT MATERIALS

TABLE 6. NITRIC ACID GROUP — Continued

% By Weight	Nitric Acid	
	Class C Materials — Continued (Materials not ordinarily considered suitable.)	
	At All Temperatures	Above Indicated Temperatures
40 to 70%	All Types 1, 2, 4, 5, 6c, 9c, 10, 12, 13A, B, 15, 16, 17, 18, 19, 21, 23, 24B, C, D, E and 25 materials; 8C, D if over 0.10% carbon	6A, B above 70° C (160° F); 7A above 90° C (195° F); 24A above 30° C (85° F)
70 to 80%	All Types 1G, 2, 4, 5, 6, 9c, 10, 12, 13A, B, 15, 16, 17, 18, 19, 21, 23, 24B, C, D, E and 25 materials; 8C, D if over 0.10% carbon	1A, B, C, E, F above 30° C (85° F); 7A above 90° C (195° F)
80 to 95%	All Types 1G, 2, 4, 5, 6, 9c, 10, 12, 13A, B, 15, 16, 17, 18, 19, 21, 23, 24B, C, D, E and 25 materials; 8C, D if over 0.10% carbon	1A, B, C, E, F (80 to 90%) above 60° C (140° F); 7A above 50° C (120° F); 7B, 8A, B, C, D and 9A, B above 70° C (160° F)
Over 95%	All Types 1G, 2, 4, 5, 6, 9c, 10, 12, 13A, B, 15, 16, 17, 18, 19, 21, 23, 24B, C, D and 25 materials; 8C, D if over 0.10% carbon	7A, B, 8A, B and 9A, B above 30° C (85° F); 8C, D above 50° C (120° F)

MIXED ACIDS — NITRATING ACIDS

Mixtures of nitric and sulfuric acid, with or without water. Ratios and concentrations may vary over the complete range of HNO_3—H_2SO_4—H_2O, from water-free mixtures of the strong acids to dilute mixtures containing approximately 0.5% total acidity. The table below is based on scattered information about specific combinations and is obviously incomplete.* It points out the general fact that the corrosivity changes as the ratio or total acidity changes, but it does not take into consideration the fact that acidity may change as a result of absorption of water from the air or the ratio change as a result of chemical reactions involving one or both acids.

	Composition of Mixed Acids (See Note 1, p. 775)		Class A and/or B Materials	Class C Materials
	% HNO_3	% H_2SO_4	See Note 2, p. 775, for materials which are Class A or B for all combinations	See Note 3, p. 775, for materials which are Class C with all combinations
1	50	50	2; 8A, B, C, D; 9A, B	1A, B, C, D, E, F, G; 2 (gray cast iron); 6; 7
2	45	53	2; 8A, B, C, D; 9A, B	1A, B, C, D, E, F, G; 2 (gray cast iron); 6; 7
3	40	58–60	2; 8A, B, C, D; 9A, B	1A, B, C, D, E, F, G; 2 (gray cast iron); 6; 7
4	30	56	2; 8B; 9A, B	1A, B, C, D, E, F, G; 2 (gray cast iron); 6; 7
5	27	73	2	1A, B, C, D, E, F, G; 2 (gray cast iron); 6; 7
6	24	68	2; 8A, B, C, D; 9A, B	1A, B, C, D, E, F, G; 2 (gray cast iron); 6; 7

* Particularly with respect to limiting temperatures. (See Note 3, p. 775.)

TABLE 6. NITRIC ACID GROUP — *Continued*

MIXED ACIDS — NITRATING ACIDS — *Continued*

	Composition of Mixed Acids (See Note 1)		Class A and/or B Materials	Class C Materials
	% HNO_3	% H_2SO_4	See Note 2, below, for materials which are Class A or B for all combinations	See Note 3, below, for materials which are Class C with all combinations
7	2	94	2; 8A, B, C, D; 9A, B	1A, B, C, D, E, F, G; 2 (gray cast iron); 6; 7
8	2	98	2; 7B; 8A, B, C, D; 9A, B	1A, B, C, D, E, F, G; 2 (gray cast iron); 6; 7
9	9	75	2; 8A, B, C, D; 9A, B	1A, B, C, D, E, F, G; 2 (gray cast iron); 6; 7
10	33⅓	33⅓	7A, B; 8A, B, C, D; 9A, B	1A, B, C, D, E, F, G; 2; 6
11	25	50	9B	1A, B, C, D, E, F, G; 2; 6; 7; 8A, C, D
12	20	60	9B	1A, B, C, D, E, F, G; 2; 6; 7; 8A, C, D
13	15	35	7A, B; 8A, B, C, D; 9A, B	1A, B, C, D, E, F, G; 2; 6
14	1	22	9B	1A, B, C, D, E, F, G; 2; 6; 7; 8
15	1	50	9B	1A, B, C, D, E, F, G; 2; 6; 7; 8
16	0.5	50	9B	1A, B, C, D, E, F, G; 2; 6; 7; 8
17	6	8	8A, B, C, D; 9A, B	1A, B, C, D, E, F, G; 2; 6; 7
18	3	7	8A, B, C, D; 9A, B	1A, B, C, D, E, F, G; 2; 6; 7
19	2	3.5	8A, B, C, D; 9A, B	1A, B, C, D, E, F, G; 2; 6; 7
20	2	5	8B; 9A, B	1A, B, C, D, E, F, G; 2; 6; 7
21	2	8	9B	1A, B, C, D, E, F, G; 2; 6; 7
22	1	9	8A, B, C, D; 9A, B	1A, B, C, D, E, F, G; 2; 6; 7
23	<1	<1	1A, B, C, E, F; 7B; 9B	
24	85	2	1A, B, C, E, F; 9A, B	

Note 1. In lines 1 to 9, inclusive, in the above table are listed mixtures which contain more than 50% sulfuric acid and less than 20% water, with which Types, 2, 6, 7, 8, and 9 materials are generally useful. Like strong sulfuric acid, these mixtures absorb water from the air or from chemicals with which they react. Equipment exposed to both the original strong acids and to the resulting weak acids should be made of material resistant to all strengths that will be encountered.

Note 2. The ratings in the above table are based in part on laboratory data which have not in all cases been confirmed in the field. For that reason, no attempt has been made to assign limiting temperatures or draw a definite line between Class A and Class B materials. Type 3 materials are Class B with all combinations up to the boiling point, and may be definitely Class A with some combinations. Materials 20A, B, C and 22A, B, C are Class A with all combinations. Among the other materials rated Class A or B, Types 6, 7, and 8 should be used with considerable caution with strong acids because of the abruptness with which they change to Class C on dilution of the mixed acids with water; and they should also be used cautiously with mixtures containing more than 20% water, especially at elevated temperatures.

Note 3. All Type 1 materials are Class C with all combinations except as indicated on the last two lines. Type 2 materials are Class C with all combinations except as indicated on the first nine lines; and with these, gray cast iron should probably be Class C because of a tendency toward intergranular oxidation and consequent breakage. Types 4, 5, 10, 12, 13A, B, 16, 17, 18, and 19 materials are Class C with all combinations.

TABLE 7. SULFURIC ACID GROUP

Types of Material
(See p. 753 for sub-groups and further details of composition.)

1. Al and Al-Base Alloys
2. Iron and Steel
3. High-Si Cast Irons
4. High-Ni Cast Irons
5. 4 to 10% Chromium Steels
6. Stainless Steels (Martensitic)
7. Stainless Steels (Ferritic)
8. Stainless Steels (Austenitic)
9. Special Fe-Cr-Ni Alloys (Austenitic)
10. Nickel
11. Ni-Cr-Fe Alloy
12. Ni-Cu Alloys (>50% Ni)
13. Ni-Mo-Fe-(Cr) Alloys
14. Ni-Cr-Cu-Mo Alloy
15. Ni-Si Alloy
16. Cu-Ni Alloys (>50% Cu)
17. Cu and Cu-Base Alloys
18. Lead
19. Silver
20. Noble Metals and Tantalum
21. Mg and Mg-Base Alloys
22. Ceramic Materials
23. Carbon and Graphite
24. Plastics
25. Rubber and Synthetic Elastomers

SULFURIC ACID

% By Weight	Class A Materials (See definition of Class A Materials on p. 748.)		
	At Room Temperature Only (up to 30° C, 85° F)	Up to Intermediate Temperature Indicated	Up to Boiling Point at Indicated Concentration
Up to 0.25%	8A, C, D; 11	3A; 13C; 15 up to 80° C (175° F); 8B up to 50° C (120° F); 18C up to 95° C (205° F); 24A, C up to 50° C (120° F); 24B, D up to 100° C (210° F); 24E and 25A-2, C-2 up to 65° C (150° F)	3B; 9A, B, C; 13A, B; 14A; 18A, B; 19; 20A, B, C; 22A, B, C; 23A, B, C, D
0.25 to 5%		3; 9A, B, C; 13C; 15 up to 80° C (175° F); 18C up to 95° C (205° F); 24A, C up to 50° C (120° F); 24B, D up to 100° C (210° F); 24E and 25A-2, C-2 up to 65° C (150° F)	3B; 13A, B; 14A; 18A, B; 19; 20A, B, C; 22A, B, C; 23A, B, C, D
5 to 10%	13A	3A; 13C; 15 up to 80° C (175° F); 9A, B, C up to 50° C (120° F); 18C up to 95° C (205° F); 24A, C up to 50° C (120° F); 24B, D up to 100° C (210° F); 24E and 25A-1, A-2, B-1, B-2, C-1, C-2, D, and E up to 65° C (150° F)	3B; 13B; 14A; 18A, B; 19; 20A, B, C; 22A, B, C; 23A, C, D
10 to 25%	13A	3A; 13C; 15 up to 80° C (175° F); 9A, B, C up to 50° C (120° F); 18C up to 95° C (205° F); 24A, C up to 50° C (120° F); 24B, D up to 100° C (210° F); 24E and 25A-1, A-2, B-1, B-2, C-1, C-2, D, and E up to 65° C (150° F)	3B; 13B; 14A; 18A, B; 19; 20A, B, C; 22A, B, C; 23A, C, D

CHEMICAL RESISTANT MATERIALS

TABLE 7. SULFURIC ACID GROUP — *Continued*

SULFURIC ACID

% By Weight	Class A Materials — *Continued* (See definition of Class A Materials on p. 748.)		
	At Room Temperature Only (up to 30° C, 85° F)	Up to Intermediate Temperature Indicated	Up to Boiling Point at Indicated Concentration
25 to 50%	13A	3A; 14A (35 to 50%); 15 up to 80° C (175° F); 9A, B, C up to 50° C (120° F); 13C up to 65° C (150° F); 18C (25 to 30%) up to 95° C (205° F); 19 up to 100° C (210° F); 23C, D up to 135° C (275° F); 24A, C, D up to 50° C (120° F); 24B up to 100° C (210° F); 24E and 25A–1, A–2, B–1, B–2, C–1, C–2, D, and E up to 65° C (150° F)	3B; 13B; 14A (25 to 35%); 18A, B; 20A, B, C; 22A, B, C; 23A
50 to 60%	9A, B, C; 13A, C; 24D, E; 25A–1, A–2, B–1, B–2, C–1, C–2, D, E	3A; 14A up to 80° C (175° F); 13B up to 95° C (205° F); 18A, B up to 120° C (250° F); 19 up to 70° C (160° F); 23C, D up to 135° C (275° F); 24A, B, C up to 50° C (120° F)	3B; 15; 20A, B, C; 22A, B, C; 23A
60 to 75%	9A, B, C; 13A, C; 24D, E	13B up to 80° C (175° F); 14A up to 60° C (140° F); 18A, B up to 120° C (250° F); 23C, D up to 135° C (275° F); 24A, B, C up to 50° C (120° F)	3A, B; 15; 20A, B, C; 22A, B, C; 23A
75 to 95%	9A, B, C; 13A; 24B	13B up to 80° C (175° F); 13C up to 65° C (150° F); 14A up to 60° C (140° F); 18A, B (75 to 80%) up to 120° C (250° F); 23C, D up to 80° C (175° F)	3A, B; 15; 22A, B, C; 23A
95 to 100%	8A, B, C, D; 13A, B, C with 96%	9A, B, C up to 50° C (120° F); 14A up to 60° C (140° F); 15 up to 150° C (300° F)	3A, B; 22A, B, C; 23A
Over 100%		8A, B, C, D; 9A, B, C; 14A up to 60° C (140° F); 23A up to 115% up to 70° C (160° F)	

CHEMICAL RESISTANT MATERIALS

TABLE 7. SULFURIC ACID GROUP — *Continued*

SULFURIC ACID

% By Weight	Class B Materials (See definition of Class B Materials on p. 749. Other materials in common use are listed under Class A Materials.)		
	At Room Temperature Only (up to 30° C, 85° F)	Up to Intermediate Temperature Indicated	Up to Boiling Point at Indicated Concentration
Up to 0.25%	1A, B, C, E, F	4A, B, C; 8A, C, D up to 50° C (120° F); (See General Note, p. 782, applying to Types 10, 12, 16 and 17 materials); 24E and 25A-2, C-2 up to 80° C (175° F)	3A; 8B; 11; 13C; 15; 24B, C, D
0.25 to 5%	1A, B, C, E, F; 4A, B, C, D; 8A, C, D; 11	8B up to 50° C (120° F); 24E and 25A-2, C-2 up to 80° C (175° F); (See General Note applying to Types 10, 12, 16 and 17 materials, p. 782)	3A; 9A, B, C; 13C; 15; 24B, C, D
5 to 10%	4A, B, C; 8B; 11	9A, B, C up to 80° C (175° F); (See General Note applying to Types 10, 12, 16, and 17 materials, p. 782); 24E and Type 25 materials up to 80° C (175° F)	3A; 13A, C; 15; 23B; 24B, C, D
10 to 25%	4A, B, C	9A, B, C up to 80° C (175° F); 24B, C, D up to 125° C (260° F); 24E and Type 25 materials up to 85° C (175° F); (See General Note applying to Types 10, 12, 16, and 17 materials, p. 782)	13A, C; 15; 23B
25 to 50%		9A, B, C up to 60° C (140° F); 13A up to 70° C (160° F); 13C up to 95° C (205° F); (See General Note applying to Types 10, 12, 16, and 17 materials, p. 782); 24B, C up to 125° C (260° F); 24E and Type 25 materials up to 80° C (175° F)	14A (35 to 50%); 15

TABLE 7. SULFURIC ACID GROUP — *Continued*

SULFURIC ACID

% By Weight	Class B Materials — *Continued* (See definition of Class B Materials on p. 749. Other materials in common use are listed under Class A Materials.)		
	At Room Temperature Only (up to 30° C, 85° F)	Up to Intermediate Temperature Indicated	Up to Boiling Point at Indicated Concentration
50 to 60%		9A, B, C up to 60° C (140° F); 13A, C up to 70° C (160° F); 13B up to 120° C (250° F); 14A up to 100° C (210° F); (See General Note applying to Types 10, 12, 16, and 17 materials, p. 782); 24B, C up to 100° C (210° F); 24E and Type 25 materials up to 80° C (175° F)	15
60 to 75%	2 above 65%; 4A, B, C	9A, B, C up to 60° C (140° F); 13A, C up to 70° C (160° F); 13B up to 120° C (250° F); 14A up to 90° C (195° F); (See General Note applying to Types 10, 12, 16, and 17 materials, p. 782); 24B, C up to 100° C (210° F); 24E up to 80° C (175° F)	
75 to 95%	8A, C, D; 11 with 93%; 18A, B (80 to 95%); 24A	2 (steel) up to 60° C (140° F)*; 2 (gray cast iron) and 4A, B, C up to 70° C (160° F); 8B; 9A, C up to 50° C (120° F); 9B up to 80° C (175° F); 13A, B, C up to 70° C (160° F); 14A up to 90° C (195° F); 18A, B (75 to 80%) up to 200° C (390° F)	
95 to 100%		2 (steel) up to 80° C (175° F); 2 (gray cast iron) and 4A, B, C up to 100° C (210° F); 8A, B, C, D; 9A, C up to 80° C (175° F); 9B up to 100° C (210° F); 13A, B, C (with 96%) up to 70° C (160° F); 14A up to 90° C (195° F)	
Over 100%	1A, B, C, E, F	2 (steel only) up to 160° C (320° F) and only with acid of strength over 101%; 8A, B, C, D; 9A, B, C up to 200° C (390° F); 14A up to 90° C (195° F)	

* See asterisk footnote on p. 782.

TABLE 7. SULFURIC ACID GROUP — Continued

SULFURIC ACID

% By Weight	Class C Materials Materials not ordinarily considered suitable.	
	At All Temperatures	Above Indicated Temperatures
Up to 0.25%	All Type 1 materials except as indicated under Class B materials at room temperature; all types 2, 5, 6, 7, 17G, 21 and all soft varieties of Type 25 materials	4A, B, C; 8A, C, D above 50° C (120° F); 24A above 50° C (120° F); 24E and 25A-2, B-2, and C-2 above 80° C (175° F)
0.25 to 5%	All Types 1, 2, 5, 6, 7, 17G, and 21 materials; all soft varieties of Type 25 materials	4A, B, C; 8A, C, D above 30° C (85° F); 8B above 50° C (120° F); 24A above 50° C (120° F); 24E and 25A-2, B-2, and C-2 above 80° C (175° F)
5 to 10%	All Types 1; 2; 5; 6; 7; 8A, C, D; 17G; and 21 materials	4A, B, C; 8B above 30° C (85° F); 9A, B, C above 80° C (175° F); 24A above 50° C (120° F); 24E and Type 25 materials above 80° C (175° F)
10 to 25%	All Types 1, 2, 5, 6, 7, 8, 17G, and 21 materials	4A, B, C above 30° C (85° F); 9A, B, C above 80° C (175° F); 11 at b.p.; 24A above 50° C (120° F); 24E and Type 25 materials above 80° C (175° F)
25 to 50%	All Types 1, 2, 4, 5, 6, 7, 8, 17G, 21 and 23B materials	9A, B, C above 60° C (140° F); 13A, C above 95° C (205° F); 13B above 120° C (250° F); 23C, D above 135° C (275° F); 24A above 50° C (120° F); 24D above 100° C (210° F); 24E and Type 25 materials above 80° C (175° F)
50 to 60%	All Types 1, 2, 4, 5, 6, 7, 8, 17G, 21, and 23B materials	9A, B, C above 60° C (140° F); 13A, C above 95° C (205° F); 23C, D above 135° C (275° F); 24A above 50° C (120° F); 24B, C above 100° C (210° F); 24D above 30° C (85° F); 24E and Type 25 materials above 80° C (175° F)
60 to 75%	All Types 1, 5, 6, 7, 8, 17G, 21, 23B, and 25 materials; Types 2 and 4 materials below 65%	2; 4A, B, C above 30° C (85° F) (65 to 75%); 9A, B, C above 60° C (140° F); 13A, C above 95° C (205° F); 13B at b.p.; 14A above 90° C (195° F); 23C, D above 135° C (275° F); 24A above 50° C (120° F); 24B, C above 100° C (210° F); 24D above 30° C (85° F); 24E above 80° C (175° F)
75 to 95%	All Types 1; 5†; 6†; 7†; 10; 12; 16; 17; 21; 23B; 24C, D, E, and 25 materials	2 (steel) above 60° C (140° F); 2 (gray cast iron) and 4A, B, C above 70° C (160° F); 8B; 9A, C above 50° C (120° F); 9B above 80° C (175° F); 13A, B, C above 95° C (205° F); 14A above 90° C (195° F); 23C, D above 80° C (175° F); 24A, B above 30° C (85° F)

† See dagger footnote on p. 782.

CHEMICAL RESISTANT MATERIALS

TABLE 7. SULFURIC ACID GROUP — *Continued*

SULFURIC ACID — *Continued*

% By Weight	Class C Materials — *Continued* Materials not ordinarily considered suitable.	
	At All Temperatures	Above Indicated Temperatures
95 to 100%	All Types 1; 5†; 6†; 7†; 10; 12; 16; 17; 21; 23B, C, D; 24; and 25 materials	2 (steel) above 80° C (175° F); 2 (gray cast iron) and 4A, B, C above 100° C (210° F); 8A, B, C, D; 9A, C above 80° C (175° F); 9B above 100° C (210° F); 13A, B at b.p.; 13C above 95° C (205° F); 14A above 90° C (195° F)
Over 100%	All Type 1 materials except as indicated under Class B materials at room temperature; 2 (gray cast iron); all Types 5†; 6†; 7†; 10; 12; 13A, B, C; 16; 17; 21; 23B, C, D; 24; and 25 materials; Type 2 with acid between 100% and 101%	2 (steel) above 160° C (320° F) with acid over 101%; 8A, B, C, D; 9A, B, C above 200° C (390° F)

ACID SALTS OF SULFURIC ACID

Non-oxidizing Salts

Aluminum potassium sulfate Aluminum sodium sulfate Aluminum sulfate Chromium potassium sulfate Ferrous sulfate Iron potassium sulfate Nickel ammonium sulfate Nickel sulfate Sodium bisulfate Stannous bisulfate Titanium sulfate Zinc sulfate	Aqueous solutions of these salts are acid and behave towards metals as do dilute solutions of sulfuric acid (less than 10%), with the following exception: crystallization, followed by retention of mother liquor beneath crystals on a metal surface may cause pitting of Type 6, 7, 8, 9, and 14 materials, especially when this is accompanied by changes in the concentration of free acid. Materials which are Class A with sulfuric acid solutions of equivalent acidity will be Class A with solutions of these salts, subject to the exception noted above. Materials which are Class B with sulfuric acid solutions of equivalent acidity may be Class A with solutions of these salts (again subject to the exception noted above) but will usually be Class B. When the salts are being used in electroplating operations, processing of paper and textiles, or for other applications which do not tolerate contamination with the respective metal salts, materials which are normally Class B may become Class C.

† See dagger footnote on p. 782.

TABLE 7. SULFURIC ACID GROUP — *Continued*

ACID SALTS OF SULFURIC ACID

	Oxidizing Salts
Copper sulfate Ferric sulfate	Aqueous solutions of these salts are acid, but in contact with most metallic materials, they do not behave as do solutions of equivalent acidity which contain sulfuric acid alone. As a general rule, the presence of these salts in solutions containing sulfuric acid accelerates the corrosion of Types 2 and 4 materials and of many non-ferrous materials. As a consequence, Types 1; 2; 4; 12; 13A, B; 16; and 17 materials are Class C with such solutions at any temperature. As a general rule, also, the presence of these salts in solutions containing free sulfuric acid inhibits the action of the acid on materials rich in chromium, provided that the concentration of the oxidizing salt is above some critical minimum value which varies with the concentration of free acid, percentage of alloyed chromium, solution composition, temperature, etc. (See Fig. 1, p. 152.) If the concentration of oxidizing salt is below the critical minimum value, the effect is to accelerate the corrosion of these materials also. Assuming a sufficiently high concentration of the oxidizing salt, Types 6 and 7 materials are usually Class A with such solutions at room temperature, and Types 8, 9, 11, 13c, and 14 materials are Class A up to substantially higher temperatures, the limiting value for which depends upon the other variables. In general, halogen ions should be absent. Types 3, 15, 18, 20, 22, and 23 materials are Class A with all concentrations at all temperatures.

* Interstate Commerce Commission Regulations permit shipment of sulfuric acid in *steel* containers if the concentration is 65° Baumé (89%) or over or when the concentration of acid is 60° to 65° Baumé (78 to 89%), and an inhibitor is added which reduces the rate of attack to at least that of 66° Baumé (93%) commercial acid.

† Types 5, 6, and 7 materials are no better than Type 2 materials and may change from Class A or B to Class C more rapidly as a result of "incidental conditions." They are therefore rated Class C for all concentrations and temperatures.

General Note. Types 10, 12, 16, and 17 materials resist corrosion by sulfuric acid over a wide range of concentrations and temperatures to a degree which varies from apparent Class A in the total absence of air or other oxidizing agents through Class B and possibly Class C with highly aerated, more concentrated solutions. Since aeration is more controlling than temperature, no attempt has been made to assign temperature limits. Under non-oxidizing conditions, the limiting temperature is probably not below 100° C (210° F) for concentrations up to 50%. Above 50%, the limiting temperature drops rapidly. Materials 17E and 17F are, in general, considered to be inferior to the others, with the corrosion resistance decreasing as the zinc content increases.

CHEMICAL RESISTANT MATERIALS

Table 8. PHOSPHORIC ACID GROUP
Types of Material
(See p. 753 for sub-groups and further details of composition.)

1. Al and Al-Base Alloys
2. Iron and Steel
3. High-Si Cast Irons
4. High-Ni Cast Irons
5. 4 to 10% Chromium Steels
6. Stainless Steels (Martensitic)
7. Stainless Steels (Ferritic)
8. Stainless Steels (Austenitic)
9. Special Fe-Cr-Ni Alloys (Austenitic)
10. Nickel
11. Ni-Cr-Fe Alloy
12. Ni-Cu Alloys (>50% Ni)
13. Ni-Mo-Fe-(Cr) Alloys
14. Ni-Cr-Cu-Mo Alloy
15. Ni-Si Alloy
16. Cu-Ni Alloys (>50% Cu)
17. Cu and Cu-Base Alloys
18. Lead
19. Silver
20. Noble Metals and Tantalum
21. Mg and Mg-Base Alloys
22. Ceramic Materials
23. Carbon and Graphite
24. Plastics
25. Rubber and Synthetic Elastomers

ORTHOPHOSPHORIC ACID*

% By Weight	Class A Materials (See definition of Class A materials on p. 748.)		
	At Room Temperature Only (up to 30° C, 85° F)	Up to Intermediate Temperature Indicated	Up to Boiling Point at Indicated Concentration
Up to 5%	8A, C, D; 11	13A up to 70° C (160° F); 18A, B, C up to 95° C (205° F); 24A up to 50° C (120° F); 24B, D up to 100° C (210° F); 24E, 25A–1, A–2, C–1, C–2, D, and E up to 65° C (150° F)	3A, B; 8B; 9A, B, C; 13B, C; 14A; 15; 19; 20A, B, C; 22A, B, C; 23A, B, C, D
5 to 25%	11	13A up to 70° C (160° F); 18A, B, C up to 95° C (205° F); 24A up to 50° C (120° F); 24B, D up to 100° C (210° F); 24E, 25A–1, A–2, C–1, C–2, D, and E up to 65° C (150° F)	3A, B; 8B; 9A, B, C; 13B, C; 14A; 15; 19; 20A, B, C; 22A, B, C; 23A, B, C, D
25 to 50%	11; 13A; 15	8B up to 95° C (205° F); 14 up to 85° C (185° F); 18A, B, C up to 95° C (205° F); 19 up to 120° C (250° F); 24A up to 50° C (120° F); 24B, D up to 100° C (210° F); 24E, 25A–1, A–2, C–1, C–2, D, and E up to 65° C (150° F)	3A, B; 9A, B, C; 13B, C; 20A, B, C; 22A, B, C; 23A, B, C, D
50 to 85%	8B; 13A, C; 15; 20A	14A up to 85° C (185° F); 18A, B, C up to 95° C (205° F); 19 up to 120° C (250° F); 24A up to 50° C (120° F); 24B up to 100° C (210° F); 24C, D up to 50° C (120° F); 24E, 25A–1, A–2, C–1, C–2, D, and E up to 65° C (150° F)	3A, B; 9A, B, C; 13B; 20B, C; 22A, B, C; 23A, B, C, D

* See General Note under Class C materials, p. 786.

TABLE 8. PHOSPHORIC ACID GROUP — *Continued*

ORTHOPHOSPHORIC ACID

% By Weight	Class B Materials (See definition of Class B materials on p. 749. Other materials in common use are listed under Class A materials.)		
	At Room Temperature Only (up to 30° C, 85° F)	Up to Intermediate Temperature Indicated	Up to Boiling Point at Indicated Concentration
Up to 5%	4A, B, C; 10	8C, D up to 60° C (140° F); 11 up to 65° C (150° F); 12 up to 70° C (160° F); 16A, B, C; 17A, B, C, D, E, F up to 60° C (140° F); 24B up to 125° C (255° F); 24C up to 100° C (210° F); 24E, 25A-1, A-2, C-1, C-2, D, and E up to 80° C (175° F)	13A; 18A, B, C
5 to 25%	8A (5 to 10%); 10	8C, D up to 60° C (140° F); 11, 12 up to 90° C (200° F); 16A, B, C; 17A, B, C, D, E, F up to 60° C (140° F); 24B up to 125° C (255° F); 24C up to 100° C (210° F)	13A; 18A, B, C
25 to 50%	8C, D; 10	11, 12 up to 90° C (200° F); 16A, B, C; 17A, B, C, D, E, F up to 60° C (140° F); 24B up to 125° C (255° F); 24C up to 100° C (210° F)	8B; 13A; 14A; 15; 18A, B, C
50 to 85%	2†; 4A, B, C; 11	8B up to 120° C (250° F); 12 up to 90° C (200° F); 13A, C up to 65° C (150° F); 14A up to 120° C (250° F); 15 up to 65° C (150° F); 16A, B, C; 17A, B, C, D, E, F up to 60° C (140° F); 18A, B up to 205° C (400° F); 18C up to 120° C (250° F); 24B up to 125° C (255° F); 24C, D up to 100° C (210° F)	
85 to 100%	24E; 25A-1, A-2, C-1, C-2, D, and E		

† Type 2 materials are Class B with solutions of crude acid containing not less than 70% phosphoric acid and, in addition sufficient arsenic to inhibit attack of the acid on iron.

Table 8. PHOSPHORIC ACID GROUP — Continued

Orthophosphoric Acid

% By Weight	Class C Materials (Materials not ordinarily considered suitable.)	
	At All Temperatures	Above Indicated Temperatures
Up to 5%	All Types 1, 2, 5, 6, 7, 17G, and 21 materials	4A, B, C; 8A above 30° C (85° F); 8C, D above 60° C (140° F); 10 above 30° C (85° F); 11 above 65° C (150° F); 12 above 70° C (160° F); Types 16 and 17 materials above 60° C (140° F); 24A above 50° C (120° F); 24B, D above 125° C (255° F); 24C above 100° C (210° F); 24E; 25A–1, A–2, C–1, C–2, D, and E above 80° C (175° F)
5 to 25%	All Types 1, 2, 4, 5, 6, 7, 17G, and 21 materials; 8A (10 to 25%)	8A (5 to 10%) above 30° C (85° F); 8C, D above 60° C (140° F); 11, 12 above 90° C (200° F); Types 16 and 17 materials above 60° C (140° F); 24A above 50° C (120° F); 24B, D above 125° C (255° F); 24C above 100° C (210° F); 24E; 25A–1, A–2, C–1, C–2, D, and E above 80° C (175° F)
25 to 50%	All Types 1, 2, 4, 5, 6, 7, 8A, 17G, and 21 materials	8C, D above 30° C (85° F); 11, 12 above 90° C (200° F); Types 16 and 17 materials above 60° C (140° F); 24A above 50° C (120° F); 24B, D above 125° C (255° F); 24C above 100° C (210° F); 24E; 25A–1, A–2, C–1, C–2, D, and E above 80° C (175° F)
50 to 85%	All Types 1; 2‡; 4; 5; 6; 7; 8A, C, D; 17G; and 21 materials	8B above 120° C (250° F); 11, 12 above 90° C (200° F); 14A above 180° C (355° F); 15 above 65° C (150° F); Types 16 and 17 materials above 60° C (140° F); 18A, B above 205° C (400° F); 18C above 120° C (250° F); 24A above 50° C (120° F); 24B, D above 125° C (255° F); 24C above 100° C (210° F); 24E; 25A–1, A–2, C–1, C–2, D, and E above 80° C (175° F)
85 to 100%		20A at 145° C (293° F); 24E; 25A–1, A–2, C–1, C–2, D, and E above 30° C (85° F)

Other Phosphoric Acids

Metaphosphoric acid	Occurs normally as a crystalline solid. Aqueous solutions tend to convert to orthophosphoric acid, especially at temperatures over 100° C (210° F). Recommendaions for orthophosphoric acid apply also to aqueous solutions of metaphosphoric acid.
Pyrophosphoric acid	More corrosive than ortho- and meta-phosphoric acids. 24E, 25A–1, A–2, C–1, C–2, D, and E are Class B up to 65° C (150° F).

‡ See footnote, p. 784.

CHEMICAL RESISTANT MATERIALS

TABLE 8. PHOSPHORIC ACID GROUP — *Continued*

OTHER PHOSPHORIC ACIDS — *Continued*

Tetraphosphoric acid	Most corrosive of the commercially available phosphoric acids.			
	Class A Materials	Class B Materials	Class C Materials	
	8B up to 60° C (140° F); 9A, B up to 120° C (250° F); 11 up to 120° C (250° F); 13c up to 180° C (355° F); 14 up to 120° C (250° F); 15 up to 60° C (140° F), 22A, B, C up to 180° C (355° F); 23A, B, C, D up to 60° C (140° F); 24B up to 60° C (140° F)	3A, B up to 60° C (140° F); 4A, B, C, up to 60° C (140° F); 8A, C, D at room temp.; 8B up to 120° C (250° F); 9A, B up to 180° C (355° F); 10 at room temp.; 11 up to 180° C (355° F); 12 up to 60° C (140° F); 14 up to 120° C (250° F); 15 up to 120° C (250° F); all Types 16 and 17 (except 17G) up to 60° C (140° F)	All materials listed in Table A above 240° C (465° F). All 1, 2, 5, 6, 7, 17G, and 21 at all concentrations and temperatures. 3A, B at 120° C (250° F); 4A, B, 8A, C, D above 30° C (85° F); 8B at 180° C (355° F); 10 at 60° C (140° F); 12 at 120° C (250° F); 14 at 180° C (355° F); all Types 16 and 17 at 120° C (250° F); 18A, B, C at 60° C (140° F)	
Monobasic or dihydrogen: Ammonium phosphate Potassium phosphate Sodium phosphate	Aqueous solutions of these salts behave like dilute solutions of orthophosphoric acid. Use recommendations for orthophosphoric acid "up to 5%."			

General Note. Types 3; 8A, C, D; 20A; and 22 materials not recommended for use with crude acid containing appreciable amounts of hydrofluoric acid.

Table 9. MISCELLANEOUS ACIDS AND ACID-FORMING GASES

Types of Material
(See p. 753 for sub-groups and further details of composition.)

1. Al and Al-Base Alloys
2. Iron and Steel
3. High-Si Cast Irons
4. High-Ni Cast Irons
5. 4 to 10% Chromium Steels
6. Stainless Steels (Martensitic)
7. Stainless Steels (Ferritic)
8. Stainless Steels (Austenitic)
9. Special Fe-Cr-Ni Alloys (Austenitic)
10. Nickel
11. Ni-Cr-Fe Alloy
12. Ni-Cu Alloys (>50% Ni)
13. Ni-Mo-Fe-(Cr) Alloys
14. Ni-Cr-Cu-Mo Alloy
15. Ni-Si Alloy
16. Cu-Ni Alloys (>50% Cu)
17. Cu and Cu-Base Alloys
18. Lead
19. Silver
20. Noble Metals and Tantalum
21. Mg and Mg-Base Alloys
22. Ceramic Materials
23. Carbon and Graphite
24. Plastics
25. Rubber and Synthetic Elastomers

Chemical	Class A Materials	Class B Materials	Class C Materials
Benzoic Acid As anhydrous vapors and as alcoholic or aqueous solution. Unless otherwise specified, ratings apply to all materials up to 100° C (210° F).	1A, B, C, E, F*; 8B; 9A, B; 11; 13C; 14A; 19; 20A, B, C; 22A, B, C; 10 with vapors only. 12 with solutions only.	Type 2 materials with anhydrous vapors. 3A, B; 4A, B, C; 16A, B, C; 17A, B, C, D.	Type 2 materials with solutions. All Class B materials with products which cannot tolerate contamination with corresponding metal salts.
Boric Acid All concentrations of aqueous solutions up to boiling point unless otherwise specified	1A, B, C, E, F‡; 3A, B; 8B at room temp.; 9A, B up to 80° C (175° F); 10; 11; 12; 19; 20A, B, C; 22A, B, C; 24A up to 50° C (120° F); 24E and all Types 25 up to 65° C (150° F)	3A, B; 4A, B, C; 10; 12; 16A, B, C; 17A, B, C, D; 18A, B, C†	2, 5, 6, 7, 21. 24A above 50° C (120° F). 24E and all Types 25 above 65° C (150° F).
Carbolic Acid	See "Phenol" below.		
Carbon Dioxide Carbonic Acid Wet gas and aqueous solutions of carbon dioxide other than carbonated beverages unless otherwise specified. Up to 100° C (210° F).	All metallic materials with dry gas up to temperature at which high temperature corrosion begins 1A, B, C, E, F; 6A, B, C; 7A, B; 8A, B, C, D; 9A, B; 10 ,11; 12; 13C; 14A; 19; 20A, B, C; 22A, B, C; 24E and all Types 25 up to 65° C (150° F)	2§; 3A, B; 4A, B, C; 5; 16A, B, C; 17A, B, C, D, E, F; 17§; 18A, B, C	2 and 17G with aerated solutions; 21; 24E and all Types 25 up to 65° C (150° F)

* Up to 60° C (140° F) only in the case of aqueous solutions.
† Most of the above materials are no worse than Class B with solutions of the pure acid and with crude liquors, and may be Class A in many situations. In the absence of specific data they are all rated Class B.
‡ In the absence of hydrochloric or sulfuric acids.
§ Corroded rapidly by aerated solutions, especially when hot.

TABLE 9. MISCELLANEOUS ACIDS AND ACID-FORMING GASES — Continued

Chemical	Class A Materials	Class B Materials	Class C Materials
Carbonated Beverages	All Class A materials with carbonic acid plus pure tin, except Types 6 and 7 materials; and provided the beverages do not contain other acids with which any of the materials is incompatible		All materials which are Class B or C with wet carbon dioxide or carbonic acid. Exception: 17A and some other Type 17 materials used with beer when circumstances permit the formation of protective "beer-stone" on metal surfaces
Chloroacetic Acid	13c; 19; 20A, B, C; 22A, B, C; 23A, B, C, D (all up to b.p.)	10; 12; 17A, B, c‖; 25A-2 at room temp. only	All Types 1, 5, 6, 7, 18, 21. All Types 25 except 25A-2. 25A-2 above 30° C (85° F)
Chlorosulfonic Acid Pure acid unless otherwise specified	1A, B, C, E, F; 8A, b; 9A, B; 10; 13c (all up to 50° C, 120° F); 8B and probably 9A, B with crude acid	2; 8A and 18A with crude acid	
Chromic Acid Up to concentrations and temperatures obtained in anodizing and electroplating baths unless otherwise specified	3A, B; 9A, B up to 60% up to 80° C (175° F); 22A, B, C; 23A, B, C, D up to 10% up to 95° C (205° F) and with 10% to 40% at room temp.; Type 8 materials with dilute solutions at moderate temperatures; 10, 11, and 12 up to 5% up to 60° C (140° F) with pure chromic acid	2¶; 8A, B, C, D; 18A, B, C; 23A, B, C, D with 10 to 40% up to 95° C (205° F). Not recommended for use with electroplating solutions. 24E up to 35% up to 55° C (130° F)	All Types 5, 6, 7, 10, 12, 13, 15, 16, 17, and 25 materials. Type 23 materials with electroplating solutions. 24E above 35% and above 55° C (130° F)
Cresylic Acid	Use recommendations for "Carbolic Acid — Phenol"		
Hydrogen Sulfide Wet gas and aqueous solutions up to 100° C (210° F) unless otherwise specified	1A, B, C, E, F; 3A, B; 9A, B; 13c; 14A	2‖; 4A, B, C; 5‖; 6‖; 7‖; 8A, B, C, D; 10; 12; 17G; 24E at room temp.; 25A-2, B-2, c-2 up to 65° C (150° F)	16A, B, C; 17A, B, C, D, E, F; 19; 24E above 30° C (85° F); 25A-1, B-1, C-1, D, E; 25A-2, B-2, C-2 above 65° C (150° F)
Oxalic Acid All concentrations of aqueous solutions up to boiling point unless otherwise specified	3A, B; 8B at room temp.; 9A, B; 12; 20A, B, C; 22A, B, C	4A, B, C; 8B; 16; 17; 19; 24E and 25A-2 up to 65° C (150° F)	All Types 1, 2, 5, 6, 7, 18, and 21

¶ Type 2 materials are subject to impingement corrosion and to local corrosion adjacent to welds.

‖ May become Class C with wet gas or aqueous solutions especially where conditions favor stress corrosion.

TABLE 9. MISCELLANEOUS ACIDS AND ACID-FORMING GASES — *Continued*

Chemical	Class A Materials	Class B Materials	Class C Materials
Phenol (Carbolic Acid) Aqueous solutions up to 100° C (210° F) unless otherwise specified	1A, B, C, E, F**; 3A, B; 8A, B, C, D††; 9A, B††; 10‡‡; 11; 13A, B, C; 14; 15; 19; 20A, B, C; 22A, B, C	4A, B, C; 12; 16A, B, C; 17A, B, C, D; 18A, B, C; Type 2, 6, and 7 materials with hot vapors up to 310° C (590° F); Types 6 and 7 better than Type 2 if sulfur compounds are present	All Types 2, 5, 6, 7, 21, 24E, and 25 materials; Type 1 materials with anhydrous vapors at 310° C (590° F); Type 2 materials above 310° C (590° F) especially if sulfur compounds are present

** Type 1 materials with hot *wet* vapors.

†† Types 8 and 9 materials with hot vapors up to 310° C (590° F); also when sulfur compounds are present.

‡‡ May become Class C if much sulfur is present.

Table 10. SULFUR DIOXIDE GROUP

Types of Material
(See p. 753 for sub-groups and further details of composition.)

1. Al and Al-Base Alloys
2. Iron and Steel
3. High-Si Cast Irons
4. High-Ni Cast Irons
5. 4 to 10% Chromium Steels
6. Stainless Steels (Martensitic)
7. Stainless Steels (Ferritic)
8. Stainless Steels (Austenitic)
9. Special Fe-Cr-Ni Alloys (Austenitic)
10. Nickel
11. Ni-Cr-Fe Alloy
12. Ni-Cu Alloys (>50% Ni)
13. Ni-Mo-Fe-(Cr) Alloys
14. Ni-Cr-Cu-Mo Alloy
15. Ni-Si Alloy
16. Cu-Ni Alloys (>50% Cu)
17. Cu and Cu-Base Alloys
18. Lead
19. Silver
20. Noble Metals and Tantalum
21. Mg and Mg-Base Alloys
22. Ceramic Materials
23. Carbon and Graphite
24. Plastics
25. Rubber and Synthetic Elastomers

Sulfur Dioxide and Sulfurous Acid

% By Weight	Class A Materials (See definition of Class A materials on p. 748.)		
	At Room Temperature Only (up to 30° C, 85° F)	Up to Intermediate Temperature Indicated	Up to 200° C (390° F) (See Note 3)
Dry sulfur dioxide (See Note 1)*	See Note 2*	See Note 2; 23c, D up to 170° C (340° F)	See Note 2; 22A, B, C; 23A, B
Moist sulfur dioxide; sulfurous acid (See Note 4)	8A, C, D; 18A, B, C; 24A	23A, B, C, D up to 100° C (210° F) or to higher temperature corresponding to steam pressure which these materials will withstand	1A, B, C, E, F (vapors only); 8B; 9A, B, C; 13C; 14A; 20A, B, C; 22A, B, C (See following section on "Sulfite Liquors")

% By Weight	Class B Materials (See definition of Class B materials on p. 749. Other materials in common use are listed under Class A materials.)		
	At Room Temperature Only (up to 30° C, 85° F)	Up to Intermediate Temperature Indicated	Up to 200° C (390° F) (See Note 3)
Dry sulfur dioxide (See Note 1)	See Note 2	See Note 2	See Note 2
Moist sulfur dioxide; sulfurous acid (See Note 4)	1A, B, C, E, F with aqueous solutions up to 3%; 17A, B, C, D; 24E	8A, C, D up to 100° C (210° F); 18C up to 120° C (250° F); 25A-2, B-2, and C-2 up to 65° C (150° F)	18A, B (See following section on "Sulfite Liquors")

* Notes are listed on p. 792.

TABLE 10. SULFUR DIOXIDE GROUP — *Continued*

Sulfur Dioxide and Sulfurous Acid

% By Weight	Class C Materials (Materials not ordinarily considered suitable.)	
	At All Temperatures	Above Indicated Temperatures
Dry sulfur dioxide (See Note 1)*	See Note 2*	See Note 2; 23c, D above 170° C (340° F)
Moist sulfur dioxide; sulfurous acid (See Note 4)	All Types 1G; 2; 3; 4; 5; 6; 7; 10; 11; 12; 13A, B; 15; 16; 19; 21; 25A–1, 25B–1, 25C–1, 25D, and 25E materials, 1A, B, C, E, F with aqueous solutions (over 3%)	8A, C, D above 100° C (210° F), especially when carbon is over 0.10%; 24E above 30° C (85° F); 25A–2, B–2, C–2 above 65° C (150° F) (See following section on "Sulfite Liquors")

Industrial Atmospheres

Waste Gases	Below the Dew Point	Above the Dew Point†
Direct products of smelting and other processes involving the combustion of sulfur-bearing materials	Class A or B Materials‡	Class A or B Materials‡
	1A, B, C, E, F; 4A, B, C; 8B; 9A, B, C; 13C; 14A; 18A, B, C; 24E up to 30° C (85° F); 25A–2, B–2, C–2 up to 65° C (150° F)	1A, B, C, E, F; 2; 4A, B, C; 5; 6; 7; 8A, B, C, D; 9A, B, C; 11; 13C; 14A; 17A, B, C, D; 18
	Class C Materials	Class C Materials
	All Types 1G; 2; 3; 4; 5; 6; 7; 10; 11; 12; 13A, B; 15; 16; 19; 21; 25A–1, B–1, C–1, D, and E materials; 24E above 30° C (85° F); 25A–2, B–2, and C–2 above 65° C (150° F)	1A, B, C, E, F above 320° C (605° F); 10; 12; 16A, B, C; 18A, B, C above 200° C (390° F)
Out-door atmospheres contaminated with above described waste gases	As above, except that some of the materials rated Class C with "Waste Gases" at temperatures below the dew point may become Class B with these same products after they have been diluted with air. Results vary with the amount of contamination and other local conditions, and are likely to be quite erratic.	

Sulfites and Bisulfites

Ammonium sulfite Calcium sulfite Magnesium sulfite Potassium sulfite Sodium sulfite	Solutions of these salts, or vapors arising from such solutions at elevated temperatures, may behave as sulfurous acid solutions or as moist sulfur dioxide vapors. In such cases the recommendations for the corresponding sulfur dioxide solutions and vapors apply here also. *Exceptions:* Copper and high-copper alloys are Class C with solutions of ammonium salts if free ammonia is present in solution or in vapors.
Ammonium bisulfite Calcium bisulfite Magnesium bisulfite Potassium bisulfite Sodium bisulfite	With slightly alkaline solutions of the normal salts Type 2 materials will be Class B and Types 6 and 7 materials will be Class A.

* Notes are listed on p. 792.
† See also *High-Temperature Corrosion*, p. 615 and *High-Temperature Resistant Materials*, p. 729.
‡ These environments vary too much to permit a sharp division between Class A and Class B materials.

792 CHEMICAL RESISTANT MATERIALS

TABLE 10. SULFUR DIOXIDE GROUP — *Continued*

SULFITE LIQUORS

Raw liquor from absorption towers Relief liquor Relief vapors Circulating liquor Cooking liquor	All these liquors contain free sulfurous acid, and the vapors arising therefrom contain moist sulfur dioxide. None of the liquors behaves exactly like sulfurous acid, however, because of contamination with sulfuric acid and constituents extracted from the wood or other material being pulped, and because they contain calcium salts and other components which deposit as scale on metal surfaces. Similarly, the vapors do not behave exactly like moist sulfur dioxide. All the materials which are Class C with moist sulfur dioxide and sulfurous acid are Class C with these liquors and vapors. Some of the materials which are Class A or B with moist sulfur dioxide and sulfurous acid take a somewhat poorer rating with these liquors and vapors. Among them are Types 1, 8, and 17 materials. Type 1 materials are Class C. Types 8A, 8C, and 8D are usually Class C. They may be Class A or B at moderate temperatures where exposed surfaces are free from scale or other deposits and in the absence of excessive amounts of sulfuric acid. Use with caution. Alloys of this group containing more than 0.10% C are always Class C. Types 8B and 9 materials are usually Class A or no worse than Class B. Types 17A, B, C, D materials are no better than Class B with hot liquors and vapors, and may be Class C in such service. They may be Class A to B with cold liquors. Type 18 materials are Class A to B up to the maximum temperature at which they are rated with sulfur dioxide and sulfurous acid. Types 22 and 23 materials are usually Class A. 24E and 25A-1, B-1, C-1, D, and E are Class B up to 65° C (150° F). Types 25A-2, B-2, and C-2 are Class A up to 65° C (150° F) and Class B up to 80° C (175° F).

Note 1. The term *dry sulfur dioxide* refers to sulfur dioxide, in either the liquid or vapor state, containing no water other than unavoidable traces, and not permitted to come in contact with air or other environment from which moisture can be absorbed.

Note 2. Theoretically, all metallic materials should be Class A with "dry" sulfur dioxide up to the limiting temperature at which high temperature corrosion occurs. Practically, it is not always possible to prevent contamination with water or to keep the temperature above the dew point at all times. Therefore, it is safer, when dealing with sulfur dioxide which is in contact with air or other environment containing moisture, to use the data for "moist" sulfur dioxide, all of which can be safely applied to the "dry" substance.

Note 3. The temperature of 200° C (390° F) was selected arbitrarily as the probable highest saturated steam pressure at which sulfur dioxide will be encountered. Most materials good up to that temperature can be used at higher temperatures.

Note 4. The term *moist sulfur dioxide* refers to atmospheres comprised substantially of sulfur dioxide and water vapor, and *sulfurous acid* refers to aqueous solutions of sulfur dioxide only. The atmospheres are assumed to be at temperatures below the dew point or in the "saturated" condition, i.e., in contact with water in the form of sulfurous acid solutions.

Table 11. ALKALIES GROUP

Types of Material
(See p. 753 for sub-groups and further details of composition.)

1. Al and Al-Base Alloys
2. Iron and Steel
3. High-Si Cast Irons
4. High-Ni Cast Irons
5. 4 to 10% Chromium Steels
6. Stainless Steels (Martensitic)
7. Stainless Steels (Ferritic)
8. Stainless Steels (Austenitic)
9. Special Fe-Cr-Ni Alloys (Austenitic)
10. Nickel
11. Ni-Cr-Fe Alloy
12. Ni-Cu Alloys (>50% Ni)
13. Ni-Mo-Fe-(Cr) Alloys
14. Ni-Cr-Cu-Mo Alloy
15. Ni-Si Alloy
16. Cu-Ni Alloys (>50% Cu
17. Cu and Cu-Base Alloys
18. Lead
19. Silver
20. Noble Metals and Tantalum
21. Mg and Mg-Base Alloys
22. Ceramic Materials
23. Carbon and Graphite
24. Plastics
25. Rubber and Synthetic Elastomers

Sodium Hydroxide

% By Weight	Class A Material (See definition of Class A Materials on p. 748.)		
	At Room Temperature Only (up to 30° C, 85° F)	Up to Intermediate Temperature Indicated	Up to Boiling Point at Indicated Concentration
Up to 10%		24A up to 50° C (120° F); 24C, D up to 100° C (210° F); 24E and all Type 25 materials up to 65° C (150° F) (Soft varieties of Type 25 materials not recommended for very dilute solutions)	4A, B, C; 8A, B, C, D; 9A, B; 10; 11; 12; 13A, B, C; 14A; 15; 16A, B, C; 19; 20B, C; 21; 23A, B, C, D
10 to 30%		8A, B, C, D up to 100° C (210° F); 24A up to 50° C (120° F); 24C, D up to 100° C (210° F); 24E and all Type 25 materials up to 65° C (150° F)	4A, B, C; 9A, B; 10; 11; 12; 13A, B, C; 14A; 16A, B, C; 19; 20B, C; 21; 23A, B, C, D
30 to 50%		4A, B, C up to 150° C (300° F); 8A, B, C, D and 16A, B, C up to 100° C (210° F); 24A up to 50° C (120° F); 24C, D up to 100° C (210° F); 24E and all Type 25 materials up to 65° C (150° F)	9A, B; 10; 11; 12; 13A, B, C; 14A; 19; 20B, C; 23A, B, C, D
50 to 70%		8A, B, C, D and 9A, B up to 100° C (210° F); 14A up to 90° C (195° F); 23C, D up to 170° C (340° F)	10; 11; 12; 19; 23A, B
70 to anhydrous up to 260° C (500° F)	10; 19; 20B, C	23A, B up to b.p.; 23C, D up to 170° C (340° F) (See Notes 1, 2, and 3)*	

*Notes are listed on p. 798.

CHEMICAL RESISTANT MATERIALS

TABLE 11. ALKALIES GROUP — *Continued*

SODIUM HYDROXIDE

% By Weight	Class A Materials — *Continued* (See definition of Class A materials on p. 748.)		
	At Room Temperature Only (up to 30° C, 85° F)	Up to Intermediate Temperature Indicated	Up to Boiling point at Indicated Concentration
Fused NaOH up to 480° C (900° F)	10; 19; 23A, B (See Notes 1, 2, and 3 on p. 798)		

% By Weight	Class B Materials (See definition of Class B materials on p. 749. Other materials in common use are listed under Class A materials.)		
	At Room Temperature Only (up to 30° C, 85° F)	Up to Intermediate Temperature Indicated	Up to Boiling Point at Indicated Concentration
Up to 10%	3A, B; 17G; 24B up to 1%	18A, B, C up to 90° C (200° F)	2; 3A, B (up to 0.5%); 5; 6; 7; 17 (all except 17G)
10 to 30%	3A, B; 5*; 6A, B, C*; 7A, B*; 17G; 18A, B, C	17 (all except 17G) up to 80° C (180° F)	2†; 8A, B, C, D; 15; 16A, B, C
30 to 50%	3A, B; 5*; 6A, B, C*; 7A, B*	2† up to 80° C (180° F)	4A, B, C; 8A, B, C, D; 15; 16A, B, C
50 to 70%	3A, B	2† up to 80° C (180° F); 4A, B, C; 8A, B, C, D and 16A, B, C up to 150° C (300° F)	10; 12; 15
70 to anhydrous up to 260° C (500° F)	2; 4A, B, C; 8A, B, C, D; 9A, B, C; 11; 12 (See Notes 1, 2, 3 on p. 798, and †)		
Fused NaOH up to 480° C (900° F)	2; 4A, B, C; 11; 12 (See Notes 1, 2, and 3 on p. 798)		

% By Weight	Class C Materials (Materials not ordinarily considered suitable.)	
	At All Temperatures	Above Indicated Temperatures
Up to 10%	All Type 1 materials; 20A; 24B above 1%; soft varieties of Type 25 materials	3A, B above 0.5% above 30° C (85° F); 17G above 30° C (85° F); 18A, B, C above 90° C (200° F); 24A above 50°C (120° F); 24C, D above 100° C (210° F); 24E and all Type 25 materials above 90° C (200° F)

* Types 5, 6, and 7 materials may be somewhat superior to some Type 2 materials with dilute solutions, but are not expected to be superior with hot concentrated solutions, and may be inferior. In the absence of specific information they are rated as indicated.

† Type 2 materials severely stressed (except gray cast iron) are subject to caustic embrittlement. (See *Boiler Corrosion*, p. 531.)

CHEMICAL RESISTANT MATERIALS

TABLE 11. ALKALIES GROUP — *Continued*

SODIUM HYDROXIDE

% By Weight	Class C Materials — *Continued* (Materials not ordinarily considered suitable.)	
	At All Temperatures	Above Indicated Temperatures
10 to 30%	All Type 1 materials; 20A; 24B	3A, B; 17G, and 18A, B, C above 30° C (85° F); 5*, 6A, B, C*, and 7A, B* above 30° C (85° F); 24A above 50° C (120° F); 24C, D above 100° C (210° F); 24E and all Type 25 materials above 90° C (200° F)
30 to 50%	All Types 1, 3, 17, and 18 materials; 20A; 24B	2 above 80° C (180° F); 5*, 6A, B, C*, and 7A, B above 30° C (85° F); 24A above 50° C (120° F); 24C, D above 100° C (210° F); 24E and all Type 25 materials above 90° C (200° F)
50 to 70%	All Types 1, 3, 17, and 18 materials; 20A; 24B	2 above 80° C (180° F); 5*, 6A, B, C*, and 7A, B above 30° C (85° F); 24A above 30° C (85° F); 24C, D above 100° C (210° F); 24E and all Type 25 materials above 90° C (200° F)
70% to anhydrous	All Types 1, 3, 5*, 6*, 7*, 15, 16, 17, 18, 20, 20A, 21, 22, 24, and 25 materials; 23C, D above 170° C (340° F)	
Fused NaOH up to 480° C (900° F)	All Types 1, 3, 5, 6, 7, 8, 9, 15, 16, 17, 18, 20, 21, 22, 23C, D, 24, and 25 materials	

OTHER HYDROXIDES

Barium hydroxide Lithium hydroxide Potassium hydroxide	The data for "Sodium Hydroxide" may be expected to apply to these hydroxides also.	Calcium hydroxide Magnesium hydroxide	All materials which are Class A with "Sodium Hydroxide" up to 10% are Class A with these hydroxides up to saturation point, and up to or above the same limiting temperature. Types 3, 6, and 7 also Class A. All Types 24 and 25 materials Class A up to temperature limitation. Type 1 materials Class C. Other materials Class B.

* Types 5, 6, and 7 materials may be somewhat superior to some Type 2 materials with dilute solutions, but are not expected to be superior with hot concentrated solutions, and may be inferior. In the absence of specific information they are rated as indicated.

TABLE 11. ALKALIES GROUP — *Continued*

ALKALINE CYANIDES AND SULFIDES			
Barium cyanide Potassium cyanide Sodium cyanide Cyanide plating solutions	The data for sodium hydroxide apply here also, subject to the following exceptions: Types 12, 16, 17, 18, 19, and 20 B, C materials which are Class A with solutions of the hydroxides are Class C with all concentrations and temperatures.	Barium sulfide Potassium sulfide Sodium sulfide	The data for sodium hydroxide apply here also, subject to the following exceptions: Types 12, 16, 17, 19, and 20B, C materials are Class C with all concentrations and temperatures. Type 18 materials are Class B with all concentrations at room temperature and with up to 25% up to 100° C (210° F). If hydrogen sulfide is liberated the soft forms of Type 25 materials become Class C and 24E becomes Class C above 30° C (85° F).

OTHER ALKALINE SALTS	
Borax Sodium aluminate Sodium bicarbonate Sodium carbonate Sodium metasilicate Sodium perborate Sodium peroxide* Sodium phosphate (tribasic) Sodium plumbite* Sodium sesquisilicate Sodium silicate Sodium stannate* Corresponding salts of other alkali metals	All materials which are Class A with sodium hydroxide up to 10% are Class A with all concentrations of these salts up to or above the same limiting temperature. Types 6 and 7 also Class A with all concentrations up to b.p. Types 4, 6, 7, 8, and 9 can be used in processing of iron-free products. Types 12 and 16 can be used in processing of copper-free products. Type 2 materials are Class B with all concentrations up to b.p., and at higher temperatures except where circumstances favor caustic embrittlement (See *Boiler Corrosion*, p. 531). Exception: Type 1 materials usually Class C. May be Class B with some solutions when these are properly inhibited. Consult manufacturers for more specific recommendations. Type 2 materials may approach Class C with sodium silicate solutions when the latter are permitted to drain and dry intermittently, with consequent cracking of the glass-like layer of dried silicate solution. Severe pitting at exposed areas may occur. Type 3 materials Class A to B with all concentrations up to b.p. Type 20B, C materials when molten Class C with those marked with asterisk. May be Class A to B with others, depending upon temperature, oxidation, and degree of conversion to hydroxide.

* Types 5, 6, and 7 materials may be somewhat superior to some Type 2 materials with dilute solutions, but are not expected to be superior with hot concentrated solutions, and may be inferior. In the absence of specific information, they are rated as indicated.

CHEMICAL RESISTANT MATERIALS

TABLE 11. ALKALIES GROUP — *Continued*

SOAP AND SOAP LIQUORS

See Table 3 for data regarding the fatty acids and fatty oils used in the manufacture of soap.
Type 1 materials are Class C with all strongly alkaline liquors and with final product containing excess alkali unless effectively inhibited. Type 2 materials are Class B except when processing iron-free products (Class C). Type 4 materials are Class B to A, useful with crude liquors containing salt. Types 8 and 9 materials are Class A except with crude liquors containing salt. Types 10, 11, and 12 materials Class A. Types 16 and 17 materials are Class A to B except Type 17 is Class C when processing copper-free products. Type 18 materials usually are Class C.

VISCOSE

(a) "Alkali-Cellulose," obtained by steeping wood pulp or cotton fiber in sodium hydroxide. (b) "Cellulose Xanthate," obtained by reaction between alkali-cellulose and carbon disulfide.‡ (c) "Viscose," obtained by dissolving cellulose xanthate in sodium hydroxide.	Type 2 materials are Class B to A throughout from standpoint of general corrosion. Widely used in cellophane and sausage casing industry. Class C in rayon industry where contamination with iron or manganese cannot be tolerated. Types 4, 8, 9, 10, 11, and 12 are Class A throughout except Type 12. They are Class C in rayon industry and Type 10 Class B in spin-bath evaporators. Types 1, 16, 17, and 18 materials are Class C. Type 25 materials are Class C where carbon disulfide is encountered.

ALKALINE PULPING LIQUORS

Soda liquors
  White
  Green
  Black
Sulfate liquors
  White
  Green
  Black
	In general, the data for "Alkaline Salts" apply to these liquors also, with sulfate liquors being somewhat more corrosive than soda liquors because they contain sulfides. Exceptions are due to the fact that some of these liquors may contain organic compounds in solution or solids in suspension, as a result of which the ordinarily protective coatings of corrosion products may be less effective, and velocity and turbulence may be more damaging. Vapors arising from these liquors may be more corrosive than the liquors themselves. Type 1 materials are Class C throughout; Type 3 materials are usually Class C. Type 2 materials are commonly employed and are Class B throughout except where they become Class C because of turbulence, abrasion, or exposure to moist vapors. Type 4 materials are Class B with sulfate liquors and Class B to A with soda liquors. Low-Ni cast irons (1 to 3% Ni) are intermediate between nickel-free and Type 4 cast irons. Low-Ni steels (3½ to 5% Ni) are superior to carbon steels. Type 5 materials are not expected to be better than carbon steel; may be inferior. Results with Types 6 and 7 materials erratic; better with vapors than with liquors. Types 8 and 9 materials are usually Class A, although Type 8 materials may become Class B where turbulence is great and temperatures are high. Types 10, 12, and 16 materials are Class B to A with soda liquors and Class B to C with sulfate liquors. Type 11

‡ Carbon disulfide is handled in Type 2 materials.

CHEMICAL RESISTANT MATERIALS

TABLE 11. ALKALIES GROUP — *Continued*

ALKALINE PULPING LIQUORS

materials are Class A throughout. Type 17 materials usually considered Class C, although 17G materials are Class B with vapors containing hydrogen sulfide. Type 18 materials are Class C.

Note 1. Corrosion by concentrated solutions and by fused sodium hydroxide varies with temperature and is aggravated by abrasion of crystals and by other factors which interfere with the formation of protective coatings. Materials rated Class A up to 260° C (500° F) are usually no worse than Class B up to 480° C (900° F). Some materials rated Class B up to 260° C (500° F) become Class C at 480° C (900° F).

Note 2. Severely stressed nickel and high-nickel alloys are susceptible to intergranular corrosion by sodium hydroxide in the concentration range where temperatures may exceed 315° C (600° F).

Note 3. The presence of oxidizable sulfur compounds in the sodium hydroxide tends to increase the corrosion rate for Types 4, 10, 11, 12, and 19 (?) materials. Types 12 (and 19?) may become Class C in such cases.

CHEMICAL RESISTANT MATERIALS

TABLE 12. AMMONIA GROUP

Types of Material

(See p. 753 for sub-groups and further details of composition.)

1. Al and Al-Base Alloys
2. Iron and Steel
3. High-Si Cast Irons
4. High-Ni Cast Irons
5. 4 to 10% Chromium Steels
6. Stainless Steels (Martensitic)
7. Stainless Steels (Ferritic)
8. Stainless Steels (Austenitic)
9. Special Fe-Cr-Ni Alloys (Austenitic)
10. Nickel
11. Ni-Cr-Fe Alloy
12. Ni-Cu Alloys (>50% Ni)
13. Ni-Mo-Fe-(Cr) Alloys
14. Ni-Cr-Cu-Mo Alloy
15. Ni-Si Alloy
16. Cu-Ni Alloys (>50% Cu)
17. Cu and Cu-Base Alloys
18. Lead
19. Silver
20. Noble Metals and Tantalum
21. Mg and Mg-Base Alloys
22. Ceramic Materials
23. Carbon and Graphite
24. Plastics
25. Rubber and Synthetic Elastomers

Ammonia and Ammoniumhydroxide, Ammonia Liquors	Class A Materials	Class B Materials	Class C Materials
Compressed liquid and gas (Note 1) anhydrous	1; 2; 3A, B; 4B, C; 5; 6; 7; 8; 9 (Note 1); 10; 11; 13A, B, C; 14A; 15; 18; 19 (Note 2); 20A, B, C; 21; 23A, B, C, D		4A; 12; 16; 17
Moist vapors Ammonium hydroxide	1; 4B*, C up to 70° C (160° F); 8; 9; 11; 13C; 14A; 20A, B, C; 21; 23A, B, C, D; 24C, D up to b.p. 24E and all Type 25 materials up to 65° C (150° F)	2; 3A, B; 5; 6; 7; 4B*, C above 70° C (160° F) up to b.p.; 19 (Note 2); 24B at room temp.	4A, B†; 10; 12; 15; 16; 17; 19, (Note 2); 24A above 30° C (85° F) with up to 10%; at all temperatures with concentrated grade. 24B above 30° C (85° F).
Ammonium carbonate Ammonium phosphate (tribasic)	In general, the ratings for ammonium hydroxide apply here. All materials which are Class A or Class C with ammonium hydroxide are similarly rated with solutions of these salts. Materials which are Class B with ammonium hydroxide may approach Class A with solutions of these salts.		
Ammonia liquors (mixtures of ammonium hydroxide, ammonium salts, and other components)	All materials which are Class C with ammonium hydroxide are Class C with ammonia liquors. The A and B ratings for ammonium hydroxide and ammonium salts will usually apply also, but they should be used cautiously, depending upon the other components.		

* Preferably Cu-free grade.
† Cu-bearing grade.
Note 1. All the ratings in this Table are based on the assumption that the temperature is below that at which dissociation of the gas and nitriding of the metal occur.
Note 2. Silver alloys containing copper are always Class C. Pure silver resists dry gas and air-free aqueous solutions well, but is attacked vigorously by aerated solutions and by hot moist vapors.

CORROSION PROTECTION

Metallic Coatings

ZINC COATINGS ON STEEL*

WILLIAM BLUM† AND ABNER BRENNER†

INTRODUCTION

The extensive use of zinc coatings on steel is largely the result of the good corrosion resistance of zinc, which under most service conditions is considerably better than that of steel. An additional advantage of zinc coatings is that under certain conditions they will protect exposed steel, for example, at cut edges, by electrochemical action (cathodic protection). The electrochemical protection is exerted when the bare steel and adjacent zinc-coated areas are in contact with a conducting aqueous solution. Under these conditions, the zinc acts as the anode of an electrolytic cell and thereby prevents corrosion of the iron which constitutes the cathode. This behavior gave rise to the term *galvanized iron*. The only other commercial metal coating on steel that yields about equal electrolytic protection is cadmium. Under conditions in which aluminum coatings remain active, they may give electrolytic protection to steel. The use of zinc for coating steel is favored because of its low cost and ease of application.

In addition to the electrolytic protection, zinc may retard corrosion of exposed steel by the fact that the solution of zinc (or of the products of its corrosion) in water may so increase the pH as to retard the corrosion of steel. In support of this hypothesis, E. A. Anderson [37]‡ has reported a pH as high as 8.5 in water saturated with the corrosion products of zinc.

One objection to the use of zinc coatings is that, especially at high temperatures and humidities, e.g., in the tropics, corrosion is rapid and results in the formation of bulky, unsightly white coatings, usually of basic zinc carbonate. One of the best methods of retarding this type of corrosion is by the application of inhibitive chromate films (p. 862).

ATMOSPHERIC CORROSION

GENERAL PRINCIPLES

Protection of the steel necessarily involves some "sacrificial" corrosion of the zinc coating. In principle, the degree of protection and the life of a given zinc coating might be evaluated by the rate at which the coating loses weight (or thickness) in given surroundings. However, these determinations are usually inconvenient, even in controlled field tests, especially on large specimens such as sheets. The results of most exposure tests of zinc-coated steel have therefore been expressed in terms of the area on which rust appeared, as a result of corrosion of exposed steel.

If a steel specimen were coated with a perfectly uniform zinc coating, and if all parts were exposed to the same corrosive influences, the zinc coating would suddenly disappear from the entire surface, after which the exposed steel would corrode at its

---

* Refer also to *Zinc*, p. 331.
† National Bureau of Standards, Washington, D. C.
‡ The numbers in brackets refer to the references at the end of this chapter (p. 816).

normal rate. Such conditions never exist in practice because (1) the zinc coatings are never uniformly distributed, even on flat sheets, (2) the different parts of the specimen do not wet or dry uniformly when exposed to rain or dew, (3) the presence of any bare steel, e.g., at cut edges, accelerates the corrosion of the adjacent zinc, and (4) in hot-dipped coatings the alloy layer may corrode at a different rate from the nearly pure zinc on the surface. The actual curves representing the progress of rust are therefore usually of the shape shown in Fig. 1, in which the fraction of the surface rusted is plotted against years of exposure.

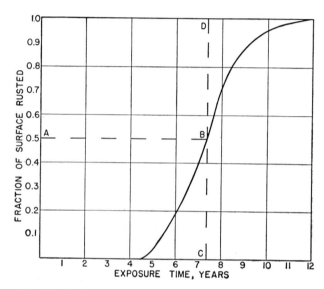

FIG. 1. Typical Curve for the Rate of Failure of Hot-Dipped Sheet Subjected to Outdoor Exposure.

Evaluation of the life of a coating from such a curve is difficult. Obviously neither the time for initial rusting nor for complete rusting is a valid criterion, especially as these points are difficult to define. The area to the left of the curve is a true index of the protection, but is not readily measured. It is numerically equal to the average abscissa, AB, of the curve, since the area $AB \times DC = AB \times 1$. The average abscissa may be called the average life of the coating because it is equal to the number of years that the coating would have given complete protection if, instead of gradually exposing the steel, it had corroded uniformly and then disappeared suddenly. If, as is true in many such tests, the curve is approximately symmetrical about a point B, which represents rusting of 50% of the surface, the number of years to produce 50% rusting may be considered as equal to the average life of the coating and taken as an arbitrary measure of the protective value of the coating.

In order to compare the protective value of different zinc coatings, in one or more locations, it is often convenient to express the results in terms of the average thickness (or weight per unit area) of zinc removed in a given period. Suggested units are milligrams per square decimeter per day (mdd), inches per year (ipy), and mils (or inches $\times 10^{-3}$) per year. The last-named unit is convenient because its reciprocal represents the life in years of a coating 1 mil (0.001 in.) thick.

ZINC COATINGS ON STEEL

ATMOSPHERIC EXPOSURE TESTS

A. S. T. M. Tests. The most extensive atmospheric tests of zinc coatings on steel are those conducted by A.S.T.M. since 1925 and still in progress. The results are summarized in Table 1. Three principal investigations were made.

HOT-DIPPED SHEETS. [1 to 19]. Corrugated sheets of hot-galvanized steel were exposed in 1926 at two industrial locations, Pittsburgh, Pa., and Altoona, Pa.; two marine locations, Sandy Hook, N. J., and Key West, Fla.; and one rural location, State College, Pa. In each location, over 100 specimens were exposed, which included variations in the type and gage of the steel and thickness of the zinc coating. The sheets were 26 by 30 in. (66 by 76 cm) and were supported on racks at 30° from horizontal. At each location there were 10 to 18 sheets having the same thickness of zinc coating.

Curves showing the typical course of failure are shown in Fig. 2. The rate of failure [13A] of the zinc coatings, in descending order, was *Altoona*, Pittsburgh, Sandy Hook, State College, and *Key West*. In contrast, uncoated steel sheets failed most rapidly at *Key West*, and least rapidly at *Altoona* and State College.

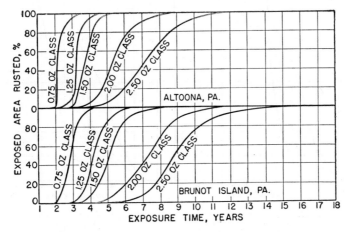

FIG. 2. Progressive Development of Rust on Steel Sheet, Coated with Hot-Dipped Zinc, Subjected to Outdoor Exposure   (Reproduced from *Proc. Am. Soc. Testing Materials*, **44**, 93, Fig. 1 (1944).)

Note that 1.18 oz Zn/sq ft of sheet (both sides) equals 1 mil thick coating.

It is difficult to correlate the behavior [15] of the galvanized steel at Key West with that in other locations. In the other locations the zinc disappeared most rapidly from the top surface where rust first appeared. At Key West, failure of the zinc and corrosion of the steel started on the lower side, and rust finally penetrated to the top surface. In the opinion of the authors, the anomalous behavior at Key West is probably the result of (1) absence there of sulfur compounds, which rapidly attack zinc, especially if they are combined with or adsorbed by solid particles; (2) rapid diurnal temperature changes and high humidity, which cause dew to condense at night and to evaporate off more slowly from the underside than from the upper surface; and (3) formation of basic carbonate on the upper surface by the salt water spray, which may retard corrosion of the zinc. If this explanation is correct, the principal attack of the zinc on the under side was caused by nearly pure condensed water. Similar corrosion has been reported inland in the tropics.

HARDWARE [1 to 19, 24]. A large variety of steel articles, such as hinges, bolts, washers, clamps, screws, and angle irons, was exposed in 1929 at the same five locations as the sheets. The coatings included hot-dipped, electroplated, and sherardized zinc, electroplated cadmium, hot-dipped aluminum and lead, and phosphate-treated articles. Because of the irregular shapes, the thickness and distribution of the coatings on each piece and lot were more variable than on the sheets. In consequence, initial rusting frequently appeared at a very early stage, and complete failure occurred much later (Fig. 3). It is therefore more difficult to draw conclusions on the life of the coatings from these tests than from those with sheets.

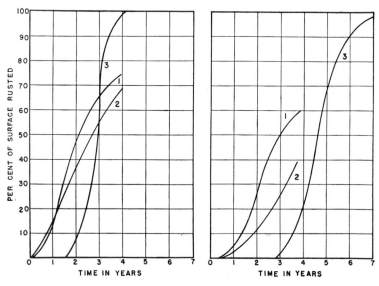

FIG. 3A. Coatings Approximately 0.8 Mil Thick.

FIG. 3B. Coatings Approximately 1.2 Mils Thick.

FIG. 3. Progressive Development of Rust on Zinc-Coated Hardware Compared with That on Hot-Dipped Sheet, Subjected to Outdoor Exposure in Pittsburgh. (Reproduced from *Symposium on the Outdoor Weathering of Metals and Metallic Coatings* of the American Society for Testing Materials, March 7, 1934, Fig. 11, p. 14.)

Curve 1: Electroplated zinc on hardware. Curve 2: Sherardized zinc on hardware. Curve 3: Hot-dipped sheet.

WIRES. In 1937, A.S.T.M. [10 to 19] started exposure tests of wires and wire fences in eleven locations, the most severe of which were Pittsburgh, Pa., Bridgeport, Conn., and Sandy Hook, N. J. Less severe corrosion occurred at State College, Pa., Lafayette, Ind., Ames, Iowa, Manhattan, Kans., Ithaca, N. Y., Santa Cruz, Calif., College Station, Texas, and Davis, Calif. The tests included steel wires of 5 gages, ranging from 6 to 14½, coated with hot-dipped and electroplated zinc, copper-clad, and hot-dipped lead; and also bare plain carbon steel and chrome-nickel steel. The unfabricated wires were 42 in. long and were welded to the racks in a horizontal position. The progress of corrosion was determined by (1) appearance, especially the extent of rust, (2) loss of weight, and (3) change in tensile strength of unfabricated wires.

Table 1 includes data on these wires for several of the severe locations. These data show that at Pittsburgh and Sandy Hook the smallest-gage wires lost zinc 10 to 20% faster than the largest-gage wires. Data not included in the table show that at severely corrosive locations the rate of loss of zinc from wires was 5 to 15% greater during the first year of exposure than during succeeding years. At the milder locations only a few specimens in the 0.2 to 0.3 oz zinc per sq ft class had rusted. At these locations, after 6 years of exposure, only the barbed wire and farm field fences with thin coat-

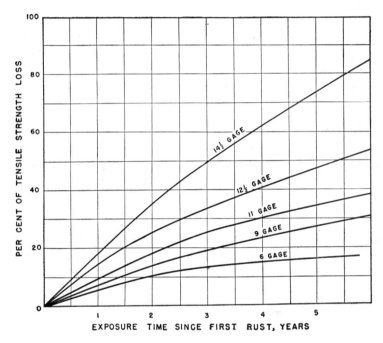

FIG. 4. Loss of Tensile Strength of Zinc-Coated Wires Subjected to Outdoor Exposure at Pittsburgh, Pa. (Reproduced from *Proc. Am. Soc. Testing Materials*, **43**, 86 [1943], Fig. 4.)

ings had rusted. The fences were not placed at the severely corrosive locations, such as Pittsburgh, Sandy Hook, and Bridgeport. In general the results on the fences agreed with those on the unfabricated wires which, however, seemed to have corroded a little faster than the fences.

Because these exposure tests are not nearly complete, no final conclusions can yet be drawn. The data thus far obtained indicate that the steel does not start to lose its tensile strength until most of the zinc has been consumed and the steel is rusted (Fig. 4).

A. E. S., A. S. T.M., N. B. S. Tests [23, 25]. In 1932, the American Electroplaters' Society, in cooperation with A. S. T. M. and National Bureau of Standards, started exposure tests of electroplated steel panels in six locations, of which four, Pittsburgh, Sandy Hook, Key West, and State College, were the same as for the earlier A. S. T. M. tests, and the other two were New York, N. Y., and Washington, D. C. (Table 1).

808 CORROSION PROTECTION

TABLE 1 PROTECTIVE VALUE OF ZINC COATINGS ON STEEL

Atmospheric Exposure

Location	Reference	Shape	Type	Thickness of Coating, mils	Av. Life, yr	Rate of Corrosion of Zinc		Average mils/yr
						mils/yr	Basis*	
New York, N. Y.	†	Sheet	Rolled zinc	...	...	0.21	cc	
	28	"	"	...	...	0.20	mc	
	33	"	"	...	...	0.19	cc	
	†	"	"	...	...	0.27	cc	0.22
						Av. sheet zinc		
	21, 28	Sheet	Hot-dipped	...	...	0.13	mc	
	21	Pipe	"	...	...	0.19	mc	
	21, 28	Sheet	Sherardized	...	...	0.13	mc	
	21	Pipe	"	...	...	0.19	mc	
	28	...	Sprayed	...	...	0.13	mc	
	28	...	Electroplated	...	...	0.17	mc	
	25	Sheet	"	0.20	1.10	0.19	al	
	25	"	"	0.50	2.30	0.22	al	
	25	"	"	1.00	3.56	0.28	al	0.22
						Av. electroplated zinc		
	25	Sheet	Hot-dipped	0.9	3.2	0.28	al	
	22	Wire diam.: in.						
	22	0.5	...	...	...	0.19	...	
	22	0.13	...	...	...	0.31	...	
	22	0.06	...	...	...	0.41	...	
	22	0.02	...	...	...	0.79	...	
Pittsburgh, Pa.	19A	Sheet	Hot-dipped	0.75	2.9	0.26	al	
	19A	"	"	1.10	4.1	0.27	al	
	19A	"	"	1.30	5.1	0.25	al	
	19A	"	"	1.71	7.4	0.23	al	
	19A	"	"	2.19	8.9	0.25	al	0.25
						Av. hot-dipped sheets		
	24	Hardware	Electroplated	0.78	2.1	0.37	al	
	24	"	"	1.22	3.0	0.41	al	
	24	"	"	1.75	3.6	0.49	al	0.42
						Av. of electroplated hardware		

* Footnotes are on p. 811.

TABLE 1. PROTECTIVE VALUE OF ZINC COATINGS ON STEEL — *Continued*

Atmospheric Exposure

Location	Reference	Shape	Type	Thickness of Coating, mils	Av. Life, yr	Rate of Corrosion of Zinc mils/yr	Basis*	Average mils/yr
Pittsburgh — *Cont.*	24	Hardware	Sherardized	0.78	2.5	0.31	al	
	24	"	"	1.22	4.7	0.26	al	
	24	"	"	1.75	5.5	0.32	al	
					Av. of sherardized hardware			0.30
	25	Sheet	Hot-dipped	0.9	3.5	0.26	al	
	25	"	Electroplated	0.20	1.1	0.18	al	
	25	"	"	0.50	2.3	0.22	al	
	25	"	"	1.00	4.0	0.25	al	
					Av. of electroplated sheet			0.22
	18C	Wire	Hot-dipped	0.47	1.0	0.47	al	
	18C	"	"	0.64	1.6	0.40	al	
	18C	"	"	0.80	2.0	0.40	al	
	18C	"	"	0.95	1.9	0.50	al	
	18C	"	"	1.10	2.4	0.46	al	
	18C	"	"	1.27	2.3	0.55	al	
	18C	"	"	1.49	2.9	0.51	al	
	18C	"	"	2.17	4.2	0.52	al	
	18C	"	"	2.98	5.0	0.60	al	
	16C	" gage 6	"	1.2 to 1.7	....	0.54	cc	
	16C	" " 9	"		....	0.57	cc	
	16C	" " 11	"		....	0.60	cc	
	16C	" " 14½	"		....	0.55	cc	
					Av. of hot-dipped wire			0.49
Altoona, Pa.	†, 33	Sheet	Rolled zinc	....	....	0.19	cc	
	19A	"	Hot-dipped	0.75	2.1	0.28	al	
	19A	"	"	1.10	3.2	0.29	al	
	19A	"	"	1.30	3.7	0.28	al	
	19A	"	"	1.71	5.3	0.31	al	
	19A	"	"	2.19	7.2	0.33	al	
					Av. of hot-dipped wire			0.57
					Av. of hot-dipped sheets			0.30

TABLE 1. PROTECTIVE VALUE OF ZINC COATINGS ON STEEL—Continued

Atmospheric Exposure

Location	Reference	Shape	Type	Thickness of Coating, mils	Av. Life, yr	Rate of Corrosion of Zinc mils/yr	Basis*	Average mils/yr
Altoona — Cont.	24	Hardware	Hot-dipped	0.78	2.3	0.34	al	
	24	"	Sherardized	0.78	3.8	0.21	al	
Bridgeport, Conn.	18C	Wire	Hot-dipped	0.47	2.5	0.19	al	
	18C	"	"	0.51	2.9	0.18	al	
	18C	"	"	0.64	3.3	0.19	al	
	18C	"	"	0.80	4.4	0.18	al	
					Av. of hot-dipped wires			0.19
Key West, Fla.	†, 33	Sheet	Rolled zinc	...	...	0.02	cc	
	†	"	"	...	...	0.14	cc	
	25‡	"	Hot-dipped	0.9	8	0.11	al	
	25§	"	Electroplated	0.20	4+	0.05—	al	
	25‖	"	"	0.50	4+	0.12—	al	
	25¶	"	"	1.00	8.5+	0.12—	al	
Sandy Hook, N. J.	†, 33	Sheet	Rolled zinc	...	...	0.06	cc	
	19A	"	Hot-dipped	0.64	6.6	0.10	al	
	19A	"	"	1.07	9.0	0.12	al	
	25	"	"	0.90	4.3	0.21	al	
	25	"	Electroplated	0.20	2.1	0.10	al	
	24	Hardware	"	0.78	5.5	0.14	al	
	24	"	Sherardized	0.78	4.1	0.19	al	
	18C	Wire	Hot-dipped	0.47	2.4	0.20	al	
	18C	"	"	0.51	2.9	0.18	al	
	18C	"	"	0.64	3.3	0.19	al	
	18C	"	"	0.80	4.1	0.20	al	
					Av. of hot-dipped wires			0.19
	18C	Wire gage 6	"	} 1.2 to 1.7		0.18	cc	
	18C	" " 9	"			0.19	cc	
	18C	" " 11	"			0.20	cc	
	18C	" " 14½	"			0.21	cc	
					Av. of hot-dipped wires			0.20

TABLE 1. PROTECTIVE VALUE OF ZINC COATING ON STEEL—*Continued*

Atmospheric Exposure

Location	Reference	Shape		Type	Thickness of Coating, mils	Av. Life, yr	Rate of Corrosion of Zinc		Average mils/yr
							mils/yr	Basis*	
State College, Pa.	33†	Sheet		Rolled zinc	....	....	0.04	cc	
	25‡	"		Hot-dipped	0.9	8	0.11	al	
	25§	"		Electroplated	0.20	4+	0.05—	al	
	25∥	"		"	0.50	4+	0.12—	al	
	25¶	"		"	1.00	10.5+	0.10—	al	
	18C	"		Hot-dipped	0.47	5.5	0.09	al	
	16C	Wire	gage 6	"	1.2 to 1.7		0.10	cc	
	16C	Wire	9	"			0.10	cc	
	16C	"	11	"			0.10	cc	
	16C	"	14½	"			0.09	cc	
Washington, D. C.	25	Sheet		Electroplated	0.20	3.7	0.05	al	
	25	"		"	0.50	6.7	0.08	al	
	25	"		"	1.00	12.3	0.08	al	0.07
						Av. of electroplated sheets	0.16		
	25	Sheet		Hot-dipped	0.9	5.6			
Lafayette, Ind.	18C	Wire		Hot-dipped	0.47	5.3	0.09	al	

* In this column the letters indicate the cleaning procedure or method used to evaluate mils/yr.
  cc = Specimens subjected to chemical cleaning.
  mc = Mechanical cleaning.
  al = The value is calculated from the average life.
† E. A. Anderson and C. E. Reinhard, *Zinc*, p. 331.
‡ Data obtained after publication of ref. 25.
§ Tests discontinued at end of 4 years.
∥ Tests discontinued at end of 8.5 years.
¶ Tests discontinued at end of 10.5 years.

These tests were conducted for only about 4 years, excepting a few specimens that were exposed over longer periods, for which data not published in reference 25 are here included. The specimens were 4 by 6 in. (10.2 by 15.2 cm) rectangles of 22-gage steel. Five specimens of each type were placed at each location on racks facing south and sloping 30° from horizontal. Cut edges of the hot galvanized sheets were protected with paint. Three different thicknesses of coating, 0.2, 0.5, and 1 mil, were exposed. There were 20 to 25 specimens of a given thickness of coating (although not all of the same type) at each location.

Bell Telephone Laboratories Tests. Exposure tests conducted by the Bell Laboratories in New York and reported in 1930 [20, 21, 28] are of interest because they included hot-dipped, electroplated, and sherardized zinc, and because efforts were made to correlate the results with accelerated tests, including simulated rainfall (Table 1). The steel sheets were 9 by 12 in. (22.8 by 30.4 cm) rectangles, $\frac{1}{16}$ in. (0.16 cm) thick, with coatings approximately 0.8 mil thick. They were hung vertically and cut edges were protected. The conclusions are based on weight losses of specimens that were brushed or scrubbed, and not cleaned chemically after exposure.

EFFECTS OF VARIOUS FACTORS

The results of various exposure tests of zinc coatings are summarized in Table 1 according to location. The basis for calculating the rate of corrosion of zinc is indicated for most of the tests.

Environment. In Table 1 the locations are listed according to the prevailing conditions in the order of industrial (New York, Pittsburgh, Altoona, and Bridgeport), marine (Key West and Sandy Hook), and suburban or rural (State College, Washington, and Lafayette). The average rates of corrosion correspond roughly to this classification. The average values in Table 1 (not weighted or with strictly comparable specimens) are, in mils penetration per year, New York, 0.26; Altoona, 0.26; Pittsburgh, 0.37; and Bridgeport, 0.19; with a general average for these four industrial locations of 0.27. At Sandy Hook, the rate is 0.15, and at Key West, 0.09; with an average for these marine exposures of 0.12. At State College, the rate is 0.08; at Lafayette, 0.09; and at Washington, 0.09; with a general average for these mild locations of 0.09.

The greater rate of corrosion of the zinc in industrial atmospheres is no doubt largely caused by the presence of SO_2 and SO_3, which increase the acidity of any condensed moisture and yield soluble zinc compounds, some of which may be washed off before the pH of the moisture is increased (by attack of the zinc) to a value at which basic zinc carbonate can form. In marine locations, protective films, usually of basic zinc carbonate, are formed and may retard corrosion. In rural or suburban areas, water is the principal corroding agent, and carbonate films are also formed.

Thickness of Coating. If the protection were directly proportional to the thickness of zinc, the corrosion rates given in Table 1, column 7, should be constant for a given location and type of specimen and coating.

For hot-dipped sheets and for zinc-coated wires, the corrosion rates are nearly constant, and the variations bear no systematic relation to the thickness. However, the electroplated sheets in New York, Pittsburgh, and Washington, including data for Washington obtained since publication of reference 25, showed an average rate of corrosion for 1-mil coatings that was 1.4 times that for 0.2-mil coatings, i.e., the thicker coatings lasted only 3.5 times as long as those that were one-fifth as thick. There is no simple explanation for this departure of the electroplated coatings from a life assumed proportional to the thickness.

Type and Purity of Zinc Coatings. Strictly comparable data on the protective value of the four types of zinc coatings, hot-dipped, electroplated, sherardized, and sprayed, are not available. Existing information indicates that, in general, a given thickness of zinc, however applied, furnishes roughly the same protection to steel. For example, the unweighted average rates of corrosion in mils per year on sheets in all comparable locations in Table 1 are hot-dipped, 0.17; electroplated, 0.14; and sheet zinc, 0.12. In the exposure tests of electroplated panels [25] a single set of hot-dipped zinc (0.9 mil) with edges protected was included. In the "mild" locations, especially Washington, D. C., the rate of failure of the hot-dipped zinc was nearly twice that of the electroplated zinc. In New York, tests in the Bell Laboratories [21, 28] showed that the electroplated zinc lost weight a little more rapidly than hot-dipped zinc. Further studies are needed to determine whether these differences are real.

TYPES OF ELECTROPLATED ZINC. No significant difference was observed in the A. E. S. tests [25] in the behavior of zinc coatings deposited from acid and cyanide baths. Previous reports [20] of the superiority of cyanide zinc deposits probably involved the better throwing power of the cyanide baths, which results in more uniform distribution of the zinc, especially on irregular shapes. In the A.E.S. study, the coatings were uniform within about 10%; hence differences in the throwing power did not enter.

COMPOSITE COATINGS INCLUDING ZINC [23]. Deposition of, e.g., 0.5 mil of nickel (with or without a final layer of chromium) over 0.5 mil of zinc on steel resulted in coatings that tended to blister, even in protected storage. On exposure, white spots appeared at pores in the nickel, and there was a tendency for subsequent peeling of the nickel. The protection afforded against rust of the steel was less than that offered by 0.5 mil of zinc alone or by 1 mil of nickel. The presence of a copper layer between the zinc and nickel slightly improved the protection of the steel, but still yielded spots and blisters.

A thin coating of chromium directly over a zinc coating on steel soon discolored and flaked and gave less protection than the zinc alone.

ALLOY COATINGS. In recent years a process* has been developed [32] in which nickel is plated on steel, followed by a layer of zinc, after which the composite coating is heated to a temperature at which the zinc and nickel form an alloy. The two metals may also be codeposited [31]. No conclusive data are available on the relative protective value of these and pure zinc coatings.

PURITY. The approximate equality of the hot-dipped and sherardized coatings, which contain Zn-Fe alloys, with the electroplated zinc, which is usually nearly pure zinc, indicates that iron is not a detrimental impurity in the zinc coating. It is reported [28] that in hot-dipped coatings lead, cadmium, and tin, up to a few tenths of a per cent, are not detrimental, but that antimony and copper are objectionable.

Shape and Position of Articles. Data in Table 1 show that the corrosion rate of zinc is higher on wires, especially on fine wires, than on panels. At Sandy Hook [17A], panels inclined about 30° from horizontal corroded more rapidly than vertical specimens. No doubt more fog, spray, and dust settled on the inclined panels (just as in the salt spray, p. 970), and thereby accelerated corrosion.

CORROSION IN AQUEOUS SOLUTIONS

No comprehensive service tests of zinc-coated steel in water or aqueous solutions have been reported. The effects of such factors as temperature and the concentrations

* Corronizing.

of dissolved oxygen, carbon dioxide, and salts upon the corrosion of the zinc coating are probably similar to those on zinc (see *Zinc*, p. 331), although the presence of other metals, such as any exposed steel, may introduce electrochemical effects.

The degree of electrochemical protection afforded to steel is directly related to the conductivity of the surrounding medium. In sea water, zinc may protect exposed steel to a distance of several inches, whereas in pure water small exposed areas, such as cut edges, will rust in a short time.

The use of galvanized steel pipe for domestic water supplies is of considerable interest. Existing data [34] indicate that in cold water the zinc exerts electrochemical (cathodic) protection to the steel. However, the Zn-Fe alloy adjacent to the steel is not so readily attacked as the pure zinc layer; hence rusting of exposed steel may occur while some alloy layer is still present. The alloy layer does, however, retard pitting of the steel. The life of the zinc coating may be appreciably increased by the ultimate deposition from the water of compounds, e.g., of calcium, inside the pipes.

In hot water, especially between 50° and 65° C (120° and 150° F), the zinc coating affords little electrochemical protection against corrosion of exposed steel, possibly because the zinc becomes electrically insulated by its corrosion products, or because under these conditions the potential between the zinc and steel becomes small, and the zinc may even become cathodic [29] to steel.

Additional data on the protective value of zinc coatings in water pipes are contained in a recent report [36] on pipes which were subjected to 15 years of service in Baltimore. With cold water, the pit depth (based on the average of the ten deepest pits in a 2-ft length of pipe) of galvanized pipe was only about one-half that of uncoated pipe. However, with hot water (40° to 65° C, 105° to 150° F) the pit depth of the galvanized pipe was one to two times as great as the pit depth of uncoated pipes.

Most impurities in the zinc coating do not affect its protective value in aqueous solutions. However, it was found [38] that 0.2% or more of tin in the zinc coating accelerated failure of the zinc and rusting of the steel when used in watering troughs. For example, normal coatings lasted as much as 15 years whereas, with tin present, they failed in a few years. No such difference was observed in the atmospheric exposure of zinc coatings with and without tin.

CORROSION IN SOILS

Corrosion of pipes in soil is complicated by the fact that the composition and texture (therefore the drainage) of the soil may vary widely, even within short distances. In addition, the water content of the soil, the concentration of dissolved salts and oxygen, and the pH may change over short periods of time. In many soils, uncoated steel pipes rust rapidly, and large pipes are usually protected with bituminous coatings. Pipes with hot-dipped zinc coatings, inside and out, are frequently used for water service pipes not more than 4 in. (10 cm) in diameter.

Extensive tests conducted by the National Bureau of Standards since 1922 [26, 27, 30, and 35] included galvanized pipes (data are given in Tables 11 and 12, p. 461-462). Conclusions were based on loss in weight (corresponding to average corrosion) and on measurements of pit depths. The two criteria give roughly parallel results, except in poorly drained soils, where there is relatively less pitting.

The behavior of zinc-coated pipe was determined on two lots of material, one of which [26] was buried in 1922 and the other [35] in 1937. The data from the two lots were not consistent, as the second lot deteriorated as much in 4 years as the first lot did in 10 years. The reason for this difference is not known. However, both sets of data are consistent in showing the decrease in rate of corrosion of the pipe brought about by the zinc coating. Most of the examples given in the following paragraphs are taken from the first set of the soil corrosion results.

It was found that galvanized steel pipes do not corrode appreciably until most of the zinc has been removed. In forty-five locations zinc coatings with 2.8 oz zinc per sq ft (4.7 mils) protected the steel from rusting or pitting for an average period of 10 years. Coatings with at least 2 oz zinc per sq ft (3.4 mils) are recommended for service pipes.

Zinc-coated pipes fail most rapidly in those soils in which steel corrodes quickly. The average weight loss per year of zinc-coated pipes in forty-five locations varied from 0.005 to 0.9 oz per sq ft, and of uncoated steel pipes from 0.17 to 2.5 oz per sq ft, i.e., the uncoated steel corroded about three times as rapidly as the galvanized pipes. The rate of loss in weight of a galvanized pipe increases when the steel is exposed (1) because the steel accelerates the corrosion of the zinc, and (2) because the steel corrodes more rapidly than the zinc.

Because pitting of the steel does not occur until most of the zinc has disappeared, the pit depths depend on the thickness of the zinc coating. In one series of tests, pipes with 2.8 oz zinc per sq ft showed, in 10 years, pits in only one of forty-six locations, whereas bare steel pitted in all locations, with an average pit depth of 0.080 in. In another series of tests, in 16 years pipes with 2.3 oz zinc per sq ft had average pit depths of 0.007 in., whereas on plain steel the pit depths averaged 0.063 in.

Reported observations [27] on galvanized pipes in service do not entirely confirm the above tests. In a 17-year period, galvanized gas service pipes in Philadelphia failed about as rapidly as uncoated pipes. This rapid failure of the zinc coatings in a large city may have resulted from (1) electrolysis from stray currents, or (2) accelerated corrosion of the zinc-coated pipes when they were used to repair systems consisting mainly of bare steel pipes. The data based on tests by the National Bureau of Standards were obtained in rural locations where electrolysis or contact with other pipes was improbable.

REFERENCES

1–19, *Proc. Am. Soc. Testing Materials*, Part I, or Committee Reports:

Ref. No.	Vol.	Year	A Hot-dipped Sheet	B Hardware	C Wire
			Page		
1	25	1925	106	115	...
2	26	1926	152	152	...
3	27	1927	191	...	...
4	28	1928	167	...	...
5	29	1929	146	149	...
6	30	1930	233	236	...
7	31	1931	181	183	...
8	32	1932	116	121	...
9	33	1933	149	154	...
10	34	1934	159	...	167
11	35	1935	88	92	...
12	37	1937	...	...	117
13	38	1938	84	90	...
14	39	1939	...	...	101
15	40	1940	106	...	...
16	41	1941	...	...	101
17	42	1942	120	...	...
18	43	1943	...	...	78
19	44	1944	92	96	109

20. C. L. Hippensteel and C. W. Borgmann, *Trans. Electrochem. Soc.*, **58**, 23 (1930).
21. C. L. Hippensteel, C. W. Borgmann, and F. F. Farnsworth, *Proc. Am. Soc. Testing Materials*, **30**, Part II, 456 (1930).
22. R. B Mears, *Bell Laboratories Record*, **11**, 141 (1933).
23. W. Blum, P. W. C. Strausser, and A. Brenner, *J. Research Natl. Bur. Standards*, **13**, 332 (1934), RP712.
24. C. D. Hocker, *Symposium on the Outdoor Weathering of Metals and Metallic Coatings*, p. 10, American Society for Testing Materials, March 7, 1934.
25. W. Blum, P. W. C. Strausser, and A. Brenner, *J. Research Natl. Bur. Standards*, **16**, 185 (1936). RP867.
26. K. H. Logan and Scott P. Ewing, *J. Research Natl. Bur. Standards*, **18**, 361 (1937). RP982.
27. Scott P. Ewing, *Soil Corrosion and Pipe Line Protection*, American Gas Association, 420 Lexington Ave., New York, N. Y., 1938.
28. R. M. Burns and A. E. Schuh, *Protective Coatings for Metals*, Reinhold Publishing Co., New York, 1939.
29. G. Schikorr, *Trans. Electrochem. Soc.*, **76**, 247 (1939).
30. K. H. Logan, *J. Research Natl. Bur. Standards*, **28**, 57 (1942). RP1446.
31. Benjamin Lustman, *Trans. Electrochem. Soc.*, **84**, 363 (1943).
32. R. Rimbach, *Metal Finishing*, **39**, 360 (1941); U. S. Patent 2,315,740 (April 6, 1943).
33. W. H. Finkeldey, Subcommittee VI. *Proc. Am. Soc. Testing Materials*, **44**, 224 (1944).
34. L. Kenworthy and M. D. Smith, *J. Inst. Metals* (London), **70**, 463 (1944).
35. K. H. Logan and M. Romanoff, *J. Research Natl. Bur. Standards*, **33**, 145 (1944). RP1602.
36. Charles F. Bonilla, *Trans. Electrochem. Soc.* **87**, 237 (1945).
37. E. A. Anderson, New Jersey Zinc Co., Palmerton, Pa., private communication to the authors.
38. Unpublished Report 540, New Jersey Zinc Company, Palmerton, Pa.

NICKEL AND CHROMIUM COATINGS

W. A. WESLEY*

By far the largest use of nickel and chromium coatings is for objects which must not only be protected from corrosion but must also present a bright and pleasing appearance. The corrosive environments in which this appearance must be maintained include all kinds of atmospheres, outdoors and indoors, as well as perspiration, soap, and other cleaning solutions, foodstuffs, and beverages. Resistance to moderate heating is required in some applications. An important reason for using nickel coatings on brass articles is to prevent season cracking. Nickel and chromium are normally passive under all these service conditions, and therefore fall into the class of cathodic or "noble" coatings when applied to such basis metals as steel, copper, brass, zinc, aluminum, or magnesium. If thin coatings could be prepared commercially, entirely continuous and free from pores, then it would be simple to calculate from the data on corrosion of the coating metal itself what minimum thickness must be specified for the intended service. Unfortunately, discontinuities are present in all commercial coatings less than about 25 microns (1 mil) thick, regardless of the method by which the films are applied. Upon exposure to corrosive environments, additional pores are developed. Since cathodic coatings do not afford galvanic protection of the basis metal laid bare at discontinuities, it is necessary to insure complete coverage or else to reduce corrosion at pores to a negligible rate.

Corrosion at Pores in Cathodic Coatings

It is generally believed that the galvanic corrosion of the basis metal at pores in a cathodic coating is under resistance control.†[1] The electrochemical relationships involved can be represented by the simple diagram of Fig. 1, from which the cathodic polarization curve for the basis metal is omitted because the local cathodic process

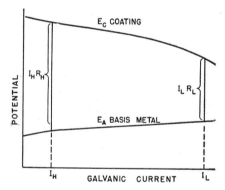

Fig. 1. Polarization Diagram for Corrosion at Pores in a Cathodic Coating.

at the base of pores is stifled by corrosion products, and the anodic curve for the coating metal is omitted upon the assumption that the latter remains substantially passive.

With light coatings the resistance of the galvanic circuit R_L is small because there

---

* Research Laboratory, The International Nickel Co., Inc., Bayonne, N. J.
† Types of control are discussed beginning p. 486.

[1] T. P. Hoar, *J. Electrodepositors' Tech. Soc.*, **14**, 42 (1938).

818   CORROSION PROTECTION

are many pores and the length of the restricted paths through which the current must flow is small. A relatively large total current (I_L) flows. Increase in thickness of the coating raises the resistance rapidly by decreasing the number of pores, decreasing their diameter, and increasing their length. A much lower total current (I_H) results from this difference in resistance.

The question of the intensity of attack, that is, the value of I over A, the anodic area, has been discussed by Evans,[2] who proposed an equation based on Ohm's Law:

$$\frac{I}{A} = \frac{E}{AR} = \frac{EA'}{\rho At} \quad (1)$$

where A' = the combined cross section of the pores,
E = the operative emf = $E_c - E_a$ (difference between polarized cathodic and anodic potentials),
ρ = the specific resistivity of the corrosive electrolyte,
t = the thickness of the coat.

According to this formula, the total corrosion of the basis metal can be reduced as much as desired by merely increasing t. The ratio A'/A, which is equal to one with very thin coatings, does not become greater than one, but less, as t is increased; therefore the *intensity* of attack I/A must diminish.

APPLICATION OF DECORATIVE COATINGS

The so-called decorative coatings of nickel and chromium are almost universally applied by electrodeposition. It is customary to plate parts after fabrication, but the use of pre-plated strip was increasing sharply before 1941 and will undoubtedly gain in popularity because of lower finishing costs and greater uniformity of coating.

Nickel Plating. The corrosion engineer need not be concerned with the many nickel plating processes in commercial use because no important influence has been demonstrated[3] of variations in plating solutions and conditions upon the protective value of sound coatings. The protective value is governed primarily by the thickness of the coat. It is true that the quality, cleanliness, and smoothness of the basis metal surface influence the porosity developed in the coat, but the effects of these variables themselves become less important as the thickness of deposit is increased. A good résumé of modern nickel plating practice is now available in convenient form.[4]

Chromium Plating. Chromium plating practice is quite well standardized.[5] The plating conditions exert some influence upon the porosity of the deposit.[6] Less porous deposits are secured at 35° C (95° F) than at higher temperatures, and the use of a high sulfate ratio is likewise beneficial. Again, however, the principal factor influencing protective value is the thickness of the deposit, which will be discussed below.

PROTECTIVE VALUE OF DECORATIVE COATINGS

The porosity of nickel electrodeposits on various basis metals as measured by conventional porosity tests* is extremely variable at thicknesses below 25 microns

* See *Tests for Metal Coatings*, p. 1032, for discussion of porosity tests.

[2] U. R. Evans, *Metallic Corrosion, Passivity and Protection*, p. 581, Edward Arnold and Co., London, 1937.
[3] W. Blum and P. W. C. Strausser, *J. Research Natl. Bur. Standards*, **24**, 443–474 (1940).
[4] W. L. Pinner, G. Soderberg, and E. M. Baker, *Modern Electroplating*, pp. 235–274, The Electrochemical Society, 1942.
[5] G. Dubpernell, *Modern Electroplating*, pp. 117–143, The Electrochemical Society, 1942.
[6] W. Blum, P. W. C. Strausser, and A. Brenner, *J. Research Natl. Bur. Standards*, **13**, 331–355 (1934).

NICKEL AND CHROMIUM COATINGS

(1 mil).[7] Most of the reported values for salt spray and ferroxyl tests lie within the shaded area of Fig. 2 and its extensions. When the best plating practice is followed and good quality basis metals are employed, the ratings will lie near the left and lower borders of the shaded zone. The fact that an increase in coating thickness above 25 microns (1 mil) tends to override other variables and guarantee low porosity is evident from the necking down of the shaded zone to the right.

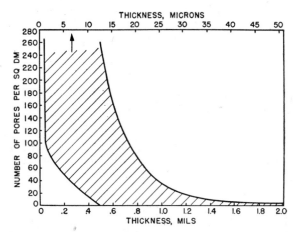

FIG. 2. Results of Conventional Porosity Tests of Nickel Coatings.

The minimum thickness of deposit required to maintain a pleasing appearance varies with the nature of service. This point is illustrated in Fig. 3, where coatings 12.5 microns (0.5 mil) thick showed approximately the same rating in suburban atmospheres as those 25 microns (1 mil) thick in marine atmospheres, whereas a thickness of 50 microns was required to equal this performance at industrial sites. It has been estimated[8] that plated coatings which maintain a satisfactory appearance for more than a year in the suburban atmospheres represented in Fig. 3 will last almost indefinitely indoors under normal household or office conditions.

Effect of Nature of Basis Metal. Occasions may arise when the choice of basis metal may be influenced by the thickness of coating required for satisfactory protection of the metal selected. Figure 4 is taken from one of the reports of the cooperative exposure tests.[9] From this figure can be judged the relative thicknesses of coating required on different basis metals to afford about the same degree of protection to each. Thus a coating thickness of 30 microns (1.2 mils) on steel was about equivalent to one of 25 microns (1 mil) on zinc alloys or 10 microns (0.4 mil) on the brasses.

Bright-finished aluminum alloys[10,11] can now be prepared with coatings of adhesion strength exceeding the tensile strength of the aluminum alloy itself.[12] Exposure data

[7] O. Bauer, H. Arndt, and W. Krause, *Chromium Plating*, pp. 128, 193, translated by E. W. Parker, Edward Arnold and Co., London, 1935.
[8] W. Blum, P. W. C. Strausser, and A. Brenner, *J. Research Natl. Bur. Standards*, **13**, 331–355 (1934).
[9] W. Blum and P. W. C. Strausser, *J. Research Natl. Bur. Standards*, **24**, 443–474 (1940).
[10] R. E. Pettit, Aluminum Company of America, private communication.
[11] R. F. Yates, *Proc. Am. Electroplaters' Soc. Convention*, p. 118, June, 1943.
[12] B. B. Knapp and W. A. Wesley, Research Laboratory, The International Nickel Co., Inc., unpublished data.

for plated aluminum alloys are not yet available. For mild outdoor exposure, the minimum thickness of nickel should be 40 microns (1.5 mils),[8,10,13] with heavier coatings indicated for more severe service.

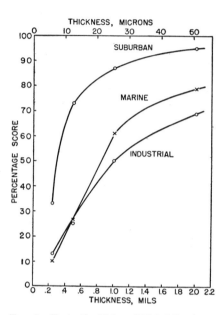

FIG. 3. Protective Value of Nickel Coatings on Steel.

18 months' exposure at:

Suburban sites: Washington, D.C., and State College, Pa.
Marine sites: Key West, Fla., and Sandy Hook, N.J.
Industrial sites: New York, N.Y., and Pittsburgh, Pa.

Method of rating specimens:

Surface Rusted, %	Rating	Corresponding Percentage Score
0	5	100
0 to 5	4	80
5 to 10	3	60
10 to 20	2	40
20 to 50	1	20
50 to 100	0	0

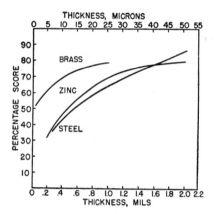

FIG. 4. Effect of Thickness of Nickel (or Nickel Plus Copper) on Percentage Score for One Year Exposure.

Method of rating specimens given in Fig. 3.
Average of six locations. All final nickel coatings covered with 0.5 micron (0.02 mil) of chromium.

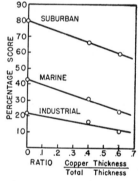

FIG. 5. Effect of Copper : Nickel Ratio on Protective Value of Thin Coatings on Zinc Alloys.

Total coating thickness: 0.5 mil (12.5 microns).
Final coating: 0.02 mil (0.5 micron) chromium.
Exposure period: 2.2 years.

Nickel plating of magnesium alloys is still a laboratory process.[14] The difference in potential between magnesium and nickel is so high that a truly impervious deposit

[13] Aluminum Company of America, *Finishes for Aluminum*, p. 54, 1938.
[14] W. S. Loose, *Trans. Electrochem. Soc.*, **81**, 213–230 (1942).

would be required for resistance to outdoor atmospheres. At the present stage of development, coatings 25 microns (1 mil) thick are satisfactory for indoor service where contact with aqueous solutions does not occur.

Effect of an Intermediate Layer of Copper. The extent to which copper can be substituted for nickel in decorative coatings in order to reduce costs and to take advantage of the greater ease of buffing copper is still a controversial subject. It is known that even a relatively thick coating of chromium applied directly to copper with no intervening nickel offers little resistance to weathering. It is also well established[15] that composite copper and nickel coatings on steel and zinc alloys are lower in protective value than nickel coatings of the same thickness. However, the importance of the effect varies with the total thickness and in different atmospheres. Representative data selected from the cooperative tests[15,16] are presented in Table 1 and Fig. 5. Definite evidence is now at hand[17] that copper corrosion products spread upon the surface of a chromium and nickel-plated steel panel cause accelerated corrosion of the coating in an industrial atmosphere. This helps explain how a copper layer in a composite coating can sometimes be harmful.

TABLE 1. EFFECT OF AN INTERMEDIATE LAYER OF COPPER ON THE PROTECTIVE VALUE OF NICKEL COATINGS

Basis Metal: Steel

Thickness in Microns and Order of Layers				Total Thickness		Duration of Exposure	Deviation in Score* from that of Nickel Controls of Same Thickness		
Ni	Cu	Ni	Cr	Microns	Mils	Years	Suburban	Industrial	Marine
1.2	2.0	3.2	...	6.4	0.25	1.5	−18	0	−6
2.5	3.8	6.5	...	12.7	0.5	1.5	+ 5	+ 6	−13
5.0	7.5	12.5	...	25.0	1.0	1.5	+ 4	0	−25
10.0	15.0	25.0	...	50.0	2.0	1.5	+ 4	− 4	−10
5.0	7.5	12.5	0.5	25.5	1.0	2.2	+11	−10	−13
5.0	12.5	7.5	0.5	25.5	1.0	2.2	+ 6	−12	−15
10.0	15.0	25.0	0.5	50.5	2.0	2.2	0	−24	+ 5

* For method of scoring, see legend of Fig. 3.

In plating on zinc alloys, a special bath must be used if nickel is to be deposited directly upon the zinc alloy. This process is more difficult to control than that of plating copper on zinc so that it is customary to coat zinc alloys with copper, followed by nickel, before chromium plating. Regardless of the total thickness of coating deemed sufficient for a given type of service, it is important to have an adequate thickness of nickel over the copper layer to prevent surface staining and attack by copper corrosion products. A final nickel layer at least 7.5 microns (0.3 mil) thick on articles intended for mild service, and at least 12.5 microns (0.5 mil) thick for outdoor exposure is recommended for this purpose.

Effect of Thickness of Chromium. The reasons why the entire thickness of 25 to 50 microns of coating for overcoming porosity is not generally applied as chromium alone are that heavy chromium plating is a more costly process than nickel plating,

[15] W. Blum and P. W. C. Strausser, *J. Research Natl. Bur. Standards*, **24**, 443–474 (1940).
[16] W. Blum, P. W. C. Strausser, and A. Brenner, *J. Research Natl. Bur. Standards*, **13**, 331–355 (1934).
[17] W. A. Wesley and B. B. Knapp, The International Nickel Co., unpublished data.

chromium is more brittle, heavy deposits of it are extremely difficult to polish, and it is much more difficult to obtain uniform distribution. It is customary to depend upon nickel to provide the resistance to corrosion and apply a thin chromium deposit whose function is to prevent tarnishing or fogging of the nickel surface.

As a chromium deposit grows in thickness, internal stresses become sufficient when a thickness of about 0.5 micron (0.02 mil) is reached to cause fine hairlike cracks. As the growth proceeds, the number of cracks increases until metal begins to deposit over the cracks first formed. The presence of these discontinuities seems to have no deleterious effect upon the total protective value except in coatings on brass where the cracks are apparently propagated through the thin nickel deposits employed. This has been suggested[18] as a reason for the perplexing differences in the effects of chromium thickness in coatings on brass, zinc, and steel as illustrated in Fig. 6. There seems to be genuine merit in the heavier coatings on a steel base.

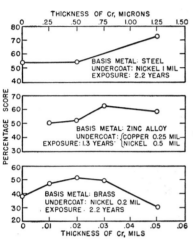

FIG. 6. Effect of Chromium Thickness on Protective Value of Composite Coatings.

Effects of Intermediate Zinc or Cadmium Layers. The cooperative exposure tests[19] led to the conclusion that the use of zinc or cadmium under nickel causes white stains and blistering without increasing the protective value of the coating.

SPECIFICATIONS FOR DECORATIVE COATINGS

The history of bright finishes has been a record of repeated increases in the thickness requirements, and there is good reason to believe that the specifications now in vogue do not represent the end of this upward trend. The present status is indicated by the representative codes in Table 2. Details of other features of the specifications, such as the selection of samples, definition of significant surfaces, and determination of thickness, can be found in the references cited in Table 2. British practice was reviewed in 1940.[20]

It is important to realize that the thickness of an electrodeposit varies considerably over the surface of most plated articles so that the average thickness actually required in order to meet specifications may be as much as several times the minimum thickness specified.

NON-DECORATIVE USES OF NICKEL COATINGS

Thick Electrodeposited Coatings. If the basis metal is one of smooth, clean surface, such as a good grade of cold-rolled steel, it is necessary to allow only about 0.075 mm (0.003 in.) thickness of nickel to block out pores and then add to this the amount of nickel estimated to be required for withstanding normal corrosion

[18] W. Blum and P. W. C. Strausser, *J. Research Natl. Bur. Standards*, **24**, 443–474 (1940).
[19] W. Blum, P. W. C. Strausser, and A. Brenner, *J. Research Natl. Bur. Standards*, **13**, 331–355 (1934).
[20] A. W. Hothersall, *Metal Finishing*, **38**, 27 (1940).

TABLE 2. EXCERPTS FROM A. S. T. M. SPECIFICATIONS FOR
ELECTRODEPOSITED COATINGS

μ = micron = 1×10^{-4} cm

A. On Steel*

Kind of service	Non-corrosive	Mild	General	Severe outdoor†
Coating designation	Q.S.	K.S.	F.S.	D.S.
Total copper plus nickel, minimum	0.4 mil (10μ)	0.75 mil (19μ)	1.25 mil (31μ)	2 mils (51μ)
Thickness final nickel, minimum	0.2 mil (5μ)	0.4 mil (10μ)	0.6 mil (15μ)	1 mil (25μ)
Chromium (if required), minimum	0.01 mil (0.25μ)	0.01 mil (0.25μ)	0.01 mil (0.25μ)	0.01 mil (0.25μ)
Salt spray test‡	16 hours	48 hours	72 hours	96 hours
No. of pores developed during the salt spray test, maximum	6 per sq ft	6 per sq ft	3 per sq ft	6 per sq ft

B. On Zinc and Zinc-Base Alloys§

Kind of service	Non-corrosive	Mild	General
Coating designation	Q. Z.	K. Z.	F. Z.
Total copper plus nickel, minimum	0.5 mil‖ (12.5μ)	0.75 mil (19μ)	1.25 mil (31μ)
Copper, minimum	0.2 mil (5μ)	0.3 mil (7.5μ)	0.4 mil (10μ)
Thickness final nickel, minimum	0.3 mil (7.5μ)	0.3 mil (7.5μ)	0.5 mil (12.5μ)
Chromium (if required), minimum	0.01 mil (0.25μ)	0.01 mil (0.25μ)	0.01 mil (0.25μ)
Salt spray test‡	16 hours	32 hours	48 hours
No. of pores developed during the salt spray test, maximum	6 per sq ft	6 per sq ft	6 per sq ft

C. On Copper and Copper-Base Alloys¶

Kind of service	Non-corrosive	Mild	General
Coating designation	Q. C.	K. C.	F. C.
Nickel, minimum	0.1 mil (2.5μ)	0.3 mil (7.5μ)	0.5 mil (12.5μ)
Chromium (if required), minimum**	0.01 mil (0.25μ)	0.01 mil (0.25μ)	0.01 mil (0.25μ)

* 1942 Book of A. S. T. M. Standards, Part I, p. 1454.
† Report of Committee B-8, A. S. T. M. Preprint, p. 5, 1945.
‡ 1944 Book of A. S. T. M. Standards, Part I, p. 1843.
§ 1942 Book of A. S. T. M. Standards, Part I, p. 1461.
‖ Total of 0.3 mil (7.5μ) if alternative of nickel only is used.
¶ 1942 Book of A. S. T. M. Standards, Part I, p. 1458.
** Chromium coatings 0.05 mil (1.25μ) or more in thickness are likely to cause cracking of nickel deposits on brass. An effort should therefore be made to obtain the required minimum thickness of chromium with as low a maximum thickness as practicable.

in the prescribed service by reference to the corrosion rates for wrought nickel given in Nickel, p. 253. If an inferior base is to be protected, an additional allowance should be made. It is commercially commonplace to deposit layers of nickel as thick as 7 mm (280 mils).

Typical applications[21] of electrodeposited nickel coatings to resist corrosion include:

1. Printing plates and rolls to resist corrosion by and avoid contamination of inks.
2. Lehr rolls in contact with hot glass.
3. Press plates and embossing rolls in the plastics industry.
4. Filter press plates for caustic soda solutions.
5. Paper mill rolls, including drier, table, wire return, paper guide, and return rolls.

[21] W. A. Wesley, Monthly Rev. Am. Electroplaters' Soc., 25, 581–603 (1938). Revised and reprinted by The International Nickel Co., Inc., 1941.

6. Alkaline storage battery containers and parts.
7. Anodes in oxygen-hydrogen electrolytic cells.
8. Stop-off in nitriding steel.
9. Sucker rods for corrosive oil wells.

A recent development is the discovery that electrodeposited nickel, because of its purity, is much more resistant to oxidation by steam at high temperatures such as 600° C (1110° F) than are the common grades of wrought nickel.[22]

Nickel-Clad Steel. This composite product is steel-protected on one side or both with a pore-free layer of wrought nickel which is bonded metallurgically to the steel base and cannot separate from it as a result of thermal or mechanical treatment. It is available in large and thick plates and is used in the construction of massive pressure vessels, evaporators, storage tanks, tank cars, and other types of equipment, where the chemical and mechanical properties of nickel are required at a lower cost than that of the solid wrought nickel. Standard claddings of 5, 10, 15, and 20%, on one or both sides, are available.

Since the nickel layer is identical with wrought nickel, its corrosion-resisting properties are not different from those of nickel described in *Nickel*, p. 253. It is important for some types of service to inspect the nickel surface for embedded particles of steel or scale. If such are found, they should be removed by pickling to avoid the danger that they might serve as nuclei for the development of pits.

Sprayed Coatings. Nickel coatings can be applied by metal spraying, but if they are intended for corrosive applications they must be made relatively thick. A minimum thickness of 0.8 mm (32 mils) is recommended because of the high porosity.

Alloy Coatings. A field of development which is very active at the present time is that of extremely thin coatings of nickel alloyed with the basis metal or with superimposed layers of other metals by diffusion. Thus nickel-zinc diffused coatings show amazing resistance to the standard salt spray,[23] and nickel-tin coatings show promise as coatings for food containers.[24] Nickel-iron coatings on steel, made by displacing nickel from a dissolved salt solution by simple immersion and subsequent diffusion of the nickel at elevated temperatures, reduce the rate of atmospheric corrosion of the steel base to the level of that of a 3% nickel steel.[25] The amount of nickel applied is only 0.5 micron (0.02 mil). It is too early to evaluate the ultimate importance of this work.

CORROSION RESISTANCE OF CHROMIUM

The behavior of thin deposits of chromium has been discussed above in connection with decorative coatings. When chromium coatings are applied for corrosive industrial service, a thickness of a much higher order is commonly specified, i.e., not less than 75 microns (3 mils). In electrodepositing thick chromium, the metal deposits over the cracks formed in the initial layer and, while additional cracks continue to form, they do not extend to the basis metal. In addition to the thickness of the coating, the rate and other conditions of deposition as well as the condition of the basis metal surface affect the porosity. Proper preparation of the original surface to eliminate pits and flaws is essential. It is the corrosion resistance of heavy electrodeposited chromium that will be considered in the following section.

[22] E. N. Skinner and J. T. Eash, *Trans. Am. Soc. Mech. Engrs.*, **66**, 289 (1944).
[23] R. Rimbach, *Metal Finishing*, **39**, 360–364 (1941).
[24] H. R. Copson and W. A. Wesley, *Trans. Electrochem. Soc.*, **84**, 211–224 (1943).
[25] W. A. Wesley and H. R. Copson, unpublished data, Research Laboratory, The International Nickel Co., Inc., New York, N.Y.

AQUEOUS CORROSION CHARACTERISTICS

Chromium in the passive condition is outstandingly corrosion-resistant. It is not employed in contact with reducing solutions or with the halogen acids or sulfuric acid solutions where passivity is lost and the corrosion rates are high. Dissolved oxygen from air is sufficiently oxidizing to maintain the passivity of chromium in neutral solutions, but in most solutions of low pH, stronger oxidizing agents must be present and the halogen acids absent in order to stabilize the passive condition. It resembles the more highly alloyed stainless steels in its general corrosion behavior.

In most of its applications, chromium is cathodic to the common metals and alloys and tends to accelerate their rates of corrosion if galvanically coupled with them. There appears to be no evidence that chromium, in media in which it normally remains passive, can be caused to corrode rapidly by galvanic contact with other metals.* It is only the possible effect of such contact upon the other member of the couple that need be considered.

Data of Table 3 represent the results of simple continuous immersion tests and are useful only as a basis for intelligent guessing as to the usefulness of chromium plate for specific corrosive environments. The authors of the data interpret them in the light of practical experience with chromium coatings as follows:

Penetration, ipy	General Classification of the Corrosive Medium
0.000 to 0.003	Non-corrosive
0.003 to 0.020	Mildly corrosive
0.020 to 0.060	Severely corrosive
Over 0.060	Prohibitively corrosive

References to other aqueous and non-aqueous corrosion data are available in convenient summaries.[26,27,28] Many of the uses of heavy chromium based primarily upon its hardness and resistance to wear also involve a certain degree of resistance to corrosion.[25,29]

HIGH-TEMPERATURE CORROSION

The rate of attack of various atmospheres on cast chromium (98.1% chromium) was determined by Hatfield[30] with the results shown in Table 4.

In the glass industry, chromium plating is applied to molds to prolong life and prevent adhesion of the molten glass. The temperatures involved are in the range of 750° to 950° C (1380° to 1740° F). At these temperatures chromium diffuses slowly into the cast iron or steel base, but the life of the part is prolonged because the high-

---

* However, in acids such as dilute sulfuric or hydrochloric, chromium often remains passive for an appreciable period of time until "activated" by (in galvanic contact with) an active metal like zinc at a single point. The entire surface of chromium thereupon progressively loses passivity starting from this point of contact, and reacts with evolution of hydrogen. EDITOR.

[26] M. Kolodney, "Chromium Plating," *Information Release* 4, War Metallurgy Committee of the National Research Council, 1943.

[27] F. Ritter, *Korrosionstabellen metallischer Werkstoffe*, Edwards Bros., Ann Arbor, Mich., 1943.

[28] R. J. McKay and R. Worthington, *Corrosion Resistance of Metals and Alloys*, pp. 331–343, Reinhold Publishing Corp., New York, 1936.

[29] W. L. Pinner, G. Soderberg, and E. M. Baker, *Modern Electroplating*, pp. 235–274, The Electrochemical Society, 1942.

[30] W. H. Hatfield, *Engineer*, **134**, 639–643 (1922).

TABLE 3. RESISTANCE OF ELECTRODEPOSITED CHROMIUM TO ACIDS, SALTS, AND VARIOUS ORGANIC SUBSTANCES*

Concentration — 10% by weight for solid solutes and 10% by volume for liquids, unless otherwise stated.

Simple Immersion Tests	Penetration, ipy	
Acids	12° C (54° F)	58° C (136° F)
Acetic	0.000	0.015
Acetic (100% glacial)	0.000	0.008
Anthranilic (satd. soln.)	0.000	0.000
Anthraquinone, 2-sulfonic	0.000	0.001
Arsenic	0.000	0.011
Benzene, sulfonic	0.000	0.002
Benzoic (satd. soln.)	0.000	0.000
Butyric	0.000	0.006
Carbolic (phenol) (satd. soln.)	0.000	0.000
Chloric	0.016	.....
Cinnamic (satd. soln.)	0.000	0.000
Citric	0.000	0.007
Dichloroacetic	0.000	0.062
Dinitrobenzoic (3,5) (satd. soln.)	0.000	0.001
Formic	0.000	1.200
Fumaric (satd. soln.)	0.000	0.000
Furoic (pyromucic)	0.000	0.000
Gluconic	0.000	0.000
Glycollic	0.000	0.023
Hydrobromic	0.001	0.186
Hydrofluoric	1.000	.....
Hydriodic	0.000	0.015
Lactic	0.000	0.006
Maleic	0.000	0.018
Malic	0.002	0.009
Malonic	0.001	0.014
Mandelic (amygdalic) (satd. soln.)	0.000	0.001
Mixed acid (36% HNO_3, 61% H_2SO_4, 3% H_2O)	0.000	0.001
Monochloroacetic (chloroacetic)	0.000	0.003
Mucic (satd. soln.)	0.000	0.000
Naphthalene, 2,7-disulfonic	0.000	0.001
Naphthionic (satd. soln.)	0.000	0.000
Nitric	0.000	0.012
Nitric (100% fuming)	0.000	0.005
Nitrobenzoic (*meta*) (satd. soln.)	0.000	0.000
Nitrocinnamic (*meta*) (satd. soln.)	0.000	0.000
Oleic (100%)	0.000	0.000
Oxalic	0.000	0.001
Palmitic (100%)	0.000	0.000
Perchloric	0.001	0.042
Phenolsulfonic (*ortho*)	0.000	0.026
Phenylacetic (satd. soln.)	0.000	0.001

NICKEL AND CHROMIUM COATINGS

TABLE 3. — *Continued*

Simple Immersion Tests	Penetration, ipy	
Acids	12° C (54° F)	58° C (136° F)
Phosphoric	0.001	0.034
Phosphoric (85%)	0.000	0.002
Phthalic (satd. soln.)	0.000	0.003
Picric (satd. soln.)	0.000	0.000
Propionic	0.000	0.005
Pyrogallic	0.000	0.000
Pyruvic	0.000	0.000
Salicylic (satd. soln.)	0.000	0.002
Stearic (100%)	0.000	0.000
Succinic	0.001	0.010
Sulfanilic (satd. soln.)	0.001	0.008
Sulfobenzoic (*ortho*)	0.000	0.129
Sulfuric	0.011	10.000
Sulfuric (100%)	0.030	0.069
Tannic	0.000	0.000
Tartaric	0.000	0.004
Toluene, sulfonic (*para*)	0.000	0.000
Trichloroacetic	0.001	0.103
Salts		
Aluminum chloride	0.000	0.003
Aluminum sulfate	0.000	0.008
Amino G salt (satd. soln.)	0.001	0.015
Ammonium chloride	0.000	0.004
Barium chloride	0.000	0.001
Calcium chloride	0.000	0.000
Calcium hypochlorite	0.002	0.035
Chromic chloride	0.000	0.003
Cupric chloride	0.015	.....
Cupric nitrate	0.002	0.007
Ferric chloride†	0.000	0.016
Ferrous chloride	0.000	0.014
Magnesium chloride	0.000	0.000
Manganese chloride	0.000	0.000
Mercuric chloride†	0.079	.....
Potassium chloride	0.000	0.000
Schaeffer salt (satd. soln.)	0.000	0.006
Sodium benzene sulfonate	0.000	0.000
Sodium chloride	0.000	0.000
Sodium formate	0.000	0.000
Sodium hydrosulfite	0.000	0.000
Sodium phenol sulfonate	0.000	0.001
Sodium sulfate	0.000	0.002
Stannous chloride	0.000	0.035
Strontium chloride	0.000	0.000
Zinc chloride	0.000	0.001

TABLE 3. — Continued

Simple Immersion Tests	Penetration, ipy	
Miscellaneous	12° C (54° F)	58° C (136° F)
Acid green	0.000	0.003
Aminophenol (meta) (satd. soln.)	0.000	0.001
Aniline hydrochloride	0.001	0.023
Benzyl chloride (satd. soln.)	0.000	0.000
Benzyl chloride (100%)	0.000	0.000
Carbon tetrachloride (satd. soln.)	0.000	0.000
Carbon tetrachloride (100%)	0.000	0.000‡
Chlorobenzene (satd. soln.)	0.000	0.000
Chlorobenzene (100%)	0.000	0.000
Chloroform (satd. soln.)	0.000	0.000
Chloroform (100%)	0.000	0.000
Chlorohydroquinone	0.000	0.001
Chlorophenol (ortho) (satd. soln.)	0.000	0.001
Nitrophenol (para) (satd. soln.	0.000	0.001
Paper pulp suspension	0.000	0.000
Phthalimide (satd. soln.)	0.000	0.001
Sodium hydroxide	0.000	0.000
Succinimide (satd. soln.)	0.000	0.001
Tartrazine	0.000	0.006
Tetrachlorobenzene (satd. soln.)	0.000	0.000

* R. E. Cleveland, M. M. Sternfels and J. M. Hosdowich, unpublished data, Chromium Corporation of America, 1931.

† These data obtained by weight loss do not indicate true penetration except where corrosion is uniform. Pitting, and hence greater actual penetration, is very likely to occur, for example, with oxidizing heavy-metal salts like $FeCl_3$ and $HgCl_2$. EDITOR.

‡ For moist carbon tetrachloride a much higher corrosion rate is expected. EDITOR.

TABLE 4. RATE OF GAIN IN WEIGHT OF CHROMIUM EXPOSED TO HOT GASES

mdd

Gas	700° C (1290° F)	800° C (1470° F)	900° C (1650° F)	1000° C (1830° F)
Oxygen	47	97	220	630
Steam	5	37	117	210
Carbon dioxide	27	34	130	308
Sulfur dioxide	16	39	320	360

chromium alloy first formed is itself resistant to oxidation.[31,32] Resistance to attack by reduced sulfur compounds, including hydrogen sulfide and organic sulfur compounds, at ordinary and elevated temperatures has led to successful applications of chromium plate in oil refining.[31,32]

[31] M. Kolodney, "Chromium Plating," *Information Release* 4, War Metallurgy Committee of the National Research Council, 1943.

[32] R. J. McKay and R. Worthington, *Corrosion Resistance of Metals and Alloys*, pp. 331-343, Reinhold Publishing Corp., New York, 1936.

Chromium Coatings Applied by Diffusion

Chromizing is a process in which an alloy of chromium and iron is formed on the surface of steel parts by diffusion. The material to be treated is packed in a closed container with chromium powder and alumina and then heated to a temperature between 1300° C (2370° F) and 1400° C (2550° F) in an atmosphere of hydrogen.[33] The chromium content of the diffused layer is said to vary between 10 and 20%. The corrosion resistance of chromized steel is somewhat similar to that of high-chromium steels. (Refer to *Chromium-Iron, Austenitic Chromium-Nickel-Iron, and Related Heat-Resistant Alloys*, p. 640.) Applications have been principally to high-temperature service in air, steam, and other oxidizing atmospheres. The modern availability of a large variety of corrosion- and oxidation-resistant alloy steels has mitigated against widespread use of this process.

Another method for impregnating the surface of steel with chromium has been developed which involves heating the parts in a tube through which an atmosphere of hydrogen and chromous chloride vapor is passed. Temperatures of 900° to 1000° C (1650° to 1830° F) are required, and a surface layer containing about 30% chromium is obtained. Corrosion resistance is said to be about what would be expected of a steel containing 30% chromium.[34,35]

TIN COATINGS

Bruce W. Gonser* and James E. Strader*

The largest, single peacetime use for tin is in coating steel sheets to make tin plate. Nearly all this is used for making containers, particularly food cans. In this way the strength and economy of steel are combined with the corrosion resistance, easy solderability, and sanitary appearance of a remarkably thin protective surface of tin.

Coatings of tin are applied also to wire and to cast or fabricated iron and steel articles, such as equipment used in the food and dairy industries. Copper wire, sheet, and utensils are likewise frequently tinned for protection or to prevent contamination of products with copper.

Methods of Tinning

The most common method of tinning is by hot dipping, whereby the cleaned and pickled base metal is dipped in flux or a grease pot, then into one or a series of pots of molten tin, and drained in air or in a grease pot.[1,2] When producing tin plate, the operation is entirely mechanized, and the coated sheet emerges from the molten tin into a bath of palm oil, passing through rolls which remove all but a very thin coating.[3,4] Normal hot dipping gives coatings of a thickness between 0.5 and 1 mil.

---

* Tin Research Institute Sponsorship, Battelle Memorial Institute, Columbus, Ohio.

[33] R. M. Burns and A. E. Schuh, *Protective Coatings for Metals*, pp. 185–186, Reinhold Publishing Corp., New York, 1939.
[34] D. W. Rudorff, *Metal Industry*, **59**, 194–195 (1941).
[35] I. Kramer and R. Hafner, "Chromizing of Steel," *Trans. Am. Inst. Mining Met. Engrs.*, **154**, 415 (1943).
[1] C. E. Homer, "Hot Tinning," Tin Research Institute *Publication* 102, pp. 1–27 (December, 1940).
[2] W. E. Hoare and H. Plummer, "Hot-Tinning 'Difficult' Mild Steels," Tin Research Institute *Publication* 107, pp. 1–28 (November, 1941).
[3] R. M. Burns and A. E. Schuh, *Protective Coatings for Metals*, pp. 130–159, Reinhold Publishing Corp., New York, 1939.
[4] "Tin Plate and Tin Cans in the United States," Tin Research Institute *Bull.* 4, pp. 1–44 (October, 1936).

When using rolls or wipers to remove excess tin, as in making tin plate or tinning wire, the thickness is usually between 0.06 and 0.11 mil. This corresponds to 1.0 to 1.8 lb per base box* of tin plate. Not all this thickness is pure tin, since in hot dipping an alloy layer is formed with the base metal. Usually about 10 to 15% of the tin is in the alloy layer, largely as the compound $FeSn_2$.

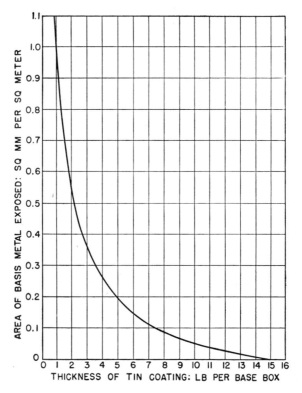

FIG. 1. Relation between Thickness of Tin Coating and the Area of Base Metal Exposed at Pinholes.

Tinning by electrodeposition is a well-established and increasingly important method, since economy in tin is effected by obtaining better thickness control and distribution than is possible by hot dipping.[5] Coating thickness may be only 0.015 mil (0.25 lb per base box), the practical minimum for good solderability on high-speed can manufacturing lines, or even less for can ends and special products. A thickness of 2 to 5 mils is recommended for dairy and food-handling equipment to give long service life. As electrodeposited, the tin coating has a matte finish and no alloy layer, but by heating the product momentarily to fuse the tin coating, a bright reflective surface and a very slight alloy layer are formed.

Replacement coatings of tin on copper-base articles give an attractive appearance and sufficient protection against tarnishing to be used commercially under mildly

---

* A base box of tin plate is 31,360 sq in. of plate (62,720 sq in. of surface) or 112 sheets, 14 by 20 in.

[5] T. W. Lippert, "Food in Cans," *Iron Age*, **149**, 29–44 (April 30, 1942).

corrosive conditions as in an indoor atmosphere and for drinking water pipes.[6] The coating is extremely thin, however, usually less than 0.01 mil. By replacement from molten salt baths containing a tin salt, a heavier alloy coating can be obtained more quickly.[7] Likewise, a tin-alloy coating on most base metals can be readily obtained by deposition of tin from volatilized stannous chloride in a hydrogen atmosphere.[8] Tin-sprayed coatings are frequently applied to equipment that cannot be readily tinned by other means, or for repair of equipment that cannot be economically dismantled for retinning. Heavy deposits can be built in this way, although the coating structure tends to be less dense and more porous than when tinning by other conventional means.

POROSITY OF COATINGS

The weakness of all tin coatings (as well as coatings of other metals) is incomplete coverage or exposure of base metal at pinholes. Figure 1 shows the calculated relationship between coating thickness and area of base metal exposed, as applied to commercial hot-dipped tin plate.[9] With heavy coatings of tin, as on "charcoal" grades of tin plate, hand-dipped articles, and hand-tinned copper sheets, there may be only a few pinholes per square decimeter. As the coating thickness decreases to one pound per base box, the number of pores are of the order of 4000 to 5000 per sq dm, and with thinner electrolytic tin coatings it becomes impractical to count the pores, although tests are used whereby comparison estimates can be made.[10] Mechanical scratches through the thin, soft coating of tin also aid in exposing the base metal locally, and potential pores exist where the outer tin layer is lacking and the brittle, intermetallic compound layer by chance has been mechanically disrupted when fabricating the tin plate. By electrodepositing from an alkaline bath over hot-dipped tin plate the number of pores has been reduced 5 to 10% of those in hot-dipped coatings of equal total thickness.[11]

Corrosion of tinned metal, then, becomes largely a problem of localized attack on the base metal, modified by the presence of an alloy layer, tin, and tin oxide. It cannot be described adequately in terms of weight loss or penetration.

CORROSION OF TIN PLATE AND TIN CANS

Under normal conditions of exposure to the atmosphere, water, and solutions, tin plate articles and the exterior of tin cans corrode at pinholes. Tin is cathodic to iron in the tin-iron couple and actually aids attack at these points, although this effect is of slight practical importance. Rust, mushrooming from these points, soon covers the tin plate, although the tin remains in place for some time and affords mechanical protection until flaking eventually occurs.

Treatment of tin plate or packed cans with a hot alkaline phosphate-dichromate

[6] J. D. Sullivan and A. E. Pavlish, "Tinning Copper and Brass by Immersion," *Metals and Alloys*, **11**, 131–134 (May, 1940).
[7] Tin Research Institute, unpublished data.
[8] B. W. Gonser and E. E. Slowter, "The Coating of Metals with Tin from the Vapor Phase," Tin Research Institute *Publication* 76 (April, 1938).
[9] W. E. Hoare, "Calculation of the Area of Basis Metal Exposed at Discontinuities in the Tin Coating of Tin Plate," *Phil. Mag.*, **26**, 1077 (December, 1938); Tin Research Institute *Publication* 86.
[10] R. Kerr, "The Testing of Continuity of Thin Tin Coatings on Steel," *J. Soc. Chem. Ind.*, **61**, 181–183 (December, 1942); Tin Research Institute *Publication* 116.
[11] A. W. Hothersall and W. N. Bradshaw, "Improvement in the Quality of Tin Plate by Superimposed Electrodeposition of Tin," *J. Soc. Chem. Ind.*, **54**, 320–326T, 1935; Tin Research Institute *Publication* 22.

solution for a few seconds has been effective in preventing rusting at pinholes over several months of outdoor exposure.[12] This method has been effective when using electrolytic tin plate of 0.5 lb per base box coating weight as well as with more heavily coated hot-dipped tin plate. The use of sodium chromate has been advocated to give a protective film over tin cans when exposed to alkaline, acid, or neutral salt solutions in the canning industry.[13] For shipment of cans to the tropics, under extremely poor wartime conditions, outside dip lacquering or baked enamel finishes have been found to be practical and effective.[14] Dipping the cans in water emulsions of some waxes also gives good protection.

Spotting and Staining of Tinned Surfaces

Black spots from localized attack on tinned surfaces by neutral or nearly neutral solutions, such as those handled in the dairy industry, are caused by electrochemical corrosion at points of weakness in the oxide film covering the tin. They can be prevented by keeping the tin in contact with a less noble metal, for example, zinc or aluminum and possibly also by anodic oxidation.[15] Only salt solutions which give no precipitate with stannous ions form black spots. Chlorides particularly aid in this type of attack.[16]

Yellow stain on tin plate and miscellaneous tinned articles is commonly caused by surface oxidation during long storage by electrochemical action of differential aeration cells on the tin surface. The stain is readily produced in the laboratory by anodic treatment in dilute sodium carbonate solution and is removed by making the tin the cathode.[17] Yellow stain on the tinned surface immediately after tinning is usually caused by too high a tinning temperature.

Corrosion inside the Can

As discussed in *Tin*, p. 323, the rate of attack of non-oxidizing acids, alkalies, and salts on tin is greatly accelerated by the presence of air. Conversely, the comparative absence of air within a food can aids in preserving the tin as well as the food. An additional factor enters to prevent attack on the base steel at pinholes in the coating. This factor consists in the reversal of potential which the tin-iron couple undergoes in the presence of many organic acids, whereby the tin galvanically protects the iron at exposed areas.[18]

Another important factor which influences the attack of foodstuffs on exposed iron in the can is the presence of stannous ions, resulting from slight attack or etching of the tin coating. As shown in Table 1, iron coupled with tin greatly reduces the rate of corrosion, but the presence of tin not in electrical contact, giving rise to stannous

---

[12] R. Kerr, "Protective Films on Tin Plate," *Tin and Its Uses*, pp. 12–13, Tin Research Institute (May, 1941).

[13] C. L. Smith and J. W. Barnet, "Control of External Corrosion of Tin Cans," *Proc. Inst. Food Tech.*, 199–204 (June 16–19, 1940).

[14] H. R. Smith, "New Developments in External Coatings as Corrosion Preventives for Canned Foods," *Proc. Inst. Food Tech.*, 26–41 (1944).

[15] S. Brennert, "Black Spots on Tin and Tinned Ware," Tin Research Institute *Publication* D-2 (August, 1935).

[16] T. P. Hoar, "The Corrosion of Tin in Nearly Neutral Solutions," *Trans. Faraday Soc.*, **33**, 1152–1167, (1937); Tin Research Institute *Publication* 63.

[17] C. E. Beynon and C. J. Leadbeater, "A Study of the Yellow Stain on Tin Plate," Tin Research Institute *Publication* D-1 (January, 1935).

[18] D. J. Macnaughtan and E. S. Hedges, "Some Recent Investigations on the Corrosion of Tin," Tin Research Institute *Publication* 34, pp. 7–13 (1936).

TABLE 1. CORROSION OF IRON IN DILUTE ORGANIC ACIDS, COUPLED
WITH TIN, AND IN PRESENCE OF STANNOUS IONS

Specimens — 2.5 × 4.0 cm (1 × 1.6 in.) of hot-rolled steel (tin-plate base) and sheet tin.
Exposed — 210 hr at room temperature in absence of air to acids of strength equivalent to 0.75% malic acid.

Acid	Corrosion Rate, mdd			
	Iron Alone	Tin Alone	Wt. Loss of Iron in Contact with Tin	Wt. Loss of Iron in Presence of Tin, No Contact
Citric	26	0.51	2.2	3.9
Malic	12	0.68	2.5	2.3
Acetic	11	0.97	7.6	5.5
Malonic	34	0.74	12.3	6.4
Succinic	15	0.62	10.6	5.8

ions, may reduce the rate even more drastically in some media.[19] Since the addition of sulfur, as 15 ppm sulfur dioxide, has been shown to increase the corrosion rate of steels in citric acid of less than about pH 4, and the addition of 15 ppm of stannous ions markedly decreases the rate of corrosion under similar conditions, one of the beneficial effects of dissolved tin is thought to be the removal of sulfur from solution.[20]

Many other materials may affect corrosion. For example, inhibition of corrosion by stannous ions may be modified by their removal as insoluble complexes with protein. Addition of apple pomace to organic acids, under the conditions listed in Table 1, has been shown to reduce drastically the rate of corrosion of iron, alone or in contact with tin.[21] Thus the rate of iron corrosion in malonic acid was reduced from 34 to 2.9 mdd by this means. Organic inhibitors, such as thiourea, have diminished the rate of attack of sour cherries and of beets, even when incorporated in the can end sealing compound, where only 2 to 4 mg become available to the contents of a No. 2 can.[22,23]

Cans for many products are coated on the inside with an enamel (a baked lacquer, not a vitreous enamel) to prevent contact of the contents with the tin plate. Until recently, this was largely to prevent darkening of sulfur-containing foods, such as peas, corn, and fish products (by formation of black iron sulfide at pinholes and ends or side seams, where iron was exposed), or to prevent bleaching of colored fruits and vegetables, such as cherries, strawberries, and beets. An inside coating has been applied to beer cans after, as well as before, the cans are formed in order to assure complete coverage, since long-continued contact with tin, as well as iron, may cause cloudiness and off-taste. Enameling, to prevent sulfide discoloration, can be eliminated in some applications by an alkaline-chromate filming treatment.[24] Since residual chromates are toxic, care is needed in thoroughly washing tin plate or unsealed cans so treated to prevent food contamination.

[19] E. F. Kohman and N. H. Sanborn, "Factors Affecting the Relative Potentials of Tin and Iron," *Ind. Eng. Chem.*, **20**, 1373 (1928).

[20] T. P. Hoar and D. Havenhand, "Factors Influencing the Rate of Attack of Mild Steels by Typical Weak Acid Media," *J. Iron Steel Inst.*, **133**, 239P (1936); Tin Research Institute *Publication* 36 (1936).

[21] E. F. Kohman and N. H. Sanborn, *loc. cit.*

[22] A. E. Stevenson and S. L. Flugge, "Container Structure with Inhibitor," U. S. Patent 2,168,107 (August 1, 1939).

[23] K. W. Brighton, "Electrolytic Tin Plate From the Can Maker's Point of View," *Trans. Electrochem. Soc.*, **84**, 227–247 (1943).

[24] R. Kerr, "A Protective Chemical Treatment for Tin Plate," *J. Soc. Chem. Ind.* (*Trans.*), **65**, No. 4 (April, 1946).

The advent of very thin electrolytic-tin coatings necessitated the use of enamel linings in such cans to prevent excessive attack by many foods. As shown in Table 2, with some foods the amount of attack merely varies with the weight of the tin coating. By using an enamel lining over the thinly tinned electrolytic plate, almost perfect protection is attained for some foods, as tuna fish or peas. However, a lining may be ineffective for a product like beets or tomato juice, although fair results are obtained when the can side seam is enamel-striped after can fabrication to reduce exposed tin plate to a minimum.[23,25]

TABLE 2. EFFECT OF TIN PLATE CONDITIONS ON STORAGE LIFE OF VARIOUS FOODS*

No. 2 cans, stored at 38° C (100° F). Vacuum loss determined by the external vacuum needed to make the can ends flip outward.

Product	Tin Coating, lb/Base Box	Enamel Lining	Days	Vacuum Loss, in.	Failures per 1000 Cans
Grapefruit	1.25	No	159	0.2	0
	0.75†	No	159	3.9	0
	0.5†	No	159	6.0	0
	0.5‡	No	159	10.1	653
Tuna fish	1.25	No	142	1.7	0
	0.5†	No	142	....	1000
	0.5†	Yes	142	1.0	0
Tomato juice	1.25	No	225	0.2	0
	0.5†	No	225	9.1	216
	0.5†,§	No	225	3.0	0
	0.5†	Yes	225	10.5	0
	0.5†,‖	Yes	225	4.2	0
Peas	1.25	Yes	262	0.4	0
	0.5†	Yes	262	0.1	0
	0.5†	No	262	8.4	0
Peaches	1.25	No	245	1.9	0
	0.5†	Yes	245	....	1000
	0.5†,‖	Yes	245	3.0	0
Beets	1.25	Yes	266	5.8	0
	0.5†	Yes	266	18.6	400
	0.5†,¶	Yes	266	12.5	40

* K. W. Brighton, "Electrolytic Tin Plate from the Can Maker's Point of View," *Trans. Electrochem. Soc.*, **84**, 227–247 (1943).
† Electrolytic tin plate, fused coating.
‡ Wire-brushed coating, unfused.
§ Tin plate given a chemical chromate treatment, exact conditions unstated.
‖ Enamel stripe made down the can side seam after fabrication.
¶ Thiourea corrosion inhibitor added to can end seam sealing compound.

EFFECT OF THE STEEL-BASE COMPOSITION

Some constituents in the steel used to make tin plate have a marked effect on the corrosion of the can. This is particularly important in enamel-lined cans, where most of the tin is covered and the chief contact of the contents with the steel and tin is along the seams where mechanical rupture of both enamel and tin coating takes place. Localized attack at these points frequently causes perforation, if the steel base favors rapid attack, since there may be insufficient tin exposed or dissolved to give protection.

[25] R. H. Lueck and K. W. Brighton, "Behavior of Foods in War-Time Cans," *Canadian Food Packer,* **15**, 17–23 (June, 1945); 18–29 (July, 1945).

Silicon in steel has been shown to be detrimental because of its effect on mechanical properties, i.e., an increased tendency for the tin plate to crack at the seams. Phosphorus actively promotes the corrosion of enamel-lined tin cans by many corrosive food products. By increasing the phosphorus in steel from 0.015 to 0.045%, the can service life was decreased from 50 to 70% in tests made with the fruits shown in Table 3.[26] The effect was less marked with unlined cans. Copper, likewise, has shown a definitely adverse effect on some fruits in enameled cans (Table 3), but on

TABLE 3. SERVICE LIFE OF FOOD CANS AS INFLUENCED BY COPPER AND PHOSPHORUS CONTENT OF THE BASE STEEL

Enameled No. 2 cans packed under identical conditions.
(a) Rimmed steel, 0.005–0.008% phosphorus.
(b) Aluminum-killed, 0.04–0.06% phosphorus.
Service life taken as days to produce 50% failure of cans tested.

		Days at Room Temperature		Days at 100° F (38° C)	
		0.02–0.04% Cu	0.18–0.20% Cu	0.02–0.04% Cu	0.18–0.20% Cu
Black cherries	(a)	1091	322	561	170
	(b)	357	288	154	90
Red, sour pitted cherries	(a)	699	191	308	83
	(b)	378	224	176	141
Italian prunes	(a)	1276	609	606	229
	(b)	549	465	309	199
Loganberries	(a)	1331	1274	535	635
	(b)	647	807	254	305

others it has little or no effect. For some fruits, such as strawberries, yellow Pershire plums, raspberries, and gooseberries, tests have shown copper to be beneficial.[27] There are some indications that nickel and, to a lesser degree, chromium are detrimental constituents in the tin plate used in packing corrosive fruits and berries.[26] Manganese, carbon, and sulfur have very little effect, or results are masked by the greater activity of other constituents. Addition of titanium to the steel is alleged to increase resistance of tin plate to perforation.[28]

Lists are available of commercial canned-food products, pH ranges, analyses, effect of packing in tin plate on vitamin content, and specifications for the type of can to be used for various products.[29,30] Many factors in handling foods affect the corrosion resistance. Storage at 38° C (100° F) usually gives a service life of only half that obtained at 27° C (80° F). For grapefruit, however, the storage life at 16° C (60° F) was 7.8 times that at 38° C (100° F), and for loganberries the service life was extended 1.8 and 3.6 times by lowering the storage temperature from 38° C (100° F) to 27° C (80° F) and 16° C (60° F), respectively.[31] Addition of sugar and of colloids, such as

[26] R. R. Hartwell, "Corrosion Resistance of Tin Plate," *Sheet Metal Ind.*, **15**, 1017–1025, 1139–1147, 1267–1268 (1941).
[27] T. P. Hoar, T. N. Morris, and W. B. Adam, "The Influence of the Steel-Base Composition on the Rate of Formation of Hydrogen Swells in Canned-Fruit Tin Plate Containers," Pt. 1, *J. Iron Steel Inst.*, **140**, 55P, (1939); Pt. 2, **144**, 133P (1941).
[28] "Canned Food Containers," National Canners Association *Bull.* 22–L, 21 (December, 1923).
[29] American Can Co., *The Canned Food Manual*, 1943.
[30] R. J. McKay and R. Worthington, *Corrosion Resistance of Metals and Alloys*, Reinhold Publishing Corp., New York, 1936.
[31] V. W. Vaurio, B. S. Clark and R. H. Lueck, *Ind. Eng. Chem.* (Anal. Ed.), **10**, 368 (1938).

gelatin, tend to inhibit corrosion under food can conditions, whereas the presence of sulfur acts as an accelerator. Fruits which often give the greatest trouble in causing perforation of cans, such as blackberries, prunes, white and black cherries, are low in acid. Greatly improved can life has been secured by adding 0.3 to 0.5% of citric acid to such packs.[32]

TIN COATINGS ON OTHER METALS

The same amount of tin 0.3 mil thick is dissolved by fresh milk from hot-tinned lead or electro-tinned lead as is dissolved from pure tin foil, according to 48-hour tests at room temperature with 20.6 sq cm exposed metal surface. With sour milk, under the same condition, the rate of corrosion is increased from 0.36 mdd to 1.7 mdd for pure tin, and slightly less tin is dissolved from the tinned-lead specimens. Insufficient lead is dissolved by milk from the tinned lead in a short time interval to be dangerously toxic, but when used as caps on milk bottles, about 1 ppm lead per pint bottle has been found, because of the greater solvent effect of the fatty acids in the cream. Using electrotinned coatings 0.02 to 0.06 mil thick on copper foil, 0.07 to 0.08 ppm copper is dissolved in 48 hours from tinned copper caps on pint bottles of fresh milk.[33]

A replacement coating of tin (between 0.01 and 0.02 mil thick) inside copper tubing decreased the amount of copper dissolved by distilled water saturated with carbon dioxide over a 40-day period from 15 ppm to less than 2 ppm. The tube was refilled ten times during this interval.[34]

As with tin plate, hot-tinned copper produces an alloy layer as well as an outer surface of substantially pure tin. In commercial coatings 0.06 to 0.11 mil thick on tinned copper wire (11 to 20 grams per sq meter), the compound amounts to about 3 grams per sq meter.[35] Such coated wire is much used in making rubber-insulated wire for electrical purposes. The tin coating, if complete, prevents sulfur in the rubber from attacking the copper, and likewise prevents rubber deterioration. Cuprous oxide inclusions in the base copper are an important source of porosity; hence the use of oxygen-free copper or of a cathodic treatment of the copper in a caustic soda solution before tinning produces a more corrosion-resistant coating.[36]

The tin coating on copper sheet and tubing, particularly that used in the dairy industry, is frequently damaged by use of corrosive alkaline cleaners. This corrosion is largely a function of the dissolved oxygen concentration; hence it can be greatly reduced by using a reducing agent, such as sodium sulfite. In 2-hour tests on tinned copper with flowing, aerated 0.5% solutions of sodium carbonate and sodium hydroxide at 100° C (212° F), the weight loss was decreased from 240 and 150 mdd, respectively, to 7 and 10 mdd by addition of 0.25% sodium sulfite.[37] Chromate inhibitors in trisodium phosphate cleaners are also effective in cleaning without excessive attack on the tin.

[32] T. N. Morris and J. M. Bryan, "The Corrosion of the Tin-Plate Container by Food Products," *Reports* 40 and 44, Food Investigation Board, Department of Scientific and Industrial Research (Cambridge, England), (1931, 1936).

[33] R. Kerr, "The Behavior of Some Metal Foils in Contact with Milk," *J. Soc. Chem. Ind.*, **61**, 128–132 (August, 1942); Tin Research Institute *Publication* 113 (1942).

[34] J. D. Sullivan and A. E. Pavlish, "Tinning Copper and Brass by Immersion," *Metals and Alloys*, **11**, 131–134 (May, 1940).

[35] P. W. Seddon, "Tin Coatings on Copper Wires," *Metal Ind.* (London), **62**, 37–38 (Jan. 15, 1943).

[36] W. D. Jones, "Influence of Surface Cuprous Oxide Inclusions on the Porosity of Hot-Tinned Coatings on Copper," *J. Inst. Metals*, **58**, 193–198 (1936).

[37] R. Kerr, "The Use of Sodium Sulfite as an Addition to Alkaline Detergents for Tinned Ware," *J. Soc. Chem. Ind.*, **54**, 217–221T (1935); Tin Research Institute *Publication* 19.

CADMIUM COATINGS

GUSTAF SODERBERG*

Cadmium coatings are applied mainly to iron and steel, to copper and copper alloys, and to a smaller extent to aluminum alloys and zinc alloys. Electrodeposition from cyanide solutions is used almost exclusively. Deposition by immersion, metal spraying, and hot dipping are much less common. Since molten cadmium does not alloy with iron and steel, hot dipping is done in Cd-Zn alloys usually of eutectic composition (82.5% cadmium). Copper wire has been hot dipped in unalloyed cadmium.

FACTORS INFLUENCING USE

The commercial use of cadmium coatings depends largely on the fact that cadmium combines to some extent the anodic behavior of zinc toward most underlying base metals with other desirable properties. Such properties are greater chemical resistance to alkalies, to wet spray of and alternate immersion in sea water (see also data in Table 19, p. 417) and sodium chloride solutions; decidedly slower formation of white corrosion products and tarnish films which interfere with mechanical and electrical functioning of moving parts and with surface conductivity for both ordinary and radio frequency currents; absence of intergranular penetration into steels at high temperatures; ease of soldering with non-corrosive fluxes; absence of gouging and seizure of moving parts; lesser hydrogen embrittlement of medium- and high-carbon steel in plating with cadmium cyanide than with zinc cyanide solutions, and greater ease in covering gray and malleable iron.[1]

The sacrificially protective action of cadmium in contact with iron and steel has been amply demonstrated both in the atmosphere and in solutions to which oxygen has access. Visual observation of rusting on cadmium-plated specimens exposed outdoors and in the salt spray test[2] shows absence of rust on bared steel nearest to the coating. Continuous immersion in $0.5 N$ NaCl of plates of electrolytic iron (2.5 × 5 cm [1 × 2 in.] and 3.8 cm [1½ in.] diameter) with cadmium plugs (1.3 cm [½ in.] and 1 cm [⅜ in.] diameter) inserted respectively in their centers caused etching of the cadmium and no corrosion of the iron.[3]

Similarly, alternate immersion for 10 days in 20% NaCl of heat-treated Duralumin sheets (2.5 × 15.3 × 0.05 cm [1 × 6 × 0.02 in.]), testing tensile strength = 42.2 kg/sq mm (60,000 psi) and elongation in 5 cm (2 in.) = 18%, with and without two cadmium-plated steel rivets across their center, lowered the tensile properties to 39.2 kg/sq mm (55,800 psi) and 16.1% and to 25.9 kg/sq mm (36,800 psi) and 3.4% respectively.[4]

A *modified electromotive series*† applicable to immersion in sea water is shown on p. 416. Cadmium protects those metals and alloys found below it in the series. However, the extent of the protection cannot be estimated from the potential because of factors not determined.

---

* Graham, Crowley and Associates, Inc., Jenkintown, Pa.
† Commonly described as *galvanic series*. EDITOR.

[1] General Preference Order M-65, War Production Board, Sept. 12, 1944.
[2] W. Blum, P. W. C. Strausser, and A. Brenner, RP 867, *J. Research Natl. Bur. Standards*, **16**, 185 (1936) and additional data supplied by W. Blum. (These results were obtained in a collaborative study undertaken by the National Bureau of Standards, the American Electroplaters' Society, and the American Society for Testing Materials.)
[3] H. S. Rawdon, *Trans. Electrochem. Soc.*, **49**, 339 (1926).
[4] M. R. Whitmore, *Ind. Eng. Chem.*, **25**, 19 (1933).

OUTDOOR CORROSION

The results of the most complete study on the outdoor corrosion of cadmium are summarized in Fig. 1.[2] If the essentially vertical portions of the curves are extended to intersect the time axis, the intercepts, as a measure of time for substantial rusting, are related to the thickness of cadmium coating as follows:

$$M = 18.1t^a + b$$

where M = time in *months* for substantial rusting,

t = thickness of cadmium coating in mils (thousandths of an inch) in the range 0.2 to 1 mil,

$a, b,$ = constants given below.

	a	b
New York City	1.00	2.2
Pittsburgh, Pa.	1.23	4.2
Sandy Hook, N. J.	1.55	15.6
Washington, D. C.	1.705	19.6

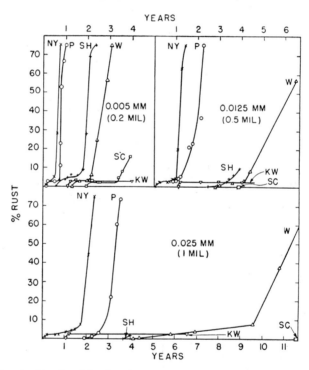

FIG. 1. Results of Outdoor Exposure Tests on Cadmium-Plated Steel for Three-Coating Thickness (0.2, 0.5, and 1 mil).

Material: Coatings from three cyanide baths with and without addition agents on 10 × 15 cm (4 × 6 in.) specimens of selected cold-rolled SAE 1010 steel.

Exposure: Panels 30° from horizontal facing south in New York City (NY), Pittsburgh, Pa. (P), Sandy Hook, N.J. (SH), Washington, D.C. (W), Key West, Fla. (KW) and State College, Pa. (SC).

These tests show that (1) there is no consistent difference between cadmium coatings produced in plating baths with or without organic addition agents and nickel at current densities of 1 or 2 amp/sq dm (9 and 19 amp/sq ft), (2) time for substantial rusting increases with, but is not a straight line function of, thickness, contrary to what has often been stated, and (3) the protective value depends greatly on the corroding atmosphere, contamination with SO_2 and SO_3 being particularly severe.

Further data on the rate of weathering of cadmium are available from substantially the same New York City location[5] and are summarized in Table 1. It is noted that

TABLE 1. WEATHERING OF CADMIUM PLATE IN NEW YORK CITY

Material — 22.9 × 30.5 × 0.16 cm (9 × 12 × 1/16 in.) steel plates with 15.5 mg/sq cm (0.02 mm, 0.8 mil) cadmium deposited at (a) 3.2 amp/sq dm (30 amp/sq ft) and (b) 1.1 amp/sq dm (10 amp/sq ft).
Exposure — Specimens mounted vertically (a) in December, 1928, and (b) in June, 1929. One half of the number of specimens was subjected to intermittent water spray for three 15-min periods daily except Sunday during March 15 to November 15, 1929, to simulate extra rain.
Determination of Weight Loss — Every other month the corroded specimens were scrubbed with a stiff bristle brush under water or lightly brushed with a soft varnish brush with no significant difference in result, weighed and re-exposed. Reported values are averages of results with triplicate specimens.

	Corrosion Rate			
Exposure Period	Normal Atmosphere		Accelerated Atmosphere	
	mdd	ipy	mdd	ipy
Cadmium (a)				
January–December	2.52	0.00042	2.33	0.00039
October–April	3.08	0.00051	2.64	0.00044
April–October	1.96	0.00031	2.02	0.00034
June–December	2.56	0.00043	2.24	0.00037
Cadmium (b)				
June–December	1.51	0.00025	1.67	0.00028

in this sea coast industrial atmosphere (1) accelerated rainfall has no great effect, (2) the rate of weathering is higher during winter than summer, and (3) starting the exposure in summertime may produce a slower rate of weathering. Item (2) is confirmed by tests in Clerkenwell, London, England, where rates of 1.61, 6.76, and 5.44 mdd (0.00027, 0.00112, and 0.00090 ipy) were found during June–October, October–December, and January–March, respectively.[6]

The *effect of porosity* of cadmium coatings on the degree of protection afforded has been studied at Woolwich, England.[7] Freshly prepared cadmium cyanide plating solutions without addition agents produced coatings which were very porous in thicknesses up to 0.0025 mm (0.1 mil). Above 0.0076 to 0.010 mm (0.3 to 0.4 mil) the coatings were substantially non-porous. Aged baths produced non-porous deposits 0.0013 mm (0.05 mil) thick. Porosity was determined by observing the number of hydrogen bubbles appearing on the surface on immersion in a 1% hydrochloric acid solution at 20° C (68° F) during 10 min. On outdoor exposure and exposure under a Stevenson screen (see footnote, p. 328), the porous coatings, depending on the degree of porosity, corroded only slightly faster than the non-porous coatings of the same thickness.

[5] C. L. Hippensteel and C. W. Borgmann, *Trans. Electrochem. Soc.* **58**, 23 (1930).
[6] W. S. Patterson, *J. Electrodepositors' Tech. Soc.*, **5**, 91 (1930).
[7] S. G. Clarke, *J. Electrodepositors' Tech. Soc.*, **8**, Paper 12 (1933).

Very *smooth and bright coatings* produced in an aged high-metal bath containing Turkey Red oil and nickel corroded more slowly than the *crystalline coatings* from an aged bath with no addition agents, possibly because of the larger actual surface present.[7]

Tests in the industrial atmosphere of Detroit, Michigan, have shown that the presence of *nickel or copper in cadmium plate* slightly decreases the time for substantial rusting.[8]

Other tests in the same location have shown that *heat treatment of cadmium-plated parts* in air used to decrease hydrogen embrittlement, may slightly decrease the protection afforded steel.[8] The maximum effect is found at the temperature at which visible oxide films are first formed, namely, about 150° C (300° F) and a treatment time of 45 to 60 minutes. Treatment for 3 hours at 250° to 300° C (480° to 570° F) has no effect on the corrosion resistance.

According to other tests in Detroit, Michigan, *bright dipping in a chromic acid-sulfuric acid bright dip* under conditions chosen to remove 0.00022 mm (0.009 mil) cadmium had no significant effect on the degree of protection provided by coatings on steel 0.0025, 0.0051, and 0.0076 mm (0.1, 0.2, and 0.3 mil) thick.[9]

LOCATIONS SHIELDED FROM RAIN

Tests in the industrial atmosphere of Woolwich, England,[7,10] have shown that in a Stevenson screen, a louvered box giving protection from rain but allowing exposure to circulating air, cadmium plate corroded only about 50% as fast as specimens freely exposed near by.

INDOOR CORROSION

Tests in London, England,[11] in a room with an average relative humidity of 55%, of 0.0165 mm (0.65 mil) cadmium plate on steel, deposited in a solution without addition agents and rubbed with No. 00 sandpaper, showed total weight increments of 27 and 31 mg/sq dm after 0.25 and 0.69 year of exposure, respectively. It is evident that the tarnish films once formed protect against corrosion under these conditions.

Where *condensation occurs*, the corrosion is more severe as shown in Table 2.[11]

TABLE 2. INDOOR CORROSION OF CADMIUM PLATE WITH FREQUENT DEW

Location — Room with large swimming pool, probably in London, England. High relative humidity, frequent dew, especially during January to March.

Elapsed time, yr	Season	Weight Increment, Total
		mg/sq dm
0.26	June–October	9.2
0.48	October–January	17.0
0.75	January–March	43.4

[8] Unpublished tests by the author.
[9] Gustaf Soderberg, *Trans. Electrochem. Soc.*, **82**, 71 (1942).
[10] S. G. Clarke, *J. Electrodepositors' Tech. Soc.*, **15**, 141 (1939).
[11] W. S. Patterson, *J. Electrodepositors' Tech. Soc.*, **5**, 91 (1930).

In another series of tests[10] in a small unheated room in Woolwich, England, with partially open window through which polluted and dusty air entered, the cadmium-plated steel specimens were suspended overnight in moist air in a closed vessel over water to cause condensation. After 6 months, 0.0013 mm (0.05 mil) coatings showed rust-colored patches and 0.013 mm (0.5 mil) coatings were dark colored with faint whitish spots, but no rust appeared except in a scratch which exposed the steel.

CORROSION BY GASES

Cadmium is attacked by *ammonia fumes* only in the presence of moisture. In warm, humid weather, moist ammonia fumes of sufficient concentration to severely discolor cadmium are formed by decomposition of cyanide residues on the surface of plated parts when kept in a container or the like.[12] Use of a chromic acid bright dip eliminates the residues and their action on the cadmium surface.[12]

When cadmium or cadmium plate is enclosed without air circulation with certain *unsaturated oils*, such as are present in certain paints and in certain impregnated fabrics or papers used for electrical insulation, very bulky white corrosion products are formed, frequently all the way to the underlying base metal, in very short time.[13] Both formate and butyrate have been identified in these products. High temperature and high humidity accelerate the attack. Treatment of cadmium in chromic acid or chromate solutions does not inhibit this effect, apparently because of the acidic nature of the corroding agents.

The *hydrogen sulfide*[14] present in the ordinary indoor atmosphere is not sufficient to cause cadmium sulfide to form on cadmium plate. High concentrations cause attack in the presence of only traces of moisture, the rate increasing with the moisture content. Practically complete removal of 0.005 mm (0.2 mil) cadmium deposits on steel has been noted after 240 hours at room temperature in moisture-saturated atmosphere.

As may be expected from its effect in the atmosphere, *sulfur dioxide*[14] is likely to attack cadmium very rapidly. Substantially dry gas in very high concentration produced only slight tarnish of cadmium plate in 48 hours at room temperature. Over water a white film was formed in 6 hours, a heavy crust after 64 hours, and rusting after 240 hours attack on 0.005 mm (0.2 mil) plate on steel.

CORROSION BY LIQUIDS

A broad picture of the effect of pH of aqueous solutions is furnished in Fig. 2.[15]

The effect of the *anion* in acid corrosion is shown in Fig. 3.[16]

A rapid non-selective attack without gas evolution by dilute nitric acid ($< 0.25\,N$) is notable because the *cadmium is brightened* in the process.[17] Similar brightening effect with solution of the cadmium is obtained with bromic acid, with certain chromic acid solutions containing also relatively low concentrations of sulfuric, nitric, hydrochloric, or formic acid, and with dilute sulfuric acid, hydrobromic acid, hydrochloric acid, formic acid, dichloracetic acid, or hydrofluosilicic acid in the presence of hydrogen peroxide.[17] In all cases the cadmium dissolves under oxidizing conditions similar to those causing anodic brightening.

Chromic acid solutions attack cadmium only very slowly and cause it to become

---

[12] Gustaf Soderberg, *Trans. Electrochem. Soc.*, **62**, 39 (1932).

[13] Gustaf Soderberg, *Monthly Rev. Am. Electroplaters' Soc.*, **17**, 30 (August, 1930).

[14] Unpublished tests by the author.

[15] B. E. Roetheli, C. J. Franz, and B. L. McKusick, *Metal Ind.*, **30**, 361 (1932).

[16] Gustaf Soderberg, *Trans. Electrochem. Soc.*, **62**, 39 (1932).

[17] See ref. in G. Soderberg and L. R. Westbrook, *Trans. Electrochem. Soc.*, **80**, 429 (1941) or *Modern Electroplating*, p. 101, Electrochemical Society, 1942.

passive. In the presence of sulfuric, nitric, hydrochloric, or formic acid in relatively high concentrations, solutions of chromic acid or chromates cause a *visible film* to form on the cadmium-plated surface, which, on exposure to high humidity, further improves its resistance to formation of white corrosion products.[17]

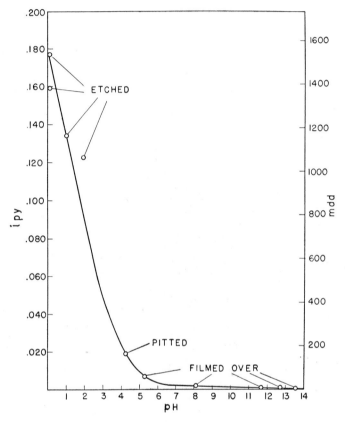

FIG. 2. Corrosion of Cadmium vs. pH in Continuously Flowing, Uniformly Agitated and Aerated Solutions of HCl or NaOH.

Material: 5 × 10 × 0.63 cm (2 × 4 × ¼ in.) cast cadmium.
Temperature: 24 ± 0.5° C (74 ± 1° F).
Time: 7 days for pH below 2; 41 days for pH above 2.

Among factors which appear to control the rate of corrosion of cadmium plate on steel in aqueous solutions is the *oxygen supply*, attack along the water line being very pronounced in distilled and tap water, and in solutions of $KAl(SO_4)_2$, $KCr(SO_4)_2$, $NaHCO_3$, ammonia, creosote, and glycerol.[18] *Temperature* has varying effects. In tap water, increasing the temperature to 65° C (150° F) causes a decrease in the rate of corrosion.[18] In dilute solutions of acetic acid, creosote, glycerol, and tannic acid, increasing temperature increases the rate of corrosion.[18]

[18] Unpublished tests by the author.

When cadmium corrodes in water or a salt solution in the presence of air, the pH of the cadmium surface rises, causing non-adhesion of subsequently applied paint coatings. The increase of alkalinity is demonstrated in Table 3.[19] Passivation

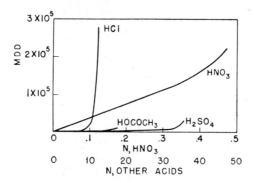

FIG. 3. Corrosion of Cadmium Plate in Acids of Various Concentrations (Normality).

Specimens: Cadmium-plated cold-rolled steel 10 × 3.2 cm (4 × 1¼ in.).
Test conditions: Specimens partially immersed for 40 sec in beakers containing acid solutions.

TABLE 3. EFFECT OF CORRODING CADMIUM ON THE pH OF THE CORROSIVE SOLUTION

Material — 1.91 × 12.7 cm (¾ × 5 in.) cadmium or cadmium-plated steel.
Apparatus — 300 ml Erlenmeyer flasks covered with watch glasses, in room free from laboratory fumes.
Solutions — Made from boiled distilled water.
Procedure — Degree of immersion and agitation not stated. pH measured at 25° C with Hellige comparator.

Medium	Time in Hours				
	0	24	48	96	400
Boiled distilled water, blank	6.7	6.7	6.8		6.9
" " " , cadmium		7.4	8.5	8.3	7.8
1 N NaCl solution, blank	6.7	6.8	6.8		7.0
" " " , cadmium		9.6	9.4	9.6	8.8
3.5 N NaCl solution, blank	6.6	6.6	6.7		7.0
" " " , cadmium		9.7	10.0	9.4	9.4

for 1 to 2 min in 3 to 5% chromic acid solutions at room temperature prevents this effect.[20]

LABORATORY TESTS

CORROSION IN SALT-SPRAY FOG

Cadmium plate is attacked so slowly in salt-spray fog that the test is not very useful for specification purposes. It is, however, being used to detect improper cleaning and pickling of the steel. No data are given because of the recent changes in salt-spray fog testing methods.

[19] M. R. Whitmore, *Ind. Eng. Chem.*, **25**, 19 (1933).
[20] Army-Navy Aeronautical Specification AN–P–61, Aug. 1, 1944

TABLE 4. APPEARANCE OF CADMIUM PLATE IN HUMIDITY TESTS

Conditions — (a) Constant 100% relative humidity (practically no condensation) at 45° C in glass vessel with humidifying liquid, closed except for small hole. (b) Varying humidity (condensation every third day) in large thermostated cabinet, automatically humidified with steam to 75% saturation at 45° C, temperature cycle 45° C for 2 days and dropping to room temperature for 1 day. Duration of test — 1 year.

Material	(a)	(b)
0.025 mm (1.0 mil) Cd	Superficial tarnish	Dark tarnished patches
0.0025 mm (0.1 mil) Cd	Tarnished, no rust	Tarnished, no rust
Bare steel	Slight rusting	More pronounced rusting

TABLE 5. SPECIFICATIONS FOR CADMIUM-PLATED COATINGS

Specification	Coating Class	Minimum Thickness		Minimum Hours in Salt Spray of 20% NaCl, 35° C (95° F)
		mm	mil	
A. S. T. M. A 165–40T (on steel)	NS	0.0127*	0.5*	....
	OS	0.0076*	0.3*	....
	TS	0.0038*	0.15*	....
U. S. Army 57–O–2C	NS	0.0127*	0.5*	72†
	OS	0.0076*	0.3*	48†
	TS	0.0038*	0.15*	24†
U. S. Navy 45P1	General	0.0127‡	0.5‡	....
	Exception§	0.0051‡	0.2‡	....
Army-Navy Aeronautical AN–P–61	General	0.0076‖	0.3‖	200‡‡
	Exception¶	0.0051‖	0.2‖	100‡‡
	Exception**	max.	max.	....
	Exception††	......	....	....

* On significant surfaces.
† Without showing white salts or iron rust.
‡ On smooth surfaces which constitute neither decided prominences nor recesses with respect to the general contour of the article.
§ On articles having integral parts which are threaded externally.
‖ On surfaces which can be touched with a 19 mm (0.75 in.) diam. sphere.
¶ On externally threaded portions of articles where the tolerances of the threads preclude deposition of 0.0003 in. and on bolts, studs, washers, nuts, and articles with major portion externally threaded.
** On parts whose dimensional tolerances will not permit a coating of 0.0003 in.
†† In holes, recesses, internal threads, and other areas where a controlled deposit cannot be obtained under normal plating conditions.
‡‡ Per specification AN–QQ–S–91 to appearance of no more than 6 corroded areas per sq ft of surface that are visible to the unaided eye or any corroded areas < 1/16 in. diam., corroded areas being exposure or corrosion of the base metal.

HUMIDITY TESTS

Laboratory humidity tests in which the contaminated air is largely excluded from the specimens appear less severe than the tests described under "indoor corrosion." Some results obtained in absence of condensation are shown in Fig. 4.[21] The appearance of cadmium coatings after one year exposure to various conditions of humidity are found in Table 4.[21] Plain steel is included for comparison.

[21] S. G. Clarke, *J. Electrodepositors' Tech. Soc.*, **15**, 141 (1939).

SPECIFICATIONS

A summary of plating specifications in force November, 1944, are given in Table 5. For detailed requirements, see the specifications themselves.

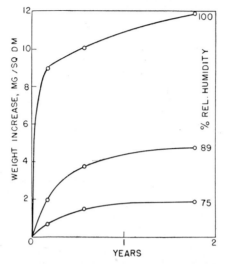

FIG. 4. Corrosion of Cadmium Coatings at Various Constant Relative Humidities.

Specimens suspended over saturated solutions of different salts kept in nearly closed vessels.

LEAD COATINGS

GUSTAF SODERBERG*

Lead coatings are used mainly on iron and steel and are applied either by "burning" on or rubbing with molten lead, producing so-called homogeneous lead coatings. They are also produced by spraying, hot dipping, or electroplating.

The *homogeneous lead coatings* are ordinarily heavy and non-porous. They are used for their resistance to chemical solutions, and their protective value is largely dependent upon the rate of corrosion of the lead itself. (See *Lead and Lead Alloys*, p. 207.)

Sprayed coatings are also heavy but tend to be very porous relative to their thickness.

Hot-dip lead coatings are seldom pure lead because of difficulties in obtaining complete coverage. Use of repeated immersions has met with only incomplete success. Better results are had by first zinc-coating the iron or steel.

The favored method of obtaining complete coverage and adhesion is to add 2 to 25% tin (alloys with 10 to 25% tin are called terne metal[†]), sometimes with 1 to 10% antimony to improve adhesion further and also to provide age-hardening characteristics. Less than ½% silver, or about 1% bismuth, or less than 2% zinc are sometimes found in low-tin alloy coatings.

* Graham, Crowley and Associates, Inc., Jenkintown, Pa.
† "Terne" meaning dull was originally used in contradistinction to "bright tin," that is tin plate. See Burns and Schuh, *Protective Coatings for Metals*, p. 205, Reinhold Publishing Corp., New York, N.Y. 1939. EDITOR.

Promising lead coatings containing 0.55% zinc as sole addition have been produced on a semi-commercial scale.[1]

Electroplated coatings are ordinarily pure lead deposited from fluoborate or sulfamate solutions. One half to 3% tin or higher have been codeposited with the lead to improve soldering and spot welding quality.

POROSITY

Both hot-dip and electrodeposited lead coatings are *porous* in commercial thicknesses; hence the base metal is a factor in the corrosion behavior. It has been shown that the early appearance of rust on relatively heavily coated steel is not significant in at least some atmospheres. In some instances, the rust disappears entirely or partly; in others, the rust development becomes substantially stationary. Since lead coatings are used largely for protection rather than appearance (however, hot-dip coatings containing silver may have a bright metallic luster), the presence of rust spots is seldom in itself a real objection.

RESISTANCE TO OUTDOOR EXPOSURE

The A. S. T. M. tests on *hot-dip lead-coated* hardware[2] have shown many instances of excellent protection over a period of 9 years. Unfortunately neither the composition nor the weight of these coatings is known. Their appearance sometimes indicates a very considerable coating thickness.

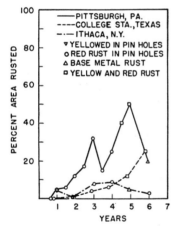

FIG. 1. Typical Progress of Rusting of 1.7 Mil Thick Zinc Plus Lead Coating on 0.188 In. Diam. Unfabricated Steel Wire.

Most of the data on hot-dip coatings are obtained on materials which were zinc-coated prior to lead coating. This is true of the 6-year A. S. T. M. wire test data[3] summarized in Table 1, the actual progress of rusting being illustrated in Fig. 1 and Table 2, and of three of the four long-term tests[4,5] shown in Table 3. Conclusions drawn from these tests do not necessarily apply to lead-alloy coatings directly on steel for which only short-time tests in unshielded locations are available (see Table 4).

The tests on lead coatings with zinc undercoatings indicate that for rather heavy coatings (1) the progress of corrosion is often irregular and interrupted and varies greatly with the exposure location, (2) the effect of corrosion is pitting of the steel without substantial spreading of the rust underneath the coating (Table 2), (3) the protective value increases with the thickness of the coating, (4) stranding of wire is detrimental to the protection afforded by the coatings, and (5) fabri-

---

[1] J. L. Bray and H. E. Zahn, Series W–37, July 12, 1943, and J. L. Bray and R. R. Dohrman, Series W–61, November 9, 1943, and W–89, May 3, 1944, Project NRC–506, War Metallurgy Committee, National Research Council, National Academy of Sciences, Washington, D. C.

[2] Reports of Committee A–5, American Society for Testing Materials, *Proc. Am. Soc. Testing Materials*, **29** (I), 149 (1929); **31** (I), 181 (1931); **33** (I), 154 (1933); **35** (I), 92 (1935); **38** (I), 90 (1938).

[3] Reports of Committee A–5, American Society for Testing Materials, *Proc. Am. Soc. Testing Materials*, **34** (I), 172 (1934); **37** (I), 122 (1937); **39**, 101 (1939); **41**, 101 (1941); **43**, 78 (1943).

[4] Private communication from W. H. Smith, Glen Dale, West Va.

[5] Private communication from Continental Steel Corporation, Kokomo, Ind.

LEAD COATINGS

TABLE 1. RESULTS OF OUTDOOR EXPOSURE OF Zn-Pb COATED WIRE

Analysis of steel — 0.04–0.10% C, 0.25–0.40% Mn, 0.24–0.29% Cu.
Analysis of coating — 0.41–0.25 oz/sq ft of total is zinc, remainder largely lead. Zinc is present as a continuous coating under the lead.
Exposure period — Approx. 6 yr.

Wire Diam., mm	4.8	3.7	3.7	3.7	3.0	3.0	2.5*	2.0
", in.	0.188	0.144	0.144	0.144	0.120	0.117	0.098	0.08
Total coating, oz/sq ft								
Aver., mm	1.4–1.6	1.1–1.5	1.5	1.75	2.0	1.7	1.8	1.1–1.9
Aver., mils	0.044	0.037	0.044	0.051	0.057	0.050	0.053	0.045
	1.7	1.5	1.7	2.0	2.2	2.0	2.1	1.8
Fabrication	None	None	Farm Field Fence	Chain Link Fence	7-Wire Strand, ⅜ in. diam.	None	Barbed	None

Location				% Area Rusted				
Pittsburgh, Pa.	20 R	20 R		15 R	10 R	20 R		20 R
Sandy Hook, N. J.	30 R + Y	30 R + Y		25 R	50 R	20 R + Y		50 R + Y
Bridgeport, Conn.	2 RP	2 RP		Y	55 R	2 RP		2 RP
State College, Pa.	15 R + Y	30 R + Y	10 R†	10 R†	15 R	15 R + Y	0	15 R + Y
Lafayette, Ind.	46 RP	43 RP	30 R + Y	20 RP	65 RP	43 RP	50 R + Y	6 RP
Ames, Iowa	25 YP	28 YP	40 YP		60 YP	26 YP	15 YP	32 YP
Manhattan, Kan.	15 YP	30 YP	30 YP		5 R	10 YP	15 YP	0
Ithaca, N. Y.	3 RP	24 RP	10 R	1 R	50 R	1 RP	3 R	2 RP
Santa Cruz, Calif.	15 SR	22 SR	30 R + Y	15 SR	30 SR	15 SR	15 SR	13 SR
College Sta., Texas	25 R + Y	51 R + Y	25 RP		50 R + Y	31 R + Y	5 R + Y	12 R + Y
Davis, Calif.	16 SR	32 SR	25 SR	15 SR	45 SR	15 SR	20 SR	8 SR

*Barbs 1 mm (0.04 in.) diam. †Approx. 5 years exposure.

Code: YP = Yellowed in pinholes.
RP = Rust in pinholes.
R = Rust of base metal.
Y = Yellow coloration.
SR = Superficial rust.

TABLE 2. RESULT OF TENSILE TESTS ON Zn-Pb COATED WIRE AFTER OUTDOOR EXPOSURE

Breaking load in kg and lb.
Elongation in 254 mm (10 in.) in per ce it.
For data, see Table 1.

Exposure	Time, yr	Wire Diameter										
		4.8 mm (0.188 in.)			3.7 mm (0.144 in.)			2.5 mm (0.098 in.)			2.0 mm (0.08 in.)		
		kg	lb	% Elong.	kg	lb	% Elong.	kg	lb	% Elong.	kg	lb	% Elong.
Unexposed	1	810	1790	9.5	485	1070	12.8	305	670	13.0	135	300	17.1
Pittsburgh, Pa. (First rust, 0.9 yr)	1.48	805	1780	9.8	480	1060	11.2	300	660	11.0	120	260	12.8
	2.93	...	...	...	...	...	...	...	...	...	125	290	11.3
	3.48	830	1835	7.6	490	1085	9.1	310	685	7.6	...	...	...
Sandy Hook, N. J. (First rust, 0.9 yr)	1.44	805	1780	8.2	470	1040	10.8	300	660	12.5	120	270	13.0
	2.90	...	...	...	...	...	...	...	...	...	135	300	14.1
	3.45	825	1820	8.0	490	1075	9.3	310	685	10.7	...	...	...
Bridgeport, Conn. (First rust, 1.4 yr)	1.45	800	1770	8.5	475	1050	11.5	300	660	10.8	120	260	13.5
	3.45	830	1830	9.0	490	1075	11.9	310	680	12.9	140	305	15.0

TABLE 3. RESULTS OF OUTDOOR EXPOSURE TESTS ON HOT-DIP LEAD COATINGS

Material	Exposure	Coating			Time, yr	Condition of Coating at End of Time Period
		oz/sq ft	mm	mils		
Corrugated roofing, unpainted*	On tin house; later hot galvanizing plant	3-3.5 incl. 0.7 Zn	0.053-0.060	2.1-2.4	28	Dark with light white incrustations, scattered rust spots
Wire, 2 mm (0.08 in.) diam.*	Near RR, mfg. plant and Ohio River	1¾ Zn + Pb	Approx. 0.053	Approx. 2.1	20	Lustrous black film over metallic lead with rusted and pitted streak along a narrow chord along the wire
Wire lawn fence,† 3.8-2.4 mm (0.15-0.096 in.) diam.§	Keg drier roof of steel plant in Kokomo, Ind.	1.32-2.32 Zn + Pb‡	0.039-0.068	1.5-2.7	10	Dull black film over metallic lead; no evidence of rust or pitted steel. Final tensile strength 40-51 kg/sq mm = 57,000-73,000 psi. Elong. in 127 mm (5 in.) = 2-6.5%
Sheet*	In open shed	1.1 incl. 0.25% Sn	0.015	0.6	8	Dark with metallic luster

* Private communication from W. H. Smith, Glen Dale, West Virginia.
† Formed and indented after coating.
‡ From 0.15 oz/sq ft (0.0065 mm, 0.25 mil) of zinc plus 1.17 oz/sq ft (0.032 mm, 1.25 mil) of lead to 0.34 oz/sq ft (0.015 mm, 0.58 mil) of zinc plus 1.98 oz/sq ft (0.053 mm, 2.1 mil) of lead.
§ Private communication from Continental Steel Corp., Kokomo, Ind.

cation into fence or barbed wire has no great effect on the protective value of the coatings.

The short-term tests on Pb-Sn alloy coatings (Table 4) show them too thin to compare with the heavier coatings referred to above. They are inferior to electroplated pure lead coatings of similar thickness.

Only relatively short-term tests are available on *electroplated coatings*. The tests of longest duration were conducted in the industrial atmospheres of Woolwich, England,[6] and Detroit, Michigan.[7] All these tests indicate that the effect of porosity is overcome in coatings 0.025 mm (1 mil) thick. Thinner coatings, 0.006 mm (0.25 mil), protect only for a short time. (See Table 5 and Fig. 2.)

Tests in Columbus, Ohio, show that the time of the year when the tests are started may determine at least the early rate of development of rust on specimens with lead coatings 0.005 mm (0.2 mil) thick.[8]

Preliminary results from approximately 0.8 year exposures at four locations are summarized in Table 4. The following *tentative* conclusions may be drawn from these tests: (1) Coatings 0.002 mm (0.08 mil) thick have little value out of doors. (2) The rate of rust formation is much more rapid in the humid sea coast atmosphere of Kure Beach, North Carolina, than in the industrial sea coast atmosphere of New York City, or the rural atmosphere of State College, or the tropical atmosphere in Tela, Honduras. (3) The presence of a copper flash under the lead is advantageous, except where the lead is no more than 0.002 mm (0.08 mil) thick.

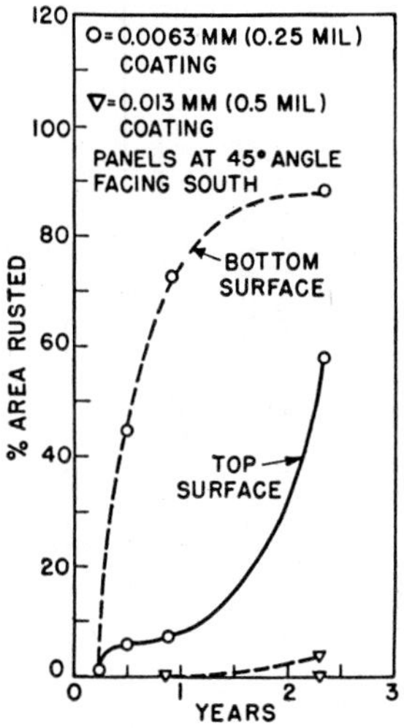

FIG. 2. Progress of Rusting o Electroplated Lead Coatings from Fluoborate Baths on Strip Steel in Detroit, Michigan.

OUTDOOR EXPOSURE IN CONTACT WITH SOIL

Hot-dip lead-coated hardware specimens which had fallen off the A.S.T.M. exposure racks[9] were badly rusted only where they were in contact with the soil.

Soil corrosion tests,[10] in which bare steel pipe, steel pipe coated by hot-dipping with about 0.025 mm (1.0 mil) lead, and lead cable sheathing were buried in a large variety of soils for about 10 years, showed that the lead coatings added to the life of the pipe in well-drained soils of high resistivity. A higher rate of corrosion of the lead coatings than of the lead sheathing, particularly in the more corrosive soils, indicates that the porosity of the coatings is a factor. The yearly average penetration of the lead-coated steel was always less than that of the bare steel in the same soil, varying between 0.0003 and 0.019 mm (0.012 to 0.75 mil) against 0.0003 to 0.068 mm (0.12 to 2.7 mils). The yearly maximum penetration by pitting was not greatly affected by the presence

[6] S. G. Clarke, *J. Electrodepositors' Technical Soc.* (England), **15**, 141 (1939).

[7] K. Gustaf Soderberg, *Proc. Am. Soc. Testing Materials*, **43**, 562 (1943). Also private communication from The Udylite Corporation, Detroit, Mich.

[8] H. A. Pray, C. L. Faust, and E. L. Combs, Series W-65, Nov. 9, 1943, W-92, March 28, 1944, and W-114, June 17, 1944, Project NRC-533, War Metallurgy Committee, National Research Council, National Academy of Sciences, Washington, D. C. See also Report of Committee B-8, American Society for Testing Materials, *Preprint* 19, 1944.

[9] Reports of Committee A-5, American Society for Testing Materials, *Proc. Am. Soc. Testing Materials*, **29** (I), 149 (1929); **31** (I), 181 (1931); **33** (I), 154 (1933); **35** (I), 29 (1935); **38** (I), 90 (1938).

[10] K. H. Logan and S. P. Ewing, *J. Research Natl. Bur. Standards*, **18**, 361 (1937).

TABLE 4. RESULTS OF OUTDOOR EXPOSURE TESTS ON LEAD-PLATED STEEL PANELS

Steel — No. 3 best bright finish, 1.0 mm (0.039 in.) thick, No. 4 temper, No. 3 edge.
Analysis of steel — 0.06% C, 0.005% Si, 0.34% Mn, 0.11% Cu, 0.005% P, 0.032% S.
Size panels — 10.0 × 15.0 cm (4 × 6 in.).
Exposure — Specimens inclined 30° from horizontal and facing south.
For each coating, first line gives results for top surface, second line for bottom surface.

| Thickness of Coating | | | | | New York City | | | State College, Pa. | | | | Kure Beach, N. C. | | | | Tela, Honduras | | |
| Pb | | Cu | | Type Pb Bath | First Rust, year | After 0.87 yr | | | First Rust, year | After 0.78 yr | | | First Rust, year | After 0.81 yr | | | First Rust, year | After 0.83 yr | |
mm	mil	mm	mil			No. Pinh.	% Stain	% Rust		No. Pinh.	% Stain	% Rust		No. Pinh.	% Stain	% Rust		No. Pinh.	% Stain	% Rust
.002	.08	...	...	f*	0.07	>25	100	20	0.30	>25	100	60	0.14	...	...	100	0.31	...	100	?
					0.18	>25	100	5	0.30	>25	93	7	0.14	...	...	100	0.42	...	95	?
.002	.08	.0004	.015	f	0.07	>25	100	75	0.07	...	...	100	0.14	...	...	100	0.31	...	100	?
					0.18	>25	80	3	0.07	...	...	100	0.14	...	...	100	0.31	...	98	?
.0063	.25	...	...	f	0.18	>25	100	1	0.30	>25	100	50	0.14	...	100	65	0.31	>25	...	...
					0.18	>25	90	2	0.78	>25	50	1	0.14	...	98	3	0.31	>25	...	...
.0063	.25	.0004	.015	f	0.87	>25	8	1	0.78	>25	1	...	0.36	>25	35	2	0.31	>25	...	...
					0.18	>25	15	1	0.78	10	45	...	0.14	>25	95	30	0.31	>25	...	...
.013	.50	...	...	f	0.18	>25	75	...	0.78	>25	...	...	0.36	>25	95	6	0.42	>25	...	...
					0.39	>25	40	1	0.78	5	...	...	0.36	>25	45	2	0.83	>25	...	...
.013	.50	.0004	.015	f	...	...	0	...	0.78	6	...	...	0.36	>25	5	...	0.31	>25	...	...
					0.87	>25	15	1	...	...	0	...	0.36	>25	12	...	0.43	>25	...	?
.013	.50	...	...	s†	...	...	30	...	...	...	40	...	0.36	>25	75	1	0.83	>25	3	...
					0.39	>25	0	1	0.78	>25	0	...	0.36	>25	35	...	0.42	>25	...	...
.013	.50	.0004	.015	s	...	...	20	...	...	...	15	...	0.36	>25	55	...	0.83	>25	...	...
					0.87	>25	45	...	0.78	>25	...	...	0.36	>25	10	...	0.42	>25	...	...
.019	.75	...	...	f	0.39	>25	40	1	0.78	5	...	...	0.36	>25	35	<1	0.42	>25	...	...
					0.39	>25	...	...	0.78	13	...	...	0.36	>25	25	...	0.58	>25	...	...
.019	.75	.0004	.015	f	...	...	3	...	0.78	1	0	...	0.36	>25	10	...	0.58	>25	5	...
					0.87	>25	...	...	...	...	...	...	0.36	>25	<1	...	0.83	10	...	...

TABLE 4. RESULTS OF OUTDOOR EXPOSURE TESTS ON LEAD-PLATED STEEL PANELS — *Continued*

Thickness of Coating					Location												
Pb		Cu		Type Pb Bath	New York City After 0.87 yr				State College, Pa. After 0.78 yr				Kure Beach, N. C. After 0.81 yr				Tela, Honduras After 0.83 yr		
mm	mil	mm	mil		First Rust, year	No. Pinh.	% Stain	% Rust	First Rust, year	No. Pinh.	% Stain	% Rust	First Rust, year	No. Pinh.	% Stain	% Rust	First Rust, year	No. Pinh.	% Stain	% Rust
.025	1.00	...	...	*f*	0.87	>25	2	...	0.78	2	...	...	0.36	>25	20	...	0.42	>25	...	...
.025	1.00	.0004	.015	*f*	0.87	>25	30	...	0.78	...	<1	...	0.36	>25	6	...	0.83	>25	...	...
.025	1.00	...	...		0.87	...	0	...	...	...	0	...	0.36	>25	<1	...	0.58	10	...	...
.025	1.00	...	...	*s*	...	...	1	...	...	...	0	...	0.36	>25	<1	...	0.83	1	...	...
.025	1.00	.0004	.015	*s*	0.87	>25	0	...	...	...	0	...	0.36	>25	<1	...	0.83	5	0	...
.025	1.00‡	...	...		...	...	1	...	...	...	0	...	0.36	>25	<1	...	...	...	...	...
.025‡	1.00‡	...	...		...	...	0	...	0.30	>25	100	60	0.14	>25	97	3	0.83	5	30	?
.025§	1.00§	...	...		0.87	>25	3	...	0.30	>25	10	1	0.14	>25	90	3	0.31	>25	30	?
.0045‖	.17‖	...	...		...	...	0	...	0.30	>25	45	7	0.14	>25	85	<1	0.42	...	...	...
.0045‖	.17‖	...	...		0.07	>25	100	7	0.07	...	0	...	0.14	...	2	<1	0.42	>25	...	...
.011¶	.44¶	...	...		0.18	>25	15	...	0.07	>25	100	70	0.14	...	100	50	0.31	>25	70	?
		...	...		0.07	>25	60	3	0.07	>25	95	20	0.14	>25	98	15	0.42	>25	60	?
		...	...		0.18	>25	15	5	0.07	>25	100	75	0.14	>25	95	35	0.31	>25	40	?
									0.07	>25	98	30	0.14	>25	90	20	0.42	...	20	?

Code: No. Pinh. = number of pinholes.
% Stain = % of area covered or stained with rust.
% Rust = % of area with flaky base metal rust.
* Panels plated in fluoborate bath marked *f*.
† Panels plated in sulfamate baths marked *s*.
‡ Panels electroplated with alloy containing 2% tin.
§ Panels electroplated with alloy containing 10% tin.
‖ Panels hot-dipped in alloy containing 20% tin.
¶ Panels hot-dipped in alloy containing 2½% tin, 2% antimony.

TABLE 5. RESULTS OF OUTDOOR CORROSION TESTS ON LEAD PLATED STEEL AT WOOLWICH, ENGLAND

Coating electrodeposited using perchlorate bath.
Panels probably exposed in vertical position.

Thickness of Lead		Period of Exposure		
mm	mil	6 months	1 yr	3 yr
0.0025	0.1	Rusted all over	....	....
0.0065	0.25	One third of surface rusted	Rusted nearly all over	Flaky rust all over
0.013	0.5	Few very small rust spots	Small rust patches	Rust all over, smooth, adherent
0.025	1.0	Few very small rust spots	Few very small rust spots	No apparent rust, good condition*

* Further comments — "The coating had become blackened and a slight degree of roughness had developed due to microscopic nodules of corrosion product. No rust was apparent, except at a scratch originally made through the coating; the rust was hard, dark in color and very localized."

of the lead coating. About one half of the soils were more severe on the bare steel pipe than on the lead-coated pipe and vice versa. The values of penetration for the lead-coated pipe varied from 0 to 0.302 mm (0 to 0.012 in.); those for the bare pipe from 0.010 to 0.308 mm (0.0004 to 0.012 in.).

Other comparative soil corrosion tests[10] of 2 years' duration indicate that the protection afforded by a lead coating takes place largely in the early stages before the coating is penetrated and before pitting sets in.

TABLE 6. RESULTS OF SEVERE INDOOR EXPOSURE OF LEAD-PLATED STEEL PANELS EXPOSED AT WOOLWICH, ENGLAND

Lead electrodeposited from perchlorate solution.
Exposure — in unheated building with window open to dusty and polluted air.
 during nights in closed vessel over water causing moisture condensation on surfaces.
Time of exposure — 6 months during winter.

Coating	Thickness of Coating			
	0.0013 mm (0.05 mil)		0.013 mm (0.5 mil) Surface	0.051 mm (2.0 mil) Surface
	In Scratch to Steel	Surface		
Lead	Rusted	Numerous rust spots	Few faint rust spots	No rust, bluish color
Nickel	Rusted	Completely covered with rust	Numerous rust spots, greenish spots on nickel	Almost free from rust spots, greenish spots on nickel
Copper	Rusted	Almost completely covered with rust	Few faint rust spots	No rust, tarnished
Zinc	Faint rust	Almost free from rust	No rust, darkened, faint whitish spots	...

INDOOR EXPOSURE

After 1½ years exposure in an office building in Washington, D. C., malleable iron fittings with *hot-dip lead coatings* 0.01 mm (0.4 mil) thick and steel stampings with *electroplated lead coatings* 0.005 mm (0.2 mil) thick showed no rust but only tarnishing of the lead itself.

Under conditions which permit frequent condensation of water, rusting occurs rapidly at pores and scratches.[6] (See Table 6.)

HUMIDITY CONDENSATION TESTS

The effect of continuous condensation of moisture on both electroplated and hot-dip coatings is shown in Figs. 3[11] and 4.[11,12] On the electroplated coatings, yellowish white deposits form almost immediately and grow thicker with time, after which

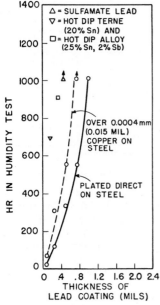

FIG. 3. Time for Development of First Rust Visible at Arm's Length in 44° C (110° F), 100% Rel. Humidity, on Lead-Coated Steel Panels Held against Copper Condensers at 33° C (90° F) and at 15° from Horizontal vs. Thickness of Coating.

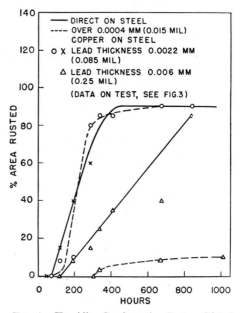

FIG. 4. Humidity Condensation Test on Plated Lead Coatings from Fluoborate Baths.

small rusted areas appear that grow in size. The hot-dipped alloy coatings containing tin do not show any white corrosion products and corrode much more slowly in this test than the electroplated lead coatings.

[11] H. A. Pray, C. L. Faust, and E. L. Combs, Series W-65, Nov. 9, 1943, W-92, March 28, 1944, and W-114, June 17, 1944, Project NRC-533, War Metallurgy Committee, National Research Council, National Academy of Sciences, Washington, D. C. See also Report of Committee B-8, American Society for Testing Materials, *Preprint* 19, 1944.

[12] Private communication from H. A. Pray, Battelle Memorial Institute, Columbus, Ohio.

TABLE 7. RESULTS OF SALT-SPRAY FOG TESTS ON LEAD AND
LEAD-ALLOY COATED SHEET

Equipment — regular National Bureau of Standards design.
Conditions — Temp. 35° C (95° F), solution 20% NaCl, air preheated and humidified at 35° C (95° F), air pressure 0.8-1.1 kg/sq cm (12-15 psi).
Maximum and minimum thickness test by Magne-Gage, average by stripping.
Specimens scratched with sharp point.

Coating	Thickness						Appearance at End of 100 Hours	
	Minimum		Maximum		Average		Rust in Scratches	White Corrosion Products on Surface
	mm	mils	mm	mils	mm	mils		
Hot-dip, 7.1% Sn, 7% Sb	...	...	...	...	0.004	0.17	Some	All over
Hot-dip, nominally 2.5% Sn, 2% Sb	0.014	0.55	0.016	0.61	...	...	Considerable	Considerable film
	0.017	0.68	0.043	1.67	0.042	1.65	Some	Considerable film
Electroplated, fluoborate bath	0.005	0.20	0.007	0.27	0.007	0.29	None	Streaks
	0.008	0.31	0.011	0.43	0.011	0.41	Slight	"
	0.022	0.83	0.025	0.95	0.028	1.11	None	"
Electroplated, sulfamate bath	0.003	0.13	0.006	0.24	0.048	0.19	None	Slight film
	0.012	0.45	0.014	0.54	0.014	0.52	Slight	"

Note. Rust always appeared around hole in specimens, but not on surface except in scratches.

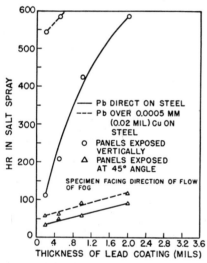

FIG. 5. Time for Development of First Rust Visible at Reading Distance in 20% NaCl, 33° C (90° F) Salt-Spray Fog Test vs. Thickness of Fluoborate Lead Coatings.

SALT-SPRAY FOG RESISTANCE

Since the salt-spray fog test has only recently been standardized, comparisons between results obtained in different laboratories cannot safely be made.

Results of tests on both hot-dipped and electrodeposited coatings on steel are found in Table 7.[13] Other tests[13] on hot-dip lead coatings (containing nominally 2.5% Sn, 2.0% Sb) on steel sheet have shown rust to develop in the bands of chatter marks appearing at intervals across the sheet. On similar sheets with thinner coatings and cold-rolled after coating, rust appeared in accidental scratches.

The effects of thickness of lead coatings from fluoborate solutions and of their angle with a vertical plane are shown in Fig. 5.[14]

The typical manner in which rust develops on lead-plated steel is indicated in Fig. 6.[15] The original small rust spots

[13] Private communication from W. Blum, National Bureau of Standards, Washington, D. C.
[14] Myron B. Diggin, *Metal Finishing*, **41**, 418 (July, 1943).
[15] Private communication from H. A. Pray, Battelle Memorial Institute, Columbus, Ohio.

increase in size and number until continuous rust areas appear. At the same time the lead itself corrodes, forming an increasingly heavy layer of white corrosion products which is replaced by rust as the test progresses.

EFFECT OF COLD ROLLING

The effect of cold rolling of steel electroplated in a fluoborate bath with 0.013 mm (0.5 mil) lead has been studied in salt-spray fog and humidity condensation tests with results shown in Table 8.[16] The electrographic test for porosity (p. 1034), contrary

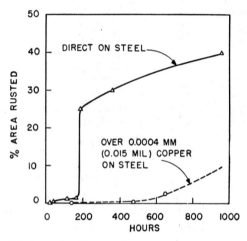

FIG. 6. Salt-Spray Fog Test (20% NaCl, 95° F) of 0.013 mm (0.5 mil) Lead Coatings from Fluoborate Bath.

TABLE 8. EFFECT OF COLD ROLLING OF LEAD-PLATED STEEL SHEET ON CORROSION IN SALT-SPRAY AND HUMIDITY CONDENSATION TEST

Data on steel panels: see Table 4.
Plated in fluoborate bath with 0.013 mm (0.5 mil) lead with and without preliminary 0.0004 mm (0.015 mil) copper flash.
Salt-spray fog test using 20% NaCl solution, 35° C (95° F).
Humidity test data: see Fig. 3.

Amount of Deformation	Hours for First Rust Visible at Arm's Length			
	Salt-Spray Fog		Humidity Condensation	
	Pb on Steel	Pb Over Cu Flash	Pb on Steel	Pb Over Cu Flash
No rolling	41	137	840	1008
Tight rolling*	137	137	840	840
0.1% Elong.	117	137	840	1008
0.5–0.7% Elong.	137	89	888	1008
1.5–2.0% Elong.	117	22	1008	1008

* No elongation.

[16] H. A. Pray, C. L. Faust, and E. L. Combs, Series W-65, Nov. 9, 1943, W-92, March 28, 1944, and W-114, June 17, 1944, Project NRC-533, War Metallurgy Committee, National Research Council, National Academy of Sciences, Washington, D. C. See also Report of Committee B-8, American Society for Testing Materials, *Preprint* 19, 1944.

TABLE 9. SPECIFICATIONS FOR LEAD AND LEAD-ALLOY COATINGS

Type of Coating	Specification	Coating Class	Requirements				
			Minimum Thickness				Minimum Hours in Salt Spray before Appearance of Rust*
			Copper		Lead		
			mm	mils	mm	mils	
Electroplated	A. S. T. M. B200–45T and U. S. Army 57-O-2C	ES	...	...	0.025	1.0	96
		EES	0.0004	0.015	0.025	1.0	96
		MS	...	...	0.013	0.5	48
		MMS	0.0004	0.015	0.013	0.5	48
		PS	...	...	0.006	0.25	24
		PPS	0.0004	0.015	0.006	0.25	24

Type of Coating	Specification	Coating Class	Lb per Double Base Box	Min. Coating Weight, oz/sq ft Sheet	Equivalent Minimum Average Thickness		Min. %	
					mm	mils	Tin	Tin and Lead
Hot-dip	Federal QQ–T–191, Long Terne	A	...	...	...	...	...	...
		B	12	0.44	0.0069	0.27	20	99
		C	15	0.55	0.0084	0.33	20	99
		D	40	1.47	0.0232	0.91	25	99
Hot-dip	Federal QQ–T–201 Terne Plate (roofing tin)		8	0.23	0.0036	0.14	25	99
			15	0.55	0.0086	0.34	25	99
			20	0.73	0.0113	0.45	25	99
			25	0.92	0.0145	0.57	25	99
			30	1.10	0.0173	0.68	25	99
			40	1.47	0.0232	0.91	25	99

Type of Coating	Specification	Type of Article				oz/sq ft Surface			Max. %	
		Weight before Coating			Centrifuged†				Tin	Antimony
		grams		oz						
Hot-dip	A. S. T. M. A267–44T On iron and steel hardware	<284		<10	Yes	0.15	0.0041	0.16	2.5	
		>284		>10	Yes	0.25	0.0069	0.27	2.5	
		...		...	No	0.51	0.0135	0.53	2.5	

Type of Coating	Specification				Requirements				
Hot-dip	A. S. T. M. B189–44T On copper wire				Certain maximum porosity requirements				20‡	6

* Unless otherwise agreed to, appearance of no more than 6 rust spots per sq ft or more than 2 rust spots on parts of <⅓ sq ft area visible to the unaided eye at normal reading distance, or any rust spots >1/16 in. diam.

† The hot-dipped articles are frequently centrifuged to throw off part of the molten coating, thereby decreasing the coating thickness.

‡ May be replaced in part with other alloying constituents.

to the salt-spray fog test, showed no significant effect of rolling on the number or size of pores on panels copper-flashed before lead plating, but indicated improvement in the specimens having the lead applied directly on the steel.[16]

Cold rolling is practiced on steel sheet coated by hot dipping with a lead alloy containing 2½% tin, 2% antimony, and < 0.5% silver.

EFFECT OF SUPPLEMENTARY OIL AND WAX COATINGS

Traces of palm oil left on hot-dip lead-alloy-coated sheet after branning retards the formation of rust spots in the salt-spray fog test.

Application of molten microcrystalline petroleum waxes to lead coatings 0.006 and 0.013 mm (0.25 and 0.5 mil) thick from fluoborate baths on steel prevented rust from appearing within 4 months of outdoor exposure in Columbus, Ohio, and retarded greatly the rusting in the salt-spray fog and humidity condensation tests.[16] Brushing with 9% ether solutions of the same waxes was ineffective or detrimental.

LEAD COATINGS ON COPPER

When lead coatings containing tin are applied to copper by hot dipping, a Cu-Sn alloy layer is formed which is cathodic to copper, at least in some atmospheres, and which causes deep pitting of the copper through pores in the coatings.

Copper sheet plated with lead coating 0.006 and 0.013 mm (0.25 and 0.5 mil) thick in fluoborate and sulfamate solutions[16] exhibited no clear-cut evidence of base-metal corrosion after 0.8 year outdoor exposure in New York City, State College, Pennsylvania, or Kure Beach, North Carolina.

SPECIFICATIONS FOR LEAD COATINGS

The numerical requirements in a number of specifications for lead and lead-alloy coatings are summarized in Table 9.

Inorganic Coatings
ANODIC TREATMENT OF METALS

JUNIUS D. EDWARDS*

Certain metals, when made the anode in suitable electrolytes, acquire a stable oxide film. Aluminum, in particular, is susceptible to this kind of treatment, and the process of anodic coating of aluminum has reached substantial proportions because of the excellent protective characteristics of the coating. Anodic coatings can also be formed on magnesium and zinc but are inherently less protective. The major part of this discussion will therefore be directed towards the anodic treatment of aluminum. In popular usage, the anodic coating of metals is sometimes referred to as anodizing.

The anodic coating of aluminum is best carried out in acid solutions. Depending on the characteristics of the electrolyte, three different conditions may prevail during anodic treatment.

First, the oxide formed is substantially insoluble in the electrolyte. A relatively impervious and thin coating is formed. This behavior is observed in electrolytes such as solutions of boric acid and borates. The principal use of this type of coating is in electrolytic condensers and rectifiers.

* Aluminum Research Laboratories, Aluminum Company of America, New Kensington, Pa.

Second, the oxide formed is slowly dissolved by the electrolyte. A porous coating, permeated by the electrolyte, is formed, and growth of the coating continues as long as the current passes. Coatings of this type are formed in electrolytes such as solutions of sulfuric, oxalic, or chromic acids.[1,2]

Third, the oxide may dissolve in the electrolyte substantially as fast as it is formed. Properly conducted, this type of anodic treatment results in electrolytic brightening or electrolytic polishing of the surface. A solution of fluoboric acid is one electrolyte employed for this purpose with aluminum.

Electrolytes of the second type are generally employed for the production of "thick" coatings for the corrosion protection of aluminum. Coatings of this type can be controlled in thickness and porosity and, through various after-treatments, can be sealed so that they offer greater protection against corrosion.

ANODIC COATING PROCEDURES FOR ALUMINUM

The natural oxide film which forms on aluminum in air is of the order of 0.01 μ (100 Å) in thickness. When aluminum is made the anode in a suitable electrolyte such as dilute sulfuric acid and a suitable voltage applied, the natural film breaks down at innumerable points and an anodic coating begins to form. This film grows by the interchange of aluminum and oxygen ions across the film layer until the film reaches a thickness of about 0.03 to 0.05 μ (300 to 500 Å). This is the so-called barrier layer.[3,4] If the film is insoluble in the electrolyte, growth stops at this point. If, however, the oxide film is slowly soluble, pores form in the outer face of the coating and growth continues. The oxide thus continues to grow in thickness as long as electrolyte can reach the barrier layer through the pores. Typical procedures for the anodic coating of aluminum are shown in Table 1.

TABLE 1. TYPICAL PROCEDURES FOR ANODIC OXIDATION OF ALUMINUM

Electrolyte	Typical Operating Characteristics	Comments
Chromic acid, 3%	40 to 50 volts, 3 amp/sq ft.	Coating gray in tint and usually about 0.1 mil (0.0025 mm) thick.
Oxalic acid, 3%	65 volts, 12 amp/sq ft.	Coating cream-colored; thickness up to about 1 mil max. (0.025 mm).
Sulfuric acid, 15%	15 volts, 12 amp/sq ft.	Coating white or transparent; thickness up to about 1 mil max. (0.025 mm).
Boric acid plus ammonium borate	50 to 500 volts.	Coating usually thin and iridescent; thickness proportional to formation voltage.

The pores are exceedingly fine, and in some types of coatings there may be as many as one trillion pores (10^{12}) per square inch.[4] It is important to note that while these pores are wide enough to permit passage of electrolyte, they are too narrow to permit the entrance of gross pigment particles, and even most colloidal particles. Because of the solvent action of the electrolyte, pore diameters near the surface of the coating

---

[1] J. D. Edwards, "Anodic Coating of Aluminum," *Monthly Rev. Am. Electroplaters' Soc.*, **26**, 513 (1939).

[2] O. F. Tarr, Marc Darrin, and L. G. Tubbs, "Anodic Treatment of Aluminum in the Chromic Acid Bath," *Ind. Eng. Chem.*, **33**, 1575 (1941).

[3] Scott Anderson, "Mechanism of Electrolytic Oxidation of Aluminum," *J. Applied Physics*, **15**, 477 (1944).

[4] J. D. Edwards and F. Keller, "The Structure of Anodic Oxide Coatings," *Trans. Am. Inst. Mining Met. Engrs.*, Institute of Metals Division, **156**, 288 (1944).

are usually larger, and, in fact, the coating near the surface may have a honeycomb structure because of the breakdown of the walls between pores. It may also be noted that the porosity, or pore volume, generally increases with increase in temperature of the electrolyte and increase in time of immersion in the electrolyte because of the increased dissolving action on the oxide.[5] The electron micrographs of a cross section and surface view of porous oxide coatings, which are shown in Figs. 1 and 2, help visualize their structure.

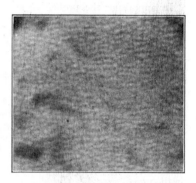

FIG. 1. Electron Micrograph of Cross Section of Portion of Thick Oxide Coating on Pure Aluminum, Made in Sulfuric Acid Electrolyte. Magnification 21,500. (Edwards and Keller.)

FIG. 2. Electron Micrograph of Surface of Extremely Thin Oxide Coating on Pure Aluminum, Made in Sulfuric Acid Electrolyte. Magnification 68,-000. (Edwards and Keller.)

On pure aluminum, the oxide coating is transparent, glassy aluminum oxide containing adsorbed ions characteristic of the electrolyte in which it was formed. However, commercial aluminum and aluminum alloys contain constituent particles whose electrolytic behavior substantially modifies the physical characteristics of the coating. These constituents can be grouped in three classes according to their behavior during anodic treatment.[6]

1. Alloying constituents which are in solid solution.
2. Constituents not in solid solution which are not appreciably dissolved or oxidized by the anodic treatment.
3. Constituents not in solid solution which are readily dissolved or oxidized by the anodic treatment.

The constituents which remain in the coating change its color and opacity and may also modify the protective characteristics. The constituents which dissolve anodically may leave voids in the coating, thus increasing the porosity and affecting its protective value.

[5] J. D. Edwards and F. Keller, "Formation of Anodic Coatings on Aluminum," *Trans Electrochem. Soc.*, **79**, 135 (1941).

[6] F. Keller, G. W. Wilcox, M. Tosterud, and C. J. Slunder, "Anodically Oxidized Aluminum Alloys —," "Anodic Coating of Al — Behavior of Constituents," *Metals and Alloys*, **10**, 187, 219 (1939).

PROTECTION BY ANODIC COATINGS

The protection afforded by anodic oxide coatings is of the same general type as that rendered by the natural oxide film on aluminum. However, the natural film is *very* thin, and anodic coatings, because of their greater thickness and abrasion resistance, offer much better protection, both against corrosive attack and mechanical injury. Aluminum oxide is relatively stable in the atmosphere and in weakly acid solutions in the range of pH 4.5 to 7.0. In strongly acid solutions and, in particular, alkaline solutions, oxide coatings on aluminum are attacked. Anodic coatings find their principal field of application, therefore, in protecting against atmospheric attack.

In appraising the protective action of anodic coatings, their structure must be considered. The protective action is fundamentally that of a continuous, stable oxide film, resistant to penetration by liquids. Emphasis should be placed on the continuous character of the oxide, for its integrity may be affected by the inclusion of certain constituents or the oxidation products of constituents or by the solution of constituents, leaving weak spots in the film. As the thickness of the film increases, the weakening effect of constituents decreases. Although the impervious barrier layer is *very* thin, it is mechanically protected by the overlying porous oxide coating. In general outdoor service, as on architectural elements, the trend is to expect greater service from the thicker coatings, other conditions being equal.

The barrier layer need not be relied upon entirely for protection since the pores can be plugged by various procedures known as sealing treatments. The protective action of the coating can also be increased by adsorbing certain corrosion inhibitors in the pores.

One of the widely used sealing processes involves treatment of the coating with hot or boiling water, resulting in the transformation of aluminum oxide into aluminum monohydrate, $Al_2O_3 \cdot H_2O$. There is no change in the appearance of the coating, but the coating is no longer adsorptive. The pores may also be sealed by treatment with hot, dilute solutions of nickel or cobalt acetate. Colloidal hydroxides of these metals formed by hydrolysis are occluded in the pores and seal them. The pores may also be sealed with oils, waxes, resins, and similarly acting materials.[7]

Impregnation of oxide coatings with chromates by adsorption from solution is a very important means of increasing the resistance of the coating to corrosion.[8] The adsorbed chromate helps repair any breaks in the coating during attack by a corrosive agent. Coatings impregnated in this manner are given increased resistance to attack by salt solutions such as sea water.

To illustrate the protective characteristics of anodic coatings in laboratory tests with salt spray and also when exposed to the weather, the results of a series of tests are presented in Table 2. In these tests it appears that the thicker coatings tend to be more protective, although the protection is not proportional to the thickness. In salt-spray environment, at least, the chromate sealing treatment gives greater protection than hot water sealing. The coatings on the copper-bearing alloy, 17S* (Duralumin) were less protective than on 52S* alloy, which contains 2.5% magnesium.

Similar panels were exposed to the weather at several locations. At New Kensington, Pennsylvania, an industrial atmosphere, panels 1, 2, 4, and 5, although somewhat stained with dirt, were virtually free from corrosion when examined after 3 years exposure. There was definite but not serious corrosion on the back of panels 3 and 6, and somewhat more substantial evidence of attack on panel 7.

In Table 3 are given data on the change in physical properties of three different coatings on alloy 17S–T, as observed in another salt-spray test.

* Code numbers for aluminum and aluminum alloys are described in Table 1, p. 39.

[7] J. D. Edwards, "Anodic Coating of Aluminum," *Monthly Rev. Am. Electroplaters' Soc.*, **26**, 513 (1939).
[8] J. D. Edwards, U. S. Patents 1,946,151; 1,946,152 (Feb. 6, 1934).

ANODIC TREATMENT OF METALS

TABLE 2. PERFORMANCE OF ANODIC COATINGS IN SALT-SPRAY TESTS*

Coatings made in 15% sulfuric acid electrolyte.
Nominal composition of 52S: 2.5% Mg. 0.25% Cr, remainder Al.
Nominal composition of 17S: 4% Cu, 0.5% Mg, 0.5% Mn, remainder Al.

Specimen No.	Alloy-Coated	Sealing Procedure	Time of Coating min	Thickness of Coating cm	Thickness of Coating mil	Weight of Coating mg per sq cm	First Attack Observed
1	52S	Hot water	20	0.00071	0.28	1.95	None†
2	52S	Hot water	30	0.00099	0.39	2.48	None†
3	52S	Dichromate	10	0.00030	0.12	0.76	18 months
4	52S	Dichromate	20	0.00074	0.29	1.80	None†
5	52S	Dichromate	30	0.00094	0.37	2.53	None†
6	17S-T	Dichromate	30	0.00104	0.41	1.91	None†
7	17S-T	Hot water	30	0.00084	0.33	1.50	600 hours

* Spray from 20% salt solution at room temperature.
† After 18 months exposure.

Note. The rating "none" includes those panels showing only a few small pinhole points of attack.

Coatings of adequate thickness made in sulfuric acid electrolyte, when properly sealed, maintain an excellent surface appearance in a clean atmosphere. In industrial atmospheres, however, a certain amount of maintenance is necessary to preserve this appearance. Dust or cinder particles, becoming attached to the surface, may adsorb moisture and sulfuric acid from the atmosphere, with resultant attack of the coating. For best service, therefore, a coating should be kept clean by occasional washing or preferably by waxing. Under these conditions, coatings have maintained a good surface appearance for periods upwards of 10 years.

TABLE 3. SALT-SPRAY EXPOSURE TEST ON ANODICALLY COATED ALUMINUM ALLOY*

Electrolyte Employed for Coating	Sealing Method	Period of Exposure to Salt Spray,‡ years	Thickness of Coating, mil	Change in Physical Properties,† %	
				Tensile Strength	Elongation
Sulfuric acid	Hot water	1	0.4	−1.7	−16.0
Sulfuric acid	Dichromate	1	0.4	−0.4	− 4.5
Chromic acid	None	1	0.1	−6.6	−32.0

* Nominal composition of 14-gage sheet 17S-T: 4% Cu, 0.5% Mg, 0.5% Mn, remainder Al.
† Average of tests on five samples.
‡ Exposed continuously to spray from a solution containing 20% sodium chloride at room temperature

One of the principal uses of anodic coatings has been as a base for paints on aircraft structures. Anodic coatings give increased adhesion and life to paint coatings because they provide an inert surface between the metal and the paint film. This minimizes the possibility that any moisture penetrating the paint coating will attack the metal and cause loss of adhesion. Anodic coatings made in chromic acid electrolyte were first used for this purpose, but there is now also a wide use of coatings made in sulfuric acid electrolyte and sealed in a chromate solution.

ANODIC COATINGS ON MAGNESIUM AND ZINC

Anodic coatings can be formed on magnesium,[9] but they are not so protective as those formed on aluminum. The oxide of magnesium is more soluble in water than is aluminum oxide, and the solubility is greatly increased by the presence of carbon dioxide. The problem of sealing is also difficult. One of the anodic treatments which has had considerable use utilizes an electrolyte containing sodium dichromate and monosodium phosphate. The coating produced therein is thin, but substantially increases the surface protection when used in combination with suitable paint coatings. A coating of substantial thickness and abrasion resistance can be produced by anodic treatment of magnesium in a solution of sodium hydroxide with or without addition agents.[10,11] (Anodic treatment can also be carried out in a chromate-sulfate-ammonium hydroxide bath. See *Chromate Coatings on Magnesium Alloys*, p. 864.) The sealing of such a coating in a chromate solution effectively increases the protective power and makes a good base for the application of protective paints.[10]

Thin anodic coatings can also be produced on zinc[9] by treatment in electrolytes containing chromic acid or chromates. Limited commercial use appears to have been made of these processes.

CHROMATE COATINGS ON ZINC

E. A. ANDERSON*

It has been pointed out in *Zinc*, p. 331, that zinc normally corrodes rather evenly in the outdoor atmosphere, with the formation of relatively thin corrosion product films. When zinc is in contact with thin, highly aerated water films for extended periods of time or in stagnant water with limited access to oxygen, the corrosion becomes uneven and rapid. Bulky corrosion products form, and, in very bad cases, pitting becomes pronounced. The most serious effect usually is that these bulky corrosion products may bind moving parts. Means to prevent this type of attack have been developed based on the known inhibitive effects of chromates.

Several procedures[1,2,3] are available by which protective films consisting essentially of chromium salts can be produced on zinc surfaces. Complete details for one process have been published.[1] In this case the film is reported to be a basic chromium chromate of the general formula $Cr_2O_3 \cdot CrO_3 \cdot XH_2O$ with a film thickness of about 0.0005 mm (0.02 mil). It is produced by simple immersion in a solution containing hexavalent chromium (e.g., $Cr_2O_7^{--}$) and sulfuric acid. When placed in contact with water, the film hydrolyzes, releasing soluble hexavalent chromium. Similar details for other procedures are not available. The data discussed below are for the process described in reference footnote 1.

The protection afforded by this type of film against corrosion in stagnant water

---

* Research Division, New Jersey Zinc Co., Palmerton, Pa.

[9] J. D. Edwards, "Anodic and Surface Conversion Coatings on Metals," *Trans. Electrochem. Soc.*, **81**, 341 (1942).

[10] R. B. Mason, "A Protective Finish for Magnesium Alloys," *Iron Age*, **157**, 48, March 21 (1946).

[11] N. H. Simpson and P. Cutter, "Anodic Process for Protecting Magnesium," *Modern Metals*, **1**, 18, March (1945).

[1] E. A. Anderson, *Proc. Am. Electroplaters' Soc.*, p. 6 (1943). Cronak process, U. S. Patent 2,035,380.
[2] A. G. Taylor, *Proc. Am. Electroplaters' Soc.*, p. 6 (1944).
[3] J. E. Starek, *Proc. Am. Electroplaters' Soc.*, p. 235 (1944).

CHROMATE COATINGS ON ZINC

TABLE 1. CORROSION OF UNTREATED AND CHROMATE-TREATED ZINC-COATED STEEL IN DISTILLED WATER*

Material — hot-dip zinc-coated steel, coating of 0.5 oz per sq ft of surface.
Specimen size — 7.6 × 10.2 × 0.061 cm (3 × 4 × 0.024 in).
Number of specimens — quadruplicate in 1 liter of water.
Water — distilled — boiled and cooled to room temperature. Changed after 1, 2, 10, 17, 24, 45, 62 days and every 2 weeks thereafter.
Specimen suspension — glass rods through holes. Holes and cut edges coated with a pitch-type enamel (Hermastic enamel).
Duration — 1768 days.
Temperature — room.

Treatment	Time in Days to Appearance of		
	Zinc Corrosion Products	Iron Rust	Specimen Perforation
None	1	90	588
Chromate Film	38	998	1768

* E. A. Anderson, Proc. Am. Electroplaters' Soc., p. 6 (1943).

is shown in Table 1. Corollary tests[4] indicate that the essential protection, particularly at discontinuities, is supplied by the soluble hexavalent chromium portion of the film. The effect is the same as that produced with untreated zinc by dichromate added to the water in the same concentration. The insoluble trivalent chromium portion of the film contributes substantially to the protection, probably by mechanical exclusion of water from the surface.

Depending upon the particular process used, the color of the chromate films is yellow, golden, iridescent, or olive green. The general tendency is for the films to lose their yellow color as the hexavalent chromium portion is leached away. The residual film usually is greenish in color.

The duration of protection in service must be determined by actual service tests since the rate of loss of the soluble hexavalent chromium portion will depend upon the moisture conditions encountered.

Tests[4] in the water layer of gasoline-water mixtures (to simulate carburetor or gasoline barrel service) showed a weight loss of only 0.0027 gram for chromate-treated rolled zinc in comparison with a loss of 0.2691 gram for untreated rolled zinc specimens of the same size in similar exposure. Chromate films have been used successfully for a number of years in the protection of die-cast zinc-alloy carburetor bowls and hot-dip-coated gasoline barrels against attack by water in the gasoline.

At the present time, chromate films are widely used in connection with thin electrogalvanized coatings on steel stampings and on zinc-alloy die castings. In most cases, severe corrosion would not have occurred even though the film were not present. However, many devices would be rendered inoperable if relatively small amounts of corrosion products were permitted to form.

Effect of Temperature

Chromate films of the present type are known to be damaged by exposure to dry heat at temperatures exceeding about 65° C (150° F). The damage takes the form of a sharp reduction in the ability of the film to protect the zinc surface at film discontinuities. The mechanism of the change appears to involve the loss of combined water.

[4] Unpublished data, the New Jersey Zinc Company.

In some cases, parts are painted in certain areas with the film left exposed in others. The adhesion of baking enamels to the chromate films appears to be adequate. The influence of temperature in the baking step has not been established clearly. Although the temperatures used exceed that known to be safe under dry heat conditions, satisfactory commercial experience has been reported.

CHROMATE COATINGS ON MAGNESIUM ALLOYS

W. S. LOOSE*

Chromate films are formed on magnesium and its alloys by immersion or anodic treatment in acid baths containing dichromates or mixtures of dichromates and other oxy-acid salts, such as the sulfates, phosphates, and nitrates. The films formed are normally amorphous mixtures composed largely of hydrated chromium and magnesium oxides. When alloying elements such as aluminum, zinc, or manganese are present in the treated alloy, they are also found in relatively large amounts in the films formed; when sulfates, phosphates, or nitrates are added to the chromate solutions, they too are found in the coating.

From the host of treatments[1,2] that have been proposed for magnesium alloys, only the corrosion results obtained from the four most commonly used are given. The application of Types I–IV treatments are described in detail in the Army-Navy Aeronautical Specification AN-M-12. Their common names and corresponding treatments are as follows:†

Type I. Chrome Pickle

Dip ½ to 2 min in a bath composed of 180 grams/liter of $Na_2Cr_2O_7 \cdot 2H_2O$ and 187 ml/liter of conc. HNO_3 maintained at room temperature. Hold 5 sec in the air after removing from bath; rinse and dry.

Type II. Sealed Chrome Pickle

Same as Type I, but followed by a sealing treatment in a boiling bath containing 100 to 150 grams/liter of $Na_2Cr_2O_7 \cdot 2H_2O$ for 30 min.

Type III. Dichromate

Step 1: Dip 5 min in an aqueous solution containing 15 to 20% by weight of HF maintained at room temperature. Rinse in cold water.

Step 2: Boil parts 45 min in a 10 to 15% $Na_2Cr_2O_7 \cdot 2H_2O$ aqueous solution, or, alternately, for 30 min in this solution saturated with MgF_2 or CaF_2. Rinse and dry.

Type IV. Galvanic Anodize

Step 1: Same as Step 1 in Type III.

Step 2: Galvanically anodize the magnesium at 1.1 amp/sq dm (10 amp/sq ft) for 10 min in a bath operated at 50° to 60° C (120° to 140° F) and composed of 30 grams/liter of $Na_2Cr_2O_7 \cdot 2H_2O$, 30 grams/liter of $(NH_4)_2SO_4$, and 10 ml/liter of NH_4OH (sp. gr. 0.88). The magnesium is electrically connected to the steel tank containing the solution to generate the necessary current.

* Metallurgical Department, Dow Chemical Co., Midland, Mich.
† The chromate treatments are applied after cleaning in alkaline or acid baths in the usual manner to remove grease and corrosion products.

[1] H. W. Schmidt, W. H. Gross, and H. K. DeLong, "The Surface Treatment of Magnesium Alloys," *The Surface Treatment of Metals*, p. 36, The American Society for Metals, Cleveland, Ohio, 1941.
[2] J. L. Bleiweis and A. J. Fusco, "Protective Treatments for Magnesium," *Metals and Alloys*, **21**, 417 (February, 1945).

CHROMATE COATINGS ON MAGNESIUM ALLOYS

The effect of these treatments in preventing corrosion or tarnish and in improving the adhesion of paints is given in the following paragraphs.

RESISTANCE — AQUEOUS MEDIA

Figure 1 shows corrosion data obtained upon testing a typical controlled and uncontrolled purity wrought alloy in a 3% NaCl solution. With the controlled purity alloy (AZ 31X), where iron and other severely cathodic impurities are maintained below the tolerance limit, the type of protective treatment has little influence on the corrosion rate. With the same alloy (A.S.T.M. AZ 31), however, when the

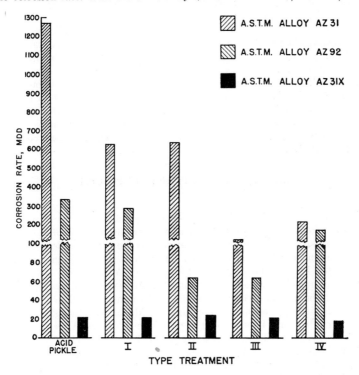

FIG. 1. Effect of Chemical Treatment on the Corrosion Rate of Magnesium Alloys in a 3% NaCl Solution.

Alloy composition — %

Designation	Alloy Form	Al	Zn	Mn	Fe	Ni
AZ 31	Sheet	3.0	0.77	0.23	0.013	<0.001
AZ 92	As cast	8.9	1.9	0.13	0.008	0.002
AZ 31X	Sheet	2.7	1.02	0.35	0.001	nil

Specimen size — 7.5 × 2.5 × 0.6 cm (3 × 1 × 0.25 in.).
Surface preparation — acid-pickled (8% HNO_3 + H_2SO_4 by vol.) prior to chemical treatment.
Temperature — 35° C (95° F).
Aeration — probably saturated.
Duration of test — 14 days.
Volume of testing solution — 100 ml.
Testing method — alternate immersion — 30 sec in medium — 2 min in air.

impurities exceed the tolerance limit, the corrosion rate is greatly affected by the chemical treatment. Types I and II chromate treatments reduce the corrosion rate, but Type III produces a maximum reduction in the rate. Probably the most important step in the use of Types III and IV treatments is the hydrofluoric acid dip. This acid is known to be especially active in removing cathodic impurities that would normally cause corrosion in accelerating media. The MgF_2 film formed on the magnesium surface also appears to be important in catalyzing or in some way promoting chromate film formation.

Also shown in Fig. 1 is the effect of the chromate treatments on the corrosion rate of a typical casting alloy (A. S. T. M. AZ 92). Although the iron content of this alloy is well above the tolerance limit, the zinc present modifies the effect of the cathodic impurities and greatly reduces the corrosion rate of the base alloy. Types II and III chromate treatments produced maximum lowering in the corrosion rate. With a controlled purity alloy of this composition, the corrosion rate would be unaffected by the chemical treatment.

When painted, then tested in 3% NaCl, both the increased corrosion resistance provided by the chemical treatment and the adhesion of the paint to the treated surface are important. Although Types I, II, and IV treatments show good corrosion resistance and paint adhesion, the Type III treatment offers maximum protection. Controlled purity alloys given these chromate treatments and then painted are much superior to similarly treated uncontrolled purity alloys.

RESISTANCE — EXTERIOR EXPOSURE

Unpainted

Figure 2 shows results obtained on a Mg-Mn alloy and two of the alloys shown in Fig. 1 when exposed to a severe industrial atmosphere for a period of 1½ years. The resistance of the coating was judged qualitatively on the general appearance, tarnishing, corrosion, and coating decomposition. The tarnish rate of magnesium alloys is greatly affected by the alloy content of the metal. Although the 3% NaCl test showed the controlled purity alloys to be much more resistant to corrosion, no effect of purity could be noted on samples exposed to the atmosphere. The chemical treatments provide little protection for the Mg-1.5% Mn alloy, although the Type IV treatment produced some improvement in resistance to change. Alloy AZ 31 was not protected by Type I or Type II treatments, but Types III and IV produced considerable resistance. The casting-type alloys exemplified by Alloy AZ 92 are much more resistant to atmospheric tarnish than the alloys of lower alloy content. When treated with Types III and IV treatments, little change in appearance was noted after 1½ years industrial exposure.

The Type III treatment also gives definite protection from corrosion even under severe coastal exposure conditions. After 2 years at Kure Beach, North Carolina, at a station 800 ft from the ocean, this coating normally shows little change in color. Although even untreated controlled purity alloys show only a 2 to 5% loss in tensile strength after 2 years exposure, the Type III treated alloys normally show less loss.

Painted

Perhaps the most important single requirement of a protective coating on magnesium alloys is that it provide a base to which paint will have good adhesion over an extended period of time. All of the four treatments listed above provide a base to which paint adheres well after 2 years industrial or coastal exposure. To obtain maximum adhesion with Types I and II, a definite etch of the metal must be produced. The build-up of magnesium nitrate in, or the addition of sulfates to, the chrome-pickle solution will inhibit the etch formation on the magnesium surface and will definitely cause a reduction in the paint adhesion characteristics. The Types III and IV coatings are somewhat thicker than the coatings produced by Types I and II, and good paint adhesion is normally assured by absorption of the paint into the relatively porous chromate film.

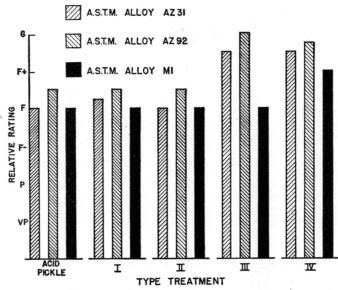

FIG. 2. Effect of Chemical Treatment in Preventing Surface Corrosion in an Industrial Atmosphere at Midland, Michigan.

	Alloy	Alloy composition —%		
Designation	Form	Al	Zn	Mn
AZ 31	Sheet	3.0	0.77	0.23
AZ 92	As cast	8.9	1.9	0.13
M1	Sheet	< 0.03	< 0.001	2.0

Specimen size — 14 × 7.5 × 0.6 cm (6 × 3 × ¼ in.).
Surface preparation — acid-pickled (8% HNO_3 + H_2SO_4 by vol.) prior to chemical treatment.
Duration of test — 1½ years.
Testing method — 45° exposure facing south.
Temperature — −6° C (21° F) winter; 22° C (71° F) summer.
P = Poor; F = Fair; G = Good.

PHOSPHATE COATINGS

V. M. DARSEY*

Phosphate coatings on metals result from chemical reaction of phosphoric acid with the metal surface to form a non-metallic coating, as contrasted with metallic or paint coatings which generally do not require chemical combination with the base

* Parker Rust Proof Co., Detroit, Mich.

metal. The most commonly used coating solutions contain zinc and iron, or manganese and iron phosphates. The solutions react chemically with the metal until the surface is converted into a crystalline phosphate coating. An oxidizing agent[1] capable of oxidizing hydrogen and thus preventing the formation of a hydrogen film on the metal surface during treatment is used to accelerate formation of the coating.

The physical characteristics of the coating such as thickness, crystalline structure, and penetration of the base metal can be controlled by the method of cleaning prior to treatment, by the manner in which the solution is applied, by time of treatment, and by modifying the composition of the solution. Advantage of these factors has been utilized in two widely used commercial processes.* One of them produces in 30- to 60-min treatment a heavy corrosion-resistant oil-absorptive phosphate coating on metal articles, and the other a thin dense crystalline phosphate coating in 1- to 5-min treatment adapted to increasing the adhesion and resultant durability of applied paint finishes.

The method of applying the phosphate coating is usually determined by the size and shape of the article to be coated. Small items such as nuts, bolts, screws, and stampings are coated by immersing in the phosphating solution, whereas large articles such as automobile bodies and refrigerator cabinets are sprayed.

Difference in Phosphoric Acid Cleaning and Phosphate-Coated Metal Surfaces

Phosphoric acid metal cleaners usually consist of phosphoric acid and a water-soluble organic grease solvent with or without a wetting agent. The purposes in the preparation of surfaces for painting with such solutions are to remove grease and rust in one operation and to provide a slight etch of the metal whereby paint adhesion is promoted. To accomplish this result, the cleaning solution must contain sufficient acid and solvent to remove rust and grease. The acid concentration is such that formation of any substantial phosphate coating is prevented, as contrasted with the use of dilute acid-phosphate solutions containing metal phosphates as described above, resulting in precipitation of a visible crystalline phosphate coating on the surface.

PHOSPHATE COATINGS AS BASE FOR PAINT

Experience has demonstrated the increase in durable life of paint finishes applied over phosphate-coated surfaces.[2] In addition to increasing the adhesion of paint films to metal surfaces, the phosphate coating retards underfilm corrosion and the spread of corrosion wherever the paint film is scratched or abraded. The comparative protection against corrosion afforded steel by two spray coats of baked synthetic enamel applied over three different methods of surface preparation was demonstrated in outdoor exposure tests of 6 years duration.[3] The three methods of metal preparation before painting consisted of (1) solvent vapor degreasing, (2) sandblasting, and (3) coating with phosphate. Observation of the panels after exposure revealed in the case of the vapor-degreased steel severe underfilm corrosion at an intentional scratch, with considerable corrosion on the dominant surface of the panel. General failure

---

* Called Parkerizing and Bonderizing.

[1] U. S. Patent 1,911,726 (1933).

[2] V. M. Darsey, "Effect of Corrosion on the Durability of Paint Films," *Ind. Eng. Chem.*, **30**, 1147–1152 (1938).

[3] Unpublished data, Parker Rust Proof Co.

occurred on the sandblasted panel at the highest points of the roughened surface, but the phosphate coating substantially improved paint protection and retarded underfilm corrosion at an intentional scratch.

PAINT RETENTION OF UNTREATED AND PHOSPHATE-COATED STEEL

The retention of paint by metal surfaces is an important factor especially where its application is by dipping or flow coating and the article contains sharp edges. During curing or drying, most paint films tend to pull away from edges because of poor adhesion, resulting thereby in reduced film thickness at such areas. In Table 1 is

TABLE 1. RETENTION OF PAINT BY UNTREATED AND PHOSPHATE-COATED STEEL

Panel size — 4 × 6 in. full-finished sheet.
Primer — U. S. Army Specification ES680A, Class 101.
Paint viscosity — 24 sec No. 4 Ford Cup at 80° F.
Dipped — withdrawal rate 5 in. per min at constant speed.

	Primer Wt. per sq ft Retained, oz	Increase in Retention over Untreated Steel, %
Steel, untreated	0.139	...
Steel, phosphate-coated (Roller application)	0.177	27.3
Steel, phosphate-coated (By spray)	0.200	43.9
Steel, phosphate-coated (By immersion [dipping])	0.242	74.0

shown the amount of paint remaining on untreated and phosphate-coated steel applied in various manners. Weight of dry primer was obtained by weighing panels before and after painting. The increased retention of paint by phosphate-coated steel in relation to untreated steel is a primary factor in preventing the spread of corrosion from edges of painted phosphate-coated surfaces.

CORROSION RESISTANCE

OIL FINISH FOR CORROSION RESISTANCE

The affinity of heavy phosphate coatings for oil, wax, stain, or coloring is one of their valuable characteristics utilized in practice for corrosion prevention. Any method for testing the corrosion resistance of phosphate coatings usually includes addition of oil or a subsequent finish.

A variety of articles such as nuts, bolts, screws, washers, and stampings are phosphate-coated and finished with oil or coloring treatment for corrosion prevention. The facts that no electric current is used and formation of the coating is solely dependent upon the processing solution contacting the metal surface make possible the production of uniform coatings on irregularly shaped articles. Various types of oil-containing corrosion inhibitors and rust-preventive finishes are used, depending upon the protection and service required. Frequently coated parts are finished by dipping in linseed oil reduced 200% with mineral spirits, followed by centrifuging to remove the excess oil. Steel sheets and strip used in the manufacture of food cans are commonly phosphate-coated and the fabricated can then finished with a lacquer or paint enamel.

SALT-SPRAY TEST OF PHOSPHATE COATING WITHOUT OIL FINISH

In Table 2 is shown the difference in corrosion resistance of phosphate-coated steel without oil or other finish and uncoated steel after various exposure periods in the salt-spray test. The extent of corrosion was determined by weighing of specimens after stripping in inhibited acid according to a method described by R. C. Ulmer.[4]

TABLE 2. CORROSION OF PHOSPHATE-COATED AND UNTREATED STEEL IN SALT-SPRAY TEST

Mg per Sq Cm

Hours in test	24	72	240	744
Steel (S. A. E. 1020), untreated	2.20	5.51	14.26	34.08
Steel, phosphate-coated	0.18	0.31	0.67	3.32

PHOSPHATE COATINGS ON ZINC

Because of the reactive nature of zinc, it is ordinarily difficult to obtain good paint adhesion without adequate chemical treatment of the surface. Phosphate coating of zinc and zinc alloys retards corrosion and increases the adhesion and resultant dura-

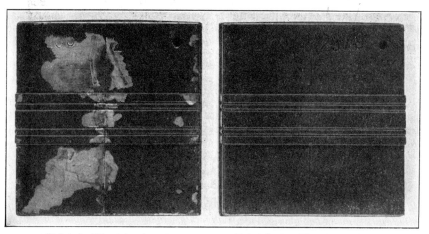

Painted without treatment.   Painted after Phosphate Treatment.

FIG. 1. Effect of Phosphate Treatment on Protection of Baked Enamel on Zinc.

Finish: 2 coats baked synthetic enamel.
Six years of outdoor exposure.

bility of applied paint coatings. In addition to the practice of phosphate coating fabricated articles made of zinc-coated steel, the hot-dipped and electroplated zinc sheets and strip are so coated at various steel mills for improved paint adherence.

In Fig. 1 is shown the difference in paint adhesion when paint is applied over an untreated zinc-base die casting and a phosphate-coated casting. Definite chipping of the paint is apparent on the untreated casting, while the finish is still intact over the phosphate-coated casting.

[4] R. C. Ulmer, "Determination of Magnetic Iron Oxide," *Ind. Eng. Chem.* (Anal. Ed.), **10**, 24 (1938).

CHEMICAL OXIDE COATINGS 871

GENERAL REFERENCES (PHOSPHATE COATINGS)

BURNS, R. M., and A. E. SCHUH, *Protective Coatings for Metals*, pp. 373–376, Reinhold Publishing Corp., New York, 1939.
FÖLDES, A., "A New Process for Cold Phosphatizing," *Korrosion u. Metallschutz*, **19**, 281 (1943).
JERNSTEDT, G., "New Phosphate Coatings with Unusual Corrosion Resistance," *Trans. Electrochem. Soc*, **83**, 361 (1943).
MACHU, W., "Phosphatizing and Its Scientific Foundations," *Metallwirtschaft*, **22**, 481–487 (1943).
ROESNER, G., L. SCHUSTER, and R. KRAUSE, "The Nature of the Different Phosphate Coating Processes," *Korrosion u. Metallschutz*, **17**, 174–179 (1941).
SCHUSTER, L., "Effect of Composition, Surface Condition and Pretreatment of Iron and Steel on Phosphate-Protective Coatings," *Korrosion u. Metallschutz*, **19**, 265–269 (1943).
SCHUSTER, L., "Present-Day Phosphate Coating Processes for Iron and Steel," *Stahl u. Eisen*, **60**, 785–790, (1940).
SCHUSTER, L., and R. KRAUSE, "Composition of Phosphating Solution, Solid Phase, Phosphate Coat and Corrosion Resistance of Coating in Known Phosphating Systems," *Korrosion u. Metallschutz*, **18**, 73–81 (1942). "Phosphating at Room Temperature," *op. cit.*, **20**, 153–161 (1944).

CHEMICAL OXIDE COATINGS

V. M. DARSEY*

Attractive black oxide coatings are obtained on steel by treatment in strong alkali solutions containing an oxidizing agent at temperatures of about 140° to 150° C (285° to 300° F), depending upon the exact composition of solution employed. For producing the coating on steel, caustic soda is usually employed in conjunction with such oxidizing agents as sodium nitrite, nitrate, or chlorate. Other methods, such as quenching in oil, are employed for producing oxide coatings on ferrous articles.

Activity of the blackening bath is controlled by means of temperature. Steels containing alloying constituents, such as nickel and chromium, require higher operating temperatures for blackening than do carbon steels.

Little dimensional change in articles results from the oxide coating, and for this reason it can be used on precision parts requiring close tolerances.

CORROSION RESISTANCE OF OXIDE COATINGS

Oxide coatings on steel are not considered a satisfactory finish for corrosion prevention without the use of some subsequent treatment with oil or wax. Salt-spray specifications for the coatings require the unoiled coating to withstand a one-half hour and the oiled coating a two-hour test without corrosion. For many applications, oxide coatings are perhaps more useful for their decorative value than for their corrosion resistance.

A satisfactory black finish is produced on copper and brass by direct oxidation at low temperatures. The coating after oiling exhibits good corrosion resistance by salt-spray test and can be used for increasing the adhesion of applied paint films. Outdoor exposure tests conducted by the author have shown that oxide-coated steel articles do not corrode uniformly but exhibit pit-type corrosion as frequently encountered with hot-rolled steel containing mill scale.

* Parker Rust Proof Co., Detroit, Mich.

VITREOUS ENAMELS

A. I. ANDREWS*

The terms *vitreous enamel, glass lining,* or *porcelain enamel* refer to a thin layer of glass fused into the surface of a metal, generally iron. Obviously, these coatings have the properties of glass, and variations in their properties are due to differences in the compositions and the physical conditions of the glasses used. These glasses must have thermal expansions compatible with the metal base; they must fuse at temperatures below the melting point of the metal and, in addition, must meet the conditions of service for which they are designed. Like other coatings, a single type of enamel glass should not be expected to meet the extremes of all kinds of conditions. For example, one should not expect to have a highly reflecting white enamel and also maximum durability toward strong acid solutions. The compositions and the physical structures of enamel glasses vary over a wide range and should be selected for the combination of properties which their service demands.

MANUFACTURE

Vitreous enamels are made by smelting (melting) together such materials as feldspar, borax, quartz, soda ash, fluorides, color oxides, and opacifiers to form a molten glass. This is then poured into a stream of water which shatters it to broken glass, which is called *frit*. The frit is ball-milled either wet or dry and applied to the properly prepared metal shape by dipping, spraying, or dusting. The ware is then placed in a furnace at about 850° C (1550° F) for sufficient time to melt the coating into a continuous layer of glass. Several coatings may be applied.

PROPERTIES

Although the properties of the enamel depend principally upon those of the glass, the serviceability of the product is also affected by the metal base and the design of the ware.

PHYSICAL PROPERTIES

Glasses in thin layers or threads have great strength and, when applied to steel, are believed to have tensile strengths in excess of 50,000 psi. In general, these coatings will withstand distortion or flexing of the steel within the elastic limit of the metal. If a permanent deformation takes place, the coating is ruptured, showing either cracks (crazes) or breaking off of the coating. All enamels which fail mechanically do so by tension. Figure 1 shows the effect of vitreous enamel of various weights per unit area on the bending strength of coated steel sheet.

The hardness of enamels is about 6 on Mohs' scale, but the actual resistance to scratching or abrasion depends greatly upon the bubble structure of the enamel. All enamel coatings contain bubbles but they should always be small as compared to the enamel thickness; and they should be uniformly distributed. Vitreous enamels in general have high resistance to scratching and abrasion.

Increased thickness of enamel coatings improve impact resistance and generally improve corrosion protection. However, thick layers lower resistance to damage by flexure and also lower resistance to thermal shock. Some physical properties are listed in Table 1.

* Department of Ceramic Engineering, University of Illinois, Urbana, Ill.

VITREOUS ENAMELS

TABLE 1. AVERAGE PHYSICAL PROPERTIES OF ENAMELS

Hardness	5 to 6 Mohs' scale
Modulus of elasticity*	$\begin{cases} 4950 \text{ to } 7480 \text{ kg per sq mm} \\ 7.05 \text{ to } 10.6 \times 10^6 \text{ psi} \end{cases}$
Tensile strength†	42,000 to 64,000 psi
Specific gravity	2.54 to 2.66; lead-bearing: 3.32 to 3.79
Thermal expansion, °C	82 to 130 × 10^{-7}
Fusibility	500° C (930° F) and up (gradual softening)

* L. D. Fetterolf and C. W. Parmelee, *J. Am. Ceram. Soc.*, **12**, 193 (1929).
† A. I. Andrews and W. W. Coffeen, *J. Am. Ceram. Soc.*, **22**, 11 (1939); C. W. Parmelee and R. G. Ehman, *J. Am. Ceram. Soc.*, **13**, 475 (1930).

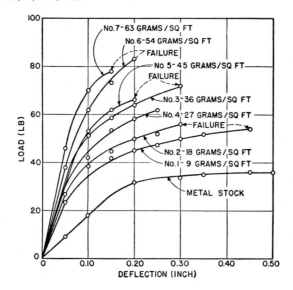

FIG. 1. Effect of Enamel on Strength of 2- by 4-in., 20-Gage Sheet Steel.

THERMAL PROPERTIES

Vitreous enamels are made for use over a wide range of temperatures from extreme cold to above red heat. Special types are used on exhaust stacks for airplanes, and others called ceramic coatings protect iron from corrosion at temperatures of 760° C (1400° F) to 815° C (1500° F). They are also made with high and low infrared emission at these higher temperatures.

Table 2 lists thermal conductivities for enameled sheet under various conditions of heat transfer.

COMPOSITION

As can be noted in Table 3, the compositions of vitreous enamels vary over a wide range. The enamels are all essentially alkali borosilicates. Fluorides and alumina are generally present to improve the processing properties of the enamel but they are a detriment to the acid solubility resistance. Such materials as antimony and zirconium compounds and the fluorides contribute the "opacity" white appearance. Cobalt, nickel, and manganese color the gloss, and in the case of ground coats, they promote

CORROSION PROTECTION

TABLE 2. OVERALL COEFFICIENTS OF HEAT TRANSFER IN ENAMELED STEELWARE*,†

Conditions	Kg cal/(hr) (sq meter) (°C)	Btu/(hr) (sq ft) (°F)
Steam, to be water-heated	400–700	80–140
Hot water, to water being heated	350	70
Steam, to boiling water	700	140
Steam, to a thick fruit product	160	32
Cooling hot water, by cold water and brine	200–600	40–120
Hot oil, to oil being heated	65–140	13–24
Hot oil, to boiling water	150–200	30–40
Steam, to water being heated in tubular heater	500–800	100–160
Steam being condensed to water in a tubular condenser jacket	700	140

* E. P. Poste, *Ind. Eng. Chem.*, **16**, 469 (1924).
† Both sides enameled. Steel normally ¼ in. thick.

adherence to the metal. Zircon improves water and alkali resistance. Titanium oxide and silica improve acid solubility resistance. In general, however, the compounding of a suitable enamel composition depends upon considerable experimentation and a balancing of the amounts of the constituents present. The mill additions, such as clay and electrolytes, are used in wet-milled enamel to control the properties of the slip and thereby facilitate the processing. Other mill additions are used for color and opacity.

DURABILITY

The resistance of enamels to corrosive environments is commonly termed durability. The durability is generally classed according to acid resistance, alkali resistance, hot-water resistance, weather resistance, and soil resistance. It is commonly determined by subjecting the surface of the ware to the solutions characteristic of service exposure and then noting either loss of gloss or loss of weight. The Porcelain Enamel Institute Test[1] utilizes the effect of the solution on the surface gloss and apparent etching. The Enameled Utensil Manufacturers Council Test[2] and many tests on chemical-enamelware base results upon loss of weight. The choice of test depends upon the requirements of appearance or corrosion protection.

ACID RESISTANCE

The acid solubility resistance of enamels varies from enamels that are durable to the attack of strong acids for long periods of time at elevated temperatures, to those that are attacked almost immediately upon exposure to weak acids in the cold. The attack varies from that on enamels in which solution is arrested after the first loss of gloss to that on enamels which are continuously dissolved away. The former type of acid-resistant enamel is obviously preferable. The order of *decreasing* resistance of different types of enamels to acids is as follows: Chemical Acid-Proof Enamel, Acid-Resisting White Cover Enamel, Cobalt Ground-Coat Enamel, Antimony White Cover Enamel, Sign Blue Cover Enamel, Fluoride Cover Enamel, Zircon White Cover Enamel. There is some overlapping in the orders given, since each type represents a range of compositions and qualities.

[1] *Test for Acid Resistance of Porcelain Enamels*, Porcelain Enamel Institute, 1010 Vermont Ave., N.W., Washington 5, D. C. Published October, 1939.
[2] "Porcelain-Enameled Steel Utensils," *Commercial Standard CS 100-44*, National Bureau of Standards, Washington, D. C. Published 1944.

TABLE 3. COMPOSITIONS OF PORCELAIN ENAMELS

	No. 1 Ground Coat		No. 2 Sign Blue Cover Enamel	No. 3 Blue Zirconium Cover Enamel	No. 4 Chemical Acid-proof Cover Enamel	No. 5 Fluoride White Cover Enamel	No. 6 Antimony White Cover Enamel	No. 7 Zirconium White Cover Enamel	No. 8 Acid-Resistant White Cover Enamel	No. 9 Titanium White Cover Enamel
	Frit A	Frit B								
Frit Batch										
Feldspar (Keystone)	20.8	21.2	38.0	28.3	...	27.0	17.6	...	9.5	4.0
Borax	34.6	34.2	30.0	33.6	2.6	28.0	22.9	...	18.0	28.6
Quartz	27.7	21.5	16.0	6.1	19.2	19.0	23.5	...	39.0	41.4
Soda ash	4.6	5.6	4.0	7.0	...	7.0	6.7	...	10.0	...
Soda niter	4.7	4.7	3.0	4.0	1.1	4.0	2.9	3.08	3.0	2.9
Fluorspar	5.1	10.3	3.0	7.0	3.0	5.0	2.5	4.40	1.5	...
Manganese dioxide	1.5	1.5	...	1.5	...	...	...	...	...	...
Cobalt oxide	0.5	0.5	2.0	1.0	1.5	...	...	...	...	...
Nickel oxide	0.5	0.5	...	...	...	...	...	...	...	...
Cryolite	...	...	4.0	...	...	10.0	8.4	12.30	...	6.5
Zircon	...	...	...	12.0	...	...	...	10.80	...	...
Zinc oxide	...	...	...	...	...	...	2.9	4.40	...	0.8
Sodium antimonate	...	...	...	...	...	...	12.6	...	9.5	0.2
Aluminum hydrate	...	...	...	...	...	...	...	3.96	...	1.0
Dehydrated borax	...	...	...	...	...	...	...	22.60	...	...
Calcite	...	...	...	...	...	...	...	7.74	...	...
Pyrophyllite	...	...	...	...	...	...	...	31.70	...	...
Sodium silico fluoride	...	...	...	...	...	...	...	...	5.0	...
Bone ash	...	...	...	...	...	...	...	...	1.5	...
Titanium dioxide	...	...	...	...	15.3	...	...	...	3.0	9.5
Sodium aluminate	...	...	...	...	5.3	...	...	...	...	...
Whiting	...	...	...	...	2.2	...	...	...	...	1.5
Sodium silicate	...	...	...	...	51.2	...	...	...	...	...
Mill Batch										
Frit	70	30	100	100	100	100	100	100	100	100
Clay (Vallendar)	7.5		...	...	...	6.0	6.0	6.0	5.0	5.0
Florida kaolin	...		6.0	6.0	1.0	...	...	...	...	...
Borax	0.75		...	...	...	...	...	...	...	...
Magnesium carbonate	...		0.25	0.25	...	0.25	0.25	0.25	...	...
Bentonite	...		...	...	0.25	...	...	...	0.25	0.25
Sodium aluminate	...		...	...	0.25	...	...	...	0.25	0.25
Barium chloride	...		...	...	0.10	...	...	...	...	...
Tin oxide	...		...	...	...	6.0	...	...	5.0	...
Treopax	...		...	...	...	...	2.0	2.0	...	...
Water	45		40	40	46	40	40	40	38	38
Firing Conditions										
Temperatures, °F.	1575		1500	1520	1520	1460	1500	1460	1530	1530
Time (minutes)	5		3	5	3	2½	2½	2½	3	3

Acids may, in some cases, be selective in their action and it is not possible to predict their activity on the basis of degree of ionization. The order of attack is not always the same for different enamels, but the following order represents common experience with 10% solutions of several acids in order of *decreasing* attack: sulfuric, hydrochloric, nitric, oxalic, tartaric, citric, malic, lactic, phosphoric, carbolic, acetic, pyrogallic. Contaminations, such as fluorides, in commercial phosphoric acid, may produce erratic results.

The rate of attack by acid solutions increases with temperature. It also increases with concentration up to a maximum, which varies with different acids and enamels, and then drops off rapidly. For sulfuric, hydrochloric, nitric, and phosphoric acids, the maximum rate generally appears close to or slightly above 10 to 20% concentration.

Alkali Resistance

Most enamels are quite resistant to alkaline solutions and, therefore, comparatively few data have been reported for these media. Enamels not resistant to acids frequently have superior resistance to alkalies. However, acid-resisting enamels may also have good alkali resistance. The addition of zirconium compounds to the enamel composition generally results in an increased resistance to alkaline solutions. The following is an approximate arrangement of the different types of enamels in order of *decreasing* resistance to a boiling 10% sodium hydroxide solution: Zirconium Blue Cover Enamel, Chemical Acid-Proof Enamels, Zirconium White Cover Enamel, Acid-Resistant White Cover Enamel, Cobalt Ground-Coat Enamel, Antimony Cover Enamel, Sign Blue Cover Enamel, Fluoride Cover Enamel.

Water, Weather, and Soil Resistance

Vitreous enamels are practically unaffected by water or clean air at atmospheric temperatures, but natural waters and atmospheres often contain acid or alkaline constituents which may affect some enamels when exposed for long periods of time.

Water Resistance. The attack by waters even when they carry large amounts of salts at atmospheric temperatures is almost imperceptible and the use of tests for enameled coatings, such as the salt-spray test, is useless. Likewise, the weatherometer does not perceptibly accelerate failure of enamel surfaces.

It is, however, important to consider the type of enamel which is to be used for lining an autoclave, or for surfaces over which hot water flows continuously or semi-continuously, as some enamels disintegrate rapidly when subjected to such conditions. Likewise, only special types of enamels can be satisfactorily used for hot water tanks, range boilers, and similar equipment. Under conditions whereby water is heated, the composition of the water is an important factor since acid waters are more corrosive than alkaline waters. Little data have been published on the subject, but it has been shown that the slight attack by boiling 10% salt solutions is related to the kind of salt present. Acid waters in general and waters containing hydrolyzable salts that produce acid conditions cause some attack of non-acid-resisting enamels.

Weather Resistance. Since vitreous enamels do not chemically react under warm moist conditions, any attack by the atmosphere must be caused largely by solution brought about by acid gases or suspended hydrolyzable salts. It is known that in industrial districts the atmosphere contains sulfurous and sulfuric acids as well as carbonic acid. For this reason, it is desirable for vitreous enamels used on signs in architectural structures, and other out-of-door applications, to have some degree of acid resistance.

Vitreous enamels do not fade or change color, but the effect of surface etching by the atmosphere sometimes gives an appearance of a white chalky film which is frequently but improperly interpreted as fading.

Soil Resistance. The resistance of vitreous enamel to soil corrosion is associated with the acidity and alkalinity of the soil and the soil waters. Deterioration is exceedingly slow if the proper enamels are used. Wherever the metal happens to be exposed to the soil (pores, edges, etc.), it corrodes and may undermine the enamel at these areas. In addition, where corrosion of the iron base of an enameled piece is taking place, atomic hydrogen produced by corrosion may diffuse through the metal. If this hydrogen comes in contact with the underside of an enamel coating, it accumulates, forms molecular hydrogen gas, and builds up sufficient gas pressure actually to force flakes of enamel, called *fishscale*, away from the iron base. These spots then act as new points where corrosion can take place.

Data on enameled iron specimens in various soils have been reported.[3]

GENERAL

Design. In addition to the chemical and physical properties of the enamels, design of the ware has a very important influence on the successful protection of the metal base. Although the expansion coefficients of the enamels are adjusted to their particular uses, some mechanical designs are much more likely to result in failure than are others.

Vitreous enamels should not be applied to small radii when corrosion resistance is critical. It is very difficult to obtain good coverage on sharp edges, and the tendency to chip or to damage by mechanical shock is greater at such points. Radii of at least 1/8 in. and preferably 1/2 in. or larger should be used. Excessive thickness of application of the enamel and especially uneven applications should be avoided.

The ware should be designed to avoid the development of stresses during firing or heating and cooling in service. Heavy and light metal sections should be kept at a minimum, and thickness of stock should be adjusted to compensate for shading in the furnace. Distortion or warping generally results in a rupture of the coating. The design must further provide for sufficient strength, particularly in tanks, to avoid stressing the steel beyond the yield point due to pressure or weight. Corrosion problems cannot be separated from these considerations of design as it is only with proper design that the coating can be expected to protect the metal completely.

Tests for Defects and Repairs. Tests for cracks, pinholes, or other defects in an enamel coating can be made, in addition to careful visual inspection, by electric resistance measurements, high-frequency discharge, or application of a cupric chloride solution.[4] The most popular method uses an electrical indicating instrument, one terminal of which is applied to the base metal and the other through a source of potential to a wet metal brush which is drawn over the surface of the enamel. Any pinhole or crack is indicated by flow of current.

The high-frequency tester is sometimes used where acidproof enamel is employed and where the most severe conditions of corrosion are to be encountered. This test is suitable only under special conditions as the high-potential current may actually rupture the enamel surface.

Repairs are difficult to make and special types must be used for different purposes. Gold leaf, silver, or other metal fillings may serve effectively to prevent progress of

---

[3] K. H. Logan and M. Romanoff, *J. Research Natl. Bur. Standards*, **33**, 145–198 (1944).

[4] "Porcelain-Enameled Tanks for Domestic Use," *Commercial Standard CS 115-44*, National Bureau of Standards (1944).

corrosion. Cements and lacquers are sometimes used satisfactorily but are generally only a temporary protection.

Use of Enamels in Cathodic Protection. Enamels are electrical insulators; hence there are many applications where this property is an important consideration. Cathodic protection of the base metal (discussed beginning p. 923) in conjunction with use of an enamel coating is a possibility that has not yet been put into extensive use, but shows promise because the protective cathodic current is focused at the small exposed areas where corrosion would otherwise occur. The current required is relatively small because of the presence of an insulating layer over the iron.

GENERAL REFERENCES

ANDREWS, ANDREW I., *Enamels*, Twin City Printing Co., Champaign, Ill.
ANDREWS, ANDREW I., "Glass-Lined Chemical Equipment," *Trans. Am. Inst. Chem. Engrs.*, **35** (No. 4), 473–488 (August 25, 1939).
ANDREWS, ANDREW I., and RALPH L. COOK, *Enamel Laboratory Manual*, the Garrard Press, Champaign, Ill.
ANDREWS, ANDREW I., and E. W. DIETTERLE, "Effect of Enamel on the Strength of Enameling Iron," *J. Am. Ceramic Soc.*, **23** (No. 1), 29–32 (1940).
COOK, RALPH L., and ANDREW I. ANDREWS, "The Chemical Durability of Porcelain Enamels," *J. Am. Ceram. Soc.*, **28**, 229–56 (1945).
LANDRUM, ROBERT D., and HERBERT D. CARTER, *Bibliography and Abstracts of Literature on Enamels*, The American Ceramic Society, Columbus, Ohio.
MCCLELLAND, E. H. (compiler), *Enamel Bibliography and Abstracts* (1928–1939), The American Ceramic Society, Columbus, Ohio.
HANSEN, J. E., *A Manual of Porcelain Enameling*, The Enamelist Publishing Co.
Porcelain Enamel Standard and Tentative Standard Tests, Porcelain Enamel Institute, Inc., 1010 Vermont Ave., N.W., Washington 5, D. C.
"Porcelain-Enameled Steel Utensils," *Commercial Standard CS 100–44*, Government Printing Office, Washington 25, D. C.
"Porcelain-Enameled Tanks for Domestic Use," *Commercial Standard CS 115–44*, Government Printing Office, Washington 25, D. C.

Organic Coatings

INTRODUCTION

G. H. YOUNG* AND COLLABORATORS

INTRODUCTION

We are here particularly interested in the matter of corrosion prevention by organic resin coatings and the considerations which control the permanency and general inertness of the protective film in corrosive environments.

We should first emphasize that, generally speaking, the resinous film is not of and in itself *corrosion inhibitive*. By this we mean that the basic corrodibility of the substrate metal is in no way altered by the overlying resin. A possible exception is the case of phenolic films, where the experimental evidence indicates that some passivation by free phenolic bodies can and does take place. On the whole, we must view the paint film as a simple mechanical barrier, interposed between the corrodible metal substrate and the corrosive environment. Accordingly, the determining properties of the barrier are those related to *continuity*, which is a measure of the ability of the resinous molecules to form a pore-free film at low coating weights; to *per-*

* Mellon Institute, Pittsburgh, Pa.

meability, which is a measure of the diffusion rates of water vapor and other gases through the "continuous" film; to *adhesion and cohesion*, which are measures of the tenacity with which the resin molecules bond to the substrate and to each other; and to *chemical inertness* against degradation by hydrolytic, oxidative, actinic, and thermal decomposition.

FILM CONTINUITY

Where the resin film is a mechanical barrier only, it is obvious that areas containing micro pores and other discontinuities will be non-protective. Thus knowledge of the factors determining continuity is of primary importance.

The relationship of film thickness, or weight per unit of area covered, to ultimate film structure and consequent film properties is of fundamental interest. With the development of quantitative methods for evaluating continuity, it became possible to study intensively the phenomena associated with coverage on active metals.[1] Of major importance is the established experimental fact that film resistivity is a discontinuous function of coating weight and that for every metal surface there exists some finite coating weight which must be exceeded before the familiar characteristics of amorphous films are encountered and before continuous film membranes are formed. This minimum weight of basic understructure is not constant for all metals but varies widely with the surface being coated.

The coating weights of a typical vinyl chloride-acetate copolymer required just to produce complete film continuity on a number of familiar metals are shown in the following table:

Metal Surface	Minimum Coating Wt.		Approximate Thickness, mil
	mg/sq cm	mg/sq in.	
Platinum	0.15–0.23	1–1.5	0.05–0.07
Copper	0.93–1.2	6–8	0.30–0.40
Tin (plate)	0.47–0.55	3–3.5	0.15–0.18
Iron	1.48–1.55	9.5–10	0.48–0.50
Zinc (galvanized iron)	2.5–2.8	16–18	0.80–0.90
Aluminum	0.85–1.0	5.5–6.5	0.28–0.33

That the experimental differences between metals cannot be attributed to differences in surface irregularities has been demonstrated. The consistency with which such data can be reproduced and the specificity of behavior on different metals suggest that the basic film structure may not be entirely amorphous or "brush heap" in make-up, as had formerly been generally supposed.

PERMEABILITY CONSIDERATIONS[2,3,4,5]

The continuity characteristics of thin resin films are related (1) to the activity of the surface on which the film is initially deposited, (2) to the composition of the

---

[1] G. H. Young and coworkers, *Ind. Eng. Chem.*, **29**, 1277 (1937); **30**, 685 (1938); **31**, 719 (1939); **33**, 72 (1941); **33**, 550 (1941).
[2] P. M. Doty, Wm. H. Aiken, and H. Mark, *Ind. Eng. Chem.* (Anal. Ed.), **16**, 686 (1944).
[3] H. F. Payne, *Ind. Eng. Chem.* (Anal. Ed.), **11**, 453 (1939).
[4] R. M. Burns and A. E. Schuh, *Protective Coatings for Metals*, Reinhold Publishing Corp., New York, 1939.
[5] J. J. Mattiello (Editor), *Protective and Decorative Coatings;* Vol. IV, *Special Studies*, John Wiley and Sons, New York, 1944.

resin and the specific nature of the polar linkages along the chain, and (3) to the average molecular weight and chain length of the polymeric molecules. It is to be borne in mind that film continuities are not directly related to *permeability* characteristics. We regard permeability as a slow or inhibited ion transfer through a *continuous* film. The mechanism may be one of adsorption, followed by gradual adsorption and diffusion, with accompanying internal swelling and perhaps even incipient degradation owing to hydrolysis and swelling stresses. It has been demonstrated, however, that only those films which are continuous in the sense of being free from gross discontinuities yield consistent data from the familiar permeability measurements — an expected but conclusive finding.

For permeability measurements, use is made of *free* or stripped films mounted on a suitable porous base material such as alundum, fine-meshed screen, paper, or cellophane. The supported films are then mounted as membranes between atmospheres maintained at different relative humidities sufficiently far apart to establish a steep diffusion gradient. The weight loss or gain of one compartment thus serves to measure the quantity of water vapor transferred through a measured area of film of a given thickness. Such studies demonstrate that the amount of moisture permeating a continuous film is inversely proportional to the thickness, provided all other variables are under strict control. It must, however, be pointed out that recorded "permeability" values strictly apply only under the experimental conditions specified. Departures from the ideal state are occasionally marked. Accordingly, it is not possible to calculate diffusion rates from one set of conditions to another on the assumption that the physical laws are obeyed quantitatively. In practice, it is better to report the permeability in terms of the experiment itself rather than to extrapolate to a state of unit parameters. Thus one preferably records "milligrams of water vapor transferred through 10 sq in. of a film 5 mils thick in 24 hours when the vapor pressure gradient across the film is 20 mm of mercury." An empirical *permeability constant* is, however, frequently employed for convenience in making close comparisons. Thus:

$$D = \frac{NX}{At\,(p_1 - p_2)}$$

where D = permeability constant,
N = wt. of water vapor transferred (in grams),
X = film thickness (in inches),
A = test area (in square inches),
t = time (in hours),
$(p_1 - p_2)$ = vapor pressure difference (in millimeters of mercury).

The relationship between film permeability as such and protective ability is not straightforward. With completely inert resinous bodies it is generally true that the lower the permeability, the better the corrosion protection afforded. There are, however, numerous complicating factors. Typical of such is the stability of the coating material to thermal effects or to actinic radiation, oxidation, or hydrolysis. Its flexibility and adherence characteristics and its general toughness and "weathering resistance" are probably of equal or greater importance than its specific permeability. On the whole, therefore, permeability measurements can only supplement orthodox performance data, not substitute for them. Some typical permeability values are listed in Table 1.

The addition of pigments can materially alter the permeability characteristics of

ORGANIC COATINGS — INTRODUCTION

a given resinous vehicle. The most profound effect is produced by "leafing" pigments, of which aluminum powder, electrolytic copper, and lead leaf are exemplary. Reductions in permeability of the order of 50% can be obtained by proper pigmentation with leafing aluminum. The other leafing metals, flake mica, etc., are less efficient.

TABLE 1. PERMEABILITY OF ORGANIC COATINGS TO MOISTURE*

Coating	Mg H_2O Passing per 24 hr per sq in. of Area per mil of Film Thickness
Cellophane (not moisture-proofed)	300
Cellulose acetate	300
Polyvinyl acetate	115
Polyvinyl chloride-acetate	28
Bodied linseed oil	80
Long oil varnishes	36–48
Short oil varnishes	17–24
Nitrocellulose-alkyd lacquers	15–87
Orange shellac	16
Asphaltic coating	5

* R. M. Burns and A. E. Schuh, *Protective Coatings for Metals*, p. 322, Reinhold Publishing Corp., New York, 1939.

The relationship between permeability and pigment binder ratio for a series of aluminum-pigmented ester gum varnishes is shown in Table 2.

TABLE 2. EFFECT OF PIGMENT TO BINDER RATIO ON PERMEABILITY TO MOISTURE

(25-gal Tung-Linseed Phenolic Resin Varnish as Test Vehicle)

Pigment/Binder Ratio	Calc'd Permeability Constant D*
0	6.30×10^{-8}
0.14	4.07
0.28	3.50
0.42	2.77
0.56	2.12
0.84	1.30

* G. H. Young and G. W. Seagren, paper presented before the American Society of Refrigerating Engineers, 1944.

Purely theoretical considerations would dictate that *any* pigmentation should profoundly influence permeability. This is because the nature and distribution of side chain substituents, determining as they do the completeness and tenacity with which the individual pigment particle is "wet" by the resin molecules, are the real controls on fundamental film properties *within the pigmented film proper*. That is, the properties of the *vehicle* have far more influence on ultimate protective ability than does the chemical make-up of the added pigment. The data in Table 3, showing permeability as a function of added pigment, clearly indicate that profoundly different pigments show much the same permeability characteristics in the same vehicle.

With the exception of a few pigments, notably red lead and zinc oxide, potentially capable of chemical *reaction* with the vehicle, all the permeabilities are of much the same order of magnitude, even though they are as chemically dissimilar as graphitic carbon and silica, for example.

TABLE 3. PERMEABILITY OF VARIOUS PIGMENTS TO MOISTURE*
Pigment Added At 25% By Volume

Pigment	Mg H_2O Passing per 24 hr per sq in. of Area per mil of Film Thickness
(Clear varnish vehicle)	72.3
Silica	40.3
Ocher	38.3
Red iron oxide	55.0
Black iron oxide	35.7
Chrome oxide	41.6
White lead	45.5
Chrome orange (lead chromate)	56.0
Lead titanate	47.0
Graphite	40.3
Lampblack	41.7
Zinc chromate	47.7
Zinc oxide	34.4
Red lead	28.8

* R. M. Burns and A. E. Schuh, *Protective Coatings for Metals*, p. 323, Reinhold Publishing Corp., New York, 1939.

ADHESION AND COHESION IN RESIN FILMS

It has been conclusively demonstrated that measured adhesion values are a function of (1) the degree of cleanliness of the metal surface and thus of its ability to be "wetted" by the resin, (2) the chemical nature of the polar groups along the macro chains (COOH, OH, $CONH_2$, etc.), (3) the average chain length up to a polymerization degree of about 4000 to 5000, and (4) the content of non-resinous polar compounds (an inverse function in this case).

TERMS AND DEFINITIONS

Paint continues to be the oldest and most widely used protective measure against corrosion. Following chapters in this section detail the basic properties and applications of the many different kinds of organic protective coatings available commercially. A brief description of the fundamental types of coating compositions may, however, be helpful at this point.

PAINTS

This term was originally strictly applied to mixtures of pigments — usually inorganic oxides — with a drying oil such as linseed oil. Oil-base paints, frequently mixed "on the job" by a master painter, are the oldest type of protective coating in general use. However, as described later, the word no longer has so restricted a meaning. Any combination of pigment and vehicle that is adapted to brush application and air-dries to a tough, adherent coating is very likely to be termed a paint.

VARNISHES

A varnish is a combination of drying oil and a fortifying resin, either natural or synthetic. The combination may or may not involve actual chemical reaction between the resin and oil. The mixture is thinned with suitable solvents to brushing or spraying viscosity and employed as a clear composition; it air-dries by oxidation of the oil component.

ORGANIC COATINGS — INTRODUCTION

ENAMELS

An enamel is a pigmented varnish, in its strictest sense. Actually, the wide use of fortifying resins in oil-base paints has resulted in the disappearance of any distinction other than an arbitrary one between paints and enamels. There is currently a tendency to term alkyd resin-base finishes *quick-dry enamels* to differentiate them from the older oleoresinous paints.

LACQUERS

Classically, a lacquer consisted of one or more selected natural resins dissolved in a rapidly volatile solvent. Such compositions were employed both as clears and pigmented. They set to very hard, glossy, non-tacky films by solvent evaporation only. Currently, the term has been broadened to designate any air drying or even baking-type clear composition usually, but not necessarily, based on nitrocellulose or similar cellulose resins.

BAKING FINISHES

This term is used broadly to describe any composition, either clear or pigmented, that requires forced drying or *baking* (in England, *stoving*) to release the solvent and *set* or *convert* the film to its ultimate hard, non-tacky, completely polymerized form.

The essential chemistry and mode of operation of these several classes of protective coatings will be more fully described in subsequent chapters.

APPLICABILITY AND SERVICE LIFE

The conditions under which paints and other coatings are applied and exposed, the almost infinite permutations to be had by combining pigment, resin, oil, and solvent, and the diversity of repaint schedules encountered in practice make it impossible to particularize on applicability and expected service life. A paint known to be in substantially perfect condition after 6 years at one location may fail in 6 months elsewhere. The weathering factors of light intensity and atmospheric contaminants have a profound influence, particularly on oleoresinous compositions. Most organic vehicles are embrittled and even decomposed at relatively low temperatures, 150° to 250° C (300° to 500° F), if maintained there for any length of time. Many finishes are excellent for normal underwater service, but are rapidly hydrolyzed and degraded by mildly alkaline or acid waters, particularly at or near the boiling point of the water.

In general, then, each condition of service must be considered as a separate problem; there is no "universal protective coating." Broadly speaking, baking finishes are more permanent than the older oxidizing-type coatings; there are, however, many applications where the baking operation is impractical or even impossible. Most *maintenance* finishes are designed for regular renewal or "touch-up." This need not be the case for an automobile body or decorated metal cabinet, can, or closure. Accordingly, one finds the maufacturer of small metal articles increasingly turning to the newer baking-type synthetics. The maintenance engineer advisedly continues to rely on regular repainting (even annually) with orthodox paints.

It is impossible to base accurate estimates of even the *comparative* service life of dissimilar resin finishes on accelerated tests alone. The majority of these — such as the standard salt-spray, humidity-cabinet, and standardized immersion exposures — were really designed to study the corroding characteristics of the substrate metal.

Most of them have little effect on the coating material *as such*. However, a corrosion process proceeding beneath a coating (perhaps initiated at a scratch, sharp edge, or other imperfection) can have a disruptive influence on adjacent intact coating. It thus comes about that a given coating on zinc, for instance, will apparently "fail" under conditions where the identical coating is completely unaffected if coated on stainless steel, silver, or platinum. For these reasons, it is imperative not to assume that a 200-hour salt-spray test means a given coating is twice as effective as one running only 100 hours to the first showing of rust. Many coatings last indefinitely if coated on glass and exposed to 100% relative humidity in the familiar *humidity cabinet*. The same coatings applied to automobile-body steel may show rusting, blistering, and even gross peeling in a matter of weeks under this test; and applied as an actual automobile finish, they frequently stand up satisfactorily for 3 to 4 years, the only gross failure being some loss of gloss.

Confronted with these anomalies as the rule rather than the exception, the experienced paint technologist cannot assign absolute values to prognosticated service life estimates. Generally speaking, salt-spray life in excess of 200 hours is considered a fair indication of reasonable protective permanence. Many finishes — notably the baking type — show no attack up to 2000 hours or more. Exposure to live steam may, however, cause the coating to peel off in a matter of hours. The average cellulose-base lacquer has a demonstrably shorter life in salt-spray and most immersion tests; but this does not necessarily mean that such lacquers are actually inferior materials. For the predominant number of applications where corrosion protection is the major goal, close attention to surface preparation prior to coating application and to the matter of insuring a uniform, sufficiently heavy and holiday-free coating — together with proper air dry time or bake — is more important than the actual choice of specific coating composition.

Each broad coating type has its place of proved applicability. The actual nature of the more important of these types is detailed in the following chapters.

PAINTS AND VARNISHES

RICHARD J. ECKART*

DEFINITION

Paints and varnishes are protective coatings which can be classified as *air drying*. This means that, upon application, the film dries essentially by oxidation and polymerization. Such coatings constitute the bulk of protective and maintenance finishes used, as contrasted with other types that dry only by solvent evaporation, forced drying, or a simple hardening process. Examples of the last are lacquers, baking enamels, bituminous coatings, rubber finishes, etc.

This discussion therefore deals with oils, resins, and pigments which, in combination, yield the protective air-drying paints. Only the more important raw materials can be mentioned. The oils, oleoresinous varnishes, and synthetic resins have infinite variety; they constitute the *vehicle* or binder for a variety of pigments.

Paint and varnish films may display extraordinary water resistance yet permit water vapor and gases to penetrate and reach the face of the metal in sufficient quantity to be damaging. It is these dissolved gases that accelerate the attack. That

* Sapolin Co., Inc., New York, N.Y.

is why added pigments play an important role, not only in improving the mechanical properties of the film but also in serving to neutralize the effects of such accelerants.

The manner in which active pigments serve to inhibit corrosion has been stated[1] to be the imparting of *alkalinity* and/or oxidizing effects. Since iron corrodes more readily as the acidity increases, basic pigments (notably the metallic oxide type such as red lead pigments) which produce an alkaline condition will retard or prevent attack. On the other hand, some strong oxidizing agents (notably the chromates) are effective because they act as passivators.

The emphasis in any paint system should properly be placed on the primer. It is better to have two good coats of a true rust-inhibitive primer followed by a mediocre top coat than the reverse. The ideal system comprising an efficient primer coat to inhibit rust with an adherent top coat that defies sun, water, and weather still remains the painter's goal.

PREPARATION OF SURFACE

Many maintenance engineers have neglected to provide a firm, solid foundation for paint systems. Paint manufacturers have failed to stress the importance of proper surface preparation and drying conditions to attain maximum service. Just as it is "penny wise and pound foolish" to use cheap paint on any structure worth saving, so is it equally foolish to apply high-quality paint to a surface on which there is mill scale, rust, grease, moisture, dirt, or old paint which is scaling and loosely attached. Premature paint failures mean more than the loss of the paint since labor and other costs in applying average 400 to 800% of the paint material itself. Furthermore, repainting becomes more expensive and the metal has been partially sacrificed.

There are adequate tools available for surface preparation, namely scrapers, wire brushes, sandpaper, and steel wool. For the larger surfaces, particularly those steel forms and structures where mill scale is still present, there are many efficient time-saving devices.[1,2,3] They may be described as:

1. Chip hammering, sandblasting, or shot blasting with air pressure.
2. Pickling in acid bath (with or without inhibitors).
3. Flame cleaning and dehydrating, preferably followed by power wire brushing.

It is important to insure the absence of mill scale since this hard crust of iron oxide, when loosened, will accelerate the eventual "letting-go" of the entire paint system down to the bare metal. Mill scale varies in its tenacity of adhesion to the metal and is brittle. It is cathodic to steel; hence, wherever there is discontinuity of scale, accelerated corrosion sets in.[1] In this connection the British Iron and Steel Institute has reported that "it is clear that complete descaling before painting is the only safe method of eliminating the danger of pitting as a result of electrochemical action, which will persist so long as any mill-scale is allowed to remain on the steel."[4]

To remove rust and mill scale, any of the above cleaning systems will suffice, depending upon the time element and the facilities on hand, i.e., air pressure, sand

---

[1] R. M. Burns and A. E. Schuh, *Protective Coatings for Metals*, Reinhold Publishing Corp., New York, 1939.
[2] J. C. Hudson, *Fifth Report of the Corrosion Committee*, Iron and Steel Institute, London, 1938.
[3] American Institute of Steel Construction, *The Preparation of Structural Steel Surfaces for Painting*, 1940.
[4] The Iron and Steel Institute (British), "First Report of the Marine Corrosion Subcommittee," **147**, Paper 9, p. 390 (1943).

and apparatus for blasting, tanks for pickling, or oxyacetylene cylinders and flame brushes for the third method. When the flame cleaning method is used, wire brushing should follow, and the surface should receive the prime coat while still hot and free from moisture.

Each cleaning system possesses advantages and disadvantages. Sometimes, however, the methods may be combined for optimum results. For example, sandblasting, followed by flame cleaning, adds the advantage of dehydration to either wet or dry sandblasting operations. In addition, the inorganic surface treatments described in *Phosphate Coatings*, p. 867, may be included so that a phosphoric acid solution might follow a pickling process or could be sandwiched between sandblasting and flame cleaning. These considerations can be weighed only in the light of the performance specifications which have been set up and in the nature of the job — shop coat vs. repaint work, ship interiors vs. bridges, etc.

KINDS OF PAINT SERVICE

For the sake of simplicity, metal protective paint systems can be classified under four usages:

1. Mild atmospheric in rural areas.
2. Severe atmospheric in industrial, tropical, or coastal areas.
3. Tidewater exposure on surfaces subject to alternate submersion and atmospheric exposure.
4. Submerged exposure: (a) in salt water; (b) in fresh water as well as water-storage tanks.

At the end of this chapter (p. 889) are listed a number of typical metal primer formulations. They will serve as illustrations to show the diversified raw materials available to paint chemists. With the formulas are designated conditions under which the type of primer is especially serviceable. It is intended that these examples of anti-corrosive primers illustrate the wide variations possible. The enlightened paint chemists of today realize that the gradual perfection of their work on protective coatings lies in the use of balanced pigment combinations and in the employment of more complex vehicles where each type of oil or resin contributes in the ultimate performance of the system. A detailed compilation[5] of raw materials lists almost 2000 pigments, 280 oils, 70 grades of fatty acids, 280 types of resins, almost 300 kinds of solvents and thinners, and 290 distinct plasticizers.

Generally speaking, there is a tendency to favor a vehicle or binder in the top coat which is similar in nature to that used in the primer so that the entire system is "welded" together as one. Hard-drying coats should not be used over soft-drying primers and vice versa. In addition, there are the practical considerations of permissible drying period, method of applying, and cost limitations. Choice of any paint system involves these factors as well as the specific service requirements.

BASIC TYPES OF VEHICLES

The non-pigment liquid portion of a paint forms the *vehicle*, *binder*, or *medium* (the British term). To the layman, a vehicle is either a varnish or an oil and, when pigmented, the varnish becomes *enamel* and the oil becomes *paint*. This is a simple and correct statement, but the line of demarcation between oils and varnishes is no

---

[5] Henry A. Gardner, *Physical and Chemical Examination of Paints, Varnishes, Lacquers and Colors*, 9th Ed., Institute of Paint and Varnish Research, Washington, D. C., 1939.

longer so sharp as it was several decades ago. At that time, linseed oil was the staple paint vehicle (and is still the most widely used drying oil today). Varnishes were made by cooking it with resin hardened with lime and zinc oxide. Natural resins such as Kauri and Congo were forerunners of the better and harder drying varnishes of today.

Then came three major developments: (1) the widespread use of China wood oil (tung oil) with its quick-drying, waterproofing properties; (2) the rapid adoption of phenolaldehyde resins as a means of further "stepping up" drying properties and improving durability; (3) the introduction and acceptance of alkyd resin vehicles, notably the glyceryl phthalate type. Though the term *synthetic* is often applied to phenolic resins, the paint industry more popularly used this designation in describing alkyd types evolved from chemical reaction of glycerin, phthalic anhydride, and fatty acids of drying oils. The alkyd paints constituted the bulk of the ultra-durable camouflage paints used on implements of war where their characteristic resistance to oxidation and film breakdown made them invaluable under the many conditions of exposure.

The widespread use of China wood oil, phenolic resins, and alkyd resins has in itself been an impetus to the creation of better finishes. Not only did they establish new standards of performance but they stimulated alternate and substitute methods for accomplishing such results.

RUST-INHIBITIVE PIGMENTS

The most successfully used rust-inhibiting pigments fall into the general class of reactive pigments and are still, by far and large, compounds of *lead* or *zinc*. A discussion of the more important pigments follows. Attention should again be drawn to the chart at the end of this chapter which illustrates their use, often in combination with other reactive pigments and/or inert, neutral pigments. These neutral pigments, such as iron oxide, magnesium silicate, and silica, have demonstrated definite advantages when used with the reactive zinc or lead pigments.[6,7] Examples are increased film strength, improved adhesion, less reactivity with the vehicle, and a "tooth" which makes for better bond between coats.

THE LEAD GROUP

Red lead is available in the usual commercial grades; 98, 97, 95, and 85%, which indicate the percentage of true red lead (Pb_3O_4). The remainder of the pigment is litharge, PbO.[8] This pigment has found very wide usage in metal primers. The reason for its effectiveness is stated as the inhibition of the anodic solution of steel or iron due to its alkaline and oxidizing nature.* This property is very desirable, particularly in the acid conditions of industrial environments.[9] In addition, it reacts with the oil to form relatively insoluble soapy formations which reinforce the film, making it more impervious to moisture.

* For further discussion of the mechanism, see *Passivity*, p. 31.

[6] R. M. Burns and A. E. Schuh, *Protective Coatings for Metals*, Reinhold Publishing Corp., New York, 1939.

[7] F. Fancutt and J. C. Hudson, "—Anti-corrosive Compositions for Ship Bottoms,—" The Iron and Steel Institute (British) **150**, 269–333 (1944).

[8] J. J. Mattiello (Editor), *Protective and Decorative Coatings*, Vol. II, John Wiley and Sons, New York, 1942.

[9] L. L. Carrick, *Eng. News-Record*, **132**, 468 (1944).

The characteristic inhibitive properties of red lead can also be used to advantage in combination with both reactive and inert pigments. Desirable improvements in elasticity are also offered by the use of vehicles and drying oils which are less susceptible to the oxidizing effect of the pigment.

Basic lead sulfate is available in two types:

White — $2PbSO_4 \cdot PbO$.

Blue — white basic lead sulfate 45% min., lead oxide 30% min., lead sulfide 12% max., lead sulfite 5% max., zinc oxide 5% max., carbon and miscellaneous 6% max.[8]

Basic lead sulfate is recently reported[10] as giving very favorable underwater performance and promises to become an increasingly important ingredient in metal primers. Being strongly basic, these sulfate pigments react with acidic ingredients in the electrolyte, neutralizing them. In addition they possess irregular particle size which connotes better penetration into surface irregularities, building a coherent, water-resisting film that protects against acid and salt solutions, oxygen, and other gases.[11]

White basic lead carbonate ($2PbCO_3 \cdot Pb[OH]_2$),[12] although a widely used pigment, has been found most valuable for use as a top coat pigment. More and more attention is being given to its use as a component pigment in primers. Success with it has also been recorded by the British.[10]

Lead chromate, broadly referred to as chrome yellow, varies in color from light yellow to deep orange. As a rust-inhibitive pigment, the basic lead chromate (60% $PbCrO_4$, 40% PbO) is generally regarded as superior to the medium or normal type (98% $PbCrO_4$) since its alkalinity and moderate reactivity with the vehicle provide good protection. It is used extensively in primers, where it differs from zinc chromate in its slower release of chromate ions, because of its very low solubility.

Other lead pigments are in the process of being evaluated.[13] Examples are basic silicate dry white lead, metallic lead flake, chromated leaded zinc, chromated blue lead, chromated white lead carbonate, lead aluminate, etc.

Zinc Pigments

Zinc pigments have become a strong contender for the dominant position held by red lead over the years as a rust-inhibitive pigment.

Zinc chromate is typified by the formula $4ZnO \cdot 4CrO_3 \cdot K_2O \cdot 3H_2O$.[12] This pigment, zinc yellow, has shown a vast increase in use due to its versatility for protecting steel as well as aluminum and magnesium. Its theoretical function is to release soluble chromates which passivate the surface. By the same token, however, it often shows a tendency to form water blisters in the presence of excessive moisture.

In highly industrial atmospheres where SO_2 is present, chromates have been reported[14] as depolarizing the cathodic hydrogen film and thereby stimulating corrosion. The incorporation of a basic pigment with zinc chromate should aid in stabilizing the latter in acid environments.

[10] F. Fancutt and J. C. Hudson "—Anti-corrosive Compositions for Ship Bottoms —," The Iron and Steel Institute (British), **150**, 269–333 (1944).

[11] L. L. Carrick, *Eng. News-Record*, **132**, 468 (1944).

[12] J. J. Mattiello (Editor), *Protective and Decorative Coatings*, Vol. II, John Wiley and Sons, New York, 1942.

[13] Henry A. Gardner and L. P. Hart, National Paint, Varnish and Lacquer Association, *S. S. Circ.* 633 (1942).

[14] R. M. Burns and A. E. Schuh, *Protective Coatings for Metals*, Reinhold Publishing Corp., New York, 1939.

ILLUSTRATIVE PAINT PRIMERS

Ingredients, lb per 100 gal paint	Specification			Specification			Non-Specification		Non-Specification
	U.S. Navy Bureau of Ships 52P18(INT)	U.S. Army Engineer Corps T-1601	U.S. Tank Automotive Center ES-680-101	Federal Specification TT-P-641*			British Ship Bottom Primer†	Special Marine Primer	General Purpose Primer
				Type I	Type III				
Zinc chromate	278	...	38	...	...		...	71	149
Red lead	...	1345(95%)	...	...	...		588(W)	391(97%)	...
Basic lead sulfate	...	...	...	...	...		...	...	447(B)
Zinc oxide	...	...	25	368	392		...	71	105
Iron oxide	25	448	257	...	...		591	50	...
Zinc dust	...	...	...	1473	1570		...	...	...
Inert pigments	127	...	197	...	...		...	128	176
Other pigments	76(T)	...	...	...	...		295(C)	...	...
Vehicle solids	236(A,V)	344(A,O)	205(A)	467(O)	250(V)		303(V)	364(A,O,V)	226(A)
Thinners and driers	370	177	393	52	240		353	269	355
Weight per gal	11.1 lb	23 lb	11.2 lb	23.5 lb	24.5 lb		21.5 lb	13.5 lb	14.6 lb
Recoating time	8 hr	16 hr	18 hr	48 hr	8 hr		12 hr	4–8 hr	12 hr
Color	Yellow	Brownish red	Red	Blue-gray	Blue-gray		Red	Brownish red	Gray
Suggested use	II, III	I, II	I	I, X	IVb		IVa	III, IVa	I, II, III

* Available in Type I (oil vehicle) Type II (alkyd vehicle), and Type III (phenolic vehicle).
† F. Fancutt and J. C. Hudson "— Anti-corrosive Compositions for Ship Bottoms —," The Iron and Steel Institute (British), **150**, 269–333 (1944).

CODE

Vehicle

(A) Alkyd resin
(O) Oil, vegetable drying
(V) Varnish, oleoresinous

Pigment

(B) Blue lead sulfate
(C) Carbonate of lead
(T) Titanium dioxide
(W) White lead sulfate

Usage

I   Mild atmospheric exposure
II   Severe atmospheric exposure
III Tidewater exposure
IVa Submerged — salt water
IVb Submerged — fresh water
X   As primer on galvanized metal

Zinc chromate has a wide compatibility with vehicles and auxiliary pigments, providing a synergistic effect in its improvement of the overall performance.

Zinc tetroxy chromate ($ZnCr_4 \cdot 4Zn[OH]_2$) is an alternate form of zinc chromate pigment that now has a background of experience when used in pigment combinations with iron oxide, zinc oxide, and inert pigments. It has been used mostly in primers for machinery, equipment, and ordnance material.

Zinc oxide (ZnO), although not usually regarded as a rust-inhibitive white pigment, has neutralizing effects that are beneficial.[15] It is commonly used with zinc dust, zinc chromate, iron oxide, and various lead pigments. As a basic pigment, zinc oxide reacts with the organic vehicle to give a "tighter" film and better resistance to both water and abrasion.

Zinc dust (Zn), a heavy metallic powder, is furnace-distilled to a finely divided powder, 97% of which passes through a No. 325 mesh sieve (0.0017-in. openings). It performs exceptionally well in paints for galvanized or other zinc surfaces and also for water-storage tanks and fresh water submersion. It is usually used with about 20% zinc oxide, to which iron oxide or chromium oxide are sometimes added. The reactivity of zinc dust with the acidic constituents of certain vehicles results in hydrogen evolution and *gassing*. A two-compartment container for zinc dust paints is therefore mandatory unless experience has demonstrated that no reaction takes place.

Leaded zinc is available in 35 and 50% grades, the percentages showing the amount of white basic lead sulfate incorporated with the zinc oxide. Since both the sulfate of lead and the oxide of zinc have proved advantageous, this composite pigment is finding a place in rust-inhibitive paints.

Other Pigments. Though of limited use or in the experimental stage, zinc aluminate,[16] chromated zinc oxide,[17] and ammonium ferrous phosphate[18] have attracted attention as have the low-solubility chromate pigments of strontium and barium.

LACQUERS AND BAKING ENAMELS

P. O. POWERS*

A variety of natural and synthetic organic materials is now available for the manufacture of lacquers and baking enamels. Each type of resin has properties that especially fit it for certain types of service; none combines all these properties.

The synthetic resins are capable of almost infinite variation. In a lacquer resin, ready solubility in low-cost solvents is desirable, but the lack of resistance to the same solvents in the finished film is not wanted. Those showing maximum adhesion to metal frequently have poor resistance to water; hardness (desirable) and brittleness (undesirable) usually accompany each other. In many cases a combination of properties is sought which are mutually exclusive. It is, however, not necessarily true that the properties of the resin as applied are the same as for the film which

---

* Battelle Memorial Institute, Columbus, Ohio.

[15] Harley A. Nelson, *Ind. Eng. Chem.*, **27**, 35 (1935).

[16] Henry A. Gardner and L. P. Hart, National Paint, Varnish and Lacquer Association, *S. S. Circ.* 633 (1942).

[17] F. Fancutt and J. C. Hudson, "— Anti-corrosive Compositions for Ship Bottoms — ," The Iron and Steel Institute (British), **150**, 269–333 (1944).

[18] A. M. Erskine, Godfrey Grimm and S. C. Horning, *Ind. Eng. Chem.*, **36**, 456 (1944).

results from it. With heat-reactive resins, for example, the basic properties change profoundly on baking.

FORMATION OF RESINOUS MATERIALS

Natural resins such as fossil gums and shellac are formed by complicated processes that have not been entirely clarified. Naturally formed large molecules like protein bodies, cellulose, and rubber are potential starting materials for coating compositions. The esters and ethers of cellulose have found wide application as lacquer ingredients. Excellent finishes have been made from zein and similar proteins. Chlorinated rubber and other rubber derivatives offer great promise in chemically resistant and underwater coatings.

One generally used method for forming purely synthetic resins is condensation polymerization. In Table 1, the methods by which a variety of commercial resins of this type are made have been outlined. Such condensation reactions can be interrupted and completed at some later time; many of the thermosetting resins are formed in this way. The molecules become larger as the condensation proceeds, but the condensation is interrupted while the resin is still soluble and fusible. There are, however, reactive groups present which, when condensed, lead to still larger molecules and in some cases to completely insoluble and infusible materials.

The other method by which polymers are formed is known as vinyl polymerization. Vinyl resins are usually permanently soluble in specific solvents, but there are exceptions. If a compound with two vinyl groups is polymerized, a highly crosslinked polymer is formed which is insoluble and infusible. Divinyl benzene is such a divinyl compound; its polymers are highly insoluble.

Vinyl compounds are often copolymerized by reacting two different monomers and forming a polymer which contains segments of both starting materials. The vinyl chloride-acetate copolymer, known by the copyrighted term Vinylite, is a well-known example. This copolymer has properties quite different from those of either polyvinyl acetate or polyvinyl chloride; the effect cannot be achieved by merely mixing the two individual polymers.

Many of the commercial plastic resins have been or can be used for protective coatings. The resins listed in Table 1 include the more important. The condensation polymers are usually heat-hardening and are supplied in soluble forms which require heat to effect a further reaction resulting in *curing* the film. Most of the vinyl-type polymers are applicable as *lacquers*, though force drying to accelerate solvent release is recommended.

Whereas it is true that much of the inherent superiority of the so-called baking finishes is attributed to this heat treatment, at the same time this places a serious limitation on their applicability to many structures, where the maximum protective action could well be afforded. Thus, for example, even were it possible to employ baked or "stoved" coatings on prefabricated structural steel bridge trusses when supplied to the contractor, damage due to handling during assembly would necessitate painting of the completed bridge; and repainting could hardly be accomplished with other than air-drying compositions. Experiments on a fairly large scale with infrared lamp banks, wherein efforts to bake out coatings in place on massive steel structures have been made, have not demonstrated this approach to be very practical. On smaller items such as reaction vessels, storage tanks, and the like where the coating serves as a *lining*, this and other types of bake-out techniques are both possible and in actual use.

On the whole, the baking-type coatings must be considered as primarily adapted to

TABLE 1*

	CONDENSATION POLYMERS				
1	Benzene + $Cl_2 \rightarrow$ Chlorobenzene $\xrightarrow{\text{NaOH or } H_2O}$ Phenol	+ Formaldehyde $\begin{smallmatrix}H\\C=O\\H\end{smallmatrix}$	$\rightarrow$	Durez, Resinox	
2	CO_2 + $NH_3 \rightarrow$ Urea $O=C\begin{smallmatrix}NH_2\\NH_2\end{smallmatrix}$	+ Formaldehyde	$\rightarrow$	Plaskon, Beetle	
3	$CaC_2 + N_2 \rightarrow C\equiv N \rightarrow$ Cyanamide; Melamine	+ Formaldehyde	$\rightarrow$	Melmac, Resimene	
4	Naphthalene + $O_2 \rightarrow$ Phthalic anhydride + Glycerol + $C_{17}H_{31}COOH$ (Linoleic acid)		$\rightarrow$	Glyptal, Rezyl, Duraplex	
5	Cyclopentadiene + Maleic anhydride → "Adduct" + Glycerol + $C_{17}H_{31}COOH$		$\rightarrow$	Carbic	
6	$C_3H_5(OOCC_{17}H_{32}OH)_3 \rightarrow$ Castor oil; Sebacic acid + Glycerol + $C_{17}H_{32}OHCOOH$ (Ricinoleic acid)		$\rightarrow$	Paraplex	
7	Castor oil $\xrightarrow{NaOH}$ $HOOC(CH_2)_8COOH$ (Sebacic acid) + $HOCH_2CH_2OH$ (Ethylene glycol)		$\rightarrow$	Paracon	
8	Phenol + $H_2 \rightarrow$ Cyclohexanol; $H_2NC(CH_2)_4CNH_2$ (Adipamide) ← $HOOC(CH_2)_4COOH$ (Adipic acid) + $H_2N(CH_2)_6NH_2$ (Hexamethylene diamine)		$\rightarrow$	Nylon	
9	$HOCH_2CH_2Cl$ (Ethylene chlorohydrin)	$\rightarrow$	Ethylene oxide	$\rightarrow$	Carbowax
10	Ethylene $C=C$ + $Cl \rightarrow$ $ClCCCl$ Ethylene dichloride	+ Na_2S_4 Sodium tetrasulfide	$\rightarrow$	Thiokol, GR-P	
	VINYL POLYMERS				
11	Cracked gases	$\rightarrow$	Ethylene $C=C$	$\rightarrow$	Polythene
12	Benzene + Ethylene $C=C$	$\rightarrow$	Ethyl benzene $HC-CH$	$\rightarrow$ Styrene $C=C$	$\rightarrow$ Styron, lustron

* Copyright, 1944, by P. O. Powers.

LACQUERS AND BAKING ENAMELS 893

TABLE 1 — *Continued*

#	Reactants / Intermediates	Products / Polymer
13	Turpentine → β-Pinene (+ HCH)	$\left[\begin{array}{c}\mathrm{CH_3\ H}\\-\mathrm{C}-\mathrm{C}-\\ \mathrm{H}\\ \mathrm{CH_3}\end{array}\right]_n$ Piccolyte
14	Coal tar distillates → Indene + Coumarone	$\left[\begin{array}{c}\mathrm{H\ H}\\-\mathrm{C}-\mathrm{C}-\mathrm{CH_2}\end{array}\right]_n$ Cumar, Nevindene, Picco
15	$\mathrm{H_3C-\underset{\mathrm{CH_3}}{\overset{\mathrm{CH_3}}{\mathrm{CH}}}}$ Isobutane → $\mathrm{\underset{H}{\overset{H}{C}}{=}\underset{CH_3}{\overset{CH_3}{C}}}$ Isobutylene	$\left[\begin{array}{c}\mathrm{H\ \ CH_3}\\-\mathrm{C}-\mathrm{C}-\\ \mathrm{H\ \ CH_3}\end{array}\right]_n$ Vistanex
16	$\underset{\mathrm{CH_3}}{\overset{\mathrm{CH_3}}{\mathrm{C}}}{=}\mathrm{O}$ + HC:N → $\mathrm{CH_3\underset{CH_3}{\overset{OH}{\underset{C\equiv N}{C}}}}$ + $\mathrm{CH_3OH}$ → $\mathrm{H\underset{OCH_3}{\overset{CH_3}{\underset{C=O}{C}}}}\overset{CH_3}{=}\mathrm{C}$ Acetone, Cyanohydrin, Methanol, Methyl methacrylate	$\left[\begin{array}{c}\mathrm{H\ \ CH_3}\\-\mathrm{C}-\mathrm{C}-\\ \mathrm{H\ \ C=O}\\ \mathrm{OCH_3}\end{array}\right]_n$ Plexiglas, Lucite
17	HC≡CH + CH₃COOH → $\mathrm{CH_3\overset{O}{\overset{\|}{C}}{-}O{-}\overset{H\ H}{\underset{H}{C}}}$ Acetylene, Acetic acid, Vinyl acetate	$\left[\begin{array}{c}\mathrm{H\ H}\\-\mathrm{C}-\mathrm{C}-\\ \mathrm{H\ O}\\ \mathrm{H_3CC{=}O}\end{array}\right]_n$ Gelva, Vinylite A
18	See 17 → $\left[\begin{array}{c}\mathrm{H\ H}\\-\mathrm{C}-\mathrm{C}-\\ \mathrm{H\ O}\\ \mathrm{O{=}C{-}CH_3}\end{array}\right]_n$ Polyvinyl acetate → $\left[\begin{array}{c}\mathrm{H\ H}\\-\mathrm{C}-\mathrm{C}-\\ \mathrm{H\ O}\\ \mathrm{H}\end{array}\right]_n$ Polyvinyl alcohol + $\mathrm{C_3H_7C{=}O}$ Butyraldehyde	$\left[\begin{array}{c}\mathrm{H\ H\ H\ H}\\-\mathrm{C}-\mathrm{C}-\mathrm{C}-\mathrm{C}-\\ \mathrm{H\ O\ H\ O}\\ \mathrm{H{-}C{-}C_3H_7}\end{array}\right]_n$ Saflex, Butacite, Vinylite X
19	$\underset{\mathrm{H\ H}}{\overset{H\ H}{C{=}C}}$ + Cl₂ → Cl$\underset{\mathrm{H\ H}}{\overset{H\ H}{C{-}C}}$Cl → $\underset{\mathrm{H\ Cl}}{\overset{H\ H}{C{=}C}}$ Ethylene, Ethylene dichloride, Vinyl chloride	$\left[\begin{array}{c}\mathrm{H\ H}\\-\mathrm{C}-\mathrm{C}-\\ \mathrm{H\ Cl}\end{array}\right]_n$ Geon, Koroseal
20	HC≡CH → $\mathrm{\underset{H}{\overset{H}{C}}{\equiv}C{-}\overset{H}{\underset{H}{C}}{=}C}$ + HCl → $\underset{\mathrm{H\ Cl\ H\ H}}{\overset{H\ H\ H}{C{=}C{-}C{=}C}}$ Acetylene, Vinyl acetylene, Chloroprene	$\left[\begin{array}{c}\mathrm{H\ \ \ \ \ H}\\-\mathrm{C}{=}\mathrm{C}-\mathrm{C}-\\ \mathrm{H\ Cl\ H\ H}\end{array}\right]_n$ Neoprene, GR-M

VINYL COPOLYMERS

#	Reactants / Intermediates	Products / Polymer
21	$\underset{\mathrm{H\ H}}{\overset{H\ H}{C{=}C}}$ + Cl₂ → HC-CH with Cl, Cl, Cl → $\underset{H\ Cl}{\overset{H\ Cl}{C{=}C}}$ + $\underset{H\ Cl}{\overset{H\ H}{C{=}C}}$ Ethylene, 1,1,2-Trichloroethane, Vinylidene chloride, Vinyl chloride	$\left[\begin{array}{c}\mathrm{H\ Cl\ H\ H}\\-\mathrm{C}-\mathrm{C}-\mathrm{C}-\mathrm{C}-\\ \mathrm{H\ Cl\ H\ Cl}\end{array}\right]_n$ Saran
22	See 17 & 19 — Vinyl chloride, Vinyl acetate	$\left[\begin{array}{c}\mathrm{H\ H\ H\ H\ H\ H}\\-\mathrm{C}-\mathrm{C}-\mathrm{C}-\mathrm{C}-\mathrm{C}-\mathrm{C}-\\ \mathrm{H\ Cl\ H\ O\ H\ Cl}\\ \mathrm{O{=}CCH_3}\end{array}\right]_n$ Vinylite V
23	$\underset{\mathrm{H\ H}}{\overset{H\ H}{HOC{-}CCl}}$ + NaCN → $\underset{\mathrm{H\ H}}{\overset{H\ H}{HOC{-}CC{\equiv}N}}$ + CH₃OH → $\mathrm{C{=}C\ \overset{H\ CH_3}{\underset{C=O}{C{=}C}}}$ Chlorohydrin, Hydracrylic nitrile, Methyl acrylate, Methyl methacrylate	$\left[\begin{array}{c}\mathrm{H\ H\ \ \ CH_3}\\-\mathrm{C}-\mathrm{C}-\mathrm{C}-\mathrm{C}-\\ \mathrm{H\ C{=}O\ H\ C{=}O}\\ \mathrm{O{-}CH_3\ \ O{-}CH_3}\end{array}\right]_n$ Acryloid
24	$\underset{\mathrm{H\ H\ H\ H}}{\overset{H\ H\ H\ H}{HC{-}C{-}C{-}CH}}$ → $\underset{\mathrm{H\ \ \ H\ H}}{\overset{H\ H}{HC{-}C{-}C{=}C}}$ → $\underset{H\ \ \ H}{\overset{H\ H}{C{=}C{-}C{=}C}}$ + Styrene Butane, Butene, Butadiene	$\left[\begin{array}{c}\mathrm{H\ H\ H\ H\ H\ H}\\-\mathrm{C}-\mathrm{C}{=}\mathrm{C}-\mathrm{C}-\mathrm{C}-\mathrm{C}-\\ \mathrm{H\ \ \ \ \ H\ H}\end{array}\right]_n$ (with phenyl) GR-S
25	$\underset{\mathrm{H}}{\overset{H}{HC}}{\overset{H}{\underset{H}{C}}}{O}$ + HCN → $\underset{\mathrm{H\ C{\equiv}N}}{\overset{H\ H}{C{=}C}}$ + $\underset{\mathrm{H\ \ \ H}}{\overset{H\ H\ H\ H}{C{=}C{-}C{=}C}}$ Ethylene oxide, Acrylonitrile, Butadiene	$\left[\begin{array}{c}\mathrm{H\ H\ H\ H\ H}\\-\mathrm{C}-\mathrm{C}{=}\mathrm{C}-\mathrm{C}-\mathrm{C}-\\ \mathrm{H\ \ \ \ \ H\ H\ C{\equiv}N}\end{array}\right]_n$ Hycar O.R., Chemigum, Perbunan, Butaprene
26	$\underset{\mathrm{CH_3\ CH_2}}{\overset{CH_3}{\underset{CH}{\bigcirc}}}$ Dipentene → $\underset{\mathrm{H\ CH_3\ H}}{\overset{H\ H\ H}{C{=}C{-}C{=}C}}$ Isoprene + $\underset{\mathrm{CH_3}}{\overset{CH_3}{C{=}C}}\overset{H}{H}$ Isobutylene	$\left[\begin{array}{c}\mathrm{H\ \ \ \ CH_3\ \ \ CH_3}\\-\mathrm{C}-\mathrm{C}{=}\mathrm{C}-\mathrm{C}-\mathrm{C}-\\ \mathrm{H\ CH_3\ \ \ CH_3\ H}\end{array}\right]_n$ Butyl, GR-I

specialty purposes. Food containers such as cans and small pails, steel drums, fabricated metal caps and closures, household and instrument cabinets, refrigerators, miscellaneous hardware, even automobile bodies, are typical successful applications where the older air-dry lacquers and varnishes or enamels are now almost entirely replaced by baked-on coatings.

The air-drying synthetics are more widely employed for decorative than for protective purposes. Metal toys, bicycles, inexpensive metal signs, ash trays, and the like consume sizeable quantities. There is a steadily increasing market for both air-dry and baking "wrinkle" finishes. They are both protective and decorative and appear on diverse metal articles such as precision instrument mountings, automobile radio cabinets, and similar familiar objects.

Where the exposure conditions are mild, such as on household articles normally used indoors, the life of these finishes, aside from mechanical damage due to scratching or marring, is practically indefinite. Outdoor exposure usually necessitates repainting after one or more years or "seasons." Most of the present automobile finishes should easily last for 3 years without touch-up or repainting. However, conditions of use and exposure vary so widely that anything other than broad qualitative estimates on service life is dangerous.

Generally speaking, the alkyd and alkyd-cellulose finishes are superior for outdoor exposure. Phenolic coatings are outstanding under conditions of water immersion and severe chemical attack; many of them will withstand even concentrated hot caustic for long times. The vinyl and acrylic coatings are unique in their oil, acid, and alkali resistance; in addition they are colorless, odorless, tasteless, and non-toxic, which accounts for their wide application as linings for sanitary food, beverage and cosmetic containers, and closures. The water resistance of all these finishes is superior to that of any but straight tung oil-phenolic varnishes.

VARIETIES OF RESINOUS MATERIALS

Phenolic Resins

The phenol-formaldehyde resins are adapted to diverse types of application. Some of the phenolic resins are used as fortifying resins with drying oils. Others are designed for use as such; they are soluble in alcohols and aromatic solvents and require baking at about 180° C (350° F) for 10 to 40 min to complete the cure. The cured film is among the most resistant to water, acids, alkalies, and solvents of all types that are available. The baked films are usually dark and cannot be pigmented to give very light colors. Since the film becomes increasingly brittle as the cure is completed, care must be exercised not to carry the bake too far. Thin films must be employed because of their inherent brittleness.

Another interesting type of phenolic resin is the so-called dispersion resin. These resins are oil-modified, and have been pre-reacted so as not to require baking. They are not soluble, but can be dispersed, in aromatic hydrocarbons. They are almost always employed in combination with other resins and pigments, as air-dry *primers*.

Alkyd Resins

Alkyd resins are essentially polyesters. They well illustrate the versatility of the synthetic resins, since they can be modified to produce hard brittle resins, soft plasticizing resins, and even polymers which resemble rubber and which can be vulcanized. A wide variety of starting materials is possible, but for the commercial coating resins, phthalic anhydride, glycerin, and the drying or semi-drying oils are most important. Maleic and succinic anhydrides are also important raw materials in the preparation of alkyds, as are terpene-maleic anhydride modifying agents.

When glycerin and phthalic anhydride are heated together with a vegetable oil or fatty acid, gelation occurs before all the acid and alcohol groups have completely reacted. The reaction is stopped short of the gel stage, and the resulting potentially reactive resinoid is then dissolved in a suitable solvent such as mineral spirits or xylene. The commercial alkyd resins vary in their content of phthalic glyceride from about 20 to 60%. The resins of low alkyd content are usually soluble in petroleum solvents and do not give particularly viscous solutions; the resins of high alkyd content must be carried in aromatic hydrocarbons and yield much more viscous solutions.

If the fatty acids which are used are derived from drying oils (linseed, perilla, oiticica, soya, tung, or dehydrated castor oil), the polyester is "unsaturated" and will dry, as the drying oils do, on exposure to air or on baking. These drying-type polyester resins are in many ways comparable to oil varnishes but usually excel them in durability and in retention of gloss and color.

Since additional hardness may be desired without loss of color, urea or melamine resins are often added to alkyd formulations. The drying-type alkyds are also widely used in combination with cellulose nitrate, where as much as four or five parts of resin may be used to one part of nitrocellulose. A relatively high resin content may be achieved in this manner. These are the familiar "lacquers" of commerce.

Maleic anhydride combines with rosin to form a tribasic acid. When this acid is esterified with glycerin in the presence of rosin, a hard resin is produced which is often used with cellulose nitrate to improve gloss and increase the solids content of lacquers.

Urea Resins

The resins from urea and formaldehyde are usually applied from alcohol solution thinned with aromatic hydrocarbons. When the coating is baked, a hard insoluble film is formed composed of highly cross-linked molecules. Since the cured film is quite brittle, it cannot be used alone to any advantage; accordingly, alkyd resins are usually blended with these resins to impart some flexibility.

Urea resins cure more rapidly in an acid environment. The acidity of the added alkyd resin is usually sufficient to cause the cure, but amine hydrochlorides, which evolve HCl on heating, are occasionally added to effect cure at lower temperatures.

Melamine Resins

The melamine resins are similar to the ureas in many of their properties. They can be modified by combination with butanol or capryl alcohol in the same way as the urea resins are modified. They are employed primarily to impart hardness to alkyd resins, where a somewhat smaller amount is required than with the urea resins for equal hardness. Their color retention is better than that of the urea resins, particularly where they are subjected to bakes at 120° C (250° F) and over.

Hydrocarbon Polymers

Since organic hydrocarbons are not usually affected by water, acids, or alkalies, they might appear to be ideal materials for protective coatings. They have not, however, been widely adopted, principally because no simple hydrocarbon polymer, with the possible exception of polyethylene, has yet been found with the desired flexibility and adhesiveness.

Polystyrene is of considerable interest as a possible raw material for coatings since styrene is now produced in large quantities. The polymer is, however, rather brittle, and plasticizers do not impart much flexibility. It cannot be blended with drying oils.

Most hydrocarbon resins are comparatively low in molecular weight; they can be blended with drying oils to improve resistance to water. They are too brittle to form satisfactory films alone. Hydrocarbon resins from cracked gases or from the polymerization of dienes are highly unsaturated and have been named *petroleum drying oils*. The color is often very dark. They polymerize further on baking to insoluble but brittle films. These resins can also be blended with drying oils, and they find their greatest utility in that way.

Acrylate Resins

The polymers of the aliphatic esters of methacrylic and acrylic acid and their copolymers offer a wide variety of resins of varying hardness, some of which are well adapted for use as coatings. The polymers become softer as the length of the carbon chain in the esterifying alcohol is increased. Thus polymethylmethacrylate is harder than polyethylmethacrylate, and n-propyl and n-butyl polymers are progressively softer. The acrylic acid polyesters are considerably softer than the corresponding methacrylic esters. The various esters copolymerize to give resins intermediate in properties between those of the polymers from single esters.

The acrylate resins are relatively expensive, but coatings from them possess many valuable properties. They yield water-white films which do not discolor on exposure to heat or sunlight. Their films are resistant to water, acids, and alkalies. Most of the acrylate resins are not affected by mineral or vegetable oils. As with many other resins, the hardest are the most resistant to chemical reagents.

Their solutions (in ketone, ester, or aromatic solvents) can be pigmented in the usual manner. The films are thermoplastic and can be fluxed on baking. Adhesion is improved by baking, but the air-dried films are often satisfactory as they are. Plasticizers may be incorporated, but are usually not required.

Polyvinyl Acetate and Acetals

Vinyl acetate polymerizes readily to a rather soft resin which is little used in protective finishes. However, if the acetate groups are removed by hydrolysis and most of the free hydroxyl groups then condensed with an aldehyde, a series of "acetal" resins is formed which have wide application in surface coatings. Thus, if formaldehyde is reacted with polyvinyl alcohol, polyvinyl formal is formed; this resin is used primarily for wire coating. Where greater hardness is required, a reactive phenol-aldehyde resin may also incorporate with the acetal resin.

Butyraldehyde yields a softer, more soluble resin. It is somewhat soft for many applications but can be hardened by fortifying with heat-hardening resins; phenolic, urea, or melamine resins can be thus employed.

Vinyl Chloride Copolymers

Many polymers of vinyl chloride have been developed as plastic materials and are widely used as extruded articles. Polyvinyl chloride, however, is not readily soluble and has not been widely used in surface coatings as such. The copolymer resulting from vinyl chloride and vinyl acetate has, however, been successfully used for many coating applications. A three-component copolymer is also available which contains

LACQUERS AND BAKING ENAMELS

a small amount of maleic anhydride in addition to the vinyl chloride and vinyl acetate.

The older vinyl chloride-acetate copolymers do not have particularly good adhesion to metal when air-dried; however, excellent adhesion is obtained when the films are properly baked. The maleic-modified copolymers, on the other hand, give satisfactory air-drying coatings.

Since all vinyl chloride resins have a tendency to evolve hydrogen chloride and darken on prolonged heating, stabilizers are usually added. Iron and zinc surfaces accelerate this breakdown, and chlorine-containing coatings applied to such surfaces must be properly stabilized. Non-vinyloid resin compositions are often used as primers. Thermal stabilizers are usually materials which combine readily with the evolved hydrogen chloride. Organic amines or amides have been used; and urea and melamine-formaldehyde resins are also reportedly effective.

The coatings are usually formulated in aliphatic ketone solvents, thinned with aromatic hydrocarbons, and are tough, flexible, non-flammable, and highly resistant to the action of moisture, alkalies, acids, oils, and fats. They may be employed as clear lacquers or pigmented in a wide variety of colors. If pigmented, a small amount of plasticizer is sometimes added with the resin. Many pigments can be used but those with a high content of iron or zinc should be avoided since they adversely affect the heat stability of the resin. Pigments containing lead and antimony, on the other hand, have a definite stabilizing effect. Vinyl chloride copolymers with acrylic esters and with vinylidene chloride also have promise in the field of surface coatings, but have not yet been offered in commercial quantities.

CELLULOSE DERIVATIVES

Cellulose Nitrate. For formulating into lacquers (Pyroxylin), the acetate esters are usually used as solvents (ethyl, butyl, or amyl acetate). Acetone and the higher ketones are also solvents. Alcohols are often added with the esters, and aromatic hydrocarbons are used as thinners. Plasticizers are usually added to increase the flexibility of the films; tricresyl phosphate and certain phthalate esters are often employed. Cellulose nitrate is nearly always blended with other resins, particularly those of the alkyd type. It is inflammable and discolors on aging.

Cellulose Acetate. Cellulose acetate is not inflammable and does not discolor on aging. It is, however, a rather difficultly soluble material and has not been used in lacquers to the extent of the nitrate.

Cellulose acetate films are characterized by considerable toughness and good color; they must be plasticized to give flexible films. The films are affected by water, but added resins and plasticizers improve the water resistance.

If part of the acetyl groups are replaced by propionyl or butyryl groups, the resulting cellulose acetate-propionate or cellulose acetate-butyrate is more readily soluble and is more readily plasticized. These mixed esters are more resistant to water than cellulose acetate. The films are tough and have excellent color.

Cellulose Ethers. A few cellulose ethers have found application in film-forming materials. The ethyl ether is the principal commercial material. The ethers are usually softer than the esters, are more readily soluble, and more easily plasticized.

Ethylcellulose solutions are increasingly employed as clear lacquers. Mixtures of aromatic hydrocarbons and aliphatic alcohols are usually employed as solvents; most of the ester-type plasticizers can be used to impart flexibility. The addition of phenolic resins has been suggested to improve the water resistance and stability to light.

BITUMINOUS COATINGS

W. F. FAIR, JR.*

Protective bituminous coatings of various compositions and properties may be divided into three classes, according to these general types of service: (1) atmospheric; (2) water immersion; (3) subsurface.

In selecting the proper type of bituminous material for a given project, consideration should be given evaluation of the flow properties of the coating in connection with the particular service conditions, temperature ranges, and stresses expected to be encountered, as well as resistance of the protective coating to moisture absorption, weathering effects, and chemical reactions. For all three of the service types mentioned above, primers are usually applied as the first step of the protection process. These primers are customarily made by "cutting back" (dissolving) the molten, unfilled, high-melting, bituminous substance subsequently to be applied hot, or a compatible equivalent material, with the proper type and amount of volatile solvent, providing a free-flowing, brushable and sprayable, quick-setting product. After application and drying, the residual film furnishes a thin adhesive coating to which the later applications of hot bituminous compositions will firmly bond. Occasionally non-bituminous primers, such as red lead, are used as undercoats for bituminous surfacings briefly described below. It should be noted that the use of red lead undercoats has not proved satisfactory for most underground services.

ATMOSPHERIC EXPOSURE

Paints

Asphalts or pitches of different softening points and penetration ranges and of different viscosity-temperature susceptibilities may be cut back with preselected compatible solvents to furnish satisfactory weather-resistant paints for tanks, bridges, structural metals, culverts, etc. Occasionally finely divided fillers, such as slate, mica, talc, and the like, are incorporated in these paints in amounts of 15 to 40%, sometimes producing thixotropic products fluid enough for easy application, yet exhibiting little or no flow for relatively thick coatings and possessing excellent weathering properties.

Another variation in bituminous paints is provided by incorporating aluminum bronze powder or paste in a properly selected bituminous vehicle or in a bituminous paint. Upon application, the aluminum leafs through the black bituminous carrier, leaving the adhesive protective bituminous film under the outer metallic aluminum surface.

Flashing cements and roofing cements, made by incorporating asbestos and other fibers or fillers in a pitch, or an asphalt, of desired weather-resistant and required rheological properties, together with a miscible solvent of coal or petroleum origin, have been used widely in recent years for application over metal flashings, downspouts, gutters, and similar metal roofing accessories. These materials afford excellent weather protection and remain soft and pliable over long periods of time, while exhibiting complete lack of flow on vertical surfaces at high atmospheric temperatures.

Built-up Membranes

Frequently it is desired to protect metal surfaces with built-up membranes. This is accomplished by priming, followed by plying alternate layers of the selected molten

---

* Koppers Company, Inc., Industrial Fellowship, Mellon Institute, Pittsburgh, Pa., and Wailes Dove-Hermiston Corp., Westfield, N.J.

pitch or asphalt and tar-impregnated or asphalt-impregnated felt until the specified number of plies has been constructed. Bituminous grouts may be used for the outer protective layer. This type of protection, properly installed, gives outstanding, long-lasting service performance.

Roofing, Siding, and Accessories

Several slightly different types of bituminous protection for metal roofing and siding, corrugated, flat, or otherwise shaped, and for roofing accessories have found acceptance in recent years. One type consists of a coating of asphalt adhesive, a ply of asphalt-impregnated asbestos felt, and an outer coating of a harder weather-resistant asphalt, the whole being integrally combined and pressed together. Another type of protection utilizes a coating application of a filled vegetable oil pitch. A third product consists of a single coat of a filled, high-melting, non-brittle asphalt, and a fourth type provides a coating of a non-brittle, non-flowing pitch compound applied by hot dipping of the preformed sheet. Mica, talc, or other finely divided inert anti-stick materials may be used with or without surface adhesives to prevent these coated sheets from sticking during stock piling and shipping.

These products, as a class, furnish excellent weather protection to metal surfaces over wide temperature ranges and under widely varying climatic and industrial-corrosive atmospheres.

WATER IMMERSION

Of extreme importance is the efficient protection of submerged structures, piers, ways, docks, hull interiors, and ships' bottoms against the action of fresh water or sea water and their contained marine-vegetable and animal life. Because of wide temperature variations in service, such as might be encountered in proceeding from tropical to cold northern waters or during rise and fall of tides in the direct sunlight, the materials selected for the protective coating must possess excellent rheological properties as well as light, water, and salt-spray resistance. Bituminous substances in themselves apparently have little toxic action against marine life, but they have been found to be excellent carriers for toxic agents, such as copper, copper oxide, and mercuric oxide.

Protection of immersed metal structures may be obtained by cold application of cut-back materials, or by coating with molten enamels.* If the enamel is used, a coating of cut-back primer is first applied to a surface which has been made as clean and dry as practical. This application should be followed by a spray-applied or daubed-on coat of preselected hot bituminous enamel, preferably containing toxic agents, followed by a cold-applied seal coat of a bituminous paint containing the recommended toxic agent.

SUBSURFACE EXPOSURE

Pipe Coatings†,‡

Many kinds of bituminous materials have been suggested and used as pipe coatings under various conditions. In an effort to establish some basic engineering principles with regard to pipe protection, the A. P. I. Three Corner Tests were instituted under

---

* Enamels as here defined are coal tar bituminous materials which contain less than 40% of filler.

† Based on data furnished by T. F. P. Kelly, Barrett Division, Allied Chemical and Dye Corp., New York, N.Y. Present address: J. E. Mavar Co., Houston, Texas.

‡ See also *Corrosion by Soils*, p. 463.

the aegis of the American Petroleum Institute, the operating pipe line companies, and the manufacturers of pipe coatings. Numerous coatings and several variations of application were included on short test nipples and main operating pipe lines, under widely varying soil conditions. Periodic examinations over the 10-year period of the test disclosed the complete inadequacy of thin and cold-applied coatings, at the same time focusing attention on the more successful, thicker, reinforced, and shielded hot-applied bituminous coatings. The results in general confirmed the findings in a similar set of tests sponsored by the American Gas Association.

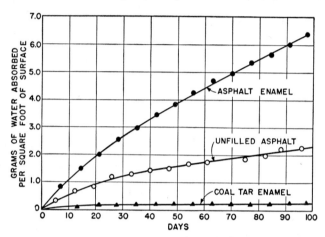

FIG. 1. Relative Water Absorption by Various Bituminous Materials. (Harry Hayes, "Service Life of Coal-Tar Enamel Protective Coatings," *J. Am. Water Works Assoc.*, **32**, No. 10, 1705–1722 [1940].)

Such cooperative tests, together with independent researches on the part of users and manufacturers of protective coatings, have resulted in great advances in application technology in the field of mechanical application and electrical inspection, as well as improvements in the bituminous materials used for coating. Figure 1 illustrates relative resistance to water absorption by various bituminous materials.

The present trend among operating engineers is to coat the pipe line throughout its entire length in contrast to the *spot* or *skip-stop coating*, which was the former practice. The use of continuous coating having good electrical resistance and providing adequate insulation against stray currents makes the use of cathodic protection economically feasible (see *Fundamentals of Cathodic Protection*, p. 923, and *Application of Cathodic Protection*, p. 935).

The accepted system of underground pipe protection entails the cold application of a suitable primer to carefully cleaned metal, followed by a hot, filled bituminous enamel coating and a wrapping of coal-tar or asphalt-saturated asbestos felt. This is the primary and basic application. In addition to the very specific physical characteristics of the bituminous primers and enamels, the combination must have definite performance characteristics, chief of which is the ability of the materials to bond to each other and to the metal surface. Figure 2 illustrates the principle of this bond.

Once the primary application is on the pipe, thicker coatings can be built up using successive layers or plies of hot-applied bituminous enamels and impregnated asbestos felts, as specifications and right-of-way conditions dictate.

Figure 3 shows how a coating system can be diversified in structure by adding to the primary simple system to take care of severe soil conditions, river crossings, swamps, etc. Coatings such as these can be applied at mills, railhead yards, or in the field over the ditch. It is fairly common practice to use a final wrapping of kraft

Fig. 2. I. Bituminous Coating Applied at Proper Temperatures over Compatible and Properly Prepared Bituminous Primers Absorbs Part of the Primer Film, and Thus Bonds to it. II. Bituminous Primer Film Applied to Clean Metal Surface Penetrates the Metal and Anchors to It. III. Metal Surface, which Must Be Thoroughly Cleaned and Free of Foreign Matter for Maximum Bond of Coatings.

paper or to white-wash the coating when the application takes place at mill or railhead yards. This serves a double purpose of keeping pipe temperatures at a minimum, which in turn reduces stresses on a welded or mechanically coupled pipe line, and facilitating final inspection of damaged coating areas before backfilling.

Fig. 3. Structure of Bituminous Coating.
A. Steel pipe surface. *B*. Bituminous primer. *C*. First coat of bituminous enamel. *D*. First wrapping of asbestos felt (reinforcement). *E*. Second coat of bituminous enamel. *F*. Second wrapping of asbestos felt (shield).

Similar types of bituminous coatings are applied to the inside of water pipes by introducing the molten enamel into the revolving pipe from troughs or retractable weirs and depending on the centrifugal force to spread the coating and hold it in place until it has cooled and bonded to the metal. Such linings, in addition to protecting the metal from corrosion and tuberculation, yield and maintain a high coefficient of flow.

The improved enamels now available exhibit much better rheological properties than their predecessors, while still retaining the excellent resistance to moisture absorption and desirable weathering characteristics of the earlier products.

Storage Tanks and Structures

The built-up membrane type of coating previously described may also be used for satisfactory protection of underground foundations, tanks, and the like. Similarly systems and combinations such as are used for pipeline coatings may be successfully applied to subsurface fuel-storage tanks and similar installations.

GENERAL REFERENCES

Primers

Abraham, H., *Asphalts and Allied Substances*, Chapter XXVII, p. 532, D. Van Nostrand Co., New York, 1938.
Ewing, Scott, *Soil Corrosion and Pipe Line Protection*, American Gas Association, 1938.
Larson, E., and G. I. Rhodes, *Pipe Corrosion and Coatings*, Chapter 21, p. 277, published by the American Gas Journal, New York, 1938.
Navy Dept., Yards and Docks Spec. 34Yb, par. 1-02.

Paints

Abraham, H., *Asphalts and Allied Substances*, Chapter XXX, p. 720, D. Van Nostrand Co., New York, 1938.
Dept. Interior, Bur. Reclamation, Spec. CA-50.
Navy Air Materials, Tent. Specs. AN-P-31.
Navy Dept., Yards and Docks Spec. 34Yb, par. 1-05.

Paints, Containing Aluminum

Aluminum Pigment for Paints, A. S. T. M. Spec. D266-41.
Aluminum Pigment Paste, A. S. T. M. Spec. D474-41.
Navy Air Materials, Tent. Specs. AN-P-31.
Navy Dept., Yards and Docks Spec. 34Yb, par. 1-06.

Built-up Membranes

Asphalt Saturated Asbestos Felt, A. S. T. M. Spec. D250-44T.
Asphalt Saturated Felt, A. S. T. M. Spec. D226-44.
Bitumized Fabric, A. S. T. M. Spec. D173-44.
Bituminous Grout, A. S. T. M. Spec. D170-41.
Bituminous Grout, A. S. T. M. Spec. D171-41.
Coal Tar Saturated Felt, A. S. T. M. Spec. D227-44.
Damp Proofing Asphalt, A. S. T. M. Spec. D449-42T.
Damp Proofing Pitch, A. S. T. M. Spec. D450-41.

Enamels

Abraham, H., *Asphalts and Allied Substances*, Chapter XXVII, p. 531, D. Van Nostrand Co., New York, 1938.
American Water Works Association, Standard Specs., Coal-Tar Enamel 7A.5-1940; 7A.6-1940.
Ewing, Scott, *Soil Corrosion and Pipe Line Protection*, American Gas Association, 1938.
Larson, E., and G. I. Rhodes, *Pipe Corrosion and Coatings*, Chapter 22, p. 298; Chapter 23, p. 309, published by The American Gas Journal, 1938.
Navy Dept., Yards and Docks Spec. 34Yb, par. 1-03.

Paint

Anti-Fouling Paint, A. S. T. M. Spec. D277-31.

Miscellaneous

Barnard, Russell, "Outline of Design Factors for Steel Water Pipelines," *J. Am. Water Works Assoc.*, **36** (No.1), 23 (1944).
Diemer, Robert B., "Removal Reconditioning and Installation of 36-Inch Steel Pipe by the Metropolitan Water District of Southern California," *J. Am. Water Works Assoc.*, **36** (No. 1), 43 (1944).
Goit, Lawrence E., "Mechanical Application of Bituminous Pipe Coatings and Linings," *J. Am. Water Works Assoc.*, **33** (No. 10), 1723 (1941).
Hayes, Harry, "Service Life of Coal-Tar Enamel Protective Coatings," *J. Am. Water Works Assoc.*, **32** (No. 10), 1705 (1940).
Hurlbut, Wm. W., "Outline of Installation Procedure for Steel Water Pipelines," *J. Am. Water Works Assoc.*, **35** (No. 10), 1281 (1943).
Logan, K. H., *A.P.I. Pipe Coating Test, Final Report*, National Bureau of Standards.

RUBBER AND RUBBER-LIKE COATINGS*

W. K. Schneider†

Rubber and rubber-like coatings find widespread employment as lining materials, where inertness to acids, alkalies, and other heavy chemicals is mandatory. This warrants their inclusion in any listing of important coating compositions to protect against corrosion.

* Refer also to *Natural and Synthetic Rubber*, p. 365.
† Stoner-Mudge, Inc., Multiple Industrial Fellowship, Mellon Institute, Pittsburgh, Pa.

RUBBER AND RUBBER-LIKE COATINGS

NATURAL RUBBER

Vulcanized rubber sheets may be cemented to metal by means of a diversity of cements or adhesives. Such sheeting is characterized by high tensile strength, resistance to abrasion, good electrical properties, resistance to acids, alkalies, salt spray, water, and many other chemicals. Its drawbacks are poor oil resistance, brittleness at low temperatures, and almost complete lack of resistance to aromatic and chlorinated hydrocarbons.

Rubber latex itself has been compounded with water dispersions of carbon black, fillers, and accelerators to give coating compositions yielding films which can be vulcanized in place with the aid of heat.

MODIFIED RUBBERS

Isomerized Rubber

Solutions of natural rubber are not particularly useful as protective coatings because of their high viscosity. Another drawback is their permanent tack. However, if rubber is treated with acid chlorides or certain salts like $SnCl_4$, a series of *isomerized* or rubber condensation products result which vary from soft chicle-like materials, through tough balata-like gums, to hard friable materials like shellac. Solutions of certain of these are useful as paints. The best are characterized by satisfactory hardness, abrasion resistance, lack of tack, pigment activity, and color. Most of the finishes oxidize to hard, glossy, but flexible films. They may be plasticized with the usual plasticizers and by raw tung oil. They are widely used in blends with other resin solutions to enhance certain properties such as hardness and water resistance.

Chlorinated Rubber

Chlorinated rubber is made by reacting rubber with chlorine until the weight percentage of chlorine approaches 67. These materials are compatible with raw or boiled drying oils, medium and long oil-modified alkyd resins, and most of the common plasticizers.

Without plasticizers or modifying resins, chlorinated rubber films are hard, brittle, non-resilient, and show poor adhesion to metals; they possess, however, excellent chemical resistance. They are not particularly stable to prolonged heat, being rather severely degraded at 120° to 150° C (250° to 300°F) in dry heat, and at 88° to 93° C (190° to 200° F) in water. Temperatures above these cause extensive decomposition with liberation of hydrogen chloride. Electrical characteristics below these temperatures are good. The moisture resistance is quite high, and abrasion and shock resistance are outstanding.

Resin-modified chlorinated rubber finishes are used effectively on machinery in the pulp and paper industry, where resistance to high humidities, acid vapors, and alkalies is of prime concern. Their alkali resistance lends itself to masonry paints; their moisture resistance, to metal finishes and marine paints generally.

Rubber hydrochloride, made by treating rubber dissolved in carbon tetrachloride with dry hydrogen chloride, can be dissolved in toluol or chlorinated solvents and used as a lacquer base. The gross properties of such finishes are not markedly different from those of chlorinated rubber compositions.

SYNTHETIC RUBBERS

BUTADIENE DERIVATIVES

Neoprene (GR–M) is a polymer of chloroprene, a butadiene derivative, and looks and handles much like crude rubber. Neoprene can be vulcanized, with or without sulfur, but not to a hard rubber. Vulcanized sheets can be cemented to metals. Neoprene cements which may be modified with other resins are often used to cement other plastic sheets to metal. In general, Neoprene compounds have the tensile, resilience, elasticity, and abrasion resistance of similar synthetic rubbers, but unlike natural rubber, maintain these properties after soaking in oils and many chemicals. They age well and do not crack in sunlight or ozone; however, they swell in aromatic hydrocarbons and chlorinated solvents.

Buna-type rubbers (GR rubbers) are copolymers of butadiene with other materials (the "S" compounds with styrene, the "N" with acrylonitrile, etc.). Sheets of these materials can be cemented to metals by the use of special cements. Buna cements can be made by dissolving the polymer in ketones and aromatic solvents. The most familiar use is in bulletproof aircraft gasoline tanks. These copolymers have fair tensile strength and good elongation. The abrasion resistance is slightly better than for natural rubber. The N copolymers resist swelling in mineral oil and low-octane gasoline. Both N and S copolymers have excellent resistance to oxidation and decomposition when exposed to heat; gas diffusion through them is lower than through natural rubber. Thermal resistance is good but both are slightly inferior to natural rubber in tear resistance and resiliency.

OTHER OLEFIN RUBBERS

Butyl rubber (GR–I) is a designation for a series of rubber-like products made by polymerizing a high percentage of a mono-olefin like isobutylene, and a small amount of a di-olefin like butadiene. The resulting products have only a fraction of the unsaturation present in natural rubber, and after vulcanization the product is essentially a cross-linked saturated hydrocarbon. Sheets of these materials may be cemented to metals. Cements and spreading doughs are made using petroleum naphthas as solvent. Since butyl rubber is essentially a paraffinic hydrocarbon, it is extremely resistant to such acids as concentrated sulfuric and nitric acids, to oxidation, and to ozone. It is swelled by aliphatic hydrocarbons and coal tar solvents, but is unattacked by vegetable and animal fats and oils. Its water absorption is about one-fourth that of natural rubber and its hydrogen and nitrogen permeability about one-tenth that of rubber.

Vistanex is a commercial product resulting from the polymerization of isobutene. The low molecular weight polymer is used in caulking compounds, hot-melt coatings and adhesives, and as a plasticizer for other resins and waxes. The medium and high molecular weight polymers are used for acid-resistant coatings, gaskets, and caulking compounds. These resins are also used as plasticizers for waxes, resins, and asphalts to lower their moisture vapor transmission and to increase their low-temperature flexibility. Vistanex cannot be vulcanized. Solutions of it can be made with hydrocarbon solvents. Vistanex is not affected by ozone and, when mixed with equal parts of rubber, the rubber is only very slowly attacked by ozone. It is quite resistant to strong mineral acids at room temperature. Moisture transmission is low, and good electrical properties are maintained even on immersion. Vistanex is superior to rubber with respect to impermeability to such gases as hydrogen, helium, nitrogen, and sulfur dioxide.

Polysulfide Rubbers

The term Thiokol is a trade name for a number of organic polysulfide rubbers (GR–P). Various materials range from "latexes" or water dispersions, through cements, curable liquid resins, uncured sheets, "cured" sheets, molding powders, and putties. Certain forms can be applied by spraying as a liquefied powder. As a class, they are practically unaffected by kerosene, fuel oils, lubricating oils, and even high aromatic gasoline. Certain of these resins can be used as coating compositions, dissolved in aromatic hydrocarbons, chlorinated solvents, alcohols, ketones, or ethers.

Some of the water dispersions can be compounded to give good film-forming compositions; the films are fairly hard. They have been used successfully in the coating of underground steel and concrete aviation gasoline-storage tanks, fuel-oil and gasoline-carrying concrete tankers, and airplane wing tanks. The gasoline permeability is quite low, but the moisture permeability is high unless the latex receives special treatment. Flexibility at very low temperatures is excellent.

"Cured" sheets and tubes are available. The sheets can be cemented to nearly any surface and have been used in tanks holding high aromatic gasolines. Tubes are available for piping benzene and other active solvents.

Some crudes are available which have remarkably little odor, and when processed have greatly decreased "permanent set," excellent water and good high octane gasoline resistance, and good low-temperature flexibility.

INHIBITORS AND PASSIVATORS

G. G. Eldredge* and J. C. Warner†

DEFINITIONS AND MECHANISM

Any substance which when added in small amounts to the corrosive environment of a metal or an alloy effectively decreases the corrosion rate is called an *inhibitor*. The various ways in which an inhibitor may function to decrease the corrosion rate may best be understood by considering the conditions for the steady state or limiting corrosion rate, i.e., the condition that the energy decrease in the corrosion process shall be equal to the sum of the energies dissipated in the various parts of the electrochemical system.[1,2] If the energy decrease in the corrosion reaction and the various energies dissipated in the process are converted into potentials ($\Delta F'_R = -NFE_R$, etc., where $\Delta F'$ is the change in free energy, N the number of equivalents taking part in the reaction, F the Faraday, and E the potential), the equation for the steady state corrosion rate may be written

$$E'_R = E_A + E_C + E_{IRi} + E_{IRe}$$

where E'_R = reversible electromotive force of the couple.

E_A = total polarization at the anode areas, which is the sum of (1) concentration polarization at the anode, (2) a possible anodic overvoltage due to some slow process in the overall anodic reaction, and (3) the IR drop through films which may cover the anode surface.

---

* Shell Development Co., Emeryville, California.
† Department of Chemistry, Carnegie Institute of Technology, Pittsburgh, Pa.

[1] R. B. Leighou and J. C. Warner, *Chemistry of Engineering Materials*, 4th Ed., pp. 424–425, McGraw-Hill Book Co., Inc., New York, 1942.
[2] J. C. Warner, *Trans. Electrochem. Soc.*, **83**, 328 (1943).

E_C = total polarization at the cathode areas, which is the sum of (1) concentration polarization, (2) cathodic overvoltage, and (3) the IR drop through films.

$E_{IRi} = IR_i$ = current flowing × resistance of electrolyte between cathode and anode areas.

$E_{IRe} = IR_e$ = current flowing × resistance of the metal between cathode and anode areas. This resistance, R_e, ordinarily is very small; hence E_{IRe} may be neglected in most cases.

It is important to note that all the dissipative terms making up the right-hand side of the above equation are functions of the current density. This fact and the significance of the limiting corrosion rate are illustrated in Fig. 1.

If the major dissipative term is E_A, the corrosion process is said to be under *anodic control*. If, on the other hand, E_C is the major term, the process is under *cathodic control*.*

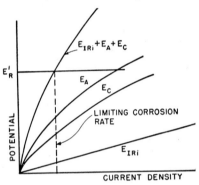

FIG. 1. The Limiting Corrosion Rate.

Inhibitors function by increasing the magnitude of one or more of these dissipative or irreversible effects. They usually have little effect on E_{IRi}. If the addition of a substance to the environment serves to increase E_A, it is known as an *anodic inhibitor*. Cathodic inhibitors, on the other hand, are those which increase the magnitude of E_C. These effects are illustrated schematically in Fig. 2, where the solid lines represent possible anodic and cathodic polarization curves in the absence of inhibitors and the dotted lines represent the corresponding curves in the presence of anodic and cathodic inhibitors. In this illustration the process in the absence of inhibitors is under cathodic control for the most part, but it is evident that an anodic inhibitor might bring the process under mixed or anodic control, depending upon the effectiveness of the inhibitor.

PROPERTIES

EXAMPLES OF INHIBITORS

Additions of soluble hydroxides, chromates, phosphates, silicates, and carbonates to decrease the corrosion rate of iron and other metals in aqueous media serve as examples of anodic inhibition. These substances increase anodic polarization, probably by helping to form or to keep in repair a protective film on the metal surface.

The addition of magnesium, zinc, or nickel salts will decrease the corrosion rate of iron and steel under conditions of partial immersion by serving as cathodic inhibitors. At the cathodic areas near the water line the alkalinity is increased as oxygen is reduced, leading to a precipitation of $Mg(OH)_2$, $Zn(OH)_2$, or $Ni(OH)_2$ over the cathodic surface as a more or less adherent porous deposit. To reach the cathodic surface, oxygen must diffuse through these deposits, and its rate of arrival is therefore decreased.

Calcium salts may act as cathodic inhibitors in waters containing CO_2 by means of the precipitation of calcium carbonate on or near the cathodic areas where the pH is high enough to give a sufficient carbonate ion concentration. Those organic

* Types of control are discussed in *Fundamental Behavior of Galvanic Couples*, p. 486.

substances which decrease the rate of acid attack on metals in such processes as pickling probably constitute another important class of cathodic inhibitors. They are discussed in more detail in a later paragraph.

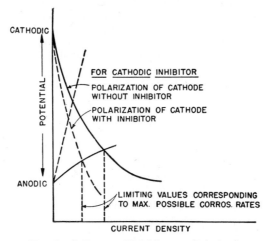

FIG. 2.   Influence of Inhibitors on Polarization.

Inhibitors useful for corrosive media in contact with magnesium are described under *Magnesium and Magnesium Alloys*, pp. 241–242.

It must be emphasized that the successful use of inhibitors frequently requires considerable knowledge of their action and a thorough understanding of the corrosion process in the system under consideration. Substances that may successfully decrease the rate of attack on a metal, or practically stop it entirely in one environment, may in another environment stimulate corrosion. In still other environments the inhibitor, although decreasing the overall corrosion rate, may bring about an increase in the intensity of attack at restricted anodic areas, leading to pitting and rapid perforation.

Although oxygen usually acts as a stimulator of corrosion, it may in some circumstances act as an inhibitor by helping to keep films in repair. It has been shown[3] that the higher the oxygen concentration at a metal surface, the lower the probability of attack on iron or steel. However, once weak points are attacked, the rate of attack will be stimulated by increasing the oxygen concentration.

SAFE AND DANGEROUS INHIBITORS

The prevention of localized corrosion or pitting is obviously equally important in bringing about a decrease in the overall corrosion rate. Failures rarely are due to moderate corrosion spread over large areas; they usually result from intense attack on small, localized areas leading to pitting and early perforation. For this reason, one must consider the influence of an inhibitor on the area attacked of equal importance with its influence upon the overall corrosion rate. Inhibitors that may increase the intensity of attack are classed as dangerous. In general, intensification of attack results when the anodic areas are very small. This situation frequently results when an insufficient concentration of an anodic inhibitor is added to a system in which the corrosion process is under cathodic control. For example, the addition of insufficient

---

[3] U. R. Evans and R. B. Mears, *Proc. Roy. Soc.* (A), **146**, 164 (1934); *Trans. Faraday Soc.*, **31**, 527 (1935).

chromate to stifle completely oxygen-type corrosion of iron, steel, zinc, and aluminum has been shown to cause serious intensification of the attack. The use of still larger amounts of chromates in these cases usually will bring the process under anodic control and give complete protection. It is important to mention that the amount of chromate necessary to eliminate pitting and give protection will depend upon the concentration of such ions as sulfate and particularly chloride in the aqueous medium. These ions interfere with the formation of passivating films on the metal.

In general, the concentration of a given inhibitor needed to give protection will depend upon a number of factors such as composition of the environment, temperature, velocity of the liquid environment past the metal, the presence or absence in the metal of internally or externally applied stresses, the composition of the metal, and the presence or absence of contact with dissimilar metals.

Whether a given inhibitor is "dangerous," "safe contractive," or "safe expansive"[4] for a given metal in a given environment may be determined by experiments involving pertinent measurements on the extent, location, and intensity of corrosive attack as a function of inhibitor concentration.

It should be emphasized that intensification of attack may occur for reasons other than the use of an insufficient concentration of a "dangerous" inhibitor. Thus small anodic areas and intensification of attack may develop under loose scale, under deposits of foreign matter, in crevices, and in similar locations relatively inaccessible to inhibitor or to oxygen.*

PASSIVATORS

Some inhibitors, when added to the environment of a metal or alloy, appreciably change the electrochemical potential to a more cathodic or noble value.[5] Such inhibitors are called *passivators*. Cathodic inhibitors are not likely to serve as passivators, but anodic inhibitors frequently do. Thus iron will stay bright indefinitely in water to which sufficient chromate has been added. Whether or not a substance will act as a passivator, as well as the concentration of the substance needed to induce passivity, depends upon the nature of the metal or alloy, the other constituents of the environment, the temperature, and probably upon still other variables. For example, higher concentrations of chromate are necessary to induce passivity in iron in solutions containing chlorides than are necessary in their absence.

Theories of passivity and passivators are discussed elsewhere in this Handbook (cf. p. 21).

APPLICATIONS

INHIBITORS IN ACID PICKLING

To prepare properly the surface of iron and steel for galvanizing, tinning, enameling, electroplating, etc., it is necessary to remove the scale that is formed in the hot-working operations. The usual method for scale removal is acid pickling. In recent

* Attack occurs predominantly at areas relatively inaccessible to the inhibitor because an inhibitor like potassium chromate is consumed at a slight but definite rate in contact with metals. If not constantly renewed by diffusion, the inhibitor is used up at these areas, which then become anodic to a large cathodic area in contact with ample amounts of inhibitor. Pitting takes place at the inaccessible area in accord with a mechanism described for passive metals and alloys. See *Pitting in Stainless Steels and Other Passive Metals*, p. 165. EDITOR.

[4] U. R. Evans, *Trans. Electrochem. Soc.*, **69**, 213 (1936). See especially R. B. Mears' discussion of this paper.

[5] R. M. Burns, *J. Applied Phys.*, **8**, 398 (1937).

years, organic-type inhibitors have been quite universally used to decrease acid attack on the basis metal, without appreciably altering the rate of scale removal.

The rate of acid attack on the basis metal is undoubtedly controlled by cathodic polarization attending hydrogen evolution. It is believed[6,7,8,9,10,11] that inhibitors function in this case by being adsorbed on the metal surface to form a fairly well-organized and oriented film which serves to increase the polarization for hydrogen evolution by increasing the hydrogen overvoltage, concentration polarization, or by introducing a high transfer resistance between the solution and the metal surface.

Many organic substances have been shown to possess inhibiting properties. A list of 112 organic substances found by Mann and others to have at least 50% protection (defined below) is given in Table 1. Some of these data have been published previously.[8,9] The conditions of the tests were all the same. Weight losses of cold-rolled steel specimens (about $0.12 \times 0.75 \times 0.75$ in.) ($0.3 \times 1.9 \times 1.9$ cm) with ground surfaces in $1 N$ (4.9%) sulfuric acid at 25° C (77° F) were measured with and without various amounts of inhibitor.[8] After 48 hours in 250 ml of acid, the percentage protection P was calculated from the overall rate of attack in grams per square centimeter per hour.

$$P = \frac{(\text{Rate uninhibited} - \text{Rate inhibited})}{\text{Rate uninhibited}} \times 100$$

Table 2 is a list of the substances tried in these same tests that in no case gave 50% protection.

These tables should not be used alone to determine the value of an inhibitor, since some of the inhibitors were erratic in individual tests or have other disadvantages. In the selection of the most suitable inhibitors for a given purpose, only plant trials will give the final answer. Others[12] have proposed laboratory tests, somewhat closer to plant pickling practice, that may be used further to select a few inhibitors worthy of a plant scale test. The properties desired in an inhibitor have been summarized by Warner.[13]

It seems evident that an organic pickling inhibitor must consist of a hydrocarbon part attached to a polar or ionizable group. In general, they contain nitrogen, oxygen, sulfur, or other elements in the fifth and sixth groups of the Periodic Table and are compounds such as amines, mercaptans, heterocyclic nitrogen compounds, substituted ureas and thioureas, sulfides, aldehydes, etc. The inhibitor may be truly soluble in the pickling acid or may be colloidally dispersed (gelatin and glue). Systematic studies[12,14,15,16] on the relation between structure and effectiveness should be consulted by the reader interested in the subject.

Effect of Inhibitor Concentration. The percentage protection generally increases with the concentration of pickling inhibitor. However, the rate of increase continually falls off, and the protection appears to approach 100% protection asymptotically.

---

[6] E. L. Chappell, B. E. Roetheli, and B. Y. McCarthy, *Ind. Eng. Chem.*, **20**, 582 (1928).
[7] J. C. Warner, *Trans. Electrochem. Soc.*, **55**, 287 (1929).
[8] C. A. Mann, *Trans. Electrochem. Soc.*, **69**, 115–129 (1936).
[9] C. A. Mann, B. E. Lauer, and C. T. Hultin, *Ind. Eng. Chem.*, **28**, 159, 1048 (1936).
[10] J. C. Warner, *Metal Cleaning and Finishing*, **10**, 104 (1938).
[11] R. M. Burns, *J. Applied Phys.*, **8**, 398 (1937). For a supplementary point of view, see H. H. Uhlig, *Ind. Eng. Chem.*, **32**, 1493 (1940).
[12] H. P. Munger, *Trans. Electrochem. Soc.*, **69**, 85 (1936).
[13] J. C. Warner, *Metal Cleaning and Finishing*, **10**, 104 (1938).
[14] C. A. Mann, *loc. cit.*
[15] C. A. Mann, B. E. Lauer, and C. T. Hultin, *loc. cit.*
[16] F. H. Rhodes and W. E. Kuhn, *Ind. Eng. Chem.*, **21**, 1066 (1929).

TABLE 1. A LIST OF INHIBITORS FOR STEEL* IN SULFURIC ACID IN THE APPROXIMATE ORDER OF DECREASING EFFECTIVENESS

The Tests Were Run 48 Hours at 25° C (77° F) in 4.9% Acid, in Stoppered Bottles or Covered Beakers

Inhibitor	Source of Data†	Wt. % of Inhibitor at Protection of	
		90%	50%
1. Butyl sulfide	E	0.003	....
2. o-Tolylthiourea	E	0.0034	....
3. p-Tolylthiourea	E	0.0046	....
4. Butyl disulfide	E	0.0036	....
5. Amyl mercaptan	E	0.006	....
6. Ethyl selenide	E	0.0062	....
7. Propyl sulfide	E	0.009	....
8. Phenylthiourea	E	0.009	....
9. Commercial inhibitor	E	0.01	....
10. Butyl methyl sulfide	E	0.01	....
11. Thiourea	E	0.011	....
12. Butyl mercaptan	E	0.011	....
13. p-Thiocresol	E	0.015	....
14. i-Butyl mercaptan	E	0.018	....
15. Triamyl amine	S	0.023	....
16. m-Thiocresol	E	0.026	....
17. Trihexyl amine	UC	0.027	....
18. Ethyl sulfide	E	0.027	....
19. Valerophenone	C	0.028	....
20. 2-Thionaphthol	E	0.032	....
21. o-Thiocresol	E	0.051	....
22. Propyl mercaptan	E	0.061	....
23. Methyl sulfide	E	0.062	....
24. Crotonaldehyde	C	0.070	....
25. Aldol (2-OH-Butraldehyde)	C	0.070	....
26. Phenylmorpholine	U	0.060	0.011
27. Formaldehyde	C	....	0.012
28. Ethyl mercaptan	E	0.093	0.013
29. Commercial wetting agent (alkylated naphthalene sodium sulfonate)	E	0.1	....
30. o-Tolualdehyde	C	0.12	0.014
31. m-Tolualdehyde	C	0.12	0.014
32. p-Tolualdehyde	C	0.12	0.014
33. Phorone (Me$_2$C:CH)$_2$CO	C	....	0.014
34. m-Butylaniline	U	0.42	0.021
35. Butrophenone	C	....	0.022
36. Thiophenol	E	0.24	0.022
37. Propionaldehyd.	C	0.17	0.023
38. Dibutylaniline	L	0.33	0.029
39. Betaine	U	....	0.030
40. Cineol	C	....	0.031
41. Diamyl amine	S	0.17	0.031
42. Commercial wetting agent (sulfonated mineral oil)	E	0.38	0.04
43. Dicyclohexyl amine	UC	1.0	0.043
44. Acetalydehyde	C	....	0.044
45. Ethanol morpholine	U	....	0.045
46. Propyl disulfide	E	....	0.045
47. α-Naphthyl amine	S	....	0.051
48. Dihexyl amine	UC	0.13	0.053
49. Propiophenone	C	....	0.054
50. Phenylhydrazine	L	....	0.054
51. Tripropyl amine	S	0.72	0.057

TABLE 1. A LIST OF INHIBITORS FOR STEEL* IN SULFURIC ACID IN THE APPROXIMATE ORDER OF DECREASING EFFECTIVENESS — Continued

Inhibitor	Source of Data†	Wt. % of Inhibitor at Protection of	
		90%	50%
52. Butraldehyde	C	0.32	0.058
53. Benzaldehyde	C	0.19	0.064
54. Morpholine	U	....	0.072
55. Cyclohexanone	C	0.24	0.079
56. Commercial wetting agent (aromatic sulfonate)	E	....	0.08
57. Butyl sulfone	E	0.41	0.082
58. Dimethyl amine	L	0.64	0.084
59. Diethylaniline	L	1.0	0.085
60. Acetophenone	C	0.29	0.098
61. p-Thiocresol	E	0.50	0.099
62. o-Toluidine	L	....	0.107
63. Dibutyl amine	S	3.0	0.11
64. Valerone (di-i-butyl ketone)	C	....	0.114
65. Anthraquinone β-sulfonic acid	C	....	0.115
66. o-Xylidine	L	0.78	0.133
67. t-Butyl mercaptan	E	....	0.135
68. Dipropyl ketone	C	....	0.15
69. Propylaniline	L	1.2	0.155
70. 3-Butylpyridine	U	....	0.16
71. Amyl methyl ketone	C	1.5	0.17
72. Quinoline	L	....	0.18
73. Lutidine	L	1.7	0.18
74. Commercial wetting agent (alcohol sulfate, from n-octyl alcohol)	E	0.4	0.2
75. Commercial inhibitor	E	0.4	....
76. Ethylmethylaniline	L	2.5	0.21
77. Acetonylacetone	C	....	0.23
78. Acridine	L	....	0.23
79. Commercial wetting agent (alkyl sodium sulfonate)	E	....	0.25
80. Triethyl amine	L	....	0.29
81. Ethyl propyl ketone	C	....	0.30
82. Hexyl amine	UC	....	0.32
83. Methylaniline	L	....	0.32
84. 3-Propylpyridine	U	....	0.34
85. Commercial wetting agent (alcohol sulfate from dodecyl alcohol)	E	....	0.35
86. Collidene (Me-Et-pyridine)	L	....	0.36
87. Ethylene diamine	L	....	0.37
88. Trimethyl amine	L	....	0.38
89. Dipropyl amine	S	....	0.38
90. Commercial wetting agent (alcohol sulfate from technical dodecyl alcohol)	E	....	0.40
91. Commercial wetting agent (alcohol sulfate)	E	....	0.40
92. Butyl methyl ketone	C	....	0.40
93. i-Butyl methyl ketone	C	....	0.40
94. t-Butyl methyl ketone	C	....	0.40
95. Pinacolone	C	1.0	0.40
96. Triethanol amine	L	0.5	0.58
97. Ethyl methyl amine	L	....	0.59
98. Picoline (4Me-Pyridine)	L	....	0.60
99. 3-Ethylpyridine	U	....	0.60
100. Amyl amine	S	....	0.61
101. Ethylaniline	L	....	0.65
102. Dimethylaniline	L	....	0.74
103. m-Propylaniline	U	2.1	0.77

TABLE 1. A LIST OF INHIBITORS FOR STEEL* IN SULFURIC ACID IN THE APPROXIMATE ORDER OF DECREASING EFFECTIVENESS — *Continued*

Inhibitor	Source of Data†	Wt. % of Inhibitor at Protection of	
		90%	50%
104. Pyridine	L	....	0.96
105. Cyclohexyl amine	UC	....	0.98
106. 3,5-Xylidine	H	....	1.0
107. Diethyl ketone	C	....	1.03
108. m-Ethylaniline	U	....	1.14
109. p-Toluidine	L	....	1.23
110. 2,6-Xylidine	H	....	1.7
111. Phenylene diamine	L	....	1.9
112. m-Toluidine	U	....	1.9

* The approximate analysis of the cold-rolled steel was C 0.15%, Si 0.06%, Mn 0.4%, and P 0.05%. The surface was ground before testing.

† The initials L, H, S, C, or E in the table refer to the theses of B. E. Lauer, 1931, C. T. Hultin, 1934, J. T. Spaniol, 1935, C. Cheh, 1939, or G. G. Eldredge, 1940, all of the University of Minnesota. The abbreviation "U" refers to unpublished data and "UC" refers to unpublished data of C. O. Chiao, in the files of the Chemical Engineering Laboratory, University of Minnesota, made available through the courtesy of C. A. Mann, who directed the thesis investigations.

TABLE 2. LIST OF SUBSTANCES TESTED AS INHIBITORS WHICH NEVER GAVE AS HIGH AS 50% PROTECTION IN ANY TEST

Substance*	Source of Data†	Substance	Source of Data†
Phenyl sulfide	E	Diphenyl amine	H
Thiophene	E	Hydrazobenzene	S
Sulfanilic acid	E	Butyl ether	C
Taurine	E	Methyl propyl ketone	C
Isethionic Acid	E	Propyl acetate	C
Commercial wetting agent (alcohol sulfate)	E	Ethyl acetate	C
		Butyl acetate	C
Commercial wetting agent (alcohol sulfate)	E	Amyl alcohol	C
		Valeric acid	C
Urea	E	Acetone	C
Guanidine	E	Methyl ethyl ketone	C
Tetramethyl ammonium ion	L	Methyl propyl ketone	C
Ammonium ion	L	Cyclopentanone	C
Methyl amine	L	Diactyl	C
Ethyl amine	L	Acetylacetone	C
n-Propyl amine	L	Benzoquinone	C
i-Propyl amine	L	Naphthoquinone	C
n-Butyl amine	L	Phenonthraquinone	C
i-Butyl amine	L	Dimenthyl pyrone	C
Diethyl amine	L	Dioxan	C
Aniline	L		

* The order is not significant.

† The initials L, H, S, C, or E in the table refer to the theses of B. E. Lauer, C. T. Hultin, J. T. Spaniol, C. Cheh, or G. G. Eldredge as listed in Table 1.

Effect of Time. Inhibitor action may be expected to become less efficient with time if the pickling inhibitor slowly decomposes. Alquist[17] found that the initially low rate of reaction of boiler steel with 15% hydrochloric acid, in the presence of a small concentration (0.5%) of several commercial inhibitors, increased with time. With the higher rates of reaction for less effective inhibitors, this tendency for increase in rate was not so noticeable.

Inhibitors in Water Systems

Inhibitors are more often used for recirculating systems than for non-recirculating systems because of the much larger amounts needed in the latter case. (However, see the section on gasoline pipe lines below.) For recirculating systems made of steel, the American Society of Refrigerating Engineers[18] has recommended about 0.2 ml of sodium silicate (water glass 40° Bé, 36 to 38%) per liter of water, or more if necessary to keep the solution basic to phenolphthalein. They also recommended 0.01% sodium dichromate ($Na_2Cr_2O_7 \cdot 2H_2O$), where this toxic compound is allowable, along with 0.0027% caustic soda.

Darrin has studied the use of sodium chromate for protection of steel.[19] He found that the rate of consumption of the chromate, highest at the start, falls off with time for 1 to 3 months and was highest at 50 ppm and lowest at 500 ppm. (The study covered the range from 25 to 1000 ppm.) The consumption of chromate increased with chloride content, especially in the first 10 days while the protecting film was being established. In a closed system a fairly high chromate content, say 500 to 1000 ppm, should be used since none will be lost, but in a system with losses this amount need be used only until films are established and then the concentration may be reduced, but not below 100 ppm.

Under conditions of little circulation in low-salinity water (20 ppm each of NaCl and Na_2SO_4) in laboratory tests, Evans[20] found that 0.3% water glass, 0.4% sodium carbonate, 0.8% disodium phosphate, or 0.2% potassium chromate was necessary for the complete protection of steel. In high-salinity water (250 ppm each of NaCl and Na_2SO_4) 0.8% water glass, 1.5% sodium carbonate, 1% disodium phosphate, or 0.4% potassium chromate was needed. Apparently more inhibitor is needed where lack of circulation allows local depletion of inhibitor.

In other laboratory tests, in recirculated water in contact with the air,[21] protection has been found with inhibitor concentrations of only 0.02%, although less complete protection was obtained than that mentioned in the paragraph above. Sodium chromate (or dichromate), sodium silicate, sodium phosphate, and sodium carbonate were of value in that order.

Organic chromate inhibitors, such as sodium chrome glucosate, have been found to be effective in industrial cooling systems.

In automobile radiator systems, it is customary to add about 0.5% soluble oil (which apparently acts by providing a thin coat of oil) or 0.2% sodium chromate. Nitrites may also be used.

---

[17] F. N. Alquist, unpublished work at Organic Research Laboratories, Dow Chemical Co., Midland, Mich.

[18] F. N. Speller, "Corrosion in Refrigeration Plant," American Society of Refrigerating Engineers, *Circ.* 10 (1930). See also *Corrosion, Causes and Prevention*, by F. N. Speller, pp. 617, 618, McGraw-Hill Book Co., New York, 1935.

[19] Marc Darrin, *Ind. Eng. Chem.*, **38**, 368 (1946).

[20] U. R. Evans, *J. Soc. Chem. Ind.*, **46**, 354T (1927).

[21] J. H. Wilson and E. C. Groesbeck, *J. Research Natl. Bur. Standards*, **24**, 665 (1940).

[22] D. W. Haering, *Ind. Eng. Chem.*, **30**, 1356 (1938).

For aluminum systems handling recirculated water, 0.1% sodium chromate or dichromate or enough to keep a permanent color should be used.[23] Chromate should be used in acidic water and dichromate in basic waters, or the waters may be neutralized first. In the presence of heavy metal salts such as copper, or when the aluminum is in contact with copper, brass, lead, or other metals as noble as these, as much as 1% sodium dichromate should be used. If a considerable content of chloride ions is present in addition to these metals, inhibition may be unavailing except at low temperatures. (See below, under "Inhibitors for Bimetallic Systems.")

Inhibitors in Gasoline Pipe Lines Containing Water

Sodium chromate has been used in gasoline pipe lines to inhibit rust and corrosion.[24] The recommended rate of inhibitor addition is about 1 lb per 160,000 gal of gasoline (0.79 ppm) added as a 12% solution in water. The amount will depend on the amount of water in the gasoline, which is appreciable but very small compared to the amount of gasoline.

Sodium nitrite[25,26] is commonly used. It may be added to an oil or gasoline pipe line at the rate of 2 to 12 lb $NaNO_2$ per day, as a 25 to 30% solution at each station and with enough sodium hydroxide to keep the pH of effluent water above 8. The amount depends on the water content and is regulated to give an effluent water with about 1% sodium nitrite. The mechanism is apparently the same as for chromate (the nitrite tends to act as an oxidizing agent and be reduced to ammonia, although it is not necessarily consumed to any practical extent in functioning as an inhibitor).

Inhibitors in Brines

For calcium chloride brines in steel equipment, the American Society of Refrigerating Engineers has recommended 1.6 grams of sodium dichromate ($Na_2Cr_2O_7 \cdot 2H_2O$) per liter[27] and enough sodium hydroxide to convert it to sodium chromate. Twice as much is recommended for sodium chloride brines or calcium magnesium chloride brines. Where chromates cannot be used, they recommend 1.6 grams of disodium phosphate ($Na_2HPO_4 \cdot 12H_2O$) per liter for sodium chloride brines. It is stated[28] that some ice plants use 2 grams per liter. The brine should first be neutralized with hydrochloric acid.

For aluminum equipment, the Aluminum Research Laboratories recommends 1% as much sodium dichromate as there is chloride ion present. For a concentrated brine, this amounts to about the same quantity of inhibitor as was recommended for steel systems above. Acid or basic brines should be neutralized, or sodium chromate used for acidic brines and sodium dichromate used for basic brines. Calcium chloride brines are commonly found to be basic.

Inhibitors in Acid Solutions

For acids in steel systems, see the section above on acid pickling. In general, for steel the same inhibitors are effective, although not to the same degree, for sulfuric, hydrochloric, or for phosphoric acids.

---

[23] R. B. Mears and G. G. Eldredge, *Trans. Electrochem. Soc.*, **83**, 403 (1943).
[24] H. G. Schad, *Proc. Am. Petroleum Inst.*, **24** (IV), 44 (1943).
[25] S. S. Smith and R. K. Schulze, *Proc. Am. Petroleum Inst.*, **24** (IV), 91 (1943).
[26] A. Wachter, and S. S. Smith, *Ind. Eng. Chem.*, **35**, 358 (1943).
[27] F. N. Speller, "Corrosion in Refrigeration Plant," American Society of Refrigerating Engineers, *Circ.* 10 (1930). See also *Corrosion, Causes and Prevention*, by F. N. Speller, pp. 617, 618, McGraw-Hill Book Co., New York, 1935.
[28] Marc Darrin, *Ind. Eng. Chem.*, **37**, 741 (1945).

Attack of copper and copper-tin alloys by dilute sulfuric acid is inhibited by benzyl thiocyanate. See Table 2, p. 99.

Aluminum in hydrochloric acid can be inhibited with inhibitors of these same types. In dilute (up to 20%) or concentrated phosphoric acid, the attack of aluminum may be very effectively inhibited by 1% of some chromate or of chromic acid. Intermediate concentrations may take more inhibitor. In sulfuric acid, the attack of aluminum is very difficult to inhibit. In concentrated nitric acid solutions (above 80%) inhibitors are not necessary. However, in more dilute nitric acid, corrosion of aluminum may be appreciable and chromates have been found to be effective.[29] In 2% HNO_3, 0.05% $K_2Cr_2O_7$ was effective; in 20% HNO_3, 1% was effective. Between 20 and 80% HNO_3, more chromate is apparently needed.

Inhibitors in Basic Solutions

Basic solutions are not ordinarily further inhibited for use with steel since, up to a fairly high pH value, the resistance to corrosion increases with the basicity. That is, the base itself is an inhibitor. However, basic solutions may be further inhibited with chromate or nitrite.

In sodium carbonate good inhibition[30] of attack of aluminum has been found for a series of sodium silicates with silica to soda ratios of 0.67 to 3.22. It was necessary to use about 6% as much silica as there was sodium carbonate present, or about 120 mg SiO_2 per 57 sq dm of aluminum surface. Similar results with sodium fluosilicate, Na_2SiF_4, were found in the same tests.

Alkaline cleaners used for aluminum alloys may contain 25 to 50% sodium disilicate, which serves as an inhibitor. Toothpastes and shaving creams may also contain a few tenths per cent of sodium silicate added to protect aluminum flexible tube containers.[31] Some basic liquids such as triethanol amine, used in aluminum equipment for the absorption of acid gases, may be inhibited by 1% sodium metasilicate or sesquisilicate.[32]

Inhibitors for Bimetallic Systems

When two or more metals with distinctly different electrolytic potentials are in contact in an electrolyte, inhibition is much more difficult.

Iron and Brass. No concentration of chromate will completely inhibit corrosion of iron in contact with brass in calcium chloride brine, but the severity of corrosion decreases with increasing amounts of chromate.[33] The attack is not serious at 2.3 grams per liter, and the results are fairly satisfactory at 4.6 grams per liter.

Iron and Aluminum. Aluminum and iron couples have been found to be protected by 2.4 grams per liter of sodium chromate in calcium chloride brine.[33] Without the chromate, aluminum protects the iron galvanically. However, in a commercial water, it has been found that addition of chromate transferred corrosion from aluminum to steel coupled to it, but did not entirely stop the attack.[34] In a calcium chloride brine in service at low temperatures, dichromate inhibitor has been found to be effective in preventing the corrosion of systems containing both aluminum and steel.[34]

[29] M. A. Whitehouse, A. Lilly, and G. G. Eldredge, unpublished work, Aluminum Research Laboratories
[30] J. F. J. Thomas, *Canadian J. Research*, **21**, Section B, 43–53 (1943).
[31] E. Blough and H. V. Churchill, U. S. Patent 1,912,175 (May 30, 1933).
[32] R. B. Mears and G. G. Eldredge, unpublished work, Aluminum Research Laboratories.
[33] Marc Darrin, *Ind. Eng. Chem.*, **37**, 741 (1945).
[34] R. B. Mears and G. G. Eldredge, *Trans. Electrochem. Soc.*, **83**, 4 (1942).

Aluminum and Copper. In laboratory tests,[32] 1% sodium chromate with 40 ppm sodium metasilicate inhibited attack of aluminum in contact with copper in 10 ppm sodium chloride solutions but not in more concentrated solutions. The more basic silicates added in this way with chromate were more effective than the less basic silicates. Smaller chromate concentrations were not effective.[35] One per cent sodium disilicate alone was sometimes effective as were some soluble oils.

TEMPORARY CORROSION-PREVENTIVE COATINGS
Slushing Compounds
W. E. Campbell*

In many industrial processes there are intervals between manufacture, assembly, shipment, and use of parts where corrosion can occur which may impair equipment or completely prevent its proper functioning. Corrosion problems during shipment and field use became acute during World War II, where apparatus was expected to arrive fit for immediate use after exposure to a varied set of atmospheric conditions. These problems have been attacked with increasing success in recent years by the use of non-drying organic coatings readily removable with available solvents, and known as *slushing compounds.*

Typical slushing compounds consist of a petroleum derivative base containing additives whose main function is to improve corrosion protection, the petroleum derivative base being selected to provide desired physical characteristics such as viscosity or consistency, melting point, pour point, flash point, etc. They range in viscosity from liquids thinner than a light spindle oil to semi-solids harder than a heavy axle grease. They are generally applied by dipping or slushing the part in a liquid compound, a solution of semi-solid in a volatile solvent or a molten semi-solid.

USES OF SLUSHING COMPOUNDS

Corrosion problems during indoor storage are generally associated with the rusting of steel and are particularly serious in the manufacture of sheet steel, ball bearings, and precision instruments. Prevention of corrosion or tarnish of other metals, such as brass or silver, is sometimes of equal importance, as, for example, where electrical contacts must be maintained at a low contact resistance. Since the general principles followed in improving the rust-preventive qualities of slushing compounds for steel apply also to other metals, this chapter will be confined to a discussion of compounds used for the prevention of rust.

Slushing compounds are employed for outdoor use when it is not feasible or economical to use methods providing more permanent protection. Examples are gears, chains, and cables on cranes or hoists, exposed bearings, farm and construction equipment, and an enormous list of parts and mechanisms in use by the Army and Navy.[1]

Frequently slushing compounds are required to serve as lubricants. This is true of compounds for protecting gears and bearings on hoists, and of oils for use in internal combustion engines, or for use on mechanisms of guns or torpedoes. They

---

* Bell Telephone Laboratories, Murray Hill, N. J.

[35] See also Marc Darrin, *Second Annual Water Conference,* Engineers Society of Western Pennsylvania, p. 33, 1941.

[1] *Army Service Forces Manual,* M 406, U. S. Government Printing Office, 1944.

may serve in a dual corrosion-prevention capacity, as in the case of airplane engines, where they protect against rusting in transit or against corrosion catalyzed by fuel combustion residues.[2] Although the problem of corrosion in steam turbines is different in many ways from that of atmospheric corrosion, some of the additives which have been successful for slushing compounds have improved the rust-preventing qualities of turbine oils.[3] Many further examples of slushing compound uses are given in the literature.[2,4,5]

TYPES OF CORROSION TO BE PREVENTED

Corrosion prevented by slushing compounds is of atmospheric origin and is caused basically by oxygen and water. Oxygen is always present so that the principal rate-determining factor is the concentration of water. There would be practically no problem in the absence of moisture, for corrosion of iron is not generally severe until a relative humidity of 50% is exceeded.[6] The principal requirement for rapid corrosion is the presence on the surface of liquid water or of hygroscopic material in a humid atmosphere.

Perhaps the most important additional accelerator of corrosion, particularly in the manufacture of precision parts such as ball bearings, is perspiration residue deposited during handling. Here chloride ion in the residue is the main accelerating agent, although fatty acids are also a factor. Additives for slushing compounds claiming to prevent corrosion due to perspiration residues are ammonia derivatives for acid neutralization, compounds such as lead soaps, which reduce the chloride ion concentration, and emulsions containing up to 10% of water. It is believed by some[7] that there is no product, which, when used in films thin enough to permit inspection and gaging of parts, will do more than delay the development of rust on fingerprinted parts in high humidity. It therefore becomes of prime importance to avoid finger printing during handling, or to remove the contaminating deposits before application of the rust preventive. A wash of 95% methanol is frequently used to remove finger prints, but it is not entirely effective and is objectionable from the standpoints of toxicity and fire hazard.

Other accelerators of corrosion in high humidity are atmospheric contaminants such as sulfur dioxide, ammonium sulfate, coke particles, dust, and salt spray. These can best be combatted by the use of a semi-solid rust preventive. Frequently corrosion is due to contaminants deposited during fabrication such as mill scale, slag inclusions, abrasives, or cutting compounds. These contaminants should be thoroughly removed before the addition of slushing compound. The Texas Company has discussed corrosion causes and has given photographic examples.[8]

TYPES OF SLUSHING COMPOUNDS

There are three main classes of compounds on a viscosity basis: (1) solvent types, which generally consist of a soft semi-solid dissolved in mineral spirits, (2) rust-inhibited oils, and (3) semi-solids. The oils are not reliable rust preventives, except

---

[2] Standard Oil Company of N. J., *The Lamp*, **26**, 17 (1943).
[3] N. B. Wilson, G. H. von Fuchs, and K. R. Edlund, *Power*, **85**, 66 (1941).
[4] The Texas Co., *Lubrication*, **29** (No. 7), 69 (1943).
[5] Penola, Inc., *Lubetext*, D 239–J 6, 1940.
[6] W. H. J. Vernon, *Trans. Faraday Soc.*, **31**, 1678 (1935).
[7] C. R. Gillette, New Departure Division, General Motors Co., paper presented at A. S. M. E. Conference, Davenport, Iowa, April, 1943.
[8] The Texas Co., *loc. cit.*

possibly for very short-term indoor protection. Unless an oil is needed for lubricating purposes, a semi-solid film should be used. The semi-solids range in consistency from that of a very soft petrolatum (penetration number < 300) to hard waxy materials used for outdoor protection (penetration number > 150).

Until recently, slushing compounds consisted largely of straight petroleum oils or petrolatums of varying viscosity sometimes compounded with small amounts of agents such as rosin, asphalt, fatty acids, and fatty acid esters or metallic soaps. Animal and vegetable fats were also sometimes used for the base compound, but the only compounds in this class which show any outstanding merit are lanolin derived from wool fat and spermaceti wax derived from sperm oil. Lanolin is still commonly and effectively used,[9] but is now generally mixed in high concentration with improved compounds of petroleum origin.

As a class, straight petroleum compounds are but fair rust preventives, their efficiency depending mainly on preventing diffusion of water to the metal surface. High-viscosity or semi-solid petrolatums thus afford limited protection for indoor use, and the heavy petrolatums are effective in very thick coatings ($\frac{1}{32}$ to $\frac{1}{4}$ in.) for outdoor exposure, but in all cases protection can be very greatly increased by the addition of suitable corrosion inhibitors to the petrolatum vehicle.

Many of the most effective additive compounds are to be found in the class of alkali and alkaline earth metal salts of sulfonated petroleum derivatives. Metallic soaps are common additives as are long chain fatty acids and their derivatives. Lead soaps and other lead organic derivatives are frequently used, and zinc naphthenate is claimed to prevent corrosion by decreasing moisture penetration.[10] Oxidation products of petroleum residues, consisting of mixtures of esters, ketones, lactones, lacto-ketones, and aldehydes (alketones) or their lead derivatives (paralketones), are the bases of many compounds. Asphalt-base, solvent-type compounds are in common use for long-time indoor and limited outdoor protection.[11]

In tests carried out in the Bell Telephone Laboratories in 90% relative humidity at 35° C (95° F) on a large number of chemically pure organic liquids, including esters such as phthalates, phosphates, tartrates, citrates, monohydric alcohol esters of straight chain fatty acids, glycols, long chain ketones, and ethers, none was found to give adequate protection and many appeared to accelerate corrosion. Aluminum, calcium, sodium, and lead soap greases were not effective. Lanolin or lanolin added to petrolatum was effective but appeared to stain the steel slightly, probably because of the presence of acidic constituents or sulfur compounds. Spermaceti wax showed some merit as an additive. The addition of rosin and of crepe rubber in small quantities improved protection. On exposure to light, rosin forms a gummy film that is difficult to remove; hence compounds containing rosin should not be employed on fine mechanisms, particularly if lubricating action is also required.

Many of the compounds in the patent literature contain inorganic inhibitors such as chromates. The value of these additives has been questioned,[12] and this view was borne out in the above series of tests.

Organic inhibitors, such as are employed as rust-preventive additives to water, also appear frequently in the patent literature, either alone or in combination with surface-active agents, as additives for slushing compounds. Typical compounds are aryl amines, amine soaps, phenol derivatives, quinones, organic-phosphorus, and

[9] C. Jakeman, *Engineering*, **120**, 123 (1925); *Special Report* 12, Department of Scientific and Industrial Research, Engineering Research, H. M. Stationery Office, 1929.
[10] M. V Borodulin and V. N. Nemchinova, *J. Chem. Ind.* (*U. S. S. R.*), **14**, 1708 (1937); **15**, 48 (1938).
[11] J. R. C. Boyer, *Corrosion and Material Protection*, **2** (No. 2), 7 (1945).
[12] P. H. Walker and L. L. Steele, U. S. Bureau of Standards, *Tech. Paper* 176 (October, 1920).

sulfur compounds. These additives may serve more than one purpose. Amino compounds, for example, may act as corrosion inhibitors, oxidation inhibitors, or as neutralizing media for perspiration or other acids. Lead soaps or leaded compounds may function as corrosion inhibitors or as agents for improving the lubricating characteristics of the oil.

THEORY

The mechanism whereby additives act is not fully understood, and there is little in the published literature on the subject. However, certain general principles can be stated. The usual practice is to incorporate in a petroleum derivative certain surface-active ingredients. These ingredients adsorb more strongly on the metal surface than does water, forming a close-packed, oriented layer through which water has difficulty in diffusing, and which it cannot displace. The surface-active agent usually consists of a long-chain molecule having a polar group at one end. The adsorbed film on the metal is probably only a few molecular layers thick. With many compounds, the free molecules outside the adsorbed film bind excess water which penetrates the outer layers of the coating, forming a creamy emulsion. In this manner a semi-rigid structure of fine droplets is formed, each consisting of a layer of polar molecules in an oil phase surrounding a water droplet. The polar compounds are oriented with their hydrophobic hydrocarbon tails extending outward, so that on continued exposure to water a semi-rigid close-packed structure is built up which may have many times the thickness of the applied film. This emulsified film has marked water-repellent qualities and offers high resistance to diffusion of water and oxygen. In some cases, protection is provided by a thin chemically combined or chemisorbed film which prevents access of corrosive ions to the surface. In other cases, adsorption of the film may take place preferentially on cathode areas, thus preventing electrochemical action.[13]

PREPARATION OF SURFACE FOR TREATMENT

The importance of thoroughly cleaning parts before application of a rust preventive cannot be too strongly stressed. The best compound may be rendered valueless if applied over a dirty surface.[14,15]

The choice of cleaners and cleaning method depends on the nature and complexity of the part and on the nature of the contaminants. Details of the methods have been given excellent treatment.[14] The principal methods used are:

1. Solvent immersion and brush cleaning.
2. Solvent spray cleaning.
3. Solvent vapor degreasing.
4. Hot soluble oil emulsion cleaning.
5. Alkaline immersion cleaning.
6. Alkaline spray cleaning.
7. Alkaline electrocleaning.
8. Emulsified solvent spray cleaning.
9. Emulsified solvent soak cleaning.
10. Methanol wash.

[13] H. P. Munger, *Trans. Electrochem. Soc.*, **69**, 85 (1936).
[14] *Army Service Forces Manual*, M 406, U. S. Government Printing Office, 1944
[15] H. Styri, *Trans. Electrochem. Soc.*, **40**, 81 (1921).

Drying of the parts may be carried out by a dried, filtered air blast or by oven treatment. In special cases vacuum-oven drying is used.

Methods used in ball bearing manufacture, where the slightest trace of rust is intolerable, may be taken as a typical example. The procedure used is briefly: alkaline spray cleaning, water rinse, soluble oil emulsion cleaning, and hot-air blast drying. For short times between operations, the oil left from the emulsion treatment provides sufficient protection; otherwise a semi-solid, solvent-type slushing compound is used. Gloves are used for handling whenever possible, but when this is impossible handling is followed by emulsion cleaning or by a methanol wash. The initial cleaning cycle follows the last operation, and a semi-solid slushing compound is added immediately after drying for storage before packaging.

METHODS OF APPLICATION

Although many parts are protected by *slushing* or *dipping*, this is not always possible, and many other methods are in common use.[16,17,18] They are:

1. Spraying or fogging.
2. Brushing.
3. Flow coating.
4. Tumbling with inert material soaked in compound.
5. Swabbing with compound-soaked cloth.

The most important considerations regarding application are that the surface is thoroughly cleaned beforehand and that the compound is applied as soon as possible after the drying step. Application by dipping is preferable for most uses, because the film is more likely to be uniform.[19] Parts should not be left immersed in an oil-type compound, particularly in a humid atmosphere. For reasons not clearly understood, breakdown of additives and rusting of the part are likely under these conditions. When application is made by dipping in a molten semi-solid, the part should be immersed long enough for the entire object to come to the temperature of the melt, or too thick a film, lacking in adhesion, may be applied. Too long an immersion may produce too thin a film. Table 1 gives an idea of the variation in thickness of film obtainable by immersion in a molten semi-solid compound, with variation in temperature of the bath and time of immersion.[20]

TABLE 1. THICKNESS OF SLUSHING COMPOUND FILM ON $\frac{1}{8}$-INCH STEEL PLATE

Temperature of Rustproof Compound Bath		Temp. of Plate: Immersion Time:	25° C (77° F) 1 sec	30° C (85° F) 30 sec
° C	° F			
65	150		39 mils	7 mils
80	175		19	5
95	200		18	4
105	225		18	4
120	250		18	3

[16] *Army Service Forces Manual*, M 406, U. S. Government Printing Office, 1944.
[17] The Texas Co., *Lubrication*, **29** (No. 7), 69 (1943).
[18] Penola, Inc., *Lubetext*, D 239-J 6, 1940.
[19] C. Jakeman, *Engineering*, **120**, 123 (1925); *Special Report* 12, Department of Scientific and Industrial Research, Engineering Research, H. M. Stationery Office, 1929.
[20] The Texas Co., *loc. cit.*

For effective rust prevention using the best type of semi-solid compound, the film thickness should lie between 5 and 50 mils for the softer compounds, and between 50 and 120 mils for the harder, high-melting grades, the thickness actually employed depending on the use and the type of protection required. Films applied from solvent-type compounds and used for indoor protection are generally relatively thin, the thickness ranging from 2 to 8 mils.

SELECTION AND TESTING

There is no generally applicable slushing compound. Selection must depend on the properties and functions of both the compound and the part to be protected. The following properties should be taken into consideration:[21,22,23,24]

1. Corrosion-preventive ability.
2. Chemical inertness towards metal part.
3. Chemical and physical stability.
4. Ease of application and removal.
5. Characteristics of film on part.
6. Fluidity.
7. Adhesiveness.
8. Inflammability.
9. Availability and cost.

In addition to the properties of the material to be tested, the nature of the packaging over the preservative must be considered.

The best test of corrosion protection is a service test, but this is often difficult or impossible to perform, so that accelerated simulated service tests have been commonly used for development of efficient rust preventives. The most common type of test consists in suspending treated specimens in a thermostat or humidity cabinet through which air is circulated at 38° C (100° F) and which is maintained at 100% relative humidity.[25,26,27,28] (See *High Humidity and Condensation Tests*, p. 1006.) In the ideal case, the temperature control should be such as to maintain continuously a condensed film on the surface but to avoid excessive condensation which will result in washing over the specimen. In practice this is difficult to achieve, but it appears that operation giving too much condensation is less likely to cause poor reproducibility than operation with undersaturation. The specimens are inspected visually at regular periods for the development of rust spots. Thorough and reproducible cleaning of the specimen is very important. The film should be of reproducible thickness, and where solvent-type compounds are used the solvent should be given adequate time to evaporate.

Another commonly used test is the well-known Salt-Spray Test,* although it is not applicable to the evaluation of fluid or soft-film coatings.

For testing coatings to be used outdoors, the above-mentioned cabinets providing

---

* For description of Salt–Spray Test, see p. 970.

[21] *Army Service Forces Manual*, M 406, U. S. Government Printing Office, 1944.
[22] U. S. Army Spec. 2–126, March 15, 1946.
[23] U. S. Army Spec. 2–120, March 25, 1944; AXS673, revision 2, October 23, 1945. U. S. Army Spec. 2–122, December 4, 1944. AXS1001, June 14, 1943.
[24] U. S. Army Spec. 2–84B, Nov. 12, 1941; 2–82C, May 14, 1940.
[25] Army-Navy Spec. AN–H–31, April 2, 1945.
[26] Penola, Inc., *Lubetext*, D 239–J 6, 1940.
[27] F. Todd, *Ind. Eng. Chem.* (Anal. Ed.), **16**, 394 (1944).
[28] H. Pray and J. L. Gregg, *Proc. Am. Soc. Testing Materials*, **41**, 758 (1941).

for alternate condensation and drying are used. Another useful test is one in which the specimens are exposed for frequent short periods to a warm water spray[29] to simulate the action of rain. Interestingly enough, coatings which may show up well in this test may show up poorly in the humidity cabinet test, indicating that both tests are advisable for outdoor-type compounds.

The weatherometer,[30] a cabinet in which the specimens are exposed to cycles of humidity, water spray, and ultraviolet light, is commonly used for testing outdoor-type compounds.

The hydrobromic acid test[31,32] is used where compounds are expected to protect against corrosion in airplane engines, but is not considered to bear much relation to service.

There is considerable difference of opinion as to the relative merits of various testing procedures. Humidity cabinet tests are difficult to reproduce from one laboratory to another, and correlation with service tests is often poor or is entirely lacking. However, great improvements in the protectiveness of slushing compounds have been made by use of one or more of these tests, and it is believed that by a proper selection of tests, compounds can be chosen with a reasonable assurance of adequate protection in the field.

TABLE 2. HUMIDITY-CABINET LIFE OF TYPICAL SLUSHING COMPOUNDS

Type	Specification	Method of Application	100° F Humidity-* Cabinet Life, hr	Use and Estimated Life
Semi-solid in volatile solvent	....	Dip, spray	50–100	Parts in process
Semi-solid in volatile solvent	....	Dip, spray	50–100	Finger print treating
Paralketone or asphaltic compound in volatile solvent	AN-C-52 (Type I) AXS-673 52-C-18 (Type I)	Dip, spray, brush	200+	Extended indoor, limited outdoor
Oil (75–200 Saybolt Seconds, viscosity at 100° F)	AXS-674	Dip, spray	200+	Machinery, indoors 3–6 months
Oil (75–200 Saybolt Seconds, viscosity at 100° F)	AXS-934 AN-VV-C-576	Spray	300+	Interior of engines
Oil-petrolatum, penetration number > 150	2-84b ANC-124 AXS-1347	Cold swab, hot dip	300+	Packing bearings for shipment, < 3 months outdoors
Petrolatum penetration number < 150	2-82c AN-C-52 (Type II)	Hot dip	400+	3–18 months indoors or outdoors

* Test on sandblasted steel. SAE 30 mineral oil gives less than one hour life.

In Table 2 are listed typical slushing compounds, with estimated information on their humidity-cabinet life. Further information on types of compound for use under various storage conditions have been given by Boyer.[33]

In testing chemical reactivity, the likelihood of contamination of associated metals or non-metals should be taken into consideration. With solvent-type compounds the purity and inertness of the solvent are very important.[34] In studies made by the Bell

[29] Penola, Inc., *Lubetext*, D 239-J6, 1940.
[30] A. S. T. M. Method, D529-39T, *Book of Standards*, II, 1449, 1944.
[31] U. S. Army Spec. 2-126, March 15, 1946.
[32] Army-Navy Spec. AN-VV-C-576b, April 28, 1945.
[33] J. R. C. Boyer, *Corrosion and Material Protection*, **2** (No. 2), 7 (1945).
[34] C. Jakeman, *Engineering*, **120**, 123 (1925); *Special Report* 12, Department of Scientific and Industrial Research, Engineering Research, H. M. Stationery Office, 1929.

Laboratories, it was found that an effective compound could actually be made to accelerate corrosion of steel and brass when deposited from a chlorinated solvent. Chemical stability is extremely important in compounds to be used for long-time storage or as lubricants. Many of those which show up best in humidity tests deteriorate over long-time periods to gums or acidic products which cause staining and pitting of the metal. Tests for chemical stability are to be found in standard government specification on lubricants.[35] Physical stability tests are described in armed service specifications.[36,37]

The protective film should be continuous, uniform in thickness, and should not check, crack, or peel. In some cases it should be abrasion resistant, and in others it should be transparent so that visual inspection will not be impeded. It should be easily removable after aging by a readily available petroleum solvent such as gasoline or kerosene.

For outdoor protection where surface temperatures as high as 65° C (150° F) are not uncommon, or for indoor protection in tropical regions where similar temperatures may be encountered for short periods, the adherence, as measured by the tendency of the film to run or slip off the specimen at elevated temperature, is important.

The viscosity becomes of importance when the compound is expected to lubricate as well as protect. Where very dusty conditions are expected, a hard, relatively non-tacky film is desirable, if other more important considerations permit its use. As a general rule, liquids and soft semi-solid solvent-type compounds are used for indoor protection, and the harder solid compounds for outdoor use.

For maximum safety, the flash point of the solvent used for solvent-type compounds should be higher than the maximum temperature of application. Where semi-solid compounds are used which must be applied molten, the flash point of the compound should be well above the temperature necessary to give the required fluidity.

The cost of application can often be reduced appreciably by determining the minimum thickness of film required to give protection for the interval over which protection is required.

FUNDAMENTALS OF CATHODIC PROTECTION

John M. Pearson*

Definition

Cathodic protection is the use of an impressed current to prevent or to reduce the rate of corrosion of a metal in an electrolyte by making the metal the cathode for the impressed current.†

THE MECHANISM OF CATHODIC PROTECTION

On the surface of a metal which is corroding in an electrolyte, anodic areas exist at which metal solution takes place, and cathodic areas at which a reducing reaction

---

* Physical Research and Development Lab., Sun Oil Co., Chester, Pa.

† The source of this impressed current is immaterial. It may derive from rectified alternating current, generated direct current, or it may be galvanic in origin as when iron and an active metal like zinc are connected together and immersed in a common electrolyte, thereby protecting the iron at the expense of the zinc.

[35] *Federal Standard Stock Catalogue*, Section IV, Part V, VV-L-791 b, Feb. 19, 1942.
[36] U. S. Army Spec. 2-84B, Nov. 12, 1941; 2-82C, May 14, 1940.
[37] Army-Navy Aeronautical Spec. AN-G-3a, April 2, 1945.

occurs not usually involving the metal. With reference to a normal hydrogen half cell as standard, we may assign a definite value to the effective potential at the anode areas (E_A) and at the cathode areas (E_C). (See Fig. 1.) The electrical network presented by a corroding surface is then a complex system of electromotive forces (emf's), resistances, and currents. The rate of corrosion as measured by weight loss per second is quite accurately proportional to the total of the currents discharged by all the anodes. The total of these currents is termed the local action current, i_L. In Fig. 1 the corroding metal is idealized by representing all anodes by the one shown symmetrically disposed between two cathodes. The half-cell potential E_M of the corroding metal must be measured at the point X, sufficiently remote from the flow lines of the local action current so that the location of the reference cell has no effect on the observed value of E_M.

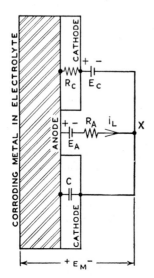

FIG. 1. Idealized Equivalent Circuit of Corroding Metal.

$$i_L = \frac{E_C - E_A}{R_A + R_C}$$
$$E_M = E_A + i_L R_A$$
$$E_M = \frac{R_C}{R_A + R_C} E_A + \frac{R_A}{R_A + R_C} E_C$$
$$E_C \geq E_M \geq E_A$$
If $R_C \gg R_A$; $E_M \cong E_A$ (cathodic control)
If $R_C \ll R_A$; $E_M \cong E_C$ (anodic control)

where C = electrostatic capacity of cathode film,
E_M = half-cell potential of corroding metal,
i_L = local action current,
E_A = effective potential of anode,
E_C = effective potential of cathode,
R_A = resistance of anode area,
R_C = resistance of cathode area.

The study of such a complex system is simplified by application of the principle of superposition of electric fields. Physically, of course, there is only one current at any one point; however, in order to arrive at its value, the principle of superposition permits us to think of it as the vector sum of independent currents. This concept is also useful in understanding what will occur when some part of the network is altered. For these purposes the principle of superposition[1] may be defined as follows:

The current in any part of a complex network of emf's and resistances is the vector sum of the individual currents which each emf would produce in that part, as if that emf were the only source of current flow.

The application of these principles in the use of cathodic protection is illustrated in Figs. 2, 3, and 4. In these figures the corroding specimen is to be cathodically protected by current impressed by the auxiliary anode. A battery is shown in the circuit to symbolize the emf producing the protective current. In some cases the source of emf is the auxiliary anode itself (zinc, for example), and the circuit external to the electrolyte contains no additional emf. In any case it is the value of the current applied to the corroding specimen which is important, not the emf which originates it.

Figure 2 represents the flow of local action current $i_L = 4^*$ from a pit on a corroding

---

* Arbitrary units.

[1] J. H. Jeans, *The Mathematical Theory of Electricity and Magnetism*, 5th Ed., pp. 90–101, 191–202, and 324–331, Cambridge University Press, 1927.

specimen to the adjacent oxide-film cathode. Figures 2, 3, and 4 represent sections through the axes of symmetry of the three-dimensional electric fields. If Fig. 2 is superposed on Fig. 3, and the two fields added vectorially, the result is given in

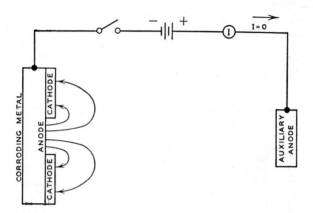

FIG. 2. Local Action Current i_L from Pit on Corroding Metal; $i_L = 4$ Units.

$$i_L = \frac{E_C - E_A}{R_C + R_A} \quad \text{(See Fig. 1.)}$$

Fig. 4. Study of the figures and captions will reveal the quantitative relations which exist at the point when the applied current is just sufficient to stop the discharge and corrosion at the anode of the protected metal.

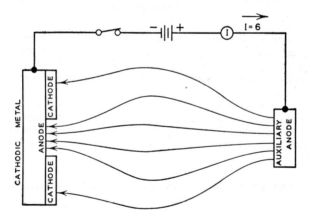

FIG. 3. Distribution of Cathodic Protection Current I Impressed from Auxiliary Anode. $I = 6$ Units.

This distribution depends on surface resistances R_A and R_C only and is independent of E_C and E_A. (See Fig. 1.) Note effect of low-resistance pit. Local action current i_L not shown but circulates as in Fig. 2.

If I in Fig. 4 is increased to twelve units, four of the additional six would enter the anode and two would enter the cathode, the distribution being as explained in Fig. 3.

The important points to keep in mind are that the current distributions are determined solely by relative resistances, independent of emf's, whereas the value of each component of current is determined solely by the emf from which it comes, if the resistances are fixed.

It is thus essential to differentiate between changes of E_M which are produced by the flow of I through R_A and R_C (in parallel) and those which are produced by changes in E_C and E_A. For the purposes of this discussion the term *polarization* as used below will mean only an alteration of the potentials E_A and E_C, excluding IR drops in R_A and R_C. (See Fig. 1.)

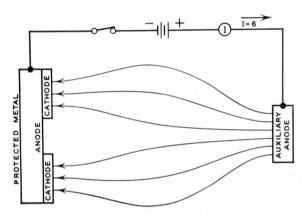

Fig. 4. Superposition of Impressed Current on Local Action Current, Showing Vector Sum of Currents in and out of Anode to Be Zero. Impressed Current I and Local Action Current i_L Are Both Present.

If the impressed current results in polarization at either the cathode or anode areas, this polarization is reflected as a change in the local action current itself, and in no way affects the distribution of the impressed current. The change in the local action current does, of course, change the amount of impressed current required to provide protection.

If, as sometimes happens, the impressed current produces a change in the film resistances at the anodes and cathodes, the distribution of the impressed current is profoundly affected, whether polarization has occurred or not. Such changes in resistance also profoundly affect the value of i_L, if the emf's remain unchanged.

CRITERIA FOR COMPLETE CATHODIC PROTECTION

For either anodic or cathodic control of corrosion rates,* complete protection occurs when the anode currents reach zero. At this point there will be no IR drops in the electrolyte near the anodes, and measurements taken there will give $E_M = E_A$. (See Fig. 1.) In many cases, the IR drop through the cathode film far outweighs the IR drop of the local action and impressed currents in the electrolyte near the cathodes. It is thus clear that, if the IR drop through the cathode film is included with its polarized potential, the potential difference measured to the electrolyte adjacent the cathode

---

* For discussion of anodic and cathodic control, see p. 486.

will nearly equal the potential of the anodes. This method of measurement has led some workers[2] to give as the criterion of the completeness of cathodic protection "the polarization of the cathodes to the open circuit potential of the anodes." (Polarization here includes IR drop with changes in potentials.) This criterion is perfectly sound.

From a study of Fig. 1, it is evident that if one knows the approximate values of E_A and E_C, the measurement of E_M will disclose whether the local action current is anodically or cathodically controlled.

Criteria for Protection of Cathodically Controlled Specimen

When a corroding specimen is to be cathodically protected, the changes in E_M measured by the remote reference cell are often constituted chiefly of IR drops through the electrolyte and through R_A and R_C in parallel. In order to observe any change due to real polarization (potential changes at the electrode), corrections must be made for the IR drops both in the electrolyte and in the surface films. This can be done by direct determination of the IR drops or by use of a null method in which they are balanced out.[3]

Measurements made in the above manner on specimens corroding under cathodic control indicate that the specimen half-cell potential remains very nearly constant until the anode current discharge is actually reversed as the impressed current is increased. (Any change at the cathodes is masked by the relatively low resistance path to the anode areas.) At the reversal of the anode current discharge, it frequently happens that another ion, such as hydrogen, is discharged rather than the metal ions. This produces a definite discontinuity in the potential, which can be observed as an indication that protection is complete.

In practice, when there are a number of anodes of varying resistance, the discontinuity is often, but not always, blurred because the anodes are not all reversed at one value of impressed current. When they have been reversed, however, the observed potential E_M varies with the logarithm of the current as it does when hydrogen is discharged.

For cathodic control, when R_C is much larger than R_A, E_M is nearly equal to E_A even before application of the protective current. It is interesting to note that, in this case, cathodic protection approaches totality without any substantial change in the half-cell potential E_M of the whole corroding specimen. The experimental evaluation of the protective current required, however, involves observations with impressed currents large enough to determine the discontinuity in E_A, which occurs when the net current at the anodes is reversed.

Criteria for Protection of Anodically Controlled Specimen

The criteria for the completeness of cathodic protection of a specimen corroding under anodic control are not so simple. In this case, the resistance of the anodes is so high that the comparatively low-resistance path to the cathode areas masks the observation of potential discontinuities at the anodes, as the impressed current is increased. When the reference cell is remote from the specimen, the cathode polarization criterion of Brown and Mears[4] is useful. In order to correct for the IR drop in the electrolyte, but not in the cathode film, the measurement of resistance should

---

[2] R. H. Brown and R. B. Mears, "Cathodic Protection" *Trans. Electrochem. Soc.*, **81**, 455–481 (1942).
[3] J. M. Pearson, " 'Null' Methods Applied to Corrosion Measurements," *Trans Electrochem. Soc.*, **81**, 485–508 (1942).
[4] R. H. Brown and R. B. Mears, *loc. cit.*

be made by the use of alternating current of frequency high enough so that the capacitative reactance of the film is negligible compared to its resistance. If a null method is used, the bridge balance should be made with alternating current for the same reason. The use of this criterion implies, of course, that the observer knows the value of E_A. This value is obtainable and may be considered generally reliable, since the anode reaction is a simple one involving the metal and its ions immediately adjacent the anode surface.

Criteria for Protection of Specimens under Mixed Control

For specimens corroding under "mixed" anodic and cathodic control (such as is encountered with some couples of dissimilar metals), the criteria for completeness of cathodic protection should be chosen according to practical circumstance. If it is possible to locate the reference half cell so that it is substantially in the field of the anode, the potential discontinuity can be used after correcting for all IR drops. If the reference half cell is remote from the specimen, and IR drops in the electrolyte alone are corrected for, the cathode polarization criterion can be used. If the reference half cell is in the field of the cathode, and the IR drops in the cathode surface are greater than those in the electrolyte, the cathode polarization criterion can be used. In the cases of some galvanic couples, however, the surface resistance of the cathode is not predominant, and the observer will be well advised to check his conclusions by more than one method.

DETERMINING MINIMUM CURRENTS FOR PROTECTION

Test Currents

The impressed current densities required for testing the completeness of cathodic protection are conventionally computed on the basis of the total area of the corroding specimen, regardless of the fact that the actual current density at the anodes when the cathode film resistance is high may be 10 to 1000 times the density at the cathodes. On the conventional basis, the testing equipment must be able to furnish current densities of 10^{-7} amp per sq cm up to 10^{-2} amp per sq cm. The lower values are associated with anodically controlled specimens, the larger values with specimens corroding under mixed or cathodic control, or corroding in acids.

The test current supply should be able to cover the desired range in geometric steps with a ratio between 1.25 and 2.00. The supply should be adjustable quickly and continuously, without interrupting the current, and it should maintain a stable selected value of current. The supply should be quickly adjustable, inasmuch as *ampere-hour-effects* (discussed below) produce a more or less steady background drift in the observation, so that it is desirable to space the latter evenly in time as well as in increment values. The maximum current needed reliably to establish the required protective current is at least four times the order of magnitude of the determined test current.

Range of Protective Currents

Figure 5 shows the orders of magnitude of the protective currents required by steel under different conditions. The current densities are based on the total area of the corroding metal, without regard to the portion which is anodic; as a result the values indicated may be from 25 to 400% of the experimentally measured values in any given case. The formation of cathode scale (see "Alkalinity as an Ampere-Hour

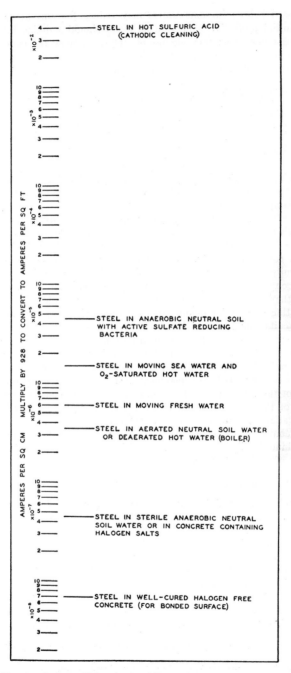

Fig. 5. Orders of Magnitude of Protective Currents for Steel.

Effect" discussed below) in some waters may reduce the current requirements to 20% of the initial protective values. In no event should a design be based on the current densities indicated; they are useful simply to show the expected orders of magnitude. Experimentally determined values of the protective current in each case are the only ones that may be considered valid.

Current Sources for Testing

For laboratory work where the maximum currents amount to less than 10 milliamperes, requirements can be met by circuits of high internal resistance and high emf for stability, and connected to supply the external circuit through an Ayrton shunt for adjustability. Radio-type potentiometer rheostats with a logarithmic taper are useful for making the adjustments in geometric steps.

Electronic circuits[5] are easily adapted to provide a high degree of current stability and to preset a number of steps of the desired ratio, for currents up to several hundred milliamperes.

When currents of greater magnitude are required, especially in the field, portable generators with separate exciters are more suitable. Some cathodic protection projects on one or more bare pipelines require currents up to several hundred amperes.

In order to correct the readings of specimen potential for IR drops, or to use a null-type circuit, the test source must be able to supply a small change of current superimposed on the steady-current value without interruption of the circuit. A small increment of the impressed direct current is generally preferable to the use of a superposed alternating current for quantitative corrections because the "skin effects" for large-scale circuits found in practice result in a distribution of alternating current which is substantially different from the distribution of the steady component of direct current. When alternating current is used in the adjustment of a null-type circuit, the observations are likely to include the IR drop in any resistive element that is bypassed by a capacitative reactance.[6] The increment of direct current is desirably about 10% of the value of the total current. It is most suitably furnished by a variation of the internal emf of the circuit by 10% if the current is to be regulated by resistance variation. A variation in the resistance of 10% can be used when the current is to be regulated by control of the circuit emf. The circuit emf can be nicely controlled by use of a continuously variable transformer with a rectifier or by suitable control of the excitation of a d-c generator.

Observing Circuit Reference Cells

Notwithstanding many statements to the contrary, all observing circuits draw slight current from the half cells connected to them. The equipment which most nearly approaches the use of zero current is a vacuum-tube-driven voltmeter. If suitable circuits are chosen, the current drawn through the connected half cell is limited to the vacuum tube grid current, which can be made very small, though finite. If any circuit draws enough current to polarize the connected reference half cell, the combination is not trustworthy for making measurements.

Calomel Cell. For laboratory and for some field work, the calomel-type half cell is very suitable. It has a very low temperature coefficient and a large current capacity without objectionable polarization. (In cathodic-polarization observations where

---

[5] J. H. Bruce and A. Hickling, "A Current Stabiliser for Electrolytic Circuits," *J. Sci. Instruments*, **14**, 367–370 (1937).

[6] J. M. Pearson, "'Null' Methods Applied to Corrosion Measurements," *Trans. Electrochem. Soc.*, **81**, 485–508 (1942).

polarizations of hundreds of millivolts are significant, an error of 2 or 3 mv in the reference cell can be tolerated.) Values of potential for the 0.1 N, 1 N, and saturated calomel electrodes are given on p. 1137.

Copper Sulfate Cell. Perhaps the most widely used reference half cell in the field is the saturated copper-copper sulfate cell. It can be made up from commercial material and is reproducible in potential to within ±0.006 volt. It has a high-temperature coefficient[7] but it is entirely satisfactory for field use, and temperature effects can be deducted if the precision of observation warrants it.

The copper-copper sulfate half cell has a tendency to reduce its effective electrode area and current capacity with age. The reduction is slow but may seriously affect the reliability of a new half cell in a few months. The change occurs as crystals of copper sulfate and copper oxide adhere to and grow on the electrode surface. The cell is objectionably polarized at a lower value of current when the electrode is cathodic than when it is anodic. The value for the half-cell potential is given on p. 1137.

INSTRUMENTS

Published reports are available covering necessary instruments[8] as well as applications to field problems.[9,10,11,12,13,14,15,16,17,18]

PHENOMENA, FAILURES, AND PITFALLS IN THE USE OF CATHODIC PROTECTION

There are a number of phenomena manifest by the anode and cathode in a practical cathodic protection circuit, some of them favorable and others decidedly unfavorable to the successful control of the corrosion rate of the cathode. Other conditions arise as the result of stray currents from the cathodic protection installation so that its design should be modified to control possible damage to contiguous structures.

ELECTRO-OSMOTIC EFFECTS

These effects occur at both the anode and cathode surfaces, to the disadvantage of the circuit. At the anode, especially if it is buried in certain clays, electro-osmosis

---

[7] Scott Ewing, *The Copper-Copper Sulfate Half-cell for Measuring Potentials in the Earth* (see also *The Cadmium-Cadmium Sulfate Half-cell for Measuring Potentials in the Earth*), Technical Section, American Gas Association Distribution Conferences, 1939.

[8] R. H. Brown and R. B. Mears, "Cathodic Protection" *Trans. Electrochem. Soc.*, **81**, 455–481 (1942).

[9] Scott Ewing, *loc. cit.* (see also *The Cadmium-Cadmium Sulfate Half-cell for Measuring Potentials in the Earth*), Technical Section, American Gas Association Distribution Conferences, 1939.

[10] Scott Ewing, *Soil Corrosion and Pipe Line Protection*, American Gas Association, 1938.

[11] J. M. Pearson, "Concepts and Methods of Cathodic Protection" *Petroleum Eng.*, **15**, Nos. 6, 7, and 8 (1944).

[12] A. V. Smith, *The Theory and Use of Cathodic Protection*, 14th American Gas Association Distribution Conference, 1937.

[13] K. H. Logan et al., "The Determination of the Current Required for Cathodic Protection," *Petroleum Eng.*, **14**, No. 10, 168–180 (1943).

[14] G. I. Rhodes, "Two Unusual Installations of Cathodic Protection of Pipe Lines," *Proc. Am. Petroleum Inst.*, **17**, 21–32 (1936).

[15] G. I. Rhodes, U. S. Patent 1,962,696, June 12, 1934.

[16] Scott Ewing, *Determination of Current Required for Cathodic Protection*, Technical Section, American Gas Association Distribution Conference, 1940.

[17] G. N. Scott, "A Rational Approach to Cathodic Protection Problems," *Petroleum Eng.*, **12**, Nos. 8, 9, 11 (1941).

[18] K. H. Logan, *The Status of Cathodic Protection of Pipe Lines in 1941*, presented to the Second Annual Water Conference of the Engineers' Society of Western Pennsylvania, Nov. 3, 1941.

of the water may dry the clay to the point where its conductivity is seriously reduced. When this point is reached, in some cases, the heat generated by the current further dries the soil. The remedy is to select a site for the anode which is under water at all times. If the anode is in an impermeable clay or dry area, treatment with salt will reduce the gradient and therefore the electro-osmosis.

At the cathode surface electroendosmosis produces the reverse effect, i.e., water is transported toward the electrode. If the cathode is protected by an insulating but permeable coating, the coating may be damaged by the water.

Ampere-Hour Effects

These effects occur at both anode and cathode surfaces. At the auxiliary anode, the principal ampere-hour effect is the electrolytic destruction of the anode. Some graphite anodes slowly disintegrate with continued electrolysis for reasons not clearly established, and ordinary metallic anodes pass into solution in proportion to the coulombs discharged. Most of the latter operate at so high a polarization emf that anions are discharged, and the efficiency of destruction of the anode is less than 100%, based on Faraday's Law. The resulting saving in anode material is offset by the greater energy consumed as the result of the higher polarization emf.

Oxygen and Ampere-Hour Effects. At the cathode (protected surface) there are a variety of ampere-hour effects. The current density at the surface to be protected is usually quite low (compared to that at the auxiliary anode), and the polarization emf is usually of the order of a few tenths of a volt. Under these conditions, the cathode reaction consists of the discharge of hydrogen ions. If dissolved oxygen is present at an interface, for example between metal oxide and electrolyte, the cathode reaction consists of a reduction of oxygen by electrons from the metal through the oxide, resulting in the formation of OH^- ions. This requires an increase in impressed current for protection. The oxidizing agent is then consumed as fast as it reaches the surface, with the result that hydrogen or another cation is discharged, and E_C (Fig. 1) is reduced or reversed.

Film Resistance and Ampere-Hour Effects. On some metals, the ampere-hour effect of continued electrolysis is reduction of the oxide film. When this occurs, the surface resistance R_C of the cathode for i_L drops abruptly. This results in a marked redistribution of the impressed current, as explained in the first part of this chapter. It would also result in an increase of local action current were it not for the simultaneous reduction of E_C.

Under some conditions the protective current is never large enough to modify the oxide cathode areas, and the local action current persists so that although the corrosion is under control, it tends to resume its original rate shortly after any interruption of current. When the oxide cathode areas are modified, the whole surface of the metal under protection becomes nearly uniformly cathodic.

Alkalinity as an Ampere-Hour Effect. If any alkali or alkaline-earth ions are present, the chief ampere-hour effect is an increase in pH. Under these conditions, an interruption of protective current does no harm in terms of the original local action circuits; however, the whole area may become the anode for a remote, unpolarized part of the same structure. This can largely be prevented by electrically insulating the cathodically protected part of the structure from the rest of it.

The increase of pH as the result of the accumulation of alkali ions at the cathode has many effects. One of the most destructive is the solution of cathodes which are amphoteric so that cathodic protection of metals such as lead, zinc, and aluminum requires caution and careful design.

Another undesirable effect is the destruction of the bond between coal tar or asphalt-type coatings and the coated structure. The bond between concrete and steel is likewise adversely affected.[19]

An interesting and useful ampere-hour effect occurs when the increase of pH results in the precipitation of insoluble carbonates on the cathode surface to form a tightly adhering, high-resistance film. The "film" is effective when barely visible, but has been observed to grow several inches in thickness. Its formation depends upon the presence of carbon dioxide in solution and the presence of calcium and magnesium ions in the catholyte.

Because of its high resistance, the formation of this film produces a profound redistribution of the impressed current, causing it to flow to more remote parts of the protected structure. This results in a spread of the area protected, and corresponds in cathodic protection to the marked "throwing power" of some electroplating solutions. Although this film is most useful in cathodic protection of ferrous materials, it can be a hazard when it forms on the more amphoteric metals because of the high pH which may accompany it.

Shielding Effects

These effects are adverse to obtaining good protection and are significant chiefly in respect to the protected metal. In general, if an electron-conducting material is located so that the current flow lines terminate on it rather than on the surface to be protected, and if there is a metallic connection from the material to the surface, the part of the surface deprived of the incoming current fails to receive protection.

Shielded Galvanic Couple. If the conducting material forms a galvanic couple as well as a shield for the corroding surface, electrical observations are unlikely to reveal the hazard, and cathodic protection fails. The prime example of this effect is the contact of a buried structure with carbonaceous cinders. The strong local action current is confined to the electrolyte between the cinder and the anodic metal. At the same time the anode on the metal is electrically shielded by the cinder so that the impressed current has no component reaching the anode; hence corrosion occurs. It is meaningless to speak of polarizing the cathode area of the cinder because the area so polarized is on the outside and is not the cathode for the local action current, which flows to the underside of the cinder. Potential observations made remote from the cinder record merely the polarization of its outside surface so that the shielded local action current is unobservable.

Shielded Cathode. Another example of shielding is that which results when a buried structure passes through a conducting conduit. If the structure contacts the conduit, all the impressed current flow lines terminate on the conduit and flow to the structure through the contact. This prevents protection of the shielded area.

When a number of corroding metal structures are separated by distances small compared to their dimensions, they are likely to act as mutual shields and are difficult to protect. Cathodic protection of the water side of tube bundles in a heat exchanger is an example of this problem. The answer is to distribute the anodes in the available space so as to apply the current where needed.

Shielding by Insulation. In the application of cathodic protection to the steel shells of large underground storage tanks, the tanks having cone or flat roofs which are very extensive compared to their depth of burial, the conductivity of the relatively thin earth layer above them reduces the delivery of protective current to the roof from any remote anode. Essentially, the electrical boundary at the earth's surface is

[19] E. B. Rosa et al., National Bureau of Standards, *Tech. Paper* 18 (March, 1913).

an insulating shield. It is effective either to coat the roof to reduce the exposed area in contact with the soil or to install distributed anodes to feed the current transversely (instead of longitudinally) through the covering soil.

Mechanical Effects

Soil Shrinkage. In some adobe soils, the shrinkage cracks which form when the soils dry out may actually open-circuit the electrolytic part of the cathodic protection circuit. With the impressed current effectively cut off from parts of a buried structure, those parts may corrode. No practical solution of this problem has been suggested.

Mechanical Depolarization. One of the most serious of mechanical hazards is so-called mechanical depolarization. This occurs, for example, if a protected structure buried under the bottom of a stream is washed out of its covering. The mechanical action of the flowing water tends to remove the alkaline products of electrolysis as well as supply more oxygen, so that a considerably increased impressed current is required for protection.

Stray-Current Hazards

Poorly engineered cathodic protection installations on one property are often a menace to adjacent properties not considered in the design. Occasionally an installation will damage the property it is intended to protect.

No proper design can omit consideration of any conductor which lies in the electrolytic circuit of the impressed current. It should be remembered that this circuit is three-dimensional and that the only sure approach is by measurement and by careful tracing of the actual paths followed by the impressed current.

One of the basic rules to follow in crowded areas is to control the field of current flow. This is best done by distributed anodes, arranged so that the current requirement of each element of cathode is exactly derived from an immediately adjacent element of anode. Another practical rule in some more simple cases is to so locate the anodes that extraneous structures are conjugate conductors in the whole protective-current circuit. Although this is usually not entirely achieved, the process greatly limits the current which must be drained from foreign structures to protect them from damage.[20]

Effect of High-Resistance Couplings

It is not an uncommon error to overlook the effect of discontinuities in the resistance of an extended structure to which cathodic protection is applied. A high-resistance joint in a pipe, for example, may cause some of the current carried by the pipe to by-pass through the earth and to damage the anodic side of the joint.

Radial Current

A poorly located anode to protect a structure which is itself of relatively poor conductivity may produce in it a so-called radial current. Part of the current picked up from the anode actually flows along the structure away from the drainage connection, discharges to earth at a remote point, and returns by a deep path to a part of the same structure near the drain point. The discharge by the structure of any part of the impressed current, except through the drainage connection, is certainly to be avoided. The remedy in most cases suggests itself when the distribution of the impressed current is determined by careful measurement.

[20] Scott Ewing, *Soil Corrosion and Pipe Line Protection*, American Gas Association, 1938.

APPLICATION OF CATHODIC PROTECTION
Robert Pope*

GENERAL

In the usual application of cathodic protection, the metal to be protected is electrically connected to the negative terminal of a source of current such as a rectifier, generator, or battery. The positive terminal is connected to an anode in

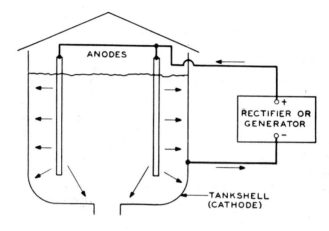

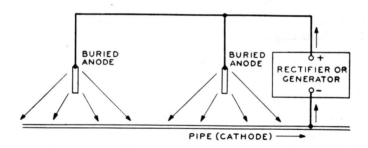

Fig. 1. Diagrams of Application of Cathodic Protection.

the corrosive electrolyte. Current† from the anode passes through the electrolyte to the protected metal, making it cathodic and reversing the current at the anodes of local cells on the protected metal. (See Fig. 1; also preceding chapter.)

Cathodic protection systems can be divided into three general classes, depending

---

* Bell Telephone Laboratories, New York, N. Y.
† Current means flow of positive electricity, opposite in direction to flow of electrons in a conductor.

upon the source of the current "drained" from the metal. *Forced drainage** uses a source of current such as a rectifier, generator, or battery provided especially for the purpose, together with an auxiliary anode. *Galvanic anode drainage* is a form of forced drainage which uses an anode of an active metal such as zinc or magnesium, the drainage current depends on the galvanic potential between the metal to be protected and the anode. *Stray-current or bus drainage* is applicable to underground pipes and cables in areas where trolleys are operated and uses part of the trolley generator potential for draining the current as explained in *Corrosion by Stray Currents*, p. 601.

PROTECTION OF UNDERGROUND PIPES AND CABLES

Probably the widest use of cathodic protection is on underground metallic pipes and cable sheaths. In addition to stray-current or bus drainage which is used wherever trolleys and d-c railroads are operated, forced drainage is used extensively on long interurban and cross-country underground pipes and telephone cables. It is also used to some extent on water and gas pipes and communication and power cables in city networks.

When a potential is applied to a long-buried metallic structure such as a pipe or cable to make it cathodic, the resultant current usually enters the structure over a considerable portion of its length and flows along the structure to the point of electrical contact. In general, such structures are electrically continuous, and the longitudinal current creates no problem. However, some pipes have joints between sections which may, by accident or design, have high electrical resistance, and longitudinal current may create an anodic condition on one side of such a joint. Before cathodic protection can be applied to such a structure, it is usually necessary to place bonds (metallic connectors) across any joints which have or may develop high resistance.

The economics of cathodic protection must take into consideration many factors in addition to the cost of making repairs. In pipe lines there is a loss of the fluid from the pipe, and in some cases this may be a serious fire hazard. In underground cables, a hole in the sheath may cause serious interruption in the services rendered.

Sources of Forced Drainage Current

Dry-type rectifiers such as copper-oxide, copper-sulfide, or selenium are most widely used as sources of current in forced drainage. They are made in a variety of output currents and adjustable voltages. Units are available in housings suitable for outside mounting, as, for example, on poles. They have proved quite satisfactory for this service, giving long life under continuous operation, even with high ambient temperature. Rectifier locations are sometimes visited only two to four times a year, the power consumption as reported on monthly power bills being used as an index of satisfactory operation.

Bulb or thermionic-type rectifiers are used to some extent where higher voltages are required to overcome unusually high resistance at the anode. Motor generator sets have been used to limited extent. Gas-driven generators are used on some long natural-gas lines, using gas from the line as fuel. Wind-driven generators have also been used where there is reasonable assurance of steady wind. Occasionally they are

---

* The term drainage has long been used by field engineers to mean the drawing of current from an underground structure by means of a metallic conductor for purposes of corrosion control. The term is not usually applied to other than underground structures.

used with a floating storage battery which supplies current when there is insufficient wind. More often, polarization of the structure is depended upon to prevent corrosion during windless intervals. Gas-driven and wind-driven generators have particular application where electric power is not available. Because of their moving parts, generators normally require more frequent inspection than rectifiers.

In trolley areas forced drainage is sometimes used to prevent corrosion resulting from stray current on a structure beyond the end of a trolley line. In such cases, the rectifier voltage can be automatically regulated to produce only as much current as is required to overcome the continuously varying trolley current.[1]

Outdoor rectifiers are usually mounted in suitable housings on poles, and windmill generators are mounted on special towers or masts. Other generators not located in available buildings usually require a small hut to be erected on the pipe line right-of-way.

Galvanic anodes require no extraneous source of current, as discussed later.

Anodes or Ground Beds

The current from the rectifier or generator is delivered to the earth through an anode or a system of anodes. These anodes (frequently referred to as ground beds), if constructed of iron, may be subject to severe corrosion as a result of the current, but since they can be relatively easily replaced (as compared to pipe or cable), they are expendable. Low-resistance anodes are desirable in order to keep down the voltage needed at the rectifier and thus reduce the power requirement.

One anode connected to a rectifier or generator is referred to as a point anode. Lower resistance and better current distribution can be obtained by using a series of anodes either in regular geometric arrangement or at the bases of poles already in place along a highway or right-of-way. Such an arrangement is called a distributed anode. Occasionally an abandoned pipe or trolley rail paralleling the structure is available and is used as a continuous anode. (See Fig. 2.) For underground cables, a special duct anode is sometimes pulled into a spare duct to minimize interference with other structures or to concentrate the effect on a short length of cable.

Pipe lines frequently use scrap steel such as old pipe for anode material. This permits installation of large quantities of metal at moderate cost to obtain low resistance and reasonably long life. Where such materials or the space required is not available, smaller anodes must be used. Steel rods driven into the ground are not practicable except as a temporary or trial arrangement, because corrosion is frequently concentrated about a foot below the ground line, and the rod becomes segregated after a relatively small portion of the total metal has corroded. A substantial increase in life can be obtained by surrounding the rod with fine coke, carbon, or graphite (Fig. 3). Carbon ground rods in various sizes are available for anodes, but they should also be surrounded by coke. In high-resistivity soils, an anode consisting of a long piece of large-size pipe or rail laid in a trench surrounded by coke may be satisfactory. Where coke, carbon, or graphite is used as environment, it not only increases the life of the anode but also reduces its electrical resistance to ground. However, carbon is electropositive (cathodic) with respect to the iron pipes or lead cables, and thus the environment produces an unfavorable galvanic potential which must be overcome in the rectifier. In selecting coke for this purpose, care must be exercised to obtain coke from coal, as petroleum coke has high electrical resistivity and may effectively insulate the anode from ground.

[1] Eric G. Carlson, "Use of Forced Drainage Systems in Stray Current Areas," *Corrosion*, **1**, 31 (March, 1945).

Coke and carbon grounds have not been found satisfactory in tidewater. Cast iron is suggested for use in such locations.

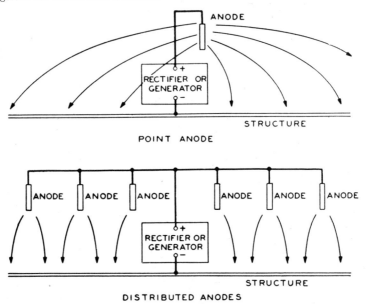

FIG. 2. Types of Forced Drainage Anodes for Underground Structures.

The resistance of a vertical ground rod or pipe to remote earth is approximately

$$R = \frac{1.64\rho}{\pi L}\left(\ln \frac{96L}{d} - 1\right) \text{ ohms}$$

where ρ is the soil resistivity in meter-ohms, L is the length of the rod below the surface of the ground in feet, and d is the diameter of the rod in inches. Where a vertical rod is surrounded by a special environment the following formula[2] can be used:

$$R = \frac{1.64\rho_1}{\pi L}\left(\ln \frac{d_1}{d}\right) + \frac{1.64\rho}{\pi L}\left(\ln \frac{96L}{d_1} - 1\right) \text{ ohms}$$

[2] H. B. Dwight, "Calculation of Resistances to Ground," *Elec. Eng.*, **55**, 1319–1328 (December, 1936).

where ρ_1 is the resistivity of the environment and d_1 is the diameter of the environment in inches. The formula assumes that the length of the rod below ground and the length of the environment are the same. (See Fig. 4.) Where ρ_1 is very small, the first term can be considered zero, and the formula becomes the same as for a rod of diameter d_1.

The resistance of a long horizontal rod or pipe to remote earth is approximately

$$R = \frac{1.64\rho}{\pi L}\left(\ln\frac{48L}{d} + \ln\frac{L}{h} - 2 + \frac{2h}{L}\right) \text{ ohms}$$

where L is the length of the rod or pipe in feet and h is the depth of the center below the surface in feet. The depth h should be at least ten times the diameter d. With environment, this becomes:

$$R = \frac{1.64\rho_1}{\pi L}\ln\frac{d_1}{d} + \frac{1.64\rho}{\pi L}\left(\ln\frac{48L}{d_1} + \ln\frac{L}{h} - 2 + \frac{2h}{L}\right) \text{ ohms}$$

and the same considerations apply as above. However, if the depth h is not so great as ten times the diameter d_1, the formula will be less accurate because of

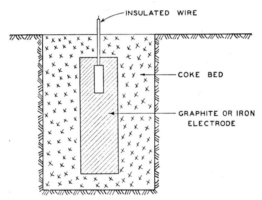

Fig. 3. Vertical-Type Anode in Coke Bed. (J. M. Pearson.)

unequal current densities at the outer surface of the environment. The presence of a bare underground metallic structure within a couple hundred feet of the ground rod or pipe will have the effect of reducing its resistance, especially in high-resistivity soil.

Effect of Coating or Conduit

Coatings applied to structures prior to their installation in the ground provide 100% protection only if they are impervious and completely cover the surface. However, a coating may contain "holidays" or pinholes or it may be damaged during installation or it may have deteriorated with time, as is frequently the case with old pipes. The surface of the pipe would then be at least partially subject to corrosion, and cathodic protection may be desirable. However, a coating, even though somewhat conductive, may materially reduce the current required for protection. Modern coatings in good condition may eliminate the need for cathodic protection at least for a period of years. However, since it is not practicable to inspect the coating to determine its condition, supplementary cathodic protection is frequently desirable.

Communication and power cables are usually installed in conduits. These conduits may be made of ceramic materials, cement, fiber, wood, or iron. Non-metallic conduit tends to provide an insulating enclosure around the cable, and the electrolytic path is largely through joints between sections of conduit. Since soil electrolyte is not excluded from the inside of the conduit, corrosion may occur on a cable sheath anywhere along its length, but cathodic protection is concentrated at the conduit joints. In metallic conduit, the iron pipe shields the cable sheath from any cathodic protection current. Although cable sheath corrosion in iron pipe conduit is not common, it cannot be controlled by cathodic protection.

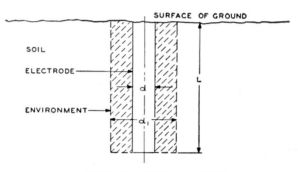

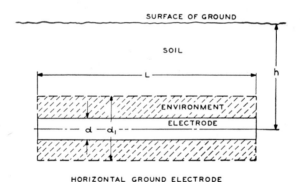

Fig. 4. Vertical and Horizontal Ground Electrodes.

Isolating a Section of Structure to Be Protected

Where a long bare structure requires protection over a relatively short length, this length is sometimes isolated from the rest of the structure by installing insulating joints at both ends. The isolated section can then be protected with relatively little current. However, the isolated section may have been acting as an anode, thus providing some protection for the structure elsewhere, and this protection would thereby be eliminated. Furthermore, by depressing the potential of the isolated section, potential differences are created across the joints which may cause anodic conditions on the structure just outside the joints. These undesirable conditions can be largely eliminated by coating the structure for several feet both sides of each joint

APPLICATION OF CATHODIC PROTECTION

and by placing a resistor across each joint so that the structure beyond the joints obtains some protection. Practically all the protective current returns to the isolated section, and the problem of cathodic current distribution along the structure is largely averted.

SPACING OF RECTIFIERS AND ANODES ON LONG STRUCTURES WITHOUT INSULATING JOINTS

On bare pipes and very poorly coated ones, close spacing of rectifiers and anodes may be required for complete protection. Pole lines are often constructed along the pipe line right-of-way to carry the power circuits, and a rectifier and anode have been installed as close as every 500 feet. On the other hand, on long-buried cables with thermoplastic coating, a fraction of an ampere every 10 miles has been sufficient for complete protection.

EFFECT OF FORCED DRAINAGE ON STRUCTURE-TO-EARTH POTENTIAL

The effect of forced drainage on the potential of the structure to near-by earth is twofold. The current raises (makes more positive or cathodic) the potential of the earth surrounding the anode (called the anode potential gradient) and lowers (makes more negative or anodic) the potential of the structure. If the anode is a considerable distance from the structure, the second effect predominates; but with an anode close to the structure the first effect may be more important, particularly on bare structures.[3]

Effect of Anode Potential Gradient. The rise in the potential of the earth surrounding an anode (anode potential gradient) is essentially the IR drop in the earth because of the anode current. The potential gradient created by a ground rod in uniform soil can be determined from the following formula:

$$E_r = \frac{1.64 I \rho}{\pi L} \ln\left(\frac{L + \sqrt{L^2 + r^2}}{r}\right) \text{ volts}$$

where E_r is the rise in earth potential (volts) at any point x, caused by current I (amperes) being delivered to the earth by the anode; ρ is soil resistivity in meter-ohms*; L is the length of the rod below the surface of the ground in feet, and r is the distance in feet from the center of the anode to point x. The value of ρ can be measured in several ways.[4]

When r is more than about ten times L, the formula becomes approximately:

$$E_r = \frac{0.525 I \rho}{r}$$

If a bare metallic structure traverses the anode potential gradient, the gradient will be distorted because of the high electrical conductivity of the structure compared with the earth. The anode current concentrates on the structure, and this effect will become more pronounced the closer the anode is to the structure. (It will also be more pronounced in high-resistivity soil than in low-resistivity soil). The structure will pick up current in the vicinity of the anode and lose it more remotely from the

---

* 1 meter-ohm = 100 ohm-cm.

[3] Robert Pope, "Attenuation of Forced Drainage Effects on Long Uniform Structures," *Corrosion*, **2**, 307–319 (December, 1946).

[4] L. M. Goldsmith, "Earth Resistivity Measurements," *Petroleum Eng.*, **9**, 68 (April, 1938).

anode. In a practical case this tendency to become anodic is overcome by the second effect discussed below.

The anode gradient effect is likely to predominate in the case of a bare structure in high-resistivity soil. When the anodes are close to such a structure, the effect on the structure-to-earth potential will be concentrated on a relatively short length of structure opposite the anodes. This is illustrated in Fig. 5, which shows the relative potentials of structure to near-by earth for close and remote anodes over a section of bare structure between two anode locations. This figure also illustrates that the critical location on the pipe is half way between the anodes.

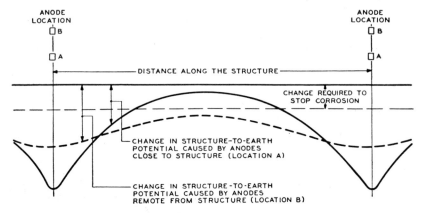

Fig. 5. Relative Effects of Close and Remote Anodes.

Effect of Depressing Structure Potential. If the anode is a considerable distance away, the structure may not traverse an appreciable portion of the potential gradient and the entire effect of the drainage may be in depressing the structure potential. At the drainage point, the change in structure potential is equal to the drainage current times the structure-to-earth resistance (R_G). Since the current and potential change can be measured, the structure-to-earth resistance can be computed

$$R_G = \frac{V_0}{\iota} \text{ ohms}$$

The change in structure potential is greatest at the drainage point and becomes less (attenuates) as the distance from the drainage point increases. For a long continuous structure in uniform soil, the change in structure-to-earth potential at any point, x (V_x), can be computed from the equation:

$$V_x = V_0 e^{\frac{-Rs}{2R_G}} \text{ volts}$$

where R is the longitudinal resistance of the structure per 1000 ft (ohm per kilofoot), s is the distance from the drainage point in thousands of feet (kilofeet), and $e = 2.718\ldots$.[5,6] The ratio $R/2R_G$, known as the attenuation constant, is indicated by the symbol γ. A typical potential attenuation curve is shown at the top of Fig. 6.

---

[5] E. D. Sunde, "Currents and Potentials along Leaky Ground-Return Conductors," *Elec. Eng.*, **55**, 1338 (December, 1936); *Bell Telephone System Monograph* B-970.

[6] H. C. Gear, "Current Density Related to Current Distribution," *Petroleum Eng.*, **14**, 182 (1943).

Rectangular coordinates are used to provide a physical concept of reduction of voltage with distance from the drainage point. The attenuation curve becomes a straight line when plotted on semi-log coordinates as indicated in Fig. 7.

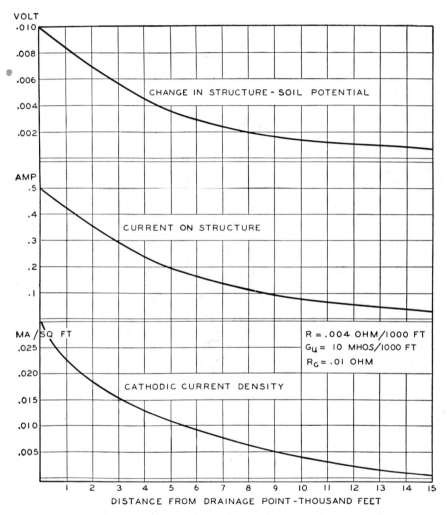

Fig. 6. Effect of Draining One Ampere from Structure.

However, on long uniform structures,

$$R_G = \frac{1}{2}\sqrt{\frac{R}{G_u}} \text{ ohms}$$

in which G_u is the leakage conductance of the structure in mhos per kilofoot and

$$\gamma = \sqrt{RG_u} \text{ per kilofoot}$$

Thus we can compute the value of γ for various values of G_u and draw a family of curves to illustrate the effects of changes in leakage conductance. The effect on structure-to-earth potential (V_x) is illustrated in Fig. 8 for a 12 in. pipe ($R = 0.004$ ohm per kilofoot). The leakage conductance of 100 ohms per kilofoot might be

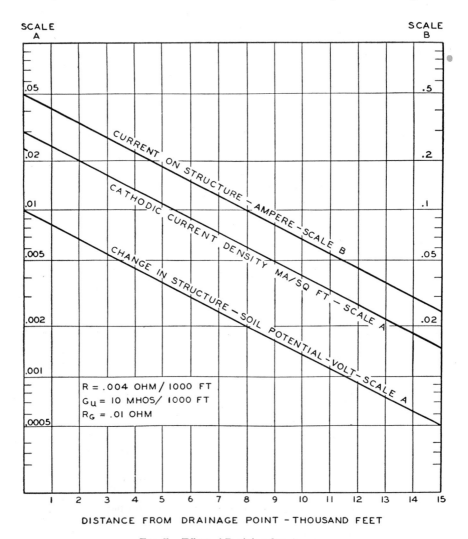

Fig. 7. Effect of Draining One Ampere.

obtained with a bare pipe in sea water ($\rho = 1.4$ meter-ohms). Ten mhos per kilofoot represents a bare pipe in very low-resistivity soil ($\rho = 14$ meter-ohms), and 1 mho per kilofoot indicates a medium-resistivity soil ($\rho = 140$ meter-ohms) or a poor coating. A leakage conductance of 0.1 mho per kilofoot represents either high-resistivity soil or a normal coating, and 0.01 mho per kilofoot can be obtained only with a very good coating.

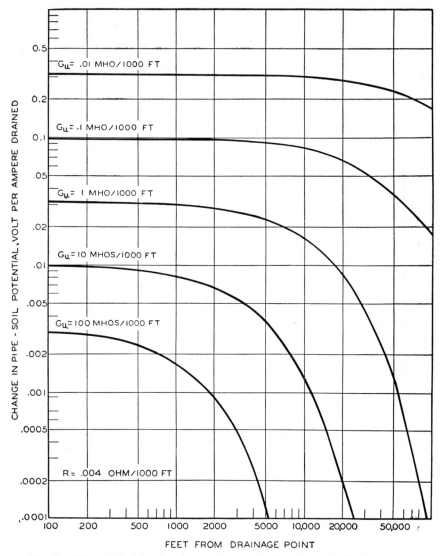

Fig. 8. Effect of Draining One Ampere from 12-in. Pipe Line with Various Leakage Conductances.

Current on Structure and Cathodic Current Density. The current on the structure follows the same attenuation formula and at any point can be determined from the equation:

$$I_x = I_0 e^{-\gamma s}$$

where I_0 is the current in the structure at the drainage point and is equal to one half the drainage current. The slope of this curve represents cathodic current per unit length, and therefore the cathodic current density can be determined from the equation:

$$i_x = \frac{12 I_0 \gamma e^{-\gamma s}}{\pi d} \quad \text{milliamperes per sq ft}$$

in which d is the outside diameter of the pipe in inches. This can also be expressed as:

$$i_x = \frac{12 \gamma I_x}{\pi d} \quad \text{milliamperes per sq ft}$$

The variation in current on the structure and cathodic current density are illustrated in Figs. 6 and 7. Figure 9 shows the effect of various values of leakage conductance on

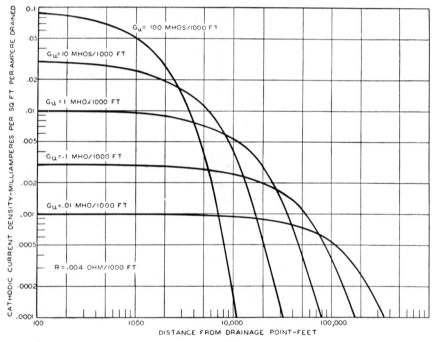

FIG. 9. Effect of Leakage Conductance on Cathodic Current Density — 12-in. Pipe.

cathodic current density. Where several drainage points affect a location, the cathodic current densities are cumulative.

The attenuation formulas and curves assume uniform condition for considerable distance both ways from the drainage point and are not strictly applicable to bare structures unless the anode is far removed from the structure.

Effect of Polarization. The attenuation formulas assume that the leakage resistance is constant. However, polarization effects tend to change the apparent structure-to-earth resistance. These effects may be small enough to be negligible on some bare structures but usually have pronounced effect especially on coated structures. Since polarization requires considerable time to become apparent in some cases, it is frequently observed that the drainage seems to increase in effectiveness for some time after its application. Polarization first reduces the cathodic current density and increases the structure-to-earth potential, in the vicinity of the drainage point, and this causes spreading of the drainage current to more remote points on the structure. This may in turn cause polarization farther from the drainage point, and in this manner polarization will extend the length of structure protected. As an extreme example, the average structure-to-earth resistance of two thermoplastic covered buried cables was 1.1 ohms for nine drainage points before applying drainage, and the average apparent resistance one month after applying drainage was 10.6 ohms.[7]

Galvanic Anodes

Galvanic anodes such as zinc, magnesium, and aluminum provide drainage without the use of a rectifier or generator.[7,8,9,10] The galvanic potential between the structure and the anode provides the current. The potential of iron or lead to zinc is in the order of 0.5 volt, and to magnesium about 1.0 volt in soil electrolytes. Because of the limited voltage, the resistance to earth of the anodes must be small to obtain appreciable current.

Zinc, in some soils, builds up a film which effectively insulates it from the surrounding earth.[11] This is particularly noticeable in soils high in carbonates. In order to overcome such effect, the zinc anodes can be installed in a special sulfate environment. Gypsum ($CaSO_4 \cdot 2H_2O$) or hydrated plaster of Paris is used for this purpose, and because of its low solubility remains in place in the soil for considerable time.[12] Magnesium uses a similar environment to provide a higher anode efficiency than would otherwise be obtained.[13]

High-purity zinc is used to keep self-corrosion (local action) to a minimum. Rolled zinc can be obtained in the form of bars or plates for this purpose. Special castings of magnesium and magnesium alloys are made for use as anodes. Both zinc and magnesium are usually installed vertically in a deep hole, and the hole is back-filled with the artificial environment. This consists of clay mixed with 25% gypsum, usually made into a slurry. Bentonite is recommended if suitable local clay is not available. For high-resistivity soils, the addition of 25% sodium sulfate helps to reduce the initial anode resistance.

For structures with low leakage resistances requiring substantial current drainage in high-resistivity soils, galvanic anodes will usually not provide sufficient current to be practicable. The limit of soil resistivity for use of magnesium is probably of the order of 80 meter-ohms (8000 ohm-cm) and for zinc about 40 meter-ohms (4000 ohm-cm). The anodes should be a sufficient distance from the structure, so that their effect is not concentrated in a very short length of the structure. Where the anodes

[7] T. J. Maitland, "Corrosion Protection for Transcontinental Cable West of Salt Lake City, Utah," *Corrosion*, **1**, 47 (June, 1945).

[8] C. L. Brockschmidt, "A Practical Application of Zinc Anode Protection to an 18-Inch Pipe Line," *Petroleum Eng.* (February, 1942).

[9] R. B. Mears and C. D. Brown, "Light Metals for Cathodic Protection of Steel Structures," *Corrosion*, **1**, 113 (September, 1945).

[10] H. A. Robinson, "Magnesium as a Galvanic Anode," *The Electrochem. Soc. Preprint* 90–4 (October, 1946).

[11] O. C. Mudd, "Experiences with Zinc Anodes," *Petroleum Industry Elec. Assoc. Elec. News* (May, 1943).

[12] H. W. Wahlquist, "Use of Zinc for Cathodic Protection," *Proc. Natl. Assoc. Corrosion Engrs.*, **1**, 61–94 (1944).

[13] Porter Hart and Yale W. Titterington, "Developing Magnesium for Cathodic Protection," *Corrosion*, **1**, 59 (June, 1945).

are close to a bare structure, close spacing of anodes is required in order that all the structure will be protected.

CRITERIA OF PROTECTION*

There are several criteria used to determine whether a structure requires protection and the adequacy of the protection. The most nearly infallible criterion is a history of corrosion failures. However, it is desirable to anticipate such failures and to take steps to prevent them.

The structure-to-earth potential is most widely used to determine the need for corrosion protection. The copper-copper sulfate half cell (discussed on p. 931) is used by many pipe companies as a reference electrode in making the potential measurement. In some cases, if the potential of the pipe to such an electrode remote from the structure is less negative† than −0.85 volt, cathodic protection is applied until this value is obtained. In other cases, iron rods are used as reference electrodes and a structure potential difference of −0.3 volt is required for a safe condition. In still other cases a pipe is considered safe from corrosion only if forced drainage is applied until the potential of the structure is depressed 0.3 volt below its original value.[14] On underground cables, a lead ground plate is used and a positive cable-to-plate potential is taken as indication of the probability of corrosion of the sheath. These criteria are based largely on experience or expediency. More scientific approaches to the problem are described in the previous chapter.

Occasionally *corrosion coupons* are used to determine the necessity for installing protection or the efficacy of existing protection.‡ They are small pieces of steel similar to the pipe to be tested and are placed on the pipe and connected to it. By comparing the weight of the coupon at the time of installation with the weight of the cleaned coupon after being in the ground for a period of time (6 months to 1 year), the rate of corrosion can be determined which is considered to be the same as that of the pipe. To avoid unnecessary errors, the coupons should be mounted flush with the pipe surface and the space between them and the pipe filled with insulating material to avoid corrosion by local action currents, should an electrolyte enter the space.

The "pattern test"[15] is adapted to this same purpose by wrapping the metal with a layer of cloth over which a tough paper cover is bound. The whole is wet with the surrounding electrolyte and covered as originally found. If any anodes on the metal discharge ferrous ions, they will migrate along the current flow lines and reach the paper. Subsequent treatment of the paper with potassium ferricyanide develops blue spots of ferrous ferricyanide where iron was corroded. The results of this test are not quantitative but are conclusive in regard to the existence or non-existence of anodes at the metal surface.

Regular but infrequent inspections of the pipe surface by excavating the pipe provide another means to check on corrosive action. Record should be made of pit locations and sizes for comparison. This method as well as the use of coupons assumes that the location selected for the test represents at least average conditions, and that

* Refer also to previous chapter, p. 923.
† "Less negative" is synonymous with "less anodic."
‡ Some information appearing in this and the next paragraph was originally supplied by J. M. Pearson.

[14] K. H. Logan et al., "The Determination of the Current Required for Cathodic Protection," *Petroleum Eng.*, **14**, 168–181 (1943).
[15] Scott Ewing and G. N. Scott, "Electrolytic Method for Determining Condition of Non-Metallic Pipe Coatings," *Am. Gas Assoc. Monthly*, **16** (No. 4), 136–139 (April, 1934).

APPLICATION OF CATHODIC PROTECTION

disturbing the soil does not affect its corrosiveness. It also assumes that the coupon, if used, is neither anodic nor cathodic with respect to the pipe. These methods are not applicable to coated pipe or to underground cables.

Effect on Other Structures

The introduction of current in the earth for forced drainage may affect other underground structures. Part of the earth current may be intercepted by such other structures and will cause corrosion where current leaves to return to the drained structure. Where the structures are approximately parallel to each other and there is no additional structure present, the location of the anode can be selected so as to minimize the effect on the intercepting structure. Where structures cross, possible damage to the intercepting structure at the crossing can be eliminated by placing a bond of proper resistance between the two structures.[16] Where drainage can be located at such crossings or where structures are closely parallel, joint drainage is frequently used. In any event, the cooperation of the various properties involved is desirable.[17]

City Networks. In communities where there is a network of underground structures, cathodic protection presents additional problems. The water and gas piping systems are usually in electrical contact with each other at customers' premises, both accidentally and at gas water heaters. The neutral conductor of the electric circuit is grounded to the water pipe, and it is usually connected to the sheath of the power cable. In this way, the water, gas, and electric power underground plants are bonded together and form an extensive network which must be drained as a unit. Communication cables and power transmission cables can be kept isolated from other plant and can be drained separately.

PROTECTION OF OTHER UNDERGROUND STRUCTURES

Cathodic protection can be applied to the outside surfaces of practically any buried metallic structures. For example, gasoline and oil tanks in so-called tank farms and refineries, together with the associated underground piping, can be protected with forced drainage.[18,19] The basic considerations and methods follow closely those applying to pipes and cables. All buried metallic structures must be electrically bonded so that they obtain protection. An isolated structure, even though it had not been corroding seriously before cathodic protection was applied, may become badly corroded as a result of intercepting current to the protected structure. Anodes must be located and the current to each anode regulated so that sufficient current reaches all parts requiring protection.

PROTECTION OF TANKS, CONDENSERS, ETC.

The use of cathodic protection on the inside surfaces of water-storage tanks, hot water tanks, condensers, water-treatment tanks, brine tanks, etc., presents interesting problems.[20] The current density required for protection depends on the material of

[16] G. R. Olson, "Cathodic Protection Coordination," *Petroleum Industry Elec. Assoc. Elec. News* (June, 1943).
[17] A. V. Smith, "Cathodic Protection Interference," *Am. Gas Assoc. Monthly*, **25**, 421 (October, 1943).
[18] D. Holsteyn, "Practical Design and Economics of Cathodic Units Applied in a Refinery," *Petroleum Industry Elec. Assoc. Elec. News* (July, 1943).
[19] R. A. Brannon, "Cathodic Protection of Tank Farms," *Petroleum Eng.*, **13**, 92 (September, 1942).
[20] L. P. Sudrabin, "Cathodic Protection of Steel Surfaces in Contact with Water," *Water Works and Sewerage* (January, February, and May, 1945).

the tank, the corrosiveness of the water, the presence of protective paint, etc. There are wide differences in corrosiveness of ordinary waters so that no set rules can be applied. Sea water, brine, sewage, etc., are relatively corrosive and require higher current densities for protection. The anodes should have long life, and therefore carbon anodes are widely used. The anodes should be so located and their respective currents so adjusted that proper protection is obtained over the entire cathode surface. This may require anodes projecting into extensions from the tanks such as stand pipes. In high-resistivity water, a greater number of anodes is needed in order to keep down the voltage required. Cathodic protection can be used to prevent corrosion of metal tanks used in chemical processes and may make possible the use of less expensive materials to handle corrosive solutions.[21]

Galvanic anodes can be used in low-resistivity waters such as brine or sea water. A galvanic anode hung in a tank should retard corrosion but will not prevent it unless there is sufficient current density at all parts of the tank.

Cathodic protection is not practicable on the inside surfaces of small pipes such as house piping.[22,23,24]

PROTECTION OF SHIPS

Protection of the hull of a ship from galvanic corrosion caused by the propeller has been limited to the attachment of a galvanic anode zinc to the bottom of the ship near the propeller.

---

[21] R. B. Mears and H. J. Fahrney, "Cathodic Protection of Aluminum Equipment," *Trans. Am. Inst. Chem. Engrs.*, **37**, 911 (1941).

[22] H. S. Warren, "Electrical Grounds on Water Pipes," *J. New England Water Works Assoc.*, **48**, 350 (September, 1934).

[23] American Research Committee on Grounding, "Interim Report of Investigations," *J. Am. Water Works Assoc.*, **36**, 383 (April, 1944).

[24] R. Eliassen and P. Goldsmith, "Effects of Grounding on Water Pipe," *J. Am. Water Works Assoc.*, **36**, 563 (May, 1944).

CORROSION TESTING

Laboratory Tests

GENERAL DISCUSSION OF LABORATORY CORROSION TESTING

CARL W. BORGMANN*

APPLICATION AND INTERPRETATION

Laboratory corrosion tests are used for a wide variety of purposes, and the type of corrosion test that will be found useful depends primarily upon the end result desired. In general, the uses to which laboratory corrosion tests are put may be divided into five classifications, namely:

1. To study the mechanism of corrosion.
2. To select the most suitable material for withstanding a definite environment.
3. To determine the environments in which a given material can be satisfactorily employed.
4. To develop new alloys.
5. To serve as a control test for the achievement of uniformity of a product.

Tests designed to study the mechanism of corrosion are of the widest variety. One must choose that type of test which will allow the simplest and easiest understanding of the processes under study. Usually it is necessary to employ quite delicate means of measurement and to use conditions that are artificially controlled within extremely close limits. In order to be able to interpret the results of such tests, it is usually simplest to control all possible variables except the one under consideration. However, the artificiality thus imposed can easily lead to results that, if not very carefully interpreted, can be most misleading.

Corrosion tests for purposes other than the study of the mechanism of corrosion have taken several forms. For many years there has been a concerted effort to find a comprehensive "accelerated corrosion test" that could be used to serve any, or at least most, of the aims previously tabulated. The salt-spray test, the simulated atmospheric corrosion test, the acid immersion test for steel, and many others were of this type. It was hoped that some form of an accelerated corrosion test could be made available which in a short time would give a rating of the behavior of the various materials that could be correlated with service exposures. For several reasons, all attempts to develop such tests have failed. These failures have occurred primarily because certain corrosive conditions have been intensified in order to cause severe corrosion in a short period of time, thus completely changing the nature of the environment. More important, perhaps, is the fact that there is such a wide variance in the behavior of metals in different types of environments that no one or even a few such accelerated tests could possibly be interpreted in terms of actual service of material. It can now be stated that attempts to develop such tests have in general been discontinued.

In corrosion studies undertaken for the purpose of finding a suitable material to withstand a particular service or of finding those environments in which a given metal

* Chemical Engineering Department, University of Colorado, Boulder, Colorado.

is satisfactory, it is best first to take advantage of the vast amount of published literature in the field of corrosion. Such a study will in general give a very good clue as to the general types of metals or alloys that should prove most satisfactory for the particular job to be done. The first laboratory tests on the selected materials should be as simple as possible, the nature of the environment in which the metal is to be used being kept in mind and at least the more important controlling factors being simulated. It is often advisable to make use of existing equipment and environments for the purpose of exposing small samples of the materials that are being considered. Often these so-called field tests will give far more information for a primary selection of materials than most laboratory tests. These simple tests should allow a further selection. The next step should be to test model-sized equipment prepared from the materials finally selected, all factors in the full-sized equipment being very carefully duplicated in the model. One word of caution should be added — in any laboratory test great care should be taken in the interpretation of the data. At best the results can only be qualitative and a great deal of common sense and experience has to be applied to such results before they become useful to the engineer.

For the development of new alloys, and particularly for control tests for alloys already in use, those laboratory tests known as "special property tests" hold the greatest promise. It has been found by experience that many metals and alloys are susceptible to attack by special chemical reagents. A typical example is the standard test for brass which consists in immersing the specimen in a solution of mercuric nitrate or chloride. If the brass is under a strained condition, it will crack quickly under such test conditions. Experience shows that the same brass under conditions of similar stress will be subject to "season" cracking when used even under fairly mild atmospheric conditions. The copper sulfate and sulfuric acid test for stainless steel (p. 1020), the sodium chloride and hydrogen peroxide tests for certain aluminum alloys, and many others illustrate the type of test being discussed. These tests do not attempt to measure the service life of the metal in a particular environment but rather are indicative of the tendency to special types of corrosion. Similarly, tests designed to measure the susceptibility of a given material to special service conditions, such as, for example, corrosion fatigue tests, stress corrosion tests, impingement tests, and many others, fall into this same category and have been found to be extremely useful in the development of special alloys. It seems unquestionable that the brightest future for laboratory corrosion tests lies in the extension of these principles to other materials and other environmental conditions.

VARIOUS METHODS FOR MEASURING CORROSION

The major methods for measuring the amount and intensity of corrosion as well as certain qualitative aids in determining the amount of corrosion are given in Table 1. Brief statements are included outlining the fields of usefulness and the major advantages and defects of the various methods. Often the use of two or more methods will remove many of the criticisms given for the individual methods.

FACTORS INVOLVING REPRODUCIBILITY

Because of the complex nature both of the mechanism of corrosion and of the diversity of environmental conditions that may be met, there are many factors of importance in determining the reproducibility of experimental results. These factors in general may be divided into two categories, namely, those involving the metal surface and those involving the environment. With regard to the metal surface itself,

TABLE 1. METHODS OF DETERMINING THE AMOUNT AND INFLUENCE OF CORROSION

Method	Fields of Usefulness	Advantages of Method	Defects of Method
Visual observation	For field tests on very large specimens. For service tests of plant equipment. To determine whether or not any attack occurs. (Inhibitors, stainless steels, etc.) To determine whether attack is uniform or localized.	1. Simplicity. 2. Valuable in conjunction with other methods.	1. Personal error. 2. A quantitative estimate of damage is impossible.
Loss in weight	For laboratory and field tests on metals or alloys not subject to special types of attack and from which the products of corrosion are easily removed.	1. Simplicity. 2. A quantitative estimate of amount of corrosion is made.	1. Error due to incomplete removal of corrosion products and to loss of uncorroded metal. 2. Special types of attack are not measured. 3. Large number of specimens necessary to determine a time-corrosion curve properly.
Gain in weight	Same as "Loss in weight" except that no loss of corrosion product must occur. Particularly applicable to indoor atmospheric corrosion and high temperature oxidation studies.	1. Same specimen may be used to determine several points on a time curve. 2. Errors due to improper removal of corrosion products eliminated.	1. An analysis of the corrosion product must be made in order to determine loss of metal. 2. Moisture in corrosion product may vary in amount and thus influence the results. 3. Same as (2) under "Loss in weight." 4. Accidental loss of corrosion products would introduce an error.
Change in electrical resistance*,†,‡	For laboratory or field tests in gaseous or poorly conducting environments or in other environments if specimen is removed for measurement. Also useful for measuring progress of intergranular attack of stainless steels by $CuSO_4$-H_2SO_4 reagent.	Produces little or no disturbance of specimen during test and hence is well adapted for measurement of time curves.	1. Impossible to distinguish between the influence of various types of attack. 2. If small wires are employed, the attack may be different in amount from that on a more massive specimen.

TABLE 1. METHODS OF DETERMINING THE AMOUNT AND INFLUENCE OF CORROSION — *Continued*

Method	Fields of Usefulness	Advantages of Method	Defects of Method
Hydrogen evolved	For laboratory tests in which corrosion takes place *solely* with the evolution of hydrogen.	Time curves are readily measured without disturbance of specimen.	Distribution of attack is not determined.
Oxygen absorbed (and hydrogen evolved) §	For laboratory tests in which the corrosion takes place *mainly* with the absorption of oxygen.	1. Time curves are readily measured without disturbance of specimen. 2. The amount of attack due to each of two different mechanisms is determinable.	1. Distribution and type of attack is not measured. 2. An analysis of the corrosion products of metals capable of existing in more than one ionic state must be made.
Depth of pitting (measurements other than microscopic)	For tests (laboratory or field) made to determine the serviceability of metals as containers of fluids (tanks, pipes, etc.)	1. Especially adapted for use with methods of determining total attack. 2. A correct measure of the penetration of a metal by corrosion except when attack is intergranular.	1. Many specimens required to determine time-penetration curves. 2. Difficulty of obtaining accurate measurement.
Microscopic	To determine kind of attack (intercrystalline, dezincification, etc.). To measure depth of pits. To determine the constituents of the metal that are specially capable of initiating attack.	Excellent tool to supplement other measures.	Not generally useful in making quantitative measurements.
Change in physical properties	Of particular use in studies on structural materials. Tensile strength measurements indicate particularly the damage by general corrosion. Pitting and intercrystalline corrosion are best measured by changes in ductility, impact resistance, or resistance to bending. The change of endurance limit indicates the susceptibility of the metal to corrosion fatigue.	Measures directly the changes in physical properties and hence is of practical value to the structural engineer.	A resultant of the several possible attacks is measured and it is not possible to evaluate the damage due to each type of attack (intercrystalline, pitting, etc.) separately.

TABLE 1. METHODS OF DETERMINING THE AMOUNT AND INFLUENCE OF CORROSION — *Continued*

Method	Fields of Usefulness	Advantages of Method	Defects of Method
Electrochemical: (a) Single electrode potential‖, ¶	To study film formation and breakdown on a metal and hence, to a limited degree, the stability of the metal in question. Important method when used in conjunction with a determination of total corrosion.	1. Determines, together with other measurements, whether anodic or cathodic control is operating. 2. Determines qualitatively the stability of surface films.	1. Does not measure amount of attack to be expected. 2. Not readily interpreted.
(b) Potential difference between two dissimilar metals	To study galvanic effects.	Determines which of two metals is likely to be more severely corroded due to electrical contact with the other in the particular solution used.	1. In no way a quantitative measurement. 2. Less valuable for purpose than a measurement of polarization characteristics of the two metals (see below)
(c) Current developed by short-circuited cell of the following type:** Metal Standard under : Environment : metal Test (Noble)	Suggested as a method to determine the relative corrosion of metals (laboratory test).	Measurement is simple.	1. The arbitrary choice of a cathode metal may distort the normal influence of such areas on the metal under study. 2. Anode and cathode are separated by a much greater distance than usual; hence resistance of solution and the unnatural formation of corrosion products are sources of error.
(d) Film resistance measurements††	To determine the penetrability of surface films by various anions.	Gives a qualitative measurement of the influence of the anion on the probability of breakdown of the film.	An arbitrary standard voltage is usually employed, and care must be taken that other reactions do not interfere. May be a measure of the decomposition voltage of an electrolyte-metal system without reference to a surface film.‡‡
(e) Anodic and cathodic polarization	To study galvanic and concentration cell corrosion (laboratory test). To determine the current necessary to polarize the two electrodes to a unipotential value.	Is relatively simple to perform and gives a semi-quantitative estimation of the corrosion rate.	Does not measure the distribution of attack.

TABLE 1. METHODS OF DETERMINING THE AMOUNT AND INFLUENCE OF CORROSION — *Continued*

Method	Fields of Usefulness	Advantages of Method	Defects of Method
Electrometric §§	To measure the thickness of surface films.	Technique is simple and quite exact.	Only useful on adherent, thin, surface films of some metals.
Optical methods ‖‖	To study the growth of tarnish and other surface films.	The formation and growth of films can be studied without disturbance of the system.	1. Requires relatively complicated apparatus. 2. The results of the simpler reflectivity methods are difficult to interpret quantitatively.
Influence of metal on environment	To determine whether the corrosion products of a metal are detrimental to the quality of a product.	In many cases, this measurement is essential to the determination of usefulness of a metal.	Naturally, such measurements tell nothing of the nature of the attack on the metal.

* J. C. Hudson, "Application of Electrical Resistance Measurements to the Study of the Atmospheric Corrosion of Metals," *Proc. Phys. Soc.* (London), **40**, 107 (1929).

† R. M. Burns and W. E. Campbell, "Electrical Resistance Method of Measuring Lead Corrosion," *Trans. Electrochem. Soc.*, **55**, 271 (1929).

‡ J. Rutherford and R. Aborn, *Trans. Am. Inst. Mining Met. Engrs.*, **100**, 293 (1932).

§ G. D. Bengough, J. M. Stuart, and A. R. Lee, "The Theory of Metallic Corrosion in the Light of Quantitative Measurements," *Proc. Roy. Soc.* (London), (A) **116**, 438 (1927); (A) **121**, 89 (1928).

‖ T. P. Hoar and U. R. Evans, "Time-Potential Curves on Iron and Steel and Their Significance," *J. Iron Steel Inst.*, **126**, 379 (1932).

¶ F. Fenwick, "Protective Films on Ferrous Alloys," *Ind. Eng. Chem.*, **27**, 1095 (1935).

** R. H. Brown and R. B. Mears, "The Electrochemistry of Corrosion," *Trans. Electrochem. Soc.*, **74**, 495–519 (1938).

†† S. C. Britton and U. R. Evans, "The Passivity of Metals. Part VI — A Comparison between the Penetrating Powers of Anions," *J. Chem. Soc.* (London), 1773 (1930).

‡‡ H. Uhlig and J. Wulff, *Trans. Am. Inst. Mining Met. Engrs.*, **135**, 494 (1939).

§§ H. A. Miley, "The Thickness of Oxide Films on Iron," Carnegie Scholarship Memoirs, *J. Iron Steel Inst.*, **25**, 200 (1936).

‖‖ L. Tronstad, "The Investigation of Thin Surface Films on Metals by Means of Reflected Polarized Light," *Trans. Faraday Soc.*, **29**, 502 (1933).

such things as cleanliness, the presence and nature of surface films, the presence of more than one phase in the alloy, the preparation of the surface for a test, and the polarization characteristics of the metal are but a few examples. Such things as composition of the environment, its homogeneity, the presence of foreign matter such as dirt, and the freedom to access of air are of importance from the standpoint of environment. In fact, anything that may be of importance to the electrochemical nature of corrosion will obviously influence the amount and nature of the corrosive attack.

In order to obtain reproducible results, therefore, the greatest care must be observed in preparing and exposing the specimens and in measuring the extent of the attack. However, even when the greatest possible pains are taken, it is not to be expected that absolute reproducibility will be achieved. There are many factors beyond the control of the experimenter.* Consequently, the absolute error of measurement can be reduced only by increasing the number of "identical" specimens tested.

* However, statistical control is often achieved. See *Total Immersion Tests*, p. 959. EDITOR.

The type of attack that occurs can greatly influence the reproducibility of the results. If the attack is uniform, the reproducibility will in general be much greater than if the attack occurs only at a few discrete points. To attain the same accuracy in this latter case, more specimens would have to be exposed.

TOTAL IMMERSION TESTS

W. A. Wesley*

The total immersion test performed with proper precautions can be made a reliable type of corrosion test. Adequate control of the important variables is in general obtainable. A state of statistical control has been demonstrated for each of the following corrosion systems: iron in potable water,[1] 70% Ni–30% Cu alloy† in aerated 5% sulfuric acid solution and in aerated 5% hydrochloric acid solution, steel in aerated 3% sodium chloride solution,[2] and zinc in aerated sodium chloride solution.[3] This and similar work encouraged formulation of recommended methods of total immersion testing for non-ferrous metals[4] by the American Society for Testing Materials which should be studied by anyone preparing to perform such tests.

METHODS OF CONTROL OF THE CORROSION VARIABLES

Figure 1 is presented as an example of the magnitude of some of the important controllable factors in total immersion corrosion, in this case the corrosion of 70% Ni–30% Cu alloy in 5% sulfuric acid solution.[5] The principal advantage of laboratory tests is that such factors can be held constant, but it is obvious that no one design of apparatus will necessarily suffice to control these variables over the full ranges encountered in practice. The physical features of the apparatus are therefore considered of importance only as they affect control of the significant variables.

Velocity

For investigating corrosion at high velocities (without impingement or cavitation which require specialized technique), simple rotation of the specimens is not ideal because the liquid tends to move with the specimen and the actual rate of movement of liquid with respect to the metal surface is not definitely known. Circulation of the solution through a tube in which the specimen is mounted[5,6,7] appears to be a better method of high-velocity control. The principle of such continuous flow apparatus is indicated in Fig. 2.

The tubular apparatus has frequently been employed also for tests at moderate velocities, but devices for moving the specimen through the solution at a uniform rate are equally satisfactory. The circular path machine sketched in Fig. 3 is useful for velocities up to 20 meters (65 ft) per min.[5,8] It offers the advantage that a vertical

---

* Research Laboratory, The International Nickel Co., Inc., Bayonne, N. J.
† Monel.

[1] R. F. Passano, *Proc. Am. Soc. Testing Materials*, **32**, Part II, 468–474 (1932).
[2] W. A. Wesley, *Proc. Am. Soc. Testing Materials*, **43**, 649–658 (1943).
[3] E. A. Anderson, unpublished data, New Jersey Zinc Co., 1944.
[4] *A. S. T. M. Standards*, 1943 Supplement, Part I, pp. 332–341, A. S. T. M. designation B185–43T.
[5] O. B. J. Fraser, D. E. Ackerman, and J. W. Sands, *Ind. Eng. Chem.*, **19**, 332–338 (1927).
[6] H. A. Trebler, W. A. Wesley, and F. L. LaQue, *Ind. Eng. Chem.*, **24**, 341–342 (1932).
[7] R. F. Passano and F. R. Nagley, *Proc. Am. Soc. Testing Materials*, **33** (II), 387–399 (1933).
[8] W. A. Wesley, *loc. cit.*

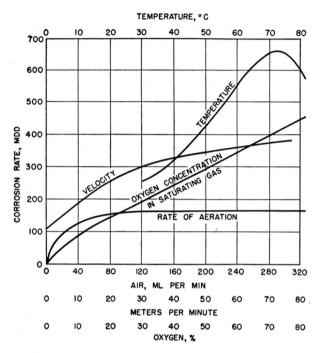

Fig. 1. Influence of Controllable Variables on Corrosion of 70% Ni–30% Cu Alloy in 5% Sulfuric Acid Solution.

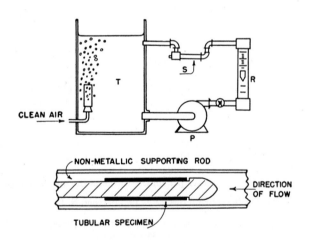

Fig. 2. Principle of a Constant Flow Apparatus.

T = Storage and solution control tank. P = Circulating pump. R = Flowmeter. S = Specimen holder.

circular motion without twist causes all parts of the specimen to move at the same velocity. It is possible to mount specimens on a rotating spindle in a way that insures substantially uniform velocity over the principal surfaces, and when this is done satisfactory velocity control is obtained.[8] A rotating spindle apparatus providing a closed container suitable for tests at low pressures is shown in Fig. 4; a similar apparatus for tests at pressures above atmospheric is illustrated in Fig. 5.

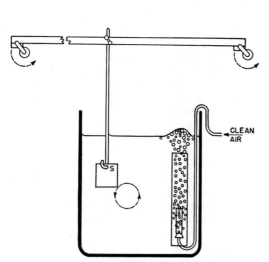

Fig. 3. Diagrammatic Sketch of Circular Path Apparatus.

Fig. 4. Rotating Spindle Apparatus — for Low Pressures.

For corrosion tests in boiling liquids it is general practice as a matter of expediency to employ stationary test pieces and depend upon the agitation provided by boiling. This is true of the standard nitric acid test for stainless steels.[9] It is probable that velocity is not a controlling factor in this test, but it might well be important in testing other combinations of metal and boiling solutions.

Zero velocity is difficult to maintain because convection currents arise from slight differences in temperature or concentration. So much care is required that a condition of zero velocity has probably been achieved only in scientific work as contrasted with industrial testing. The classic example of the former is the work of Bengough, Stuart, Lee, and Wormwell.[10]

It is not safe to predict corrosion rates at one velocity from rates observed at a different velocity. Protective films which form and adhere at low velocities may not withstand higher rates of flow. On the other hand, instances are known of alloys resistant to corrosion at high velocities which fail under conditions of stagnation. For example, brasses that dezincify at low velocity are free from such effects at high velocity, and stainless steels which retain an unblemished surface in salt water at high rates of flow may be severely pitted under stagnant conditions.

[9] *A.S.T.M. Standards*, 1943 Supplement, Part I, pp. 205–208, A.S.T.M. designation A262–43T.
[10] Summarized by U. R. Evans, *Metallic Corrosion, Passivity and Protection*, pp. 183 and 635, Edward Arnold and Co., London, 1937.

Temperature

For temperatures in the range between room temperature and 100° C, thermostatic control at the desired temperature plus or minus 1° C is usually considered satisfactory. A somewhat greater variation is often permissible at test temperatures above 100° C.

Specimen Supports

Supports for the specimens should be designed so as to insulate specimens from each other and from any metallic container or supporting device used with the apparatus. The supporting device and container should not be affected by the corroding agent to an extent that might cause contamination of the testing solution so as to change its corrosivity. The shape and form of the specimen support should be such as to avoid, as much as possible, any interference with free contact of the specimen with the corroding solution. When it is desired to set up conditions favoring contact corrosion, deposit attack, or other forms of concentration cell action, the means by which these types of attack are favored should be such as to insure exact reproducibility from specimen to specimen and test to test.[11] Examples of methods of support which have been satisfactory for the particular combination of metal and solution under study are available in the references already cited.

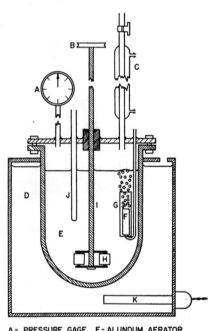

A - PRESSURE GAGE   F - ALUNDUM AERATOR
B - DRIVE PULLEY   G - GLASS CHIMNEY
C - CONDENSER   H - TEST SPECIMEN
D - OIL BATH   I - ROTATING SPINDLE
E - TEST SOLUTION   J - THERMOMETER WELL
                    K - ELECTRIC HEATER

Fig. 5. Rotating Spindle Apparatus — for Pressures above Atmospheric.

Aeration

The guiding principle in control of the gas content of the corroding solution is this: it is easier to maintain a solution at the saturation point with respect to the gas phase than at any other concentration. The desired concentration of any one component of the gas phase, such as oxygen, in the solution is then determined by how much is present in the aerating gas. Pure oxygen would be used only if it is desired to maintain the solution saturated with oxygen. If air is employed, the solution is commonly maintained saturated with air, and the oxygen concentration in the solution is thus held at about one-fifth that of an oxygen-saturated solution. Air diluted with nitrogen can be employed for still lower oxygen concentrations.

The means adopted for saturating the solution with gas depend upon the type of apparatus employed. Thus in a continuous flow apparatus it is feasible to maintain the solution saturated with air by using an air lift as part of the circulating system.[12]

[11] *A.S.T.M. Standards*, 1943 Supplement, Part I, pp. 332–341, A.S.T.M. designation B185–43T.
[12] H. A. Trebler, W. A. Wesley, and F. L. LaQue, *Ind. Eng. Chem.*, **24**, 341–342 (1932).

The most reliable method is to blow air through the solution, using an Alundum thimble or a sintered glass diffusion disk of medium porosity to break the air stream into small bubbles. Such devices should introduce the air at the base of a glass chimney over each aerator so as to prevent the impingement of the bubbles upon any test specimen.

When good reproducibility is desired, the rate of flow of saturating gas should be measured with a flowmeter. Suggestions as to the actual rates of flow necessary are available.[13,14] If it is certain that the practical problem involves an oxygen-free solution, this condition is more reliably reproduced in the laboratory test by saturation with an inert gas such as purified nitrogen* than by omitting aeration.

Test Solutions

The recommended ratio of volume of test solution to area of specimens is 250 ml per sq in. (4 liters per sq dm) unless it is desirable to reproduce the ratio that exists in the practical case. An instance of the latter choice is a test, the object of which is to determine the effect of metals upon the characteristics of a dye solution.

Changes in the Solution. Perhaps the most common danger of mistaken interpretation of laboratory immersion test results arises from exhaustion of ingredients in the original solution which control the rate or type of corrosion, or from accumulation of products which are corrosion inhibitors or accelerators. An example of the first is related to the great difficulty encountered in reproducing in the laboratory the type and rate of corrosion of almost any metal in natural sea water. The gross composition of sea water is readily reproduced, or natural sea water can be carried to the laboratory; but the delicate balance of oxidizing and reducing agents, of carbon dioxide and oxygen, maintained by the living matter in sea water is difficult or impossible to sustain indoors. Another important example is the corrosion behavior of metals in tap water, which differs so greatly when a continuous flow of fresh water is supplied and when the test is run in a restricted volume.

Copper in solution is a classic example of a corrosion product which accelerates corrosion, as in the corrosion of 70% Ni–30% Cu alloy in aerated hydrochloric acid solution. As little as 0.5 gram per liter of dissolved copper can increase the rate of corrosion about ten times its normal rate. It is interesting to note also that dissolved copper can act as an inhibitor for the corrosion of stainless steel in sulfuric acid.

Effects of the above kinds can usually be detected by changes in corrosion rates with time.

Test Specimens

Sheared edges, and even machined edges, of corrosion specimens are often less resistant to attack than the normal surface. If the ratio of edge area to total area is much larger than would be utilized in practical equipment, as is usually the case, the corrosion rate calculated from the laboratory test may be misleadingly high.

The usual precautions regarding treatment of sheared edges, surface preparation, and selection of specimens apply in total immersion testing as in other types of test.[14] If localized corrosion attack is encountered or expected, small specimens should not be used because the probability of occurrence of this type of corrosion increases

---

* Passed over a sufficient length and area of copper at approximately 450° C to remove oxygen. Carbon dioxide can be removed using soda lime or a liquid air trap. EDITOR.

[13] W. A. Wesley, *Proc. Am. Soc. Testing Materials*, **43**, 649–658 (1943).

[14] *A.S.T.M. Standards*, 1943 Supplement, Part I, pp. 332–341, A.S.T.M. designation B185–43T.

with increase in the area of metal surface tested.[15] Methods for cleaning specimens before and after test are discussed in *Preparation and Cleaning of Specimens*, p. 1078.

Data obtained with annealed metals may not be representative of the corrosion behavior of the same materials after other heat treatments, such as for hardening or tempering, or those associated with welding or forging. The corrosion resistance of welds may vary with the thickness of the section welded. The condition of the surface of the test specimens may not be representative of that commonly supplied. This is particularly important if corrosion is of the pitting type. It is not important if corrosion is uniformly distributed and the proper duration of test has been employed.

Number of Specimens per Test Container. It has been found that the practice of testing duplicate specimens of a material in the same container at the same time occasionally gives misleadingly good agreement between them and may encourage a false estimate of the reliability of the test.[16] This can be avoided by placing duplicate specimens in separate containers or testing them consecutively. In contrast to this, the same reasoning leads to the conclusion that it may be advantageous to immerse two different specimens (not duplicates) in the same container in comparing corrosion rates, provided that the corrosion products of one material have no effect upon the rate of corrosion of the other.

Duration of Test

A constant rate of corrosion is encountered much more often in total immersion than in other types of corrosion. Nevertheless it is not safe to assume that this relationship holds without the supporting evidence of a corrosion time curve. If the test is of too short duration, some materials which build up protective corrosion product films slowly may be ruled out as unsatisfactory. The corrosion of cast iron in many waters is of this type. If a test is too long, the effects of exhaustion of ingredients or accumulation of corrosion products may be pronounced.

Measurement of Corrosion Rate

Methods of measuring the extent of damage of the specimens by corrosion or otherwise determining the rate of corrosion are discussed in the preceding chapter, p. 953.

REPORTING RESULTS

The standard unit for expressing weight loss data is milligrams per square decimeter per day (mdd). If the distribution of corrosion has been uniform, the mdd values are often converted to inches penetration per year (ipy). Average and maximum depths of pits are reported, with a distinction made between pitting on surfaces freely exposed to the testing solution, edge attack, and localized corrosion underneath contacts with supporting devices. It is not advisable to convert pit depths to a unit such as inches penetration per year since the rate of pitting is usually not proportional to time. Pitting should be expressed as maximum depth for the actual test period.

Whenever possible, results should be presented in the form of time curves, whether the rates were measured as weight loss, loss in tensile strength, depth of intergranular

---

[15] R. B. Mears and R. H. Brown, *Ind. Eng. Chem.*, **29**, 1088 (1937).
[16] W. A. Wesley, *Proc. Am. Soc. Testing Materials*, **43**, 649–658 (1943).

attack, depth of pits, or otherwise. It is now generally agreed[17] that reports of laboratory total immersion tests should include the following information:

1. The chemical compositions of the metals and alloys tested.
2. The exact size, shape, and area of the specimens.
3. The fabrication and metallurgical history of the specimens.
4. The surface treatment used to prepare specimens for test.
5. The number of specimens of each material tested, whether each specimen was tested in a separate container, or which specimens were tested in the same container.
6. The chemical composition and volume of the testing solution, and information as to how and to what extent the composition was held constant or how frequently the solution was replaced.
7. The temperature of the testing solution and the maximum variation in temperature during the test.
8. The degree of aeration of the solution in terms of milliliters (ml) of air per liter of solution per minute and the maximum variation in this flow, or similar information for the gas or mixture of gases employed. The type of aerator should also be described.
9. The velocity of relative movement between the test specimens and the solution and a description of how this movement was effected and controlled.
10. The nature of the apparatus used for the test.
11. The duration of the test or of each part of it if made in more than one stage.
12. The method used to clean specimens after exposure and the extent of any error introduced by this treatment.
13. The actual weight losses of the several specimens, depths of pits (plus notes on their size, shape, and distribution, as by sketch), data on mechanical properties before and after exposure if determined, and results of microscopic examination or qualitative bend tests.
14. Corrosion rates for individual specimens calculated in milligrams per square decimeter per day (mdd).

Complete Examination of Specimens

Unless experience has shown it to be unnecessary for the particular combination of metal and corrosive medium being studied, the report should include the reassurance that no damage has occurred to the test specimens other than that shown by the data. Examination of the test specimens after test should have been complete enough to indicate the presence or absence of such effects as dezincification, intergranular corrosion, embrittlement, and stress corrosion. If any such damage is detected, special tests to evaluate the susceptibility of the metal to these types of attack are recommended.

ALTERNATE IMMERSION TESTS

H. L. Burghoff*

The alternate immersion test has long been a favorite method of conducting laboratory corrosion experiments. As the name implies, the test material is cyclically

---

* Chase Brass and Copper Co., Waterbury, Conn.

[17] *A.S.T.M. Standards*, 1943 Supplement, Part I, pp. 332–341, A.S.T.M. Designation B185-43T.

displaced with respect to the corroding solution, so that it is alternately suspended in the solution and in the atmosphere above the solution. The test involves consideration of, and control of, several factors as follows:

Preparation of suitable specimens.
Concentration and volatility of solution.
Volume of solution per unit surface of specimen.
Temperature.
Duration of test.
Alternate immersion time cycle and proportioning of time in solution and time in atmosphere.
Rate of drying of specimens when in atmosphere above solution.
Laboratory atmosphere.
Qualitative and quantitative evaluation of the corrosion experienced.

The test is most commonly used as a means of comparing the corrosion resistance of various metals and alloys in a wide variety of corrosive media. It is also used as a control test to check the quality of successive lots of certain alloys in commercial production and as so used must be very rigidly controlled. Common test procedures have been described,[1,2] and at the present time tentative recommended procedure has been formulated by Committee B-3 of the American Society for Testing Materials.[3]

EXPERIMENTAL ARRANGEMENT

The arrangement of one version of the test as used for copper alloys is shown in Fig. 1. In this the corroding solution is contained in glass museum or battery jars, which stand in a circulating water bath of controlled temperature. The specimens consist of strip tensile specimens prepared in accordance with A. S. T. M. requirements.[4] In each jar there is mounted on glass supports one group of five or six specimens, separated and insulated from each other by glass beads. The group of specimens is suspended by strong nylon fishline, which runs over a glass rod above the jar and thence to a motor-driven crank, which imparts the necessary raising and lowering motion. In this particular test the total cycle of specimen movement is of one minute's duration, and the solution level is so adjusted that the specimens are immersed during 50% of the cycle. As the machine is provided with a rather tightly fitting insulating hood, the humidity in the chamber is high and the specimens remain completely wet at all times.

In another version of the test as applied to copper alloys the specimens are suspended from a rack or frame. The entire frame is alternately lowered and raised by means of a crank, the specimens being immersed in the solution for 1½ min and in the atmosphere for the same time in each cycle. The equipment is not provided with a hood. The specimens at no time become completely dried, although some evaporation obviously occurs.

In the simplest form of the test the corrosive conditions are uniformly distributed

---

[1] D. K. Crampton and N. W. Mitchell, *Alternate-Immersion and Water-Line Tests*, Symposium on Corrosion Testing Procedures, p. 74, Chicago Regional Meeting, American Society for Testing Materials, 1937.

[2] R. B. Mears, C. J. Walton, and G. G. Eldredge, "An Alternate Immersion Test for Aluminum-Copper Alloys," *Proc. American Society for Testing Materials*, **44**, 639 (1944).

[3] "Alternate Immersion Corrosion Test of Non-ferrous Metals, "*A.S.T.M. Standards* I-B (B192-44T), p. 780 (1946).

[4] "Standard Methods of Tension Testing of Metallic Materials (E8-42), *A.S.T.M. Standards*,' Part I, p. 899 (1942).

over the entire specimen, and there is no tendency for strongly localized attack except as peculiarities of some alloys may dictate. The assembly of specimens may readily be modified, however, so that the action of deposit attack may be simulated. This modification is accomplished, as shown in Fig. 2, by placing a short length of heavy cotton cord between the specimens of the group and maintaining this composite assembly in position by means of a metal clip. An oxygen concentration cell is thus developed in the alternate immersion cycle, and the attack may be localized under or immediately adjacent to the cord. This is a good laboratory approach to the problem of estimating corrosion resistance of material as affected by surface deposits in the presence of corrosive media.

FIG. 1. View of an Alternate Immersion Corrosion Testing Machine Used in Investigating Copper Alloys.

Specimens are suspended above the solution in the glass jars from nylon lines attached to the crank arm overhead. Glass jars stand in water bath. Insulating hood is thrown back.

An alternate immersion test has been described[5] (Fig. 3), the purpose of which is to serve as a routine check upon the quality of a certain aluminum-base alloy. In this test, strip specimens are used, some unstressed and some stressed, as simple beams. The test solution consists of a distilled water solution of sodium chloride and hydrogen peroxide and is contained in Pyrex glass breadpans. The alternate immersion is

[5] R. B. Mears, C. J. Walton, and G. G. Eldredge, "An Alternate Immersion Test for Aluminum-Copper Alloys," Proc. American Society for Testing Materials, **44**, 639 (1944).

accomplished by raising and lowering the pans rather than the specimens. In the usual cycle the solution pans are rapidly raised and held in the raised position for 1½ min and then are rapidly lowered and held in this position for 1½ min. Tem-

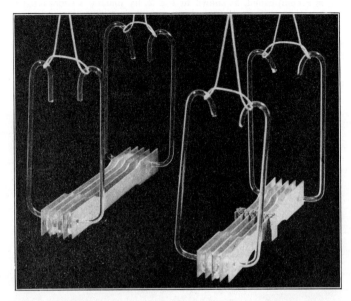

FIG. 2. Typical Arrangements of Strip Tensile Specimens on Glass Supports, Nylon Line Suspension, for Alternate Immersion Testing.

The group at the left is the usual, simple arrangement. In the group at the right, short lengths of cotton cord are held in contact with central portions of specimens in order to simulate deposit attack.

perature is normally controlled at 30° C ± 1° (85° F). The equipment is enclosed in a large insulated cabinet. Each test is run for 48 hours, and the solution is changed after 24 hours. The test has been conducted regularly for several years, and statistically consistent results are reported to be obtained from it. Material of known characteristics is used as a standard of comparison in each run.

SPECIMENS

Specimens are best prepared from strip metal, so that the ratio of the surface to mass is large and that of edge area to total area small. The conditions may be such that the specimens are completely dried between each period of immersion, or they may remain completely wet. In either case there is opportunity for aeration of the liquid film adhering to the specimen during the period when it is out of the solution proper. It is desirable to have as high a ratio of quantity of solution to area of metal surface as is possible. The tentative specification of the American Society for Testing Materials[3] states that about 250 ml per sq in (40 ml per sq cm) of specimen area is desired, but that as little as 20 ml per sq in. (3 ml per sq cm) may be sufficient. The solution should be completely changed at reasonable intervals during the testing in order to minimize the effect of contamination by corrosion products.

The value of any test is naturally very greatly enhanced if the service behavior of at least one of the test materials is known. Furthermore, in running successive tests under the same conditions for a wide variety of alloys, it is advisable to include in each test at least one or two materials that can serve as standards of comparison.

NUMBER OF SPECIMENS

Five specimens form a satisfactory number for an alternate immersion test. These are best scattered among five jars of solution along with the specimens of other materials that are being compared, rather than placed as a single group in one jar.*

FIG. 3. An Arrangement for Alternate Immersion Testing of Al-Cu Alloys. (R. B. Mears, C. J. Walton, and G. G. Eldredge.)

The comparative results appear to be more certain when this plan is followed. Of course there may be instances where a material is so badly corroded or is so markedly different that its corrosion product will unduly affect the solution and its action upon other test materials. In such an event it is well to segregate specimens of the offending material.

TEMPERATURE

Attention must be given to testing temperature, temperature control, and distribution of specimens. The lowest controllable temperatures commonly used are 30° to 35° C (85° to 95° F), and 60° C (140° F) is a convenient elevated temperature. At

---

* This practice is also followed in total immersion testing. See p. 964. EDITOR.

this latter temperature, loss of water by evaporation of the test solution is within reasonable bounds and may be offset by addition of distilled water once or twice daily. Temperature is desirably and readily controlled to $\pm 1°$ C.

CORROSIVE MEDIA

The solutions that may be used are almost unlimited. As sea water is perhaps the most widely encountered highly corrosive medium, tests using it or, more often, solutions of sodium chloride, of sea salt, or of several of the salts found in sea water are frequently conducted. Acids, bases, and salts may be used in various concentrations. The important point is to choose the medium, its concentration, and its temperature in such a way as to approach as closely as possible the actual service condition for which information is sought and thereby develop the attack that is most likely to be suffered in service.

DURATION OF TEST

Duration of test is an important consideration. The value of having more than one time period is unquestioned. If, as sometimes happens, it is impossible to arrange this, experience is helpful in determining a proper period. For the testing of copper alloys in sea water or dilute sodium chloride at 60° C (140° F), 100 days will provide a good comparison, based on loss of tensile strength of strip specimens which are 0.032 in. (0.081 cm) in original thickness. A shorter time would suffice in a more corrosive medium, for example 30 days in dilute (1 to 5%) sulfuric acid at 30° C (85° F).

EVALUATING CORROSION

The manner and extent of corrosion may be evaluated in several ways, none of which is peculiar to this type of test. Weight loss, depth of pitting, loss of tensile strength, and microscopic observation may be applied individually or in combination. Simple bending may also reveal intercrystalline corrosion or so-called selective attack such as dezincification.

SALT-SPRAY TEST

WILLIAM BLUM* AND LEO J. WALDRON†

PURPOSE AND FUNCTIONS

The demand for an accelerated corrosion test and the frequent occurrence of corrosion in marine environments no doubt account for the original proposal[1] of the salt-spray test, in which the articles are subjected to a fine spray or fog of a salt solution. Extensive experience has shown that although this test yields results somewhat similar to those in marine exposure, it does not reproduce all the factors that may be involved in marine service, much less under widely different conditions of exposure. The salt-spray test should therefore be considered as an arbitrary per-

* National Bureau of Standards, Washington, D. C.
† Naval Research Laboratory, Washington, D. C.

[1] J. A. Capp, "A Rational Test for Metal Protective Coatings," *Proc. Am. Soc. Testing Materials*, **14** (II), 474 (1914).

formance test for metals, with or without protective coatings. The validity of this test for a given metal or coating depends upon the extent to which a correlation has been established between the results of the test and the behavior under expected conditions of service.[2]

PRINCIPLES

As usually specified, the test involves exposure of the specimens to a fine spray or mist of a solution of sodium chloride at a specified temperature. The fog particles settle upon the test surfaces (which are preferably inclined) and thereby constantly replenish and replace the film of solution on the surface. The extent and nature of the corrosion of the metal or coating after a specified period of exposure serve as a measure of quality.

In order to accelerate corrosion it is advantageous to increase the temperature or, with less effect, the concentration (or activity) of the corroding medium. The customary temperature of 35° C (95° F) is not much above the temperature prevailing in many outdoor exposures. The use of a 20% solution of sodium chloride does not accelerate corrosion much above that with a 3% solution (e.g., sea water).

It is therefore probable that much of the acceleration of corrosion in the salt spray results from the replenishment of oxygen in the film of solution on the metal surface. If that film were stationary, the portion immediately adjacent to the metal (e.g., zinc) would become depleted in oxygen if fairly rapid initial corrosion occurred. Continued corrosion would then depend upon the rate at which oxygen from the surrounding air diffused through the film to the metal surface. If, however, as occurs in the operation of the salt spray, the fog particles settle continually on the surface and thereby replace the existent film, the oxygen content throughout the film will approach more closely that of a solution saturated with respect to the air.

Approximate computations show that the amount of oxygen dissolved in the volume of salt solution condensed is less than is required for the observed rate of corrosion of zinc. If appreciable amounts of oxygen were adsorbed on the surface of the fog particles in addition to that which is dissolved in the liquid of the droplets, condensation of the fog would result in a solution that is momentarily supersaturated with oxygen, whereby corrosion might be still further accelerated.

This hypothesis is supported by the fact that oxygen reduces the surface tension of water and hence is adsorbed on the liquid particles. Because the ratio of total surface of particles to their volume increases as the particle size decreases, the adsorption of gases on small particles is relatively large. However, approximate computations indicate that on fog particles one micron in diameter, the amount of oxygen adsorbed on the surface is only a small percentage of that dissolved in the droplet. It is doubtful whether this small increase in the concentration of oxygen on fog particles will account for the accelerated corrosion observed in the salt spray or in natural fogs.

APPARATUS

Chamber

Although it is not necessary to specify the exact apparatus to be employed, certain features are essential. The specimens are exposed to the salt spray in a closed chamber, constructed of any corrosion-resistant material, such as albarene stone, slate, pottery,

---

[2] C. H. Sample, "Use and Misuse of the Salt-Spray Test as Applied to Electrodeposited Metallic Finishes," American Society for Testing Materials, *Bull.* 123, p. 19 (August, 1943); discussion, p. 22 (August, 1943).

glass, wood, or metal covered with rubber or similar material. The door or cover is sealed with a gasket or water seal to prevent leakage of fog. According to the size and number of test specimens, the boxes may vary in volume from 0.3 to 23 cu meters (10 to 800 cu ft), those of the largest size being small rooms. One purpose of recent specifications[3,4] is to insure similar results regardless of the size or design of the boxes. A typical small box is shown in Fig. 1.

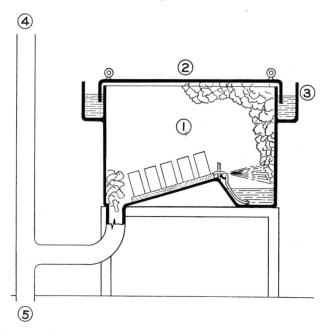

FIG. 1. Diagrammatic Sketch of Salt-Spray Box Used at National Bureau of Standards. 1. Exposure chamber. 2. Removable cover. 3. Water seal. 4. Vent. 5. Sewer outlet.

SOLUTION RESERVOIR

A few days' supply of the salt solution is usually kept within the box in a compartment, thus preventing contamination by the drip of solution that has condensed on the specimens. Occasionally an external container, such as a carboy, is used to supply solution to the nozzles.

NOZZLES

The spray or fog is generally produced by blowing humidified air through an atomizer so that the solution is converted into a stream of fine particles. The nozzles may be made of glass, hard rubber, plastic, 70% Ni–30% Cu alloy,* or other corrosion-

* Monel.

[3] "Tentative Method of Salt Spray (Fog) Testing," *A.S.T.M. Standards*, B117–44T, Part IB, p. 773 (1946).

[4] Federal Specification TT–P–141a, "Paint, Varnish, Lacquer, and Related Materials; General Specification for Sampling and Test Methods."

resistant metals. Sketches of typical nozzles are shown in Fig. 2. As far as known, no critical study has been made of the effects of design of the nozzles on the amount or size of the fog particles.

BAFFLES

If the spray as produced were allowed to impinge directly on the specimens, marked localized effects in the corrosion might result from convection currents or from the presence of relatively coarse and more rapidly settling particles in the spray. Greater uniformity is obtained by directing the spray against baffles, on which the larger particles condense. In Fig. 1 the end of the box serves as a baffle and at the same time returns to the reservoir about 75% of the sprayed liquid, in an uncontaminated state.

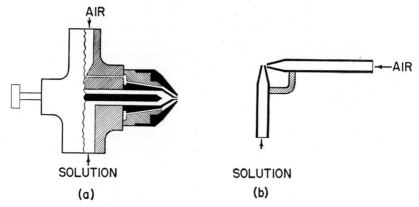

FIG. 2. Types of Nozzles Used for the Salt Spray.

(a) High-pressure Monel metal nozzle (10–25 lb air pressure). (b) Low-pressure glass nozzle (10–15 lb air pressure).

EXHAUST AND DRAIN

It is necessary to remove from the box practically the total volume of air introduced plus the solution that has settled on the specimens or within the test portion of the box. This may be done readily as in Fig. 1 by passing the air and spent solution into a pipe connected both with the sewer and with a ventilating stack. The exhaust gases should be carried away from the building because they contain corrosive fog.

OPERATION

During the past few years efforts[5,6,7] have been made to obtain more reproducible results. The definition and range of each factor in the proposed specifications[8,9,10]

[5] "Summary of Important Factors in Salt Spray Testing," *Circ.* 670, National Paint, Varnish, and Lacquer Association, Scientific Section, February, 1944.
[6] E. H. Dix, Jr., and J. J. Bowman, "Salt Spray Testing," *Symposium on Corrosion Testing Procedures*, Chicago Regional Meeting, American Society for Testing Materials, p. 57 (1937).
[7] V. M. Darsey, "The Salt Spray Test," American Society for Testing Materials, *Bull.* 128, p. 31, May, 1944.
[8] "Salt Spray Testing of Non-Ferrous Metals," *A. S. T. M. Standards*, B117-44T, Part I, p. 1843 (1944).
[9] Federal Specification TT-P-141a, "Paint, Varnish, Lacquer, and Related Materials; General Specification for Sampling and Test Methods."
[10] Federal Specification QQ-M-151a, "Metals; General Specification for Inspection of."

SALT SOLUTION

The test is now most generally applied with a 20% solution of sodium chloride, although 3, 5, or 10% solutions are sometimes used. Available evidence indicates that the salt concentration is not an important variable, but that with some metals the 20% solution is somewhat more corrosive. The oxygen solubility of the 20% solution is less affected by temperature than a solution of lower salt concentration.[11] One objection to the higher concentration of salt is a greater tendency to clog the nozzles, but this is avoided if the air is properly humidified.

Chemically pure salt is desirable but expensive so that "dairy" salt, containing at least 99.8% NaCl and not more than 0.1% NaI, is often used. Distilled water is preferable, but natural water containing not more than 200 ppm of total solids is acceptable. The solution should be neutral, i.e., the pH should be between 6.5 and 7.2, measured electrometrically or colorimetrically, and adjusted by dilute HCl or NaOH.

In the effort to simulate marine exposure, natural or "synthetic" sea water has sometimes been used.* It is probable that its content of other salts, especially of magnesium chloride, makes the sea water somewhat more corrosive than pure salt. Heussner[11] reported that sea water was more corrosive than 20% NaCl solution on steel plated with nickel and chromium at 35° C (95° F) but about equal at 22° C (72° F). Since, as mentioned, the salt-spray test does not usually approximate actual marine exposure, the extra trouble and expense in using sea water are hardly justified.

TEMPERATURE

Because corrosion, like other chemical reactions, is markedly affected by temperature, operation at a constant temperature is essential for reproducible results. "Room temperature" varies widely at different times and places, so that temperature control is necessary. Because heating is more easily applied and controlled than cooling, an elevated temperature is preferable, and it also accelerates the corrosion. Practically, it has been found that a temperature of 35° C (95° F) can be readily maintained, and it is now specified[12,13,14] with a permissible range within the box from 33.5° to 36° C (92° to 97° F).

Temperature control may be accomplished by (1) placing the apparatus in a constant temperature room, e.g., at 37° C (99° F); (2) applying heat through a surrounding water jacket; or (3) heating the incoming air. A fourth method, use of a heater immersed in the reservoir, is undesirable because it leads to large local variations in temperature and humidity.

The temperature of the air within the box will tend to be lower than the surroundings because of (1) cooling of the air by expansion (the Joule-Thompson effect), which is usually not greater than about 0.2° C (0.4° F); and (2) cooling by evaporation of water. The second effect is small if, as is advisable, the air delivered to the nozzles is nearly saturated with water. The temperature should be measured and controlled in the portion of the box where the test samples are placed.

---

* Recently a formula for synthetic ocean water, based on all available data, was published. (T. P. May and C. E. Black, "Synthetic Ocean Water," *Naval Research Laboratory Report* P-2909, August, 1946.)

[11] C. E. Heussner, "Comparison of Salt Spray and Ocean Spray Testing," Educational Sessions, *Proc Am. Electroplaters' Soc.*, pp. 75-88 (June 8, 1942).

[12] "Tentative Method of Salt Spray (Fog) Testing," *A. S. T. M. Standards*, B117-44T, Part IB, p. 773 (1946).

[13] Federal Specification TT-P-141a, "Paint, Varnish, Lacquer, and Related Materials; General Specification for Sampling and Test Methods."

[14] Federal Specification QQ-M-151a, "Metals; General Specification for Inspection of."

The Air

The air used for producing the spray should be free from dust, oil, an excessive amount of liquid-water particles, and any foreign gases. It should contain sufficient water vapor to be approximately in equilibrium with the vapor pressure of the salt solution at that temperature.

The removal of impurities from the large volumes of compressed air employed requires efficient filters that must be checked and cleaned at intervals. Passage through water towers will remove such harmful gases as sulfur dioxide, and will humidify the air.

To be in equilibrium with a 20% NaCl solution at 35° C (95° F), the air (at atmospheric pressure) should have a relative humidity of 84% (with respect to pure water). This humidification of the air is rendered difficult by the fact that the saturation with water must take place while the air is compressed, but must be adequate when the air is released to atmospheric pressure.[15] If, e.g., the air has 100% humidity at 35° C (95° F) and a pressure of 0.8 kg/sq cm (12 psi) above atmospheric, when it is expanded to atmospheric pressure at 35° C it will have a relative humidity of only about 55%. In order to introduce sufficient water vapor into the air at 0.8 kg/sq cm (12 psi) to yield 84% humidity at 35° C and atmospheric pressure, it is necessary to preheat the compressed air (or the water) to about 43° C (110° F). Although this preheating is advantageous, it is not always necessary, because the air that passes through water towers usually carries some liquid-water particles (fog) with it. When the air is saturated at the same temperature as the room, 35° to 37° C (95° to 99° F), it may even carry over water sufficient to slightly dilute the salt solution.

Atomization

Method. Use of the nozzles described above leads to the formation of a finely divided fog of salt solution, which moves through the box from the baffles to the exit at an average velocity determined by the rate at which air is introduced and the cross section of the box. Although much work has been done recently upon the character and concentration of the fog, these results have not been directly correlated with the design or operation of the nozzles, or with the number required for a box of given size. On general principles it is an advantage to use at least two nozzles in a box, as then the failure of one nozzle will not completely interrupt operations.

Fog Characteristics. Although it is known that the fog particles must be extremely small to be carried along by slowly moving air, no measurements of their size have been reported. Computations based on the amount of liquid in a liter of fog and the rate of settling indicate that the particles are of the order of 5 microns in diameter. Observations in a box with a volume of about 12 cu ft (0.35 cu meter), operated to meet the present specifications indicate that each liter of fog contains a total of about 0.1 ml of salt solution, or 20 mg of NaCl.

The concentration of salt in the fog particles depends not only on the concentration of the solution in the reservoir but also on the humidity of the air delivered to the nozzles. If this is less than 84% at 35° C (95° F), the fog particles from a 20% salt solution will rapidly lose water and become more concentrated, and vice versa. The most serious objections to an increase in the concentration of the fog are (1) tendency to clog the nozzles and (2) tendency to form "dry" fog, which does not wet the

---

[15] V. M. Darsey, "The Salt Spray Test," American Society for Testing Materials, *Bull.* 128, p. 31, May, 1944.

specimens (possibly because the outer surface of each particle may then consist of solid NaCl). To avoid either concentration or dilution of the fog, the specifications provide that the liquid condensed from the fog must have a salt content between 18 and 22%, i.e., a sp. gr. of 1.126 to 1.157 when measured at a temperature of 33.5° to 36° C (92° to 97° F).

As suggested above, the most important factor in the corrosion is the rate at which the fog settles out on the specimens. Darsey[16] recognized this factor and devised an ingenious method for collecting the condensed liquid, the volume and concentration of which can then be measured. He employed a circular glass funnel, 10 cm (4 in.) in diameter, into which the fog settled and then drained into a graduated cylinder. A still simpler form employs a circular crystallizing dish, of any convenient diameter, which may be placed in any part of the box.

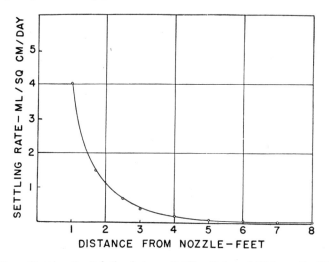

Fig. 3. Curve Showing the Relation between Settling Rate and Distance the Mist Travels from a Glass Nozzle, 10 to 15 lb Air Pressure.

With this device many observations have recently been made on the rate of settling in salt-spray boxes of different types and sizes. On this basis, tentative limits have been set of 0.5 to 3 ml of solution per hour condensed on a horizontal circular surface 10 cm (4 in.) in diameter (about 80 sq cm or 12 sq in.). These limits can be more conveniently expressed as 0.15 to 0.9 ml/sq cm/day, in order to make use of collectors of any size. The corresponding depths of condensed liquid, in any container with parallel sides, are 0.15 to 0.9 cm in 24 hours.

The rate of settling of fog at a given point is determined by (1) the average size of the fog particles, (2) the number present in unit volume, (3) the distance from the source, and (4) their horizontal velocity. In Fig. 3 it is shown that 60 to 80% of the liquid in the fog condenses within 1 meter (3 ft) of the source, under the conditions used in this box.

The effect of the rate of settling upon the rate of corrosion of zinc is shown in Fig. 4. This confirms for zinc the belief that at least 0.5 ml/80 sq cm/hour (0.15

---

[16] V. M. Darsey, "The Salt Spray Test," American Society for Testing Materials, *Bull.* 128, p. 31, May, 1944.

ml/sq cm/day) is necessary for satisfactory operation of the salt spray, but indicates no sharply defined upper limit. Further studies with other metals are desirable.

SPECIMENS

Size and Shape. The articles may either be actual devices, large or small, to be subjected to acceptance tests, or specimens prepared for the purpose. In the latter case, several relatively small flat pieces, e.g., 10 × 15 cm (4 × 6 in.) are desirable, with which duplicate results are obtainable. When tensile tests are to be made, the size and shape of the samples are usually specified.

Preparation. Unless specimens are to be tested as received, it is desirable to clean the surface to remove any grease or foreign materials that might affect the corrosion. Metal surfaces should be cleaned by organic solvents, soap and water, or magnesium

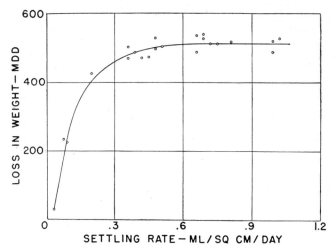

FIG. 4. Relation between Settling Rate of the Mist and Loss in weight of Sheet Zinc Exposed for 24 Hours to Salt Spray.

oxide and water, until the surface is completely wet by water, i.e., there is no "water-break." Paints and other organic films may be wiped with clean dry cloth or paper.

Mounting. Because corrosion is accelerated by the settling of the fog, it is desirable to have the principal surfaces inclined so as to have a horizontal component and to permit drainage and fairly close spacing without masking each other. Flat specimens are conveniently held at about 15° from vertical and parallel to the principal direction of flow of the fog.* Vertical suspension, formerly used, results in less severe and less uniform corrosion. The specimens should be supported by non-metallic racks or hooks, made e.g., of wood, glass, or waxed string. They should not be in contact with each other or with the box, and should not permit solution condensed on one specimen to drip on another.

* In general, when specimens are used solely for the salt-spray tests, the condition of the under surface is disregarded in rating the results, since usually very little change occurs there. However, if the actual behavior of such material in service is involved, any changes in the condition of the under surface should be reported. In extreme cases, it may be desirable to wax or paint the lower surface, but usually this is not done if the two surfaces are in essentially the same condition at the beginning of the test.

Inspection. The boxes are generally operated continuously, and inspections are made at specified intervals, first shorter, and then intervals of 24 hours after the first day. At the conclusion of the period specified for the test, the specimens are removed, rinsed in warm water, and immediately dried. In specific cases, corrosion products may be brushed off to reveal the condition of the underlying metal.

RESULTS

Expression. The type of inspection will depend upon the materials involved, but usually includes qualitative or semi-quantitative observations on the character and extent of corrosion or other surface changes. The change in tensile strength or elongation may serve as a measure of the corrosion. In any case, the results are usually expressed (1) as the relative behavior of two or more materials, especially in research work, or (2) as to whether the material meets the pertinent specification.

Interpretation. In the final analysis, the results of the salt-spray test merely represent the behavior of the material under those conditions. For specification purposes, this performance test is justifiable if research and experience have shown that there is at least an approximate relation between this test and the expected conditions of service.

This test is also of value in research, especially to compare the behavior of similar coatings or materials, but not of widely different materials. Comparisons of similar products do not justify predictions of their actual life, especially under widely different conditions, but they may indicate the relative life of the articles, especially under marine conditions.

GENERAL REFERENCES

"Electrodeposited Coatings of Nickel and Chromium on Steel," *A. S. T. M. Standards*, A166-41T, Part I, p. 1455 (1942).

FINN, A. N., "Method of Making the Salt Spray Corrosion Test," *Proc. Am. Soc. Testing Materials*, **18** (I), 237 (1918).

GROESBECK, E. C., and W. A. TUCKER, "Accelerated Laboratory Corrosion Test Methods for Zinc-Coated Steel," *J. Research Natl. Bur. Standards*, **21**, 272 (1928); RP-10.

MUTCHLER, W. H., R. W. BUZZARD, and P. W. C. STRAUSSER, "Salt Spray Test," *Letter Circ.* 530, National Bureau of Standards, July 1, 1938.

RAWDON, H. S., A. I. KRYNITSKY, and W. H. FINKELDEY, "Types of Apparatus Used in Testing the Corrodibility of Metals," *Proc. Am. Soc. Testing Materials*, **24** (II), 717 (1924).

Report of Subcommittee III of Committee B-8, *Proc. Am. Soc. Testing Materials*, **43**, 187 (1943).

STRAUSSER, P. W. C., A. BRENNER, and W. BLUM, "Accelerated Tests of Nickel and Chromium Plating on Steel," *J. Research Natl. Bur. Standards*, **13**, 519 (1934); RP-724.

WALDRON, L. J., "Basic Requirements in the Standardization of the Salt Spray Corrosion Test," *Proc. Am. Soc. Testing Materials*, **44**, 654 (1944).

POTENTIAL MEASUREMENTS AND ELECTROCHEMICAL TESTS

H. E. HARING*

The electrochemical theory of corrosion is relatively new and far from complete, although remarkable progress has been achieved in its development within the past years. In consequence, electrochemical techniques based on this theory are of still more recent origin. Most of these techniques are in a state of flux and development and are not sufficiently fixed to be recognized as standard procedures. In other words, they are methods for study rather than methods for test. For this reason,

* Bell Telephone Laboratories, Murray Hill, N. J.

in this chapter emphasis will be placed upon the scope, status, and future possibilities of electrochemical methods for corrosion testing rather than upon a detailed description of testing procedures, which in many cases have been used only by the original investigators. For a detailed description of these procedures, the original sources should be consulted.

DIRECT AND INDIRECT METHODS

For purposes of discussion, electrochemical methods for corrosion testing may be divided into two classes, which will be termed "direct" and "indirect." A direct method may be defined as one based upon the electrical expression for the corrosion reaction, viz., Ohm's Law, $I = E/R$. In this case I is the current flowing between the anodes and cathodes that exist on the surface of a corroding metal, and therefore by virtue of Faraday's Law is a direct measure of the rate of corrosion; E is the difference between the working potentials of these anodes and cathodes; and R is the resistance of the electrolyte between them. An indirect electrochemical method may be defined as any electrochemical method not based directly upon the laws of Ohm and Faraday.

Theoretically I may be *measured* directly or *calculated* from a knowledge of E and R. However, it usually is impracticable, in the light of present knowledge, to isolate the anodes and cathodes and subject them to measurement. Nevertheless, the resultant of anodic and cathodic potentials that exist upon the surface of a metal in contact with a corrosive environment can be measured and plotted against time. For a detailed discussion of these potentials, their effect on corrosion, and the manner in which they combine to produce an overall resultant and measurable potential, the reader is urged to consult Evans,[1] and Burns and Schuh,[2] as well as *Fundamental Behavior of Galvanic Couples*, p. 481.

If it were possible to measure I, the corrosion current, as the corrosion reaction proceeds, there would be little need for any other type of corrosion test, except those that provide information on corrosion distribution, because according to Faraday's Law the amount of current that flows in a given time is a quantitative measure of the amount of metal that has corroded in that time. Unfortunately the measurement of this current has been possible to date in only a limited number of tests that permit isolation of the anodic and cathodic areas.[3] The corrosion current usually can be determined more readily by measuring the quantity of hydrogen discharged (or oxygen consumed) by the corrosion reaction or the actual amount of metal that has reacted, e.g., by means of the polarograph.

THE POLAROGRAPH

The *polarograph*[4] is an electrochemical device admirably suited to the determination of minute quantities of dissolved metals. The very definite value of the instrument for this purpose is demonstrated by Fig. 1. This is a graphic summary of a study of the rates of solution of two grades of lead, known to the trade as

---

[1] U. R. Evans, *Metallic Corrosion, Passivity and Protection*, 2nd Ed., Edward Arnold and Co., London, 1937.

[2] R. M. Burns and A. E. Schuh, *Protective Coatings for Metals*, Reinhold Publishing Corp., New York, 1939.

[3] R. H. Brown and R. B. Mears, *Trans. Electrochem. Soc.*, **74**, 495 (1938). R. H. Brown and R. B. Mears, *Trans. Faraday Soc.*, **35**, 467 (1939). L. J. Benson, R. H. Brown, and R. B. Mears, *Trans. Electrochem Soc.*, **76**, 259 (1939).

[4] I. M. Kolthoff and J. J. Lingane, *Polarography*, Interscience Publishers, Inc., New York, 1941.

(1) "chemical" lead and (2) "corroding" lead, in distilled water at room temperature.[5] The results of duplicate runs on a standard lead solution are shown in the graph at the top of Fig. 1. The length of the current "step" is a measure of the lead content. It will be noted that the chemical lead dissolves more rapidly than the corroding lead for the first hour or two (as indicated by the current), but that thereafter they behave similarly.

It is possible to determine concentrations of two or more ions in the same measurement with the polarograph, provided that their deposition potentials on mercury differ by 0.2 volt or more. This fact suggests the use of the instrument for determining the corrosion rates of the constituents of complex alloys. The cathode ray oscillograph has been used in conjunction with the dropping-mercury cathode to provide similar results much more rapidly.[6] This is but one of many types of problems in the field of corrosion research to which the oscillograph might be applied to advantage.

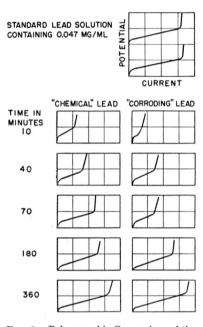

FIG. 1. Polarographic Comparison of the Rates of Solution of Two Grades of Lead in Distilled Water at Room Temperature.

POTENTIAL-TIME CURVES

Potential-time curves, so called, frequently have been used to determine whether a metal is active or passive. This possibility is illustrated by Fig. 2, in which the potential-time behavior of iron immersed in water and a variety of water solutions is compared with the time-potential behavior of 18-8 chromium-nickel stainless steel immersed in tap water. It will be observed that in sulfuric acid, which is known to corrode iron, the potential rapidly becomes strongly electronegative (anodic) with time. In water, the potential of iron first becomes more electropositive (cathodic), thus suggesting the presence of an air-formed film, and then moves in the electronegative (anodic) direction, indicating that this film is breaking down and that active corrosion is taking place. Careful inspection of the specimen after the test confirms this theory. Stainless steel, which does not corrode appreciably in tap water, is strongly electropositive (cathodic) when immersed in it. As might be anticipated, iron immersed in a chromate solution behaves similarly.

The potential-time method for determining whether a metal is or is not passive has been applied to a sufficiently wide variety of metals and environments to be considered by some investigators as a reasonably reliable method for test. Others question its value for this purpose because in certain instances they have observed corrosion notwithstanding the fact that the potential-time curves indicated passivity. Probably this apparent discrepancy is due to the fact that the potential measured was the

[5] R. M. Burns, *J. Applied Phys.*, **8**, 398 (1937).

[6] L. A. Matheson and N. Nichols, *Trans. Electrochem. Soc.*, **73**, 193 (1938). R. H. Müller, R. L. Garman, M. E. Droz, and J. Petras, *Ind. Eng. Chem.*, Anal. Ed., **10**, 339 (1938).

potential of the entire electrode, while the corrosion observed was the corrosion of isolated areas. Obviously the potential-time and visual methods for detecting corrosion should be made equally discriminating if the results obtained are to be compared. If an isolated or pitting type of corrosive attack is anticipated, the potential-time measurement should be made upon a large number of small areas and not upon one large area.

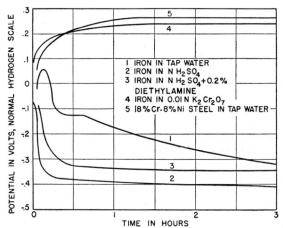

FIG. 2. Potentials of Iron and Stainless Steel in Water and Aqueous Solutions.

It should be emphasized that potential-time curves provide little if any information on the rate of corrosion; they should be regarded primarily as a means for determining corrosion tendency, as is indicated in Fig. 2 by curve 3 for iron immersed in sulfuric acid containing diethylamine, an inhibitor. The corrosion rate is known to be markedly reduced by the inhibitor, but this fact is not reflected by the curve.

The use of potential-time curves is not restricted to metals in direct contact with an aqueous environment. It has been demonstrated (refer to *Testing of Organic Coatings*, p. 1023) that the corrosion behavior of painted iron can be studied[7] by using a vacuum tube electrometer[8] in conjunction with the usual reference half cell and a recording potentiometer. Much valuable information concerning the protective value of paints and the nature of their protective mechanism can be obtained in this manner, and it is anticipated that relatively "foolproof" methods for electrochemically testing paints eventually will be developed along these lines.

RESISTANCE MEASUREMENTS

Measurements of the resistance of the corrosive environment have not been widely employed in corrosion testing. However, such measurements have served a very useful purpose in estimating the corrosivity of soils (p. 1038), ground water, and other dilute electrolytes particularly when used in conjunction with a knowledge of the nature of the anions present and the solubility of the corrosion product that they form with

[7] R. M. Burns and H. E. Haring, *Trans. Electrochem. Soc.*, **69**, 169 (1936). H. E. Haring and R. B. Gibney, *Trans. Electrochem. Soc.*, **76**, 287 (1939).

[8] K. G. Compton and H. E. Haring, *Trans. Electrochem. Soc.*, **62**, 345 (1932). D. B. Penick, *Rev. Sci. Instruments*, **6**, 115 (1935).

the metal in question.[9] The indications are that this type of measurement should be employed much more extensively, particularly for those types of corrosion that are known to be "resistance controlled."

SURFACE ANALYSIS

The preceding discussion indicates that the so-called direct methods for corrosion testing are not well developed and that the indirect methods are the most useful at the present time. Probably the primary reason for the undeveloped state of the direct methods is a lack of knowledge of the physical and chemical nature of metal surfaces and of their electrochemical behavior in various electrolytes. Actually such knowledge frequently is all that is required to permit the prediction of a corrosion result or the explanation of a corrosion phenomenon. The remainder of the chapter will be devoted to a discussion of methods which may be employed to obtain this basic information.

Well-known electrochemical methods that have been used for qualitatively analyzing metal surfaces are the ferroxyl test (p. 1033) and a wide variety of other so-called electrographic methods.[10] Such methods, which are based upon the chemical reactions and corresponding color changes that can be made to take place at metal surfaces as the result of an inherent galvanic potential or an externally applied electromotive force, are proving to be an effective method for mapping metal surfaces, particularly if the dissimilar areas are macroscopic in size.

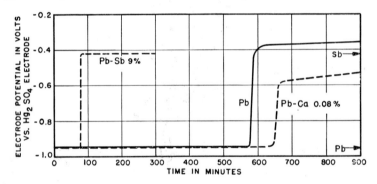

Fig. 3. Typical Sulfation Curves for Lead, Pb-Sb, and Pb-Ca in 7 N H_2SO_4.

Other electrochemical methods with distinct quantitative possibilities are available. These methods are based upon observation of the electrochemical behavior of a metal as induced either by an inherent galvanic couple or by an externally applied electromotive force. They differ from electrographic methods in that they are electrometric in character rather than colorimetric, and quantitative rather than qualitative. For example, when a Pb-Sb alloy is immersed in sulfuric acid, a potential-time or "sulfation" curve such as that shown in Fig. 3 is obtained.[11] Initially the potential measured is practically that of lead, but within a short time, because of couple action between the portions of the surface that are essentially lead (the anodic areas) and the portions that are essentially antimony (the cathodic areas),

[9] R. M. Burns, *Bell System Tech. J.*, **15**, 603 (1936).
[10] H. W. Hermance, *Bell Lab. Record*, **18**, 269 (1940). M. S. Hunter, J. R. Churchill, and R. B. Mears, *Metal Progress*, **42**, 1070 (1942).
[11] H. E. Haring and U. B. Thomas, *Trans. Electrochem Soc.*, **68**, 293 (1935).

the anodic areas become coated with an insulating film of lead sulfate and the potential changes abruptly to that of the residual antimony areas. The time required for this change to take place is a measure of the antimony content of the surface, as

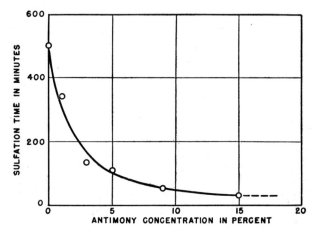

FIG. 4. Effect of Antimony Concentration on the Sulfation Time of Pb-Sb Alloys.

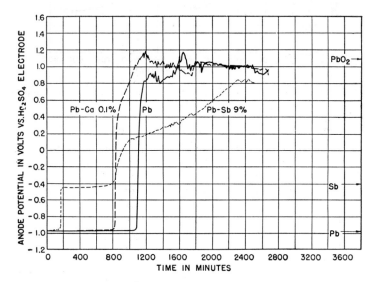

FIG. 5. Anode Potential-Time Curves for Lead, Pb-Sb, and Pb-Ca in 7 N H_2SO_4 at a Current Density of 0.001 amp per sq dm.

is shown by Fig. 4, in which the time required for sulfation, i.e., insulation of the anodic areas by lead sulfate, is plotted against the antimony content of the alloy.[11]

Similar results are obtained if the Pb-Sb alloy is made anode in a sulfuric acid solution.[11] These are shown by Fig. 5, in which are plotted the anode potential-time curves for lead, Pb-Sb, and Pb-Ca alloys in 7 N H_2SO_4 at a current density of 0.001 amp/sq dm. It will be noted that there are three steps in the curve for Pb-Sb alloy, due to (1) the sulfation of the lead areas, (2) the solution of the antimony areas, and (3) the oxidation of the sulfate-coated areas to lead peroxide. The potential of the step identifies the reaction, and its duration at a given current density indicates the quantity of reactant. It is conceivable that from such data obtained in properly selected electrolytes the anodic and cathodic areas on the surfaces of a variety of alloys could be identified and their total areas could be computed.

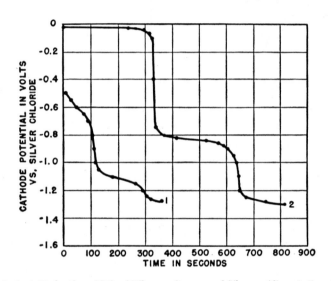

Fig. 6. Cathode Reduction of Mixed Films on Copper and Silver. (Campbell and Thomas.)

Film Composition	Current Density	Film Thickness
1. Cuprous oxide–cuprous sulfide.	0.05 milliamp/sq cm	69 Å — Cu_2O 130 Å — Cu_2S
2. Silver oxide–silver sulfide.	0.09 milliamp/sq cm	470 Å — Ag_2O 480 Å — Ag_2S

This method may be used in reverse as an electrochemical method for qualitatively and quantitatively analyzing tarnish films on metal surfaces.[12,13] The tarnished specimen is made cathode in a suitable electrolyte at a constant current density, and the resulting cathode potential is plotted against time. The potential at which reduction takes place identifies the compound being reduced, and by virtue of Faraday's Law the duration of reduction at a given current density indicates the quantity present. The possibilities of this method of analysis are illustrated by Fig. 6, in which are plotted the cathode reduction curves for mixed films of (1) cuprous oxide and cuprous sulfide on copper and (2) silver oxide and silver sulfide on silver.[13] This method has been limited to the reductions that may be accomplished in aqueous solution at potentials more cathodic than the discharge potential for hydrogen, but the indications are that by resorting to a non-aqueous electrolyte it may be extended

[12] U. R. Evans and H. A. Miley, *Nature*, **139**, 283 (1937).
[13] W. E. Campbell and U. B. Thomas, *Trans. Electrochem. Soc.*, **76**, 303 (1939).

to include the more anodic reductions as well.[14] If it can be so extended, analysis of the tarnish or corrosion films on any metal may be within its scope.

CURRENT DENSITY CURVES

Current density-time curves resulting from the application of a constant voltage may sometimes be employed advantageously. For example, the surface of aluminum and some of its alloys can be "analyzed" by anodizing in a suitable film-forming electrolyte at a fixed potential. Initially the current that flows is high, but it rapidly decreases and in a short time becomes constant at a "leakage current," which is a measure of the soluble constituents in the surface of the alloy as well as of the porosity of the anodized film. The fact that leakage current is an excellent measure of porosity suggests a method that might be applied in studies of the porosity of those metallic coatings which can be filmed over and insulated in a suitable electrolyte. Lead and aluminum are obvious examples of such metals.

Passive films on ferrous metals have been studied by chemical titration with a chloride solution,[15] and also by anodic treatment in halide solutions.[16] A critical survey of these methods has been published.[17]

Potential-current density curves have been used in corrosion studies as a means for determining the electrochemical behavior of a number of metals when anodic and also when cathodic.[18] Anode potential-current density measurements provide information on the film-forming and film-breakdown characteristics of metals in various electrolytes;[16] and cathode potential-current density measurements indicate the presence or absence of cathodic areas, which may promote corrosion by facilitating hydrogen discharge. For example, the cathode potential-current density curves in Fig. 7 show quite clearly that areas of low hydrogen overvoltage (antimony) exist on the surface of the Pb-Sb alloy.[19] The extent of depolarization of the hydrogen discharge curve in this particular case is a measure of the antimony content of the alloy. A frequent error in the application of potential-current density curves in corrosion studies has been the use of current densities far in excess of the current densities that actually prevail in the corrosion process. Conclusions drawn from such data should take this factor into account.

HINTS ON PROCEDURE

Most of the apparatus required for potential measurements and electrochemical tests of the type described is standard electrical and chemical laboratory equipment. The measurement of the potential of individual electrodes must be accomplished with a minimum of current flow in the measuring circuit, so that neither the electrode being measured nor the reference electrode will be appreciably polarized. A potentiometer, a vacuum tube voltmeter, or some other practically zero-current instrument may be employed for the purpose. For convenience these devices should preferably, though not necessarily, be of the continuous recording variety. If the potential must

---

[14] R. B. Gibney and H. E. Haring, unpublished experiments, Bell Telephone Laboratories, Murray Hill, N. J.

[15] F. Fenwick, *Ind. Eng. Chem.*, **27**, 1095 (1935).

[16] H. J. Donker and K. A. Dengg, *Korrosion u. Metallschutz*, **3**, 217, 241 (1927). S. Brennert, *J. Iron Steel Inst.* (London), **135**, 101 (1937).

[17] H. H. Uhlig and John Wulff, *Trans. Am. Inst. Mining Met. Engrs.*, Iron and Steel Division, **135**, 494 (1939).

[18] W. Blum and H. S. Rawdon, *Trans. Electrochem. Soc.*, **52**, 403 (1927).

[19] H. E. Haring and U. B. Thomas, *Trans. Electrochem. Soc.*, **68**, 304 (1935).

be measured through a high resistance, e.g., through a paint film, the ordinary potentiometer will not suffice unless used in conjunction with a vacuum tube circuit such as that previously mentioned.[20] The vacuum tube instruments now commercially available for glass electrode work are also suitable for this purpose.

The reference electrode is preferably, but not necessarily, selected to match the electrolyte; e.g., a calomel electrode or a silver chloride electrode is preferred for chloride solutions, a mercurous sulfate electrode for sulfate solutions, and a mercuric oxide electrode for alkaline solutions. Special precautions should be taken to prevent contamination of the electrolyte in the test cell by diffusion of electrolyte from the

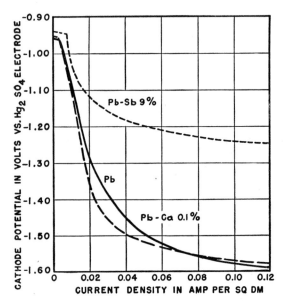

FIG. 7. Cathode Potential-Current Density Curves for Lead, Pb-Sb, and Pb-Ca in 7 N H_2SO_4.

reference half cell, unless the two are identical. This precaution is particularly important if the tip of the reference electrode is in actual contact with the electrode being studied. A convenient method for preventing such contamination is to effect a junction between the two electrolytes at some distance from the point of measurement, e.g., by using an auxiliary tip on the reference electrode and filling it with the electrolyte being studied.[21]

An extremely simple but effective type of reference half cell is shown in Fig. 8. A silver wire coated with a film of silver chloride serves as a silver chloride reference electrode, and the electrolyte, common to both test cell and reference half cell, is a known chloride solution. The tip of the reference electrode is brought into direct contact with the electrode being measured if current is flowing to or from the specimen under observation; otherwise IR drop is included in the measurement. Such

[20] K. G. Compton and H. E. Haring, *Trans. Electrochem. Soc.*, **62**, 345 (1932). D. B. Penick, *Rev. Sci Instruments*, **6**, 115 (1935).
[21] H. E. Haring, *Trans. Electrochem. Soc.*, **49**, 417 (1926).

contact is also necessary if it is desired to measure the potential of a particular small area upon any electrode.

Potential measurements are usually made with the test specimen completely immersed. Sometimes, however, it is desirable to wet an extremely small area only; in such measurements a drop of the desired electrolyte may effect a junction between the test specimen and the tip of the reference half cell. The tip of the reference half cell is upturned to prevent trapping air or gas bubbles, which might break the continuity of the measuring circuit.

Experimental conditions should coincide as closely as possible with intended service conditions unless such precautions have been demonstrated to be unnecessary. In this connection, the presence or absence of atmospheric oxygen may be of considerable importance. In certain instances the potential of an electrode is influenced by the concentration of dissolved oxygen and by the depth of immersion of the electrode in the electrolyte. The latter effect is due to the difference in oxygen concentrations at various levels. Sometimes the higher oxygen concentration at the water line causes this section of the electrode to be cathodic to the more deeply submerged portion. On the other hand, the water line region may be anodic to the more deeply submerged region because a downward flow of heavy, corrosion product-laden electrolyte causes an influx of fresh and therefore more corrosive electrolyte at the water line. Whether or not the potential of a given electrode is adversely affected by these influences and the manner in which these or other undesirable effects may be avoided is another matter, which at this stage of development must be left to the judgment of the investigator. Insulation of the water line has solved the problem in certain instances; a nitrogen atmosphere has been used to advantage in others. Obviously any undesirable effect due to oxygen or any other cause must be detected and eliminated, or at least properly interpreted, before a method for study can be utilized as a reliable method for test.

FIG. 8. A Simple Method for Measuring Electrode Potentials.

Methods for measuring electrode potentials that do not require the use of reference half-cells have been suggested[21] and in certain instances may be applied to advantage, but, in general, for corrosion study a recognized reference half cell should be considered as essential and standard.

TESTS FOR CORROSION FATIGUE

Blaine B. Wescott*

GENERAL CONSIDERATIONS

Methods and apparatus used for corrosion fatigue testing are patterned after those used for endurance testing in air. Suitable alterations are incorporated to provide for exposure of the specimen to the corrodent while it is undergoing cyclic stressing.

* Gulf Research and Development Co., Pittsburgh, Pa.

The types of testing machines most commonly used apply cycles of completely reversed bending, axial, or torsional stresses. Less commonly used are machines which apply one of these three types of stress in a unidirectional manner between maximum and minimum limits. Occasionally, machines which combine two types of stress loading are employed as, for instance, reversed torsion under constant tension load. For a complete discussion of fatigue testing machines, reference should be made to books on fatigue of metals.[1,2]

TESTING IN AIR

Practically any endurance-testing apparatus can be used for making corrosion fatigue tests in air merely by the addition of a device to subject the specimen to the action of the corrodent in the manner desired.

McAdam[3] used a rotating cantilever beam fatigue machine in this manner, applying the water diagonally so that the conically tapered specimen was surrounded with water from the outer to the inner fillet. Speller, McCorkle, and Mumma[4] used the R. R. Moore rotating beam machine fitted with a tube to flow water on the specimen by gravity and having a receiver underneath the specimen so that the water could

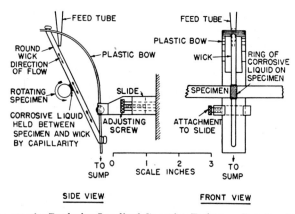

Fig. 1. Apparatus for Producing Localized Corrosion Fatigue on Rotating Beam Specimen. (Huddle and Evans.)

be recirculated. Air was continuously bubbled through the water in the reservoir to keep it saturated. Gould,[5] using a rotating beam machine, suspended a cotton tape just under the specimen but without touching it. The tape was kept saturated by dripping the corrodent on to it, and a liquid film of constant width was kept on the specimen by capillarity. This arrangement permits study of the effect of corrosion fatigue at any chosen spot of the uniformly stressed specimen. Huddle and Evans[6]

[1] H. F. Moore and J. B. Kommers, *The Fatigue of Metals*, McGraw-Hill Book Co., Inc., New York, 1927.
[2] H. J. Gough, *The Fatigue of Metals*, Ernest Benn, Limited, 1926.
[3] D. J. McAdam, Jr., "Stress-Strain-Cycle Relationship and Corrosion Fatigue of Metals," *Proc. Am. Soc. Testing Materials*, **26** (2), 224 (1926).
[4] F. N. Speller, I. B. McCorkle, and P. F. Mumma, "Influence of Corrosion Accelerators and Inhibitors on Fatigue of Ferrous Metals," *Proc. Am. Soc. Testing Materials*, **28** (2), 159 (1928).
[5] A. J. Gould, *Engineering*, **136**, 453 (1933).
[6] A. U. Huddle and U. R. Evans, "Some Measurements of Corrosion Fatigue Made with a New Feeding Arrangement," *J. Iron Steel Inst.*, **149**, 109 (1944).

used a modification of Gould's method shown schematically in Fig. 1 to study the protective value of paints as well as certain aspects of corrosion fatigue.

Kenyon[7] used a machine for corrosion fatigue testing of wire in which the wire, held in the shape of an arc, was rotated about its own curved axis, thus applying tension compression stresses. The specimen was immersed in the corrodent.

Gough and Sopwith used a Haigh axial stress machine fitted with chambers enclosing the specimen which could be exhausted to high vacuum[8] or in which the specimen could be immersed completely in a liquid[9] (Fig. 2).

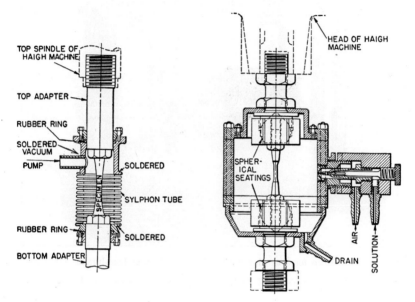

FIG. 2. Methods for Conversion of Haigh Axial Fatigue Machine to Corrosion Fatigue. (Gough and Sopwith.)

Left: Sylphon tube chamber for fatigue testing in vacuum. *Right:* Chamber for fatigue testing in liquid spray. (A slight modification would provide for testing with specimen completely immersed.)

The apparatus shown in Fig. 3 was used by Rawdon[10] to test sheet Duralumin by repeated flexure. The strip specimen, S, bolted to an upright, U, at each end, was flexed by rotation of the uprights at a rate of 75 cycles per min. A tank, T, containing the corrodent was raised to immerse the specimen for one minute at intervals of 15 min.

---

[7] J. N. Kenyon, "The Rotating-Wire Arc Fatigue Machine for Testing Small-Diameter Wire," *Proc. Am. Soc. Testing Materials,* **35** (2), 156 (1935). J. N. Kenyon, "A Corrosion-Fatigue Test to Determine the Protective Qualities of Metallic Platings," *Proc. Am. Soc. Testing Materials,* **40** (2), 705 (1940).

[8] H. J. Gough and D. G. Sopwith, "Atmospheric Action as a Factor in Fatigue of Metals," *J. Inst. Metals,* **49**, 93 (1932).

[9] H. J. Gough and D. G. Sopwith, "Some Comparative Corrosion-Fatigue Tests Employing Two Types of Stressing Action," *Engineering,* **136**, 75 (July 21, 1933). H. J. Gough and D. G. Sopwith, "Atmospheric Action as a Factor in Fatigue of Metals," *Engineering,* **134**, 694 (December 9, 1932).

[10] H. S. Rawdon, "The Effect of Corrosion, Accompanied by Stress, on the Tensile Properties of Sheet Duralumin," *Proc. Am. Soc. Testing Materials,* **29** (2), 314 (1929).

990   CORROSION TESTING (LABORATORY TESTS)

FIG. 3. Apparatus for Corrosion Fatigue Testing of Sheet Duralumin. (Rawdon.)

S. Specimen of sheet Duralumin.   U. Oscillating shafts to flex specimen.   T. Tank containing corrodent, raised at intervals.

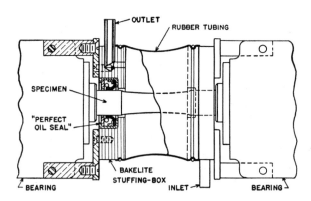

FIG. 4. Modification of R. R. Moore Rotating Beam Machine for Corrosion Fatigue Testing with Sample Immersed in Absence of Air. (Wescott and Bowers.)

TESTING BY IMMERSION IN ABSENCE OF AIR

Wescott and Bowers[11] adapted the R. R. Moore rotating beam machine so that the specimen was immersed in air-free oil well brine saturated with natural gas while subjected to cyclic stresses. A sectional view of the cell is shown in Fig. 4. The brine flowed by gravity through the cell into a receiver, from which it was lifted to the reservoir by natural gas lift. Natural gas was also bubbled continuously through the brine in the reservoir, the pressure of which served to exclude air. The same equipment was also used to determine the corrosion fatigue limits of steels in brine saturated with hydrogen sulfide by bubbling H_2S continuously through the reservoir brine. This procedure has been adopted as a code for corrosion fatigue testing of sucker rod materials by the American Petroleum Institute.[12]

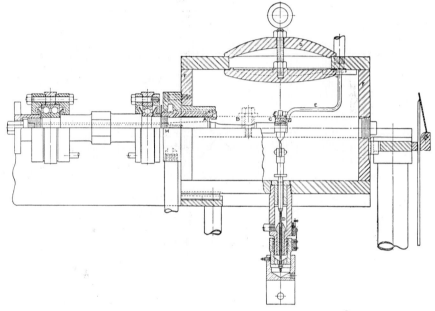

Fig. 5. Rotating Cantilever Beam Machine for Corrosion Fatigue Testing in Steam. (Fuller.)

A. Specimen holder. *B.* Specimen. *C.* Bearing on specimen for load application. *D.* Steam packing around load application rod. *E.* Pipe for lubrication of bearing *C*. *F.* Steam box. *G.* Steam box cover. *H.* Steam packing around rotating specimen shaft.

TESTING IN STEAM

A sectional view of the apparatus devised by Fuller[13] for corrosion fatigue testing in steam with a rotating cantilever-type machine is shown in Fig. 5. Steam packings

---

[11] B. B. Wescott and C. N. Bowers, "Corrosion Fatigue and Sucker-Rod Failures," 14th Annual Meeting, *Proc. Am. Petroleum Inst.*, **14**, Sect. IV, 29–42 (Oct. 24, 1933).

[12] *A. P. I. Code No. 30 for Corrosion Fatigue Testing of Sucker Rod Materials*, American Petroleum Institute, Division of Production, Dallas, Texas.

[13] T. S. Fuller, "Endurance Properties of Steel in Steam," *Trans. Am. Inst. Mining Met. Engrs.*, Iron and Steel Div., **90**, 280 (1930).

were used around the rotating shaft and the rod for application of the load. Water was drained from these packings by vacuum. Tests were made with steam pressures and temperatures as high as 220 psi and 370° C (700° F), respectively.

METHOD FOR REPORTING TEST RESULTS

The most concise and informative method for reporting the results of corrosion fatigue tests is by the use of curves plotted on semi-log paper, using the ordinate for the imposed stress and the log scale as abscissa for the number of reversals to failure. These are the usual S-N type of charts used to report results of air endurance tests. The curves for endurance limit and corrosion fatigue limit may be plotted on the same chart to facilitate comparison. Such a chart as used for cold-rolled and annealed

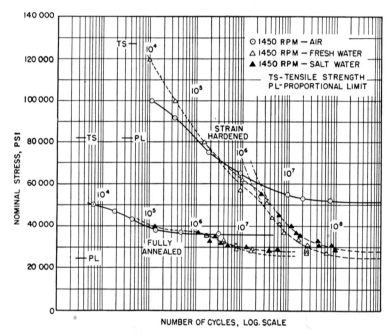

Fig. 6. S-N Curves for Reporting Results of Fatigue and Corrosion Fatigue Tests. (Results for 70% Ni – 30% Cu Alloy by D. J. McAdam, Jr., *Proc. Am. Soc. Testing Materials*, **27**, 110 [1927].)

70% Ni–30% Cu alloy* by McAdam is shown in Fig. 6. When plotted in this manner, the *corrosion fatigue limit* is clearly indicated as the stress at which the slope changes sharply, and the curve becomes approximately asymptotic to the abscissa. It should be borne in mind that if a metal is eventually damaged by corrosion fatigue the S-N curve will never become truly asymptotic. For most metals, however, there will be a marked change in slope, and the corresponding stress is considered to be the corrosion fatigue limit. This value can be obtained for steels and some nonferrous metals by 10 million cycles of stress, but aluminum and its alloys and certain other alloys require 100 million or more cycles for reliable results.

* Monel.

Charts to express specific aspects of damage to the endurance strength by prior corrosion and corrosion fatigue, such as total, net, and constant damage, have been described in the numerous papers by McAdam.

In some cases, it is desirable to compare the resistance of metals to corrosion fatigue by determining the number of cycles required to cause failure at some constant stress higher than the corrosion fatigue limit. Data derived in this manner are most reliable when the stress chosen permits a large number of cycles before failure.

PRECAUTIONS TO BE OBSERVED

Since it has been established that the corrosion resistance of a metal is the property having the greatest influence upon its resistance to corrosion fatigue, it is important that the medium chosen for testing duplicates in all essential respects that encountered in service.

The rate of stressing has a direct influence upon the value of the corrosion fatigue limit. The value is higher for rapid rates of stressing because the specimen is exposed to corrosion for a shorter period of time. Therefore, when the tests are intended to provide a basis for selecting a metal for a specific service, it is desirable to employ a rate of stressing that is not greatly disproportionate to that actually prevailing.

It is essential that the apparatus be so designed that concentration cells or electrochemical action from extraneous causes do not influence the results. It has been shown that the former can completely nullify the protective action of inhibitors.[14] Furthermore, it was found that when stuffing boxes of steel similar to the specimens were used in the apparatus shown in Fig. 4, the corrosion fatigue limit of S. A. E. 1040 steel was 21,600 psi whereas, when aluminum stuffing boxes were used, the corrosion fatigue limit was increased to 57,000 psi.[15] The air endurance limit of this steel was 63,000 psi.

TESTS FOR DAMAGE BY CAVITATION

S. Logan Kerr*

MEASURE OF DAMAGE

Several methods for measurement of damage by cavitation have been proposed, but as yet there is no one generally accepted unit. Either depth and area of the damage or weight loss is used, weight loss being feasible in the measurement of laboratory specimens where they can be weighed before and after exposure. Weight loss is not suitable to the evaluation of damage to turbine runners or pump impellers, however, where the damage is frequently determined by the nature of critical hydraulic dimensions.

The best criterion of relative damage by cavitation is the performance of the full-scale installation of the turbine, pump, valve, or other device. Service tests are the final and conclusive evidence, and even they can be evaluated only with some degree of accuracy by relating the surface finish and the setting with regard to tail water, to

---

* Consulting Engineer, Philadelphia, Pa.

[14] F. N. Speller, I. B. McCorkle, and P. F. Mumma, "Influence of Corrosion Accelerators and Inhibitors on Fatigue of Ferrous Metals," *Proc. Am. Soc. Testing Materials*, **28** (2), 159 (1928).

[15] B. B. Wescott and C. N. Bowers, "Corrosion Fatigue and Sucker-Rod Failures," 14th Annual Meeting, *Proc. Am. Petroleum Inst.*, **14**, Sect. IV, 29–42 (Oct. 24, 1933).

suction conditions, and to other operating conditions of the unit. The development of cavitation test laboratories for turbines and pumps has contributed much to the knowledge of operation directed to avoid the ill effects of general low pressure and the loss of capacity and efficiency.

TYPES OF TESTS

Venturi Tests

As described by Mousson,[1,2] use is made of the subnormal pressure effect beyond the Venturi throat as a cavitation-producing medium. Tests require 16 hours or longer to get measurable losses. Longer times may be needed if the test set-up cannot produce the relative level of cavitation intensity necessary to produce measurable damage.

Jet Tests

Tests have been made to demonstrate that a free jet,[3] impinging on a plane or irregular surface, can produce erosion. Such tests have shown that cavitation damage is not restricted to turbine runners or pump impellers, but occurs wherever the formation of local areas of extreme low pressures cause the formation of cavities in the fluid.

Recently a high-pressure jet test was utilized to compare various metals for a large hydroelectric installation. The jet was directed at an angle and impinged upon a recess in the metal plate. Cavitation erosion occurred at a critical point beyond the hole. Results were obtained in a matter of a few days.

Vibratory Tests

The basic characteristic of cavitation, consisting of pressures alternating between subzero and high values, has resulted in the artificial development of such conditions through the use of vibratory apparatus. Two types have been employed, one consisting of a reciprocating piston within a cylinder and the other making use of the vibrating action produced by an oscillatory electromagnetic circuit, causing magnetostriction of a pure nickel tube.

Extensive tests[4,5] were made using this method, whereby a ⅝-in. diameter specimen was attached at the end of the oscillating rod (6700 cycles per sec) and immersed in fresh or sea water. It was possible to produce measurable weight loss in 30 min and substantial weight loss within 90 min. The surface appearance of the specimens after exposure was identical with the characteristics found in rotating machinery where cavitation erosion was experienced.

CORRELATION OF TEST RESULTS

Time factors in the various test methods range from a matter of years in actual installation to a matter of hours with Venturi test, and minutes with vibratory

[1] J. M. Mousson, "Practical Aspects of Cavitation and Pitting," *Edison Elec. Inst. Bull.* **5**, 373 (September-October, 1937).

[2] J. M. Mousson, "Pitting Resistance of Metals under Cavitation Conditions," *Trans. Am. Soc. Mech. Engrs.*, **59** (No. 5), 399 (July, 1937).

[3] Unpublished tests of Shawinigan Water and Power Co., Shawinigan Falls, P. Q., Canada.

[4] S. L. Kerr, "Determination of the Relative Resistance to Cavitation Erosion by the Vibratory Method," *Trans. Am. Soc. Mech. Engrs.*, **59** (No. 5), 373 (July, 1937).

[5] R. Beeching, "Resistance to Cavitation Erosion," *J. Inst. of Engrs. and Shipbuilders in Scotland*, pp. 210–276 (1942).

TABLE 1. RELATIVE RESISTANCE OF METALS TO CAVITATION DAMAGE BY THE VIBRATORY TEST METHOD*

Non-ferrous	Form	\multicolumn{9}{c	}{Material}	Weight Loss at 25° C for Last 60-min. Exposure, mg/hr	
		Cu	Sn	Zn	Mn	Si	Ni	Fe	Pb	Al	Fresh Water	Sea Water†
Bronze (Cu, Zn, Sn)	Rolled	60	1	39							69.5	65.2
Brass (Cu, Zn)	Rolled	60		40							77.8	68.7
Brass (Cu, Zn)	Rolled	85		15							115.2	101.3
Brass (Cu, Zn)	Rolled	90		10							134.9	122.8
Bronze (Cu, Al)	Cast	80						Fe		10	15.3	14.5
Bronze (Cu, Sn, Ni)	Cast	87.5	11				Ni				54.6	62.4
Bronze (Cu, Sn, Pb)	Cast	88	10				1.5		2		60.4	48.5
Bronze (Cu, Si)	Cast	92–94									42.6	40.4
Bronze (Cu, Si, Mn)	Cast	94		Zn	1			Fe		Al	52.4	54.5
Bronze (Cu, Zn, Al, Mn)	Forged	60–70		20–30						Al	19.2	19.9
Bronze (Cu, Zn, Fe, Mn)	Cast	58		40				Fe		Al	53.0	55.4
Bronze (Cu, Sn, Zn)	Cast	88	10	2				1			65.8	57.4
Nickel (Cu, Fe, Si)	Cast	32–33				4	62–63	2			20.0	21.4
Nickel (Cu, Fe, Mn)	Drawn	29			1		68	1			53.3	53.2
Nickel (Cu)	Rolled	70					30				86.2	87.6

Ferrous	Form	C	Si	Cu	Mo	S	P	Mn	Cr	Ni	Fresh Water	Sea Water†
Iron	Cast	3.1	2.3			0.12	0.07	0.75			50.1	80.9
Iron	Cast	3.4	1.3			0.08	0.25	0.75			69.8	115.3
Iron	Cast	3.4	2.3					0.59			89.7	100.2
Iron (Cu, Ni, Cr, Si)	Cast	3.0	1.9	6.0					4.0	14.4	41.6	51.4
Iron (Mo)	Cast	3.3	1.3		0.40			0.51			54.1	63.9
Iron (Mn, Cu, Ni, Cr)	Cast	3.0	1–2	6.0		0.10	0.04	1.0	1–3	12–15	85.3	95.3
Steel	Rolled	0.35					0.45	0.67			34.2	39.6
Steel	Rolled	0.27				0.40	0.45	0.48			68.3	77.8
Steel	Rolled	0.20				0.03	0.02	0.50			78.2	82.4
Steel	Cast	0.37	0.31			0.04	0.04	1.10			44.8	53.6
Steel	Cast	0.26	0.32			0.04	0.04	0.60			72.9	80.9
Steel (Ni, Cr)	Rolled	0.34	0.20			0.03	0.02	0.52	0.60	1.18	20.0	22.0
Steel (Ni)		0.19				0.02	0.02	0.60		2.2	61.3	64.0
Stainless steel (Cr)	Rolled	0.08	0.57			0.02	0.03	0.47	17.2	0.34	11.8	10.8
Stainless steel (Cr)	Rolled	0.09	0.38			0.02	0.02	0.43	12.2	0.32	20.6	23.0
Stainless steel (Cr, Ni)	Cast	0.15	0.50					0.50	16–20	8–12	13.5	13.4
Stainless steel (Cr, Ni)	Rolled	0.07	0.37			0.14	0.19	0.48	18.4	8.7	16.1	15.3

* S. L. Kerr, "Determination of the Relative Resistance to Cavitation Erosion by the Vibratory Method," *Trans. Am. Soc. Mech. Engrs.*, **59** (No.5), 373 (July, 1937).
† Natural Sea Water from Cobscook Bay, Eastport, Me.

TABLE 2. RELATIVE RESISTANCE OF METALS TO CAVITATION EROSION BY THE VENTURI METHOD*,†

Fresh Water

No.	Type of Material	Analysis							Condition	Cavitation Loss in 16 Hours, Cubic mm of Metal at 20° C
		C	Mn	P	S	Si	Ni	Cr		
1	Cast iron	3.2	0.5	....	....	2.1	....	....	As cast	636.0
2	Cast iron with casting skin	3.2	0.5	....	....	2.0	....	....	As cast	396.0
3	Nitrided 0.34% steel	0.34	0.53	0.017	0.020	....	....	....	As cast	109
4	0.33% C steel of turbine grade	0.33	0.71	0.03	0.03	....	0.3	(0.2 Mo, 1.0 Al)	As cast	62.4
5	Stainless steel, 14% Cr, 1% Ni	0.05	0.3	0.02	0.02	0.35	1.1	14.5	As cast	32.0
6	", 14% Cr, 2% Ni	0.05	0.4	0.02	0.02	0.31	2.0	14.3	As cast	12.9
7	", 14% Cr, 3% Ni	0.07	0.4	0.02	0.02	0.40	3.4	14.3	As cast	9.8
8	", 18% Cr, 8% Ni	0.08	....	....	....	....	8.3	18.5	As cast	8.8
9	Stainless 18-8 clad carbon steel	....	....	....	....	....	....	....	Rolled	11.2
10	Stainless 12% Cr-5% Ni Steel	....	....	....	....	....	....	....	Two layers welded on wrought iron	8.4
11	Stainless 12% Cr-5% Ni Steel	....	....	....	....	....	....	....	Two layers welded on boiler plate	8.1
12	Stainless 18-8 Cr-Ni Steel	....	....	....	....	....	....	....	Two layers welded on wrought iron	8.2
13	Stainless 17-7 Cr-Ni Steel	....	....	....	....	....	....	....	Two layers welded on wrought iron	1.3

* J. M. Mousson, "Practical Aspects of Cavitation and Pitting," *Edison Elec. Inst. Bull.*, **5**, 373 (September-October, 1937).
† J. M. Mousson, "Pitting Resistance of Metals under Cavitation Conditions," *Trans. Am. Soc. Mech. Engrs.*, **59** (No. 5), 399 (July, 1937).

apparatus. Comparison of the various results shows that there is the same general relative order of cavitation damage for different materials, irrespective of the test methods employed. (See Tables 1 and 2.)

When large-scale specimens are tested by the Venturi method, coupled with microscopic examination of the surface and a section of the material,[6] the damage is of the same general type as found in specimens taken from full-scale installations. It is highly probable that similar confirmation could be obtained from microscopic analysis of specimens from the vibratory tests.

The effect of temperature, vapor pressure, and absolute pressure conditions in the laboratory test procedure are equally as important as the effect of corresponding conditions in full-scale equipment.

HIGH-TEMPERATURE TESTS

F. E. BASH*

Alloys used at high temperatures may be called upon to withstand various environments, among which are:

1. Air and other gases.
2. Furnace atmospheres.
3. Carburizing or nitriding atmospheres.
4. Molten and solid salts.
5. Molten metals.

In determining the relative values of alloys for high-temperature service, it is essential that a method of test be available which is representative of the conditions under which the material is to be used. The other alternative is to test the material in actual service. It is more desirable, of course, to have an accelerated test that serves as a measure of service performance.

Test methods have been worked out by engineering society committees and other groups for the study of the behavior of metals and alloys at high temperatures in air. None of these methods is completely representative of conditions in service, but some approach them with a fair degree of accuracy. The primary objective of a corrosion test is to determine the rate at which the metal is destroyed by any particular set of conditions.

INTERNAL HEATING OF TEST SPECIMENS

A method of life test[1] has been worked out by the American Society for Testing Materials to determine the relative life of metals used for electrical heating. Briefly, this test consists of passing a current through a wire 0.025 in. (0.064 cm) in diameter, suspended from a binding post with a lower connection in a cup of mercury. A current is passed through the wire until it has reached the temperature of test as determined by use of an optical pyrometer. A rheostat in series with the wire used for varying the current. The temperature is readjusted at intervals and may be controlled at a constant value; or the impressed voltage across the sample may be maintained

---

* Driver-Harris Co., Harrison, N. J.

[6] H. N. Boetcher, "Failure of Metals Due to Cavitation under Experimental Conditions," *Trans. Am. Soc. Mech. Engrs.*, **58**, Paper HYE-58-1, p. 355 (1936).

[1] "Accelerated Life Test for Metallic Materials for Electrical Heating," Test B76-39, *A. S. T. M. Standards*, Part I, p. 915 (1944).

constant with the temperature dropping off as the sample under test decreases in cross section due to oxidation. A record of the variations of resistance of the wire with time may be plotted, as an indication of the change of the cross-sectional area.

This method may utilize an intermittent heating and cooling cycle which provides a greatly accelerated life test; or the temperature may be maintained constant with no interruptions in heating. It has been found that a cycle of 2 min heating and 2 min cooling for Ni-Cr and Ni-Cr-Fe alloy wires 0.025 in. (0.064 cm) in diameter gives the shortest life. This is one eighth to one tenth the life of a wire maintained at constant temperature without any cooling. It is obvious that with intermittent heating and cooling a differential expansion of the metal and protective oxide scale will result in much more rapid scaling than if the material is maintained at a constant temperature.*

It is usually found that after the electrical resistance of the sample under test has increased by 10% or more, hot spots form at places where the cross-sectional area of the wire has been reduced more than at other points. After the hot spot or hot spots form, oxidation increases at a rapid rate, and the wire fails at that point. It is not desirable, therefore, to use the time to burn out as a measure of the resistance of a wire to oxidation at any test temperature. For this reason, *useful life* has been specified as the *time required for the resistance of the specimen under test to increase by 10%* and is employed as a measure of the rate of oxidation of any metal under specified conditions.

With some alloys there may be a change in the specific resistance with time independent of oxide formation. This change is not usually important, and it may be determined, in any event, and compensated for.

The relationship between life of a wire in hours as described above and the temperature of the test is given by the following equation:

$$\log L = K - St$$

in which $L =$ hours of test life,

$S =$ slope of the temperature-life curve plotted on semi-logarithmic paper,

$t =$ temperature,

$K =$ a constant.

Examples are given in Fig. 1 of the resulting straight lines obtained within working range of the alloys. Constants of the above equation for the three alloys included in Fig. 1 are:

Alloy				
Cr	Ni	Fe	S	K
20	80		0.0047	12.265
16	60	Bal.	0.0033	9.062
14	80	6	0.0030	7.622

EXTERNAL HEATING OF TEST SPECIMENS

WEIGHT CHANGE METHOD

Tests have been devised where the sample under test is heated externally in a given environment at a known temperature for a specified period of time and then cooled

---

* Hessenbruch (*Zunderfeste Legierungen*, p. 60, Berlin, 1940) points out that the surface oxide of cooled metal may continue to spall appreciably for some minutes, and at a minor rate for 10 to 20 hours. The period over which a metal remains at room temperature in a cycle of heating and cooling is of some importance, therefore, to the measured corrosion rate. It is stated that for some alloys a wire heated in a cycle of 2 min on and 6 *min off* may actually have 20 to 30% shorter life than a wire heated 2 min on and 2 *min off*. EDITOR.

to room temperature. This cycle is repeated for a definite length of time, or number of cycles, and the loss or gain in weight of the specimen is reported. When no provision is made to remove all the oxide scale before weighing, this test method is of value in comparing the relative corrosion rates of metals or alloys of approximately the same general analysis. Results, however, may be extremely misleading if comparisons are made between alloys of considerably varying analyses. If a metal is free scaling, the

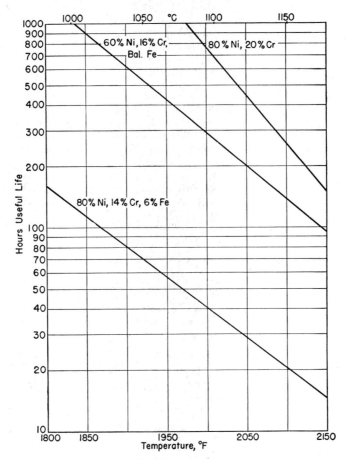

FIG. 1. Relation between Life of Heat-Resistant Wires and Temperature of Operation.

loss in weight may be a reasonably accurate measure of the reduction in cross section of the metal, but, on the other hand, a metal may form a spongy oxide which does not detach itself readily from the surface. For the latter an increase in weight will be noted, although the reduction in cross section may be as great as that of a metal which is free scaling. Again, it is possible for a metal to lose some scale and to build up an adherent oxide at such rates that one will balance the other, and the weight of the test sample will remain approximately constant. The test is only an accurate measure of the rate of corrosion provided the oxide formed at high-temperature exposure is totally removed without removing the metal. (See *Preparation and Cleaning of Specimens*, p. 1077.) It should be noted in making oxidation tests of this kind, and where the sample is weighed at intervals, that the test is usually continued until a change in weight of more than 1% has occurred.

When the extent of reaction is not detectable by the ordinary analytical balance, a more sensitive means is available. A very sensitive microbalance has been described[2] in which a specimen is suspended as part of the balance system within a quartz tube. The specimen is heated by induction and any weight change produced by reaction with a given gas environment can be followed. The balance is sensitive to 1×10^{-6} gram or the order of magnitude of a monolayer of oxygen atoms on the specimen surface (15 sq cm area).

Microscopic Examination*

The weight-change method of determining corrosion rate has the disadvantage of failing to indicate the condition of metal remaining. It is sometimes advisable, therefore, to examine the specimen cross section under the microscope, using common metallographic techniques. The rate of corrosion can be determined at the same time.

After high-temperature exposure, selected cross sections of the specimens are examined in the etched and unetched condition. The thickness of corrosion product is noted, and the remaining metal examined for intergranular attack, internal oxidation, or other change. The rate of attack is calculated from the thickness of sound metal before and after a given time of exposure.

Bend Test*

Corrosion and phase changes may have a marked effect upon the residual ductility after exposure. This effect can be evaluated by a simple bend test in a tensile machine. Comparative ductility is expressed as the load or the angle of bend required to produce the first sign of cracking of the specimen.

Other Methods†

A convenient method for determining oxidation or reaction rate makes use of optical measurement of film thickness.[3] This is accomplished by noting either interference colors or the state of polarized light reflected from the surface.[4] The method is restricted to temperature ranges corresponding to relatively low rates of attack. Data obtained for oxidation of iron and nickel by observing interference colors are reproduced in Figs. 1 and 2, p. 12.

Another method sensitive especially at low pressures involves measurement of gas consumed by the metal-gas reaction within a given time. The metal reacts at constant temperature in a closed system to which is attached either a gas burette[5,6] or a pressure gage,[7] depending upon whether the rate is measured at constant or

---

* This discussion is by G. F. Geiger, International Nickel Co., Inc.
† This discussion is by the Editor.

[2] E. A. Gulbransen, *Trans. Electrochem. Soc.*, **81**, 327 (1942).
[3] G. Tammann, *Z. Anorg. Chem.*, **111**, 78, 1920; (with W. Köster) **123**, 196 (1922).
[4] The two methods are reviewed in detail by U. R. Evans, *Metallic Corrosion, Passivity and Protection*, pp. 672–682, Edward Arnold and Co., London, 1937.
[5] J. S. Dunn, *Proc. Royal Soc.* (A), **111**, 203–209 (1926).
[6] A. Portevin, E. Pretet, and H. Jolivet, *Rev. Met.*, **31**, 101, 186, 219 (1934).
[7] C. N. Hinshelwood, *Proc. Roy. Soc.* (A), **102**, 318 (1922).

changing gas pressure. With low pressure gases especially, but also at atmospheric pressures, it is important that the temperature of the specimen be known in addition to that of the reacting gas. The heat of reaction may raise the specimen temperature appreciably above the surrounding temperature.

An interesting pressure change method employs a differential water or oil manometer between a cell holding the specimen and an identical dummy cell at the same pressure.* This has been used effectively in the study of oxidation rates over a wide range of pressures.

The amount of film substance for some metal compounds (e.g., Cu_2O) can be accurately measured electrometrically by reducing the film cathodically in a suitable electrolyte.[8,9] The amount of electricity that has passed is noted at the time all the surface compound is reduced, detected by a change in potential of the specimen. The equivalent weight of substance involved is calculated by applying Faraday's Law.

CORROSION IN FURNACE ATMOSPHERES

The A. S. T. M. life test method quoted above can be used with modification by enclosing the test piece and circulating the gases under investigation through a suitable chamber. This method gives a quantitative measure of the comparative quality of alloys in a specified atmosphere at a particular temperature.

Another A. S. T. M. method[10] has been devised in which a number of pieces of metal strip, approximately one foot long, are suspended in a furnace in which the temperature varies from approximately 1100° C (2000° F) in the center to approximately 550° C (1000° F) at the ends. The test gas is circulated slowly through the furnace, and in this way the reaction of eight or ten samples to a particular atmosphere at varying temperatures may be determined. It is not easy, however, to determine the rate of corrosion quantitatively. Examination of the surface and cross section of pieces of strip approximately 0.375×0.030 in. $(1.0 \times 0.08$ cm$)$ is a qualitative measure of corrosive attack.

MOLTEN AND SOLID SALTS

There is no standard test method devised for containers of molten salts. The only known method is to suspend samples of material for test in the molten salt, making sure that only part of the specimen is immersed, since the heaviest corrosion is usually at the line marking the surface of the salt, where air comes in contact.

For solid salts such as carburizing compounds contained in alloy boxes, it appears that the best method of test at present is to make a box from the alloy under consideration and observe it in service.

MOLTEN METALS

Conditions similar to those employing molten salts are encountered with molten metals such as lead, tin, and solder in metal containers. Test procedures are conducted similarly.

* W. E. Campbell, private communication (The Electrochemical Society *Preprint* 91-24 [1947]).

[8] H. A. Miley, *J. Am. Chem. Soc.*, **59**, 2626 (1937).

[9] W. E. Campbell and U. B. Thomas, *Trans. Electrochem. Soc.*, **76**, 303 (1939).

[10] "Effect of Controlled Atmospheres upon Alloys in Electric Furnaces," Test B 181-43T, *A. S. T. M. Standards*, Part I, p. 1799 (1944).

GALVANIC COUPLE TESTS

C. H. Sample*

The primary purpose of galvanic couple tests is to obtain useful information in predicting or controlling the extent to which galvanic couple action occurs when dissimilar metals are in contact under given conditions of service. Such tests may include studies of galvanic protection, as well as of the deterioration resulting from galvanic corrosion.

In order for galvanic couple action to occur, the two or more metallic elements of the couple must be in metallic contact while exposed to an electrolyte; and a difference in potential must exist between them. When these basic requirements are satisfied, galvanic couple action can and will occur. The extent to which it occurs, however, can only be determined by actual test. It has been shown[1] that the amount of corrosion resulting from galvanic couple action is proportional to the total current flowing in the cell. This in turn is dependent primarily on the relation between the open-circuit potential of the cell, less all counter potentials resulting from the flow of current, and the total resistance of the circuit, the latter including primarily the resistance of the electrolyte.

Since these factors are profoundly affected by the dimensions of the electrodes and the physical arrangement of the cell and, in addition, by the composition, temperature, and movement of the electrolyte, it will be obvious that there can be no one single design of galvanic corrosion test specimen or apparatus that will yield results applicable to all problems. It should be emphasized that if the test conditions are not identical with conditions of service, the results can never be considered to be more than semi-quantitative.

The principal means for studying the galvanic behavior of metals and alloys has been summarized as follows:[2]

1. Measurement of the open-circuit potential difference between metals or alloys forming a couple.
2. Measurement of the potential of a metal or alloy in the testing solution with reference to a standard of potential such as a calomel half cell.
3. Measurement of the direction and magnitude of the current generated by a couple throughout a test period.
4. Comparison of the extent of corrosion of coupled and uncoupled specimens exposed under identical conditions. Such measurements of the extent of corrosion may be made by means of weight-loss determinations, observations of changes in mechanical properties (strength and ductility), and measurements of the extent and distribution of local attack (pitting).
5. Determination of the polarization characteristics of the specimens in the testing solution.

Since contact of electrolyte between the two metallic elements of a galvanic couple is a primary requisite, it is assumed that galvanic action is most pronounced under conditions of continuous or intermittent immersion in relatively strong electrolytes. Tests embodying these or comparable conditions, therefore, are probably of primary

---

* Bell Telephone Laboratories, New York, N. Y. Present Address: The International Nickel Co., New York, N. Y.

[1] R. H. Brown and R. B. Mears, *Trans. Electrochem. Soc.*, **74**, 495–517 (1938). W. A. Wesley, *Proc. Am. Soc. Testing Materials*, **40**, 690–704 (1940). U. R. Evans and T. P. Hoar, *Proc. Roy. Soc.* (A), **137**, 343–365 (1932).

[2] F. L. LaQue and G. L. Cox, *Proc. Am. Soc. Testing Materials*, **40**, 670–689 (1940).

importance. Such tests are illustrated by the work of LaQue and Cox[2] dealing with the action of flowing sea water on various metallic couple combinations in the form

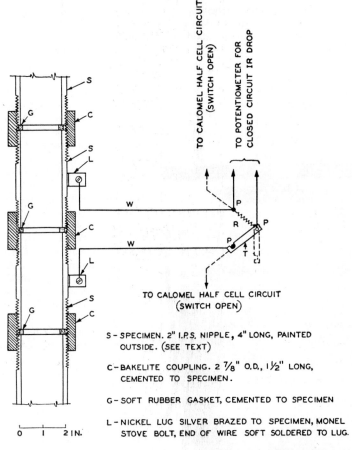

S – SPECIMEN. 2" I.P.S. NIPPLE, 4" LONG, PAINTED OUTSIDE. (SEE TEXT)

C – BAKELITE COUPLING. 2 7/8" O.D., 1 1/2" LONG, CEMENTED TO SPECIMEN.

G – SOFT RUBBER GASKET, CEMENTED TO SPECIMEN

L – NICKEL LUG SILVER BRAZED TO SPECIMEN, MONEL STOVE BOLT, END OF WIRE SOFT SOLDERED TO LUG.

W – RUBBER-INSULATED #16 GA. STANDARD COPPER WIRE.

P – NICKEL PLATED BRASS BINDING POSTS, NICKEL SILVER NUTS PERMANENTLY FASTENED TO A CENTRAL BOARD.

R – RESISTANCE, 1 OHM

T – NICKEL TERMINAL SWITCH (SLOTTED)

FIG. 1. Method of Coupling Galvanic Test Specimens in the Form of Pipe. (LaQue and Cox.)

of pipe. Their method of coupling the test specimens is shown in Fig. 1. Another type of specimen used in conducting galvanic corrosion tests in sea water[3] is the spool type shown on page 1054.

[3] H. E. Searle and F. L. LaQue, *Proc. Am. Soc. Testing Materials*, **35** (II), 249–260 (1935).

A convenient laboratory set-up for conducting galvanic corrosion tests under controlled conditions of total immersion has been described by Wesley.[4] Typical test specimens and the manner of coupling them through a known low resistance during the test are illustrated in Fig. 2. It will be noted that in addition to the coupled specimens, the assembly includes an identical pair of uncoupled test pieces. The latter provide information on the normal rate of corrosion of the metal under the same corrosive conditions. During a test run the whole assembly is placed in a suitable water bath in order to control the temperature of the corroding medium. Suitable apparatus for obtaining pertinent electrochemical data relative to galvanic corrosion has also been described by Mears and Brown.[5]

For use in studying the galvanic couple behavior of different metal and alloy combinations exposed to various outdoor atmospheres, Subcommittee VIII of the American Society for Testing Materials Committee B-3 employed[6] the type of specimen shown in Fig 3. Although these tests provided useful data, experience with this design of specimen was not entirely satisfactory and led to the design of the test specimen shown in Fig. 4 in which only the two center dissimilar disks are weighed. The latter design is currently being used by that committee in outdoor exposure tests of various metals and alloys coupled to stainless steels. It eliminates one serious fault of the old design, namely, the variations in normal corrosion on the sides of the disk resulting from variable failure of the paint. In addition to eliminating the need for paint protection and minimizing the area subjected to normal weathering, the new specimen provides greater contact area between the two metallic elements of the couple where galvanic action will be promoted. The two specimens have the one common fault of possible variation in contact resistance between the two metallic elements of the couple, particularly after corrosion products have accumulated.

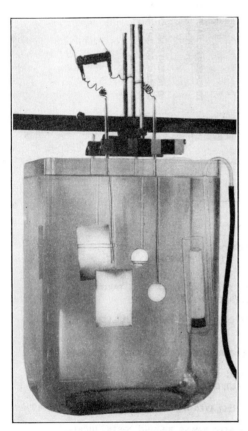

FIG. 2. Laboratory Set-up for Conducting Galvanic Corrosion Tests. (Courtesy of International Nickel Co.)

Figure 5 shows a test specimen designed by the author primarily for use in atmos-

---

[4] W. A. Wesley, *Trans. Electrochem. Soc.*, **73**, 539 (1938).
[5] R. Mears and R. Brown, *Trans. Electrochem. Soc.*, **74**, 522 (1938).
[6] C. L. Hippensteel, *Proc. Am. Soc. Testing Materials*, **32** (I), 248–251, (1932).

pheric exposures. Its principal advantage is that significant and reproducible effects of couple corrosion can be detected after relatively short exposure periods. These are made possible by the favorable ratio of surface area to weight of the anodic element and by a relatively large area of close contact between the anode and cathode where galvanic action is encouraged by the capillary retention of electrolyte between the

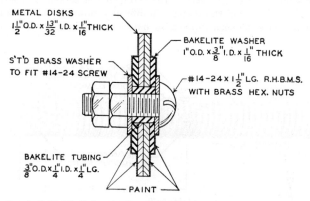

Fig. 3. A. S. T. M. Galvanic Couple for Outdoor Exposure (Early Design).

two surfaces. This type of specimen has the further advantage of assuring a low-resistance contact between the two metallic elements of the couple for the duration of the test.

With slight modification, this type of test specimen has also been used successfully to obtain open-circuit time-potential curves between the elements of the couple when

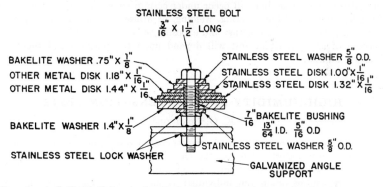

Fig. 4. A. S. T. M. Galvanic Couple for Stainless Steel Tests, Outdoor Exposure.

exposed to various atmospheric conditions. For this purpose the wire is insulated from the bolt by means of a thin layer of electrolyte-free, long-fiber paper, with non-conducting inserts in the retaining holes for the wire and set screws. The two metallic elements are then connected directly to a high-resistance recording potentiometer, which provides a continuous record of changes in the potential, which may occur, for example, as a result of variations in weather conditions.

A disadvantage of this type of specimen is that only significant weight changes of the anodic element can be obtained, whereas in the disk-type specimen of Fig. 4 the confirming evidence of galvanic protection may be obtained simultaneously from the same specimen by weight-change measurements of the more noble metal. Other limiting features are that materials may be tested in wire form only and also that a protective agent must be depended upon to confine the couple action to the area under test and prevent its occurrence at the two points of permanent contact between the two metallic elements of the couple. No difficulty, however, has been experienced from this factor on specimens exposed to outdoor environments for periods up to 4 months.

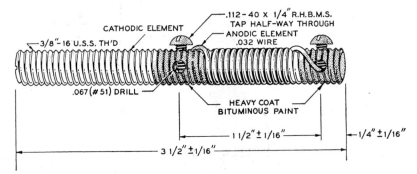

FIG. 5. Bolt and Wire Type Galvanic Couple Test Specimen.

The above tests were designed primarily for yielding information of a quantitative nature. When only rough qualitative information is desired, the tests and specimen design need not be so elaborate. For example, useful information as to whether or not serious galvanic couple action will occur under certain conditions of service has been obtained by simply bolting, riveting, welding, or otherwise joining strips, plates, or washers of dissimilar metals together and exposing them to the conditions in question. The choice of any particular test will depend upon the purpose and the type of information desired.

HIGH-HUMIDITY AND CONDENSATION TESTS

K. G. COMPTON*

Humidity chambers are widely used for accelerated testing of the indoor corrosion resistance of many materials. In size the chambers vary from large rooms to small battery jars. Similarly the temperatures and humidities employed have covered a wide range depending upon the individual views of the persons concerned. Constructional details of some cabinets have been published.[1,2,3,4]

Tests in very small sealed enclosures do not reproduce the results obtained in large humidity rooms, probably because of depletion of contaminants in the early stages

---

* Bell Telephone Laboratories, New York, N. Y.

[1] F. Todd, *Ind. Eng. Chem.*, Anal Ed., **16**, 394 (1944).
[2] H. Pray and J. Gregg, *Proc. Am. Soc. Testing Materials*, **41**, 758 (1941).
[3] R. H. W. Burkett and H. K. Henisch, *J. Sci. Instruments*, **21**, 106 (1944).
[4] *Alternate Performance of the Standard Cycle for Moisture Resistance Tests of Component Parts*, Performance Test Section, Squier Laboratory, Signal Corps, U. S. Army, November 8, 1944.

of the corrosion process. There is little detailed correlation between the results obtained in a constant temperature and humidity chamber and those obtained by indoor exposure, but experience indicates that if certain materials or finishes resist deterioration for a given period in a humidity chamber, they are not likely to give

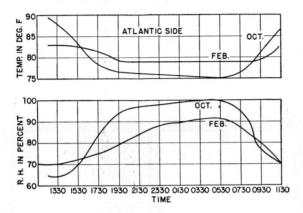

Fig. 1. Representative Relative Humidity at Various Times during the Day at Canal Zone, Panama.

trouble in service in temperate climates. In general, the length of the exposure period must be determined for each set of conditions.

For simulated tropical indoor corrosion, it is necessary to add controlled periods of condensation to the humidity test, inasmuch as condensation occurs nearly every night in most sections of the tropics. It must be borne in mind that condensation occurs on objects as they lag in temperature behind the ambient air temperature during the daily cycle. This tropical test cycle can be greatly accelerated by obtaining

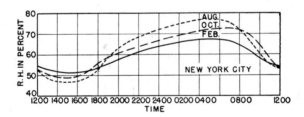

Fig. 2. Representative Relative Humidity at Various Times during the Day in New York City.

two condensations per day and by maintaining the steady state conditions at 48° C (120° F) and 95% relative humidity. A higher temperature makes operation difficult and is likely to produce excessive acceleration of the corrosion of certain types of material such as zinc or cadmium.

Two methods of obtaining the condensation cycle are in general use. For small-scale testing, the specimens are removed from the test chamber, chilled to some temperature well below the dew point, and returned to the humidity chamber. Condensation forms on the surface and tends to persist as a stagnant film of moisture. For large-scale tests, the chamber is cycled by dropping the temperature rapidly to some temperature in the neighborhood of 23° C (70° F), holding at this temperature for a short period until the specimens become chilled, and then raising the temperature rapidly to the steady-state condition of 48° C (120° F) and 95% relative humidity. It is important that the relative humidity be kept as high as possible during the period of rising temperature.

In Fig. 1 is given the daily march of humidity on representative days in cleared areas during the first week of October and of February for the Atlantic side of the Isthmus of Panama, near the Canal Zone. In the jungle the humidity will not vary greatly from these values but will depend upon the distance above ground and upon the density of the foliage. The variation in temperature and humidity will not be as

Condensation Twice Daily.   No Condensation.

FIG. 3. Zinc Plate (0.2 mil) on Steel Exposed Three Weeks at 54° C (130° F) and 85% R.H.

great near the ground in heavy growth as it will be at some distance above ground because of radiation from the earth during the cool period at night. Similarly in Fig. 2 is shown the daily march of humidity in New York City for the first weeks of August, October, and February. From these data it is apparent that equipment intended for indoor use in temperate climates can be tested in humidity chambers without condensation, but that intended for tropical climates requires a condensation cycle.

The behavior of metals and of protective coatings varies considerably in the two environments. Zinc and its alloys corrode badly in the condensation cycle but give satisfactory service in chambers without condensation. The principal use of the straight humidity chamber is to test the efficacy of porous or permeable coatings on ferrous materials. Figure 3 shows the relative corrosion of zinc-plated steel panels at 54° C (130° F) and 85% relative humidity with and without condensation.

The efficacy of temporary preservatives (slushing compounds) is usually determined by humidity testing as the tests closely simulate service conditions.

TESTS FOR STRESS CORROSION CRACKING

H. R. Copson*

The discussion of stress corrosion cracking on p. 569 shows that tensile stress, environment, time, and internal structure of the alloy are important factors, and that their importance varies. It follows that there can be no universal test of susceptibility to stress corrosion cracking. By ingenious selection of test conditions most metals can be made to crack. For a laboratory test to be of practical value, test conditions must be selected with due regard to service conditions. Special test conditions may be needed for a new problem.

Correlation between laboratory tests and service performance is difficult to obtain. Exact correlation should not be expected because service conditions are frequently complex. It may be impossible to reproduce the environment or state of stress. Substitution of some other conditions may lead to different results. Thus, with magnesium alloys, laboratory tests in salt and in salt chromate solutions showed opposite trends for the effect of precipitation heat treatments.[1]

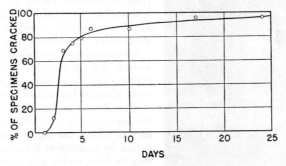

FIG. 1. Cracking of 18-8 Stainless Steel (Type 302) Specimens in a Boiling Concentrated Calcium Magnesium Chloride Brine.

Laboratory tests are restricted as to time and to specimen size. Figure 1 shows the effect of time on the percentage of specimens cracked in a particular laboratory test. Failure in an accelerated laboratory test, however, does not necessarily mean that failures are to be expected in practice. Experience has shown that in the absence of high tensile stress (internal or applied) stress corrosion cracking is not encountered in service.

Laboratory tests are useful as a means of control to insure that proper fabrication procedure is consistently employed. Thus a mercurous nitrate test is used as a control to insure that internal stresses are low in brass parts.[2]

Laboratory tests are useful in developing new alloys. They show which alloy compositions are markedly susceptible to stress corrosion cracking. For compositions found to be susceptible the conditions required to reduce this tendency can be studied. Comparisons can be made with alloys in commercial use.

Laboratory tests are useful in isolating the specific components of the environment that are responsible for stress corrosion cracking. Thus they showed that hydrogen

---

* Research Laboratory, The International Nickel Co., Inc., Bayonne, N. J.

[1] W. S. Loose, Dow Chemical Co., private communication.
[2] H. Rosenthal and A. L. Jamieson, *Proc. Am. Soc. Testing Materials*, **41**, 897 (1941).

cyanide and moisture impurities in town gas were responsible for the stress corrosion cracking of high-pressure steel cylinders,[3] and that the hydrochloric acid decomposition product of ethyl chloride was responsible for the cracking of stainless steel thermowells.[4]

Results are usually based on the observation that either a specimen cracks or it does not. Cracking may be obvious but sometimes micro-examination is necessary. Tensile test data can be used to indicate partial failure.

FIG. 2. Stress Corrosion Specimen, Direct Load. (Courtesy Dow Chemical Co.)

STRESS

Laboratory tests may be run with internal or applied stress. Internal stress may be estimated by mechanical, X-ray, or other methods.[5] With a suitable set-up, applied tensile stress may be measured directly, or from stress-strain curves by means of extensometers or electric strain gages. To avoid complications it may be preferable to start with stress-relieved material.[6]

A constant load applied directly (Fig. 2), through a lever (Fig. 3), or by means of a spring (Fig. 4) permits accurate measurement of applied stress and gives quantitative results. Precautions must be taken to keep the specimens in line, and it may be necessary to protect the highly stressed bolt hole region by painting or other means. Weights may be hung on the specimens so as to produce bending (Fig. 5), but eccentric loading (Fig. 6) seems to give more consistent results. Such tests are inexpensive, but with materials having relatively high creep rates such as magnesium they must be used with discretion.

To obtain stresses more similar to service conditions, constant deformation is sometimes preferred. U-bend (Fig. 7) and C-bend[6] specimens and other modifications[7] have been used. Care must be taken never to back up on nuts holding the specimens or stress will be relieved. Constant deformation may be applied in other ways, as by pure tensile stress without any bending or twisting.[8] Again the possibility of creep should be considered.

[3] H. Buchholtz and R. Pusch, *Stahl u. Eisen*, **62**, 21 (1942).
[4] J. C. Hodge and J. L. Miller, *Trans. Am. Soc. Metals*, **28**, 25 (1940).
[5] C. S. Barrett, *Metals and Alloys*, **5**, 131, 154, 170, 196, 224 (1934). C. A. Wilkins, *Metallurgia*, **27**, 115-117, 120 (1943).
[6] M. A. Scheil, O. Zmeskal, J. Waber, and F. Stockhausen, *Am. Welding Soc. J.* (Supplement), **22**, 493-s, 1943.
[7] W. C. Schroeder and A. A. Berk, U. S. Department of the Interior, Bureau of Mines, *Bull.* **443** (1941).
[8] E. H. Dix, Jr., *Trans. Am. Inst. Mining Met. Engrs.*, **137**, 11 (1940).

With constant deformation, cracks relieve the stress, and, if many small cracks form, growth of cracks may cease. With constant load, cracks increase the stress on remaining sound metal and hasten failure. Stress is constant only so long as dimensions do not change. Small eccentricities that throw excessive loads on certain parts of the test pieces must be avoided for consistent results. Even at high loads plastic flow will not distribute unequal distortion.

In other than atmospheric tests it is necessary to surround the specimen with the corrosive medium. One means of doing this is shown in Fig. 3. The center part of bent specimens may be immersed in the medium. Other methods are described in the literature.[8,9] It may be necessary to protect the specimen at the water line.

ENVIRONMENT

For each metal, stress corrosion cracking is associated with specific environments. These environments have to be reproduced in laboratory tests. The more common environments are described below.

ALUMINUM

Aluminum alloys high in zinc or magnesium have cracked in the atmosphere and in tap water. Stress corrosion cracking tests may be run in these environments. However, cracking is more rapid in salt solutions. Accordingly it is more customary to use solutions containing 3 to 5% sodium chloride. Complete immersion, alternate immersion, and spray have been used. To accelerate the tests, about 3 grams per liter of hydrogen peroxide* or 1% of hydrochloric acid may be added to the salt solution. To accelerate corrosion still further an anodic current has been applied to the specimen.[10] The details of many tests are given by Sager, Brown, and Mears.[11]

FIG. 3. Stress Corrosion Specimen Surrounded with Aqueous Medium, and with Load Applied through a Lever. (Courtesy Dow Chemical Co.)

* 9.1 ml of 30% H_2O_2 per liter of solution.

[9] A. Morris, *Trans. Am. Inst. Mining Met. Engrs.*, **89**, 256 (1930).
[10] E. H. Dix, Jr., *Trans. Am. Inst. Mining Met. Engrs.*, **137**, 11 (1940).
[11] *Symposium on Stress-Corrosion Cracking (1944)*, p. 255, American Society for Testing Materials and American Institute of Mining and Metallurgical Engineers, Philadelphia and New York, 1945.

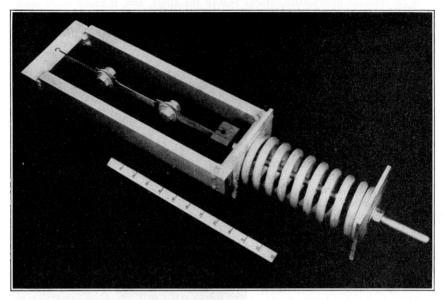

Fig. 4. Stress Corrosion Specimen, Load Applied by a Spring. (Courtesy Dow Chemical Co.)

Fig. 5. Stress Corrosion Specimens, Weights Applied at Ends. (Courtesy Dow Chemical Co.)

Brass and Other Copper Alloys

There are two outstanding accelerated stress corrosion cracking tests for copper alloys. The first is the ammonia cracking test. This endeavors to reproduce conditions causing service failures. Essentially the test involves exposing specimens to a gas phase containing ammonia, air, water vapor, and carbon dioxide. If desired, stress may be applied to the specimens. Several modifications of this test have been published.[11,12] For reproducible results the temperature must be controlled and also the composition of the gas phase, particularly the carbon dioxide content. With this test there seems to be no threshold stress below which cracking will not occur. Time to failure is a function of the stress.

The other test is the mercury cracking test,[11,13,14] which is widely used. A solution containing 1% mercurous nitrate and 1% nitric acid is recommended.[13] The specimens are degreased,

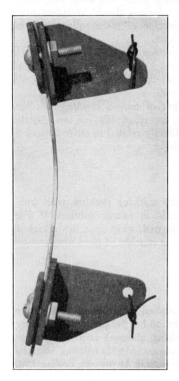

Fig. 6. Stress Corrosion Specimen, Eccentric Load. (Courtesy Dow Chemical Co.)

Fig. 7. Stress Corrosion U-Bend Specimen.

pickled in 40% nitric acid for 30 sec, and immersed in this solution for 60 min at 21° ± 5° C (70° ± 10° F). For each 6.5 sq cm (1 sq in.) of surface area there should be 10 ml of solution. The specimens must be carefully cleaned, but results give considerable scatter. Cracks may be visible, but if not, the deposited mercury should be volatilized and the specimens examined at 10 to 20 magnifications. Cracking may occur after the specimens have stood in air for some hours.

An alternative solution containing 10% mercurous nitrate and 1% nitric acid is sometimes specified.[15] With this solution an immersion of 15 min is considered

---

[12] A. Morris, *Trans. Am. Inst. Mining Met. Engrs.*, **89**, 256 (1930).

[13] H. Rosenthal and A. L. Jamieson, *Proc. Am. Soc. Testing Materials*, **41**, 897 (1941).

[14] H. P. Croft and G. Sachs, *Iron Age*, **151**, (No. 10), 47 (1943); (No. 11), 62 (1943). H. S. Rawdon, *Proc. Am. Soc. Testing Materials*, **18** (II), 189 (1918). H. Gisser, *Metals and Alloys*, **16**, 238 (1942).

[15] *A. S. T. M. Standards*, Part I, B111-42, p. 772 (1942).

sufficient. Cracking time decreases with increasing concentration of mercurous nitrate, whereas the rate of mercury deposition increases. Excessive mercury deposition tends to hide the cracks.

Small variations in acid concentration have little influence on cracking time, if the specimens have been thoroughly pickled. Cracks tend to start at defects, and the fractures are invariably wetted with mercury. Polishing delays cracking. Mercurous salts are more active than mercuric, wherefore test solutions should be stored over mercury. Cracking time is independent of contamination of the solution up to a concentration of 25 grams of copper and/or zinc per liter.

The mercury cracking test seems suitable as an inspection test. It is fast and easy to control and reveals high internal stress, but it requires a minimum threshold stress. Service cracking has been known to occur in material that has passed the mercury test. It can hardly be called a true stress corrosion test.

The action of mercury is probably similar to that of molten metals.[16] Molten metals penetrate stressed material intergranularly, an effect that is not strictly classed as corrosion. The ammonia cracking test is directly related to stress corrosion cracking and is more suitable for experimental studies.

IRON*

Caustic embrittlement is the chief stress corrosion cracking problem with iron. Accordingly laboratory tests have usually been made in caustic solutions.[17] For field tests, an embrittlement detector has been devised which can be attached directly to a steam boiler to determine if the water is capable of producing caustic embrittlement. This is described on p. 532.

STAINLESS STEEL†

Intergranular corrosion of stainless steel has been studied by means of 10% copper sulfate plus 10% sulfuric acid solution.‡ No single solution has yet proved satisfactory for the study of transcrystalline stress corrosion cracking, although acidified chloride solutions seem indicated. Boiling 60% hydrated magnesium chloride solution acidified with hydrochloric acid to pH 4 was selected for some tests.[18] At present, boiling 42% magnesium chloride (154° C, 309° F) is preferred.

MAGNESIUM

Stress corrosion cracking tests on magnesium alloys are usually made in 3% sodium chloride solution, in a solution containing 3.5% sodium chloride and 2% potassium chromate, or in the atmosphere, both inland and marine. The salt chromate solution is said to correlate fairly well with exterior tests, but not with tests in the salt solution.[19]

---

* Refer also to discussion of embrittlement in boilers, p. 531.
† Further discussion of stress corrosion cracking in stainless alloys appears on p. 174.
‡ In this country sometimes called "Strauss solution."

[16] L. J. G. van Ewijk, *J. Inst. Met.*, **56** (I), 241 (1935).
[17] W. C. Schroeder and A. A. Berk, U. S. Department of the Interior, Bureau of Mines, *Bull.* 443 (1941).
[18] M. A. Scheil, O. Zmeskal, J. Waber, and F. Stockhausen, *Am. Welding Soc. J.* (Supplement), **22**, 493-s, 1943.
[19] W. S. Loose, Dow Chemical Co., private communication.

SPECIAL METHODS
Tests for Brass
C. L. Bulow[*]

DEZINCIFICATION

Solutions of cupric chloride (cupric sulfate and ferric chloride) have been used in accelerated laboratory corrosion tests in studying the effect of alloy elements on the susceptibility of the Cu-Zn alloys to dezincification. Generally the tests are conducted with simple equipment under the following conditions:

Temperature: 20° to 80° C (68° to 175° F).
Specimens: Flat tensile specimens (0.050 in. gage; approximately 70 sq cm of surface) made to A. S. T. M. Designation E8–42.
Velocity: Specimens completely submerged, either motionless or moving at a rate of 40 cm per min (16 in. per min).
Concentration: 10 grams of cupric chloride per liter of solution. (Sometimes hydrochloric acid is added.)
Ratio of Solution Volume to specimen surface area: 2000 ml of solution per A. S. T. M. tension specimen.
Degree of Aeration: None, or numerous fine bubbles of air produced by passing one liter of air per minute through a No. 3 Pyrex porous glass plate.
Duration of Tests: Five to 30 weeks, depending upon alloy and temperature.

The ratio of depth of penetration based upon loss in tensile strength to the depth of penetration based on loss in weight gives a measure of any dezincification. A ratio of approximately 1.3 to 1.6 indicates uniform corrosion whereas a ratio of 3.0 to 6.0 or more indicates dezincification. This assumes uniform and careful cleaning. The nature of the corrosive attack should be checked by microscopic examination to detect cracking, intergranular corrosion, or pitting. This method of testing has given some information concerning the possible effectiveness of various dezincification inhibitors. Like many accelerated laboratory tests, the results obtained require the confirmation of field and service tests.

IMPINGEMENT CORROSION TEST

Various types of impingement tests have been used in the laboratory to study the effect of air-water impingement upon the copper-base alloys. This resulted in the discovery that entrained air in a turbulent stream of water entering a condenser tube may lead to the rapid pitting of the tube. The laboratory test that simulates service conditions in the best manner is the one that consists of a stream of natural water (fresh, brackish, or sea water coming from a lake, river, harbor, or the ocean), containing a known or fixed amount of air, directed against a small specimen of the alloy in question. It is advisable to conduct these tests through all seasons of the year. The essential parts of such a piece of testing equipment are shown in Fig. 1.

Tests conducted with such equipment have revealed:

1. That small quantities of air added to a stream of water may not be damaging, but beyond a critical quantity, the increase in rate of pitting is approximately proportional to the quantity of air.

[*] Bridgeport Brass Co., Bridgeport, Conn.

2. That the rate of pitting during the winter season differs quite markedly from the rate obtained during the summer season.

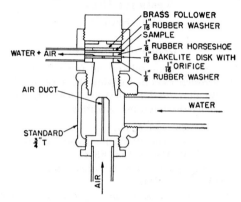

FIG. 1. Design of Jet for Impingement Tests. (Alan Morris "Seasonal Variation in Rate of Impingement Corrosion," *Trans. Am. Inst. Mining Met. Engrs.*, **99**, 274 [1932].)

SEASON CRACKING TESTS

Refer to *Tests for Stress Corrosion Cracking*, p. 1013.

Tests for Stainless Steels

H. A. PRAY*

A number of special laboratory test methods for determining the corrosion characteristics of the stainless steels are in current use. Only the more significant of them will be described here. In general, their function is to give information relating to certain particular properties of the material, such as resistance to oxidizing acid attack, susceptibility to intergranular corrosion, or resistance to pitting. The data obtained from such tests ordinarily bear no direct relation to behavior in a particular service, and care and judgment must be exercised in their interpretation and application.

BOILING NITRIC ACID TEST

This is the most commonly used acceptance and quality test for the Fe-Cr-Ni and the Fe-Cr alloys. The effects of poorly balanced alloy composition and of improper heat treatment are revealed by the test. It should not be used as the basis for prediction of behavior in other than nitric acid service, although the test results can be so used if correlated with previous experience with similar alloys in the particular service in question. The test has proved very useful in the development of improved alloys and heat treatments.

First described by W. R. Huey,[1] the details of testing procedure have changed somewhat since the original description. The following procedure is representative of

---

* Battelle Memorial Institute, Columbus, Ohio.

[1] *Trans. Am. Soc. Steel Treating*, **18**, 1126 (1930).

present practice and is in conformance with current (unpublished) A.S.T.M. recommendations.

Corrosive Medium

The solution used is C.P. analytical reagent grade nitric acid, adjusted with distilled water to 65% ± 0.2% by weight as determined by analysis.

Apparatus

The nitric acid solution is maintained at boiling temperature, usually in a 1-liter Erlenmeyer flask provided with either a 30-in. reflux condenser (with ground glass joint) or a bulb-type condenser to prevent loss of acid. The flask is fitted with glass hooks or cradles for supporting the test specimen. The ordinary laboratory electric hot plate is satisfactory for heating. A typical set-up is shown in Fig. 1.

Fig. 1. Laboratory Set-up for Boiling Nitric Acid Test.

Test Specimen

The size and shape of the specimens are determined by the type of product being tested and the capacity of the apparatus available. The surface area of the specimen is limited in accordance with the volume of nitric acid solution used. In cutting the specimen, care is taken to avoid an unnecessarily large cross-sectional area, or, in the case of a casting, a large area from its interior. With the 1-liter flask described above, it is usually possible to secure an adequate test specimen that can be weighed on an analytical balance.

Except where it is necessary to test a particular surface finish, all surfaces of the specimens are finished with No. 120 grit abrasive paper. The specimen is then measured and the total surface area calculated. It is then degreased in acetone or other suitable solvent, dried, and weighed.

Procedure

The specimen is completely immersed in a quantity of the nitric acid solution sufficient to provide at least 125 ml of solution per sq in. (19 ml/sq cm) of surface area. The solution is heated to boiling and maintained at boiling for a period of 48 hours. The specimen is removed, rinsed, freed of any adherent corrosion products, dried, solvent degreased, and weighed. The test is repeated to a total of five 48-hour periods, weight loss being determined, and fresh acid provided for each test period.

A separate flask is ordinarily used for each specimen, although a number may be tested in the same container provided (1) the total area does not exceed the 125 ml per sq in. requirement, (2) the materials are of the same type, and (3) the corrosion rate of none of the specimens is excessive.

In many cases, as much information is secured at the end of three test periods as at the end of five. However, most specifications require that the five-period test be completed, and it is considered to be the safer practice. It is usually unnecessary to make duplicate tests.

Results

Data are most commonly reported as inches penetration per month (30 days) for each test period and for the entire 240 hours. This may be calculated from the weight loss using the following equation:

$$\text{Penetration (in./month)} = \frac{43.9 \times \text{Wt. loss (grams)}}{\text{Area (sq in.)} \times \text{Density (grams/cc)} \times \text{Time of test (hr)}}$$

Examples of data obtained from the test are given in Table 1. Alloys A, B, and C are compositions that differ in average corrosion rate, but show satisfactory constancy of rate from period to period. This constancy of rate indicates that the alloys have been properly heat-treated. Alloy D shows about the same average rate as alloy C. The individual period rates, however, increase progressively. The heat treatment of this material was the same as for A, B, and C. The fact that the corrosion rate increases from period to period indicates in this case that the composition is outside

TABLE 1. EXAMPLES OF CORROSION DATA FOR CAST 19% CR–9% Ni ALLOYS IN BOILING 65% HNO₃

Alloy	Penetration in./month from Weight Loss — 48-Hour Test Periods					Average
	1	2	3	4	5	
A	0.00087	0.00072	0.00075	0.00073	0.00077	0.00077
B	0.00115	0.00117	0.00114	0.00113	0.00115	0.00113
C	0.00194	0.00186	0.00165	0.00173	0.00172	0.00178
D	0.00142	0.00152	0.00185	0.00195	0.00206	0.00177
E	0.00538	0.01348	....	....	....	....

the range amenable to the heat treatment to which A, B, and C respond. The alloy, in that condition, would be considered unsuitable for many uses. Alloy E is identical in composition with A but differs in heat treatment. The very high corrosion rate and the rapid increase shown in the second test period are indicative of the severe intergranular attack which was evident on subsequent microscopic examination.

INTERGRANULAR CORROSION TESTS

The austenitic stainless steels in certain environments and with certain thermal histories are subject to intergranular attack associated with the precipitation of carbides at the grain boundaries. (Discussed on p. 161.) The susceptibility of a given material to intergranular corrosion can be detected by microscopic examination for the presence of precipitated intergranular carbides, and the relative susceptibility can be determined by special corrosion tests.

Boiling Nitric Acid Test

As indicated in Table 1 (alloy E), a high corrosion rate and a rapidly increasing rate during successive periods of the boiling 65% nitric acid test are evidence of intergranular susceptibility. The test procedure can be modified to give a more nearly quantitative idea of the relative susceptibility of materials, particularly useful for cast alloys.

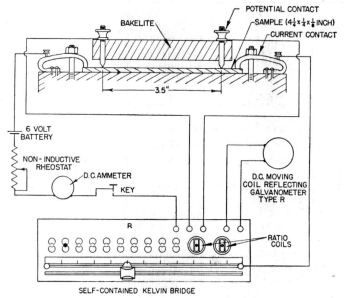

Fig. 2. Diagram of Apparatus for Measuring the Change in Electrical Resistance Caused by Intergranular Corrosion.

The modification consists in measurement of the change in electrical resistance, as well as weight loss, of the test specimen. Figure 2 describes diagrammatically a method of making the electrical resistance measurements. Dimensions of the test specimen determine the precision with which they can be made. Intergranular attack decreases the effective cross section of the sample for current flow, and the depth of intergranular penetration is readily calculated from the resulting increase in electrical resistance.[2]

[2] J. J. B. Rutherford and R. H. Aborn, *Trans. Am. Inst. Mining Met. Engrs.*, **100**, 293 (1932).

In the *absence* of intergranular attack, the penetration calculated from weight loss and that calculated from the increase in electrical resistance are the same or nearly so, the agreement between the two values being dependent upon the precision of the resistance measurements. Obviously, if intergranular corrosion occurs, the value determined by the resistance method is the greater of the two. The ratio of the electrical resistance penetration to the weight loss penetration is an index to the character of the attack. When this ratio is small, corrosion is relatively uniform and confined to the metal surface, whereas high values indicate the occurrence of intergranular attack.

Examples of data obtained from a single alloy in various conditions of heat treatment are shown in Table 2. The specimen of example F was not attacked intergranularly, whereas the specimens of examples G, H, and I were. The severity of intergranular attack is, at least, semi-quantitatively defined by the magnitude of the ratio.

TABLE 2. EXAMPLES OF CORROSION DATA FOR A CAST 19% Cr–9% Ni ALLOY IN BOILING 65% NITRIC ACID

Example	Average Penetration (in./month)		
	From Wt. Loss (b)	From Electrical Resistance (a)	Ratio a/b
F	0.00095	0.0009	1.0
G	0.00244	0.0042	1.7
H	0.00549	0.0207	3.8
I	0.00608	0.0580	9.5

COPPER SULFATE AND SULFURIC ACID TEST

This test[3] is at present in use to a limited extent for determining relative intergranular susceptibility. It depends upon the highly selective attack of the boiling $CuSO_4$–H_2SO_4 mixture on the susceptible grain-boundary material. Fully quench-annealed austenite is attacked at an extremely slow rate. The corrosive solution now generally used contains 47 ml per liter of concentrated C.P. sulfuric acid and 13 grams/liter of $CuSO_4 \cdot 5H_2O$. The specimen is totally immersed in the boiling solution in an apparatus similar to that used for the boiling nitric acid test. The test is ordinarily conducted for an indefinite number of 72-hour periods, with fresh solution provided for each period. A specimen is considered to be susceptible if cracks appear when it is subjected to a bend of 180° over a radius of half the specimen thickness. The time on test before the appearance of these cracks defines the relative susceptibility of different specimens.

More nearly quantitative information can be obtained by use of electrical resistance measurements.[4]

The test has proved useful in the study of intergranular corrosion of stainless steels, but it is considerably less sensitive in revealing susceptibility than boiling nitric acid. This is shown by the examples listed in Table 3. Examples J and K show no evidence of intergranular attack in either reagent. In the absence of intergranular corrosion, it can be seen that the $CuSO_4$–H_2SO_4 solution corrodes the alloy very slowly, whereas appreciable rates are observed for an identical specimen in the nitric

[3] B. Strauss, H. Schottky, and J. Hinnüber, *Z. anorg. allgem. Chem.*, **188**, 309 (1930).
[4] J. J. B. Rutherford and R. H. Aborn, *Trans. Am. Inst. Mining Met. Engrs.*, **100**, 293 (1932).

TABLE 3. COMPARISON OF THE RATES OF PENETRATION (in./month) OF CAST 19% Cr–9% Ni ALLOYS IN BOILING 65% HNO₃ AND BOILING CuSO₄-H₂SO₄ SOLUTIONS

Example	Nitric Acid*		$CuSO_4$-H_2SO_4		Total Time of Test (hr)
	From Weight Loss	From Electrical Resistance	From Weight Loss†	From Electrical Resistance†	
J	0.0013	0.0016	0.00002	0.0000	288
K	0.0018	0.0018	0.00003	0.0001	288
L	0.0024	0.0235	0.00003	0.0002	864
M	0.0066	0.0770	0.00105	0.0451	288

* Average of five 48-hr periods.
† Calculated penetration for examples J, K, and L using $CuSO_4$-H_2SO_4 is of the same order as the experimental error; hence interpretation is uncertain.

acid test. The specimen of example L is clearly shown to be susceptible by the nitric acid test, but the similar effect of the $CuSO_4$-H_2SO_4 reagent, even after prolonged exposure, does not reveal this fact. The higher carbon, and therefore more susceptible, alloy M is shown to be definitely subject to intergranular corrosion by either test.

The principal difficulties in the use of the $CuSO_4$-H_2SO_4 test lie in its lack of sensitivity and in the uncertainty regarding the duration of test that is necessary to reveal carbide precipitation. This is particularly true in cases of low or intermediate susceptibility.

ELECTROLYTIC TEST

A rapid test for determining the susceptibilty of the austenitic Cr-Ni-Fe alloys to intergranular corrosion has been developed[5] whereby precipitated carbides are revealed by anodic surface treatment.

A diagram of the cell and of the electrical circuit used are shown in Fig. 3. The electrolyte is placed in the small lead cup, C, which serves as the cathode, is insulated from the test specimen, D, by a rubber gasket, E, and is held firmly in place by an insulated clamp. The size of the spot can be changed to fit conditions. An area of contact between specimen and electrolyte ⅜ in. in diameter is a convenient size. A 6- to 9-volt battery, a variable resistance, and a 10-amp ammeter are connected in the circuit as shown.

The area to be tested is given a fairly smooth finish with several grades of emery paper to correspond to the first stage of a metallographic preparation. All scratches, however, need not be removed.

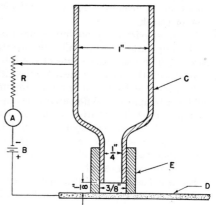

FIG. 3. Electrolytic Cell Used in Rapid Test for Intergranular Susceptibility of Austenitic Stainless Steels.

[5] H. W. Russell, H. A. Pray, and P. D. Miller, *Trans. Am. Inst. Mining Met. Engrs.*, **145**, 357 (1941); *Trans. Am. Foundrymen's Assoc.*, **50**, 918 (1942).

The electrolyte is a 60% (by weight) solution of a sulfuric acid, containing 5 ml/liter of Glycyrrhiza Extract, U. S. P. (licorice extract).

To conduct the test, 2 ml of electrolyte are placed in the cell, and the current adjusted to an anode current density of 14 amp per sq in. With a spot 3/8 in. in diameter, this amounts to 1.5 amp. As the treatment progresses, some heating of the electrolyte occurs, but the current is maintained by adjustment of the variable resistance. The time required for 19% Cr–9% Ni alloys is 1.5 min, and for molybdenum-containing alloys 3 min.

After the specimen is rinsed and dried, the treated spot is visually examined or photographed under oblique light. In the absence of intergranular precipitated carbide, the test surface will have acquired a high specular reflectivity so that it appears black under oblique illumination. However, precipitated carbides if present are selectively attacked and impart a rough frosty or grainy appearance to the spot under the same illuminating condition. The angle of illumination should be 20° or less.

The sensitivity of the test is essentially the same as for the boiling nitric acid test. The principal difficulty is interpreting the appearance of the spot. With a little experience in its use, the operator is able to reveal as much information on the existence of carbide precipitation and susceptibility to intergranular attack by this test as by the more laborious and time-consuming nitric acid test.

PIT CORROSION TESTS

Under some conditions of exposure the stainless steels are subject to a severe pitting type attack, particularly in the presence of chlorides. (Discussed on p. 165.) Various chloride solutions, particularly ferric chloride, have been used in laboratory tests to evaluate the resistance of materials to this tendency. Ferric chloride, and mixtures of ferric with other chlorides, in solution will cause pitting of these alloys. The phenomenon is more or less erratic, and the pitting tests have been most successful in demonstrating gross differences between materials susceptible to this type of attack.

The most effective procedure is described in detail by Smith.[6]

The apparatus is essentially an inverted U-tube, the legs of which are water-jacketed. The lower, open ends of the tube fit into a glass disk. When the apparatus is filled with the corrosive medium, the disk is put into loose contact with the sample to be tested, and solution is caused to circulate over the specimen surface by maintaining a temperature differential between the two legs of the tube. The test solution contains 10.8% $FeCl_3 \cdot 6H_2O$ in $0.05 N$ HCl, although the more resistant materials require a more concentrated solution. The duration of the test is ordinarily 4 hr. Both weight loss and the number of visible pits are recorded. It is necessary to probe the surface with a needle (dental probe) to reveal the depth of pitting since undercutting is prevalent.

The test has been useful in development work, although correlation between test results and service behavior has not been fully established.*

STRESS CORROSION CRACKING

Refer to *Stress Corrosion Cracking in Stainless Alloys*, p. 174.

* Some comparative data are listed in Table 2, p. 172.

[6] H. A. Smith, *Metal Progress*, **33**, 596 (1938).

Steam Test for Zinc

E. A. Anderson*

Zinc die casting alloys deficient in magnesium or exceeding the A.S.T.M. limits (see specification B86–43) for impurity content are subject in service to subsurface corrosion, generally of an intercrystalline type. This occurs most rapidly in warm, moist atmospheres. A laboratory test to simulate service failures of this type has been in widespread use. It consists of exposing samples in an atmosphere of air saturated with water at 95° C (203° F) under conditions permitting condensation of hot water on the metal. This condensation is accomplished by suspending the samples in an insulated air chamber over a water bath containing immersed heating units. The water temperature is adjusted to keep the atmosphere temperature at 95° C (203° F). To obtain results comparable with other installations, it is often necessary to use an average temperature slightly above or below 95° C.

Experience has shown that castings whose properties (e.g., impact strength) and dimensions are not significantly altered by a 10-day exposure in this test will not suffer intergranular attack in atmospheric service. Conversely, castings damaged in this test also fail in service.

Although the 95° C steam test has been used to some extent as an acceptance or control test, it is best used as a laboratory tool in the evaluation of new alloys. Control and acceptance requirements are adequately covered by analytical determinations of the composition and the impurity contents.

Testing of Organic Coatings†

K. G. Compton‡ and R. J. Phair‡

Two general procedures may be followed in establishing the corrosion protection afforded by organic finish systems. One method is based on exposure studies in various corrosive environments for sufficient intervals of time to develop visible deterioration of the coating or corrosion of the base. The second procedure consists in making measurements of the physical properties of the organic coatings at periodic intervals during exposure and in estimating the amount of expected protection from the rates of degradation.

Nearly all of the exposure tests made upon bare metals or metallic coatings are utilized in evaluating the protective level of organic coatings. In general, these include the natural environments such as indoor and outdoor atmospheric exposure, sea water immersion, and burial in the soil; and the various laboratory media such as humidity rooms, salt-spray chambers, cycling devices, and exposure to various chemicals.

In making comparisons of the corrosion protection afforded by organic coatings, it is usually advisable to employ panel stock of the same material as that to which the coatings will be applied in use, and it is extremely important that the surfaces of the test specimen be properly prepared. Improper surface preparation may seriously

---

* Research Division, New Jersey Zinc Co., Palmerton, Pa.
† Refer also to *Field Testing of Paints*, p. 1071.
‡ Bell Telephone Laboratories, New York, N. Y.

affect adherence[1,2,3] of the organic coatings and vitiate the test results. For exposure testing of organic coatings on steel surfaces, a tentative method[4] covering the type of steel, its surface condition, and methods for surface preparation is available.

APPLICATION OF COATINGS AND MEASUREMENT OF FILM THICKNESS

Since it has been repeatedly demonstrated that the thickness of the organic coating has decided effects on its protective level and physical characteristics, much attention has been devoted to methods for applying films of controlled and uniform thickness. The development of automatic spray machines,[5] controlled dipping devices,[6] and doctor blade application methods[7] has made possible the control of films to within 0.0025 mm (0.0001 in.) of the desired thickness. Comparative studies[7] of the various methods of controlled application have been undertaken, and A. S. T. M. Method D–823–45T covering the different application techniques has been prepared.

Measurement of dry film thickness may be accomplished by several means. These include (1) accurate micrometer measurements of a spot on the panel before and after removal of the film, (2) needle penetration devices[8] with electrical indication of contact with the base surfaces, and (3) magnetic methods measuring in arbitrary units[9] the change in attractive force (with variations of film thickness) between a magnet and the base, or being influenced by the reluctance of a magnetic circuit produced by interposition of a non-magnetic film between the measuring device[10] and a magnetic base. A method[11] based on a change of inductance is reputed to give reliable results when measuring non-conducting films on non-magnetic bases.

EVALUATING PROTECTION

The method for determining corrosion protection based on exposure to the point of failure is estimated by measuring, as a function of exposure time, the depth of pitting[12] or weight loss in the base, and by visual estimates of the degree of deterioration of the coating or the amount and type of corrosion covering the surface. Estimates of the type and degree of paint deterioration and of the amount of corrosion developed are usually based on arbitrary numerical or descriptive scales originated by the separate investigators.[13] Efforts at standardization of visual methods have resulted in the establishment of photographic guides, one of which indexes and illustrates six degrees of corrosion without blistering and four degrees with blistering[14]; another represents degrees of paint deterioration.[15]

---

[1] National Paint, Varnish, and Lacquer Association, Scientific Section, *Circ.* 604, pp. 501–528 (1941).

[2] L. A. Jordan, *Chemistry and Industry*, **56**, 361–371 (1937).

[3] R. M. Burns and A. E. Schuh, *Protective Coatings for Metals*, pp. 312–321, Reinhold Publishing Corp., New York, 1939.

[4] *A. S. T. M. Standards*, D–609–45T, Part II, Supplement, p. 192 (1945).

[5] H. G. Arlt, *Metal Ind.* (New York), **38**, 123–124 (1940).

[6] H. F. Payne, *Ind. Eng. Chem.*, Anal. Ed., **15**, 48–56 (1943).

[7] National Paint, Varnish and Lacquer Others, Scientific Section, *Circ.* 604, pp. 415–436. (1941)

[8] R. E. Wheeler, *Products Finishing*, **5**, 54–56, 58 (April, 1941).

[9] A. Brenner, *J. Research Natl. Bur. Standards*, **20**, 357–368 (1938).

[10] E. S. Gallagher. *Am. Machinist*, **86**, 1155–1157 (1942).

[11] A. L. Alexander and Others, *Ind. Eng. Chem.*, Anal. Ed., **17**, 389–393 (1945).

[12] G. H. Young and Others, *Ind. Eng. Chem.*, **36**, 341–344 (1944).

[13] H. A. Gardner, *Physical and Chemical Examination of Paints, Varnishes, Lacquers, and Colors*, 9th Ed., pp. 158–159, Institute of Paint and Varnish Research, Washington, D. C., 1939

[14] *A. S. T. M. Standards*, D–610–43, Part II, pp. 1109–1115 (1944).

[15] National Paint, Varnish and Lacquer Association, Scientific Section, *Fed. Papers*, pp. 79–92 (1940)

A modification of the usual visual method of evaluating corrosion resistance involves application of the finish to a thin iron foil (0.127 mm [0.005 in.]) secured to a transparent backing.[16] Because of the thinness of the metal base the occurrence of corrosive attack is almost immediately visible through the transparent backing, permitting evaluation before evidence of attack becomes visible at the finish surface.

In the second method of evaluation based on measurements of the physical characteristics of the coating and the rate of change of these properties during exposure, selection of the tests depends in large measure on the proposed usage of the finish system. However, since adequacy of the coating as a protective medium is dependent in large measure on its adhesional level and distensibility, physical tests showing changes in these properties are basic in indicating rates of degradation. Associated with these tests, initial measurements of moisture permeability of the film and of the electrochemical behavior as influenced by pigmentation are valuable in determining the probable protective level. In addition to corrosion protection, the selection of protective coatings is influenced by the amount and kind of physical abuse to be expected during their service life; hence tests that evaluate impact and abrasion resistances and hardness are frequently employed.

Since temperature and humidity have decided effects[17] on the physical characteristics of organic coatings, equilibration to some standard conditions at each testing period is extremely important, if valid results are to be obtained.

ADHESIONAL MEASUREMENTS

In the actual measurement of the adherence of a coating to its base, the results are modified and influenced by such other characteristics of the film as its cohesion, plasticity, and brittleness. For instance, in the direct tensile method of measurement, unless coherence of the film structure is greater than the adhesion between coating and base it will be the former rather than the latter property that will be evaluated. From a theoretical standpoint, therefore, the various methods for determining adherence may not be wholly adequate, but in practice reasonable correlation is obtained between the test results and performance on exposure.

The methods commonly employed may be grouped in three general classifications; those dependent (1) on direct tensile pull, (2) on scratching or cutting action, and (3) on deformation of the base. Only a limited number of instruments representative of the various groups[18] will be mentioned here.

Tensile Pull Test

A critical survey[19] of available methods undertaken by the New York Paint and Varnish Production Club to establish a satisfactory adhesion tester led to the adoption of a technique based on tensile pull. In this method[20] the adhesional level is expressed as the force, applied parallel to the plane of the film, required to strip the coating from the base. The painted panel is secured in one jaw of a tensile test machine, and a wooden block secured under pressure to the paint surface with a thermoplastic cement so as to cover a standardized area of film, is held in the other.

[16] G. D. Patterson and C. K. Sloan, *Ind. Eng. Chem.*, Anal. Ed., **16**, 234–238 (1944).

[17] R. M. Burns and A. E. Schuh, *Protective Coatings for Metals*, p. 337, Reinhold Publishing Corp., New York, 1939.

[18] H. A. Gardner, *Physical and Chemical Examination of Paints, Varnishes, Lacquers, and Colors*, 9th Ed., pp. 125–128, Institute of Paint and Varnish Research, Washington, D. C., 1939.

[19] National Paint, Varnish and Lacquer Association, Scientific Section, *Fed. Papers*, pp. 123–146 (1940).

[20] National Paint, Varnish and Lacquer Association, Scientific Section, *Circ.* 604, pp. 439–482 (1941).

A modification[21] of this technique reported in a study of adhesion of lacquers to metal involves small areas of contact and therefore permits bonding of the two previously painted sections of the tensile specimen by application of the same material as that under test.

SCRATCHING METHOD

Adhesion evaluation by the scratching or cutting method is accomplished by measuring the force that must be applied to a cutting or scratching tool to remove the finish from its base. The tool may be relatively blunt[22,23] and thus tend to push the film from the substrate as it is drawn over the finish, or it may be sharpened to a knife edge and introduced to the finish surface in such a manner as to cut[19,24] the film from the base or to act as a wedge[25] forced between the two elements. In these methods gradually increasing pressures are applied until stripping of the film occurs, or the reactive force is transmitted to calibrated spring or lever arrangements, which indicate the pressures developed. Stresses developed in an attached film as it is cut in successive parallel lines will be sufficient to displace the coating if the area between such cuts is small enough. The spacing between lines to produce displacement of the film will be decreased as the adhesional level increases. A machine based on this principle has been announced.[26] Reasonable degrees of precision have been reported, but the ductility and cohesion of the film and any non-uniformity of the substrate surface influence quantitative interpretations of the results obtained by these methods.

DEFORMATION OF THE BASE

Estimations of adhesional level based on the behavior of the applied paint during deformation of the base, as with the Erickson tester,[27] have been advocated.[26,28] In operation, a hemispherical plunger is slowly forced down on the back of the test panel, which is supported on a firm base having a round opening beneath the plunger. This action is continued until rupture of the film occurs. It is apparent however, that the distensibility or elongation characteristics of the coating will affect the results obtained by this method.

DISTENSIBILITY

Determinations of the distensibility or elongation characteristics of attached organic films have served as useful criteria[29] in estimating their aging behavior. Measurements of this property are made by direct extension of the base to which the coating is applied or by bending the coated specimen over a mandrel. In the former method[29] a coated specimen is slowly stretched in a modified tensile machine until the coating ruptures. From the increase in distance between lines scratched in the metal before the extension, the percentage of elongation may be determined. In the latter method[29] coated metal strips are bent over a series of cylindrical mandrels,

---

[21] R. L. Savage and Others, *Am. Paint J.*, **27**, 13–18 (Oct. 27, 1942).
[22] E. Rossmann, *Farbe u. Lack*, pp. 485–497 (1938).
[23] R. J. Phair, *Bell Lab. Record*, **23** (8), 284–286 (1945).
[24] P. Koole, *Philips Tech. Rev.*, **8** (5), 147–148 (1946).
[25] W. König, *Farben-Ztg*, **44**, 1230–1232 (1939).
[26] K. N. Kathju, *Ind. Eng. Chem.*, Anal. Ed., **16**, 144 (1944).
[27] H. A. Gardner, *Physical and Chemical Examination of Paint, Varnishes, Lacquers, and Colors*, 9th Ed., p. 123, Institute of Paint and Varnish Research, Washington, D. C., 1939.
[28] A. V. Blom, *Paint Oil Chem. Rev.*, **101**, 9–10 (1939).
[29] A. E. Schuh and H. C. Theuerer, *Ind. Eng. Chem.*, Anal. Ed., **29**, 9–12 (1937).

which vary in diameter. The degree to which the film may be distended is determined by the smallest diameter mandrel over which bending occurs without rupture of the coating. The amount of elongation at the surface of $\frac{1}{32}$-in. panels when bent over the various mandrels has been experimentally determined for three different metals, and the results may therefore be expressed as percentage of elongation.

This method, however, is discontinuous and groups the coating in one of six levels of distensibility. To overcome this drawback, a conical mandrel[30] apparatus, which would give a continuous rather than a stepwise evaluation, has been developed. The mandrel, which is 8 in. long, varies in diameter from 0.317 cm (0.125 in.) at the smaller end to 3.8 cm (1.5 in.) at the larger end. A drawbar wraps the coated panel closely around the mandrel, and the distensibility is determined from the maximum diameter at which rupture of the film occurs. Evaluation is facilitated by charts giving percentage of elongation in terms of distance from the base of the cone. The method is approved under A.S.T.M. designation D-522-41.

The Erickson tester previously mentioned is also employed in determining distensibility.

IMPACT AND ABRASION RESISTANCE

Testers giving a measure of the resistance that may be expected of a coating when subjected to impact shock and to abrasive action are useful in determining the degree to which a coating will preserve continuity of coverage in service.

Impact Resistance

Impact resistance may be measured by the free falling ball method or by hammer arrangements. In one of the simplest devices a steel ball is dropped from a fixed height on to the test panel. The number of falls necessary to produce rupture of the finish is a measure of the impact resistance. A somewhat similar principle is utilized in a falling hammer device,[31] which is loosely pivoted at the handle end and is dropped from a vertical position on to the panel. The height of the hammer is adjustable, permitting direct or glancing blows, and the striking surface of the hammer may be either ball-like or chisel-shaped.

A further modification of impact testing is realized in a machine[32] in which a loosely pivoted arm rounded to a ball on the impacting end is rotated at gradually increasing speed and strikes glancing blows on the coated surface as the test specimen is moved beneath it at a rate synchronized to the hammer rotation. Impact resistance is expressed as the speed of rotation of the hammer required to cause rupture of the coating through to the base.

Abrasion Resistance

Devices that measure abrasion resistance may be separated into two groups. In one group, the abrading is accomplished by the rubbing action of the abrasive medium, whereas in the other, abrasive particles impinge on the paint surface. The rubbing technique is exemplified in a well-known abrading device[33] in which the test panel is mounted on a table rotating at a definite speed beneath two arms carrying abrasive

[30] H. G. Arlt, American Society for Testing Materials, *Bull.* 89, pp. 5–9 (December, 1937).
[31] H. A. Gardner, *Physical and Chemical Examination of Paints, Varnishes, Lacquers, and Colors*, 9th Ed., p. 125, Institute of Paint and Varnish Research, Washington, D. C., 1939.
[32] H. G. Arlt, *Bell Lab. Record*, **15** (4), 108–110 (1936).
[33] H. A. Gardner, *Physical and Chemical Examination of Paints, Varnishes, Lacquers, and Colors*, 9th Ed., p. 131, Institute of Paint and Varnish Research, Washington, D. C., 1939.

wheels. The arms are freely pivoted, permitting the wheels to rest on the paint surface in such a position that a shearing action is developed. Mounting the panel to the table requires that a hole be drilled in the center of the specimen. The precision is dependent on the preservation of uniformity of the abrading surface during test and on the temperature developed during abrasion. Abrasion resistance is expressed as the number of revolutions of the specimen beneath the wheels required to expose the base.

Abrasion by impingement is accomplished in a tester[34] employing a finely divided abrasive, which is driven against the paint surface under a controlled air blast. The abrasive, falling at a controlled rate through an orifice, enters a special nozzle in which it is picked up by the air stream and strikes the specimen mounted beneath it. Abrasion resistance is expressed as the number of grams of abrasive required to wear through the coating and expose a definite area of the base surface. The results are influenced by such factors as abrasive particle size, nozzle dimensions, air pressure, and angle of impingement. By careful control of these variables[35] a high degree of reproducibility is realizable. The method has been tentatively approved under A. S. T. M. Standards, designation D–658–44.

Another test device[36] employs a falling stream of sand as the abrasive medium.

ABSORPTION AND PERMEABILITY[37]

Ascertainment of the moisture resistance of coatings requires measurement of both absorption and diffusion characteristics. Absorption levels may be determined by equilibrating the coating specimen to two conditions of humidity and measuring the weight increase due to absorption at the higher level. Similar measurements may be taken before and after immersion in water for various time intervals. By suspending[38] the specimen on a calibrated spring, the rate of absorption upon exposure to a specific humidity or the amounts of water absorbed at increasing vapor pressures may be determined.

Moisture diffusion is usually measured by interposing the organic film between two atmospheres differing in water-vapor pressure and determining the amount of moisture passing through the material in a given time. Numerous modifications[39,40,41] in technique have been studied in measuring diffusion constants, but a method commonly employed consists in carefully sealing the material either as a free film or as a coating applied to unplasticized cellophane, parchmentized paper,[40] or glass cloth[42] over the top of a cuplike cell. The test may be conducted by placing a water-absorbent substance such as phosphorus pentoxide in the cell prior to sealing, exposing the specimen to a controlled humid atmosphere, and recording the gain in weight at periodic intervals. The measurement may also be made by noting the loss in weight when the cell is partially filled with water and is exposed to a desiccated atmosphere. The diffusion constant is calculated from the following relationship based on Fick's linear diffusion law.

[34] A. E. Schuh and E. W. Kern, *Ind. Eng. Chem.*, Anal. Ed., **3**, 72–76 (1931).
[35] H. G. Arlt, *Proc. Am. Soc. Testing Materials*, **40**, 967–975 (1940).
[36] C. C. Hipkins and R. J. Phair, *Am. Soc. Testing Materials Bull.*, **143**, 18–22 (December, 1946).
[37] *Protective and Decorative Coatings*, J. J. Mattiello (Editor), Vol. IV, pp. 194–222, John Wiley and Sons, New York, 1944.
[38] P. T. Newsome, *Ind. Eng. Chem.*, **20**, 827 (1928).
[39] R. L. Taylor and Others, *Ind. Eng. Chem.*, **28**, 1255–1263 (1936).
[40] National Paint, Varnish and Lacquer Association, Scientific Section, *Circ.* 546, pp. 1–81 (1938).
[41] A. S. T. M. Standards, D–697–42T, Part III, pp. 1280–1283 (1942).
[42] H. F. Payne, *Ind. Eng. Chem.*, Anal. Ed., **11**, 453–458 (1939).

$$D = \frac{G \cdot X}{A \cdot T \, (P_1 - P_2)}$$

where
D = Diffusion constant,
G = Amount of water gained or lost,
X = Film thickness of coating,
T = Time of exposure,
A = Area of film exposed,
$P_1 - P_2$ = Difference in partial pressures of water vapor.

The diffusion constant is determined after a constant rate of gain or loss of moisture has been attained.

POROSITY

Films may be tested for porosity and other discontinuities by such methods as application of copper sulfate solution to a coated steel panel and examination for localized copper precipitation. An electrographic method[43] reputed to give a greater resolution of such imperfections has been reported. In use on iron or steel, a paper saturated with potassium ferrocyanide is placed over the coated panel, which is made the anode in a low-voltage circuit. Almost immediate evidence of porosity is indicated by the development of the characteristic blue coloration of the iron salt.

It has been shown[44] that film electrical resistivity is a discontinuous function of coating weight and that a definite threshold thickness must be attained before continuous films are formed. Furthermore, the minimum coating weight is not a constant for the film-forming material but varies considerably depending on the metal surface to which the coating is applied.

The quantitative determination[44] of this characteristic of films is possible by a measurement of the current instantly flowing when the coated panel is employed as the anode in a single cell circuit. As actual current measurement is unsatisfactory under the conditions of test, it is calculated from the potentials developed and the internal cell resistance.

ELECTROCHEMICAL TESTS

It has been shown[45] that a knowledge of the corrosion behavior of painted iron and of the mechanism of the corrosion protection afforded by various paints may be obtained by a study of the time-potential curves obtained with a vacuum tube electrometer for the cell:

Fe; paint film, tap water : sat'd KCl, HgCl; Hg

Coated iron rods may be immersed in the water, or flat coated panels may be tested by employing a glass U-tube connection. In the latter method[46] it has been found desirable to employ a nitrogen atmosphere around the electrode. Curves having a time axis of a few hours or more are usually sufficiently indicative of corrosion behavior.

The time-potential curves so obtained that show iron indefinitely cathodic to the normal hydrogen electrode indicate, as with red lead, corrosion inhibition; whereas those that in time show iron anodic, as with iron oxide, indicate physical protection.

[43] A. Glazunov and L. Jenicek, *Korrosion u. Metallschutz*, **16**, 341–344 (1940).
[44] G. H. Young and G. W. Gerhardt, *Ind. Eng. Chem.*, **29**, 1277–1280 (1937).
[45] R. M. Burns and H. E. Haring, *Trans. Electrochem. Soc.*, **69**, 169–179 (1936).
[46] H. E. Haring and R. B. Gibney, *Trans. Electrochem. Soc.*, **76**, 287–302 (1939).

The method has been applied to a study[46] of a large number of pigmentations with evidence of its usefulness in predetermining probable exposure behavior of paints.

Tests for Metal Coatings

WILLIAM BLUM*

INTRODUCTION

In the laboratory, the protective value of metal coatings may be estimated directly by some accelerated test, or indirectly by measuring those properties that are known or believed to be directly related to the protective value, e.g., the thickness, adhesion, hardness, and porosity of the coatings.

THICKNESS

In most cases the protection furnished by a metallic coating depends upon the minimum thickness of the coating. This is frequently specified. Determination of the average thickness by appropriate stripping methods is still employed with metals such as gold and silver whose intrinsic cost is significant; and also for some coatings whose local thickness is not easily measured.

Experience has shown that the protective value of zinc and cadmium coatings on steel is nearly proportional to their thickness. With cathodic metal coatings such as copper, nickel, chromium, and silver on steel, the protective value increases with the coating thickness (but not linearly) principally because the porosity of the coatings decreases with an increase in thickness. When such coatings are thick enough to be nearly impervious, further increase in thickness has no direct effect upon their protective value.

The local thickness of a metal coating may be determined by the following methods:

1. *Microscopic*.[1] This is the most generally applicable procedure and is commonly used as the primary or umpire method. It is accurate to about ±5% for coatings more than .0025 mm (0.1 mil) thick. It requires special technique with soft metal coatings such as tin and zinc. The specimens are destroyed.

2. *Chord*.[2,3] This method involves cutting through the coating on a curved surface with a flat file, or on a flat surface with a precision grinding wheel, and measuring the width of the exposed chord. The thickness is $C^2/8R$, where C is the chord width and R is the radius of the surface or wheel. Its accuracy is about ±10% for coatings down to .005 mm (0.2 mil).

3. *Immersion*. In the Preece test for zinc coatings, the article is immersed in neutral copper sulfate solution for one minute intervals until adherent copper is deposited. This test[4] measures the relative distribution rather than the actual thickness at a given point.

* National Bureau of Standards, Washington, D.C.

[1] C. E. Heussner, *Monthly Rev. Am. Electroplaters' Soc.*, **23**, 5 (1936). *A. S. T. M. Standards*, Part I, p. 1830 (1944).

[2] F. C. Mesle, *Metal Ind.* (New York), **33**, 283 (1935).

[3] W. Blum and A. Brenner, *J. Research Natl. Bur. Standards*, **16**, 171 (1936).

[4] E. C. Groesbeck and H. H. Wallkup, *J. Research Natl. Bur. Standards*, **12**, 785 (1934). *A. S. T. M. Standards*, Part I, p. 561 (1944).

4. *Spot Tests.* The thickness of very thin chromium coatings[5] is measured (to ±10%) by the time required for one drop of concentrated hydrochloric acid completely to dissolve the coating at that point.

5. *Dropping Method.*[6,7,8] In this method an appropriate reagent is allowed to drop on the surface at a controlled rate, and the time for penetration of the coating is noted. The accuracy is about ±10%. This method is used principally for zinc and cadmium coatings on steel, but is also applicable to tin and copper coatings on steel.

6. *Jet Method.*[9] This is similar to the dropping test, except that the reagent is applied in a continuous fine stream. This method has been applied to nickel, copper, and silver coatings on steel and non-ferrous metals.

7. *Electrochemical.*[10] The period in which a coating of, e.g., chromium, dissolves anodically at a specified current density is used as a measure of thickness to an accuracy of ±5%.

8. *Magnetic.* All the above methods destroy the coatings. The magnetic methods are non-destructive. In one of the magnetic methods,[11] the change in attraction of a permanent magnet is used as a measure of the thickness of a non-magnetic (or less magnetic) coating on steel. The attraction of the magnet is also used to measure the thickness of nickel (or other magnetic metals) on a non-magnetic base. The accuracy is ±10%.

In other types[12] an electromagnet is used to measure the thickness of the coatings.

9. *X-ray.* Methods have been suggested[13] for measuring the thickness of metal coatings but have not been widely adopted. Proposals for use of electron diffraction have also received consideration.

10. *Density.*[14] When the base metal and the coating have quite different specific gravities, measurement of the average specific gravity indicates the proportion of each and hence the average (but not the local) thickness of the coating.

ADHESION

The protective value of any metal coating depends partly upon its adhesion. If the adhesion is poor, the coating is easily detached by severe abrasion or distortion of the surface and may blister as a result of temperature changes or of local corrosion.

There are no very satisfactory methods for quantitative measurement of the adhesion of coatings, especially on finished articles. For them, it is customary to bend, break, stretch, twist, hammer, or otherwise deform the article and the coating, and note whether any separation occurs, especially at a point of fracture. Separation may be facilitated by using a knife point or a chisel edge. At best, such methods are not quantitative and usually indicate merely "good" or "poor" adhesion of the coating.

Recently[15] the use of a rapidly vibrating hammer for testing the adhesion of

---

[5] W. Blum and W. A. Olson, *Proc. Am. Electroplaters' Soc.*, **28**, 25 (1940). *A. S. T. M. Standards*, Part I, p. 1831 (1944).

[6] S. G. Clarke, *J. Electrodepositors' Tech. Soc.*, **8**, No. 11, 1–10 (1933).

[7] R. O. Hull and P. W. C. Strausser, *Monthly Rev. Am. Electroplaters' Soc.*, **22**, 9–14 (March, 1935).

[8] A. Brenner, *J. Research Natl. Bur. Standards*, **23**, 387 (1939).

[9] S. G. Clarke, *J. Electrodepositors' Tech. Soc.*, **12**, 1, 157 (1936).

[10] S. Anderson and R. W. Manuel, *Trans. Electrochem. Soc.*, **78**, 373 (1940).

[11] A. Brenner, *J. Research Natl. Bur. Standards*, **18**, 565 (1937); **20**, 357 (1938).

[12] W. H. Tait, *J. Electrodepositors' Tech. Soc.*, **14**, 108 (1938).

[13] H. R. Isenberger, *Proc. Am. Electroplaters' Soc.*, **27**, 77 (1939).

[14] L. S. Larkin, *J. Electrodepositors' Tech. Soc.*, **18**, 48 (1942).

[15] A. W. Hothersall and C. J. Leadbeater, *J. Electrodepositors' Tech. Soc.*, **19**, 49 (1944).

metal coatings has been suggested. The evidence there presented and also some unpublished data indicate that this method serves only to detect coatings that have very poor adhesion.

Methods are now under investigation in which the reflection of X-rays or of supersonic rays is used to detect a poor bond at an interface. Such methods may prove useful for specific items but are not likely to be generally applicable.

Another method depends upon the application of a fluorescent oil to a surface at which a boundary is exposed, removing the excess of oil, and examining with ultraviolet or "black" light. This serves to detect a crack or gap at the exposed edge but gives no certain evidence of the adhesion at some distance from the edge.

The two quantitative adhesion tests that have been proposed involve the preparation of special specimens and are therefore not applicable to plated products.

In Jacquet's method,[16] and as modified by Mesle,[17] the force necessary to pull the coating from a strip is measured.

In Ollard's method[18] the force required to pull an annular ring of metal from the end of a cylinder is determined. Roehl[19] has described the precautions necessary for reliable results by this method. Recently a "micro-Ollard" test has been described in which smaller specimens are used.

HARDNESS

Although the hardness of metal coatings is chiefly of interest when good resistance to abrasion is required, the latter property may influence the protective value, especially of coatings cathodic to the base metal. Scratches through the coating permit corrosion of the base metal. Actually the wear resistance of a metal is dependent upon other factors in addition to hardness, including brittleness and toughness. No general quantitative methods for measuring wear resistance are available.

One difficulty in measuring the hardness of relatively thin metal coatings is that the apparent hardness is affected by that of the base metal. To avoid or reduce this error, the depth of the indentations should be only a small fraction of the coating thickness. The most commonly used methods are the micro-Vickers and the Knoop indentor.[20] Scratch hardness measurements are usually less reproducible.

POROSITY

If a metal coating is cathodic to the base metal in the prevailing environment, the coating corrodes less readily than the base metal and does not protect it at pores or discontinuities. In such cases, the protective value depends primarily upon the porosity of the coating. With plain carbon steel as the base metal, practically all presently applied metal coatings except zinc and cadmium act as noble metals. (Laboratory tests indicate that aluminum coatings on steel act as noble metals in pure or nearly pure water but as "sacrificial" metals in salt solutions.)

Because zinc and cadmium are less noble than steel, they protect appreciable exposed areas of steel; hence the porosity of the coatings is not very important. Exposure of the steel in pores may slightly accelerate corrosion of the adjacent zinc or cadmium so that the presence of many pores, especially in thin coatings, may be

[16] P. A. Jacquet, *Trans. Electrochem. Soc.*, **66**, 393 (1934).
[17] F. C. Mesle, *Proc. Am. Electroplaters' Soc.*, **27**, 152 (1939).
[18] E. A. Ollard, *Trans. Faraday Soc.*, **21**, 81 (1926).
[19] E. J. Roehl, *Iron Age*, **146**, 17 (Sept. 26) and 30 (Oct. 3, 1940).
[20] C. G. Peters and F. Knoop, *Metals and Alloys*, **12**, 292 (1940).

detrimental. No satisfactory method has been devised for detecting pores in zinc or cadmium coatings.

The detection of pores in any cathodic metal coating, e.g., on steel, depends upon the application of a reagent that will attack and thereby reveal any exposed steel and that will at the same time not attack the coating to any appreciable extent.

1. *Ferroxyl Test.* The best known of these methods is the Ferroxyl Test, in which a corroding reagent, usually sodium chloride, is mixed wih sodium ferricyanide and is applied in an aqueous solution containing a gel such as agar.* This mixture may be absorbed in paper that is moistened and applied to the coating. Blue spots appear wherever steel is exposed.

Although this test is useful for routine inspection, the results are usually not sufficiently reproducible to warrant its inclusion in specifications. If the concentration of ferricyanide is too high, coatings such as nickel may be attacked and pores may form.

2. *Hot-Water Test.* Immersion in hot distilled water[21] has been used to detect pores in tin or nickel coatings on steel.

3. *Salt-Spray Test* (described on p. 970). For cathodic coatings on steel, the salt-spray test[22] is essentially a measure of porosity. The time required for the first appreciable corrosion is usually less significant than the number of rust spots at the end of a specified period, e.g., 50 or 100 hours. Many specifications[23] provide that in a given period not more than six visible rust spots per square foot shall appear or any spots more than $\frac{1}{16}$ in. (1.6 mm) in diameter.

It has been found that prolonged exposure of, e.g., nickel coatings to the salt spray sometimes greatly increases the number of pores, apparently by local attack of the coating. Although it is possible that the resultant corrosion corresponds to the behavior of the coating that might take place in marine exposure, the results of very long exposures of coatings to salt spray are of doubtful validity.

For sacrificial metals like zinc and cadmium, the salt-spray test is a rough measure of the thickness of coating. Zinc coatings are attacked to form white corrosion products (basic zinc carbonate) and, when the zinc is removed from an appreciable area, the steel rusts. The rate of attack of cadmium in the salt spray is much slower than that of zinc. The protective value of zinc and cadmium depends mostly on the coating thickness, which can be rapidly and accurately measured. The salt-spray test is, therefore, not generally specified for zinc or cadmium coatings, although it is often employed to test the chromate finishes on zinc.

4. *Humidity Test* (described on p. 1006). Exposure to air with high humidity at high temperatures may also serve to detect pores in noble metal coatings. Intermittent cooling and condensation of moisture usually accelerate the attack, but at best this procedure is a very slow and not a very reproducible method of detecting pores.

5. *Intermittent Immersion.* Periodic immersion of the specimens in a solution of sodium chloride or other reagent, with intermediate drying in the air, is sometimes recommended. To secure reproducible results it is necessary to control not only the temperature of the solution and of the air but also the humidity and circulation of the air. The latter determines the rate of drying and may thereby affect corrosion.

6. *Hydrogen Evolution.*[24] This highly specific test is used to estimate the relative

---

* Dissolve 10 grams of agar in one liter of hot water. Add one gram of NaCl and one gram of $K_3Fe(CN)_6$. To detect cathodic areas as well, add a few drops of phenolphthalein indicator.

[21] A. W. Hothersall and R. A. F. Hammond, *Trans. Electrochem. Soc.*, **73**, 449 (1938).
[22] *A. S. T. M. Standards*, B117–44T, Part I, p. 1843 (1944).
[23] *A. S. T. M. Standards*, Part I, p. 1820 (1944).
[24] V. Vaurio, B. Clark, and R. Lueck, *Ind. Eng. Chem.* (Anal. Ed.), **10**, 368 (1938).

porosities of tin coatings on steel, as used in the canning industry. The time required to evolve 5 ml of gas from a specified area when treated with $1 N$ HCl at $57°$ C ($135°$ F) is taken as a measure of porosity.

7. *Electrographic Test.** In this test a flat surface of the coated article is made the anode for a few seconds in a "sandwich" type of cell under pressure (500 psi) in which the cathode consists of aluminum or other metal not yielding colored compounds. A double layer of filter paper moistened with 5% Na_2CO_3 and 1% NaCl solution serves as the intervening electrolyte.

Sufficient potential is applied to the cell to overcome couple action between the coating and exposed basis metal, and the current is permitted to flow for a controlled period of 5 to 15 sec. In general, with most coatings of practical thickness, the probability of creating pinholes during the test is not great since the quantity of current passed is relatively small.

The paper member of the cell sandwich is removed and immersed in a solution containing an anion that, with the cations of the basis metal, precipitates an identifiable colored compound. The procedure may be simplified by having the indicating agent in the paper of the cell before test.

For some coatings on steel, a suitable impregnant is potassium ferrocyanide. For tin-coated brass, a reagent paper containing precipitated antimony sulfide may be used. It is immersed in dilute phosphoric acid before use. In the first case, the pinholes are revealed by the appearance of spots of Turnbull's blue, and in the second case by the precipitation of copper sulfide. For lead coatings on steel an electrolyte of 0.05% sodium chloride 2.5% potassium ferrocyanide has proved satisfactory.

GENERAL REFERENCES

Among the American Electroplaters' Society research projects, the following relate to this chapter:
1. *Methods of Testing Adhesion*, on which *Am. Electroplaters' Soc. Research Reports* 1 and 2, by A. L. Ferguson and Associates, have been published.
2. *Porosity of Electrodeposits*, directed by N. Thon.
3. *Methods for Testing Thickness of Electrodeposits*, directed by H. J. Read.
4. *Physical Properties of Electrodeposited Metals*, directed by Abner Brenner.

Tests for Corrosion by Lubricants

C. M. Loane†

BEARING CORROSION

The weight losses of bearings in full-scale engine block tests and of bearing test strips in laboratory beaker tests are used as a measure of the bearing corrosion characteristics of lubricating oils. The engine tests are well standardized and are the performance criteria (other measurements besides bearing corrosion are made in these tests) in U. S. Army Specification 2–104–B for heavy-duty motor oils. A number of laboratory tests have been used but, so far, none is generally accepted as predicting performance in service.

---

* Described by C. L. Sample, The International Nickel Co., New York, N. Y. See also Burns and Schuh, *Protective Coatings for Metals*, p. 252, Reinhold Publishing Corp., New York, 1939.

† Research Department, Standard Oil Co., Whiting, Ind.

TABLE 1. ENGINE BEARING CORROSION TEST METHODS

Name of Test	Engine Used*	Test Bearings	RPM	Brake Horse-power	Jacket Temp., °C	Oil Temp., °C	Fuel	Test Duration, Hours	Max. Cu-Pb Weight Loss Tolerated	Reference to Test Procedure
Chevrolet 36-hour Test	Conventional Chevrolet engine	Two new Cu-Pb bearings	3150	30	93.3	138 or 130†	2.5 to 3 ml PbEt₄ per gal.	36	Not specified (0.35 gram or less per whole bearing is OK)	C. R. C. Designation L-4-545
Caterpillar "Hot Box" Test	Specially equipped 4-cylinder Cat. Diesel (run in "hot box" to maintain high temperatures)	At least two new Cu-Pb bearings	1400	37	93.3	100	Clear (no lead tetraethyl)	120	0.100 gram per whole bearing	C. R. C. Designation L-3-545
G. M. C. 500-Hour Test	Model 71 Diesel Engine	Two new Cu-Pb bearings	2000	72	82.2	110	Clear (no lead tetraethyl)	500	Not specified (0.40 gram or less per whole rod bearing is OK)	C. R. C. Designation L-5-545

* In all tests the engines are mounted on stationary beds and are coupled to dynamometers or fan brakes to absorb the load.
† 130° C — 10 grade oils; 138° C — 30 and 50 grade oils.

TABLE 2. LABORATORY BEARING

Test Name	Sample Container	Sample Size, ml	Rate of Stirring	Air Blowing, liters per hour	Oil Temp., °C
Stirring Tests					
Caterpillar	500-ml tall form beaker	400	60 rpm	....	132
Indiana Stirring*	500-ml tall form beaker	250	1300 rpm	....	149
MacCoull	500-ml tall form beaker	125	3000 rpm	....	177
Shell Corrosion and Stability	1-liter beaker	225	1310 rpm	36	184
Shell Thrust Corrosion	160-steel cup	35	2400 rpm	3	Varied 100 to 170
Air Blowing Tests					
Continental	Large test tube	160	....	10	172
Sohio	500-ml test tube	160	....	70	138
Shell Existent Corrosivity	20 + ml test tube	20	....	0.38	156
Circulating Oil Tests					
Underwood	Large metal sump	1500	(Oil sprayed through air)	....	163

* Bearing segments usually Cu-Pb.

Engine Tests[1]

The three engine test methods are presented in summarized form in Table 1. Detailed test procedures are described in the Cooperative Research Council Test Procedures L–3, 4, and 5–545.

The Chevrolet Test is the most widely used test for evaluating corrosion characteristics of oils, and extensive cooperative test work has been carried out on the method.[2]

Laboratory "Beaker" Tests

A number of the more prominent laboratory bearing corrosion tests are summarized in Table 2. The Underwood Test has been used more extensively than any one of

[1] W. B. Bassett, *Natl. Petroleum News*, **36**, R–450 (1944).
[2] H. C. Mougey and S. A. Moller, *S. A. E. Journal*, **50**, 417 (1942).

CORROSION TEST METHODS

Catalysts	Test Duration, hours	Bearing Test Strip	Special Features	Reference
\multicolumn{5}{c}{Stirring Tests}				
Cu, Pb	65	Pb	Only mild agitation provided, sufficient to keep oil circulating	....
Cu, Pb Cu-Pb PbO Powder	(Until 3300 mg/sq dm Pb loss)	Pb	Life-type test; a number of Pb strip weighings are made	Ind. Eng. Chem. Anal. Ed., **17**, 89 (1945)
Cu, bearing segment*	10	Bearing segment*	Oil forced at high rate of shear over bearing surface	S. A. E. Trans., **50**, 338 (1942)
Bearing segment*	3 or 5	Bearing segment*	Bearing segment wiped by felt pad	Ind. Eng. Chem. (Anal. Ed), **15**, 550 (1943)
Bearing segment*	20	Bearing segment*	Positive washing of bearing surface by oil shear	Ind. Eng. Chem. (Anal. Ed.), **17**, 168 (1945)
\multicolumn{5}{c}{Air Blowing Tests}				
Iron wire	48 or 72	Bearing segment	Indiana oxidation-type test, modified by the use of catalysts	S. A. E. Preprint, June, 1940
Fe, Cu, Cu-Pb Fe (Et hexoate) PbBr$_2$	36	Cu-Pb, Cu	Attempt to provide all types of catalytic conditions prevalent in Chevrolet 36-Hour Test	Ind. Eng. Chem., Anal. Ed., **17**, 302 (1945)
Pb or Cd	0.33	Pb or Cd	Attempt to measure existent corrosivity of oil without further oxidation	Ind. Eng. Chem., **36**, 263, 1944
\multicolumn{5}{c}{Circulating Oil Tests}				
Cu, bearing segments,* metal naphthenates	5 or 10	Cu-Pb or Cd-Ag	Oil sprayed through air against bearing surface and circulated by pump	S. A. E. Trans., **43**, 385 (1938)

the other tests, although its limitations in predicting engine performance have been recognized.

COPPER CORROSION IN LUBRICANTS

Copper Strip Test for Sulfur Corrosion[3]

A 1.3 by 7.6 cm (0.5 by 3 in.) clean copper strip is immersed in a sample of oil held in a test tube fitted with a vented stopper. The tube is heated at 100° C (212° F) for 3 hours. The strip is removed, rinsed free of oil, and examined. Any discoloration or pitting indicates corrosion.

[3] *Federal Standard Stock Catalogue*, Section IV, Part 5, p. 259, May 1945, Method 530.31.

TEST FOR TENDENCY OF OIL TO DISSOLVE COPPER[4]

Fifty ml of oil in a test tube are heated at 93° C (200° F) in contact with 56 cm (22 in.) of clean No. 22 copper wire. One-ml samples of oil are taken periodically and mixed with an equal volume of dithizone reagent. If the green color of the reagent is changed to purple, the oil contains dissolved copper. The number of hours elapsed before a positive copper test is obtained is a measure of the resistance of the oil toward dissolving copper. (The test shows only fair correlation with copper plating in refrigerator systems.)

RUST-PREVENTING CHARACTERISTICS OF STEAM TURBINE OILS[5]

A polished, cylindrical steel specimen, 12.7 by 140 mm (0.5 by 5.5 in.) is suspended in 300 ml of oil held in a 400-ml tall form beaker. Thirty ml of distilled water are added. The oil-water mixture is held at 60° C (140° F) while being stirred at 1000 rpm. After 48 hours, the steel specimen is removed, washed with naphtha, and examined. If the steel shows no rust spots (in duplicate tests), the oil is considered to have passed the test.

Soil Corrosion Tests

I. A. DENISON*

For the study of soil corrosion phenomena in the laboratory, a corrosion cell has been described[1] with which the behavior of various soils and metals can be investigated under uniform conditions of moisture, aeration, etc. The cell consists essentially of two electrodes of the same metal separated by a layer of moist soil constituting the electrolyte. Differential aeration of the two electrodes is brought about by making one of them more accessible to air than the other so that the cell develops an electromotive force.

The essential parts of the corrosion cell, illustrated in Fig. 1, are a Bakelite ring 44 mm (1.75 in.) in internal diameter and 20 mm (0.79 in.) high; a metal disk 38 mm (1.50 in.) in diameter, which serves as the anode; and another disk of the same material, which serves as the cathode and is perforated with 51 holes per sq cm (330 per sq in.), the diameter of each being 0.84 mm (0.033 in.). The metal disks are cut from the same sheet material with a punch and die. A narrow tab or projection is provided on the cathode to facilitate electrical connection. The total area of the perforated cathode is almost twice the area of one side of the anode.

The cell is packed with soil saturated with water by capillarity. The excess water is then removed by suction, the pressure difference being 59 cm of mercury. This pressure applied for one hour reduces the moisture content of the soil to that normally contained by soils in nature when gravitational and capillary forces have become established after saturation of the soil by rainfall. After adjustment of the moisture content, the anode, previously prepared by rubbing one face with 1G French emery paper and cleaning with carbon tetrachloride, is placed in position on

* National Bureau of Standards, Washington, D. C.

[4] C. E. Waring (Frigidaire Corp.), private communication.
[5] *A. S. T. M. Standards on Petroleum Products and Lubricants*, p. 369 (1945), Method D665–46T.
[1] K. H. Logan, S. P. Ewing, and I. A. Denison, *Symposium on Corrosion Testing Procedures*, p. 97. American Society for Testing Materials, 1937.

the surface of the soil, and a connection is made with the cathode by means of a No. 24 copper wire soldered to the reverse side of the anode near the edge. A No. 7 rubber

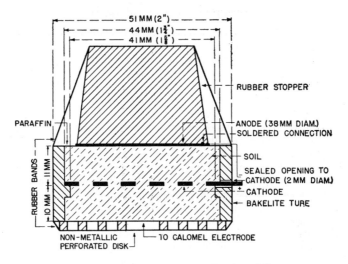

FIG. 1. Cross Section of the Denison Cell.

stopper is placed on the anode, a perforated insulating disk is placed over the lower end of the cylinder, and the whole assembly is fastened together by rubber bands. The cells are placed in individual one-pint friction-top cans containing water to provide a saturated atmosphere. The details of packing, adjustment of the moisture content, and assembly of the cells have been previously given.[1]

Short-circuit currents of the cell are measured by means of a "zero resistance" ammeter,[2] the circuit for which is given in Fig. 2. This kind of instrument is used because the introduction of even a low-resistance milliammeter in a low-resistance circuit may alter the total current by an appreciable amount.

Measurements of the separate electrode potentials in the cell while current is flowing are made by the method of Hickling[3] as adapted to the Denison cell by Darnielle.[4] In this method the current is periodically interrupted for very short intervals of time and the potentials are measured during the period of inter-

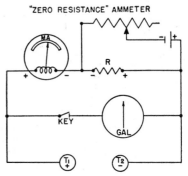

FIG. 2. A Drop in Voltage across the Terminals T_1 and T_2 Is Opposed by an Equal Voltage Drop across Resistance R, Such That the Net Drop across T_1 and T_2 Is Reduced to Zero.

[2] Sensitive Research Instrument Corp., *Catalog* 40, New York, N. Y.
[3] A. Hickling, "Studies in Electrode Polarization. Part 1 — The Accurate Measurement of the Potential of a Polarized Electrode," *Trans. Faraday Soc.*, **33**, 1540 (1937).
[4] R. B. Darnielle, "Measurement of Electrode Potentials and Polarization in Soil Corrosion Cells," *J. Research Natl. Bur. Standards*, **25**, 421 (1940).

ruption, which can be made very short (of the order of 10^{-5} sec) so that errors due to depolarization are small.

INTERPRETATION OF DATA

From data obtained with the corrosion cell, various relationships and critical values can be determined that indicate the relative corrosiveness of different soils, the relative resistance to corrosion of different metals, and the type of control of the corrosion reaction.

If it were not for the high internal resistance of the cell, which limits the value of the current on closed circuit, the rate of corrosion could be determined simply by measuring the cell current at a suitable stage of corrosion. It is preferable, however, to eliminate the effect of resistance by making use of current-potential curves. As the current density is increased, the potentials of the anode and cathode approach the same value. At the point where the two potentials are the same, the cell is completely polarized, the internal potential difference is zero, and the applied electromotive force just balances the internal IR drop. The potential of the electrodes at this point is known as the *corrosion potential*. The associated current density is the limiting value for the particular combination of soil and electrodes and corresponds to the current for the maximum rate of corrosion.[5]

The results of the various electrical measurements made on the corrosion cell with steel electrodes are illustrated in Table 1 for a group of soils differing widely in

TABLE 1. CORROSIVENESS OF SOILS AS INDICATED BY ELECTROLYTIC MEASUREMENTS AND LOSS OF WEIGHT OF THE ANODES*

Soil No.	Soil Type	Maximum Open-Circuit Potential of Cell, volt (H_2 Scale)	Corrosion Potential,† volt (H_2 Scale)	Current at Corrosion Potential, ma	Maximum Short-Circuit Current, ma	Loss of Weight,‡ mdd
64	Docas clay	0.31	0.34	2.72	2.19	404
103	Billings silt loam	0.30	0.32	3.86	2.85	337
45	Unidentified alkali soil	0.38	0.29	2.55	2.20	337
56	Lake Charles clay	0.37	0.29	2.92	1.64	316
113	Imperial clay	0.38	0.31	2.90	1.70	310
57	Merced clay adobe	0.30	0.35	2.00	1.70	285
23	Merced silt loam	0.34	0.22	1.88	1.05	198
51	Acadia clay	0.12	0.31	1.20	0.80	194
8	Fargo clay loam	0.12	0.37	0.78	0.55	159
111	Fresno fine sandy loam	0.16	0.21	1.20	0.60	146
2	Bell clay	0.14	0.33	0.90	0.54	138
7	Unidentified soil	0.14	0.24	0.50	0.42	103
1	Allis silt loam	0.14	0.22	0.60	0.23	67
41	Summit silt loam	0.07	0.26	0.40	0.20	59
20	Mahoning silt loam	0.20	0.30	0.67	0.32	55
25	Miami clay loam	0.01	0.32	0.005	0.003	4

* I. A. Denison and R. B. Darnielle, "Observations on the Behavior of Steel Corroding under Cathodic Control in Soils," *Trans. Electrochem. Soc.*, **76**, 199 (1939).
† Potential at intersection of current density-potential curves.
‡ Calculated from results in a 2-week run.

[5] I. A. Denison and R. B. Darnielle, "Observations on the Behavior of Steel Corroding under Cathodic Control in Soils," *Trans. Electrochem. Soc.*, **76**, 199 (1939).

corrosiveness. The data are arranged in descending order of corrosiveness as indicated by the loss in weight of the anodes during a 2-week test period. It can be seen that there is a definite correlation between the loss in weight and the current at the corrosion potential or the maximum short-circuit current. The ratio of these two currents is fairly constant in soils of low resistivity and tends to increase in soils of high resistivity, as would be expected.

Metals such as stainless steel, lead, copper, and copper alloys, which tend to form protective films in soils, develop only small potential differences, and, since the above-described cell has a high resistance, only a negligible current flows between electrodes of these metals. Under these conditions the corrosion is due almost entirely to local action currents, and obviously the short-circuit current would not be sufficient to account for the observed corrosion. In order to measure the rate of corrosion under such conditions, use is made of the fact that where the rate of corrosion is determined by the cathode reaction, the local action current and the current required to protect the metal cathodically are identical.

The "protective" current is determined by causing increasing currents to flow from the perforated disk to the solid disk and measuring the potential of the solid disk corresponding to each value of current. By plotting the logarithm of the current with the potential, curves are obtained, consisting usually of a horizontal or nearly horizontal part and a section having a negative slope (Fig. 3). The protective current is taken as the value of current at the intersection of the two curves.* For materials corroding under anodic control the protective current can be considered only an approximation to the true corrosion rate because under such conditions the current required to protect against corrosion is always greater than the local action current.

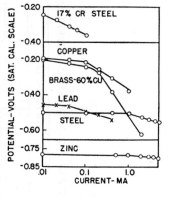

FIG. 3. Current-Potential Curves for Solid Disks of Various Metals When Made Cathodic in Docas Clay.

Comparisons have been made between the results of laboratory tests with a variety of metals and those of long-time field tests.[6] The rates of corrosion measured electrolytically were in agreement with the field tests to the extent that the laboratory measurements indicated whether or not the materials were seriously corroded in the different soil environments.

To investigate the mechanism of corrosion in soils as is done in other environments it is often desirable to determine the type of control of the corrosion reaction, that is, at which electrode the rate of corrosion is chiefly controlled.† From inspection of current-potential curves for various metals corroding under a variety of soil conditions, general conclusions have been drawn about the type of control for different metals in various soil environments. The results of this study are given in Table 2.

The corrosion of zinc and steel is under cathodic control in all of the soil conditions studied because in the presence of chloride, sulfate, carbonate, and bicarbonate ions in an environment deficient in oxygen, the primary corrosion products of zinc and

---

* The theory is discussed in *Fundamentals of Cathodic Protection*, p. 923.
† For discussion of types of corrosion control, refer to p. 486.

[6] I. A. Denison, "Electrolytic Behavior of Ferrous and Non-Ferrous Metals in Soil-Corrosion Circuits," *Trans. Electrochem. Soc.*, **81**, 435 (1942).

TABLE 2. TYPE OF CORROSION RATE CONTROL OF METALS IN DIFFERENT SOIL ENVIRONMENTS*,†

Metal	Soil Environment							
	Alkali Carbonate		Chloride	Sulfate	Chloride and Sulfate	Reducing, with Organic Acids	Reducing, with Organic Acids and Sulfate	Reducing, with Organic Acids, Chloride and Sulfate
	Good Aeration	Poor Aeration						
Low-carbon steel	Cathodic‡	Cathodic	Cathodic	Cathodic	Cathodic	Cathodic	Cathodic	Cathodic
Steel with 17% Cr	Anodic	Anodic	Anodic	Anodic	Anodic	Anodic	Anodic	Cathodic
18-8 steel	Anodic	Anodic	Anodic	Anodic	Anodic	Anodic	Anodic	Cathodic
Copper	Anodic	Anodic	Anodic	Anodic	Anodic	Cathodic	Cathodic	Cathodic
Brass (60–40)	Anodic	Anodic	Anodic	Anodic	Anodic	Cathodic	Cathodic	Cathodic
Zinc	Cathodic	Cathodic	Cathodic	Cathodic	Cathodic	Cathodic	Cathodic	Cathodic
Lead	Anodic	Anodic	Anodic	Anodic	Anodic	Cathodic	Anodic	Anodic

* I. A. Denison, "Electrolytic Behavior of Ferrous and Non-ferrous Metals in Soil Corrosion Circuits," *Trans. Electrochem. Soc.*, **81**, 435 (1942).
† A third type of control is "mixed control," in which both the anode and cathode appreciably polarize upon flow of electric current. This classification specifies only the electrode reaction that is *predominantly* controlling.
‡ Anodic with exceptionally good aeration.

steel are soluble and consequently diffuse away from the anode surface readily. Under these conditions a fairly large potential can be maintained at the anode even at relatively high current densities. However, in very porous soils that are either strongly alkaline or deficient in soluble salts the type of control of steel and zinc shifts from cathodic to anodic. The fact that the corrosion of zinc is generally under cathodic control has naturally an important bearing on its use as a protective coating for steel and as an anode in cathodic protection installations.

Field and Service Tests

ATMOSPHERIC EXPOSURE TESTS

C. P. LARRABEE*

Many costly investigations have been made in an attempt to develop an accelerated laboratory test that enables one to find the comparative atmospheric corrosion resistance of materials. Some are described by Rawdon,[1] but none is entirely reliable. The comparative protective qualities of the rusts that form on plain and low-alloy steels in outdoor exposure determine to a large extent their subsequent corrosion rates. As the oxide films produced under accelerated conditions may not be so protective, the results of such tests are usually misleading. Therefore, in order to obtain results comparable with service, steels must be exposed where a rust typical of that produced under service conditions will be formed. There is no adequate substitute for the service test.

The desirability of obtaining information on the resistance of ferrous materials to corrosion by various atmospheres has led to the adoption of substantially similar methods for comparing steels and arriving at certain definite conclusions as to their relative merits. Hereafter, steels will be used as examples, although other materials such as lead, copper, or zinc can be exposed in like manner.

The following factors must be considered in arriving at the optimum conditions for testing.

TYPES OF ATMOSPHERES

Considerable evidence is available[2] to prove that the ratio of corrosion rates for plain and copper steels is about the same whether the test is made in an industrial atmosphere (Pittsburgh, Pa.), at a semi-rural location (Fort Sheridan, Ill.), or where there is less industrial pollution (Annapolis, Md.), if the two kinds of steels are exposed at a given location at the same time. This similarity extends to the low-alloy steel classes. Therefore, a test made in a very severe industrial atmosphere will give the relative merits of materials at locations much less corrosive in character but will not necessarily place them in the same numerical ratios. In marine atmospheres the effect of certain constituents of steel on the corrosion rate may be greater than in industrial or semi-rural atmospheres, but frequently is of lower magnitude (p. 120). Therefore, for the most reliable indication, tests should be made in the type of atmosphere in which the product is most likely to be used.

* Research Laboratory, Carnegie-Illinois Steel Corp., Pittsburgh, Pa.

[1] H. S. Rawdon, "Atmospheric Corrosion Testing," *Symposium on Corrosion Testing Procedures*, pp. 36–56, American Society for Testing Materials, Philadelphia, Pa., 1937.

[2] "Reports of Committee A-5, Subcommittee III," *Proc. Am. Soc. Testing Materials*, **16-44**.

Nature of Test Data Desired

The kind of information required determines the relative importance of (1) discoloration, (2) pitting, (3) loss of weight, (4) visible perforation, (5) loss of strength, and (6) changes in other physical properties. For (1), only visual inspection may be necessary. For (2) and (3), removal of the corrosion products as described in *Preparation and Cleaning of Specimens*, p. 1077, is essential. The loss of strength can be determined by preparing tensile test specimens from the corroded sample without further treatment, or by exposure of the tensile test specimens themselves. On certain materials fatigue tests are employed as a measure of performance. In the tests of A.S.T.M. Committee A-5, Subcommittees III* and VIII,* visible perforation is the measure of performance. However, with the more resistant materials, the number of years before results are available may be considerable, and, furthermore, there is always a question of the rust clogging actual perforations in the metal itself for an indeterminate length of time.

Number of Different Materials to Be Tested

The variety of materials also depends upon the purpose of the investigation. Available materials for many applications are numerous, but previous practice and economical considerations narrow the field so that those most likely to give satisfactory service with reasonable overall cost can be selected.

The number of duplicate specimens to be used depends upon whether any are to be removed at stated intervals. For visual inspection, two samples are usually sufficient. For loss of weight, the desired accuracy and type of information are the controlling factors. Some investigators consider that four samples of each material should be removed at one time. Others prefer a single specimen of each material but use two or more individual sources for each type of product. For plain and low-alloy steels, five removal periods, each longer than the preceding one, are considered necessary for conclusive results. Specimens removed and cleaned should not be re-exposed, as the surface is different from that of the original test piece.

Size and Preparation of Specimens

The type of test to be employed determines the nature of the specimens. If visual inspection is to be the criterion, as is usually the case with electroplated and hot-dipped coatings, the specimens should receive the same treatment as is accorded similar finishes before being put into actual service.

For coated specimens in loss of weight and pitting tests, 4- by 6-in. (approximately 10- by 15-cm) specimens seem to be a good compromise. This size has been adopted by Subcommittee II of A.S.T.M. Committee B-8[3] for its 1944 exposure test of steels plated with non-ferrous metals. With smaller samples there is danger of error due to non-uniformity and intensified edge effect. With larger samples, the necessary rack space makes a test of many materials prohibitive and increases the difficulties of weighing and handling.

The nature of the rolling scale on steels varies greatly and may or may not be protective. In order that the effect of the composition of the steel on corrosion will not be masked by other variables, and to give more comparable initial results, it is advisable to use cleaned surfaces. They can be cleaned chemically (pickling) or mechanically (sand or shot blasting, or machining). Pickling is recommended as more nearly approximating the treatment of surfaces used in actual service.

---

[3] "Reports of Committee B-8, Subcommittee II," *Proc. Am. Soc. Testing Materials*, **44**.
* Now Subcommittee XIV.

A suitable means of identification of specimens should be used. Steel stencils are satisfactory for materials such as stainless steels, but for those that corrode more severely, one of two methods is employed, i.e., drilling holes according to a numbered template or notching the edges at certain distances from the corners.

When tensile tests are to be made, the specimens should be large enough so that several test pieces can be cut after exposure, unless the tensile specimens themselves are exposed.

Method of Exposure

Probably the most frequently used rack for the exposure of 4- by 6-in. specimens is that shown in Fig. 1, which is the same as described by Rawdon.[4] It consists of a frame made of 14-gage metal strips formed for rigidity into channels and angles. Porcelain insulators mounted with machine screws and nuts support the specimens. Seventy specimens are accommodated in an area roughly 68 by 36 in. (173 by 91 cm).

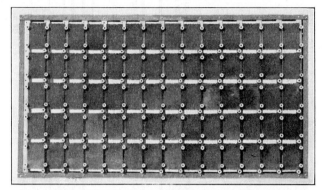

Fig. 1. View of Rack Holding Seventy Specimens Measuring 4 by 6 Inches.

Fig. 2. View of Racks Illustrating Method of Their Support on Pipe Frames.

Most corrosion investigators in this country follow the procedure of A.S.T.M. Committee A–5, and expose specimens at an angle of 30° to the horizontal and facing south. A convenient method of supporting the rack is by a frame made of galvanized pipe, as shown in Fig. 2. By varying the length of the uprights any desired angle of exposure can be secured. Detailed plans for the construction of both rack and pipe frames are given in Fig. 3.

[4] H. S. Rawdon, "Atmospheric Corrosion Testing," *Symposium on Corrosion Testing Procedures*, pp. 36–56, American Society for Testing Materials, Philadelphia, Pa., 1937.

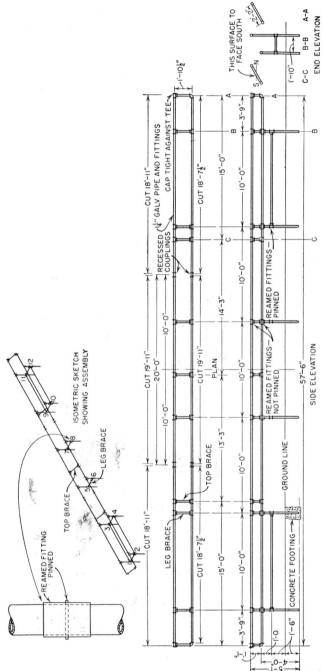

Fig. 3. (*Part 1*). Rack for Atmospheric Exposure Tests. (Carnegie-Illinois Steel Corp.)

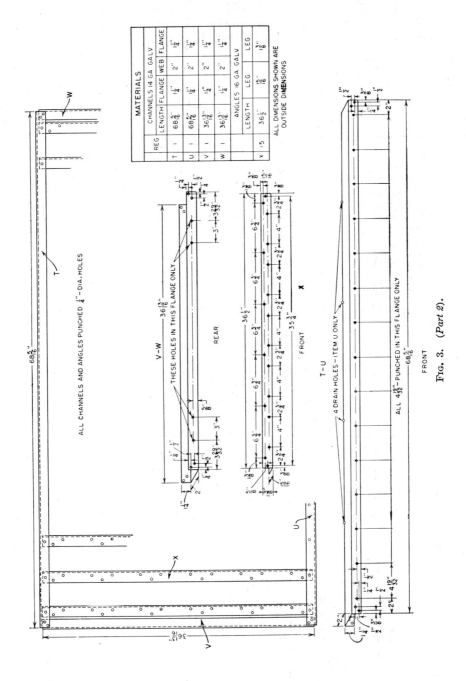

Fig. 3. (Part 2).

Non-ferrous materials are exposed vertically by A. S. T. M. Committee B–3, although the test of Committee B–8, referred to earlier, is at 30°. Many tests of painted specimens are exposed at a 45° slope.

Length of Exposure

This factor varies, depending upon the type of test. Semi-annual visual inspections are made on exposure tests of Subcommittees III* and VIII* of A. S. T. M. Committee A–5. Loss-of-weight determinations of plain and low-alloy steels in severe industrial or marine atmospheres should follow a sequence of exposures such as 1, 2, 4, 8, and 16 years. On account of the parabolic form of the time-corrosion curves for certain of these alloys (see Figs. 2 and 3, pp. 123–124), some believe that specimens should be removed after 0.5, 1.5, 3.5, 7.5, and 15.5 years in order to locate the "break" in the curve more accurately. In highly polluted atmospheres, exposures of 0.5 year will *indicate* major differences in relative corrosion ratios; after 3.5 years conclusions from the data are considered *tentative;* the longer periods are necessary for *conclusive* results. In less polluted atmospheres the losses of weights are less, and the periods of exposure could well be 2, 4, 8, and 16 years. The nature of the materials being tested and the severity of the atmosphere are the most important factors to be considered in arriving at a decision on the optimum length of exposures.

Treatment of Exposed Specimens if Loss of Weight Is Desired

This subject is covered in *Preparation and Cleaning of Specimens*, p. 1077.

SOIL EXPOSURE TESTS

I. A. Denison†

The test must include a sufficient number of specimens to yield a reliable average, the standard deviation and standard error of which can be calculated. The greater the diversity of soil conditions, the greater should be the number of specimens. The Bureau of Standards data show that if the conditions at the test site are uniform, twelve specimens will yield an average having a standard deviation not greater than 10%. A *minimum of four specimens* should be provided.

For preparation of specimens for exposure, and removal of corrosion products after exposure, the usual accepted techniques should be employed. (See *Preparation and Cleaning of Specimens*, p. 1077.)

Since corrosion data obtained from one soil may not apply to another, tests must be made in each class of soil for which data are desired. Care should be exercised in the selection of the test site, which should be typical of the type of soil in which the test is made.

Since most soils consist of several horizons that may differ widely in their physical and chemical characteristics, specimens should be buried at the depth that will expose them to the soil horizon to be encountered by the structure for which corrosion data are desired.

All points along the bottom of the trench should have the same elevation and the same depth of soil above them.

* Now Subcommittee XIV.
† National Bureau of Standards, Washington, D. C.

Details of Specimens

Specimens should be so spaced that they will not affect each other either through the migration of corrosion products or through a change in the electrical conductivity of the paths traversed by the corrosion currents. The minimum spacing will therefore depend to some extent upon the size and conductivity of the specimens and possibly upon the conductivity of the soil. The rule adopted in the Bureau of Standards test is that the distance between the specimens should be at least twice the diameter of the specimens.

Although cylindrical specimens may corrode more uniformly when placed on end, it is probable that the corrosion on specimens placed horizontally more nearly represents the corrosion which would occur in service. For this reason it is preferable to place the specimens horizontally.

Sheet specimens should be placed on edge in order that both sides receive the same aeration.

If specimens of pipe are tested the ends should be closed to prevent corrosion on the inside of the pipe.

Specimens of different areas should be provided or a sufficient number of one size should be exposed to permit the treating of two or more specimens as one. The purpose of this requirement is to allow the pit depth-area relation to be determined for the specific conditions of the test. It can be shown that a small change in a constant of the pit depth-area equation (to be described) results in a large change in the estimated pit depth if the associated area is much greater than that of the specimen.

Specimens should be designed so that a part of the original surface will be preserved. It is from this uncorroded surface that the depths of pits are measured.

Duration of Exposure

Specimens should be removed after several periods of exposure in order that the relation between corrosion and time may be determined. This procedure will render it unnecessary to test specimens to destruction, which is often impracticable as well as of uncertain value.

Recording of Data

The depths of a limited number of pits in addition to the deepest should be recorded in order that the distribution of pit depths can be studied.

The data should be reduced to the amount of corrosion (loss of weight and depth of the deepest pit) for a unit area and a unit period of exposure, say for one square foot for one year. These data must be accompanied by certain constants (to be described) to permit adjustment of the observed pit depths for area and time.

Tests for loss of strength are desirable for such materials as cast iron and yellow brass, in which the corrosion products are retained to some extent within the pits.

Interpretation of Data

The corrosiveness of a soil toward a given material is most completely defined in terms of three constants. A full discussion of these relations and the methods used in calculating the constants have been published.[1]

It is useful to know the value for the deepest pit to be encountered in a given area

---

[1] K. H. Logan, S. P. Ewing, and I. A. Denison, "Soil Corrosion Testing," *Symposium on Corrosion Testing Procedures*, American Society for Testing Materials, 1937.

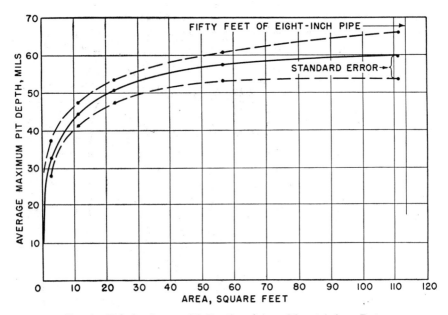

Fig. 1. Relation between Pit Depth and Area, Mount Auburn Data.

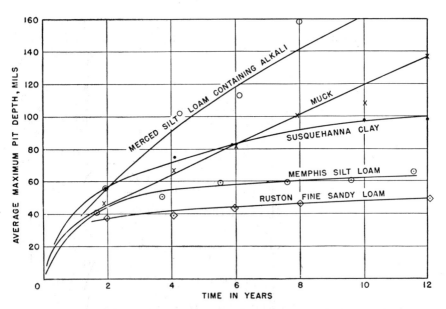

Fig. 2. Relation between Pit Depth and Time of Exposure in Various Soils.

of steel, knowing the deepest pits that occur within a smaller specimen area. Over a limited range of area, the relation between pit depth and area is expressed by the equation:[2]

$$P = bA^a$$

where $P =$ the depth of the deepest pit in the area A,
$a =$ constant depending on the conditions of exposure,
$b =$ constant equal to the average depth of the deepest pits in unit area.

This relation is illustrated in Fig. 1.

When the logarithms of the maximum pit depth are plotted with the logarithms of the corresponding areas associated with the recorded depths, a curve that is very nearly a straight line is obtained. Its slope gives the value of a and its intercept with the ordinate at $\log A = 0$ gives the logarithm of b.

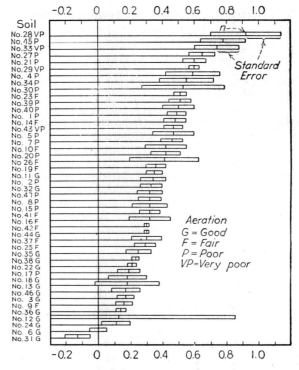

FIG. 3. Values of n for Soils in the National Bureau of Standards Soil Corrosion Test Sites.

The relation between pit depths and time for a variety of soils is illustrated in Fig. 2. This relation can be expressed mathematically by the equation:

$$P = kT^n$$

where $P =$ depth of the deepest pit at time T.

[2] G. N. Scott, "Adjustment of Soil-Corrosion Pit-Depth Measurements for Size of Sample," *Proc. Am. Petroleum Inst.*, **14**, Section IV, 204 (1933).

In plotting the logarithm of P with the logarithm of T, n equals the slope of the line, and log k is equal to the intercept on the log P axis.

The values of n arranged in order of decreasing magnitude for a group of soils are shown in Fig. 3. It is evident that the aeration of the soil is the chief factor determining the value of n, the poorly drained and poorly aerated soils being at the top of the list and the light-textured, well-aerated soils at the bottom.

The constants having been evaluated, the experimental data may be adjusted for any desired area and time by means of the equation:

$$P = kT^n A^a$$

where k is now the average value of the depth of the deepest pit in unit area at unit time.

CHEMICAL PLANT EQUIPMENT TESTS

O. B. J. Fraser*

REASONS FOR PLANT TESTS

Corrosion testing in operating equipment is the most convincing and frequently the most reliable procedure for solving specific plant corrosion problems. The difficulty, and often the impracticability, of reproducing actual service conditions in the laboratory often dictates the plant test for evaluating metals and alloys.

In a laboratory program one variable at a time can be studied and evaluated, but such study is rarely feasible in operating equipment. The effects of changes in concentrations, substances present, and solution characteristics, such as viscosity, that may occur in evaporation, distillation, digestion, polymerization, sulfonation, and other unit processes are integrated automatically in a plant test. The influence on corrosion rates of natural substances not certainly identified, and difficult or impossible to maintain at constant concentration in the small and not continuously renewed volumes of fluid used in laboratory testing, is assured in a plant test.

The volume of solution in an operating unit makes it possible to test a large number of specimens at the same time. Other advantages of the plant test are that very little technical supervision is required and it is not essential that the investigator know the precise composition of the solutions. This last is a convenient factor for the operator who either does not have complete information on solution compositions or, for some reason, does not wish to disclose such information.

OBJECTIVES OF PLANT TESTS

The objective of a plant test is not always the same. It may be that some portion of the equipment is corroding more rapidly than seems desirable. A properly conducted test will include test pieces of the materials already being used in the equipment so that a direct comparison may be made with operating experience. In the occasional instance where rapid corrosion of large areas of metal is occurring, the possibly disturbing influence of relatively large concentrations of dissolved metals upon normal corrosion rates of other materials being tested must be considered.

A second problem arises when process conditions are being changed, or when an entirely different process is introduced into existing equipment, and it is desired

* Development and Research Division, The International Nickel Co., Inc., New York, N. Y.

without interruption of production to evaluate the new hazard to which the equipment is exposed. A test under the new conditions will provide information not only on how rapidly the equipment is being dissolved, by reference to weight losses of test pieces of the material constituting the equipment, but also about the most suitable materials for replacements.

In a third problem it may be necessary to select materials for equipment of a newly developed chemical process not yet brought beyond the laboratory scale of operation. When, as is often true, a pilot plant stage built of materials selected on the basis of first cost, previous experience, or laboratory test results is available, it will usually be possible to conduct the plant type test simultaneously with the working-out of process operating procedure. The results will either confirm the choice of materials for the pilot plant equipment or indicate what other materials should be used for the large-scale plant.

TEST PROCEDURE

The simplest form of plant test, but one not usually recommended, consists of simply hanging a piece of metal in the liquid or vapor contents of operating equipment. The chances that unreliable results will be obtained are great. Contact with support wires of different composition from the test piece or with metallic parts of the equipment may result in either galvanic protection or accelerated corrosion, depending on whether the test-piece material is more noble or less so than the other material. There is also danger of mechanical damage or loss of a loosely suspended specimen.

The mechanical essentials of a satisfactory plant test procedure are:

1. Well-prepared and weighed test pieces of suitable size.
2. A means of supporting the test pieces, rigidly and with no direct contact, except when galvanic effects are being studied, with other test pieces or with the metal parts of the equipment in which the test is being run.

Spool-Type Specimen Holder

A procedure that has given reliable results over many years makes use of a specimen holder in which a number of test pieces are supported on insulating materials

Table 1. BILL OF MATERIAL FOR A TYPICAL SPECIMEN HOLDER

Number Required	Name of Parts	Material	Size
5	Rods	Monel‡	$3/16$ in. diam. by 12 in. long; ends threaded $1\frac{1}{4}$ in.; $19/32$ threads
20	Hex nuts	Monel‡ machine screw nuts, No. $19/32$	
2	Brackets	Monel‡	11 U. S. Standard gage, $3/4$ in. by $3\frac{3}{4}$ in., having 3 holes* $7/32$ in. diam.
1	Insulator	Molded Bakelite tubing	$5/16$ in. O.D. by $13/64$ in. I.D. by $10\frac{3}{4}$ in. long
2	End disks	Molded Bakelite sheet	4 in. diam. by $1/4$ in. thick with 5 holes†
15	Spacers	Molded Bakelite tubing	$7/16$ in. O.D. by $21/64$ in. I.D. by $5/8$ in. long
1	Short spacer	Molded Bakelite tubing	$7/16$ in. O.D. by $21/64$ in. I.D. by $5/16$ in. long

* Center of end hole located $1/2$ in. from end on long axis; other hole centers are, respectively, $1\frac{3}{4}$ in and $2\frac{3}{4}$ in. from center of first hole.

† Hole $21/64$ in. diam. drilled at disk center; centers of other holes, $7/32$ in. diam., located at four points 90° apart and at a radial distance of $1\frac{3}{4}$ in. from the center of the disk.

‡ Or type 316 stainless steel.

so that galvanic contacts are obviated.[1] Standardized details of such a test are available.[2] Preparation, cleaning, and weighing of specimens before and after testing should be carried out in accordance with recommendations outlined in *Preparation and Cleaning of Specimens*, p. 1077.

The spool-type specimen holder with its component parts is illustrated in Fig. 1. Specimens are machined to an exact diameter of 5.672 cm (2.233 in.), with a 0.833 cm

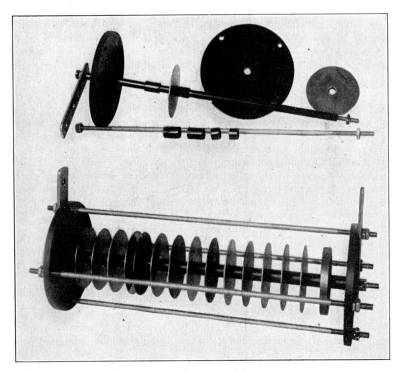

Fig. 1. Spool-Type Specimen Holder.

($21/64$ in.) diameter hole in the center, and are of a standard thickness of 0.079 cm ($1/32$ in.). A specimen when mounted on the holder has an exposed area of 0.5 sq dm (7.75 sq in.). For thicker specimens, the area may be taken from Table 2.

Cast specimens may be 0.476 cm ($3/16$ in.) thick and 5.291 cm (2.083 in.) in diameter, so that the exposed area, when installed on the specimen holder, will again be exactly 0.5 sq dm (7.75 sq in.).

Specimens can be weighed readily on an analytical balance. Because of their uniform size, the time involved in changing weights is minimized. For certain purposes special specimens may be prepared, as, for example, by welding or soldering pieces of metal together.

[1] H. E. Searle and F. L. LaQue, "Corrosion Testing Methods," *Proc. Am. Soc. Testing Materials*, **35** (II), 249–260 (1935).

[2] "Recommended Practice for Conducting Plant Corrosion Tests," *A. S. T. M. Standards*, Part IA, pp. 491–499 (1946).

TABLE 2. AREAS OF 5.672 CM (2.233 IN.) DIAMETER SPECIMENS ACCORDING TO THICKNESS

U. S. Standard Gage	Thickness		Area Exposed	
	cm	in.	sq dm	sq in.
3	0.6350	0.2500	0.599	9.286
7	0.4763	0.1875	0.571	8.847
11	0.3175	0.1250	0.543	8.409
12	0.2779	0.1094	0.535	8.299
14	0.1934	0.0781	0.521	8.080
16	0.1588	0.0625	0.514	7.970
18	0.1270	0.0500	0.509	7.883
20	0.0953	0.0375	0.503	7.795
21	0.0874	0.0344	0.502	7.773
22	0.0795	0.0313	0.500	7.752
23	0.0714	0.0281	0.499	7.729
24	0.0635	0.0250	0.497	7.707
25	0.0556	0.0219	0.496	7.686
26	0.0478	0.0188	0.494	7.664

Usually at least two specimens of each metal or alloy are used for checking purposes.

The four rods of the assembly serve two purposes: they afford protection to the specimens against mechanical abuse, and they make the assembly very strong. One rod passes through holes in the supporting brackets to hold them firmly.

Spacing of specimens is regulated by the length of the spacer; usually it is 1.588 cm (⅝ in.). Short spacers may be used to provide for any variations in thickness of specimens.

When galvanic effects are investigated, an insulating spacer may be replaced by a spacer of the normal, or a shorter length made from one of the materials of the galvanic couple. Ordinarily, the metal spacer is exposed to the corrosive attack and is considered part of the metal specimen comprising one half the couple; but, if desirable, the exposed surface of the metal spacer may be coated with an insulating lacquer or wax. Similarly, corrosion of the specimens forming the couple may be confined to adjacent faces by lacquering the backs. When a galvanic arrangement of this kind is used, it is common to use a compression spring, made of a corrosion-resisting material, between a spacer and the end disc; the spring takes up any slack that might result from thinning of specimens by corrosion.

The choice of materials from which to make the holder is important. These need not be durable enough to last indefinitely, but they must last long enough to insure satisfactory completion of the test. Metallic parts may be made of nickel, of a 70% Ni–30% Cu alloy, or of stainless steel or other corrosion-resistant metals, depending upon the corrosive environment. Insulating materials used are phenolic resins, porcelain, neoprene, and glass. The phenolic resins answer most purposes; their principal limitations are unsuitability for use at temperatures over 150° C (300° F) and lack of adequate resistance to concentrated alkalies and certain organic compounds, such as coal tar products.

The method of support during the test period is important. The preferred position is with the long axis of the holder horizontal; it avoids dripping of corrosion products from one specimen to another. The holder must be located so as to comply with conditions of exposure to be studied. It may be submerged, exposed only to the vapors, or located at liquid level. Sometimes all three locations may be utilized.

Liquid-level exposure is accomplished most readily by building a suitable float, usually of wood, around the holder so that it will float on the surface. Various means have been utilized for supporting the holders in liquids or in vapors. The simplest is to suspend the holder by means of a heavy wire or light metal chain.

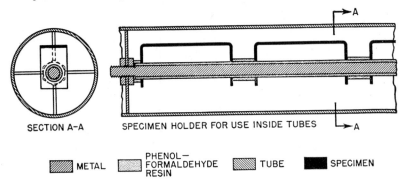

FIG. 2. Specimen Holder for Measuring Corrosion Inside Pipe Lines or Tubes.

SPECIAL CASES

Occasionally the standard spool-type holder is not applicable. A suitable holder for measuring corrosion rates inside pipe lines or tubes, for example, is illustrated in Fig. 2. The specimens are strips 1.27 cm (½ in.) wide by 7.62 cm (3 in.) long, bent

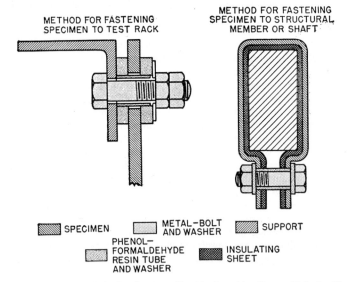

FIG. 3. Methods for Attaching Specimens to Test Racks and to Parts of Moving Equipment.

at each end, and provided with 0.833 cm (21/64 in.) diameter holes in the bent portions. They may be assembled and spaced on a rod of the same diameter as used in the standard holder. End disks are replaced by twisted wire spiders to locate specimens in the center of the tube. This device does not interfere seriously with the flow of liquid through tubes 4 cm (1.5 in.) inside diameter or larger.

Another way to study corrosion in pipe lines is to install in the line short sections of pipe of the materials to be tested. These should be insulated from each other and from the rest of the piping system by means of non-metallic couplings. Insulating

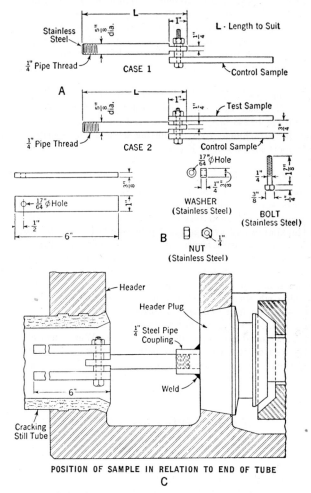

FIG. 4. Test Specimen for Determination of Corrosion Resistance in Petroleum Refinery Equipment, and Suggested Methods of Support.

gaskets should be placed between the ends of the pipe specimens where they meet inside the couplings. Such joints may be sealed with various types of "dope" or cement. It is desirable to paint the outside of the specimens so as to confine corrosion to the inner surfaces.

It is occasionally desirable to expose specimens in operating equipment without the previously described holders. The specimens can be attached directly to some part of the operating equipment, and the necessary insulation against galvanic effects can be provided as shown in Fig. 3.

OIL REFINERY EQUIPMENT

The Committee on Corrosion of Refinery Equipment of the American Petroleum Institute (A.P.I.) drew up a code in 1930 to cover corrosion testing of metals and alloys in petroleum refinery equipment. The recommended procedure, designed for exposure of test pieces in petroleum hydrocarbon liquids and vapors at relatively high temperatures, is adequate for that purpose, but since no provision is made for insulating the specimens electrically it is not suited for use in electrically conducting liquids.

The A.P.I. code calls for test pieces 15.239 ± 0.102 cm (6 ± 0.04 in.) long, 2.54 ± 0.025 cm (1 ± 0.01 in.) wide, and 0.953 ± 0.025 cm (0.375 ± 0.01 in.) thick, with a hole 0.676 cm (0.266 in.) diameter located on the center line 1.27 cm (0.5 in.) from one end, for attachment to a supporting device. The specimen and suggested methods for support are shown in Fig. 4.

EXAMINATION OF SPECIMENS

Although loss in weight is the principal measure of corrosion, this calculation should be always supplemented by an examination of each specimen, before it is cleaned, for tarnishing and the nature of any thicker corrosion products. If pitting has occurred, the maximum and average pit depths should be determined by means of a calibrated microscope or, with larger pits, a needle-pointed depth gage. Distinction should be made between pits on parts of the surfaces that were exposed freely and those on parts under the spacers or materials making contact with the metal. A corroded specimen may be subjected also to bending tests and microscopic examination to determine whether intergranular deterioration or other less usual forms of corrosion, such as dezincification, have occurred.

The standard disk specimens are not well adapted to the quantitative measurement of changes in mechanical properties due to corrosion. When such changes are to be studied, test pieces of sufficient size to permit cutting into standard sheet-metal, tensile-test specimens should be used; or precut tensile-test specimens may be employed, provided that preferential corrosion at edges is not prominent.

WATER PIPE SERVICE TESTS

V. V. KENDALL[*]

The pipe selected should be "run-of-mill" product. Aside from cleaning and degreasing, it should receive no preliminary treatment. If it is desired to determine the effect of special surface treatment, such as removal of mill scale by pickling, sandblasting, or other means, such pipe should be included as a separate material.

The test line should be installed in a by-pass to permit inspection and removal without interfering with the normal operation of the system. If possible, the test line should be straight to eliminate variable turbulence effects due to change in direction of flow. All the specimens should be in the same line, as adjacent lines will have different rates of corrosion.

[*] National Tube Co., Pittsburgh, Pa.

Pipe specimens generally used are 2 to 3 ft in length and 1 to 2 in. in diameter. Not less than three specimens of each material should be installed, and they should be taken from different lengths of pipe. They should be placed in alternate positions, e.g., material A, B, C, A, B, C, etc.

Galvanic or dissimilar metal effects must be considered. If the materials are significantly different, such as brass, copper, and steel, they should be separated by insulating couplings. If the galvanic behavior of any material is unknown, insulating couplings should be used. Experience with many service tests has not indicated a

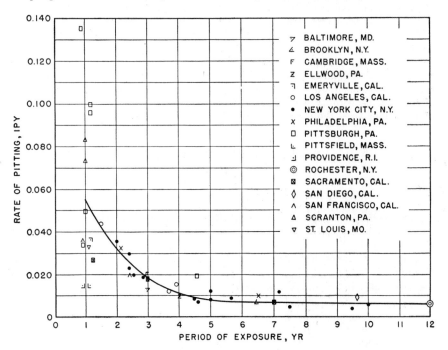

FIG. 1. The Effect of Time on Rate of Pitting of Iron Pipe in Hot Water Lines at Various Locations.

significant galvanic effect in combinations of the following rolled materials: open hearth steel, Bessemer steel, wrought iron and copper steel.

It is desirable to install a water meter and a recording thermometer in the test line. If this is not practicable, then some alternate means should be used to determine periodically, the maximum, minimum, and average temperature of the water over a 24-hour period. An analysis of the water should be made monthly as there is considerable variation in the composition of a water supply at different seasons of the year. This analysis should be complete but, in any case, it must include the pH, total dissolved solids, alkalinity, and the calcium content to permit the calculation of the saturation or the stability index. (See p. 502.)

At the end of the test, the pipe specimens should be removed, split longitudinally, and cleaned. Cleaning can be done by hot ammonium citrate, wire brushing, or sand- or shotblasting. (Refer to *Preparation and Cleaning of Specimens*, p. 1077.)

The life of pipe in hot water lines is dependent on the depth of pitting, therefore this measure is the significant one. On a 2- or 3-ft specimen, the ten deepest pits are measured and the maximum, minimum, and average reported. In measuring the depth of pitting, 2 in. on each side of the coupling are ignored to allow for any possible dissimilar metal effect. (*Note:* The writer has previously recommended ignoring 6 in. on each side of the coupling, but a careful check of a large number of service tests does not indicate any necessity for this length.) In one type of steam return line corrosion there is a general thinning of the pipe with very little pitting. In this case, the wall thickness of the specimen should be measured and the percentage decrease in wall thickness calculated, based on the original nominal wall thickness.

The test can run as long as desired. Figure 1 presents[1] the data from forty-two service tests in hot water lines in seventeen cities. It indicates that the initial attack on the metal, as measured by the results within the first year, is a function of the corrosivity of the water. After the initial period of rust formation, the rate of corrosion is slowed down. The attack on the metal is then dominated by the character of the rust, its porosity, its ability to permit diffusion of oxygen, etc. By the end of about four years the rate of corrosion has become approximately constant.

SEA-WATER CORROSION TESTS

F. L. LaQue*

PRECAUTIONS

The essential requirement of any method of testing specimens in sea water is to insure that the method of test does not influence the normal behavior of the test pieces to any significant extent. Where corrosion is the principal consideration of the test, means of supporting specimens should be such as to avoid all the following disturbing factors:

1. Galvanic action between different specimens or between the specimens and the rack.
2. Local shielding of any appreciable area of a test piece, so as to provide an opportunity for concentration-cell corrosion within, or alongside, the shielded zone with coincident electrochemical protection of adjacent areas.
3. Non-uniform flow of water past the surfaces of different specimens.
4. Corrosion-accelerating or -inhibiting effect of corrosion products from the rack material or from chemicals that may be leached from treated wood or other non-metallic racks.
5. Abrasion of loose specimens due to rubbing against their supports.
6. Mechanical damage by floating objects.
7. Loss of specimens due to rack failure.

Materials to be compared directly should be exposed at the same time and for the same length of time. Specimens exposed first in April may not corrode the same as specimens exposed first in November. There is greater effect of fouling organisms on the corrosion of specimens exposed in the spring and summer months when fouling organisms become attached very quickly.

In studying the effect of depth of immersion on corrosion, it is not sufficient

---

* Development and Research Division, The International Nickel Co., Inc., New York, N. Y.

[1] V. V. Kendall, "A Review of Data on the Relationship of Corrosivity of Water to Its Chemical Analysis," *Proc. Am. Soc. Testing Materials*, **40**, Pt. II, 1317–1322 (1940).

simply to expose a series of individual specimens immersed at different depths. It would be desirable to provide some electrical contact between such specimens so as to take into account any electrochemical action associated with differences in the concentration of dissolved oxygen or other constituents of sea water that may vary with depth.

When the anti-fouling properties of coatings are to be observed, the following additional factors should be considered:

1. All panels to be compared should be exposed with equal access to light; fouling in the shade is often different from fouling with more exposure to light. In this hemisphere there is a marked difference in the fouling of surfaces facing north as compared with identical surfaces facing south.
2. All panels to be compared should be exposed at the same depth. There is a variation in fouling organisms with depth so that those exposed at one depth may escape fouling organisms that would come into contact with specimens exposed either nearer the surface or deeper.
3. All panels should be subjected to the same water currents that carry embryonic organisms and food for growing organisms.
4. Metal racks should be made of highly corrosion-resisting material, since metals like zinc and iron can inactivate copper and mercury in anti-fouling paints either by direct electrochemical effect or by interactions of their corrosion products with the paint constituents.
5. The racks used to expose anti-fouling coating test panels should not be coated with anti-fouling paints, since a variable factor may be introduced by such paint.
6. As a general rule, specimens should be submerged in a vertical position so as to avoid disturbances from silt or other matter that might accumulate on horizontal surfaces.

In studying the corrosion of *steel piling*, probably the best practice is to drive test piles and observe the progress of corrosion by periodic micrometer measurements of the remaining thickness of unattacked metal at different points above and below the water.

RACK CONSTRUCTION

Although the use of wood for racks provides a convenient means of insulating specimens from each other, wooden racks have the serious disadvantage of being subject to rapid attack by teredos and other marine borers at many locations. This damage can be avoided by using wood thoroughly impregenated with creosote after all cutting and drilling have been completed. However, creosote tends to bleed from the wood with possible effects on corrosion of specimens that may become contaminated with it. In addition, creosote-treated wood is rather messy to handle in warm weather.

One way to avoid contamination of specimens by creosote-treated wood is to use Bakelite, Micarta, or other plastic inserts in creosoted wood racks as indicated by Fig. 1, based on a design developed by G. C. Quelch, Jr., of the North Florida Test Service, Daytona Beach, Florida.

A design of treated-wood rack used by the Battelle Memorial Institute that is well suited to the testing of painted panels is shown in Fig. 2. An important feature of this construction is that each specimen is held in place by a wooden stop or plate cap at the bottom so that a single panel can be removed without disturbing the others.

A rack used by the Woods Hole Oceanographic Institution for exposing wood panels is shown in Fig. 3.

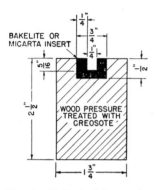

FIG. 1. Method of Preventing Contact of Test Panels with Creosote-Treated Wood Racks.

A novel method of supporting panels for testing anti-fouling coatings was developed by R. Eckart of the Sapolin Company; it is shown in Fig. 4. The advantages of this method of racking include:

1. Complete insulation with no danger of electrolytic effects.
2. A freely swinging panel that eliminates in-shore and off-shore variations on front and back surfaces.
3. The bottom flap — when used — permits observations of phototropic and geotropic influences on fouling organisms.
4. The device may be exposed quickly at any dock, float, or other location without elaborate preparation or provision of supporting structures.
5. The parts are durable and may be used over again.

The difficulties with wooden racks are avoided by using metal racks which, for reasons mentioned previously, should be sufficiently resistant to corrosion to avoid any effects of soluble corrosion products on the behavior of test panels. The author

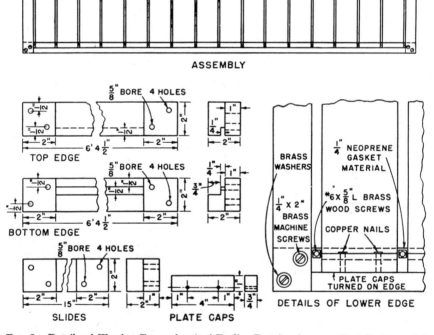

FIG. 2. Details of Wooden Frame for Anti-Fouling Test Specimens. (Battelle Memorial Institute.)

FIG. 3. Panel Rack for 3/8-in. Plywood Panels. (Woods Hole Oceanographic Institution.)

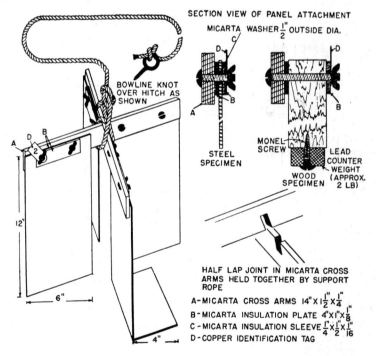

FIG. 4. Submersion Test Rack. (Sapolin Co., Inc.)

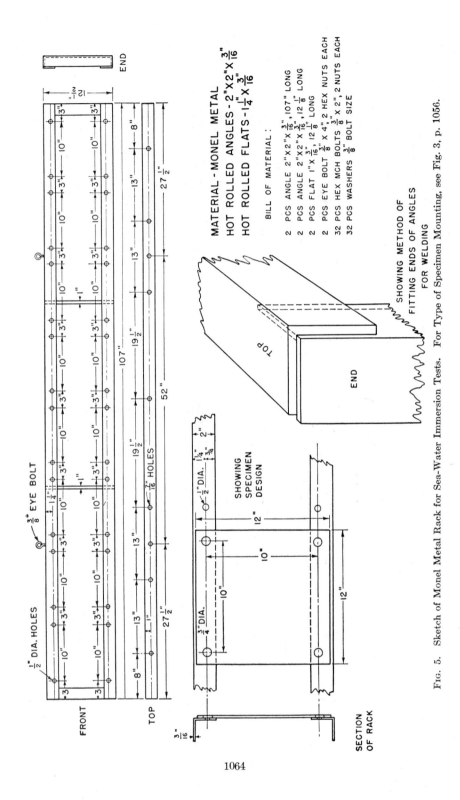

Fig. 5. Sketch of Monel Metal Rack for Sea-Water Immersion Tests. For Type of Specimen Mounting, see Fig. 3, p. 1056.

has found 70% Ni–30% Cu alloy* to be very satisfactory for racks constructed as shown in Fig. 5. Adequate insulation is provided by Bakelite sleeves and washers used with the alloy bolt and nut assemblies by which specimens are fastened to the racks as shown by the detail of the fastening assembly in Fig. 3, p. 1056 (left). For example, magnesium and zinc specimens have been so exposed without any evidence of galvanic action.

Although the method of insulation of specimens from metal racks illustrated by Fig. 5 is generally satisfactory, some difficulties may be encountered with steel panels coated with anti-fouling paints from which copper is leached, and with copper

FIG. 6. Lowering Metal Rack (Details in Fig. 5) Containing Test Panels into Sea Water.

and high-copper alloys yielding substantial amounts of copper in corrosion products. Under such circumstances, conditions within and around the fastening assembly may permit the building up of a relatively high copper ion concentration in the water inside the fastening assembly. This, in turn, may be followed by deposition of copper on the metal under the Bakelite washers and the growth of this deposited copper in the form of a "tree" between the insulating tube and washers. The deposited metal may serve to bridge the gap between the specimen and the rack or the fastening,

* Monel.

and thus provide metallic contact between a specimen and the rack and any other specimens that come into similar accidental galvanic contact with the rack.*

For this reason, it is unwise to use bare metal racks, as in Figs. 5 and 6, to support steel panels coated with anti-fouling paints containing copper or cuprous oxide. Either non-metallic racks or metallic racks covered with an insulating coating are preferred. The latter racks also would be more reliable for the exposure of copper and high-copper alloys. For the same reason, the copper-base metals or other metals coated with anti-fouling paints containing copper would not be well suited to the fabrication of racks for the exposure of metallic specimens such as illustrated in Figs. 5 and 6.

Alloys demonstrating passivity in sea water may suffer localized corrosion beneath the Bakelite washers used for insulation. Unless it is desired to determine susceptibility to such action, it can be reduced, and frequently eliminated, by coating the under surfaces of the Bakelite washers with petrolatum at the time the specimens are being fastened to the racks.

When metallic racks are supported by metallic hangers from a metal structure, they should be insulated from the structure so as to prevent galvanic effects between racks of different compositions or between specimens that may come into accidental galvanic contact with different racks. A rack of this type and the structure for supporting a number of racks in a sea water channel are shown in Fig. 6.

FIG. 7A. Marine Corrosion Subcommittee Large Exposure Raft at Caernarvon. (Courtesy of J. C. Hudson.)

Figures 7A and 7B illustrate the method used by investigators for the Marine Corrosion Subcommittee of the British Iron and Steel Institute. Figure 7A shows a raft from which the test specimens are supported on the steel frames shown in Fig. 7B. The usual specimen size is 15 by 10 by 0.25 in. (38 × 25 × 0.64 cm). The frames are raised and lowered by means of a movable hoisting gear.

The specimens are secured to the frames by steel bolts, washers, and nuts and are insulated by means of laminated plastic sleeves, bushes, and washers in much the

---

* It is also remotely possible that dissimilar metal complex cells may be established. These do not involve metallic contact, hence are not dependent on the presence of copper, and arise only because of differences in electrolyte composition over portions of the specimen surface because of marine organisms or because of stagnant pockets of liquid at the insulated support or for other reasons. See *Fundamental Behavior of Galvanic Couples*, p. 493. EDITOR.

same way as illustrated in Fig. 5. It has been found desirable to use fairweathers (false bows appearing in Fig. 7A) rigidly secured to each end of the rafts so as to prevent accumulation of floating weeds in the raft.

Another type of rack that provides positive insulation between specimens was designed by C. P. Larrabee of the Carnegie-Illinois Steel Corporation. The details of

FIG. 7B. Marine Corrosion Subcommittee Double Exposure Frame in Use at Caernarvon Lifted Ready for Inspection. (Courtesy of J. C. Hudson.)

such a rack to hold 12 by 12 in. (30 by 30 cm) specimens are shown in Figs. 8 and 9. It will be seen that each specimen rests in grooves in large porcelain insulators strung on two rods at the bottom of the rack and between similar insulators strung on two more rods at each side of the specimens. A keeper rod across the top of the rack prevents loss of specimens in case the rack should be tipped over accidentally. The racks and fastenings may be made of 70% Ni–30% Cu alloy or other suitable corrosion-resisting alloy, such as Type 316 stainless steel. The minimum spacing between specimens in such a rack should be 3 in. (7.5 cm) so that the growth of marine organisms will not entirely clog the space between specimens.

Whatever kind of rack is used, it should be supported so that the flat surfaces of all specimens are parallel to the usual water currents.

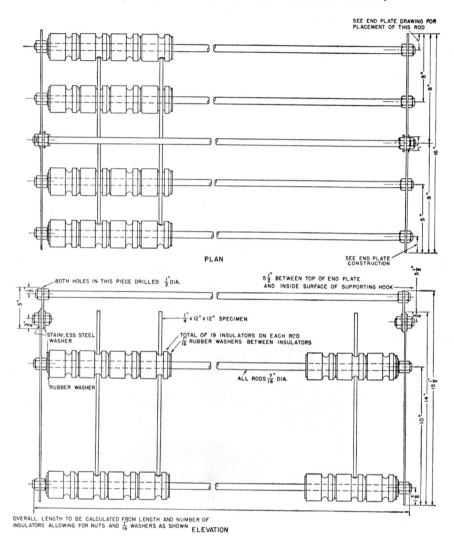

Fig. 8A. Assembly Sea-Water Rack for Specimens $\frac{1}{4} \times 12 \times 12$ inches (Carnegie-Illinois Steel Corporation.)

HALF-TIDE EXPOSURES

In studying corrosion at half-tide level, allowance must be made for the protective effect of oils and greases from polluted harbor waters that may coat the alternately wet and dry surfaces and thus indicate much less corrosion in this critical region than would be encountered by the same metals exposed to clean salt water where no protective oily films are present. It is difficult to expose specimens partially immersed

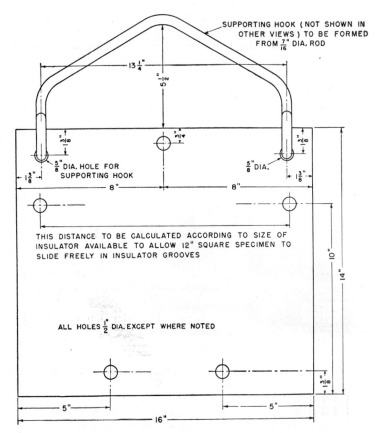

Fig. 8B. End Plate Sea-Water Rack for Specimens 1/8 × 12 × 12 inches or 1/4 × 12 × 12 inches. (Carnegie-Illinois Steel Corporation.)

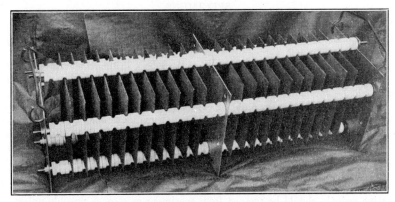

Fig. 9. Rack for Exposure of Panels to Sea Water. (Carnegie-Illinois Steel Corp.) For details see Fig. 8A and 8B.

with a constant water line on account of the action of tides. This condition may be approached by fastening specimens to a raft, but the gradual increase in the weight of the raft and the specimens due to the accumulation of marine organisms will cause the raft to sink gradually and change the location of the original water line. The William F. Clapp Laboratories at Duxbury, Massachusetts, use a float for supporting racks and test specimens as illustrated in Fig. 10.

FIG. 10. Raft for Partially Immersed Panels in Sea Water. (William F. Clapp Laboratories.)

Where there is little or no tidal action, water-line effects may be observed on specimens mounted on racks similar to those of Figs. 5 and 6. The racks should be made larger to accommodate specimens long enough so that part will always be in the water and part always out.

SPECIMENS

Precautions in the preparation of specimens for sea-water corrosion tests are the same as for any other kind of corrosion test. (See *Preparation and Cleaning of Specimens*, p. 1077.)

Specimens may be identified by notching the edges, drilling holes to a template, or simply by stamping a number or symbol. The location of the means of identification should be specific so as to afford information at any time after the test as to top and bottom of the specimen during exposure, and its relative position in the rack.

Because of the possibility of losing identification marks through corrosion, a record should be kept of the relative positions of specimens when they are placed on racks so that they can later be identified by position if necessary. Identification tags, if used, should be attached as soon as the specimens are taken from the racks and should be kept on the specimens until the presence of other means of identification has been established.

Marine growths should be scraped from specimens as soon as possible unless they are to be preserved in situ as part of the record of the test. Further cleaning of specimens should then be undertaken using appropriate methods as described in *Preparation and Cleaning of Specimens*, p. 1078. If panels are used for paint tests, there are some additional desirable features:

1. Steel panels should have rounded edges and corners. In many instances, it is desirable to provide thicker or supplemental coatings on edges and corners.
2. A single panel should not be used to test more than one formulation of protective coating or anti-fouling composition.
3. The absence of holes is desirable so that racks that do not require holes in the panels are preferred.
4. Wood panels should be sanded and the edges rounded prior to painting. Mahogany, cedar, oak, and white pine are satisfactory. Supplemental edge protection is desirable in testing coatings on wood that has been treated with some preservative to prevent interference by leaching of the preservative at such points.

FIELD TESTING OF PAINTS*

Harley A. Nelson†

Aside from providing information on product quality, an important reason for interest in field testing today is the need for data with which to calibrate laboratory tests. Field tests may be conducted on test surfaces ranging from panels to large structures, which may be exposed to outdoor weathering or placed in direct contact with corrosive agents under a variety of service conditions. Few large-scale field tests with paint systems on structures have been fully recorded. The most ambitious and interesting of them is the Havre de Grace Bridge test conducted by the American Society for Testing Materials[1] from 1906 to 1913. This record deserves study by any one planning extensive field tests on structures.

One advantage of field testing on properly selected structures is the assurance that service conditions are reproduced. A serious disadvantage is the lack of control of atmospheric conditions during the three important steps in a paint job — surface preparation, application, and drying. Panel testing, on the other hand, offers opportunities for almost any desired degree of control during these steps but often presents difficulties in obtaining faithful reproduction of service conditions. Consequently, panel testing is still looked upon as an exploratory procedure, and, if the two do not agree because of the many variables involved, first consideration is given to the results on structures. However, the trend created by improved laboratory technique is to stress the advantages of panel testing. In recognition of this trend, the necessarily limited discussion that follows will emphasize panel testing on steel. This is a good compromise because steel is a difficult test base, and technical control is required to obtain significant and reproducible results with it in field tests of any kind.

Size of Test Areas

Reduced edge effects and better averaging of the effects of surface condition or composition of the metal are arguments in favor of large panels. Panels 24 by 36 in. (61 by 92 cm) are recommended and used by the American Society for Testing

---

* For discussion of laboratory testing of organic coatings see p. 1023.
† Technical Department, The New Jersey Zinc Company (of Pennsylvania), Palmerton, Pa.

[1] "Reports on Inspection of Havre de Grace Bridge," *Proc. Am. Soc. Testing Materials*, **15**, Part I, 189 (1915) and previous volumes back to **6**.

Materials.[1,2,3,4] Regardless of this, many field exposure tests are made on relatively small panels ranging from about 4 by 12 in. (10 × 30 cm) to 12 by 18 in. (30 × 46 cm). Thickness is usually proportionate to size in order to avoid excessive weaving or fluttering; that is, $\frac{1}{16}$ to $\frac{1}{8}$ in. (0.16 to 0.32 cm) for the smaller and $\frac{1}{8}$ to $\frac{1}{4}$ in. (0.32 to 0.64 cm) for the larger sizes recommended by the American Society for Testing Materials. The smallest size mentioned is about the minimum that can be used with any degree of accuracy, and such small panels are generally used only for preliminary or pilot tests. Sections of representative structural shapes are also extensively used, for example, 18 to 24 in. (46 to 61 cm) lengths of I-beams or angle sections. The dimensions of these should be sufficient (4 in. [10 cm] or more) to avoid a too high proportion of edges to surface area.

DUPLICATION OF TESTS

A common practice which is subject to criticism is the acceptance of tests on one panel, or one specimen, or one localized area on a structure. Theoretically, the use of large panels and the control of other variables reduce the need for multiple tests. If all these variables can be controlled to a known degree and the metal is thoroughly cleaned, one panel might suffice. Considering the number of variables that exist, however, it is logical that standard practice should require more than one test area for each system. As the amount of mill scale or degree of corrosion present at the time of painting increases, there is increasing need for multiple tests. At least three or preferably more should be used. The variables that enter into the planning of exposure tests have been more thoroughly discussed[5,6] in connection with paint tests on wood, but with some exceptions the same variables are important in tests on metal surfaces.

PREPARATION OF TEST AREA FOR PAINTING

The preparation of a test area is often dictated by the accepted practice for the service conditions, but where comparisons of different coatings are involved, the trend is to favor some procedure that will, in so far as possible, insure uniformity of surface. For example, in the case of steel, it is still argued that the presence of mill scale is a prevailing condition in service and that test surfaces should either be left as is or have the scale removed to some specified extent by preweathering. If the latter practice is followed, the effect of the more variable surface produced by patches of scale must be recognized by using a greater number of panels or test sections. The current view is that if shop coating cannot be done at the mill when the scale is in good condition, it should be removed entirely by pickling or sandblasting in order to obtain reproducibility of paint tests.[7,8,9,10] This is true to an increasing degree as

---

[2] "Report of Subcommittee on Standard Methods of Conducting Service Tests," *Proc. Am. Soc. Testing Materials*, **5**, 83 (1905).

[3] "Reports on Atlantic City Steel Plate Paint Tests," *Proc. Am. Soc. Testing Materials*, **15**, Part I, 214 (1915) and previous volumes back to **8**.

[4] "Report of Subcommittee XXIX, on Preparation and Painting of Structural Iron and Steel," *Proc. Am. Soc. Testing Materials* (Committee Reports), **41**, 322 (1941).

[5] G. W. Ashman, "Panel Test Evaluation of Exterior House Paints," *Ind. Eng. Chem.*, **28**, 934 (1936).

[6] E. W. McMullen and E. J. Ritchie, "Organization of a Systematic Test Fence Program," *Ind. Eng. Chem.*, **35**, 161 (1943).

[7] H. A. Gardner, *Physical and Chemical Examination of Paints, Varnishes, Lacquers and Colors*, 9th Ed., Chapters 9, 12, p. 155, Institute of Paint and Varnish Research, Washington, D. C., 1939.

[8] J. C. Hudson, *The Corrosion of Iron and Steel*, Chapter 5, p. 95, Chapman and Hall, Ltd., London, 1940.

[9] "Statement on Principles of Testing Paints and Combinations of Paints on Steel Panels," *Proc. Am. Soc. Testing Materials*, **35**, Part I, 322 (1935).

[10] L. A. Jordan, "The Preparation of Metal Surfaces for Painting," *Chem. Ind.*, **56**, 361 (1937).

the severity of the exposure increases. For cleaning in the field, sandblasting is the prevailing method, but extensive tests have been made with flame cleaning. The importance of painting as soon as possible after sandblasting is generally accepted.

Phosphate- and chromate-type pretreatments have been advocated for producing more uniform testing surfaces on steel, but their use for this specific purpose is very limited. However, surface pretreatments applied at the source are normal for commercial non-ferrous alloys of aluminum and magnesium, and these generally require only a solvent cleaning prior to paint application. Zinc-coated (galvanized) surfaces may or may not be pretreated by etching or phosphate coating, but in any case these usually require only standard solvent cleaning before painting, or as recommended for steel surfaces.

No standardized cleaning procedures have been adopted for field tests but, in addition to the references just mentioned, a number of methods designed for laboratory tests are found in Army, Navy, and Federal Specifications. Most of them try to minimize the surface preparation and cleaning problem by using cold-rolled steel. In a recently adopted tentative method for preparing panels for accelerated laboratory tests, the American Society for Testing Materials[11] specifies cold-rolled steel of a grade known to be relatively uniform and reproducible so that no sanding or cleaning other than with a suitable solvent is necessary. The advantages of surface uniformity and reproducibility could justify the use of these specially prepared grades of metal for comparing paint systems in field tests if it is known that their corrosion characteristics are typical of the steels used commercially.

VARIABLES IN APPLICATION

Variables encountered in application of organic protective coatings[12] cannot be fully discussed, but those of major importance in corrosion-resistance tests will be briefly covered.

Aging of freshly prepared paint is important because the normal reactions between pigment and vehicle should attain a balance and because the activity of the drier may change. One week at room temperature, 25° C (77° F), should be the minimum, and two weeks or one month would be better.

Temperature of the paint and of the surface influences the consistency of the paint and results in variations of film thickness. If temperatures cannot be controlled, they should at least be recorded. Tests should include the range that is likely to be encountered in service.

Film thickness and number of coats are important factors in protection against corrosion, but the film thickness that should be used in comparing paints is still debated. Some insist that paints should be applied at the spreading rates normally obtained by an experienced operator with the individual test products. It is difficult to refute this argument in field tests on structures, although it is probable that in most cases either the paints or the application procedure could be adjusted to approximate equal film thicknesses with the test systems. Unless the products are of very different types, comparison of paint systems at equal dry-film thicknesses is accepted as a desirable goal. There is less excuse for ignoring this where smaller articles are coated in the laboratory because, in addition to weighing the amount of

---

[11] "Tentative Method for Preparation of Steel Panels for Testing Organic Protective Coatings," A. S. T. M. Standards, D 609–46 T, Part II, p. 1601 (1946).

[12] G. W. Ashman, "Panel Test Evaluation of Exterior House Paints," Ind. Eng. Chem., **28**, 934 (1936).

paint applied, a number of practical schemes have been developed for approximating equal and more uniform film thicknesses, particularly on panels.[13,14] Also, instruments[15,16,17,18,19] are commercially available for checking both wet and dry film thicknesses of paint systems on any metal surface. Dry-film thickness data or their equivalent should appear in all test records.

Method of application on panels (brush, spray, or dip) should normally follow the practice that prevails in service but with attention to control of film thickness. There is a tendency to ignore brushing and standardize on spraying[20] or dipping,[21] which are most easily controlled and eliminate non-uniformity due to brush marks. Some maintain that brushing results in better adhesion. However, there is no consistent evidence that any method is generally superior. Paint systems may differ so widely in application properties that it is often wise to include more than one method.

Number of Coats in Test Systems. Except when tested as shop coats, one-coat systems of priming paints should not be compared because many priming paints are not designed for such use. When intended for a specific service, the system should include the finish coats normally recommended; or, if the purpose is to compare a group of priming paints, one or two coats of a finish coat of recognized quality should be included. Even paints that can be used effectively as both primers and finish coats should be tested as multicoat systems unless the recommendations state otherwise.

Drying schedules and maturing of films before exposure depend upon the products that are being tested but, where practicable, an attempt should be made to obtain some degree of control of the drying conditions. The ideal scheme is indoor painting and conditioning under uniform room temperature and humidity conditions (25° C [77° F] at 50% relative humidity), a practice which is still not universally approved or followed although it eliminates important variables. Unless a specific schedule must be met, aging of the film before exposure should be adjusted to the time required by the slowest drying product to attain proper maturity. One week after application of the final coat is commonly allowed for systems applied indoors. Large-scale outdoor tests should be applied during a season of average drying weather except when the objective is to test the effect of adverse drying conditions. In immersion and chemical resistance tests, it is particularly important that the paints should be given equal opportunity to reach a proper stage of maturity before the test is started.

Effects of conditions near the test area are receiving attention because of recent evidence[22] that localized corrosion can influence the life of the coating on other parts of a test area. This justifies questioning a number of common practices in making paint tests, such as leaving the backs of panels unpainted or relying on one or two

[13] H. G. Arlt, "Paint Films of Controlled Thickness," *Steel*, **98**, 42 (June 15, 1936); *Metal Finishing* (N. Y.), **38**, 123 (1940).
[14] H. F. Payne, "... An Instrument. . for Uniform Films," *Ind. Eng. Chem.* (Anal. Ed.), **15**, 48 (1943); *Official Digest*, No. 238, p. 400, Federation of Paint and Varnish Production Clubs, September, 1944.
[15] E. W. McMullen and E. J. Ritchie, "Organization of a Systematic Test Fence Program," *Ind. Eng. Chem.*, **35**, 161 (1943).
[16] A. Brenner, "Film Thickness Measurements," *J. Research Natl. Bur. Standards*, **20**, 357 (1938).
[17] R. E. Wheeler, "Measurements of Paint Film Thickness," *Products Finishing*, **5**, 54 (April, 1941).
[18] A. L. Alexander, P. King, and J. E. Dinger, *An Instrument for Measuring the Thickness of Nonconducting Films over Nonmagnetic Metals*, Preprint of Division of Paint, Varnish and Plastics Chemistry, American Chemical Society, September, 1944.
[19] A. H. Pfund, "The Thickness of Wet Paint Films," *J. Franklin Inst.*, **191**, 517 (April, 1921).
[20] H. G. Arlt, *loc. cit.*
[21] H. F. Payne, "Application of Uniform Films," *Ind. Eng. Chem.* (Anal. Ed.), **15**, 48 (1943); *Official Digest*, No. 238, p. 400, Federation of Paint and Varnish Production Clubs, September, 1944.
[22] W. W. Kittelberger, "Water Immersion Testing of Metal Protective Paints," *Ind. Eng. Chem.*, **34**, 943 (1942).

coats of any paint that may be at hand, applying more than one test system on a panel, and comparing systems with different numbers of coats on a panel. The use of large panels may minimize these effects, but the safest procedure is to paint all sides of the panels with the test system and add an extra coat on the edges. A similar question arises in planning tests on structures where contiguous areas may increase electrolytic effects due to the presence of bare steel or of more noble metals which are unpainted. To minimize such effects, each test system should be allowed as much area as circumstances permit and be tested in both favorable and unfavorable locations.

MOUNTING OF TEST SPECIMENS

In atmospheric testing, the results of vertical, angular, and horizontal exposures may not agree[23] as to amount or order of failure because of the differences in light and heat intensity and the extent to which the water accumulates on the surface. Common practice is to expose panels at 45° facing south with the idea that this represents average conditions. Recently, 30° exposure has been proposed for subtropical locations in order to take full advantage of the maximum average exposure to the sun, although many believe that moisture conditions are more important. Some investigators give the test panels a 135° bend so that either vertical and 45° or horizontal and 45° exposures can be obtained with the same panel.

Continuous water immersion tests are generally exposed in a vertical position, but the racks may be adjusted to catch the impact of currents at some specific angle. Whether panels in so-called tidewater or alternate-immersion tests should be mounted vertically or 45° facing south apparently is still being debated. Tests on the exterior of large structures should be distributed on both sunny and shady areas, with consideration also being given to moisture conditions.

One advantage of using heavy I-beams or angle sections on test racks is that they require no anchorage. Insulation of test specimens from each other and from the metal rack is usually desirable under any condition, but it is very necessary where moisture and salt water conditions prevail. One easy way to obtain sufficient insulation for most exposure conditions is to attach the panels to boards in such a manner that the panels and their wood supports can be mounted as units. Galvanized screws and washers should be used and space allowed to avoid tendencies to hold water at any point between the back of the panel and the support. In general, the construction of the rack can follow the plan for a metal corrosion rack illustrated by Figs. 1 and 2, p. 1045. In all severe tests involving direct contact with water or chemicals, complete electrical insulation of the test specimen should be provided by means as effective as those recommended by the American Society for Testing Materials[24] for corrosion resistance tests of metals.

TIMING OF EXPOSURE, INSPECTION, AND RECORDING

For complete information, it is often desirable to start duplicate tests at different seasons because freshly applied paint films are more sensitive to unusual conditions. Otherwise, outdoor test painting or panel exposure should be done under conditions normally considered acceptable for applying paint.

[23] S. E. Beck, "Laboratory Testing of Metal Finishes," *Am. Paint J.*, **28**, 54 (Nov. 1, 1943).
[24] "Recommended Practice for Conducting Plant Corrosion Tests," *A. S. T. M. Standards*, A 224–41, Part I, 492 (1942).

Plans should be made for inspections at stated intervals, but the time interval must of necessity vary wtih the severity of the exposure. For moderate atmospheric conditions, a cursory inspection at 3 or 6 months to see if anything unusual is happening, with a more careful inspection after one year and annually thereafter, may suffice. Severe industrial conditions and salt water immersion demand more frequent inspections, ranging from monthly to three or four times a year. Frequent inspections are important if statistical methods of data analysis are to be used.

Although the inclusion of a comparison standard is helpful in interpreting results, pictorial standards for degrees of paint failures are available which offer promise of putting inspection data on a more useful basis. These standards for rusting[25] and blistering[26] were developed for use in panel tests, but experience indicates that they will be equally useful for larger-scale field tests.[27] Standards are also available for other common failures of paint systems which can accelerate the appearance of corrosion, such as chalking, erosion, checking, and cracking.[28]

The commonly accepted rating schemes are based on a numerical scale, but there are differences of opinion about the scale. Many commercial laboratories still prefer a 0-10 scale with $0 =$ absence of failure and $10 =$ complete failure, whereas the American Society for Testing Materials[29] and the Federation of Paint and Varnish Production Clubs[30] have agreed on 10-0 with $10 =$ perfect or absence of failure and $0 =$ complete failure or poorest condition conceivable.

SELECTION OF TEST LOCATIONS

Weather conditions at test locations are important because organic coatings are susceptible to climate and weather. Assuming that atmospheric conditions are not unusual, essentially the same degree and type of surface failure are obtained on paints in widely separated locations in the same latitude where the climate may be considered comparable or average. Examples of exceptions due to abnormal local conditions that may exist in the same latitude are such locations as Denver and Pittsburgh. Also typically different in character as well as degree in their effects on paint systems are locations in southern latitudes, such as Miami and the Gulf Coast, where severe surface chalking and erosion occur, and northern latitudes where temperature changes promote cracking and peeling.

The amount of moisture and the manner in which it contacts the protective coating are important. For example, ordinary intermittent rain is less damaging than fog or condensation (dew).[31] The net result is that tests in humid southern locations, particularly those near the seashore, are essentially accelerated tests when compared with most of the country. Such locations are extensively used for this reason, although they may be misleading if the paint system happens to be susceptible to the effects of temperature changes. Seashore exposures are usually accelerated tests when compared with inland exposures, as are highly industralized locations when compared with

---

[25] *A. S. T. M. Pictorial Rusting Standards*, Method D 610–43, Part II, p. 1601 (1946).

[26] *A. S. T. M. Tentative Method for Evaluating Degree of Blistering*, D 714–45, Part II, p. 1171 (1946).

[27] W. F. Singleton, "The Interpretation of Visual Rusting Standards," *Proc. Am. Soc. Testing Materials*, **44** (1944).

[28] *A. S. T. M. Standards*, Part II (1946) — *Pictorial Standards or Paint Failures:* "Chalking," Method D 659–44, p. 1175; "Checking," Method D 660–44, p. 1177; "Cracking," Method D 661–44, p. 1180 "Erosion," Method D 662–44, p. 1183.

[29] *A. S. T. M. Pictorial Rusting Standards*, Method D 610–43, Part II, p. 1186 (1946).

[30] Report of Exposure Test Committee, Federation of Paint and Varnish Production Clubs, 1933; American Paint and Varnish Manufacturers Association, *Circ.* 445 (November, 1933).

[31] R. J. Wirshing, "Some Causes of Paint-Film Failure," *Ind. Eng. Chem.*, **33**, 234 (1941).

rural communities, etc. The safe course is to include locations that represent the extremes that may be encountered in service. Experience indicates that the following typical atmospheric exposure conditions should be considered for field testing of paints:

Inland industrial location of low mean altitude near latitude 40° north.
Moist, hot (semi-tropical), seashore location (Florida or Gulf Coast).
Cold seashore location (New England or West Coast).
Dry, normally hot, inland location where rapid temperature changes are encountered (West Texas or Colorado).
Dry, normally cold, inland location where rapid temperature changes are encountered (North Central States or Canada).

However, some investigators are not convinced that the last two exposure conditions are essential for testing paints on metals, but that two types of inland exposure conditions should be used, namely, industrial and suburban. The question is under discussion in A.S.T.M. subcommittees, and the reader is advised to refer to reports that will appear from this source.

PREPARATION AND CLEANING OF SPECIMENS

B. B. Knapp*

PREPARATION OF SPECIMENS

SIZE

The corrosion specimen will vary in size and shape with the type and purpose of the test and the testing apparatus or equipment. In general, the size may be limited by maintaining a proper ratio of specimen area to volume of testing solution (preferred minimum ratio — 5 liters/sq dm). Where the corrosion rate is measured only by weight loss, the ratio of area to mass and total area to edge area should be large. These conditions are especially important where the weight losses are small, such as in short-time laboratory tests or in mildly corrosive media. When practicable this condition can be met by the use of flat disks or squares of a size that will permit weighing on an analytical balance. Where other measurements are used to measure corrosion, the size and shape of specimen will be governed by the particular test.

SURFACE PREPARATION

The surface preparation will also vary with the type of test. In general the surface should be representative of what would be used in practice. If there is no preferred surface condition, a standard one should be used. Edges should be finished by machining to remove the sheared metal. Pickling is necessary for some materials to obtain a reproducible surface, but most materials can be finished by simply resurfacing with No. 120 abrasive paper. Pickling or stress relief annealing are sometimes used to avoid effect of cold-worked surfaces.

Preliminary degreasing can be done in a solvent or in vapor of a solvent. This should be followed by wet scrubbing with pumice until the surface is free of water breaks, rinsing in water, and drying. Immediate and complete drying is necessary to prevent oxidation of some materials. Rinsing in a water-miscible solvent, such as

* Research Laboratory, The International Nickel Co., Inc., Bayonne, N. J.

alcohol or acetone, before drying, will aid rapid drying, but it should be done with care to avoid depositing a grease film on the surface. Prepared specimens should be stored in a desiccator until tested.

IDENTIFICATION

Corrosion specimens are usually and most conveniently identified by stenciling. However, other methods are used where stencil marks are likely to become obliterated. Such methods as notching, drilling holes, and various means of protecting the stencils are used. In all tests a record should be made of the relative positions of the specimens in order to maintain their identity.

CLEANING OF SPECIMENS

The cleaning of corrosion specimens at the end of the test depends upon the method used for measuring the corrosion. Specimens that are used for weight loss determinations must be thoroughly cleaned of all corrosion products without loss of any sound metal. There are many satisfactory methods of cleaning specimens but, *whatever the method, its effect in removing base metal should be determined for each material.* An appraisal of the amount of attack should determine the cleaning method, as drastic methods should not be used when there are indications of only a small weight loss resulting from the corrosion test.

Methods for cleaning corroded specimens vary with the metal and to some extent with the nature of the corrosion product. The removal of the corrosion product may not be a simple procedure as no one method can be relied upon. No hard and fast rules can be laid down; consequently it is necessary for the individual to study the problem and select the most suitable procedure.

The various cleaning methods may be classified as follows:

I. *Mechanical Treatment:*
   A. Scrubbing with bristle brush.
   B. Scraping.
   C. Wire brushing.
   D. Shocking and sandblasting.

II. *Chemical Treatment:*
   A. Organic solvents.
   B. Chemical reagents.

III. *Electrolytic Treatment as Cathode:*
   A. Sulfuric acid.
   B. Citric acid.
   C. Potassium cyanide.
   D. Caustic soda.

MECHANICAL TREATMENT

Wet scrubbing with a bristle brush and some abrasive and detergent is a good method for removing light non-adherent films. This is also an exceptionally good method for the last step in any cleaning procedure. A brush suitable for this purpose is a flat, stiff-bristled plater's brush. Sapolio soap is a good abrasive and detergent. This method should be used with a certain degree of care as some metal can be removed by abrasion from excessive scrubbing. (See Table 1.)

Table 1. Weight Losses of Clean Metals Scrubbed Vigorously with Brush,* Sapolio Soap, and Water for One Minute

140 Strokes per Minute

Material	Weight Loss ($mg/sq\ dm$)
Aluminum	57
Cast iron	4.5
Copper	95.3
Inconel	32.3
Lead	121
Monel	38.7
Mild steel	21.2
Nickel	27.4
Ni-Resist	4.5
Stainless steel	17.0
Tin	183

* Brush used was a stiff-bristled plater's brush.

Specimens with a heavy scale, such as commonly found on soil or atmospheric corrosion specimens, are usually subjected to a rather harsh mechanical treatment to remove the outer layers of the deposit. Such treatment may include scraping with a dull chisel or knife, wire brushing, and mechanical shocking such as hammering or dropping on a hard surface. The one danger with such drastic treatment is that it might mar the metal surface and thus hinder its examination if it is carried too far. Such methods should be used only as a preliminary cleaning and should be followed by chemical or electrolytic treatment to remove the last traces of corrosion product.

Sandblasting is another mechanical method which has been used with success, especially on steels. The amount of metal it removes appears to be a function of the hardness. The following results were obtained on three materials* treated for the same length of time.[1]

Metal	Weight loss, $mg/sq\ dm$
Hot-rolled steel	298
Cold-reduced steel	16.1
Galvanized	114 (no zinc visible)

It is evident that the method can only be used on heavily corroded specimens. It also has the disadvantage that it renders the surface useless for examination.

Chemical Treatment

Corrosion specimens coated with a gummy resinous product or greases and oils are best treated with some organic solvent before attempting other cleaning methods. Acetone is usually tried first, but the choice of solvent is best left to the discretion of the operator. Other solvents that have been used are carbon tetrachloride, alcohol, ether, gasoline, benzene, and kerosene.

Various chemical reagents are used, but they are more specific for certain materials than for general use. These special treatments are listed on p. 1081.

Electrochemical Treatment

Several solutions have been used for the electrolytic treatment with the specimen as cathode. The best one appears to be sulfuric acid solution with organic inhibitor.

* Time necessary for sandblasting depends on the corrosion products. For exposed paint panels, 6¼ min. sufficed for the removal of paint and corrosion products.

[1] Unpublished data from Mellon Institute.

This treatment is used under the following conditions:

Solution: Sulfuric acid, 5% by weight.
Anodes: Lead or carbon.
Cathode: Corrosion specimen.
Cathode current density: 20 amp per sq dm (1.3 amp per sq in.).
Inhibitor: 2 ml organic inhibitor per liter of solution.*
Temperature: 75° C (165° F).
Exposure: 3 to 5 min.

Table 2 lists the weight losses obtained on a number of materials when the clean metals were subjected to this cleaning treatment. The metals that are likely to be attacked using this method are magnesium, zinc, and possibly lead.

As clear cut as this method appears, it still has its pitfalls for the unwary operator. A poor contact will not make the specimen cathodic and will allow it to corrode in the acid solution. This can usually be detected by a low current or no visible gassing from the specimen. The best remedy is to devise some holder for the specimens that will insure good electrical contact. Another source of trouble is contamination of the solution with an easily reducible metal ion, such as copper, silver, or tin, which will plate out as a metal deposit on the specimen and thus lower the apparent weight

TABLE 2. WEIGHT LOSSES OF CLEAN METALS SUBJECTED TO ELECTROLYTIC CLEANING TREATMENT

Inhibited 5% sulfuric acid bath

Time of Exposure — 3 min.

Material	Weight Loss, mg/sq dm
Aluminum (commercially pure)	1.6
Brass (admiralty)	0.2
Brass (red)	0.0
Brass (yellow)	0.4
Bronze (phosphor, 5% tin)	0.0
Bronze (silicon)	0.4
Bronze (cast, 85-5-5-5)	2.0
Copper	0.2
Copper-nickel (70-30)	0.0
" " (55-45)	0.2
" " (30-70)	0.0
Copper-nickel-zinc (75-20-5)	0.0
Iron (mild steel)	0.8
" (cast)	4.4
" (alloy cast)	1.2
" (18-8 stainless steel)	0.0
Lead (chemical)	6.0
Nickel	2.2
Nickel-molybdenum-iron (60-20-20)	0.8
Nickel-chromium-iron (80-13-7)	0.0
Tin	0.6
Zinc	Too high to be useful

loss. Errors from this source may be serious. A good plan is to clean such alloys in a separate solution or else clean them last before discarding the solution. Copper deposits acquired in the cathodic treatment can be removed easily in an aerated 2%

* E.g., Rodine. Any of the following may also be used in concentration of 0.5 gram per liter: diorthotolyl thiourea, quinolin ethiodide, or betanaphthol.

ammonia solution provided such a solution does not attack the base metal. A third source of trouble is decomposition of the inhibitor, which will then allow attack of metal under the corrosion products. This can be overcome by avoiding prolonged use of the cleaning solution.*

A similar electrolytic cleaning method was described for the removal of scale formed in high-temperature steam on various steels and alloys.[2] The acid and inhibitor (quinoline ethiodide) concentration was 10% and 1%, respectively, contained in a lead jar serving as anode, and the cathodic current density was 15 amp per sq dm (1 amp per sq in.). The time of cleaning required for four oxidized steels was:

Steel	Time, hours
S.A.E. 1010	½
4–6% Cr-Moly	1
12% Cr	2
18–8	4

No attack of the base metal took place below 45° C (110° F).

Other solutions used for the cathodic cleaning of corrosion specimens, especially rusted steel, include a saturated solution of citric acid, a 10% solution of sodium or potassium cyanide, and a 10% solution of caustic soda. The corrosion committee of the Iron and Steel Institute[3,4] investigated these three solutions. It was reported that the citric acid solution was not too efficient on rusts that had aged. At high current densities the solution becomes hot and above 80° C (175° F) it will attack steel. The cyanide solution was found to be superior to the citric acid for lightly rusted steels. The caustic soda solution was found to be of little value for removing rust from steel. For additional methods see items 6 and 8 below.

SPECIAL METHODS FOR CLEANING SPECIFIC MATERIALS

1. *Copper and Nickel Alloys.* Dip for 2 to 3 min in 18% hydrochloric acid or 10% sulfuric acid at room temperature and then scrub with bristle brush.
2. *Aluminum Alloys.* Dip for 2 to 3 min in concentrated nitric acid at room temperature. Scrub lightly with bristle brush as some aluminum alloys in an annealed condition may be abraded. In some environments an oxide film is developed which resists the nitric acid. In such cases an alternative treatment is usually effective. This consists of about a 10-min dip in a solution containing 2% chromic acid and 5% phosphoric acid at a temperature of 80° to 85° C (175° to 185° F). This treatment can be followed by the nitric acid dip.
3. *Tin Alloys.* Dip for 10 min in a boiling solution of 15% trisodium phosphate. Scrub lightly with bristle brush.
4. *Lead Alloys.*
    (a) *For removal of PbO:* Dip for 10 min in boiling solution of 1% acetic acid. Scrub lightly.
    (b) *For removal of PbO and/or $PbSO_4$:* Dip in hot solution of 25% ammonium acetate. Scrub lightly.

---

* A fourth possible error occurs when the corrosion product can be cathodically reduced to metal, thereby giving rise to low corrosion weight losses. No data are available to indicate the importance of this factor, but it is likely that it need be considered only when dealing with thin coatings such as tarnish films.

[2] H. L. Solberg, G. A. Hawkins, and A. A. Potter, *Trans. Am. Soc. Mech. Engrs.*, **64**, 303–316 (1942).
[3] *First Corrosion Report of Iron and Steel Institute*, p. 207, 1931.
[4] *Third Corrosion Report of Iron and Steel Institute*, p. 59, 1935.

(c) *For removal of PbO_2:* The closely adhering lead peroxide on battery grids can be removed with the following solution:

Sodium hydroxide — 80 grams per liter
Mannite — 50 grams per liter
Hydrazine sulfate — 0.62 gram per liter

After brushing to remove loose corrosion product the specimens are placed in the boiling solution for 30 min or until the metal surface is clean. Wash in warm running water and scrub lightly to remove a possible film of oxide on surface.

5. *Zinc.* Immersion for 5 min in a 10% ammonium chloride solution followed by a light scrub should remove the corrosion products. Then immerse for 20 sec in a boiling solution containing 5% chromic acid and 1% silver nitrate. This procedure gives a blank loss of about 4.6 mg per sq dm. In making up the chromic acid solution, it is advisable to dissolve the silver nitrate separately and add it to the boiling chromic acid to prevent excessive crystallization of the silver chromate. The chromic acid must be free from sulfates and chlorides to avoid attack on the zinc.

6. *Zinc-Coated Steel After Rusting.* Several methods have been used for such specimens. They are:
    (a) Treat cathodically at 1.1 amp/sq dm (10 amp/sq ft) in the following solution at 70° C (160° F): trisodium phosphate, 30 grams per liter; sodium carbonate, 15 grams per liter; sodium hydroxide, 10 grams per liter.
    (b) Soak in a solution of 10% trisodium phosphate for 2 to 15 hours, depending upon the amount of corrosion product.
    (c) Immerse in a 20% ammonium hydroxide solution for 2 to 5 min.
    (d) Treat cathodically in 5% disodium phosphate for 5 min.
    (e) Immerse in a 10% ammonium persulfate solution.

7. *Magnesium Alloys.* Dip for approximately 1 min in 20% chromic acid solution to which has been added with agitation 1% $AgNO_3$ in solution form. It is important to use C.P. chromic acid, as sulfates and chlorides present will attack the magnesium. This method was recently modified to a 15-min dip in a boiling solution of 15% chromic anhydride and 1% silver chromate for the reason that nitrates may cause some attack.[5]

8. *Corroded Iron.* Several methods in addition to those described have been used for rusted iron specimens.
    (a) Cathodic treatment in 10% ammonium citrate at 10 amp/sq ft (1.1 amp/sq dm).
    (b) Cathodic treatment in 10% sodium cyanide at 20° C (68° F), 15 amp/sq ft (1.6 amp/sq dm) for 20 min.
    (c) A 5-min immersion in a boiling solution of 20% sodium hydroxide to which has been added 200 grams per liter zinc dust is effective in loosening a deposit.

9. *Stainless Steel Alloys.* A 10% nitric acid solution at 60° C (140° F) is an effective cleaning solution. Contamination with chlorides should be avoided or attack of the base metal may result.*

---

* A pickle suitable for some purposes in preparing stainless test specimens is given in item 2 under "The Alloy," p. 173. EDITOR.

[5] F. A. Fox and C. J. Bushrod. *J. Inst. Metals,* **70,** 325 (1944).

GENERAL REFERENCES
(PREPARATION AND CLEANING OF SPECIMENS)

A. S. T. M. *Standards*, Designation B 185–43T, Part I, Supplement, p. 332 (1943).
A. S. T. M. *Standards*, Designation A 224–41, Part I, Supplement, p. 72 (1941).
Symposium on Corrosion Testing Procedures, American Society for Testing Materials, 1937.

Statistical Methods

G. G. Eldredge*

INTRODUCTION

The primary purposes of this chapter are: (1) to draw attention to statistical concepts, (2) to provide a minimum reference list, and (3) to show how statistical reasoning can contribute to studies on corrosion in the planning of corrosion experiments and in identifying causes of corrosion.

In corrosion testing, as in experimental data of any kind, in order to measure the effect of any cause it is necessary to take into account all the other causes, known and unknown. The results of unknown causes are called error and may be divided into assignable error and residual error. The interpretation of data usually involves the comparison of the magnitude of the results with a measure or estimate of the experimental error, even if this comparison takes no more definite form than an opinion that the results are significant. This chapter describes various elementary statistical techniques used in comparing results with errors. Definitions, nomenclature, and calculation short cuts are given at the end of the chapter.

In order that data may be interpreted efficiently, the first and most important requirements are that the experiment be properly designed to give the information wanted and carefully controlled in the important respects. If information is being obtained from a process beyond the control of the investigator, it is all the more necessary to set up the sampling procedure to insure getting data without unnecessary bias. Statistical design requires taking into account variation attributable to causes under study and known unwanted causes, and it also must provide enough information to measure the overall effect of unknown causes which make up the residual error.

After the best has been done in designing the test, the first step in interpreting the data is to arrange and present them. This should be done in a way to show:

(1) The measures of the effects being studied (averages, differences, rates, coefficients, etc.), in addition to
(2) The quantity of data represented.
(3) If possible, the quality of data (evidence of *control*†).
(4) Measures of the spread or precision of the data.
(5) How the data are distributed around the averages or other measures.

A discussion of the subject of presentation of data (from which this list is adapted) is given in a pamphlet of the American Society for Testing Materials.[1] For presenting a set of observations of a single variable obtained under the same essential conditions, the average (a), the *standard deviation* (d), and the number of observations (b) are

---
* Shell Development Company, Emeryville, California.
† Words in italics are defined at the end of the chapter.

[1] American Society for Testing Materials, *Manual on Presentation of Data*, 1940.

the minimum specifications. The pamphlet referred to recommends the form:

Number of Tests	Average $\bar{X}$	Limits for $\bar{X}$ Chances 9 in 10	Standard Deviation
10	575.2	±5.1	8.26

The limit quantity is calculated from the other three quantities. Its calculation and meaning, as well as that of the standard deviation, will be discussed later.

If enough observations are available, the shape of the distribution (e) can be indicated by the showing of a grouped frequency distribution or, in terms of statistics, measuring *skewness* and *kurtosis*. It must be admitted that this standard, as logical as it sounds, is not met by most published data on corrosion.

An extensive example of the presentation of measures of the spread of corrosion data may be examined in an A. S. T. M. report on 10-year atmospheric corrosion tests of non-ferrous metals.[2] In this example, tensile strengths and standard deviations of tensile strengths are given in tables (points a and d mentioned above), but the quantity of data represented by each average is given only in the text (point b). With regard to point c, evidence concerning control is presented as a set of modified control charts of the uncorroded material (see below for a discussion of control charts) with plotted standard deviations of the distribution at each data point instead of control limits.

Number of Places to be Retained in Calculation and Presentation of Data

During calculation one or two or more places than will be retained should be kept to prevent compounding of error. For the presentation of data, the number of places to be retained will depend on what use is to be made of the results. A general rule, therefore, cannot be laid down. If the arrangement of data suggested above is followed, there will be nothing very misleading about extra digits carried in an average. It would appear usually unnecessary to carry more than one or two doubtful figures, that is, the plus and minus limits given above may be expressed to two figures and the averages may be expressed to the same decimal as the limits.

A conservative rule generally equivalent to retaining at least two doubtful places would be to express measurement to at least the nearest sixth of the standard deviation and a statistic, such as an average or standard deviation, to at least the nearest sixth of its expected standard deviation.* With any given standard of retention of digits to express individual values, one extra place may be justified for their averages, up to 100 values. For larger samples, more places should be carried.†

Carrying standard deviations to three significant figures is recommended.[3] When *probable errors* are used in the literature (see "Confidence Ranges or Confidence Intervals," p. 1087) one or two places are more common.

TESTING CONTROL

The most used method of testing for statistical control (statistical uniformity, absence of unknown *assignable causes*) is the control chart technique. The control

---

* This will give an expected range of thirty-six or more times the least interval indicated. Since an average is more precisely known than an individual value, it may be expressed to more places.

† This results from the form of the expression for the expected standard deviation of an average $\sigma_{\bar{x}} = \sigma_x/\sqrt{n}$ (correction for sample size not important here).

[2] Subcommittee VI of Committee B-3, *Proc. Am. Soc. Testing Materials*, **44**, 224 (1944).

[3] American Society for Testing Materials, *Manual on Presentation of Data*, 1940.

chart technique is a procedure widely used in controlling industrial production and is being applied more and more in laboratory research. There is considerable literature on control charts.[3,4,5,6,7] Examples of the application of this technique to corrosion problems have been published.[8,9,10,11,12]

A control chart may be made from any sufficiently complete series of values in which it is desired to find if there are non-random causes of variation worth trying to find. Control charts may be made by plotting averages or any other statistic, such as standard deviation, range, or fraction rejected, along with objective control limits, which indicates values suspected of not being random. Only the description of a control chart for averages will be given here along with some comments on the interpretation. Discussion of control charts for the other statistics mentioned may be found in the references cited above.

Construction of a Control Chart for Averages

The following directions parallel the procedure given by the American Society for Testing Materials and the American Standards Association.[13,14]

The first step is to divide all the data into small subgroups (usually four or five values to a subgroup) and plot their averages consecutively. (The subgroups are usually chronological but may have some other logical basis. It is necessary that causes of lack of control should affect differences between subgroups more than differences within subgroups.)

The second step is to plot the central average and the control limits at plus and minus three times (or some other multiple of) the expected standard deviation of the averages above and below the central average. This triple expected standard deviation of the averages is obtained by multiplying a standard deviation assumed from past experience by a factor $A = 3/\sqrt{n}$ or by multiplying the arithmetic average standard deviation of the individual subgroups by a similar factor A_1, which contains a correction for the bias of the averaging of σ's and for the bias due to small sample size. Tables of these factors, A and A_1, are given in works on this subject.[13,14] The control limits may also be estimated from the ranges of the subgroups with similar factors. This latter method is easier and but little less efficient for samples of four and five, although markedly less efficient for considerably larger samples. Another method[15] of getting the height of the control limit is to use $\sqrt{(\sigma^2)_{\text{ave.}}/(n-1)}$.

[4] L. E. Simon, *An Engineers' Manual of Statistical Methods*, John Wiley and Sons, New York, 1941.

[5] American Standards Association, New York, "Guide for Quality Control," A. S. A. War Standard Z1.1 (1941); "Control Chart Method for Analyzing Data," Z1.2 (1942); "Control Chart Method of Controlling Quality during Production," Z1.3 (1942).

[6] E. S. Pearson, British Standards Institution, London, "The Application of Statistical Methods to Industrial Standardization and Quality Control," 600 (1935); B. P. Dudding and W. J. Jennett, "Quality Control Charts," 600R (1942).

[7] W. A. Shewhart, *Economic Control of Quality of Manufactured Products*, D. Van Nostrand Co., New York, 1931.

[8] R. F. Passano, *Proc. Am. Soc. Testing Materials*, **32**, (II), 468 (1932).

[9] R. F. Passano and F. R. Nagley, *Proc. Am. Soc. Testing Materials*, **33** (II), 387 (1933).

[10] W. A. Wesley, *Proc. Am. Soc. Testing Materials*, **43**, 655 (1943).

[11] R. B. Mears, C. J. Walton, and G. G. Eldredge, *Proc. Am. Soc. Testing Materials*, **44**, 639 (1944).

[12] W. E. Campbell, *Trans. Electrochem. Soc.*, **81**, 379–390 (1942).

[13] American Society for Testing Materials, *Manual on Presentation of Data*, 1940.

[14] American Standards Association, New York, "Guide for Quality Control," A. S. A. War Standard Z1.1 (1941); "Control Chart Method for Analyzing Data," Z1.2 (1942); "Control Chart Method of Controlling Quality during Production," Z1.3 (1942).

[15] E. S. Pearson, British Standards Institution, London, "The Application of Statistical Methods to Industrial Standardization and Quality Control," 600 (1935); B. P. Dudding and W. J. Jennett, "Quality Control Charts," 600R (1942).

Interpretation of a Control Chart

Points falling outside the control limits are indications of non-random causes or lack of *control*. Random variation does not interfere with this indication, since it is present in both the within-subgroup and among-subgroup variations compared. A change in the subgroup size will change the spread of the control limits by changing the factor A or A_1.

When points fall outside the limits the causes may not be evident, but experience shows that an investigation into the source of the data will often reveal (enough to justify the search) these assignable causes to the engineer or researcher, whose knowledge of the process or experiment will suggest possibilities. An improvement of the experimental conditions can be made to remove these causes one at a time until no more non-random causes appear to be present. After the system is in statistical control, predictions may be made of future results. In particular, other statistical tests may be used with assurance.

Other criteria of control may be used besides the finding of a point outside the control limits. A long run of successive points above or below the average (or more logically above or below the median) is an indication of lack of control. Table 1 shows the length of a run in a series of a given number of points, which is an indication of lack of control at the two significance levels 95% and 99%.*[16]

TABLE 1. LENGTH OF RUNS ABOVE AND BELOW THE MEDIAN SIGNIFICANT AT THE INDICATED LEVEL OF PROBABILITY*

Runs as long or longer are significant

Length of Series	Runs Above (or Below) Alone		Runs Either Above or Below	
	$P = .95$	$P = .99$	$P = .95$	$P = .99$
10	5	..	5	..
20	7	8	7	8
30	8	9	8	9
40	8	9	9	10
50	8	10	10	11

* F. Mosteller, *Annals Math. Statistics*, **12**, 229 (1941). In the same place is a table giving probabilities of the runs of different lengths.

Still another criterion of control is the occurrence of a run of points all in ascending or descending order. A table similar to Table 1, giving significant lengths of runs ascending or descending or the probability of the occurrence of an arrangement with at least one run ascending or descending of a critical length or more, is not available, although the equations for the expected number of runs of various lengths and their standard deviations are in the statistical literature.[17]

However, the total number of runs above and below the median or runs ascending and descending may be tested to see if there are more or less than would be expected

---

* These levels are lower than the 99.73% levels used for the falling out of individual points, but a little reflection will show that this is a logical reduction of limits. The chance of "looking for trouble" in a controlled system, because the first point falls out of the 99.73% limits, is 0.27%, but the chance increases with the number of points. Actually in the same numbers of points shown in the table, in a system actually in control, the chance of at least one point's falling out of control is somewhat greater than the chance of a run of one of the lengths shown.

[16] F. Mosteller, *Annals Math. Statistics*, **12**, 229 (1941).
[17] P. S. Olmstead, *J. Am. Statistical Assoc.*, **37**, 152 (1942). See also H. Levene and J. Wolfowitz, *Annals Math. Statistics*, **15**, 58 (1944).

from a random distribution. For N observations the expected number of runs with a range to include probabilities of 95% for runs above and below the median is*

$$E = \frac{N}{2} + 1 \pm \left(2\sqrt{\frac{N^2 - 2N}{4(N-1)}} + \frac{1}{2}\right)$$

and of runs ascending and descending

$$E = \frac{2N-1}{3} \pm \left(2\sqrt{\frac{16N - 29}{90}} + \frac{1}{2}\right)$$

If the total number of runs of either sort falls out of the ranges given, it is an indication of lack of control. If the number of runs is excessive, the cause leading to lack of control may be periodic.

Campbell has described an application of both types of runs to corrosion work.[18]

TESTS OF SIGNIFICANCE

When there is assurance that statistical control exists, data may be interpreted with assurance. When the relation of results to the amount of error is not sufficiently obvious, tests of significance may be used.

CONFIDENCE RANGES OR CONFIDENCE INTERVALS

INTRODUCTION TO THE t-TEST

Averages are frequently expressed with a confidence interval in the form $\bar{X} + L$. An example was given on p. 1084. The limit quantity like the 5.1 in this example can be separated into two parts, in several equivalent ways, either Limit $= a\sigma = d(\sigma/\sqrt{n})$, or Limit $= t(s/\sqrt{n})$. The two alternative expressions involve the symbol $\sigma = \sqrt{\Sigma(X - \bar{X})^2/n}$ usually employed in engineering literature or the quantity $s = \sqrt{\Sigma(X - \bar{X})^2/n - 1}$ usually used in the literature on tests of significance. It may be seen that a, d, or t will vary with the probability that it is desired to indicate by the limits. In addition because σ and s are calculated from the limited samples for a given probability, t and d as well as a are functions of the size of the samples or the number of tests. In older work it has been common to neglect the effect of sample size on d and use values of $d = 0.6745$, 1, 2, or 3 giving limits which have been called "probable errors," "1-sigma limits," "2-sigma limits," or "3-sigma limits," respectively, and enclosing for very large samples from a normal population the respective probabilities .5000, .6826, .9555, and .9973. When applied to smaller samples, however, these intervals enclose the "true" mean less frequently.

However, it is now recommended[19,20] to use values of a, d, or t to give limits which have a constant probability of including the "true" mean of normal populations. Here "true" mean is the mean that would be approached by using more or larger samples. The A. S. T. M. pamphlet[19] gives a table of values of a from which to get limits for .9 or .99 probability. Table 2 of this chapter gives values of t which can be used for limits of .95, .99, or .999 probability.

* The radical is the standard deviation; the ½ term is a theoretical adjustment for non-normality of distribution. These equations were adapted from equations given in mimeographed sheet form by F. C. Mosteller.

[18] W. E. Campbell, *Trans. Electrochem. Soc.*, **81**, 379–390 (1942).
[19] American Society for Testing Materials, *Manual on Presentation of Data*, 1940.
[20] L. E. Simon, *An Engineers' Manual of Statistical Methods*, John Wiley and Sons, New York, 1941.

TABLE 2. VALUES OF t EXPECTED TO BE EXCEEDED (IN EITHER DIRECTION) BY CHANCE ALONE WITH THE PROBABILITY GIVEN*

Size of a Single Sample	Degrees of Freedom	Probability		
n	$n-1$	.05	.01	.001
2	1	12.706	63.657	636.619
3	2	4.303	9.925	31.598
4	3	3.182	5.841	12.941
5	4	2.776	4.604	8.610
6	5	2.571	4.032	6.859
7	6	2.447	3.707	5.959
8	7	2.365	3.499	5.405
9	8	2.306	3.355	5.041
10	9	2.262	3.250	4.781
11	10	2.228	3.169	4.587
12	11	2.201	3.106	4.437
13	12	2.179	3.055	4.318
14	13	2.160	3.012	4.221
15	14	2.145	2.977	4.140
16	15	2.131	2.947	4.073
17	16	2.120	2.921	4.015
18	17	2.110	2.898	3.965
19	18	2.101	2.878	3.922
20	19	2.093	2.861	3.883
21	20	2.086	2.845	3.850
22	21	2.080	2.831	3.819
23	22	2.074	2.819	3.792
24	23	2.069	2.807	3.767
25	24	2.064	2.797	3.745
26	25	2.060	2.787	3.725
27	26	2.056	2.779	3.707
28	27	2.052	2.771	3.690
29	28	2.048	2.763	3.674
30	29	2.045	2.756	3.659
31	30	2.042	2.750	3.646
41	40	2.021	2.704	3.551
61	60	2.000	2.660	3.460
121	120	1.980	2.617	3.373
∞	∞	1.960	2.576	3.291

*Reprinted from R. A. Fisher and F. Yates, *Statistical Tables for Biological, Agricultural, and Medical Research*, p. 30, Oliver and Boyd, Ltd., London, 1943. By permission of the Author and Publisher.

The number n is the number of comparisons or degrees of freedom from which t was calculated.

t-Test Directions

A t test is a way to compare the difference between a mean and an arbitrary value, or the difference between two means, with the experimental error. The difference being tested is divided by its expected standard deviation and the quotient compared with the appropriate quantity t, which may be taken from Table 2. For significance of the difference between an average and a known value, X_0, usually zero, use:

$$t = \frac{\bar{X}_1 - X_0}{s\sqrt{1/n}} > t \text{ (from table)}$$

where $s = \sqrt{\Sigma(X - \bar{X})^2/n - 1}$, and the table for t is entered with $n - 1$ *degrees of freedom*. For the difference between two averages

$$t = \frac{\bar{X}_1 - \bar{X}_2}{s\sqrt{\dfrac{1}{n_1} + \dfrac{1}{n_2}}} > t \text{ (from table)}$$

In this latter case s is estimated from both samples as $s = \sqrt{\dfrac{\Sigma(X_1 - \bar{X}_1)^2 + \Sigma(X_2 - \bar{X}_2)^2}{n_1 + n_2 - 2}}$
and the table for t is entered from the column giving the number of degrees of freedom, $n_1 + n_2 - 2$.

INTERPRETATION OF t TEST

What is done by the t test is to compare the difference found from the data, with the difference which might be expected because of experimental error in an experiment where no real difference existed. The difference is divided by its expected standard deviation $s\sqrt{1/n}$ or $s\sqrt{1/n_1 + 1/n_2}$ in order to eliminate the effect of sample size and units of measurements as well as to compare with experimental error. The t test tells how often such a difference might be caused by chance. A difference may be real but not significant by the t test. When engineering experience suggests that a difference is real, it is often suggested that a new experiment be planned with sufficiently larger samples to show that the difference is significant. When this may not be done, it is good practice to take the observed difference as an estimate of the difference expected.

EXAMPLE OF A t TEST

Specimens of a certain alloy were intermittently exposed to $NaCl-H_2O_2$ by a test method described elsewhere.[21] Losses in tensile strength were compared between unstressed specimens and specimens stressed in bending to a stress of 75% of the yield strength.

Table 3 shows the calculation of the t test. Columns 2 and 5 give experimental data. In columns 3 and 6 a number near the mean is subtracted from each value to simplify the work without changing the results. In columns 4 and 7 are the squares of the values in 3 and 6. The sums of squares are adjusted by subtracting the square of the sum of the quantities, divided by their number n as explained at the end of the chapter under technical short cuts. At the bottom of this table, t is calculated. It turns out to be 3.76, which is definitely larger than the value (from Table 2) required for significance at a 95% probability level, which is 2.07. This means that a value of t as large as that found in this test would not be expected to arise as often as once in 20, if there were no differences between the two types of specimens. In fact, since Table 2 shows a value of 3.79 as compared to 3.76 for 0.001 probability, it may be seen that so large a value of t would be expected to arise by chance only about once in a thousand times.

Since the stressed and unstressed specimens were tested at the same time, it might be expected that some of the causes of error that might affect them could be eliminated from the comparisons, by comparing the values by pairs. This is done in Table 4, where the pair differences of the same data are analyzed according to this method. As expected, the variance in the denominator of this equation is smaller than before and t is therefore larger. However, the value of t required for significance is larger since only eleven comparison or *degrees of freedom* were available for

[21] R. B. Mears, C. J. Walton, and G. G. Eldredge, *Proc. Am. Soc. Testing Materials*, **44**, 639 (1944).

TABLE 3. THE t TEST APPLIED TO AN ALTERNATE IMMERSION CORROSION TEST

Comparison of Percentage Losses in Tensile Strength of Specimens Exposed Unstressed and Stressed to Salt Peroxide Solution by Alternate Immersion

(1) Test	(2)	(3) Unstressed	(4)	(5)	(6) Stressed	(7)
	X_1	$(X_1 - 10)$	$(X_1 - 10)^2$	X_2	$(X_2 - 10)$	$(X_2 - 10)^2$
1	11.4	1.4	1.96	14.2	4.2	17.64
2	9.6	−0.4	0.16	12.9	2.9	8.41
3	9.6	−0.4	0.16	12.3	2.3	5.29
4	11.4	1.4	1.96	13.0	3.0	9.00
5	8.2	−1.8	3.24	10.7	0.7	0.49
6	10.2	0.2	0.40	12.6	2.6	6.76
7	11.5	1.5	2.25	10.7	0.7	0.49
8	9.9	−0.1	0.10	9.1	−0.9	0.81
9	9.3	−0.7	0.49	13.1	3.1	9.61
10	10.6	0.6	0.36	11.5	1.5	2.25
11	8.7	−1.3	1.69	11.9	1.9	3.61
12	9.6	−0.4	0.16	11.0	1.0	1.00
Sum	120.0	0.0	12.93	143.0	23.0	65.36
Adjustment = (Sum)^2/12		....	0.00	....	....	44.08
Adjusted sum of squares		....	12.93	....	....	21.28

Significance of difference of two Means

$$t = \frac{X_2 - X_1}{s\sqrt{1/n_1 + 1/n_2}} = \frac{11.92 - 10.0}{\sqrt{\frac{21.28 + 12.93}{12 + 12 - 2}\left(\frac{2}{12}\right)}} = 3.76$$

t required for two samples of 12, that is with 22 degrees of freedom (Table 2), is

        2.07 for 5% Probability
        2.82 for 1%   "
        3.79 for 0.1% "

estimating the standard deviation instead of twenty-two for the other method. For this example, it turns out that the probability is about the same for the two methods of comparison. One method of comparison has been about as sensitive as the other.

ANALYSIS OF VARIANCE

Introduction

The t test just described is a simple case of the analysis of *variance*. For cases where comparisons are made among more than two means, the t test may be generalized to take care of the necessarily more complicated measures of the cause being studied and of the unassignable causes.* Equations corresponding to those given for the t test are not commonly written since very convenient general tabular forms have been worked out for making comparisons.

Design of Experiment

In designing an experiment from which more than one kind of comparison is to be made or which will be affected by more than one cause, whether or not a formal analysis of variance is to be made, the test should be arranged so that averages showing the effects of each cause are not biased by the other causes. Common sense

---
* See definition of assignable cause.

TABLE 4. THE t TEST BY PAIRING APPLIED TO AN ALTERNATE IMMERSION TEST

Difference in Percentage Losses in Tensile Strength of Specimens Exposed Unstressed and Stressed to Salt Peroxide Solution by Alternate Immersion. Same Corrosion Test Results in Table 5.

Test	$X_2 - X_1$	$(X_2 - X_1)^2$
1	2.8	7.84
2	3.3	10.89
3	2.7	7.29
4	1.6	2.56
5	2.5	6.25
6	2.4	5.76
7	−0.8	0.64
8	−0.8	0.64
9	3.8	14.44
10	0.9	0.81
11	3.2	10.24
12	1.4	1.96
Sum	23.0	69.32
Adjustment = (Sum)^2/12	....	44.08
Adjusted Sum of Squares	....	25.24

Significance of Mean of Pair Difference

$$t = \frac{X_2 - X_1}{s\sqrt{\frac{1}{12}}} = \frac{11.93 - 10}{\sqrt{\frac{25.24}{12-1}\left(\frac{1}{12}\right)}} = 4.38$$

t required for 11 degrees of freedom equals
2.20 for 5% Probability
2.82 for 1%   "
3.79 for 0.1%   "

and visualization of all factors are required. Plans for efficient statistical design of experiments and discussion of this subject are given in various books on statistics.[22,23]

Procedure for Analysis of Variance

In carrying out an analysis of variance one starts with a table of values showing the apparent effects of the several variables. The steps to be carried through are listed below. Tables 5 to 7 illustrate the calculations on a sample problem. Procedures for different forms of analysis of variance are explained in various books on statistics.[23,24]

An Example of Analysis of Variance

Table 5 gives the losses in yield strength of machined tensile specimens of an alloy exposed by the same test method as in the previous example.[25] The data are subdivided by blocks, sheets, and specimens. The blocks were groups of four sheets held 2 in. apart and quenched simultaneously after the heat treatment. Then the specimens were cut out and exposed to the corrosion test.

For this example the causes of variance analyzed are differences among blocks quenched at different times, difference among the four positions, top to bottom, in

---

[22] R. A. Fisher, *The Design of Experiments*, Oliver and Boyd, Ltd., London, 1942.
[23] G. W. Snedecor, *Statistical Methods*, Iowa State College Press, 1940.
[24] R. A. Fisher, *Statistical Methods for Research Workers*, Oliver and Boyd, Ltd., London, 1944
[25] R. B. Mears, C. J. Walton, and G. G. Eldredge, *Proc. Am. Soc. Testing Materials*, **44**, 639 (1944).

TABLE 5. YIELD STRENGTH LOSSES IN PSI OF SPECIMENS EXPOSED TO SALT-HYDROGEN PEROXIDE SOLUTION BY ALTERNATE IMMERSION

Block	Specimen Position	Sheet 1	Sheet 2	Sheet 3	Sheet 4	Ave.
1	1	1900	2100	2000	2400	
	2	2100	2500	1900	3300	
	3	2800	2100	3200	4300	
	4	1900	2000	3400	3900	
	Ave.	2175	2175	2625	3475	*2612*
2	1	5200	2600	3700	3700	
	2	3100	2600	3000	3800	
	3	3200	3100	3200	3600	
	4	2800	2800	2800	3200	
	Ave.	3575	2775	3175	3575	*3275*
3	1	2900	3000	2700	3400	
	2	2500	3500	2700	3700	
	3	2400	3300	3100	3500	
	4	2800	3000	2800	3000	
	Ave.	2650	3200	2825	3400	*3019*
4	1	3800	2300	3200	3400	
	2	3100	4400	3600	3400	
	3	3600	4500	3600	3300	
	4	3700	3400	2800	3000	
	Ave.	3550	3650	3300	3275	*3444*
5	1	3200	3700	3700	4100	
	2	3100	3400	4200	2800	
	3	3500	3700	3200	2800	
	4	2700	3300	3800	3100	
	Ave.	3125	3525	3725	3200	*3394*

Grand Average *3129*

a sheet as quenched, and differences due to inner or outer positions of the sheet in the block. A further breakdown could be made of the variance (3 degrees of freedom) due to position of the specimens in the sheet. By analysis of *covariance* or by regression analysis (see discussion below) the part due to linear variation with sheet position (one degree of freedom) could be isolated from the remainder (two degrees of freedom), and the analysis of variance extended.

STEPS IN AN ANALYSIS OF VARIANCE

A. A convenient working mean may be subtracted from all values to reduce their size and number of digits. This is analogous to the use of a working mean in calculating a standard deviation. See the calculation short cut at end of chapter. All further work is done with the simplified values (Table 6). The choice of the working mean does not affect the final results. This step cuts down the amount of work, but, if calculating machines are available, the original values may be used and a possible source of mistakes avoided.

B. The table of values as thus simplified is then broken down according to all classifications of interest and totaled and subtotaled (Table 6).

Table 6

		Yield Strength Losses in Excess of 3000 psi Expressed in Hundreds.* Data from Table 5.					Squares of Corresponding Values and Sums Given at the Left†				
Block	Specimen Position	Sheet 1	Sheet 2	Sheet 3	Sheet 4	Sum					
1	1	−11	− 9	−10	− 6		121	81	100	36	
	2	− 9	− 5	−11	+ 3		81	25	121	9	
	3	− 2	− 9	+ 2	+13		4	81	4	169	
	4	−11	−10	+ 4	+ 9		121	100	16	81	
	Sum	−33	−33	−15	+19	−62	1089	1089	225	361	3844
2	1	+22	− 4	+ 7	+ 7		484	16	49	49	
	2	+ 1	− 4	0	+ 8		1	16	0	64	
	3	+ 2	+ 1	+ 2	+ 6		4	1	4	36	
	4	− 2	− 2	− 2	+ 2		4	4	4	4	
	Sum	+23	− 9	+ 7	+23	+44	529	81	49	529	1936
3	1	− 1	0	− 3	+ 4		1	0	9	16	
	2	− 5	+ 5	− 3	+ 7		25	25	9	49	
	3	− 6	+ 3	+ 1	+ 5		36	9	1	25	
	4	− 2	0	− 2	0		4	0	4	0	
	Sum	−14	+ 8	− 7	+16	+ 3	196	64	49	256	9
4	1	+ 8	− 7	+ 2	+ 4		64	49	4	16	
	2	+ 1	+14	+ 6	+ 4		1	196	36	16	
	3	+ 6	+15	+ 6	+ 3		36	225	36	9	
	4	+ 7	+ 4	− 2	0		49	16	4	0	
	Sum	+22	+26	+12	+11	+71	484	676	144	121	5041
5	1	+ 2	+ 7	+ 7	+11		4	49	49	121	
	2	+ 1	+ 4	+12	− 2		1	16	144	4	
	3	+ 5	+ 7	+ 2	− 2		25	49	4	4	
	4	− 3	+ 3	+ 8	+ 1		9	9	64	1	
	Sum	+ 5	+21	+29	+ 8	+63	25	441	841	64	3969
Sum for all Block	1					30					900
	2					27					729
	3					60					3600
	4					2					4
	Sum	+ 3	+13	+26	+77	119	9	169	676	5929	14,161

Outside sheets (1 and 4) Sum = 80   6400
Inside sheets (2 and 3) Sum = 39   1521

Quantities Needed for Table 7

All causes sum of squares = Sum of 80 values at right = 3,413
Blocks sum of squares = 3844 + 1936 + 9 + 5041 + 3969 = 14,799
Specimen position sum of squares = 9 + 169 + 679 + 5929 = 5,233
Sheet position sum of squares = 6400 + 1521 = 7,921
Square of grand sum = 14,161

* In other words, a working Mean of 3000 psi = 30 hundred psi has been used.
† These values are not totaled on this side of the table, that is, 1089 is $(-33)^2$ and not a sum of other values.

TABLE 7. ANALYSIS OF VARIANCE DUE TO BLOCKS, SPECIMEN POSITIONS, AND SHEET POSITIONS

Cause of Variance (1)	Sums of Squares of Total (2)	Sample Sizes (3)	Un-corrected Sum of Squares (4)	Corrected Sum of Squares (5)	Degrees Freedom (6)	Variance (7)	Ratio of Variance (8)	Ratio Expected Once in Twenty (9)	Estimated Standard Deviation (10)
All causes	3,413	1	3,413.0	3,236.0	79	(40.9)	....	....	6.3
Blocks	14,799	16	924.9	748.0	4	187	6.0	2.50	....
Specimen positions	5,233	20	261.65	84.6	3	28.2	0.91	2.74	....
Sheet positions	7,921	40	198.0	21.0	1	21.0	0.68	3.98	....
(Sum)	14,161	80	177.0	....	....	....	....	....	....
Error (residue)	....	....	....	2,194.0	71	30.9	1.0	....	5.6

C. The individual values are all squared as are the totals and subtotals (Table 6). This completes the work preliminary to forming the table for the analysis of variance (Table 7), which is made up as follows:

(1) In column one are listed all the causes of variance starting with "total" or the sum of all causes.

(2) Column two of Table 7 gives the sums of the squares of the individual values and subtotals of the various kinds. The final number in column two is the square of the grand sum of all the individual values.

(3) The sample sizes of the samples on which each line is based are listed in column three of Table 7 and used as divisors to get column four. The sample size of the grand sum is, of course, the total number of values in the whole test.

(4) To get column five of Table 7, the final number in column four is then subtracted from each of the other numbers in that column. This adjustment step removes the effect of the arbitrary choice of working mean. If the mean of all the observations had been used as working mean this adjustment would have been zero.

(5) In column five the effects of all causes of variance are additive.* Subtracting all the accounted-for items from the "total" (first) item measures the effects unaccounted for, or error.

(6) These effects in column five of Table 7 are made up each of an intensive and an extensive factor. The extensive factor is the number of comparisons or *degrees of freedom* involved. It is ordinarily one less than the number of values compared. This number is written in column six and divided into column five to give column seven. The values in column seven are called *variances*. The usual estimate of error variance is the variance found by dividing the residue of column five by the residual degrees of freedom of column six.

(7) Since the variances in column seven of Table 7 are independent in a probability sense, their ratios to the error variance are taken in column eight to be compared to values expected by chance one time in twenty (95% probability) as obtained from Table 8 or from similar tables for other probabilities.

* An unfortunate paradox in nomenclature should be pointed out in the term *analysis of variance*. *Variances* are defined as the quantities given in column seven and are the quotients after dividing the adjusted sum of squares by the number of degrees of freedom. However, the quantities that are "analytic" or additive are in columns five and six.

TABLE 8. VARIANCE RATIO*

5% Points of Ratio

n_2 \ n_1	\multicolumn{10}{c}{Degrees of Freedom of Larger Variance}
	1	2	3	4	5	6	8	12	24	∞
1	161.4	199.5	215.7	224.6	230.2	234.0	238.9	243.9	249.0	254.3
2	18.51	19.00	19.16	19.25	19.30	19.33	19.37	19.41	19.45	19.50
3	10.13	9.55	9.28	9.12	9.01	8.94	8.84	8.74	8.64	8.53
4	7.71	6.94	6.59	6.39	6.26	6.16	6.04	5.91	5.77	5.63
5	6.61	5.79	5.41	5.19	5.05	4.95	4.82	4.68	4.53	4.36
6	5.99	5.14	4.76	4.53	4.39	4.28	4.15	4.00	3.84	3.67
7	5.59	4.74	4.35	4.12	3.97	3.87	3.73	3.57	3.41	3.23
8	5.32	4.46	4.07	3.84	3.69	3.58	3.44	3.28	3.12	2.93
9	5.12	4.26	3.86	3.63	3.48	3.37	3.23	3.07	2.90	2.71
10	4.96	4.10	3.71	3.48	3.33	3.22	3.07	2.91	2.74	2.54
11	4.84	3.98	3.59	3.36	3.20	3.09	2.95	2.79	2.61	2.40
12	4.75	3.88	3.49	3.26	3.11	3.00	2.85	2.69	2.50	2.30
13	4.67	3.80	3.41	3.18	3.02	2.92	2.77	2.60	2.42	2.21
14	4.60	3.74	3.34	3.11	2.96	2.85	2.70	2.53	2.35	2.13
15	4.54	3.68	3.29	3.06	2.90	2.79	2.64	2.48	2.29	2.07
16	4.49	3.63	3.24	3.01	2.85	2.74	2.59	2.42	2.24	2.01
17	4.45	3.59	3.20	2.96	2.81	2.70	2.55	2.38	2.19	1.96
18	4.41	3.55	3.16	2.93	2.77	2.66	2.51	2.34	2.15	1.92
19	4.38	3.52	3.13	2.90	2.74	2.63	2.48	2.31	2.11	1.88
20	4.35	3.49	3.10	2.87	2.71	2.60	2.45	2.28	2.08	1.84
21	4.32	3.47	3.07	2.84	2.68	2.57	2.42	2.25	2.05	1.81
22	4.30	3.44	3.05	2.82	2.66	2.55	2.40	2.23	2.03	1.78
23	4.28	3.42	3.03	2.80	2.64	2.53	2.38	2.20	2.00	1.76
24	4.26	3.40	3.01	2.78	2.62	2.51	2.36	2.18	1.98	1.73
25	4.24	3.38	2.99	2.76	2.60	2.49	2.34	2.16	1.96	1.71
26	4.22	3.37	2.98	2.74	2.59	2.47	2.32	2.15	1.95	1.69
27	4.21	3.35	2.96	2.73	2.57	2.46	2.30	2.13	1.93	1.67
28	4.20	3.34	2.95	2.71	2.56	2.44	2.29	2.12	1.91	1.65
29	4.18	3.33	2.93	2.70	2.54	2.43	2.28	2.10	1.90	1.64
30	4.17	3.32	2.92	2.69	2.53	2.42	2.27	2.09	1.89	1.62
40	4.08	3.23	2.84	2.61	2.45	2.34	2.18	2.00	1.79	1.51
60	4.00	3.15	2.76	2.52	2.37	2.25	2.10	1.92	1.70	1.39
120	3.92	3.07	2.68	2.45	2.29	2.17	2.02	1.83	1.61	1.25
∞	3.84	2.99	2.60	2.37	2.21	2.09	1.94	1.75	1.52	1.00

Degrees of Freedom of Smaller Variance

Lower 5% points are found by interchange of n_1 and n_2, i.e., n_1 must always correspond with the greater mean square.

* Reprinted from R. A. Fisher and F. Yates, *Statistical Tables for Biological, Agricultural and Medical Research*, p. 30, Oliver and Boyd, Ltd., London, 1943. By permission of the Author and Publisher.

INTERPRETATION OF THE EXAMPLE OF ANALYSIS OF VARIANCE

The only statistically significant cause of variance revealed in the analysis of this particular example appears to be the differences of the blocks. Its ratio of 6.0 in column eight is the only one which is larger than the value expected once in twenty by chance. In this case, the value of the ratio expected by chance is 2.50 (column nine). Quenching in different blocks has affected the resistance to corrosion. The position of the sheet in the block has made no consistent difference, although it was thought that outer as opposed to inner sheets would be quenched at different rates, and in advance this cause was thought as likely to cause a difference as difference between blocks. If the interactions of various effects are to be included, as they frequently are, the analysis becomes a little more complicated.[26,27]

LINEAR CORRELATION

Correlation is a non-random relation of one set of values with corresponding values from another set. For example, values measuring concentration and corrosion in sulfuric acid solutions are ordinarily correlated. The linear or proportional part of the correlation is measured by the correlation coefficient. Where correlation is curvilinear it may be only partly indicated by the correlation coefficient. For example, around a maximum in the curve of corrosiveness of sulfuric acid, linear correlation may not exist although the corrosiveness does depend upon the concentration.

Correlation may be taken into account in an analysis of variance by including covariance[26,27], or correlation coefficients may be calculated.

A correlation coefficient between two quantities X and Y is defined in the following identical ways:

$$r = \frac{\Sigma xy/n}{\sigma_x \sigma_y} \quad \text{or} \quad \frac{\Sigma yx/(n-1)}{s_x s_y} \quad \text{or} \quad \frac{\Sigma xy}{\sqrt{\Sigma x^2 \Sigma y^2}}$$

where $x = X - \bar{X}$ and $y = Y - \bar{Y}$. Shorter methods of calculation are available, developed in the manner of the method of the note at the end of the chapter.

The significance of a linear correlation coefficient depends on the sample size. Large samples give more definite information. Even a random distribution of data with no real correlation will give a non-zero coefficient, being smaller for a large sample. From the number of pairs, using tables that are available,[28,29,30] it can be determined whether a correlation coefficient is large enough so that it is improbable it is due to chance.

SIMPLE REGRESSION ANALYSIS

Some corrosion data may be best analyzed by expressing the results as functions of the test variables in the form of equations. Linear dependence on one or more variables, that is, dependence expressible by a plotted straight line, is conveniently handled by the method of least squares. By the expression "method of least squares" is meant that the unaccounted-for variance in the dependent variable is made as small as possible by the arrangement of the line. Where the relationships are precise or where the highest efficiency is not required, the visually best straight lines may be used. For non-linear relationships, equations higher than first degree may be used[31] or methods may be used[30] in which the fitness of a visually best line is

[26] G. W. Snedecor, *Statistical Methods*, Iowa State College Press, 1940.
[27] R. A. Fisher, *Statistical Methods for Research Workers*, Oliver and Boyd, Ltd., London, 1944.
[28] R. A. Fisher and F. Yates, *Statistical Tables for Biological, Agricultural and Medical Research*, p. 30, Oliver and Boyd, Ltd., London, 1943.
[29] R. A. Fisher, *loc. cit.*
[30] Mordecai Ezekiel, *Methods of Correlation Analysis*, Second Edition. John Wiley and Sons, New York, 1941.
[31] W. Edwards Deming, *Statistical Adjustment of Data*, John Wiley and Sons, New York, 1943.

verified by analyzing its residual variance, and if necessary after taking into account other variables.

For the two variables X and Y, if Y is a function of X and the experimental errors being minimized are errors in Y, the constants a and b in the equation $Y = a + bX$ are given as follows:

$$b = \Sigma xy/\Sigma x^2 \quad ; \quad a = \bar{Y} - b\bar{X}$$

where $x = X - \bar{X}$ and $y = Y - \bar{Y}$.

The most convenient method of calculation of b is usually

$$b = \frac{\Sigma xy - \Sigma x \Sigma y/n}{\Sigma x^2 - (\Sigma x)^2/n}$$

where, this time, x and y are deviations from rounded-off averages. With two independent variables the analysis becomes more complicated. For more variables or for curvilinear relationships, books on statistics should be consulted.[31,32]

CALCULATION SHORT CUTS

The standard deviation is defined as one of the following identical quantities:

$$(A) \quad \sigma = \sqrt{\frac{\Sigma(X - \bar{X})^2}{n}}$$

$$(B) \quad \sigma = \sqrt{\frac{\Sigma x^2 - (\Sigma x)^2/n}{n}} \quad \text{or} \quad \sqrt{\frac{\Sigma x^2 - n\bar{x}^2}{n}}$$

where $x = X$ minus a convenient rounded-off mean value.

$$(C) \quad \sigma = \sqrt{\frac{\Sigma X^2 - (\Sigma X)^2/n}{n}} \quad \text{or} \quad \sqrt{\frac{\Sigma X^2 - n\bar{X}^2}{n}}$$

In the straightforward method (A) the steps in getting the quantity under the radical sign are: (1) calculating the average $\bar{X}$, (2) subtracting it from each value, and (3) adding them all up. See the left of Table 9. However, $\bar{X}$ will ordinarily have several more decimal places than the individual values of X, and squares will have still more. These extra places cause considerable unnecessary work. Using fewer decimal places in the average will usually cause only a small error, and the error can be eliminated by means of an adjustment. See equation B and the middle of Table 9. That is, any convenient number, usually a rounded-off average, may be taken from all values of X and an adjustment made proportional to the square of the difference between this number and the average.

In fact, if it is wished, no subtraction need be made and the original values may be used. See equation B and the right of Table 9. Equations B and C are the result of simple algebra, and their derivations from A will provide considerable insight into the meaning of the quantities involved.

These same short cuts or similar ones also apply to calculation of other statistics, such as correlation coefficients, variances, and skewness.

[32] Mordecai Ezekiel, *Methods of Correlation Analysis*, Second Edition, John Wiley and Sons, New York, 1941.

TABLE 9. SEVERAL METHODS OF CALCULATING σ OR s

(A)	(B)*	(C)
$\sigma = \sqrt{\dfrac{\Sigma(X-\bar{X})^2}{n}}$	$\sigma = \sqrt{\dfrac{\Sigma x^2 - (\Sigma x)^2/n}{n}}$	$\sigma = \sqrt{\dfrac{\Sigma X^2 - (\Sigma X)^2/n}{n}}$

X psi	x or $X - \bar{X}$ psi	x^2 or $(X-\bar{X})^2$	X hundred psi	x or $X - 20$ hundred psi	x^2 or $(X-20)^2$	X hundred psi	X^2
1,900	− 712	506,944	19	−1	1	19	361
1,900	− 712	506,944	19	−1	1	19	361
2,100	− 512	262,144	21	1	1	21	441
1,900	− 712	506,944	19	−1	1	19	361
2,400	− 212	44,944	24	4	16	24	576
3,900	+1,288	1,658,944	39	19	361	39	1,521
2,100	− 512	262,144	21	1	1	21	441
2,100	− 512	262,144	21	1	1	21	441
2,000	− 612	374,544	20	0	0	20	400
3,200	+ 588	345,744	32	12	144	32	1,024
3,300	+ 688	473,344	33	13	169	33	1,089
2,800	+ 188	35,344	28	8	64	28	784
2,500	− 112	12,544	25	5	25	25	625
2,000	− 612	374,544	20	0	0	20	400
3,400	+ 788	620,944	34	14	196	34	1,156
4,300	+ 688	2,849,344	43	23	529	43	1,899
41,800	912	9,097,504	418	98	1,510	418	11,830

Average = 2,612 psi
Sum of squares = 9,097,504

$98^2/16 = 600.25$
Adjusted sum of squares
$1500 - 600.25 = 909.75$

$418^2/16 = 10,920.25$
Adjusted sum of squares
$11830 - 10,920.25 = 909.75$

$$\sigma = \sqrt{\dfrac{9097504}{16}} = 754 \text{ psi} \quad \text{or} \quad \sigma = \sqrt{\dfrac{909.75}{16}} = 7.54 \text{ hundred psi}$$

$$s = \sqrt{\dfrac{9097504}{15}} = 779 \text{ psi} \quad \text{or} \quad s = \sqrt{\dfrac{909.75}{15}} = 7.79 \text{ hundred psi}$$

*$x = X$ minus a rounded-off mean

DEFINITIONS OF TERMS USED IN THIS CHAPTER

Assignable Causes are causes that lead to significant variation. If a suitable test for control indicates lack of control, the causes of the lack of control are called assignable, because it is expected that they can be searched for with a reasonable (economic) chance of finding them.

Average and Mean. In this chapter and in most use of statistics these two terms are used as synonyms for the common arithmetic mean.

Bias. An error or source of an error that does not tend to diminish with larger and larger samples.

Control. Controlled data, or data obtained under conditions of statistical control, are data that contain no evidence of known or unknown assignable causes for variation when tested statistically in such a way that causes would be expected to be revealed.

Covariance. A joint variance of two quantities.

$$\text{Covariance} = \dfrac{\Sigma(X-\bar{X})(Y-\bar{Y})}{n-1} = \dfrac{\Sigma XY - \Sigma X \Sigma Y/n}{n-1}$$

Degrees of Freedom. The number of degrees of freedom is the number of independent comparisons. In an ordinary sample it is one less than the number of values. Thus, when comparing two values, there is only one comparison. In comparing three values, A, B, and C, there are only two independent comparisons. Of the three comparisons, A vs. B, B vs. C, and C vs. A, whichever one is taken third, adds no information not already known from the other two.

Error Variance. The variance not depending on assigned or assignable causes.

Expected Standard Deviation in a Population Sampled. This quantity is also called the "Student" standard deviation, and is represented by the letter s,

$$s = \sqrt{\Sigma(X - \bar{X})^2/(n-1)}, \quad s = \sigma\sqrt{n/(n-1)}.$$

It is a better estimate, from a sample, of the standard deviation of the population from which the sample comes than the sample standard deviation. Both s and the standard deviation are included here since both are used in literature the corrosion engineer will use. In this literature the equations will sometimes have to be recognized by their forms since the nomenclature is not standardized.

Kurtosis. The flatness of a distribution. Even symmetrical distributions with the same standard deviations may differ in shape. A measure of kurtosis is

$$\beta_2 = \frac{\Sigma(X - \bar{X})^4}{n\sigma^4} = \frac{n\Sigma(X - \bar{X})^4}{(\Sigma[X - \bar{X}]^2)^2}$$

Kurtosis and skewness are ordinarily usefully calculated only for large samples (at least 250). Their significances vary with the sample size.[33,34]

Probable Error. A name commonly used for 0.6745 times the standard deviation. See the section on ranges or confidence intervals.

Skewness. The departure from symmetry of a distribution. A measure of skewness is k,

$$k = \frac{\Sigma(X - \bar{X})^3}{n\sigma^3} = \frac{n\Sigma(X - \bar{X})^3}{\Sigma(X - \bar{X})^{3/2}}$$

Standard Deviation. The standard measure of the spread of a distribution. It is ordinarily somewhat larger than the average of the deviations from the average. It is defined as the square root of the average of the squares of the deviation from the mean and is represented by the small Greek letter σ (sigma); thus, $\sigma = \sqrt{\Sigma(X - \bar{X})^2/n}$, where X is in turn each of the values of a sample or group of values, $\bar{X}$ is the mean of the values, n is the number of values, and Σ (capital sigma) means that all the quantities are added together. See also the calculation short cuts on p. 1097.

Variance. While it is used for variability in general, the term variance refers specifically to quantities which are sums of squares of deviations divided by the number of degrees of freedom. For example, in one sample

$$\text{Variance} = s^2 = \frac{\Sigma(X - \bar{X})^2}{n - 1},$$

or in two samples

$$\text{Variance} = s^2 = \frac{\Sigma(X_1 - \bar{X}_1)^2 + \Sigma(X_2 - \bar{X}_2)^2}{n_1 + n_2 - 2}$$

(See the footnote in the text under the discussion of analysis of variance.)

---

[33] W. A. Shewhart, *Economic Control of Quality of Manufactured Products*, D. Van Nostrand Co., New York, 1931.

[34] R. A. Fisher, *Statistical Methods for Research Workers*, Oliver and Boyd, Ltd., London, 1944.

MISCELLANEOUS INFORMATION

ILLUSTRATIONS OF TYPICAL FORMS OF CORROSION

R. F. STEARN*

FIG. 1. Pitting of Passive Alloy Exposed to Sea Water. $\times 1/3$. Note Contact Corrosion (Crevice Corrosion) Produced under Washers (Not Shown) at Two Places of Support. (Courtesy of F.L. LaQue.)

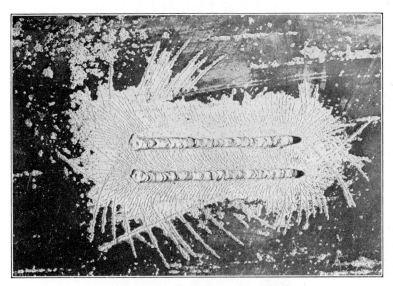

FIG. 2. Concentration of Corrosion in a Lüder's Line Pattern around a Weld Bead on Scaled Steel Exposed to Salt Water. $\times 3/8$. (Courtesy of Paul Ffield.)

* Research Laboratory, The International Nickel Co., Bayonne, N. J.

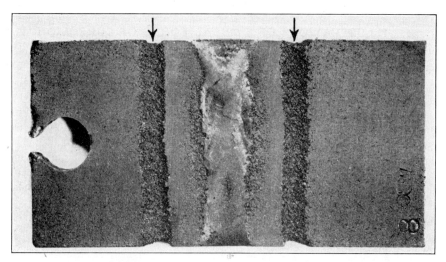

Fig. 3. Typical Weld Decay. Specimen Exposed to 25% HNO$_3$. ×2.

Fig. 4. Effect of Velocity. "Wire Drawing" of Valve Seat by Wet Steam. Two Views ×20. (Courtesy of E. N. Skinner).

Fig. 5. Tuberculation on the Inside of a Steel Water Main after 50 Years of Service. (Courtesy of J. E. Garratt).

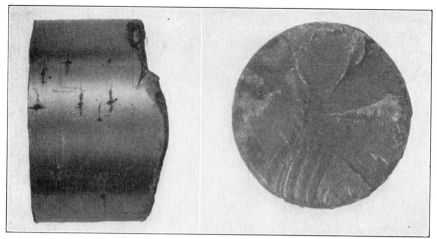

Fig. 6. Corrosion Fatigue Cracking and Fracture of Propeller Shaft. ×1. (Courtesy of E. N. Skinner.)

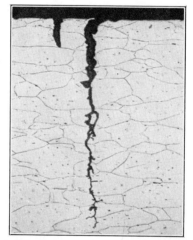

Fig. 7. Transcrystalline Corrosion Fatigue Crack. ×50.

Fig. 8. Deposit Attack (Contact Corrosion) of Condenser Tube. (Courtesy of C. A. Gleason).

Fig. 9A. Longitudinal Cross Section of Under-Cut Pits Associated with Impingement Attack by Salt Water. Flow of Solution from Left to Right. ×7.

Fig. 9B. Appearance of Surface of Condenser Tube Exposed to Impingement Attack by Salt Water. Flow of Solution from Left to Right. ×1. (Courtesy of F. L. LaQue.)

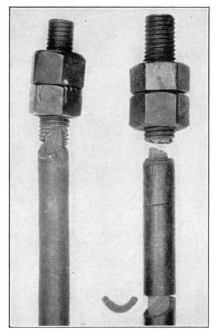

Fig. 10. Dezincification Type of Attack (Layer Type). ×½. (Courtesy of F.L. LaQue.)

Fig. 11A. Flattened Piece of Brass Pipe Subject to Layer Type Dezincification. (Courtesy of H. L. Shuldener.) Approx. ×1.

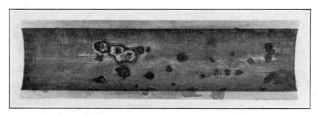

Fig. 11B. Section of Brass Pipe That Has Suffered Plug-Type Dezincification (66% Cu, 33.5% Zn). Approx. ×1. (Courtesy of H. L. Burghoff.)

Fig. 12A. Typical Transcrystalline Season Cracking. Hard-Drawn Tube after Alternate Immersion in 15 N NH$_4$OH for 24 Hours. ×125. (Courtesy of C. L. Bulow.)

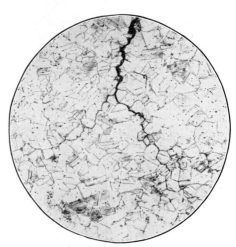

Fig. 12B. Typical Intergranular Season Cracking. Specimen Stored One Year. ×45. (Courtesy of C. L. Bulow.)

Fig. 13. Under-Film Corrosion on Lacquered Tin Can Section after Exposure to Indoor Atmosphere. ×1.

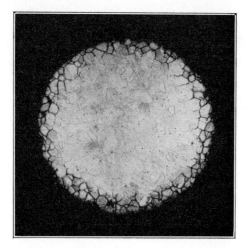

Fig. 14. High-Temperature Intergranular Attack of Nickel by Sulfur. ×100. (Courtesy of E. N. Skinner.)

Fig. 15. Internal Oxidation of a 0.1% Zn-Cu Alloy after 192 Hours at 600° C. Etched. ×500. (Courtesy of F. N. Rhines.)

CHARACTERISTICS OF SEA WATER

Alfred C. Redfield*

DEFINITIONS

The proportions of the major constituents of natural sea water obtained from the open oceans are very uniform, the ratio of water to total salt content being the principal variant. Many properties of sea water can consequently be determined from tables if the total salt content and the temperature are defined. The quantity of any constituent may be estimated if the chlorinity (defined below) of a sample is measured. (See Tables 1 and 2.)

Table 1. THE MAJOR CONSTITUENTS OF SEA WATER*

Chlorinity = 19.00 0/00†

Ion	Parts per Million	Equivalents per Million	Parts per Million per Unit Chlorinity
Chloride, Cl^-	18,980.0	535.3	998.90
Sulfate, SO_4^{--}	2,649.0	55.1	139.40
Bicarbonate, HCO_3^-	139.7	2.3	7.35
Bromine, Br^-	64.6	0.8	3.40
Fluoride, F^-	1.3	0.1	0.07
Boric acid, H_3BO_3	26.0	....‡	1.37
Total		593.6	
Sodium, Na^+	10,556.1	159.0	555.60
Magnesium, Mg^{++}	1,272.0	104.6	66.95
Calcium, Ca^{++}	400.1	20.0	21.06
Potassium, K^+	380.0	9.7	20.00
Strontium, Sr^{++}	13.3	0.3	0.70
Total		593.6	

* H. U. Sverdrup, M. W. Johnson, and R. H. Fleming, *The Oceans*, Prentice-Hall, Inc., New York, 1942. J. Lyman and R. H. Fleming, *J. Marine Research*, **3**, 134–146, 1940.

† 0/00 is used to denote grams per kilogram or parts per thousand.

‡ Undissociated at usual pH.

The total salt content of sea water is frequently expressed as *salinity*. Salinity is defined as the total amount of solid material in grams contained in one kilogram of sea water when all carbonate has been converted to oxide, the bromine and iodine replaced by chlorine, and all organic matter is completely oxidized.

Because of the convenience of chlorine analysis by titration with silver nitrate, potassium dichromate being used as an indicator, the salt content of sea water is frequently expressed as *chlorinity*. Chlorinity is defined approximately as the total amount of chlorine, bromine, and iodine in grams contained in one kilogram of sea water, assuming that the bromine and the iodine have been replaced by chlorine. Since this definition depends upon the accepted values of atomic weights, which have changed since the definition was established by an international commission, in exact oceanographic work chlorinity is established by comparison with a standard "Normal Water" now prepared at the Woods Hole Oceanographic Institution.

* Woods Hole Oceanographic Institution, Woods Hole, Mass., and Harvard University, Cambridge, Mass.

TABLE 2. ELEMENTS PRESENT IN SOLUTION IN OCEANIC SEA WATER EXCLUSIVE OF DISSOLVED GASES*

Chlorinity = 19.00 0/00

Element	Parts per Million	Element	Parts per Million
Chlorine	18980	Copper	0.001–0.01
Sodium	10561	Zinc	0.005
Magnesium	1272	Lead	0.004
Sulfur	884	Selenium	0.004
Calcium	400	Cesium	0.002
Potassium	380	Uranium	0.0015
Bromine	65	Molybdenum	0.0005
Carbon	28	Thorium	0.0005
Strontium	13	Cerium	0.0004
Boron	4.6	Silver	0.0003
Silicon	0.02–4.0	Vanadium	0.0003
Fluorine	1.4	Lanthanum	0.0003
Nitrogen†	0.006–0.7	Yttrium	0.0003
Aluminum	0.5	Nickel	0.0001
Rubidium	0.2	Scandium	0.00004
Lithium	0.1	Mercury	0.00003
Phosphorus	0.001–0.10	Gold	0.000006
Barium	0.05	Radium	$0.2\text{--}3 \times 10^{-10}$
Iodine	0.05	Cadmium	present
Arsenic	0.002–0.02	Cobalt	present
Manganese	0.001–0.01	Tin	present

* H. U. Sverdrup, M. W. Johnson, and R. H. Fleming, *The Oceans*, Prentice-Hall, Inc., New York, 1942.
† Nitrogen in combined forms.

TABLE 3. CHLORINITY, SALINITY, AND SPECIFIC GRAVITY OF SEA WATER*

Chlorinity, 0/00	Salinity, 0/00	Specific Gravity at 17.5° C	Chlorinity, 0/00	Salinity, 0/00	Specific Gravity at 17.5° C
0.00	0.00	1.00 000			
1.00	1.84	0 144	12.00	21.69	1.01 658
2.00	3.64	0 283	13.00	23.50	1 795
3.00	5.45	0 421	14.00	25.30	1 933
4.00	7.25	0 558	15.00	27.11	1.02 070
5.00	9.06	1.00 696	16.00	28.91	2 208
6.00	10.86	0 834	17.00	30.72	2 346
7.00	12.67	0 971	18.00	32.52	2 484
8.00	14.47	1 109	19.00	34.33	2 622
9.00	16.28	1 246	20.00	36.13	1.02 760
10.00	18.08	1.01 383	21.00	37.94	2 899
11.00	19.89	1 521	22.00	39.74	3 037

* For more complete tables see M. Knudsen, *Hydrographical Tables*, 63 pp., Copenhagen, 1901, or N. N. Subow, *Oceanographical Tables*, 208 pp., Moscow, 1931.

Salinity is related to chlorinity as follows:

$$\text{Salinity} = 0.03 + 1.805 \times \text{chlorinity} \tag{1}$$

Where great precision is not required, chlorinity or salinity may be obtained conveniently from a measurement of specific gravity (Table 3) or density (Table 4), using a hydrometer. Measurements of refractive index or conductivity are applicable where special circumstances warrant the use of more elaborate apparatus.

In the open sea, the salinity varies between 32 and 36 0/00 (chlorinity 18 and 20 0/00), depending upon geographical position. In order to state rough working rules relative to its properties, *natural sea water* may be considered to have a chlorinity of approximately 19 0/00 (salt content 3.4%). Slightly higher values are found in the oceanic waters of subtropical regions. Extreme values of salinity are found in enclosed seas subject to high evaporation (Mediterranean, 38.6 0/00; Red Sea, 41 0/00). Coastal waters and tide-swept harbors are usually characterized by slightly lower salinities. Enclosed bodies of water with extensive land drainage, such as the Baltic Sea and Chesapeake Bay, have very much lower salinities, and all degrees of dilution are to be expected in the tidal estuaries of rivers.

Table 4. DENSITY OF SEA WATER*

As a Function of Temperature and Salinity

Salinity, Parts per Thousand	Temperature, °C							
	0	5	10	15	17.5	20	25	30
0	0.99987	1.00000	0.99973	0.99913	0.99870	0.99813	0.99707	0.99567
5	1.00397	0401	1.00367	1.00301	1.00257	1.00207	1.00086	9943
10	0801	0796	0756	0684	0638	0585	0461	1.00314
15	1204	1191	1144	1067	1019	0964	0836	0686
20	1.01606	1.01585	1.01532	1.01449	1.01399	1.01341	1.01210	1.01057
25	2008	1980	1920	1832	1779	1720	1585	1429
30	2410	2374	2308	2215	2160	2099	1960	1801
35	2813	2770	2698	2600	2542	2479	2337	2176
40	1.03216	1.03166	1.03088	1.02985	1.02925	1.02860	1.02715	1.02551

* For more complete tables see M. Knudsen, *Hydrographical Tables*, 63 pp., Copenhagen, 1901, or N. N. Subow, *Oceanographical Tables*, 208 pp., Moscow, 1931.

PHYSICAL PROPERTIES RELATED TO CHLORINITY

Freezing Point (Table 5)

The freezing point is given as the temperature of the initial formation of ice. As soon as any ice is formed, the concentration of dissolved solids increases, and the formation of additional ice takes place only at a lower temperature.

Temperature of Maximum Density

Unlike fresh water, natural sea water continues to increase in density as it cools and freezes at temperatures above that of maximum density (Table 5). The transition in behavior comes at a salinity of 24.7 0/00 (chlorinity = 13.7 0/00). At lower salinities, the temperature of maximum density occurs above the freezing point, and the behavior resembles fresh water.

Osmotic Pressure, OP_0

It may be estimated from the freezing point depression, $\Delta\theta_f$ (Table 5), by the equation:

$$OP_0 = 12.08\Delta\theta_f$$

TABLE 5. PROPERTIES OF SEA WATER OF DIFFERENT SALINITY*

Salinity, 0/00	Freezing Point, °C	Temperature of Maximum Density, °C	Osmotic Pressure, Atmospheres	Specific Heat
0	0.00	3.95	0	1.00
5	−0.27	2.93	3.23	0.982
10	−0.53	1.86	6.44	0.968
15	−0.80	0.77	9.69	0.958
20	−1.07	−0.31	12.98	0.951
25	−1.35	−1.40	16.32	0.945
30	−1.63	−2.47	19.67	0.939
35	−1.91	−3.52	23.12	0.932
40	−2.20	−4.54	26.59	0.926

* N. N. Subow, *Oceanographical Tables*, 208 pp., Moscow, 1931. T. G. Thompson, "The Physical Properties of Sea Water," *Bull.* 85, "Physics of the Earth. V. Oceanography," p. 63, National Research Council of the National Academy of Sciences, Washington, 1932.

The osmotic pressure at any temperature, θ, is given by

$$OP = OP_0 \times \frac{273 + \theta}{273} \qquad (2)$$

The osmotic pressure of natural sea water is of the order of 20 atmospheres.

Vapor Pressure

The vapor pressure, ϵ, of sea water of any chlorinity may be estimated from the vapor pressure of distilled water, ϵ_0, by the equation:

$$\frac{\epsilon}{\epsilon_0} = 1 - 0.00969 \times \text{chlorinity} \qquad (3)$$

Roughly speaking, the vapor pressure is reduced 0.1% per 1 0/00 chlorinity. The vapor pressure of natural sea water is about 2% lower than that of fresh water.

Electrical Conductivity

For a rough working rule, the resistance of natural sea water may be taken as one ohm per foot cubed. Exact values of the specific conductance of sea water as a function of temperature and chlorinity are given in Table 6.

TABLE 6. SPECIFIC CONDUCTANCE OF SEA WATER AS A FUNCTION OF TEMPERATURE AND CHLORINITY*

Conductance in reciprocal ohm-centimeters

Chlorinity, Parts per Thousand	Temperature, °C					
	0	5	10	15	20	25
1	0.001839	0.002134	0.002439	0.002763	0.003091	0.003431
2	0.003556	0.004125	0.004714	0.005338	0.005971	0.006628
3	0.005187	0.006016	0.006872	0.007778	0.008702	0.009658
4	0.006758	0.007845	0.008958	0.010133	0.011337	0.012583
5	0.008327	0.009653	0.011019	0.012459	0.013939	0.015471
6	0.009878	0.011444	0.013063	0.014758	0.016512	0.018324
7	0.011404	0.013203	0.015069	0.017015	0.019035	0.021121
8	0.012905	0.014934	0.017042	0.019235	0.021514	0.023868
9	0.014388	0.016641	0.018986	0.021423	0.023957	0.026573
10	0.015852	0.018329	0.020906	0.023584	0.026367	0.029242
11	0.017304	0.020000	0.022804	0.025722	0.028749	0.031879
12	0.018741	0.021655	0.024684	0.027841	0.031109	0.034489
13	0.020167	0.023297	0.026548	0.029940	0.033447	0.037075
14	0.021585	0.024929	0.028397	0.032024	0.035765	0.039638
15	0.022993	0.026548	0.030231	0.034090	0.038065	0.042180
16	0.024393	0.028156	0.032050	0.036138	0.040345	0.044701
17	0.025783	0.029753	0.033855	0.038168	0.042606	0.047201
18	0.027162	0.031336	0.035644	0.040176	0.044844	0.049677
19	0.028530	0.032903	0.037415	0.042158	0.047058	0.052127
20	0.029885	0.034454	0.039167	0.044114	0.049248	0.054551
21	0.031227	0.035989	0.040900	0.046044	0.051414	0.056949
22	0.032556	0.037508	0.042614	0.047948	0.053556	0.059321

* B. D. Thomas, T. G. Thompson, and C. L. Utterback, *J. du conseil*, **9**, 28–35 (1934).

CHEMICAL PROPERTIES RELATED TO CHLORINITY

SOLUBILITY OF GASES

The concentration of a gas dissolved in a liquid is directly proportional to the partial pressure of the gas in the saturating gas phase. Table 7 gives the coefficients of saturation of oxygen, nitrogen, and carbon dioxide in sea water at different temperatures and chlorinities. The coefficients are defined as the concentrations of gas in equilibrium with one atmosphere (760 mm) of the pure gas. The values for carbon dioxide represent the amount of free CO_2 and H_2CO_3 dissolved and do not include considerable quantities of CO_2 present as bicarbonate.

The composition of the normal atmosphere with which surface waters are approximately saturated is given in Table 8.

TABLE 7. COEFFICIENTS OF SATURATION OF ATMOSPHERIC GASES IN WATER*

Concentrations of Oxygen, Nitrogen, and Carbon Dioxide in Equilibrium with 1 Atmosphere (760 mm) of Designated Gas

Gas	Chlorinity, 0/00	Temperature, °C	Concentration	
			Milliliters per Liter	Parts per Million
Oxygen	0	0	49.2†	70.4
		12	36.8	52.5
		24	29.4	42.1
	16	0	40.1	56.0
		12	30.6	42.9
		24	24.8	34.8
	20	0	38.0	52.8
		12	29.1	40.4
		24	23.6	32.9
Nitrogen	0	0	23.0†	28.8
		12	17.8	22.7
		24	14.6	18.3
	16	0	15.0	18.4
		12	11.6	14.2
		24	9.36	11.5
	20	0	14.2	17.3
		12	11.0	13.4
		24	8.96	10.9
Carbon dioxide‡	0	0	1715†	3370
		12	1118	2198
		24	782	1541
	16	0	1489	2860
		12	980	1888
		24	695	1342
	20	0	1438	2746
		12	947	1814
		24	677	1299

* Calculated from data in H. U. Sverdrup, M. W. Johnson, and R. H. Fleming, *The Oceans*, Prentice-Hall, Inc., New York, 1942.

† These values differ slightly from those given in Table 12, p. 1146, for fresh water.

‡ Includes CO_2 present as H_2CO_3 but not as HCO_3^- or CO_3^{--}

Atmospheric gases are present in ocean water in approximately the following quantities:

	Milliliters per liter	Parts per million
Oxygen	0–9	0–12
Nitrogen	8–15	10–18
Carbon dioxide*	33–56	64–107
Argon	0.2–0.4	0.4–0.7
Helium and neon	1.7×10^{-4}	0.3×10^{-4}†

* Includes CO_2 present as H_2CO_3, HCO_3^- and CO_3^{--}.

† Estimated as helium.

Hydrogen sulfide is normally absent but may reach concentrations of 22 ml per liter or 33 ppm under exceptional conditions.

TABLE 8. COMPOSITION OF NORMAL ATMOSPHERE*

Gas	Volume or Pressure, %	Partial Pressure, mm Hg, 1 atm Total Pressure
Nitrogen	78.03	593.02
Oxygen	20.99	159.52
Argon	0.94	7.144
Carbon dioxide	0.03	0.228
Hydrogen, neon, helium	0.01	0.088
	100.00	760.000

* From H. U. Sverdrup, M. W. Johnson and R. H. Fleming, *The Oceans*, Prentice-Hall, Inc., New York, 1942.

Table 9 gives the quantity of oxygen dissolved in sea water at different temperatures and chlorinities when in equilibrium with a normal atmosphere saturated with water vapor. It thus represents the condition approached by the surface water when biological activity is not excessive.

TABLE 9. OXYGEN DISSOLVED IN SEA WATER IN EQUILIBRIUM WITH A NORMAL ATMOSPHERE (760 MM) OF AIR SATURATED WITH WATER VAPOR*

	Parts per Million				
Chlorinity 0/00	0	5	10	15	20
Salinity 0/00	0	9.06	18.08	27.11	36.11
Temperature °C					
0	14.62†	13.70	12.78	11.89	11.00
5	12.79	12.02	11.24	10.49	9.74
10	11.32	10.66	10.01	9.37	8.72
15	10.16	9.67	9.02	8.46	7.92
20	9.19	8.70	8.21	7.77	7.23
25	8.39	7.93	7.48	7.04	6.57
30	7.67	7.25	6.80	6.41	5.37

* Based on data of C. J. J. Fox, Conseil permanent international pour l'exploration de la mer, Copenhagen, *Publication de circonstance* 41 (1907).

† The values for solubility in water of zero chlorinity differ slightly from those given in Table 12, p. 1146, for fresh water.

Detailed tables on the solubility of nitrogen in sea water are available.[1,2]

ALKALINITY

In oceanographic chemistry, alkalinity is defined as the number of milliequivalents of hydrogen ions necessary to combine with the ions of weak acids in a quantity of

[1] C. J. J. Fox, Conseil permanent international pour l'exploration de la mer, Copenhagen, *Publication de circonstance* 41 (1907).

[2] N. W. Rakestraw and V. M. Emmel, *J. Phys. Chem.*, **42**, 1211 (1938).

water which at 20° C has a volume of one liter. Thus it represents approximately the concentration of bicarbonate and carbonate ions present.

The alkalinity of ocean water is related fairly closely to chlorinity by the expression:

$$\text{Alkalinity} = 0.120 \times \text{chlorinity} \times \rho_{20}$$

where ρ_{20} is the density of the water at 20° C. Departures from this ratio are usually attributed to changes in the calcium carbonate content of sea water, such as result from biological activity or the dilution of sea water with river water rich in calcium carbonate. Alkalinity may also be used as an index of pollution by acid or basic materials. There are several methods of determining alkalinity.[3,4,5]

Hydrogen-Ion Concentration

Owing to the excess of strong basic ions, pure sea water is slightly alkaline. The pH of sea water in equilibrium with the air varies between 8.1 and 8.3. The removal of carbon dioxide by photosynthetic processes of marine plants increases the pH somewhat, an exceptional limiting value observed being as great as 9.7. The decomposition of organic matter in water removed from the influence of the atmosphere decreases the pH. Values as low as 7.5 are observed at certain depths in the Pacific, where decomposition has almost entirely removed the dissolved oxygen. In stagnant basins where large amounts of H_2S are present, the pH may approach 7.0.

The carbonates constitute the principal buffer system of sea water. Boric acid is also present in small amounts (0.43 mg moles/liter) but is almost completely undissociated at commonly existing pH. The buffer capacity of sea water is equal to the *alkalinity*. When strong base is added, magnesium hydrate is precipitated at pH values between 10 and 11, and higher pH values are not obtained until the considerable quantity of magnesium present is exhausted. Calcium is precipitated in a similar way at higher pH values.

There are tables giving the pH and total CO_2 content of sea water in equilibrium with various pressures of CO_2 at different temperatures and chlorinities.[6]

Solubility of Salts

Because of the influence of other ions upon the activity of a given ion, the solubility product of salts in distilled water cannot be applied to sea water. The solubility products of several salts in distilled water and in sea water of salinity 35 0/00 at 20° C are compared in Table 10.

The solubility product of calcium carbonate given in Table 10 is less than the product of the concentration of calcium and carbonate ions in natural sea water (270×10^{-8} at pH 8.2), which indicates that sea water is supersaturated in respect to this salt. Because some uncertainty exists in the matter, the original studies should be consulted.[7,8,9]

[3] D. M. Greenberg, E. G. Moberg, and E. C. Allen, *Ind. Eng. Chem.*, Anal, Ed., **4**, 309 (1932).

[4] P. H. Mitchell and N. W. Rakestraw, *Biol. Bull.*, **65**, 437 (1933).

[5] S. Grippenberg, International Association of Physical Oceanography (Association d'Océanographic physique, Union Géodesie et Géophysique, internationale, *Rapport et procès-verbal*, No. 2, p. 150, Liverpool (1937).

[6] K. Buck, H. W. Harvey, H. Wattenberg, and S. Grippenberg," Über das Kohlensauresystem in Meerwasser," Conseil permanent international pour l'exploration de la mer, *Rapport et procès-verbal*, **79** (1932).

[7] H. Wattenberg and E. Timmermann, *Ann. Hydrog. u. Mar. Meteor.*, **64**, 23-31 (1936) and *Kieler Meeresforschung*, **2**, 81-94 (1938).

[8] R. Revelle and R. H. Fleming, Fifth Pacific Scientific Congress, Canada, 1933, Proc., **3**, 2089-2092 (1934).

[9] R. Revelle, *J. Sedim. Petrol.*, **4**, 103-110 (1934).

TABLE 10. SOLUBILITY PRODUCTS OF CERTAIN SALTS IN DISTILLED WATER AND IN SEA WATER OF SALINITY 35 0/00 AT 20° C*

Salt	Distilled Water	Sea Water
$CaCO_3$	0.5×10^{-8}	50×10^{-8}
$MgCO_3 \cdot 3H_2O$	0.1×10^{-4}	3.1×10^{-4}
$SrCO_3$	0.3×10^{-9}	500×10^{-9}
$Mg(OH)_2$	1×10^{-11}	5×10^{-11}

* From H. U. Sverdrup, M. W. Johnson, and R. H. Fleming, *The Oceans*, Prentice-Hall, Inc., New York, 1942.

Calcium carbonate becomes less soluble under conditions which increase the temperature or pH of sea water. Calcium and magnesium carbonates and hydroxides are precipitated on cathodic surfaces, as in cathodic protection and in galvanic couples, as the result of induced increases in the ionic products at the electrode.

The presence of calcium and magnesium introduces special difficulties when sea water is treated with many substances owing to the tendency of these ions to form relatively insoluble compounds.

CHEMICAL SUBSTANCES INFLUENCED BY ORGANIC ACTIVITY

A number of substances present in sea water are absorbed by growing plants and are set free by the decomposition of organic matter. Their concentrations are consequently widely variable and are related to the seasonal cycle of biological activity. Pollution with domestic sewage or organic wastes, and the resulting bacterial activity, increases the concentration of these materials in sea water.

The growth of marine plants may almost completely remove the phosphates and nitrates from sea water during the late spring and summer. The silicate concentration may be greatly reduced, and because of the absorption of CO_2 in photosynthesis, the pH is increased slightly. Photosynthesis may produce sufficient oxygen to increase the oxygen content in the subsurface water 10 to 20% above the saturation value. During the fall and winter in the surface layers of the sea, processes of decomposition predominate, higher values of nitrate, phosphate, and silicon are to be expected, and the oxygen content of the water is very nearly in equilibrium with the air.

In deep water this is also true at all seasons except that the oxygen content is reduced by the oxidation of organic matter and falls below the saturation value.

In tropical waters the seasonal cycle is less pronounced, and the surface waters remain depleted of phosphate and nitrate throughout the year.

The range in concentration of the various inorganic forms of nitrogen and of phosphate phosphorus to be encountered in natural (unpolluted) sea water are:

Nitrogen as NO_3	0.001 –0.6 ppm
Nitrogen as NO_2	0.0001–0.05 ppm
Nitrogen as NH_3	0.005 –0.05 ppm
Phosphorus as PO_4	0.001 –0.10 ppm

Pollution with domestic sewage or organic wastes increases the concentration of organic derivatives, i.e., nitrogen compounds and phosphates, in the sea water, and may result in a marked deficiency in the oxygen content. If the pollution is sufficiently intense to result in anaerobic conditions, the activity of bacteria may result in the production of hydrogen sulfide consequent to the reduction of sulfates. Conditions of this kind occur commonly in the muds of polluted regions where organic debris tends to accumulate.

In the decomposition of nitrogenous matter, ammonia is liberated initially. It is first oxidized to nitrite and subsequently to nitrate. The presence of unusual quantities of ammonia and nitrite may consequently be taken as signs of recent active pollution.

WATER OF ESTUARIES

The dissolved solids of river water differ in composition from those of sea water (Table 11). Consequently where sea water is mixed with considerable quantities of river water, the concentration of various constituents cannot be estimated precisely

TABLE 11. PERCENTAGE COMPOSITION OF DISSOLVED SOLIDS IN SEA WATER AND RIVER WATER*

Substance	Sea Water	River Water Weighted Average	Mississippi River	Columbia River	Colorado River
CO_3^{--}	0.41 (HCO_3^-)	35.15	34.98	36.15	13.02
SO_4^{--}	7.68	12.14	15.37	13.52	28.61
Cl^-	55.04	5.68	6.21	2.82	19.92
NO_3^-	....	0.90	1.60	0.49	....
Ca^{++}	1.15	20.39	20.50	17.87	10.35
Mg^{++}	3.69	3.41	5.38	4.38	3.14
Na^{++}	30.62	5.79	8.33	8.12	19.75
K^+	1.10	2.12	....	1.95	2.17
$(Fe,Al)_2O_3$	....	2.72	0.58	0.08	....
SiO_2	....	11.67	7.05	14.62	3.04
Sr^{++}, H_3BO_3, Br^-	0.31	....	....	....	....
Salt content, ppm	35,000	....	166	92.4	702

* H. U. Sverdrup, M. W. Johnson, and R. H. Fleming, *The Oceans*, Prentice-Hall, Inc., New York, 1942. F. W. Clarke, U. S. Geological Survey, *Bull.* 770, 5th Ed., 841 pp. (Washington, D. C., 1924).

from the chlorinity. River water is usually relatively richer in calcium carbonate in particular, and also in magnesium, potassium, sulfate, iron, and silicon. It is for this reason that the salinity of sea water is not exactly proportional to the chlorinity, as shown by relation 1 (p. 1113). In harbors and estuaries, the proportions of the constituents may be further disturbed by pollution. When this is of domestic origin or contains organic wastes, the tendency is toward a decrease in oxygen content and pH and an increase in nitrogenous compounds, particularly ammonia, and of sulfur as sulfide.

Where river water mixes with sea water there is frequently a tendency for stratification to develop, the fresh water flowing out over the salt water which lies at greater depth. Under such circumstances examination of samples collected from the surface will give incomplete information, and marked changes in salinity may accompany the tidal cycle.

ARTIFICIAL SEA WATER

For experimental work where the physical properties of sea water, such as osmotic pressure or electrical conductivity, are at issue a 3.4% solution of sodium chloride may be used. Where the action of the water to be examined is of a chemical nature a more exact reproduction of sea water is desirable, depending upon the nature of the problem. Formulas for artificial sea water are given in Table 12. Preparations of natural sea salt may also be employed.

Table 12. FORMULAS FOR ARTIFICIAL SEA WATER

Chlorinity = 19.00 0/00

Naval Aircraft Factory Process Specification PS-1 for synthetic sea water, for use in testing corrosion-resisting steel tubing (Navy Department Specification 44T27b, dated July 1, 1940), is as follows:

Stock Solution

Potassium chloride	10 grams
Potassium bromide	45 grams
Magnesium chloride	550 grams
Calcium chloride	110 grams
Sterile distilled water to make 1 liter	

This stock solution is used with other chemicals to make the synthetic sea water as follows:

Sodium chloride — NaCl	23 grams
Sodium sulfate — $Na_2SO_4 \cdot 10H_2O$	8 grams
Stock solution	20 ml
Sterile distilled water to make 1 liter	

Other recommended compositions are as follows:

McClendon et al (1917)*		Brujewicz (Subow, 1931)†		Lyman and Fleming (1940)‡	
Salt	grams/kg	Salt	grams/kg	Salt	grams/kg
NaCl	26.726	NaCl	26.518	NaCl	23.476
$MgCl_2$	2.260	$MgCl_2$	2.447	$MgCl_2$	4.981
$MgSO_4$	3.248	$MgSO_4$	3.305	Na_2SO_4	3.917
$CaCl_2$	1.153	$CaCl_2$	1.141	$CaCl_2$	1.102
KCl	0.721	KCl	0.725	KCl	0.664
$NaHCO_3$	0.198	$NaHCO_3$	0.202	$NaHCO_3$	0.192
NaBr	0.058	NaBr	0.083	KBr	0.096
H_3BO_3	0.058			H_3BO_3	0.026
Na_2SiO_3	0.0024			$SrCl_2$	0.024
$Na_2Si_4O_9$	0.0015			NaF	0.003
H_3PO_4	0.0002				
Al_2Cl_6	0.013				
NH_3	0.002				
$LiNO_3$	0.0013				
Total:	34.4406		34.421		34.481
Water to:	1,000.0000		1,000.000		1,000.000

* J. F. McClendon, C. C. Gault, and S. Mulholland, Carnegie Institution of Washington, *Publication* 251 (Papers from Dept. of Marine Biology), pp. 21–69 (1917).

† N. N. Subow, *Oceanographical Tables*, U. S. S. R. Oceanographic Institute Hydro-Meteoral Com. 208 pp. Moscow, 1931.

‡ J. Lyman and R. H. Fleming, *J. Marine Research*, **3**, 134–146 (1940).

It should not be assumed that the results of corrosion tests in any synthetic sea water may be applied directly to an estimation of the performance of materials under natural conditions of exposure to sea water.

GENERAL REFERENCES

"The Determination of Chlorinity by the Knudsen Method," a translation of "Chloruration par la Méthode de Knudsen," by M. Oxner, and a reprint of "Hydrographical Tables," by M. Knudsen. Woods Hole Oceanographic Institution, Woods Hole, 1946.

HARVEY, H. W., *Recent Advances in the Chemistry and Biology of Sea Water*, The University Press, Cambridge, 1945.

"Physics of the Earth — V. Oceanography," National Research Council of the National Academy of Sciences, Washington, *Bull.* 85 (1932).

SVERDRUP, H. U., M. W. JOHNSON, and R. H. FLEMING, *The Oceans, Their Physics, Chemistry, and General Biology*, Prentice-Hall, Inc., New York, 1942.

HIGH-TEMPERATURE EQUILIBRIA FOR OXIDATION AND CARBURIZATION OF IRON

J. B. AUSTIN*

The data presented in Figs. 1 to 3 show the variation with temperature of the equilibrium constant for each of the four reactions chiefly responsible for the oxidation, carburization, or decarburization of iron or steel at elevated temperature, namely:

$$Fe + H_2O = FeO + H_2 \qquad K = pH_2/pH_2O \qquad (1)$$
$$Fe + CO_2 = FeO + CO \qquad K = pCO/pCO_2 \qquad (2)$$
$$Fe_3C + CO_2 = 3Fe + 2CO \qquad K = p^2CO/pCO_2 \qquad (3)$$
$$3Fe + CH_4 = Fe_3C + 2H_2 \qquad K = p^2H_2/pCH_4 \qquad (4)$$

The constants are given in terms of the partial pressure of each gas, that is, the total pressure multiplied by the volume fraction (not weight percentage) of the gas in question. For example, in a gas mixture containing 45% CO, 15% CO_2, and 40% N_2 at a total pressure of one atmosphere, the partial pressure of CO is 0.45, that of CO_2 is 0.15.

The curves presented in the several figures enable one to predict the way in which a mixture containing CO and CO_2, or H_2 and H_2O *tends* to react with iron or iron oxide. (They do not indicate the rate at which this tendency may be followed.) As an illustration, for the gas mixture described above, the ratio pCO/pCO_2 is 0.45/0.15 or 3. Figure 1 shows that this ratio is equal to the equilibrium constant at 1130° C (2065° F); hence at this temperature this mixture is inert to iron or iron oxide, at a higher temperature it cannot reduce iron oxide and tends to oxidize iron, and at a lower temperature it tends to reduce iron oxide and cannot oxidize iron.

To ascertain whether this mixture tends to carburize or to decarburize steel at 800° C, the value of the equilibrium constant for reaction 3 is computed; it is $p^2CO/pCO_2 = (0.45)^2/0.15 = 1.35$. Figure 2 shows that at 800° C (1470° F) this gas is decarburizing toward steel of any carbon content. A ratio of 6 instead of 1.35 would indicate (see Fig. 2) that at 800° C (1470° F) the gas mixture tends to decarburize steel containing less than 0.8% carbon and to carburize steel containing a higher concentration of carbon.

* Research Laboratory, U. S. Steel Corporation, Kearny, N. J.

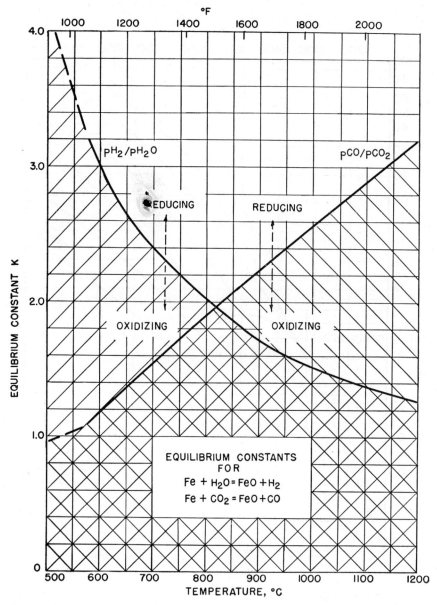

FIG. 1. Chart Illustrating the Variation with Temperature of the Equilibrium Constant for the Reaction between Iron and Water Vapor or Carbon Dioxide.

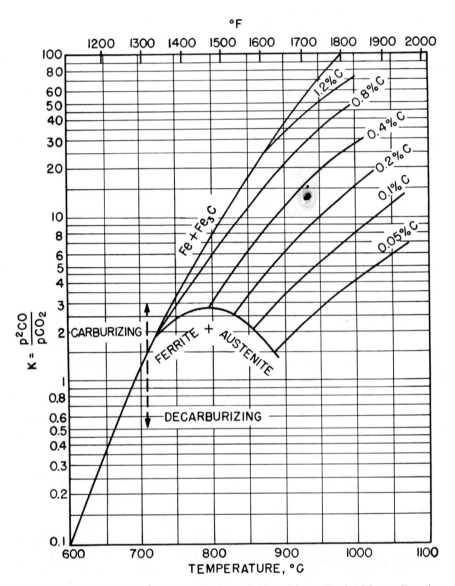

Fig. 2. Equilibrium Constants for the Reaction $Fe_3C + CO_2 = 3Fe + 2CO$, as a Function of Temperature and Carbon Content.

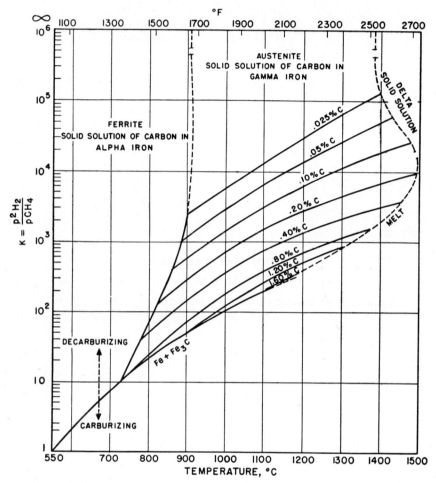

FIG. 3. Equilibrium Constants for the Reaction $3Fe + CH_4 = Fe_3C + 2H_2$, as a Function of Temperature and Carbon Content.

The use of these charts for mixtures of hydrogen and methane (Fig. 3) is analogous to that for carbon monoxide and carbon dioxide.

Note that at high temperatures the oxide which first forms is FeO, and if the partial pressure of oxygen is low this is the only oxide that forms. At somewhat higher oxygen pressures a layer of Fe_3O_4 forms on top of FeO, and at still higher pressures a layer of Fe_2O_3 forms over Fe_3O_4.

Below about 600° C (1100° F) FeO is unstable and decomposes into iron and Fe_3O_4 so that high-temperature scales on iron examined at room temperatures show only iron, Fe_3O_4 and possibly Fe_2O_3.

GENERAL TABLES

Prepared by J. C. Warner, D. S. McKinney, and J. P. Fugassi

Department of Chemistry, Carnegie Institute of Technology, Pittsburgh, Pa.

TABLE 1. PHYSICAL CONSTANTS OF THE ELEMENTS

Element	Symbol	Atomic Weight (1941)	Density, grams/cc (20 °C)	Melting Point, °C	Boiling Point, °C at 1 atm	Specific Heat, gram cal /gram /° C, Room Temperature (18°–25° C)	Heat of Fusion, gram cal / gram atom
Actinium	Ac	227	....	....	....	....	....
Aluminum	Al	26.97	2.70	660.2	2057	0.2259	2500
Antimony	Sb	121.76	6.62	630.5	1440	0.0493	4770
Argon	A	39.944	1.663×10^{-3}	−189.3	−185.8	0.1252	290
Arsenic	As	74.91	5.73	814	610	0.0822	6620
Barium	Ba	137.36	3.5	704	1638	0.0680	1400
Beryllium	Be	9.02	1.85	1284	2780	0.425	3092
Bismuth	Bi	209.00	9.80	271.0	1450	0.0290	2505
Boron	B	10.82	2.3	2300	2550	0.3091	....
Bromine	Br	79.916	3.12	−7.2	58.0	0.0703	2580
Cadmium	Cd	112.41	8.65	320.9	766	0.0547	1460
Calcium	Ca	40.08	1.55	850	1487	0.157	2230
Carbon	C	12.010	2.22	3600	4830	0.165	11000
Cerium	Ce	140.13	6.9	780	1400	0.05	2120
Cesium	Cs	132.91	1.9	28.4	690	0.0521	500
Chlorine	Cl	35.457	....	−101.0	−34.1	0.226	1531
Chromium	Cr	52.01	7.14	1550	2482	0.12	3930
Cobalt	Co	58.94	8.9	1490	2906	0.0989	3660
Columbium	Cb	92.91	8.57	1950	3300	....	....
Copper	Cu	63.57	8.94	1083	2595	0.0918	3110
Dysprosium	Dy	162.46	....	....	....	....	....
Erbium	Er	167.2	....	....	....	....	....
Europium	Eu	152.0	....	....	....	....	....
Fluorine	F	19.00	....	−223	−188.2	....	....
Gadolinium	Gd	156.9	....	....	....	....	....
Gallium	Ga	69.72	5.91	29.8	2071	0.0788	1336
Germanium	Ge	72.60	5.36	959	2700	0.0733	8300
Gold	Au	197.2	19.3	1063.0	2970	0.0308	3030
Hafnium	Hf	178.6	11.4	1700	5390	....	....
Helium	He	4.003	0.1664×10^{-3}	−271.4	−268.4	1.25	....
Holmium	Ho	164.94	....	....	....	....	....
Hydrogen	H	1.0080	0.08375×10^{-3}	−259.2	−252.7	3.415	28
Indium	In	114.76	7.31	156.4	1450	0.0568	781
Iodine	I	126.92	4.93	113.0	183.0	0.0523	3650
Iridium	Ir	193.1	22.4	2409	4910	0.0322	....
Iron	Fe	55.85	7.87	1535	3000	0.1075	3560
Krypton	Kr	83.7	....	−157	−152.9	....	360

TABLE 1. PHYSICAL CONSTANTS OF THE ELEMENTS — *Continued*

Element	Symbol	Atomic Weight (1941)	Density, grams/cc (20°C)	Melting Point, °C	Boiling Point, °C at 1 atm	Specific Heat, gram cal / gram /°C, Room Temperature (18°–25° C)	Heat of fusion, gram cal / gram atom
Lanthanum	La	138.92	6.15	826	1780	0.0446	....
Lead	Pb	207.21	11.34	327.4	1744	0.030	1224
Lithium	Li	6.940	0.53	186	1372	0.79	1100
Lutecium	Lu	174.99	....	....	....	....	....
Magnesium	Mg	24.32	1.74	651	1109	0.249	2160
Manganese	Mn	54.93	7.44	1242	2151	0.107	3450
Mercury	Hg	200.61	13.55	−38.9	357	0.0332	557
Molybdenum	Mo	95.95	10.2	2622	4800	0.0647	6660
Neodymium	Nd	144.27	7.05	840	....	0.447	....
Neon	Ne	20.183	0.8387×10^{-3}	−248.5	−246.0	....	77
Nickel	Ni	58.69	8.9	1452	2900	0.112	4200
Nitrogen	N	14.008	1.1649×10^{-3}	−210.0	−195.8	0.247	172
Osmium	Os	190.2	22.5	2700	5490	0.031	....
Oxygen	O	16.0000	1.3318×10^{-3}	−118.9	−183.0	0.2184	106
Palladium	Pd	106.7	12.0	1554	3980	0.0587	4120
Phosphorus	P	30.98	1.82	44.2	280	0.177	615
Platinum	Pt	195.23	21.45	1773	4390	0.0319	4700
Potassium	K	39.096	0.86	63.5	776	0.177	574
Praeseodymium	Pr	140.92	6.63	940	....	0.458	2700
Protactinium	Pa	231	....	....	....	....	....
Radium	Ra	226.05	5.0	960	1140	....	....
Radon	Rn	222	....	−71	−61.8	....	....
Rhenium	Re	186.31	20	3000	5870	0.0346	....
Rhodium	Rh	102.91	12.44	1966	4500	0.0598	....
Rubidium	Rb	85.48	1.53	39.1	679	0.0802	525
Ruthenium	Ru	101.7	12.2	2450	4900	0.061	....
Samarium	Sm	150.43	7.7	1300	....	....	....
Scandium	Sc	45.10	2.5	1200	2430	....	....
Selenium	Se	78.96	4.81	217	753	0.084	1220
Silicon	Si	28.06	2.4	1427	2290	0.1762	9470
Silver	Ag	107.880	10.5	960.5	2212	0.0558	2700
Sodium	Na	22.997	0.97	97.7	892	0.295	630
Strontium	Sr	87.63	2.6	770	1384	....	2190
Sulfur	S	32.06	2.07	112.8	444.6	0.175	298
Tantalum	Ta	180.88	16.6	2850	5870	0.0356	....
Tellurium	Te	127.61	6.24	453	1090	0.0468	3230
Terbium	Tb	159.2	....	310	....	....	....
Thorium	Th	232.12	11.5	1845	5200	0.0276	....
Thulium	Tm	169.4	....	....	....	....	....

TABLE 1. PHYSICAL CONSTANTS OF THE ELEMENTS — *Continued*

Element	Symbol	Atomic Weight (1941)	Density, grams/cc (20°C)	Melting Point, °C	Boiling Point, °C at 1 atm	Specific Heat, gram cal / gram /°C, Room Temperature (18°–25°C)	Heat of fusion, gram cal / gram atom
Tin	Sn	118.70	7.30	231.8	2270	0.054	1720
Titanium	Ti	47.90	4.5	1800	5100	0.142	....
Tungsten	W	183.92	19.3	3390	5900	0.034	8400
Thallium	Tl	204.39	11.85	302.5	1457	0.0311	1030
Uranium	U	238.07	18.7	1150	3500	0.0276	....
Vanadium	V	50.95	5.68	1710	3000	0.1153	....
Xenon	Xe	131.3	5.495×10^{-3}	−111.5	−108.0	....	740
Ytterbium	Yb	173.04	....	....	....	....	....
Yttrium	Y	88.92	5.51	1480	4600	....	....
Zinc	Zn	65.38	7.14	419.5	907	0.09	1595
Zirconium	Zr	91.22	6.4	1700	5050	0.066	....

Element	Heat of Vaporization, gram cal /gram atom	Resistivity, microhm-cm, Room Temperature	Linear Coefficient Thermal Expanson per °C, Room Temperature (18°–25° C)
Actinium	....	....	....
Aluminum	61,020	2.655	24×10^{-6}
Antimony	46,670	39	11.29
Argon	1,590	....	....
Arsenic	31,000	35	3.86
Barium	35,670	....	....
Beryllium	....	18.5	12.3
Bismuth	....	115	13.45
Boron	....	1.8×10^{12}	2
Bromine	7,420	....	....
Cadmium	23,870	7.59	29.8
Calcium	36,580	4.6	25
Carbon	....	1000	1.2
Cerium	....	78	....
Cesium	16,320	20	97
Chlorine	4,878	....	....
Chromium	....	13.1	8.1
Cobalt	....	9.7	12.08
Columbiun	....	20	7.2
Copper	72,810	1.682	16.42
Dysprosium	....	....	....

TABLE 1. PHYSICAL CONSTANTS OF THE ELEMENTS — *Continued*

Element	Heat of Vaporization, gram cal/gram atom	Resistivity, microhm-cm, Room Temperature	Linear Coefficient Thermal Expansion per °C, Room Temperature (18°–25° C)
Erbium	....	....	....× 10^{-6}
Europium	....	....	....
Fluorine	1,640	....	....
Gadolinium	....	....	....
Gallium	....	57.1	18.3
Germanium	....	....	....
Gold	81,800	2.42	14.4
Hafnium	....	....	....
Helium	22	....	....
Holmium	....	....	....
Hydrogen	216	....	....
Indium	....	9	33
Iodine	10,390	1.3×10^{15}	93
Iridium	....	6.08	6.41
Iron	84,600	9.8	11.9
Krypton	2,310	....	....
Lanthanum	....	59	....
Lead	42,060	20.65	29.5
Lithium	32,250	8.5	56
Lutecium	....	....	....
Magnesium	32,520	4.46	25.7
Manganese	55,150	....	23
Mercury	13,980	95.8	....
Molybdenum	128,000	4.77	5.49
Neodymium	....	79	....
Neon	440	....	....
Nickel	87,300	6.9	13.7
Nitrogen	1,336	....	....
Osmium	....	9	5.70
Oxygen	1,629	....	....
Palladium	....	10	11.60
Phosphorus	12,520	10^{17}	125
Platinum	107,000	9.83	8.8
Potassium	18,920	7.0	83
Praseodymium	....	88	....
Protactinium	....	....	....
Radium	....	....	....
Radon	4,010	....	....

TABLE 1. PHYSICAL CONSTANTS OF THE ELEMENTS — *Continued*

Element	Heat of Vaporization, gram cal/gram atom	Resistivity, microhm-cm, Room Temperature	Linear Coefficient Thermal Expansion per °C, Room Temperature (18°–25° C)
Rhenium	....	21	12.5×10^{-6}
Rhodium	....	4.93	8.9
Rubidium	18,110	12.5	90.0
Ruthenium	....	10	8.5
Samarium	....	....	....
Scandium	....	....	....
Selenium	25,490	12	37
Silicon	....	85×10^3	....
Silver	60,720	1.62	18.9
Sodium	23,120	4.6	71
Strontium	33,610	....	....
Sulfur	20,200	1.9×10^{17}	67.48
Tantalum	....	15.5	6.5
Tellurium	....	....	16.8
Terbium	....	....	....
Thorium	....	18	12.3
Thulium	....	....	....
Tin	68,000	11.5	22.4
Titanium	....	....	7.14
Tungsten	176,000	5.48	4.0
Thallium	38,810	18.1	28.0
Uranium	....	60	....
Vanadium	....	26	....
Xenon	3,110	....	....
Ytterbium	....	....	....
Yttrium	....	....	....
Zinc	27,430	6.2	32.5
Zirconium	....	41	6.3

TABLE 2. MISCELLANEOUS PHYSICAL CONSTANTS*

Constant	Symbol	Value
Standard atmosphere	A_0	1.01325×10^6 dyne·cm^{-2}
Standard atmosphere (Lat. 45°)	A_{45}	1.01320×10^6 dyne·cm^{-2}
Liter (1000 ml)	l	1000.028 cm^3
Joule equivalent	J_{15}	4.1855 Absolute joules·cal$_{15}^{-1}$
Ice point	T_0	273.16° K
Velocity of light	c	2.99776×10^{10} cm·sec^{-1}
Electronic charge	e^-	4.80×10^{-10} Absolute e.s.u.
Electron volt		1.602×10^{-12} ergs
Electron volt		23,052 cal$_{15}$·mole^{-1}
Faraday's constant	F	96,501 International coulomb·gram-equiv.$^{-1}$
Volume of ideal gas (0° C, A_0)	V_0	22.4140 liter-atm·mole^{-1}
Volume of ideal gas (0° C, A_{45})	V_0'	22.4151 liter-atm·mole^{-1}
Avogadro's number	N_0	6.0228×10^{23} mole^{-1}
Gas constant	R_0	8.3144×10^7 erg·deg^{-1}·mole^{-1}
Gas constant	R_0'	1.9865 cal$_{15}$·deg^{-1}·mole^{-1}
Gas constant	R_0''	8.2054×10^{-2} liter·atm·deg^{-1}·mole^{-1}
Boltzmann's constant	k	1.3805×10^{-16} erg·deg^{-1}
Planck's constant	h	6.624×10^{-27} erg·sec

* R. T. Birge, *Rev. Modern Phys.*, **13**, 233 (1941).

TABLE 3. ELECTROCHEMICAL EQUIVALENTS

Electrode Reaction	Electrochemical Equivalent* (grams/coulomb)
$K = K^+ + e^-$	4.0513×10^{-4}
$Ca = Ca^{++} + 2e^-$	2.0767×10^{-4}
$Na = Na^+ + e^-$	2.3830×10^{-4}
$Mg = Mg^{++} + 2e^-$	1.2600×10^{-4}
$Al = Al^{+++} + 3e^-$	0.9316×10^{-4}
$Mn = Mn^{++} + 2e^-$	2.8461×10^{-4}
$Zn = Zn^{++} + 2e^-$	3.3875×10^{-4}
$Cr = Cr^{+++} + 3e^-$	1.7965×10^{-4}
$Fe = Fe^{++} + 2e^-$	2.8938×10^{-4}
$Cd = Cd^{++} + 2e^-$	5.8243×10^{-4}
$Co = Co^{++} + 2e^-$	3.0539×10^{-4}
$Ni = Ni^{++} + 2e^-$	3.0409×10^{-4}
$Sn = Sn^{++} + 2e^-$	6.1502×10^{-4}
$Pb = Pb^{++} + 2e^-$	10.736×10^{-4}
$Fe = Fe^{+++} + 3e^-$	1.9292×10^{-4}
$H_2 = 2H^+ + 2e^-$	0.1045×10^{-4}
$Cu = Cu^{++} + 2e^-$	3.2937×10^{-4}
$Cu = Cu^+ + e^-$	6.5875×10^{-4}
$2Hg = Hg_2^{++} + 2e^-$	20.788×10^{-4}
$Ag = Ag^+ + e^-$	11.180×10^{-4}
$Hg = Hg^{++} + 2e^-$	10.394×10^{-4}
$Pt = Pt^{++} + 2e^-$	10.115×10^{-4}
$Au = Au^{+++} + 3e^-$	6.8117×10^{-4}
$Au = Au^+ + e^-$	20.435×10^{-4}
$2I^- = I_2 + 2e^-$	13.150×10^{-4}
$2Br^- = Br_2(l) + 2e^-$	8.2814×10^{-4}
$Cl^- = \frac{1}{2}Cl_2 + e^-$	3.6743×10^{-4}
$2F^- = F_2 + 2e^-$	1.9689×10^{-4}
$S^{--} = S + 2e^-$	1.6611×10^{-4}

* These values are based on the International Atomic Weights for 1941 and the value of 96,501 international coulombs per gram-equivalent for Faraday's constant.

TABLE 4. ELECTROMOTIVE FORCE SERIES*

Electrode Reaction	Standard Electrode Potential, E^0 (volts), 25° C
$K = K^+ + e^-$	−2.922
$Ca = Ca^{++} + 2e^-$	−2.87
$Na = Na^+ + e^-$	−2.712
$Mg = Mg^{++} + 2e^-$	−2.34
$Be = Be^{++} + 2e^-$	−1.70
$Al = Al^{+++} + 3e^-$	−1.67
$Mn = Mn^{++} + 2e^-$	−1.05
$Zn = Zn^{++} + 2e^-$	−0.762
$Cr = Cr^{+++} + 3e^-$	−0.71
$Ga = Ga^{+++} + 3e^-$	−0.52
$Fe = Fe^{++} + 2e^-$	−0.440
$Cd = Cd^{++} + 2e^-$	−0.402
$In = In^{+++} + 3e^-$	−0.340
$Tl = Tl^+ + e^-$	−0.336
$Co = Co^{++} + 2e^-$	−0.277
$Ni = Ni^{++} + 2e^-$	−0.250
$Sn = Sn^{++} + 2e^-$	−0.136
$Pb = Pb^{++} + 2e^-$	−0.126
$H_2 = 2H^+ + 2e^-$	0.000
$Cu = Cu^{++} + 2e^-$	0.345
$Cu = Cu^+ + e^-$	0.522
$2Hg = Hg_2^{++} + 2e^-$	0.799
$Ag = Ag^+ + e^-$	0.800
$Pd = Pd^{++} + 2e^-$	0.83
$Hg = Hg^{++} + 2e^-$	0.854
$Pt = Pt^{++} + 2e^-$	ca 1.2
$Au = Au^{+++} + 3e^-$	1.42
$Au = Au^+ + e^-$	1.68

* Signs of potential employed by the American Chem. Soc. are opposite to those of this table

Table 5. STANDARD OXIDATION-REDUCTION POTENTIALS AT 25° C*,†,‡

Couple	E^0
$Cr^{++} = Cr^{+++} + e^-$	-0.41
$Pb + SO_4^{--} = PbSO_4 + 2e^-$	-0.355
$Pb + 2Cl^- = PbCl_2 + 2e^-$	-0.268
$Cl^- + Cu = CuCl + e^-$	0.124
$H_2S = S + 2H^+ + 2e^-$	0.141
$Sn^{++} = Sn^{++++} + 2e^-$	0.15
$Cu^+ = Cu^{++} + e^-$	0.167
$Ag + Cl^- = AgCl + e^-$	0.222
$2Hg + 2Cl^- = Hg_2Cl_2 + 2e^-$	0.268
$Fe(CN)_6^{----} = Fe(CN)_6^{---} + e^-$	0.36
$2I^- = I_2 + 2e^-$	0.535
$2Hg + SO_4^{--} = Hg_2SO_4 + 2e^-$	0.615
$H_2O_2 = O_2 + 2H^+ + 2e^-$	0.682
$Fe^{++} = Fe^{+++} + e^-$	0.771
$2H_2O = O_2 + 4H^+(10^{-7}M) + 4e^-$	0.815
$Hg_2^{++} = 2Hg^{++} + 2e^-$	0.910
$HNO_2 + H_2O = NO_3^- + 3H^+ + 2e^-$	0.94
$NO + H_2O = HNO_2 + H^+ + e^-$	0.99
$2Br^- = Br_2(l) + 2e^-$	1.065
$2H_2O = O_2 + 4H^+ + 4e^-$	1.229
$Tl^+ = Tl^{+++} + 2e^-$	1.25
$Mn^{++} + 2H_2O = MnO_2 + 4H^+ + 2e^-$	1.28
$Au^+ = Au^{+++} + 2e^-$	ca 1.29
$Cl^- = \tfrac{1}{2}Cl_2 + e^-$	1.358
$2Cr^{+++} + 7H_2O = Cr_2O_7^{--} + 14H^+ + 6e^-$	1.36
$Pb^{++} + 2H_2O = PbO_2 + 4H^+ + 2e^-$	1.456
$Cl^- + H_2O = HClO + H^+ + 2e^-$	1.49
$Mn^{++} = Mn^{+++} + e^-$	1.51
$Mn^{++} + 4H_2O = MnO_4^- + 8H^+ + 5e^-$	1.52
$Ce^{+++} = Ce^{++++} + e^-$	1.61
$\tfrac{1}{2}Cl_2 + H_2O = H^+ + HClO + e^-$	1.63
$MnO_2 + 2H_2O = MnO_4^- + 4H^+ + 3e^-$	1.67
$PbSO_4 + 2H_2O = PbO_2 + SO_4^{--} + 4H^+ + 2e^-$	1.685
$Pb^{++} = Pb^{++++} + 2e^-$	1.69
$Ni^{++} + 2H_2O = NiO_2 + 4H^+ + 2e^-$	1.75
$2H_2O = H_2O_2 + 2H^+ + 2e^-$	1.77
$Co^{++} = Co^{+++} + e^-$	1.84
$Ag^+ = Ag^{++} + e^-$	1.98
$2SO_4^{--} = S_2O_8^{--} + 2e^-$	2.05
$O_2 + H_2O = O_3 + 2H^+ + 2e^-$	2.07
$H_2O = O(g) + 2H^+ + 2e^-$	2.42
$2F^- = F_2 + 2e^-$	2.85
$Mg + 2OH^- = Mg(OH)_2 + 2e^-$	-2.67
$Al + 4OH^- = H_2AlO_3^- + H_2O + 3e^-$	-2.35

‡ Signs of potential employed by the American Chem. Soc. are opposite to those of this table.

TABLE. 5. STANDARD OXIDATION-REDUCTION POTENTIALS AT 25° C*,† — Cont.

Couple	E^0
$Al + 3OH^- = Al(OH)_3 + 3e^-$	-2.31
$Mn + 2OH^- = Mn(OH)_2 + 2e^-$	-1.47
$Zn + S^{--} = ZnS + 2e^-$	-1.44
$Mn + CO_3^{--} = MnCO_3 + 2e^-$	-1.35
$Cr + 3OH^- = Cr(OH)_3 + 3e^-$	-1.3
$Zn + 2OH^- = Zn(OH)_2 + 2e^-$	-1.245
$Cd + S^{--} = CdS + 2e^-$	-1.23
$Zn + 4OH^- = ZnO_2^{--} + 2H_2O + 2e^-$	-1.216
$Cr + 4OH^- = CrO_2^- + 2H_2O + 3e^-$	-1.2
$Ni + S^{--} = NiS(\gamma) + 2e^-$	-1.07
$Zn + CO_3^{--} = ZnCO_3 + 2e^-$	-1.07
$Fe + S^{--} = FeS + 2e^-$	-1.00
$Pb + S^{--} = PbS + 2e^-$	-0.98
$2Cu + S^{--} = Cu_2S + 2e^-$	-0.95
$Co + S^{--} = CoS(\alpha) + 2e^-$	-0.93
$Fe + 8OH^- = Fe_3O_4 + 4H_2O + 8e^-$	-0.91
$Fe + 2OH^- = Fe(OH)_2 + 2e^-$	-0.877
$Ni + S^{--} = NiS(\alpha) + 2e^-$	-0.86
$H_2 + 2OH^- = 2H_2O + 2e^-$	-0.828
$Cd + 2OH^- = Cd(OH)_2 + 2e^-$	-0.815
$Cd + CO_3^{--} = CdCO_3 + 2e^-$	-0.80
$Sn + 3OH^- = HSnO_2^- + H_2O + 2e^-$	-0.79
$Cu + S^{--} = CuS + 2e^-$	-0.76
$Fe + CO_3^{--} = FeCO_3 + 2e^-$	-0.755
$Co + 2OH^- = Co(OH)_2 + 2e^-$	-0.73
$Fe + 3OH^- = Fe(OH)_3 + 3e^-$	ca -0.73
$2Ag + S^{--} = Ag_2S + 2e^-$	-0.71
$Hg + S^{--} = HgS + 2e^-$	-0.70
$Ni + 2OH^- = Ni(OH)_2 + 2e^-$	-0.66
$Co + CO_3^{--} = CoCO_3 + 2e^-$	-0.632
$S^{--} + 6OH^- = SO_3^{--} + 3H_2O + 6e^-$	-0.61
$Pb + 2OH^- = PbO(r) + H_2O + 2e^-$	-0.578
$Fe(OH)_2 + OH^- = Fe(OH)_3 + e^-$	-0.56
$Pb + 3OH^- = HPbO_2^- + H_2O + 2e^-$	-0.54
$Pb + CO_3^{--} = PbCO_3 + 2e^-$	-0.506
$Ni + CO_3^{--} = NiCO_3 + 2e^-$	-0.45
$Mn(OH)_2 + OH^- = Mn(OH)_3 + e^-$	-0.40
$2Cu + 2OH^- = Cu_2O + H_2O + 2e^-$	-0.361
$2CN^- + Ag = Ag(CN)_2^- + e^-$	-0.29
$Cu + 2OH^- = Cu(OH)_2 + 2e^-$	-0.224
$Cr(OH)_3 + 5OH^- = CrO_4^{--} + 4H_2O + 3e^-$	-0.12
$OH^- + HO_2^- = O_2 + H_2O + 2e^-$	-0.076
$4NH_3 + Cu = Cu(NH_3)_4^{++} + 2e^-$	-0.05
$Cu + CO_3^{--} = CuCO_3 + 2e^-$	0.053
$Hg + 2OH^- = HgO(r) + H_2O + 2e^-$	0.098

Table 5. Standard Oxidation-Reduction Potentials at 25° C*,† — Cont.

Couple	E^0
$2Hg + 2OH^- = Hg_2O + H_2O + 2e^-$	0.123
$2Hg + CO_3^{--} = Hg_2CO_3 + 2e^-$	0.32
$2Ag + 2OH^- = Ag_2O + H_2O + 2e^-$	0.344
$2NH_3(Aq) + Ag = Ag(NH_3)_2^+ + e^-$	0.373
$4OH^- = O_2 + 2H_2O + 4e^-$	0.401
$2Ag + CrO_4^{--} = Ag_2CrO_4 + 2e^-$	0.446
$2Ag + CO_3^{--} = Ag_2CO_3 + 2e^-$	0.47
$Ni(OH)_2 + 2OH^- = NiO_2 + 2H_2O + 2e^-$	0.49
$MnO_4^{--} = MnO_4^- + e^-$	0.54
$MnO_2 + 4OH^- = MnO_4^- + 2H_2O + 3e^-$	0.57
$Cl^- + 2OH^- = ClO^- + H_2O + 2e^-$	0.94
$O_2 + 2OH^- = O_3 + H_2O + 2e^-$	1.24

* W. M. Latimer, *Oxidation Potentials*, Prentice-Hall, Inc., New York, 1938. Note that the American Chemical Society uses a convention of sign opposite to that employed here.

† For metal electrode potentials, refer also to Table 4, p. 1134.

Table 6. Standard Reference Electrodes

$$E = E_{25°} + \frac{dE}{dt} t$$

Name	Cell	$E_{25°}$ (volts)	$E_{25°}$ (strong acids)	$E_{25°}$ (buffers)	$\frac{dE}{dt}$ (volts/deg. C)
0.1 N calomel	$Hg/Hg_2Cl_2/KCl(0.1\ N)$	0.3337*	0.3386†	0.336 †	-0.7×10^{-4}
1.0 N calomel	$Hg/Hg_2Cl_2/KCl\ (1.0\ N)$	0.2800*	0.2848†	0.2828†	-2.4×10^{-4}
Saturated calomel‡	$Hg/Hg_2Cl_2/KCl(saturated)$	0.2415*	0.2457†	0.2434†	-7.6×10^{-4}
Silver chloride	$Ag/AgCl/KCl\ (0.1\ N)$	0.2881*	....	....	-6.5×10^{-4}
Copper sulfate‡§	$Cu/CuSO_4/CuSO_4\ (sat'd)$	0.316	....	....	9.0×10^{-4}

* The E values so marked are true values in that they do not include liquid junction potentials.

† The E values so marked include an approximate value for the liquid junction potentials. See W. J. Hamer, *Trans. Electrochem. Soc.*, **72**, 62 (1937).

‡ The E of this electrode is subject to pronounced hysteresis effects and should be used only at constant temperature.

§ S. Ewing, *The Copper-Copper Sulfate Half Cell for Measuring Potentials in the Earth*, p. 10, Committee Report, American Gas Association, 1939.

TABLE 7. SOLUBILITY PRODUCT CONSTANTS AT 25° C

Reaction	Solubility Product
$Al(OH)_3 = Al^{+++} + 3OH^-$	1.9×10^{-33}
$Al(OH)_3 = H^+ + H_2AlO_3^-$	4×10^{-13}
$Cd(OH)_2 = Cd^{++} + 2OH^-$	1.2×10^{-14}
$CdS = Cd^{++} + S^{--}$	1.4×10^{-28}
$CdCO_3 = Cd^{++} + CO_3^{--}$	2.5×10^{-14}
$Ca(OH)_2 = Ca^{++} + 2OH^-$	7.9×10^{-6}
$CaCO_3 = Ca^{++} + CO_3^{--}$	4.82×10^{-9}*
$CaSO_4 \cdot 2H_2O = Ca^{++} + SO_4^{--} + 2H_2O$	2.4×10^{-5}
$Ca_3(PO_4)_2 = 3Ca^{++} + 2PO_4^{---}$	1×10^{-25}
$CaHPO_4 = Ca^{++} + HPO_4^{--}$	ca 5×10^{-6}
$Cr(OH)_3 = Cr^{+++} + 3OH^-$	6.7×10^{-31}
$Cr(OH)_3 = H^+ + CrO_2^- + H_2O$	9×10^{-17}
$Co(OH)_2 = Co^{++} + 2OH^-$	2×10^{-16}
$CoCO_3 = Co^{++} + CO_3^{--}$	1.0×10^{-12}
$½Cu_2O + ½H_2O = Cu^+ + OH^-$	1.2×10^{-15}
$Cu_2S = 2Cu^+ + S^{--}$	2.5×10^{-50}
$Cu(OH)_2 = Cu^{++} + 2OH^-$	5.6×10^{-20}
$CuS = Cu^{++} + S^{--}$	4×10^{-38}
$CuCO_3 = Cu^{++} + CO_3^{--}$	1.37×10^{-10}
$Fe(OH)_2 = Fe^{++} + 2OH^-$	1.65×10^{-15}
$FeCO_3 = Fe^{++} + CO_3^{--}$	2.11×10^{-11}
$FeS = Fe^{++} + S^{--}$	1×10^{-19}
$Fe(OH)_3 = Fe^{+++} + 3OH^-$	4×10^{-38}
$Pb(OH)_2 = Pb^{++} + 2OH^-$	2.8×10^{-16}
$PbO(r) + H_2O = Pb^{++} + 2OH^-$	5.5×10^{-16}
$Pb(OH)_2 = H^+ + HPbO_2^-$	2.1×10^{-16}
$PbO(y) + OH^- = HPbO_2^-$	5.31×10^{-2}
$PbO(r) + OH^- = HPbO_2^-$	4.03×10^{-2}
$PbCl_2 = Pb^{++} + 2Cl^-$	1.7×10^{-5}
$PbCO_3 = Pb^{++} + CO_3^{--}$	1.5×10^{-13}
$PbS = Pb^{++} + S^{--}$	1.0×10^{-29}
$PbCrO_4 = Pb^{++} + CrO_4^{--}$	1.8×10^{-14}
$PbSO_4 = Pb^{++} + SO_4^{--}$	1.8×10^{-8}
$PbO_2 + 2H_2O = Pb^{++} + 4OH^-$	10^{-64}
$Mg(OH)_2 = Mg^{++} + 2OH^-$	5.5×10^{-12}
$MgCO_3 \cdot 3H_2O = Mg^{++} + CO_3^{--} + 3H_2O$	ca 1×10^{-5}*
$Mn(OH)_2 = Mn^{++} + 2OH^-$	7.1×10^{-15}
$MnCO_3 = Mn^{++} + CO_3^{--}$	8.8×10^{-11}
$MnS = Mn^{++} + S^{--}$	5.6×10^{-16}
$Hg_2O + H_2O = Hg_2^{++} + 2OH^-$	1.6×10^{-23}
$Hg_2Cl_2 = Hg_2^{++} + 2Cl^-$	1.1×10^{-18}
$Hg_2CO_3 = Hg_2^{++} + CO_3^{--}$	9×10^{-17}
$Hg_2SO_4 = Hg_2^{++} + SO_4^{--}$	6.2×10^{-7}
$Hg_2S = Hg_2^{++} + S^{--}$	1×10^{-45}
$HgO + H_2O = Hg^{++} + 2OH^-$	1.7×10^{-26}

TABLE 7. SOLUBILITY PRODUCT CONSTANTS AT 25° C — *Continued*

Reaction	Solubility Product
$HgS = Hg^{++} + S^{--}$	3×10^{-53}
$Ni(OH)_2 = Ni^{++} + 2OH^-$	1.6×10^{-14}
$NiCO_3 = Ni^{++} + CO_3^{--}$	1.36×10^{-7}
$NiS(\alpha) = Ni^{++} + S^{--}$	3×10^{-21}
$NiS(\beta) = Ni^{++} + S^{--}$	1×10^{-26}
$NiS(\gamma) = Ni^{++} + S^{--}$	2×10^{-28}
$Pt(OH)_2 = Pt^{++} + 2OH^-$	ca 1×10^{-35}
$\tfrac{1}{2}Ag_2O + \tfrac{1}{2}H_2O = Ag^+ + OH^-$	2.0×10^{-8}
$AgCl = Ag^+ + Cl^-$	1.7×10^{-10}
$Ag_2S = 2Ag^+ + S^{--}$	1.0×10^{-51}
$Ag_2CO_3 = 2Ag^+ + CO_3^{--}$	8.2×10^{-12}
$Ag_2SO_4 = 2Ag^+ + SO_4^{--}$	1.18×10^{-5}
$Sn(OH)_2 \rightleftharpoons Sn^{++} + 2OH^-$	5×10^{-26}
$H_2SnO_2 \rightleftharpoons H^+ + HSnO_2^-$	6×10^{-18}
$SnS = Sn^{++} + S^{--}$	8×10^{-29}
$Sn(OH)_4 = Sn^{++++} + 4OH^-$	ca 1×10^{-56}
$Sn(OH)_4 + 2OH^- = Sn(OH)_6^{--}$	2.1×10^{-4}
$Zn(OH)_2 = Zn^{++} + 2OH^-$	4.5×10^{-17}
$ZnCO_3 = Zn^{++} + CO_3^{--}$	6×10^{-11}
$ZnS = Zn^{++} + S^{--}$	4.5×10^{-24}

* For solubility product in sea water, see Table 10, p. 1119.

TABLE 8. DISSOCIATION CONSTANTS OF WEAK ELECTROLYTES AT 25° C

Reaction	K
$H_3BO_3 = H^+ + H_2BO_3^-$	5.79×10^{-10}
$H_2CO_3 = H^+ + HCO_3^-$	4.45×10^{-7}
$HCO_3^- = H^+ + CO_3^{--}$	4.69×10^{-11}
$HCOOH = H^+ + HCOO^-$	1.77×10^{-4}
$H_2C_2O_4 = H^+ + HC_2O_4^-$	5.9×10^{-2}
$HC_2O_4^- = H^+ + C_2O_4^{--}$	5.18×10^{-5}
$HCN = H^+ + CN^-$	4×10^{-10}
$HClO = H^+ + ClO^-$	5.6×10^{-8}
$HCrO_4^- = H^+ + CrO_4^{--}$	3.2×10^{-7}
$Cr_2O_7^{--} + H_2O = 2HCrO_4^-$	2.3×10^{-2}
$HNO_2 = H^+ + NO_2^-$	4.5×10^{-4}
$NH_4OH = NH_4^+ + OH^-$	1.81×10^{-5}
$H_2O = H^+ + OH^-$	1.008×10^{-14}
$H_2O_2 = H^+ + HOO^-$	2.4×10^{-12}
$H_3PO_4 = H^+ + H_2PO_4^-$	7.52×10^{-3}
$H_2PO_4^- = H^+ + HPO_4^{--}$	6.22×10^{-8}
$HPO_4^{--} = H^+ + PO_4^{---}$	1×10^{-12}
$H_2SiO_3 = H^+ + HSiO_3^-$	1×10^{-10}
$HSiO_3^- = H^+ + SiO_3^{--}$	1×10^{-12}
$H_2S = H^+ + HS^-$	1.15×10^{-7}
$HS^- = H^+ + S^{--}$	1.0×10^{-15}
$H_2SO_3 = H^+ + HSO_3^-$	1.72×10^{-2}
$HSO_3^- = H^+ + SO_3^{--}$	6.24×10^{-8}
$HSO_4^- = H^+ + SO_4^{--}$	1.01×10^{-2}
$CH_3COOH = H^+ + CH_3COO^-$	1.76×10^{-5}

TABLE 9. DISSOCIATION CONSTANTS OF SOME WEAK ELECTROLYTES AS A FUNCTION OF TEMPERATURE

Reaction	Experimental Temperature Range, °C	A	B	C
$H_3BO_3 = H^+ + H_2BO_3^-$	5 to 50	−2193.55	3.0395	−0.016499
$HCO_3^- = H^+ + CO_3^{--}$	0 to 50	−2902.39	6.4980	−0.02379
$HCOOH = H^+ + HCOO^-$	0 to 60	−1342.85	5.2744	−0.0151682
$HC_2O_4^- = H^+ + C_2O_4^{--}$	0 to 50	−1539.31	7.1966	−0.021200
$H_2O = H^+ + OH^-$	0 to 300	−4470.99	6.0875	−0.017060
$H_3PO_4 = H^+ + H_2PO_4^-$	0 to 50	−1264.51	7.6601	−0.018590
$H_2PO_4^- = H^+ + HPO_4^{--}$	5 to 50	−1648.88	3.2542	−0.016534
$CH_3COOH = H^+ + CH_3COO^-$	0 to 60	−1170.48	3.1649	−0.013399

The constants A, B, and C, on being inserted in the equation

$$\log_{10} K = \frac{A}{T} + B + CT$$

will give K as a function of T, where T = deg. Kelvin (deg. C + 273.1°). From H. S. Harned and B. B. Owen, *The Physical Chemistry of Electrolytic Solutions*, American Chemical Society Monograph 95, Reinhold Publishing Corp., New York, 1943.

t °C	$H_2CO_3 = H^+ + HCO_3^-$ * K	$HSO_4^- = H^+ + SO_4^{--}$ † K	t °C	$NH_4OH = OH^- + NH_4^+$ ‡ K
0	2.647×10^{-7}	....	18	1.72×10^{-5}
5	3.040×10^{-7}	1.80×10^{-2}	25	1.81×10^{-5}
10	3.430×10^{-7}	....	51	1.81×10^{-5}
15	3.802×10^{-7}	1.36×10^{-2}	75.2	1.64×10^{-5}
20	4.147×10^{-7}	....	100	1.35×10^{-5}
25	4.452×10^{-7}	1.01×10^{-2}	124.8	1.04×10^{-5}
30	4.710×10^{-7}	....	156	0.63×10^{-5}
35	4.914×10^{-7}	0.75×10^{-2}	218	0.18×10^{-5}
40	5.058×10^{-7}	....	306	0.0093×10^{-5}
45	5.139×10^{-7}	0.56×10^{-2}		
50	5.161×10^{-7}	....		
55	....	0.41×10^{-2}		

* H. S. Harned and R. Davis, Jr., *J. Am. Chem. Soc.*, **65**, 2030 (1943).

† H. S. Harned and B. B. Owen, *The Physical Chemistry of Electrolytic Solutions*, American Chemical Society Monograph 95, Reinhold Publishing Corp., New York, 1943.

‡ A. Noyes and Y. Kato, *Carnegie Inst., Publ.* **63**, 178 (1907). R. B. Sosman, *op. cit.*, p. 228

TABLE 10. SPONTANEITY OF COMMON CORROSION REACTIONS*

Metal	Solid Product†	Hydrogen-Type‡ Corrosion, p.H$_2$ = 1.0 atm.		Oxygen-Type§ Corrosion, p.O$_2$ = 0.21 atm.	
		E (volt)	ΔF per gram atom Metal, gram cal	E (volt)	ΔF per gram atom Metal, gram cal
Mg	Mg(OH)$_2$	+1.823	−84,000	+3.042	−140,000
Al	Al(OH)$_3$ (?)	+1.48	−102,600	+2.70	−180,700
Mn	Mn(OH)$_2$	+0.60	−27,600	+1.81	−83,200
	Mn(OH)$_3$	+0.256	−17,700	+1.50	−103,000
	MnO$_2$	−0.14	+12,700	+1.11	−101,000
Cr	Cr(OH)$_3$	+0.47	−32,500	+1.69	−117,000
Zn	Zn(OH)$_2$ (?)	+0.417	−19,200	+1.636	−75,200
Fe	Fe$_3$O$_4$	+0.082	−5,000	+1.30	−80,000
	Fe(OH)$_2$	+0.049	−2,300	+1.27	−58,500
	Fe(OH)$_3$	−0.07	+4,700	+1.15	−80,000
Cd	Cd(OH)$_2$	−0.013	+600	+1.1206	−55,600
Co	Co(OH)$_2$	−0.098	+4,500	+1.12	−51,700
Ni	Ni(OH)$_2$	−0.17	+7,800	+1.05	−48,500
Pb	PbO(red)	−0.250	+11,500	+0.97	−44,600
Cu	Cu$_2$O	−0.413	+9,500	+0.80	−18,600
	Cu(OH)$_2$	−0.604	+27,800	+0.615	−28,300
	CuO	−0.537	+24,800	+0.680	−31,500
Hg	HgO	−0.926	+42,600	+0.293	−13,600
	Hg$_2$O	−0.951	+21,970	+0.268	−6,200
Ag	Ag$_2$O	−1.172	+27,000	+0.047	−1,100

* R. Brown, B. Roetheli, and H. Forrest, *Ind. Eng. Chem.*, **23**, 350 (1931) probably were the first investigators to attempt a summary of this type. They also used the terms *Hydrogen Type* and *Oxygen Type* in the same sense (cf. definitions below). For a more detailed discussion of these data, see J. C. Warner, *Trans. Electrochem. Soc.*, **83**, 319 (1943).

† Except when the formula is followed by (?), the data are quite certainly for the formation of the solid corrosion product as indicated. The formulas given do not, in any case, attempt to indicate the extent of hydration of the solid phase.

‡ The data in the column under "Hydrogen-Type Corrosion" are for reactions of the type

$$M(s) + 2H_2O(l) = M(OH)_2(s) + H_2(g, 1 \text{ atm.})$$

or

$$M(s) + H_2O(l) = MO(s) + H_2(g, 1 \text{ atm.})$$

where s = solid, l = liquid, and g = gas.

The potentials given are the reversible potentials for the galvanic corrosion cells, in which the anode and cathode reactions add up to the overall corrosion reactions, thus:

$$\text{Anode: } Zn(s) + 2OH^- = Zn(OH)_2(s) + 2e^-$$
$$\text{Cathode: } 2H_2O(l) + 2e^- = 2OH^- + H_2(g, 1 \text{ atm.})$$
$$\overline{\text{Total: } Zn(s) + 2H_2O(l) + Zn(OH)_2(s) + H_2(g, 1 \text{ atm.})}$$

A positive value for E or a negative value for ΔF corresponds to a spontaneous reaction. This follows the convention

$$\Delta F = -nFE$$

where F = Faraday (96,501 coulombs), E = emf, and n = number of electrons (equivalents) taking part in the reaction.

§ In the columns under "Oxygen-Type Corrosion" the data are for reactions of the type

$$M(s) + H_2O(l) + \tfrac{1}{2}O_2(g, p = 0.21 \text{ atm.}) = M(OH)_2(s)$$

or
$$M(s) + \tfrac{1}{2}O_2(g, p = 0.21 \text{ atm.}) = MO(s)$$

The data have been calculated for these reactions with O_2 at a partial pressure of 0.21 atmosphere because this is approximately the partial pressure of O_2 in dry air under a total pressure of one atmosphere.

TABLE 11. VALUES OF HYDROGEN OVERVOLTAGE

A. Interrupter Method*

Material	Overvoltage (volts at 16° C ±1° in 1 N HCl)			
C. D. $\frac{\text{amp}}{\text{sq cm}} \rightarrow$	10^{-3}	10^{-2}	10^{-1}	1
Mercury	1.04	1.15	1.21	1.24
Cadmium	0.99	1.20	1.25	1.23
Lead†	0.91	1.24	1.26	1.22
Tin	0.85	0.98	0.99	0.98
Bismuth	0.69	0.83	0.91	1.01
Lead	0.67	0.97	1.12	1.08
Copper	0.60	0.75	0.82	0.84
Aluminum	0.58	0.71	0.74	0.78
Copper†	0.50	0.62	0.74	0.80
Silver	0.46	0.66	0.76	....
Chromium	....	....	0.67	0.77
Iron	0.40	0.53	0.64	0.77
Nickel	0.33	0.42	0.51	0.59
Tungsten	0.27	0.35	0.47	0.54
Platinum†	0.25	0.35	0.40	0.40
Gold†	0.17	0.25	0.32	0.42
Platinum	0.09	0.39	0.50	0.44
Rhodium	0.08	0.22	0.33	0.34
Platinum (platinized)	0.01	0.03	0.05	0.07
Carbon (filament)	0.95	1.13	1.18	1.17
Carbon (graphite)	0.47	0.76	0.99	1.03
Carbon (arc)	0.27	0.34	0.41	0.41

* These values were obtained using an electronic interrupter. For details, see A. Hickling and F. W. Salt, *Trans. Faraday Soc.*, **33**, 1540 (1937); **36**, 1226 (1940); **37**, 333 (1941). The voltages obtained are presumed to be essentially free from IR drop and concentration polarization. They should not be used for any steady state calculation unless appropriate corrections are made for these two effects.

† Specimens were in the form of electrodeposited films on copper except for copper which was deposited on silver.

TABLE 11. VALUES OF HYDROGEN OVERVOLTAGE — *Continued*

B. Direct Method‡

Material	Overvoltage (volts at 25° C in 2 N H$_2$SO$_4$)			
C. D. $\dfrac{\text{amp}}{\text{sq cm}} \longrightarrow$	10^{-3}	10^{-2}	10^{-1}	1
Cadmium	0.98	1.13	1.22	1.25
Mercury	0.90	1.04	1.07	1.12
Tin	0.86	1.08	1.22	1.23
Bismuth	0.78	1.05	1.14	1.23
Zinc	0.72	0.75	1.06	1.23
Graphite	0.60	0.78	0.98	1.22
Aluminum	0.56	0.83	1.0	1.29
Nickel	0.56	0.75	1.05	1.21
Lead	0.52	1.09	1.18	1.26
Brass	0.50	0.65	0.91	1.25
Copper	0.48	0.58	0.8	1.25
Silver	0.47	0.76	0.88	1.09
Iron	0.40	0.56	0.82	1.29
Monel	0.28	0.38	0.62	1.07
Gold	0.24	0.39	0.59	0.80
Duriron	0.20	0.29	0.61	1.02
Palladium	0.12	0.3	0.7	1.0
Platinum	0.024	0.07	0.29	0.68
Platinum (platinized)	0.015	0.03	0.04	0.05

‡ The values listed were abstracted from the table prepared by M. Knobel, *International Critical Tables*, **6**, 339. Overvoltages obtained by the direct method necessarily include the IR drop and concentration polarization.

TABLE 12. SOLUBILITY OF GASES IN WATER

$t(°C)$	O_2* $\alpha \times 1000$	N_2* $\alpha \times 1000$	CO_2† $\alpha \times 1000$	H_2S* $\alpha \times 1000$	Air* ml/liter at Total Pressure‡ = 760 mm	Air* O_2 in Dissolved Air, Volume %
		ml/liter§				
0	48.89	23.54	1710	4670	29.18	34.91
10	38.02	18.61	1195	3399	22.84	34.47
20	31.02	15.45	875	2582	18.68	34.03
25	28.31	14.34	759	2282	17.08	33.82
30	26.08	13.42	669	2037	15.64	33.60
40	23.06	11.84	530	1660	....	....
50	20.90	10.88	435	1392	....	....
60	19.46	10.23	359	1190	....	....
70	18.33	9.77	....	1022	....	....
75	....	....	308∥	....	....	....
80	17.61	9.58	....	917	....	....
90	17.2	9.5	....	840	....	....
100	17.0	9.5	231∥	810	....	....

$t(°C)$	SO_2 Total Pressure‡ = 760 mm, grams/100 grams water	$t(°C)$	NH_3 Total Pressure‡ = 760 mm, grams/100 grams Solution	NH_3 Vapor
10	15.39	0	47.3	....
20	10.64	10	40.6	99.2
29.9	7.58	20	34.6	98.5
40	5.54	30	29.1	97.0
48.15	4.39	40	24.0	94.5
60	3.25	50	19.0	90.2
70	2.61	60	14.4	82.5
80	2.13	70	10.0	70.0
90	1.81	80	6.1	52.6
		90	2.9	31.8

Notes:

α = Bunsen absorption coefficient = volume of dissolved gas (calculated to 0° C and 760 mm pressure) per volume of solvent (the volume of the solvent being measured at the temperature considered) when the partial pressure of gas is 760 mm.

1000α = milliliters of gas (standard conditions) per liter of solvent, when the partial pressure of gas is 760 mm. To calculate the solubility of a gas at a given total pressure, when the total pressure is low:

$$S_t (\text{ml/liter solvent}) = 1000\alpha_t \frac{P - P_s}{P_{\text{std}}}$$

P = observed total pressure
P_s = vapor pressure of solvent at temperature t
P_{std} = one standard atmosphere
α_t = Bunsen coefficient at temperature t

The solubility of oxygen from air may be calculated from the data for oxygen by the following formula:

$$S_t (\text{oxygen from air, ml/liter}) = 1000\alpha_t \frac{(P - P_s) \times 0.208}{P_{\text{std}}}$$

where 0.208 is the mole fraction of O_2 in air.

Notes (Cont.):

The solubility of a gas in parts per million, at a partial pressure of 760 mm, may be calculated from α by the following equation:

$$S(\text{ppm})(P_A = 1 \text{ atm.}) = \frac{1000\alpha_t \cdot d_a}{d_s}$$

where d_a = density of gas at 0° C and 760 mm in grams per liter,
d_s = density of solution at temperature t in grams per cc.

For O_2, CO_2, and H_2S, d_s can be taken as the density of water with sufficient accuracy for most purposes.

Gases such as O_2 and N_2 obey Henry's Law at low and moderate pressures. CO_2, H_2S, and SO_2 obey Henry's Law at low pressures only when ionization is inappreciable. When ionization is appreciable, Henry's Law is obeyed by the un-ionized fraction of these gases.

The data given above may be used to calculate the un-ionized dissolved CO_2, H_2S, or SO_2. At 25° C, Table 14, "Ion Activities of Weak Electrolytes as a Function of pH," may then be used to calculate the total gas dissolved when the pH of the solution is known. The un-ionized portion of NH_3 obeys Henry's Law only at very low pressures. For the total solubility of SO_2 and NH_3 in pure water at low pressures, see R. T. Haslam, R. L. Hershey, and R. H. Keen, *Ind. Eng. Chem.*, **16**, 1224 (1924), Figs. 4 and 5.

* *Handbook of Chemistry and Physics*, 28th Edition, Chemical Rubber Publishing Co., 1944–1945.

† *International Critical Tables*, **3**, 260.

‡ Partial pressure of gas plus aqueous tension at stated temperature equals 760 mm.

§ Milliliter of gas (0° C, 760 mm) per liter of water when partial pressure of gas is 760 mm.

‖ Zelvenskii, *J. Chem. ind.* (U. S. S. R.), **14**, 1250 (1937); *Chem. Abstracts*, **32**, 852 (1938).

TABLE 13. SOLUBILITY OF OXYGEN IN CERTAIN ELECTROLYTE SOLUTIONS

(See Tables 7 and 9, pp. 1116 and 1117, for solubility of oxygen in sea water.)

Electrolyte	Conc. of electrolyte, moles/liter: 0.5	1.0	2.0	
	Solubility, ml /liter*			
HNO_3	27.67	27.03	26.03	⎫
HCl	27.13	26.30	24.47	⎪
H_2SO	25.20	23.00	19.15	⎬ 25° C
NaCl	24.01	20.44	14.48	⎪
KOH	23.09	18.88	....	⎪
NaOH	22.91	18.69	12.19	⎭
HNO_3	33.00	31.86	29.87	⎫
HCl	32.62	31.01	28.35	⎪
H_2SO_4	32.05	31.76	22.09	⎬ 15° C
NaCl	29.20	24.65	17.26	⎪
KOH	27.59	22.19	....	⎪
NaOH	27.31	21.90	14.41	⎭

* Solubility given in milliliters of gas (°C, 1 atm.) dissolved in 1 liter of solution when partial pressure of the gas equals one atmosphere.

TABLE 14. ION ACTIVITIES OF WEAK ELEC

pH	H$_2$O		NH$_3$		CO$_2$			H$_2$S		
	a_{H^+} ×10^3	a_{OH^-} ×10^3	$\dfrac{a_{NH_4^+}}{a_N}$	$\dfrac{a_{NH_3}}{a_N}$	$\dfrac{a_{CO_2}}{a_C}$	$\dfrac{a_{HCO_3^-}}{a_C}$	$\dfrac{a_{CO_3^{--}}}{a_C}$	$\dfrac{a_{H_2S}}{a_S}$	$\dfrac{a_{HS^-}}{a_S}$	$\dfrac{a_{S^{--}}}{a_S}$
1	100.0	....	....	....	....	....	....	....	....	....
1.5	31.62	....	....	....	....	....	....	....	....	....
2	10.00	....	....	....	1.0000	....	....	....	....	....
2.5	3.162	....	....	....	0.9999	0.0001	....	1.0000	....	....
3	1.000	....	....	....	0.9996	0.0004	....	0.9999	0.0001	....
3.5	0.3162	....	....	....	0.9986	0.0014	....	0.9996	0.0004	....
4	0.1000	....	....	....	0.9957	0.0043	....	0.9988	0.0012	....
4.5	0.0316	....	1.0000	....	0.9866	0.0134	....	0.9964	0.0036	....
5	0.0100	....	0.9999	0.0001	0.9587	0.0413	....	0.9886	0.0114	....
5.5	0.0032	....	0.9998	0.0002	0.8800	0.1200	....	0.9649	0.0351	....
6	0.0010	....	0.9994	0.0006	0.6988	0.3012	0.0000	0.8969	0.1031	....
6.5	0.0003	....	0.9982	0.0018	0.4232	0.5767	0.0001	0.7333	0.2667	....
7	0.0001	0.0001	0.9944	0.0056	0.1883	0.8113	0.0004	0.4651	0.5349	....
7.5	....	0.0003	0.9872	0.0173	0.0683	0.9303	0.0014	0.2157	0.7843	....
8	....	0.0010	0.9472	0.0528	0.0226	0.9728	0.0046	0.0800	0.9200	....
8.5	....	0.0032	0.8503	0.1497	0.0072	0.9783	0.0145	0.0268	0.9732	....
9	....	0.0101	0.6423	0.3577	0.0022	0.9530	0.0448	0.0086	0.9914	....
9.5	....	0.0319	0.3622	0.6378	0.0006	0.8701	0.1293	0.0027	0.9973	....
10	....	0.1008	0.1522	0.8478	0.0002	0.6801	0.3197	0.0009	0.9991	....
10.5	....	0.3188	0.0537	0.9463	0.0000	0.4022	0.5978	0.0003	0.9997	0.0000
11	....	1.008	0.0176	0.9824	....	0.1754	0.8246	0.0001	0.9998	0.0001
11.5	....	3.188	0.0056	0.9944	....	0.0630	0.9370	0.0000	0.9997	0.0003
12	....	10.08	0.0018	0.9982	....	0.0208	0.9792	....	0.9990	0.0010
12.5	....	31.18	0.0006	0.9994	....	0.0067	0.9933	....	0.9968	0.0032
13	....	100.8	0.0002	0.9998	....	0.0021	0.9979	....	0.9901	0.0099

Note. The symbol a as used above indicates activity of a given ion. The fractional activities are best explained by an example: $\dfrac{a_{CO_2}}{a_C}$ means $\dfrac{a_{CO_2}}{a_{CO_2} + a_{HCO_3^-} + a_{CO_3^{--}}}$. At infinite dilution the activities become equal to concentrations in moles per liter.

TROLYTES AS A FUNCTION OF pH AT 25° C*

H₂SO₃†			H₃PO₄			
$a_{H_2SO_3}$	$a_{HSO_3^-}$	$a_{SO_3^{--}}$	$a_{H_3PO_4}$	$a_{H_2PO_4^-}$	$a_{HPO_4^{--}}$	$a_{PO_4^{---}}$
$a_{S^{+4}}$	$a_{S^{+4}}$	$a_{S^{+4}}$	a_P	a_P	a_P	a_P
0.8532	0.1468	....	0.9302	0.0698	....	....
0.6477	0.3523	....	0.8083	0.1917	....	....
0.3676	0.6324	....	0.5714	0.4286	....	....
0.1553	0.8447	....	0.2966	0.7034	0.0000	....
0.0549	0.9450	0.00006	0.1176	0.8823	0.0001	....
0.0180	0.9818	0.0002	0.0405	0.9593	0.0002	....
0.0058	0.9936	0.0006	0.0132	0.9862	0.0006	....
0.0018	0.9962	0.0020	0.0042	0.9939	0.0019	....
0.0006	0.9932	0.0062	0.0013	0.9925	0.0062	....
0.0002	0.9804	0.0193	0.0004	0.9804	0.0192	....
0.0001	0.9412	0.0587	0.0001	0.9415	0.0584	....
0.0000	0.8352	0.1648	0.0000	0.8361	0.1639	....
....	0.6157	0.3842	....	0.6173	0.3827	....
....	0.3363	0.6637	....	0.3378	0.6622	0.0000
....	0.1381	0.8619	....	0.1390	0.8609	0.0001
....	0.0482	0.9518	....	0.0485	0.9512	0.0003
....	0.0158	0.9842	....	0.0159	0.9831	0.0010
....	0.0050	0.9950	....	0.0051	0.9918	0.0031
....	0.0016	0.9984	....	0.0016	0.9885	0.0099
....	0.0005	0.9995	....	0.0005	0.9689	0.0306
....	0.0002	0.9998	....	0.0001	0.9090	0.0909
....	0.00005	0.99995	....	0.0000	0.7598	0.2402
....	....	1.00	....	....	0.5000	0.5000
....	....	....	....	....	0.2403	0.7597
....	....	....	....	....	0.0909	0.9091

* Reprinted from D. S. McKinney, "Calculation of Corrections to Conductivity Measurements for Dissolved Gases," *Proc. Am. Soc. Testing Materials*, **41**, 1290 (1941).

† From $K_1 = 1.72 \cdot 10^{-2}$; $K_2 = 6.24 \cdot 10^{-8}$. H. V. Tartar and H. H. Garretson, "The Thermodynamic Ionization Constants of Sulfurous Acid at 25°," *J. Am. Chem. Soc.*, **63**, 808 (1941). All other values from dissociation constants given in W. M. Latimer's *Oxidation Potentials*, Prentice-Hall, Inc., New York, 1938.

TABLE 15. AMERICAN IRON AND STEEL INSTITUTE (A.I.S.I.) TYPE NUMBERS FOR STAINLESS STEELS

Type Number	Carbon, %	Manganese, % Max.	Silicon, % Max.	Phosphorus, % Max.	Sulfur, % Max.	Chromium, %	Nickel, %	Other Elements, %
301	Over 0.08–0.20	2.00	1.00	0.04	0.04	16.00–18.00	6.00–8.00	
302	Over 0.08–0.20	2.00	1.00	0.04	0.04	17.00–19.00	8.00–10.00	
302B	Over 0.08–0.20	2.00	2.00–3.00	0.04	0.04	17.00–19.00	8.00–10.00	
303	0.20 max.	2.00	1.00	...	...	17.00–19.00	8.00–10.00	P or S or Se min. 0.07 Zr or Mo max. 0.60
304	0.08 max.	2.00	1.00	0.04	0.04	18.00–20.00	8.00–10.00	
308	0.08 max.	2.00	1.00	0.04	0.04	19.00–21.00	10.00–12.00	
309	0.20 max.	2.00	1.00	0.04	0.04	22.00–24.00	12.00–15.00	
310	0.25 max.	2.00	1.50	0.04	0.04	24.00–26.00	19.00–22.00	
316	0.10 max.	2.00	1.00	0.04	0.04	16.00–18.00	10.00–14.00	Mo 1.75–2.50
321	0.10 max.	2.00	1.00	0.04	0.04	17.00–19.00	8.00–11.00	Ti 4 × C min.
347	0.10 max.	2.00	1.00	0.04	0.04	17.00–19.00	9.00–12.00	Cb 8 × C min.
403	0.15 max.	1.00	1.00	0.04	0.04	11.50–13.00	...	Turbine Quality
405	0.08 max.	1.00	1.00	0.04	0.04	11.50–13.50	...	Al 0.10–0.30
406	0.15 max.	1.00	1.00	0.04	0.04	12.00–14.00	...	Al 3.50–4.50
410	0.15 max.	1.00	1.00	0.04	0.04	11.50–13.50	...	
414	0.15 max.	1.00	1.00	0.04	0.04	11.50–13.50	1.25–2.50	
416	0.15 max.	1.00	1.00	0.04	...	12.00–14.00	...	P or S or Se min. 0.07 Zr or Mo max. 0.60
420	Over 0.15	1.00	1.00	0.04	0.04	12.00–14.00	...	
430	0.12 max.	1.00	1.00	0.04	0.04	14.00–18.00	...	
430F	0.12 max.	1.00	1.00	...	...	14.00–18.00	...	P or S or Se min. 0.07 Zr or Mo max. 0.60
431	0.20 max.	1.00	1.00	0.04	0.04	15.00–17.00	1.25–2.50	
440A	0.60–0.75	1.00	1.00	0.04	0.04	16.00–18.00	...	
440B	0.75–0.95	1.00	1.00	0.04	0.04	16.00–18.00	...	
440C	0.95–1.20	1.00	1.00	0.04	0.04	16.00–18.00	...	
442	0.35 max.	1.00	1.00	0.04	0.04	18.00–23.00	...	
443	0.20 max.	1.00	1.00	0.04	0.04	18.00–23.00	...	0.90–1.25 Cu
446	0.35 max.	1.00	1.00	0.04	0.04	23.00–27.00	...	
501	Over 0.10	1.00	1.00	0.04	0.04	4.00–6.00	...	
502	0.10 max.	1.00	1.00	0.04	0.04	4.00–6.00	...	

Note: No specific composition limits within the above range should be placed on Types 301, 302, 302B, and 303 except that carbon may be specified to a 4-point range within the above limits.

Table 16. STANDARD DESIGNATIONS OF ALLOY COMPOSITIONS FOR HEAT AND CORROSION-RESISTANT CASTINGS

Alloy Casting Institute

Designation	% Nickel	% Chromium	% Carbon	% Other Elements
CA-14	1 max.	11-14	.14 max.	....
CA-40	1 max.	11-14	.20-.40	....
CB-30	2 max.	18-22	.30 max.	....
CC-35	3 max.	27-30	.35 max.	....
CD-10 M	3-6	27-30	.10 max.	Molybdenum 2.00 max
CE-30	8-11	27-30	.30 max.	....
CF-7	8-10	18-20	.07 max.	....
CF-10	8-10	18-20	.10 max.	....
CF-16	8-10	18-20	.16 max.	....
CF-20	8-10	18-20	.20 max.	....
CF-7 Se	8-10	18-20	.07 max.	Selenium 0.20-0.35
CF-7 C	8-10	18-20	.07 max.	Columbium 10 × Carbon
CF-7 M	8-10	18-20	.07 max.	Molybdenum 2.5-3.5
CF-10 M	8-10	18-20	.10 max.	Molybdenum 2.5-3.5
CF-16 M	8-10	18-20	.16 max.	Molybdenum 2.5-3.5
CF-7 MC	8-10	18-20	.07 max.	Molybdenum 2.5-3.5 Columbium 10 × Carbon
CG-7	10-12	20-22	.07 max.	....
CG-10	10-12	20-22	.10 max.	....
CG-16	10-12	20-22	.16 max.	....
CG-16 Se	10-12	20-22	.16 max.	Selenium 0.20-0.35
CG-7 C	10-12	20-22	.07 max.	Columbium 10 × Carbon
CG-7 M	10-12	20-22	.07 max.	Molybdenum 2.5-3.5
CG-10 M	10-12	20-22	.10 max.	Molybdenum 2.5-3.5
CG-16 M	10-12	20-22	.16 max.	Molybdenum 2.5-3.5
CG-7 MC	10-12	20-22	.07 max.	Molybdenum 2.5-3.5 Columbium 10 × Carbon
CH-10	10-12	23-26	.10 max.	....
CH-20	10-12	23-26	.20 max.	....
CH-10 C	10-12	23-26	.10 max.	Columbium 10 × Carbon
CH-10 M	10-12	23-26	.10 max.	Molybdenum 2.5-3.5
CH-20 M	10-12	23-26	.20 max.	Molybdenum 2.5-3.5
CH-10 MC	10-12	23-26	.10 max.	Molybdenum 2.5-3.5 Columbium 10 × Carbon
CK-25	19-21	23-26	.25 max.	....
CM-25	19-22	8-11	.25 max.	....
CN-25	23-26	18-22	.25 max.	....
CS-25	29-32	8-12	.25 max.	....
CT-25	34-37	13-17	.25 max.	....
HB	2 max.	18-22	....	....
HC	3 max.	27-30	....	....
HD	3-6	27-30	....	....
HE	8-11	27-30	....	....
HF	8-11	18-23	....	....
HH	10 13	23-27	....	....
HI	13-16	26-30	....	....
HK	19-21	23-26	....	....
HL	19-21	28-32	....	....
HN	23-26	18-22	....	....

TABLE 16. STANDARD DESIGNATIONS OF ALLOY COMPOSITIONS FOR HEAT AND CORROSION-RESISTANT CASTINGS — *Continued*

Designation	% Nickel	% Chromium	% Carbon	% Other Elements
HP	29–31	28–32	....	....
HS	29– 2	8–12	....	....
HT	34–37	13–17	....	....
HU	37–40	17–21	....	....
HW	59–62	10–14	....	....
HX	65–68	15–19	....	....

Designations with the initial letter "C" indicate alloys generally used to resist corrosive attack at temperatures less than 650° C (1200° F). Designations with the initial letter "H" indicate alloys generally used under conditions where the metal temperature is in excess of 650° C (1200° F).

All the above designations apply to type compositions, and no attempt is made to cover elements such as manganese, silicon, and, in the case of heat-resistant alloys, carbon.

Table 17. TEMPERATURE CONVERSION TABLE

Interpolation Factors

°C	°F	°C	°F
0.56	1	1.8	3.33	6	10.8
1.11	2	3.6	3.89	7	12.6
1.67	3	5.4	4.44	8	14.4
2.22	4	7.2	5.00	9	16.2
2.78	5	9.0	5.56	10	18.0

−459.4 to −60			−60 to 35			35 to 76		
°C	°F		°C		°F	°C		°F
−273	−459.4		−51.1	−60	−76	1.67	35	95.0
−268	−450		−45.6	−50	−58	2.22	36	96.8
−262	−440		−40.0	−40	−40	2.78	37	98.6
−257	−430		−34.4	−30	−22	3.33	38	100.4
−251	−420		−28.9	−20	−4	3.89	39	102.2
−246	−410		−23.3	−10	14	4.44	40	104.0
−240	−400		−17.8	0	32	5.00	41	105.8
−234	−390		−17.2	1	33.8	5.56	42	107.6
−229	−380		−16.7	2	35.6	6.11	43	109.4
−223	−370		−16.1	3	37.4	6.67	44	111.2
−218	−360		−15.6	4	39.2	7.22	45	113.0
−212	−350		−15.0	5	41.0	7.78	46	114.8
−207	−340		−14.4	6	42.8	8.33	47	116.6
−201	−330		−13.9	7	44.6	8.89	48	118.4
−196	−320		−13.3	8	46.4	9.44	49	120.2
−190	−310		−12.8	9	48.2	10.0	50	122.0
−184	−300		−12.2	10	50.0	10.6	51	123.8
−179	−290		−11.7	11	51.8	11.1	52	125.6
−173	−280		−11.1	12	53.6	11.7	53	127.4
−169	−273	−459.4	−10.6	13	55.4	12.2	54	129.2
−168	−270	−454	−10.0	14	57.2	12.8	55	131.0
−162	−260	−436	−9.44	15	59.0	13.3	56	132.8
−157	−250	−418	−8.89	16	60.8	13.9	57	134.6
−151	−240	−400	−8.33	17	62.6	14.4	58	136.4
−146	−230	−382	−7.78	18	64.4	15.0	59	138.2
−140	−220	−364	−7.22	19	66.2	15.6	60	140.0
−134	−210	−346	−6.67	20	68.0	16.1	61	141.8
−129	−200	−328	−6.11	21	69.8	16.7	62	143.6
−123	−190	−310	−5.56	22	71.6	17.2	63	145.4
−118	−180	−292	−5.00	23	73.4	17.8	64	147.2
−112	−170	−274	−4.44	24	75.2	18.3	65	149.0
−107	−160	−256	−3.89	25	77.0	18.9	66	150.8
−101	−150	−238	−3.33	26	78.8	19.4	67	152.6
−95.6	−140	−220	−2.78	27	80.6	20.0	68	154.4
−90.0	−130	−202	−2.22	28	82.4	20.6	69	156.2
−84.4	−120	−184	−1.67	29	84.2	21.1	70	158.0
−78.9	−110	−166	−1.11	30	86.0	21.7	71	159.8
−73.3	−100	−148	−0.56	31	87.8	22.2	72	161.6
−67.8	−90	−130	0	32	89.6	22.8	73	163.4
−62.2	−80	−112	0.56	33	91.4	23.3	74	165.2
−56.7	−70	−94	1.11	34	93.2	23.9	75	167.0
−51.1	−60	−76	1.67	35	95.0	24.4	76	168.8

MISCELLANEOUS INFORMATION

TABLE 17. TEMPERATURE CONVERSION TABLE — *Continued*

76 to 340			340 to 830			830 to 1320		
°C		°F	°C		°F	°C		°F
24.4	76	168.8	171	340	644	443	830	1526
25.0	77	170.6	177	350	662	449	840	1544
25.6	78	172.4	182	360	680	454	850	1562
26.1	79	174.2	188	370	698	460	860	1580
26.7	80	176.0	193	380	716	466	870	1598
27.2	81	177.8	199	390	734	471	880	1616
27.8	82	179.6	204	400	752	477	890	1634
28.3	83	181.4	210	410	770	482	900	1652
28.9	84	183.2	216	420	788	488	910	1670
29.4	85	185.0	221	430	806	493	920	1688
30.0	86	186.8	227	440	824	499	930	1706
30.6	87	188.6	232	450	842	504	940	1724
31.1	88	190.4	238	460	860	510	950	1742
31.7	89	192.2	243	470	878	516	960	1760
32.2	90	194.0	249	480	896	521	970	1778
32.8	91	195.8	254	490	914	527	980	1796
33.3	92	197.6	260	500	932	532	990	1814
33.9	93	199.4	266	510	950	538	1000	1832
34.4	94	201.2	271	520	968	543	1010	1850
35.0	95	203.0	277	530	986	549	1020	1868
35.6	96	204.8	282	540	1004	554	1030	1886
36.1	97	206.6	288	550	1022	560	1040	1904
36.7	98	208.4	293	560	1040	566	1050	1922
37.2	99	210.2	299	570	1058	571	1060	1940
37.8	100	212.0	304	580	1076	577	1070	1958
43	110	230	310	590	1094	582	1080	1976
49	120	248	316	600	1112	588	1090	1994
54	130	266	321	610	1130	593	1100	2012
60	140	284	327	620	1148	599	1110	2030
66	150	302	332	630	1166	604	1120	2048
71	160	320	338	640	1184	610	1130	2066
77	170	338	343	650	1202	616	1140	2084
82	180	356	349	660	1220	621	1150	2102
88	190	374	354	670	1238	627	1160	2120
93	200	392	360	680	1256	632	1170	2138
99	210	410	366	690	1274	638	1180	2156
100	212	413	371	700	1292	643	1190	2174
104	220	428	377	710	1310	649	1200	2192
110	230	446	382	720	1328	654	1210	2210
116	240	464	388	730	1346	660	1220	2228
121	250	482	393	740	1364	666	1230	2246
127	260	500	399	750	1382	671	1240	2264
132	270	518	404	760	1400	677	1250	2282
138	280	536	410	770	1418	682	1260	2300
143	290	554	416	780	1436	688	1270	2318
149	300	572	421	790	1454	693	1280	2336
154	310	590	427	800	1472	699	1290	2354
160	320	608	432	810	1490	704	1300	2372
166	330	626	438	820	1508	710	1310	2390
171	340	644	443	830	1526	716	1320	2408

TABLE 17. TEMPERATURE CONVERSION TABLE — *Continued*

1320 to 1810			1810 to 2300			2300 to 2790		
°C		°F	°C		°F	°C		°F
716	**1320**	2408	988	**1810**	3290	1260	**2300**	4172
721	**1330**	2426	993	**1820**	3308	1266	**2310**	4190
727	**1340**	2444	999	**1830**	3326	1271	**2320**	4208
732	**1350**	2462	1004	**1840**	3344	1277	**2330**	4226
738	**1360**	2480	1010	**1850**	3362	1282	**2340**	4244
743	**1370**	2498	1016	**1860**	3380	1288	**2350**	4262
749	**1380**	2516	1021	**1870**	3398	1293	**2360**	4280
754	**1390**	2534	1027	**1880**	3416	1299	**2370**	4298
760	**1400**	2552	1032	**1890**	3434	1304	**2380**	4316
766	**1410**	2570	1038	**1900**	3452	1310	**2390**	4334
771	**1420**	2588	1043	**1910**	3470	1316	**2400**	4352
777	**1430**	2606	1049	**1920**	3488	1321	**2410**	4370
782	**1440**	2624	1054	**1930**	3506	1327	**2420**	4388
788	**1450**	2642	1060	**1940**	3524	1332	**2430**	4406
793	**1460**	2660	1066	**1950**	3542	1338	**2440**	4424
799	**1470**	2678	1071	**1960**	3560	1343	**2450**	4442
804	**1480**	2696	1077	**1970**	3578	1349	**2460**	4460
810	**1490**	2714	1082	**1980**	3596	1354	**2470**	4478
816	**1500**	2732	1088	**1990**	3614	1360	**2480**	4496
821	**1510**	2750	1093	**2000**	3632	1366	**2490**	4514
827	**1520**	2768	1099	**2010**	3650	1371	**2500**	4532
832	**1530**	2786	1104	**2020**	3668	1377	**2510**	4550
838	**1540**	2804	1110	**2030**	3686	1382	**2520**	4568
843	**1550**	2822	1116	**2040**	3704	1388	**2530**	4586
849	**1560**	2840	1121	**2050**	3722	1393	**2540**	4604
854	**1570**	2858	1127	**2060**	3740	1399	**2550**	4622
860	**1580**	2876	1132	**2070**	3758	1404	**2560**	4640
866	**1590**	2894	1138	**2080**	3776	1410	**2570**	4658
871	**1600**	2912	1143	**2090**	3794	1416	**2580**	4676
877	**1610**	2930	1149	**2100**	3812	1421	**2590**	4694
882	**1620**	2948	1154	**2110**	3830	1427	**2600**	4712
888	**1630**	2966	1160	**2120**	3848	1432	**2610**	4730
893	**1640**	2984	1166	**2130**	3866	1438	**2620**	4748
899	**1650**	3002	1171	**2140**	3884	1443	**2630**	4766
904	**1660**	3020	1177	**2150**	3902	1449	**2640**	4784
910	**1670**	3038	1182	**2160**	3920	1454	**2650**	4802
916	**1680**	3056	1188	**2170**	3938	1460	**2660**	4820
921	**1690**	3074	1193	**2180**	3956	1466	**2670**	4838
927	**1700**	3092	1199	**2190**	3974	1471	**2680**	4856
932	**1710**	3110	1204	**2200**	3992	1477	**2690**	4874
938	**1720**	3128	1210	**2210**	4010	1482	**2700**	4892
943	**1730**	3146	1216	**2220**	4028	1488	**2710**	4910
949	**1740**	3164	1221	**2230**	4046	1493	**2720**	4928
954	**1750**	3182	1227	**2240**	4064	1499	**2730**	4946
960	**1760**	3200	1232	**2250**	4082	1504	**2740**	4964
966	**1770**	3218	1238	**2260**	4100	1510	**2750**	4982
971	**1780**	3236	1243	**2270**	4118	1516	**2760**	5000
977	**1790**	3254	1249	**2280**	4136	1521	**2770**	5018
982	**1800**	3272	1254	**2290**	4154	1527	**2780**	5036
988	**1810**	3290	1260	**2300**	4172	1532	**2790**	5054

TABLE 17. TEMPERATURE CONVERSION TABLE — *Continued*

2790 to 2860			2860 to 2930			2930 to 3000		
°C	°F		°C	°F		°C	°F	
1532	2790	5054	1571	2860	5180	1610	2930	5306
1538	2800	5072	1577	2870	5198	1616	2940	5324
1543	2810	5090	1582	2880	5216	1621	2950	5342
1549	2820	5108	1588	2890	5234	1627	2960	5360
1554	2830	5126	1593	2900	5252	1632	2970	5378
1560	2840	5144	1599	2910	5270	1638	2980	5396
1566	2850	5162	1604	2920	5288	1643	2990	5414
1571	2860	5180	1610	2930	5306	1649	3000	5432

TABLE 18. GENERAL CONVERSION FACTORS
(Prepared by the Editor)

Multiply	by	to *Obtain*
ampere-hours (abs.)	3600.0	coulomb (abs.)
ampere-hours (abs.)	0.03731	Faradays
Angstrom units	3.937×10^{-9}	inches
Angstrom units	1×10^{-10}	meters
Angstrom units	1×10^{-4}	microns
atmospheres	1.0133	bars
atmospheres	29.921	inches of mercury at 32° F
barrels, oil	42	gallons (U.S.)
bars	0.9869	atmospheres
Btu	252	calories, gram
Btu	1055	joules
Btu	0.2930	watt-hours
Btu /(hr) (sq ft) (°F /ft)	0.00413	gram cal /(sec) (sq cm) /(°C /cm)
calories, gram (mean)	4.131×10^{-2}	liter atmospheres
calories, gram	3.968×10^{-3}	Btu
calories, gram (mean)	4.186	joules (abs.)
calories, gram (mean)	0.0011628	watt-hours
gram-cal /(sec) (sq cm) (°C /cm)	242	Btu /(hr) (sq ft) (°F /ft)
centimeters	1×10^{8}	Angstrom units
centimeters	0.3937	inches
cubic centimeters	0.06102	cubic inches
cubic centimeters	9.9997×10^{-4}	liters
cubic centimeters	0.03381	ounces (U.S., fluid)
cubic feet (U.S.)	1728	cubic inches (U.S.)
cubic feet (U.S.)	7.481	gallons (U.S.)
cubic feet of water (60° F)	62.37	pounds
cubic inches (U.S.)	16.387	cubic centimeters
cubic inches (U.S.)	5.787×10^{-4}	cubic feet (U.S.)
cubic inches (U.S.)	0.5541	ounces (U.S., fluid)
days (mean solar)	1440	minutes
days (mean solar)	86400	seconds
equivalents per million	0.001*	normality
equivalents per million	50.0	parts per million as $CaCO_3$
equivalents per million	2.92*	grains per gallon (U.S.) as $CaCO_3$
ergs	2.389×10^{-8}	calories, gram (mean)
ergs	1×10^{-7}	joules
Faradays	26.80	ampere-hours (abs.)
feet	0.3048	meters
gallons (U.S.)	0.1337	cubic feet
gallons (U.S.)	0.003785	cubic meters
gallons (U.S.)	3.785	liters
grains	0.06480	grams
grains per gallon (U.S.)	17.12†	parts per million

* These values are exact when multiplied by the specific gravity of the solution.
† These values are exact when divided by the specific gravity of the solution.

TABLE 18. GENERAL CONVERSION FACTORS — *Continued*

Multiply	by	to Obtain
grains per gallon (U.S.) as $CaCO_3$	0.3424†	equivalents per million
grams	0.03527	ounces (avoirdupois)
grams	0.03215	ounces (troy)
grams per liter	1000†	parts per million
inches	2.540	centimeters
joules (abs.)	0.2389	calories, gram (mean)
joules (abs.)	1×10^7	ergs
joules	0.009869	liter atmospheres
kilograms	2.205	pounds (avoirdupois)
kilograms per sq cm	14.223	pounds per sq. in.
kilometers	0.6214	miles (U.S.)
liter atmospheres	24.21	calories, gram (mean)
liter atmospheres	101.33	joules (abs.)
liters	1.0567	quarts (U.S., liquid)
meters	3.281	feet
micrograms	1×10^{-6}	grams
microhms	1×10^{-6}	ohms
microns	1×10^{-4}	centimeters
microns	0.03937	mils
miles (U.S.)	1.6094	kilometers
milliliters (ml)	1.000027	cubic centimeter
milliliters dissolved oxygen (N. T. P.) per liter	1.429†	parts per million
millimeters	39.37	mils
mils	0.002540	centimeters
normality	1000†	equivalents per million
ounces (avoirdupois)	437.5	grains
ounces (avoirdupois)	28.350	grams
ounces (avoirdupois)	0.9115	ounces (troy)
ounces (troy)	31.104	grams
ounces (U.S., fluid)	29.57	cubic centimeters
ounces (U.S., fluid)	0.03125	quarts (U.S., liquid)
parts per million	0.0584*	grains per gallon (U.S.)
parts per million	0.001*	grams per liter
parts per million as $CaCO_3$	0.02	equivalents per million
parts per million of dissolved oxygen	0.70*	milliliters dissolved oxygen (N. T. P.) per liter
pounds (avoirdupois)	0.4536	kilograms
pounds (troy)	0.3732	kilograms
pounds per square inch	0.07031	kilograms per sq cm

* These values are exact when multiplied by the specific gravity of the solution.
† These values are exact when divided by the specific gravity of the solution.

GENERAL TABLES

TABLE 18. GENERAL CONVERSION FACTORS — *Concluded*

Multiply	*by*	*to Obtain*
quarts (U.S., liquid)	0.9463	liters
square centimeters	0.1550	square inches
square decimeters	100	square centimeters
square feet (U.S.)	0.09290	square meters
square inches (U.S.)	6.452	square centimeters
square kilometers	0.3861	square miles (U.S.)
square meters	10.764	square feet (U.S.)
watt-hours	860.01	calories, (gram mean)
watt-hours	3600	joules
weeks	168	hours
weeks	10080	minutes
weeks	604800	seconds
years (mean solar)	8765.8	hours (mean solar)
years (mean solar)	3.156×10^7	seconds (mean solar)

TABLE 19. CONVERSION FACTORS FOR CORROSION UNITS

(Prepared by the Editor)

Multiply	by	to Obtain
in. penetration per yr (ipy)	696 × density	mg per sq dm per day (mdd)
cm penetration per yr (cmpy)	274 × density	"
oz per sq ft per day	3050	"
grams per sq meter per yr	0.0274	"
grams per sq meter per day	10	"
grams per sq meter per hour	240	"
mg per sq dm per day (mdd)	0.00144/density	in. penetration per yr (ipy)
cm penetration per yr (cmpy)	0.394	"
oz per sq ft per day	4.39/density	"
grams per sq meter per yr	0.0000394/density	"
grams per sq meter per day	0.0144/density	"
grams per sq meter per hour	0.346/density	"

Metal	Density (grams per cc)	$\dfrac{0.00144}{\text{Density}}$	696 × Density
Aluminum	2.72	0.000529	1890
Aluminum-magnesium-silicon (98, 0.5, 1.0)	2.69	0.000535	1870
Aluminum-manganese (99, 1)	2.73	0.000528	1901
Aluminum-copper (duralumin) (Cu 4, Mn 0.5, Mg 0.5, Al remainder)	2.79	0.000516	1940
Brass (Admiralty)	8.54	0.000168	5950
Brass (red)	8.75	0.000164	6100
Brass (yellow)	8.47	0.000170	5880
Bronze (10 % tin)	8.77	0.000164	6100
Bronze (18 % tin)	8.80	0.000163	6135
Bronze (phosphorus, 5 % tin)	8.86	0.000162	6170
Bronze (silicon)	8.54	0.000168	5950
Bronze (manganese) (Cu 66.5, Zn 19, Al 6, Fe 4.5, Mn 4)	7.89	0.000182	5500
Cadmium	8.65	0.000167	6020
Columbium	8.4	0.000171	5850
Copper	8.92	0.000161	6210
Copper-nickel (70, 30)	8.95	0.000161	6210
Copper-nickel-zinc (75, 20, 5)	8.86	0.000162	6170
Copper-nickel-zinc (65, 18, 17)	8.75	0.000164	6100
Copper-silicon-manganese (96, 3, 1)	8.53	0.000168	5950
Copper-nickel-tin (70, 29, 1)	8.87	0.000162	6170
Iron	7.87	0.000183	5480
Iron-chromium (86–88, 12–14, 0.3 C)	7.7	0.000187	5360
Iron-chromium (82–84, 16–18, .12 C)	7.7	0.000187	5360
Iron-chromium (73–77, 23–27, .35 C)	7.6	0.000189	5290
Iron-chromium-nickel [18–8] (75–79, 17–19, 8–10, .1 C)	7.9	0.000182	5500
Iron-silicon (Duriron) (84, 14.5, 0.9 C, 0.4 Mn)	7.0	0.000205	4870

CONVERSION FACTORS — *Continued*

Metal	Density (grams per cc)	$\dfrac{0.00144}{\text{Density}}$	$696 \times$ Density
Lead (chemical)	11.35	0.000127	7900
Lead (antimony) (99, 1)	11.30	0.000127	7870
Magnesium	1.74	0.000826	1210
Nickel	8.89	0.000162	6180
Nickel-copper (Monel) (70, 30)	8.84	0.000163	6140
Nickel-chromium-iron (Inconel) (80, 13, 7)	8.51	0.000169	5920
Nickel-molybdenum-iron (Hastelloy A) (60, 20, 20)	8.80	0.000163	6120
Silver	10.50	0.000137	7300
Tantalum	16.6	0.0000868	11550
Tin	7.29	0.000197	5080
Zinc	7.14	0.000201	4970

INDEX

Key: **T** refers to tables. **F** refers to figures. **FN** refers to footnotes.

Acetic acid, resistance of, chromium-nickel iron alloys, 154*T*, 156*F*
  copper, 65*F*
  copper-nickel alloys, 88-9
  copper-silicon alloys, 108*T*
  copper-tin alloys, 99
  molybdenum, 253*T*
  nickel, 260*T*
  nickel-chromium alloys, 289*T*
  nickel-copper alloys, 274-5
  nickel-molybdenum alloys, 298*T*
  tin, 325
Acetic acid group chemicals materials resistant to, **757-9*T***
Acids (refer also to specific acid)
  inhibitors for, 914-5
  mechanism of attack by, 8-9
  non-oxidizing, theory of attack by, 8
  oxidizing, theory of attack by, 8-9
  resistance of, chromium-iron alloys, 145-8, 148*T*
    columbium, 60
    copper, 64-6
    copper-beryllium alloys, 112
    copper-nickel alloys, 87-9
    copper-silicon alloys, 107
    copper-tin alloys, 97-9
    copper-zinc alloys, 74-5
    indium alloys, 120*T*
    lead alloys, 211-2
    magnesium alloys, 225
    molybdenum, 253*T*
    nickel, 265-9
    nickel-chromium alloys, 283-7
    nickel-copper alloys, 270-4
    nickel-iron alloys, 194-6
    nickel-molybdenum alloys, 295-6
    nickel-molybdenum-iron alloys, 297*T*
    silver, 315, 316*T*
    steel, 132-5
    tantalum, 321, 322*T*
    tin, 324-5, 324*T*
    zinc, 336-8
Acids and acid-forming gases, materials resistant to, **787-9*T***
Activities, ion, 1148-9*T*
Adhesives, for wood, **376-8*T***

Admiralty metal (*see* Copper-zinc alloys)
Aeration, effect in total immersion tests, 962-3
Alclad, 53-4
Alkalies (refer also to specific alkalies)
  inhibitors for, 915
  materials resistant to, **793-8*T***
  resistance of, chromium-iron alloys, 148*FN*
    copper, 64-6
    copper-nickel alloys, 89-90
    copper-silicon alloys, 108
    copper-tin alloys, 99
    copper-zinc alloys, 75-6
    indium alloys, 120*T*
    lead alloys, 213
    magnesium alloys, 225-6
    nickel, 262
    nickel-chromium alloys, 290-1
    nickel-copper alloys, 276
    nickel-iron alloys, 198
    nickel-molybdenum alloys, 296
    silver, 315-8
    steel, 134-5, 636-8
    tantalum, 321
    tin, 325-6
    zinc, 336-8
Alumel (*see* Nickel alloys)
Aluminum (*see* Aluminum alloys)
Aluminum alloys, **39-56**
  anodic treatment of, **857-61**
  cathodic corrosion of, 53
  clad, 53-4
  code numbers, 39-40*T*
  compositions of wrought and cast, 39-41*T*
  corrosion fatigue of, 48
  effect of, cold work, 48
    heat treatment, 53
    pH, 42
    precipitation in alloys, 48-9
    scratches, 7
    stress, 48, 570
    structure of alloys, 48-9
    temperature, 41-2
    velocity, 47
  effect on oxidation of magnesium, 672*F*
  galvanic corrosion, 490*T*
  intergranular corrosion of, 49, 54-5

1164   INDEX

Aluminum alloys, mechanical properties of, 39–40T
  oxidation, internal, 617
  passivity of, 32
  protection of, 49–51
  resistance to, acids, 43–5
    organic, 45
    alkalies, 46
    amines, 46
    ammonia, 46, 49
    atmospheric corrosion, 51–6
      galvanic, 53
      indoor, 52–3
      outdoor, 51–2
      protection against, 55–6
    fluorinated hydrocarbons, 49
    gases, 49
      dissolved, 41T, 42T
      high-temperature, 617
    hydrogen chloride, 49
    hydrogen fluoride, 49
    hydrogen sulfide, 49
      high-temperature, 617
    lead, molten, 617
    mercury, 618
    mercury salts, 47FN
    metals, molten, 617–8
    methyl alcohol, 49
    organic compounds, 49
    oxygen, 41
    phenol, 49
    salt solutions, 46–7, 47T
    salts, molten, 617–8
    soil corrosion, 50–1
    steam, 617
    steam condensate, 43
    sulfur dioxide, 49
      high-temperature, 617
    tin, molten, 617
    water, distilled, 42
      fresh, 42
      mine, 43
      sea, 43, 406, 410–1T
    zinc, molten, 617
Aluminum brass (see Copper-zinc alloys)
Aluminum-iron alloys, high-temperature corrosion, **638–40**
Ammeter, zero resistance, 1039F
Ammonia solubility in water, 1146T
Ammonia at high temperature, resistance of, cobalt alloys, 619
  copper alloys, 629–30
  heat-resistant alloys, 657–8, 681, 735T
Ammonia group chemicals, materials resistant to, 799T

Ammonium hyroxide, resistance of, aluminum alloys, 46
  copper-nickel alloys, 89–90
  copper-tin alloys, 99
  copper-zinc alloys, 76
  steel, 131–2, 135
Anodes, corrosion-resistant, **606–9**
  applications, 606–8
  mechanism of, 608–9
  palladium, 306–7
  platinum, 303–4
Anodic control, 486
Anodic corrosion of gold, 117
Anodic treatment, **857–62**
Anti-fouling measures, **441–6** (see also Fouling)
  behavior of alloys, 427–9T
  biocidal agents, 445–6
  copper alloys, 401–2
  electric currents for, 445
  fresh water anchorage, 445
  paint (see Paint, anti-fouling)
  tests, racks for, 1062F
  toxics for, 445
Arrhenius equation, 16
Arsenic, effect in copper, 67
  on dezincification, 551
Asbestos cement, soil resistance, 454
Asbestos pipe, soil resistance, 454
A.S.T.M. galvanic couple test, 1005
A.S.T.M. life test, high-temperature, 997–8
Atmosphere, composition of, 1117T
  furnace, alloys for, **730–2T**
  types of, 1043
Atmospheric corrosion, resistance of, aluminum alloys, 51–6
    galvanic, 53
    indoor, 52–3
    outdoor, 51–2
    protection, 55–6
  cast iron, 124–5
  chromium-iron alloys, 149–50
  chromium-manganese-iron alloys, 192–3
  chromium-nickel-iron alloys, 164–5
  coatings, bituminous, 898–9
    metal, 803–57
  cobalt alloys, 59
  columbium, 61
  copper, 67–8
  copper-nickel alloys, 95
  copper-silicon alloys, 109 1)
  copper-tin alloys, 102–3
    corrosion products of, 103
  copper-zinc alloys, 84–5
  indium alloys, 119–20

INDEX 1165

Atmospheric corrosion, resistance of, lead
    alloys, 215–6
  magnesium alloys, 243–52
    corrosion fatigue of, 249
    effect of humidity, 245
    effect of pollution, 245–6
    effect of stress, 247–9
    galvanic, 246—7
    indoor, 243
    marine atmosphere, 246
    metallurgical factors, 250–1
    outdoor, 244
    protection, 251–2
  nickel, 265–6
  nickel-chromium alloys, 293
  nickel-copper alloys, 280–1
  nickel-iron alloys, 200
  palladium, 308–9
  steel, 124–5
    effect of copper, 121–2
    effect of phosphorus, 122
    factors affecting, 120
    low alloys, 122–4
  tantalum, 323
  tin, 327–9
  zinc, 342–6
    effect of stress, 346
    galvanic, 346
    indoor, 342–3
    outdoor, 344–6
    protection, 347
Atmospheric tests (see Tests [field], atmospheric)

Bacteria (see Micro-organisms)
Bases (see Alkalies)
Bearing corrosion, **559–68**
  effect of, catalysts, 562–3
    oxidizing characteristics of lubricating oil, 563
    temperature, 561–2
  mechanism of, 564–5
  prevention, 565–7
Bearings, indium-plated, 119
  surface destruction by fretting corrosion, 594
Beryllium, **56–7**
Beryllium-copper (see Copper-beryllium alloys)
Boiler-corrosion (see Water, boiler)
Boiler embrittlement, 531
Boron, **378–9**
Brass (see Copper-zinc alloys)
Brines, inhibitors for, 498–9, 914
Brinelling, false, 594
Bromine (see Gases, and Halogen gases, high-temperature)

Bronze (see Copper-tin alloys)
Buna N (see Rubber)
Buna S (see Rubber)
Butyl rubber (see Rubber)

Cable sheathing, **609–13**
Cadmium (see Coatings, cadmium)
Calcium chloride, resistance of, cast iron, 132T
  nickel, 263T
  nickel-copper alloys, 277T
  nickel-iron alloys, 196T
  steel, 132T
Calorizing, 638–40
Cans, tin, 831–6
Carbon, **348–52**
  characteristics, 348–9
  effect in, chromium-manganese-iron alloys, 190
    chromium-nickel-iron alloys, 160
    nickel, 679
    steel, 140
  electrical conductivity, 351
  forms, 349
  high-temperature uses, **723**
  impregnated, 351
  physical properties, 350T
  porosity, 349
  purity, 349
  thermal conductivity, 351
  thermal shock, resistance to, 351
Carbon dioxide, dissolved, effect on, copper, 62–4
  copper-nickel alloys, 86T
  magnesium alloys, 218
  nickel, 256T
  nickel-copper alloys, 269T
  nickel-iron alloys, 199T
  steel, 127
  high-temperature resistance of, aluminum alloys, 617
    magnesium alloys, 673
    nickel alloys, 679
    silver, 719
  solubility, in water, 1146T
    in sea water, 1116T
Carbon tetrachloride (see Chlorinated solvents)
Carbonic acid (see Carbon dioxide, dissolved)
Carburizing containers, 735T
Cast iron, chromium, high-temperature corrosion, 662–4
  graphitic corrosion (graphitization), 139
  nickel cast iron, 196–8
    resistance to, sulfuric acid, 196–8
    various media, 197–8T

1166 INDEX

Cast iron, resistance to, atmospheric corrosion, 124–5
  water, sea, 392, 384–7T
  silicon (*see* Silicon-iron alloys)
  versus steel and wrought iron, 139
Cathodic control, 486
Cathodic corrosion of aluminum alloys, 53
Cathodic protection, **923–50**
  ampere-hour effects, 932–3
  anode arrangement, 935F, 938–40F, 942F
  applications, **935–50**
    in fresh water, 500–1
    tanks, 949–50
    underground pipes and cables, 936–49
    underground structures, 949
  criteria for protection, 926–8
  current sources for testing, 930
  definition, 923 (*see also* Glossary)
  effect of soil shrinkage, 934
  electro-osmotic effects, 931–2
  high-resistance couplings, effect of, 934
  local action current, 923
  mechanical effects, 934
  mechanism of, 923–6
  minimum currents, determination of, 928–30
  radial currents, 934
  reference cells, 930–1
  shielding effects, 933–4
  stray-current hazards, 934
Cavitation, **597–601** (*see also* Tests, laboratory, cavitation damage)
  damage, occurrence of, 500
  definition, 597 (*see also* Glossary)
  effect of corrosion, 597–8
    temperature, 598
  metals resistant to, 600, 995–6T
  remedies for damage by, 600–1
Cells, complex, 493–4
  Denison, 1038–9
  oxidation, tarnish, 17F
  passive-active, 165 (*see also* Glossary)
Cellulose ethers, 897
Cellulose nitrate, 897
Cement mortar, soil resistance, 462
Cementite, as cathode, 8
Ceramics (*see* Stoneware)
Chemicals, materials resistant to, **747–99**T
Chlorinated solvents, resistance of, copper, 66
  copper-nickel alloys, 92–3T
  copper-silicon alloys, 109
  copper-zinc alloys, 76
  nickel, 264, 265T
  stainless steels, 157T
  steel, 141

Chlorinated solvents, resistance of, zinc. 339
Chlorine (*see* Gases, and Halogen gases, high-temperature)
Chlorinity, definition of, 1111
Chromates, as inhibitors, corrosion fatigue, 589
  in paints, 430, 888–90
  limitations of, 500
  theory of action, 9, 31
  uses, 498–500, 913–15
  effect on corrosion of, copper-nickel alloys, 92T
  iron in CaCl$_2$, 132T
  zinc, 341T
  potential of iron in, 981F
Chromel (*see* Nickel-chromium alloys)
Chromic acid, resistance of, aluminum alloys, 45
  copper-nickel alloys, 87–8
  magnesium alloys, 225
  materials resistant to, 788T
  steel, 133
Chromium, **824–9**
  coatings, 817
  effect in steel, 140
  high-temperature corrosion, **825–8**
Chromium-iron alloys, **143–150**
  compositions, heat treatment and properties, 146–7T
  creep strength, 643T
  effects of, alloy additions, 148–9
    silicon, 664–7
  fabrication and heat treatment, 144
  high-temperature corrosion, **640–64**
  high-temperature strength, 642–3
  oxidation, 646
  passivity of, 24–5
  physical properties, 144
  pitting of, 165
  potentials of in NaCl, 27F
  resistance to, acids, 145–8, 148T
    air, high-temperature, 644–6, 662–4
    alkalies, 148FN
    ammonia, high-temperature, 657–8
    atmospheric corrosion, 149–50
    gases, exhaust, 659–61
      flue, 661–2
      hot carbonaceous, 655–7
      hot sulfur compounds, 647–54
    hydrogen, high-temperature, 657–8
    metals, molten, 658–9
    nitric acid, 144T
    oxygen, high-temperature, 646
    petroleum products, high-temperature, 654–5

Chromium-iron alloys, resistance to, salts, molten, 658–9
  stress corrosion cracking, 178–81
  water spray, 28F
  uses, cutlery, 145
  tool steels, 145
Chromium-aluminum-iron alloys, high-temperature corrosion, 667–70
Chromium-manganese-iron alloys, **188–93**
  compositions, 188
  effect of, carbon content, 190
    nitrogen content, 190
  general corrosion properties, 188
  high-temperature corrosion, **662**
  intergranular corrosion, 191
    effect of columbium and titanium, 191–2
  mechanical properties, 189T
  resistance to, aqueous media, 189–91
    atmospheric corrosion, 192–3
    nitric acid, 189T, 190T, 191T
Chromium-manganese-nickel-iron alloys (*see* Chromium-manganese-iron alloys)
Chromium-nickel-iron alloys, including 18-8, **150–65** (*see also* Nickel-chromium, and Nickel-chromium [-iron] alloys, high-temperature)
  carburizing tests, 657T
  compositions, 150–2, 1150–2T
  creep strength, 643T
  crevice (contact) corrosion, 155–9
  design strength, high-temperature, 643F
  effect of, alloying elements, 160T
    carbon content, 160
    cold work, 159–60
    manganese content, 160, 662
    pH, 152
    silicon content, 160, 662
  galvanic corrosion of, 159
  high-temperature corrosion, **640–64**, 683–99
  high-temperature strength, 642–3
  intergranular corrosion, 161–4
    effect of columbium and titanium in preventing, 163–4
  mechanical properties, 151T
  oxidation, 646
  oxide layer, characteristics of, 646
  passivation of, 158
  pitting of, 165–73
  properties, high-temperature, 640
  resistance to, air, high-temperature, 644–6
    ammonia, high-temperature, 657–8
    atmospheric corrosion, 164–5
    carbon dioxide, high-temperature, 646, 648T

Chromium-nickel-iron alloys, resistance to, gases, 155, 157T
    exhaust, 659–61
    flue, 661–2, 691–5
    hot carbonaceous, 655–7
    hot sulfur compounds, 647–54
    hydrogen, high-temperature, 657–8
    metals, molten, 658–9
    non-aqueous media, 157T
    organic compounds, 157T
    oxygen, high temperature, 646
    petroleum products, high temperature, 654–5
    salts, molten, 658–9
    steam, 511–20
    various media, 154T
  stress corrosion cracking of, 174–82, 571
  types and designations, 1150–2T
  16-2 alloy, **182–8**
    general corrosion properties, 183–5
    nitrogen content, 188
    resistance to, nitric acid, 183T
      salt spray, 184, 186–7T
      steam, 185
      water, fresh, 184
      sea 184–5, 184T
    specifications, 183T
    tempering, 185
Chromium-silicon-iron alloys, high-temperature corrosion, **664–7**
Chromizing (*see* Coatings, Cr, diffusion)
Citric acid group chemicals, materials resistant to, **760–61**T
Clad steel, 824 (*see also* Coatings)
Clad aluminum (*see* Aluminum and Aluminum alloys)
Cleaning corrosion specimens, 1077–83
Coatings, aluminum, high-temperature corrosion (*see* Aluminum-iron alloys)
  anodic, for aluminum alloys, 55, 857–61
    for magnesium alloys, 862
    for zinc, 862
  baking finishes, definition, 883
  bituminous, **898–902**
    resistance to, atmospheric corrosion, 898–9
      soil corrosion, 463–5, 899–901
      water corrosion, 899
    uses, roofing, 899
      tanks, 901–2
  cadmium, **837–45**
    resistance to, atmospheric corrosion, 838–41
      gases, 841
      humidity tests, 844
      liquids, 841–3

Coatings, cadmium, resistance to, salt spray, 843–4
  specifications, 845
 cathodic, corrosion at pores, 817–8
 concrete (cement), for fresh water, 505
 chromate, on magnesium alloys, 864–7
  on zinc, 862–4
 chromium, **817–29**
  application of, 818
  corrosion resistance, 824–5
  diffusion coatings, 829
  effect of, basis metal, 819–20
   copper layer, 821
   thickness, 821–2
  high-temperature corrosion, 825–8
  protective value, 818–9
 copper, effect on nickel and chromium plate, 821
 decorative, protective value, 818–9
  specifications, 822–4
 enamel, vitreous, **872–8**
 grease, soil corrosion, 463
  slushing compounds, 916–23
 inorganic, **857–78**
 lead, **845–57**
  application of, 845–6
  effect of cold rolling, 855–7
  humidity tests, 853–4
  oil and wax coatings for, 857
  on copper, 857
  porosity, 846
  resistance to, atmospheric corrosion, 846–9
   salt spray, 854–5
   soil corrosion, 849–52
  specifications, 857
 metal spray, 56, 824
 metallic, **803–57**
  soil corrosion, 460–2
 nickel, **817–29**
  alloy, 824
  application of, 818
  cladding on steel, 824
  effect of, basis metal, 819–20
   copper layer, 821
  protective value, 818–9
  sprayed, 824
 organic, **878–905** (*see also* Paints, Lacquers, Enamels)
  adhesion and cohesion, 882
  applicability and service life, 883–4
  film continuity, 879
  for aluminum alloys, 55
  permeability, 879–81
  resistance to sea water, 430–33
 oxide, 871 (*see also* Coatings, anodic)

Coatings, phosphate, **867–71**
  as base for paints, 868–9
 rubber, **902–5**
  butadiene derivatives, 904
  butyl, 904
  chlorinated, 903
  isomerized, 903
  natural, 903
  polysulfide, 905
 temporary (slushing compounds), **916–23**
  application of, 920–2
  preparation of surface, 919–20
  testing, 921–3
  theory of additives, 918
  types, 917–9
  uses, 916–7
 tin, **829–36**
  application, 829–31
  porosity, 831
  uses, cans, 831–6
 zinc, **803–16**
  protection for steel in atmosphere, 808–11T
  resistance to, aqueous solutions, 813–14
   atmospheric corrosion, 803–13
   soil corrosion, 460–2, 814–16
Cobalt alloys, **57–9**
 compositions, 57T
 effect of velocity, 59
 erosion resistance, 58–9
 resistance to, air, high-temperature, 618–9
  ammonia, high-temperature, 619
  aqueous solutions, 58T
  atmospheric corrosion, 59
  hydrogen, high-temperature, 619
  salts, molten, 619–20
  steam, 619
 uses in human body, 59
 valves, 620
Cold work, effect on corrosion of, aluminum alloys, 48
 chromium-nickel-iron alloys, 159–60, 577F
 copper, 67
 magnesium alloys, 239–41
 nickel, 577F
 nickel-copper alloys, 577F
 steel, 138
Colmonoy (*see* Nickel-chromium alloys)
Columbium, **61–2**
 effect on intergranular corrosion, 163–4, 191–2
 resistance to, air, high-temperature, 620–1 620T
  acids, 60
  atmospheric corrosion, 61

INDEX

Columbium, resistance to, gases, high-temperature, 621T
    various media, 60T
Concentration cell corrosion, in condensers, [549–51
Condenser-tube alloys, physical and mechanical properties, **556–7**T
    specifications, 546T
Condenser-tube corrosion, **545–59** (*see also* Copper-zinc alloys)
    cold wall effect, 554–5
    concentration cell corrosion, 549–51
    dezincification, 551
    effect of, stress, 555
        temperature, 551–5
        velocity, 545–9
    galvanic effects, 555
    hot-wall effect, 551–4
    impingement attack, 545–9
        designs to avoid, 547F
Condensers, materials for, **555–9**
Constantan (*see* Copper-nickel alloys)
Constants, miscellaneous, 1132T
    physical, of elements, 1127–31T
Contact corrosion (*see* Crevice corrosion)
Conversion factors, corrosion units, 1160–61T
    general, 1157–60T
    temperature, 1153–6T
Copper, **61–9**
    composition, 62T
        effect on tarnishing, 14FN
    corrosion fatigue, 67, 586T
    effect of, arsenic content, 67
        cold work, 67
    effect on, magnesium alloys, 235–9
        steel, 140
            atmospheric corrosion, 121–2
    embrittlement by hydrogen, 629
    galvanic corrosion, 66–7
    impingement attack, 66
    lead coatings for, 857
    oxidation, 621–4
        cyclic, and reduction, 630 [621–2
        dissociation pressures for copper oxides,
        effect of, crystal habit, 623
            crystal orientation, 34
        rate, 623–4
            effect of moisture, 15
        rate constant, 18
    patina, 67
    potential behavior, 65
        in hydrochloric acid, 66F
    properties, 62T
    red stain, 630
    resistance to, acids, inorganic, 64–6
        alkalies, 64–6

Copper, resistance to, ammonia, high-temperature, 629
    atmospheric corrosion, 67–8
    carbon dioxide, 63–4
        dissolved, 62–4
    hydrogen, high-temperature, 629
    lubricants and oils, 567
    organic compounds, 66
    oxygen, 62–4
        dissolved, 62–3, 65F
        high-temperature, 623–4
    refrigerants, 66
    salts, 64–6
    soils, 67
    steam, 520, **627**
        impingement attack by, 83–4
    steam condensate, 64
    sulfur, 66
    sulfur compounds, 66, 567
        high-temperature, 628–9
        in lubricants, 567
    water, fresh, 62–4
        sea, 64, 394T
    tarnishing, effect of, composition, 14FN
        surface, 15
Copper alloys, effect of stress, 571
    oxidation, 624–7
    protection, 627
    resistance to, ammonia, high-temperature, 629–30
        gases, hot carbon compounds, 628
        hot halogens, 630
        hot sulfur compounds, 628
        hydrogen, high-temperature, 629
        phosphorus, molten, 630
        soil corrosion, 452–3, 457T
        steam, 520
        water, sea 393–402, 405T
Copper-beryllium alloys, **110–2**
    composition, 110
    impingement attack, 111
    intergranular attack, 110
    physical constants, 110, 111T
    potential in sea water, 112
    resistance to, acids, 112
        water, sea, 111
    tensile properties, 111T
Copper-nickel alloys, **85–95** (*see also* Nickel-copper alloys)
    compositions, 85
    corrosion fatigue, 94
    effect of stress, 93–4
        velocity, 93
    galvanic corrosion, 93
    impingement attack, effect of alloying additions, 94

Copper-nickel alloys, metallurgical factors, 94
  oxidation rate, 676F
  passivity of, 30
  resistance to, acids, 87-9
    organic, 88-9
  alkalies, 89-90
  atmospheric corrosion, 95
  black (paper) liquor, 90
  chlorinated solvents, 92-3T
  chromates, 87-8, 91-2
  cupric salts, 87, 91
  ethyl acetate, 93
  ferric salts, 87, 91
  gases, 94-5
  hydrogen sulfide, 85
  mercaptides, 90
  metals, molten, 95
  organic acids, 88-9, 89T
  organic compounds, 92-3
  salts, 90-2, 91T
  steam, 95
  steam condensate, 86T
  sulfides, 90
  water, boiler, 86
    fresh, 85-6
    sea, 86-7, 399T
  uses, condensers, 558-9
Copper-silicon alloys, **104-10**
  compositions, 105
  effect of silicon content, 106T
  physical constants, 104T
  resistance to, acids, 107-8
    acetic, 108T
  alkalies, 108
  atmospheric corrosion, 109-10
  gases, 109
  organic compounds, 109
  salts, 108-9
  steam, 109
  water, fresh, 105
    mine, 106
    sea, 105-6, 395T
  stress corrosion cracking, 110
  tensile strength, 105T
Copper-strip test, 1037
Copper-tin alloys, **96-103**
  compositions, 96
  corrosion fatigue, 101
  corrosion products, periodic formation of, 103
  effect of stress, 101
    velocity, 100
  galvanic corrosion, 100-1
  resistance to, acids, 97-9
    alkalies, 99
    atmospheric corrosion, 102-3

Copper-tin alloys, resistance to, atmospheric corrosion, corrosion products, 103
  black (paper) liquor, 99
  chromates, 98
  fatty acids, 99
  gases, 101-2
  organic compounds, 101
  salts, 99-100
  steam, 102
  sulfite process liquor, 98
  water, fresh, 96-7
    sea, 96-7, 398T, 400T
  uses, 96
Copper-zinc alloys, **69-85**
  composition, 69T
  crevice corrosion, 82
  dezincification, 69-70, 551
    effect of alloying, 70
    in sea water, 71-2, 393
  effect of, mineral scales, 71
    stress, 77-81, 555, 570-1
    velocity, 76-7, 545
  galvanic corrosion, 77
  impingement attack, 76-7, 545
    by steam, 83
  metallurgical factors, 81-2
  patina, 84
  protection, 82
  resistance to, acids, 74-5
    organic, 75
  alkalies, 75-6
  atmospheric corrosion, 84-5
  gases, 82-4
    dissolved, 70-1
  organic compounds, 76
  salts, 74
  soil, 82
  steam, 83-4
  sulfur dioxide, 82-3
  water, boiler, 73-4
    fresh, 71
    mine, 74
    sea, 71-3, 393-7
      pitting in, 73, 393-7
      soft, 73-4
  season cracking, 77
  uses, condensers, 556-8
Corronizing (*see* Coatings, nickel alloy, 813, 824)
Corrosion, definition, 3 (*see also* Glossary)
  mechanism of, **3-11**
    high-temperature, 11-20
  reactions, free energy of, 1142-3T
  types of, illustrations, 1103-10
Corrosion fatigue, **578-91** (*see also* Tests, laboratory, corrosion fatigue)

Corrosion fatigue, effect of crystal orientation, 35
  evidence of, 581–2
  illustration, 1106F
  inhibitors to prevent, 589–90
  in steam, 588
  limits, 582–7
    definition, 992
    of non-ferrous materials, 586T
    of steel in waters, 584T, 585T
  nature of, **578–81**
  notch effect from, 587–8
  of aluminum alloys, 48, 586T
  of copper, 67, 586T
  of copper-nickel alloys, 94, 586T
  of copper-tin alloys, 107
  of magnesium alloys, 234–5
    in atmosphere, 249
  of nickel, 586T
  of nickel-copper alloys, 586T
  of stainless steels, 584T
  of steel, 583–90
  of zinc, 339
  prevention of, 588–90
Corrosion fissuring, of copper-nickel alloys, 95
Corrosion potential, definition, 482, 484T, 1040
Corrosion reactions, spontaneity (free energy) of, 1142–3T
  types of control, 486
Corrosion-resistant anodes (*see* Anodes)
Corrosion units (*see also* Preface)
  conversion factors, 1160–1T
Corrosiron (*see* Silicon-iron alloys)
Cracking of oxide films, 13
  (*see also* Stress corrosion, and Tests, laboratory, stress corrosion cracking)
  season cracking, 77, 1108F
Crevice (contact) corrosion, 82, 172, 550
  effect of inhibitors, 10
  of chromium-nickel-iron alloys, 155–9
  of copper-zinc alloys, 82
  prevention, 158
Crystals, corrosion of single, 34
Crystal orientation, effect on corrosion fatigue, 35
  variation of corrosion with crystal face, 34–5
Cutlery, chromium-iron alloys for, 145

Deactivation, 506
Deaeration, 507
Denison cell, 1039F
Deposit attack, 550, 1106F
Dezincification, **69–70**

Dezincification, effect of alloying, 70
  in sea water, 71–2
  of condensers, 551
  plug-type, 70, 1108F
  tests (*see* Tests, laboratory, brass)
  uniform layer type, 69–70, 1108F
Dichromates (*see* Chromates)
Difference effect, 482
Differential aeration, effect on, condenser tubes, 549–51
  steel, 126–7
Diffusion coatings, aluminum, 638–40
  chromium, 829
  nickel-tin, 824
  nickel-zinc, 813, 824
Dissociation constants of electrolytes, 1140–1T
Durichlor (*see* Silicon-iron alloys)
Duriron (*see* Silicon-iron alloys)

Economizer corrosion, 524–6
Electrical heating alloys, 694–8
Electrochemical attack, 4–6
  definition, 4–5
  mechanism of, 5
  proof of, 6–7
Electrochemical equivalents, 1133T
Electrodes, standard reference, 930–1, 1137T
Electrographic test (coatings), 1034
Electrolytes, dissociation constants of weak, 1140–1T
  ion activities of weak, 1148–9T
Electromotive force (Emf) series, 1134T
Elements, physical constants of, 1127–31T
Embrittlement, caustic (*see* Embrittlement in boilers)
  detector, 532–3
  in boilers, 531–5
  of columbium, 60, 621T
  of tantalum, 321, 721–2
  of zirconium, 348FN
Enamels, baking, **890–97**
  definition 883
  vitreous, 872–8
Erosion (*see also* Cavitation)
  definition, 3
Everdur (*see* Copper-silicon alloys)

Fatigue (*see also* Corrosion fatigue)
  failures due to fretting corrosion, 590–2
  of bolts, 593
Fatty acid group chemicals, materials resistant to, **762–3T**
Fatty acids, resistance of, copper-nickel alloys, 89
  copper-tin alloys, 99

Fatty acids, resistance of, copper-zinc alloys, 75
  nickel, 260
  nickel-copper alloys, 275
  nickel-chromium alloys, 287
  steel, 141T
  tin, 325
Ferric salts, resistance of, copper-nickel alloys, 87
  copper-tin alloys, 98
  copper-zinc alloys, 75
  steel, 132
    stainless, 153
Ferroxyl test, 1033
Films, and passivity, 20–33 [19–20
  equations for porous and non-porous, oxide, growth of, 11–3
  oxide and tarnish, permeability of, 11–2
  thickness of, by cathodic reduction, 984
Fineness of gold, definition, 113
Flue gases (see Gases, flue)
Fluorine (see Halogen gases, high-temperature, and Gases)
Forms of corrosion, illustrations, 1103–10F
Fouling, **433–41** (see also Anti-fouling)
  attachment, 439–40
  effect of, color on, 439
    copper ion concentration, 419
    corrosion products, 439
    ocean currents, 437–9
    protective coatings, 440–1
    seasons, 437–9
    surface, 439
    temperature, 437
    velocity, 439
  effect on corrosion, 440–1
  organisms, distribution, 435–6
    film-forming, 435
    motile, 434
    seasons and attachment, 438F
    semi-motile, 434
    sessile, 433–4
  water, sea, 419–29
Free energy of corrosion reactions, 1142T
Fretting corrosion, **590–7**
  and hardness, 596
  and lubrication, 596
  and stress, 596–7
  and welding, 595
  bearing surface destruction due to, 594
  bolt fatigue failures due to, 593
  destruction of protective films by, 594–5
  fatigue failures due to, 590–2
  loss of dimensional accuracy due to, 592–3
  slipping, 596
  susceptibility to, 595

Fruit juices, resistance of, nickel, 261T
  nickel-copper alloys, 276T
  steels, stainless, 154T

Galvanic corrosion, **481–96** (see also Tests, laboratory, galvanic)
  anodic control, 486
  cathodic control, 486
  definition, 481–2
  difference effect, 493
  diffusion control, 487
  distribution, 491–3
  effect of, cathode material, 489–91
    relative areas, 487–9
  fundamentals, **481–96**
  galvanic current and cathode area, 493
  graphic estimation of currents, 483–5
  inhibitors for, 915–6
  in sea water, 418–9, 420–6T
    with paints, 432
  internal resistance, calculation of, 485
  Kirchhoff's law, 482–3
  limiting galvanic current, 483
  limiting rules, 493
  mixed control, 486
  of aluminum, 48, 490T
    in atmosphere, 53
  of chromium-nickel-iron alloys, 159
  of condensers, 555
  of copper, 66–7
  of copper-nickel alloys, 93
  of copper-tin alloys, 100–1
  of copper-zinc alloys, 77
  of lead, 213
  of magnesium, 228T, 490T, 228–31
    in atmosphere, 246–7
    inhibitors, 231
  of silicon-iron alloys, 204
  of steel, 136–7
  of steel, stainless, 159
    in sea water, 415–8
  of zinc, 339
    in atmosphere, 346
  polarization, 483
  potentials of corroding metals, 482
  protection against, 495–6
    cathodic protection, 496
    inhibitors, 495–6
    insulation, 495
    painting, 495
  relation of galvanic current to, 485–7
  resistance control, 487
Galvanic series, definition of (see Glossary)
  in sea water, 416T
Galvanized coatings (see Coatings, zinc)

Galvanized pipe, soil corrosion, 461–2T, 814–5
Gases, acid-forming, materials resistant to, **787–89T**
Gases (refer also to specific gas)
  resistance of, aluminum alloys, 49
    cadmium coatings, 841
    chromium-nickel-iron alloys, 155
    copper-nickel alloys, 94–5
    copper-silicon alloys, 109
    copper-tin alloys, 101–2
    copper-zinc alloys, 82–4
    lead alloys, 213
    magnesium alloys, 242–3
    nickel, 265
    nickel-chromium alloys, 293
    nickel-copper alloys, 280
    nickel-molybdenum alloys, 298
    silicon-iron alloys, 204
    silver, 319
    steel, 142–3
    tantalum, 323
    zinc, 340
  solubility in water, 1146–7T
Gases (dissolved), resistance of, aluminum alloys, 41T, 42T
  copper-zinc alloys, 70–4
  zinc, 332–5
Gases (exhaust), resistance of heat-resisting alloys, 659–61, 661–2
  valve alloys, 739
Gases (flue), resistance of, nickel-chromium-iron alloys, 661, 691–4
  typical composition, 693T
Gases (high-temperature), resistance of, aluminum alloys, 617
  columbium, 621T
  copper alloys, 628
  gold, 715
  heat-resisting alloys, 647–54, 655–7
  magnesium alloys, 674–5
  nickel alloys, 681–2, 679
  nickel-chromium and nickel-chromium-iron alloy wires, 698–9
  platinum group metals, 715
  silver, 719–20
  steel, 635
    effect of velocity, 635
Gasoline lines, inhibitors for, 501, 914
Glass, **354–9**
  compositions, 354T
  phosphorus, 357
  physical properties, 356T
  resistance to, aqueous solutions, 355T
    sodium carbonate, 355T
    sodium hydroxide, 355T

Glass, resistance to, steam, 355T
  silica, 357–8
  vitreous, 358–9
  weight losses on heating, 357T
Glass-working machinery, platinum group metals for, 715–6
Gold, **112–9** (*see also* Gold alloys)
  anodic corrosion, 117
  corrosion characteristics, **114–7**
  passivity, 117
  resistance to, gases, hot carbonaceous, 715
    halogens, high-temperature, 710-4
    hydrochloric acid, high-temperature, 714–5
    hydrogen, high-temperature, 709-10
    phosphoric acid, high-temperature, 715
    sulfuric acid, high-temperature, 707–9
    sulfur dioxide, high-temperature, 708
    various media, 115–7T
  uses, 112–3
Gold alloys, **112–9**
  corrosion characteristics, 117–9
  effect of, copper, 113–4
    nickel, 114
    palladium and platinum, 113
    silver, 114
  fineness, definition, 113FN
  nitric acid spot test, 118FN
  ordering, 113–4
  oxidation, 699–700
  parting, 118
  reaction limits, 118T
  tarnishing, 118
  uses, 112–4
Graphite (*see* Carbon)
Graphitic corrosion, 139, 198
  definition, (*see* Glossary)

Halogen gases, high-temperature (*see also* Gases)
  resistance of, copper alloys, 630
    gold, 710–14
    nickel-alloys, 681–2
    platinum, 711–4
    steel, 142–3
    various metals, 682T
Hastelloys (*see* Nickel-molybdenum alloys)
Heat-resistant alloys (*see also* Chromium-iron, Chromium-nickel-iron, and Nickel-chromium (-iron) alloys, high-temperature
  applications, 738–9T
  compositions, 740–3T
  designations, 1151–2T
  high-temperature corrosion, **640–664**, 729–43T

Heat-resistant alloys, high-temperature strength, 642–3
  oxidation, 646, 730–2T
  tests, 645T
  resistance to, air, high-temperature, 644–6
    ammonia, high-temperature, 657–8
    gases, exhaust, 659–61
      flue, 661–2, 691
      hot carbonaceous, 655–7, 666F, 669T, 681, 699F
      hot sulfur compounds, 647–54, 669T, 689
    hydrogen, high-temperature, 657–8, 669T 699F
    metals, molten, 658–9, 735T
    oxygen, high-temperature, 646, 669F, 684T, 699F
    petroleum products, high-temperature, 654–5, 666F
    salts, molten, 658–9, 736–7T
    steam, 511–20, 733-4T
  uses, exhaust valves, 660T
Heat treatment, effect on, aluminum alloys, 54
  magnesium alloys, 239–41
  nickel-molybdenum alloys, 298
  steel, 139–40
  steel, stainless, **161**, 169
High-temperature corrosion, 615–725 (*see also* Tests, laboratory, high-temperature)
  materials resistant to, **729–43T**
Human body, use of cobalt alloys, 59
  tantalum, 321–2
Humidity, at Canal Zone and New York City, 1007F
  and corrosion of iron, 917
  magnesium, 245
Humidity tests (*see* Tests, laboratory, humidity and condensation)
Hydrochloric acid, resistance of aluminum alloys, 43, 41T, 45F
  copper-silicon alloys, 107T
  copper-tin alloys, 97–8
  copper-zinc alloys, 75
  molybdenum, 253T
  nickel, 257–9, 258F
  nickel-chromium alloys, 283, 285T
  nickel-copper alloys, 272T
  nickel-molybdenum alloys, 294F, 295, 297T
  silicon-iron alloys, 203F
  silicon-iron-molybdenum alloys, 205F, 206T, 207T
  steel, 132–3
  tungsten, 330T

Hydrochloric acid group chemicals, materials resistant to, **764–68T**
Hydrofluoric acid, resistance of aluminum alloys, 43
  copper-nickel alloys, 87, 88T
  magnesium alloys, 225T
  nickel-chromium alloys, 283
  nickel-copper alloys, 273T, 272–3
  nickel-iron alloys, 194
  steel, 133
Hydrofluoric acid group chemicals, materials resistant to, **769–71T**
Hydrogen, high-temperature resistance of, cobalt alloys, 619
  copper alloys, 629
  gold alloys, 709–10
  heat-resisting alloys, 657–8
  nickel alloys, 681
  palladium, 709–10
  silver, 719
  interstitial, effect on passivity, 24
  overvoltage, values of, 1144–5T
Hydrogen chloride, resistance of, aluminum alloys, 49
  gold, 715
  nickel alloys, 681
  platinum, 714–5
  various metals, 682T
Hydrogen sulfide, high-temperature resistance of, aluminum alloys, 617
  nickel-chromium and nickel-chromium-iron alloys, 651–5, 689–91
  resistance of, aluminum alloys, 49
  copper-nickel alloys, 85
  nickel, 259
  nickel-chromium alloys, 283
  solubility in water, 1146T

Immersion (*see also* Tests, laboratory, alternate or total immersion)
  partial, electrochemical effects, 5
  total, mechanisms of attack by, 3–11
Impingement attack, 76–7, 111
  by steam, 83
  designs to avoid, 547F
  illustration, 1107F
  in condensers, 545–9
  inlet end corrosion, 76
  of copper, 66
  of copper-beryllium alloys, 111
  of copper-zinc alloys, 76–7
Inconel (*see* Nickel-chromium alloys)
Indium and indium alloys, **119–20**
  place in Emf series, 119
  resistance to, acids, 120T
    alkalies, 120T

INDEX

Indium and indium alloys, resistance to, atmospheric corrosion, 119–20
    lubricating oils, 119
    salt spray, 120
Inhibitors, **905–16**
    anodic, 9
    cathodic, 10
    chromates, theory of action, 9, 31
    definition, 905–6 (see also Glossary)
    effect at crevices, 10
    for acid pickling, 908–13
    for acids, 914–5
    for alkalies, 915
    for automobile radiator, 499
    for brines, 498–9, 914
    for diesel and gas engine systems, 499
    for fire protection systems, 498
    for galvanic couples, 231 (Mg alloys), 915–6
    for gasoline lines, 501, 914
    for prevention of corrosion fatigue, 589–90
    for water systems, 499, 913–4
    in pigments, 10–11, 887–90
    mechanism of action, 905–6
    properties, 906–7
    safety of, 907–8
    soluble, action of, 9–10
Insoluble anodes (see Anodes)
Intergranular corrosion, of aluminum alloys, 54–5
    of chromium-manganese-iron alloys, 191–2
    of chromium-nickel-iron alloys, 161–4
    of copper, 629
    of copper-beryllium, 110
    of nickel alloys, 679, 681
    of silver, 719
    prevention with columbium and titanium, 163–4, 191–2
Invar (see Nickel-iron alloys)
Iridium, **309–10**
    corrosion properties, 309T
    oxidation, 705
    resistance to sulfuric acid at high temperature, 707–9
Iron (see also Steel)
    carburization, equilibria for high temperature, 1122–5
    effect in magnesium alloys, 235–9
    effect of, mill scale, 5
    sulfur content, 8
    oxidation, equilibria for high temperature, 1122–5
    oxidation rate, effect of moisture on, 15
    passivity of, 21, 26, 127, 130, 132
    potential in chromates, 981F
        pickling inhibitors, 21, 981F
        sodium chloride solutions, 27F

Iron, resistance to, salt solutions, 7–8
    water, 7–8

Karbate (see Carbon)
Kirchhoff's law, 482–3

Lacquers, **890–7**
    definition, 883
Langelier index, 502
Lead and Lead Alloys, **207–18** (see also Cable sheathing)
    chemical, 209
    common, 209
    corroding, 208–9
    effect of, pH, 211
        stress, 571
    galvanic corrosion, 213
    mechanical factors, 213–4
    metallurgical factors, 213–4
    molten, alloys for pots, 735T
        resistance of aluminum alloys, 617
    permanent fiber stresses, 208F
    potential-time curves, 982–3
    protection, 214–5
    resistance to, acid and alkali solutions, 211F
        acids, 211–2
            organic, 212–3
        alkalies, 213
        atmospheric corrosion, 215–6
        gases, 213
        salts, 213
        soil, 454, 459T, 612
        water, distilled, 210F
            fresh, 4, 209–11
    specifications, 209T
    with antimony, **217**, 982F, 983F, 986F
    with calcium, **217–8**, 982F, 983F, 986F
    with copper, 209
    with silver, 217
    with tin, 216–7
Lead pigments, 887–90
    inhibiting action of, 10–11
Local corrosion, definition, 481–2
"L" Nickel (see Nickel)
Lubricants, **559–68** (see also Bearing corrosion)
    and fretting corrosion, 596
    resistance of, copper, 567
        steel, 568
Lubricating oils, effect of oxidizing characteristics, 563

Macro-organisms (see Fouling, organisms)
Magnesium (see Magnesium alloys)
Magnesium alloys, **218–52**
    anodic treatment, 862

INDEX

Magnesium alloys, chromate coatings, 864–7
  compositions amd properties, 219T
  corrosion fatigue, 234–5
  designations, 219T
  effect of, aluminum content on oxidation, 672F
    cold work, 239–41
    copper content, 235–9
    heat treatment, 239–41
    iron content, 235–9
    nickel content, 235–9
    stress, 231–4, 232F, 233F, 234F, 235F, 572–3
    temperature, 218–21
      on oxidation, 671F
    time of exposure, 218
    velocity, 228
    zinc on oxidation, 673F
  galvanic corrosion, 228–31, 228T, 490T
    effect of inhibitors, 231
  high-temperature corrosion, **670–5**
  protection of, 241–2
  resistance to, acids, 225
      organic, 225, 226T
    alkalies, 225–6
    atmospheric corrosion, 243–52
      corrosion fatigue, 249
      effect of humidity, 245
      effect of pollution, 245–6
      galvanic effects, 246–7
      high-temperature, 670–3
      indoors, 243
      marine atmosphere, 246
      metallurgical factors, 250–1
      outdoors, 244
      protection, 251–2
    carbon dioxide, dissolved, 218
      high-temperature, 673
    gases, 242–3
      high-temperature, 674–5
      protection, 674
    organic compounds, 226–7, 227T
      non-aqueous, 242
    oxygen, dissolved, 218
      high-temperature, 670–3
    salts, 221–3, 223T
    steam, 674
    sulfur dioxide, high-temperature, 673
    water, fresh, 222F
      sea, 223, 406–13, 412T
        effect of heat treatment and painting, 224T
    tolerance limit, 221FN
Manganese, effect in, chromium-nickel-iron alloys, 160
  nickel alloys, 675–82

Manganese, effect in, steel, 140
Mercury, resistance of, aluminum, 618
  nickel, 264
  nickel-copper alloys, 279
  test for brass, 1013
Metals, molten, pots for, 735T
  resistance of, aluminum alloys, 617–8
    heat-resisting alloys, 658–9
    nickel alloys, 682
Methyl alcohol, resistance of, aluminum alloys, 49
  magnesium, 242
Microbiological corrosion, 466–81, 497
  definition, 466
Micro-organisms, **466–81**
  aerobic, 468–9
    sulfur oxidizing, 478–80
  anaerobic, 467–8
    methane fermentation, 478
    nitrate-reducing, 477–8
    sulfate-reducing, 467, 469–77
  effect on, corrosion, 466–7
    steel, 126, 497
  iron bacteria, 480–1
  miscellaneous, 481
  sulfur bacteria, 480
Mill scale, effect on, iron corrosion, 5
  steel in sea water, 388–9
Molybdenum, **252–3**
  resistance to acids, 253T
Molybdenum-iron alloys, passivity of, 28–9
Molybdenum-nickel-iron alloys (*see also* Nickel-molybdenum alloys)
  passivity of, 28–30
Monel (*see* Nickel-copper alloys)
Muntz metal (*see* Copper-zinc alloys)

N.D.H.A. Corrosion Tester, steam condensate, 544
Neoprene (*see* Rubber)
Nickel, **253–66**
  effect in magnesium alloys, 235–9
  effect of stress, 264–5, 577F
  high-temperature corrosion, 675–83
    by sulfur, illustration of, 1109F
  oxidation, effect of addition elements on, 677T
  resistance to, acids, 256–9
      organic, 260–1
    alkalies, 262
    atmospheric corrosion, 265–6
    chlorinated solvents, 264, 265T
    food products, 261
    gases, 265
    hydrogen sulfide, 259
    mercury, 264

INDEX 1177

Nickel, resistance to, salts, 262–4, 263*T*
  steam, 520, 679, 824
  steam condensate, 254
  water, carbonated, 254
    distilled, 256*T*
    fresh, 254
    mine, 254, 256*T*
    sea, 254–6, 402, 407*T*
Nickel alloys, high-temperature (*see also* Nickel-chromium (-iron) alloys, high-temperature)
  compositions, 676*T*
  cyclic oxidation and reduction, 682
  effect of stress, 573–4
  high-temperature corrosion, **675–83**
  limiting service temperatures, chlorine and hydrogen chloride, 682*T*
  sulfurous and sulfur-free atmospheres, 680*T*
  oxide, spalling of, 678
  resistance to (high-temperature), air, 676–9, 696*F*
    carbon dioxide, 679
    carbon monoxide, 679
    gases, 681–2
    hot sulfur compounds, 679
    metals, molten, 682
    oxygen, 676–9
    salts, fused, 682–3
    steam, 679
  uses: thermocouples, 678–9
Nickel-chromium alloys, **281–94** (*see also* Nickel-chromium (-iron) alloys, high-temperature)
  compositions and properties, 281*T*
  effect of stress, 291–2, 577*F*
  resistance to, acetic acid, 289*T*
    acids, 283–90
      organic, 287–90
    alkalies, 290–1
    atmospheric corrosion, 293
    gases, 293
    hydrogen sulfide 283
    salt solutions, 291–3, 292*T*, 293*T*
    steam condensate, 282
    water, boiler, 282
      fresh, 282
      mine, 282–3
      sea, 283, 402, 409*T*
Nickel-chromium (-iron) alloys, high-temperature, 683–99 (*see also* Chromium-nickel-iron alloys [including 18-8])
  effect of, alloy additions, 695–8
    theory of, 697
    chromium, 696*F*

Nickel-chromium (-iron) alloys, effect of, silicon, 696*F*
  resistance to, air (oxidation), 683–91, 730–2*T*
    gases, flue, 661, 691–5
    hydrogen sulfide, 689–91
    sulfur dioxide, 689
  uses, electrical, 694–8
  wires, effect of hot gases, 698–9
Nickel-copper alloys, 266–81
  compositions and properties, 267*T*
  effect of stress, 279–80, 577*F*
  high-temperature corrosion, 675–83
  passivity of, 30, 31*F*
  resistance to, acids, 270–4
    organic, 274–6, 275*T*
    sulfuric, **270–1**, 960*F*
  alkalies, 276
  atmospheric corrosion, 280–1
  carbonic acid, 269*T*
  fruit juices, 276*T*
  gases, 280
  mercury, 279
  salts, 276–9, 277–8*T*
  steam condensate, 268–9
  water, boiler, 268
    carbonated, 268
    distilled, 268
    fresh, 268
    mine, 269
    sea, 269–70, 402–3, 404*T*, 408*T*
Nickel-iron alloys, 194–200
  resistance to, acids, 194–8
    sulfuric, 30*F*, 194–6
  alkalies, 198–9
  atmospheric corrosion, 200
  salts, 194–6
  various media, 197–8*T*
  water, well, 199*T*
  rust characteristics, 200*T*
  stress corrosion cracking, 199
Nickel-manganese alloys, high-temperature corrosion, 675–83
Nickel-molybdenum alloys, 294–8
  composition of commercial alloys, 294*T*
  intergranular corrosion, 298
  metallurgical and mechanical factors, 296–8
  passivity of, 28–30
  resistance to, acids, mineral, 295–6, 297*T*
    organic, 298*T*
    sulfuric, 29*F*, 296*F*
  alkalies, 296
  gases, 298
  salt solutions, 296
  sea water, 403, 409*T*

Ni-Resist (*see* Cast iron, nickel)
$n/8$ rule, 27
Nitric acid, resistance of, aluminum alloys, 43–4
    chromium-iron alloys, 144T, 148T
    chromium-manganese-iron alloys, 189T, 190T, 191T
    chromium-nickel-iron alloys, 16-2 alloy, 183T
        18-8 alloy, 159T, 160–1T, 1018T, (cast alloy) 1020T
    copper-nickel alloys, 88
    copper-tin alloys, 97
    copper-zinc alloys, 74–5
    molybdenum, 253T
    nickel-chromium alloys, 287, 288T
    nickel-molybdenum alloys, 297T
    steel, 133–4
    tungsten, 330T
    spot test for gold alloys, 118FN
    theory of attack by, 8
    test, *see* Tests, laboratory, stainless steels
Nitric acid group chemicals, materials resistant to, **772–5T**
Nitriding containers, 735T
Nitrites, as inhibitor in petroleum lines, 501, 914
    passivating action of, 31
Nitrogen, effect in, chromium-manganese-iron alloys, 190
    chromium-nickel-iron alloys, 16-2 alloy, 188
        18-8 alloy, 160, 169
    steel, 140–1
    solubility in, sea water, 1116T
        water, 1146T
Notch effect, from corrosion, 587–8

Oils, lubricating, resistance of indium alloys, 119
Open-circuit potential, 482
Ordering, of gold-copper alloys, 113–4
Organic acids, resistance of, aluminum alloys, 46F, 45–6, 49
    copper-nickel alloys, 88–9, 89T
    copper-tin alloys, 99
    copper-zinc alloys, 75
    lead alloys, 212–3
    magnesium alloys, 225, 226T, 260–1
    nickel-chromium alloys, 287–90, 290T
    nickel-copper alloys, 275T, 274–6
    nickel-molybdenum-iron alloys, 298T
    steel, 134
    tin, 325
Organic compounds, resistance of, aluminum alloys, 45–6, 49
    chromium-nickel-iron alloys, 157T
    copper, 66
    copper-nickel alloys, 92–3
    copper-silicon alloys, 109
    copper-tin alloys, 101
    copper-zinc alloys, 76
    magnesium alloys, 226–7, 227T
        in non-aqueous, 242
    silver, 317T
    steel, 141–2
    tin, 327
    zinc, 338
        in non-aqueous, 339–40
Osmium, **311–2**
    general corrosion resistance, 312T
    oxidation, 705–6
    resistance to, phosphoric acid, high-temperature, 715
    sulfuric acid, high-temperature, 707–9
Overvoltage, values for hydrogen, 1144–5T
Oxidation, effect of, environment, 15
    lattice structure, 13
    metal composition, 14
    metal orientation, 13
    metal properties, 13–4
    metal surface, 14–5
    moisture, 15
    temperature, 16
    equations for, 15–20
    fundamentals of, **11–20**
    internal, illustration of, 1110F
    of aluminum alloys, internal, 617
    of aluminum-iron alloys, 638–9
    of chromium, 825–8
    of chromium-aluminum-iron alloys, 668–70
    of chromium-silicon-iron alloys, 664–7
    of cobalt alloys, 618–9
    of columbium, 620–1
    of copper, 621–4
        effect of crystal habit, 623
        rate, 623–4
        rate constant, 18
        temperature and pressure limits for copper oxides 621–2
    of copper alloys, 624–7
        protection against, 627
    of copper-nickel alloys, rate, 676F
    of gold, 699–700
    of magnesium alloys, 671–3
        effect of, aluminum content, 672F
        temperature, 671F
        zinc content, 673F
    of nickel, effect of addition elements, 677T
    of nickel alloys, 676–9

Oxidation, of nickel-chromium and nickel-chromium-iron alloys, 683–99, 730–2T
of platinum and platinum alloys, 700–5
of silver, 718–9
of steel, 632–5, 730–2T
of tantalum, 720–1
of zinc, 722
protection against, 20
Oxidation rule of Pilling and Bedworth, 11
Oxidation theory, electrolytic, 16–20
Wagner's, 16–9
Oxidation-reduction potentials, 1135-7T
and pitting of 18-8, 167T
Oxide coatings (*see* Coatings, oxide)
Oxide films, growth of, 11–3
permeability of, 11–2
spalling, 13
effect of surface, 13FN
nickel alloys, 678
thickness of, by cathodic reduction, 984
Oxygen, adsorbed, effect on passivity, 24
dissolved, effect on, aluminum alloys, 41
copper, 62–3
lead, 4
magnesium alloys, 218
steel, 125–6
zinc, 4
maximum allowable in various waters, 506
high-temperature, resistance of, heat-resistant alloys, 646, 684T, 699T
magnesium alloys, 670–3
nickel alloys, 676–9
steel, 634F, 646T
solubility in, electrolytes, 1147T
water, 1146T
sea water, 1116T, 1117T

Paint, **884–90**
anti-fouling, 442–5
corrosion by, 444
evaluation of, 444–5
formulation, 442–3
inhibition of, 443
toxics, 443
vehicle, 443–4
bituminous (*see* Coatings, bituminous)
definition, 882
for sea water, 430–3
blistering, 430
surface preparation, 431–2
phosphate coatings as base for, 868–9
pigments, rust-inhibiting, 887–90
preparation of surface, 505, 895–6
primer formulations, 889T

Paint, uses, 886
vehicles, 886–7
Palladium, **304–9**
general corrosion behavior, 305, 307T
oxidation, 705
properties, 304–5
resistance to, atmospheric corrosion, 308–9
hydrogen, high-temperature, 709–10
sulfur dioxide, high-temperature, 707
sulfuric acid, high-temperature, 707–9
tarnishing, 308–9
uses, 305–7
anodes, 306–7
Parkerizing (*see* Coatings, phosphate)
Parting, 118
definition (*see* Glossary)
Passivators, **905–16**
definition, 908 (*see also* Glossary)
theory of, **30–2**
Passive-active cell, 165
Passive metals and alloys, properties of, 22–3
Passivity, **20–32**
and sublimation energy, 25–6
and work function, 25
by red lead, 31
by sodium nitrite, 31
by zinc chromate, 30–1
definition of, 20–2
destruction by, chloride ions, 32
halogen ions, 22–3
determination, 22, 980–1
effect of adsorbed oxygen and interstitial hydrogen on, 24
history, 22
mechanism, 27–30
of aluminum, 32
of chromium-iron alloys, 24–5
of chromium-nickel-iron alloys, 28
of copper-nickel alloys, 30
of iron and steel, 21, 26, 127, 130, 132
of molybdenum-iron alloys, 28
of molybdenum-nickel-iron alloys, 28
of platinum, 304
theories, 9, 23–6
electron configuration, 23–6
general discussion of, 32
generalized film, 23
in alloys, 27–30
oxide film, 23
Patina, definition (*see* Glossary)
on copper, 67
on copper-zinc alloys, 84
Perbunan (*see* Rubber)
Permalloy (*see* Nickel-iron alloys)
pH, effect on, aluminum alloys, 42
cadmium, 842F

1180   INDEX

pH, effect on, chromium-nickel-iron alloys, 152
    iron, 129
    lead alloys, 211
    silicon-iron alloys, 204
    zinc, 335
    of water in tubercles, 497
Phosphate coatings (*see* Coatings, phosphate)
Phosphor bronze, (*see* Copper-tin alloys)
Phosphoric acid, resistance of, aluminum alloys, 43–4
    copper-nickel alloys, 88
    copper-tin alloys, 98–9
    copper-zinc alloys, 75
    molybdenum, $253T$
    nickel, 259
    nickel-chromium alloys, $286T$, 287
    nickel-copper alloys, 273, $274T$
    nickel-molybdenum alloys, $297T$
Phosphoric acid, high-temperature, resistance of, gold, 715
    osmium, 715
    platinum, 715
    rhodium, 715
Phosphoric acid group chemicals, materials resistant to, **783–7T**
Phosphorus, effect in, brass, 70, 551–2
    steel, 140
    effect on atmospheric corrosion, 121–2
    molten, resistance of copper alloys, 630
Physical constants of elements, **1127–32T**
    miscellaneous constants, $1132T$
Pickling acid, inhibitors for, 908–13
Pigments, inhibitive, 887–90
    theory of, 10–11, 30
    lead, 888
    zinc, 888–90
Pilling and Bedworth, oxidation rule of, 11
Pipe, bituminous coatings for, 463–5, 899–901
    tests, 1058
Pitting, in soils, 451–2; calculation of max. pit depth, 1050–2
    of boilers, 521–4
    of copper-zinc alloys in sea water, 73, 393–7
    of steel in sea water, 388–9
    of stainless steels, 165–73
        and contact (crevice) corrosion, 172
        avoidance, 173
        conditions affecting, 166–8
        effect of, alkalinity, 167
            environment, 166–8
            mechanical work, 171–2
            metal structure, 169–71
            oxidation-reduction potential, $167T$
            pH, $169F$

Pitting, in soils, of stainless steels, effect of, temperature, 167
    tests, $172T$, 1022
    theory of, 165–6
Plastics, **359–64**
    properties, $362T$–$3T$
    resistance to various chemicals, 360–$2T$, $364T$
Platinum, **299–304**
    general corrosion behavior, 302–3, $303T$
    oxidation, 700–5
    passivity, 304
    properties, 299
    resistance to, gases, hot carbonaceous, 715
        halogens, hot, 711–4
        hydrochloric acid, high-temperature, 714–5
        phosphoric acid, high-temperature, 715
        salts, molten, 716–8
        sulfur dioxide, high-temperature, 707
        sulfuric acid, high-temperature, 707–9
        various chemicals, 300–$1T$
    uses, 299
        anodes, 303–4
        glass-working machinery, 715–6
Platinum group metals, **299–314** (*see also* individual metals)
Polarization, 22, 483
    definition (*see* Glossary)
Polarograph, 979–80
Porcelain, **365**
    compositions, $365T$
Potassium hydroxide, resistance of, aluminum alloys, 46
    copper-nickel alloys, 90
    steel, 134
Potential, corrosion, 482, $484T$, 1040
    measurement of, 978–87
    open-circuit, 482
    standard, $1134T$, 1135–$37T$
    -time curves, 980, $981F$, $983F$
    of paint coatings, 1029–30
Protection against corrosion, **803–950** (*see also* type of protection in question)

Reaction limits, $118T$
Red brass (*see* Copper-zinc alloys and Condenser corrosion)
Refractories, high-temperature uses, **723–5**
Resinous materials, **890–7**
Resins, acrylate, 896
    alkyd, 894–5
    melamine, 895
    phenolic, 894
    urea, 895
Rhenium, **314**

Rhodium, **310–1**
 general corrosion resistance, 311T
 oxidation, 705
 resistance to, phosphoric acid, high-temperature, 715
  sulfuric acid, high-temperature, 707–9
Roofing, bituminous coatings for, 899
Rubber, **365–71**
 chemical resistance, 370–1T
 properties, 366–7T
 uses, linings, 369T
Ruthenium, **312–3**
 general corrosion resistance, 313T
 oxidation, 706

Salinity, definition of, 1111
Salts, molten, resistance of, alloys for pots, 736–7T
 aluminum alloys, 617–8
 cobalt alloys, 619–20
 heat-resisting alloys, 658–9
 nickel alloys, 682–3
 platinum group metals, 716–8
 silver, 720
 steel, 636
 solutions (*see also* Water, sea)
  resistance of, aluminum alloys, 46–7, 47T
   chromium-manganese-iron alloys, 189
   chromium-nickel-iron, 16-2 alloy, 184
   copper, 64–6
   copper-nickel alloys, 90–2
   copper-silicon alloys, 108–9
   copper-tin alloys, 99–100
   copper-zinc alloys, 74
   iron, 7–8
   lead alloys, 213, 854–5
   magnesium alloys, 221–3, 223T
   nickel, 262–4, 263T
   nickel-chromium alloys, 291, 292T
   nickel-copper alloys, 276–9, 277–8T
   nickel-iron alloys, 194–6
   nickel-molybdenum alloys, 296
   silicon-iron alloys, 204
   silver, 315–6
   steel, 130–2, 132T, 636–8
   tantalum, 321–2
   tin, 326–7
   zinc, 336–8, 337T, 977F
Salt-spray test (*see* Tests, laboratory, salt spray)
Saturation Index, 502
 relation to copper-zinc alloys, 71
Sea water (*see* Water, sea)
Season cracking, 77, 1108F
Silica, vitreous (*see* Glass)

Silicon, **380**
 effect on, chromium-iron alloys, high-temperature, 664–7
  chromium-nickel-iron alloys, 160
   high-temperature, 662
  copper, 104–10
  steel, 140
Silicon bronze (*see* Copper-silicon alloys)
Silicon-iron alloys, **201–7**
 compositions, 201
 effect of, pH, 204
  temperature, 202–3
 galvanic corrosion, 204
 general corrosion resistance, 201–2
 metallurgical factors, 204–5
 physical characteristics, 201T
 resistance to, gases, 204
  hydrochloric acid, 203F
  salts, 204
  sulfuric acid, 202F
 uses, 207
Silicon-molybdenum-iron alloys, **205–7**
Silicone rubber (*see* Rubber)
Silver and silver alloys, **314–20**
 embrittlement, high temperatures, 719
 high-temperature corrosion, **718–20**
 oxidation, 718–9
 resistance to, acids, 315, 316T
  alkalies, 315–8
  atmosphere, high-temperature, 718–9
  carbon monoxide, high-temperature, 719
  gases, 319
   high-temperature, 719–20
  organic compounds, 317T
  salts, 315–8
   molten, 720
  sodium hydroxide, 317T, 720
 tarnish rate constant for, 18
 tarnishing, 14FN, 15, 319–20
  effect of moisture, 15
  protection against, 20, 319–20
 uses, 318–9
Slushing compounds (*see* Coatings, temporary, slushing compounds)
S-N curves, 992
Sodium hydroxide, resistance of, aluminum alloys, 46
 copper-nickel alloys, 90
 copper-silicon alloys, 108T
 copper-tin alloys, 99
 glass, 355T
 magnesium, 226T
 nickel-copper alloys, 277T
 nickel-iron alloys, 199T
 silver, 317T
 steel, 134–5

Sodium nitrite (*see* Nitrites)
Soil corrosion, **446–66**
 definition, 446
 effect of soil, physical properties, 451T
 effect of soil acidity on repairs, 450T
 protection against, 460–6 (*see also* Cathodic protection)
  with bituminous coatings, 463–5, 899–901
  with cement mortar, 462
 resistance of, aluminum alloys, 50–1
  asbestos cement, 454
  asbestos pipe, 454
  coatings, bituminous, 463–5, 899–901
   grease, 463
   lead, 849–52
   metallic, 460–2
   zinc, 461–2T, 814–6
  copper, 67
  copper and copper alloys, 452, 457–8T
  copper-zinc alloys, 82, 457–8T
  galvanized pipe, 461–2T
  lead, 454, 459T, 612–3
  steel, 451–2, 455T
  zinc, 454, 460T
Soils, classification, Whitney diagram, 448F
 compositions, 449T
 corrosive properties, **446–51**
 groups, 446–51
 treatment, 465–6
Solder (*see* Lead-tin alloys)
Solubility product constants, 1138–9T
 in sea water, 1119T
Spalling, definition (*see* Glossary)
 effect on wire life, 998–1000
 of oxide and tarnish films, 13
  effect of surface curvature, 13FN, 518
  on nickel alloys, 670
Specifications, for cadmium coatings, 844–5
 for coatings, decorative, 822–4
 for coatings, lead, 856–7
 for condenser tube alloys, 546T
 for stainless steel, 16-2 alloy, 183T
Specimens, cleaning and preparation, 1077–83
Spool-type specimen holder, 1053–6
Spot test, nitric acid, for gold alloys, 118FN
Statistical methods, 1083–99
Steam, **511–20**
 corrosion fatigue in, 588
 corrosion rates of alloys, 511–20, **733–4T**
 impingement attack in, 83
  wire drawing, 84
 in boilers, 535–7
 resistance of, aluminum alloys, 617
  chromium-iron and chromium-nickel-iron alloys, 511–20
   16-2 alloy, 185

Steam, resistance of, cobalt alloys, 619
 copper and copper alloys, 520, 627
 copper-nickel alloys, 95
 copper-silicon alloys, 109
 copper-tin alloys, 112
 copper-zinc alloys, 83–4
 glass, 355T
 magnesium alloys, 674
 nickel and nickel alloys, 520, 679, 824
 steels, 511–20, 733–4T
  effect of, pressure, 511
   stress, 518
   surface, 518
   temperature, 512–5
   time, 511
  types of scales, 515–7
Steam condensate, **538–44**
 carbon dioxide attack, 538–9
 gases, solution of deleterious, 539–40
 measuring corrosion by, 544–5
 metals resistant to, 542–3
 N.D.H.A. corrosion tester, 544
 oxygen attack, 538–9
 prevention of corrosion by, design of equipment, 541
  treatment of, 541–2
  treatment of boiler feed water, 540–1
 removal of insoluble deposits, 543–4
 resistance of, aluminum alloys, 42–3, 542F
  brass, 542F
  copper, 64, 542F
  copper-nickel alloys, 86T, 542F
  nickel, 254
  nickel-copper alloys, 268–9, 282
  steel, 542F
Steel, 125–43 (*see also* Iron)
 corrosion fatigue, in steam, 588
  limits in, oil well brine, 585T
   fresh water, 584T
 effect of, bacteria, 126, 466
  carbon content, 140
  chromium content, 140, 391–2
  cold work, 138
  copper content, 121–4, 140
  differential aeration, 126–7
  heat treatment, 139–40
  manganese content, 140
  nickel content, 121–4, 140–1
  nitrogen content, 140
  pH, 129
  phosphorus content, 121–3, 140
  silicon content, 140
  stress, 137–8, 571
  sulfur content, 140
  temperature, 128

INDEX 1183

Steel, effect of, velocity, 135-6
    in sea water, 390-1
  galvanic corrosion, 136-7
  high-temperature corrosion, **630-8**
    effect of, alloying elements on, 635
      surface on, 631-5
  nickel-clad, 824
    welds in, 198-9
  open hearth vs. Bessemer, 139
  passivation, 127, 132 (see also Iron)
  resistance to, acids, 132-5
    organic, 134
    alkalies, 134-5
      high-temperature, 636-8
    ammonium hydroxide, 135
    ammonium salts, 131-2
    atmospheric corrosion, 120-5
    carbon dioxide, 127
    gases, 142-3
      high-temperature, 635
    lubricants, 568
    organic compounds, 141-2
    oxygen, dissolved, 125-6
    salt solutions, 130-2, 132T
      high-temperature, 636-8
    salts, molten, 636
    soil corrosion, 451-2, 455T
    steam, 511-20
    steam condensate, 538-44
    water, boiler, 520-37
      fresh, 130
      mine, 132
      sea, 383-92, 384-7T
  scaling, 636
  stress corrosion cracking, 137-8 (see also Embrittlement, in boilers)
  vs. cast and wrought iron, 139
  water-line attack, 126-7
Steels, high-alloy (see Chromium-iron, Chromium-nickel-iron, and Nickel-chromium [-iron] alloys)
  soil corrosion, 452
  low-alloy
    atmospheric corrosion, 122-124
      effect of copper content, 121-122
      effect of phosphorus content, 122
    sea-water corrosion, 383-92
    soil corrosion, 451
Steels, stainless (see also Chromium-iron alloys, Chromium-manganese-iron alloys, Chromium-nickel-iron alloys, and Nickel-chromium [-iron] alloys)
  Alloy Casting Institute designations, 1151-2T
  A.I.S.I. type numbers, 1150T
  effect of stress, 571

Steels, stainless, effect of stress, in magnesium chloride, 180-1T
  intergranular corrosion, 161-4,191
  pitting of, 165-173
  resistance to sea water, 172T,413-8, 413T
  stress corrosion of, 174-82, 571
    effect of heat treatment, 178T
    in chlorides, 174, 180-1T
Stellites (see Cobalt alloys)
Stoneware, chemical, **353-4**
  physical properties, 353T
Stray-current corrosion, 601-6
  definition, 601 (see also Glossary)
  detection, 604-6
  effects, 602-3
  lead cable sheathing, 613
  prevention, 603-4
  sources, 601-2
Stress, and fretting corrosion, 596-7
  effect of static stress, 576-7
    corrosion by steam, 518
    embrittlement in boiler water, 531
    permissible fiber stress in lead alloys, 208F
    threshold for stress corrosion cracking of stainless alloys, 176
Stress corrosion, **569-78** (refer also to Season cracking)
  magnesium chloride for test solution, 174
  of aluminum alloys, 48, 570
    condensers, 555
    copper alloys, 571
    copper-nickel alloys, 93-4
    copper-silicon alloys, 110
    copper-tin alloys, 101
    copper-zinc alloys, 77-81, 570-1
    lead, 571
    magnesium alloys, 231-4, 232F, 233F, 234F, 235F, 572-3
    in atmosphere, 247-9
    nickel, 264-5
    nickel alloys, 573-4
    nickel-chromium alloys, 291-2
    nickel-copper alloys, 279-80
    nickel-iron alloys, 199-200
    stainless alloys, **174-82**
    steel, 137-8, 571
      stainless, 174-182, 571
    zinc, 339, 574
    in atmosphere 346
  remedies, 574-5
  theories, 575-6
  tests (see Tests, laboratory, stress corrosion cracking)
Stress relief temperatures, various metals, 574
  A.S.M.E. boiler code, 164

1184　　　　　　　　　INDEX

Sublimation energy and passivity, 25–6
Sulfur (*see also* Gases, hot sulfur compounds)
　effect in, iron, 8
　　steel, 140
Sulfur dioxide, high-temperature, resistance of, aluminum alloys, 617
　　gold, 706
　　magnesium alloys, 673
　　nickel-chromium (-iron) alloys, 689
　　palladium, 707
　　platinum, 707
　resistance of, aluminum alloys, 49
　　copper-zinc alloys, 82–3
　solubility in water, 1146*T*
Sulfur dioxide group chemicals, materials resistant to, **790–2 *T***
Sulfuric acid, resistance of, aluminum alloys, 43–4
　cast iron-nickel alloys, 196–8
　chromium-iron alloys, 148*T*
　chromium-nickel-iron alloys, 29*F*, 152–5, 156*F*
　copper-nickel alloys, 87
　copper-silicon alloys, 107*T*
　copper-tin alloys, 98
　copper-zinc alloys, 75
　lead, 212*F*
　molybdenum, 253*T*
　nickel, 256–7, 257*T*
　nickel-chromium alloys, 283
　nickel-copper alloys, 270–1,960*F*
　nickel-iron alloys, 194, 195*F*, 196*F*
　nickel-molybdenum alloys, 296*F*, 297*T*
　silicon-iron alloys, 202*F*
　steel, 133
　tungsten, 330*T*
　high-temperature, resistance of, gold, 707–9
　　iridium, 707–9
　　osmium, 707–9
　　palladium, 707–9
　　platinum, 707–9
　　rhodium, 707–9
Sulfuric acid group chemicals, materials resistant to, **776–82 *T***
Sulfurous acid, resistance of, chromium-nickel-iron alloys, 155
　copper-nickel alloys, 88
　copper-tin alloys, 98
　copper-zinc alloys, 75
　nickel, 259
　nickel-chromium alloys, 287
　nickel-copper alloys, 273–4
Superheater tube corrosion 535–7

Tanks, storage, bituminous coatings for, 901–2

Tantalum, **320–3**
　embrittlement, 321, 721–2
　high-temperature corrosion, 720–2
　resistance to, acids, 321
　　air, high-temperature, 720–1
　　alkalies, 321
　　atmospheric corrosion, 323
　　gases, 323
　　halogens, 722
　　nitrogen, high-temperature, 721*T*
　　salts, 321–2
　　steam, 721
　uses in human body, 321–2
Tarnish, effect of environment, 15
　effect of moisture, 15
　equations for, 15–20
　fundamentals, 11–20
　of copper, 14*FN*
　　gold alloys, 118
　　palladium, 308–9
　　silver, 14*FN*, 15, 319–20
　　　rate constant for, 18
　protection against, 20, 319–20
　theories of, electrolytic, 16–20
　　Wagner's, 16–19
Temperature, conversion units, 1153–6*T*
　effect on, aluminum alloy corrosion, 41–2
　　alternate immersion testing, 969–70
　　bearing corrosion, 561–2
　　cavitation, 598
　　chromium-nickel-iron, 16-2 alloy, 185
　　condenser corrosion, 551–5
　　fouling, 437
　　magnesium alloy corrosion, 218–21
　　pitting, 167
　　salt-spray tests, 974
　　sea-water corrosion of, copper-zinc alloys, 72–3
　　　copper-nickel alloys, 86–7
　　　steel, 391
　　silicon-iron alloy corrosion, 202–3
　　steam corrosion, 512–5
　　steel corrosion, 128
　　total immersion tests, 962
　　zinc corrosion, 335
Terne plate (*see* Lead-tin alloys and Lead coatings)
Tests, **953–1099**
　electrochemical, 6–7
　embrittlement in boiler waters, 532–5
Tests (field), **1043–99**
　atmospheric, 1043–8
　　atmosphere types, 1043
　　data desired, 1044
　　duration, 1048
　　exposure methods, 1045–8

INDEX

Tests (field), atmospheric, galvanic couple tests, A.S.T.M., 1005
  racks for, 1045–7F
  specimens, 1044–5
 chemical plant, **1052–8**
  objectives, 1052–5
  oil-refinery equipment, 1058
  procedure, 1053–8
  specimen examination, 1058
  spool test, 1053–6
 paints, 1071–7
  application variables, 1073–5
  duplication, 1072
  duration of test, 1075–6
  mounting of specimens, 1075
  reporting results, 1075–6
  test areas, 1071–2
   preparation of, 1072–3
  test locations, 1076–7
 soil, 1048–52
  calculation of pit depth, 1050–2
  duration of test, 1049
  reporting results, 1049–52
  specimens, 1049
 steam, service tests, 518–20
 steam condensate, 544–5
 water, sea, 1060–71
  half tide, 1068–70
  precautions, 1060–1
  rack construction, 1061–8
  specimens, 1070–1
 water pipe, 1058–60
Tests (laboratory), **953–1042**
 alternate immersion, 965–70
  apparatus, 966–8
  duration of test, 970
  effect of temperature, 969–70
  reproducibility of results, 970
  solutions, 970
  specimens, 968–9
 brass, 1015–6
  dezincification, 1015
  impingement, 1015–6
  season cracking, 1013
 cavitation damage, 993–7
  jet test, 994
  measure of damage, 993–4
  venturi test, 994
  vibratory test, 994
 corrosion fatigue, 987–93
  in air, 988–91
  in liquids, 991
  in steam, 991–2
  precautions, 993
  reporting results, 992–3
 electrochemical, 978–87

Tests (laboratory), electrochemical, current density curves, 985
  methods, 979
  paints, 1029
  polarograph, 979–80
  potential-time curves, 980–1
  procedure, 985–7
  resistance measurements, 981–2
 galvanic couple, 1002–6
 high-temperature, 997–1001
  A.S.T.M. life test, 997–8
  external heating of specimens, 998–1000
  furnace atmospheres, 1001
  internal heating of specimens, 997–8
  microscopic examination, 1000
  molten metals, 1001
  salts, molten and solid, 1001
 humidity and condensation, 1006–8
  of coatings, Cd, 844
  of coatings, Pb, 853–4
 lubricant corrosion, 1034–8
  beaker tests, 1036–7
  bearing corrosion, 1034–7
  copper-strip test, 1037–8
  engine tests, 1036
 metal coatings, 1030–4
  adhesion, 1031–2
  hardness, 1032
  porosity, 1032–4
  thickness, 1030–31
 methods, various, 953–9
 organic coatings, **1023–30**
  abrasion resistance, 1027–8
  absorption and permeability, 1028–9
  adhesion, 1025–6
  distensibility, 1026–7
  electrochemical, 1029
  film thickness, 1024
  impact resistance, 1027
  porosity, 1029
  potential-time measurements, 1029–30
 outline and interpretation, various tests, 953–9
 salt spray, 970–8
  air for, 975
  apparatus, 971–3
  functions, 970–1
  reporting results, 978
  solutions, 974
  specifications, Cd coatings, 844
   Pb coatings, 856
  specimens, 977–8
  temperature, 974
 soil, 1038–43
  Denison cell, 1039
  interpretation, 1040–3

Tests (laboratory), stainless steels, 1016–22
   copper sulfate-sulfuric acid test, 1020–1
   intergranular corrosion, 1019–22
   nitric acid, boiling, 1016–20
   pitting, 1022
   Strauss test (see Copper sulfate-sulfuric acid test)
   stress corrosion cracking, 174
  steam, test for zinc, 1023
   tests in (see references in *Corrosion by High-Temperature Steam*, p. 511)
  stress corrosion cracking, 1009–14
   environment, 1011–4
   environment for, aluminum alloys, 1011–2
    copper alloys, 1013–4
     mercury test for brass, 1013–4
    iron, 137, 1014
    magnesium alloys, 1014
    stainless steels, 174, 1014
   stress, methods for applying, 1010–1
  total immersion, 959–65
   duration, 964
   effect of, aeration, 962–3
    temperature, 962
    velocity, 959–61
   reporting results, 964–5
   solutions, 963
   specimen supports, 962
   specimens, 963–4
  slushing compounds, 921–3
  specimens, cleaning of, 1077–82
   preparation, 1077–82
  statistical methods, 1083–99
  stray-current corrosion, 604–6
  zinc, steam test, 1023
Theory of, action of additives to slushing compounds, 919
  action of chromates as inhibitors, 9, 31
  attack by acids, 8–9
  electrochemical corrosion, 3–11
  electrolytic oxidation and tarnish, 16–20
  electron configuration of passivity, 24–6
  generalized film of passivity, 23–4
  oxide film of passivity, 23
  passivators, 30–2
  passivity, 23–6
   general discussion of, 32
   in alloys, 27–30
   pitting, 165–6
   stress corrosion, 575–6
   Wagner's of tarnish, 16–9
Thermocouple alloys, 678–9
Thiokol (see Rubber)
Tin, 323–9
  as alloy in indium, 120

Tin, coatings (see Coatings, tin)
  molten, resistance of aluminum alloys to, 617
  resistance to, acids, 324–5
   organic, 325
   alkalies, 325–6
   atmospheric corrosion, 327–9
   organic compounds, 327
   salts, 326–7
   water, fresh, 323–4
   sea, 324, 417T
Tinning, methods of, 829–31
Titanium, 329
  additions to prevent intergranular corrosion, 163–4, 191–2
Tolerance limit, 221FN
Tool steels, chromium-iron alloys for, 145
Tubercles, pH of water in, 497
  theory of formation, 497
  illustration, 1105F
Tungsten, 330
Twoscore (see Chromium-nickel-iron, 16-2 alloy)

Uranium, 331

Valves, cobalt alloys for, 620
  exhaust, heat-resisting, properties of, 660T
Varnishes, 884–90
  definition, 882
Vehicles for paints, 886–7
Velocity, effect on, aluminum alloys, 47
  cobalt alloys, 58–9
  condenser corrosion, 545–9
  copper alloys in sea water, 393–401
  copper-nickel alloys, 93
  copper-tin alloys, 100
  copper-zinc alloys, 76–7
  fouling, 439
  lead, 212
  magnesium alloys, 228
  stainless steel in sea water, 414–5
  steel, 135–6
   in hot gases, 635
   in sea water, 390–1
   total immersion testing, 959–61
  zinc, 338
Vibration (see Corrosion fatigue)
Vinyl acetate and acetals, 896
Vinyl chloride, 896–7
Vistanex (see Rubber)
Vitallium (see Cobalt alloys)

Wagner's theory, 16–9
Water (boiler), **520–37**
  concentrated, 526–31

INDEX 1187

Water (boiler), corrosive characteristics, 520–1
  embrittlement. 531–5
    cracks, 533–5
    protection against, 535
    stress for, 531
    test for, 532–3
  pitting, 521–4
  resistance of, copper-nickel alloys, 86
    copper-zinc alloys, 73–4
    nickel-chromium alloys, 282
    nickel-copper alloys, 268
  treatment, 540
Water (carbonated) resistance of, nickel, 254
  nickel-copper alloys, 268
Water (distilled), resistance of, aluminum alloys, 42
  lead, 210F
  nickel, 256T
  nickel-copper alloys, 268
  tin, 323
  zinc, 4
Water (fresh), **496–506**
  cathodic protection in, 501, 949–50
  coatings for, 504–6 (see also Coatings)
    bituminous, 899
    cement, 505
  cooling, 499
    non-recirculating, 500
  deactivation, 506–7
  deaeration, 507–10
  domestic, 501–3
    carbonate balance system, 502
    hexametaphosphate treatment, 503
    silicate of soda for, 502–3
    treatment, 501–3
  inhibitors for, 499 (see also Inhibitors and Passivators, p. 905)
  Langelier Index, 502
  paints for, 505–6 (see also topic Organic Coatings, 878–905)
  pipe materials for, 503–4
  resistance of, aluminum alloys, 42
    chromium-manganese-iron alloys, 190
    chromium-nickel-iron, 16-2 alloy, 184
    copper, 62–4
    copper-nickel alloys, 85–6
    copper-silicon alloys, 105
    copper-tin alloys, 96–7
    copper-zinc alloys, 71, 73–4
    iron and steel, 7–8, 130
    lead alloys, 209–11
    magnesium alloys, 222F
    nickel, 254
    nickel-chromium alloys, 282
    nickel-copper alloys, 268

Water (fresh), resistance of, nickel-iron alloys, 199T
  tantalum, 321
  tin, 323–4
  zinc, 335–6
  tuberculation, 497–8
Water (mine), resistance of, aluminum alloys, 43
  copper-silicon alloys, 106
  copper-zinc alloys, 74
  nickel, 254, 256T
  nickel-chromium alloys, 282–3
  nickel-copper alloys, 269
  steel, 132
Water (river), dissolved solids in, 1120T
Water (sea), **383–429**
  alkalinity, 1117–8
  artificial, formulas for 1121T
  characteristics, 1111–22T
  coatings, organic, for, 430–3
  conductance, 1114, 1115T
  constituents, 1111T
  density, 1113T
    maximum, 1113
  dissolved oxygen content, 1117T
  dissolved solids, 1120T
  effect of pollution, 1119–20
  fouling in, 419–29, 433–46
  fouling tendencies of alloys in, 427–9T
  freezing point, 1113, 1114T
  galvanic corrosion, 418–9, 420–6T
  galvanic series in, 416T
  gases, solubility in, 1115–17T
  osmotic pressure, 1114
  paints for, 430–3 (see also Paints, and Coatings, organic)
    blistering, 430
    dry docking, 431
    galvanic effects of, 432
    surface preparation for, 431–2
    vehicles, 431
  pH, 1118
  piling, 390
  racks for tests, 1064F, 1068–9F
  resistance of, aluminum alloys, 43, 406, 410–11T
    cast iron, 384–7T, 392
    chromium-manganese-iron alloys, 190
    chromium-nickel-iron, 16-2 alloys, 184–5
    chromium-nickel-iron alloys, 184T, 413–8, 413T
    copper, 64, 394T
    copper alloys, 393–402, 405T
    copper-beryllium, 111
    copper-nickel alloys, 86–7, 399T
    copper-silicon alloys, 105–6, 395T

Water (sea), resistance of,
  copper-tin alloys, 96–7, 300T, 398T
  copper-zinc alloys, 71–3, 396–7T
  galvanic corrosion, 415–8
  magnesium alloys, 223, 224T, 406–13, 412T
  miscellaneous materials, 417T
  nickel, 254–6, 407T
  nickel alloys, 402–6, 409T
  nickel-chromium alloys, 283, 409T
  nickel-copper alloys, 269–70, 408T
  nickel-molybdenum alloys, 403, 409T
  steel, 383–92, 384–7T
    effect of, composition, 391–2
      mill scale, 388–9
      temperature, 391
      velocity, 390–1
      welding, 391
    continuous immersion, 383–8
    half-tide immersion, 389–90
    mud line corrosion, 390
    pitting, 388–9
  steels, stainless, 172T, 413–8, 413T
  tin, 324
  solubility products of salts in, 1118–9
  specific gravity, 1113T
  specific heat, 1114T
  vapor pressure, 1114
Water-line attack, 5–6
  of steel, 126
Weld decay, 162, 1104F
Welding, and fretting corrosion, 595
Whitney diagram of soil classification, 448F
Wire drawing, in steam impingement attack, 84, 1104F
Wires, heat-resisting, life of, 999F
  life test for, 997–8
  nickel-chromium and nickel-chromium-iron, behavior in gases, 698–9
Wood, **372–8**
  adhesives for, properties, 376–8T

Wood, in contact with lead, 216
  properties, 374–5T
  resistance to various solutions, 373T
Work function and passivity, 25

Zinc, **331–47**
  anodic coatings for, 862
  chromate coatings for, 862–4
  coatings (see Coatings, zinc)
  compositions, 332T
  corrosion fatigue, 331
  effect of, pH, 335
    stress, 339, 574
    temperature, 335
    velocity, 338
  effect on oxidation of magnesium, 673F
  galvanic corrosion, 339
  high-temperature corrosion, 722
  metallurgical factors, 339
  molten, resistance of aluminum to, 617
  pigments, 888–90
  protection of, 340–2
  resistance to, acids, 336–8
    alkalies, 336–8
    atmospheric corrosion, 342–7
      effect of stress, 346
      galvanic, 346
      indoor, 342–3
      outdoor, 344–6
      protection from, 347
    gases, 340
      dissolved, 332–5
    organic compounds, 338
    non-aqueous, 339–40
  salts, 336–8, 337T, 977F
  soil, 454, 460T (see also Coatings, zinc)
  water, distilled, 4
    fresh, 335–6
Zirconium, 347–8

— FIRST —

AN INTRODUCTION TO THE FEATURE BY SERGIO PONCHIONE

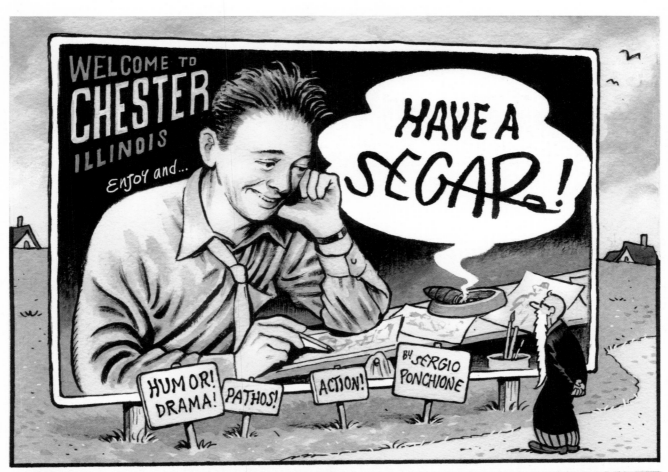

— AND NOW —
"OOMPH"
A POPEYE SHORT
BY CATHY MALKASIAN

DESIGN: Jacob Covey
EDITORIAL: Conrad Groth & Gary Groth
PUBLICITY: Jacq Cohen
ASSOCIATE PUBLISHER: Eric Reynolds
PUBLISHER: Gary Groth

THIS EDITION OF *Popeye Volume 1: Olive Oyl and Her Sweety* is copyright © 2021 Fantagraphics Books, Inc. All Segar comics and drawings © 2021 King Features Syndicate, Inc. / ™Hearst Holdings, Inc. Strips provided by Bill Blackbeard and his San Francisco Academy of Comic Art. "Have a Segar!" is copyright © 2021 Sergio Ponchione. Translation is copyright © Jamie Richards. "Oomph" is copyright © 2021 Cathy Malkasian. All rights reserved. Permission to reproduce content must be obtained from the publisher.

FANTAGRAPHICS BOOKS, INC.
7563 Lake City Way NE
Seattle, WA 98115

www.fantagraphics.com
facebook.com/fantagraphics
@fantagraphics

ISBN: 978-1-68396-462-9
LOC: 2021935360
First Fantagraphics Books edition: September 2021
Printed in China

OUR FEATURE: POPEYE

THE E.C. SEGAR POPEYE SUNDAYS

STARRING IN:
"Olive Oyl & Her Sweety"
VOLUME ONE OF A SERIES
March 1930 – February 1932

GROTH & SON, IN ASSOCIATION WITH KING FEATURES SYNDICATE

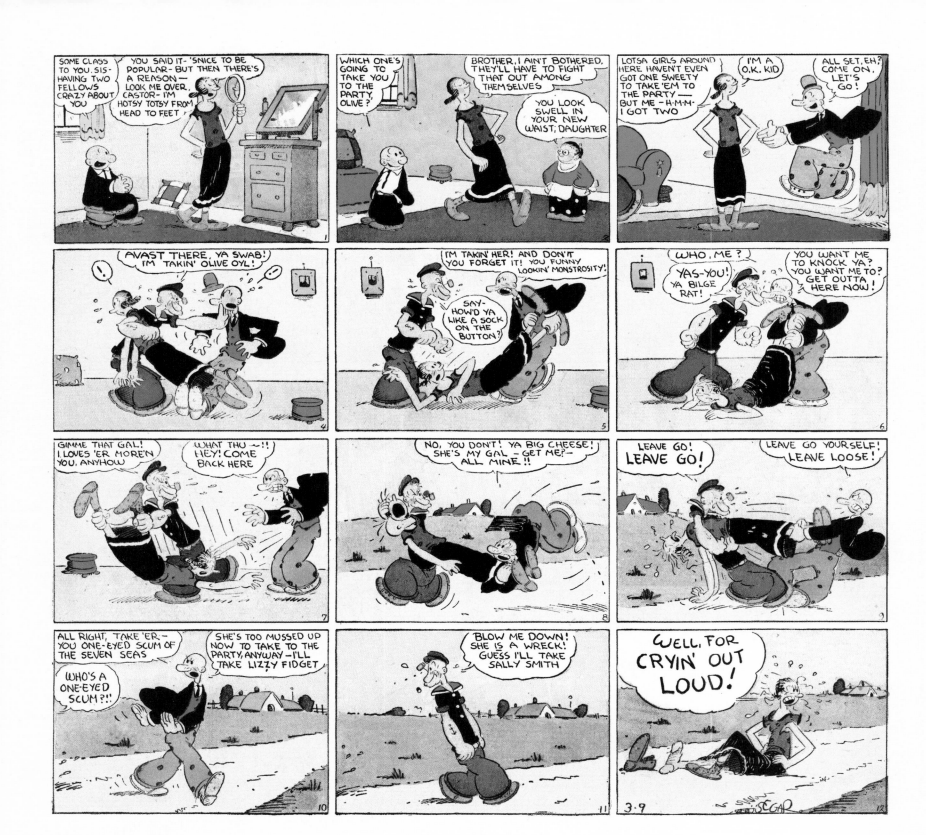

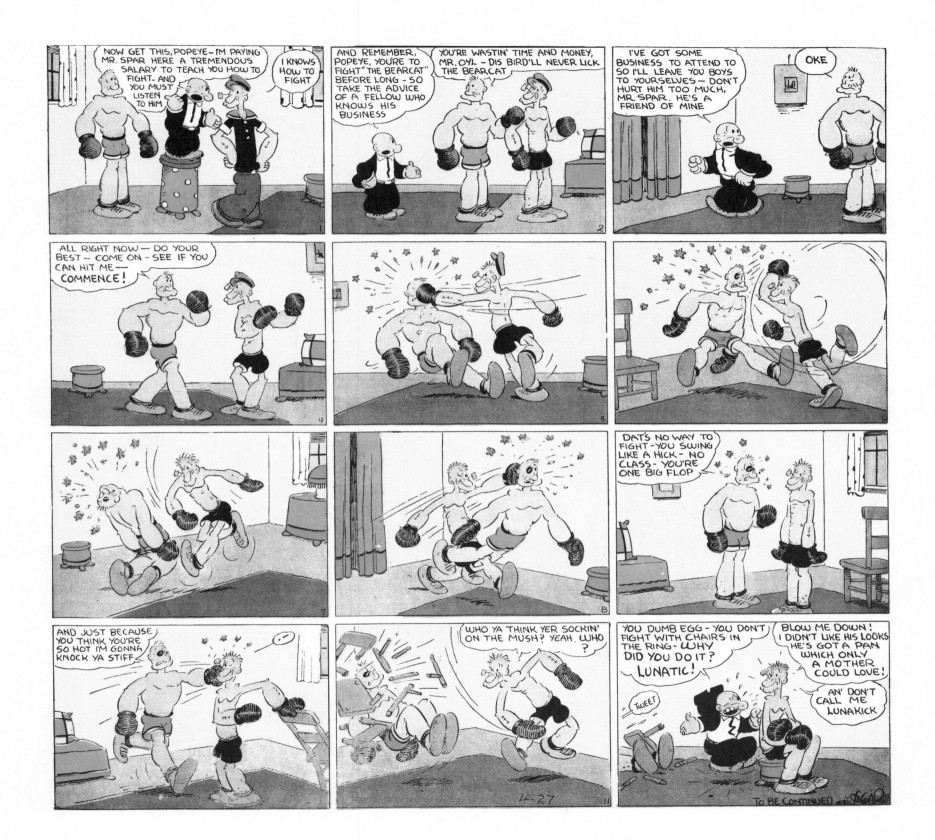

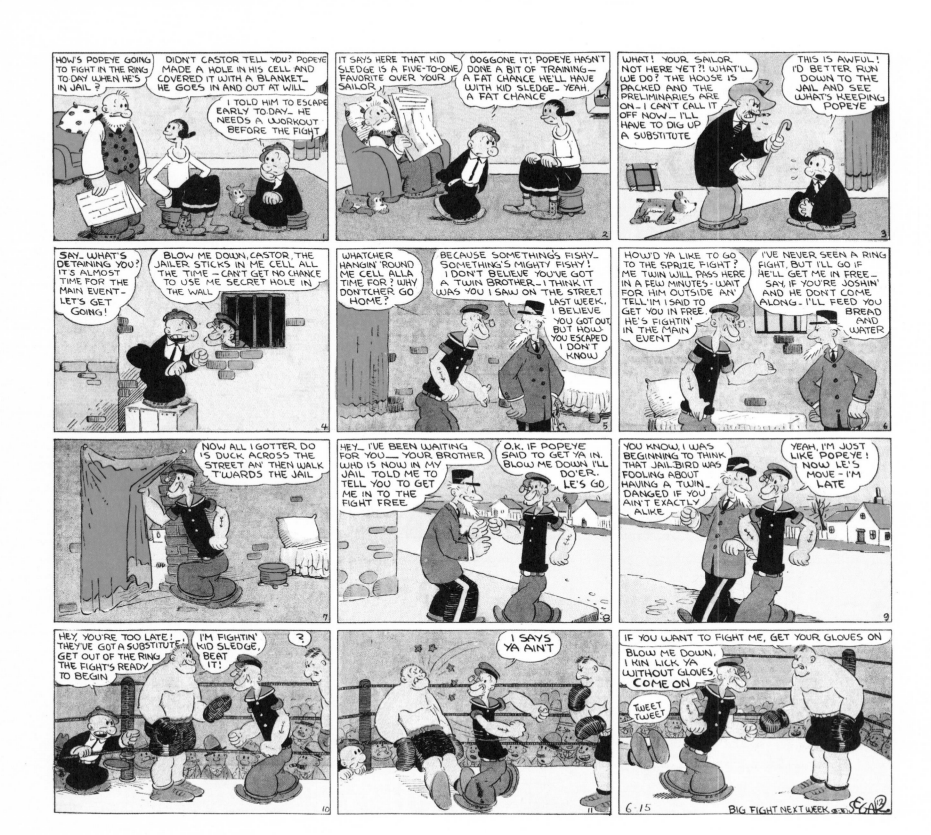

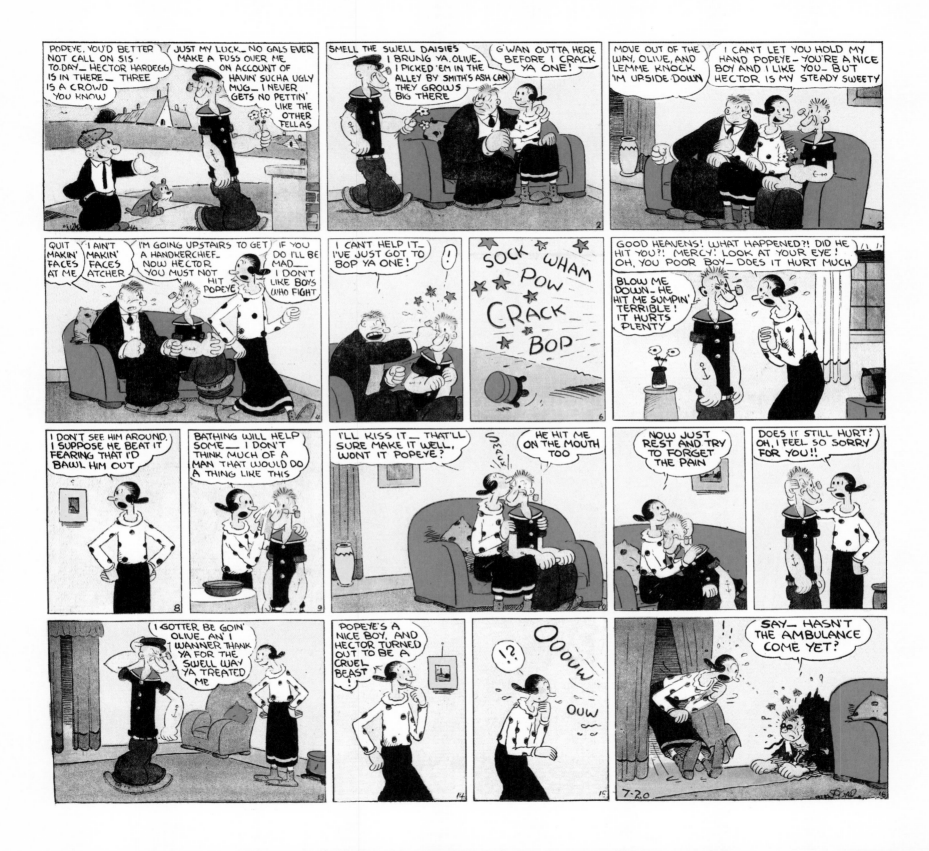

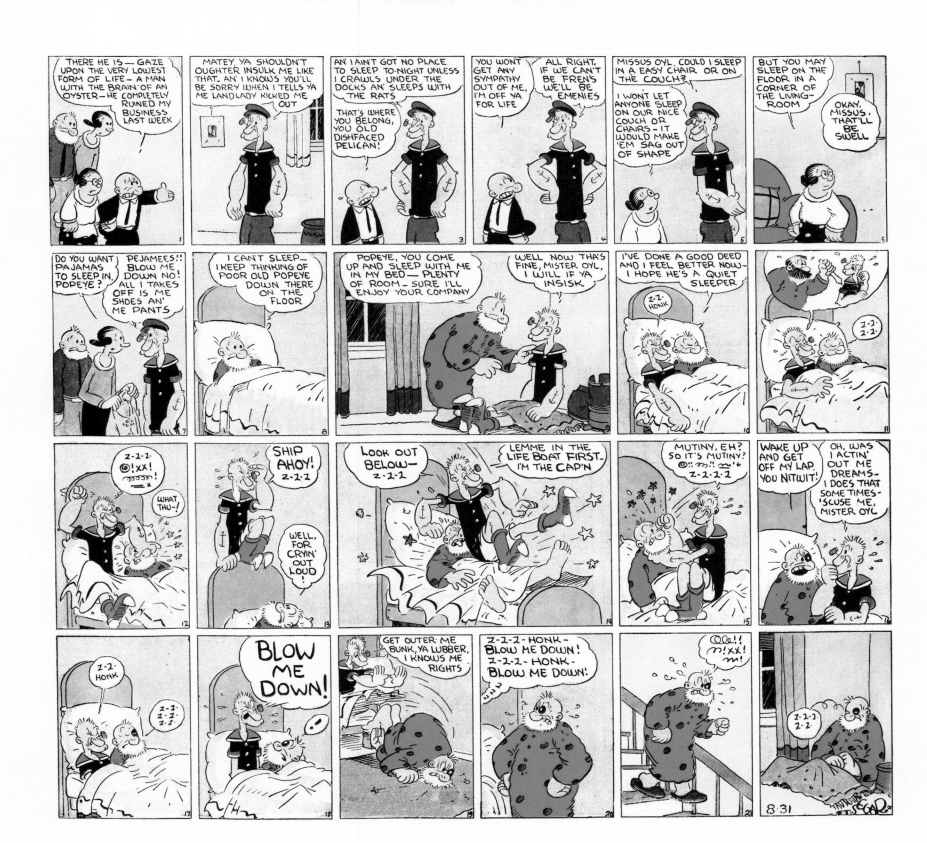

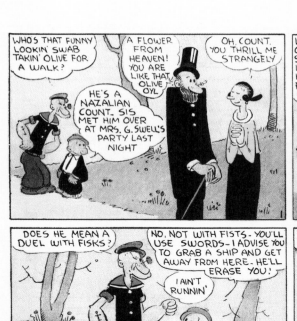

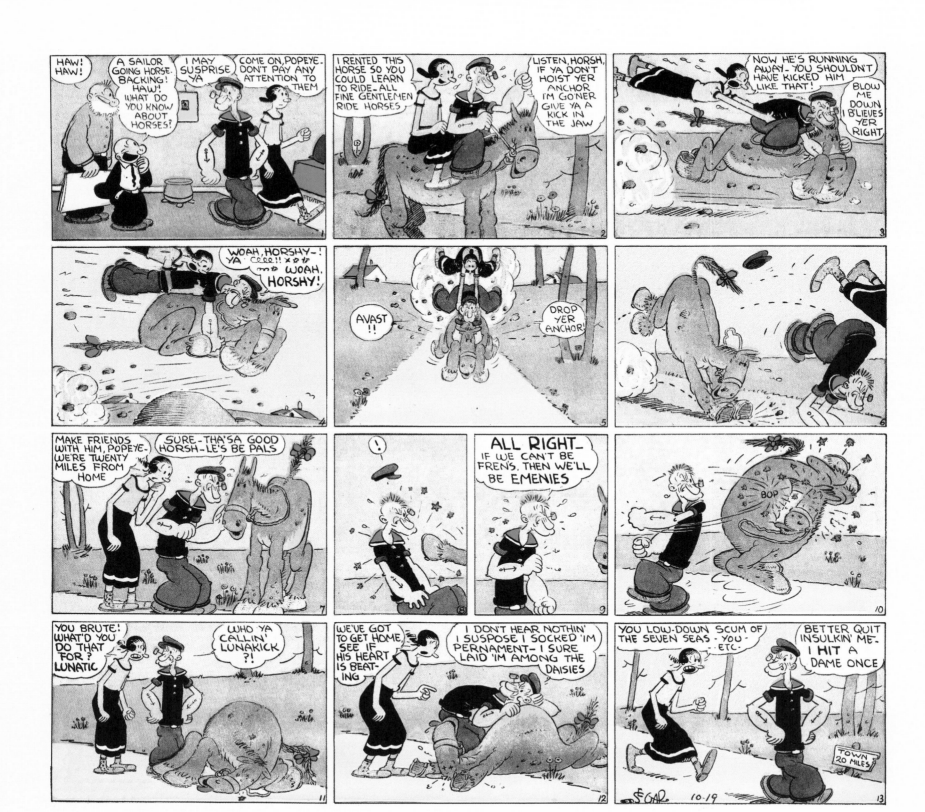

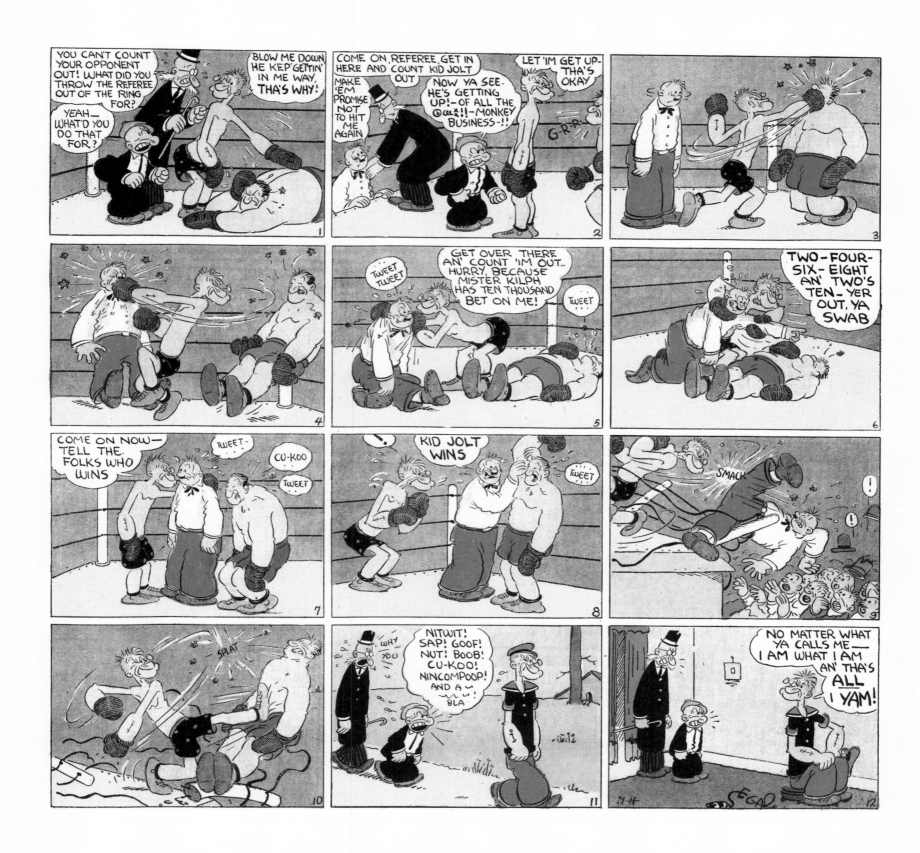

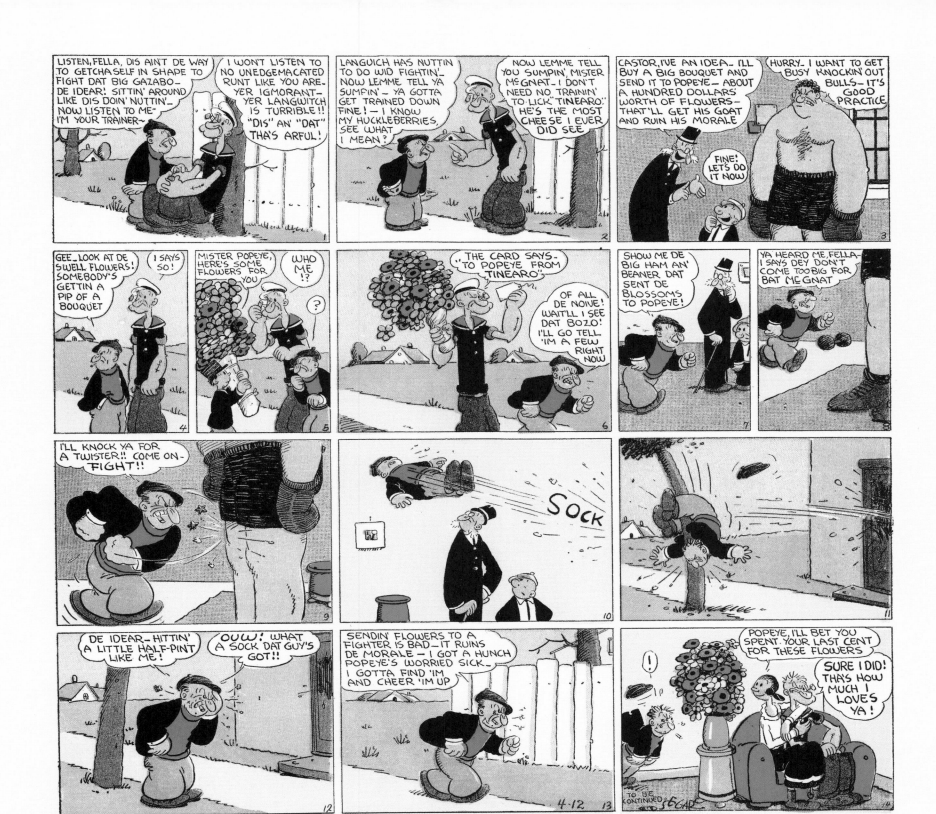

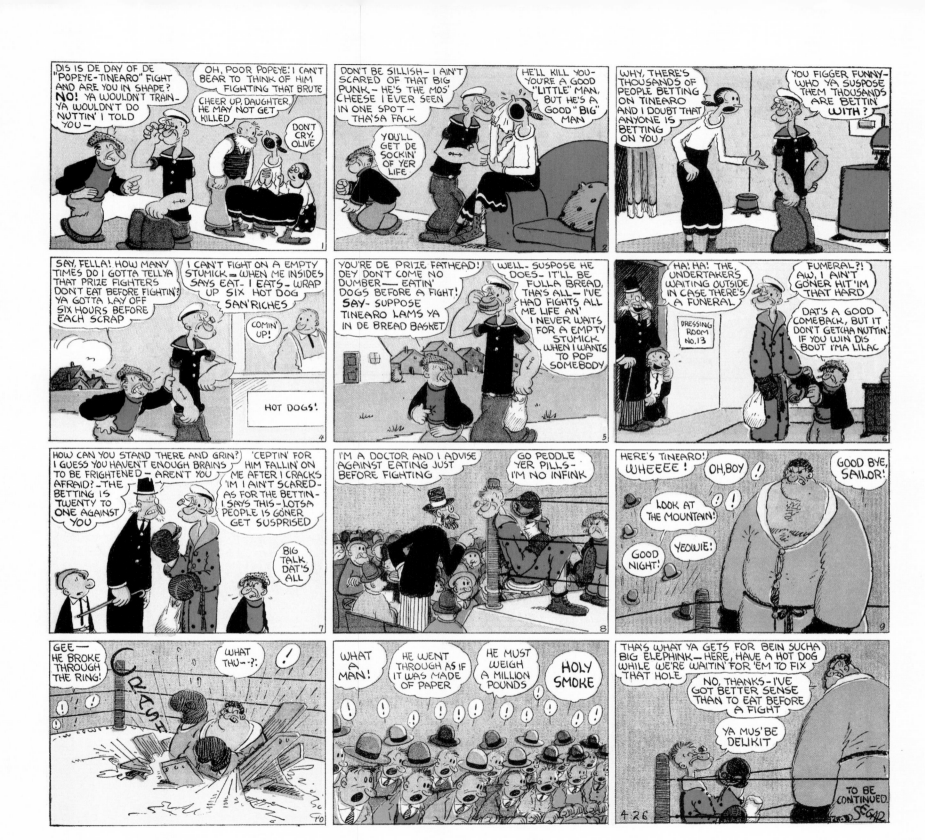

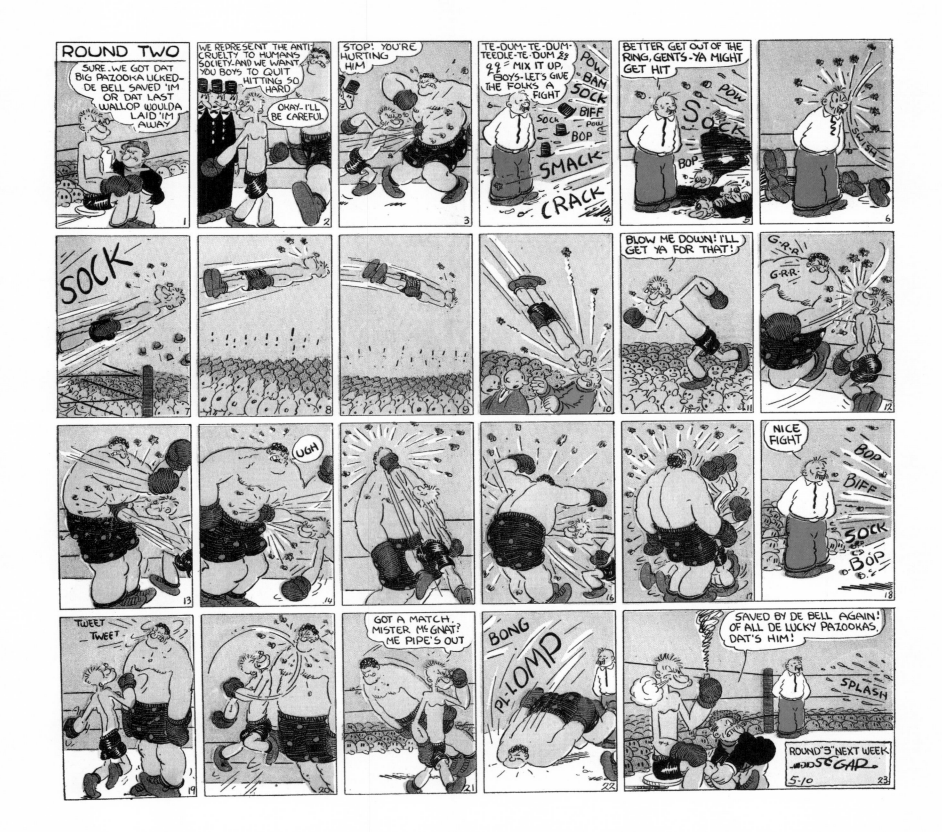

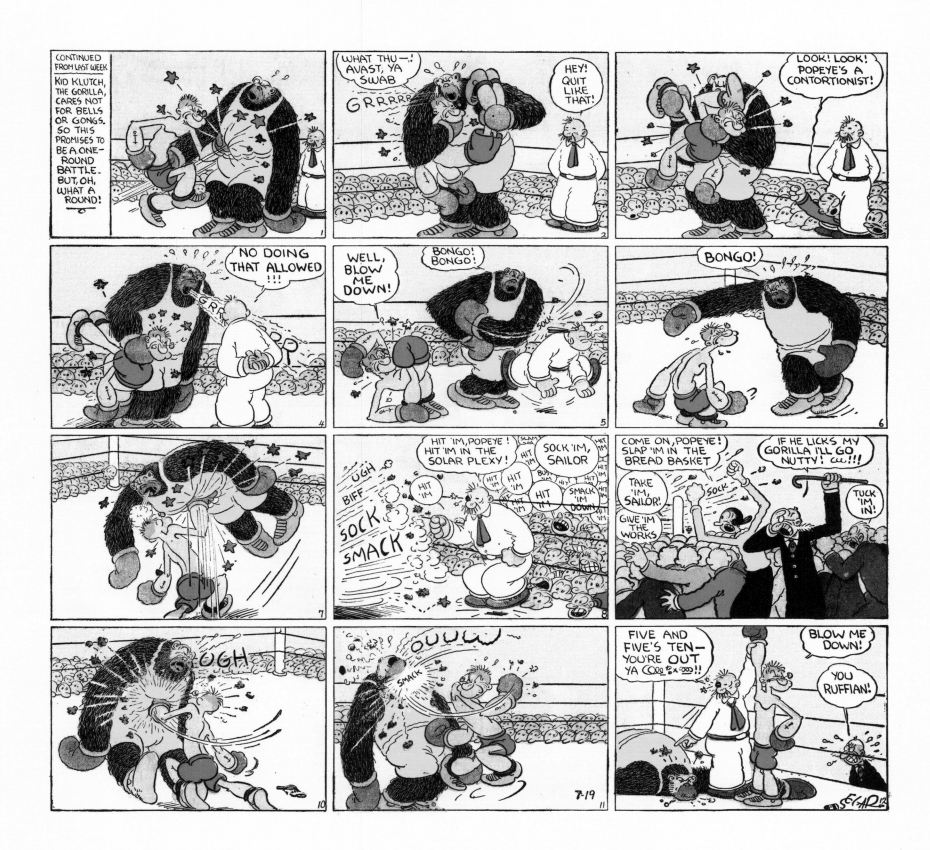

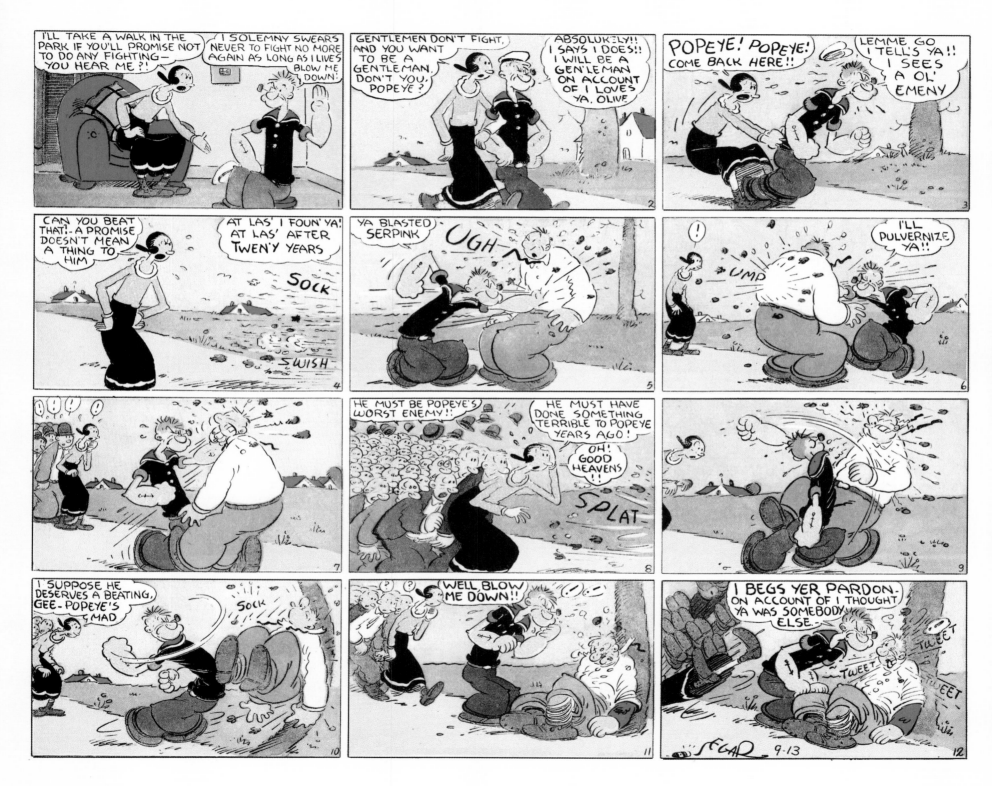

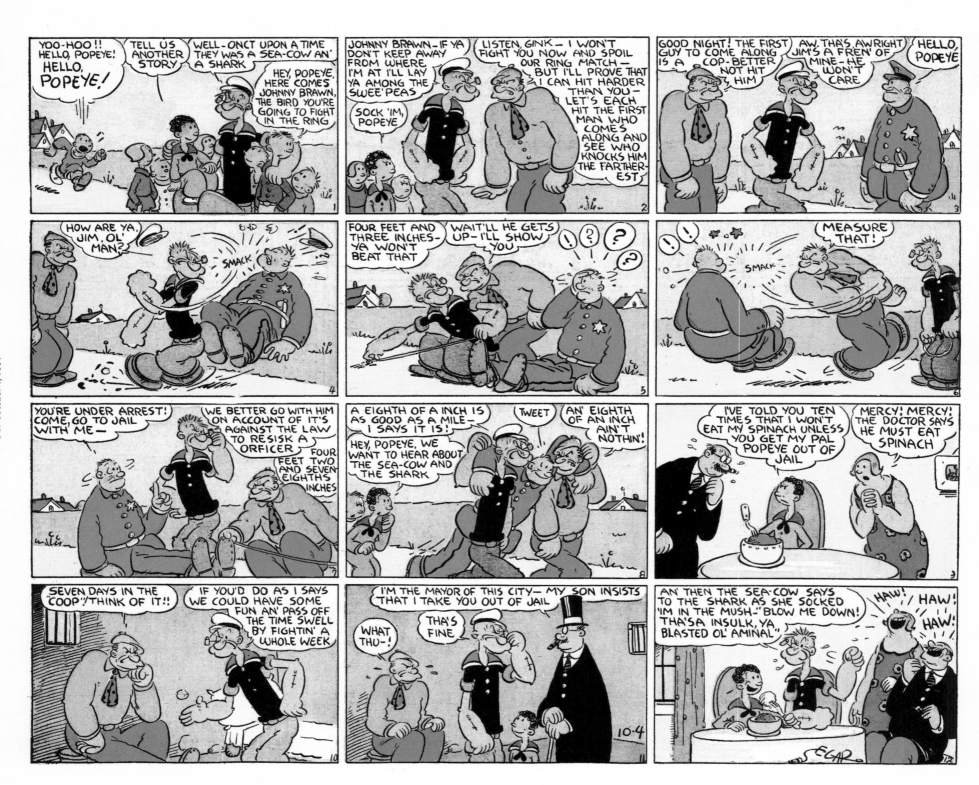

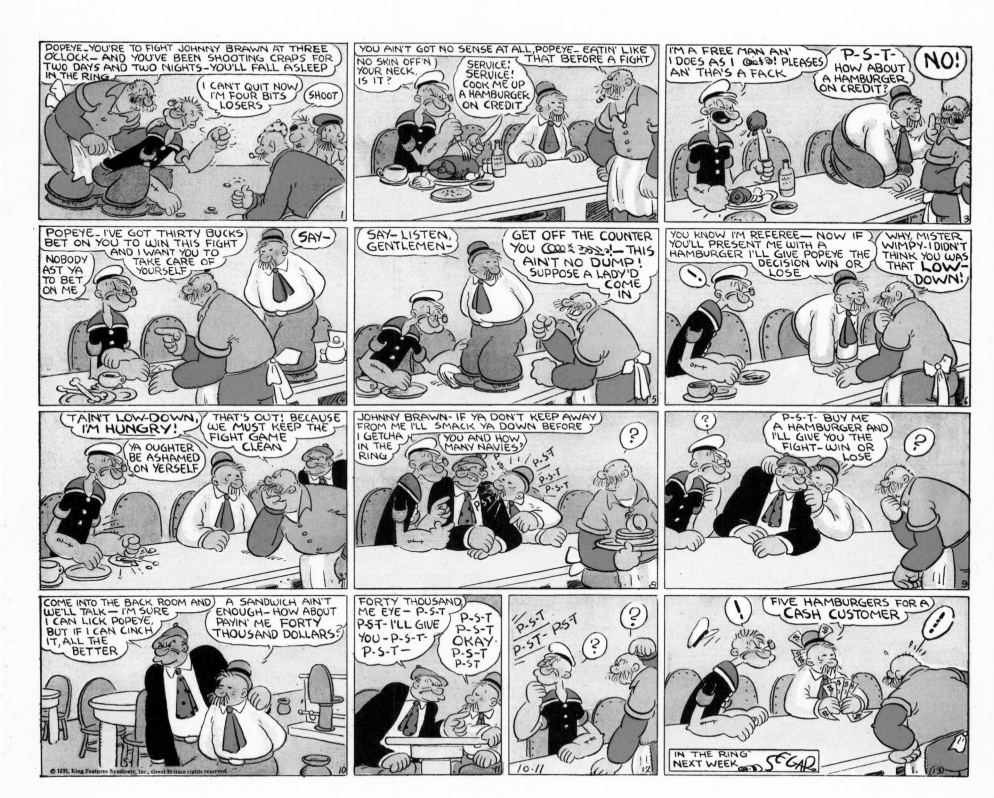

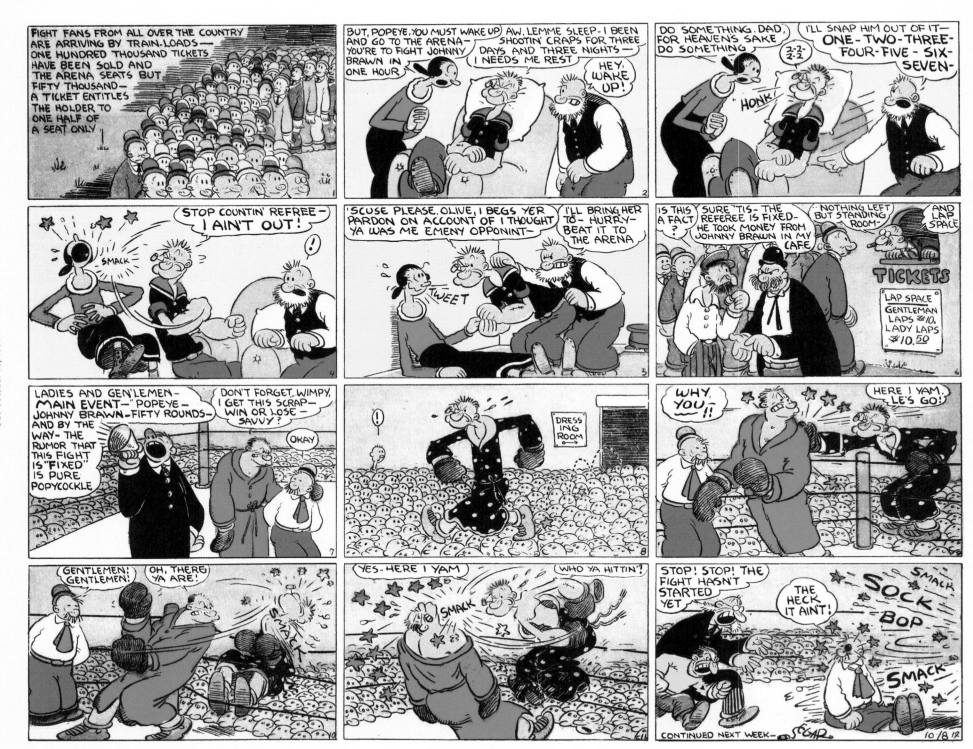

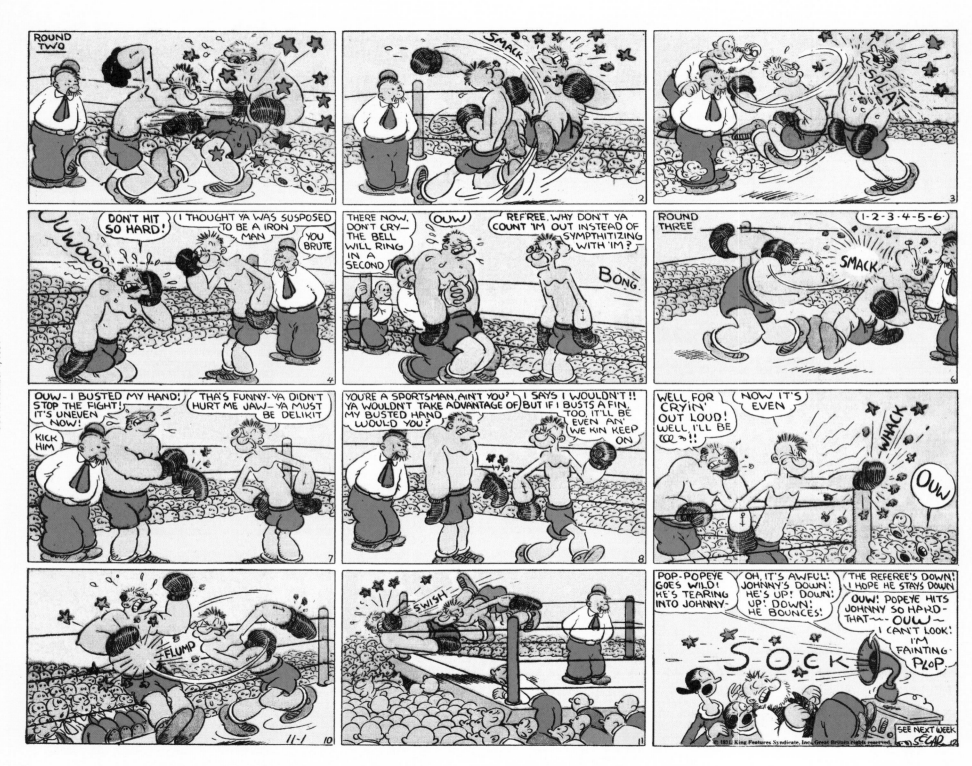

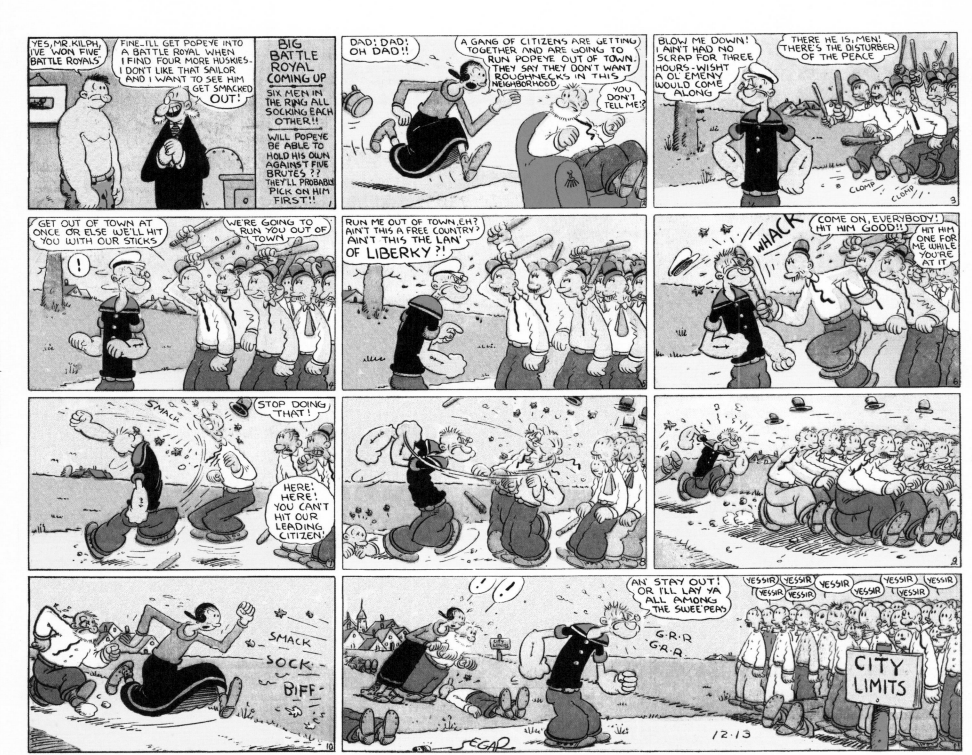

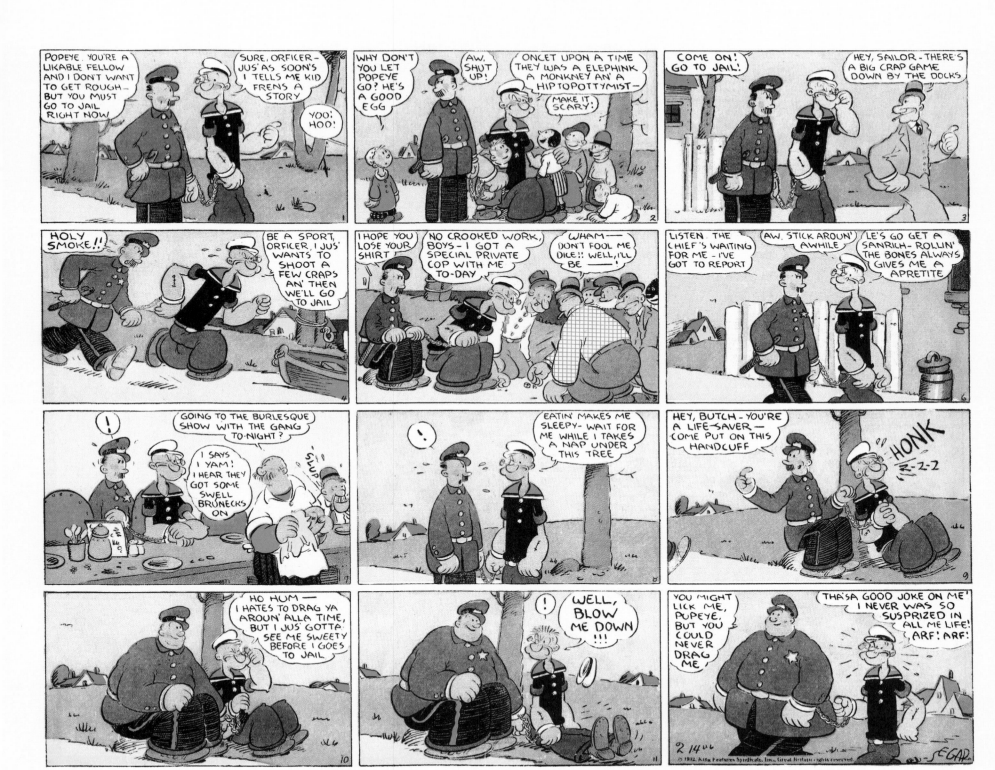

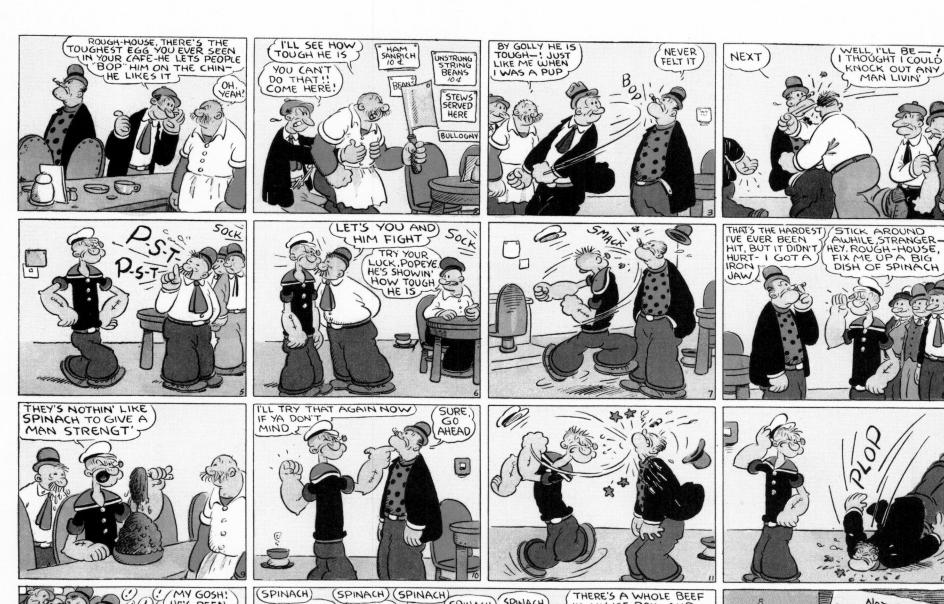

ELZIE CRISLER SEGAR was born in Chester, IL, on December 8, 1894. At the age of 12, he got a job at the Chester Opera House, a motion-picture theater, where he drew show cards, played drums in the orchestra, and ran the projectors. This early interest in sequential storytelling led him to take a correspondence course in cartooning, which honed his drawing skills to the extent that in 1916 he was hired as a staff cartoonist at the Chicago *Herald*. There, he worked on the series *Charlie Chaplin's Comic Capers* for about a year, before he was able to create his own strip, *Barry the Boob*. In 1918, Segar joined the staff of the *Chicago Evening American*, where he created *Looping the Loop*. This strip proved popular enough that he was summoned to New York in 1919 to create a new series for the *Evening Journal* — *Thimble Theatre*, which was to become his magnum opus. Initially a parody of vaudeville shows and film serials, *Thimble Theatre* soon developed into an original story chronicling the ill-conceived schemes and romantic escapades of Olive Oyl, her brother Castor Oyl, and her boyfriend Ham Gravy. In 1920, Segar also created the long-running gag comic *Sappo*, which followed the everyday misadventures of inventor John Sappo, his wife Myrtle, and their mad scientist boarder, Professor Wotasnozzle.

Thimble Theatre's readership kept growing when, on January 17, 1929, Segar offhandedly introduced a new character to the strip — Popeye. This idiosyncratic sailor captivated readers and soon took center stage; by 1931, the series was retitled *Thimble Theatre Starring Popeye* and Segar had a hit on his hands. As the strip went on, Segar would add many memorable characters to the ensemble, including J. Wellington Wimpy, the Sea Hag and the Goon, Oscar, Bluto, Swee'pea, and Eugene the Jeep — but Popeye remained the star of the show. At the height of his cartooning career, Segar died of leukemia and liver disease on October 13, 1938, at the age of 43. However, his signature creation has lived on in print, in film, and as a fixture of pop culture today.